Fusillo
Nature Classification System
Spore Plant Species Checklist
1st Edition

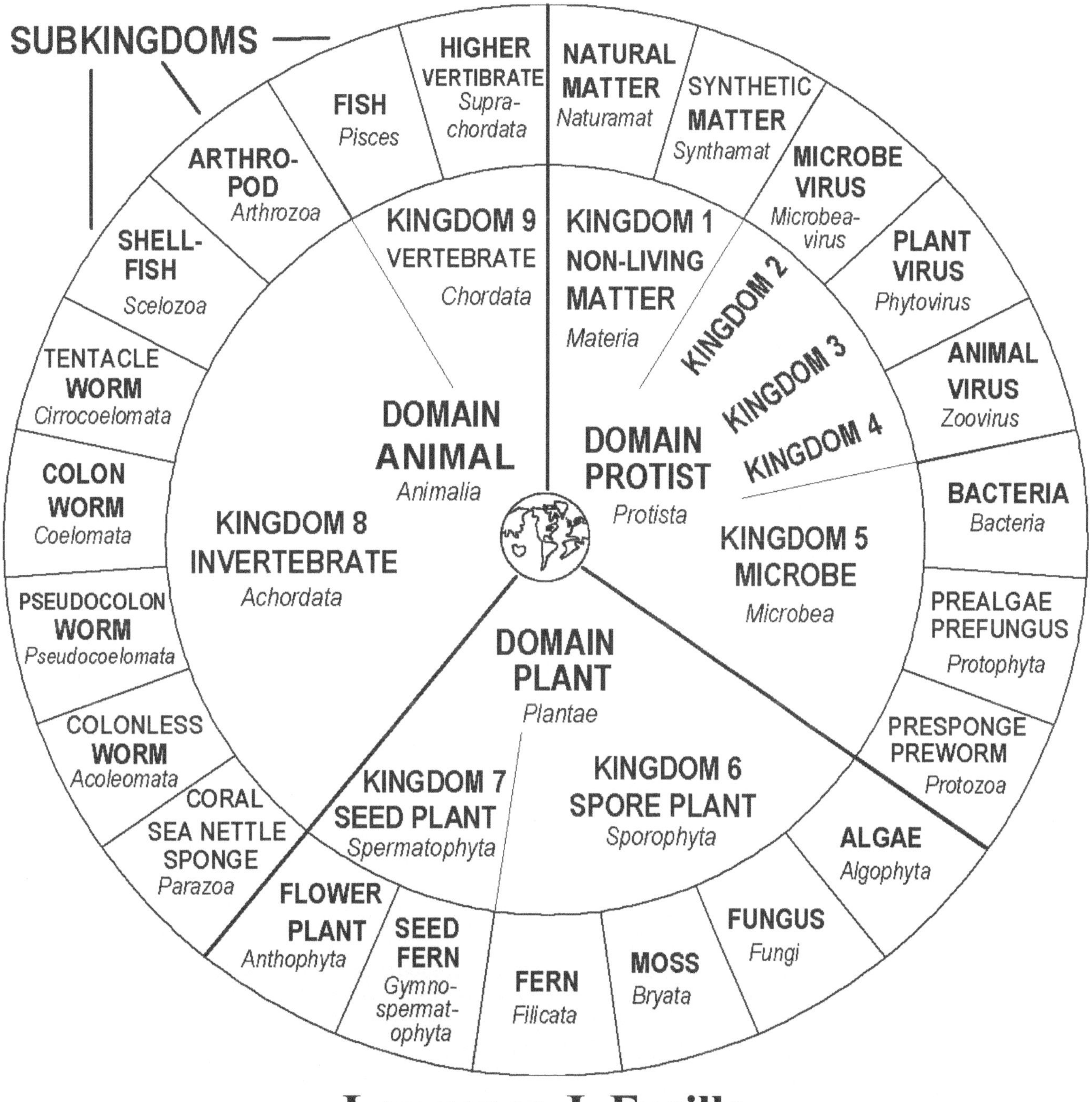

Lawrence J. Fusillo

Trademarks

Trademark names are used. Rather than put a trademark symbol after every occurrence of the trademarked name, the author used the names in an editorial fashion only, and to the benefit of the trademark owner, with no intention of infringement of the trademark. Where such designations appear in this book, they have been put in initial caps.

Warning and Disclaimer

The author of this work has added value to the underlying factual information herein by unique and original selection, by unique and original arrangement, and by unique and original classification of the information. Every effort has been made to make this book as complete and accurate as possible. However, no warranty of fitness is implied. The information provided is on an as-is basis.

Quantity Discounts

Check the publisher's web site, www.buybooksontheweb.com for sales on small quantities of books. The publisher offers discounts on this book when ordered in quantity for bulk purchases and special sales. For bulk sales within the United States please call Toll Free: 877-BUY-BOOK (877-289-2665), Local 610-941-9999, Fax 610-941-9959.

ISBN 0-7414-5878-0

Published by:

INFINITY
PUBLISHING.COM

1094 New DeHaven Street, Suite 100
West Conshohocken, PA 19428-2713

Printed in the United States of America

The paper used in this book meets the minimum requirements of the American National Standard for Information Sciences Permanence of Paper for Printed Library Materials ANSI Z39.48-1984

Published 1 May 2010

Bibliographic Data

LIBRARY OF CONGRESS BIBLIOGRAPHIC DATA	1. INTL. STD. BOOK NO. (ISBN) ISBN: 0-7414-5878-0	2. CIP Accession No.	3. NTIS Accession No.

4. TITLE AND SUBTITLE
Fusillo Nature Classification System
Spore Plant Species Checklist 1st Edition

5. Publication Date 1 May 2010
6. Performing Org. Code SCS

7. AUTHOR(S)
Lawrence J. Fusillo

8. Performing Org. Rep. No. SCS-FNCS 005-1

9. PERFORMING ORGANIZATION NAME AND ADDRESS
Lawrence J. Fusillo
280 Piccadilly Square, Apt E2
Athens, GA, 30605

10. Project/Task/Work No. WDI-WBS 005-1
11. Contract/Grant No. None

12. Sponsoring Organization Name and Complete Address (Street, City, State, ZIP)
This work was performed without a contract, a sponsor, or a grant.

13. Period Covered Not Applicable
14. Sponsoring Agency Code Not Applicable

15. SUPPLEMENTARY NOTES

This book *supercedes* and *replaces* the book "Systema Naturae, Ed. X. (Systema naturae per regna tria naturae, secundum classes, ordines, genera, species, cum characteribus, differentiis, synonymis, locis. Tomus I. Editio ecimal, reformata.)", by Carl Linnaeus, Holmiae publisher, 1758.
SCS – Scientific Computer Software, FNCS – Fusillo Nature Classification System. WDI – World Data Infrastructure, WBS – World Basic Standard.

16. ABSTRACT (A 200 word or less summary of most significant information If document includes a significant bibliography or literature survey, mention it here)

The Fusillo Nature Classification System (FNCS) is the first major revision to classification in 250 years. This is the first in human history classification of all of the higher taxa of spore plants and all of the species of spore plants in one book. Spore plants include Algae, Fungus, Lichen, Moss, and Ferns. The following numbering system is used for spore plants.

Fusillo Nature Classification System (FNCS)

Domain *Plantae*	Plant
Kingdom 6 *Sporophyta*	Spore Plant
Subkingdom 1 *Algophyta*	Algae
Phylum 1 *Bacillariophyta*	Diatom Brown Algae
Phylum 2 *Chlorophyta*	Olive Green Algae
Phylum 3 *Chrysophyta*	Golden Ocher Algae
Phylum 4 *Cyanidophyta*	Blue Green Algae
Phylum 5 *Ochrophyta*	Ocher Algae
Phylum 6 *Prasinophyta*	Green Algae
Phylum 7 *Rhodophyta*	Red Algae
Subkingdom 2 *Fungi*	Fungus
Phylum 1 *Ascomycota*	Sac Fungus
Phylum 2 *Basidiomycota*	Club Fungus
Phylum 3 *Zygomycota*	Multinucleate Fungus
Subkingdom 3 *Bryata*	Moss
Phylum 1 *Bryophyta*	Moss
Phylum 2 *Anthocerophyta*	Hornwort
Phylum 3 *Hepatophyta*	Liverwort
Subkingdom 4 *Filicata*	Fern
Phylum 1 *Lycopodiophyta*	Moss Fern
Phylum 2 *Pteridophyta*	Leaf Fernl

KEY WORDS (six to twelve entries; alphabetical order; capitalize only the first letter of the first key word unless a proper name; separated by semicolons)
Algae, Fungus; Lichen; Moss; Fern; Taxonomy

18. AVAILABILITY | X | Unlimited
| X | For Commercial Distribution Do not release to NTIS
| X | Order from Infinity Publishing, www.buybooksontheweb.com
| X | Order from Barnes and Noble
| X | Order from Borders
| X | Order from Amazon.com

19. Data CLASS WORLD DATA INFRASTRUCTURE
20. Standards CLASS WORLD BASIC STANDARD
21. NO. OF PAGES
22. Price $49.95

Preface

(1) The Fusillo Nature Classification System is a bilingual "Latin/English System"

This is the first in the history of mankind, original publication of an **English Language Classification System**. This book is used by environmental protection organizations, teachers, & students as an English Language "Classification Decoding Key" to decoding Latin based Classification Systems from Latin into English. (All Catalogue of Life CoL species are in Latin.)

(2) The Fusillo Nature Classification System has no "Missing Categories"

The Fusillo Nature Classification is a Decanomial system for "variety, species, Genus, Family, Order, Class, Phylum, Subkingdom, Kingdom, & Domain". A complete system with each species in the Physical Universe properly assigned to all 8 higher taxonomic categories has never been published before. (CoL has missing categories.)

(3) The Fusillo Nature Classification System has no "Duplicate Families"

Single Genus Families are recognized when a Genus is truly distinct from all other Genera in the same Order. But when a Genus in a Single Genus Family is closely related to Genera in another Family in the same Order, that Genus has been moved to the closely related Family & the Single Genus Family has been deleted. (CoL has single Genus Families.)

(4) The Fusillo Nature Classification System is "Backward Compatible"

While scientists currently disagree on Family, Order, Class, Phylum, & Subkingdom assignments, there is universal agreement on the assignment of Latin *Genus* & *species* names. The Fusillo Nature Classification System has been designed to be 100% backward compatible with Latin *Genus* & *species* names. A few Genera have been split in two, retaining the old Latin *Genus* name & adding a new Latin *Genus* name. In these cases, the Latin *species* name provides 100% backward compatibility.

(5) The Fusillo Nature Classification System is an "Open Standard"

Fusillo Nature Classification System (FNCS) categories are on CD are in universally readable MS Excel spreadsheet format (.xls). Purchase of a CD includes a personal use license for FNCS spreadsheet files for use on three sites; 1 home desktop, 1 mobile laptop, & 1 work. Import into database software for personal use is encouraged.

Latin Names for viruses, microbes, plants, & animals were obtained from the following books & CDs;

Title	Author/Publisher		Year
"*Systema Naturae per regna tria naturae, secundum classes, ordines, genera, species, cum characteribus,differentiis, synonymis, locis. 10th Edition*"	Carl Linnaeus/Holmiae		1758
"International Code of Nomenclature for Cultivated Plants"	Seattle Botanical Congress		1969
"A World Checklist of Mammalian Species 2nd Edition"	Corbet & Hill/Facts on File Publisher		1986
"Plant Families & Their Relationship" extracted from page 958 of "Botanica"	Dahlgren & Cronquist	ISBN:3-8331-1253-0	2004
"*Systema Naturae 2008 Edition*" on Catalog of Life (CoL) CD	UK Species 2000 Org.	(www.sp2000.org)	2008

English Names for viruses, microbes, plants, & animals were obtained from the following books;

Title	Author/Publisher		Year
"A World Checklist of Birds"	Monroe & Sibley	ISBN:0-300-07083-7	1993
"Latin Names Explained; Birds, Reptiles, Mammals	Gotch, A.F.	ISBN:0-8160-3377-3	1995
"Stearns Dictionary of Plant Names, Revised"	Stearn, W.	ISBN:0-88192-556-X	1996
"Latin Concise Dictionary" (Latin to English)	Harper Collins	ISBN:0-06-053690-X	1997
"The World of Primates"	Hook,P.	ISBN:0-517-16204-0	2000
"Encyclopedia of Aquatic Life"	Bannister & Campbell	ISBN:0-8160-1257-1	2002
"Stedmen's Medical Dictionary 28th Ed.	Stedman	ISBN:0-7817-6192-1	2005
"Virus Taxonomy 8th Report on Taxonomy of Viruses"	Fauquet,C./Elsevier	ISBN:0-12-249551-X	2005
"Encyclopedia of Virology 3rd Edition"	Mahy&Regenmortel	ISBN:0-12-373735-7	2008

The Fusillo Nature Classification System (FNCS) containing all living organisms was published in;

	Fusillo Nature Classification System (FNCS) Title	Author/Publisher		Year
1	"Fusillo Nature Classification System Genus & Higher Taxa 1st Edition"	Fusillo, L.J./Infinity	ISBN:0-7414-4495-3	Jan 2008
2	"FNCS Nonliving Matter Species Checklist 1st Edition"	Fusillo, L.J./Infinity	ISBN:0-7414-4459-3	Mar 2008
3	"FNCS Virus Species Checklist 1st Edition"	Fusillo, L.J./Infinity	ISBN:0-7414-5214-6	Mar 2009
4	"FNCS Microbe Species Checklist 1st Edition"	Fusillo, L.J./Infinity	ISBN:0-7414-5475-0	Aug 2009
5	"FNCS Spore Plant Species Checklist 1st Edition"	Fusillo, L.J./Infinity	ISBN:0-7414-5878-0	May 2010
6	"FNCS Seed Plant Species Checklist 1st Edition"	Fusillo, L.J./Infinity		
7	"FNCS Orchid Species Checklist 1st Edition"	Fusillo, L.J./Infinity	ISBN: 0-7414-5668-0	Dec 2009
8	"FNCS Herb, Vegetable, & Fruit variety Checklist 1st Edition"	Fusillo, L.J./Infinity		
9	"FNCS Rose Species Checklist 1st Edition"	Fusillo, L.J./Infinity		
10	"FNCS Anna Belle Nature Trail Guide 1st Edition"	Fusillo, L.J./Infinity	ISBN:0-7414-4201-9	Jul 2007
11	"FNCS Invertebrate Parazoan, & Worm Species Checklist 1st Edition"	Fusillo, L.J./Infinity		
12	"FNCS Invertebrate Arthropod Species Checklist 1st Edition"	Fusillo, L.J./Infinity		
13	"FNCS Invertebrate Shellfish & Vertebrate Fish Species Checklist"	Fusillo, L.J./Infinity		
14	"FNCS Vertebrate Amphibian, Reptile, Bird, & Mammal Species Checklist 1st Editon"	Fusillo, L.J./Infinity		

Lawrence J. Fusillo
280 Piccadilly Sq. Apt. E-2
Athens, GA 30605

TABLE OF CONTENTS

KINGDOM 1 *Materia* Nonliving Matter
KINGDOM 2 *Micobeavirus* Microbe Virus
KINGDOM 3 *Phytovirus* Plant Virus
KINGDOM 4 *Animaliavirus* Animal Virus
KINGDOM 5 *Microbea* Microbe

KINGDOM 6 *Sporophyta* Spore Plant

Latin: *Spora* spore, reproduced from spores
Latin: *Phyta* phyton, shoot, sole of the foot
a shoot from the ground, plant

SUBKINGDOMS		Species
61 *Algophyta*	Algae	33,500 [M&S]
62 *Fungi*	Fungus	53,600 [M&S]
63 *Bryata*	Moss	16,100 [M&S]
64 *Filicata*	Fern	12,117 [M&S]

The male pollen single chromosome (haploid) cell gamete (→) fertilizes the female pistil single nucleus (haploid) egg cell gamete (†) forming a double nucleus cell (cytogamy zygote) or a double chromosome single nucleus cell (karogamy zygote). The fertilized egg (zygote) grows by cell division which halves the number of chromosomes (meiosis). The evolution of zygote embryo case (a polyp case for mosses and ferns, a seed case for conifers and flower plants) protected the zygote embryo from drying and other environmental hazards was a major factor in the spread of plants from oceans to dry land.
This is the original publication of English Names for Algae, Fungus, Moss, and Fern Phylums, Classes, Orders, and Families.

SUBKINGDOM 61 *Algophyta* Algae

Latin: *Alga* algae
Latin: *Phyta* phyton, plant
33,500 species [Marg. & Schw.]
22,692 species [Guiry, Dec. 2004]

PHYLUMS		Species
611 *Bacillariophyta*	Diatom Brown Algae	10000 [M&S]
612 *Chlorophyta*	Olive Green Algae	16000 [M&S]
613 *Chrysophyta*	Golden Ocher Algae	
614 *Cyanidiophyta*	Blue Green Algae	
615 *Ochrophyta*	Ocher Algae	
616 *Prasinophyta*	Unicell Green Algae	147
617 *Rhodophyta*	Red Algae	

PHYLUM 611 *Bacillariophyta* Diatom Brown Algae

Latin: *Bacca* berry
Latin: *Phyta* phyton, plant
10,523 species [Guiry, Dec. 2004]
10,000 species [Margulis & Schwartz, 1998]

CLASSES		Species
611.1 *Bacillariophycaceae*	Feather Diatom Brown Algae	8,336[G]
611.2 *Coscinodiscophyceae*	Radial Diatom Brown Algae	2,187[G]

Phylum *Bacillariophyta* Diatom Brown Algae species are single celled and sometimes come together to form filament colonies. They have a shell (test) composed of pectin organic material impregnated with silicon.

Class *Bacillariophycaceae* Feather Diatom Brown Algae species are axially symmetric and are shaped like a feather, boat, or needle.

Class *Coscinodiscophyceae* Cos Disc Radial Diatom Brown Algae are radially symmetric and are shaped like a coin or short cylinder. They inhabit both freshwater and saltwater. Diatom Brown Algae species decompose to form diatomaceous sediment which is used for abrasives, filters, chalk, and talc.

CLASS 611.1 *Bacillariophycaceae* Feather Diatom Brown Algae

ORDER 1 *Bacillariales* Feather Diatom Brown Algae

FAMILY 1 *Bacillariaceae* Feather Diatom Brown Algae
GENUS *Bacillaria* Feather Diatom Brown Algae
SPECIES *cistula, cursoria, paxillifer, phoenicenteron, socialis, viridis, vulgaris*
VARIETIES *paxillifer tumidula*
GENUS *Algites* Algite Feather Diatom Brown Algae
SPECIES *samantae*
GENUS *Allogonium* Allogon Feather Diatom Brown Algae
SPECIES *depressum*
GENUS *Alveolophora* Pit Bearing Feather Diatom Brown Algae
SPECIES *tscheremissinovae*
GENUS *Alveovallum* Pitted Feather Diatom Brown Algae
SPECIES *beyensii*
GENUS *Alveus* Pit Feather Diatom Brown Algae
SPECIES *marinus*
GENUS *Amblypyrgus* Ambling Pyrgi Feather Diatom Brown Algae
SPECIES *campanellus*
GENUS *Amicula* Amicable Feather Diatom Brown Algae
SPECIES *specululum*
GENUS *Amphipentas* Dual Pentad Feather Diatom Brown Algae
SPECIES *antillarum, boryana, campechiana, godeffroyi, grovei, hungarica, jimboi, juncta, laeta, loczyi, nobilis, pentacrius, quinquelobata*
VARIETIES *juncta fossilis*
GENUS *Amphitropis* Dual Feather Diatom Brown Algae
SPECIES *ornata*
GENUS *Ancylopyrgus* Ancyclic Pyrgi Feather Diatom Brown Algae
SPECIES *reticulatus*
VARIETIES *reticulatus elongatus*
GENUS *Annellus* Ring Feather Diatom Brown Algae
SPECIES *californicus*
VARIETIES *californicus hannae*
GENUS *Archaegladiopsis* Archaic Sword Feather Diatom Brown Algae
SPECIES *davidii, etheridgei, jackii*
GENUS *Archepyrgus* Archaic Pyrgi Feather Diatom Brown Algae
SPECIES *melosiroides*
GENUS *Arctyocha* Arc Pyrgi Feather Diatom Brown Algae
SPECIES *bukryi*
GENUS *Astartiella* Astarte Feather Diatom Brown Algae
SPECIES *bahusiensis, bahusiensoides, bremeyeri, glacialis, ignota, iks, modica, producta, punctifera, tenuis, wellsiae*
VARIETIES *bremeyeri rostrata*
GENUS *Asteroplanus* Star Flat Feather Diatom Brown Algae
SPECIES *karianus*
GENUS *Auriculopsis* Ear Feather Diatom Brown Algae
SPECIES *sparsipunctata*
GENUS *Austariella* Southern Feather Diatom Brown Algae
SPECIES *admissa, jamalinensis*
VARIETIES *jamalinensis subcircularis*
GENUS *Azpeitiopsis* Azpeiti Feather Diatom Brown Algae
SPECIES *morenoensis*
GENUS *Basilicostephanus* Basilica Stephan Feather Diatom Brown Algae
SPECIES *mirabilis, ornatus*
GENUS *Benetorus* Benet Feather Diatom Brown Algae
SPECIES *craspedodiscoides, morenoensis*
GENUS *Bergonia* Bergon Feather Diatom Brown Algae
SPECIES *angelica*
GENUS *Biblarium* Biblical Feather Diatom Brown Algae
SPECIES *ellipticum, emarginatum, glans, leptostauron, lineare*
GENUS *Bilingua* Two Lip Feather Diatom Brown Algae
SPECIES *rossii*
GENUS *Bogorovia* Bogorov Feather Diatom Brown Algae
SPECIES *barronii, curvata, gombosii, lancettula, mediopunctata, moholensis, paleacea, praepaleacea, puncticulata, rostrata, tatsunokuchiensis, veniamini*
VARIETIES *paleacea elongata*

GENUS *Brasiliella* Brazil Feather Diatom Brown Algae
SPECIES *helix*
GENUS *Calyptosporium* Calypter Spore Feather Diatom Brown Algae
SPECIES *carinatum, cyclacaenum, effalum, exasperatum, galeolum, lentiginosum, maximum, perexiguum, pustulatum*
GENUS *Campylostylus* Field Feather Diatom Brown Algae
SPECIES *normanianus*
GENUS *Catillus* Catill Feather Diatom Brown Algae
SPECIES *subimpletus*
GENUS *Cavernosa* Cavern Feather Diatom Brown Algae
SPECIES *kapitiana*
GENUS *Cavitatus* Cavity Feather Diatom Brown Algae
SPECIES *jouseanus*
GENUS *Cayeuxia* Cayeux Feather Diatom Brown Algae
SPECIES *americana, anae, atanasiui, elliottii, fruticulosa, kurdistanensis, lemaitrae, mediterranea, nodosa, piae*
GENUS *Climacosphenia* Climac Wedge Feather Diatom Brown Algae
SPECIES *robusta*
GENUS *Cometes* Comet Feather Diatom Brown Algae
SPECIES *globosa*
GENUS *Corona* Crown Feather Diatom Brown Algae
SPECIES *temperei, tetragona*
GENUS *Coxia* Cox Feather Diatom Brown Algae
SPECIES *guillauminii*
GENUS *Coxliella* Little Cox Feather Diatom Brown Algae
SPECIES *ampla, helix, pseudannulata*
GENUS *Craspedopleura* Pleural Fringed Feather Diatom Brown Algae
SPECIES *kryophila*
GENUS *Craspedoporus* Fringed Feather Diatom Brown Algae
SPECIES *briggeri, johnsonianus, ralfsianus, reticulatus*
GENUS *Craspedostauros* Star Fringed Feather Diatom Brown Algae
SPECIES *amphoroides, australis, britannicus, capensis, decipiens, indubitabilis, neoconstrictus*
GENUS *Creswellia* Creswell Feather Diatom Brown Algae
SPECIES *barbadensis, ferox, palmeriana, superba*
GENUS *Crossophialus* Saucer Fringed Feather Diatom Brown Algae
SPECIES *glabrus, gyroscolus, paterus*
GENUS *Crucidenticula* Tortured Tooth Feather Diatom Brown Algae
SPECIES *ikebei, kanayae, nicobarica, paranicobarica, punctata*
GENUS *Cussia* Cuss Feather Diatom Brown Algae
SPECIES *lancettula, moholensis*
GENUS *Cyclophora* Cyclophore Feather Diatom Brown Algae
SPECIES *tenuis*
GENUS *Cylindrospira* Cylindric Spire Feather Diatom Brown Algae
SPECIES *simsiae*
VARIETIES *simsiae homannii*
GENUS *Cylindrotheca* Cylindric Case Feather Diatom Brown Algae
SPECIES *closterium, fusiformis, gracilis*
GENUS *Cymbellopsis* Little Cyme Feather Diatom Brown Algae
SPECIES *apiculata, guyana, krammeri, lanceolata, metzeltinii, mirabilis, neocaledonica, pumila, terminis*
GENUS *Cymbopleura* Pleural Cyme Feather Diatom Brown Algae
SPECIES *aequistriata, crassinervia, declivis, delicatula, fluminea, heinii, kanakarum, madagascariensis, michelcostei, neocaledonica, semigibbosa*
VARIETIES *delicatula judaica*
GENUS *Cypellachaetes* Goblet Bristle Feather Diatom Brown Algae
SPECIES *intricatus*
GENUS *Dasyangea* Daisy Feather Diatom Brown Algae
SPECIES *dactylethra*
GENUS *Decussata* Wicker Feather Diatom Brown Algae
SPECIES *hexagona, placenta*
VARIETIES *placenta obtusa*
GENUS *Denticella* Small Toothed Feather Diatom Brown Algae
SPECIES *tridens*
GENUS *Denticula* Little Toothed Feather Diatom Brown Algae
SPECIES *crassula, creticola, dimorpha, elegans, erifuga, eximia, frigida, hyalina, inflata, interrupta, kuetzingii, laevis, lauta, miocenica, nicobarica, norwegica, ocellata, ornata, parva, praedimorpha, punctata, rainerensis, sinuata, subtilis, tabellaria, tenuis, thermalis, valida, watanabei*

VARIETIES *elegans elegans, elegans constricta, elegans ocellata, elegans valida, kuetzingii rumrichae, punctata hustedtii, tenuis tenuis, tenuis crassula, tenuis frigida*
GENUS *Denticulopsis* Lesser Little Toothed Feather Diatom Brown Algae
SPECIES *hustedtii, meridionalis, praedimorpha, praehyalina, praelauta*
VARIETIES *hustedtii aspera*
GENUS *Dickensoniaforma* Dickenson Feather Diatom Brown Algae
SPECIES *arctica, navicularia*
GENUS *Dickieia* Dickie Feather Diatom Brown Algae
SPECIES *dansei, excepta, pinnata, resistans, soodensis, subinflata, wajdae*
GENUS *Dicladia* Dual Clad Feather Diatom Brown Algae
SPECIES *elliptica, norwegica*
GENUS *Diomphala* Dual Navel Feather Diatom Brown Algae
SPECIES *clavaherculis*
GENUS *Diprora* Dual Prow Feather Diatom Brown Algae
SPECIES *haenaensis*
GENUS *Drewsandria* Drewsander Feather Diatom Brown Algae
SPECIES *nodulifera*
GENUS *Ectodictyon* External Net Feather Diatom Brown Algae
SPECIES *varians*
GENUS *Ehrenbergia* Ehrenberg Feather Diatom Brown Algae
SPECIES *granulosa, hauckii*
GENUS *Entopyla* Entic Pyle Feather Diatom Brown Algae
SPECIES *australis*
GENUS *Eolimna* Marsh Feather Diatom Brown Algae
SPECIES *adnata, inconspicua, martinii, minima, neocaledonica, rhombelliptica, rhombica, subadnata, subminuscula, thioense, verecundaeformis, verecundaeformis, zalokariae*
GENUS *Epiphalaina* Perching Moth Feather Diatom Brown Algae
SPECIES *aleutica, radiata*
VARIETIES *aleutica lineata*
GENUS *Eunophora* Eunophore Feather Diatom Brown Algae
SPECIES *indistincta, oberonica, tasmanica*
GENUS *Eurossia* Ross Feather Diatom Brown Algae
SPECIES *irregularis*
VARIETIES *irregularis incurvatus*
GENUS *Exilaria* Meager Feather Diatom Brown Algae
SPECIES *curvata, fasciculata, vaucheriae*
VARIETIES *fasciculata minor*
GENUS *Florella* Foral Feather Diatom Brown Algae
SPECIES *portoricensis*
GENUS *Flosculosphaera* Flos Sphere Feather Diatom Brown Algae
SPECIES calceolariopsis, sacculus
GENUS *Fogedia* Foged Feather Diatom Brown Algae
SPECIES *acuta, christensenii, finmarchia, finmarchica, geisslerae, giffeniana, heterovalvata*
VARIETIES *giffeniana yemenitica*
GENUS *Fontigonium* Fontigon Feather Diatom Brown Algae
SPECIES *rectangulare*
GENUS *Fromentella* Froment Feather Diatom Brown Algae
SPECIES *granulosa*
GENUS *Frustalia* Frustule Feather Diatom Brown Algae
SPECIES *krammeri*
GENUS *Gephyria* Gephyre Feather Diatiom Algae
SPECIES *media, telfairiae*
GENUS *Giffenia* Giffen Feather Diatom Brown Algae
SPECIES *cocconeiformis*
GENUS *Giraudyopsis* Giraudy Feather Diatom Brown Algae
SPECIES *stellifera*
GENUS *Gladiolithus* Rock Sword Feather Diatom Brown Algae
SPECIES *striatus*
GENUS *Gladiopsis* Sword Feather Diatom Brown Algae
SPECIES *antarctica, antarctica, ellipsoidea, jouseana, lagenoides, maxima, modica, pacifica, pacifica, speciosa*
GENUS *Goniothecium* Gonad Case Feather Diatom Brown Algae
SPECIES *coronatum, flexuosum, loricatum, prolongatum, tenue*
GENUS *Groentvedia* Groentved Feather Diatom Brown Algae
SPECIES *Groentvedia elliptica*
GENUS *Halurina* Halurine Feather Diatom Brown Algae
SPECIES *Halurina itaparicana*

GENUS *Hamatusia* Hooked Feather Diatom Brown Algae
SPECIES *Hamatusia reediana*
GENUS *Hantzschia* Hantzsch Feather Diatom Brown Algae
SPECIES *abundans, achroma, alkaliphila, amphilepta, amphioxys, anguis, angusta, barckhausenii, bardii, brunneri, compacta, compactoides, considerata, distinctepunctata, doigiana, dorgaliensis, elegantula, elongata, eximia, fenestrata, fusiformis, graciosa, hyalina, hyperborea, kaiseri, longa, longiareolata, marina, mossae, nitzschioides, obtusa, pannonica, paracompacta, psammicola, pseudomarina, rhaetica, rigida, sardiniensis, spectabilis, subrupestris, subvivacior, virgata, vivacior, vivax, weyprechtii, wittii, xerophila*
GENUS *Henshawia* Henshaw Feather Diatom Brown Algae
SPECIES *Henshawia biddulphioides*
GENUS *Himantidium* Hemantic Feather Diatom Brown Algae
SPECIES *denticulatum, exiguum, faba, gracile, guianense, majus, marinum, minus, papilio, polydentula, strictum, veneris*
VARIETIES *majus bidens*
GENUS *Homoeocladia* Homoclad Feather Diatom Brown Algae
SPECIES *conferta, elongata, filiformis, germanica, martiana, sigmoidea, subcohaerens*
VARIETIES *subcohaerens chinensis, subcohaerens scotica*
GENUS *Hustedtia* Greater Hustedt Feather Diatom Brown Algae
SPECIES *mirabilis*
GENUS *Hustedtiella* Lesser Hustedt Feather Diatom Brown Algae
SPECIES *baltica, sinuata*
GENUS *Huttonia* Hutton Feather Diatom Brown Algae
SPECIES *norwegica*
GENUS *Hyalinella* Transparent Feather Diatom Brown Algae
SPECIES *lateripunctata*
GENUS *Hyalotrochus* Transparent Wheel Feather Diatom Brown Algae
SPECIES *incompositus, radiatus*
GENUS *Hygropetra* Wet Rock Feather Diatom Brown Algae
SPECIES *balfouriana, elongata*
GENUS *Ikebea* Ikebe Feather Diatom Brown Algae
SPECIES *amphistriata, bifurcata, clavata, coronata, robusta, tenuis*
GENUS *Inoderma* Inner Dermal Feather Diatom Brown Algae
SPECIES *lamellosum*
GENUS *Katahiraia* Katahir Feather Diatom Brown Algae
SPECIES *aspera, oblonga, pauperata*
GENUS *Kathablepharis* Kathar Feather Diatom Brown Algae
SPECIES *remigera*
GENUS *Kerkis* Kerk Feather Diatom Brown Algae
SPECIES *bispinosa*
GENUS *Kidoa* Kido Feather Diatom Brown Algae
SPECIES *graviarmata*
GENUS *Kisseleviella* Kisselev Feather Diatom Brown Algae
SPECIES *ezoensis, magnaareolata*
GENUS *Koizumia* Koizum Feather Diatom Brown Algae
SPECIES *adaroi, akibae, tatsunokuchiensis*
GENUS *Kozloviella* Kozlov Feather Diatom Brown Algae
SPECIES *subrotunda*
GENUS *Kreagra* Kreger Feather Diatom Brown Algae
SPECIES *bifalcata, forfex, tripartita*
GENUS *Kurpiszia* Kurpisz Feather Diatom Brown Algae
SPECIES *kryokonites, semiaperta, subprotracta*
GENUS *Lacunicula* Lake Feather Diatom Brown Algae
SPECIES *bella, sardiniensis*
GENUS *Lecohuia* Lecohu Feather Diatom Brown Algae
SPECIES *corrugata, geniculata, martinii, riverae*
GENUS *Liparogyra* Glossy Gyre Feather Diatom Brown Algae
SPECIES *circularis, dendroteres, spiralis*
GENUS *Liriogramma* Lily Grass Feather Diatom Brown Algae
SPECIES *petterssonii*
GENUS *Lisitzinia* Lisitzin Feather Diatom Brown Algae
SPECIES *ornata*
GENUS *Lunella* Moon Feather Diatom Brown Algae
SPECIES *bisecta, ghalebii*
GENUS *Lyngbyea* Lyngby Feather Diatom Brown Algae
SPECIES *cuneata*
GENUS *Lysigonium* Loose Gonad Feather Diatom Brown Algae
SPECIES *granulata*

GENUS *Macrora* Large Tail Feather Diatom Brown Algae
SPECIES *barbadensis, najae*
GENUS *Matteia* Matte Feather Diatom Brown Algae
SPECIES *conchicola*
GENUS *Mayamaea* Maya Feather Diatom Brown Algae
SPECIES *agrestis, aliena, arida, asellus, atomus, crassistriata, destrica, destricta, elongata, excelsa, fossalis, fossaloides, lacunolaciniata, mediterranea, muraliformis, nolensoides, pseudopermitis, recondita*
GENUS *Mediaria* Media Feather Diatom Brown Algae
SPECIES *magna, splendida*
GENUS *Meristosolen* Meristo Solitary Feather Diatom Brown Algae
SPECIES *cylindratus*
GENUS *Microorbis* Tiny Orb Feather Diatom Brown Algae
SPECIES *convexus, undulatus*
GENUS *Microrhabdulinus* Tiny Ribbon Feather Diatom Brown Algae
SPECIES *antihelicoideus, ceratorhabdulinus, helicoidites*
GENUS *Monobrachia* Monad Branched Feather Diatom Brown Algae
SPECIES *simplex, unicornuta*
GENUS *Myxochrysis* Slime Feather Diatom Brown Algae
SPECIES *paradoxa*
GENUS *Navigiolum* Navel Feather Diatom Brown Algae
SPECIES *canoris, exiliforme, sardiniense, spineum, spinosissimum*
GENUS *Neodenticula* Modern Toothed Feather Diatom Brown Algae
SPECIES *kamtschatica, koizumii, seminae*
GENUS *Odontidium* Toothed Feather Diatom Brown Algae
SPECIES *elegans, hyemale, hyemale, hyemale, mesodon, mesodon, mutabile, sinuatum, tabellaria*
GENUS *Okedenia* Okeden Feather Diatom Brown Algae
SPECIES *inflexa*
GENUS *Paulinella* Paulin Feather Diatom Brown Algae
SPECIES *chromatophora, ovalis, taeniata*
GENUS *Peltomonas* Pelt Monad Feather Diatom Brown Algae
SPECIES *volitans*
GENUS *Perrya* Perry Feather Diatom Brown Algae
SPECIES *pulcherrima*
GENUS *Phaeodactylum* Dusky Finger Feather Diatom Brown Algae
SPECIES *tricornutum*
GENUS *Picophagus* Picophage Feather Diatom Brown Algae
SPECIES *flagellatus*
GENUS *Pinnulariosigma* Pinnate Sigma Feather Diatom Brown Algae
SPECIES *raeanum*
GENUS *Pinnunavis* Pinnate Naval Feather Diatom Brown Algae
SPECIES *armoricana, elegans, genustriata, yarrensis*
GENUS *Pleurocyclos* Pleural Cyclic Feather Diatom Brown Algae
SPECIES *stechlinensis*
GENUS *Pleurostaurum* Pleural Star Feather Diatom Brown Algae
SPECIES *javanicum*
GENUS *Pleurostaurum* Pleural Cross Feather Diatom Brown Algae
SPECIES *acutum, janischii, legumen, lineare, obtusum, parvulum, sagitta, scaphulaeforme, smithii, spicula, tumidulum*
GENUS *Podochrysis* Pod Golden Feather Diatom Brown Algae
SPECIES *roscoffensis*
GENUS *Podosphenia* Pod Wedge Feather Diatom Brown Algae
SPECIES *communis, communis, cuneata, debilis, hyalina, juergensii, ovata, paradoxa, tenuis*
GENUS *Poretzkia* Poretzk Feather Diatom Brown Algae
SPECIES *circularis, mirabilis, ovalis*
GENUS *Porocyclia* Pore Cyclic Feather Diatom Brown Algae
SPECIES *dendrophila*
GENUS *Pravifusus* Pravifuse Feather Diatom Brown Algae
SPECIES *hyalinus*
GENUS *Protoraphis* Needle Feather Diatom Brown Algae
SPECIES *hustedtii*
GENUS *Psammodictyon* Sand Net Feather Diatom Brown Algae
SPECIES *areolatum, bisculptum, bombiforme, constrictum, corpulentum, ferox, inductum, mediterraneum, molle, panduriforme, roridum, rudum*
GENUS *Pseudodimerogramma* Lesser Dimer Grass Feather Diatom Brown Algae
SPECIES *elegans, ellipticum, elongatum, filiforme, oligocenicum*
GENUS *Pseudoguinardia* Lesser Guinard Feather Diatom Brown Algae
SPECIES *recta*

GENUS *Pseudohimantidium* Lesser Hamantic Feather Diatom Brown Algae
SPECIES *pacificum*
GENUS *Pseudoleyanella* Lesser Leyan Feather Diatom Brown Algae
SPECIES *lunata*
GENUS *Rattrayella* Rattray Feather Diatom Brown Algae
SPECIES *californica*
GENUS *Rebecca* Rebecca Feather Diatom Brown Algae
SPECIES *helicata*
GENUS *Rhabdium* Rod Feather Diatom Brown Algae
SPECIES *cuneatum*
GENUS *Rhamnochrysis* Buckthorn Gold Feather Diatom Brown Algae
SPECIES *aestuarinae*
GENUS *Rhinozeta* Snout Feather Diatom Brown Algae
SPECIES *addoensis, caecalis, cristata, multiplata, rhinozeta, triciliata, unilaminata*
GENUS *Rhipidophora* Needle Bearing Feather Diatom Brown Algae
SPECIES *anglica, dalmatica, nubecula, oedipus, paradoxa, paradoxum*
GENUS *Rosaria* Rosary Feather Diatom Brown Algae
SPECIES *ramosa*
GENUS *Rossiella* Lesser Ross Feather Diatom Brown Algae
SPECIES *elongata, fennerae, fourtanierae, gombosii, moholensis, symmetrica*
GENUS *Rouxia* Roux Feather Diatom Brown Algae
SPECIES *californica, diploneides, grandis, naviculoides, obesa*
GENUS *Rutilariopsis* Rutile Feather Diatom Brown Algae
SPECIES *ventricosa*
GENUS *Sawamuraia* Sawamur Feather Diatom Brown Algae
SPECIES *biseriata, multibullata, quadriseriata*
GENUS *Schizostauron* Split Wicker Feather Diatom Brown Algae
SPECIES *andicola*
GENUS *Sieminskia* Sieminsky Feather Diatom Brown Algae
SPECIES *wohlenbergii, zeta*
GENUS *Sigmatella* Sigma Feather Diatom Brown Algae
SPECIES *nitzschii*
GENUS *Simonsenia* Greater Simonsen Feather Diatom Brown Algae
SPECIES *delogneii*
GENUS *Simonseniella* Lesser Simonsen Feather Diatom Brown Algae
SPECIES *barboi, praebarboi*
GENUS *Smithiella* Smith Feather Diatom Brown Algae
SPECIES *marina*
GENUS *Sphaeraspis* Round Asp Feather Diatom Brown Algae
SPECIES *pascheri*
GENUS *Sphenella* Wedge Feather Diatom Brown Algae
SPECIES *angustata, obtusata, parvula*
GENUS *Sphinctocystis* Sphincter Cyst Feather Diatom Brown Algae
SPECIES *elliptica, librilis*
GENUS *Stauronella* Wicker Feather Diatom Brown Algae
SPECIES *amphoroides, arctica, decipiens, indubitabilis*
GENUS *Stephanogonia* Stephan Gonad Feather Diatom Brown Algae
SPECIES *horrida, mutabilis*
GENUS *Stoermeria* Stoermer Feather Diatom Brown Algae
SPECIES *californica, trifoliata*
GENUS *Stomatochone* Mouth Chone Feather Diatom Brown Algae
SPECIES *infundibuliformis*
GENUS *Stoschia* Greater Stosch Feather Diatom Brown Algae
SPECIES *mirabilis, paleacea*
GENUS *Stoschiella* Lesser Stosch Feather Diatom Brown Algae
SPECIES *hebetata*
GENUS *Styllaria* Styllar Feather Diatom Brown Algae
SPECIES *cuneata*
GENUS *Stylobiblium* Biblical Style Feather Diatom Brown Algae
SPECIES *excentricum*
GENUS *Subathua* Lesser Athuan Feather Diatom Brown Algae
SPECIES *balcanica, sahnii, spinosa*
GENUS *Sulcochrysis* Furrowed Golden Feather Diatom Brown Algae
SPECIES *biplastida*
GENUS *Symbolophora* Symbol Feather Diatom Brown Algae
SPECIES *amblyoceros*
GENUS *Syringidium* Syringe Feather Diatom Brown Algae

SPECIES *bicorne*
GENUS *Syringogyra* — Syringe Gyre Feather Diatom Brown Algae
SPECIES *viridis*
GENUS *Systephania* — Systematic Stephan Feather Diatom Brown Algae
SPECIES *corona*
GENUS *Tenuocharta* — Tenuous Chart Feather Diatom Brown Algae
SPECIES *cloudii*
GENUS *Tertiarius* — Tertiary Feather Diatom Brown Algae
SPECIES *pygmaeus, transylvanicus*
GENUS *Thumia* — Thume Feather Diatom Brown Algae
SPECIES *elegans*
GENUS *Toxarium* — Toxic Feather Diatom Brown Algae
SPECIES *hennedyanum, undulatum*
GENUS *Tripterion* — Three Wing Feather Diatom Brown Algae
SPECIES *kalamense, philoderma*
GENUS *Trochus* — Wheel Feather Diatom Brown Algae
SPECIES *elegantulus*
GENUS *Tubaformis* — Tube Feather Diatom Brown Algae
SPECIES *unicornis*
GENUS *Tubularia* — Tubule Feather Diatom Brown Algae
SPECIES *compressa, fragilis*
GENUS *Tursiocola* — Turse Feather Diatom Brown Algae
SPECIES *olympica, omurae, staurolineata*
GENUS *Vanheurckia* — Vanheurck Feather Diatom Brown Algae
SPECIES *cuspidata, cuspidata, hitchcockii, lewisiana, neglecta, rhomboides, viridula, vulgaris*
GENUS *Veigaludwigia* — Viega Ludwig Feather Diatom Brown Algae
SPECIES *urbana, willerii*
GENUS *Vikingea* — Viking Feather Diatom Brown Algae
SPECIES *gibbocalyx, promunturii*
GENUS *Zonosphaeridium* — Zoned Sphere Feather Diatom Brown Algae
SPECIES *densum, foveolatum, limpaticum, punctatum*
GENUS *Zygoceros* — Zygote Wax Feather Diatom Brown Algae
SPECIES *pelagicum*

FAMILY 2 *Neisaceae* — Neid Feather Diatom Brown Algae
GENUS *Cocconeiopsis* — Berry Nied Feather Diatom Brown Algae
SPECIES *breviata, fraudulenta, girondica, kantsinensis, orthoneoides, patrickae, pseudoorthoneoides, pullus, regularis, wrightii*
GENUS *Desikaneis* — Desik Neid Feather Diatom Brown Algae
SPECIES *gessneri, howellii, riesgoviensis, stauroneiformis*
GENUS *Dictyoneis* — Netted Neid Feather Diatom Brown Algae
SPECIES *hungarica, marginata*
GENUS *Lancineis* — Lance Neid Feather Diatom Brown Algae
SPECIES *clavata, fatula, fusus, lancettula, parilis, petropolitana, praeparilis*
GENUS *Lapideacassis* — Lapid Case Neid Feather Diatom Brown Algae
SPECIES *blackii, cornuta, multispinata, tricornus, trispina*
GENUS *Membraneis* — Membrane Neid Feather Diatom Brown Algae
SPECIES *challengeri, impostor*
GENUS *Microneis* — Tiny Neid Feather Diatom Brown Algae
SPECIES *trinodis*
GENUS *Nanoneis* — Dwarf Neid Feather Diatom Brown Algae
SPECIES *hasleae*
GENUS *Nephroneis* — Liver Neid Feather Diatom Brown Algae
SPECIES, *macintirei*
GENUS *Orthoneis* — Straight Neid Feather Diatom Brown Algae
SPECIES *binotata, clevei, wrightii*
GENUS *Pleuroneis* — Pleural Neid Feather Diatom Brown Algae
SPECIES *costata*
GENUS *Pogoneis* — Beard Neid Feather Diatom Brown Algae
SPECIES *bahrainii*
GENUS *Pseudodictyoneis* — Lesser Netted Neid Feather Diatom Brown Algae
SPECIES *hungarica*
GENUS *Scrinocassis* — Scrin Case Neid Feather Diatom Brown Algae
SPECIES *downiei*
GENUS *Tropidoneis* — Tropic Neid Feather Diatom Brown Algae
SPECIES *confusa, elegans, gibberula, lepidoptera, maxima, phantasma, primoris, pusilla, recta, semistriata, seriata, seychellensis, subulata, vanheurckii, vitrea*

FAMILY 3 *Mastogloiaceae* — Mast Glow Feather Diatom Brown Algae

GENUS *Mastogloia* Mast Glow Feather Diatom Brown Algae

SPECIES *angulata, angustata, apiculata, baldjikiana, baltica, binotata, biocellata, bisulcata, braunii, canalicula, closei, corsicana, corsicana, dansei, elliptica, erythraea, exigua, fallax, gallica, grevillei, halophila, ignorata, lacustris, lanceolata, manguinii, maxima, multiloculata, ovata, ovulum, portierana, pseudobalteata, pseudoexigua, pseudolacrimata, pseudolatecostata, pumila, pusilla, recta, rectirobusta, reimeri, smithii, splendida, stephensiana, streptoraphe, urveae, vasta*

VARIETIES *bisulcata corsicana, dansei elliptica, dansei streptoraphe, elliptica australis, elliptica dansei, elliptica oblonga, erythraea biocellata, grevillei subconstricta, pusilla subcapitata, smithii smithii, smithii amphicephala, smithii lacustris*

GENUS *Aneumastus* Aneue Mast Feather Diatom Brown Algae

SPECIES *albanicus, apiculatus, aratus, balticus, carolinianus, humboldtianus, laetus, maculosus, minor, rosettae, rostratus, stroesei, tuscula, tusculoides*

VARIETIES *stroesei stroesei, stroesei lanceolata, tuscula tuscula, tuscula angulata, tusculus capitata, tuscula minor, tuscula obtusa, tuscula rostrata, tuscula schinzii*

GENUS *Animikiea* Animikie Mast Feather Diatom Brown Algae

SPECIES *beltensis, crassiuscula, indica, punctata, septa*

FAMILY 4 *Nemabacillariaceae* Thread Feather Diatom Brown Algae

GENUS *Chrysonema* Golden Thread Feather Diatom Brown Algae

SPECIES *litoralis*

GENUS *Colletonema* Collet Thread Feather Diatom Brown Algae

SPECIES *bullosum, eximium, lucustre, subcohaerens, viridulum, vulgare*

GENUS *Gloeonema* Glow Thread Feather Diatom Brown Algae

SPECIES *paradoxum, triangulatum*

GENUS *Monema* Monad Thread Feather Diatom Brown Algae

SPECIES *prostratum*

GENUS *Rhabdonema* Striped Thread Feather Diatom Brown Algae

SPECIES *adriaticum, arcuatum, crassum, minutum, robustum*

VARIETIES *arcuatum robustum*

GENUS *Schizonema* Split Thread Feather Diatom Brown Algae

SPECIES *amplius, crucigera, ehrenbergii, eximium, kobbense, laciniatum, neglectum, obtusum, parvulum, rutilans, spicula, viridulum, vulgare*

FAMILY 5 *Nitzschiaceae* Nitzsch Feather Diatom Brown Algae

GENUS *Nitzschia* Nitzsch Feather Diatom Brown Algae

SPECIES *acicularioides, acicularis, acidoclinata, actinastroides, acula, adamata, adamatoides, adapta, admissa, aequora, aequorea, aerophiloides, affinis, agnewii, agnita, albicostalis, alexandrina, alpina, alpinobacillum, amphibia, amphibioides, amphicephala, amphioxoides, amplectens, andicola, angularis, angustata, angustatula, angustiforaminata, aniae, archibaldii, arcula, arenosa, armoricana, aurariae, aurica, auricula, austriaca, bacata, bacillariaeformis, bacilliformis, bacillum, barronii, bavarica, behrei, bergii, beyeri, bicapitata, bilobata, blankaartensis, bombiformis, braarudii, brebissonii, bremenensis, bremensis, brevior, brevirostris, brevissima, brunoi, bryophila, bulnheimiana, buschbeckii, calcicola, californica, capitata, capitellata, challengeri, chandolensis, chasei, chauhanii, chungara, circumsuta, clarissima, clausii, clementei, clementia, clevei, communis, commutata, commutatoides, compressa, concordia, confinis, congolensis, constricta, coutei, creticola, cursoria, curvilineata, curvula, dealpina, debilis, debliis, delauneyi, delognei, denticula, desertorum, diserta, disputata, dissipata, distans, diversa, draveillensis, dubia, dubiformis, dubiiformis, ebroicensis, eglei, elegantula, elegens, elongata, epiphytica, epiphyticoides, epithemoides, fasciculata, fasciculata, fibula-fissa, fibulafissa, filiformis, flexa, flexoides, fluminenesis, fluorescens, fogedii, fonticola, fonticoloides, fontifuga, fossilis, frequens, frigida, frustulum, fruticosa, gaertnerae, gandersheimiensis, geitleri, gessneri, gisela, gotlandica, graciliformis, gracilis, graciloides, guttula, habirshawii, hadriatica, hamburgiensis, hantzschiana, harderi, heidenii, heteropolica, heufleriana, hierosolymitana, holastica, hollerupensis, holsatica, homburgiensis, hustedtiana, hyalina, hybrida, impressa, improvisa, incognita, inconspicua, incrustans, incurva, indica, innominata, insecta, insignis, intermedia, interrupta, invicta, invisa, irregularis, johnmartinii, juba, jugata, kahlii, kanakarum, kanayae, kittlii, kolaczeckii, kuetzingiana, kuetzingii, kuetzingioides, labella, lacrima, lacunarum, lacuum, laevis, lagunensis, lamprocampa, lanceolata, latiuscula, lauenbergiana, leehyi, leistikowii, leucosigma, levidensis, liebetruthii, ligowskii, linearis, linearis, linkei, littorea, longissima, lorenziana, lucisensibilis, lunaris, lunata, luzonensis, macaronesica, machardyae, macilenta, mancoides, marina, martiana, maxima, media, mediocris, metzeltinii, microcephala, minuta, minutissima, minutula, modesta, moissacensis, monachorum, nana, natans, navicularis, navis-varingica, neotropica, nienhuisii, normannii, notabilis, novae-guineaensis, oberheimiana, obsidialis, obtusa, ovalis, pacifica, palacea, palea, paleacea, paleaeformis, paleoides, palustris, pararostrata, peisonis, pelagica, pellucida, perminuta, permitis, perpusilla, perspicua, petitiana, philippinarum, pilum, plana, planctonica, plioveterana, polaris, praecurta, praefossilis, praereinholdii, prolongata, pseudoatomus, pseudocapitata, pseudocommunis, pseudocylindrica, pseudofonticola, pseudohybrida, pseudosigma, pumila, punctata, pura, pusilla, quadrangula, radicula, recta, rectiformis, rectilonga, rectirobusta, rectissima, reimeri, reimersenii, retusa, rigida, rolandii, romana, rosenstockii, rostellata, rostrata, salinarum, salinicola, scalaris, scalpelliformis, schwabei, schweikertii, scutellum, sellingii, semirobusta, serpentiraphe, sicula, sigma, sigmatella, sigmoidea, silica, siliqua, similis, sinuata, smithii, sociabilis, socialis, solgensis, solita, spathulata, speciosa, spectabilis, sphaerophora, spiculoides, spiculum, spinarum, stompsii, strelnikovae, stricta, subaccommodata, subacicularis, subamphioxoides, subapiculata, subcapitellata, subcohaerens, subconstricta, subfrustulum, subgraciloides, subinvicta, sublaevis, sublica, sublinearis, sublongirostris, subromana, subrostrata, subrostratoides, subrostroides, subtilioides, subtilis, subtubicola, suchlandtii, supralitorea, tabellaria, taeniiformis, tantata, tarda, temperei, tenerifa, tenuis, tenuissima, tergestina, terrestris, thermalis, thermaloides, tropica, tryblionella, tsarenkoi, tubicola, umaoiensis, umbonata, vacua, valdecostata, valdestriata, valida, vasta, vermicularis, vermicularoides, vexans, vildaryana, villarealii, vitrea, vivax, vixnegligenda, vulga, weaveri, witkowskii, wodensis, woltereckii, woltereckoides, wuellerstorfii, yunchengensis*

VARIETIES *acicularis closterioides, amphibia abbreviata, amphibia acutiuscula, amphibia amphibia, amphibia fossilis, amphibia frauenfeldii, amphibia intermedia, amphibia pelagica, amphibia rostrata, amphibia umbrosa, angularis affinis, angustata acuta, angustata minuta, archibaldii crassistriata, bacata linearis, bicapitata faeroensis, bilobata hybrida, bilobata minor, bremensis brunsvigensis, communis communis, communis abbreviata, communis hyalina, compressa minor, compressa pararostrata, compressa vexans, constricta parva, curvula subcapitata, denticula delognei, dissipata acula, dissipata borneensis, dissipata media, dissipata oligotraphenta, dissipata undulata, distans subsigmoidea, epithemoides disputata, filiformis conferta, fonticola capitata, fonticola pelagica, frustulum bryophila, frustulum bulnheimiana, frustulum minutula, frustulum perminuta, frustulum perpusilla, frustulum subsalina, frustulum tenella, gandersheimiensis tenuirostris, gracilis acicularoides, hantzschiana subserians, incurva incurva, incurva lorenziana, incurva subtilis, insignis insignis, insignis notabilis, insignis smithii, invicta lanceolata, jugata patrickiae, kuetzingiana exilis, lanceolata lanceolata, lanceolata incrustans, lanceolata minor, levidensis salinarum, liebetruthii major, linearis subtilis, linearis tenuis, longissima closterium, longissima parva, lorenziana incerta, lorenziana incurva, lorenziana major, lorenziana muscicola, lorenziana subtilis, moissacensis heidenii, nana ignorata, nana rousselinii, obtusa brevissima, obtusa nana, obtusa scalpelliformis, obtusa schweinfurthii, palea barodensis, palea debilis, palea fonticola, palea minuta, palea palea, palea sumatrana, palea tenuirostris, plana fennica, punctata apiculata, punctata coarctata, punctata constricta, punctata curta, pungens multiseries, pusilla monoensis, recta robusta, sicula bicuneata, sicula migrans, sicula rostrata, sigma amoricana, sigma amphioxys, sigma brevior, sigma diminuta, sigma fasciculata, sigma fonticola, sigma intercedens, sigma rigida, sigma rigidula, sigma sigmatella, sigmoidea armoricana, sinuata sinuata, sinuata delognei, sinuata tabellaria, socialis baltica, socialis kariana, spathulata hyalina, subtilis palacea, tenuis parva, thermalis capitata, thermalis intermedia, thermalis minor, tryblionella levidensis, tryblionella maxima, tryblionella salinarum, tryblionella subsalina, tryblionella victoriae, valida longissima, vermicularis minuta, vermicularis vialis, vitrea major, vitrea salinarum, vitrea subvitrea, vivax elongata, vivax minor*

GENUS *Cymatonitzschia* — Cyme Nitzsch Feather Diatom Brown Algae
SPECIES *marina*

GENUS *Cymbellonitzschia* — Boat Nitzsch Feather Diatom Brown Algae
SPECIES *diluviana, hossamedinii, manguini, minima, szulczewskii*

GENUS *Gomphonitzschia* — Bolt Nitzsch Feather Diatom Brown Algae
SPECIES *ungeriana*

GENUS *Pseudo-nitzschia* — Lesser Nitzsch Feather Diatom Brown Algae
SPECIES *americana, australis, brasiliana, cuspidata, delicatissima, fossilis, fraudulenta, granii, heimii, inflatula, linea, lineola, multiseries, prolongatoides, pseudodelicatissima, pungens, pungiformis, seriata, subcurvata, subfraudulenta, subpacifica, turgidula, turgiduloides*
VARIETIES *granii curvata, seriata obtusa*

GENUS *Tryblionella* — Shell Nitzsch Feather Diatom Brown Algae
SPECIES *acuminata, acuta, adducta, aerophila, angustata, ardua, balatonis, bathurstensis, bicuneata, brightwellii, calida, campechiana, chutteri, circumsuta, coarctata, cocconeiformis, conformata, cruciata, davidsonii, debilis, debilis, didyma, divergens, graeffii, granulata, halteriformis, hungarica, jelineckii, lanceola, limicola, littoralis, magnacarina, marginulata, maxima, nicobarica, obscurepunctata, partita, perversa, plana, plicatula, ponciensis, princeps, pseudohungarica, rabenhorstii, salvadoriana, scutellum, sibula, sulcata, umbilicata, victoriae, visurgis, zebuana*
VARIETIES *acuminata subconstricta, calida salinarum, chutteri aestuarii, hungarica linearis, littoralis slesvicensis, littoralis tergestina, marginulata didyma, nicobarica continua, nicobarica minor, nicobarica panduriformis*

FAMILY 6 *Pyxillaceae* — Pyxie Feather Diatom Brown Algae

GENUS *Pyxilla* — Pyxie Feather Diatom Brown Algae
SPECIES *johnsoniana*

GENUS *Cylindropyxis* — Cylindric Pyxie Feather Diatom Brown Algae
SPECIES *profunda, tremulans*

GENUS *Gladius* — Sword Pyxie Feather Diatom Brown Algae
SPECIES *antarcticus, antiquus, jouseanus, maximus, pacificus, perfectus, rhizophorus, speciosus, trispinosus*
VARIETIES *antarcticus altus, antiquus nodosus, antiquus tenuis, pacificus minor, speciosus aculeolatus, speciosus ellipsoideus, speciosus poratus, trispinosus tenuis*

GENUS *Mastogonia* — Mast Pyxie Feather Diatom Brown Algae
SPECIES *crux*

GENUS *Pseudopyxilla* — Lesser Pyxie Feather Diatom Brown Algae
SPECIES *furcata, sistena*

GENUS *Pyxidicula* — Little Pyxie Feather Diatom Brown Algae
SPECIES *aculeata, cruciata, mediterranea*

GENUS *Rhynchopyxis* — Beaked Pyxie Feather Diatom Brown Algae
SPECIES *siphonoides*

GENUS *Xanthiopyxis* — Yellow Pyxie Feather Diatom Brown Algae
SPECIES *leguminiformis, ornata, structuralis*

FAMILY 7 *Rhophalodiaceae* — Club Feather Diatom Brown Algae

GENUS *Rhopalodia* — Club Feather Diatom Brown Algae
SPECIES *acuminata, argus, brebissonii, constricta, contorta, dubia, gibba, gibberula, globosa, iriomotensis, musculus, novae-zelandiae, operculata, pacifica, parallela,rumrichiae, rupestris, sculpta, sterrenburgii, tholulata*

VARIETIES *acuminata protracta, gibba minuta, gibba parallela, gibba ventricosa, gibberula globosa, gibberula minuens, gibberula minuta, gibberula perpusilla, gibberula producta, gibberula protracta, gibberula rupestris, gibberula vanheurckii, musculus constricta, musculus mirabilis, operculata constricta, parallela distorta*

GENUS *Cystopleura* — Pleural Cyst Club Feather Diatom Brown Algae

SPECIES *turgida, zebra*

VARIETIES *turgida turgida, turgida capitata, turgida granulata, turgida plicata, turgida vertagus, turgida westermannii, turgida zebrina, zebra intermedia, zebra porcellus, zebra saxonica*

GENUS *Epithemia* — Epithemian Club Feather Diatom Brown Algae

SPECIES *adnata, alpestris, andrewsii, argus, cistula, constricta, emarginata, frickei, gibba, goeppertiana, gracilis, hyndmanni, illuminata, intermedia, krammeri, miranda, muelleri, ocellata, porcellus, proboscidea, reichelti, reticulata, rupestris, saxonica, smithii, sorex, westermannii, zebra*

VARIETIES *adnata adnata, adnata intermedia, adnata minor, adnata porcellus, adnata proboscidea, adnata saxonica, argus argus, argus alpestris, argus amphicephala, argus longicornis, argus protracta, argus testudo, gibba genuina, gibba parallela, sorex gracilis*

GENUS *Protokeelia* — Keel Club Feather Diatom Brown Algae

SPECIES *aculeata, bassonii, cholnokyi, hottingeri, quinquelobata, spinifera*

GENUS *Yoshidaia* — Yoshida Club Feather Diatom Brown Algae

SPECIES *constricta, densicostata, divergens, loculata, purpurifera*

FAMILY 8 *Siraceae* — Sirus Feather Diatom Brown Algae

GENUS *Bacterosira* — Rod Sirus Feather Diatom Brown Algae

SPECIES *bathyomphala*

GENUS *Brevisira* — Short Sirus Feather Diatom Brown Algae

SPECIES *arentii*

GENUS *Hyalosira* — Transparent Sirus Feather Diatom Brown Algae

SPECIES *aberrans, delicatula, hustedtiana, interrupta, minutissima, obtusangula, parietina, rectangula, tropicalis*

GENUS *Miosira* — Pupil Sirus Feather Diatom Brown Algae

SPECIES *rhoenana*

GENUS *Pseudoaulacosira* — Furrowed Sirus Feather Diatom Brown Algae

SPECIES *moisseeviae*

GENUS *Pseudopodosira* — Pod Sirus Feather Diatom Brown Algae

SPECIES *areolata, aspera, kosugii, modesta, pileiformis, reticulata, simplex, westii*

GENUS *Skeletonemopsis* — Skeleton Sirus Feather Diatom Brown Algae

SPECIES *barbadensis, crawfordii, mahoodii, morenoensis, rugosa, simbirskiana*

GENUS *Stephanosira* — Stephan Sirus Feather Diatom Brown Algae

SPECIES *epidendron, hamadryas*

GENUS *Trochosira* — Wheel Sirus Feather Diatom Brown Algae

SPECIES *coronata*

FAMILY 9 *Striatellaceae* — Striped Feather Diatom Brown Algae

GENUS *Striatella* — Striped Feather Diatom Brown Algae

SPECIES *aberrans, delicatula, fenestrata, hustedtiana, interrupta, unipunctata*

GENUS *Grammatophora* — Grass Striped Feather Diatom Brown Algae

SPECIES *angulosa, balfouriana, gibberula, hamulifera, macilenta, marina, oceanica, serpentina*

VARIETIES *oceanica macilenta, oceanica subtilissima*

GENUS *Microtabella* — Tab Striped Feather Diatom Brown Algae

SPECIES *delicatula, interrupta*

ORDER 2 *Achnanthales* Achnanth Feather Diatom Brown Algae

FAMILY 1 *Achnanthaceae* — Achnanth Feather Diatom Brown Algae

GENUS *Achnanthes* — Achnanth Feather Diatom Brown Algae

SPECIES *abundans, adnata, altaica, alteragracillima, amoena, amphicephala, angustata, arcus, aretasii, atacamae, atomus, aubertii, austriaca, bahusiensis, baicalensis, baltalensis, bergiani, biasolettiana, bicapitata, bijuga, bioretii, bistriata, blancheana, borealis, borneolanceolata, boudoui, bremeyeri, brevicostata, brevipes, buccola, calcar, carissima, carmichaelii, cataractarum, catenata, ceramii, childanos, clevei, coarctata, cognita, conspicua, convergens, costata, cotteriensis, crassa, crenulata, curtissima, danica, daonensis, delicatula, depressa, didyma, dispar, divergens, dornii, duthiei, elegans, exigua, exilioides, exilis, expressa, fimbriata, flexella, fluviatilis, fogedii, fontana, frequentissima, frigida, fugei, germainii, gracillima, grana, gregoriana, grishunum, groenlandica, hallandica, hasta, hauckii, helvetica, herteri, hettensis, hintzii, holsatica, hudsonis, hungarica, hyperboreoides, imperfecta, impexa, incognita, inflata, ingratiformis, interrupta, investians, japonica, johncarteri, joursacense, kolbei, koshovii, kranzii, kriegeri, kryophila, kuelbsii, kuwaitensis, lacunarum, lacus-baicali, lacus-vulcani, laevis, lanceolata, lapidosa, laterostrata, latissima, lauenburgiana, lemmermannii, leonardii, levanderi, lilljeborgei, leariformis, linearis, linkei, lintonensis, longipes, lorenziana, lutheri, manguinii, manifera, marginulata, maxima, medioconvexa, mercurii, mercuryi, mesoconstricta, metaffinis, metakryophila, meyerii, microcephala, minima, minuscula, minuta, minutissima, modesta, modestiformis, mollis, montana, multiarticulata, neocaledonica, ninckei, nitidiformis, nodosa, nollii, obliqua, oblongella, obscura, oestrupii, okunoi, onegensis, ostenii, parexigua, parvula, pennaeformis, peragalli, percava, perfida, pericava, petersenii, pirogueana, ploenensis, polaris, profunda, pseudoaffinis, pseudobliqua, pseudobrevipes, pseudogroenlandica, pseudopunctulata, pseudosuchlandtii, pseudoswazi, punctata, punctulata, pusilla, quadripunctata, rechtensis, reidensis, renei, reversa, ricula, rosenstockii, rossii, rostellata, rostrata, rupestoides, rupestris, saccula, saxophila, scotica, semiaptera, seriata, siamexilis, siamlinearis, similis, skvortzowii, stauroneiformis, stauroneioides, stewartii, stolida, striata, subatomoides, subcrassa, subexigua, subinflata, submarina, subsaloides, subsalsa, suchlandtii, suspecta, taeniata, tenuistriata, therezienii, thermalis, trinodis, valpestris, ventralis, ventraloconfusa, vistulana, woodlawnensis, ziegleri*

VARIETIES *altaica minor, austriaca austriaca, austriaca minor, austriaca parallela, baltalensis ovalis, biasolettiana subatomus, biasolettiana thienemannii, borealis baicalensis, bremeyeri rostrata, brevipes brevipes, brevipes angustata, brevipes intermedia, calcar australis, calcar baicalensis, calcar pulcherrima, clevei clevei, clevei balcanica, clevei bottnica, clevei rostrata, coarctata sinaensis, conspicua conspicua, conspicua brevistriata, delicatula engelbrechtii, delicatula elliptica, delicatula hauckiana, delicatula robusta, delicatula rostrata, delicatula septentrionalis, delicatula subcapitata, dispar angulata, dispar fontellii, exigua angustirostrata, exigua baicalensis, exigua elliptica, exigua semiaperta, flexella alpestris, flexella arctica, helvetica alpina, helvetica minor, inflata elata, joursacense elongata, laevis aratasii, laevis austriaca, laevis diluviana, laevis fennica, laevis ninckei, laevis quadratarea, lanceolata abbreviata, lanceolata biporoma, lanceolata boyei, lanceolata capitata, lanceolata dubia, lanceolata elliptica, lanceolata elliptico-lanceolata, lanceolata faeroensis, lanceolata haynaldii, lanceolata inflatum, lanceolata lanceolatoides, lanceolata magna, lanceolata maxima, lanceolata minimum, lanceolata minor, lanceolata pura, lanceolata rhomboidealis, lanceolata robusta, lanceolata rostrata, lanceolata rusticum, lanceolata semipura, lanceolata ventricosa, lanceolata vulgaris, lapidosa appalachiana, lapidosa chiliensis, lapidosa lanceolata, lapidosa subrostrata, lemmermannii lineata, lemmermannii obtusa, linearis curta, linearis nipponica, linearis pusilla, longipes carmichaelii, minutissima minutissima, minutissima affinis, minutissima cryptocephala, minutissima gracillima, minutissima inconspicua, minutissima jackii, minutissima macrocephala, minutissima saprophila, minutissima scotica, oestrupii aperta, oestrupii lanceolata, oestrupii parvula, oestrupii pungens, ostenii mesogongyla, ostenii parva, peragalli fossilis, peragalli parvula, petersenii densestriata, petersenii kryophila, petersenii subhudsonis, pirogueana capitata, ploenensis gessneri, ploenensis woldstedtii, pseudoaffinis capitata, pusilla pusilla, pusilla procera, rosenstockii sinepuncta, striata rostrata, thermalis rumrichorum, ventralis crassa, ventraloconfusa simplex*

GENUS *Achnantheiopsis* — Least Achnanth Feather Diatom Brown Algae

SPECIES *delicatula, distincta, fragilarioides, frequentissima, lanceolata, peragalli*

FAMILY 2 *Achnanthidiaceae* — Lesser Achnanth Feather Diatom Brown Algae

GENUS *Achnanthidium* — Lesser Achnanth Feather Diatom Brown Algae

SPECIES *affine, alteragracillimum, archibaldianum, biporomum, brevipes, clevei, coarctatum, contrarea, convergens, deflexum, duthii, eutrophilum, exiguum, exilis, flexellum, hauckianum, hungaricum, indicatrix, japonicum, koghisense, kranzii, latecephalum, levanderi, macrocephalum, maximum, minutissimum, minutum, modestiforme, peridotiticum, pusillum, pyrenaicum, saprophila, semiapertum, straubianum, subatomus, trinodis, ventralis, woltereckii*

VARIETIES *brevipes intermedia, clevei rostratum, coarctatum coarctata, exiguum constrictum, exiguum heterovalvum, koghisense angustifasciatum, minutissimum gracillima*

GENUS *Karyevia* — Karyev Achnanth Feather Diatom Brown Algae

SPECIES *clevei,laterostrata*

VARIETIES *clevei rostrata*

GENUS *Kolbesia* — Kolbes Achnanth Feather Diatom Brown Algae

SPECIES *amoena, kolbei, ploenensis, suchlandtii*

VARIETIES *ploenensis gessneri*

GENUS *Lemnicola* — Marsh Achnanth Feather Diatom Brown Algae

SPECIES *hungarica*

GENUS *Planothidium* — Flat Achnanth Feather Diatom Brown Algae

SPECIES *apiculatum, aurerii, baicalense, biporomum, borneolanceolatum, calcar, chilense, daui, delicatulum, deperditum, diplopunctatum, dispar, distinctum, dubium, ellipticum, engelbrechtii, fonticolanceolatum, fragilarioides, frequentissimum, granum, hauckianum, haynaldii, heidenii, heteromorphum, holstii, infrequens, jacobii, jan-marcinii, joursacense, lanceolatoides, lanceolatum, lilljeborgei, linkei, magnificum, marinum, minutissimum, miotum, neglectum, oculatum, oestrupii, peragalloi, pericavum, polaris, pseudotanense, pungens, quarnerensis, rhombicuneatum, robustius, robustum, rostratum, salvadorianum, schwabei, septentrionalis, slupskii*

VARIETIES *frequentissimum magnum, frequentissimum minus, hauckianum rostratum, haynaldii elliptico-lanceolata, lanceolatum bimaculatum, lanceolatum genuinum, lanceolatum omissum, peragalloi fossilum, peragalloi parvulum*

GENUS *Psammothidium* — Sand Achnanth Feather Diatom Brown Algae

SPECIES *abundans, acidoclinatum, altaicum, bioretii, bristolicum, buccolum, chlidanos, confusiforme, daonense, didymum, frigidum, germainii, grischunum, helveticum, incognitum, investians, kuelbsii, lacus-vulcani, laurenbergianum, levanderi, manguinii, marginulatum, oblongellum, perpusillum, pseudoswazi, punctulatum, rechtensis, reversum, rosenstockii, rossii, sacculum, scoticum, stauroneioidum, strenzkei, subatomoides, ventralis*

VARIETIES *abundans rosenstockii, grischunum daonense*

GENUS *Rossithidium* — Ross Achnanth Feather Diatom Brown Algae

SPECIES *duthiei, linearis, petersennii, pusillum*

FAMILY 3 *Cocconeidaceae* — Berry Neid Feather Diatom Brown Algae

GENUS *Cocconeis* — Berry Neid Feather Diatom Brown Algae

SPECIES *amygdalina, archeri, biflexa, brevicostata, britannica, californica, capensis, caribensis, cataractarum, ceticola, clandestina, confusa, costata, cyclophora, danica, depressa, diaphana, dirupta, diruptoides, disculus, distans, diversa, egena, egregia, finnica, flexella, fluminensis, fluviatilis, granulifera, grata, guttata, hauniensis, irregularis, japonica, krammeri, lineolata, lyra, magnifica, maxima, minuta, molesta, multiperforata, neocaledonica, neodiminuta, neothumensis, nitida, norvegica, notata, ornata, pacifica, parthenopaea, pediculus, pellucida, pelta, peltoides, pinnata, placentula, placentuloides, pseudomarginata, pseudothumensis, quarnerensis, regalis, rhombus, rugosa, sancti-paulii, scutelliformis, scutellum, sedimentaria, skvortzowii, speciosa, stauroneiformis, sublittoralis, tenuis, theinemannii, therezienii, thwaitesii, transversalis, utermohlii, wrightii*

VARIETIES *amygdalina major, amygdalina minor, brevicostata obscura, californica ambigua, costata hexagona, costata pacifica, diaphana amygdalina, dirupta antarctica, dirupta flexella, dirupta parva, disculus minor, fluminensis subimpleta, molesta molesta, molesta amygdalina, molesta angusta, molesta crucifera, molesta major, molesta minor, neodiminuta aegagropilae, placentula placentula, placentula australica, placentula baicalensis, placentula euglypta, placentula intermedia, placentula klinoraphis, placentula lineata, placentula pseudolineata, placentula rouxii, placentula tenuistriata, scutellum scutellum, scutellum ampliata, scutellum distans, scutellum japonica, scutellum minutissima, scutellum ornata, scutellum parva, scutellum posidoniae, scutellum speciosa, scutellum stauroneiformis, thwaitesii arctica*

GENUS *Amphicocconeis* — Dual Berry Neid Feather Diatom Brown Algae

SPECIES *disculoides*

GENUS *Anorthoneis* — Straight Berry Neid Feather Diatom Brown Algae

SPECIES *dulcis, eurystoma, excentrica, hummii, hyalina*

GENUS *Bennettella* — Bennett Berry Neid Feather Diatom Brown Algae

SPECIES *ceticola*

GENUS *Campyloneis* — Field Berry Neid Feather Diatom Brown Algae

SPECIES *grevillei, regalis*

VARIETIES *regalis minuta*

GENUS *Campylonema* — Field Thread Neid Feather Diatom Brown Algae

SPECIES *indicum, lahorense*

GENUS *Canistrolithus* — Canister Rock Neid Feather Diatom Brown Algae

SPECIES *valliformis*

GENUS *Epipellis* — Perching Pell Berry Neid Feather Diatom Brown Algae

SPECIES *oiketis*

GENUS *Eucocconeis* — Good Berry Neid Feather Diatom Brown Algae

SPECIES *alpestris, aretasii, austriaca, depressa, diluviana, dorogostaisky, flexella, laevis, lapponica, leptostriata, minuta, ninckei, onegensis, poretzkyi, quadratarea*

VARIETIES *flexella alpestris, poretzkyi gracilis, quadratarea fennica*

GENUS *Psammaococconeis* — Sand Berry Neid Feather Diatom Brown Algae

SPECIES *brasilensis, disculoides*

ORDER 3 *Cymbellales* — Boat Feather Diatom Brown Algae

FAMILY 1 *Cymbellaceae* — Boat Feather Diatom Brown Algae

GENUS *Cymbella* — Boat Feather Diatom Brown Algae

SPECIES *abnormalis, acutiuscula, aequalis, affiniformis, affinis, alpestris, amphicephala, amphioxys, ancyli, anglica, angustata, angustissima, arctissima, aspera, austriaca, avenacea, bengalensis, bengaliformis, bernensis, blancheana, boeckii, borealis, bourrellyi, brasiliana, breweriana, buechleri, cantonatii, carenageana, cesatii, cistula, cistuliformis, citrus, claasseniae, compacta, cornuta, costei, cucumis, curta, cuspidata, delicatula, descripta, designata, diminuta, directa, disparestriata, disposita, diversa, diversiformis, diversistigmata, dorse-notata, dorsirostrata, dubitabilis, dumbeana, duplopunctata, ehrenbergii, elginsis, elizabethana, elliptica, excisa, excisiformis, exigua, falaisensis, flexella, fluminea, fonticola, frigida, gadjiana, gaeumannii, gallica, geddiana, gemeinhardtii, giluwensis, gondwana, graciliformis, gracilis, guettingeri, halophila, hantzschiana, hauckii, hebridica, hedinii, helvetica, hercynica, heteropleura, hofmanniae, hungarica, hustedtii, hybrida, inaequalis, inca, incerta, incertastriata, incurvata, inducta, islandica, italica, jordani, kappii, kemiana, kolbei, krasskei, lacustre, lacustris, laevis, lanceolata, lancettuliformis, langii, lapponica, lata, latarea, laterostrata, lembus, leptoceros, linearis, longa, lunata, maculata, maggiana, magna, marina, metzeltinii, mexicana, microcephala, minutissima, modicepunctata, moelleriana, mongolica, naviculacea, naviculiformis, neglecta, nekliaiana, neocistula, neoleptoceros, neuquina, norvegica, novazeelandiana, oahuensis, obtusiformis, obtusiuscula, octensis, oestrupii, omaniana, orientalis, pachycephala, palpebralis, pantocsekii, parva, parviformis, parvula, pediculus, peraffinis, peraffinis, peraspera, percapitata, percymbiformis, perfossilis, perjaponica, pernodensis, perparva, pervarians, pisciculus, praerupta, prostrata, proxima, pseudonorvegica, pusilla, rabenhorstii, rainierensis, reinhardtii, reviersiana, robertii, rossii, rotundata, rumrichae, runcina, rupicola, schilleri, schubartoides, scotica, scotlandica, scutariana, semicircularis, semisymmetrica, similis, simonsenii, sinensis, sinuata, sphaerophora, spuria, stauroneiformis, stomatophora, strontiana, stuxbergii, subaequalis, subantarctica, subarctica, subaspera, subcistula, subcuspidata, subdirecta, subhelvetica, subkolbei, subleptoceros, subtruncata, subturgidula, subventricosa, suecica, sumatrensis, superparva, symmetrica, tasmaniensis, terrafuegiana, thumensis, ticinensis, towoetensis, transsilvanica, tropica, tsonka, tumida, tumidula, turgida, turgidula, turgiduliformis, uebelackeri, vanoyei, variabilis, variostriata, vasta, ventricosa, volkii, vulgata, weslawskii, yateana, zambesiana*

VARIETIES *abnormalis antiqua, acutiuscula acutiuscula, acutiuscula minor, aequalis diminuta, aequalis oblonga, aequalis obtusa, aequalis pisciculus, affinis nonpunctata, affinis parva, affinis procera, amphicephala citrus, aspera aspera, aspera gastroides, aspera gigas, aspera incerta, aspera minor, austriaca erdoebenyiana, blancheana directa, cesatii cesatii, cesatii capitata, cesatii geitleri, cistula cistula, cistula convexa, cistula fastigiata, cistula gibbosa, cistula gracilis, cistula maculata, cistula minor, cistula truncata, cuspidata anglica, cuspidata elliptica, cuspidata impressa, cuspidata naviculiformis, cuspidata schulzii, delicatula costei, ehrenbergii subcuspidata, excisa angusta, excisa procera, excisa subcapitata, excisiformis nonprotracta, fonticola dumbeana, gadjiana acuminata, gracilis scotica, hantzschiana borealis, hantzschiana strontiana, helvetica helvetica, helvetica balatonis, helvetica compacta, helvetica gracilis, helvetica punctata, heteropleura minor, hustedtii compacta, hustedtii crassipunctata, hustedtii neocaledonicum, hustedtii rhombica, hybrida lanceolata, incerta crassipunctata, kolbei angusta, lacustris inflata, laevis lata, lanceolata aspera, lanceolata bottnica, lata minor, leptoceros angusta, leptoceros excisa, leptoceros rostrata, leptoceros sublanceolata, mexicana lata, microcephala obtusa, microcephala ruttneri,neocistula islandica, neocistula lunata, neoleptoceros tenuistriata, neuquina fastigata, parva hungarica, peraspera colombiana, pernodensis delicata, pisciculus incerta, prostrata auerswaldii, prostrata caespitosa, proxima borealis, proxima bottnica, rupicola rupicola, rupicola rostrata, scotica scotica, scotica naviculacea, sinuata sinuata, sinuata antiqua, sinuata ovata, stuxbergii baikalensis, subaequalis krasskei, subaequalis oblonga, subaspera salina, subtruncata similis, tropica tenuipunctata, tumida tumida, tumida australica, tumida borealis, tumida fossilis, tumida gracilis, tumidula lancettula, tumidula latiuscula, turgida excisa, turgida pseudogracilis, turgidula bengalensis, turgidula venezolana, variostriata granulata, ventricosa auerswaldii, vulgata plivicensis*

GENUS *Brebissonia* Brebisson Boat Feather Diatom Brown Algae

SPECIES *boeckii*

GENUS *Didymosphenia* Dual Wedge Boat Feather Diatom Brown Algae

SPECIES *clava-herculis, clavaherculis, curvata, geminata, pumila*

GENUS *Encyonema* Ency Boat Feather Diatom Brown Algae

SPECIES *accedens, aequilongum, affine, africanum, alpiniforme, alpinum, altiplanense, amanianum, ampliatum, angustecapitatum, apiculatum, auerswaldii, biarcuatum, braunii, brehmiforme, brehmii, brevicapitatum, caespitosum, canadense, carina, caronianum, cascadense, chilense, directiforme, distinctepunctatum, distinctum, donatoi, dubium, elginense, evergladianum, fogedii, formosum, friedelhinzae, gaeumananii, geisslerae, gracile, graciloides, grossestriatum, guatemalense, guettingeri, hebridica, hebridicum, hebridiforme, hilliardii, hintzii, hophense, hungaricum, inarense, incertum, incurvatum, indistinctum, insulsum, javanicum, jemtlandicum, jordaniforme, juriljii, kamtschaticum, kuelbsii, kukenanum, lacustre, lancettulum, lange-bertalotii, latareolatum, latecapitatum, latens, latissimum, latum, limula,, lineolatum, lunatum, magnum, malaysianum, marginestriatum, melidense, menisculus, mesianum, metzeltinii, mexicanum, minutiformae, minutiforme, minutum, muelleri, neocaledonicum, neogracile, neomuelleri, norvegica, obscuriforme, obscurum, obtusum, ochridanum, oestrupii, pankowii, paradoxum, parvum, paucistriatum, perelginense, pergracile, perminutum, perpusillum, persilesiacum, polystigmoideum, ponteanum, procerum, prostratum, pseudomuelleri, reichardtii, rifelanum, riotecense, rostratum, rugosiforme, rugosum, rumrichae, schimanskii, schneideri, schoemanii, semilanceolatum, sibericum, silesiacum, simile, sinicum, sparsipunctatum, spiculum, spitsbergense, stigmoideum, subalpinum, subbrevistriatum, subelginense, subgracile, subjavanicum, subkukenanum, sublanceolatum, subminutum, subnorvegicum, subobscurum, subperpusillum, subrostratum, subtenerum, subtenuissimum, subtriste, subturgidum, supergracile, tapajoz, tasmaniense, tenerum, tenuicephalum, tenuissimum, thermale, thioense, triangulatum, trianguliforme, triangulum, tropicum, truncatum, turgidum, vanoyei, venezolanum, ventricosa, ventricosum, vulgare, wallaceanum, wojtalae, yarrense, yellowstonianum, yucatanense*

VARIETIES, *caespitosum comensis, caespitosum maxima, elginense stigmoideum, jemtlandicum stigmoidea, jemtlandicum venezolana, kamtschaticum parva, lange-bertalotii inarensis, lange-bertalotii obscuriformis, lineolatum major, lunatum borealis, minutiforme apicodeclinatum, minutum minutum, minutum latens, minutum pseudogracile, muelleri ventricosum, neocaledonicum leve, neogracile tenuipunctata, norvegica alpinum, norvegica capitatum, norvegica crassipunctatum, norvegica lapponica, obscurum alpina, obtusum truncatum, perpusillum moreirae, perpusillum striator, persilesiacum ata-tau, silesiacum altensis, silesiacum distigmata, silesiacum distinctepunctata, silesiacum elegans, silesiacum excisa, silesiacum lata, silesiacum latarea, silesiacum latistriata, silesiacum ventriformis, subturgidum neocaledonica, subturgidum wallaceana, turgidum ventricosum, vanoyei grande, venezolanum similis, vulgare chilensis, vulgare tasmanensis, wallaceanum neocaledonicum*

GENUS *Encyonopsis* Lesser Ency Boat Feather Diatom Brown Algae

SPECIES *alpina, angusta, blanchensis, buedeliformis, buedelii, costaricana, degenerata, descripta, difficiliformis, eifelana, fararama, floridana, fragilis, frequentiformis, glacialis, gomphonemiformis, grunowii, kergueliformis, krammerioides, kriegeri, kurtii, lanceo-elliptica, mantasoana, moseri, namibiana, neoamphioxis, opima, patagonica, perborealis, perfloridana, radialis, recta, speciosa, spiculiformis, subcapitata, subfonticola, subruttneri, subspicula, substodderi, tavirana, thumensis, tiroliana, vandamii*

VARIETIES *blanchensis neocaledonicum, descripta asymmetrica, difficiliformis capitata, kriegeri fossilis*

GENUS *Gomphocymbella* Bolt Boat Feather Diatom Brown Algae

SPECIES *ancyli, brunii, muelleri, rossii, tanganyikae*

GENUS *Placoneis* Placid Boat Feather Diatom Brown Algae

SPECIES *abiskoensis, apicalicostata, centropunctata, chilensis, clementioides, clementis, conveniens, dacostae, demeraroides, dicephala, disparalis, disparilis, elginensis, exigua, explanata, gastrum, ignorata, jatobensis, madagascariensis, molesta, neotropica, paraelginensis, pellaifa, placentula, pseudanglica, pseudanglica signata, pseudodemerarae, santaremensis, scharfii, surinamensis, tersa, undulata, vicina, zimmermannii*

VARIETIES *clementis linearis, elginensis cuneata, elginensis neglecta, elginensis rostrata, elginensis undulata, exigua signata, gastrum signata, jatobensis capitata, placentula apiculata, placentula jenisseyensis, placentula latiuscula, placentula maculata, placentula rostrata*

FAMILY 2 *Anomoeoneidaceae* Anomalous Neid Feather Diatom Brown Algae

GENUS *Anomoeneis* Anomalous Neid Feather Diatom Brown Algae

SPECIES *bohemica, brachysira, costata, dumbeana, exellii, exilis, follis, garrensis, gomphonemoides, neocaledonica, paludigena, palustris, rhetica, sculpta, sphaerophora, starmuehlneri, variabilis, zellensis*

VARIETIES *exilis gomphonemacea, exilis lanceolata, exilis thermalis, neocaledonica angusta, sphaerophora sphaerophora, sphaerophora costata, sphaerophora magnifica, sphaerophora sculpta, zellensis difficilis*

GENUS *Staurophora* Wicker Neid Feather Diatom Brown Algae

SPECIES *amphioxys, anuschkae, elata, gregoryi, rossii, salina, wislouchii*
VARIETIES *amphioxys obtusa, salina latior*

FAMILY 3 *Gomophonemataceae* Bolt Feather Diatom Brown Algae
GENUS *Gomphonema* Bolt Feather Diatom Brown Algae

SPECIES *abbreviatum, acidoclinatiforme, aciforme, acuminatum, acutiusculum, aequale, aequirostrum, affine, alineae, americanum, americobtusatum, amoenum, angustatum, angusticephalum, angustivalva, anoenum, apicatum, apuncto, archaevibrio, astridiae, augur, auritum, bacillum, bavaricum, berggrenii, bipunctatum, blancheanum, bourbonense, bozenae, brasiliense, brebissonii, calcareum, calcifugum, camburnii, capitatum, carlsenii, carolinense, cathedralis, cholnokyi, chubichuense, clava, clavatoides, clavatulum, clavatum, clevei, consector, constrictum, contraturris, coronatum, costei, crassicingulatum, crenulata, crocodilei, cuneatum, cuneolus, curtum, cymbelliclinum, daphnoides, delicatula, designatum, dichotomum, distans, drutelingense, dumbeanum, duplipunctatum, ellipsis, elongatum, equadorianum, eriense, exilissimum, extentum, fanensis, fibula, firma, frigidum, geisslerae, genkalii, germainii, globiferum, gracile, grasmueckii, grovei, halbachii, hasta, hastatum, hatherlianum, hebridense, helveticum, holmquistiae, imprime, inkadoriae, innata, innocens, insigne, insigniforme, interpositum, intricatum, inversum, jadwigiae, javanicum, kaweckana, kobayashiae, lacunicola, lacus-vulcani, lagerheimei, lagerheimii, lanceolatum, lateripunctatum, laticeps, laticollum, latipes, latistigmatum, latius, leptocampum, leptoproductum, lindsayi, lingulatiforme, lingulatum, lippertii, littorale, longiceps, longilineare, longipes, lowei, maclaughlinii, makarovae, malayense, manguinii, manifestum, manubrium, marinum, mexicanum, micropumilum, micropus, minusculum, minutiforme, minutum, montanum, multiforme, mustela, nasutum, naviculoides, neoapiculatum, neobourrellyi, neonasutum, netriviale, occidentale, occultum, okavangoanum, olivaceum, opimum, ortalii, oxycephalum, pala, paludosum, parallelistriatum, parvulum, patrickiae, perdurans, periintricatum, pertusum, pfannkucheae, pierrebourrellyi, platypus, potaroense, pratense, procerum, productum, pseudoaugur, pseudobohemicum, pseudoclavatum, pseudopusillum, pseudotenellum, puiggarianum, pulvinatum, pumilum, punae, pygmaeum, quadripunctatum, quasicrocodilei, ravidum, reichardtii, reicheltii, rhombiciforme, rhombicum, ricardii, riotecense, robustum, roehringeri, rosenstockianum, sacrophagus, sagitta, salae, salinarum, sarcophagus, scharfii, scoticollegarum, semiapertum, separatipunctatum, septum, signyense, spauldingae, spectabilissimum, sphaerophorum, sphenovertex, staurophorum, stonei, subangustatum, subarcticum, submehleri, subtile, superiorense, supersedens, tackei, tenellum, tenerrimum, tenue, tergestinum, tetrastigmatum, tholus, triangulare, trigonocephalum, tristigmatum, trulliforme, truncatum, tumens, tumidum, turgidum, turris, undulatum, utae, variostriatum, variscohercynicum, ventricosum, venustum, vibrioides, vitreum, wiltschkorum, wulasiensis, yucatanense, zairense, zellense*

VARIETIES *acuminatum brebissonii, acuminatum coronatum, acuminatum intermedia, acuminatum malayensis, acuminatum trigonocephala, affine elongatum, affine insigne, affine major, affine rhombicum, angustatum bohemicum, angustatum linearis, angustatum producta, angustatum sarcophagus, angustatum undulatum, archaevibrio cuneatum, augur angulatum, augur gautieri, augur rotundatum, augur sphaerophorum, augur turris, capitatum gracile, capitatum macilentum, clavatoides rostratum, clevei javanica, constrictum australe, constrictum capitatum, constrictum capitatum-curta, constrictum curta, dumbeanum distans, dumbeanum octense, gracile aurita, gracile cymbelloides, gracile lanceolata, gracile naviculacea, gracile turris, grovei lingulatum, helveticum tenuis, holmquistiae alaskaense, intricatum bohemicum, intricatum dichotoma, intricatum fossile, intricatum pulvinatum, intricatum pumila, intricatum pusillum, intricatum subcapitatum, intricatum vibrio, lanceolatum insignis, lanceolatum turris, longiceps gracilis, longiceps montana, longiceps subclavatum, longiceps suecica, longipes subclavatum, malayense melanesicum, micropus aequale, minutum curtum, minutum lemanense, minutum pachypus, minutum syriacum, montanum montanum, montanum suecicum, naviculoides navicella, neobourrellyi strictum, neobourrellyi prolongatum, occidentale abbreviata, olivaceum balticum, olivaceum calcareum, olivaceum fonticola, olivaceum minutissimum, olivaceum olivaceoides, olivaceum quadripunctatum, olivaceum staurophorum, olivaceum tenellum, oxycephalum acuta, oxycephalum subacuta, parvulum exilis, parvulum exilissimum, parvulum lagenula, parvulum lanceolatum, parvulum micropus, parvulum parvulius, parvulum saprophilum, parvulum subcapitatum, parvulum subelliptica, parvulum undulatum, puiggarianum aequatorialis, pumilum elegans, pumilum rigidum, ricardii pachypus, salinarum staurophorum, semiapertum tergestinum, subtile subtile, subtile malayensis, subtile sagitta, truncatum truncatum, truncatum capitatum, truncatum elongatum, truncatum macilentum*

GENUS *Gomphogramma* Grass Bolt Feather Diatom Brown Algae
SPECIES *rupestris*
GENUS *Gomphoneis* Neid Bolt Feather Diatom Brown Algae
SPECIES *elegans, eriense, geitleri, hastata, herculeana, linearis, mammilla, mesta, quadripunctatum, rhombica, septa, trullata*
VARIETIES *herculeana robusta*
GENUS *Gomphopleura* Pleural Bolt Feather Diatom Brown Algae
SPECIES *noblis*
GENUS *Pseudogomphonema* Lesser Bolt Feather Diatom Brown Algae
SPECIES *kamtschaticum, monikae, plinskii*
GENUS *Reimeria* Reimer Bolt Feather Diatom Brown Algae
SPECIES *lacus-idahoensis, sinuata*
VARIETIES *sinuata antiqua, sinuata ovata, uniseriata*
GENUS *Reimerothrix* Reimer Thread Bolt Feather Diatom Brown Algae
SPECIES *floridensis*

FAMILY 4 *Rhoicospheniaceae* Sumac Wedge Feather Diatom Brown Algae
GENUS *Rhoicosphenia* Sumac Wedge Feather Diatom Brown Algae
SPECIES *abbreviata, flexa, marina, pullus, stauroneiformis, vanheurckii*
VARIETIES *abbreviata curvatum, abbreviata marinum*
GENUS *Campylopyxis* Pyxie Sumac Wedge Feather Diatom Brown Algae
SPECIES *garkeana*
GENUS *Cuneolus* Cuniform Sumac Wedge Feather Diatom Brown Algae
SPECIES *minutus*
GENUS *Gomphonemopsis* Bolt Thread Sumac Wedge Feather Diatom Brown Algae

SPECIES *exigua, lindae, obscura*
VARIETIES *exigua minutissimum, exigua platypus*
GENUS *Gomphoseptatum* Bolt Septic Sumac Wedge Feather Diatom Brown Algae
SPECIES *aestuarii, pseudoseptatum*
GENUS *Gomphosphenia* Bolt Sumac Wedge Feather Diatom Brown Algae
SPECIES *grovei, holmquistiae, lingulatiformis, paradoxa, pfannkucheae, reicheltii, tackei, tenerrima*
VARIETIES *grovei lingulata, holmquistiae alaskaensis*
GENUS *Gondisphaeridium* Gond Sphere Wedge Feather Diatom Brown Algae
SPECIES *leve*

ORDER 4 *Eunotiales* Marked Feather Diatom Brown Algae

FAMILY 1 *Eunotiaceae* Marked Feather Diatom Brown Algae
GENUS *Eunotia* Marked Feather Diatom Brown Algae
SPECIES *abbottiae, acicularis, acmocephala, actuariola, acus, acutinasuta, aequalis, ambigua, amphioxys, anamargaritae, andinofrequens, angusta, antiqua, arctica, arcuatissima, arcubus, arculus, arcuoides, arcus, astricleveae, australominor, bactriana, baicalensis, ballinaensis, bicapitata, bicornigera, bidens, bidentula, bidentuloides, bigibba, bilunaris, boreoalpina, boreotenuis, botellus, botuliformis, brevis, brinkmannii, cameliopsis, camelus, cancellata, cantica, chelonia, circumborealis, clevei, coccinea, compacta, coringii, costaricana, crassispinata, crassula, crispula, cristagalli, cuneiformis, curtagrunowii, curvata, dacostae, damasii, deformis, densistriata, denticulata, depressa, desmogonioides, diadema, didyma, diodon, disproportionata, distinguenda, ehrenbergii, elegans, elongata, eruca, eurycephala, eurycephaloides, excelsa, exgracilis, exigua, exsecta, faba, falcata, falcifera, fallax, filum, flexuosa, formica, frickie, fukushimae, genufalcata, genuflexa, georgei, georgii, germainii, gibberula, gibbosa, glacialifalsa, glacialis, gracilis, gracillima, granulata, grunowii, hebridica, hemicyclus, hepaticola, hexaglyphis, hyperborea, iatriaensis, implicata, impressa, incisa, incisatula, indigenarum, inflata, inspectabilis, intermedia, itapuana, jarensis, jemtlandica, kanakarum, kocheliensis, krammeri, krasskei, lacus-baicali, lapponica, lenis, limax, lineolata, longicollis, longicornis, lunula, marginistriata, maxima, media, meisteri, melanogaster, meridiana, michaelii, microcephala, minor, minutissima, monodon, monodontiforma, montana, multiplastidica, muscicola, naegelii, napoleonica, neofallax, neomundana, nodosa, noerpeliana, novaecaledonica, nymanniana, ocellata, odebrechtiana, paludosa, papilio, parallela, parapectinalis, parasiolii, pectinalis, pentaglyphis, perpusilla, pileus, pirarucu, pirla, pocosinensis, polydentula, polyglyphis, ponspendens, praerupta, praestans, pseudolunaris, pseudopapilio, pseudopectinalis, pseudoserra, pseudovalida, punctistriata, pyramidata, pyramidatoides, quaternaria, quinaria, rabenhorstiana, reichardtii, repens, rhomboidea, rhynchocephala, robusta, rostellata, sahayadrensis, sardiniensis, sarekensis, sarraceniae, satelles, scandinavica, schneideri, schwabei, seminulum, sennae, septentrionalis, serra, serra serra, siberica, sicula, silvae, silvahercynia, siolii, sioliopsis, smithiae, smithii, soleirolii, soleri, sparsiornata, sparsistriata, sphagnophila, steineckei, stevensonii, subaequalis, subarcuatoides, subgibba, subtilissima, sudetica, suecica, tambaquina, tapacumae, tapacuminor, tapacumopsis, tenella, tenelloides, tenuis, tetraglyphis, tetraodon, thumii, tibia, torganiae, transfuga, transylvanica, triconfusa, tridentula, trigibba, triodon, trivialiforma, tschirchiana, uncinata, ursamajoris, valida, vanheurckii, varioundulata, veneris, ventralis, ventriosa, vitiosa, weisingii, woleirotii, yanomami, yoshitakeae, zasuminuensis, zebra, zebrina, zygodon*
VARIETIES *arcus arcus, arcus bidens, arcus curvata, arcus fallax, arcus minor, arcus plicata, arcus tenella, arcus uncinata, bigibba bigibba, bigibba pumila, bilunaris biceps, bilunaris capitata, bilunaris excisa, bilunaris mucrophila, bilunaris sicula, bilunaris subarcuata, camelus denticulata, camelus kaarverensis, camelus ventricosa, cancellata mutabilis, clevei baicalensis, clevei hispida, curvata attenuata, curvata bergii, curvata capitata, curvata curvata, curvata linearis, curvata sicula, curvata subarcuata, damasii melanesica, denticulata denticulata, denticulata fennica, diadema tetraodon, didyma amazonensis, didyma claviculata, didyma cuyabenensis, didyma elongata, didyma maxima, didyma media, didyma recta, diodon diminuta, exigua exigua, exigua bidens, exigua compacta, exigua linearis, exigua tenella, exigua tridentula, exigua undulata, faba densestriata, faba intermedia, faba pachycephala, faba parvula, faba rhomboidea, fallax gracillima, fallax groenlandica, flexuosa flexuosa, flexuosa eurycephala, flexuosa linearis, flexuosa pachycephala, flexuosa platycephala, formica elongata, formica richmondiae, frickie elongata, hebridica bergii, media jemtlandica, meisteri meisteri, meisteri bidens, microcephala tridentata, monodon monodon, monodon bidens, monodon compacta, monodon constricta, monodon curta, monodon impressa, monodon inelegans, monodon major, monodon neocaledonica, monodon scandica, monodon tropica, monodon ventricosa, muscicola perminuta, muscicola tridentula, odebrechtiana essequiboensis, paludosa arculus, paludosa groenlandica, paludosa trinacria, paludosa undulata, parallela parallela, parallela angusta, parallela carenagea, parallela carenageana, parallela densestriata, pectinalis biconstricta, pectinalis compacta, pectinalis crassa, pectinalis impressa, pectinalis intermedia, pectinalis minor, pectinalis rostrata, pectinalis stricta, pectinalis undulata, pectinalis undulata-fossilis, pectinalis ventralis, perpusilla perminuta, perpusilla tridentata, pileus guianensis, polydentula perpusilla, praerupta africana, praerupta bidens, praerupta curta, praerupta excelsa, praerupta inflata, praerupta maxima, praerupta minor, praerupta monodon, praerupta monas, praerupta muscicola, praerupta nana, praerupta triodon, pyramidata ventralis, rabenhorstiana elongatum, rhynchocephala satelles, robusta abrupta, robusta diadema, robusta octodon, robusta tetraodon, serra abrupta, serra australis, serra diadema, serra tetraodon, sudetica australis, tapacumae sphagnophila, tibia tibia, tibia bidens, tridentula perminuta, tridentula perpusilla, vanheurckii densestriata, vanheurckii intermedia, vanheurckii obtusa, ventriosa brevis, zygodon arcuatissima, zygodon depressa, zygodon elongata, zygodon emarginata, zygodon lata*
GENUS *Actinella* Radial Marked Feather Diatom Brown Algae
SPECIES *australis, bourrellyi, brasiliensis, candelabrum, guianensoides, madagascariensis, manguinii, modesta, pararobusta, pseudohantzschia, punctata, reviersii, tasmaniensis*
VARIETIES *punctata alakaiensis, punctata australis*
GENUS *Amphicampa* Dual Marked Feather Diatom Brown Algae
SPECIES *hemicyclus, mirabilis*
VARIETIES *mirabilis eskdalensis, mirabilis transilvanica*
GENUS *Desmogonium* Fiber Marked Feather Diatom Brown Algae
SPECIES *guinense, rabenhorstianum*
GENUS *Semiorbis* Orb Marked Feather Diatom Brown Algae

SPECIES *hemicyclus*

FAMILY 2 *Peroniaceae* — Peron Marked Feather Diatom Brown Algae

GENUS *Peronia* — Peron Marked Feather Diatom Brown Algae

SPECIES *erinacea, fibula, heribaudii*

VARIETIES *erinacea lysebuensis*

GENUS *Peronyella* — Lesser Peron Marked Feather Diatom Brown Algae

SPECIES *pandorinae*

ORDER 5 *Naviculales* — Navicule Feather Diatom Brown Algae

FAMILY 1 *Naviculaceae* — Navicule Feather Diatom Brown Algae

GENUS *Navicula* — Navicule Feather Diatom Brown Algae

SPECIES *abaujensis, aberrans, abiskoensis, aboensis, abrupta, absoluta, abstrusa, accedens, acceptata, acidobionta, acuta, adhaerens, adminii, adnata, advena, affinis, agatkae, aggerica, agnita, agrestiformis, agrestis, aikensis, ajajensis, aleemii, algida, alineae, alpestris, altiplanensis, americana, ammophila, amphibola, amphiceropsis, amphigomphus, amphirhychus, amphisbaena, ampliata, amygdalina, anatis, anca, ancilla, angulosa, angusta, angustata, anhuiensis, anna, antigua, antiqua, antonii, apiculata, appendiculata, approximatella, aquaedurae, archeriana, arcta, arctotenelloides, arcuata, arcus, arenariaeformis, arenicola, argens, arida, arkona, armata, armoricana, arvensis, asellus, aspergilla, associata, atacamae, athenae, atlantica, atomus, aurora, australomediocris, austrocollegarum, avenacea, avonensis, bachmannii, bacillaris, bacillifera, bacilliformis, bacilloides, bacillum, bahusiensis, baicalensis, baileyana, balatonis, balcanica, barclayana, barodensis, bastowii, begeri, begerii, bella, bendaensis, bergerii, bicapitellata, bicephala, bicontracta, bicuneolus, bidentula, biflexa, bifrons, biglobosa, binodis, bipectinalis, biskanterae, biskanteri, bistriata, bisulcata, bjoernoeyaensis, bogotensis, bohemica, borealis, borowkae, bottnica, bourrellyivera, brachysira, brasiliana, brauniana, brebissonii, breitenbuchii, bremensis, brevissima, britannica, britannica, brockmannii, broetzii, bryophila, bryophiloides, buderi, buderiformis, caenosus, calida, campylogramma, canalis, canariana, cancellata, cantonatii, capitata, capitatoradiata, capitoradiata, capsa, cari, carinifera, cariocincta, carteri, cascadensis, catalanogermanica, cataractarheni, caterva, celinei, centropunctata, certa, cesatii, charlatii, chauhanii, chiarae, chilena, cincta, cingens, cistella, citrus, clagesii, clamans, clarensiana, clathrata, claytonii, clementioides, clementis, clementoides, clevei, cluthensis, cocconioides, coffeiformis, collersonii, commutata, comoides, complanata, completa, compositestriata, concentrica, confecta, confervacea, congolensis, consentanea, constans, constantinii, contempta, contraria, conveniens, coraliana, cornubiensis, corrugata, corymbosa, costata, costulata, costuloides, crassa, crassinervia, crassulexigua, creuzbergensis, crucicula, cruciculoides, crucifera, crucigera, crucigeroides, cryptocefalsa, cryptocephala, cryptocephaloides, cryptofallax, cryptonella, cryptostriata, cryptotenella, crystallina, cumbriensis, curta, curtestriata, curtisterna, cuspidata, cuspis, cymbelliformis, cymbula, damasii, dapaloides, dealpina, debilissima, declivis, decussis, defluens, demeraroides, densa, denselineolata, densilineolata, depressa, destricta, detenta, detersa, diabolica, diahotana, dicephala, diengensis, difficilima, difficilis, difficillima, difficillimoides, digito-radiata, digitulus, digna, dilata, dilatata, dilucida, diluviana, dimidiata, directa, diserta, disjuncta, disparalis, disparata, dispersa, disphenia, disputans, dissipata, dissociata, distans, distinctastriata, divaricata, divergentissima, diversa, doehleri, dolomitica, dorogostaiskyi, dubia, dubitanda, duerrenbergiana, dulcis, dunstonii, dutoitiana, dvoracheckii, ehrenbergii, ehrlichiae, eichhorniophila, eidrigeana, eidrigiana, elata, electrolytifuga, elegans, elginensis, elkab, elliptica, elongata, ergadensis, erifuga, escorialis, eta, evanida, excentrica, excepta, exigua, exiguiformis, exiguoides, exiliformis, exilis, exilissima, expansa, expectata, expleta, farta, fasciata, favus, fennica, fenzlii, festiva, feuerborni, finitima, finnmarchica, firma, flagellifera, flanatica, florentina, florinae, floriniae, flotowii, fluens, follis, fonticola, formenterae, fortis, fossalis, fossaloides, fraudulenta, frigida, fromenterae, frugalis, fuchsii, fuscata, fusiformis, fusioides, galikii, gandrupii, gastriformis, gastrum, geinitzi, geisslerae, geitleri, gelida, gentilis, genustriata, gerloffii, germainii, germanopolonica, gibbosa, gibbula, giffeniana, glaciei, globiceps, globosa, globulifera, globuliferiformis, glomus, goeppertiana, gondwana, gorjanovicii, gotlandica, gracillima, graeffii, granii, granulata, grasmueckii, gregaria, gregoriana, gregorii, grevilleana, grimmei, grimmeioides, grippii, groschopfii, halinae, halophila, halophiloides, hambergii, hamiltonii, hamulifera, hanseatica, hanseni, harderi, hasta, hastatula, hawaiiensis, hebes, heimansii, heimansioides, hemiptera, herbstiae, heufleriana, hilseana, hintzii, hitchcockii, hoeflerii, hofmanniae, holmenii, holmiensis, horstii, humboldtiana, humilis, hustedtii, hyalosira, ignota, ilopangoensis, imperfecta, impexa, importuna, inaequalis, incarum, incerta, incertata, incongruens, incurva, incus, indica, inexpectans, inflata, inflatoides, inflexa, ingapirca, ingenua, ingrata, inornata, inscriptura, insignita, instabilis, insulsa, integra, interglacialis, intermedia, intractata, iranensis, irmengardis, iserentantii, islandica, jaagii, jasnitskyi, jatobensis, jeffreyae, jentzschii, jogensis, johnsoniana, johnsonii, joubaudii, jungii, kariana, kohlenbachii, kohlmaieri, korzeniewskii, kotschyi, kraeuselii, krammerae, krasskei, kriegeri, krookii, kryophila, kuelbsii, kuseliana, kutzingiana, lacunarum, lacunolaciniata, ladogensis, laevissima, lagerstedtii, lambda, lanceolata, lange-bertalotii, lapidosa, lapila, lapsa, latens, lateropunctata, laterostrata, laticeps, latiuscula, lauca, legumen, leistikowii, lemmermannii, lentzii, lepida, leptogongyla, leptosoma, leptostriata, lesmonensis, lestikowii, levanderi, libellus, liber, libonensis, librile, liburnica, limosa, lineolata, litoris, lohmannii, loibl-sittlerii, longa, longiceps, luciae, lucidula, ludloviana, lunatapicalis, lundii, lunulifera, luxuriosa, luzonensis, maceria, macilenta, macrogongyla, macula, maculata, maculosa, madeirensis, mahoodii, maidanae, major, margalithii, margaritiana, margino-nodularis, margino-ornata, marnieri, mascarenae, maxima, medioconvexa, mediocostata, mediocris, mediterranea, melanesica, menaiana, menda, menisculoides, menisculus, meniscus, mesoleiae, mesolepta, microcari, microcephala, microdigitoradiata, micropunctata, micropupula, minima, miniscula, minisculoides, minnewaukonensis, minthe, minutissima, minutula, misella, modica, moenofranconica, molaris, molesta, molestiformis, mollicula, mollis, moniliformis, mormonorum, moskalii, mournei, multicostata, muraliformis, muraloides, muscora, mutata, namibica, naumanii, neglecta, neoventricosa, nipponica, nivaloides, nobilis, nodosa, nolensoides, northumbrica, notanda, notata, notha, novasiberica, nuda, nyassensis, oahuensis, obdurata, oblonga, oblongella, oblongolinearis, obsidialis, obsoleta, obtecta, obtusa, obtusangula, occulta, octavosignata, oculata, oculus, odiosa, okunoi, oligotraphenta, opima, opportuna, oppugnata, orbis, ostenfeldii, ovata, paanaensis, paca, pagophila, palpebralis, paludosa, paracari, parahasta, paranipponica, parasubtilissima, parasura, parca, pargemina, parinacota, parsura, parva, parvula, patula, paucivisitata, paul-schulzii, paulina, paupercula, pavillardii, peguana, pelagica, pelliculosa, peltoensis, pennata, peraustralis, peregrina, peregrinopsis, perelegans, perfidissima, perhibita, perlepida, perminuta, perparva, perplexa, perrotettii, petersenii, phylleptosoma, placenta, placentula, placentula, plathii, platycephala, platystoma, plicata, podzorskii, polyonca, ponticulus, poretzkiae, porifera, potzgeri, pozzuoliensis, praeterita, producta, protracta, pseudoanglica, pseudoarvensis, pseudobacillum, pseudobryophila, pseudoclamans, pseudoclementis, pseudoconcamerata, pseudocrassirostris, pseudocryptocephala, pseudodemerarae, pseudofrugalis, pseudohalophila, pseudoislandica, pseudokotschyi, pseudolagerheimii, pseudolanceolata, pseudolitoricola, pseudomenisculus, pseudopalpebralis, pseudoppugnata, pseudopupula, pseudoreinhardtii, pseudosilicula, pseudosubtilissima, pseudosydowii, pseudothienemannii, pseudotuscula, pseudoventralis, puella, pulchripora, pusio, quadripartita, quasidisjuncta, quechua, quinquenodis, rabenhorstii, radiosa, radiosiola, radiostriata, rajmundii, rakowskae, ramosissima, ranomafanensis, rattrayi, recens, recondita, recurva, reichardtiana, reimeri, reinhardtii, reinickeana, reissii, relicta, resecta, restituta, retrocurvata, retusa, rhombica, rhombicula, rhynchocephala, rhynchotella, ricardae, ricardii, riediana, riojae, riotecensis, riparia, rivularis, rivulorum, rolandii, rossii, rostellata, rostrata, rotaeana, rotula, rotunda, rudiformis, rupicola, sabiniana, sacrophagus, salinarum, salinicola, sansegana, saprophila, saugerii, saxicola, scandinavica, schadei, schassmanii, schmidtii, schoenfeldii, schonkenii, schroeterii, schumanniana, schwabei, scirpus, scoliopleuroides, scotica, scutelloides, scutum, secreta, secura, seductilis, seibigiana, seibigii, sejuncta, semen, semenoides, semihyalina, seminoides, seminuloides, seminulum, semivirgata, septentrionalis, serians, shiloi, siamexilis, siamlinearis, sieminskiae, sigma, sigmoidea, silicula, silvahercynia, similis, simplex, simplexoides, simsii, simulans, singularis, siofokensis, skabitschewskyi, skuae, slesvicensis, soehrensis, solaris, soodensis, sovereignae, sphaerophora, spicula, spinifera, splendicula, splendida, splendidula, sponsa, spuria, stachurae, stankovicii, starmachii, starmachioides, staurifera, streckerae, strenzkii, striolata, stroemii, structa, sub-bacillum, subadnata, subalpina, subbacillum, subcarinata, subclementis, subconcentrica, subconstricta, subcostulata, subdeliberata, subelata, subfortis, subfossalis, subfraudulenta, subgastriformis, subgrimmei, subhalophila, subinflata, subinflatoides, sublanceolata, subminuscula, submolesta, submuralis, subocculata, subplacentula, subretusa, subrhynchocephala, subrotundata, subsolaris, subsulcata, subtilissima, sudora, suecorum, sulcifera, supergregaria, supralitoralis, suprinii, surinamensis, suspica, symmetrica, syvertsenii, tabellaria, tabellariaeformis, taishanica, tau, taylori, tecta, tectissima, tenella, tenelloides, tenuicephala, tenuipunctata, tenuis, termes, terminata, termitina, terricula, tersa, testata, texana, theinemannii, thoroddsenii, thuringiaca, thurstonensis, torellii, torneensis, toulae, toxa, transistantioides, transitans, tranversa, tridentula, tripunctata, trivialis, trivialis oligotraphenta, trochus, trophicatrix, tropicoidea, turgida, twymanniana, udintsevii, ultratenelloides, ulvacea, umbonata, undosa, unilaterarea, utermoehlii, vacillans, valdestriata, validicostata, vandamii, vanei, vanhoeffenii, variabilis, varians, variolinea, vaucheriae, vekhovii, veneta, ventosa, ventraloides, ventricosa, verecunda, verecundaeformis, vetita, vicina, vilaplanii, viridula, viridulacalcis, virosa, vitabunda, vitiosa, vixvisibilis, vula, vulgaris, vulpina, wasmundii, weissflogii, wellneri, westermannii, wiesneri, wiktoriae, wildii, wisei, witkowskii, wittrockii, woltereckii, wrightiana, wrightii, wuestii, wulffii, wunsamiae, wygaschii, yarrensis, zachariasii, zanoni, zebra, zellensis, zeta, zohdyi, zostereti*

VARIETIES *abrupta rattrayi, affinis undulata, americana alastis, americana bacillaris, americana minor, americana parallela, ammophila flanatica, amphibola baicalensis, amphibola gracilis, amphibola perrieri, amphibola polymorpha, arvensis major, atomus alcimonica, atomus recondita, bacillaris thermalis, bacilliformis cruciata, bacillum bacillum, bacillum gregoryana, bacillum lepida, bacillum major-baicalensis, bacillum rostrata, bahusiensis arctica, bahusiensis istriana, bicephala manguinii, bottnica acuta, brebissonii diminuta, brockmannii undulata, bryophila bryophila, bryophila lapponica, bryophila paucistriata, bryophiloides linearis, cancellata ammophila, cancellata gregorii, cancellata retusa, cancellata subretusa, capitata capitata, capitata hungarica, capitata linearis, capitata lueneburgensis, capitata sarolata, cari angusta, cari barodensis, cari divergentissima, cari recens, cincta leptocephala, cincta minuta, cincta rostrata, clementis linearis, clementis quadristigmata, cluthensis cluthensis, cluthensis rostrata, coffeiformis subcircularis, complanata subinflata, compositestriata rostrata, confervacea baicalensis, confervacea laterostrata, confervacea peregrina, confervacea tenuis, constans symmetrica, crucicula crucicula, crucicula obtusata, crucifera apiculata, cryptocephala exilis, cryptocephala intermedia, cumbriensis minor, cuspidata cuspidata, cuspidata ambigua, cuspidata craticula, cuspidata curta, cuspidata heribaudi, cuspidata major, dicephala abiskonensis, dicephala elginensis, dicephala minor, dicephala neglecta, dicephala rostrata, dicephala subcapitata, dicephala undulata, digito-radiata convergens, digito-radiata linearis, digito-radiata minima, digito-radiata rostrata, digito-radiata striolata, directa incus, directa remota, directa subtilis, eichhorniophila peracuta, elegans cuspidata, elginensis elginensis, elginensis cuneata, elginensis lata, elginensis minor, elginensis minuta, elginensis neglecta, elginensis rostrata, elginensis signata, elginensis subsalsa, elliptica minutissima, ergadensis minor, exigua capitata, exigua signata, expleta domblittensis, firma subampliata, firma subundulata, flanatica ammophila, fortis opima, fossalis obsidialis, gastrum exigua, gastrum galikii, gastrum jenisseyensis, gastrum latiuscula, gastrum minor, gastrum signata, geitleri asymmetrica, gelida asymmetrica, gerloffii lanceolata, globiceps krockii, globulifera robusta, gotlandica helvetica, gotlandica woltericki, granulata maculosa, grevilleana comoides, halophila fogedii, halophila robusta, halophila subcapitata, halophila tenuirostris, hanseatica circumarctica, heufleriana minor, hustedtii obtusa, ignota ignota, ignota acceptata, ignota anglica, ignota palustris, inflata laterostrata, inflata rostrata, integra helvetica, integra truncata, jatobensis capitata, kariana kariana, kariana curta, kariana frigida, kariana minor, kariana perpusilla, kotschyi kotschyi, kotschyi hueckii, kotschyi robusta, kotschyi undulata, lanceolata arenaria, lanceolata curta, lanceolata cymbula, lanceolata phyllepta, lanceolata rostellata, lanceolata tenuirostris, laterostrata rostrata, legumen decrescens, legumen parva, levanderi capitata, levanderi tatrensis, limosa gibberula, limosa subinflata, limosa undulata, maculata acuta, maculata gigantea, maculata inflata, major crassa, maxima bicuneata, maxima umbilicata, mediocris atomus, mediocris barodensis, menisculus menisculus, menisculus grunowii, menisculus krenneri, menisculus obtusa, menisculus tenuis, menisculus upsaliensis, minima minima, minima okamurae, miniscula minuscula, miniscula alpestris, miniscula istriana, nobilis dactylus, nodosa curta, nodosa denudata, nyassensis minor, oblonga oblonga, oblonga lanceolata, oblonga subcapitata, palpebralis angulosa, palpebralis barclayana, parvula producta, parvula prominula, peregrina peregrina, peregrina angusta, peregrina asiatica, peregrina kefvingensis, peregrina meniscus, peregrina minor, peregrina minusculus, perrotettii enervis, placentula apiculata, placentula jenisseyensis, placentula lanceolata, placentula latiuscula, placentula maculata, placentula rostrata, platystoma pantocsekii, porifera opportuna, protracta protracta, protracta elliptica, protracta subcapitata, pseudoanglica signata, pseudobacillum rostrata, radiosa radiosa, radiosa acuta, radiosa fallax, radiosa intermedia, radiosa nipponica, radiosa parva, ramosissima amplia, ramosissima amplior, ramosissima mollis, ramosissima mucosa, ramosissima torquata, recondita recondita, reichardtiana crassa, reinhardtii reinhardtii, reinhardtii elliptica, reinhardtii gracilior, retusa cancellata, retusa gregorii, rhynchocephala amphiceros, rhynchocephala elongata, rhynchocephala germanii, rotaeana oblongella, salinarum salinarum, salinarum capitata, salinarum intermedia, salinarum minima, schmidtii skagerakiensis, schroeterii schroeterii, schroeterii escambia, schumanniana biconstricta, seminulum hustedtii, seminulum radiosa, serians minima, serians minor, serians thermalis, soehrensis soehrensis, soehrensis capitata, soehrensis hassiaca, soehrensis muscicola, torneensis aboensis, toulae capitata, transitans derasa, transitans derasa-delicatula, tridentula tridentula, tridentula indica, tridentula parallela, tridentula tenuis, tripunctata tripunctata, tripunctata cuneata, tripunctata schizomenoides, trochus biconstricta, vacillans minuta, vandamii mertensiae, ventricosa minuta, ventricosa truncatula, vilaplanii longicephala, viridula abbreviata, viridula avenacea, viridula capitata, viridula denselineolata, viridula ehrenbergii, viridula germanica, viridula linearis, viridula slesvicensis, viridula vulpina, viridulacalcis neomundana, vulgaris denudata, vulpina avenacea, wittrockii directarea*

GENUS *Adlafia* — Adlaf Navicule Feather Diatom Brown Algae

SPECIES *aquaeductae, bryophila, coringii, drouetiana, minuscula, muscora, parabryophila, suchlandtii*

VARIETIES *minuscula muralis*

GENUS *Brachysira* — Short Siris Navicule Feather Diatom Brown Algae

SPECIES *acutirhomboides, acutissima, amoena, angusta, aponina, archibaldii atacamae, australofollis, babuschka, bouletiana, brebissonii, calcicola, calligraphica, coraliana, estoniarum, exilis, feickertiae, follis, garrensis, guttiformis, hofmanniae, inamoena, incognita, inflata, insolita, insulsa, irawanoides, julio, krammeri, kuntzei, labrata, lange-bertalotii, lecohui, lehmanniae, liliana, longirostris, macroserians, maillardii, manfredii, metzeltinii, microcephala, microclava, microserians, minor, nanoclava, neoacuta, neocaledonica, neoexilis, nubigena, obesa, palustris, procera, pseudoacuta, pseudoexilis, pulchra, pumila, rhomboides, rostrata, rumrichiae, ruppeliana, seippii, serians, silvicola, simplex, speluncola, sphaerophora, spicula, staurophora, steindorfiana, steinitziae, styriaca, sublinearis, subrostrata, superba, supriniana, tenuiserians, trinodis, venter, vitrea, wygaschii, zellensis*

VARIETIES *aponina brachysira, australofollis angustior, brebissonii superba, neocaledonica angusta, serians serians, serians acuta, serians bacillifer, serians brachysira, serians modesta, serians rostrata, serians thermalis, vitrea lanceolata, zellensis difficilis*

GENUS *Capartogramma* — Caport Seaweed Navicule Feather Diatom Brown Algae

SPECIES *crucicula, pumila*

GENUS *Catacombus* — Catacomb Navicule Feather Diatom Brown Algae

SPECIES *gaillonii*

GENUS *Cavinula* — Cavin Navicule Feather Diatom Brown Algae

SPECIES *cocconeiformis, intractata, jaernefeltii, lacustris, lapidosa, mollicula, pseudoscutiformis, scutelloides, scutiformis, thoroddsenii, variostriata, weinzierlii*

VARIETIES *cocconeiformis capitata, cocconeiformis parva*

GENUS *Cocconema* — Round Navicule Feather Diatom Brown Algae

SPECIES *australasicum, boeckii, cornutum, cuspidatum, cymbiforme, delicatulum, fusidium, gastroides, gracile, laeve, leptoceros, parva, scotica, sinuatum, stomatophorum, tumidum*

VARIETIES *stomatophorum boreale, stomatophorum fossile*

GENUS *Craticula* — Wicker Navicule Feather Diatom Brown Algae

SPECIES *accomoda, accomodiformis, acidoclinata, ambigua, angustilancea, buderi, citroides, citrus, cuspidata, dissociata, fumantii, germainii, guatemalensis, halopannonica, halophila, halophilioides, lanceola, minusculoides, molesta, molestiformis, neglecta, nonambigua, pampeana, paramolesta, perrotettii, riparia, subhalophila, submolesta, sverirschopkae, vixnegligenda, vixvisibilis*

VARIETIES *ambigua craticula, cuspidata heribaudi, cuspidata major, halophila robusta, halophila subcapitata, halophila tenuirostris, riparia mollenhaueri, vixvisibilis distinctior*

GENUS *Fischerites* — Fischer Navicule Feather Diatom Brown Algae

SPECIES *coreanicus*

GENUS *Fistularia* — Tube Navicule Feather Diatom Brown Algae

SPECIES *compressa, erecta, intestinalis*

VARIETIES *, intestinalis compressa*

GENUS *Fistulifera* — Tubule Navicule Feather Diatom Brown Algae

SPECIES *iranensis, pelliculosa, saprophila*

GENUS *Geissieria* — Geiss Navicule Feather Diatom Brown Algae

SPECIES *annulata, arkonensis, boreosiberica, cummerowi, declivis, decussis, dolomitica, ignota, kriegeri, lange-bertalotii, mirabilis, moseri, neotropica, paludosa, perelegans, schmidiae, schoenfeldii, similis, subdecussis, tectissima, thingvallae*

VARIETIES *annulata baicalensis*

GENUS *Haslea* — Hasle Navicule Feather Diatom Brown Algae

SPECIES *britannica, crucigera, crucigeroides, crystallina, gigantea, ostrearia, spicula, trompii, wawrikae*

GENUS *Hippodonta* — Horse Tooth Navicule Feather Diatom Brown Algae

SPECIES *arctica, arkonensis, avittata, caotica, capitata, costulata, costulatiformis, coxiae, geocollegarum, germainii, grunowii, hungarica, kaiseri, kissii, kociolekii, lesmonensis, linearis, luneburgensis, neglecta, pseudacceptata, pseudopinnularia, rostrata, ruthnielseniae, subcostulata, subtilissima*

GENUS *Kobayasia* — Kobayasi Navicule Feather Diatom Brown Algae

SPECIES *bicuneus, grandis, jaagii, krasskei, madumensis, micropunctata, neocaledonica, okadae, parasubtilissima, pseudostauron, saxicola, subtilissima*

GENUS *Kobayasiella* — Lesser Kobayasi Navicule Feather Diatom Brown Algae

SPECIES *mirabilis, pseudosubtilissima*

GENUS *Krasskella* — Krassk Navicule Feather Diatom Brown Algae

SPECIES *hyalina, kriegerana, manguinii*

GENUS *Lyrella* — Lyre Navicule Feather Diatom Brown Algae

SPECIES *abrupta, abruptoides, amphoroides, approximata, approximatella, approximatoides, atlantica, baldjickiensis, barbara, barbitos, californica, circumsecta, clavata, concilians, copiosa, crebra, david-mannii, diffluens, discreta, durandii, excavata, exsul, fogedii, genifera, hennedyella, hennedyi, hennedyoides, illustrioides, illustris, imitans, inaurata, inhalata, investigata, irrorata, irroratoides, lyra, lyroides, mediopartita, miranda, nebulosa, novaeseelandiae, oamaruensis, opuntioides, perfecta, perplexoides, praetexta, pseudoapproximata, robertsiana, sandriana, schaarschmidtii, semiapproximata, signatula, spectabilis, subirroratoides, sulcifera, variolata, venusta, zanzibarica*

VARIETIES *approximata approximata, approximata morei, approximata nicaensis, approximata substauroneiformis, clavata indica, clavata wrightii, hennedyi bacillifera, hennedyi manca, lyra asymmetrica, lyra constricta, lyra dilatata, lyra elliptica, lyra granulata, lyra intermedia, lyra robertsiana, lyra robertsoniana, lyra seductilis, praetexta abundans, spectabilis constricta, spectabilis punctata*

GENUS *Melonavicula* Melon Navicule Feather Diatom Brown Algae

SPECIES *marylandica*

GENUS *Mueniera* Muenier Navicule Feather Diatom Brown Algae

SPECIES *membranacea*

GENUS *Microcostatus* Ribbed Navicule Feather Diatom Brown Algae

SPECIES *andinus, egregius, krasskei, kuelbsii, maceria, naumannii, ornicollegarum, polonarum, pseudomuralis, stapputana*

GENUS *Muelleria* Mueller Navicule Feather Diatom Brown Algae

SPECIES *algida, bachmannii, dubitanda, gibbula, holmenii, islandica, luculenta, meridionalis, peraustralis, portomontana, splendidula, terrestris, variolata, varipunctata*

GENUS *Naviculadicta* Net Navicule Feather Diatom Brown Algae

SPECIES *ambiguissima, angustata, brasiliana, bremensiformis, bryophila, chilensis, circumborealis, cosmopolitana, digituloides, digitulus, elorantana, geissleriae, ivatoensis, kaiseri, lepida, linearis, moratoria, multiconfusa, nagumoi, nana, nanogomphonema, obdurata, parasemen, pseudofallacia, pseudostauron, raederae, rionautensis, sassiana, schneckenburgeri, stauroneioides, tomesii, vaucheriae, vekhovii, ventraloconfusa*

VARIETIES *tomesii clagesii, ventraloconfusa chilensis*

GENUS *Nupela* Nupel Navicule Feather Diatom Brown Algae

SPECIES *astartiella, brachysiroides, butterfassiana, chilensis, comperei, cymbelloidea, deformis, encyonopsis, giluwensis, gracillima, imperfecta, impexiformis, jahniae-reginae, lapidosa, lesothensis, neotropica, obliqua, pallavicinii, paludigena, praecipua, rhetica, rumrichorum, schoemaniana, silvahercynia, subinvicta, subpallavicinii, tenuicephala, tenuistriata, tristis, tropica, vyvermanii, wellneri, zizkae*

GENUS *Obliquipithonella* Oblique Pith Navicule Feather Diatom Brown Algae

SPECIES *echinosa, globosa, krasheninnikovii, operculata, parva*

GENUS *Obrachevella* Obrachev Navicule Feather Diatom Brown Algae

SPECIES *condensata*

GENUS *Orthopithonella* Straight Pith Navicule Feather Diatom Brown Algae

SPECIES *aequilamellata, antarctica, aspera, congruens, cookii, flora, krasheninnikovii, minuta, veeversii, williambensonii*

GENUS *Pithonella* Pith Navicule Feather Diatom Brown Algae

SPECIES *antarctica, carteri, cookii, echinosa, edgarii, francadecimae, globosa, gustafsonii, heirtzleri, helentappaniae, johnstonei, krasheninnikova, krasheninnikovii, loeblichii, mcknightii, nonarenziae, parva, patriciagreeleyae, quiltyi, robinsonii, rockeri, sheilasantawiae, thayeri, veeversii*

GENUS *Playaensis* Beach Navicule Feather Diatom Brown Algae

SPECIES *circumfimbria, furtiva*

GENUS *Proschkinia* Proschkin Navicule Feather Diatom Brown Algae

SPECIES *bulnheimii, complanata, complanatoides, complanatula, hyalosirella, longirostris, poretzkajae*

VARIETIES *bulnheimii bulnheimii, bulnheimii belgica*

GENUS *Seminavis* Lesser Navicule Feather Diatom Brown Algae

SPECIES *barbara, cymbelloides, gracilenta*

GENUS *Stauroptera* Crosshatch Wing Navicule Feather Diatom Brown Algae

SPECIES *amphioxys, isostauron, legumen, microstauron, parva, platycephala, scalaris, semicruciata*

FAMILY 2 *Amphipleuraceae* Dual Pleural Navicule Feather Diatom Brown Algae

GENUS *Amphipleura* Dual Pleural Navicule Feather Diatom Brown Algae

SPECIES *arctica, chiapasensis, delicatissima, kriegeriana, lindheimeri, micans, pellucida, rigida, rutilans, sigmoidea*

VARIETIES *micans fragilis*

GENUS *Amphiprora* Dual Proran Navicule Feather Diatom Brown Algae

SPECIES *alata, angustata, brebissoniana, calumetica, chinensis, cholnokyi, clepsydra, conspicua, constricta, costata, crenulata, decussata, delicatula, didyma, duplex, eximia, hustedtii, hyalina, hyperborea, jolisianakjellmanii, kuetzingiana, lata, lepida, longa, meneghiniana, nitida, oblonga, ornata, paradoxa, plicata, pulchra, rabenhurstiana, rectangularis, reediana, robusta, sulcata, superba, surirelloides, thwaitesiana, undulata, venusta*

VARIETIES *alata holdererii, decussata septentrionalis, kjellmanii subtilissima, lata lata, lata angustior, pulchra pulchella*

GENUS *Cistula* Cist Navicule Feather Diatom Brown Algae

SPECIES *lorenziana*

GENUS *Frustulia* Frustrum Navicule Feather Diatom Brown Algae

SPECIES *acidophilissima, acrosphaeria, acuminata, adnata, aequalis, alfinitosilviae, altimontana, amosseana, amphipleuroides, angustata, angustior, appendiculata, australis, avenacea, bahlsii, baicalensis, bisulcata, blancheana, chilensis, corneliae, crassinervia, crassipunctata, crassiundosa, creuzbergensis, creuzburgensis, cuneata, depressa, dilatata, disjuncta, dispersata, ellipticoides, elongata, elongatissima, entrancensis, erifuga, gaertnerae, gastroides, guayanensis, huberi, hustedtii, interposita, jajaensis, kosmolliana, krammeri, lacustris, lakatosii, lanceolata, lange-bertalotii, lata, leptocephala, linkei, longinqua, lyngbyei, lyrata, maculata, madagascariensis, magaliesmontana, magna, major, mamillaris, mediocris, modesta, nana, neocaledonica, neofrenguellii, neomundana, nitzschii, oblonga, operculata, ovalis, pangaea, pangaeopsis, pararhomboides, pelliculosa, pellucida, peridotitica, pseudomagaliesmontana, pumilio, pusilla, quadrisinuata, quinquepunctata, rhodesica, rhomboides, rostrata, rostratiformis, scalaris, septentrionalis, spicula, splendens, stagnalis, submarina, tenuirostris, tenuissima, trepanata, undosa, ventricosa, vermicularis, vermiculata, vulgaris, weinholdii, zizkae*

VARIETIES *blancheana sulcata, neocaledonica minima, pararhomboides pararhomboides, pararhomboides aff, pararhomboides subundosa, quadrisinuata meridiana, rhomboides amphipleuroides, rhomboides capitata, rhomboides crassinervia, rhomboides dispersata, rhomboides elongatissima, rhomboides hustedtii, rhomboides lacustris, rhomboides rostrata, rhomboides saxonica, rhomboides saxonica-capitata, rhomboides saxonica-undulata, rhomboides undulata, rhomboides viridula, rostrata angustior, spicula alpina, spicula judaica, vulgaris vulgaris, vulgaris angusta, vulgaris brevis, vulgaris capitata, vulgaris elliptica*

GENUS *Pseudoamphiprora* Lesser Dual Proran Feather Diatom Brown Algae

SPECIES *australis, stauroptera*

FAMILY 3 *Berkeleyaceae* Berkeley Navicule Feather Diatom Brown Algae

GENUS *Berkeleya* Berkeley Navicule Feather Diatom Brown Algae

SPECIES *interrupta, linearis, obtusa, pellucida, rutilans, scopulorum*

VARIETIES *scopulorum belgica, scopulorum perlonga*

GENUS *Berkella* Berk Navicule Feather Diatom Brown Algae

SPECIES *alpina, linearis, spicula*

GENUS *Climaconeis* Climax Neid Navicule Feather Diatom Brown Algae

SPECIES *colemaniae, inflexa, koenigii, lorenzii, scalaris, stromatolitis*

GENUS *Parlibellus* Parlib Navicule Feather Diatom Brown Algae

SPECIES *adnatus, bennikei, calvus, coxiae, crucicula, cruciculoides, delognei, ellipticus, hendeyi, panduriformis, perytii, phoebeae, pozzuoliensis, protracta, protractoides, rhombicula*

GENUS *Stenoneis* Stencil Neid Navicule Feather Diatom Brown Algae

SPECIES *inconspicua, obtuserostrata, wojtek-kowalskii*

VARIETIES *inconspicua baculus*

FAMILY 4 *Diadesmidaceae* Fiber Navicule Feather Diatom Brown Algae

GENUS *Diadesmis* Fiber Navicule Feather Diatom Brown Algae

SPECIES *aerophila, andegavensis, arctica, arcuata, bigibba, biscutella, brekkaensis, brekkaensoides, confervacea, confervaceoides, contenta, corrugata, discordabilis, dissimilis, elegans, fukushimae, gallica, implicata, lacunosa, laevissima, pantropica, paracontenta, peregrina, perpusilla, pseudocontenta, tabellariaeformis, tahitiensis, templiniana, virginiana, williamsonii*

VARIETIES *brekkaensis andegavensis, contenta biceps, contenta elliptica, contenta parallela, contenta undulata, gallica laevissima, gallica montana, gallica perpusilla*

GENUS *Diademoides* Lesser Fiber Navicule Feather Diatom Brown Algae

SPECIES *luxuriosa*

GENUS *Luticola* Lute Fiber Navicule Feather Diatom Brown Algae

SPECIES *acidoclinata, argutula, binodis, clagesii, cohnii, costei, dapaliformis, dapalis, dapaloides, dismutica, frenguellii, gaussii, goeppertiana, heufleriana, incoacta, incoactoides, inserata, kraeuselii, lacertosa, lagerheimii, mitigata, mollis, monita, murrayi, mutica, muticoides, muticopsis, nivalis, obligata, palaearctica, paramutica, peguana, plausibilis, pseudomutica, saxophila, seposita, spinifera, tomesii, undulata, ventriconfusa, ventricosa, voigtii*

VARIETIES *inserata undulata, lagerheimii intermedia, mutica mutica, mutica binodis, mutica cohnii, mutica gibbosa, mutica hyalina, mutica intermedia, mutica lanceolata, mutica ornata, mutica ovalis, mutica rhomboidea, mutica sinica, mutica subcircularis, mutica subhexagona, mutica undulata, mutica ventricosa, nivalis distinguenda*

FAMILY 5 *Neidiaceae* Neid Navicule Feather Diatom Brown Algae

GENUS *Neidium* Neid Navicule Feather Diatom Brown Algae

SPECIES *affine, alpinum, ampliatum, andinum, angustirostratum, apiculatum, arenarzense, bergii, binode, binodeforme, bisulcatum, calvum, carterii, catarinense, coralieae, cuneatum, curtihamatum, decoratum, demerariense, densestriatum, dilatatum, distincte-punctatum, dubiforme, dubium, elongatum, essequiboanum, excisum, falklandicum, firmum, glaberrimum, globiceps, gracile, hamatum, hankensis, hercynicum, herrmannii, hitchcockii, hustedtii, improvisum, inconspicuum, incurvum, infirmum, iridis, javanicum, juba, ladogensis, lanceolata, latum, levanderi, longiceps, magellanica, meadeense, minutissimum, miopsa, mirum, obliquum, odamii, opulentum, paraffine, perforatum, perminutum, polygibbum, popeiense, productum, quadriundulatum, reimeri, sacoense, sardiniense, schmidtii, septentrionalis, subamphigomphus, subdubium, tabletta, temperei, tenuissimum, testa, vekhovii*

VARIETIES *affine affine, affine capitata, affine ceylonicum, affine globiceps, affine hankense, affine hercynicum, affine humerus, affine longiceps, affine tenuirostris, affine undulatum, bisulcatum, bisulcatum baicalense, bisulcatum subampliatum, bisulcatum subundulatum, bisulcatum undulatum, dilatatum jacutica, dubium dubium, dubium constrictum, globiceps triundulata, hitchcockii sauramoi, iridis iridis, iridis affinis, iridis amphigomphus, iridis amphirhynchus, iridis ampilata, iridis ampliatum, iridis bryophilum, iridis conspicua, iridis firmum, iridis firmum-subundulatum, iridis genuinum, iridis intermedia, iridis intermedium, iridis intermedium-subundulatum, iridis maxima, iridis maximum, iridis parallela, iridis producta, iridis robusta, iridis subampliatum, iridis subundulatum, iridis vernalis, kozlowi, kozlowi baicalensis-robusta, kozlowi ellipticum, kozlowi parvum, productum constricta*

GENUS *Diploneis* Diploid Neid Navicule Feather Diatom Brown Algae

SPECIES *advena, aestiva, aestuari, alpina, baicalensis, biremiformis, boldtiana, bombiformis, bomboides, bombus, bombus bombiformis, bombus egena, bombus gemina, burgitensis, caffra, chersonensis, chilensis, coffaeiformis, colae, confecta, constantinii, constricta, contigua, crabro, cynthia, decipiens, didyma, divergens, domblittensis, elliptica, entomon, eugenia, expleta, finnica, fusca, gorjanovicii, graeffii, gruendleri, hyalina, incurvata, interrupta, jasnitski, kahlii, kawabatae, krammeri, late-elliptica, lineata, littoralis, major, marginestriata, mauleri, mediterranea, meyerii, minuta, modica, nitescens, nitescens candida, notabilis, oamaruensis, oblongella, ocellata, oculata, ovalis, papula, parca, parma, petersenii, pfitzeriana, pneumatica, praesejuncta, pseudovalis, puella, rhombica, skvortzowii, smithii, solea, splendida, stroemii, subadvena, subcincta, suborbicularis, subovalis, turgida, vacillans, vetula, weissflogii*

VARIETIES *advena sansegana, aestiva aestiva, aestiva delicata, aestiva fusca, aestiva major, boldtiana acuminata, boldtiana australica, boldtiana ovalis, coffaeiformis subcircularis, constricta alpina, contigua eudoxia, contigua eugenia, crabro excavata, crabro expleta, crabro multicostata, crabro pandura, didyma linearis, divergens bacillaris, divergens elliptica, divergens parallela, divergens undulata, domblittensis domblittensis, domblittensis incisa, domblittensis subconstricta, elliptica elliptica, elliptica ladogensis, elliptica ostracodorum, elliptica pygmaea, fusca aestiva, fusca delicata, fusca gregorii, fusca major, fusca norvegica, fusca tenuipunctata, incurvata stricta, interrupta clancula, littoralis littoralis, littoralis clathrata, mediterranea gemmata, oblongella oblongella, oblongella gibbosa, ovalis arctica, ovalis oblongella-gibbosa, ovalis pumila, papula constricta, skvortzowii punctata, smithii aestiva, smithii nitescens, smithii pumila, smithii recta, smithii rhombica, suborbicularis constricta, subovalis baicalensis, vacillans delicatula, vacillans minuta*

GENUS *Anomoeoneis* Anomalous Neid Navicule Feather Diatom Brown Algae

SPECIES *exilis, follis, garrensis, serians, serians sphaerophora, starmuehlneri, styriaca, zellensis*

VARIETIES *exilis gomphonemacea, exilis lanceolata, exilis subacuminata, acuta, serians brachysira, serians brachysira-thermalis, serians thermalis, sphaerophora guntheri, sphaerophora polygramma, sphaerophora sculpta, styriaca ventricosa, zellensis linearis*

GENUS *Caloneis* Callisto Neid Navicule Feather Diatom Brown Algae

SPECIES *aequatorialis, aerophila, africana, alpestris, amphisbaena, angustivalva, arctica, bacillaris, bacillum, backmanii, bicuneata, borealis, branderii, brevis, budensis, callosa, clevei, constans, continua, crassa, dorogostaiskyi, dumbeana, fenzlii, fontinalis, fossilis, fusoides, gjeddeana, glaberrima, gracilis, hendeyi, hitoyosiensis, hustedtii, hyalina, inliberalis, intermedia, kristinae, lacunarum, ladogensis, lamella, lancettula, latiuscula, lauta, lepidula, leptosoma, lewisii, liber, limosa, linearis, macedonica, molaris, nubicola, obesa, obtusa, odiosa, ostenii, patagonica, perlepida, permagna, pulchra, schroederi, schumanniana, silicula, stauroneiformis, subclevei, sublinearis, subsalina, tenuis, thermalis, trinodis, undulata, undulatiformis, vaucheriae, westii, zachariasii*

VARIETIES *alpestris inflata, amphisbaena aequata, amphisbaena fenzlii, amphisbaena fuscata, amphisbaena undulata, bacillaris bacillaris, bacillaris chandolensis, bacillaris thermalis, bacillum aemula, bacillum fonticola, bacillum lancettula, backmanii fasciata, brevis africana, clevei undulata, clevei uruguayensis, ladogensis densistriata, ladogensis latestriata, latiuscula rhombica, latiuscula subholstei, liber bicuneata, liber excentrica, liber umbilicata, limosa gibberula, limosa lancettula, limosa subinflata, limosa undulata, permagna elongata, pulchra interrupta, schumanniana biconstricta, schumanniana lancettula, silicula alpina, silicula arctica, silicula blancheana, silicula curta, silicula gibberula, silicula inflata, silicula intermedia, silicula limosa, silicula longissima, silicula minuta, silicula peisonis, silicula robusta, silicula subundulata, silicula truncatula, silicula tumida, silicula undulata, silicula ventricosa, westii cuyabenensis, westii holmiensis*

GENUS *Cosmineis* Cosmic Neid Navicule Feather Diatom Brown Algae

SPECIES *delawarensis, eta, grossepunctata, indifferens, lundstroemii, pusilla*

VARIETIES *lundstroemii frieseana, pusilla lanceolata*

GENUS *Cymatoneis* Cyme Neid Navicule Feather Diatom Brown Algae

SPECIES *margarita, sulcata*

GENUS *Neidiopsis* Lesser Neid Navicule Feather Diatom Brown Algae

SPECIES *levanderi, vekhovii, wulffii*

VARIETIES *levanderi capitata, levanderi tatrensis*

GENUS *Petroneis* Winged Neid Navicule Feather Diatom Brown Algae

SPECIES *arabica, deltoides, glacialis, granulata, groveoides, humerosa, japonica, latissima, marina, monilifera, plagiostoma, punctigera, subdiffusa, transfuga, xi*

VARIETIES *humerosa arabica, humerosa constricta, humerosa minor, monilifera heterosticta*

GENUS *Pexisperma* Pex Disc Navicule Feather Diatom Brown Algae

SPECIES *amplectens, dichrosperma, lutescens, sputo, truncata*

GENUS *Raphidodiscus* Radish Disc Navicule Feather Diatom Brown Algae

SPECIES *christianii, febigeri, marylandicus*
GENUS *Rhoikoneis* Rhoiko Neid Navicule Feather Diatom Brown Algae
SPECIES *trinodis*
GENUS *Scolioneis* Twisted Neid Navicule Feather Diatom Brown Algae
SPECIES *brunkseiensis, tumida*
VARIETIES *tumida adriatica, tumida anglica, tumida subsalsa*
GENUS *Stauroneis* Crosshatch Neid Navicule Feather Diatom Brown Algae
SPECIES *accedens, acidobionta, acuta, acutiuscula, aerophila, africana, agrestis, alabamae, aleutica, alpina, amica, amicula, amphibia, amphicephala, amphilepta, amphioxys, amphoroides, amplipora, anceps, ancepsopsis, angustevittata, angustilancea, arctica, aspera, aueri, bacillum, baileyi, biformis, biundulata, borgei, borrichii, boudetii, branderii, brevis, brunii, capula, cataractae, circumborealis, conspicua, constricta, decipiens, didacta, dilatata, dubia, dubitabilis, elisa, exilis, finmarchica, fluminea, frauenfeldiana, fulmen, goeppertiana, graciliopsis, gracilior, gracilis, gracillima, gregoryii, heinii, hyperborea, impenda, inflata, italica, janischii, jarensis, javanica, joignerezii, koeltzii, koniamboensis, kriegeri, lanceolata, lapidicola, laterostrata, lauenburgiana, legleri, legumen, leguminiformis, leguminopsis, linearis, livingstonii, lundii, madagascariensis, mangelsdorffii, margaritorae, maunakeaensis, membranacea, mirabilis, muriella, nana, nebulosa, neofossilis, neohyalina, nobilis, obesa, obtusa, olympica, omurae, pachycephala, palustris, parajavanica, parasubgracilis, parviporis, parvula, parvulissima, perminuta, phoenicenteron, phylloides, producta, prominula, pseudosuboptusoides, punensis, pygmaea, quadripedis, recondita, rectangularis, reichardtii, resoluta, respectabilis, sagitta, salina, scandinavica, scaphulaeformis, schimanskii, schroederi, semen, semiaperta, septrionalis, siberica, simulans, singula, smithii, spicula, stellae, stodderi, subalpina, subgracilis, tackei, thermicola, truncata, tumidula, tyrrhenica, undata, valderostrata, ventriosus, wittrockii, zairensis*
VARIETIES *africana acuminata, alabamae angulata, amphioxys obtusa, anceps anceps, anceps americana, anceps amphicephala, anceps birostris, anceps elliptica, anceps elongata, anceps extrema, anceps gracilis, anceps hyalina, anceps javanica, anceps lata, anceps linearis, anceps obtusa, anceps recta, anceps rhomboidalis, anceps siberica, anceps subrostrata, biformis australis, borrichii borrichii, borrichii subcapitata, dilatata baicalensis, exilis constricta, exilis exigua, fulmen capitata, gracilis bicapitellata, javanica lapponica, kriegeri undulata, legumen balatonis, legumen parva, nobilis baconiana, obtusa catarinensis, obtusa lapponica, obtusa rostrata, parvula capitata, parvula major, phoenicenteron amphilepta, phoenicenteron brevis, phoenicenteron brunii, phoenicenteron gracilis, phoenicenteron hattorii, phoenicenteron nobilis, salina latior, salina wislouchii, smithii smithii, smithii borgei, smithii incisa, smithii minima, smithii sagitta*
GENUS *Trachyneis* Rough Neid Navicule Feather Diatom Brown Algae
SPECIES *aspera, clepsydra, clepsydra scotica*
VARIETIES *aspera elliptica, aspera intermedia, aspera pulchella*

FAMILY 6 *Pinnulariaceae* Pin Navicule Feather Diatom Brown Algae
GENUS *Pinnularia* Pin Navicule Feather Diatom Brown Algae
SPECIES *abaujensis, abbreviata, acidophila, acorica, acoricola, acrosphaeria, acuminata, acuta, aestuarii, alabamae, alfinitoana, alpina, alta, altiplanensis, amabilis, amazonica, ambigua, ambositrana, amphicephala, amphioxys, anglica, angulata, angusta, angustiborealis, angustistriata, angustivalva, aperta, apiculata, appendiculata, araucanensis, asanzae, bacilla, bacilliformis, balatonis, balfouriana, baltica, bardii, bertrandii, biceps, biclavata, bigibba, biglobosa, bilobata, bipectinalis, bistriata, blasii, bleischii, boecheri, bogotensis, borealis, boyeriformis, brandeliformis, brandelii, brauniana, brevicostata, britannica, bullacostae, burkii, canadensis, cardinaliculus, cardinalis, carlsonii, carlssonii, carminata, carteri, castraregina, cavacinii, certa, cherryfieldiana, cincta, circumducta, claasseniae, claviculiformis, claviculus, cleveana, clevei, cleveiformis, cocconeis, commutata, complexa, confirma, copulata, costata, crabro, cruciformis, cruxarea, cuneata, cuneatiformis, cuneola, curticostata, curtispinulosa, curtistriata, cuyabenensis, dactylus, debesii, debilis, denotata, denticulata, discifera, discrepantia, dispar, disphenia, distinguenda, divergens, divergentissima, diversarea, domingensis, domii, dubitabilis, egregia, ehrenbergii, elegans, elginensis, episcopalis, ergadensis, esox, falaiseana, fasciata, ferroindulgentissima, ferrophila, fistuca, flexuosa, fontellii, fossilis, frauenbergiana, frequentis, fumantii, garganica, gasseana, gastrum, genkalii, gentilis, gibba, gibbiformis, gigaformis, globiceps, gracillima, graciloides, graphica, grunowii, halophila, hartleyana, hemiptera, hemiptera, hemiptera hemipteriformis, heteropleura, higoensis, hilseana, humilis, iatriaensis, ignobilis, ilka, imperfecta, inconstans, incrassata, infirma, inflata, insignis, instabiliformis, instabilis, intermedia, interrupta, iridis, islandica, isostauron, ivaloensis, jarensis, julma, jungii, kanakarum, karelica, kefvingensis, kittelii, kneuckerii, kociolekii, kolbei, krammeri, krasskei, krookii, krookiformis, kuetzingii, kuolensis, kwacksii, lacunicola, lagerheimii, lagerstedtii, lanceolata, lange-bertalotii, lapponica, lata, latarea, latestriata, latevittata, laticeps, laucensis, legumen, legumiformis, lenticula, linearis, lokana, luculenta, lunata, lundii, macilenta, maculata, magnaconvexa, magnacuneata, major, martinii, mayeri, meridiana, mesogongyla, mesolepta, metzeltinii, microstauron, minuta, minutiformis, molaris, mondimia, montgomeryana, mormonorum, murdeguensis, neglecta, neglectiformis, neomajor, nitzschiophila, nobilefasciata, nobilis, nodosa, nodosiformis, normandiana, notabilis, notata, novazealandiae, oblongolinearis, obscura, obtusicuneata, oriunda, oriundiformis, ovalis, panduraelongata, parallela, parva, parvula, parvulissima, paschalis, patagonensis, paulina, percuneata, perinarensis, perinstabilis, perinterrupta, perirrorata, permontana, perspicua, persudetica, petersenii, placentula, platycephala, pluvianiformis, poasana, podzorskii, polyonca, problematica, procera, pseudacuminata, pseudocardinalis, pseudogibba, pseudomesogongyla, pseudoparva, pseudosimilis, pulchra, quadratarea, raã«ana, rabenhorstii, radiosa, rangoonensis, rathsbergiana, rectangulata, reichardtii, renata, renatiformis, retusa, rhenohassiaca, rhombarea, rhombica, rhomboelliptica, rhombofasciata, rhynchocephala, rotundicapitata, rumrichae, rupestris, ruttneri, sakulensis, santafiorana, saprophila, sardoelliptica, savona, saxicola, scapha, schimanskii, schoenfelderi, schroederii, schroeterae, schwabei, scotica, semicruciata, semizebra, septentrionalis, siebenkittlii, silesiaca, similiformis, similis, simplex, sinistra, siona, sistassa, socialis, spauldingae, spinea, spinosissima, spitsbergensis, stauntonii, staurocuneata, stauroneiformis, stauroptera, stidolphii, stomatophora, streckerae, streptoraphe, styliformis, subacuminata, subanglica, subbacillum, subboyeri, subcapitata, subcommutata, subconstricta, subdivergens, subfalaiseana, subflamma, subgibba, subgigas, subinterrupta, sublanceolata, sublinearis, submicrostauron, subnodosa, subrhombica, subrostrata, subrupestris, subsilvatica, subsolaris, substomatophora,*

substreptoraphe, subtriumvirorum, subumbrosa, subviridis, suchlandtii, sudetica, sulcata, sundaensis, superdivergentissima, sylvatica, tabellaria, tagliaventiae, tenuirugosa, tenuis, termes, termitina, tibetana, tolottiana, torta, toscana, transversa, transversiformis, trevelyana, triangulata, trigonocephala, trivialiformis, trudis, tsonka, tumescens, turfosiphila, turnerae, tynnii, tyrrhenica, umbrosa, undulata, vacua, vaetterana, vangariana, vasta, vastoides, viridiformis, viridis, viridoides, weningeri, westii, wisconsinensis, woerthensis

VARIETIES *abaujensis subundulata, acoricola elongatirhombica, acoricola linearis, acoricola pumilio, acrosphaeria maxima, acrosphaeria turgida, acrosphaeria undulata, acuminata cuyabenensis, acuminata novazealandiae, acuminata steinitzii, angusta pisciculus, angusta rostrata, aperta doelle-jardii, appendiculata budensis, bertrandii angustefasciata, bertrandii orleansiana, biceps biceps, biceps amphicephala, biceps petersenii, biceps pusilla, biclavata abbreviata, borealis barodensis, borealis brevicostata, borealis islandica, borealis lanceolata, borealis rectangularis, borealis scalaris, borealis subislandica, borealis sublinearis, borealis tenuistriata, borealis thuringiaca, borealis undulata, brauniana brauniana, brauniana amphicephala, brevicostata sumatrana, brevicostata ventricosa, canadensis pulchella, cleveana cleveana, cleveana stauroneiformis, clevei angusta, cleveiformis ventricosa, complexa minor, cruciformis faeroensis, cuneata reducta, cuneata subacuta, cuyabenensis hyalina, divergens bacillaris, divergens elliptica, divergens fontellii, divergens ignorata, divergens linearis, divergens mesoleptiformis, divergens parallela, divergens rostrata, divergens sublinearis, divergens truncata, divergens undulata, divergentissima minor, ergadensis minor, esox esox, esox eifeliana, esox gracilescens, esox ventricosa, falaiseana lanceola, fasciata doloma, flexuosa cuneata, flexuosa neglecta, fossilis erratica, fumantii pumila, gibba brevistriata, gibba linearis, gibba mesogongyla, gibba parva, gibba sancta, gibba subundulata, gibbiformis floralensis, globiceps globiceps, globiceps angustefasciata, globiceps crassior, globiceps krookei, globiceps linearis, graciliodes latecapitata, hartleyana orientalis, hemiptera, hemiptera brevistriata, hemiptera debilis, hemiptera inconstans, hemiptera interrupta, ilka marchica, interrupta minutissima, julma tirolensis, krasskei ventricosa, lapponica alpina, lapponica constricta, lata lata, lata costata, lata latestriata, lata rabenhorstii, lata thuringiaca, latarea thermophila, latevittata domingensis, legumen legumen, legumen cuneata, legumen florentina, legumen undulata, linearis macrostauron, lundii baltica, lundii linearis, major major, major convergens, major horrida, major hyalina, major linearis, major neglecta, major paludosa, major pulchella, major woerthensis, martinii capitata, mayeri similis, meridiana meridiana, meridiana concava, meridiana parallela, mesolepta mesolepta, mesolepta angusta, mesolepta gibberula, mesolepta intermedia, mesolepta stauroneiformis, mesolepta tenuis, microstauron ambigua, microstauron biundulata, microstauron diminuta, neomajor cuneata, neomajor frequentis, neomajor inflata, nobilis gentilis, nobilis inarensis, nobilis regularis, nodosa nodosa, nodosa capitata, nodosa constricta, nodosa percapitata, notabilis inariensis, parva minuta, parva rostratissima, parva ventricosa, percuneata constricta, percuneata minor, persudetica silvatica, polyonca stidolphii, polyonca sumatrana, problematica problematica, problematica subundulata, pseudogibba rostrata, pulchra angusta, pulchra subtilis, quadratarea minor, quadratarea soederlundii, quadratarea subproducta, rectangulata subundulata, rhombarea brevicapitata, rhombarea decrescens, rhombarea halophila, rhombarea undulata, rhombarea variarea, rhombica ovata, rhomboelliptica inflata, rhomboelliptica novazealandiae, rhomboelliptica rhombica, rhomboelliptica zuricana, rupestris cuneata, schroeterae elliptica, septentrionalis americana, sinistra sublanceolata, stauroptera interrupta, stauroptera minuta, stauroptera parva, stomatophora recta, streptoraphe gibbosa, subcapitata diversa, subcapitata elongata, subcapitata hilseana, subcapitata paucistriata, subcapitata stauroneiformis, subcapitata subrostrata, subcapitata undulata, subgibba angustarea, subgibba capitata, subgibba gracilis, subgibba hustedtii, subgibba lanceolata, subgibba sublinearis, subgibba tirolensis, subgibba undulata, subrhombica angusta, subrostrata marchica, subrostrata obscuriformis, subrupestris parva, substreptoraphe bornholmiana, sudetica sudetica, sudetica brittanica, sudetica commutata, sudetica leptogongyla, sudetica rupestris, tenuis interrupta, umbrosa tropica, undulata mesoleptiformis, undulata stauroneiformis, undulata subundulata, viridis commutata, viridis diminuta, viridis distinguenda, viridis elliptica, viridis fallax, viridis intermedia, viridis leptogongyla, viridis mayeri, viridis minor, viridis rupestris, viridis semicruciata, viridis sudetica*

GENUS *Chamaepinnularia* — Chamise Pin Navicule Feather Diatom Brown Algae

SPECIES *alexandrowiczii, begeri, brasiliana, brasilianopsis, bremensis, calida, circumborealis, clamans, distantepunctata, evanida, ferrarioana, gandrupii, justa, krasskei, krookiformis, krookii, margaritacea, margaritiana, mediocris, oculus, reinventa, resalutata, rhombelliptica, rupicola, schauppiana, soehrensis, submusciola, tongatensis, truncata, vyvermanii, wiktoriae*

VARIETIES *reinventa urtroquarea*

GENUS *Diatomella* — Lesser Pin Navicule Feather Diatom Brown Algae

SPECIES *balfouriana, lecohui, ouenkohana*

VARIETIES *balfouriana constricta*

GENUS *Oestrupia* — Ostrup Pin Navicule Feather Diatom Brown Algae

SPECIES *bicontracta, ergadensis, musca, supergradata, zachariasii*

VARIETIES *zachariasii undulata*

FAMILY 7 *Plagiotropidaceae* — Deep Sea Navicule Feather Diatom Brown Algae

GENUS *Plagiotropis* — Deep Sea Navicule Feather Diatom Brown Algae

SPECIES *antarctica, arizonica, australis, baltica, cochinvitreopsis, delicatula, elegans, eximia, fasciata, gaussii, gibberula, grevillei, lepidoptera, longa, maxima, neapolitana, neosamoensis, neovitrea, oblonga, oviformis, primoris, recta, semihyalina, semistriata, subalata, tayrecta, vitrea*

VARIETIES *lepidoptera proboscidea*

GENUS *Banquisia* — Banquis Navicule Feather Diatom Brown Algae

SPECIES *belgicae*

GENUS *Entosphaeroides* — Entic Sphere Navicule Feather Diatom Brown Algae

SPECIES *bilinearis, irregularis*

GENUS *Eolimma* — Limm Navicule Feather Diatom Brown Algae

SPECIES zalokariae

GENUS *Ephemera* — Ephemeral Navicule Feather Diatom Brown Algae

SPECIES *bjoernoeyaensis, planamembranacea*

GENUS *Manguinea* — Manguin Navicule Feather Diatom Brown Algae

SPECIES *fusiformis, rigida*

GENUS *Mannsia* — Manns Navicule Feather Diatom Brown Algae

SPECIES *phantasma*
GENUS *Pachyneis* Thick Neid Navicule Feather Diatom Brown Algae
SPECIES *gerlachii*
GENUS *Plagiolemma* Lemma Navicule Feather Diatom Brown Algae
SPECIES *confusum, inflatum*
GENUS *Stauropis* Crosshatch Navicule Feather Diatom Brown Algae
SPECIES *granii*
GENUS *Staurotropis* Crosshatch Tropic Navicule Feather Diatom Brown Algae
SPECIES *seychellensis*
GENUS *Stilis* Stile Navicule Feather Diatom Brown Algae
SPECIES *subulatus*

FAMILY 8 *Pleurosigmataceae* Pleural Sigma Navicule Feather Diatom Brown Algae
GENUS *Pleurosigma* Pleural Sigma Navicule Feather Diatom Brown Algae
SPECIES *acutum, amarum, arcuatum, chandolensis, curiosum, cuspidatum, densestriatum, directum, elongatum, falcatum, finnmarchicum, lanceolatum, latum, longum, macilentum, majus, marinum, maroccanum, minutum, naviculaceum, normanii, planktonicum, rhombeum, salinarum, siberica, simonsenii, subhyalinum, subrigidum, subsalsum, tenuiforme*
VARIETIES *lanceolatum transversale-beta, longum subrigidum, siberica siberica, siberica manitounukensis*
GENUS *Donkinia* Donkin Sigma Navicule Feather Diatom Brown Algae
SPECIES *angusta, carinata*
GENUS *Gyrosigma* Gyre Sigma Navicule Feather Diatom Brown Algae
SPECIES *acuminatum, aestuarii, algoris, angulatum, attenuatum, balticum, cali, decorum, delicatulum, diaphanum, diminutum, diminutum, dissimile, distortum, elongatum, euhippus, exilis, eximium, fasciola, fogedii, fonticolum, formosum, gibbyi, gondwana, hippocampus, intermedium, kuetzingii, lacustre, littorale, macrum, mediterraneum, moresbyana, murphyi, nodiferum, obliquum, obscurum, obtusatum, parkerii, peisonis, pensacolae, prolongatum, quadratum, rectum, reimeri, reversum, rigidum, scalproides, scalprum, sciotoense, speciosum, spencerii, sterrenburgii, stidolphii, strigile, strigilis, strigosum, subsalina, subsalinum, temperei, tenuissimum, wansbeckii, wormleyi*
VARIETIES *acuminatum brebissonii, acuminatum curta, acuminatum gallica, acuminatum scalprum, angulatum angulatum, angulatum quadratum, angulatum robustum, attenuatum scalprum, balticum diminutum, delicatulum delicatulum, delicatulum gracile, delicatulum salinarum, distortum diaphanum, distortum parkeri, distortum stauroneoides, distortum undulatum, fasciola arcuatum, fasciola sulcatum, formosum formosum, formosum balearica, intermedium nubecula, prolongatum closterioides, rectum minutum, rigidum rigidum, rigidum giganteum, scalproides eximia, speciosum pulchrum, spencerii spencerii, spencerii curvula, spencerii exilis, spencerii minor, spencerii nodifera, spencerii subsalina*
GENUS *Hadubrandia* Hadubrand Sigma Navicule Feather Diatom Brown Algae
SPECIES *decipiens*
GENUS *Rhoicosigma* Rhoic Sigma Navicule Feather Diatom Brown Algae
SPECIES *arcticum, compactum, reichardtianum*
GENUS *Toxonidea* Toxic Sigma Navicule Feather Diatom Brown Algae
SPECIES *gregoriana, insignis, undulata*
VARIETIES *insignis undulata*

FAMILY 9 *Scoliotropidaceae* Twisted Navicule Feather Diatom Brown Algae
GENUS *Scoliotropis* Twisted Navicule Feather Diatom Brown Algae
SPECIES *latestriata*
GENUS *Biremis* Birem Twisted Navicule Feather Diatom Brown Algae
SPECIES *ambigua, baculus, circumtexta, clarensis, digitus, fritschii, hartzii, kriegeriana, leeawuliana, lucens, osbornei, ridicula, solitaria, tasmanica*
VARIETIES *fritschii lata*
GENUS *Scoliopleura* Pleural Twisted Navicule Feather Diatom Brown Algae
SPECIES *adriatica, campylogramma, peisonis, tumida, westii*

FAMILY A *Sellaphoraceae* Sellaphore Navicule Feather Diatom Brown Algae
GENUS *Sellaphora* Sellaphore Navicule Feather Diatom Brown Algae
SPECIES *alastos, americana, amoena, auldreekie, bacillarioides, bacillum, blackfordensis, capitata, costei, densistriata, disjuncta, fusticulus, gibbula, gregoriana, guyanensis, illustris, julma, laevissima, lambda, lanceolata, lange-bertalotii, laterostellata, laterostrata, madagascariensis, mantasoana, minima, mutata, mutatoides, nana, nanoides, nyassensis, obesa, omuelleri, pupula, rectangularis, rectilinearis, reviersii, riojae, sardiniensis, seminulum, subdeliberata, sublinearis, tapajosensis, tau, tenerrima, triundulata, ventraloconfusa, verecundiae, vitabunda*
VARIETIES *bacillum gregoriana, bacillum lepida, lambda densistriata, nyassensis minor, pupula pupula, pupula aquaeductae, pupula bacillarioides, pupula capitata, pupula convergens, pupula densistriata, pupula elliptica, pupula major, pupula mutata, pupula nyassensis, pupula paenei, pupula parapupula, pupula pseudopupula, pupula rectangularis, pupula rostrata, pupula undulata*
GENUS *Caponea* Cap Sellaphore Navicule Feather Diatom Brown Algae
SPECIES *caribbea*
GENUS *Fallacia* Fallacy Sellaphore Navicule Feather Diatom Brown Algae

SPECIES *aequoria, amphipleuroides, arenaria, auriculata, balnearis, bioculata, brachium, carpentariae, cassubiae, clepsidroides, clipeiformis, costericardii, cryptolyra, cunoniae, diploneoides, dithmarsica, ecuadoriana, egregia, escorialis, fenestrella, florinae, forcipata, fracta, frustulum, gemmifera, helensis, hudsonis, hummii, hyalinula, inattingens, indifferens, inscriptura, insociabilis, lange-bertalotii, litoricola, lucens, margino-punctata, marnieri, melanocephala, minima, mitis, monoculata, naumannii, nicobarica, nummularia, ny, nyella, oculiformis, omissa, perlucida, pseudoforcipata, pseudomitis, pseudomuralis, pseudony, pseudosemilyrata, pulchella, pygmaea, reichardtii, reissii, scaldensis, schaeferae, schoemaniana, semilyrata, solutepunctata, spatiata, standeriella, subforcipata, subhamulata, sublucidula, submitis, subsulcatoides, suspirii, tenera, teneroides, tenerrima, umpatica, versicolor, veterana, vitrea, vittata, wuestii, zonata*

VARIETIES *forcipata densistriata, forcipata punctata, pygmaea linearis, pygmaea producta, pygmaea rhomboidea, pygmaea subpygmaea, subhamulata undulata*

GENUS *Rossia* — Ross Sellaphore Navicule Feather Diatom Brown Algae

SPECIES *elliptica*

ORDER 6 *Rhaponeidales* Rhapid Feather Diatom Brown Algae

FAMILY 1 *Raphoneidaceae* — Radish Neid Feather Diatom Brown Algae

GENUS *Rhaponeis* — Radish Neid Feather Diatom Brown Algae

SPECIES *amphiceros, angulata, belgicaclavata, cocconeides, elliptica, fatula, fluminensis, fusus, krusadaiensis, lancettula, liburnica, mediopunctata, minutissima, miocenica, nitida, ossiformis, parallelica, parilis, petropolitana, praeparilis, robustata, schulzii, scutelloides, simonsenii, tatsunokuchiensis*

VARIETIES *belgica intermedia*

GENUS *Delphineis* — Delphi Neid Feather Diatom Brown Algae

SPECIES *angustata, kamenooensis, karstenii, sheshukovae, simonsenii, surirella, surirelloides*

GENUS *Diplomenora* — Menora Neid Feather Diatom Brown Algae

SPECIES *cocconeiformis*

GENUS *Neodelphineis* — Modern Delphi Neid Feather Diatom Brown Algae

SPECIES *indica, pelagica*

GENUS *Perissonoa* — Perissono Neid Feather Diatom Brown Algae

SPECIES *cruciata, parvula*

GENUS *Sceptroneis* — Sceptre Neid Feather Diatom Brown Algae

SPECIES *caduceus, dimorpha, facialis, gemmata, grunowii, humuncia, ligulata, marina, mayenica, mutabilis, ossiformis, pesplanus, propinqua, pupa, talwanii, tenuis, vermiformis*

VARIETIES *humuncia tridens, marina parva, marina perminuta*

FAMILY 2 *Psammodiscaceae* — Sand Disc Neid Feather Diatom Brown Algae

GENUS *Psammodiscus* — Sand Disc Neid Feather Diatom Brown Algae

SPECIES *nitidus*

ORDER 7 *Surirellales* Surir Feather Diatom Brown Algae

FAMILY 1 *Surirellaceae* — Surir Feather Diatom Brown Algae

GENUS *Surirella* — Surir Feather Diatom Brown Algae

SPECIES *amphioxys, anceps, angusta, apiculata, arachnoidea, archibaldii, armoricana, astridae, asymmetrica, atomus, barrowcliffia, bifrons, birostrata, biseriata, bohemica, brebissonii, brebissonii brebissonii, brebissonii kuetzingii, brebissonii punctata, bryophila, calcarata, capronii, cardinaliskitton, caspia, celtica, citrus, comis, conifera, constricta, costei, cruciata, crumena, cuspidata, delicatissima, desikacharyi, didyma, elegans, elliptica, elongata, engleri, fastuosa, fluminensis, gemma, gotohiana, gracilis, granulata, guatamalensis, hastata, helvetica, hispida, hoefleri, hybrida, intermedia, iowensis, islandica, jenneri, lacus-vulcani, ladogensis, lagerheimii, lapponica, lata, lepida, librile, linearis, marina, microcora, minima, minuta, moelleriana, muscicola, nana, neocaledonica, nervosa, nobilis, norvegica, oregonica, ovalis, pandura, panduriformis, patella, patrimonii, peisonis, pinnata, pseudotenuissima, pygmaea, pyriformis, rattrayi, rauhii, regina, regula, rhombica, roba, robusta, robustior, rottii, rumrichorum, saxonica, schuwerkii, sigmoidea, simplex, smithii, solea, spinifera, spiralis, splendida, striata, striatula, strigosa, subsalsa, suecica, suevica, susanae, sutilis, tenera, tenerisilex, tenuis, tenuissima, terricola, thienemanni, thumensis, tibetica, turgida, turkanensis, undulata, variabilis, ventricosa, verrucosa, visurgis*

VARIETIES *angusta angusta, angusta apiculata, angusta constricta, angusta panduriformis, biseriata biseriata, biseriata amphioxys, biseriata bicuspidata, biseriata bifrons, biseriata constricta, biseriata diminuta, biseriata minor, biseriata orientalis, biseriata punctata, biseriata rostrata, biseriata subparallela, capronii capronii, capronii calcarata, caspia cardaria, celtica constricta, crumena baltica, crumena brightwellii, delicatissima longirostris, elegans constricta, fastuosa fastuosa, fastuosa cuneata, fastuosa hohenackeri, helvetica helvetica, helvetica tibetica, hybrida contracta, lata contracta, linearis linearis, linearis amphioxys, linearis constricta, linearis helvetica, linearis pseudolinearis, neocaledonica punctata, ovalis ovalis, ovalis aequalis, ovalis africana, ovalis ambigua, ovalis angusta, ovalis apiculata, ovalis baltica, ovalis crumena, ovalis lewisii, ovalis maxima, ovalis minuta, ovalis pinnata, ovalis salina, peisonis pyriformis, pinnata minuta, pinnata panduriformis, robusta robusta, robusta armata, robusta constricta, robusta hustedtiana, robusta lata, robusta obtusa, robusta obtusa-plana, robusta punctata, robusta splendida, splendida nobilis, splendida punctata, striatula algensis, tenera tenera, tenera nervosa, tenera splendidula, tenera subconstricta, turgida turgida, turgida margarita*

GENUS *Campylodiscus* Field Disc Surir Feather Diatom Brown Algae

SPECIES *angularis, biangulatus, bicostatus, birostratus, centralis, clypeus, costatus, daemelianus, decorus, echineis, eximius, fastuosus, hibernicus, hodgsonii, horologium, hypodromus, impressus, indianorum, innominatus, lacusbaikali, latus, levanderi, limbatus, lorenzianus, noricus, pervulsus, ralfsii, robertsianus, simulans, spiralis*

VARIETIES *biangulatus biangulatus, biangulatus lorenzianus, clypeus bicostatus, clypeus ovatus, decorus ralfsii, echineis echineis, echineis cribrosus, fastuosus parvulus, fastuosus thuretii, hypodromus gregorii, noricus hibernica, noricus tenera*

GENUS *Cymatopleura* Cyme Surir Feather Diatom Brown Algae

SPECIES *angulata, apiculata, brunii, cochlea, elliptica, hibernica, librile, nobilis, solea, spiralis*

VARIETIES *elliptica apiculata, elliptica constricta, elliptica hibernica, elliptica nobilis, librile apiculata, solea apiculata, solea clavata, solea gracilis, solea graciloides, solea pfuhlii, solea regula, solea subconstricta*

GENUS *Hydrosilicon* Hydric Silicon Surir Feather Diatom Brown Algae

SPECIES *mitra*

GENUS *Petasaria* Petas Surir Feather Diatom Brown Algae

SPECIES *heterolepis*

GENUS *Petrodictyon* Wing Net Surir Feather Diatom Brown Algae

SPECIES *contiguum, foliatum, gemma, gemmoides, patrimonii*

GENUS *Plagiodiscus* Pelagic Disc Surir Feather Diatom Brown Algae

SPECIES *enervatus*

GENUS *Stenopterobia* Stencil Wing Surir Feather Diatom Brown Algae

SPECIES *alpina, arctica, capitata, crassior, curvula, cuspidata, delicatissima, densestriata, eximia, heribaudii, intermedia, krammeri, minuscula, muscicola, planctonica, pumila, ruppeliana, scolopendra, sigmatella*

VARIETIES *cuspidata giluwensis, heribaudii detrita, intermedia capitata, intermedia densestriata, intermedia subacuta*

FAMILY 2 *Auriculaceae* Ear Surir Feather Diatom Brown Algae

GENUS *Auricula* Ear Surir Feather Diatom Brown Algae

SPECIES *amphitritis, cholnokyi, complexa, decipiens, densestriata, dubia, insecta, intermedia, japonica, minuta, mucronata, ostrea, punctata, quinquelobata, staurophora*

FAMILY 3 *Entomoneidaceae* Insect Neid Surir Feather Diatom Brown Algae

GENUS *Entomoneis* Insect Neid Surir Feather Diatom Brown Algae

SPECIES *aequabilis, alata, corrugata, costata, crenulata, decussata, densistriata, kjellmanii, oestrupii, ornata, paludosa, pseudoduplex, punctulata, robusta, vertebralis*

VARIETIES *paludosa paludosa, paludosa borealis, paludosa densistriata, paludosa duplex, paludosa hyperborea, paludosa punctulata, paludosa subsalina*

ORDER 8 *Thalassiophysales* Thalassic Feather Diatom Brown Algae

FAMILY 1 *Thallasiophysaceae* Thalassic Feather Diatom Brown Algae

GENUS *Thalassiophysa* Thalassic Feather Diatom Brown Algae

SPECIES *hyalina, rhipidis*

FAMILY 2 *Catenulaceae* Chain Thalassic Feather Diatom Brown Algae

GENUS *Catenula* Chain Thalassic Feather Diatom Brown Algae

SPECIES *adhaerens, pelagica, robusta*

GENUS *Amphora* Amphore Chain Thalassic Feather Diatom Brown Algae

SPECIES *acuta, acutiscula, aequalis, ajajensis, angularis, angusta, arcuata, arcus, arenaria, arenicola, argus, atomus, ayensuensis, bacillaris, bigibba, binodis, bioculata, borealis, bullatoides, bullosa, calumetica, cingulata, coffeaeformis, commutata, copulata, costata, costulata, crassa, crucifera, cruciferoides, cuneata, cymbamphora, cymbifera, decipiens, decussata, delicatissima, dissimilis, donkinii, dubia, dusenii, eatoniana, elegans, elliptica, elongata, ergadensis, eunotia, excisa, exigua, eximia, extensa, flexuosa, fluminensis, fogediana, fonticola, fontinalis, fusca, gigantea, gigas, globosa, gracilis, graeffeana, granulata, gregorii, grevilleana, hassatica, hemicycla, hendeyi, holsatica, holsaticoides, huronensis, hyalina, inariensis, insecta, jostesorum, laevis, laevissima, lagerheimii, lanceolata, leighsmithiana, levenensis, libyca, linearis, lineolata, longa, lyrata, macilenta, magnifica, margalefi, marina, membranacea, mexicana, michiganensis, micrometra, milesiana, minutissima, monglica, monilifera, montana, mucronata, neglecta, neomarina, nobilis, normanii, oblonga, obscura, obtusa, ocellata, oligotraphenta, ostrearia, ovalis, parallelistriata, paraveneta, pediculus, perpusilla, praelata, proboscidea, proteus, pseudoholsatica, pseudohyalina, pulchella, pulchra, pusilla, pusio, quadrata, recta, rhombica, robusta, rotunda, sabyii, salina, sibirica, sinuata, spectabilis, spitzbergensis, streckerae, strigosa, subcapitata, subcostulata, subholsatica, subinflata, subturgida, sulcata, tenera, tenerrima, tenuistriata, terroris, thromboliticola, thumensis, tomiakae, towutensis, transcaspica, triundulata, truncata, turgida, undulata, valida, veneta, ventricosa, vitrea*

VARIETIES *acuta arcuata, arcus sulcata, arenaria donkinii, arenaria rattrayi, arenicola minor, arenicola oculata, coffeaeformis acutiuscula, coffeaeformis angularis, coffeaeformis borealis, coffeaeformis perpusilla, coffeaeformis thumensis, copulata epiphytica, gigantea gigantea, gigantea fusca, globosa perpusilla, graeffeana minor, laevis laevissima, lagerheimii minuta, macilenta macilenta, macilenta ergadensis, margalefi lacustris, obscura diminuta, obtusa rectangularis, ocellata cingulata, ostrearia lineata, ostrearia vitrea, ovalis affinis, ovalis constricta, ovalis gracilis, ovalis libyca, ovalis minor, ovalis minutissima, ovalis pediculus, proteus oculata, veneta veneta, veneta angularis, veneta capitata, veneta grossestriata*

GENUS *Amphoropsis* — Lesser Amphore Chain Bladder Feather Diatom Brown Algae

SPECIES *decipiens, vanheurckii*

GENUS *Undatella* — Wavy Chain Thalassic Feather Diatom Brown Algae

SPECIES *lineata, quadrata*

FAMILY 3 *Thalassionemataceae* — Thread Thalassic Feather Diatom Brown Algae

GENUS *Thalassionema* — Thread Thalassic Feather Diatom Brown Algae

SPECIES *antiquum, bacillare, claviforme, frauenfeldii, hirosakiense, javanicum, nitzschioides, pseudonitzschioides, robustum, synedriforme*

VARIETIES *nitzschioides lanceolata, nitzschioides obtusa*

GENUS *Lioloma* — Liolome Thread Thalassic Feather Diatom Brown Algae

SPECIES *delicatulum, elongatum, pacificum*

GENUS *Thalassiothrix* — Lesser Thread Thalassic Feather Diatom Brown Algae

SPECIES *fruaenfeldii, gibberula, longissima, miocenica, monospina, robusta, spathulata*

GENUS *Trichotoxon* — Toxic Thread Thalassic Feather Diatom Brown Algae

SPECIES *reinboldii*

FAMILY 4 *Thalassiphoraceae* — Least Thread Thalassic Feather Diatom Brown Algae

GENUS *Thalassiphora* — Least Thread Thalassic Feather Diatom Brown Algae

SPECIES *maxima*

GENUS *Thalassomyxa* — Mycote Thread Thalassic Feather Diatom Brown Algae

SPECIES *canariensis*

GENUS *Subathua* — Athuan Thread Thalassic Feather Diatom Brown Algae

SPECIES *balcanica, sahnii, spinosa*

CLASS 611.2 *Coscinodiscophyceae* Disc Diatom Brown Algae

ORDER 1 *Coscinodiscales* Disc Diatom Brown Algae

FAMILY 1 *Coscinodiscaceae* — Disc Diatom Brown Algae

GENUS *Coscinodiscus* — Disc Diatom Brown Algae

SPECIES *alboranii, angstii, apiculatus, aralensis, argus, asteromphalus, australiensis, bellum, bipartitus, bouvet, centralis, commutatus, concinniformis, concinnus, curvatulus, debilis, decrescens, demergitus, devius, dissoniformis, dissonus, distinctus, divisus, fallax, fimbriatus, flexuosus, gigas, granii, gregorii, hajosiae, janischii, joergensenii, jonesianus, kolbei, kurzii, lacunosus, lacustris, lanceolatum, lewisianus, marginatus, minutus, miocenicus, moelleri, moholensis, morsianus, mutabilis, nitidus, nobilis, norwegicus, oblongus, obovatus, obscurus, oculus-iridis, ovalis, pavillardii, perforatus, planiusculus, plicatoides, plicatus, polymorphus, praenitidus, praepaleaceus, praeyabei, princeps, radiatus, rothii, sellatus, septentrionalis, steinyi, subbulliens, subtilis, temperei, thorii, tuberculatus, uralensis, velatus, vulnificus, wailesii, wittianus, woodwardii*

VARIETIES *apiculatus apiculatus, apiculatus ambiguus, asteromphalus hybrida, lacustris denticulatus, lacustris iris, lacustris papillatus, lacustris pellucidus, lacustris stellatus, lacustris tympaniformis, lewisianus concavus, lewisianus levis, perforatus pavillardi, rothii subsalsum, subtilis subtilis, subtilis fluviatilis, subtilis normanii, temperei aspinosus, temperei delicatus*

GENUS *Actinodictyon* — Ray Net Disc Diatom Brown Algae

SPECIES *atlanticum, elegans, fergusonii*

GENUS *Actinodiscus* — Ray Disc Diatom Brown Algae

SPECIES *barbadensis, contabulatus, corolla, italicus, javanicus, mascarenensis, mutabilis, philippinarum*

GENUS *Actinophaenia* — Dusky Ray Disc Diatom Brown Algae

SPECIES *splendens*

GENUS *Actinostephanos* — Ray Stephan Disc Diatom Brown Algae

SPECIES *matrensis, rudis*

GENUS *Anisodiscus* — Anise Disc Diatom Brown Algae

SPECIES *pantocsekii*

GENUS *Arachnoidiscus* — Spider Disc Diatom Brown Algae

SPECIES *ehrenbergii, japonicus*

GENUS *Aulacodiscus* — Aulac Disc Diatom Brown Algae

SPECIES *argus, breviprocessus, kittonii, kryshtofovichii, mastoideus, petersii, septus, sollitianus, tenuis*

VARIETIES *septus pentagonus, septus tetragonus*

GENUS *Australodiscus* — Australia Disc Diatom Brown Algae

SPECIES *peruvianus*

GENUS *Cestodiscus* — Girdle Disc Diatom Brown Algae

SPECIES *antarcticus, ovalis, reticulatus*

GENUS *Corethron* — Corethron Disc Diatom Brown Algae

SPECIES *columna, criophilum, hispidum, inerme, murrayanum, pelagicus, pennatum, valdiviae*

VARIETIES *criophilum inerme*
GENUS *Cosmiodiscus* Cosmic Disc Diatom Brown Algae
SPECIES *insignis*
VARIETIES *insignis triangula*
GENUS *Craspedodiscus* Fringed Disc Diatom Brown Algae
SPECIES *coscinodiscus, excavatus, microdiscus*
GENUS *Cymatodiscus* Cyme Disc Diatom Brown Algae
SPECIES *planetophorus*
GENUS *Discodiscus* Lesser Disc Diatom Brown Algae
SPECIES *tetrapora*
GENUS *Discoplea* Pleated Disc Diatom Brown Algae
SPECIES *astrea, comta, oligactis, oregonica*
GENUS *Discus* Least Disc Diatom Brown Algae
SPECIES *porcelaineus*
GENUS *Ethmodiscus* Ethmon Disc Diatom Brown Algae
SPECIES *gazellae*
GENUS *Glyphodiscus* Glyph Disc Diatom Brown Algae
SPECIES *intermedius*
GENUS *Gnathum* Gnath Disc Diatom Brown Algae
SPECIES *leve*
GENUS *Gossleriella* Gossler Disc Diatom Brown Algae
SPECIES *tropica*
GENUS *Hyalodiscopsis* Hyaline Disc Diatom Brown Algae
SPECIES *plana*
GENUS *Klinodiscus* Inclined Disc Diatom Brown Algae
SPECIES *obliquus*
GENUS *Palmeria* Palmer Disc Diatom Brown Algae
SPECIES *hardmaniana*
GENUS *Phacodiscus* Lentil Disc Diatom Brown Algae
SPECIES *punctulatus*
GENUS *Pontodiscus* Black Sea Disc Diatom Brown Algae
SPECIES *makarovae*
GENUS *Porodiscus* Pore Disc Diatom Brown Algae
SPECIES *calyciflos, minor, venkataramanii*
GENUS *Rocella* Rocelle Disc Diatom Brown Algae
SPECIES *gelida, gemma, princeps, schraderi, vigilans*
GENUS *Stellarima* Star Disc Diatom Brown Algae
SPECIES *distincta, microtrias, primalabiata, steinyi, stellaris*
GENUS *Stictocyclus* Stick Cyclic Disc Diatom Brown Algae
SPECIES *varicus*
GENUS *Stictodiscus* Stick Disc Diatom Brown Algae
SPECIES *gelidus, hardmanianus, parallelus, punctatus, rota, triangularis*
VARIETIES *hardmanianus coronatus, parallelus madagascarensis*

FAMILY 2 *Anaulaceae* Anaul Disc Diatom Brown Algae
GENUS *Anaulus* Anaul Disc Diatom Brown Algae
SPECIES *australis, mediterraneus, scalaris, sibiricus, simonsenii, uniseptatus*
VARIETIES *mediterraneus intermedius*
GENUS *Eunotogramma* Seaweed Anaul Disc Diatom Brown Algae
SPECIES *laevis, leve, marinum, novemloculatum, quinqueloculatum, rectum, septemloculatum, triloculatum*

FAMILY 3 *Asterolampraceae* Star Lamp Disc Diatom Brown Algae
GENUS *Asterolampra* Star Lamp Disc Diatom Brown Algae
SPECIES *centraster, gradiata, marylandica, praeacutiloba, vulgaris*
VARIETIES *vulgaris hyalina*
GENUS *Asteromphalus* Navel Star Lamp Disc Diatom Brown Algae
SPECIES *arachne, brookei, centraster, concavus, darwinii, elegans, flabellatus, heptactis, hookeri, hyalinus, inaequabilis, kennettii, oligocenicus, parvulus, roperianus, roundii, sarcophagus, symmetricus*

FAMILY 4 *Hemidiscaceae* Half Disc Diatom Brown Algae
GENUS *Hemidiscus* Half Disc Diatom Brown Algae
SPECIES *cuneiformis, triangularis, barkleyi*
VARIETIES *cuneiformis gibba, cuneiformis inornata, cuneiformis radiata*
GENUS *Actinocyclus* Ray Half Disc Diatom Brown Algae
SPECIES *actinochilus, barkleyi, cholnokyi, circellus, curvatulus, delicatus, dimorphus, exiguus, fasciculatus, fistulis, fryxelliae, fulvus, ingens, katzingii, kozyrenkoae, levis, ligularis, maccollumii, makarovae, niagarae, normanii, octonarius, roperii, sagittulus, spiritus, subtilis, tenuissimus, tesselatus, tunkaensis, vestigulus*
VARIETIES *ingens irregularis, ingens nodus, ingens ovalis, ingens planus, normanii subsalsus, octonarius crassus, octonarius ralfsii, octonarius sparsus, octonarius tenellus*

GENUS *Azpeitia* Azpeit Half Disc Diatom Brown Algae
SPECIES *africana, antiqua, barronii, biannulata, gombosii, neocrenulata, nodulifera, obscura, tabularis*
GENUS *Cymatotheca* Cyme Case Disc Diatom Brown Algae
SPECIES *weissflogii*
GENUS *Euodia* Euode Disc Diatom Brown Algae
SPECIES *udiensis*
GENUS *Euodiella* Lesser Euode Disc Diatom Brown Algae
SPECIES *bicornigera, biddulphioides, cretacea, eocenica, hendeyi, perplexa, semicircularis, tristictia*
GENUS *Roperia* Roper Half Disc Diatom Brown Algae
SPECIES *marginata, praetesselata, tesselata*

FAMILY 5 *Hemipeltaceae* Pelt Disc Diatom Brown Algae
GENUS *Hemipeltus* Pelt Disc Diatom Brown Algae
SPECIES *communis*
GENUS *Actinoptychus* Ray Fold Pelt Disc Diatom Brown Algae
SPECIES *amblyoceros, minutus, senarius, splendens, tenuis, tuberculatus, vulgaris*
GENUS *Glorioptychus* Glory Fold Pelt Disc Diatom Brown Algae
SPECIES *callidus*
GENUS *Lepidodiscus* Thin Pelt Disc Diatom Brown Algae
SPECIES *elegans*

FAMILY 6 *Lithodesmiaceae* Rock Fiber Disc Diatom Brown Algae
GENUS *Lithodesmium* Rock Fiber Disc Diatom Brown Algae
SPECIES *cornigerum, pliocenicum, rotundum, undulatum*
GENUS *Ditylum* Dity Rock Fiber Disc Diatom Brown Algae
SPECIES *brightwellii*
GENUS *Lithodemsoides* Lesser Rock Fiber Disc Diatom Brown Algae
SPECIES *polymorpha*

FAMILY 7 *Paraliaceae* Paral Disc Diatom Brown Algae
GENUS *Paralia* Paral Disc Diatom Brown Algae
SPECIES *sulcata, thybergii*
VARIETIES *sulcata biseriata, sulcata coronata, sulcata radiata*
GENUS *Ellerbeckia* Ellerbeck Paral Disc Diatom Brown Algae
SPECIES *kochii*
VARIETIES *kochii antiqua*
GENUS *Psammoparalia* Sand Paral Disc Diatom Brown Algae
SPECIES *elliptica*

FAMILY 8 *Rhizosoleniaceae* Rhizid Disc Diatom Brown Algae
GENUS *Rhizosolenia* Rhizid Disc Diatom Brown Algae
SPECIES *acicularis, acuminata, antarctica, antennata, arafurensis, barboi, bergonii, bezrukovae, borealis, bulbosa, calcar-avis, castracanei, chunii, clevei, costata, crassa, cretacea, curvata, curvirostris, debyana, decipiens, drafurensis, faeroensis, fallax, formosa, gracilis, gravida, hebetata, hyalina, imbricata, interposita, longiseta, massiva, minima, miocenica, morsa, norwegica, ostenfeldii, palliola, pokrovskajae, polydactyla, praealata, praebarboi, pungens, robusta, setigera, sigma, sima, similis, simplex, striata, styliformis, temperei, truncata, twistata*
VARIETIES *antennata semispina, castracanei neglecta, clevei communis, hebetata semispina, hebetata subacuta, polydactyla squamosa, sima silcea*
GENUS *Calyptrella* Calypter Rhizid Disc Diatom Brown Algae
SPECIES *robusta*
GENUS *Carciosolenia* Carcion Rhizid Disc Diatom Brown Algae
SPECIES *murrayi*
GENUS *Dactyliosolen* Finger Rhizid Disc Diatom Brown Algae
SPECIES *antarcticus, blavyanus, fragilissimus, phuketensis, tenuijunctus*
GENUS *Guinardia* Guinard Rhizid Disc Diatom Brown Algae
SPECIES *cylindrus, delicatula, flaccida, striata, tubiformis*
VARIETIES *striata maxima*
GENUS *Gunflintia* Gunflint Rhizid Disc Diatom Brown Algae
SPECIES *virtex*
GENUS *Gyrodiscus* Gyre Rhizid Disc Diatom Brown Algae
SPECIES *barghoornii, magna*
GENUS *Neocalyptrella* Modern Calypter Rhizid Disc Diatom Brown Algae
SPECIES *robusta*
GENUS *Probiscia* Snout Rhizid Disc Diatom Brown Algae
SPECIES *alata, barboi, cretacea, curvirostris, eumorpha, indica, inermis, interposita, praebarboi, subarctica, truncata*
VARIETIES *alata curvirostris, alata gracillima, alata indica*
GENUS *Pseudosolenia* Lesser Rhizid Disc Diatom Brown Algae
SPECIES *calcar-avis*

GENUS *Urosolenia* Tail Rhizid Disc Diatom Brown Algae
SPECIES *eriensis, gracilis, longiseta*
VARIETIES *eriensis gracilis, eriensis morsa*

ORDER 2 *Biddulphiales* Box Disc Diatom Brown Algae

FAMILY 1 *Biddulphiaceae* Box Disc Diatom Brown Algae
GENUS *Biddulphia* Box Disc Diatom Brown Algae
SPECIES *alternans, antediluviana, aurita, balearica, biddulphiana, calamus, cornuta, echinulata, ecornuta, granulata, moronensis, obtusa, pentacrinus, petrolia, radiata, regina, reticulata, reticulum, rhombus, sinensis, spinosa, striolata, tenera, tridentata, tridentula, tuomeyi*
VARIETIES *antediluviana pentagona, moronensis nicobarica, rhombus tetragona, rhombus trigona, tuomeyi charkoviana, tuomeyi tridentata*
GENUS *Biddulphiopsis* Lesser Box Disc Diatom Brown Algae
SPECIES *titiana*
VARIETIES *titiana arlyniana*
GENUS *Flexibiddulphia* Flexible Box Disc Diatom Brown Algae
SPECIES *semicircularis, udiensis*
GENUS *Hydrosera* Hydric Box Disc Diatom Brown Algae
SPECIES *triquetra, whampoensis*
GENUS *Isthmia* Isthmus Box Disc Diatom Brown Algae
SPECIES *enervis, nervosa*
VARIETIES *enervis subcylindica*
GENUS *Neohuttonia* Modern Hutton Box Disc Diatom Brown Algae
SPECIES *reichardtii*
GENUS *Terpsinoa* Terpsin Box Disc Diatom Brown Algae
SPECIES *musica*
GENUS *Trigonum* Trigonal Box Disc Diatom Brown Algae
SPECIES *alternans, arcticum, latum, moronense, tetragonum*
VARIETIES *arcticum quadratum, arcticum sexangulatum, latum dissimile, moronense nicobaricum, , tetragonum reticulum, tetragonum arcticum, tetragonum formosum*

ORDER 3 *Chaetocerotales* Bristle Disc Diatom Brown Algae

FAMILY 1 *Chaetocerotaceae* Bristle Disc Diatom Brown Algae
GENUS *Chaetoceros* Bristle Disc Diatom Brown Algae
SPECIES *aequatorialis, affine, anastomosans, angularis, armatum, atlanticum, bermejense, boreale, borealoides, breve, buceros, bungei, ceratosporum, cinctus, coarctatum, concavicorne, constrictum, convolutum, coronatum, costatum, crinitus, criophilus, curvisetum, dadayi, danicus, debile, decipiens, densum, diadema, didymum, difficile, diversicurvatum, diversum, eibenii, elmorei, elongatum, exospermum, externum, fallax, filiforme, flexuosum, fragile, furcellatum, galvestonense, gaussii, glandazii, gracile, hohnii, holsaticum, imbricatum, ingolfianum, janischianum, karianum, komarovii, laciniosum, lauderi, lorenzianum, mertensii, messanense, minimum, mitra, muellerii, neapolitanum, neogracile, paradoxum, pelagicum, perpusillum, peruvianus, phuketense, polygonum, pseudocurvisetum, radians, radicans, rostratus, saltans, scolopendra, seiracanthus, sessile, simile, simplex, skeleton, sociale, subsalsum, subsecundum, subtile, tenuisimum, teres, tetrachaeta, tetrastichon, thienemannii, throndsenii, tortissimum, transisetum, vistulae, volans, wighamii, willei, zachariasii*
VARIETIES *ceratosporum brachysetus, concavicorne volans, convolutum trisetosa, muellerii subsalsum, simplex calcitrans, simplex pumilus, throndsenii throndsenii, throndsenii trisetosus*
GENUS *Baeteriastrum* Baeter Star Bristle Disc Diatom Brown Algae
SPECIES *biconicum, delicatulum, elongatum, furcatum, hyalinum, parallelum, solitarium*

FAMILY 2 *Acanthocerataceae* Spiny Bristle Disc Diatom Brown Algae
GENUS *Acanthoceras* Spiny Bristle Disc Diatom Brown Algae
SPECIES *madeburgensis, zachariasii*

FAMILY 3 *Attheyaceae* Atthey Bristle Disc Diatom Brown Algae
GENUS *Attheya* Atthey Bristle Disc Diatom Brown Algae
SPECIES *arenicola, armata, decora, flexuosa, gaussii, lata, longicornis, septentrionalis, zachariasii*
VARIETIES *flexuosa enodulifer, zachariasii curvata*

FAMILY 4 *Leptocylindraceae* Thin Cylinder Bristle Disc Diatom Brown Algae
GENUS *Leptocylindrus* Thin Cylinder Bristle Disc Diatom Brown Algae
SPECIES *danicus, mediterraneus, minimus*

ORDER 4 *Cymatosirales* Cyme Disc Diatom Brown Algae

FAMILY 1 *Cymatosiraceae* Cyme Disc Diatom Brown Algae
GENUS *Cymatosira* Cyme Disc Diatom Brown Algae
SPECIES *adaroi, belgica, compacta, cornuta, coronata, elliptica, fossilis, lorenziana, praecompacta, robusta*

GENUS *Arcocellulus* — Arc Cyme Disc Diatom Brown Algae
SPECIES *cornucervis, mammifer*
GENUS *Brockmanniella* — Brockmann Cyme Disc Diatom Brown Algae
SPECIES brockmannii
GENUS *Campylosira* — Field Cyme Disc Diatom Brown Algae
SPECIES *alexandrica, cymbelliformis*
GENUS *Extubocellulus* — Tube Cyme Disc Diatom Brown Algae
SPECIES *cribriger, spinifer*
GENUS *Lennoxia* — Lennox Cyme Disc Diatom Brown Algae
SPECIES *faveolata*
GENUS *Leyanella* — Leyan Cyme Disc Diatom Brown Algae
SPECIES *arenaria*
GENUS *Minutocellus* — Minute Cyme Disc Diatom Brown Algae
SPECIES *polymorphus, scriptus*
GENUS *Papiliocellus* — Butterfly Cyme Disc Diatom Brown Algae
SPECIES *elegans, simplex*
GENUS *Plagiogrammopsis* — Pelagic Seaweed Cyme Disc Diatom Brown Algae
SPECIES *crawfordii, mediaequata, panduriformis, vanheurckii*

FAMILY 2 *Rutilariaceae* — Rutile Cyme Disc Diatom Brown Algae
GENUS *Rutilaria* — Rutile Cyme Disc Diatom Brown Algae
SPECIES *amplectans, areolata, attenuata, australis, barbadensis, briggeri, cantabrigiensis, craspedostigma, delicatula, epsilon, flabellifera, hannae, hustedtii, hyalina, kamyshlovensis, longirostris, minima, schenkii, tenuicornis, tenuis*
VARIETIES *areolata tenior, hannae diomedeae, minima danica, tenuicornis paleogena*
GENUS *Syndetocystis* — Syndet Cyst Rutile Cyme Disc Diatom Brown Algae
SPECIES *barbadensis, grevilleana, uralensis*
VARIETIES *grevilleana grevilleana, grevilleana monocellata*

ORDER 5 *Fragilariales* — Fragile Disc Diatom Brown Algae

FAMILY 1 *Fragilariaceae* — Fragile Disc Diatom Brown Algae
GENUS *Fragilaria* — Fragile Disc Diatom Brown Algae
SPECIES *acidoclinata, acuta, acutirostrata, aequalis, alpestris, amicorum, amphicephala, anceps, arcus, austriaca, barbararum, barbatula, berolinensis, biceps, bidens, binodis, bipunctata, bronkei, californica, canariensis, capitata, capucina, capucinadesm, cassubica, chilensis, constricta, costata, crassa, crotonensis, cyclopum, danica, delicatissima, demerarae, dorsiventralis, eichhornii, entomon, exiguiformis, fonticola, fossilis, fragilarioides, gaillonii, gedanensis, geocollegarum, germainii, gessneri, gracillima, grandis, guenter-grassii, hantzschiana, harrisonii, heidenii, henryi, hirosakiensis, hungarica, hyalina, improbula, incisa, incognita, inflata, intermedia, islandica, karelica, kriegeriana, labei, lacus-baikalensis, laevigata, lanceolata, lancettula, lapponica, lata, lenzii, leptostauron, linearis, loetschertii, longiceps, maillardii, martyi, mazamaensis, magocsyi, mesodon, mesolepta, microstriata, miniscula, montana, mutabilis, nanana, nanoides, neoproducta, nitzschioides, northumbrica, oldenburgiana, opacolineata, pectinalis, perminuta, polonica, polygonata, producta, pseudoconstruens, pulchella, punctata, punctiformis, quadrata, quebecensis, radians, reicheltii, rhomboides, riesgoviensis, robusta, rolandschmidtii, rotundissima, santaremensis, schulzii, shiloi, similis, socia, sopotensis, spinarum, strangulata, striatula, submicroscopica, suboldenburgiana, subsalina, synegrotesca, tabulata, tenera, tenuicollis, trigibba, ulna, undata, ungeriana, utermoehlii, vaucheriae, venter, viereckiana, vitrea, voeringia, zeilleri*
VARIETIES *aequalis producta, binodis pantocsekii, capitata chandolensis, capitata fossilis, capucina capucina, capucina acuta, capucina amphicephala, capucina austriaca, capucina capitellata, capucina gracilis, capucina lanceolata, capucina mesolepta, capucina perminuta, capucina radians, capucina rumpens, capucina subsalina, constricta stricta, constricta tetranodis, constricta trinodis, delicatissima angustissima, germainii acostata, harrisonii harrisonii, harrisonii amphitetras, harrisonii dubia, harrisonii rhomboides, heidenii heidenii, heidenii istvanffyi, hungarica tumida, intermedia continua, intermedia fallax, leptostauron leptostauron, leptostauron dubia, leptostauron martyi, linearis curta, mutabilis intercedens, pectinalis undulata, producta bohemica, pulchella constricta, pulchella lanceolata, tabulata australis, tabulata obtusa, tabulata tenuistriata, tabulata truncata, ulna acus, ulna amphirhynchus, ulna goulardi, ulna ungeriana, undata quadrata, undata stricta, undata tetranodis, vitrea minima, zeilleri elliptica*
GENUS *Adoneis* — Adonis Neid Fragile Disc Diatom Brown Algae
SPECIES *pacifica*
GENUS *Amphicampha* — Dual Camphor Fragile Disc Diatom Brown Algae
SPECIES *hemicyclus*
GENUS *Andrewsiella* — Andrews Fragile Disc Diatom Brown Algae
SPECIES *fossilis*
GENUS *Asterionella* — Aster Fragile Disc Diatom Brown Algae
SPECIES *bleakeleyi, candelabrum, cleveana, fibula, formosa, frauenfeldii, gracillima, inflata, madagascariensis, notata, ralfsii, synedriformis, thomassonii, zasuminensis*

VARIETIES *formosa formosa, formosa acaroides, formosa bleakeleyi, formosa gracillima, formosa ralfsii, formosa subtilis, formosa subtilissima, madagascariensis minor, madagascariensis tibiiformis, ralfsii americana, ralfsii hustedtiana*

GENUS *Asterionellopsis* Lesser Aster Fragile Disc Diatom Brown Algae

SPECIES *glacialis, socialis*

GENUS *Bleakeleya* Bleakeley Fragile Disc Diatom Brown Algae

SPECIES *notata*

GENUS *Catacombas* Catacomb Fragile Disc Diatom Brown Algae

SPECIES *gaillonii, obtusa*

VARIETIES *gaillonii gaillonii, gaillonii macilenta*

GENUS *Centronella* Central Fragile Disc Diatom Brown Algae

SPECIES *reicheltii*

GENUS *Ceratoneis* Horn Neid Fragile Disc Diatom Brown Algae

SPECIES *arcus, chilensis*

VARIETIES *arcus amphioxys, arcus linearis*

GENUS *Ctenophora* Comb Fragile Disc Diatom Brown Algae

SPECIES *pulchella*

VARIETIES *pulchella lanceolata*

GENUS *Diatoma* Disc Diatom Brown Algae

SPECIES *anceps, anomalum, costata, elongatum, fenestratum, gracillima, grande, hiemale, hyemale, minimum, moniliformis, obliquatum, ochridana, problematica, tenue, vitrea, vulgaris*

VARIETIES *anceps linearis, anomalum curtum, anomalum longissimum, costata curta, elongatum actinastroides, elongatum curta, elongatum hybridum, elongatum minor, elongatum subsalsa, elongatum tenuis, hyemale hyemale, hyemale mesodon, hyemale quadratum, moniliformis ovalis, tenue tenue, tenue elongatum, tenue pachycephala, vulgaris vulgaris, vulgaris brevis, vulgaris capitulatum, vulgaris constrictum, vulgaris grande, vulgaris lineare, vulgaris ovalis, vulgaris productum*

GENUS *Distrionella* Distrion Fragile Disc Diatom Brown Algae

SPECIES *asterionelloides, incognita*

GENUS *Falcula* Falcule Fragile Disc Diatom Brown Algae

SPECIES *rogallii*

GENUS *Fossula* Foss Fragile Disc Diatom Brown Algae

SPECIES *arctica*

GENUS *Fragilariforma* Lesser Fragile Disc Diatom Brown Algae

SPECIES *acidobiontica, bicapitata, constricta, floridana, hungarica, platensis, virescens*

VARIETIES *constricta quadrata, constricta stricta, constricta tetranodis, constricta trinodis, hungarica tumida, virescens capitata, virescens diatomacea, virescens elliptica, virescens exigua, virescens lata, virescens mesolepta, virescens oblongella, virescens producta, virescens subsalina*

GENUS *Fragilariopsis* Least Fragile Disc Diatom Brown Algae

SPECIES *atlantica, bjoernoeyaensis, curta, cylindriformis, cylindrus, doliolus, humboldtianorum, kerguelensis, oceanica, peragallii, pseudonana, reginae-jahniae, rhombica, ritscheri, separanda, sublinearis*

GENUS *Frankophila* Frank Fragile Disc Diatom Brown Algae

SPECIES *horstii, loetschertii, maillardii, similoides*

GENUS *Hannaea* Hanna Fragile Disc Diatom Brown Algae

SPECIES *arcus*

VARIETIES *arcus arcus, arcus amphioxys, arcus linearis*

GENUS *Hyalosynedra* Haline Synergetic Fragile Disc Diatom Brown Algae

SPECIES *laevigata*

GENUS *Martyana* Marty Fragile Disc Diatom Brown Algae

SPECIES *atomus, martyi, schulzii*

VARIETIES *martyi baicalensis*

GENUS *Meridion* Meridian Fragile Disc Diatom Brown Algae

SPECIES *anceps, circulare, constrictum, hohii, intermedium, vernale, zinckenii*

VARIETIES *circulare constricta, circulare zinckenii, intermedium constrictum*

GENUS *Neosynedra* Modern Synergetic Fragile Disc Diatom Brown Algae

SPECIES *provincialis*

GENUS *Opephora* Opephore Fragile Disc Diatom Brown Algae

SPECIES *ansata, burchardtiae, gemmata, guenter-grassii, horstiana, krumbeinii, marina, minuta, mutabilis, naveana, neoelliptica, olsenii, pacifica, parva, pinnata, schulzii, schwartzii*

VARIETIES *pacifica trigona*

GENUS *Podocystis* Pod Cyst Fragile Disc Diatom Brown Algae

SPECIES *adriatica, spathulata*

GENUS *Porannulus* Pore Ring Fragile Disc Diatom Brown Algae

SPECIES *contentus*

VARIETIES *contentus circulus*

GENUS *Psammosynedra* Sand Synergetic Fragile Disc Diatom Brown Algae

SPECIES *closterioides*

GENUS *Pseudoeunotia* Lesser Notable Fragile Disc Diatom Brown Algae

SPECIES *flexuosa, lunaris*
GENUS *Pseudostaurosira* Lesser Crosswise Pit Disc Diatom Brown Algae
SPECIES *brevistriata, naveana, pseudoconstruens, robusta, zeilleri*
VARIETIES *brevistriata inflata, brevistriata trigibba, brevistriata turgida*
GENUS *Pteroncola* Rock Fragile Disc Diatom Brown Algae
SPECIES *marina*
GENUS *Punctastriata* Dotted Stripe Fragile Disc Diatom Brown Algae
SPECIES *linearis, ovalis*
GENUS *Stauroforma* Crosswise Fragile Disc Diatom Brown Algae
SPECIES *enermis, exiguiformis, inermis*
GENUS *Staurosira* Crosswise Siren Fragile Disc Diatom Brown Algae
SPECIES *altiplanensis, aventralis, berolinensis, capucina, construens, elliptica, entomon, harrisonii amphitetras, harrisonii fossilis, lapponica, laucensis, laucensis vulpina, mercedes, microstriata, mutabilis, oldenburgiana, proboscoidea, pseudoconstruens, punctiformis, robusta, subsalina*
VARIETIES *construens binodis, construens exigua, construens pumila, construens pumila-subrotunda, construens pusilla, construens rhomboidalis, construens subsalina, construens triundulata, construens venter*
GENUS *Staurosirella* Lesser Crosswise Siren Fragile Disc Diatom Brown Algae
SPECIES *africana, ansata, berolinensis, lapponica, leptostauron, pinnata, spinosa*
VARIETIES *leptostauron dubia, pinnata acuminata, pinnata baicalensis, pinnata capitata, pinnata elliptica, pinnata inflata, pinnata intercedens, pinnata lancettula, pinnata minutissima, pinnata parallela, pinnata sublitoralis, pinnata subrotunda, pinnata trigona*
GENUS *Synedra* Synergetic Fragile Disc Diatom Brown Algae
SPECIES *acicularis, actinastroides, acula, acus, aggregata, alpina, amoricana, amphicephala, amphirhyncha, amphirhynchus, arctica, arcuata, austriaca, bacillaris, baculus, balatonis, berolinensis, biceps, bicurvata, bilunaris, campyla, capilla, capitellata, chaseii, closterioides, constricta, crotonensis, crystallina, cuneata, cyclopum, danica, debilis, delicatissima, demerarae, dissipata, falcata, famelica, familiaris, fasciculata, filiformis, formosa, fragilaroides, fulgens, goulardi, gracilis, hamata, hartii, hemicyclus, inaequalis, incisa, inflata, joursacensis, jouseana, lanceolata, lata, limnetica, lismorensis, longa, longiceps, longissima, lunaris, miniscula, minutissima, montana, nana, netronoides, nitzschii, nyansae, obtusa, ostenfeldii, oxyrhyncha, palea, parva, parvula, perpusilla, pulchella, radians, rostrata, rumpens, scalaris, simalongis, smithii, socia, spathulifera, spectabilis, splendens, stela, subaequalis, subarcuata, subtilis, subula, superba, tabulata, tenera, tenuis, tenuissima, ulna, utermoehlii, vaucheriae, vermicularis, vitrea*
VARIETIES *acus acus, acus acula, acus angustissima, acus delicatissima, acus neogena, acus ostenfeldii, acus radians, acus tenuissima, acus varipunctata, amphicephala austriaca, amphicephala densestriata, amphicephala fallax, amphicephala intermedia, bilunaris elongata, crotonensis prolongata, crystallina bacillaris, familiaris major, familiaris parva, fasciculata fasciculata, fasciculata affinis, fasciculata obtusa, fasciculata parva, fasciculata truncata, filiformis exilis, goulardi fluviatilis, inaequalis jumlensis, lanceolata curvata, lunaris capitata, minutissima pelliculosa, pulchella flexella, pulchella lacerata, pulchella minuta, pulchella naviculacea, pulchella subconstricta, radians tenuissima, rumpens rumpens, rumpens constricta, rumpens familiaris, rumpens fragilarioides, rumpens meneghiana, rumpens scotica, superba robusta, tabulata fasciculata, tabulata grandis, tenera utermoehlii, ulna ulna, ulna acus, ulna aequalis, ulna amphirhynchus, ulna biceps, ulna capitata, ulna chaseana, ulna chaseii, ulna claviceps, ulna clavipes, ulna constricta, ulna contracta, ulna curta, ulna danica, ulna delicatissima, ulna delicatula, ulna goulardi, ulna impressa, ulna lanceolata, ulna lanceolata-undulata, ulna longissima, ulna notata, ulna obtusa, ulna oxyrhynchus, ulna oxyryhnchus-mediocontracta, ulna pygmaea, ulna ramesi, ulna rhynchocephala, ulna rostrata, ulna spathulifera, ulna splendens, ulna subaequalis, ulna thalheimii, ulna undulata, ulna undulata, ulna ungeriana, ulna vitrea, vaucheriae capitellata, vaucheriae truncata, vitrea distorta*
GENUS *Synedrella* Lesser Synergetic Fragile Disc Diatom Brown Algae
SPECIES *parasitica*
VARIETIES *parasitica parasitica, parasitica constricta, parasitica lata, parasitica rhombica, parasitica rhomboidalis, parasitica subconstricta*
GENUS *Synedropsis* Least Synergetic Fragile Disc Diatom Brown Algae
SPECIES *fragilis, hyperborea, hyperboreoides, laevis, lata, recta*
VARIETIES *lata angustata*
GENUS *Tabularia* Tabular Fragile Disc Diatom Brown Algae
SPECIES *affinis, barbatula, investiens, ktenoeides, tabulata, waernii*
VARIETIES *waernii famelica, waernii littoralis*
GENUS *Thalassioneis* Thalassic Neid Fragile Disc Diatom Brown Algae
SPECIES *signyensis*
GENUS *Trachysphenia* Rough Sphene Fragile Disc Diatom Brown Algae
SPECIES *australis*

ORDER 6 *Hemiaulales* Half Furrowed Disc Diatom Brown Algae

FAMILY 1 *Hemiaulaceae* Half Furrowed Disc Diatom Brown Algae
GENUS *Hemiaulus* Half Furrowed Disc Diatom Brown Algae

SPECIES *alatus, altar, amplectans, claviger, echinulatus, fragilis, giganteus, giganteus, gondolaformis, grassus, hastatus, hauckii, inaequilaterus, latus, longicornis, lyriformis, orthoceras, pacificus, peripterus, polycystinorum, proteus, pyxilloides, rectus, sinensis, stabilus, taurus, tenuicornis, unicornutus, vitreus*

VARIETIES *polycystinorum dissonus, rectus twista*

GENUS *Abas* — Aba Half Furrowed Disc Diatom Brown Algae

SPECIES *wittii*

GENUS *Ailuretta* — Cat Half Furrowed Disc Diatom Brown Algae

SPECIES *cerataulinoides*

GENUS *Baxteriopsis* — Baxter Half Furrowed Disc Diatom Brown Algae

SPECIES *brunii*

GENUS *Briggera* — Brigger Half Furrowed Disc Diatom Brown Algae

SPECIES *ornithocephala*

GENUS *Cerataulina* — Horned Half Furrowed Disc Diatom Brown Algae

SPECIES *bicornis, pelagica*

GENUS *Climacodium* — Climactic Half Furrowed Disc Diatom Brown Algae

SPECIES *frauenfeldianum*

GENUS *Eucampia* — Field Half Furrowed Disc Diatom Brown Algae

SPECIES *antarctica, cornuta, groenlandica, zodiacus*

VARIETIES *antarctica recta*

GENUS *Keratophora* — Horn Half Furrowed Disc Diatom Brown Algae

SPECIES *nitida*

GENUS *Pseudorutilaria* — Rutile Half Furrowed Disc Diatom Brown Algae

SPECIES *monile, monomembranacea*

GENUS *Riedelia* — Reid Half Furrowed Disc Diatom Brown Algae

SPECIES *alata, altar, clavigera, longicornis, lyriformis, mirabilis, sublevata, tenuicornis*

GENUS *Sphynctolethus* — Sphyncter Half Furrowed Disc Diatom Brown Algae

SPECIES *monstrosus, pacificus*

GENUS *Spinasphaera* — Spine Sphere Half Furrowed Disc Diatom Brown Algae

SPECIES *robusta*

GENUS *Strelnikova* — Strelnikov Half Furrowed Disc Diatom Brown Algae

SPECIES *antiqua*

GENUS *Trilobatum* — Three Lobe Half Furrowed Disc Diatom Brown Algae

SPECIES *yunnanense*

GENUS *Trinacria* — Three Point Half Furrowed Disc Diatom Brown Algae

SPECIES *yunnanense, anisimovae, cancellata, caudata, cornuta, coronata, ecostata, excavata, fimbriata, indefinita, regina, senta, tessela, tristictia*

VARIETIES *barronii barronii, barronii tetragona, cornuta pentagona, excavata inflata*

FAMILY 2 *Bellerocheaceae* — Beller Disc Diatom Brown Algae

GENUS *Bellerochea* — Beller Disc Diatom Brown Algae

SPECIES *malleus*

VARIETIES *malleus biangulata*

GENUS *Neostreptotheca* — Twisted Beller Disc Diatom Brown Algae

SPECIES *subindica*

GENUS *Subsilicea* — Silicon Beller Disc Diatom Brown Algae

SPECIES *fragilarioides*

ORDER 7 *Melosirales* — Mellow Disc Diatom Brown Algae

FAMILY 1 *Melosiraceae* — Mellow Disc Diatom Brown Algae

GENUS *Melosira* — Mellow Disc Diatom Brown Algae

SPECIES *aequalis, arctica, arenaria, arentii, binderana, borreri, catenata, crenulata, crotonensis, dickiei, dozyana, elegans, excurrens, franklinii, granulata, granulosa, helvetica, herzogi, hustedtii, hyperborea, ikapoensis, juergensii,kochii, lineata, longispina, marina, mikkelsenii, moisseeviae, moniliformis, nivalis, normanii, nummuloides, octogona, orichalcea, papilio, punctata, roeseana, salina, solida, subflexilis, subsalsa, tenuissima, undulata, varians, westii*

VARIETIES *arenaria teres, crenulata ambigua, crenulata binderana, crenulata laevis, crenulata tenuis, crenulata valida, granulata angustissima, granulata angustissima-curvata, granulata circinalis, granulata curvata, granulata muzzanensis, granulata sparsipunctata, granulata spinosa, granulata spiralis, granulata valida, granulata valida-robusta, ikapoensis minor, juergensii juergensii, juergensii octogona, juergensii subangularis, roeseana dendroteres, roeseana dendroteres-porocyclia, roeseana dendroteres-spiralis, roeseana elongata, roeseana epidendron, roeseana epidendron-porocyclia, roeseana hamadryas, roeseana normalis, roeseana porocyclia, roeseana spiralis, roeseana tropica, undulata undulata, undulata normanii, varians moniliformis*

GENUS *Druridgea* — Druridge Mellow Disc Diatom Brown Algae

SPECIES *compressa*

GENUS *Orthoseira* — Straight Mellow Disc Diatom Brown Algae

SPECIES *americana, circularis, crenulata, dendrophila, dendroteres, epidendron, punctata*

VARIETIES *dendroteres roeseana*

FAMILY 2 *Aulacoseiraceae* Aulac Mellow Disc Diatom Brown Algae
GENUS *Aulacoseira* Aulac Mellow Disc Diatom Brown Algae
SPECIES *agassizii, alpigena, ambigua robusta, baicalensis, canadensis, ceretana, crassipunctata, crenulata, dendrophila, dendroteres, dianchiensis, distans, epidendron, fennoscandica, islandica, italica, lacustris, laevissima, lyrata, muzzanensis, nygaardii, perglabra, pfaffiana, pseudoamericana, pseudoislandica, robusta, skvortzowii, solida, subborealis, tenella, tenuior, tenuis, tethera, valida*
VARIETIES *agassizii malayensis, ambigua distans distans, distans africana, distans alpigena, distans blelhamensis, distans humilis, distans lacustris, distans laevissima, distans limnetica, distans lyrata, distans nivalis, distans perglabra, distans pfaffiana, distans septentrionalis, distans seriata, distans tenella, islandica ambigua, islandica curvata, islandica helvetica, italica italica, italica binderana, italica crenulata, italica curvata, italica pachyderma, italica recta, italica subarctica, italica subborealis, italica tenuissima, italica valida, lyrata lyrata, lyrata alpigena, lyrata biseriata, lyrata lacustris, lyrata perglabra, tenuiores perglabra florinae*

FAMILY 3 *Endictyaceae* Net Mellow Disc Diatom Brown Algae
GENUS *Endictya* Net Mellow Disc Diatom Brown Algae
SPECIES *hendeyi, oceanica*

FAMILY 4 *Hyalodiscaceae* Hyaline Mellow Disc Diatom Brown Algae
GENUS *Hyalodiscus* Hyaline Mellow Disc Diatom Brown Algae
SPECIES *ambiguus, cervinus, elegans, kryshtofovichii, laevis, planus, plicatus, pustulatus, radiatus, scoticus, subtilis*
VARIETIES *ambiguus kamtschaticus*
GENUS *Podosira* Pod Mellow Disc Diatom Brown Algae
SPECIES *aspera, hyalina, modesta, montagnei, simplex, stelligera*

FAMILY 5 *Stephanopyxidaceae* Stephan Pyxie Mellow Disc Diatom Brown Algae
GENUS *Stephanopyxis* Stephan Pyxie Mellow Disc Diatom Brown Algae
SPECIES *biseriata, californica, charkoviana, crenata, defecta, dimorpha, edita, grunowii, horridus, inermis, joynsonii, kulmii, lavrenkoae, lavrenkoi, marginata, megapora, nipponica, palmeriana, punctata, schenckii, schulzii, similis, spinosa, spinossima, superba, turris, turris arctica, turris campanulata, turris polaris, uralensis, weissflogii*
VARIETIES *schulzii cretacea, schulzii nuda, superba bispinosa, superba trispinosa*

ORDER 8 *Thalassiosirophytales* Greenish Disc Diatom Brown Algae

FAMILY 1 *Thallassiosiraceae* Greenish Disc Diatom Brown Algae
GENUS *Thalassiosira* Greenish Disc Diatom Brown Algae
SPECIES *aestivalis, allenii, andamanica, angulata, anguste-lineata, antarctica, antiqua, australiensis, australis, baldaufii, baltica, bingensis, bioculata, bipora, bramaputrae, brunii, bulbosa, burckliana, californica, cardiophora, cedarkeyensis, centra, complicata, concaviuscula, conferta, convexa, curviseriata, delicata, delicatula, depressa, dichotomica, diporocyclus, dubiosa, duostra, eccentrica, echinata, elliptipora, endoseriata, exceptiuncula, exigua, fasciculata, flexuosa, fraga, fryxelliae, gersondei, gracilis, gravida, grunowii, guillardii, hendeyi, hibernalis, hyalina, hydra, hyperborea, incerta, insigna, inura, irregulata, japonica, jouseae, kanayae, kolbei, kushirensis, labimarginata, lacustris, lentiginosa, leptopus, levanderi, lineata, lusca, mahoodii, makarovae, mala, mediaconvexa, mediterranea, minima, minuscula, miocenica, mioplicata, moholensis, multipora, nana, nidulus, nordenskioeldii, oceanica, oestrupii, oliverana, oranica, ornata, pacifica, partheneia, patagonica, perispinosa, perpusilla, planiuscula, plicata, plicatoides, poroseriata, praenidulus, praeyabei, primalabiata, profunda, proschkinae, pseudonana, punctigera, ritscheri, rotula, rudolfi, sancettae, scotia, simonsenii, simplex, strelnikovae, striata, subsalina, subtilis, symbolophora, tealata, temperei, tenera, tetraoestrupii, transitoria, trifulta, tumida, umaoiensis, urahoroensis, variantia, visurgis, webbii, weissflogii, yabei*
VARIETIES *antiqua septata, bioculata exigua, convexa aspinosa, eccentrica eccentrica, eccentrica decipiens, flexuosa tenella, gracilis expecta, lentiginosa obovatus, leptopus elliptica, nidulus delicata, oliverana sparsa, tetraoestrupii reimeri*
GENUS *Bacteriosira* Bacteria Greenish Disc Diatom Brown Algae
SPECIES *fragilis*
GENUS *Leioplanktona* Smooth Plankton Greenish Disc Diatom Brown Algae
SPECIES *madhensis, verrucosa*
GENUS *Microsiphonia* Siphon Greenish Disc Diatom Brown Algae
SPECIES *potamos*
GENUS *Minidiscus* Mini Greenish Disc Diatom Brown Algae
SPECIES *chilensis, comicus, decoratus, trioculatus*
GENUS *Neosolenopora* Modern Pore Greenish Disc Diatom Brown Algae
SPECIES *ramaraoi*
GENUS *Parthaea* (Synonym for *Planktoniella*)
GENUS *Planktoniella* Plankton Greenish Disc Diatom Brown Algae
SPECIES blanda, sol
GENUS *Porosira* Pore Greenish Disc Diatom Brown Algae
SPECIES *antarctica, glacialis, pseudodelicatula, pseudodenticulata*

VARIETIES *glacialis antarctica*
GENUS *Praesolenopora* Primitive Pore Greenish Disc Diatom Brown Algae
SPECIES *formosa, furcata, liaoningensis*
GENUS *Praethalassiosiropsis* Primitive Greenish Disc Diatom Brown Algae
SPECIES *hasleae*
GENUS *Rotiramulus* Branched Greenish Disc Diatom Brown Algae
SPECIES *pilifer*
GENUS *Roundia* Round Greenish Disc Diatom Brown Algae
SPECIES *cardiophora*
GENUS *Stephanocostis* Stephan Greenish Disc Diatom Brown Algae
SPECIES *chantaicus, lucens*
GENUS *Takanoa* Takano Greenish Disc Diatom Brown Algae
SPECIES *bingensis*
GENUS *Thalassiocyclus* (Synonym for *Stephanocostis*)
GENUS *Thalassiosiropsis* Lesser Greenish Disc Diatom Brown Algae
SPECIES *Thalassiosiropsis wittiana*

FAMILY 2 *Lauderiaceae* Lauder Greenish Disc Diatom Brown Algae
GENUS *Lauderia* Lauder Greenish Disc Diatom Brown Algae
SPECIES *annulata, mediterranea*

FAMILY 3 *Skeletonemaceae* Skeleton Greenish Disc Diatom Brown Algae
GENUS *Skeletonema* Skeleton Greenish Disc Diatom Brown Algae
SPECIES *barbadense, costatum, denticulatum, menzellii, mirabile, potamos, probabile, pseudocostatum, simbirskianum, subsalsum, tropicum, ventricosum*
GENUS *Cyclotubicoalitus* Tube Skeleton Greenish Disc Diatom Brown Algae
SPECIES *undatus*
GENUS *Detonula* Deton Skeleton Greenish Disc Diatom Brown Algae
SPECIES *confervacea, pumila*

FAMILY 4 *Stephanodiscaceae* Stephan Greenish Disc Diatom Brown Algae
GENUS *Stephanodiscus* Stephan Greenish Disc Diatom Brown Algae
SPECIES *aegyptiacus, agassizensis, alpinus, asteroides astraea, atmosphaerica, berolinensis, binderanus, bramaputrae, carconensis, conspicueporus, costatilimbus, damasii, delicatus, dubius, dzhilindus, excentricus, galileensis, gravitoides, hantzschii, heterostylus, inconspicuus, invisitatus, irregularis, khurseviczae, komoroensis, kutzingiana,, lucens, makarovae, marginatus, matrensis, medius, minor, minutulus, muelleri, nativus, nemanensis, neoastraea, niagarae, novaezeelandiae, omarensis, oregonicus, pantocsekii, parvus, perforatus, ponticus, prohantzschii, pseudosuzukii, punctatus, radiatus, rotula, rugosus, skabitschevskyi, subsalsus, subtilis, subtransilvanicus, superiorensis, suzukii, tenuis, transsylvanicus, transylvanicus, vestibulis, yukonensis*
VARIETIES *asteroides intermedia, astraea intermedia, astraea minutula-minor, carconensis pusilla, hantzschii tenuis, niagarae magnifica, tenuis minor*
GENUS *Crateriportula* Crater Stephan Greenish Disc Diatom Brown Algae
SPECIES *inconspicuus*
GENUS *Cyclostephanos* Cyclic Stephan Greenish Disc Diatom Brown Algae
SPECIES *costatilimbus, damasi, delicatus, dubius, fenestratus, guatamalae, invisitatus, lacrimis, malawiensis, marginatus, novaezeelandiae, omarensis, pantocsekii, pliocenicus, ponticus, servant-vildaryi, sibiricus, stelliformis, tholiformis, undatus*
GENUS *Cyclotella* Cycle Stephan Greenish Disc Diatom Brown Algae
SPECIES *aegaea, agassizensis, ambigua, americana, antiqua, arentii, areolata, atomus, austriaca, azigzensis, baikalensis, bastowii, bodanica, bornata, catenata, cavita,, chaetoceras, choctawhatcheeana, comensis, corona, cretica, cryptica, cyclopuncta, delicatissima, delicatula, distinguenda, elgeri, elymaea, exigua, facetia, fottii, gamma, glabriuscula, glomerata, gordonensis, gracilis, hellae, iris, jonesii, kisselevii, krammeri, kuetzingiana, kurdica, lucernensis, mascarenica, meduanae, melosiroides, meneghiniana, michiganiana, minuta, minutula, multipunctata, ocellata, operculata, orientalis, ornata, papillosa, planctonica, plitvicensis, polymorpha, praekuetzingiana, praetermissa, proshkinae, pseudocomensis, pseudonana, pseudostelligera, pumila, punctata, pygmaea, quadriiuncta, quillensis, radiosa, rossii, schroeteri, schumannii, servant-vildaryi, sevilleana, sexnotata, socialis, spinosa, stelligera, stelligeroides, striata, strisignata, stylorum, styriaca, substelligera, tasmanica, temperei, texta, thienemannii, transylvanica, trichonidea, tripartita, unipunctata, vorticosa, wolterecki, wuerthrichiana, wulfiae*
VARIETIES *aegaea pythagorea, bodanica affinis, bodanica comta, bodanica intermedia, bodanica lemanensis, bodanica michiganensis, bodanica stellata, distinguenda mesoleia, distinguenda unipunctata, glomerata stelligera, iris baltalensis, kuetzingiana parva, kuetzingiana planetophora, kuetzingiana radiosa, meneghiniana brevistriata, meneghiniana convexa, meneghiniana fluviatilis, meneghiniana laevissiina, meneghiniana major, meneghiniana minutissima, meneghiniana plana, meneghiniana quadrata, ocellata kuetzingiana, operculata mesoleia, operculata unipunctata, radiosa bodanica, radiosa comta, radiosa glabriuscula, radiosa oligactis, radiosa paucipunctata, radiosa quadrijuncta, servant-vildaryi elegans, servant-vildaryi robusta, striata striata, striata ambigua, striata bipunctata, thienemannii minuscula, transylvanica disseminatepunctata, trichonidea parva*
GENUS *Mesodictyon* Middle Net Stephan Greenish Disc Diatom Brown Algae
SPECIES *cyclotelloides, fovis, gasseae, hyalinum, magnum, nemanense, parvulum, plicatum, sectorale, undulatum*
GENUS *Pelagodictyon* Pelagic Net Stephan Greenish Disc Diatom Brown Algae

SPECIES *fritzii, spinosum, tenue*
GENUS *Pleurocyclus* Pleural Cycle Stephan Greenish Disc Diatom Brown Algae
SPECIES *stechlinensis*
GENUS *Pilocaenicus* Pile Mud Stephan Greenish Disc Diatom Brown Algae
SPECIES *costatus, harzii, hercynicus, omarensis, pantocsekii, pygmaeus, undulatus*
VARIETIES *costatus leprindus, costatus sibiricus*

ORDER 9 *Triceratiales* Three Horn Disc Diatom Brown Algae

FAMILY 1 *Triceratiaceae* Three Horn Disc Diatom Brown Algae
GENUS *Triceratium* Three Horn Disc Diatom Brown Algae
SPECIES *acutangulum, amblyoceros, anissimovae, balearicum, bicornigerum, campeachianum, cancellatum, caudatum, cornutum, coronatum, dictyotum, dissimile, divisum, dubium, excavatum, exiguum, favus, gombosii, groningense, inconspicuum, indefinitum, irregulare, kanayae, laeve, latipes, mesoleium, moronense, nancoorense, obscurum, parallelum, pentacrinus, perplexum, planum, plicatum, polymorphum, quinqueguttatum, rectangulare, reticulum, robertsianum, sacrophagus, schmidtii, schulzii, semicirculare, sentum, sparsimpunctatum, subundosum, tessela, trifoliatum, unguiculatum, variabile, venustum, whampoensis*
VARIETIES *favus octogonum, groningense quadratum, inconspicuum trilobatum, kanayae quadriloba, laeve annuliferum, moronense nicobaricum, parallelum madagascarense, schulzii quadrilobum, unguiculatum quinquangulum, unguiculatum triangulum, venustum costata*
GENUS *Amphitetras* Amphitetrad Three Horn Disc Diatom Brown Algae
SPECIES *acutangula, alternans, anderssonii, atlantica, balearica, bicornis, blandita, canalifera, coerulescens, conciliata, cornuta, dutertrei, inconspicua, ingloria, intermedia, kusnetzkiana, lanceolata, lata, margaritifera, minuta, oculata, perryana, pethoei, pulvillus, quadrangularis, quadricornis, rectangularis, sturtii, tetragona, unguiculata, zonatulata*
VARIETIES *conciliata validior*
GENUS *Auliscus* Furrowed Three Horn Disc Diatom Brown Algae
SPECIES *cylindricus, intercedens, lucidus, nebulosus, sculptus, spinosus*
VARIETIES *sculptus major, sculptus nankoorensis*
GENUS *Cerataulus* Lesser Furrowed Three Horn Disc Diatom Brown Algae
SPECIES *laevis, pacificus, radiatus, smithii, turgidus*
GENUS *Eupodiscus* Eupolis Three Horn Disc Diatom Brown Algae
SPECIES *gregorianus, inconspicuus, kamischevensis, paracaensis, radiatus*
GENUS *Fryxelliella* Fryxell Three Horn Disc Diatom Brown Algae
SPECIES *floridana, inconspicua*
GENUS *Lampriscus* Lamp Three Horn Disc Diatom Brown Algae
SPECIES *kittonii*
GENUS *Medlinia* Medlin Three Horn Disc Diatom Brown Algae
SPECIES *pseudoweissei, simbirskiana*
GENUS *Odontella* Tooth Three Horn Disc Diatom Brown Algae
SPECIES *aurita, calamus, cornuta, granulata, hastata, litigiosa, longicruris, mobiliensis, regia, rhombus, septentrionalis, sinensis, weissflogii*
VARIETIES *rhombus rhombus, rhombus trigona*
GENUS *Pleurosira* Pleural Three Horn Disc Diatom Brown Algae
SPECIES *laevis*
VARIETIES *laevis polymorpha*
GENUS *Porguenia* Porguen Three Horn Disc Diatom Brown Algae
SPECIES *peruviana*
GENUS *Pseudauliscus* Least Furrowed Three Horn Disc Diatom Brown Algae
SPECIES *grevillei*
GENUS *Pseudotriceratium* Lesser Three Horn Disc Diatom Brown Algae
SPECIES *fallax*
GENUS *Sextiputeus* Six Pute Three Horn Disc Diatom Brown Algae
SPECIES *inornatus, reticulatus*
GENUS *Sheshukovia* Sheshukov Three Horn Disc Diatom Brown Algae
SPECIES *cellulosa, duplicata, fenestrata, sundbyensis, synica, uralensis, weissei*
VARIETIES *cellulosa simbirskiana*

FAMILY 2 *Plagiogrammaceae* Pelagic Gramm Disc Diatom Brown Algae
GENUS *Plagiogramma* Pelagic Gramm Disc Diatom Brown Algae
SPECIES *gregorianum, grevillei, interruptum, laeve, minimum, ornatum, parallelum, sigmoideum, staurophorum, van-heurckii*
GENUS *Dimeregramma* Dimer Pelagic Gramm Disc Diatom Brown Algae
SPECIES *lanceolatum, marinum, minor*
VARIETIES *marinum lanceolatum, minor nanum*
GENUS *Glyphodesmis* Glyph Pelagic Gramm Disc Diatom Brown Algae
SPECIES *distans, eximia, williamsonii*

PHYLUM 612 *Chlorophyta* Olive Green Algae

Latin: *Chlora* chloric, green
Latin: *Phyta* phyton, plant

16,000 species 1998]
[Margulis & Schwartz,
5,857 species [Guiry, Dec. 2004]

CLASSES		Species
612.1 *Chlorophyceae*	Olive Green Algae	
612.2 *Bryopsidophyceae*	Mossy Olive Green Algae	
612.3 *Charophyceae*	Elegant Olive Green Algae	
612.4 *Klebsormidiophyceae*	Klebs Olive Green Algae	
612.5 *Mesostigmatophyceae*	Stigma Olive Green Algae	
612.6 *Pedinophycea*	Pedin Olive Green Algae	
612.7 *Trebouxiophyceae*	Treboux Olive Green Algae	
612.8 *Ulvophyceae*	Filament Olive Green Algae	
612.9 *Zygonematophyceae*	Conjugating Olive Green Algae	

Phylum *Chlorophyta* Olive Green Algae species contain chlorophyll a and chlorophyll be as well as carotene pigments astaxanthin, canthoxanthin, flavoxanthin, loraxanthin, neoxanthin, violaxanthin, and zeathanthin. They are a major component of freshwater and ocean phytoplankton. They remove one billion tons of carbon from the environment per year and convert it to organic compounds. Their cell walls are made of cellulose, pectin, xylose, and mannose.

Class 1 *Chlorophyceae* Olive Green Algae species include the common tree scum algae *Chlorococcum*. Symbiote and free living *Chorella* species are grown in the laboratory and used to study photosynthesis.

Class 2 *Bryopsidophyceae* Mossy Olive Green Algae includes the famous *Caulerpa* species which invaded the Mediterranean Sea from aquarium water dumped into the Mediterranean by the staff of the Jacque Custeau Institute.

Class 3 *Charophyceae* Elegant Olive Green Algae are multi-celled and live in freshwater and brackish water. Genera *Nitella* and *Chara* are delicate pond weeds.

Class 4 *Klebsormidiophyceae* Klebs Olive Green Algae includes the Genus *Klebsmorbidium*.

Class 8 *Ulvophyceae* Filament Olive Green Algae species include the sea lettuce algae *Ulva* and the estuary lettuce algae *Enteromorpha*. Green Seaweeds such as *Codium* and *Acetabularia* are in this class.

CLASS 612.1 *Chlorophyceae* Olive Green Algae

ORDER 1 *Chlorophytales* Olive Green Algae

FAMILY 1 *Chlorphytaceae* — Olive Green Algae
GENUS *Asphaltinella* — Asphalt Olive Green Algae
SPECIES *horowitzii, peratrovichensis*
GENUS *Botryella* — Berry Olive Green Algae
SPECIES *spinosa*
GENUS *Boueina* — Bouein Olive Green Algae
SPECIES *hochstetteri, marondei*
GENUS *Chloremys* — Mouse Olive Green Algae
SPECIES *sessilis*
GENUS *Chloronomala* — Anomalous Olive Green Algae
SPECIES *cuprecola, palmelloides*
GENUS *Chlorosphaeropsis* — Sphere Olive Green Algae
SPECIES *alveolata, epidemica*
GENUS *Crassikamaena* — Crass Olive Green Algae
SPECIES *foraminosa*
GENUS *Cribrokamaena* — Sieve Olive Green Algae
SPECIES *citrosa, furcillata*
GENUS *Entransia* — Entrance Olive Green Algae
SPECIES *fimbriata*
GENUS *Gymnozyga* — Nude Olive Green Algae
SPECIES *moniliformis*
VARIETIES *moniliformis gracilescens, moniliformis minima*
GENUS *Halochlorococcum* — Halogen Round Olive Green Algae
SPECIES *dilatatum, marinum, moorei, operculatum, porphyrae, saccatum*
GENUS *Hofmania* — Hofman Olive Green Algae
SPECIES *africana*
GENUS *Hortobagyiella* — Hortobagy Olive Green Algae
SPECIES *verrucosa*

GENUS *Hsiungia* — Hsiung Olive Green Algae
SPECIES *triciliata*
GENUS *Hyalococcus* — Lesser Halogen Round Olive Green Algae
SPECIES *dermatocarponis*
GENUS *Hyalophycus* — Halogen Olive Green Algae
SPECIES *ocellatus*
GENUS *Hydraea* — Hydric Olive Green Algae
SPECIES *calcarata*
GENUS *Lacrymorphus* — Weeping Olive Green Algae
SPECIES *catenaeformis*
VARIETIES *catenaeformis sardus*
GENUS *Lobosphaeropsis* — Lobe Sphere Olive Green Algae
SPECIES *lobophora, pyrenoidosa*
GENUS *Mateola* — Mate Olive Green Algae
SPECIES *curvata*
GENUS *Microspora* — Microspore Olive Green Algae
SPECIES *abbreviata, aequabilis, amoena, crassior, ficulinae, floccosa, flocculosa, fugacissima, irregularis, loefgrenii, membranacea, pachyderma, palustris, quadrata, stagnorum, tenuiderma, tumidula, willeana, wittrockii*
VARIETIES *amoena gracilis*
GENUS *Neodesmus* — Modern Fiber Olive Green Algae
SPECIES *pupukensis*
GENUS *Neosphaerochloris* — Modern Sphere Olive Green Algae
SPECIES *asymmetrica*
GENUS *Octacanthium* — Eight Spine Olive Green Algae
SPECIES *bifidum, borgei, controversum, japonicum, longispinum, mucronulatum, octocorne, sachlanii, trispinatum*
GENUS *Oltmannsiellopsis* — Oltmanns Olive Green Algae
SPECIES *geminata, unicellularis, viridis*
GENUS *Ortonella* — Orton Olive Green Algae
SPECIES *cipitensis*
GENUS *Ovillaria* — Ovillar Olive Green Algae
SPECIES *catenata*
GENUS *Parallela* — Parallel Olive Green Algae
SPECIES *novae-zelandiae*
GENUS *Phaeophila* — Dusky Olive Green Algae
SPECIES *bulbochaete, dendroides, endophytum, filiformis, hirsuta, polymorpha, ramulosa, vagans*
VARIETIES *dendroides calcicola, dendroides pachyderma*
GENUS *Phycastrum* — Star Olive Green Algae
SPECIES *pecten*
GENUS *Physactis* — Radial Olive Green Algae
SPECIES *bullata*
GENUS *Physidrum* — Bladder Olive Green Algae
SPECIES *aggregatum, hyalinum, pisiforme, rubescens*
GENUS *Physotris* — Ear Bladder Olive Green Algae
SPECIES *glomerata*
GENUS *Phytelis* — Telose Olive Green Algae
SPECIES *atra, granulata, macrocarpa, radiata, sulcata, tuberculata*
GENUS *Phythelios* — Theliose Olive Green Algae
SPECIES *de-tonii*
GENUS *Pseudocarteria* — Lesser Carter Olive Green Algae
SPECIES *polychloris*
GENUS *Pseudodidymocystis* — Twin Cyst Olive Green Algae
SPECIES *fina, inconspicua, lineata, planctonica*
GENUS *Pseudoprostratum* — Lesser Prostrate Olive Green Algae
SPECIES *indicum*
GENUS *Sangirellum* — Sangir Olive Green Algae
SPECIES *taeniforme*
GENUS *Sarfatiella* — Sarfat Olive Green Algae
SPECIES *sarda*
GENUS *Scotlandella* — Scotland Olive Green Algae
SPECIES *alpina, nodosa*
GENUS *Selenodictyon* — Netted Olive Green Algae
SPECIES *brasiliense, elongatum, naviculiforme*
GENUS *Seletonella* — Seleton Olive Green Algae
SPECIES *sajanica*
GENUS *Tasmanites* — Tasman Olive Green Algae
SPECIES *marshalliae, pelagica, vindhyanensis*

GENUS *Tethysicodium* Sea Goddess Olive Green Algae
SPECIES *wrayi*
GENUS *Urnella* Urn Olive Green Algae
SPECIES *terrestris*
VARIETIES *terrestris abnormis*
GENUS *Uvella* Bunched Olive Green Algae
SPECIES *chamaemora, uva*

FAMILY 2 *Chlorosarcinaceae* Packed Olive Green Algae
GENUS *Chlorosarcina* Packed Olive Green Algae
SPECIES *brevispinosa, consociata, longispinosa, sempevirens, superba*
GENUS *Chlorosarcinopsis* Lesser Packed Olive Green Algae
SPECIES *aggregata, amylophila, arenicola, auxotrophica, bastropiensis, caeca, communis, deficiens, delicata, dissociata, eremi, gelatinosa, minuta, sempervirens, superba, variabilis*
GENUS *Desmochloris* Fiber Packed Olive Green Algae
SPECIES *halophila*
GENUS *Fasciculochloris* Fascicule Packed Olive Green Algae
SPECIES boldii, mexicana
GENUS *Ignatius* Volcanic Packed Olive Green Algae
SPECIES *tetrasporus*
GENUS *Neochlorosarcina* Modern Packed Olive Green Algae
SPECIES *minor, negevensis, pseudominor*
VARIETIES *negevensis ferruginea*
GENUS *Polysphaera* Polysphere Packed Olive Green Algae
SPECIES *composita*

FAMILY 3 *Oedogoniaceae* Neurotic Olive Green Algae
GENUS *Oedogonium* Neurotic Olive Green Algae
SPECIES *acrosporum, ahlstrandii, alpinum, alternans, angustistomum, anomalum, arcyosporum, areschougii, bengalense, birmanicum, borisianum, boscii, bourellyanum, brasiliense, braunii, calcereum, calliandrum, candollei, capillaceum, capillare, capilliforme, capitellatum, cardiacum, ciliatum, cleveanum, concatenatum, contortum, crassidens, crassiusculum, crassum, crassum, crenulato-costatum, crispum, croasdalea, cryptoporum, curtum, curvum, cyathigerum, cyclostomum, cymatosporum, de-tonii, decipiens, donnellii, echinospermum, echinosporum, excisum, excitans, flavescens, foveolatum, fragile, franklinianum, fuscolutescens, geniculatum, giganteum, globosum, gracile, grande, gunnii, hians, hirnii, howardii, hystrix, idioandrosporum, infirmum, intermedium, itzogsohnii, landsboroughii, lautumniarum, longicolle, macrandrium, macrospermum, microgonium, minus, monile, multisporum, nanum, nobile, nodulosum, oblongum, obsoletum, obtruncatum, oelandicum, pachydermatosporum, parthasarathii, petri, pisanum, plagiostomum, platygynum, pluviale, princeps, pringsheimii, pseudomitratum, punctato-striatum, punctatostriatum, punctatum, pusillum, pyriforme, pyrulum, reinschii, rivulare, rothii, rufescens, sexangulare, spechtii, spiripennatum, stagnale, suecicum, szymanskae, tapeinosporum, tenellum, tumidulum, undulatum, urbicum, vaucherii, vernale, vesicatum, viaticum, victoriense, wissmanii, wolleanum, zigzag*
VARIETIES *acrosporum bathmidosporum, acrosporum majusculum, acrosporum puntatum, alternans brevicellulare, arcyosporum burmense, brasiliense seriatum, cardiacum carbonicum, concatenatum hutchinsiae, concatenatum rectangulare, crispum hawaiense, crispum obesum, cyathigerum perfectum, fragile robustum, howardii minus, itzogsohnii minus, longicolle senegalense, monile victoriense, multisporum cylindricum, platygynum ambiceps, pringsheimii nordstedtii, pringsheimii pachydermatosporum, rufescens lundelllii, sexangulare ochratum, suecicum australis, tumidulum parasitica, undulatum moebiusii*
GENUS *Bulbochaete* Bulb Bristle Neurotic Olive Green Algae
SPECIES *angulosa, annularis, borealis, brebissonii, crassiuscula, crenulata, elachistandra, elatior, ellipsospora, gigantea, hiloensis, imperialis, insignis, intermedia, iyengarii, minor, mirabilis, monile, nana, obliqua, pygmaea, quadrata, rectangularis, repanda, rhadinospora, sessilis, setigera, subintermedia, varians*
VARIETIES *imperialis regalis, obliqua subreticulata, rectangularis hiloensis, varians antiqua, varians hawaiiensis, varians subsimplex*
GENUS *Oedocladium* Branched Neurotic Olive Green Algae
SPECIES *albermalense, carolinianum, cirratum, eriense, himalayense, indicum, lewisii, medium, operculatum, prescottii, protonema, terrestre, tiffanyanum, wettsteinii*

FAMILY 4 *Prasiolaceae* Leek Olive Green Algae
GENUS *Prasiola* Leek Olive Green Algae
SPECIES *borealis, calophylla, crispa, cristata, delicata, filiformis, fluviatilis, formosana, furfuracea, japonica, linearis, mauritiana, meridionalis, mexicana, nevadensis, sinica, stipitata, subareolata, tessellata, velutina, yunnanica*
VARIETIES *crispa antarctica, crispa muralis, crispa sudetica, filiformis minuta*
GENUS *Filoprotococcus* Filament Round Leek Olive Green Algae
SPECIES *polymorphus*
GENUS *Hormidium* Chain Leek Olive Green Algae
SPECIES *crassum, gaditanum, lubricum, neesi, pseudomirabile, subsalsum, tribonematoideum, velutinum*
GENUS *Prasiococcus* Round Leek Olive Green Algae

SPECIES *calcarius*
GENUS *Prasiolopsis* Lesser Leek Olive Green Algae
SPECIES *ramosa*
GENUS *Rosenvingiella* Rosenving Leek Olive Green Algae
SPECIES *constricta, polyrhiza, radicans*
VARIETIES *radicans tenuior*

FAMILY 5 *Tetracystaceae* Ten Cyst Olive Green Algae
GENUS *Tetracystis* Ten Cyst Olive Green Algae
SPECIES *aeria, aggregata, aplanospora, compacta, diplobionticoidea, dissociata, elliptica, excentrica, fissurata, illinoisensis, intermedia, isobilateralis, macrostigmata, micropyrenoides, nagasakiensis, pampae, pulchra, sarcinalis, tetraspora, texensis, vinatzeri*
GENUS *Axilosphaera* Axial Ten Cyst Olive Green Algae
SPECIES *vegetata*
GENUS *Desmotetra* Fiber Ten Cyst Olive Green Algae
SPECIES *antarctica, aureospora, stigmatica*
GENUS *Heterotetracystis* Variable Ten Cyst Olive Green Algae
SPECIES *akinetos, intermedia*
GENUS *Spongiococcum* Sponge Ten Cyst Olive Green Algae
SPECIES *tetrasporum*

ORDER 2 *Chaetophorales* Bristle Olive Green Algae

FAMILY 1 *Chaetophoraceae* Bristle Olive Green Algae
GENUS *Chaetophora* Bristle Olive Green Algae
SPECIES *attenuata, elegans,incrassata, pisiformis, punctiformis, tuberculata, tuberculosa*
VARIETIES *attenuata claytonii, elegans minor*
GENUS *Aphanochaete* Inconspicuous Bristle Olive Green Algae
SPECIES *magna, pilosissima, polychaete, repens, vermiculoides*
GENUS *Arthrochaete* Straight Bristle Olive Green Algae
SPECIES *penetrans*
GENUS *Bolbocoleon* Bulb Bristle Olive Green Algae
SPECIES *endophytica, jolyi, piliferum*
GENUS *Bolboforma* Bulb Form Bristle Olive Green Algae
SPECIES *antarctica*
GENUS *Cedercreutziella* Cedercreutz Bristle Olive Green Algae
SPECIES *savoniensis*
GENUS *Chaetomnion* Omni Bristle Olive Green Algae
SPECIES *pyriferum*
GENUS *Chaetonema* Thread Bristle Olive Green Algae
SPECIES *irregulare, ornatum*
GENUS *Chaetonemopsis* Lesser Thread Bristle Olive Green Algae
SPECIES *pseudobulbochaete*
GENUS *Chaetosphaera* (Synonym for *Chaetophora*)
GENUS *Chamaetrichon* Hairy Bristle Olive Green Algae
SPECIES *capsulatum*
GENUS *Chlorofilum* Filament Bristle Olive Green Algae
SPECIES *ephemerum*
GENUS *Choreoclonium* Chore Clone Bristle Olive Green Algae
SPECIES *procumbens*
GENUS *Cloniophora* Clone Bristle Olive Green Algae
SPECIES *macrocladia, paihiae, plumosa, shanxiensis, spicata*
GENUS *Coccobotrys* Berry Bristle Olive Green Algae
SPECIES *mucosus, verrucariae*
GENUS *Coleonema* Thread Bristle Olive Green Algae
SPECIES *arenifera*
GENUS *Crenacantha* Scalloped Bristle Olive Green Algae
SPECIES *orientalis*
GENUS *Dermatophyton* Dermal Bristle Olive Green Algae
SPECIES radians
GENUS *Desmococcus* Fiber Round Bristle Olive Green Algae
SPECIES *apatococcus, endolithicus, olivaceus, spinocystis, viridis*
GENUS *Diaphragma* Diaphragm Bristle Olive Green Algae
SPECIES *radiosum*
GENUS *Dicranochaete* Dicran Bristle Olive Green Algae
SPECIES *bohemica, britannica, quadriseta, reniformis*
GENUS *Didymosporangium* Dual Spore Bristle Olive Green Algae
SPECIES *repens*

GENUS *Dilabifilum* Dual Lip Bristle Olive Green Algae
SPECIES *arthropyreniae, incrustans, prostratum*
GENUS *Diplosphaera* Dual Sphere Bristle Olive Green Algae
SPECIES *chodati, mucosa*
GENUS *Draparnaldia* Draparnald Bristle Olive Green Algae
SPECIES *desikacharyi, elongata, judayi, mutabilis, pusilla*
GENUS *Draparnaldiana* Least Draparnald Bristle Olive Green Algae
SPECIES *ciliata*
GENUS *Draparnaldiopsis* Lesser Draparnald Bristle Olive Green Algae
SPECIES *alpinis, krishnamurthyi, taylorii*
GENUS *Ectochaete* Ectophyte Bristle Olive Green Algae
SPECIES *perforans*
GENUS *Elaterodiscus* Disc Bristle Olive Green Algae
SPECIES *appendiculatus*
GENUS *Endoclonium* Endic Clone Bristle Olive Green Algae
SPECIES *marinum*
GENUS *Endoderma* Endic Dermal Bristle Olive Green Algae
SPECIES maius
GENUS *Endophyton* Endophyte Bristle Olive Green Algae
SPECIES *atroviridis, laurenciae, ramosum*
GENUS *Entocladia* Entic Branched Bristle Olive Green Algae
SPECIES *anandii, cingens, cladophorae, codicola, condensata, endolithica, endophytica, endophyticum, endozoica, hypoglossiae, leptochaete, maculans, major, majus, mexicana, moewusae, pennata, polymorpha, polysiphoniae, reticulata, robusta, taylorii, testarum, vagans, velutina, ventriculosa, viridis, wittrockii*
VARIETIES *robusta valoniopsidis, viridis majus*
GENUS *Entodictyon* Entic Net Bristle Olive Green Algae
SPECIES *schilleri*
GENUS *Fritschiella* Fritsch Bristle Olive Green Algae
SPECIES *tuberosa*
GENUS *Gongrosira* Gonger Bristle Olive Green Algae
SPECIES *calcifera, cavernicola, chapmanii, debaryana, fluminensis, gaditana, incrustans, jollyi, malardii, onusta, papuasica, pseudoprostrata, pseudorostrata, schmidlei, sclerococcus, scourfieldii, stagnalis*
VARIETIES *papuasica viridis*
GENUS *Gongrosirella* Lesser Gonger Bristle Olive Green Algae
SPECIES *vermiformis*
GENUS *Hazenia* Hazen Bristle Olive Green Algae
SPECIES *mirabilis*
GENUS *Helicodictyon* Helix Net Bristle Olive Green Algae
SPECIES *planctonicum*
GENUS *Helicotheca* Helix Bristle Olive Green Algae
SPECIES *tamesis*
GENUS *Herposteiron* Herpetic Bristle Olive Green Algae
SPECIES *confervicolum*
VARIETIES *confervicolum bicellularis*
GENUS *Internoretia* Internoret Bristle Olive Green Algae
SPECIES *fryeana*
GENUS *Ireksokonia* Ireksokon Bristle Olive Green Algae
SPECIES *formosa*
GENUS *Iwanoffia* Ivanoff Bristle Olive Green Algae
SPECIES *terrestris*
GENUS *Jaagiella* Jaag Bristle Olive Green Algae
SPECIES *alpicola*
GENUS *Jaoa* Jao Bristle Olive Green Algae
SPECIES *prasina*
GENUS *Klebahniella* Klebahn Bristle Olive Green Algae
SPECIES *elegans*
GENUS *Kymatotrichon* Kymat Bristle Olive Green Algae
SPECIES *armatum*
GENUS *Leptochaete* Thin Bristle Olive Green Algae
SPECIES *capsosirae, crustacea, fonticola, gracilis, hansgirgi, marina, nidulans, parasitica, rivularis, stagnalis*
VARIETIES *rivularis tenuior*
GENUS *Leptosira* Thin Pit Bristle Olive Green Algae
SPECIES *erumpens, mediciana, obovata, polychloris, thrombii, vischeri*
GENUS *Leptosiropsis* Lesser Thin Pit Bristle Olive Green Algae
SPECIES torulosa
GENUS *Micropoa* Grass Bristle Olive Green Algae
SPECIES *leptochaete*

GENUS *Myxonema* Thread Bristle Olive Green Algae
SPECIES *nanum*
GENUS *Myxonemopsis* Lesser Thread Bristle Olive Green Algae
SPECIES *crassimembranacea*
GENUS *Nayalia* Nayal Bristle Olive Green Algae
SPECIES *terrestris*
GENUS *Periplegmatium* Plegma Bristle Olive Green Algae
SPECIES *ceramii*
GENUS *Pilinella* Pilin Bristle Olive Green Algae
SPECIES *californica*
GENUS *Pleurangium* Many Angled Bristle Olive Green Algae
SPECIES *amphibium*
GENUS *Pleurococcus* Pleural Round Bristle Olive Green Algae
SPECIES *angulosus, bradypodis, mucosus, murorum, vulgaris*
GENUS *Pringsheimia* Pringsheim Bristle Olive Green Algae
SPECIES *marchantae*
GENUS *Pringsheimiella* Lesser Pringsheim Bristle Olive Green Algae
SPECIES *marchantae, conchyliophila, gratulans, mauritiana, sanctae-luciae, scutata, striata*
GENUS *Pseudendoclonium* Lesser Endic Clone Bristle Olive Green Algae
SPECIES *basiliense, conchiliophila, dynamenae, fucicola, laxum, marinum, printzii, prostratum, submarinum*
VARIETIES *basiliense brandii, laxum deminutum, laxum laterale, prostratum pseudoprostratum, prostratum radiatum*
GENUS *Pseudochaete* Lesser Bristle Olive Green Algae
SPECIES *crassisetum, gracilis*
GENUS *Pseudodictyon* Lesser Net Bristle Olive Green Algae
SPECIES *reticulatum*
GENUS *Pseudopleurococcus* Lesser Pleural Round Bristle Olive Green Algae
SPECIES *botryoides, printzii*
GENUS *Pseudopringsheimia* Lesser Pringsheim Bristle Olive Green Algae
SPECIES *apiculata, confluens, reinkei*
GENUS *Pseudulvella* Lesser Bunched Bristle Olive Green Algae
SPECIES *consociata*
GENUS *Rhexinema* Rhex Thread Bristle Olive Green Algae
SPECIES *paucicellularis*
GENUS *Schizomeris* Split Sea Bristle Olive Green Algae
SPECIES *leibleinii*
GENUS *Skvortzoviothrix* Skvortzov Bristle Olive Green Algae
SPECIES *terrestris*
GENUS *Stigeoclonium* Stige Clone Bristle Olive Green Algae
SPECIES *aestivale, amoenum, askenasyi, attenuatum, australense, biasolettianum, elongatum, falklandicum, farctum, fasciculare, flagelliferum, helveticum, islamii, longipilum, lubricum, nanum, nudiusculum, ovisporum, pachydermum, pascheri, protensum, pusillum, segarare, setigerum, simplex, stagnatile, subsecundum, subspinosum, subuligerum, tenue, thermale, uniforme, variabile*
VARIETIES *amoenum novizelandicum, farctum anglicum, fasciculare amrutii, fasciculare henriquesii, lubricum nairii, lubricum nathanii, nudiusculum lamii, subsecundum tenuis, subspinosum falklandicum, tenue lubricum, tenue uniforme*
GENUS *Streptochlora* Twisted Bristle Olive Green Algae
SPECIES *moniliformis*
GENUS *Stromatella* Stroma Bristle Olive Green Algae
SPECIES *monostromatica, papillosa*
GENUS *Syncoryne* Club Bristle Olive Green Algae
SPECIES *reinkei*
GENUS *Tellamia* Tellam Bristle Olive Green Algae
SPECIES *contorta*
GENUS *Thamniochaete* Thamnid Bristle Olive Green Algae
SPECIES *aculeata, huberii*
GENUS *Thamniochloris* Lesser Thamnid Bristle Olive Green Algae
SPECIES *atroviridis, ochlochaetoides, variabilis*
GENUS *Trichodiscus* Hairy Disc Bristle Olive Green Algae
SPECIES *elegans*
GENUS *Trichophilus* Hairy Bristle Olive Green Algae
SPECIES *welckeri*
GENUS *Trichosarcina* Hairy Sarcin Bristle Olive Green Algae
SPECIES *mucosa, polymorpha*
GENUS *Tumulofilum* Tumor Bristle Olive Green Algae
SPECIES *terrestre*
GENUS *Uronema* Tailed Bristle Olive Green Algae

SPECIES *acuminatum, brasiliense, confervicola, curvata, elongatum, marina, terrestre, trentonense*
GENUS *Zoddaea* Zodd Bristle Olive Green Algae
SPECIES *viridis*
GENUS *Zygomitus* Miter Bristle Olive Green Algae
SPECIES *reticulatus*

FAMILY 2 *Pleurastraceae* Pleural Star Olive Green Algae
GENUS *Pleurastrum* Pleural Star Olive Green Algae
SPECIES *insigne, photoheterotrophicum, terricola*
GENUS *Parietochloris* Pariet Pleural Star Olive Green Algae
SPECIES *bilobata, incisa*
GENUS *Schizogonium* Split Angled Pleural Star Olive Green Algae
SPECIES *boryanum, contortum, delicatulum, laetevirens, murale, neesii, nodosum, pallidum, parietinum, percursum, thermale*
VARIETIES *laetevirens majus, laetevirens crispatum, murale uniseriatum*

ORDER 3 *Chlorococcales* Round Olive Green Algae

FAMILY 01 *Chlorococcaceae* Round Olive Green Algae
GENUS *Chlorococcum* Round Olive Green Algae
SPECIES *acidum, aegyptiacum, botryoides, choloepodis, citriforme, costatozygotum, diplobionticum, echinozygotum, elbense, elkhartiense, ellipsoideum, humicolum, hypnosporum, infusionum, isabeliense, littorale, lobatum, macrostigmatum, minimum, minutum, murorum, novae-angliae, oleofaciens, pamirum, pinguideum, pseudodictyosphaerium, pyrenoidosum, salinum, schizochlamys, schwarzii, submarinum, tatrense, vacuolatum, viride*
VARIETIES *infusionum macrostigmatica*
GENUS *Ankyra* Ankara Round Olive Green Algae
SPECIES *ancora, inermis, judayi, paradoxioides, starrii*
GENUS *Apodochloris* Apodal Round Olive Green Algae
SPECIES *dinobryonis, irregularis, polymorpha, simplicissima*
GENUS *Axilococcus* Axile Round Olive Green Algae
SPECIES *clingmanii*
GENUS *Botryosphaerella* Berry Round Olive Green Algae
SPECIES sudetica
GENUS *Bracteacoccus* Bract Round Olive Green Algae
SPECIES *aerius, aggregatus, anomalus, cinnabarinus, cohaerens, engadinensis, giganteus, grandis, helveticus, medionucleatus, minor, minutus, pseudominor*
VARIETIES *minor desertorum, minor glacialis*
GENUS *Chlorococcus* Lesser Round Olive Green Algae
SPECIES *endozoicum*
GENUS *Chlorohippotes* Horse Round Olive Green Algae
SPECIES *tenax*
GENUS *Chlorotetraedron* Tetrahedron Round Olive Green Algae
SPECIES *polymorphum, bitridens*
GENUS *Closteridium* Spindle Round Olive Green Algae
SPECIES *bengalicum, hebetatum*
GENUS *Coccochlorum* Least Round Olive Green Algae
SPECIES *murorum*
GENUS *Coccothrix* Hairy Round Olive Green Algae
SPECIES *chlorolobata*
GENUS *Coenochloris* Mud Round Olive Green Algae
SPECIES *asymmetrica, bilobata, fottii, granulata, mucolamellata, pyrenoidosa, signiensis*
GENUS *Coenococcus* Lesser Mud Round Olive Green Algae
SPECIES *apyrenoidosus, planctonicus, polycoccus*
GENUS *Coronastrum* Crown Star Round Olive Green Algae
SPECIES *ellipsoideum*
GENUS *Cystomonas* Cyst Monad Round Olive Green Algae
SPECIES *actinosphaerii, indica, starrii*
GENUS *Desmatractum* Fiber Dark Round Olive Green Algae
SPECIES *bipyramidatum, plicatum*
GENUS *Dicellula* Dicell Round Olive Green Algae
SPECIES *geminata*
GENUS *Didymogenes* Double Round Olive Green Algae
SPECIES *anomala, palatina*
GENUS *Ecballocystis* Ecball Cyst Round Olive Green Algae
SPECIES *fluitans, japonica*
GENUS *Ectogeron* Ectophyte Round Olive Green Algae
SPECIES *elodeae*

GENUS *Elliptochloris* Elliptic Round Olive Green Algae
SPECIES *bilobata, reniformis, subsphaerica*
VARIETIES *subsphaerica antarctica*
GENUS *Emergococcus* Emerging Round Olive Green Algae
SPECIES *lucens*
GENUS *Emergosphaera* Lesser Emerging Round Olive Green Algae
SPECIES *superficialis*
GENUS *Ettlia* Ettl Round Olive Green Algae
SPECIES *alveolaris, bilobata, carotinosa, cohaerens, fusispora, minuta, oleoabundans, pseudoalveolaris, terrestris, texensis*
GENUS *Euastropsis* Star Round Olive Green Algae
SPECIES *richteri*
GENUS *Fernandinella* Fernando Round Olive Green Algae
SPECIES *alpina*
VARIETIES *alpina semiglobosa*
GENUS *Ferricystis* Ferric Cyst Round Olive Green Algae
SPECIES *mucicola*
GENUS *Follicularia* Follicle Round Olive Green Algae
SPECIES *botryoides, ettlii, komarekii, paradoxalis, skujae, starrii, texensis*
GENUS *Glenobotrydion* Glen Berry Round Olive Green Algae
SPECIES *kanshiense, tetragonalum*
GENUS *Granulocystopsis* Granule Cyst Round Olive Green Algae
SPECIES *coronata, decorata, pseudocoronata*
GENUS *Hydrianum* Hydric Round Olive Green Algae
SPECIES *coronatum, heteromorphum, ovale*
GENUS *Keratococcus* Horn Round Olive Green Algae
SPECIES *bicaudatus, dispar, dissociatus, raphidioides*
GENUS *Korshikoviella* Korshikov Round Olive Green Algae
SPECIES *michailovskoensis, rotundata, setosa*
GENUS *Lobocystis* Lobe Cyst Round Olive Green Algae
SPECIES *neodichotoma, planctonica*
VARIETIES *planctonica mucosa*
GENUS *Macrochloris* Large Round Olive Green Algae
SPECIES *chlorococcoides, cohaerens, multinucleata, radiosa*
GENUS *Nautococcus* Nauticle Round Olive Green Algae
SPECIES *caudatus, emersus, minutus, pyriformis, solutus, terrestris, verrucosus*
GENUS *Neglectella* Neglected Round Olive Green Algae
SPECIES *asterifera, chodatii, eremosphaerophila, ovalis, paludensis, rotula*
GENUS *Neglectellopsis* Lesser Neglected Round Olive Green Algae
SPECIES *skujae*
GENUS *Neochloris* Modern Round Olive Green Algae
SPECIES *aquatica, cohaerens, conjuncta, gelatinosa, pseudostigmatica, pyrenoidosa, terrestris, texensis, vigensis, wimmeri*
GENUS *Neospongiococcum* Modern Sponge Round Olive Green Algae
SPECIES *alabamense, bipapillatum, commatiforme, concentricum, excentricum, gelatinosum, macropyrenoidosum, mobile, munimentiferum, polymorphum, vacuolatum*
GENUS *Nephrochlamys* Kidney Cloak Round Olive Green Algae
SPECIES *rostrata, rotunda*
GENUS *Octogoniella* Octogon Round Olive Green Algae
SPECIES *sphagnicola*
GENUS *Palaeoanacystis* Paleoic Anacyst Round Olive Green Algae
SPECIES *psilata, punctata, reticulata, suketensis, verrucosa*
GENUS *Phaseolaria* Bean Round Olive Green Algae
SPECIES *obliqua*
VARIETIES *obliqua major*
GENUS *Podohedra* Podohedron Round Olive Green Algae
SPECIES *bicaudata, falcata, longipes*
GENUS *Poloidion* Polo Round Olive Green Algae
SPECIES *didymos*
GENUS *Pseudocharacium* Lesser Characin Round Olive Green Algae
SPECIES *acuminatum, americanum, obtusum*
GENUS *Pseudodictyochloris* Lesser Net Round Olive Green Algae
SPECIES *dissecta, multinucleata*
GENUS *Pseudoplanophila* Lesser Flat Round Olive Green Algae
SPECIES *sphagnothermalis*
GENUS *Pseudoquadrigula* Lesser Quad Round Olive Green Algae
SPECIES *britannica, lagoensis*

GENUS *Pseudosiderocelopsis* Lesser Siderite Round Olive Green Algae
SPECIES *antarctica*
GENUS *Pseudospongiococcum* Lesser Sponge Round Olive Green Algae
SPECIES *protococcoides*
GENUS *Pseudotetracystis* Lesser Tetrad Cyst Round Olive Green Algae
SPECIES *aurescens, terrestris*
GENUS *Pseudotetrastrum* Lesser Tetrad Star Round Olive Green Algae
SPECIES *punctatum*
GENUS *Pseudotrochiscia* Lesser Wheel Round Olive Green Algae
SPECIES *areolata*
GENUS *Quadricoccus* Quad Round Olive Green Algae
SPECIES *euryhalinicus, laevis, verrucosus*
GENUS *Quadrodiscus* Quad Disc Round Olive Green Algae
SPECIES *gombosii, indicus, minutus*
GENUS *Rhodoplax* Red Plaque Round Olive Green Algae
SPECIES *schinzii*
GENUS *Schroederia* Schroeder Round Olive Green Algae
SPECIES *jadayi, planktonica, robusta, setigera*
GENUS *Sclernax* Hard Round Olive Green Algae
SPECIES *lutescens, truncata*
GENUS *Sclerococcus* Hardened Round Olive Green Algae
SPECIES *jutoghensis*
GENUS *Scotiellopsis* Scot Round Olive Green Algae
SPECIES *levicostata, reticulata, rubescens*
GENUS *Siderocelis* Siderite Round Olive Green Algae
SPECIES *minor, ornata*
GENUS *Siderocystopsis* Siderite Cyst Round Olive Green Algae
SPECIES *punctifera*
GENUS *Skujaster* Skuja Star Round Olive Green Algae
SPECIES *asteriferus*
GENUS *Sphaerobotrys* Berry Round Olive Green Algae
SPECIES *fluviatilis*
GENUS *Spongiochloris* Sponge Round Olive Green Algae
SPECIES *excentrica, gigantea, incrassata, lamellata, llanoensis, minor, spongiosa, typica*
GENUS *Spongioplastidium* Plastid Sponge Round Olive Green Algae
SPECIES *proliferum*
GENUS *Sporotetras* Spore Tetrad Round Olive Green Algae
SPECIES *fluviatilis, pyriformis*
GENUS *Sykidion* Sycon Round Olive Green Algae
SPECIES *dilitatum, droebakense, dyeri*
GENUS *Symbiococcum* Symbiote Round Olive Green Algae
SPECIES *hydrae*
GENUS *Tetrachlorella* Tetrad Chloric Round Olive Green Algae
SPECIES *alternans*
GENUS *Tetraedron* Tetrahedron Round Olive Green Algae
SPECIES *acutum, bifurcatum, caudatum, conicum, constrictum, cruciforme, gigas, gracile, granulosum, hortense, incus, limneticum, lobulatum, longispinum, minimum, muticum, obtusum, octaedricum, planctonicum, rectangulare, regulare, sexconicum, striatum, trigonum, tumidulum, verrucosum, victorieae, vulgare, wasteneysii*
VARIETIES *acutum rectilineare, caudatum australe, gigas tetraedricum, limneticum gracile, lobulatum crassum, lobulatum decussatum, lobulatum polyfurcatum, lobulatum sydneyense, lobulatum triangulare, minimum minimum, minimum australe, minimum excavata, minimum morsum, minimum pinacidium, minimum quadratum, minimum rectilineare, minimum scrobiculatum, minimum trigonum, muticum hispanica, regulare granulata, regulare incus, regulare inerme, regulare major, regulare majus, regulare octaedricum, regulare papilliferum, trigonum gracile, trigonum mammillatum, victorieae major, vulgare major*
GENUS *Trochisciopsis* Wheel Round Olive Green Algae
SPECIES *insignis, tetraspora*

FAMILY 02 *Actinochloridaceae* Radial Round Olive Green Algae
GENUS *Actinochloris* Radial Round Olive Green Algae
SPECIES *sphaerica, terrestris*
GENUS *Deasonia* Deason Radial Round Olive Green Algae
SPECIES *bispora, granata, irregularis, multinucleata, prolifera, punctata, saccata*

FAMILY 03 *Botryococcaceae* Berry Round Olive Green Algae
GENUS *Botryococcus* Berry Round Olive Green Algae
SPECIES *australis, braunii, calcareus, canadensis, comperei, fernandoi, neglectus, pila, protuberans, terribilis*
GENUS *Caulodendron* Wood Berry Round Olive Green Algae
SPECIES *dichotomococcoides*

FAMILY 04 *Characiochloridaceae* Charming Round Olive Green Algae
GENUS *Characiochloris* Charming Round Olive Green Algae
SPECIES *acuminata, apiculata, characioides, sasae, sessilis*
GENUS *Chlamydopodium* Cloak Pod Charming Round Olive Green Algae
SPECIES *fusiforme, perforatum, pluricoccum, starrii, vacuolatum*
GENUS *Chledripole* Claw Pole Charming Round Olive Green Algae
SPECIES *lobata, tubulosa*

FAMILY 02 *Chlorosiphonaceae* Siphon Round Olive Green Algae
GENUS *Characiosiphon* Charming Siphon Round Olive Green Algae
SPECIES *rivularis*
GENUS *Lobocharacium* Lobed Charming Siphon Round Olive Green Algae
SPECIES *coloradoense*
GENUS *Protosiphon* Primitive Siphon Round Olive Green Algae
SPECIES *botryoides*
GENUS *Schizosiphon* Split Siphon Round Olive Green Algae
SPECIES *fucicola*
GENUS *Siphorus* Siphon Round Olive Green Algae
SPECIES *alternus, fasciculatus*
GENUS *Siphytus* Lesser Siphon Round Olive Green Algae
SPECIES *filiformis, hexodon, obconicus*

FAMILY 03 *Coccomyxaceae* Mycote Round Olive Green Algae
GENUS *Coccomyxa* Round Olive Green Algae
SPECIES *arvernensis, astericola, brevis, chodatii, confluens, corbierei, curvata, elongata, gloeobotrydiformis, litoralis, mucigena, olivacea, parasitica, peltigerae, pringsheimii, rayssiae, solorinae, subellipsoida, subglobosa, viridis*
VARIETIES *peltigerae variolosae, solorinae solorinae, solarinae bisporae, solarinae croceae, solorinae saccatae*
GENUS *Dactylothece* Finger Mycote Round Olive Green Algae
SPECIES *arcuatum, braunii*
GENUS *Dispora* Dispore Mycote Round Olive Green Algae
SPECIES *crucigenioides*
GENUS *Lusitania* Portugal Mycote Round Olive Green Algae
SPECIES *henriquesii*
GENUS *Nannochloris* Dwarf Mycote Round Olive Green Algae
SPECIES *atomus, bacillaris, coccoides, eucaryotum, normandinae, oculata*
GENUS *Ourococcus* Tailed Mycote Round Olive Green Algae
SPECIES *bicaudatus*
GENUS *Palmogloea* Palm Glow Mycote Round Olive Green Algae
SPECIES *protuberans*
GENUS *Pseudococcomyxa* Lesser Mycote Round Olive Green Algae
SPECIES *gloeoglossa, simplex*

FAMILY 04 *Coelastraceae* Hollow Star Round Olive Green Algae
GENUS *Coelastrum* Hollow Star Round Olive Green Algae
SPECIES *astroideum, crenatum, howardii, indicum, intermedium, microporum, obtusum, polychordum, printzii, probiscideum, pseudomicroporum, pulchellum, pulchrum, quadricellulare, reticulatum, schizodermaticum, sphaericum, triangulare, verrucosum*
VARIETIES *crenatum cubicum, pulchrum intermedium, sphaericum compacta, sphaericum subpulchrum, verrucosum acutiverrucosum*
GENUS *Asterarcys* Arcy Hollow Star Olive Green Algae
SPECIES *quadricellulare*
GENUS *Coelastropsis* Lesser Hollow Star Olive Green Algae
SPECIES *costata*
GENUS *Eutetramorus* Tetrad Hollow Star Olive Green Algae
SPECIES *globosus*
GENUS *Soropediastrum* Soroped Hollow Star Olive Green Algae
SPECIES *howardii, printzii*

FAMILY 05 *Dictyosphaeriaceae* Netted Round Olive Green Algae
GENUS *Dictyosphaerium* Netted Round Olive Green Algae
SPECIES *chlorelloides, coacervatum, dichotomum, ehrenbergianum, elongatum, granulatum, pulchellum, regulare, terrestre, tetrachotomum*
GENUS *Dactylosphaerium* Finger Netted Round Olive Green Algae
SPECIES *sociale*
GENUS *Dictyochloris* Chloric Netted Round Olive Green Algae
SPECIES *fragrans, pulchra*
GENUS *Dictyococcus* Lesser Netted Round Olive Green Algae
SPECIES *bradypodis, schumacherensis, varians*

GENUS *Dimorphococcus* — Dimorphic Netted Round Olive Green Algae
SPECIES *lunatus*
GENUS *Gilbertsmithia* — Gilbert Smith Netted Round Olive Green Algae
SPECIES *grandis*
GENUS *Pseudodictyosphaerium* — Least Netted Round Olive Green Algae
SPECIES *scoticum*
GENUS *Tetracoccus* — Tetrad Netted Round Olive Green Algae
SPECIES *lacustris*
GENUS *Westella* — West Netted Round Olive Green Algae
SPECIES *botryoides, linearis*

FAMILY 05 *Endosphaeraceae* — Endic Round Olive Green Algae
GENUS *Chlorochytrium* — Chytrid Endic Round Olive Green Algae
SPECIES *bristolae, cohnii, crassum, dermatocolax, facciolaae, grande, inclusum, lemnae, porphyrae, sarcophyci, schmitzii*
VARIETIES *facciolaae minor*
GENUS *Codiolum* — Bell Endic Round Olive Green Algae
SPECIES *dudresnayi, gregarium, pusillum*
VARIETIES *gregarium intermedium, pusillum americanum, pusillum longipes, pusillum subsessile*
GENUS *Kentrosphaera* — Kent Endic Round Olive Green Algae
SPECIES *austriaca, gibberosa, lemnae*
VARIETIES *gibberosa polymorpha*
GENUS *Phyllobium* — Phyllode Endic Round Olive Green Algae
SPECIES *dimorphum*
GENUS *Rhodochytrium* — Red Chytrid Endic Round Olive Green Algae
SPECIES *spilanthidis*

FAMILY 06 *Gomontiaceae* — Gomont Round Olive Green Algae
GENUS *Gomontia* — Gomont Round Olive Green Algae
SPECIES *aegagropila, arhiza, caudata, codiolifera, lignicola, polyrhiza*
GENUS *Chlorojackia* — Jack Gomont Round Olive Green Algae
SPECIES *pachyclados*

FAMILY 06 *Hormotilaceae* — Hormotile Round Olive Green Algae
GENUS *Hormotila* — Hormotile Round Olive Green Algae
SPECIES *blennista, mucigena*
GENUS *Dendrocystis* — Wood Cyst Hormotile Round Olive Green Algae
SPECIES *raoi*
GENUS *Heleococcus* — Sun Hormotile Round Olive Green Algae
SPECIES *mucicolus*
GENUS *Palmodactylon* — Palm Finger Hormotile Round Olive Green Algae
SPECIES *simplex*
GENUS *Palmodictyon* — Palm Net Hormotile Round Olive Green Algae
SPECIES *resendeanum, varium, viride*
GENUS *Planochloris* — Flat Hormotile Round Olive Green Algae
SPECIES *pyrenoidifera*

FAMILY 07 *Hydrodictyaceae* — Hydric Netted Round Olive Green Algae
GENUS *Hydrodictyon* — Hydric Netted Round Olive Green Algae
SPECIES *africanum, majus, patenaeforme, reticulatum*
VARIETIES *reticulatum bernardii, reticulatum minimum, reticulatum nodosum*
GENUS *Paradoxia* — Paradox Hydric Netted Round Olive Green Algae
SPECIES *multiseta, paradoxioides*
GENUS *Pediastrum* — Pedal Star Hydric Netted Round Olive Green Algae
SPECIES *angulosum, asymmetricum, bareilliense, bijuga, biradiatum, boryanum, braunii, conchense, contiguum, duplexellipticum, gracillimum, integrum, kawraiskyi, longecornutum, musterii, orbitale, patagonicum, privum, simplex, tetras, tricornutum, tricyclium*
VARIETIES *angulosum asperum, angulosum rugulosum, bijuga capitata, biradiatum braunii, boryanum australe, boryanum capitatum, boryanum cornutum, boryanum granulatum, boryanum haynaldii, boryanum longicorne, boryanum ornatum, boryanum productum, boryanum rugulosum, duplex asperum, duplex convexum, duplex denticulatum, duplex gracillimum, duplex irregulare, duplex longicorne, duplex reticulatum, duplex rugulosum, duplex subgranulatum, integrum braunianum, integrum privum, simplex clathrata, simplex ovata, tetras apiculatum, tetras cuspidatum, tetras fluviatile, tetras integrum, tetras longicornutum, tetras octogonis, tetras tetraodon*
GENUS *Sorastrum* — Soral Star Hydric Netted Round Olive Green Algae
SPECIES *americanum, hathoris, spinulosum*
GENUS *Tetrapedia* — Tetraped Hydric Netted Round Olive Green Algae
SPECIES *reinschiana*

FAMILY 08 *Hypnomonadaceae* — Mossy Round Olive Green Algae
GENUS *Hypnomonas* — Mossy Round Olive Green Algae

SPECIES *pleiopyrenigerum*
GENUS *Glaucosphaera* Glaucous Mossy Round Olive Green Algae
SPECIES *vacuolata*
GENUS *Kremastochloris* Kremast Mossy Round Olive Green Algae
SPECIES *conus*
GENUS *Sphaerellocystis* Cyst Mossy Round Olive Green Algae
SPECIES *aplanosporum, ellipsoidea*

FAMILY 09 *Micractiniaceae* Tiny Round Olive Green Algae
GENUS *Micractinium* Tiny Round Olive Green Algae
SPECIES *bornhemiense, conococcoides, pusillum, quadrisetum, strigoniense*
GENUS *Acanthosphaera* Spiny Tiny Round Olive Green Algae
SPECIES *zacharaisi*
GENUS *Echinosphaeridium* Urchin Tiny Round Olive Green Algae
SPECIES *nordstedtii*
GENUS *Golenkinia* Golenkin Tiny Round Olive Green Algae
SPECIES *aculeata, longispicula, paucispina, radiata*
VARIETIES *radiata australis, radiata paucispina*
GENUS *Golenkiniopsis* Lesser Golenkin Tiny Round Olive Green Algae
SPECIES *chlorelloides, gujaratensis, longispina, parvula*

FAMILY 10 *Oocystaceae* Oocyst Round Olive Green Algae
GENUS *Oocystis* Oocyst Round Olive Green Algae
SPECIES *alpina, apiculata, asymmetrica, australiensis, borgei, chodati, elliptica, eremosphaeria, granulata, hunanensis, lacustris, marssonii, minuta, naegelii, natans, nodulosa, nordstedtiana, novaesemliae, ovalis, panduriformis, parva, polymorpha, pusilla, rifeum, rotula, rupestris, solitaria, striata, subhexagona, submarina, subsphaerica*
VARIETIES *apiculata asymmetrica, apiculata major, apiculata obesa, apiculata simplicior, apiculata splendida, hunanensis calida, lacustris natans, lacustris paludensis, minuta ellipsoidea, minuta vacuolata, naegelii curta, naegelii obesa, nodulosa australis, nodulosa crassa, nordstedtiana rotunda, novaesemliae novaesemliae, novaesemliae australica, novaesemliae major, novaesemliae maxima, ovalis cylindracea, ovalis subtruncata, panduriformis pachyderma, solitaria elongata, solitaria granulosa, solitaria ostenfeldii, solitaria terrestris, submarina major*
GENUS *Amphikrikos* Amphibian Krik Oocyst Round Olive Green Algae
SPECIES *heynigii, minutissimus, nanus, variabilis*
VARIETIES *variabilis krienitz*
GENUS *Bernardia* Bernard Oocyst Round Olive Green Algae
SPECIES *chodati, tetraedrica, wratislavensis*
VARIETIES *chodati cruciata*
GENUS *Catenocystis* Chain Oocyst Round Olive Green Algae
SPECIES *cohaerens*
GENUS *Cerasterias* Cherry Oocyst Round Olive Green Algae
SPECIES *raphidioides*
GENUS *Chloroidium* Chloric Oocyst Round Olive Green Algae
SPECIES *kruegeri*
GENUS *Chlorolobion* Chloric Lobe Oocyst Round Olive Green Algae
SPECIES *braunii, lunulatum, obtusum, tatrae*
VARIETIES *braunii pusilla, braunii viride*
GENUS *Chloropteridella* Chloric Wing Oocyst Round Olive Green Algae
SPECIES *tetragona*
GENUS *Chodatella* Chodat Oocyst Round Olive Green Algae
SPECIES *brevispina, chodati, circumfilata, pilosa, subsalsa, wratislaviensis*
VARIETIES *subsalsa longiseta*
GENUS *Chodatellopsis* Lesser Chodat Oocyst Round Olive Green Algae
SPECIES *elliptica*
GENUS *Chondrosphaera* Cartilage Oocyst Round Olive Green Algae
SPECIES *lapponica*
GENUS *Coenolamellus* Mud Lamella Oocyst Round Olive Green Algae
SPECIES *botryoideus*
GENUS *Conradia* Conrad Oocyst Round Olive Green Algae
SPECIES *incrustans*
GENUS *Dactylococcus* Finger Oocyst Round Olive Green Algae
SPECIES *dispar, infusionum, lacustris*
VARIETIES *infusionum minor*
GENUS *Diacanthos* Two Spine Oocyst Round Olive Green Algae
SPECIES *belenophorus*
GENUS *Diogenes* Diogene Oocyst Round Olive Green Algae
SPECIES *bacillaris, rotundus*
GENUS *Ecdysichlamys* Cloak Oocyst Round Olive Green Algae

SPECIES *obliqua*
GENUS *Echinocoleum* Urchin Oocyst Round Olive Green Algae
SPECIES *elegans*
GENUS *Eremosphaera* Sand Oocyst Round Olive Green Algae
SPECIES *antillana, gigas, viridis*
VARIETIES *viridis acuminata, viridis chodatii, viridis doliformis, viridis nodosa, viridis ovalis, viridis tetraedrica*
GENUS *Ettliella* Ettl Oocyst Round Olive Green Algae
SPECIES *tetraspora*
GENUS *Excentrosphaera* Excentric Oocyst Round Olive Green Algae
SPECIES *viridis*
GENUS *Fotterella* Fotter Oocyst Round Olive Green Algae
SPECIES *tetrachlorelloides*
GENUS *Franceia* France Oocyst Round Olive Green Algae
SPECIES *amphitricha, breviseta, oblonga, ovalis*
GENUS *Fusola* Fused Oocyst Round Olive Green Algae
SPECIES *viridis*
GENUS *Glaucocystopsis* Glaucous Oocyst Round Olive Green Algae
SPECIES *africana*
GENUS *Glochiococcus* Gloch Oocyst Round Olive Green Algae
SPECIES *aciculiferus*
GENUS *Granulocystis* Granule Oocyst Round Olive Green Algae
SPECIES *verrucosa*
GENUS *Halochlorella* Halogen Oocyst Round Olive Green Algae
SPECIES *rubescens*
GENUS *Hemichloris* Half Oocyst Round Olive Green Algae
SPECIES *antarctica*
GENUS *Hyalochlorella* Hyaline Oocyst Round Olive Green Algae
SPECIES *marina*
GENUS *Hyaloraphidium* Hyaline Needle Oocyst Round Olive Green Algae
SPECIES *contortum, curvatum*
GENUS *Juranyiella* Jurany Oocyst Round Olive Green Algae
SPECIES *javorkae*
GENUS *Keriochlamys* Kery Cloak Oocyst Round Olive Green Algae
SPECIES *styriaca*
GENUS *Kirchneriella* Kirchner Oocyst Round Olive Green Algae
SPECIES *aperta, arcuata, cornuta, elegans, elongata, gracillima, incurvata, irregularis, lunaris, major, microscopica, obesa, phaseoliformis, roselata, secktiana*
VARIETIES *lunaris acuta, lunaris aperta, lunaris approximata, obesa aperta*
GENUS *Kirchneriellosaccus* Kirchner Sac Oocyst Round Olive Green Algae
SPECIES *lunatus*
VARIETIES *lunatus indicus*
GENUS *Lagerheimia* Lagerheim Oocyst Round Olive Green Algae
SPECIES *acuminata, chodatii, ciliata, circumfilata, citriformis, comosa, coronata, cristata, elliptica, genevensis, globosa, hindakii, longiseta, subsalsa, wratislaviensis*
VARIETIES *ciliata amphitricha, ciliata inermis, coronata inflata, coronata striolata, genevensis gracilis, genevensis subglobosa, subsalsa pilosa*
GENUS *Lithraphidites* Lithic Needle Oocyst Round Olive Green Algae
SPECIES *praequadratus*
GENUS *Makinoella* Makino Oocyst Round Olive Green Algae
SPECIES *tosaensis*
GENUS *Monoposthium* Neglected Oocyst Round Olive Green Algae
SPECIES *bracchium, latum, vulgare*
GENUS *Monoraphidium* Monad Needle Oocyst Round Olive Green Algae
SPECIES *arcuatum, contortum, convolutum, griffithii, irregulare, komarkovae, litorale, lunare, minutum, mirabile, nanum, obtusum, pusillum, skujae, terrestre, tortile*
VARIETIES *convolutum pseudosabulosum*
GENUS *Muriella* Muriel Oocyst Round Olive Green Algae
SPECIES *australis, decolor, magna, terrestris, zofingiensis*
GENUS *Muriellopsis* Lesser Muriel Oocyst Round Olive Green Algae
SPECIES *geosphaera, pyrenigera, sphaerica*
GENUS *Mycacanthococcus* Spined Oocyst Round Olive Green Algae
SPECIES *cellaris*
GENUS *Mychonastes* Mychonaste Oocyst Round Olive Green Algae
SPECIES *desiccatus, ruminatus*
GENUS *Mycotetraedron* Tetrahedron Oocyst Round Olive Green Algae
SPECIES *cellare*

GENUS *Nephrocytium* Kidney Oocyst Round Olive Green Algae
SPECIES *agardhianum, alantoideum, limneticum, lunaris, lunatum*
VARIETIES *agardhianum allantoideum, agardhianum planktonicum*
GENUS *Oocystaenium* Ribbon Oocyst Round Olive Green Algae
SPECIES *elegans*
GENUS *Oocystidium* Lesser Oocyst Round Olive Green Algae
SPECIES *ovale*
GENUS *Oocystopsis* Least Oocyst Round Olive Green Algae
SPECIES *granulata*
GENUS *Oogamochlamys* Cloak Oocyst Round Olive Green Algae
SPECIES *ettlii*
GENUS *Oonephris* Lesser Kidney Oocyst Round Olive Green Algae
SPECIES *obesa*
GENUS *Pachycladella* Thick Clad Oocyst Round Olive Green Algae
SPECIES *umbrina*
GENUS *Palmellococcus* Palm Oocyst Round Olive Green Algae
SPECIES *marinus, protothecoides*
GENUS *Picocystis* Pico Oocyst Round Olive Green Algae
SPECIES *salinarum*
GENUS *Pilidiocystis* Pile Oocyst Round Olive Green Algae
SPECIES *endophytica, multispora*
GENUS *Planktosphaerella* Lesser Plankton Oocyst Round Olive Green Algae
SPECIES *terrestris*
GENUS *Planktosphaeria* Plankton Oocyst Round Olive Green Algae
SPECIES *gelatinosa*
GENUS *Polyedriopsis* Polyhedron Oocyst Round Olive Green Algae
SPECIES *spinulosa*
GENUS *Pseudochlorococcum* Lesser Chloric Oocyst Round Olive Green Algae
SPECIES *polymorphum, typicum*
GENUS *Pseudokirchneriella* Lesser Kirchner Oocyst Round Olive Green Algae
SPECIES *arcuata, contorta, danubiana, elongata, extensa, gracillima, inclinata, irregularis, major, mayorii, microscopica, mucosa, phaseoliformis, roselata, rotunda, subcapitata, turfosa, valida, van-goorii*
VARIETIES *van-goorii decussata*
GENUS *Pseudomuriella* Least Muriel Oocyst Round Olive Green Algae
SPECIES *aurantiaca*
GENUS *Quadrigula* Quad Bristle Oocyst Round Olive Green Algae
SPECIES *chodatii, closterioides, korshikovii, pfitzer*
GENUS *Rayssiella* Rayss Oocyst Round Olive Green Algae
SPECIES *hemisphaerica*
GENUS *Reinschiella* Reinsch Oocyst Round Olive Green Algae
SPECIES *longispina*
GENUS *Rhombocystis* Rhombic Oocyst Round Olive Green Algae
SPECIES *complanata*
GENUS *Saturnella* Saturn Oocyst Round Olive Green Algae
SPECIES *corticola, elegans*
GENUS *Scotiella* Scot Oocyst Round Olive Green Algae
SPECIES *fritschii, muscicola, oboviformis, spinosa, tuberculata, turficola*
GENUS *Selenastrum* Moon Star Oocyst Round Olive Green Algae
SPECIES *bibraianum, capricornutum, gracile*
VARIETIES *gracile minutum*
GENUS *Selenoderma* Moon Dermal Oocyst Round Olive Green Algae
SPECIES *malmeana*
GENUS *Sestosoma* Sestosome Oocyst Round Olive Green Algae
SPECIES *villosum*
GENUS *Thelesphaera* Nipple Oocyst Round Olive Green Algae
SPECIES *alpina*
GENUS *Thorakochloris* Thorak Oocyst Round Olive Green Algae
SPECIES *lacustris, nygaardii, planktonica, tetras*
GENUS *Treubaria* Treubar Oocyst Round Olive Green Algae
SPECIES *appendiculata, crassicornuta, schmidlei, setigera, triappendiculata*
GENUS *Trigonidiella* Trigon Oocyst Round Olive Green Algae
SPECIES *galea*
GENUS *Zoochlorella* Parasite Oocyst Round Olive Green Algae
SPECIES *parasitica*

FAMILY 6 *Radiococcaceae* Disc Round Olive Green Algae
GENUS *Radiococcus* Disc Round Olive Green Algae
SPECIES *nimbatus, planktonicus*

GENUS *Catenococcus* Chain Disc Round Olive Green Algae
SPECIES *tortuosus*
GENUS *Coenocystis* Mud Disc Round Olive Green Algae
SPECIES *obtusa, planctonica, subcylindrica*
GENUS *Crucigloea* Cross Glow Disc Round Olive Green Algae
SPECIES *crassiseta*
GENUS *Dictyochlorella* Netted Disc Round Olive Green Algae
SPECIES *globosa*
GENUS *Diplochloris* Diploid Disc Round Olive Green Algae
SPECIES *decussata, lunata*
GENUS *Hindakochloris* Hindak Disc Round Olive Green Algae
SPECIES *insularis, nygaardii, planktonica*
GENUS *Myurococcus* Tailed Disc Round Olive Green Algae
SPECIES *urococcus*
GENUS *Neocystis* Modern Cyst Disc Round Olive Green Algae
SPECIES *ovalis*
GENUS *Palmellosphaerium* Palm Disc Round Olive Green Algae
SPECIES *planktonicum*
GENUS *Pectodictyon* Pectin Net Disc Round Olive Green Algae
SPECIES *cubicum*
GENUS *Tomaculum* Tomacule Disc Round Olive Green Algae
SPECIES *catenatum*

FAMILY 11 *Scenedesmaceae* Scene Fiber Round Olive Green Algae
GENUS *Scenedesmus* Scene Fiber Round Olive Green Algae
SPECIES *aciculatus, aculeolatus, acuminatus, acutiformis, acutus, aldavei, anhuiensis, apiculatus, arcuatus, bernardii, bicaudatus, bicellularis, bidentatus, bijugus, brasiliensis, breviaculeatus, caudato-aculeolatus, chlorelloides, circumfusus, costatus, dimorphus, dissociatus, distentus, ecornis, ellipticus, falcatus, fenestratus, flavescens, fuscus, granulatus, gujaratensis, hanleyi, heimii, heteracanthus, hindakii, hirsutus, huangshanensis, incrassatulus, jovais, longispina, magnus, microspina, multistriatus, naegelii, obliquus, obtusus, obtusus graevenitzii, oocystiformis, parisiensis, pecsensis, pectinatus, planctonicus, platydiscus, prismaticus, prismaticus spinosus, pseudobernardii, quadrialatus, quadricaudus, raciborskii, reginae, reniformis, securiformis, semipulcher, serrato-perforatus, smithii, soli, terrestris, tricostatus, tschudyi, verrucosus, wuhuensis*
VARIETIES *acuminatus inermius, acuminatus tortuosus, acutiformis brasiliensis, acutiformis costatus, acutus alternans, arcuatus gracilis, arcuatus platydiscus, bijugus disciformis, bijugus duplex, bijugus inermis, bijugus obtusiusculus, bijugus ralfsii, circumfusus bicaudatus, ecornis disciformis, flavescens breviaculeatus, fuscus peruvianus, granulatus salina, heimii longispina, magnus bidigitatus, magnus carpetana, obliquus antennatus, obliquus arcuatus, pectinatus distentus, pseudobernardii globosus, pseudobernardii procerus, quadricaudus crassiaculeatus, quadricaudus horridus, quadricaudus incurvus, quadricaudus inermius, quadricaudus longispinus, quadricaudus maximus, quadricaudus minus, quadricaudus opoliensis, quadricaudus quadrispinus, smithii spinulosus*
GENUS *Astrocladium* Star Clad Scene Fiber Round Olive Green Algae
SPECIES *cerastioides*
GENUS *Bilgramia* Bilgram Scene Fiber Round Olive Green Algae
SPECIES *indica*
GENUS *Crucigenia* Cross Scene Fiber Round Olive Green Algae
SPECIES *australis, cordata, cruxmichaeli, fenestrata, floralis, lauterbornii, quadrata, tetrapedia, triangularis, truncata, unicellularis, variabilis*
VARIETIES *quadrata secta, tetrapedia ornata*
GENUS *Crucigeniella* Lesser Cross Scene Fiber Round Olive Green Algae
SPECIES *apiculata, crucifera, irregularis, rectangularis*
GENUS *Danubia* Nubian Scene Fiber Round Olive Green Algae
SPECIES *africana*
GENUS *Desmodesmus* Lesser Scene Fiber Round Olive Green Algae
SPECIES *abundans, ambuehlii, armatus, arthrodesmiformis, asymmetricus, bicellularis, brasiliensis, communis, costato-granulatus, cuneatus, curvatocornis, denticulatus, dispar, flavescens, grahneisii, hystrix, insignis, intermedius, kissii, komarekii, komarekii hirsutus, lefevrei, lunatus, maximus, multiformis, opoliensis, pannonicus, perforatus, pleiomorphus, polyspinosus, protuberans, pseudodenticulatus, schnepfii, serratus, spinosus, spinulatus, subspicatus, tropicus*
VARIETIES *abundans longicauda, abundans peruvianus, armatus bicaudatus, armatus longispina, armatus microspinosus, armatus pluricostatus, armatus spinosus, armatus subalternans, brasiliensis serrato-perforatus, communis rectangularis, costato-granulatus elegans, denticulatus australis, denticulatus brevispinus, denticulatus disciformis, denticulatus fenestratus, denticulatus gracilis, denticulatus granulatus, denticulatus linearis, denticulatus lunatus, dispar denticulatus, dispar elegans, dispar rabae, dispar robustus, dispar semidenticulatus, flavescens breviaculeatus, flavescens longicaudatus, intermedius acutispinus, intermedius balatonicus, intermedius inflatus, lefevrei muzzanensis, maximus peruviensis, opoliensis alatus, opoliensis carinatus, opoliensis mononensis, perforatus bicaudatus, perforatus spinosus, spinosus microspinosus*

GENUS *Dicloster* — Dicloster Scene Fiber Round Olive Green Algae
SPECIES *acutus*
GENUS *Dictyastrum* — Netted Star Scene Fiber Round Olive Green Algae
SPECIES *mirabile*
GENUS *Didymocystis* — Dual Cyst Scene Fiber Round Olive Green Algae
SPECIES *fina, inconspicua, inermis, lineata, planktonica*
GENUS *Elakatothrix* — Elakat Scene Fiber Round Olive Green Algae
SPECIES *acuta, gelatinosa, inflexa, obtusata, parvula, viridis*
VARIETIES *gelatinosa aplanospora*
GENUS *Enallax* — Enallac Scene Fiber Round Olive Green Algae
SPECIES *acutiformis, coelastroides*
GENUS *Komarekia* — Komarek Scene Fiber Round Olive Green Algae
SPECIES *appendiculata, rotundata*
GENUS *Lauterborniella* — Lauterborn Scene Fiber Round Olive Green Algae
SPECIES *elegantissima*
GENUS *Sampaionema* — Sampa Scene Fiber Round Olive Green Algae
SPECIES *lacerdana*
GENUS *Schroederiella* — Schroeder Scene Fiber Round Olive Green Algae
SPECIES *africana*
GENUS *Tetradesmus* — Tetrad Scene Fiber Round Olive Green Algae
SPECIES *cumbricus, wisconsinensis*
VARIETIES *wisconsinensis reginae*
GENUS *Tetrallantos* — Tetrad Allantos Scene Fiber Round Olive Green Algae
SPECIES *lagerheimii, novae-geronae*
GENUS *Tetranephris* — Tetrad Kidney Scene Fiber Round Olive Green Algae
SPECIES *brasiliensis*
GENUS *Tetrastrum* — Tetrad Star Scene Fiber Round Olive Green Algae
SPECIES *asymmetricum, delicatospinosum, elegans, glabrum, heteracanthum, komarekii, staurogeniaeforme, triangulare*
VARIETIES *asymmetricum heterospinum, elegans dentatum, glabrum hispanica*

FAMILY 12 *Sphaeropleaceae* — Plebe Round Olive Green Algae
GENUS *Sphaeroplea* — Plebe Round Olive Green Algae
SPECIES *annulina, fragilis, robusta, soleirolii, wilmani*
GENUS *Atractomorpha* — Dark Plebe Round Olive Green Algae
SPECIES *echinata, porcata*

ORDER 4 *Tetrasporales* Ten Spore Olive Green Algae

FAMILY 1 *Tetrasporaceae* — Ten Spore Olive Green Algae
GENUS *Tetraspora* — Ten Spore Olive Green Algae
SPECIES *apiocystioides, explanata, gelatinosa, giraudii, hexanematoidea, intestinalis, intricata, lacustris, lemmermannii, lubrica*
GENUS *Apiocystis* — Bee Cyst Ten Spore Olive Green Algae
SPECIES *brauniana*
GENUS *Gloeodendron* — Glow Wood Ten Spore Olive Green Algae
SPECIES *catenatum, ramosum*
GENUS *Heleochloris* — Sun Ten Spore Olive Green Algae
SPECIES *pallida*
GENUS *Octosporiella* — Octogon Ten Spore Olive Green Algae
SPECIES *coloradoensis*
GENUS *Paulschulzia* — Paul Schulz Ten Spore Olive Green Algae
SPECIES *elegans, pseudovolvox, tenera*
GENUS *Phacomyxa* — Bundle Ten Spore Olive Green Algae
SPECIES *submontana*
GENUS *Placosphaera* — Plate Ten Spore Olive Green Algae
SPECIES *cavanillesiana*
GENUS *Planktogloea* — Plankton Glow Ten Spore Olive Green Algae
SPECIES *lundii*
GENUS *Pseudosphaerocystis* — Lesser Sphere Cyst Ten Spore Olive Green Algae
SPECIES *lacustris*
GENUS *Radiosphaera* — Radio Sphere Ten Spore Olive Green Algae
SPECIES *minuta, negevensis*
VARIETIES *negevensis minor*
GENUS *Schizochlamys* — Split Cloak Ten Spore Olive Green Algae
SPECIES *gelatinosa*
GENUS *Schulziella* — Schulz Ten Spore Olive Green Algae
SPECIES *pseudovolvox*

GENUS *Stylosphaeridium* — Sphere Ten Spore Olive Green Algae
SPECIES *stipitatum*
GENUS *Tetrasporella* — (Synonym for *Tetraspora*)

FAMILY 2 *Chlorangiellaceae* — Angled Ten Spore Olive Green Algae
GENUS *Chlorangiella* — Angled Ten Spore Olive Green Algae
SPECIES *pygmaea*
GENUS *Chlorepithema* — Pith Angled Ten Spore Olive Green Algae
SPECIES *hemisphaericum*
GENUS *Chlorophysema* — Bladder Angled Ten Spore Olive Green Algae
SPECIES *apiocystiforme, inertis*
GENUS *Gloeochloris* — Glow Angled Ten Spore Olive Green Algae
SPECIES *planctonica*
GENUS *Malleochloris* — Malle Angled Ten Spore Olive Green Algae
SPECIES *sessilis*
GENUS *Swarchewskiella* — Swarchewsk Angled Ten Spore Olive Green Algae
SPECIES *rotans*

FAMILY 3 *Chlorangiopsidaceae* — Lesser Angled Ten Spore Olive Green Algae
GENUS *Chlorangiopsis* — Lesser Angled Ten Spore Olive Green Algae
SPECIES *epizootica*
GENUS *Cecidochloris* — Cut Lesser Angled Ten Spore Olive Green Algae
SPECIES *adnata*
GENUS *Chlamydomonadopsis* — Cloak Lesser Angled Ten Spore Olive Green Algae
SPECIES *elliptica*
GENUS *Metapolytoma* — Polytome Lesser Angled Ten Spore Olive Green Algae
SPECIES *bacteriferum*
GENUS *Physocytium* — Bladder Lesser Angled Ten Spore Olive Green Algae
SPECIES *confervicola*
GENUS *Stylosphaeridiella* — Sphere Lesser Angled Ten Spore Olive Green Algae
SPECIES *crucigeniae*

FAMILY 4 *Gloeocystaceae* — Glow Cyst Ten Spore Olive Green Algae
GENUS *Gloeocystis* — Glow Cyst Ten Spore Olive Green Algae
SPECIES *adnata, ampla, australis, gigas, papuana, polydermatica, scopulorum, vesiculosa*
VARIETIES *australis ampla, gigas maxima, vesiculosa alpina*
GENUS *Asterococcus* — Star Glow Cyst Ten Spore Olive Green Algae
SPECIES *limneticus, superbus*
GENUS *Chlamydocapsa* — Cloak Capsule Glow Cyst Ten Spore Olive Green Algae
SPECIES *ampla, lobata, planctonica, retrospectans*
GENUS *Gloeococcus* — Round Glow Cyst Ten Spore Olive Green Algae
SPECIES *maximus, minor, minutissimus, schroeteri, tetrasporus*
GENUS *Gloeocystopsis* — Lesser Glow Cyst Ten Spore Olive Green Algae
SPECIES *limneticum*
GENUS *Tetrasporidium* — Least Glow Cyst Ten Spore Olive Green Algae
SPECIES *javanicum*

FAMILY 5 *Palmellaceae* — Palm Ten Spore Olive Green Algae
GENUS *Palmella* — Palm Ten Spore Olive Green Algae
SPECIES *alpicola, hyalina, mediterranea, microscopica, mucosa, oceanica*
GENUS *Askenasyella* — Askenasy Palm Ten Spore Olive Green Algae
SPECIES *chlamydopus*
GENUS *Chalarodora* — Chalar Palm Ten Spore Olive Green Algae
SPECIES *azurea*
GENUS *Chloranomala* — Anomalous Palm Ten Spore Olive Green Algae
SPECIES *cuprecola, palmelloides*
GENUS *Gloiodictyon* — Glow Net Palm Ten Spore Olive Green Algae
SPECIES *varium, viride*
GENUS *Gorffyana* — Gyorffy Palm Ten Spore Olive Green Algae
SPECIES *humicola*
GENUS *Hormotilopsis* — Hormotile Palm Ten Spore Olive Green Algae
SPECIES *gelatinosa, tetravacuolaris*
GENUS *Korschpalmella* — Korsch Palm Ten Spore Olive Green Algae
SPECIES *apyrenoidosa, microscopica*
GENUS *Oncosaccus* — Onc Sac Palm Ten Spore Olive Green Algae
SPECIES *tetrasporoides*
GENUS *Palmellocystis* — Cyst Palm Ten Spore Olive Green Algae
SPECIES *planktonica*
GENUS *Palmoclathrus* — Cage Palm Ten Spore Olive Green Algae
SPECIES *stipitatus*

GENUS *Planctococcus* Plankton Palm Ten Spore Olive Green Algae
SPECIES *sphaerocystiformis*
GENUS *Sphaerocystis* Sphere Cyst Palm Ten Spore Olive Green Algae
SPECIES *oleifera, planctonica, schroeteri, gelatinosa*
VARIETIES *oleifera antarctica, schroeteri tetraedrica*

FAMILY 6 *Palmellopsidaceae* Lesser Palm Ten Spore Olive Green Algae
GENUS *Palmellopsis* Lesser Palm Ten Spore Olive Green Algae
SPECIES *gelatinosa, muralis, texensis*
VARIETIES *texensis occidentalis*
GENUS *Nautocapsa* Capsule Lesser Palm Ten Spore Olive Green Algae
SPECIES *neustophila*
GENUS *Palmophyllum* Phyllode Lesser Palm Ten Spore Olive Green Algae
SPECIES *crassum, umbracola*
VARIETIES *crassum gestroi, crassum orbiculare*
GENUS *Ploeotila* Ploeotile Lesser Palm Ten Spore Olive Green Algae
SPECIES *ramosa*
GENUS *Pseudotetraspora* Mock Lesser Palm Ten Spore Olive Green Algae
SPECIES *marina*
GENUS *Verdigellas* Verdig Lesser Palm Ten Spore Olive Green Algae
SPECIES *fimbriata, nektongammea, peltata*

ORDER 5 *Volvocales* Coil Olive Green Algae

FAMILY 1 *Volvocaceae* Coil Olive Green Algae
GENUS *Volvox* Coil Olive Green Algae
SPECIES *africanus, aureus, capensis, carteri, gigas, globator, lismorensis, maderaspatensis, powersii, rousseletii, tertius*
VARIETIES *aureus hemisphaericus, carteri hazenii, carteri iyengarii, carteri kawasakiensis, carteri nagariensis, carteri weismannia, globator australis, globator maderaspatiensis, lismorensis globulifera, tertius godwardensis, tertius guttulosa, tertius ovalis, tertius tessellata*
GENUS *Chlamydonephris* Cloak Kidney Coil Olive Green Algae
SPECIES *pomiformis*
GENUS *Eudorina* Dorine Coil Olive Green Algae
SPECIES *cylindrica, elegans, illinoisensis, indica, minodii, unicocca*
VARIETIES *elegans richmondiae, elegans wallichii, unicocca peripheralis*
GENUS *Furcilla* Furco Coil Olive Green Algae
SPECIES *stigmatophora*
GENUS *Lundiella* Lundy Coil Olive Green Algae
SPECIES *indica*
GENUS *Mastigosphaera* Mastigote Sphere Coil Olive Green Algae
SPECIES *gobii*
GENUS *Pandorina* Pandora Coil Olive Green Algae
SPECIES *bengalensis, charkowiensis, colemaniae, elegans, minodii, morum, unicocca*
VARIETIES *morum tropica*
GENUS *Platydorina* Flat Dorine Coil Olive Green Algae
SPECIES *caudata*
GENUS *Pleodorina* Pleon Dorine Coil Olive Green Algae
SPECIES *californica, illinoisensis, indica, japonica*
GENUS *Stephanoon* Stephan Coil Olive Green Algae
SPECIES akenasii
GENUS *Volvulina* Lesser Coil Olive Green Algae
SPECIES *boldii, compacta, pringsheimii, steinii*
VARIETIES *steinii lenticularis, steinii parvicellula, steinii subreniformis*
GENUS *Vulcanella* Volcano Coil Olive Green Algae
SPECIES *hannae*
GENUS *Yalisphaeridium* Yale Sphere Coil Olive Green Algae
SPECIES *baculatum, xizangensis*
GENUS *Yamagishiella* Yamagishi Coil Olive Green Algae
SPECIES *unicocca*

FAMILY 2 *Chlamydomonadaceae* Cloak Coil Olive Green Algae
GENUS *Chlamydomonas* Cloak Coil Olive Green Algae
SPECIES *acidophila, actinochloris, acuta, acutata, akimovii, allensworthii, alpina, amici-mei, angulosa, anticontata, apiocystiformis, appendiculata, applanata, asymmetrica, augustae, baca, bacillaris, ballenyana, bicaudata, bicocca, bipapillata, boldii, bolyaiana, braunii, britannica, calcicola, callosa, callunae, carrizoensis, caudata, cavanillesiana, chlamydogama, chlorastera, chlorococcoides, cingulata, coccoides, concinna, corrosa, crassa, cristata, culleus, cylindrica, dalecarlica, debaryana, desmidii, dinobryonis, dorsoventralis, dysosmos, ehrenbergii, elliptica, epsilon, eumaculata, euryale, eustigma, fimbriata, foraminata, fottii, frankii, fungicola,*

gelatinosa, gerloffii, gigantea, globosa, globulosa, gloeocystiformis, gloeogama, gloeopara, gloeophila, gracilis, grandistigma, grovei, gyrus, hemigymnos, heterogama, hydra, incisa, inepta, inertis, inflexa, intermedia, irregularis, isabeliensis, leiostraca, leptochaete, lismorensis, lundii, macroplastida, macrostellata, maculata, mantonii, media, melanospora, meslinii, metastigma, minuta, minutissima, mirabilis, moewusii, monadina, monticola, mucicola, multitaeniata, muriella, mutabilis, nasuta, nekrassovii, nivalis, noctigama, obesa, oblonga, oblongella, oogama, orbicularis, ovata, oviformis, parkeae, parvula, paupercula, penium, perigranulata, perpusilla, pertusa, pertyi, peterfii, pila, pisiformis, planoconvexa, platyrhyncha, platystigma, plethora, proboscigera, proteus, pseudagloe, pseudintermedia, pseudocaeca, pseudococcum, pseudoelegans, pseudomacrostigma, pseudomicrosphaera, pulsatilla, pulvinata, pulvisculus, pumilio, pusilla, pygmaea, pyramidalis, radiata, rapa, reginae, reinhardtii, reisiglii, reniformis, retronatans, rima, rotula, sajao, segnis, sestinensis, siderogloea, silvicola, similis, simplex, smithii, snowiae, speciosa, sphagnicola, sphagnophila, sphagnophila dysosmos, spreta, stellata, stercoraria, stipitata, subangulosa, subehrenbergii, subfusiformis, subreticulata, subtilis, surtseyiensis, terrestris, tetragonia, tetras, toveli, typica, ucrainica, ulvaensis, uva-maris, varians, zebra, zimbabwiensis

VARIETIES *angulosa obesa, cingulata globulifera, corrosa oculata, dalecarlica minor, globosa maculata, globulosa pertusa, gloeocystiformis australis, gloeophila irregularis, hydra micropapillata, lismorensis gracilis, lismorensis ovata, maculata oblonga, maculata planktonica, moewusii microstigmata, monadina ovalis, mucicola rotundata, ovata pulchra, pila major, pisiformis attenuata, pisiformis cylindracea, pisiformis gloecystiformis, pisiformis obesa, pisiformis ocellata, pisiformis ovalis, pseudoelegans minor, pseudoelegans symmetrica, pseudoelegans wuhanensis, rapa vasta, similis curta, snowiae grandis, snowiae ovata, snowiae palmelloides, snowiae pluristigma, stagnophila, steinii, subtilis australis*

GENUS *Brachiomonas* — Short Cloak Coil Olive Green Algae

SPECIES *simplex, submarina*

GENUS *Carteria* — Carter Cloak Coil Olive Green Algae

SPECIES *acidicola, arenicola, australis, bullulina, cerasiformis, cruciata, crucifera, eugametos, globosa, granulosa, incisa, iyengari, klebsii, micronucleolata, multifissa, obtusa, olivieri, peterhofiensis, polychloris, radiosa, rugulosa, scrobiculata, sphaerelloides, wuhanensis*

VARIETIES *australis ovata, crucifera polonica, eugametos contaminans, rugulosa angulata, rugulosa inaequalis, scrobiculata ovalis*

GENUS *Chlorogonium* — Angled Cloak Coil Olive Green Algae

SPECIES *capillatum, elegans, elongatum, euchlorum, fusiforme, kasakii, minimum, neglectum, tetragamum, tetragonum*

GENUS *Chloromonas* — Monad Cloak Coil Olive Green Algae

SPECIES *angustissima, brevispina, clathrata, gelatinosa, gerloffii, mirabilis, multipes, orbicularis, palmelloides, paradoxa, perforata, platystigma, polychloris, pseudoparadoxa, rosae, rubrifilum, rubroleosa, serbinowi, variabilis*

GENUS *Diplostauron* — Diploid Crosshatch Cloak Coil Olive Green Algae

SPECIES *angulosum, elegans, elephas, panduriforme, pentagonium, platyrhynchum*

VARIETIES *angulosum trigonum*

GENUS *Gigantochloris* — Giant Cloak Coil Olive Green Algae

SPECIES *permaxima*

GENUS *Gloeomonas* — Glow Monad Cloak Coil Olive Green Algae

SPECIES *kupfferi, ovalis*

GENUS *Heterochlamydomonas* — Variable Cloak Coil Olive Green Algae

SPECIES *inaequalis, lobata, rugosa*

GENUS *Hyalobrachion* — Hyaline Short Cloak Coil Olive Green Algae

SPECIES *omphalotus*

GENUS *Hyalogonium* — Hyaline Angled Cloak Coil Olive Green Algae

SPECIES *fusiforme*

GENUS *Lobomonas* — Lobe Monad Cloak Coil Olive Green Algae

SPECIES *ampla, denticulata, francei, piriformis, rostrata, sphaerica*

VARIETIES *ampla okensis*

GENUS *Oltmannsiella* — Oltmanns Cloak Coil Olive Green Algae

SPECIES *viridis*

GENUS *Parapolytoma* — Lesser Polytome Cloak Coil Olive Green Algae

SPECIES *satura*

GENUS *Peterfiella* — Peterfi Cloak Coil Olive Green Algae

SPECIES *alata*

GENUS *Phyllariochloris* — Phyllode Cloak Coil Olive Green Algae

SPECIES *torta*

GENUS *Polytoma* — Polytome Cloak Coil Olive Green Algae

SPECIES *caudatum, coimbrigense, cylindraceum, fusciforme, ulvella, uva, uvella*

GENUS *Protococcus* — Round Cloak Coil Olive Green Algae

SPECIES *consociatus, infusionum, minor, monas, pluvialis, verrucosus, wimmeri*

GENUS *Provasoliella* — Provasol Cloak Coil Olive Green Algae

SPECIES *ovata*

GENUS *Pyramichlamys* — Pyram Cloak Coil Olive Green Algae

SPECIES *quadrangulata*
GENUS *Sphaerella* Sphere Cloak Coil Olive Green Algae
SPECIES *lacustris*
GENUS *Sphaerellopsis* Lesser Sphere Cloak Coil Olive Green Algae
SPECIES *aulata, fluviatilis, gloeocystiformis, lefevrei, montana, mucosa, spiralis, velata*
VARIETIES *gloeocystiformis ferruginosa*
GENUS *Spirogonium* Spire Angled Cloak Coil Olive Green Algae
SPECIES *chlorogonioides*
GENUS *Tetrablepharis* Tetrad Fringed Cloak Coil Olive Green Algae
SPECIES *cyamos, globulus*
GENUS *Tetratoma* Tetrad Polytome Cloak Coil Olive Green Algae
SPECIES *archerii*
GENUS *Tussetia* Tusset Cloak Coil Olive Green Algae
SPECIES *elaiogenetes, polytomoides*
GENUS *Vitreochlamys* Vitreous Cloak Coil Olive Green Algae
SPECIES *gloeocystiformis, incisa, nekrassovii, ordinata, pinguis*

FAMILY 3 *Goniaceae* Angled Coil Olive Green Algae
GENUS *Gonium* Angled Coil Olive Green Algae
SPECIES *dispersum, formosum, formosum, formosum, indicum, lacustris, multicoccum, octonarium, pectorale, quadratum, sacculiferum, viridistellatum*
GENUS *Astrephomene* Star Angled Coil Olive Green Algae
SPECIES *gubernaculifera, perforata*

FAMILY 4 *Haematococcaceae* Blood Round Coil Olive Green Algae
GENUS *Haematococcus* Blood Round Coil Olive Green Algae
SPECIES *capensis, carocellus, lacustris, murorum, pluvialis, thermalis, zimbabwiensis*
GENUS *Balticola* Baltic Blood Round Coil Olive Green Algae
SPECIES *buetschlii, droebakensis*
VARIETIES *droebakensis fastigatus*
GENUS *Stephanosphaera* Stephan Blood Round Coil Olive Green Algae
SPECIES *pluvialis*

FAMILY 5 *Phacotaceae* Bundle Coil Olive Green Algae
GENUS *Phacotus* Bundle Coil Olive Green Algae
SPECIES *australis, bullatus, crassus, glaber, lendneri, lenticularis, rectangularis, reticulatus, sphaericus*
VARIETIES *bullatus conicus*
GENUS *Coccomonas* Monad Round Bundle Coil Olive Green Algae
SPECIES *orbicularis, platyformis*
GENUS *Dysmorphococcus* Dysmorphic Round Bundle Coil Olive Green Algae
SPECIES *coccifer, globosus, sphaericus, variabilis*
GENUS *Hemitoma* Half Polytome Bundle Coil Olive Green Algae
SPECIES *meandrocystis*
GENUS *Pedinopera* Pedin Bundle Coil Olive Green Algae
SPECIES *granulosa*
GENUS *Pteromonas* Wing Monad Bundle Coil Olive Green Algae
SPECIES *aculeata, aequiciliata, angulosa, cruciata, protracta, spinosa, tenuis, varians*
VARIETIES *angulosa australis, angulosa scutiformis, angulosa takedena, angulosa vexilliformis, cruciata pulchra*
GENUS *Thoracomonas* Thorax Monad Bundle Coil Olive Green Algae
SPECIES *irregularis, laminata*

FAMILY 6 *Sphondylomoraceae* Sphondylid Coil Olive Green Algae
GENUS *Spondylomorum* Sphondylid Coil Olive Green Algae
SPECIES *quaternarium*
VARIETIES *quaternarium rostratum*
GENUS *Chlamydobotrys* Cloak Berry Sphondylid Coil Olive Green Algae
SPECIES *elegans, gracilis*
GENUS *Pascherina* Pascher Sphondylid Coil Olive Green Algae
SPECIES *tetras*
GENUS *Pyrobotrys* Fire Berry Sphondylid Coil Olive Green Algae
SPECIES *casinoensis, stellata*
GENUS *Uva* (Synonym for *Pyrobotrys*)

FAMILY 7 *Tetrabaenaceae* Ten Ribbon Coil Olive Green Algae
GENUS *Tetrabaena* Ten Ribbon Coil Olive Green Algae
SPECIES *socialis*
VARIETIES *socialis saccula*
GENUS *Basichlamys* Cloak Ten Ribbon Coil Olive Green Algae
SPECIES *sacculifera*

CLASS 612.2 *Bryopsidophyceae* Mossy Olive Green Algae

ORDER 1 *Bryopsidales* Mossy Olive Green Algae

FAMILY 1 *Bryopsidaceae* Mossy Olive Green Algae
GENUS *Bryopsis* Mossy Olive Green Algae
SPECIES *adriatica, africana, aishae, arbuscula, australis, cespitosa, composita, corticulans, corymbosa, cupressina, cupressoides, dasyphylla, derbesioides, dichotoma, duchassaignii, duplex, eckloniae, feldmannii, flagellata, foliosa, galapagensis, gemellipara, halliae, harveyana, hypnoides, implexa, indica, lubrica, lyngbyei, macraildii, macrocarpa, magellenica, manorensis, maxima, minor, muscosa, myosuroides, pakistanica, palliolatis, penicillata, penicillum, pennata, pennulata, peruviana, plumosa, pottsii, pusilla, qasimensis, ramulosa, rhizophora, rosea, ryukyuensis, salvadoreana, secunda, spinescens, stenoptera, stolonifera, tenuis, thuyoides, triploramosa, vestita*
VARIETIES *duchassaignii filicina, duplex nuda, duplex prolifera, duplex simpliciuscula, duplex pseudoderbesia, halliae filicina, hypnoides arbuscula, hypnoides lagunarum, hypnoides praelongata, hypnoides subnuda, indica triseriata, indica unilaterialis, pennata leprieurii, pennata minor, pennata secunda, plumosa arbuscula, plumosa hypnoides, vestita brevis*
GENUS *Bryopsidella* Lesser Mossy Olive Green Algae
SPECIES *neglecta, ostreobiformis*
GENUS *Halimeda* Halim Mossy Olive Green Algae
SPECIES *bikiniensis, borneensis, copiosa, cryptica, cuneata, cylindracea, discoidea, distorta, favulosa, fragilis, gigas, goreaui, gracilis, hederacea, howensis, hummii, incrassata, lacrimosa, lacunalis, macroloba, macrophysa, magnidisca, melanesica, micronesica, minima, monile, opuntia, papyracea, peltata,, rectangularis, renschii, scabra, simulans, stuposa, taenicola, tuna, velasquezii, xishaensis*
VARIETIES *copiosa elongata, cryptica acerifolia, cuneata digitata, cuneata undulata, goreaui compacta, gracilis opuntioides, incrassata gracilis, incrassata tripartita, lacrimosa globosa, lacunalis lata, monile cylindrica, monile robusta, opuntia macrocarpa, opuntia minor, tuna triloba, velasquezii robusta*
GENUS *Lambia* Lamb Mossy Olive Green Algae
SPECIES *antarctica*
GENUS *Ostreobium* Oyster Mossy Olive Green Algae
SPECIES *brabantium, constrictum, quekettii*
GENUS *Pseudobryopsis* Lesser Mossy Olive Green Algae
SPECIES *gracilis, oahuensis, parva, plantonica, venezolana*
GENUS *Smithsoniella* Smithson Mossy Olive Green Algae
SPECIES *earleae*
GENUS *Trichosolen* Hairy Mossy Olive Green Algae
SPECIES *blomquistii, duchassaingii, hainanensis, longipedicellata, mauritianus, molassensis, mucronatus, myura, pambanensis, retrorsa, solomonensis, thikkodiensis, venezolana, venezoleanus*

FAMILY 2 *Caulerpaceae* Caulerpa Olive Green Algae
GENUS *Caulerpa* Caulerpa Olive Green Algae
SPECIES *agardhii, alternans, annulata, antoensis, articulata, ashmeadii, bartoniae, bikinensis, biserrulata, brachypus, brownii, buginense, buginensis, cactoides, cliftonii, constricta, cupressoides, dichotoma, diligulata, distichophylla, ellistoniae, elongata, faridii, fastigiata, fergusonii, filicoides, filiformis, flexilis, floridana, hedleyi, heterophylla, holmesiana, imbricata, juniperoides, kempfii, lagara, lanuginosa, lentillifera, lessonii, longifolia, manorensis, mexicana, microphysa, murrayi, nummularia, obscura, okamurae, oligophylla, ollivieri, opposita, papillosa, parvula, paspaloides, peltata, pinnata, plumulifera, pressoides, prolifera, pusilla, qureshii, racemosa, remotifolia, reniformis, reyesii, scalpelliformis, sedoides, selago, serrulata, sertularioides, seuratii, simpliciuscula, subserrata, taxifolia, trifaria, urvilleana, vanbossea, veravalensis, verticillata, vesiculifera, webbiana, zeyheri*
VARIETIES *brachypus brasiliana, brachypus nordestina, brachypus parvifolia, cupressoides amicorum, cupressoides denudata, cupressoides disticha, cupressoides elegans, cupressoides ericifolia, cupressoides flabellata, cupressoides gardineri, cupressoides intermedia, cupressoides lycopodium, cupressoides lycopodium-alternifolia, cupressoides mamillosa, cupressoides nuda, cupressoides plumarioides, cupressoides serrata, cupressoides turneri, elongata disticha, fastigiata confervoides, fastigiata delicatula, fastigiata minor, filicoides andamanensis, filiformis latior, filiformis palmata, flexilis muelleri, imbricata minor, imbricata mixta, lanuginosa delicatula, lentillifera compacta, lessonii tuticorinensis, longifolia crispata, mexicana exposita, mexicana laxior, mexicana pectinata, mexicana pluriseriata, mexicana vietnamica, okamurae oligophylla, paspaloides compressa, paspaloides flabellata, paspaloides laxa, paspaloides minima, paspaloides phleoides, paspaloides phyllaphlaston, paspaloides wurdemanni, peltata exigua, peltata macrodisca, peltata stellata, pressoides denudata, prolifera obovata, prolifera zosterifolia, pusilla mucronata, racemosa caespitosa, racemosa clavifera, racemosa complanata, racemosa compressa, racemosa condensata, racemosa corynephora, racemosa cylindracea, racemosa depauperata, racemosa exigua, racemosa gracilis, racemosa imbricata, racemosa intermedia, racemosa intermediata, racemosa lamourouxii, racemosa lamourouxii-requienii, racemosa laxa, racemosa macra, racemosa macrophysa, racemosa occidentalis, racemosa planiuscula, racemosa reducta, racemosa remota, racemosa semifalcata, racemosa simplicissima, racemosa turbinata, scalpelliformis denticulata, scalpelliformis dwarkensis, scalpelliformis intermedia, sedoides crassicaulis, sedoides geminata, sedoides mixta, sedoides tasmanica, sedoides tasmanica, serrulata angusta,*

serrulata boryana, serrulata boryana-longifolia, serrulata hummii, serrulata lata, serrulata longidentata, serrulata occidentalis, serrulata pectinata, serrulata serrulata-torulosa, serrulata spiralis, sertularioides brevipes, sertularioides corymbosa, sertularioides farlowii, sertularioides flagellata, sertularioides longipes, sertularioides longiseta, sertularioides umbellata, taxifolia asplenoides, taxifolia interrupta, taxifolia tristichophylla, urvilleana tristicha, urvilleana vitiensis, verticillata charoides, webbiana disticha, webbiana divergens, webbiana elegans, webbiana pickeringii, webbiana tomentella

GENUS *Caulerpella* — Lesser Caulerpa Olive Green Algae
SPECIES *ambigua*
GENUS *Herpochaeta* — Serpent Bristle Caulerpa Olive Green Algae
SPECIES *charoides*

FAMILY 3 *Chaetosiphonaceae* — Bristle Siphon Olive Green Algae
GENUS *Chaetosiphon* — Bristle Siphon Olive Green Algae
SPECIES *moniliformis*
GENUS *Blastophysa* — Bladder Bristle Siphon Olive Green Algae
SPECIES *rhizopus*
GENUS *Dichotomosiphon* — Dichotome Mossy Olive Green Algae
SPECIES *marinus, salinus*

FAMILY 4 *Codiaceae* — Seaweed Olive Green Algae
GENUS *Codium* — Seaweed Olive Green Algae
SPECIES *acuminatum, adhaerens, amplivesiculatum, anastomosans, arabicum, australicum, barbatum, bartlettii, brandegeei, bulbopilum, bursa, campanulatum, capitatum, capitulatum, carolinianum, cattaniae, cervicorne, cicatrix, coactum, conjunctum, contractum, convolutum, coralloides, cranwelliae, cuneatum, cylindricum, dawsonii, decorticatum, desultorium, dichotomum, distichium, duthieae, dwarkense, edule, effusum, elisabethae, extricatum, fernandezianum, flabellatum, flabelliforne, formosanum, foveolatum, fragile, galeatum, geppiorum, giraffa, globosum, gracile, guineã«nse, harveyi, hawaiiense, hubbsii, incognitum, indicum, intermedium, intertextum, intricatum, isaacii, isabelae, isthmocladum, johnstonei, laminarioides, latum, lineare, longiramosum, lucasii, macdougalii, madagascariense, mamillosum, megalophysum, minus, minutissimum, mozambiquense, muelleri, ovale, papenfussii, papillatum, pelliculare, perriniae, phasmaticum, picturatum, platyclados, platylobium, pocockiae, pomoides, prostratum, pugniforme, reductum, reediae, repens, ritteri, saccatum, saifullahii, santamariae, schmiederi, setchellii, shameelii, silvae, simulans, spinescens, spongiosum, stephensiae, subantarcticum, subtubulosum, sursum, taylorii, tenue, tobrukense, tomentosum, tunue, unilaterale, vermilara, yezoense*
VARIETIES *decorticatum mucronatum, dichotomum proliferum, fragile atlanticum, fragile capense, fragile novae-zelandiae, fragile scandinavicum, fragile tasmanicum, fragile tomentosoides, isthmocladum clavatum, lucasii capense, papillatum hainanense, tobrukense elmasirense, tomentosum mucronatum, tomentosum proliferum*
GENUS *Anthocodium* — Flower Seaweed Olive Green Algae
SPECIES *fragile*
GENUS *Appeninocodium* — Appenine Seaweed Olive Green Algae
SPECIES *convolvens*
GENUS *Arabicodium* — Arabian Seaweed Olive Green Algae
SPECIES *bicazense, hansii*
GENUS *Bacinellacodium* — Bacin Seaweed Olive Green Algae
SPECIES *calcareum*
GENUS *Bevocastria* — Bevocaster Seaweed Olive Green Algae
SPECIES *elliottii, toomeyi*
GENUS *Callipsygma* — Callipso Seaweed Olive Green Algae
SPECIES *wilsonis*
GENUS *Carpathocodium* — Carpathian Seaweed Olive Green Algae
SPECIES *anae*
GENUS *Flabellia* — Fan Seaweed Olive Green Algae
SPECIES *petiolata*
GENUS *Garhundacystis* — Garhund Cyst Seaweed Olive Green Algae
SPECIES *spadensis*
GENUS *Garwoodia* — Garwood Seaweed Olive Green Algae
SPECIES *bardosi, fluegelii*
GENUS *Geppella* — Gepp Seaweed Olive Green Algae
SPECIES *prolifera, yaeyamensis*
GENUS *Hedstroemia* — Hedstrom Seaweed Olive Green Algae
SPECIES *moldavica*
GENUS *Johnsonicodium* — Johnson Seaweed Olive Green Algae
SPECIES *stratiforme*
GENUS *Kakabekia* — Kakabek Seaweed Olive Green Algae
SPECIES *flabelliformis, rarea*
GENUS *Pseudocodium* — Seaweed Olive Green Algae
SPECIES *australasicum, convolvens, de-vriesii, floridanum*

FAMILY 5 *Derbesiaceae* — Derbes Olive Green Algae
GENUS *Derbesia* — Derbes Olive Green Algae
SPECIES *attenuata, boergesenii, cervicornis, corallicola, corallina, fastigiata, furcata, hollenbergii, indica, longifructa, marina, minima, novae-zelandiae, osterhoutii, pacifica, padinae, prolifica, rhizophora, sirenarum, souriei, tenuissima, turbinata, vaucheriaeformis*
GENUS *Halicystis* — Haline Cyst Derbes Olive Green Algae
SPECIES *ovalis, pyriformis*
GENUS *Pedobesia* — Pedal Derbes Olive Green Algae
SPECIES *clavaeformis, feldmannii, ryukyuensis, simplex, solieri*
GENUS *Pseudoderbesia* — Lesser Derbes Olive Green Algae
SPECIES *arbuscula*

FAMILY 6 *Udoteaceae* — Udote Olive Green Algae
GENUS *Udotea* — Udote Olive Green Algae
SPECIES *abbottiorum, argentea, caribaea, conglutinata, cyathiformis, dixonii, dotyi, fibrosa, flabellum, fragilifolia, geppii, glaucescens, goreaui, indica, kuetzingii, looensis, luna, minima, norrisii, occidentalis, orientalis, palmetta, papillosa, reniformiis, spinulosa, tenax, tenuifolia, unistratea, velutina, verticillosa, wilsonii, xishaensis, yamadae*
VARIETIES *argentea spumosa, cyathiformis flabellifolia, cyathiformis infundibulum, cyathiformis sublittoralis, flabellum longifolia, papillosa subpapillata, spinulosa palmettoidea*
GENUS *Avrainvillea* — Avrainville Udote Olive Green Algae
SPECIES *amadelpha, asarifolia, calathina, canariensis, clavatiramea, digitata, elliottii, erecta, fenicalii, fulva, gardineri, geppii, hayi, hollenbergii, lacerata, levis, longicaulis, mazei, nigricans, obscura, rawsonii, ridleyi, riukiuensis, rotumensis, silvana, sylvearleae, tentaculosa*
VARIETIES *amadelpha submersa, asarifolia olivacea, levis translucens, longicaulis laxa, nigricans floridana, nigricans fulva, nigricans parva, nigricans spongiosa*
GENUS *Boodleopsis* — Boodle Udote Olive Green Algae
SPECIES *carolinensis, hawaiiensis, indica, pusilla, siphonacea, sundarbanensis, vaucherioidea, verticillata*
GENUS *Chlorodesmis* — Fiber Udote Olive Green Algae
SPECIES *baculifera, caespitosa, dotyi, fastigiata, haterumana, hildebrandtii, major, mexicana, papenfussii, sinensis, torresiensis*
GENUS *Cladocephalus* — Clad Head Udote Olive Green Algae
SPECIES *excentricus, luteofuscus, scoparius*
GENUS *Penicillus* — Penicillin Udote Olive Green Algae
SPECIES *capitatus, comosus, dumentosus, dumetosus, lamourouxii, manaarensis, nodulosus, oblongus, pyriformis, sibogae*
VARIETIES *capitatus elongatus, capitatus laxus, dumentosus expansus, lamourouxii gracilis, pyriformis expanatus*
GENUS *Pseudochlorodesmis* — Lesser Fiber Udote Olive Green Algae
SPECIES *australis, furcellata, tenuis*
VARIETIES *furcellata canariensis*
GENUS *Pseudopenicillus* — Lesser Penicillin Udote Olive Green Algae
SPECIES *aegaeicus*
GENUS *Rhipidosiphon* — Wicker Siphon Udote Olive Green Algae
SPECIES *floridensis, javensis*
GENUS *Rhipilia* — Wicker Udote Olive Green Algae
SPECIES *amamiensis, crassa, diaphana, fungiformis, micronesica, millarii, multiplex, nigrescens, orientalis, penicilloides, pusilla, sinuosa, tenaculosa, tomentosa*
VARIETIES *tomentosa zonata*
GENUS *Rhipiliella* — Least Wicker Udote Olive Green Algae
SPECIES *verticillata*
GENUS *Rhipiliopsis* — Lesser Wicker Udote Olive Green Algae
SPECIES *carolyniae, corticata, echinocaulos, gracilis, howensis, madagascariensis, millarii, mortensenii, multiplex, papuensis, peltata, profunda, reticulata, robusta, stri, yaeyamensis*
GENUS *Rhipocephalus* — Wicker Head Udote Olive Green Algae
SPECIES *oblongus, phoenix*
VARIETIES *phoenix brevifolius, phoenix longifolius*
GENUS *Siphonogramen* — Siphon Seaweed Udote Olive Green Algae
SPECIES *abbreviata, parva*
GENUS *Tydemania* — Tydeman Udote Olive Green Algae
SPECIES *expeditionis*

CLASS 612.3 *Charophyceae* Elegant Olive Green Algae

ORDER 1 *Charales* — Elegant Olive Green Algae

FAMILY 1 *Characeae* — Elegant Olive Green Algae
GENUS *Chara* — Elegant Olive Green Algae

SPECIES *aculeolata, alopecuroides, angarica, aequistriata, aspera, baltica, baueri, bharadwajae, biformis, braunii, bulbillifera, canescens, chidamuensis, columnaria, connivens, contraria, corallina, crinita, curta, denudata, eboliangensis, ecklonii, fibrosa, fragilifera, galioides, globularis, halina, hispida, hookeri, horrida, huangii, hydropytis, imperfecta, intermedia, leei, leptopitys, montagnei, muscosa, palaeofragilis, polycarpica, pouzolsii, rudis, rusbyana, sadleri, setosa, spinescens, stoechadum, tomentosa, virgata, visianii, vulgaris, wallrothii, zeylanica*

VARIETIES *alopecuroides montagnei, alopecuroides papulosa, alopecuroides pouzolsii, alopecuroides wallrothii, angarica tenera, aequistriata micracantha, aequistriata polyphylla, aequistriata tylacantha, aspera lacustris, aspera pulchella, baueri muelleri, braunii brasiliensis, braunii oahuensis, corallina lucida-tenerior, corallina nobilis, corallina nobilis-inflata, corallina nobilis-lucida, corallina nobilis-nobilis, corallina nobilis-simplicissima, corallina nobilis-stuartiana, corallina nobilis-vieillardii, corallina stuartiana, crinita toletana, ecklonii albaniensis, ecklonii mollusca, fibrosa fibrosa, fibrosa acanthopitys, fibrosa bancroftii, fibrosa benthamii, fibrosa brachyphylla, fibrosa hookeri, fibrosa humilis, fibrosa major, fibrosa microptilis, fibrosa microteles, fibrosa myriophylla, fibrosa polyphylla, globularis annulata, globularis laxa, globularis leptosperma, globularis macrophylla, globularis microptila, globularis verrucosa, hispida baltica, hispida dalmatica, hispida major, hydropytis rororiverensis, leptopitys subebracteata, palaeofragilis ornata, visianii dalmatica, vulgaris vulgaris, vulgaris crassicaulis, vulgaris denudata, vulgaris gallocantae, vulgaris gymnophylla, vulgaris longibracteata, vulgaris papillata, zeylanica diaphana-armata*

GENUS *Amblyochara* — Ambly Elegant Olive Green Algae

SPECIES *brevis, producta*

GENUS *Amphorochara* — Amphor Elegant Olive Green Algae

SPECIES *grambastii*

GENUS *Charites* — Lesser Elegant Olive Green Algae

SPECIES *angustior, bangueshanensis, chidamuensis, columnaria, daxiaensis, eboliangensis, gigantofusiformis, guanpingensis, hejiapensis, leei, longiconica, parvula, siwalikus, tenuis, xiagouensis, yangtzensis*

GENUS *Croftiella* — Croft Elegant Olive Green Algae

SPECIES *lii*

GENUS *Gobichara* — Gobi Desert Elegant Olive Green Algae

SPECIES *deserta*

GENUS *Grambastichara* — Seaweed Stitched Elegant Olive Green Algae

SPECIES *atopa, breviconica, communis, fusiformis, qinghaiensis, yuntaishanensis*

VARIETIES *yuntaishanensis acuta*

GENUS *Grovesichara* — Groves Elegant Olive Green Algae

SPECIES *changzhouensis, minheensis, yangii*

GENUS *Gyrogona* — Gyre Elegant Olive Green Algae

SPECIES *crassinodosa, grambasti, huajiazhuangensis, hubeiensis, problematica, qianjiangica, xindianensis*

GENUS *Harrisichara* — Harris Elegant Olive Green Algae

SPECIES *bressonii, denticulata, diastemosa, honghuensis, margaritata, monspeliensis, paracylindrica, saportae, stipitata*

VARIETIES *bressonii major*

GENUS *Hornichara* — Horn Elegant Olive Green Algae

SPECIES *leduensis, yiduensis*

GENUS *Lamprothamnium* — Lamp Thamnid Elegant Olive Green Algae

SPECIES *aragonense, macropogon,papulosum, succinctum, toletanum*

VARIETIES *macropogon heterophylla, macropogon tasmanica*

GENUS *Lamprothamnion* — Least Lamp Thamnid Elegant Olive Green Algae

SPECIES *hansenii*

GENUS *Lamprothamnus* — Lesser Lamp Thamnid Elegant Olive Green Algae

SPECIES *alopecuroides, aragonensis, papulosus, toletanus*

VARIETIES *alopecuroides montagnei, alopecuroides pouzolsii, alopecuroides wallrothii, papulosus montagnei, papulosus pouzolsii*

GENUS *Latochara* — Wide Elegant Olive Green Algae

SPECIES *columelaria, curtula, cylindrica, qinshuiheensis, shiheziensis, sphaerica, yuananensis*

GENUS *Lychnothamnus* — Phyllode Thamnid Elegant Olive Green Algae

SPECIES *barbatus, wallrothii*

GENUS *Maedleriella* — Medler Elegant Olive Green Algae

SPECIES *angusta, cumulata, mangenotii*

VARIETIES *mangenotii meridionalis*

GENUS *Maedlerisphaera* — Medler Sphere Elegant Olive Green Algae

SPECIES *chinensis, globula, grambastorum, jiadianensis, leduensis*

GENUS *Mesochara* — Middle Elegant Olive Green Algae

SPECIES *gumudiensis, latiovata, producta, stipitata, taxiheensis, xiagouensis, xuanziensis*

VARIETIES *xuanziensis minor*

GENUS *Microchara* — Tiny Elegant Olive Green Algae

SPECIES *cristata, deserta, leiocarpa*

GENUS *Multispirochara* — Multispire Elegant Olive Green Algae

SPECIES *subovalis*
GENUS *Nemegtichara* Nemegt Elegant Olive Green Algae
SPECIES *brevicylindrica, brevis, latiovalis, microcylindrica, sadleri*
GENUS *Neochara* Modern Elegant Olive Green Algae
SPECIES *dangyangica, gaochunensis, huananensis*
GENUS *Nitella* Shiny Elegant Olive Green Algae
SPECIES *acuminata, australiensis, axillaris, capillaris, confervacea, confusa, congesta, cristata, diffusa, flexilis, furcata, gelatinifera, gracilens, haagenii, heterophylla, horikawae, hyalina, ignescens, imahorii, imperialis, inversa, iyengarii, leonhardii, leptostachys, lhotzkyi, mariammai, moniliformis, monopodiata, moriokae, mucronata, myriotrichia, nilgherensis, opaca, orientalis, penicillata, polycarpa, polycephala, praelonga, pseudoflabellata, pulchella, sonderi, spanioclema, spiciformis, stuartii, subtilissima, superba, tasmanica, tenuissima, thillaiammai, tolypelloides, translucens, tumida, ungula, verticillata, wahlbergiana, woodii*
VARIETIES *cristata cristata, cristata ambigua, cristata diffusa, cristata microteles, cristata remota, cristata reticulata, flexilis spanioclena, furcata glaziovi, furcata australiensis, gelatinifera gelatinifera, gelatinifera afoliolata, gelatinifera beaugleholii, gelatinifera cladostachya, gelatinifera cryptostachya, gelatinifera microcephala, gelatinifera microstachya, gelatinifera opaca, gelatinifera podostachya, hyalina athanurensis, hyalina brachyactis, hyalina chelliammai, hyalina puduchattirensis, leonhardii leonhardii, leonhardii intermedia, leptostachys fruticosa, leptostachys leonhardii, leptostachys viscosa, mucronata gracillima, myriotrichia acuminata, myriotrichia incurvata, orientalis orientalis, orientalis phauloteles, polycarpa gujaratensis, pseudoflabellata confusa, pseudoflabellata congesta, pseudoflabellata hawaiensis, pseudoflabellata heterothalia, pseudoflabellata homothalia, pseudoflabellata imperalis, pseudoflabellata leptosoma, pseudoflabellata mucosa, pseudoflabellata sathamangalensis, sonderi aemula, subtilissima subtilissima, subtilissima biformis, subtilissima gloeostachys, subtilissima major, subtilissima microphylla, tasmanica arthroglochin, tasmanica attenuata, tasmanica heteromorpha, verticillata kodaikanalensis*
GENUS *Nitellopsis* Lesser Shiny Elegant Olive Green Algae
SPECIES *bulbillifera, morulosa, obtusa, ovalis, stelligera*
GENUS *Nodosochara* Nodose Elegant Olive Green Algae
SPECIES *heilongjiangensis, specialis*
GENUS *Obtusochara* Obtuse Elegant Olive Green Algae
SPECIES *honghuensis, jianglingensis, luodianensis, songheensis, sphericovalis, subparallelos, tenuiconica*
GENUS *Peckichara* Peck Elegant Olive Green Algae
SPECIES *cancellata, dangyangensis, gejiajiensis, nanzhangensis, paomagangensis, pectinata, rudongensis, sertulata, zhijiangensis, zoumalingensis*
GENUS *Peckisphaera* Peck Sphere Elegant Olive Green Algae
SPECIES *macrocarpa*
GENUS *Piriformachara* Piriform Elegant Olive Green Algae
SPECIES *gumudiensis, tuguluensis*
GENUS *Platychara* Flat Elegant Olive Green Algae
SPECIES *caudata, grambasti*
GENUS *Raskyaechara* Rasky Elegant Olive Green Algae
SPECIES *bellatula, elegana, jucunda, purniagiriensis, subcylindrica, tangyinensis*
GENUS *Sphaerochara* Sphere Elegant Olive Green Algae
SPECIES *biambonaria, bicarinata, convexa, intricata, kuitunensis, magna, minheensis, minuscula, rugulosa, subquadrata, yangii*
GENUS *Stephanochara* Stephan Elegant Olive Green Algae
SPECIES *acris, fortis, jingmenensis*
GENUS *Tectochara* Tecton Elegant Olive Green Algae
SPECIES *doliumis, fusiformis, lii, morulosa, thaleri, wangdianensis*
VARIETIES *thaleri pradensis*
GENUS *Tolypella* Tolyp Elegant Olive Green Algae
SPECIES *giennensis, glomerata, hispanica, intricata, nidifica, prolifera, stipitata*
VARIETIES *intricata dispersa, nidifica glomerata*
GENUS *Tolypellopsis* Lesser Tolyp Elegant Olive Green Algae
SPECIES *stelligera*
GENUS *Vorticella* Whorled Elegant Olive Green Algae
SPECIES *microstoma*
GENUS *Wallidinellum* Wallidin Elegant Olive Green Algae
SPECIES *dalei*
GENUS *Wangichara* Wang Elegant Olive Green Algae
SPECIES *chanjiensis, tanshanensis*
GENUS *Watanabea* Wallidin Elegant Olive Green Algae
SPECIES *reniformis*
GENUS *Wattsia* Watts Elegant Olive Green Algae
SPECIES *nodosa*

FAMILY 2 *Aclistocharaceae* Aclist Elegant Olive Green Algae
GENUS *Aclistochara* Aclist Elegant Olive Green Algae

SPECIES *brevis, datongheensis, faceta, firma, fortis, gulistanica, hongguensis, huangshuiensis, jimusaerensis, lianmuxinensis, longiconica, longiformis, lufengensis, maxima, nuguishanensis, platyglobata, poculiformis, shanchengensis, shiwanensis, stellerides, subquadrularia, truncata, umbonata, urniformis, usitata, wangii, xiangtangensis, yunnanensis*

GENUS *Qinghaichara* — Qinghai Aclist Elegant Olive Green Algae

SPECIES *longiconica, ovalis*

FAMILY 3 *Characiaceae* — Characin Elegant Olive Green Algae

GENUS *Characium* — Characin Elegant Olive Green Algae

SPECIES *acuminatum, acutum, angustum, astipitatum, braunii, bulgariense, cerassiforme, cylindricum, groenlandicum, guttula, heteromorphum, longipes, marinum, obtusum, ornithocephalum, ovale, polymorphum, pseudopyriforme, pyriforme, rostratum, saccatum, sieboldii, strictum, subulatum, tenue, westianum*

VARIETIES *cerassiforme minimum, ornithocephalum harpochytriforme, ornithocephalum pringsheimii*

GENUS *Actidesmium* — Radial Fiber Characin Elegant Olive Green Algae

SPECIES *hookeri*

GENUS *Bicuspidella* — Bicuspid Characin Elegant Olive Green Algae

SPECIES *incus*

GENUS *Characiella* — Lesser Characin Elegant Olive Green Algae

SPECIES *rukwae*

GENUS *Characiellopsis* — Least Characin Elegant Olive Green Algae

SPECIES *anophelesii*

GENUS *Characiopodium* — Pod Characin Elegant Olive Green Algae

SPECIES *californicum, hindakii, oviforme, pseudopolymorphum, typicum*

GENUS *Coleochlamys* — Cloak Characin Elegant Olive Green Algae

SPECIES *apoda, cucumis, maxima, oleifera, perforata*

GENUS *Deuterocharacium* — Second Characin Elegant Olive Green Algae

SPECIES *polyplastidicum*

GENUS *Lanceola* — Lance Characin Elegant Olive Green Algae

SPECIES *spatulifera*

GENUS *Marthea* — Marthe Characin Elegant Olive Green Algae

SPECIES *tetras*

GENUS *Pseudochlorothecium* — Theca Characin Elegant Olive Green Algae

SPECIES *mucigenum*

GENUS *Pseudoschroederia* — Schroeder Characin Elegant Olive Green Algae

SPECIES *punctata, robusta*

GENUS *Rhopalosolen* — Rhopal Characin Elegant Olive Green Algae

SPECIES *cylindricus, saccatus*

FAMILY 3 *Clavatoraceae* — Club Elegant Olive Green Algae

GENUS *Flabellochara* — Fan Club Elegant Olive Green Algae

SPECIES *hangzhouensis, jurongica, niwanensis*

GENUS *Flosculosphaera* — Flosc Sphere Club Elegant Olive Green Algae

SPECIES *calceolariopsis, sacculus*

GENUS *Globator* — Glob Club Elegant Olive Green Algae

SPECIES *steinhauseri*

VARIETIES *steinhauseri maillardii*

GENUS *Nodosoclavator* — Nodose Club Elegant Olive Green Algae

SPECIES *qinghaiensis*

GENUS *Retusochara* — Retuse Club Elegant Olive Green Algae

SPECIES *macrocarpa*

GENUS *Septorella* — Septor Club Elegant Olive Green Algae

SPECIES *ultima*

FAMILY 4 *Karpinskyaceae* — Karpinsky Elegant Olive Green Algae

GENUS *Gemmichara* — Gem Karpinsky Elegant Olive Green Algae

SPECIES *sinensis*

GENUS *Pseudomoellerina* — Moeller Karpinsky Elegant Olive Green Algae

SPECIES *altaiensis, maslovii, sexcostata*

FAMILY 5 *Porocharaceae* — Pore Elegant Olive Green Algae

GENUS *Porochara* — Pore Elegant Olive Green Algae

SPECIES *aksuensis, anluensis, gonganzhaiensis, huzhuensis, jingshanensis, kucheensis, maedleri, moyuensis, parasphaerica, sardiniae, stipitata*

GENUS *Cuneatochara* — Wedge Pore Elegant Olive Green Algae

SPECIES *bellatula, elliptica, jinxiensis, nanpiaoensis, xinjiangensis*

GENUS *Leonardosia* — Leonard Pore Elegant Olive Green Algae

SPECIES *jinxiensis, turpanensis*

GENUS *Minhechara* — Minhe Pore Elegant Olive Green Algae

SPECIES *columelaria, sphaerica, xiaoxiaensis*

GENUS *Ministeria* — Minister Pore Elegant Olive Green Algae

SPECIES *marisola*
GENUS *Musacchiella* Mucilage Pore Elegant Olive Green Algae
SPECIES *maxima*
GENUS *Paracuneatochara* Lesser Wedge Pore Elegant Olive Green Algae
SPECIES *elliptica, gansuensis, jimsarensis, jinxiensis, nanpiaoensis, xinjiangensis*
GENUS *Pseudolatochara* Lesser Wide Pore Elegant Olive Green Algae
SPECIES *gracilis, jianghanensis, rhombica*
GENUS *Stomochara* Mouth Pore Elegant Olive Green Algae
SPECIES *bellatula, luii, moyuensis*
VARIETIES *bellatula longa*
GENUS *Turbochara* Turbo Pore Elegant Olive Green Algae
SPECIES *specialis*
GENUS *Turrilithus* Turret Lithic Pore Elegant Olive Green Algae
SPECIES *latericioides*

FAMILY 6 *Stellatocharaceae* Star Elegant Olive Green Algae
GENUS *Stellatochara* Star Elegant Olive Green Algae
SPECIES *minor*
GENUS *Acutochara* Acute Star Elegant Olive Green Algae
SPECIES *chinensis, gansuensis, jimsarensis, turpanensis*

FAMILY 7 *Sycidiaceae* Syce Elegant Olive Green Algae
GENUS *Sycidium* Syce Elegant Olive Green Algae
SPECIES *beiliuense, beiliuense bipyramidum, beiliuense longum, miniglobosum, sipaiense, volborthii, volborthii sipaiense*

FAMILY 8 *Trochiliscaceae* Wheel Elegant Olive Green Algae
GENUS *Trochiliscus* Wheel Elegant Olive Green Algae
SPECIES *altaiensis, lipuensis, maslovii, sexcostatus*

CLASS 612.4 *Klebsormidiophyceae* Klebs Olive Green Algae

ORDER 1 *Klebsormidiales* Klebs Olive Green Algae

FAMILY 1 *Klebsormidiaceae* Klebs Olive Green Algae
GENUS *Klebsormidium* Klebs Olive Green Algae
SPECIES *bilatum, crenulatum, dissectum, elegans, flaccidum, fluitans, klebsii, marinum, montanum, mucosum, nitens, pseudostichococcus, rivulare, scopulinum, sterile, subtile, subtilissimum*
VARIETIES *flaccidum minor, flaccidum tumidum*
GENUS *Chlorokybus* Kybus Klebs Olive Green Algae
SPECIES *atmophyticus*
GENUS *Heterothrichopsis* Variable Hairy Klebs Olive Green Algae
SPECIES *viridis*
GENUS *Raphidonema* Needle Filament Klebs Olive Green Algae
SPECIES *nivale, pyrenoidifera, sempervirens*
VARIETIES *pyrenoidifera elongata*

FAMILY 2 *Coleochaetaceae* Bristle Klebs Olive Green Algae
GENUS *Coleochaete* Bristle Klebs Olive Green Algae
SPECIES *andreae, baileyi, circularis, conchata, divergens, echinocystis, irregularis, leve, nitellarum, orbicularis, pulvinata, reptans, scutata, sieminskiana, soluta*
VARIETIES *divergens minor, pulvinata baileyi, scutata minor, scutata pinguis*
GENUS *Phyllactidium* Radial Bristle Klebs Olive Green Algae
SPECIES *pulchellum*

CLASS 612.5 *Mesostigmatophyceae* Stigma Olive Green Algae

ORDER 1 *Mesostigmatales* Stigma Olive Green Algae

FAMILY 1 *Mesostigmataceae* Stigma Olive Green Algae
GENUS *Mesostigma* Stigma Olive Green Algae
SPECIES *viride*

FAMILY 2 *Chaetosphaeridiaceae* Bristle Stigma Round Olive Green Algae
GENUS *Chaetosphaeridium* Bristle Stigma Round Olive Green Algae
SPECIES *globosum, huberi, pringsheimii*
VARIETIES *globosum depressum, globosum microscopicum, pringsheimii conferta*
GENUS *Chaetotheke* Theke Bristle Stigma Round Olive Green Algae
SPECIES *reptans*
GENUS *Conochaete* Conical Bristle Stigma Round Olive Green Algae

SPECIES *comosa, polytricha*
GENUS *Dicoleon* — Dual Bristle Stigma Round Olive Green Algae
SPECIES *nordstedtii*
GENUS *Diplochaete* — Double Bristle Stigma Round Olive Green Algae
SPECIES *solitaria*
GENUS *Oligochaetophora* — Few Bristle Stigma Round Olive Green Algae
SPECIES *simplex*
GENUS *Polychaetophora* — Many Bristle Stigma Round Olive Green Algae
SPECIES *lamellosa*

CLASS 612.6 *Pedinophyceae* Pedin Olive Green Algae

ORDER 1 *Pedinomonadales* Pedin Olive Green Algae

FAMILY 1 *Pedinomonadaceae* — Pedin Olive Green Algae
GENUS *Marsupiomonas* — Marsupial Pedin Olive Green Algae
SPECIES *pelliculata*
GENUS *Resultor* — Result Pedin Olive Green Algae
SPECIES *mikron*

CLASS 612.7 *Trebouxiophyceae* Treboux Olive Green Algae

ORDER 1 *Trebouxiales* Treboux Olive Green Algae

FAMILY 1 *Trebouxiaceae* — Treboux Olive Green Algae
GENUS *Trebouxia* — Treboux Olive Green Algae
SPECIES *aggregata, anticipata, arboricola, asymmetrica, corticola, crenulata, decolorans, erici, excentrica, flava, galapagensis, gelatinosa, gigantea, glomerata, higginsiae, impressa, incrustata, irregularis, italiana, jamesii, magna, phycobiontica, potteri, pyriformis, showmanii, simplex, usneae*
VARIETIES *decolorans gelatinosa*
GENUS *Actinastrum* — Radial Treboux Olive Green Algae
SPECIES *aciculare, bacillare, cerastioides, gracillimum, guttula, hantzschii*
GENUS *Picochlorum* — Pico Treboux Olive Green Algae
SPECIES *oklahomensis*
GENUS *Pseudotrebouxia* — (Synonym for *Trebouxia*)

FAMILY 2 *Chlorellaceae* — Chlorella Olive Green Algae
GENUS *Chlorella* — Chlorella Olive Green Algae
SPECIES *angustoellipsoidea, anitrata, antarctica, autotrophica, botryoides, capsulata, conglomerata, desiccata, ellipsoidea, emersonii, faginea, fusca, glucotropha, homosphaera, kessleri, luteoviridis, marina, miniata, minutissima, mirabilis, nocturna, ovalis, parasitica, parva, peruviana, protothecoides, pyrenoidosa, regularis, rugosa, saccharophila, salina, sorokiniana, spaerckii, sphaerica, stigmatophora, subsphaerica, trebouxioides, vulgaris*
VARIETIES *anitrata minor, autotrophica atypica, ellipsoidea minor, emersonii globosa, emersonii rubescens, pyrenoidosa tumidus, regularis minima, vulgaris globosa, vulgaris minuscula, vulgaris suboblonga, vulgaris tertia, vulgaris viridis*
GENUS *Ankistrodesmopsis* — Lesser Ancist Fiber Chlorella Olive Green Algae
SPECIES *gabretae-silvae*
GENUS *Ankistrodesmus* — Ancist Fiber Chlorella Olive Green Algae
SPECIES *acicularis, antarcticus, braunii, convolutus, cucumiformis, densus, dulcis, extensus, falcatus, fasciculatus, fusiformis, gelifactum, marinus, mayorii, mucosus, nannoselene, quaternus, spiralis, stipitatus*
VARIETIES *acicularis mirabilis, braunii minutus, convolutus minutum, dulcis cingulum, falcatus biplex, falcatus contortus, falcatus setiformis, falcatus setigerus, falcatus spiralis, falcatus stipitatus*
GENUS *Apatococcus* — Apatal Round Chlorella Olive Green Algae
SPECIES *constipatus, lobatus, minor, perspicuus*
GENUS *Apodococcus* — Apodal Round Chlorella Olive Green Algae
SPECIES *characioides*
GENUS *Auxenochlorella* — Auxen Chlorella Olive Green Algae
SPECIES *protothecoides*
GENUS *Catena* — Chain Chlorella Olive Green Algae
SPECIES *de-toniana, viridis*
GENUS *Choricystis* — Cyst Chlorella Olive Green Algae
SPECIES *minor, skujae*
GENUS *Closteriopsis* — Closter Chlorella Olive Green Algae
SPECIES *acicularis, longissima, scolia*
VARIETIES *acicularis africana, longissima tropica*
GENUS *Coelastrella* — Celestial Star Chlorella Olive Green Algae
SPECIES *corconica, multistriata, oocystiformis, striolata, terrestris, vacuolata*

VARIETIES *multistriata corcontica, striolata multistriata*
GENUS *Graesiella* (Synonym for *Coelastrella vacuolata*)
GENUS *Kermatia* Kermat Chlorella Olive Green Algae
SPECIES *pupukensis*
GENUS *Lobosphaera* Lobe Sphere Chlorella Olive Green Algae
SPECIES *tirolensis, undulata*
GENUS *Marvania* Marvan Chlorella Olive Green Algae
SPECIES *aerophytica, geminata*
GENUS *Nanochlorum* Dwarf Chlorella Olive Green Algae
SPECIES *eucaryotum*
GENUS *Prototheca* Primitive Theca Chlorella Olive Green Algae
SPECIES *chlorelloides, kruegeri, moriformis, portoricensis, zopfii*
VARIETIES *portoricensis trisporus*
GENUS *Pseudochlorella* Lesser Chlorella Olive Green Algae
SPECIES *pyrenoidosa*
GENUS *Viridiella* Virid Chlorella Olive Green Algae
SPECIES *fridericiana*

FAMILY 3 *Microthamniaceae* Thamnid Chlorella Olive Green Algae
GENUS *Dictyochloropsis* Netted Thamnid Chlorella Olive Green Algae
SPECIES *irregularis, reticulata, splendida, symbiotica*
GENUS *Fusochloris* Fused Thamnid Chlorella Olive Green Algae
SPECIES *perforata*
GENUS *Koliella* Kolie Thamnid Chlorella Olive Green Algae
SPECIES *antarctica, longiscta, nivalis, spiculiformis*
GENUS *Koliellopsis* Lesser Kolie Thamnid Chlorella Olive Green Algae
SPECIES *inundata*
GENUS *Lobococcus* (Synonym for *Myrmecia*)
GENUS *Myrmecia* Ant Thamnid Chlorella Olive Green Algae
SPECIES *astigmatica, biatorellae, bisecta, globosa, incisa, irregularis, israeliensis, macronucleata, pyriformis*
VARIETIES *irregularis antarcticus*
GENUS *Stichococcus* Stitch Round Thamnid Chlorella Olive Green Algae
SPECIES *allas, bacillaris, castellanus, chlorelloides, contortus, cylindricus, exiguus, fragilis, minutissimus, minutus, mirabilis, pelagicus, sequoieti, subtilis, undulatus, variabilis*
VARIETIES *pelagicus spiralis*

CLASS 612.8 *Ulvophyceae* Seaweed Olive Green Algae

ORDER 1 *Ulvales* Seaweed Olive Green Algae

FAMILY 1 *Ulvaceae* Sea Lettuce Seaweed Olive Green Algae
GENUS *Ulva* Sea Lettuce Seaweed Olive Green Algae
SPECIES *acanthophora, anandii, arasakii, armoricana, atroviridis, attenuata, beytensis, bifrons, brevistipitata, burmanica, californica, chaetomorphoides, coccinea, conglobata, cornuta, covelongensis, crassa, crassimembrana, curvata, dactylifera, denticulata, elegans, elminthoides, erecta, expansa, fasciata, fenestrata, gelatinosa, geminoidea, gigantea, grandis, hopkirkii, indica, intestinaloides, intricata, intybacea, javanica, kylinii, lactuca, lactucaefolia, laetevirens, laingii, linearis, linkiana, lippii, litoralis, lobata, lubrica, myriotrema, neapolitana, nematoidea, ohnoi, olivacea, pacifica, papenfussii, paradoxa,parva, parvula, patengensis, percursa, pertusa, phyllosa, popenguinensis, porrifolia, profunda, pseudocurvata, pseudolinza, pulchra, purpurascens, quilonensis, radiata, ralfsii, ranunculata, reticulata, rhacodes, rigida, rotundata, rubens, scagelii, scandinavica, sericea, serrata, simplex, sorensenii, spinulosa, stenophylla, sublittoralis, taeniata, tenera, tetragona, tuberosa, uncialis, uncinata, utriculosa, uvoides, ventricosa*
VARIETIES *erecta plumosa, lactuca bullosa, lactuca crispa, lactuca laciniata, lactuca lacinulata, lactuca macrogyna, lactuca mesenteriformis, lactuca myriotrema, lactuca pulvinata, lactuca rosulans, paradoxa tenuissima, rigida densa, rigida fimbriata, rigida laciniata, rigida tropica*
GENUS *Alcyonidium* Alcyon Sea Lettuce Seaweed Olive Green Algae
SPECIES *bullatum, nostoch*
GENUS *Blidingia* Bliding Sea Lettuce Seaweed Olive Green Algae
SPECIES *chadefaudii, groenlandica, marginata, minima, nana, ramifera, subsalsa, tuberculosa*
VARIETIES *minima stolonifera, minima subsalsa, minima vexata*
GENUS *Capsosiphon* Cap Siphon Sea Lettuce Seaweed Olive Green Algae
SPECIES *aurea, fulvescens, groenlandicus*
GENUS *Chloropelta* Pelt Sea Lettuce Seaweed Olive Green Algae
SPECIES *caespitosa*
GENUS *Diplonema* Dual Thread Sea Lettuce Seaweed Olive Green Algae
SPECIES *confervoideum, rupicola*
GENUS *Enteromorpha* Enteromorph Sea Lettuce Seaweed Olive Green Algae

SPECIES *adriatica, angusta, aragoensis, arctica, basiramosa, bayonnensis, bulbosa, bullosa, byssoides, caerulescens, canaliculata, capillaris, chartacea, clathrata, clavata, comosa, complanata, compressa, confervacea, confervicola, constricta, corniculata, cornucopiae, coziana, crinita, crispa, crispata, cruciata, cylindracea, dangeardii, denudata, echinata, ectocarpoidea, falcata, fascia, fasciculata, firma, flabellata, flexuosa, fucicola, fulvescens, gayraliae, gracillima, gujaratensis, gunniana, hendayensis, hirsuta, hookeriana, howensis, intestinalis, juergensii, jugoslavica, lanceolata, latifolia, limosa, lingulata, linza, littorea, livida, longissima, maeotica, marchantae, marginata, maxima, micrococca, microphylla, minima, multiramosa, musciformis, nana, nizamuddinii, opposita, ovata, pacifica, pallescens, perestenkoae, peruviana, plana, procera, prolifera, pulcherrima, quaternaria, ramellosa, rivularis, robertii-lamii, rugosa, saifullahii, sancti-joannis, saxicola, scopulorum, spermatoidea, stipitata, subulata, szegediensis, torta, undulata, usneoides, utricularis, vexata, vulgaris*

VARIETIES *bulbosa africana, bulbosa australasica, bulbosa dilatata, bulbosa elongata, bulbosa foliacea, bulbosa peruana, chartacea elongata, chartacea minor, clathrata abbreviata, clathrata agardhiana, clathrata ambigua, clathrata angusta, clathrata angustimembrana, clathrata asiatica, clathrata confervacea, clathrata confervoides, clathrata erecta, clathrata foeniculacea, clathrata leptopoda, clathrata nudiuscula, clathrata plumosa, clathrata pumila, clathrata robusta, clathrata rothiana, clathrata tenuis, clathrata tenuissima, clathrata uncinata, complanata crispa, complanata ramulosa, compressa abbreviata, compressa caespitosa, compressa crinita, compressa crispata, compressa gracillis, compressa nana, compressa prolifera, compressa subsimplex, coziana arcassonensis, crinita grossa, crinita latior, crinita procera-ramulifera, crinita rothiana, denudata minor, denudata simplex, flexuosa biflagellata, flexuosa linziformis, flexuosa pilifera, flexuosa submarina, fucicola ramosissima, gracillima breviramosa, intestinalis asexualis, intestinalis bullosa, intestinalis clava, intestinalis compressa, intestinalis cornucopiae, intestinalis crispa, intestinalis cylindracea, intestinalis filiformis, intestinalis flagelliformis, intestinalis intestiniformis, intestinalis maxima, intestinalis mesenteriformis, intestinalis nana, intestinalis rivularis, intestinalis suhrii, intestinalis tenuis, lanceolata angusta, lanceolata crispata, lanceolata latifolia, lanceolata plana, lanceolata ramifera, lanceolata smaragdina, lanceolata undulata, linza angusta, linza bicornuta, linza cornucopiae, linza crispata, linza flexicaulis, linza interrupta, linza lanceolata, linza lata, linza latior, linza linearis, linza minor, linza oblanceolata, linza ramifera, linza spiralis, minima fastigiata, minima glacialis, minima rivularis, nana bullosa, nana glacialis, nana gunniana, nana minima-bullosa, nana subsalsa, procera eramosa, procera novae-zelandiae, procera opposita, procera roscoffensis, procera subundulata, prolifera capillaris, prolifera crispatissima, prolifera eramosa, prolifera flexuosa, prolifera gullmariensis, prolifera guttulata, prolifera ramosa, prolifera setosa, prolifera tenuis, prolifera tenuis-setulosa, prolifera trichoclada, quaternaria aureola, quaternaria ochracea, quaternaria ramosa, stipitata linzoides, torta percursum, vulgaris capillaris, vulgaris cornucopiae, vulgaris lacustris*

GENUS *Enteronia* — Enteric Sea Lettuce Seaweed Olive Green Algae

SPECIES *confervoides, rigidula*

GENUS *Gayralia* — Gayral Sea Lettuce Seaweed Olive Green Algae

SPECIES *oxysperma*

VARIETIES *oxysperma laceratum, oxysperma wittrockii*

GENUS *Gemina* — Gemini Sea Lettuce Seaweed Olive Green Algae

SPECIES *clavata, enteromorphoidea, letterstedtoidea, ulvoidea*

GENUS *Hydrosolen* — Hydric Sea Lettuce Seaweed Olive Green Algae

SPECIES *compressus, intestinalis*

GENUS *Kallonema* — Callisto Sea Lettuce Seaweed Olive Green Algae

SPECIES *caespitosum, ralfsii*

GENUS *Kornmannia* — Kornmann Sea Lettuce Seaweed Olive Green Algae

SPECIES *leptoderma*

GENUS *Letterstedtia* — Letterstedt Sea Lettuce Seaweed Olive Green Algae

SPECIES *insignis, stipitata, ulvoidea*

GENUS *Lobata* — Lobed Sea Lettuce Seaweed Olive Green Algae

SPECIES *foliosa*

GENUS *Percursaria* — Percurse Sea Lettuce Seaweed Olive Green Algae

SPECIES *confervoidea, dawsonii, fucicola, percursa, rigens*

GENUS *Phycoseris* — Seris Sea Lettuce Seaweed Olive Green Algae

SPECIES *bertolonii, smaragdina*

VARIETIES *smaragdina angustifolia, smaragdina latifolia*

GENUS *Rhizenteron* — Rhizid Enteric Sea Lettuce Seaweed Olive Green Algae

SPECIES *saxatile*

GENUS *Solenia* — (Synonym for *Enteromorpha*)

GENUS *Tetranema* — (Synonym for *Ulva percursa*)

GENUS *Ulvophys* — (Synonym for *Ulva*)

GENUS *Ulvaria* — Lesser Sea Lettuce Seaweed Olive Green Algae

SPECIES *fusca, shepherdii*

VARIETIES *fusca blyttii*

GENUS *Umbraulva* — Umbrella Sea Lettuce Seaweed Olive Green Algae

SPECIES *amamiensis, japonica, olivascens*

GENUS *Uncinisphaera* — Hook Sphere Sea Lettuce Seaweed Olive Green Algae

SPECIES *imaguncula, tribulosa*

GENUS *Zignoa* Zigno Sea Lettuce Seaweed Olive Green Algae
SPECIES *muricata*
FAMILY 2 *Ulvellaceae* Seaweed Olive Green Algae
GENUS *Ulvella* Seaweed Olive Green Algae
SPECIES *acervus, calcicola, confluens, lens, nadsonii, peltata, prostrata, protuberans, setchellii, stellata*
GENUS *Acrochaete* Bristle Seaweed Olive Green Algae
SPECIES *cladophorae, endostraca, endozoica, geniculata, heteroclada, inflata, leptochaete, marchantiae, operculata, polymorpha, repens, wittrockii*
GENUS *Epicladia* Clad Seaweed Olive Green Algae
SPECIES *cyclostoma, flustrae, halimedae, heterotricha, perforans, phillipsii, pontica, testarum*
GENUS *Ochlochaete* Few Bristle Seaweed Olive Green Algae
SPECIES *hystrix*
VARIETIES *hystrix ferox*
GENUS *Protoderma* Primitive Seaweed Olive Green Algae
SPECIES *amorphum, beesleyi, cavanillesiana, cohaerens, concharum, frequens, pax-augustana, polyrhizum, sarcinoideum, viride*
GENUS *Pseudodendoclonium* Wood Clone Seaweed Olive Green Algae
SPECIES *brasiliense, dynamenae, informe, prostratum*
VARIETIES *brasiliense brandii*
GENUS *Ruthnielsenia* Ruth Nielsen Seaweed Olive Green Algae
SPECIES *tenuis*
GENUS *Trichothyra* Hairy Thyre Seaweed Olive Green Algae
SPECIES *irregularis*

ORDER 2 *Cladophorales* Branched Olive Green Algae

FAMILY 1 *Cladophoraceae* Branched Olive Green Algae
GENUS *Cladophora* Branched Olive Green Algae
SPECIES *aculeata, aegagropila, aegagropiloidea, aequalis, afra, albida, amphibia, anisogona, aokii, arbuscula, arenaria, aucklandica, australis, bainesii, battersii, bicolor, blomquistii, bombayensis, brasiliana, canalicularis, capensis, catenata, ceylanica, chlorocontracta, clavuligera, coelothrix, colabensis, colensoi, columbiana, columbina, conchopheria, conferta, confusa, conglomerata, constricta, contexta, corallicola, corallinicola, cornuta, crinalis, crispula, cristata, crouani, crucigera, crystallina, cymopoliae, dalmatica, daviesii, debilis, demissa, dichotomo-divaricata, diffusa, divergens, dotyana, echinus, enomotoi, falklandica, feredayi, feredayoides, flagelliformis, flavescens, flexuosa, fluviatilis, fracta, fritschii, fusca, globifera, globulina, glomerata, goweri, graminea, hauckii, hawaiiana, hesperia, horii, hospita, hutchinsiae, hutchinsioides, inclusa, incompta, intertexta, isaacii, japonica, jongiorum, kamerunica, kemariensis, koeiei, kosterae, laetevirens, lehmanniana, liebetruthii, liniformis, longicellulata, luxurians, macdougalii, magellanica, michaelensis, microcladioides, minisakaii, mirabilis, montagneana, nigrescens, nitellopsis, nordstedtii, nudiuscula, ohkuboana, okamurai, oligoclada, oligocladoidea, opaca, pachyderma, pachyliebetruthii, panamensis, papenfussii, parriaudii, parvula, patentiramea, pectinella, pellucida, pellucidoidea, penicillioides, perpusilla, peruviana, piscinae, prehendrens, prolifera, pseudobainesii, pseudorupestris, pseudostruveoides, pura, pusilla, pygmaea, quisumbingii, radiosa, rechingeri, refracta, retroflexa, rhizoclonioidea, rigidula, rivularis, ruchingeri, rupestris, ryukyuensis, sakaii, samoensis, saracenica, saviniana, savoeana, saxatilis, senegalensis, sericea, sericioides, sibogae, sivashensis, socialis, speciosa, stimpsonii, stolonifera, submarina, subnitida, subsimplex, surera, teusmannii, theotonii, thwaitesii, tiburonensis, timorensis, tomentosa, traillii, tranquebariensis, trinitatis, tropica, uncinella, vadorum, vagabunda, valonioides, vandenhoeki, vaughanii, verticillata, virgatula, viridifusca, weizenbauii, willeelloides, woolsii, wrightiana*
VARIETIES *coelothrix antarctica, diffusa diffusa, diffusa virgata, flexuosa australis, flexuosa densa, flexuosa expansa, flexuosa subflexuosa, flexuosa tenuis, fracta flavescens, fracta intricata, fracta normalis, fracta rigidula, fritschii kemariensis, globifera occidentalis, glomerata crassior, glomerata marina, japonica kajimurae, lehmanniana valida, luxurians radicans, okamurai adnata, pellucida australis, pellucida tenuissima, rupestris distorta*
GENUS *Aegagropila* Shield Branched Olive Green Algae
SPECIES *kannoi*
GENUS *Arnoldiella* Arnold Branched Olive Green Algae
SPECIES *conchophila*
GENUS *Basicladia* Basal Branched Olive Green Algae
SPECIES *chelonum, crassa, ramulosa*
GENUS *Bryobesia* Mossy Branched Olive Green Algae
SPECIES *johannae*
GENUS *Chaetomorpha* Bristle Branched Olive Green Algae
SPECIES *adriani, aerea, akineta, antennina, auricoma, basiretrorsa, billardierii, brachygona, breviarticulata, californica, cannabina, chelonum, clavata, coliformis, crassa, dubyana, exposita, fibrosa, firma, geniculata, gracilis, implexa, implicata, indica, inflata, intestinalis, javanica, kellersii, kerguelensis, linoides, linum, litorea, longiarticulata, mawsonii, melagonium, minima, moniligera, monilina, natalensis, nodosa, obscura, olneyi,*

pachynema, pacifica, picquotiana, recurva, restricta, robusta, saccata, septentrionalis, solitaria, spiralis, stricta, stuhlmannii, tenuissima, valida, zernovii
VARIETIES *antennina stolonifera, brachygona crassipellita, linum brachyarthra*
GENUS *Chaetonella* Least Branched Olive Green Algae
SPECIES *goetzii*
GENUS *Cladogonium* Gonad Branched Olive Green Algae
SPECIES *ogishimae*
GENUS *Cladophorella* Lesser Branched Olive Green Algae
SPECIES *calcicola, netzahualpillii, sundarbanensis*
GENUS *Conferva* Conferva Branched Olive Green Algae
SPECIES *albida, ansonii, antillarum, atrovirens, bangioides, barbata, bombycinum, canalicularis, ceratophyllum, compressa, confragosa, decomposita, densissima, dilatata, distorta, forcipata, fruticulosa, fugacissima, fulva, gracilis, hyemalis, intricata, lineata, lucens, lutescens, minor, monilina, muralis, myriophyllum, nidifica, nitida, orichalcea, pachynema, pinnata, pisum, polymorpha, reticulosa, rugosa, sandvicensis, seticulosa, setosa, subdivisa, subulata, supradecomposita, tenax, tumidulum, utriculosa, virgata*
VARIETIES *compressa flexuosa, dilatata globulifera, forcipata glabella*
GENUS *Gemmiphora* Gem Branched Olive Green Algae
SPECIES *compacta*
GENUS *Hormiscia* Hormisc Branched Olive Green Algae
SPECIES *delicatula, moniliforme, sphaerulifera, wormskioldii*
GENUS *Hormotrichum* Hormisc Hairy Branched Olive Green Algae
SPECIES *delicatulum, vermiculare*
GENUS *Lola* Lola Branched Olive Green Algae
SPECIES *irregularis, tortusa*
GENUS *Pithophora* Pith Branched Olive Green Algae
SPECIES *affinis, macrospora, oedogonia*
GENUS *Prolifera* Prolific Branched Olive Green Algae
SPECIES *rothii*
GENUS *Rama* Filament Branched Olive Green Algae
SPECIES *antarctica*
GENUS *Rhizoclonium* Rhizid Clone Branched Olive Green Algae
SPECIES *africanum, ambiguum, antillarum, arboreum, bolbogoneum, crassipellitum, curvatum, erectum, fontanum, fontinale, fractiflexum, grande, hieroglyphicum, , implexum, kerneri, minutissimum, rhizophilum, riparium, samoense, sandvicense, setaceum, tortuosum*
VARIETIES *crassipellitum robustum, hieroglyphicum crispum, hieroglyphicum hosfordii riparium polyrhizum, riparium validum, tortuosum crispa*
GENUS *Vaughaniella* Vaughan Branched Olive Green Algae
SPECIES *rupicola*
GENUS *Wittrockiella* Witrock Branched Olive Green Algae
SPECIES *lyallii, paradoxa, salina*

FAMILY 2 *Anadyomenaceae* Anady Branched Olive Green Algae
GENUS *Anadyomene* Anady Branched Olive Green Algae
SPECIES *brownii, eseptata, howei, lacerata, leclancheri, linkiana, menziesii, pavonina, plicata, rhizoidifera, saldanhae, stellata, wrightii*
GENUS *Macrodictyon* Large Net Anady Branched Olive Green Algae
SPECIES *clathratum*
GENUS *Microdictyon* Small Net Anady Branched Olive Green Algae
SPECIES *boergesenii, calodictyon, curtissiae, kraussii, laxereticulatum, marinum, mokilensis, nigrescens, obscurum, okamurae, palmeri, pseudohapteron, schmitzii, setchellianum, tenuius, thiebautii, umbilicatum, vanbossea*
VARIETIES *pseudohapteron luciparense*
GENUS *Rhipidiphyllon* Wicker Anady Branched Olive Green Algae
SPECIES *reticulatum*
GENUS *Valoniopsis* Valon Anady Branched Olive Green Algae
SPECIES *hancockii, pachynema*
GENUS *Willeella* Wille Anady Branched Olive Green Algae
SPECIES *japonica, ordinata*

ORDER 3 *Codiolales* Cody Olive Green Algae

FAMILY 1 *Acrosiphoniaceae* Tall Siphon Cody Olive Green Algae
GENUS *Acrosiphonia* Tall Siphon Cody Olive Green Algae
SPECIES *arcta, arctiuscula, coalita, effusua, flagellata, grandis, hemisphaerica, incurva, orientalis, saxatilis, sonderi, spinescens*
VARIETIES *arcta prescettii, arcta zosterae, saxatilis chamissonis*
GENUS *Chlorothrix* Hairy Tall Siphon Cody Olive Green Algae
SPECIES *gracilis, intermedia, kornmannii*

GENUS *Spongomorpha* Sponge Tall Siphon Cody Olive Green Algae
SPECIES *aeruginosa, cincinnata, coalita, conjuncta, duriuscula, glacialis, heterocladia, hystrix, mertensii, minima, oxyclada, pacifica, spiralis, tomentosa, vernalis*
GENUS *Urospora* Tail Spore Tall Siphon Cody Olive Green Algae
SPECIES *bangioides, curvata, doliifera, elongata, grandis, hartzii, laeta, media, microscopica, neglecta, penicillatum, penicilliformis, sphaerulifera, tetraciliata, vancouveriana, wormskioldii*

FAMILY 2 *Monostromataceae* Stroma Cody Olive Green Algae
GENUS *Monostroma* Stroma Cody Olive Green Algae
SPECIES *alittoralis, angicava, antarcticum, arcticum, balticum, bullosum, crassidermum, crassissimum, dactyliferum, ecuadoreanum, expansa, fractum, grevillei, hariotii, lactuca, lindaueri, lubricum, membranaceum, moorei, nitidum, obscurum, pacificum, parvum, quaternarium, saccodeum, sandei, zostericola*
VARIETIES *grevillei arctica, grevillei cornucopiae, grevillei lactuca, grevillei lubricum, grevillei vahlii*
GENUS *Protomonostroma* Primitive Stroma Cody Olive Green Algae
SPECIES *undulatum*
VARIETIES *undulatum pulchrum*

ORDER 4 *Dasycladales* Daisy Olive Green Algae

FAMILY 1 *Dasycladaceae* Daisy Olive Green Algae
GENUS *Dasycladus* Daisy Olive Green Algae
SPECIES *densus, ramosus, vermicularis*
GENUS *Batophora* Bat Daisy Olive Green Algae
SPECIES *largoensis, occidentalis, oerstedii*
VARIETIES *occidentalis largoensis, oerstedii occidentalis*
GENUS *Bornetella* Bornet Daisy Olive Green Algae
SPECIES *capitata, clavellina, nitida, oligospora, ovalis, sphaerica, vitileviana*
VARIETIES *nitida minor*
GENUS *Chloroclados* Branched Daisy Olive Green Algae
SPECIES *australasicus*
VARIETIES *australasicus philippinensis*
GENUS *Chlorocladus* (Synonym for *Chloroclados*)
GENUS *Cymopolia* Cyme Daisy Olive Green Algae
SPECIES *barbata, vanbosseae*
GENUS *Epimastopora* Mast Pore Daisy Olive Green Algae
SPECIES *yoshimurae*
GENUS *Halicoryne* Haline Club Daisy Olive Green Algae
SPECIES *spicata, wrightii*
GENUS *Myrsidrum* Myrtle Daisy Olive Green Algae
SPECIES *bursa, clavatum, dilatatum, ramosum*
GENUS *Neomeris* Modern Sea Daisy Olive Green Algae
SPECIES *annulata, bilimbata, cokeri, dumetosa, mucosa, stipitata, vanbosseae*
GENUS *Xainzanella* Xainzan Daisy Olive Green Algae
SPECIES *sinensis*

FAMILY 2 *Polyphysaceae* Bladder Seaweed Olive Green Algae
GENUS *Polyphysa* Bladder Seaweed Olive Green Algae
SPECIES *exigera, myriospora, polyphysoides, pusilla*
GENUS *Acetabularia* Acetic Bladder Seaweed Olive Green Algae
SPECIES *antillana, calyculus, crenulata, dentata, farlowii, kilneri, major, peniculus, ryukyuensis, schenckii, toxasii*
VARIETIES *crenulata monodisca, ryukyuensis philippinensis*
GENUS *Acicularia* Pointed Bladder Seaweed Olive Green Algae
SPECIES *shenckii*
GENUS *Chalmasia* Chalmas Bladder Seaweed Olive Green Algae
SPECIES *antillana*
GENUS *Parvocaulis* Short Stalk Bladder Seaweed Olive Green Algae
SPECIES *clavata, exigua, parvula, polyphysoides, pusilla*

ORDER 5 *Siphonocladales* Siphon Olive Green Algae

FAMILY 1 *Siphonocladaceae* Siphon Olive Green Algae
GENUS *Siphonocladus* Siphon Olive Green Algae
SPECIES *concrescens, feldmannii, filiformis, forsskalii, pusillus, rhodensis, rigidus, tropicus, xishaensis*
GENUS *Apjohnia* Apjohn Siphon Olive Green Algae
SPECIES *laetevirens*
GENUS *Boergesenia* Boergesen Siphon Olive Green Algae
SPECIES *forbesii*
GENUS *Chamaedoris* Chamaedor Siphon Olive Green Algae

SPECIES *auriculata, delphinii, orientalis, peniculum*
GENUS *Cladophoropsis* Branched Siphon Olive Green Algae
SPECIES *adhaerens, brachyartra, carolinensis, corallinicola, dichotoma, gerloffii, gracillima, herpestica, infestans, javanica, limicola, macromeres, magna, membranacea, modonensis, neocaledonica, palauensis, pallida, peruviana, philippinensis, psyttaliensis, sundanensis, voluticola*
GENUS *Nereodictyon* Nere Net Siphon Olive Green Algae
SPECIES *imitans*
GENUS *Spongocladia* Sponge Siphon Olive Green Algae
SPECIES *vaucheriaeformis*
GENUS *Struveopsis* Struve Siphon Olive Green Algae
SPECIES *chagoensis, covalamensis, robusta, siamensis*
GENUS *Ventricaria* Ventricle Siphon Olive Green Algae
SPECIES *nutrix, ventricosa*

FAMILY 2 *Valoniaceae* Valon Siphon Olive Green Algae
GENUS *Valonia* Valon Siphon Olive Green Algae
SPECIES *aegagropila, barbadensis, caespitosa, cladophora, fastigiata, macrophysa, oblongata, trabeculata, utricularis*
VARIETIES *aegagropila australis, utricularis crustacea*
GENUS *Ascothamnion* Sac Thamnid Valon Siphon Olive Green Algae
SPECIES *intricatum*
GENUS *Dictyosphaeria* Net Sphere Valon Siphon Olive Green Algae
SPECIES *australis, cavernosa, enteromorpha, intermedia, ocellata, sericea, spinifera, versluysii*
VARIETIES *cavernosa bullata, intermedia solida*
GENUS *Ernodesmis* Fiber Valon Siphon Olive Green Algae
SPECIES *verticillata*
GENUS *Petrosiphon* Rock Valon Siphon Olive Green Algae
SPECIES *adhaerens*

ORDER 6 *Trentepohliales* Trent Pohl Olive Green Algae

FAMILY 1 *Trentepholiaceae* Trent Pohl Olive Green Algae
GENUS *Trentepohlia* Trent Pohl Olive Green Algae
SPECIES *abietina, allorgei, angadickalensis, annulata, arborum, aurea, calamicola, chinensis, cochunesis, cucullata, cyanea, depressa, dialepta, diffracta, ellipsiocarpa, elongata, flintii, fusco-atra, hunanensis, iolithus, jucunda, leprieurii, lichenicola, minima, monilia, negeri, odorata, pathanamthittaensis, peruana, prolifera, randhawaiana, rigidula, tentaculata, thevalliensis, treubiana, umbrina, uncinata, villosa, willeana, willei*
VARIETIES *abietina crassisepta, abietina minor, abietina tenue, aurea acutata, cucullata sandvicensis, diffracta colorata, diffracta sandvicensis, ellipsiocarpa africana, iolithus anthonyi, iolithus bovina, odorata compacta, odorata elongata, odorata unilateralis, rigidula lynchii*
GENUS *Acroblaste* Blastid Trent Pohl Olive Green Algae
SPECIES *reinschii*
GENUS *Byssus* Byssus Trent Pohl Olive Green Algae
SPECIES *botryoides*
GENUS *Cephaleuros* Cap Trent Pohl Olive Green Algae
SPECIES *biolophus, diffusus, drouetii, endophyticus, expansa, henningsii, karstenii, lagerheimii, minimus, parasiticus, pilosa, solutus, tumidae-setae, virescens*
VARIETIES *parasiticus nana*
GENUS *Chlorotylium* Tyle Trent Pohl Olive Green Algae
SPECIES *cataractum*
GENUS *Ctenocladus* Comb Trent Pohl Olive Green Algae
SPECIES *circinnatus*
GENUS *Epibolium* Bowl Trent Pohl Olive Green Algae
SPECIES *dermaticola*
GENUS *Friedaea* Frieda Trent Pohl Olive Green Algae
SPECIES *torrenticola*
GENUS *Gloeoplax* Glow Plaque Trent Pohl Olive Green Algae
SPECIES *weberi*
GENUS *Lochmium* Loch Trent Pohl Olive Green Algae
SPECIES *piluliferum*
GENUS *Mycoidea* Mycoid Trent Pohl Olive Green Algae
SPECIES *parasitica*
GENUS *Phycopeltis* Pelt Trent Pohl Olive Green Algae
SPECIES *amboinensis, arundinacea, aurea, costaricensis, dorsopapillosa, epiphyton, expansa, flabellata, flabelligera, irregularis, kosteriana, maritima, microcystis, minuta, nigra, novae-zelandiae, parva, pilosa, prostrata, pseudotreubii, terminopapillosa, theaensis, treubii, treubioides, vaga*
GENUS *Printzina* Printz Trent Pohl Olive Green Algae
SPECIES *bosseae, diffusa, dusenii, effusa, lagenifera, lagerheimii, luteofusca, santurcensis*

VARIETIES *bosseae brevicellulis, bosseae crassimura, bosseae samoensis, lagenifera africana, lagenifera rugulosa*

GENUS *Pseudendocloniopsis* Wood Clone Trent Pohl Olive Green Algae
SPECIES *botryoides*
GENUS *Rhizothallus* Rhizid Thallus Trent Pohl Olive Green Algae
SPECIES *islandicus*
GENUS *Sporocladus* Spore Branched Trent Pohl Olive Green Algae
SPECIES *fragilis*
GENUS *Stomatochroon* Stoma Trent Pohl Olive Green Algae
SPECIES *coalitus, consociatus, lagerheimii, reniformis*

ORDER 7 *Ulotrichales* Hairy Olive Green Algae

FAMILY 1 *Ulotrichaceae* Hairy Olive Green Algae
GENUS *Ulothrix* Hairy Olive Green Algae
SPECIES *aequalis, australica, australis, belkae, brunnthaleri, confervicola, cylindricum, cylindricus, discifera, dissecta, endospongialis, fimbriata, flacca, flexuosa, gigas, idiospora, implexa, juergensii, laetevirens, longicauda, minulata, minuta, minutula, moniliformis, mucosa, nitens, oscillatorioides, palusalsa, quaternaria, speciosa, subflaccida, tenerrima, tenuissima, trentonense, verrucosa, zonata*
VARIETIES *aequalis cataeniformis, flacca geniculata, flacca maxima, flacca minor, flacca roscoffensis, quaternaria angustissima, zonata valida*
GENUS *Binuclearia* Binuclear Hairy Olive Green Algae
SPECIES *palhinhae, tectorum*
VARIETIES *tectorum minor*
GENUS *Chlorhormidium* Hormid Hairy Olive Green Algae
SPECIES *catenatum, subtilis*
GENUS *Chodatia* Chodat Hairy Olive Green Algae
SPECIES *tetrallantoidea*
GENUS *Eugomontia* Gomont Hairy Olive Green Algae
SPECIES *sacculata, stelligera*
GENUS *Fottea* Fotte Hairy Olive Green Algae
SPECIES *cylindrica, pyrenoidosa, sphaeroides, stichococcoides*
GENUS *Geminella* Gem Hairy Olive Green Algae
SPECIES *amphigranulata, crenulatocollis, interrupta, marina, minor, mutabilis, ordinata, scalariformis, terricola*
VARIETIES *crenulatocollis filamentosa, interrupta cylindrica, interrupta mutabilis, scalariformis marina*
GENUS *Gloeotila* Glow Tile Hairy Olive Green Algae
SPECIES *cateniformis, monospora, pallida, protogenita, subconstricta*
GENUS *Gloeotilopsis* Lesser Glow Tile Hairy Olive Green Algae
SPECIES *planctonica, sterilis*
GENUS *Hormidiella* Lesser Hormid Hairy Olive Green Algae
SPECIES *attenuata, parvula*
GENUS *Hormidiopsis* Least Hormid Hairy Olive Green Algae
SPECIES *crenulata*
GENUS *Hormidiospora* Spore Hormid Hairy Olive Green Algae
SPECIES *verrucosa*
GENUS *Ingenhouzella* Ingenhouz Hairy Olive Green Algae
SPECIES *fugacissima, priestleyi*
GENUS *Pirula* Pirule Hairy Olive Green Algae
SPECIES *salina*
GENUS *Planctonema* Plankton Hairy Olive Green Algae
SPECIES *aenigmaticum, lauterbornii*
GENUS *Psephotaxus* Psephon Hairy Olive Green Algae
SPECIES *lamellosus*
GENUS *Pseudoneochloris* Modern Hairy Olive Green Algae
SPECIES *marina*
GENUS *Pseudoschizomeris* Split Sea Hairy Olive Green Algae
SPECIES *caudata, mucosa*
GENUS *Radiofilum* Radial Hairy Olive Green Algae
SPECIES *conjunctivum, flavescens, irregulare, paradoxum, scalariforme, transversalis*
VARIETIES *paradoxum reticulatum, scalariforme argentinense*
GENUS *Raphidonemopsis* Needle Hairy Olive Green Algae
SPECIES *perifitica, sessilis*
GENUS *Ulotrichopsis* Lesser Hairy Olive Green Algae
SPECIES *cylindrica*

FAMILY 2 *Borodinellaceae* Borod Hairy Olive Green Algae
GENUS *Borodinellopsis* Borod Hairy Olive Green Algae

SPECIES *oleifera, texensis*
GENUS *Botryokoryne* — Berry Club Borod Hairy Olive Green Algae
SPECIES simplex
GENUS *Chloroplana* — Lesser Flat Borod Hairy Olive Green Algae
SPECIES *terricola*
GENUS *Planophila* — Flat Borod Hairy Olive Green Algae
SPECIES *asymmetrica, bipyrenoidosa, laetevirens, microcystis*

FAMILY 3 *Chaetopeltidaceae* — Pelt Hairy Olive Green Algae
GENUS *Chaetopeltis* — Pelt Hairy Olive Green Algae
SPECIES *barbata, minor, orbicularis*
GENUS *Floydiella* — Floyd Pelt Hairy Olive Green Algae
SPECIES *terrestris*
GENUS *Pabia* — Pab Pelt Hairy Olive Green Algae
SPECIES *signiensis*

FAMILY 4 *Collinsiellaceae* — Collins Hairy Olive Green Algae
GENUS *Collinsiella* — Collins Hairy Olive Green Algae
SPECIES *cava, japonica, tuberculata*
GENUS *Collinsiellopsis* — Lesser Collins Hairy Olive Green Algae
SPECIES *expansa*

FAMILY 5 *Cylindrocapsaceae* — Cylinder Capsule Hairy Olive Green Algae
GENUS *Cylindrocapsa* — Cylinder Capsule Hairy Olive Green Algae
SPECIES *conferta, geminella, involuta, oedogoniales*
GENUS *Cylindrocapsopsis* — Lesser Cylinder Capsule Hairy Olive Green Algae
SPECIES *indica*

CLASS 612.9 *Zygonematophyceae* Conjugating Olive Green Algae

ORDER 1 *Zygnematales* Conjugating Olive Green Algae

FAMILY 1 *Zygnemataceae* — Conjugating Olive Green Algae
GENUS *Zygnema* — Conjugating Olive Green Algae
SPECIES *aequale, anomalum, atrocoeruleum, binuclearioides, carinthiacum, carteri, circumcarinatum, coeruleum, conspicuum, cruciatum, cylindricum, cylindrospermum, extenue, gorakhporense, guininum, insigne, irregulare, lacustre, leiospermum, margalefianum, melanosporum, momoniense, montanum, oveidanum, parvulum, pectinatum, peliosporum, ralfsii, ramosum, rhynchonema, rivulare, spontaneum, stagnale, stellinum, sterile, subcylindricum, subtile, tenue, tenuissimum, vaginatum, vaucherii*
VARIETIES *leiospermum minor, peliosporum maximum*
GENUS *Conjugata* — (Synonym for *Zygnema pectinatum*)
GENUS *Debarya* — Debary Conjugating Olive Green Algae
SPECIES *glyptosperma, hardyi*
GENUS *Hallasia* — Hallas Conjugating Olive Green Algae
SPECIES *reticulata*
GENUS *Mesocarpus* — (Synonym for *Mougeotia*)
GENUS *Mougeotia* — Mougeot Conjugating Olive Green Algae
SPECIES *abnormis, acadiana, angusta, boodlei, calcarea, capucina, cotopaxiensis, cyanea, decussata, depressa, elegantula, ellipsoidea, gelatinosa, genuflexa, gracillima, laetevirens, laevis, microverrucosa, nummuloides, oblongata, ovalis, ovalispora, padulosa, paludosa, parvula, poinciana, pulchella, quadrangulata, recurva, scalaris, sestertisignifera, smithii, subcrassa, tenuis, transeaui, varians, ventricosa, victoriensis, violacea, virescens, viridis*
VARIETIES *cotopaxiensis parallela, ellipsoidea tenuissima, parvula angusta, parvula tenuissimus, recurva scotica*
GENUS *Mougeotiopsis* — Lesser Mougeot Conjugating Olive Green Algae
SPECIES *calospora*
GENUS *Sirocladium* — Branched Pit Conjugating Olive Green Algae
SPECIES *kumaoense*
GENUS *Sirogonium* — Gonad Pit Conjugating Olive Green Algae
SPECIES *khoriense, sticticum, tenuis*
VARIETIES *sticticum tenuis*
GENUS *Sphaerocarpus* — Round Carpel Conjugating Olive Green Algae
SPECIES *microcarpus*
GENUS *Spirogyra* — Spire Gyre Conjugating Olive Green Algae
SPECIES *affinis, alpinum, alternata, australiensis, baileyi, bellis, bichromatophora, biharensis, borgeana, borysthenica, britannica, calchaquiesiae, calospora, catenaeformis, chunkingensis, circumlineata, colligata, columbiana, communis, condensata, coumbiana, crassa, crassispina, crassiuscula, cupientis, cylindrica,*

cylindrospora, decima, decimina, dictyospora, diversizygotica, dubia, elegantissima, ellipsospora, elongata, fallax, farlowii, favosa, fennica, flavescens, flavicans, fluviatilis, frigida, gracilis, grevilleana, gujaratensis, hassallii, heeriana, inflata, insignis, irregularis, juergensii, jugalis, liana, lismorensis, longata, lubrica, lutetiana, macrospora, majuscula, marchica, maxima, meinningensis, mirabilis, mithalaensis, moebii, neglecta, nitida, notabilis, occidentalis, paradoxa, parva, parvula, patliputri, peipingensis, pellucida, plena, portacalis, pratensis, protecta, pseudoneglecta, punctata, punctulata, quadrata, quadrilaminata, quinina, rectangularis, reflexa, regularis, reticulata, rhizobrachialis, rivularis, sarmae, scrobiculata, setiformis, silvicola, singularis, spreeiana, subechinata, submargaritata, submaxima, subsalsa, suomaniana, tenuior, tenuissima, teodoresci, ternata, tolosana, transeauiana, tucumaniae, turfosa, tuwensis, varians, velata, verrucogranulata, weberii, westii

VARIETIES *decima cylindrosperma, grevilleana australis, gujaratensis oryzae, lutetiana minor, maxima cupientis, mirabilis gracilis, mirabilis major, neglecta columbiana, peipingensis occidentalis, punctata tenuior, punctulata cylindrospora, punctulata macrospora, quinina inaequalis, reticulata regularis, setiformis diversizygotica, subsalsa macrospora*

GENUS *Spirophycus* — Spire Conjugating Olive Green Algae

SPECIES *acicularis*

GENUS *Staurocarpus* — Crosshatch Carpel Conjugating Olive Green Algae

SPECIES *gracilis*

GENUS *Temnogametum* — Temnon Gamete Conjugating Olive Green Algae

SPECIES *sinense, thaxteri*

GENUS *Temnogyra* — Temnon Gyre Conjugating Olive Green Algae

SPECIES *madhuensis, tuwensis*

GENUS *Trigonum* — Trigonal Conjugating Olive Green Algae

SPECIES *indicum*

GENUS *Tyndaridea* — Tyndar Conjugating Olive Green Algae

SPECIES *cruciata, lutescens*

GENUS *Zygnemopsis* — Lesser Conjugating Olive Green Algae

SPECIES *desmidoides, tiffiana, vermae*

GENUS *Zygogonium* — Gonad Conjugating Olive Green Algae

SPECIES *affine, ericetorum, kumaoense, tenue*

FAMILY 2 *Closteriaceae* — Closter Conjugating Olive Green Algae

GENUS *Closterium* — Closter Conjugating Olive Green Algae

SPECIES *abruptum, acerosum, aciculare, acutum, anguineum, angustatum, archerianum, arcus, attenuatum, baillyanum, balmacarense, biclavatum, braunii, calosporum, cancer, ceratium, closteroides, cornu, costatosporum, costatum, decorum, dianae, didymotocum, directum, eboracense, ehrenbergii, evesiculatum, exiguum, exile, forte, gracile, groenbladii, guyanense, hibernicum, idiosporum, incurvum, infractum, intermedium, jenneri, juncidum, kuetzingii, lagoense, laterale, leibleinii, limneticum, lineatum, littorale, lundellii, lunula, magnificum, malinvernianiforme, molle, moniliferum, navicula, naviculoideum, nematodes, nordstedtii, nylandicum, pakistanicum, parvulum, peracerosum, porrectum, praelongum, pritchardianum, pronum, pseudokuetzingii, pseudolunula, pseudopygmaeum, pusillum, pygmaeum, ralfsii, rectimarginatum, regulare, rostratum, scoticum, semicirculare, setaceum, sinuosum, spetsbergense, sphaerosporum, strigosum, striolatum, subcoronulatum, subjuncidiforme, sublaterale, subscoticum, subulatum, tortum, toxon, tumidulum, tumidum, turgidum, variable, venus*

VARIETIES *abruptum majus, abruptum nilssonii, acerosum acerosum, acerosum angolense, acerosum borgei, acerosum casinoensis, acerosum coloratum, acerosum elongatum, acerosum kriegeri, acerosum lanceolatum, acerosum minus, acerosum rectum, acutum latius, acutum linea, acutum variabile, angustatum gracilius, angustatum sculptum, baillyanum alpinum, baillyanum multinucleatum, biclavatum australiense, calosporum brasiliense, closteroides closterioides, closteroides elongatum, closteroides intermedium, cornu brasiliense, cornu lundellii, cornu upsaliense, costatum borgei, costatum multipyrenoidum, costatum westii, decorum elongatum, decorum nordstedtii, decorum rectum, dianae dianae, dianae arcuatum, dianae brevius, dianae cingulum, dianae compressum, dianae minus, dianae pseudodianae, didymotocum asperulatum, directum minor, directum oligocampylum, directum recurvatum, ehrenbergii malinvernianum, gracile bicurvatum, gracile japonicum, gracile substrigosum, gracile tenue, incurvum majus, intermedium hibernicum, jenneri jenneri, jenneri curvatissimum, jenneri hibernicum, jenneri minutum, jenneri punctatum, jenneri robustum, kuetzingii capense, kuetzingii onchyosporum, kuetzingii procerum, kuetzingii vittatum, lagoense glabrum, leibleinii cornutum, leibleinii minimum, limneticum tenue, lineatum costatum, lineatum elongatum, lineatum guyanense, littorale crassum, lunula biconvexum, lunula intermedium, lunula lunula, lunula massartii, lunula massartii-nasutum, lunula maximum, lunula minus, lunula mourense, lunula nasutum, lunula submoniliferum, navicula crassum, nematodes proboscideum, nematodes tumidum, nordstedtii polystictum, parvulum angustum, parvulum cornutum, parvulum majus, parvulum minor, porrectum angustatum, praelongum brevior, praelongum brevius, pritchardianum alpinum, pronum hyalina, pusillum laticeps, pusillum maius, pusillum monolithum, ralfsii hybridum, ralfsii pterosporum, ralfsii typicum, rostratum brevirostratum, setaceum elongatum, setaceum vittatum, spetzbergense australe, spetzbergense palustre, strigosum elegans, striolatum borgei, striolatum recta, striolatum typicum, tumidum nylandicum, tumidum sphaerosporum, turgidum glabra, venus inflatum, venus westii*

GENUS *Spinoclosterium* — Spined Closter Conjugating Olive Green Algae

SPECIES *curvatum, cuspidatum*

FAMILY 3 *Desmidiaceae* — Fiber Conjugating Olive Green Algae

GENUS *Desmidium* Fiber Conjugating Olive Green Algae

SPECIES *aequale, aptogonum, baileyi, bengalicum, coarctatum, curvatum, graciliceps, grevillei, occidentale, pseudostreptonema, quadrangulare, quadrangulatum, quadratum, suboccidentale, swamii, swartzii*

VARIETIES *aptogonum acutius, aptogonum ehrenbergii, aptogonum quinquagulare, aptogonum tetragonum, baileyi gemima, baileyi longiprocessum, baileyi tetragonum, baileyi undulatum, coarctatum cambricum, swamii quadrangulatum, swartzii amblyodon, swartzii gracile, swartzii quadrangulatum*

GENUS *Actinotaenium* Radial Fiber Conjugating Olive Green Algae

SPECIES *adelochondrum, australe, borgeanum, capax, clevei, colpopelta, crassiusculum, cruciferum, cucurbita, cucurbitinum, curtum, didymocarpum, diplosporum, elongatum, gelidum, geniculatum, globosum, habeebense, hallei, hibernicum, inconspicuum, incurvum, lagenarioides, minutissimum, mooreanum, obcuneatum, palangula, perminutum, phymatosporum, pyramidatum, rufescens, silvae-nigrae, spinospermum, subglobosum, subpalangula, subtile, truncatum, turgidum, wollei*

VARIETIES *australe crassius, capax minus, clevei crassum, cruciferum latius, cucurbitinum crassior, cucurbitinum major, cucurbitinum minor, cucurbitinum roduntatum, cucurbitinum subpolymorphum, cucurbitinum subpolymorphum-crassior, curtum attenuatum, curtum obtusum, elongatum africanum, elongatum lanceolatum, lagenarioides sydneyense, perminutum australe, silvae-nigrae parallelum, turgidum turgidum, turgidum westii, wollei longius*

GENUS *Allorgeia* Allorge Fiber Conjugating Olive Green Algae

SPECIES *valiae*

GENUS *Amscottia* Amscott Fiber Conjugating Olive Green Algae

SPECIES *gulungulana, mira*

GENUS *Arthrodesmus* Straight Fiber Conjugating Olive Green Algae

SPECIES *bifidus, borgei, bulnheimii, constrictus, , controversus, curvatus, ellipticus, extensus, impar, indentatus, longispinus, mucronulatus, quiriferus, sachlanii, spechtii, tenuissimus, trispinatus*

VARIETIES *bifidus latidivergens, bifidus truncatus, bulnheimii subincus, constrictus longispinus extensus retusus, tenuissimus longispina*

GENUS *Bambusina* Bamboo Fiber Conjugating Olive Green Algae

SPECIES *borreri,sphaerospora*

VARIETIES *borreri brasiliensis, borreri constricta, borreri gracilescens*

GENUS *Bourellyodesmus* Bourelly Fiber Conjugating Olive Green Algae

SPECIES *heimii*

GENUS *Brachytheca* Short Theca Fiber Conjugating Olive Green Algae

SPECIES *inopinata, sulcata*

GENUS *Cosmarium* Cosmo Fiber Conjugating Olive Green Algae

SPECIES *abbreviatum, abruptum, achondroides, acutigonum, adoxum, affine, alpestre, amoenum, amplum, anceps, andinum, angulare, angulatum, angulosum, angustatum, anisochondrum, annulatum, arctoum, arnellii, askenasyi, asphaerosporum, asterosporum, attenuatum, auriculatum, australe, basilicum, berryense, bicuneatum, binerve, binum, bioculatum, bipapillatum, bipunctatum, bireme, biretiforme, biretum, bisphaericum, bitriangulum, blyttii, boeckii, boitierense, botrytis, brasiliense, braunii, brebissonii, broomei, burkillii, caelatum, calcareum, callosissimum, cambricum, canaliculatum, candianum, capitulum, carinthiacum, centiculatum, ceylanicum, circulare, ciskeium, clepsydra, coarctatum, collectorense, commissurale, concinnum, confusum, connatum, conspersum, constrictum, contractiforme, contractum, controversum, corbula, coronatum, corribense, corriense, corsavaggense, cosmarioides, costatum, crassipelle, crenatum, crenulatum, creperum, cristatum, cruciferum, cucumis, cuneatum, cupiforme, cyclicum, cyclops, cylindricoideum, cylindricum, cymatonotophorum, cymatopleurum, debaryii, decedens, decoratum, decussiferum, denotarisii, denticulatum, dentiferum, depauperatum, depressum, dickii, didymochondrum, didymoprotupsum, difficile, difficillimum, difficiloides, dilatatum, diplodesmium, distichum, dorsitruncatum, dovrense, dubium, dybowskii, eductum, eichlerianum, elegantissimum, entochondrum, etchachanense, exasperatum, excavatum, exiguum, fastidiosum, favum, feldmannii, fictopraemorsum, flavum, fluviatile, fontigenum, formulosum, furcatospermum, fusum, galeritum, garrolense, gayanum, geminatum, geometricum, giganteum, globosum, glyptodermum, gonoides, granatoides, granatum, grantii, granulatum, granuluscum, hammeri, hardyi, helcangulare, hexalobum, holmiense, humile, impressulum, incognitum, inconspicuum, incurvum, intermedium, isthmium, isthmochondrum, jaoi, jenisejense, kjellmani, klebsii, kostei, kotilainense, laeve, lagenarioides, latereprotractum, laticollum, latifrons, latum, lepidum, levinotabile, logiense, lundellii, lutetianum, luxuriosum, magnificum, mansangense, margaritatum, margaritiferum, maskellii, mateyucense, mbabanicum, medioscrobiculatum, melanosporum, meneghinii, mesoleium, messikommeri, microsphinctum, minimum, minor, minutum, moniliforme, monochondrum, monomazum, morsum, multigranulatum, multiordinatum, multituberculatum, murrayi, naegelianum, nagelianum, nasutum, neapolitanum, nitidulum, norimbergense, notabile, notatum, novae-semliae, nudum, nutallii, nymannianum, obcuneatum, obgoletum, obliquum, oblongum, obsoletum, obsoletum obtusatum, ocellatum, ochthodes, oenpelliense, oligogongrus, oliveri, ophelii, orbiculatum, ordinatum, ornatum, orthogonum, orthopunctulatum, orthostichum, ovale, pachydermum, palustre, panduriforme, papilliferum, paragranatoides, pardalis, paucigranulatum, perfissum, perforatum, pericymatium, perigranulatum, perpusillum, petsamoense, phaseolus, plicatum, pokornyanum, polygonum, porrectum, portianum, praegrande, praemorsum, prominulum, promontorium, protomphalum, protractum, protuberans, pseudamoenum, pseudamoenum basilare, pseudarctoides, pseudatlantoideum, pseudoamoenum, pseudoarmatum, pseudobiremum, pseudobroomei, pseudoconnatum, pseudoexiguum, pseudonitidulum, pseudoornatum, pseudopachydermum, pseudophaseolus, pseudoprotuberans, pseudopyramidatum, pseudoquadratulum, pseudoretusum, pseudoscenedesmus, pseudospeciosum, pseudotaxichondrum, pseudovariolatum, pseudowembarense, pulcherrimum, punctulatum, pusillum, pycnochondrum, pygmaeum, pyramidatum, pyriforme, quadratulum, quadratum, quadridentatum, quadrimamillatum, quadrum, quasillus, quaternarium, quinarium, raciborskii, radiosum, ralfsii, rarissimum, rectangulare, rectangulum, regnellii, regnesii, reinschii, reniforme, repandum, retusiforme, retusum, rotundum, saxicolum, scopularum, scoticum, scottii, securiforme, seelyanum, septentrionale, sexangulare, sexnotatum, simii, sinostegos, slewdrumense, smolandicum, sniatyniense, solidum, spechtii, speciosissimum, speciosum, sphaeroideum, sphagnicolum, sphalerostichum, sphyrelatum, spinuliferum, spirale, sportella, striolatum, subalatum, subarctoum, subaversum, subbinale, subbroomei, subcapitulum, subcontractum, subcostatum, subcrenulatum, subcucumis, subcylindricum, subdanicum, subdepressum, subdistichum, subexcavatum, subgranatum, subimpressulum, submammilliferum, subnotabile, suborthostichum, subpalangula, subphaseolus, subprotumidum, subquadrans, subquadratum, subraciborskii, subreniforme, subretusiforme, subspeciosum, subtholiforme, subtrinodulum, subtriordinatum, subtumidum, subturgidum, subundulatum, succissum, synthlibomenum, tatricum, taxichondriforme, taxichondrum, taxillus, tenue, tetrachondrum, tetragonum, tetraophthalmum, tholiforme, thwaitesii, tinctum, tjibenongense, trachydermum, trachypleurum, trafalgaricum, trilobulatum, triplicatum, truncatellum, truncatum, tuberculatum, tumens, tumidum, turgidum, turneri, turpinii, tyrolicum, uliginosum, umbilicatum, undulatum, ungerianum, variolatum, varsoviense, venustum, vexatum, vicenistriatum, viride, vogesiacum, willeanum, wittrockii, wollei, woodlawnense, zonatum*

VARIETIES *abbreviatum minor, abbreviatum planctonicum, abruptum granulatum, affine excavatum, amoenum intumescens, amoenum mediolaeve, amplum ciastoni, anceps crispulum, anceps tatricoides, angulare octangulare, angulatum conicum, angulatum major, angulosum concinnum, anisochondrum confusum, annulatum elegans, arctoum minor, arctoum tatricum, arnellii compressa, arnellii notatum, arnellii subrotundatum, askenasyi latum, askenasyi undulatum, asphaerosporum strigosum, australe crassius, bicuneatum tatricum, binum australiensis, binum fontense, bioculatum depressum, bioculatum hians, bipunctatum subrectangularis, biretum trigibberum, bitriangulum bitrapezoidale, blyttii blyttii, blyttii australica, blyttii bipunctatum, blyttii bipunctatum, blyttii casinoense, blyttii lismorense, blyttii novae-sylvae, blyttii quaternigranulatum, blyttii richmondiae, boeckii imperfectum, boeckii trifoliatum, boitierense inambitiosum, botrytis dayense, botrytis depressum, botrytis emarginatum, botrytis gemmiferum, botrytis mediolaeve, botrytis paxillosporum, botrytis subtumidum, botrytis tumidum, brasiliense contractum, brasiliense dimaziforme, brasiliense kanitzii, brasiliense mamilliferum, brasiliense moerlianum, brasiliense pseudoincrassatum, broomei minus, burkillii depressum, caelatum hexagonum, caelatum spectabile, calcareum spetsbergense, capitulum australe, capitulum groenlandicum, capitulum rectangula, ceylanicum minus, commissurale acutum, commissurale crassum, concinnum laeve, connatum truncatum, conspersum latum, conspersum rotundatum, conspersum scottii, constrictum minor, contractum cracoviense, contractum ellipsoideum, contractum ellipsoideum-interme, contractum ellipsoideum-minor, contractum ellipticum, contractum gartanense, contractum incrassatum, contractum indentatum, contractum intermedia, contractum jacobsenii, contractum minutum, contractum notatum, contractum pachydermum, contractum rotundatum, contractum subellipticum, contractum subfoveatum, crassipelle champesianum, crassipelle crassangulatum, crassipelle laterale, crassipelle monochrondrum, crassipelle noduliferum, crenatum bicrenatum, cucumis helveticum, cucumis magnum, cuneatum verrucosum, cyclicum arcticum, cyclicum nordstedianum, cymatopleurum cymatopleurum, cymatopleurum minus, decussiferum medionudum, dentiferum quadrum, dentiferum sublatum, depressum depressum, depressum achondrum, depressum dorsitruncatum, depressum elevatum, depressum mediopunctatum, depressum minutum, depressum planctonicum, depressum reniforme, difficile laeve, difficile minutissimae, difficile rectangulum, difficile sublaeve, dilatatum concavum, distichum heterochondrum, dorsitruncatum pseudoscenedesmus, exasperatum spinatum, excavatum duplo-majus, exiguum pressum, exiguum subrectangulum, feldmannii nasiferum, formulosum minor, formosulum nathorstii, garrolense crassum, gayanum eboracense, geminatum ornatum, globosum wollei, glyptodermum tuberculatum, gonoides subturgidum, gonoides variolatum, granatum concavum, granatum elongatum, granatum gibbosum, granatum grunowii, granatum nordstedii, granatum pentagonum, granatum pyramidal, granatum scrobiculatum, granatum subangulare, granatum subgranatum, hammeri africanum, hammeri homalodermum, hammeri protuberans, hammeri queenslandica, hammeri sublaeve, hexalobum minus, holmiense attenuatum, holmiense hibernicum, holmiense integrum, holmiense undatum, humile danicum, humile glabrum, humile striatum, humile substriatum, impressulum suborthogonum, isthmium hibernica, isthmium horizontale, isthmochondrum decussiferum, isthmochondrum pergranulatum, jenisejense australe, kjellmani grande, kjellmani ornatum, laeve cymatium, laeve depressum, laeve lacustre, laeve octangulare, laeve septentrionale, laeve westii, laticollum minutum, levinotabile heterocrenatum, logiense subcirculare, lundellii corruptum, lundellii ellipticum, magnificum fluviatile, magnificum italicum, magnificum minus, margaritatum pyramidatum, margaritatum quadrum, margaritiferum regularius, meneghinii meneghinii, meneghinii concinnum, meneghinii granatoides, meneghinii nanum, meneghinii regnellii, meneghinii reinschii, mesoleium adulta, mesoleium arcuatum, mesoleium asperrima, mesoleium australis, mesoleium rhomboideum, mesoleium rhomboideum-australis, microsphinctum majus, microsphinctum parvulum, minor australis, moniliforme moniliforme, moniliforme ellipticum, moniliforme elongatum, moniliforme indentatum, moniliforme limneticum, moniliforme panduriforme, moniliforme pseudarctoum, moniliforme pulcherrimum, moniliforme subpyriforme, moniliforme subquadratum, monomazum polymazum, multiordinatum burmense, nasutum asperum, nasutum subnasutum, neapolitanum australicum, nitidulum gaditana, norimbergense depressa, notabile transiens, novae-semliae granulatum, novae-semliae sibiricum, obliquum trigonum, obsoletum, obsoletum dentata, obsoletum major, obsoletum punctatum, obsoletum sitvense, obsoletum sitvense-dentata, obtusatum beanlandii, ocellatum incrassatum, ochthodes amoebum, ochthodes subcirculare, ornatum perornatum, ovale subglabrum, pachydermum pacydermum, pachydermum aethiopicum, pachydermum minus, pachydermum parvum, pachydermum transitoria, palangula, palustre circulare, palustre ovale, pericymatium corrugatum, pericymatium eboracense, perpusillum nanum, phaseolus phaseolus, phaseolus elevatum, phaseolus minor, phaseolus punctulata, phaseolus skujae, plicatum hibernicum, plicatum majus, portianum nephroideum, portianum orthostichum, prominulum subundulatum, protomphalum angulare, protuberans padulosa, protuberans guyanense, pseudoamoenum inornatum, pseudophaseolus tithophoroides, pseudoprotuberans australe, pseudoprotuberans tumidum, pseudopyramidatum pseudopyramidatum, pseudopyramidatum extensum, pseudopyramidatum minor, pseudopyramidatum raciborskii, pseudopyramidatum spirale, pseudopyramidatum stenonotum-minor, pseudoretusum africanum, pseudotaxichondrum asymmetricum, pseudovariolatum major, punctulatum seriatum, punctulatum subpunctulatum, pygmaeum apertum, pyramidatum stenonotum, pyriforme reductum, quadratum angustatum, quadratum willei, quadrifarium, quadrifarium quadrifarium, quadrifarium gemmulatum, quadrifarium hexastichum, quadrifarium major, quadrifarium octastichum, quadrifarium rosacea, quadrifarium trifidum, quadrum minus, quadrum sublatum, ralfsii montanum, ralfsii rotundatum, rectangulare altius, rectangulare angustior, rectangulare angustius, rectangulare australe, rectangulare australe-multigranulatum, rectangulare biretum, rectangulare biretum-angustior, rectangulare biretum-rectilinea, rectangulare boldtii, rectangulare boldtii-minor, rectangulare cambrense, rectangulare caraibense, rectangulare cyclopeum, rectangulare dentatum, rectangulare finmarkiae, rectangulare glabrum, rectangulare granulatum, rectangulare hexagonum, rectangulare hexagonum-major, rectangulare latior, rectangulare major, rectangulare minor, rectangulare minus, rectangulare multigranulatum, rectangulare nodulatum, rectangulare nodulatum-minor, rectangulare notatum, rectangulare ocellatum, rectangulare papillatum, rectangulare patereforme, rectangulare quadrigeminatum, rectangulare quadrigeminatum-latior, rectangulare rectilinearis, rectangulare reniforme, rectangulare repandum, rectangulare rotundatum, rectangulare saxonicum, rectangulare subhexagonum, rectangulare subhexagonum-granulatum, rectangulare subreniforme, rectangulare sulcatum, regnellii minimum, regnellii pseudoregnellii, regnesii montanum, regnesii productum, regnesii tritum, reinschii eboracense, reniforme apertum, reniforme compressum, reniforme elevatum, repandum minor, repandum minus, repandum papillatum, retusiforme major, retusum angustatum, scottii profundum, seelyanum elegans, sexangulare minima, sexangulare minus, sexnotatum bipunctatum, sexnotatum tristriatum, sinostegos obtusus, smolandicum angustatum, sniatyniense sydneyensis, speciosissimum validius, speciosum australianum, speciosum biforme, speciosum difficilis, speciosum incrassatum, speciosum rostafinskii, speciosum simplex, sportella subnudum, striolatum nordstedtii, subarctoum punctatum, subbinale euastroideum, subbroomei isthmochondrum, subcapitulum minus, subcontractum detritum, subcostatum australe, subcostatum beckii, subcostatum biturberculatum, subcostatum minor, subexcavatum ordinatum, submammilliferum quadrigemme, subspeciosum simplex, subspeciosum validius, subtholiforme badense, subtriordinatum rotundatum, subtumidum groenbladii, subtumidum klebsii, subtumidum pachydermum, subturgidum minus, succissum hians, tatricum novizelandicum, tatricum sphaeruliferum, taxichondrum undulatum, taxichondrum yamagishii, tetragonum davidsonii, tetragonum elegans, tetragonum intermedium, tetragonum lundellii, tetragonum ornatum, thwaitesii penioides, tinctum excisum, tinctum intermedium, tinctum subretusum, tinctum tumidum, tjibenongense minor, trachypleurum minus, trachypleurum spinosum, trachypleurum subglabrum, trilobulatum depressum, triplicatum paucius, tumidum subrectangulare, turgidum alpestre, turgidum ovatum, turgidum subrotundatum, turpinii eximium, turpinii podolicum, tyrolicum beanlandii, undulatum minutum, undulatum wollei, ungerianum minus, ungerianum subtriplicatum, variolatum cataractarum, variolatum extensum, variolatum rotundatum, variolatum skujae, variolatum tortum, varsoviense tumidum, venustum venustum, venustum borgei, venustum excavatum, venustum excavatum-latum, venustum hypohexagonum, venustum induratum, venustum latum, venustum majus, venustum minor, venustum minus, vexatum rotundatum, viride minor, wollei longius, zonatum subcirculare*

GENUS *Cosmoastrum* — Star Cosmo Fiber Conjugating Olive Green Algae

SPECIES *hystrix, lapponicum, muricatiforme, orbiculare, teliferum, turgescens, wellipticum*

VARIETIES *hystrix pannonicum, teliferum pecten, turgescens sparsigranulatum*

GENUS *Cosmocladium* — Branched Cosmo Fiber Conjugating Olive Green Algae

SPECIES *constrictum, perissum, pulchellum, saxonicum, subramosum, tuberculatum, tumidum*

VARIETIES *tumidum evolutum, tumidum hallei*

GENUS *Cosmoeca* — Eca Cosmo Fiber Conjugating Olive Green Algae

SPECIES *ventricosa*

GENUS *Croasdalea* — Croasdale Fiber Conjugating Olive Green Algae

SPECIES *marthae*

GENUS *Cruciangulum* — Cross Fiber Conjugating Olive Green Algae

SPECIES *lateroprominens*

GENUS *Crucicribrum* — Cross Crib Fiber Conjugating Olive Green Algae

SPECIES *striatum*

VARIETIES *striatum constansii*

GENUS *Disphinctium* — Disphincter Fiber Conjugating Olive Green Algae

SPECIES *connactum, cucurbita,cylindrus, palangula, ralfsii, speciosum, subglobosum*

VARIETIES *cucurbita inflatum, speciosum simplex*

GENUS *Docidium* — Docid Fiber Conjugating Olive Green Algae

SPECIES *basiundatum, brefeldii, cylindricum, elongatum, gracile, horridum, nobile, pygmaeum, rectum, robustum, subundulatum, tessellatum, trabecula, tridentulum, undulatum*

VARIETIES *brefeldii gracilior, gracile aculeatum, rectum rectum, rectum australis, robustum cristatum, trabecula trabecula, trabecula baculoides, trabecula baculum, trabecula contricta, trabecula delpontei, trabecula delpontei-constricta, trabecula diadematum, trabecula ehrenbergii, trabecula elongatum, trabecula farquharsonii, trabecula maximum, trabecula mediolaeve, trabecula ovatum, trabecula rectissimum, trabecula rectum, trabecula truncatum, undulatum noble, undulatum perundulatum*

GENUS *Dysphinctium* — (Synonym for *Disphinctium*)

GENUS *Echinella* — Urchin Fiber Conjugating Olive Green Algae

SPECIES *cuneata, vitrea*

GENUS *Euastrum* — Star Fiber Conjugating Olive Green Algae

SPECIES *aboense, acanthophorum, affine, ampullaceum, angustatum, ansatum, asperum, attenuatum, bainsii, bidentatum, bimorsum, binale, biverrucosum, boldtii, bombayense, borgeanum, brevisinuosum, bullatum, carduetum, circulare, conicum, coralloides, cornubiense, cosmarioides, crassangulatum, crassicolle, crassum, crispulum, croasdaleae, cruciforme, cuneatum, cuspidatum, degii, deminutum, denticulatum, didelta, diplostauron, divaricatum, divergens, dubium, elegans, elobatum, elongatum, engleri, erosum, evolutum, exile, fittkaui, gayanum, gemmatum, groenbladii, guyanense, horikawae, humerosum, hypochondrum, inerme, inermius, insigne, insulare, intermedium, jenneri, labrys, lacustre, laticolle, latipediforme, longicolle, luetkemuelleri, mirum, moebii, montanum, nasiferum, obesum, oblongum, octogibbosum, oculatum, panamense, pectinatum, pictum, pinnatum, pinque, plesiocoralloides, praemorsum, pseudojenneri, pseudotuddalense, pulchellum, pulcherrimum, quadratum, quadriceps, quadrilobum, sachlanii, serratum, sibiricum, sinuosum, spetbergense, sphyroides, spinulosum, subalpinum, subincisum, sublobatum, subrostratum, triangulare, turgidum, turneri, umbonatum, undulatum, validum, ventricosum, verrucosum, vinnulum*

VARIETIES *acanthophorum acanthophorum, acanthophorum asymmetricum, acanthophorum latius, acanthophorum minus, angustatum australis, ansatum attentuatum, ansatum australianum, ansatum biscrobiculata, ansatum compactum, ansatum concavum, ansatum dideltiforme, ansatum dideltiforme-australianum, ansatum javanicum, ansatum laeve, ansatum maxima, ansatum pyxidatum, ansatum simplex, ansatum suprapositum, ansatum triporum, attenuatum saitoi, bidentatum bidentatum, bidentatum bioculatum, bidentatum oculatum, bidentatum praemorsum, bidentatum speciosum, binale borgei, binale cosmarioides, binale elobatum, binale gutwinskii, binale hians, binale minus, binale retusum, binale subelobatum, bombayense gokakense, bombayense simplex, brevisinuosum dissimile, carduetum antillarum, conicum imperforatum, coralloides reductum, coralloides subintegrum-reductum, coralloides trigibberum, cornubiense medianum, crassangulatum ornatum, crassum crispulum, crassum microcephalum, crassum scrobiculatum, crassum taturnii, croasdaleae pumicosum, cruciforme schizostauron, cuneatum basiventricosum, cuneatum punctatum, cuneatum solum, cuneatum solum-punctatum, cuspidatum goyazense, denticulatum elongatum, denticulatum incisum, denticulatum quadrifarium, denticulatum quadrifarium-incisum, denticulatum rectangulare, denticulatum stictum, didelta bengalicum, didelta bengalicum-minus, didelta minus, didelta quadriceps, diplostauron yirrakallense, divergens coronulum, divergens rhodesiense, divergens rhodesiense-coronulum, dubium anglicanum, dubium cambrense, dubium glabrum, dubium ornatum, dubium snowdoniense, dubium tumidum, elegans compactum, elegans medianum, elegans novaesemliae, elegans ornatum, elegans ornithocephalum, elegans pseudelegans, elegans spiniferum, elegans spinosum, elobatum elobatum, elobatum subelobatum, engleri madagascariense, erosum granulosum, evolutum pseudoglaziovii, exile triporum, humerosum affine, hypochondrum divergens, insulare basichondrum, insulare silesiacum, insulare silesiacum-brasiliense, intermedium poriferum, intermedium speciosum, longicolle australicum, longicolle capitatum, longicolle capitatum-minus, longicolle minus, luetkemuelleri carniolicum, luetkemuelleri laterepunctatum, moebii burmese, moebii diplocanthylum, moebii extensum, moebii insolitum, moebii javanica, moebii latum, moebii tetrachastriforme, obesum extensum, obesum tetmemoroides, oculatum gronbladii, pectinatum brachylobum, pseudotuddalense perforatum, pulchellum ceylanicum, pulchellum protrusum, pulchellum retusum, pulcherrimum menggalense, quadratum intermedium, quadratum javanicum, quadratum perornatum, quadriceps dideltoides, quadriceps javanicum, quadriceps triangulum, quadrilobum guyanense, sinuosum campanulatum, sinuosum ceylanicum, sinuosum germanica, sinuosum oculum, sinuosum parallelum, sinuosum polonica, sinuosum reductum, sinuosum scrobiculatum-oculum, sinuosum subjenneri, spinulosum borgei, spinulosum burmense, spinulosum emarginatum, spinulosum inermius, spinulosum lindae, spinulosum octogonum, spinulosum ornatum, spinulosum protrudens, spinulosum sudanense, subalpinum crassum, sublobatum brevisinuosum, sublobatum dissimile, sublobatum incrassatum, sublobatum obtusatum, sublobatum subdissimile, turgidum auburnense, turgidum cruxaustralis, turgidum grunovii, turgidum simplex, turneri australiense, turneri australiense-extremum, turneri extremum, turneri minor, turneri stictum, validum subvalidum, verrucosum verrucosum, verrucosum alatum, verrucosum coarctatum, verrucosum cruxaustralis-evoluta, verrucosum cruxaustralis-intermedia, verrucosum planctonicum, verrucosum reductum, verrucosum wallichianum*

GENUS *Groenbladia* — Groenblad Fiber Conjugating Olive Green Algae

SPECIES *bourrellyi, neglecta, undulata*

VARIETIES *bourrellyi elongata, neglecta elongata, neglecta major, neglecta minor, undulata kriegeri, undulata perundulata*

GENUS *Haplotaenium* — Ribbon Fiber Conjugating Olive Green Algae

SPECIES *minutum, rectum*

VARIETIES *minutum crassum, minutum elongatum, minutum gracile, minutum polonicum, rectum foesteri, rectum foesteri-elongatum*

GENUS *Heimansia* — Heiman Fiber Conjugating Olive Green Algae

SPECIES *pusilla, tumida*

VARIETIES *tumida evoluta*

GENUS *Hyalotheca* — Hyaline Theca Fiber Conjugating Olive Green Algae

SPECIES *burmensis, dissiliens, dubia, hians, indica, mucosa*

VARIETIES *dissiliens dissiliens, dissiliens hians, dissiliens tatrica, dubia subconstricta, indica javanica, mucosa laevincta, mucosa minor*

GENUS *Ichthyocercus* — Fish Tail Fiber Conjugating Olive Green Algae

SPECIES *angloensis, australiensis, longispinus*

GENUS *Ichthyodontum* — Fish Tooth Fiber Conjugating Olive Green Algae

SPECIES *sachlanii*

GENUS *Micrasterias* — Star Fiber Conjugating Olive Green Algae

SPECIES *adscendens, alata, americana, anomala, apiculata, bourrellyana, brachyptera, ceratofera, conferta, crenata, crux-melitensis, decemdentata, denticulata, doveri, fimbriata, foersteri, foliacea, furcata, hardyi, hieronymusii, incisa, indica, jenneri, laticeps, lux, mahabuleshwarensis, mohii, murrayi, nordstedtii, oscitans, papillifera, pinnatifida, radians, radiata, radiosa, rotata, schweinfurthii, subdenticulata, suboblonga, subtruncata, thomasiana, torreyi, tropica, truncata, verrucosa, zeylanica*

VARIETIES *alata alata, alata gracillior, alata parallela, anomala bifurcata, apiculata brachyptera, apiculata fimbriata-spinosa, apiculata lacerata, apiculata lacerata-australiensis, ceratofera maxima, conferta harmata, decemdentata intermedia, denticulata angulosa, doveri curvata, fimbriata brasiliensis, foliacea ornata, furcata dichotoma, furcata evoluta, furcata pseudo-crux, hieronymusii crameri, incisa wallichiana, indica wallichiana, jenneri braziliensis, jenneri simplex, laticeps acuminata, laticeps javanica, laticeps minor, lux longibracchiata, lux spinosa, mahabuleshwarensis ampullacea, mahabuleshwarensis dichotoma, mahabuleshwarensis intermedia, mahabuleshwarensis novizelandica, mahabuleshwarensis reducta, mahabuleshwarensis surculifera, mahabuleshwarensis tetradonta, mahabuleshwarensis wallichii, murrayi triquetra, nordstedtii australiensis, oscitans decemdentata, oscitans laticeps, oscitans pinnatifida, pinnatifida incudiformis, radians bogoriensis, radians gracilior, radiata gracillima, radiosa aculeata, radiosa elegantior, radiosa evoluta, radiosa extensa, radiosa murrayi, radiosa ornata, radiosa ornata-aculeata, rotata granulata, rotata major, rotata simplicior, suboblonga australis, thomasiana evoluta, thomasiana major, thomasiana notata, thomasiana ornata, thomasiana pulcherrima, thomasiana schweinfurthii, thomasiana torneensis, torreyi crameri, torreyi curvata, tropica australiensis, tropica crassa, tropica indivisa, tropica polonica, tropica polonica-australiensis, truncata australica, truncata crenata, truncata laticipiformis, truncata punctata, truncata upsaliensis, zeylanica biconvexa, zeylanica wallichiana*

GENUS *Onychodromus* Claw Drome Fiber Conjugating Olive Green Algae

SPECIES *quadricornutus*

GENUS *Onychonema* Claw Fiber Conjugating Olive Green Algae

SPECIES *uncinatum*

GENUS *Oocardium* Heart Fiber Conjugating Olive Green Algae

SPECIES *stratum*

GENUS *Phymatodocis* Docin Fiber Conjugating Olive Green Algae

SPECIES *alternans, irregulare, nordstedtiana*

VARIETIES *irregulare intermedia*

GENUS *Pleurotaeniopsis* Lesser Ribbon Fiber Conjugating Olive Green Algae

SPECIES *ampla, bigibbum, caldense, debaryi, indicum, javanicum, rectum, simplicissimum, subcoronulatum, subturgida, turgida, undatum*

VARIETIES *caldense cristatum, simplicissimum insigne, subcoronulatum africanum, subcoronulatum detum, subturgida minor, turgida ovata*

GENUS *Pleurotaenium* Ribbon Fiber Conjugating Olive Green Algae

SPECIES *archeri, australianum, baculoides, baculum, basiundulatum, burmense, clavatum, coronatum, coroniferum, coronulatum, cosmarioides, cristatum, cuyabense, ehrenbergii, elatum, engleri, eugenium, excelsum, gloriosum, hutchinsonii, hypocymatium, kayei, mamillatum, maximum, mediolaeve, minutum, multitaeniatum, nodosum, nodulosum, ovatum, raciborskii, sceptrum, stuhlmannii tridentulum, truncatum, undatum, verrucosum*

VARIETIES *archeri elongatum, baculum baculum, baculum australe, baculum inflatum, burmense curtum, burmense longissimum, coronatum constrictum, coronatum fluctuatum, coronatum jorgensenum, coronatum nodulosum, coronatum nodulosum-constri, coronatum robustum, coronatum undulatum, coroniferum multinodosum, cuyabense inornatum, ehrenbergii borgei, ehrenbergii constrictum, ehrenbergii crenulatum, ehrenbergii elongatum, ehrenbergii granulatum, ehrenbergii indicum, ehrenbergii quantillum, ehrenbergii undulatum, eugenium eugenium, eugenium scotica, excelsum rhomphaeum, hypocymatium angustum, mediolaeve gracilior, minutum excavatum, minutum latum, multitaeniatum coronatum, nodosum australica, nodosum borgei, nodosum constrictum, nodosum dentatum-australica, nodosum gutwinskii, nodosum mammillatum, ovatum inermis, ovatum tumidum, raciborskii engleri, sceptrum hexacanthum, truncatum farquharsonii*

GENUS *Prescottiella* Prescott Fiber Conjugating Olive Green Algae

SPECIES *sudanensis*

GENUS *Sphaerozosma* Sphere Fiber Conjugating Olive Green Algae

SPECIES *aubertianum, compressum, cosmarioides, elegans, formosum, laeve, quadrispinatum, vertebratum, wallichii*

VARIETIES *aubertianum aubertianum, aubertianum archerii, laeve constrictum, laeve elongatum, laeve latum, laeve micracantha, laeve oscitans, laeve sumatranum, quadrispinatum evolutum, wallichii anglicum, wallichii australis*

GENUS *Spinocosmarium* Spine Fiber Conjugating Olive Green Algae

SPECIES *quadridens*

GENUS *Spondylosium* Spondylose Fiber Conjugating Olive Green Algae

SPECIES *compressum, ellipticum, javanicum, moniliforme, nitens, panduriforme, papillosum, planum, pulchellum, pulchrum, pygmaeum, secedens, tetragonum*

VARIETIES *nitens major, nitens minor, nitens triangulare, panduriforme limneticum, pygmaeum compressum, pygmaeum monile*

GENUS *Staurastrum* Crosshatch Star Fiber Conjugating Olive Green Algae

SPECIES *acanthastrum, acarides, acestophorum, aciculiferum, aculeatum, acutum, affine, aggeratum, agumbeyense, alternans, amphidoxon, anatinoides, anatinum, anisacanthum, approximatum, arachne, archerii, arctiscon, arcuatum, armatum, armigerum, arnellii, asperum, assurgens, asterias, asteroideum, auburnense, aureolatum, avicula, bacillare, barbaricum, barbulae, bibrachiatum, bicorne, bicoronatum, bienneanum, bifidum, bifurcum, bilobum, bitrapezicum, biwaense, bloklandiae, boergesenii, bohlinianum, boldtii, boreale, borgeanum, botanense, botrophilum, brachiatum, brachioprominens, brachycerum, brasiliense, brebissonii, campanulatum, capitulum, centenniale, cerastes, ceratophorum, ceylanicum, chaetoceros, chavesii, cingulum, clepsydra, clevei, columbetoides, connatum, contectum, corniculatum, cornutum, cosmarioides, cosmospinosum, crenulatum, cristatum, cruciforme, cumbricum, cuneatum, cuniculosum, cyathipes, cyclacanthum, cyrtocerum, denticulatum, diabolo, diacanthum, digitatum, dilatatum, dispar, disputatum, distentum, donardense, dorsidentiferum, duacense, dubium, eboracense, echinodermum, elegans, elegantissimum, ellipticum, elongatum, emaciatum, ensiferum, erasum, erostellum, eskolense, excavatum, exporrectum, fennicum, floriferum, foersteri, fontense, forficulatum, formosum, franconicum, freemanii, furcatum, furcigerum, fusoideum, galeatum, gatniense, glabrum, gladiosum, gracile, granulosum, gutwinskii, haaboeliense, habeebense, hantzschii, hardyi, heimerlianum, hexacerum, hirsutum, hollandicum, horametrum, hypacanthum, hystrix, inconspicuum, inflatum, inflexum, iotanum, irregulare, iversenii, javanicum, johnsonii, kaiseri, kouwetsii, kurilense, laeve, laevispinus, lapponicum, latecurvatum, latiusculum, lepidum, leptacanthum, leptopus, libeltii, limneticum, longebrachiatum, longipes, longiradiatum, longispinum, luetkemuelleri, lunatum, maamense, manfeldtii, margaritaceum, maskellii, megalonotum, meriani, micron, micronoides, minimum, moniliferum, monticulosum, muricatiforme, muricatum, muticum, natator, navigiolum, neglectum, nigrae-silvae, nodiferum, nodosum, nodulosum, obscurum, octodontum, oenpelliense, oligacanthum, ophiura, orbiculare, orbiculispiniferum, osceolense, oxyacanthum, paradoxum, patens, paxilliferum, pelagicum, pendulum, pentacerum, pentacladum, pentacrinum, peristephes, perundulatum, petsamoense, picum, pileolatum, pilosellum, pilosum, pingue, pinnatum, planctonicum, platycerum, playfairii, polonicum, polymorphum, polytrichum, proboscideum, protectum, pseudobiretum, pseudofurcigerum, pseudoiotanum, pseudolevanderi, pseudopelagicum, pseudosebaldi, pseudosuecicum, pseudotetracerum, pseudowillsii, punctulatum, pungens, pyramidatum, quadrangulare, quadrispinatum, raphidicanthum, ravenellii, rectangulare, renardii, renardii cornifrons, retusum, rhabdophorum, rosei, rostellum, rotula, rugulosum, sagittarium, saltans, saxonicum, scabrum, scottii, sebaldi, senarium, setigerum, sexangulare, sexcostatum, simonyi, sinuatum, sirnonyi, smithii, sonthalianum, spiniceps, spiniferum, spinipendens, spinosum, spongiosum, stelliferum, striolatum, subarmigerum, subavicula, subcruciatum, subdenticulatum, subgemmulatum, subgracillimum, subgrande, sublaevispinum, submonticulosum, subnudibrachiatum, suborbiculare, subscabrum, tangaroae, tauphorum, teliferum, tenuissimum, tetracerum, thienemannii, tiara, tohopekaligense, tonsum, tortum, trachygonum, trachytithophorum, tridentulum, triprocessum, tsukubicum, turgescens, uhtuense, ungeri, urinator, valdiviense, varians, ventricosum, verticillatum, vestitum, victoriense, volans, wildemanii, xanthium, yirrkallense, zahlbruckneri, zonatum*

VARIETIES *acarides eboracense, acutum badense, acutum paxilliferum, aggeratum aculeatum, aggeratum granulosum, aggeratum leve, anatinum aculeatum, anatinum biradiatum, anatinum controversum, anatinum grande, anatinum longibrachiatum, anatinum nodulosum, anatinum parvum, anatinum simplicius, anatinum truncatum, anatinum vestitum, arachne arachnoides, arachne curvatum, arctiscon glabrum, armigerum furcigerum, arnellii inornatum, arnellii spiniferum, assurgens australe, assurgens victoriense, asteroideum munitum, asteroideum salebrosum, auburnense minor, avicula aciculiferum, avicula subarcuatum, bacillare obesum, bibrachiatum cymatium, bicorne evolutum, bicorne longebrachiatum, bicorne quadrifidum, bicoronatum ornatum, bienneanum ellipticum, bienneanum spetsbergensis, bitrapezicum apertum, biwaense sorabanum, boergesenii gracilius, bohlinianum subpygmaeum, brachioprominens inflatum, brasiliense lundellii, brebissonii brevispinum, capitulum spetsbergense, cerastes coronatum, cerastes coronatum-inflatum, cerastes inflatum, cerastes pulchrum, cerastes triradiatum, ceratophorum multiplicatum, cingulum cingulum, cingulum affine, cingulum obesum, claviferum brasiliense, clepsydra minimum, clepsydra obtusum, clepsydra sibiricum, clepsydra trigona, columbetoides intermedium, connatum muticum, connatum rectangulum, connatum spencerianum, contectum inevolutum, corniculatum australis, corniculatum australis-granulata, corniculatum granulata, corniculatum pseudoconnatum, corniculatum spinigerum, corniculatum variabile, crenulatum continentale, crenulatum continentale-mammiferum, crenulatum mammiferum, cyclacanthum armigerum, cyclacanthum armigerum-longipes, cyclacanthum gracilius, cyclacanthum longipes, cyclacanthum subacanthum, cyclacanthum submanfeldtoides, dilatatum australica, dilatatum insigne, dilatatum obtusilobum, dispar semicirculare, elegantissimum brasiliense, forficulatum africanum, forficulatum aristeron, forficulatum eximium, freemanii evolutum, freemanii evolutum-trispinatum, freemanii nudiceps, freemanii tridentatum, freemanii triquetrum, freemanii trispinatum, furcatum aciculiferum, furcatum renardi, furcatum senarium, gracile bulbosum, gracile elongatum, gracile kriegeri, gracile longipes, gracile minimum, gracile uniseriatum, granulosum acutum, gutwinskii reductum, hantzschii minutum, hexacerum aversum, hirsutum muricatum, hystrix pannonicum, inflexum brachycerum, iversenii americanum, javanicum apiculiferum, johnsonii altius, johnsonii cornutum, johnsonii elegans, laeve latidivergens, laevispinus subbrachiatum, laevispinus sydneyensis, lepidum latecurvatum, leptopus variabile, limneticum aculeatum, limneticum burmense, longebrachiatum longebrachiatum, longebrachiatum australe, longebrachiatum kriegeri, longebrachiatum pseudoanchora, longipes contractum, longiradiatum majus, longiradiatum subnudum, longispinum bidentatum, longispinum depressum, lunatum ovale, lunatum planctonicum, manfeldtii eichleri, manfeldtii planctonicum, manfeldtii submanfeldtii, margaritaceum cruciatum, margaritaceum hirtum, margaritaceum ornatum, margaritaceum subtile, meriani minutum, micron perpendiculatum, micron spinulosum, monticulosum bifarium, monticulosum duplex, muticum convexum, muticum depressum, muticum exeftensum, muticum hibernicum, muticum horii, muticum minus, muticum ralfsii, muticum subcurtum, muticum tetragona, muticum victoriense, nigrae-silvae trispinatum, oligacanthum incisum, ophiura cambricum, ophiura horizontale, orbiculare denticulatum, orbiculare denticulatum-minor, orbiculare depressum, orbiculare forficulatum, orbiculare germinosum, orbiculare germinosum-spinosa, orbiculare major, orbiculare minor, orbiculare muticum, orbiculare planktonicum, orbiculare protractum, orbiculare ralfsii, orbiculare spinosa, orbiculare subangulatum, orbiculare verrucosum, osceolense fennicum, oxyacanthum polycanthum, paradoxum nodulosum, paradoxum parva, paradoxum perornatum, patens patens, patens coronatum, patens minutum, patens obesa, patens planctonicum, patens planctonicum-australica, pelagicum bifidum, pendulum pinguiforme, pentacerum curvatum, petsamoense trispinosum, pileolatum cristatum, pinnatum robustum, pinnatum subpinnatum, pinnatum subpinnatum-robustum, planctonicum bulbosum, planctonicum ornatum, polymorphum divergens, protectum rangoonense, pseudoiotanum ornatum, pseudosebaldi bicornis, pseudosebaldi biradiata, pseudosebaldi coralloideum, pseudosebaldi coralloideum-biradiata, pseudosebaldi elongatum, pseudosebaldi latum, pseudosebaldi pectinatum, pseudosebaldi planctonicum, pseudosebaldi tonsum, punctulatum kjellmanii, punctulatum obtusum, punctulatum pygmaeum, punctulatum pygmaeum-obtusum, punctulatum striatum, punctulatum subglabrum, pyramidatum hirsutum, quadrangulare contectum, quadrispinatum furcatum, quadrispinatum spicatum, quadrispinatum spicatum-furcatum, rectangulare verrucosum, retusum boreale, retusum granulatum, retusum punctulatum, rosei elongatum, rostellum erostellum, rotula smithii, sagittarium longispinum, sagittarium scottii, saltans orientale, saltans polycharax, saltans sumatranum, sebaldi africanum, sebaldi altum, sebaldi bifidum, sebaldi gracile, sebaldi minus, sebaldi novizelandica, sebaldi ornatum, sebaldi ornatum-bifidum, sebaldi ornatum-novizelandica, sebaldi ornatum-planctonicum, sebaldi planctonica, sebaldi productum, sebaldi pusillum, setigerum subvillosum, sexangulare asperum, sexangulare australica, sexangulare dentatum, sexangulare gemmescens, sexangulare incurvum, sexangulare platycerum, sexangulare productum, sexangulare stellinum, sexangulare subglabrum, sexcostatum productum, spiniceps trifidum, spongiosum griffithsianum, spongiosum perbifidum, stelliferum laeve, subgracillimum curvatum, subgracillimum tortum, subgrande convexum, sublaevispinum aureolatum, sublaevispinum digitatum, subscabrum scabrior, tetracerum cameloides, tetracerum evolutum, tetracerum undulatum, tohopekaligense brevispinum, tohopekaligense insigne, tohopekaligense minus, tohopekaligense trifurcatum, tortum trigona, urinator ornatum, valdiviense ornatum, varians badense, varians subglabrum, vestitum semivestitum, victoriense tasmanicum, volans elegans, volans trigonum, wildemanii majus, wildemanii unispiniferum, zahlbruckneri africanum, zahlbruckneri granulatum, zonatum ceylanicum, zonatum majus, zonatum verrucicinctum*

GENUS *Staurodesmus* Crosshatch Fiber Conjugating Olive Green Algae

SPECIES *angulatus, apiculatus, arcuatus, aristiferus, asiaticus, aversus, brevispinus, bulnheimii, conspicuus, controversus, convergens, crassus, crispus, curvatus, cuspidatus, dejectus, dickiei, extensus, furcatospermus, gibberulus, glaber, grandis, hallei, incus, indentatus, inelegans, isthmosus, jaculiferus, lanceolatus, leptodermus, lobatus, lunatus, mamillatus, megacanthus, minutissimus, mucronatus, multinodulosum, nudus, omearii, obesus, obsoletus, orientalis, pachyrhynchus, phimus, psilosporus, pterosporus, quadratus, quirificus, ralfsii, reginae, rhomboideum, sellatus, sibiricus, skujae, smolandicus, spencerianus, spetsbergensis, subhexagonus, subpygmaeus, subtriangularis, subulatus, triangularis, tumidus, unicornis, validus*

VARIETIES *angulatus planctonicum, arcuatus octospinatus, aristiferus protuberans, brevispinus altum, brevispinus altus, brevispinus boldtii, brevispinus obversum, brevispinus obversus, brevispinus retusum, bulnheimii huitfeldtii, bulnheimii subincus, convergens ralfsii, convergens ralfsii-curtus, crispus americanus, crispus obesus, curvatus borgei, curvatus brevispinum, curvatus depressus, curvatus kalimantanus, curvatus latus, curvatus latus-depressus, curvatus variabile, cuspidatus cuspidatus, cuspidatus coronulatus, cuspidatus curvatus, cuspidatus divergens, cuspidatus maximus, cuspidatus tricuspidatus, dejectus dejectus, dejectus apiculatus, dejectus inflatum, dejectus patens, dejectus subapplanatum, dickiei bifidus, dickiei circularis, dickiei circularis-bifidus, dickiei denticulatus, dickiei forcipatus, dickiei grande, dickiei maximus, dickiei maximus-grande, dickiei minor, dickiei rhomboideus, gibberulus bimucronatus, gibberulus ellipticus, gibberulus mucronatus, gibberulus trimucronatus, glaber debaryanus, glaber hirudinella, grandis grandis, grandis parvus, incus incus, incus juressi, incus latiusculum, incus minor, incus quadrata, incus ralfsii, incus ralfsii-subhexagona, incus subhexagonus, incus validus, lanceolatus compressus, leptodermus corniculatus, leptodermus subcorniculatus, lobatus giganteus, mamillatus maximus, megacanthus australis, megacanthus giganteus, megacanthus orientalis, megacanthus rhomboideus, megacanthus scoticus, megacanthus subcurvatus, megacanthus triangularis, mucronatus bimucronatus, mucronatus delicatulus, mucronatus subtriangularis, omearii omearii, omearii minutus, obsoletus sitvensis, pachyrhynchus convergens, pachyrhynchus ellipticus, phimus phimus, phimus convexus, phimus hebridarus, phimus minimissimus, phimus occidentalis, phimus occidentalis-minimissimus, quirificus minor, sibiricus occidentalis, smolandicus pyramidalis, spencerianus gracilis, spencerianus spencerianus, spencerianus triangulatus, spetsbergensis florinae, subpygmaeus subangulatus, subtriangularis inflatus, subulatus americana, subulatus nordstedtii, subulatus subaequalis, triangularis triangularis, triangularis inflatus, triangularis latodivergens, triangularis limneticus, triangularis minor, triangularis subparallelus, unicornis gracilis, unicornis gracilis-longicollis, unicornis longicollis, unicornis nodiferum, unicornis subscolopacinus, validus apertus, validus gibbosus, validus subapertus, validus subapertus-gibbosus*

GENUS *Streptonema* Twisted Fiber Conjugating Olive Green Algae

SPECIES *trilobatum*

GENUS *Teilingia* Teiling Fiber Conjugating Olive Green Algae

SPECIES *excavata, granulata, wallachi*

VARIETIES *wallachi anglica, wallachi borgei*

GENUS *Tetmemorus* Tetmemor Fiber Conjugating Olive Green Algae

SPECIES *brebissonii, euastroides, granulatus, laevis, minutus*

VARIETIES *brebissonii brebissonii, brebissonii attenuatus, brebissonii immanis, brebissonii minor, brebissonii tenuissima, brebissonii turgidus, brebissonii varians, laevis minutus*

GENUS *Triploceras* Triple Horn Fiber Conjugating Olive Green Algae

SPECIES *denticulatum, gracile, serratum, splendens, verticillatum*

VARIETIES *gracile aculeatum, gracile aculeatum-australica, gracile australica, gracile bidentatum, gracile bidentatum-spinosum, gracile bilobatum, gracile curvatum, gracile denticulatum, gracile elegans, gracile genuinum, gracile robustum, gracile septundulatum, gracile spinosum, gracile undulatum, gracile undulatum-septundulatum, verticillatum verticillatum, verticillatum ornatum, verticillatum superbum*

GENUS *Xanthidium* Yellow Fiber Conjugating Olive Green Algae

SPECIES *acanthophorum, aculeatum, antilopaeum, apiculatum, apiculiferum, armatum, basidentatum, bengalicum, bifidum, bifurcatum, brevinasutum, burkilii, calcarato-aculeatum, caulfieldii, ceylanicum, concinnum, controversum, cristatum, decemdenticulatum, fasciculatum, gerloffii, hastiferum, hastiferum hexagonum, hirsutum, inchoatum, japonicum, kalimantanum, multicorne, octocorne, octonarium, orcadense, perissacanthum, pilosum, regulare, robinsonianum, scottii, sexmamillatum, simplicius, smithii, subhastiferum, submicracanthum, subtrilobum, superbum, tetracentrotum, triangulare, variabile, zhejiangense*

VARIETIES *aculeatum basidentatum, antilopaeum depauperatum, antilopaeum hebridarum, antilopaeum hebridarum-triquetrum, antilopaeum javanica, antilopaeum laeve, antilopaeum laeve-longispinum, antilopaeum longispinum, antilopaeum polymazum, antilopaeum triquetrum, armatum anguliferum, armatum basidentatum, armatum cervicorne, armatum cervicorne-triquetrum, armatum coogeeanum, armatum depressum, armatum fissum, armatum irregularis, bifurcatum simplex, concinnum boltianum, cristatum angulatum, cristatum delpontii, cristatum leiodermum, cristatum spinuliferum, cristatum uncinatum, fasciculatum oronense, fasciculatum perornatum, hastiferum, hastiferum inevolutum, hastiferum javanicum, inchoatum alpinum, inchoatum cracoviense, inchoatum mammillatum, inchoatum reductum, octocorne depressum, octocorne impar, octocorne majus-hawaiensis, octonarium cornutum, octonarium latum, regulare novangulare, sexmamillatum pulneyense, simplicius botanicum, simplicius intermedium, simplicius pseudosmithii, smithii collum, smithii majus, subhastiferum murrayi, subtrilobum inornatum, tetracentrotum protuberans, tetracentrotum quadricornutum*

FAMILY 4 *Mesotaeniaceae* — Ribbon Conjugating Olive Green Algae

GENUS *Mesotaenium* — Ribbon Conjugating Olive Green Algae

SPECIES *caldariorum, chlamydosporum, degreyi, endlicherianum, kramstai, macrococcum, mirificum, scrobiculatum, truncatum, violascens*

VARIETIES *chlamydosporum minus, chlamydosporum violascens, degreyi breve, degreyi exiguum, degreyi major, degreyi tenuis, endlicherianum grande, macrococcum nigrescens*

GENUS *Ancylonema* — Ancyclic Ribbon Conjugating Olive Green Algae

SPECIES *nordenskioeldii*

GENUS *Cylindrocystis* — Cylindric Cyst Ribbon Conjugating Olive Green Algae

SPECIES *acanthospora, brebissonii, crassa, cushleckae, debaryi, gracilis, jenneri, minutissima, obesa, ovalis, pyramidata*

VARIETIES *brebissonii brebissonii, brebissonii jenneri, brebissonii minor, brebissonii turgida*

GENUS *Geniculus* — Genicule Ribbon Conjugating Olive Green Algae

SPECIES *gatunensis*

GENUS *Netrium* — Net Ribbon Conjugating Olive Green Algae

SPECIES *digitus, elongatum, interruptum, lanceolatum, minus, minutum, naegelii, nepalense, oblongum, scottii*

VARIETIES *digitus constrictum, digitus curtum, digitus elegans, digitus lamellosum, digitus latum, digitus minus, digitus naegelii, digitus parvum, digitus rectum, digitus rhomboideum, digitus scottii, interruptum minor, interruptum scottii, interruptum sectum, naegelii elongatum, oblongum brevius, oblongum cylindricum*

GENUS *Polytaenia* — Many Ribbon Conjugating Olive Green Algae

SPECIES *cambrica, obscura, trabeculata*

GENUS *Roya* — Roy Ribbon Conjugating Olive Green Algae

SPECIES *anglica, cambrica, obtusa, pseudoclosterium*

VARIETIES *cambrica limnetica, obtusa montana, pseudoclosterium minor*

GENUS *Spirotaenia* — Spire Ribbon Conjugating Olive Green Algae

SPECIES *acuta, bispiralis, bryophila, closteridia, condensata, diplohelica, eboracensis, endospira, erythrocephala, kirchneri, minuta, truncata*

VARIETIES *bispiralis fusiformis, kirchneri erythropunctata, minuta eboracencis, minuta minutissima*

GENUS *Tortitaenia* — Twisted Ribbon Conjugating Olive Green Algae

SPECIES alpina, *luetkemuelleri*

FAMILY 5 *Peniaceae* — Pen Conjugating Olive Green Algae

GENUS *Penium* — Pen Conjugating Olive Green Algae

SPECIES *adelochondrum, australe, clevei, crassiusculum, cruciferum, cylindrus, didymocarpum, exiguum, globosum, inconspicuum, jenneri, lanceolatum, libellula, margaritaceum, minutissimum, mooreanum, naegelii, phymatospermum, polymorphum, rufescens, spinulosum, spirostriolatiforme, spirostriolatum, suboctangulare, subtile, truncatum*

VARIETIES *australe crassior, clevei crassum, cylindrus cylindrus, cylindrus cuticulare, exiguum expansum, globosum maxima, globosum wollei-maxima, libellula elongatum, libellula granulatum, libellula intermedium, libellula interruptum, libellula punctatum, margaritaceum indivisum, margaritaceum irregulare, margaritaceum pulverulentum, polymorphum cylindraceum, polymorphum minutissimum, polymorphum mooreanum, polymorphum turneri, spirostriolatiforme rectispirum, spirostriolatum spirostriolatum, spirostriolatum gracillimum*

GENUS *Genicularia* — Genicule Pen Conjugating Olive Green Algae

SPECIES *elegans, spirotaenia*

GENUS *Gonatozygon* — Gonad Pen Conjugating Olive Green Algae

SPECIES *aculeatum, brebissonii, chadefaudii, dicranoaculeatum, kinahani, kjellmanni, monotaenium, pilosum, tenuissimum*

VARIETIES *aculeatum gracile, aculeatum groenbladii, brebissonii laeve, brebissonii minutum, brebissonii vulgare, kinahani kjellmanni, monotaenium gracile, monotaenium pilosellum, monotaenium tenue*

PHYLUM 613 *Chrysophyta* Golden Ocher Algae

Latin: *Chryso* gold, yellow
Latin: *Phyta* plant
1231 species
[Guiry, Dec. 2004]

CLASSES		Species
613.1 *Chrysophyceae*	Golden Ocher Algae	88 [G]
613.2 *Dictyochophyceae*	Netted Golden Ocher Algae	100 [G]
613.3 *Eustigmatophyceae*	Eyespot Green Yellow Ocher Algae	17 [G]
613.4 *Synurophyceae*	Unicell Golden Ocher Algae	621 [G]
613.5 *Xanthophyceae*	Green Yellow Golden Ocher Algae	405 [G]

Phylum *Chrysophyta* Golden Ocher Algae species live in saltwater and freshwater. Chrysophytes (Golden Ocher Algae) form a group of algae whose plastids contain golden yellow pigments. They are widespread in freshwater temperate lakes and ponds. Freshwater chrysophytes survive desiccation by making membrane outer coat structures called statocysts. The statocysts become encrusted with silicon and iron.

Only one group of chrysophytes, called the silicoflagellates, are marine.The marine planktonic silicoflagellates interact with silica and iron to form shells called tests. The test shells have beautiful patterns that are used to distinguish genera. Some tests have been perserved as fossils dating to 500 million years ago.

Because the pigments are golden yellow or ocher yellow, these algae are considered separate and distinct from algae that have red and orange ocher pigments. Algae with ocher colors other than ocher yellow, are placed into the Phylum *Ochrophyta*.

Other Classifications

Scientists Margulis and Schwartz previously recognized Class *Eustigmantophyceae* and Class *Xanthophyceae* as Phylum *Eustigmatophyta* and Phylum *Xanthophyta*, respectively. Scientist M. Guiry created Super-Phylum *Ochrophyta* on 14 December 2004 and lowered *Chrysophyta*, *Eustigmatophyta and Xanthophyta* to Classes within Super-Phylum *Ochrophyta*. **Phylum *Chrysophyta* Golden Ocher Algae** was previously recognized by scientists Margulis and Schwartz as containing Class *Chrysophyceae* and Class *Dictyophyaceae* and Class *Pedinellophyceae*. In the Fusillo Classification System, Class *Pedinellophyceae* has been lowered to Order *Pedinellales* under Class *Dictyophyceae*. The M. Guiry *Ochrophyta* Super-Phylum has been reduced to regular Phylum status.

Reclassification of Phylum *Chrysophyta*

		AlgaeBase
PHYLUM	*Chrysophyta*	~~*Ochrophyta*~~
CLASS 1	***Chrysophyceae***	***Chrysophyceae***
ORDER 1	*Chrysapiales*	*Chrysapiales*
FAMILY 1	*Chrysapiaceae*	~~*Not Assigned*~~
FAMILY 2	*Sarcinochrysidaceae*	*Sarcinochrysidaceae*
ORDER 2	*Chrysocapsales*	*Chrysocapsales*
FAMILY 1	*Chrysocapsaceae*	*Chrysocapsaceae*
FAMILY 2	*Hydruraceae*	*Hydruraceae*
ORDER 3	*Phaeothamniales*	*Phaeothamniales*
FAMILY 1	*Phaeothamniaceae*	*Phaeothamniaceae*
FAMILY 2	*Phaeosaccionaceae*	*Phaeosaccionaceae*
ORDER 4	*Thallochrysidales*	*Thallochrysidales*
FAMILY 1	*Thallochrysidaceae*	*Thallochrysidaceae*
FAMILY 2	*Chrysomeridaceae*	*Chrysomeridaceae*
CLASS 2	***Dictyochophyceae***	***Dictyochophyceae***
ORDER 1	*Dictyochales*	*Dictyochales*
FAMILY 1	*Dictyochaceae*	*Dictyochaceae*
		~~*Vallacertaceae*~~
FAMILY 2	*Cornuaceae*	*Cornuaceae*
FAMILY 3	*Distephanacea*	*Distephanacea*
FAMILY 4	*Mesocenaceae*	*Mesocenaceae*
ORDER 2	*Pedinellales*	~~*Chromulinales*~~
FAMILY 1	*Pedinellaceae*	*Pedinellaceae*
ORDER		*Pedinellales*
FAMILY 2	*Cyrtophoraceae*	*Cyrtophoraceae*
CLASS 3	***Eustigmatophyceae***	***Eustigmatophyceae***
ORDER 1	*Eustigmatales*	~~*Not Assigned*~~
FAMILY 1	*Eustigmataceae*	*Eustigmataceae*
FAMILY 2	*Monodopsidaceae*	*Monodopsidaceae*
FAMILY 3	*Pleurochloridaceae*	*Pleurochloridaceae*
FAMILY 4	*Pseudocharaciopsidaceae*	*Pseudocharaciopsidaceae*
CLASS 4	***Synurophyceae***	***Synurophyceae***
ORDER 1	*Synurales*	~~*Ochromonadales*~~
FAMILY 1	*Synuraceae*	*Synuraceae*
ORDER		~~*Hibberdiales*~~
FAMILY 4	*Hibberdiaceae*	*Hibberdiaceae*
ORDER		~~*Parmales*~~
FAMILY 7	*Triparmaceae*	*Triparmaceae*
		~~*Pentalaminaceae*~~
ORDER	*Synurales*	*Synurales*
FAMILY 5	*Mallomonadaceae*	*Mallomonadaceae*
ORDER 1	*Synurales*	~~*Not Assigned*~~
FAMILY 2	*Archaeomonadaceae*	*Archaeomonadaceae*
FAMILY 1	*Synuraceae*	~~*Aurosphaeraceae*~~
FAMILY 1	*Synuraceae*	~~*Bicoecaceae*~~
FAMILY 3	*Chrysosaccaceae*	*Chrysosaccaceae*
FAMILY 6	*Monosigaceae*	*Monosigaceae*
FAMILY 1	*Synuraceae*	~~*Phalansteriaceae*~~
ORDER 2	*Chromulinales*	*Chromulinales*
FAMILY 1	*Chromulinaceae*	*Chromulinaceae*

		AlgaeBase
		~~*Chrysothallaceae*~~
		~~*Paraphysomonadaceae*~~
		~~*Rhizochromulinaceae*~~
FAMILY 2	*Chrysococcaceae*	*Chrysococcaceae*
FAMILY 3	*Chrysolepidomonadaceae*	*Chrysolepidomonadaceae*
FAMILY 4	*Chrysosphaeraceae*	*Chrysosphaeraceae*
ORDER		~~*Chrysosphaerales*~~
FAMILY 5	*Stichogloeaceae*	*Stichogloeaceae*
ORDER 3	*Chrysamoebidales*	*Chrysamoebidales*
FAMILY 1	*Chrysamoebaceae*	*Chrysamoebaceae*
FAMILY 2	*Stylococcaceae*	*Stylococcaceae*
ORDER 4	*Ochromonadales*	*Ochromonadales*
FAMILY 1	*Ochromonadaceae*	*Ochromonadaceae*
FAMILY 2	*Dinobryaceae*	*Dinobryaceae*
CLASS 5	***Xanthophyceae***	***Xanthophyceae***
ORDER 1	*Xanthophytales*	~~*Not Assigned*~~
FAMILY 1	*Xanthophytaceae*	~~*Not Assigned*~~
		~~*Phyllosiphonaceae*~~
		~~*Pseudochloridaceae*~~
ORDER		~~*Heterogloeales*~~
FAMILY 2	*Heterogloeaceae*	*Heterogloeaceae*
		~~*Pleurochloridellaceae*~~
ORDER 2	*Mischococcales*	*Mischococcales*
FAMILY 1	*Mischococcaceae*	*Mischococcaceae*
		~~*Chloropediaceae*~~
		~~*Gloeopodiaceae*~~
		~~*Ophiocytaceae*~~
		~~*Trypanochloridaceae*~~
FAMILY 2	*Botrydiopsidaceae*	*Botrydiopsidaceae*
FAMILY 3	*Botryochloridaceae*	*Botryochloridaceae*
FAMILY 4	*Centritactaceae*	*Centritactaceae*
FAMILY 5	*Characiopsidaceae*	*Characiopsidaceae*
FAMILY 6	*Gloeobotrydaceae*	*Gloeobotrydaceae*
ORDER 3	*Rhizochloridales*	*Rhizochloridales*
ORDER		~~*Chloramoebales*~~
FAMILY 1	*Rhizochloridaceae*	*Rhizochloridaceae*
		~~*Myxochloridaceae*~~
FAMILY 2	*Chloramoebaceae*	*Chloramoebaceae*
ORDER		*Rhizochloridales*
FAMILY 3	*Rhizogranulochloridaceae*	*Rhizogranulochloridaceae*
FAMILY 4	*Rhizounochloridaceae*	*Rhizounochloridaceae*
FAMILY 5	*Stipitococcaceae*	*Stipitococcaceae*
ORDER 4	*Tribonematales*	*Tribonematales*
FAMILY 1	*Tribonemataceae*	*Tribonemataceae*
		~~*Neonemataceae*~~
FAMILY 2	*Xanthonemataceae*	*Xanthonemataceae*
FAMILY 3	*Heteropediaceae*	*Heteropediaceae*
ORDER 5	*Vaucheriales*	*Vaucheriales*
FAMILY 1	*Vaucheriaceae*	*Vaucheriaceae*
FAMILY 2	*Botrydiaceae*	*Botrydiaceae*

CLASS 613.1 *Chrysophyceae* Golden Ocher Algae

ORDER 1 *Chrysapiales* Parsley Golden Ocher Algae

FAMILY 1 *Chrysapiaceae*	Parsley Golden Ocher Algae
GENUS *Chrysapion*	Parsley Golden Ocher Algae
SPECIES *rhigophilos*	
GENUS *Chrysapsis*	Lesser Parsley Golden Ocher Algae
SPECIES *fenestrata*	
GENUS *Chrysopora*	(Synonym for *Chrysapsis fenestrata*)
FAMILY 2 *Sarcinochrysidaceae*	Pack Parsley Golden Ocher Algae
GENUS *Sarcinochrysis*	Pack Parsley Golden Ocher Algae
SPECIES *marina*	
VARIETIES *marina filamentosa*	
GENUS *Ankylochrysis*	Ankylose Pack Parsley Golden Ocher Algae
SPECIES *lutea*	
GENUS *Aureoumbra*	Umbrella Pack Parsley Golden Ocher Algae
SPECIES *lagunensis*	
GENUS *Chrysoderma*	Dermal Pack Parsley Golden Ocher Algae
SPECIES *mucosum*	
GENUS *Chrysocystis*	Cyst Pack Parsley Golden Ocher Algae
SPECIES *fragilis*	
GENUS *Chrysonephos*	Kidney Pack Parsley Golden Ocher Algae
SPECIES *lewisii*	
GENUS *Chrysoreinhardia*	Reinhard Pack Parsley Golden Ocher Algae
SPECIES *algicola, feldmannii, giraudii*	
GENUS *Nematochrysopsis*	Thread Pack Parsley Golden Ocher Algae
SPECIES *marina*	
VARIETIES *marina giganteus*	

ORDER 2 *Chrysocapsales* Capsule Golden Ocher Algae

FAMILY 1 *Chrysocapsaceae*	Capsule Golden Ocher Algae
GENUS *Chrysocapsa*	Capsule Golden Ocher Algae
SPECIES *epiphytica, planktonica, vernalis*	
GENUS *Arthrogloea*	Straight Glow Capsule Golden Ocher Algae
SPECIES *annelidiformis*	
GENUS *Bourrellia*	Bourrell Capsule Golden Ocher Algae
SPECIES *skujae*	
GENUS *Chalkopyxis*	Chalk Pyxie Capsule Golden Ocher Algae
SPECIES *tetrasporoides*	
GENUS *Chrysocapsella*	Lesser Capsule Golden Ocher Algae
SPECIES *paludosa, planctonica*	
GENUS *Chrysocapsopsis*	Least Capsule Golden Ocher Algae
SPECIES *rupicola*	
GENUS *Chrysotilos*	Chrysotile Capsule Golden Ocher Algae
SPECIES *ferrea*	
GENUS *Dermatochrysis*	Dermal Capsule Golden Ocher Algae
SPECIES *pelagica, pseudofenestrata, reticulata*	
GENUS *Eirmodesmus*	Fiber Capsule Golden Ocher Algae
SPECIES *phaeotilus*	
GENUS *Geochrysis*	(Synonym for *Gloeochrysis turfosa*)
GENUS *Gloeochrysis*	Glow Capsule Golden Ocher Algae
SPECIES *pyrenigera, turfosa*	
GENUS *Heimiochrysis*	Heim Capsule Golden Ocher Algae
SPECIES *actinotrichia*	
GENUS *Kremastochrysis*	Cremast Capsule Golden Ocher Algae
SPECIES *minor, pendens*	
GENUS *Kremastochrysopsis*	Lesser Cremast Capsule Golden Ocher Algae
SPECIES *ocellata*	
GENUS *Kussiella*	Kuss Capsule Golden Ocher Algae
SPECIES *tuanshanziensis*	
GENUS *Naegeliella*	Naegel Capsule Golden Ocher Algae
SPECIES *britannica, flagellifera*	
GENUS *Pascherella*	Pascher Capsule Golden Ocher Algae
SPECIES *yserensis*	
GENUS *Phaeaster*	Dusky Star Capsule Golden Ocher Algae

SPECIES *pascheri*
GENUS *Phaeococcus* — Dusky Round Capsule Golden Ocher Algae
SPECIES *adnatus, clementii*
GENUS *Tetrasporopsis* — Tetraspore Capsule Golden Ocher Algae
SPECIES *fuscescens*

FAMILY 2 *Hydruraceae* — Freshwater Tailed Golden Ocher Algae
GENUS *Hydrurus* — Freshwater Tailed Golden Ocher Algae
SPECIES *foetidus, irregularis*
VARIETIES *irregularis causta, irregularis crassa, irregularis moravicus*
GENUS *Celloniella* — Cellon Freshwater Tailed Golden Ocher Algae
SPECIES *palensis*
GENUS *Chrysonebula* — Nebula Freshwater Tailed Golden Ocher Algae
SPECIES *holmesii*
GENUS *Nanurus* — Dwarf Freshwater Tailed Golden Ocher Algae
SPECIES *flaccidus*

ORDER 3 *Phaeothamniales* Dusky Golden Ocher Algae

FAMILY 1 *Phaeothamniaceae* — Dusky Golden Ocher Algae
GENUS *Phaeothamnion* — Dusky Golden Ocher Algae
SPECIES *confervicola*
VARIETIES *confervicola britannica*
GENUS *Apistonema* — Parsley Thread Dusky Golden Ocher Algae
SPECIES *aestuarii, carterae, commutatum, pyrenigerum, submarinum*
GENUS *Chrysoclonium* — Clone Dusky Golden Ocher Algae
SPECIES *ramosum*
GENUS *Chrysodesmis* — Fiber Dusky Golden Ocher Algae
SPECIES *gloeophila*
GENUS *Chrysophaeum* — Lesser Dusky Golden Ocher Algae
SPECIES *taylorii*
GENUS *Koinopodion* — Coin Pod Dusky Golden Ocher Algae
SPECIES *aggregatum*
GENUS *Nematochrysis* — Thread Dusky Golden Ocher Algae
SPECIES *hieroglyphica, sessilis*
GENUS *Phaeogloea* — Glow Dusky Golden Ocher Algae
SPECIES *mucosa*
GENUS *Phaeoschizochlamys* — Lesser Split Cloak Dusky Golden Ocher Algae
SPECIES *mucosa*
GENUS *Schizochlamydella* — Split Cloak Dusky Golden Ocher Algae
SPECIES *capsulata, delicatula, minutissima, sphaerica*
GENUS *Selenophaea* — Selene Dusky Golden Ocher Algae
SPECIES *granulosa*
GENUS *Sphaeridiothrix* — Sphere Hair Dusky Golden Ocher Algae
SPECIES *compressa*
GENUS *Tetrachrysis* — Tetrad Dusky Golden Ocher Algae
SPECIES *dendroides*
GENUS *Tetrapion* — Tetrad Pion Dusky Golden Ocher Algae
SPECIES *gloeocystiforme*

FAMILY 2 *Phaeosaccionaceae* — Sac Dusky Golden Ocher Algae
GENUS *Phaeosaccion* — Sac Dusky Golden Ocher Algae
SPECIES *collinsii*
GENUS *Antarctosaccion* — Antarctic Sac Dusky Golden Ocher Algae
SPECIES *applanatum*

ORDER 4 *Thallochrysidales* Thallus Golden Ocher Algae

FAMILY 1 *Thallochrysidaceae* — Thallus Golden Ocher Algae
GENUS *Thallochrysis* — Thallus Golden Ocher Algae
SPECIES *litoralis, pascheri*
GENUS *Phaeodermatium* — Dusky Thallus Golden Ocher Algae
SPECIES *rivulare*
GENUS *Phaeoplaca* — Lesser Dusky Thallus Golden Ocher Algae
SPECIES *thallosa*

FAMILY 2 *Chrysomeridaceae* — Seaweed Thallus Golden Ocher Algae
GENUS *Chrysomeris* — Seaweed Thallus Golden Ocher Algae
SPECIES *articulata, ramosa*
GENUS *Chrysowaernella* — Waern Seaweed Thallus Golden Ocher Algae

SPECIES *hieroglyphica*

CLASS 613.2 *Dictyochophyceae* Netted Golden Ocher Algae

ORDER 1 *Dictyochales* Netted Golden Ocher Algae

FAMILY 1 *Dictyochaceae* Netted Golden Ocher Algae
GENUS *Dictyocha* Netted Golden Ocher Algae
SPECIES *aculeata, aegea, alta, anguinea, angulata, aperta, apiculata, arbutusensis, arctios, ausonia, biapiculata, binoculus, brachyacantha, brevispina, byronalis, challengeri, circulus, concavata, concinna, curta, deflandrei, delicata, diommata, elata, eocaenica, epiodon, extensa, fallacia, fibula, grandis, haliomma, helix, hessii, katharinae, lamellifera, lingii, lockeri, longa, mesophthalma, messanensis, navicula, ornata, paradistephanus, perfecta, perlaevis, pulchella, pumila, pygmaea, pygmaea, quadrangula, quadria, rhombica, rotundata, septenaria, stapedia, stelliformis, subaculeata, subclinata, tamarae, torta, transenna, triacantha, vanandelii, varia, vexativa*
VARIETIES *aculeata binoculus, aculeata subaculeata, apiculata inermis, apiculata monolineata, biapiculata lata, brevispina ausonia, deflandrei lobata, deflandrei producta, delicata bisecta, elata reducta, epiodon subaculeata, fibula aculeata, fibula asymmetrica, fibula brevispina, fibula cannopilea-hexacantha, fibula eocaenica, fibula octagona, fibula pumila, fibula rhombica, fibula speculum, lockeri pentagona, longa asymmetrica, longa paxilla, messanensis aculeata, messanensis aspinosa, messanensis aspinosa, messanensis speculum, messanensis stapedia, navicula aspera, navicula minor, ornata africana, perlaevis delicata, perlaevis ornata, perlaevis pentaradiata, pulchella inflata, rotundata secta, stapedia aspinosa, triacantha inermis, varia extensa*
GENUS *Bachmannocena* Bachmann Netted Golden Ocher Algae
SPECIES *occidentalis*
GENUS *Caryocha* Cary Netted Golden Ocher Algae
SPECIES *depressa, ernestinae, ichikawae, jouseae, latifenestrata, picassoi*
GENUS *Eunaviculopsis* (Synonym for *Naviculopsis obtusarca*)
GENUS *Halicalyptra* Haline Calypter Netted Golden Ocher Algae
SPECIES *depressa, virginica*
GENUS *Hannaites* (Synonym for *Dictyocha quadria*)
GENUS *Naviculopsis* Navicule Netted Golden Ocher Algae
SPECIES *americana, aspera, contraria, foliacea, lacrima, minor, nordica, obtusarca, pacifica, ponticula, ponticula spinosa, sicca, vemae*
VARIETIES *aspera clinata, aspera pygmaea, foliacea tumida, nordica hyalina, obtusarca acicula, pacifica pansa*
GENUS *Neonaviculopsis* Modern Navicule Netted Golden Ocher Algae
SPECIES *danica, eobiapiculata, lata, neonautica, punctilia, robusta, spinosa*
VARIETIES *lata obliqua, neonautica cocosensis, neonautica praenautica, punctilia taenia*
GENUS *Octactis* Octal Netted Golden Ocher Algae
SPECIES *octonaria, pulchra, tenuis*
GENUS *Paradictyocha* (Synonym for *Paramesocena dumitricae*)
GENUS *Pseudorocella* Rocelle Netted Golden Ocher Algae
SPECIES *barbadensis*
GENUS *Stylochromonas* Monad Netted Golden Ocher Algae
SPECIES *minuta*
GENUS *Vallacerta* Vallacert Netted Golden Ocher Algae
SPECIES *hannae*
GENUS *Variramus* Variable Branched Netted Golden Ocher Algae
SPECIES *loperi*

FAMILY 2 *Cornuaceae* Horned Golden Ocher Algae
GENUS *Cornua* Horned Golden Ocher Algae
SPECIES *aculeifera*
GENUS *Corbellia* Lesser Corbis Horned Golden Ocher Algae
SPECIES *contorta*
GENUS *Corbisema* Corbis Horned Golden Ocher Algae
SPECIES *angularis, bimucronata, bukryi, constricta, convexa, cuspis, elata, exilis, falklandensis, globulata, hastata, inermis, katharinae, lamelliferum, media, navicula, naviculoidea, ovalis, panda, perch-nielseniae, regina, schulzii, toxeuma, triacantha, xenicum*
VARIETIES *bimucronata rotatoria, hastata globulata, hastata miranda, inermis crenulatum, inermis ballantina, inermis disymmetrica, media reducta, navicula constricta, triacantha convexa, triacantha mediana*
GENUS *Crassicorbisema* Crass Corbis Horned Golden Ocher Algae
SPECIES *disymmetrica*
VARIETIES *dumitricae, crenulatum, naviculoidea*
GENUS *Lyramula* Lyre Horned Golden Ocher Algae
SPECIES *bipedicellata*
GENUS *Phyllodictyocha* Net Horned Golden Ocher Algae

SPECIES *schulzii*

FAMILY 3 *Distephanaceae* Distephan Golden Ocher Algae

GENUS *Distephanus* Distephan Netted Golden Ocher Algae

SPECIES *acanthicus, boliviensis, bukryi, frugalis, irregularis, jimlingii, major, norvegiensis, paradistephanus, polyactis, pseudofibula, pulcher, quintus, quintus, rosae, speculum, sulcatus, trioctus, variabilis, varians, xenus*

VARIETIES *boliviensis frugalis, boliviensis jimlingii, boliviensis major, boliviensis spinosus, polyactis crassus, polyactis literatus, speculum bispicatus, speculum cannopilea-hexacantha, speculum diommata, speculum elongatus, speculum fenestratus, speculum giganteus, speculum haliomma, speculum minutus, speculum nonarius, speculum notabilis, speculum octonarius, speculum patulus, speculum pentagonus, speculum polyommata, speculum pseudocrux, speculum pseudofibula, speculum pseudopentagonus, speculum septenarius, speculum spiralis, speculum tenuis, speculum varians*

GENUS *Cannopilus* Pilose Distephan Golden Ocher Algae

SPECIES *antarcticus, desikacharyi, diommata, ernestinae, haliomma, hemisphaericus, heptacanthus, ichikawae, jimlingii, jouseae, latifenestratus, literatus, quintus, sulcatus, tholus, trioctus*

VARIETIES *hemisphaericus polyommata, heptacanthus antarcticus*

GENUS *Distephanopsis* Lesser Distephan Netted Golden Ocher Algae

SPECIES *bachmannii, bipartita, contraria, crux, hannae, longispina, macilenta, octangulata, pseudocrux, schauinslandii, slavnicii, soljanii, stauracanthus, stauracanthus octagonus, staurodon, stradneri*

VARIETIES *crux apiculata, crux aspera, crux bispinosa, crux carolae, crux darwinii, crux fenestrata, crux hannae, crux loeblichii, crux longispina, crux mesophthalma, crux parva, crux paulii, crux scutulata, schauinslandii rotundata, schauinslandii stradneri, slavnicii naviculoidea, slavnicii orbicularis, staurodon pentagona, staurodon quadrangula, stradneri grandis*

GENUS *Nothyocha* False Distephan Golden Ocher Algae

SPECIES *insolita*

VARIETIES *insolita quadrangula*

GENUS *Paracannopilus* Lesser Pilose Distephan Golden Ocher Algae

SPECIES *miltiadei, picassoi*

FAMILY 4 *Mesocenaceae* Mesocene Golden Ocher Algae

GENUS *Mesocena* Mesocene Golden Ocher Algae

SPECIES *bispicata, concava, connudata, convexa, diodon, elliptica, hexalitha, nodulifera, oamaruensis, oamaruensis quadrangula, occidentalis, ovata, polyactis, polymorpha, quadrangula, tsumurae, venusta*

VARIETIES *convexa triangula, diodon borderlandensis, diodon nodosa, elliptica quadrangula, hexalitha heptalitha, nodulifera triangula, polymorpha biseptenaria, polymorpha fossilis, polymorpha glabra, polymorpha quadrangula*

GENUS *Bukryella* Bukry Mesocene Golden Ocher Algae

SPECIES *eocaenica*

GENUS *Paramesocena* Lesser Mesocene Golden Ocher Algae

SPECIES *apiculata, bioctonaria, biseptenaria, circulus, polyactis*

VARIETIES *apiculata curvata, apiculata evexa, apiculata glabra, apiculata inflata, apiculata monolineata, bioctonaria fossilis, biseptenaria fossilis, circulus apiculata*

GENUS *Septamesocena* Septad Mesocene Golden Ocher Algae

SPECIES *pappii, schulzii*

ORDER 2 *Pedinellales* Ribbon Golden Ocher Algae

FAMILY 1 *Pedinellaceae* Ribbon Golden Ocher Algae

GENUS *Pedinella* Ribbon Golden Ocher Algae

SPECIES *hexacostata, marina, tricostata*

GENUS *Actinomonas* Radial Monad Ribbon Golden Ocher Algae

SPECIES *mirabilis*

GENUS *Apedinella* Mock Ribbon Golden Ocher Algae

SPECIES *spinifera*

GENUS *Ciliophrys* Ciliate Ribbon Golden Ocher Algae

SPECIES *infusionum*

GENUS *Mesopedinella* Mesocene Ribbon Golden Ocher Algae

SPECIES *arctica*

GENUS *Parapedinella* Least Ribbon Golden Ocher Algae

SPECIES *reticulata*

GENUS *Pseudopedinella* Lesser Ribbon Golden Ocher Algae

SPECIES *elastica, pyriforme, tricostata*

GENUS *Pteridomonas* Winged Monad Ribbon Golden Ocher Algae

SPECIES *danica, pulex*

FAMILY 2 *Cyrtophoraceae* Arc Ribbon Golden Ocher Algae

GENUS *Cyrtophora* Arc Ribbon Golden Ocher Algae

SPECIES *pedicellata*

GENUS *Palatinella* Palatin Arc Ribbon Golden Ocher Algae

SPECIES *cyrtophora*

CLASS 613.3 *Eustigmatophyceae* Eyspot Green Yellow Golden Ocher Algae

ORDER 1 *Eustigmatales* Eyespot Green Yellow Ocher Algae

FAMILY 1 *Eustigmataceae* Eyespot Green Yellow Ocher Algae
GENUS *Eustigmatos* Eyespot Green Yellow Ocher Algae
SPECIES *magnus, polyphema, vischeri*
GENUS *Polyhedriella* (Synonym for *Eustigmatos*)
GENUS *Vischeria* Vischer Eyespot Green Yellow Ocher Algae
SPECIES *helvetica, punctata, stellata, torta*

FAMILY 2 *Monodopsidaceae* Monad Eyespot Green Yellow Ocher Algae
GENUS *Monodopsis* Monad Eyespot Green Yellow Ocher Algae
SPECIES *subterranea*
GENUS *Nannochloropsis* Nannite Monad Eyespot Green Yellow Ocher Algae
SPECIES *gaditana, granulata, limnetica, oceanica, oculata, salina*

FAMILY 3 *Pleurochloridaceae* Pleural Eyespot Green Yellow Ocher Algae
GENUS *Pleurochloris* Pleural Eyespot Green Yellow Ocher Algae
SPECIES *acidophila, cavernicola, commutata, imitans, inaequalis, lobata, meiringensis, nanella, polychloris, pyrenoidosa, terrestris*
GENUS *Actinellipsoidion* Radial Pleural Eyespot Green Yellow Ocher Algae
SPECIES *sol*
GENUS *Arachnochloris* Spider Pleural Eyespot Green Yellow Ocher Algae
SPECIES *major*
GENUS *Bracchiogonium* Arm Pleural Eyespot Green Yellow Ocher Algae
SPECIES *ophiaster*
GENUS *Chlorapion* Parsley Pleural Eyespot Green Yellow Ocher Algae
SPECIES *grande*
GENUS *Chlorarkys* Ark Pleural Eyespot Green Yellow Ocher Algae
SPECIES *reticulata*
GENUS *Chloridella* Lesser Pleural Eyespot Green Yellow Ocher Algae
SPECIES *cystiformis, ferruginea, glacialis, major, minuta, neglecta, polychloris, simplex*
GENUS *Chlorocloster* Closter Pleural Eyespot Green Yellow Ocher Algae
SPECIES *engadinensis, inaequalis, minimus, minor, pyreniger, raphidioides, simplex, terrestris*
VARIETIES *pyreniger minor*
GENUS *Diachros* Diachrome Pleural Eyespot Green Yellow Ocher Algae
SPECIES *pleiochloris, simplex*
GENUS *Ellipsoidion* Ellipsoid Pleural Eyespot Green Yellow Ocher Algae
SPECIES *acuminatum, anulatum, cryophilum, oocystoides, ovoideum, pachydermum, parrae, parvum, perminimum, pulchrum, regulare, salinum, simplex, solitare, stellatum, stichococcoides*
VARIETIES *perminimum cryophilum*
GENUS *Endochloridion* Endic Pleural Eyespot Green Yellow Ocher Algae
SPECIES *simplex*
GENUS *Goniochloris* Gonad Pleural Eyespot Green Yellow Ocher Algae
SPECIES *contorta, fallax, mutica, parvula, sculpta, smithii, stellata, tripus*
VARIETIES *fallax bifida*
GENUS *Isthmochloron* Isthmus Pleural Eyespot Green Yellow Ocher Algae
SPECIES *gracile, lobulatum*
GENUS *Keriosphaera* Ker Sphere Pleural Eyespot Green Yellow Ocher Algae
SPECIES *gemma*
GENUS *Meringosphaera* Mering Sphere Pleural Eyespot Green Yellow Ocher Algae
SPECIES *aculeata, brevispina, mediterranea, spinosa*
GENUS *Monallantus* Monallant Pleural Eyespot Green Yellow Ocher Algae
SPECIES *brevicylindrus, monochloris, pyreniger, salina, stichococcoides*
GENUS *Monodus* Monad Pleural Eyespot Green Yellow Ocher Algae
SPECIES *acuminata, amicimei, chodatii, coccomyxa, coccomyxoides, cystiformis, fusiformis, guttula, pyreniger, subglobosa, unipapilla*
GENUS *Nephrodiella* Kidney Pleural Eyespot Green Yellow Ocher Algae
SPECIES *acuta, lunaris, minor, phaseolus, semilunaris, tingitana*
GENUS *Pleurogaster* Gastric Pleural Eyespot Green Yellow Ocher Algae
SPECIES *anisodus, lunaris*
GENUS *Polygoniochloris* Polygon Pleural Eyespot Green Yellow Ocher Algae
SPECIES *circularis, polygonia*
GENUS *Prismatella* Prism Pleural Eyespot Green Yellow Ocher Algae
SPECIES *hexagona*

GENUS *Pseudopolyedriopsis* Polyhedron Pleural Eyespot Green Yellow Ocher Algae
SPECIES *skujae*
GENUS *Pseudostaurastrum* Crosshatch Pleural Eyespot Green Yellow Ocher Algae
SPECIES *enorme, gallicum, hastatum, limneticum*
VARIETIES *hastatum elegans, hastatum palatinum, hastatum simplex, hastatum triangulare*
GENUS *Rhomboidella* Rhomboid Pleural Eyespot Green Yellow Ocher Algae
SPECIES *obliqua*
GENUS *Schilleriella* Schiller Pleural Eyespot Green Yellow Ocher Algae
SPECIES anurea, *trieseta*
GENUS *Skiadosphaera* Skiad Sphere Pleural Eyespot Green Yellow Ocher Algae
SPECIES *divergens*
GENUS *Sklerochlamys* Hard Cloak Pleural Eyespot Green Yellow Ocher Algae
SPECIES *fragilis, pachyderma*
GENUS *Tetraedriella* Tetrahedral Pleural Eyespot Green Yellow Ocher Algae
SPECIES *gallica, jovetti, limbata, polychloris, regularis, spinigera, tumidula, verrucosa*
GENUS *Tetraplektron* Tetrad Pleural Eyespot Green Yellow Ocher Algae
SPECIES *deniseae, torsum, tribulus*
VARIETIES *torsum longispinum*
GENUS *Trachydiscus* Rough Disc Pleural Eyespot Green Yellow Ocher Algae
SPECIES *lenticularis*

FAMILY 4 *Pseudocharaciopsidaceae* Characin Eyespot Green Yellow Ocher Algae
GENUS *Pseudocharaciopsis* Characin Eyespot Green Yellow Ocher Algae
SPECIES *minuta, ovalis*
GENUS *Botryochloropsis* Berry Characin Eyespot Green Yellow Ocher Algae
SPECIES *similis*

CLASS 613.4 *Synurophyceae* Unicell Golden Ocher Algae

ORDER 1 *Synurales* Unicell Golden Ocher Algae

FAMILY 1 *Synuraceae* Unicell Golden Ocher Algae
GENUS *Synura* Unicell Golden Ocher Algae
SPECIES *adamsii, australiensis, curtispina, echinulata, favus, granulosa, mamillosa, mollispina, nygaardii, petersenii, sphagnicola, spinosa, uvella, virescens*
VARIETIES *granulosa pusilla, petersenii columnata, spinosa nygaardii, spinosa striata*
GENUS *Aurosphaera* Sphere Unicell Golden Ocher Algae
SPECIES *ovalis*
GENUS *Bicoeca* Bicoecid Unicell Golden Ocher Algae
SPECIES *petiolatum*
GENUS *Chlorodesmos* Fiber Unicell Golden Ocher Algae
SPECIES *hispida*
GENUS *Chromophysomonas* Chromic Bladder Unicell Golden Ocher Algae
SPECIES *involuta, trioralis*
GENUS *Chrysodidymus* Dual Unicell Golden Ocher Algae
SPECIES *synuroideus*
GENUS *Jaoniella* Jaon Unicell Golden Ocher Algae
SPECIES *planctonica*
GENUS *Microglena* Glen Unicell Golden Ocher Algae
SPECIES *arenicola, butcheri, monadina*
GENUS *Monochrysis* Monad Unicell Golden Ocher Algae
SPECIES *lutheri, major*
GENUS *Paraphysomonas* Lesser Bladder Unicell Golden Ocher Algae
SPECIES *butcheri, caelifrica, corbidifera, coronata, cribosa, cylicophora, erinacea, faveolata, formanifera, imperforata, manubriata, pileata, planus, porosa, poteriophora, runciniferopsis, sideriophora, vestita*
VARIETIES *coronata poteriophora, poteriophora manubriata*
GENUS *Phalansterium* Phalanster Unicell Golden Ocher Algae
SPECIES *consociatum, intestinum*
GENUS *Physomonas* Bladder Unicell Golden Ocher Algae
SPECIES *elongata, socialis*
GENUS *Pseudosyncrypta* Cryptic Unicell Golden Ocher Algae
SPECIES *volvox*
GENUS *Sollasites* Sollas Unicell Golden Ocher Algae
SPECIES *falklandensis, thiersteinii*
GENUS *Spiniferomonas* Spine Unicell Golden Ocher Algae
SPECIES *abei, abrupta, bourrellyi, breakneckii, cornuta, coronacircumspina, enigmata, genuiformis, hamata, involuta, nichollsii, pectinata, septispina, triangularis, trioralis*
GENUS *Stichochrysis* Stitch Unicell Golden Ocher Algae

SPECIES *immobilis*
GENUS *Synuropsis* — Lesser Unicell Golden Ocher Algae
SPECIES *danubiensis*
GENUS *Tessellaria* — Tessellate Unicell Golden Ocher Algae
SPECIES *volvocina*

FAMILY 2 *Archaeomonadaceae* — Archaic Unicell Golden Ocher Algae
GENUS *Archaeomonas* — Archaic Unicell Golden Ocher Algae
SPECIES *karinae, ninae, norvegiensis*
GENUS *Litharchaeocystis* — Rock Cyst Archaic Unicell Golden Ocher Algae
SPECIES *talwanii, udintsevii*
GENUS *Pararchaeomonas* — Lesser Archaic Unicell Golden Ocher Algae
SPECIES *challengeri*

FAMILY 3 *Chrysosaccaceae* — Sac Unicell Golden Ocher Algae
GENUS *Chrysosaccus* — Sac Unicell Golden Ocher Algae
SPECIES *incompletus, membraniger*
GENUS *Phaeosphaera* — Dusky Sphere Sac Unicell Golden Ocher Algae
SPECIES *gelatinosa*
GENUS *Saccochrysis* — Lesser Sac Unicell Golden Ocher Algae
SPECIES *colligera, piriformis, rotundata*

FAMILY 4 *Hibberdiaceae* — Hibberd Golden Ocher Algae
GENUS *Hibberdia* — Hibberd Golden Ocher Algae
SPECIES *magna*
GENUS *Chromophyton* — Chromic Hibberd Golden Ocher Algae
SPECIES *rosanoffii*

FAMILY 5 *Mallomonadaceae* — Soft Unicell Golden Ocher Algae
GENUS *Mallomonas* — Soft Unicell Golden Ocher Algae
SPECIES *acaroides, adamas, akrokomos, allantoides, allorgei, alphaphora, alpina, annulata, areolata, asmundae, australica, calceolus, caudata, ceylanica, clavus, connensis, corymbosa, crassisquama, cratis, cylindracea, doignonii, duerrschmidtiae, eoa, epithalattia, favosa, flora, galeiformis, grossa, guttata, hamata, heterospina, hexagonis, inornata, insignis, intermedia, leboimii, lelymene, limnicola, litomesa, liturata, lychenensis, mangofera, marsupialis, matvienkoae, monograptus, morrisonensis, multisetigera, multiunca, nuussuaqensis, ouradion, oviformis, palaestrica, papillosa, parvula, paucispinosa, paxillata, perfossa, perpusilla, phasma, pillula, plumosa, producta, pseudocratis, pugio, pumilio, punctifera, rasilis, retifera, retrorsa, rhombica, sabulosa, sexangularis, soleatus, sphagniphila, splendens, stellata, striata, subsalina, tasmanica, teilingii, tolerans, tonsurata, torquata, variabilis, wujekii*
VARIETIES *acaroides muskokana, acaroides obtusa, australica gracillima, australica subglobosa, corymbosa interrupta, crassisquama papillosa, doignonii robusticostis, guttata simplex, litomesa curta, mangofera foveata, mangofera reticulata, papillosa ellipsoida, papillosa monilifera, parvula nichollsii, pumilio silvicola, splendens arnhemensis, splendens biceps, splendens pusilla, striata serrata, tonsurata tonsurata, tonsurata etortisetifera, torquata simplex*
GENUS *Conradiella* — Conrad Soft Unicell Golden Ocher Algae
SPECIES *calva*
GENUS *Lepidoton* — Thin Soft Unicell Golden Ocher Algae
SPECIES *dubium*
GENUS *Mallomonopsis* — Lesser Soft Unicell Golden Ocher Algae
SPECIES *bangladeshica, elliptica, peronoides*
VARIETIES *elliptica oviformis, elliptica salina, peronoides bangladeshica*
GENUS *Ochromallomonas* — Least Soft Unicell Golden Ocher Algae
SPECIES *pelophila*
GENUS *Pseudomallomonas* — Mock Soft Unicell Golden Ocher Algae
SPECIES *anglica, bernardinensis*

FAMILY 6 *Monosigaceae* — Siga Unicell Golden Ocher Algae
GENUS *Monosiga* — Siga Unicell Golden Ocher Algae
SPECIES *marina, skvortzovii*
GENUS *Desmarella* — (Synonym for *Monosiga*)
GENUS *Sphaeroeca* — Sphere Siga Unicell Golden Ocher Algae
SPECIES *volvox*

FAMILY 7 *Triparmaceae* — Parmal Unicell Golden Ocher Algae
GENUS *Triparma* — Triad Parmal Unicell Golden Ocher Algae
SPECIES *columacea, laevis, retinervis, strigata, verrucosa*
VARIETIES *columacea alata, laevis mexicana, laevis pinnatilobata, laevis ramispina, retinervis crenata*
GENUS *Pentalamina* — Pentad Parmal Unicell Golden Ocher Algae
SPECIES *corona*
GENUS *Tetraparma* — Tetrad Parmal Unicell Golden Ocher Algae
SPECIES *insecta, pelagica*

ORDER 2 *Chromulinales* Chromic Golden Ocher Algae

FAMILY 1 *Chromulinaceae* — Chromic Golden Ocher Algae
GENUS *Chromulina* — Chromic Golden Ocher Algae
SPECIES *aerophila, chionophila, cuneata, ferrea, flavicans, gelatinosa, lunaris, nebulosa, nevadensis, ochracea, ochromonoides, ovalis, pallida, parvula, placentula, pleiades, pyriformis, sporangifera, wislouchiana*
GENUS *Amphichrysis* — Dual Chromic Golden Ocher Algae
SPECIES *compressa*
GENUS *Anthophysa* — Flower Bladder Chromic Golden Ocher Algae
SPECIES *vegetans*
GENUS *Atraktochrysis* — Atract Chromic Golden Ocher Algae
SPECIES *rotans*
GENUS *Chrysobotriella* — Lesser Berry Chromic Golden Ocher Algae
SPECIES *spondylomorum*
GENUS *Chrysobotrys* — Berry Chromic Golden Ocher Algae
SPECIES *feldmannii*
GENUS *Chrysodendron* — Lesser Wood Chromic Golden Ocher Algae
SPECIES *ramosum*
GENUS *Chrysoglena* — Glen Chromic Golden Ocher Algae
SPECIES *verrucosa*
GENUS *Chrysonephele* — Kidney Chromic Golden Ocher Algae
SPECIES *palustris*
GENUS *Chrysoxys* — Sharp Chromic Golden Ocher Algae
SPECIES *major*
GENUS *Cladonema* — Branched Thread Chromic Golden Ocher Algae
SPECIES *laxum*
GENUS *Cyclonexis* — Cyclic Chromic Golden Ocher Algae
SPECIES *annularis, erinus, uraliensis*
GENUS *Dendromonas* — Wood Chromic Golden Ocher Algae
SPECIES *compacta, virgaria*
GENUS *Didymochrysis* — Dual Chromic Golden Ocher Algae
SPECIES *paradoxa*
GENUS *Eusphaerella* — Sphere Chromic Golden Ocher Algae
SPECIES *turfosa*
GENUS *Monas* — Monad Chromic Golden Ocher Algae
SPECIES *flavicans, lens, mica, nephrodes, uva*
GENUS *Mucosphaera* — Mucus Sphere Chromic Golden Ocher Algae
SPECIES *geophila*
GENUS *Mycochrysis* — Mycote Chromic Golden Ocher Algae
SPECIES *oligothiophila*
GENUS *Ochrostylon* — Stylon Chromic Golden Ocher Algae
SPECIES *epiplankton*
GENUS *Oikomonas* — Oiko Chromic Golden Ocher Algae
SPECIES *mutabilis*
GENUS *Pyramidochrysis* — Pyramid Chromic Golden Ocher Algae
SPECIES *splendens*
GENUS *Rhizochromulina* — Rhizome Chromic Golden Ocher Algae
SPECIES *marina*
GENUS *Siderodendron* — Iron Wood Chromic Golden Ocher Algae
SPECIES *manganiferum*
GENUS *Siphomonas* — Siphon Chromic Golden Ocher Algae
SPECIES *fritschii*
GENUS *Sphaleromantis* — Slippery Chromic Golden Ocher Algae
SPECIES *ochracea, subsalus, tetragona*
GENUS *Spumella* — Frothing Chromic Golden Ocher Algae
SPECIES *elongata, vulgaris*
GENUS *Stipitochrysis* — Stippled Chromic Golden Ocher Algae
SPECIES *monorhiza*
GENUS *Uroglena* — Tailed Glen Chromic Golden Ocher Algae
SPECIES *americana, articulata, dendracantha, eustylis, lindii, soniaca, volvox*
GENUS *Uroglenopsis* — Lesser Tailed Glen Chromic Golden Ocher Algae
SPECIES *americana*
GENUS *Wellheimia* — Wellheim Chromic Golden Ocher Algae
SPECIES *pfeifferi*

FAMILY 2 *Chrysococcaceae* — Round Chromic Golden Ocher Algae
GENUS *Chrysococcus* — Round Chromic Golden Ocher Algae

SPECIES *cordiformis, cystophorus, diaphanus, minutus, pascheri, punctiformis, radians, rufescens, tessellatus*
VARIETIES *cystophorus astigmata, rufescens tripora*
GENUS *Calycomonas* Calico Round Chromic Golden Ocher Algae
SPECIES *cylindrica, dilatata, globosa, gracilis, wulffii*

FAMILY 3 *Chrysolepidomonadaceae* Thin Chromic Golden Ocher Algae
GENUS *Chrysolepidomonas* Thin Chromic Golden Ocher Algae
SPECIES *anglica, dendrolepidota, marina*
GENUS *Lepidochrysis* Lesser Thin Chromic Golden Ocher Algae
SPECIES *glomerifera*
GENUS *Polylepidomonas* Poly Thin Chromic Golden Ocher Algae
SPECIES *vacuolata*

FAMILY 4 *Chrysosphaeraceae* Sphere Chromic Golden Ocher Algae
GENUS *Chrysosphaera* Sphere Chromic Golden Ocher Algae
SPECIES *magna, nitens*
GENUS *Chrysosphaerella* Lesser Sphere Chromic Golden Ocher Algae
SPECIES *annulata, brevispina, enigmata, longispina, salina, septispina*
GENUS *Epichrysis* Epiphyte Sphere Chromic Golden Ocher Algae
SPECIES *paludosa*
GENUS *Epicystis* Epiphyte Cyst Sphere Chromic Golden Ocher Algae
SPECIES *peridinearum*

FAMILY 5 *Stichogloeaaceae* Stitch Glow Chromic Golden Ocher Algae
GENUS *Stichogloea* Stitch Glow Chromic Golden Ocher Algae
SPECIES *doederleinii, olivacea*
GENUS *Phaeobotrys* Dusky Berry Stitch Glow Chromic Golden Ocher Algae
SPECIES *solitaria*

ORDER 3 *Chrysamoebidales* Ameba Golden Ocher Algae

FAMILY 1 *Chrysamoebaceae* Ameba Golden Ocher Algae
GENUS *Chrysamoeba* Ameba Golden Ocher Algae
SPECIES *mikrokonta, pyrenoidifera, radians, scherffelii*
GENUS *Amphirhiza* Dual Rhizid Ameba Golden Ocher Algae
SPECIES *epizootica*
GENUS *Brehmiella* Brehm Ameba Golden Ocher Algae
SPECIES *chrysohydra, endoloricata*
GENUS *Chrysarachnion* Spider Ameba Golden Ocher Algae
SPECIES *insidians*
GENUS *Chrysidiastrum* Star Ameba Golden Ocher Algae
SPECIES *catenatum*
GENUS *Chrysostephanosphaera* Crown Sphere Ameba Golden Ocher Algae
SPECIES *globulifera*
GENUS *Leukapsis* Light Parsley Ameba Golden Ocher Algae
SPECIES *vorax*
GENUS *Leukochrysis* Light Ameba Golden Ocher Algae
SPECIES *pascheri*
GENUS *Platychrysella* Flat Ameba Golden Ocher Algae
SPECIES *vorax*
GENUS *Rhizoochromonas* Rhizid Ameba Golden Ocher Algae
SPECIES *endoloricata*

FAMILY 2 *Stylococcaceae* Round Ameba Golden Ocher Algae
GENUS *Stylococcus* Round Ameba Golden Ocher Algae
SPECIES *aureus*
GENUS *Bitrichia* Hairy Round Ameba Golden Ocher Algae
SPECIES *chodatii, longispina, wolhynica*
GENUS *Chrysoamphipyxis* Dual Pyxie Round Ameba Golden Ocher Algae
SPECIES *canadensis*
GENUS *Chrysoamphitrema* Dual Tremor Round Ameba Golden Ocher Algae
SPECIES *brunnea*
GENUS *Chrysocrinus* Lily Round Ameba Golden Ocher Algae
SPECIES *hydra*
GENUS *Chrysopodocystis* Pod Cyst Round Ameba Golden Ocher Algae
SPECIES *socialis*
GENUS *Chrysopyxis* Pyxie Round Ameba Golden Ocher Algae
SPECIES *bipes, globosa, stenostoma*
GENUS *Chrysothecopsis* Theca Round Ameba Golden Ocher Algae

SPECIES *epiphyticus*
GENUS *Chrysothylakion* — Thylacion Round Ameba Golden Ocher Algae
SPECIES *vorax*
GENUS *Derepyxis* — Dere Pyxie Round Ameba Golden Ocher Algae
SPECIES *amphora*
GENUS *Heliapsis* — Sun Parsley Round Ameba Golden Ocher Algae
SPECIES *achromatica, mutabilis*
GENUS *Heliochrysis* — Sun Round Ameba Golden Ocher Algae
SPECIES *eradians*
GENUS *Heterolagynion* — Variable Lagynion Round Ameba Golden Ocher Algae
SPECIES *oedogonii*
GENUS *Hyalocylix* — Hyaline Round Ameba Golden Ocher Algae
SPECIES *stipitata*
GENUS *Kybotion* — Cybot Round Ameba Golden Ocher Algae
SPECIES *eremita*
GENUS *Lagynion* — Lagynion Round Ameba Golden Ocher Algae
SPECIES *ampullaceum, macrotrachelum, scherffelii*
GENUS *Platytheca* — Flat Theca Round Ameba Golden Ocher Algae
SPECIES *micropora*
GENUS *Porostylon* — Pore Round Ameba Golden Ocher Algae
SPECIES *transversale*
GENUS *Rhizaster* — Rhizid Star Round Ameba Golden Ocher Algae
SPECIES *crinoides*
GENUS *Stephanoporos* — Crown Pore Round Ameba Golden Ocher Algae
SPECIES *epiphyticus*
GENUS *Styloceras* — Horn Round Ameba Golden Ocher Algae
SPECIES *longissimus*

ORDER 4 *Ochromonadales* Monad Golden Ocher Algae

FAMILY 1 *Ochromonadaceae* — Monad Golden Ocher Algae
GENUS *Ochromonas* — Monad Golden Ocher Algae
SPECIES *allorgei, aspera, bourrellyi, cosmopolita, crenata, cylindracea, danica, minima, minuscula, minuta, oblonga, ostreaeformis, pyriformis, reducta, resistens, spherocystis, triangulata, tuberculata, variabilis, verrucosa, villosa*
GENUS *Boekelovia* — Boekelov Monad Golden Ocher Algae
SPECIES *gallica, hooglandii*
GENUS *Entodesmis* — Entic Fiber Monad Golden Ocher Algae
SPECIES *litoralis, maritima, scenedesmoides*
GENUS *Mitcheldeania* — Mitcheld Monad Golden Ocher Algae
SPECIES *americana*
GENUS *Monadodendron* — Wood Monad Golden Ocher Algae
SPECIES *compactum*
GENUS *Olisthodiscus* — Olisthon Disc Monad Golden Ocher Algae
SPECIES *luteus*
GENUS *Syncrypta* — Syne Cryptic Monad Golden Ocher Algae
SPECIES *spongiarum*
GENUS *Synochromonas* — Syne Monad Golden Ocher Algae
SPECIES *elaeochrus, janei, pallida*

FAMILY 2 *Dinobryaceae* — Mossy Monad Golden Ocher Algae
GENUS *Dinobryon* — Mossy Monad Golden Ocher Algae
SPECIES *balticum, bavaricum, borgei, brucei, calyciforme, condensatum, crenulatum, cylindricum, dillonii, divergens, emoriens, faculiferum, pediforme, petiolatum, praecambrianum, protuberans, sertularia, sociale, suecicum, unguentariforme, utriculus*
VARIETIES *bavaricum undulatum, bavaricum vanhoeffenii, cylindricum alpinum, cylindricum divergens, cylindricum palustre, divergens schauinslandii, sertularia angulatum, sertularia protruberans, sociale americanum, sociale stipitatum*
GENUS *Angulochrysis* — Angular Mossy Monad Golden Ocher Algae
SPECIES *erratica*
GENUS *Arthrochrysis* — Straight Mossy Monad Golden Ocher Algae
SPECIES *leptopus*
GENUS *Arthropyxis* — Straight Pyxie Mossy Monad Golden Ocher Algae
SPECIES *annulata*
GENUS *Chrysolykos* — Wolf Mossy Monad Golden Ocher Algae
SPECIES *gracilis, planctonicus, skujae*
GENUS *Codonobotrys* — Codon Berry Mossy Monad Golden Ocher Algae
SPECIES *physalis*

GENUS *Codonodendron* Codon Wood Mossy Monad Golden Ocher Algae
SPECIES *ocellatum*
GENUS *Conradocystis* Conrad Cyst Mossy Monad Golden Ocher Algae
SPECIES *dinobryonis*
GENUS *Epipyxis* Epiphyte Pyxie Mossy Monad Golden Ocher Algae
SPECIES *anglica, aureus, condensata, diplostoma, gracilis, kenaiensis, kodiakensis, lauterbornii, polymorpha, proteus, pulchra, ramosa, singoalikensis, sinuosa, tabellariae, tubulosa, turgida, utriculus*
VARIETIES *lauterbornii mucicola, utriculus acuta, utriculus tabellaria*
GENUS *Erythrolobas* Red Lobe Mossy Monad Golden Ocher Algae
SPECIES *fottii*
GENUS *Hyalobryon* Hyaline Mossy Monad Golden Ocher Algae
SPECIES *decipiens, diplostomum, lauterbornei, mucicola, polymorphum, ramosum*
GENUS *Kephyrion* Kephyrion Mossy Monad Golden Ocher Algae
SPECIES *littorale, moniliferum, ovum, petasatum, rubri-claustri, sitta*
VARIETIES *littorale constricta*
GENUS *Lepochromulina* Thin Chromic Mossy Monad Golden Ocher Algae
SPECIES *bursa*
GENUS *Ollicola* Ollie Mossy Monad Golden Ocher Algae
SPECIES *vangoorii*
GENUS *Porochrysis* Pore Mossy Monad Golden Ocher Algae
SPECIES *aspergillus*
GENUS *Poteriochroomonas* Poter Mossy Monad Golden Ocher Algae
SPECIES *malhamensis, nutans, stipitata*
GENUS *Pseudokephyrion* Lesser Kephyrion Mossy Monad Golden Ocher Algae
SPECIES *anglica, ellipsoideum, formosisimum, klarnetii, ovum, pseudospirale, ruttnerii, tortosaccatum, undulatum*
GENUS *Sphaerobryon* Sphere Mossy Monad Golden Ocher Algae
SPECIES *fimbriatum*
GENUS *Stenocodon* Thin Mossy Monad Golden Ocher Algae
SPECIES *epiplankton*
GENUS *Stokesiella* Stokes Mossy Monad Golden Ocher Algae
SPECIES *acuminata, cavanillesiana*
GENUS *Stylochrysalis* Style Mossy Monad Golden Ocher Algae
SPECIES *aurea, parasita*
GENUS *Stylopyxis* Pyxie Style Mossy Monad Golden Ocher Algae
SPECIES *bachmannii*
GENUS *Woronichiniella* Woronichin Mossy Monad Golden Ocher Algae
SPECIES *pentagona*

CLASS 613.5 *Xanthophyceae* Yellow Green Ocher Algae

ORDER 1 *Xanthophytales* Yellow Green Ocher Algae

FAMILY 1 *Xanthophytaceae* Yellow Green Ocher Algae
GENUS *Aeronema* Aero Yellow Green Ocher Algae
SPECIES *polymorphum*
GENUS *Hydrogastrum* Gastric Yellow Green Ocher Algae
SPECIES *utriculare*
GENUS *Pelagocystis* Cyst Yellow Green Ocher Algae
SPECIES *oceanica*
GENUS *Phyllosiphon* Siphon Yellow Green Ocher Algae
SPECIES *arisari*
GENUS *Polyedrium* Polyhedron Yellow Green Ocher Algae
SPECIES *hastatum, tetraedricum, tumidulum*
VARIETIES *hastatum palatinum*
GENUS *Pseudogoniochloris* Gonad Yellow Green Ocher Algae
SPECIES *tripus*
GENUS *Pseudopleurochloris* Pleural Yellow Green Ocher Algae
SPECIES *antarctica*

FAMILY 2 *Heterogloeaceae* Glow Yellow Green Ocher Algae
GENUS *Heterogloea* Glow Yellow Green Ocher Algae
SPECIES *endochloris*
GENUS *Helminthogloea* Helminth Glow Yellow Green Ocher Algae
SPECIES *ramosa*
GENUS *Heterodendron* (Synonym for *Heterogloea*)
GENUS *Malleodendron* Soft Wood Glow Yellow Green Ocher Algae
SPECIES *gloeopus*

GENUS *Pleurochloridella* Pleural Glow Yellow Green Ocher Algae
SPECIES *botryidopsis, vacuolata*
GENUS *Pseudogloea* Lesser Glow Yellow Green Ocher Algae
SPECIES *crucifera, micronucleolata, multifissa, polychloris*

ORDER 2 *Mischococcales* Round Yellow Green Ocher Algae

FAMILY 1 *Mischococcaceae* Round Yellow Green Ocher Algae
GENUS *Mischococcus* Round Yellow Green Ocher Algae
SPECIES *confervicola, sphaerocephalus*
GENUS *Chloropedia* Pedal Round Yellow Green Ocher Algae
SPECIES *plana*
GENUS *Gloeopodium* Glow Pod Round Yellow Green Ocher Algae
SPECIES *elephantipes*
GENUS *Ophiocytium* Ophiocyte Round Yellow Green Ocher Algae
SPECIES *apiculatum, arbuscula, bicuspidatum, capitatum, cochleare, gracilipes, majus, parvulum, tenellum, terrestre*
VARIETIES *capitatum longispinum, cochleare arbuscula, cochleare bicuspidatum, tenellum apiculatum*
GENUS *Trypanochloris* Trypan Round Yellow Green Ocher Algae
SPECIES *clausiliae*

FAMILY 2 *Botrydiopsidaceae* Lesser Berry Round Yellow Green Ocher Algae
GENUS *Botrydiopsis* Lesser Berry Round Yellow Green Ocher Algae
SPECIES *anglica, arhiza, callosa, constricta, eriensis, intercedens, pyrenoidosa*
GENUS *Excentrochloris* Excentric Lesser Berry Round Yellow Green Ocher Algae
SPECIES *gigas*
GENUS *Perone* Perone Lesser Berry Round Yellow Green Ocher Algae
SPECIES *dimorpha*

FAMILY 3 *Botryochloridaceae* Berry Round Yellow Green Ocher Algae
GENUS *Botryochloris* Berry Round Yellow Green Ocher Algae
SPECIES *cumulata, minima, simplex*
GENUS *Chlorellidiopsis* Lesser Chloric Berry Round Yellow Green Ocher Algae
SPECIES *separabilis*
GENUS *Chlorellidium* Chloric Berry Round Yellow Green Ocher Algae
SPECIES *astigmatum, tetrabotrys*
GENUS *Dichotomococcus* Dichotomy Berry Round Yellow Green Ocher Algae
SPECIES *curvatus*
GENUS *Heterodesmus* Fiber Berry Round Yellow Green Ocher Algae
SPECIES *bichloris*
GENUS *Ilsteria* Ilster Berry Round Yellow Green Ocher Algae
SPECIES *tetracoccus*
GENUS *Raphidiella* Needle Berry Round Yellow Green Ocher Algae
SPECIES *fascicularis*
GENUS *Sphaerosorus* Sorus Berry Round Yellow Green Ocher Algae
SPECIES *coelastroides*
GENUS *Tetraktis* Tetrad Berry Round Yellow Green Ocher Algae
SPECIES *aktinastroides*

FAMILY 4 *Centritactaceae* Centric Round Yellow Green Ocher Algae
GENUS *Centritractus* Centric Round Yellow Green Ocher Algae
SPECIES *belenophorus, dubius, ellipsoideus*
VARIETIES *ellipsoideus excentricus*
GENUS *Bumilleriopsis* Bumiller Centric Round Yellow Green Ocher Algae
SPECIES *biverruca, brevis, filiformis, megacystis, peterseniana, pyrenoidosa, terricola*

FAMILY 5 *Characiopsidaceae* Characin Round Yellow Green Ocher Algae
GENUS *Characiopsis* Characin Round Yellow Green Ocher Algae
SPECIES *acuta, aquilonaris, armata, avis, bourrellyi, columnaris, curvata, ettliana, frutosiae, heeringiana, lilloensis, longipes, microcysticola, minor, naegelii, obovoidea, pyriformis, saccata, spinifer, subulata, tuba, turgida*
VARIETIES *aquilonaris chompi, armata claricis, columnaris pusilla, longipes minor, microcysticola phacuscola, spinifer correntinensis, tuba minor*
GENUS *Chlorokoryne* Club Characin Round Yellow Green Ocher Algae
SPECIES *petrovae*
GENUS *Chlorothecium* Theca Characin Round Yellow Green Ocher Algae
SPECIES *africanum, pirittae*
GENUS *Dioxys* Diad Sharp Characin Round Yellow Green Ocher Algae
SPECIES *incus, polyceras*
GENUS *Harpochytrium* Harp Chytrid Characin Round Yellow Green Ocher Algae

SPECIES *hyalothecae*
GENUS *Hemisphaerella* — Hemisphere Characin Round Yellow Green Ocher Algae
SPECIES *operculata*
GENUS *Lutherella* — Luther Characin Round Yellow Green Ocher Algae
SPECIES *adhaerens*
GENUS *Peroniella* — Peron Characin Round Yellow Green Ocher Algae
SPECIES *hyalothecae*

FAMILY 6 *Gloeobotrydaceae* — Glow Berry Round Yellow Green Ocher Algae
GENUS *Gloeobotrys* — Glow Berry Round Yellow Green Ocher Algae
SPECIES *arborum, chlorinus, ellipsoideus, gelatinosus, limneticus, ovalis, phaseolus, pyriformis, terrestris*
VARIETIES *chlorinus major*
GENUS *Asterogloea* — Star Glow Berry Round Yellow Green Ocher Algae
SPECIES *gelatinosa*
GENUS *Chlorobotrys* — Lesser Glow Berry Round Yellow Green Ocher Algae
SPECIES *gloeothece, regularis, simplex, terrestris*
GENUS *Chlorosaccus* — Sac Glow Berry Round Yellow Green Ocher Algae
SPECIES *fluidus*
GENUS *Gloeoskene* — Skene Glow Berry Round Yellow Green Ocher Algae
SPECIES *turfosa*
GENUS *Gloeosphaeridium* — Least Glow Berry Round Yellow Green Ocher Algae
SPECIES *firmum*

ORDER 3 *Rhizochloridales* Rhizid Yellow Green Ocher Algae

FAMILY 1 *Rhizochloridaceae* — Rhizid Yellow Green Ocher Algae
GENUS *Rhizochloris* — Rhizid Yellow Green Ocher Algae
SPECIES *arachnoides, lilloensis, mirabilis*
GENUS *Aldavemyxa* — Aldave Rhizid Yellow Green Ocher Algae
SPECIES *coeruleo-viridis*
GENUS *Chlamydomyxa* — Cloak Rhizid Yellow Green Ocher Algae
SPECIES *labrynthuloides, montana*
GENUS *Heterocalycina* — Calycine Rhizid Yellow Green Ocher Algae
SPECIES *levantinoides*

FAMILY 2 *Chloramoebaceae* — Ameba Yellow Green Ocher Algae
GENUS *Chloramoeba* — Ameba Yellow Green Ocher Algae
SPECIES *heteromorpha*
GENUS *Ankylonoton* — Ankylon Ameba Yellow Green Ocher Algae
SPECIES *pyrenigerum*
GENUS *Anotropis* — Anotropic Ameba Yellow Green Ocher Algae
SPECIES *subsalina*
GENUS *Beckingia* — Becking Ameba Yellow Green Ocher Algae
SPECIES *salina*
GENUS *Bothrochloris* — Bother Ameba Yellow Green Ocher Algae
SPECIES *longeciliata*
GENUS *Chlorokardion* — Heart Ameba Yellow Green Ocher Algae
SPECIES *pleurochloron, subsalsum*
GENUS *Chloromeson* — Meson Ameba Yellow Green Ocher Algae
SPECIES *agile, luteoviride, parvum*
GENUS *Heterochloris* — Variable Ameba Yellow Green Ocher Algae
SPECIES *mutabilis*
GENUS *Nephrochloris* — (Synonym for *Chloramoeba*)
GENUS *Phacomonas* — Dusky Ameba Yellow Green Ocher Algae
SPECIES *pelagica*
GENUS *Polykyrtos* — Polycyrtose Ameba Yellow Green Ocher Algae
SPECIES *vitreus*
GENUS *Xanthophycomonas* — Lesser Dusky Ameba Yellow Green Ocher Algae
SPECIES *thalassoides*

FAMILY 3 *Rhizogranulochloridaceae* — Granular Rhizid Yellow Green Ocher Algae
GENUS *Garciamyxa* — Garcia Granular Rhizid Yellow Green Ocher Algae
SPECIES *rhizaspiformis*
GENUS *Herreramyxa* — Herrera Granular Rhizid Yellow Green Ocher Algae
SPECIES *stagnalis*

FAMILY 4 *Rhizounochloridaceae* — Uniary Rhizid Yellow Green Ocher Algae
GENUS *Lopezmyxa* — Lopez Uniary Rhizid Yellow Green Ocher Algae
SPECIES *viridis-flavescens*
GENUS *Requejomyxa* — Requejo Uniary Rhizid Yellow Green Ocher Algae

SPECIES *armata*

FAMILY 5 *Stipitococcaceae* Stipe Rhizid Yellow Green Ocher Algae
GENUS *Stipitococcus* Stipe Rhizid Yellow Green Ocher Algae
SPECIES *urceolatus*
GENUS *Rhizolekane* Lekane Stipe Rhizid Yellow Green Ocher Algae
SPECIES *campanuliformis, sessilis*
GENUS *Stipitochloris* Lesser Stipe Rhizid Yellow Green Ocher Algae
SPECIES *vas*
GENUS *Stipitoporos* Pore Stipe Rhizid Yellow Green Ocher Algae
SPECIES *polychloris*

ORDER 5 *Tribonematales* Thread Yellow Green Ocher Algae

FAMILY 1 *Tribonemataceae* Thread Yellow Green Ocher Algae
GENUS *Tribonema* Thread Yellow Green Ocher Algae
SPECIES *aequale, affine, angustissimum, crassum, elegans, fonticola, minus, missouriense, ulotrichoides, utriculosa, viride, vulgare*
VARIETIES *affine abbreviata, vulgare pallida, vulgare tenue*
GENUS *Bumilleria* Bumiller Thread Yellow Green Ocher Algae
SPECIES *fernandesii, klebsiana, sicula*
GENUS *Neonema* Modern Thread Yellow Green Ocher Algae
SPECIES *pumilum*

FAMILY 2 *Heteropediaceae* Pedal Thread Yellow Green Ocher Algae
GENUS *Heteropedia* Pedal Thread Yellow Green Ocher Algae
SPECIES *polychloris, simplex*
GENUS *Capitulariella* Cap Pedal Thread Yellow Green Ocher Algae
SPECIES *radians*
GENUS *Heterococcus* Round Pedal Thread Yellow Green Ocher Algae
SPECIES *africanus, akinetus, anguinus, botrys, brevicellularis, caespitosus, canadensis, capitatus, clavatus, conicus, corniculatus, crassulus, curvatus, curvirostris, dissociatus, endolithicus, erectus, filiformis, fontanus, fuornensis, furcatus, geniculatus, granulatus, implexus, leptosiroides, longicellularis, mainxii, marietanii, mastigophorus, maximus, moniliformis, nepalensis, papillosus, plectenchymaticus, pleurococcoides, presolanensis, protonematoides, quadratus, ramosissimus, stellatus, stigeoclonioides, subterrestris, tectiformis, teleutosporoides, thermalis, tiroliensis, undulatus, unguis, vesiculosus, virginis, viridis, zonatus*

FAMILY 3 *Xanthonemataceae* Lesser Thread Yellow Green Ocher Algae
GENUS *Xanthonema* Lesser Thread Yellow Green Ocher Algae
SPECIES *antarcticum, bristolianum, debile, exile, hormidioides, monochloron, montanum, quadratum, sessile, solidum, stichococcoides, tribonematoides, ulotrichoides*
GENUS *Heterotrichella* Hairy Lesser Thread Yellow Golden Ocher Algae
SPECIES *gracilis*

ORDER 5 *Vaucheriales* Vaucher Yellow Green Ocher Algae

FAMILY 1 *Vaucheriaceae* Vaucher Yellow Green Ocher Algae
GENUS *Vaucheria* Vaucher Yellow Green Ocher Algae
SPECIES *acrandra, adela, alaskana, arcassonensis, arrhyncha, aversa, bermudensis, bicornigera, borealis, bursata, caloundrensis, canalicularis, compacta, coronata, cruciata, debaryana, dichotoma, dillwynii, disperma, erythrospora, fontinalis, frigida, gardneri, geminata, glomerata, gyrogyna, hercyniana, incurva, intermedia, javanica, karachiensis, lii, litorea, longata, longicaulis, mayyanadensis, medusa, micranthera, minuta, nanandra, nasuta, pachyderma, patagonica, piloboloides, prescottii, prolifera, prona, pronosperma, pseudogeminata, racemosa, schleicheri, sescuplicaria, sessilis, simplex, submarina, subsimplex, synandra, taylorii, terrestris, undulata, velutina, verticillata, vipera, walzii, woroniniana*
VARIETIES *compacta dulcis, dichotoma marina, longicaulis macounii, longicaulis orientalis, mayyanadensis sundarbanensis, pachyderma cerebrina, subsimplex synoica, undulata ornata, velutina separata*
GENUS *Pseudodichotomosiphon* Dichotomy Siphon Vaucher Yellow Green Ocher Algae
SPECIES *constricta*
VARIETIES *constricta minor*
GENUS *Wolea* Wole Vaucher Yellow Green Ocher Algae
SPECIES *lemnae*

FAMILY 2 *Botrydiaceae* Berry Vaucher Yellow Green Ocher Algae
GENUS *Botrydium* Berry Vaucher Yellow Green Ocher Algae
SPECIES *argillaceum, becherianum, corniforme, cystosum, divisum, granulatum, milleri, pachydermum, stoloniferum, tuberosum, wallrothii*
VARIETIES *granulatum kolkwitzianum*
GENUS *Asterosiphon* Star Siphon Berry Vaucher Yellow Green Ocher Algae
SPECIES *dichotomus*

PHYLUM 614 *Cyanidiophyta* Blue Green Algae

Latin: *Cyanea* cyan, blue green
Latin: *Phyta* phyton, plant
18 species [Guiry, Dec. 2004]

CLASSES		Species
614.1 *Cyanidiophyceae*	Blue Green Algae	14 [G]
614.2 *Glaucophyceae*	Cyan Gray Algae	4 [G]

Phylum *Cyanidophyta* Blue Green Algae species are the direct descendents of Phylum *Cyanobacteria*. They are the direct ancestors of Phylum *Bryophyta* Moss and Phylum *Pteridophyta* Fern. Phylum *Glaucophyta*, created by Scientist M. Guiry on 14 December 2004, has been lowered to the status of Class by the author of this book and placed in the *Cyanidophyta* Phylum. They are the genetic link between bacteria and plants.

Photosynthetic Bacteria → Blue Green Algae → Green Algae → Moss, Fern → Seed Plants

CLASS 614.1 *Cyanidiophyceae* Blue Green Algae

ORDER 1 *Cyanidiales* Blue Green Algae

FAMILY 1 *Cyanidiaceae* — Blue Green Algae
GENUS *Cyanidium* — Blue Green Algae
SPECIES *caldarium, daedalum, maximum, partitum, rumpens, sulphurarium*
VARIETIES *caldarium rumpens*
GENUS *Cyanidioschyzon* — Schyzon Blue Green Algae
SPECIES *merolae*
GENUS *Galdieria* — Galdier Blue Green Algae
SPECIES *daedala, maxima, partita, sulphuraria*
GENUS *Rhodococcus* — Reddish Round Blue Green Algae
SPECIES *caldariorum*
GENUS *Rhodospora* — Reddish Spore Blue Green Algae
SPECIES *sordida*

CLASS 614.2 *Glaucophyceae* Cyan Algae

ORDER 1 *Glaucophytales* Cyan Algae

FAMILY 1 *Glaucophytaceae* — Cyan Algae
GENUS *Cyanophora* — Cyan Algae
SPECIES *biloba, paradoxa*
GENUS *Gloeochaete* — Glow Bristle Cyan Algae
SPECIES *wittrockiana*

FAMILY 2 *Glaucocystaceae* — Cyst Cyan Algae
GENUS *Glaucocystis*
SPECIES *nostochinearum*

PHYLUM 615 *Ochrophyta* Ocher Algae

Latin: *Ochra* ochre, i.e. earth mineral, tan, orange brown
Latin: *Phyta* phyton, plant
103 species [Guiry, Dec. 2004]

CLASSES		Species
615.1 *Bodonophyceae*	Bodo Ocher Algae	12 [G]
615.2 *Bolidophyceae*	Bolid Ocher Algae	2 [G]
615.3 *Craspedophyceae*	Fringed Ocher Algae	25 [G]
615.4 *Fragilariophyceae*	Fragile Ocher Algae	37 [G]
615.5 *Pelagophyceae*	Pelagic Ocher Algae	3 [G]
615.6 *Pinguiophyceae*	Greasy Ocher Algae	5 [G]
615.7 *Phaeophyceae*	Dusky Brown Ocher Algae	2,049
615.8 *Raphidophyceae*	Needle Ocher Algae	19 [G]

Phylum *Ochrophyta* Ocher Algae includes all algae that have an ocher color other than golden ocher.

Class *Phaeophyta* Dusky Brown Ocher Algae, phaeophyte, species include brown seaweeds. Giant Kelp can reach sizes up to 30 feet in length. Most species live along the rocky coastlines in temperate regions. They form great seaweed beds in the inter-tidal zone between the surf and the beach. They contain the pigments chlorophyll a, chlorophyll c, and fucoxanthin. The rapidly growing seaweeds of this Phylum are used for food, drugs, and fuel. Species of the well known *Sargassum* Genus are found in the Sargasso Sea. Sargasso seaweed form huge floating colonies which serve as food and shelter for colonies of marine worms and protozoans. The rock seaweed *Fucus* is common on temperate rocky coastlines. Phaeophytes reproduce in alternating haploid sporophyte and diploid gametophyte generations.

Other Classifications
Phylum *Chrysophyta* recognized by Scientists Margulis and Schwartz in 1998, was merged into Phylum *Ochrophyta* by scientist M. Guiry custodian of AlgaeBase www.algaebase.org on 3 Dec. 2004. Phylum *Chrysophyta* Golden Ocher Algae has been restored to Phylum status.

CLASS 615.1 *Bodonophyceae* Bodo Ocher Algae

ORDER 1 *Bodonales* Bodo Ocher Algae

FAMILY 1 *Bodonaceae*	Bodo Ocher Algae
GENUS *Bodo*	Bodo Ocher Algae
SPECIES *curvifilus, edax, saltans*	
GENUS *Heteromastix*	Mastigote Bodo Ocher Algae
SPECIES *angulata, minuta, rotunda*	
GENUS *Kamera*	Kamer Bodo Ocher Algae
SPECIES *lens*	
GENUS *Rhynchobodo*	Beak Bodo Ocher Algae
SPECIES *taeniata*	
GENUS *Rhynchomonas*	Monad Beak Bodo Ocher Algae
SPECIES *nasuta*	
GENUS *Rhynchophus*	Phuse Beak Bodo Ocher Algae
SPECIES *amitus*	
FAMILY 2 *Cryptobiaceae*	Cryptic Bodo Ocher Algae
GENUS *Cryptobia*	Cryptic Bodo Ocher Algae
SPECIES *libera*	
GENUS *Crustomastix*	Crust Cryptic Bodo Ocher Algae
SPECIES *didyma, stigmatica*	

CLASS 615.2 *Bolidophyceae* Bolid Ocher Algae

ORDER 1 *Bolidomonadales* Bolid Ocher Algae

FAMILY 1 *Bolidomonadaceae*	Monad Bolid Ocher Algae
GENUS *Bolidomonas*	Monad Bolid Ocher Algae
SPECIES *mediterranea, pacifica*	

CLASS 615.3 *Craspedophyceae* Fringed Ocher Algae

ORDER 1 *Craspedomonadales* Monad Fringed Ocher Algae

FAMILY 1 *Acanthoecaceae* — Spiny Monad Fringed Ocher Algae
GENUS *Crinolina* — Lily Spiny Monad Fringed Ocher Algae
SPECIES *isefiordensis*
GENUS *Crucispina* — Cross Spiny Monad Fringed Ocher Algae
SPECIES *cruciformis*
GENUS *Diaphanoeca* — Diaphanous Spiny Monad Fringed Ocher Algae
SPECIES *grandis, pedicellata*
GENUS *Parvicorbicula* — Corbic Spiny Monad Fringed Ocher Algae
SPECIES *aculeata, manubriata, pedunculata, quadricostata, socialis, superpositus*
GENUS *Polyfibula* — Fibula Siga Monad Fringed Ocher Algae
SPECIES *sphyrelata*
GENUS *Stephanoeca* — Crown Spiny Monad Fringed Ocher Algae
SPECIES *apheles, campanula, diplocostata*

FAMILY 2 *Codosigaceae* — Siga Monad Fringed Ocher Algae
GENUS *Codosiga* — Codon Siga Monad Fringed Ocher Algae
SPECIES *botrytis*
GENUS *Kentrosiga* — Kent Siga Monad Fringed Ocher Algae
SPECIES *echina*
GENUS *Pleurasiga* — Pleural Siga Monad Fringed Ocher Algae
SPECIES *minima, reynoldsii, tricaudata*

FAMILY 3 *Salpingoecaceae* — Salping Monad Fringed Ocher Algae
GENUS *Salpingoeca* — Salping Monad Fringed Ocher Algae
SPECIES *ampullacea*
GENUS *Bicosta* — Two Rib Salping Monad Fringed Ocher Algae
SPECIES *antennigera, minor, spinifera*
GENUS *Calliacantha* — Callisto Spiny Salping Monad Fringed Ocher Algae
SPECIES *longicaudata, natans, simplex*
GENUS *Conion* — Cone Salping Monad Fringed Ocher Algae
SPECIES *groenlandicum*
GENUS *Salpingella* — Lesser Salping Monad Fringed Ocher Algae
SPECIES *laminata, secata*

CLASS 615.4 *Fragilariophyceae* Fragile Ocher Algae

ORDER 1 *Licmophorales* Licmo Fragile Ocher Algae

FAMILY 1 *Licmophoraceae* — Licmo Fragile Ocher Algae
GENUS *Licmophora* — Licmo Fragile Ocher Algae
SPECIES *anglica, argenescens, communis, dalmatica, debilis, ehrenbergii, flabellata, gracilis, hyalina, juergensii, lyngbyei, minuta, nubecula, oedipus, ovata, ovulum, paradoxa, splendida, tenuis*
VARIETIES *gracilis aegyptiaca, gracilis anglica, gracilis elongata*
GENUS *Licmosphaenia* — Sphene Licmo Fragile Ocher Algae
SPECIES *clevei*
GENUS *Ovoidites* — Oval Licmo Fragile Ocher Algae
SPECIES *ellipsoideus*

ORDER 2 *Tabellariales* Tabular Fragile OcherAlgae

FAMILY 1 *Tabellariaceae* — Tabular Fragile Ocher Algae
GENUS *Tabellaria* — Tabular Fragile Ocher Algae
SPECIES *fenestrata, flocculosa, grandis, kuetzingiana, quadriseptata, trinodis, ventricosa*
GENUS *Oxyneis* — Sharp Neid Tabular Fragile Ocher Algae
SPECIES *binalis*
VARIETIES *binalis elliptica*
GENUS *Tetracyclus* — Tetrad Cyclic Tabular Fragile Ocher Algae
SPECIES *ellipticus, emarginatus, excentricus, glans, lacustris, pseudocastellum, rupestris, subclypeus, subdivisium*
VARIETIES *lacustris elongata, lacustris rhombica*

CLASS 615.5 *Pelagophyceae* Pelagic Ocher Algae

ORDER 1 *Pelagomonadales* Pelagic Ocher Algae

FAMILY 1 *Pelagomonadaceae* — Monad Pelagic Ocher Algae
GENUS *Pelagomonas* — Monad Pelagic Ocher Algae
SPECIES *calceolata*
GENUS *Aureococcus* — Golden Round Monad Pelagic Ocher Algae
SPECIES *anophagefferens*
GENUS *Pelagococcus* — Round Monad Pelagic Ocher Algae
SPECIES *subviridis*

CLASS 615.6 *Phaeophyceae* Dusky Brown Ocher Algae

ORDER 1 *Phaeophytales* Dusky Brown Ocher Algae

FAMILY 1 *Phaeophytaceae* — Dusky Brown Ocher Algae
GENUS *Asterocladon* — Clad Star Dusky Brown Ocher Algae
SPECIES *lobatum*
GENUS *Australofilum* — Australia Dusky Brown Ocher Algae
SPECIES *incommodum*
GENUS *Desmithamnium* — Fiber Thamnid Dusky Brown Ocher Algae
SPECIES *tergestinum*
GENUS *Melastictis* — Dark Stick Dusky Brown Ocher Algae
SPECIES *desmarestiae*
GENUS *Phaeosiphonia* — Siphon Dusky Brown Ocher Algae
SPECIES *cryophila*
GENUS *Phasgonon* — Phasgon Dusky Brown Ocher Algae
SPECIES *bulbosum*
GENUS *Phaxantha* — Yellowish Dusky Brown Ocher Algae
SPECIES *lichenoides*
GENUS *Pilinia* — Pilin Dusky Brown Ocher Algae
SPECIES *endophytica, lunatiae, minor, morsei, novae-zelandiae, reinschii, rimosa*
VARIETIES *lunatiae simplex*
GENUS *Stegastrum* — Stegid Star Dusky Brown Ocher Algae
SPECIES *porphyrae*
GENUS *Syringoderma* — Syringe Dermal Dusky Brown Ocher Algae
SPECIES *abyssicola, australe, floridana, japonicum, phinneyi*

FAMILY 2 *Ascoseiraceae* — Sac Dusky Brown Ocher Algae
GENUS *Ascoseira* — Sac Dusky Brown Ocher Algae
SPECIES *mirabilis*
GENUS *Ascoseirophila* — Lesser Sac Dusky Brown Ocher Algae
SPECIES *violodora*

FAMILY 3 *Choristocarpaceae* — Choris Dusky Brown Ocher Algae
GENUS *Choristocarpus* — Choris Dusky Brown Ocher Algae
SPECIES *tenellus*
GENUS *Discosporangium* — Disc Choris Dusky Brown Ocher Algae
SPECIES *mesarthrocarpum*
GENUS *Onslowia* — Onslow Choris Dusky Brown Ocher Algae
SPECIES *bahamensis, endophytica*
GENUS *Verosphacela* — Vero Choris Dusky Brown Ocher Algae
SPECIES *ebrachia*

FAMILY 4 *Cutleriaceae* — Cutler Dusky Brown Ocher Algae
GENUS *Cutleria* — Cutler Dusky Brown Ocher Algae
SPECIES *adspersa, chilosa, compressa, cyclindrica, cylindrica, hancockii, irregularis, kraftii, mollis, multifida*
VARIETIES *multifida pacifica*
GENUS *Microzonia* — Zoned Cutler Dusky Brown Ocher Algae
SPECIES *velutina*
GENUS *Stifftia* — Stifft Cutler Dusky Brown Ocher Algae
SPECIES *nardi, squamaria*
GENUS *Zanardinia* — Zanardini Cutler Dusky Brown Ocher Algae
SPECIES *typus*

FAMILY 5 *Desmarestiaceae* — Desmarest Dusky Brown Ocher Algae
GENUS *Desmarestia* — Desmarest Dusky Brown Ocher Algae

SPECIES *aculeata, anceps, antarctica, chordalis, confervoides, distans, dresnayi, farcta, firma, foliacea, herbacea, inanis, intermedia, kurilensis, latifrons, latissima, ligulata, media, menziesii, muelleri, munda, pinnatinervia, pteroides, rossii, sivertsenii, tabacoides, tortuosa, tropica, viridis*

VARIETIES *aculeata attenuata, aculeata complanata, aculeata media, aculeata subnervis, distans tenuis, intermedia fuscescens, intermedia teretifolia, ligulata firma, media tenuis, viridis major*

GENUS *Arthrocladia* Exoskin Desmarest Dusky Brown Ocher Algae

SPECIES *septentrionalis, villosa*

VARIETIES *villosa australis*

GENUS *Himantothallus* Himan Thallus Desmarest Dusky Brown Ocher Algae

SPECIES *grandifolius*

GENUS *Iridea* Iridescent Desmarest Dusky Brown Ocher Algae

SPECIES *clavellosa, fissa*

GENUS *Krybolopteris* Krybol Wing Desmarest Dusky Brown Ocher Algae

SPECIES *oltmannsi*

GENUS *Phaeurus* Tailed Desmarest Dusky Brown Ocher Algae

SPECIES *antarcticus*

GENUS *Spinularius* Spiny Desmarest Dusky Brown Ocher Algae

SPECIES *latifrons*

FAMILY 6 *Dictyotaceae* Netted Dusky Brown Ocher Algae

GENUS *Dictyota* Netted Dusky Brown Ocher Algae

SPECIES *acuminata, acutiloba, adhaerens, adnata, alternans, alternifida, angusta, bartayresiana, beccariana, bidentata, bifurca, binghamiae, canaliculata, caribaea, cervicornis, ceylanica, ciliolata, concrescens, crenulata, crinita, crispata, crux, cuspidata, decumbens, dichotoma, diemensis, dilatata, dilitata, divaricata, dumosa, fasciola, fastigiata, fenestrata, flabellata, friabilis, furcellata, grossedentata, guajirae, guineensis, gunniana, hamifera, hauckiana, humifusa, implexa, intermedia, jamaicensis, johnstonii, kohlmeyeri, lata, linearis, littoralis, liturata, magneana, major, marginata, marginispora, masonii, maxima, mediterranea, menstrualis, mertensii, moniliformis, naevosa, ocellata, okamurae, papenfussii, patens, pfaffii, pinnata, pinnatifida, prolifera, pulchella, rigida, robusta, rotundata, sandvicensis, serrulata, spathulata, spinulosa, spiralis, squamaria, squamata, stolonifera, submaritima, suhrii, tenera, tournefortiana, tournefortii, virellus, volubilis, zonaria*

VARIETIES *bartayresiana denticulata, cervicornis pseudobartayresii, cervicornis pseudohamata, ceylanica anastomosans, ciliolata bermudensis, crenulata canariensis, crux parva, dichotoma elongata, dichotoma intricata, dichotoma jedanensis, fasciola repens, linearis intricata, linearis minor, mediterranea crassus*

GENUS *Chlanidophora* Chlanid Netted Dusky Brown Ocher Algae

SPECIES *madagascariensis, microphylla*

GENUS *Dictyopteris* Wing Netted Dusky Brown Ocher Algae

SPECIES *acrostichoides, australis, camiguinensis, cokeri, crassinervia, delicatula, diaphana, dichotoma, divaricata, fucoides, gracilis, hoytii, iyengarii, jamaicensis, johnstonei, jolyana, justii, latiuscula, ligulata, longifolia, macrocarpa, muelleri, nigricans, papenfussii, plagiogramma, polypodioides, prolifera, repens, serrata, tripolitana, undulata, woodwardia*

VARIETIES *australis karachiensis*

GENUS *Dictyotopsis* Lesser Netted Dusky Brown Ocher Algae

SPECIES *propagulifera*

GENUS *Dilophus* Two Crest Netted Dusky Brown Ocher Algae

SPECIES *angustus, gunnianus, marginatus, okamurae, pinnatus, radicans, tener*

GENUS *Distromium* Distrom Netted Dusky Brown Ocher Algae

SPECIES *decumbens, didymothrix, flabellatum, multifidum, skottsbergii*

GENUS *Exallosorus* Exall Sorus Netted Dusky Brown Ocher Algae

SPECIES *harveyanus, olsenii*

GENUS *Glossophora* Gloss Netted Dusky Brown Ocher Algae

SPECIES *coriacea, galapagensis, kunthii, nigricans*

GENUS *Glossophorella* Lesser Gloss Netted Dusky Brown Ocher Algae

SPECIES *coriacea, dhofarensis*

GENUS *Gymnosorus* Nude Sorus Netted Dusky Brown Ocher Algae

SPECIES *collaris*

GENUS *Homoeostrichus* Strick Netted Dusky Brown Ocher Algae

SPECIES *canaliculatus, flabellatus, multifidus, olsenii, sinclairii*

GENUS *Lobophora* Lobe Netted Dusky Brown Ocher Algae

SPECIES *dichotoma, indica, minima, papenfussii, variegata*

VARIETIES *variegata indica, variegata minima*

GENUS *Lobospira* Lobe Spire Netted Dusky Brown Ocher Algae

SPECIES *bicuspidata*

GENUS *Newhousia* Newhouse Netted Dusky Brown Ocher Algae

SPECIES *imbricata*

GENUS *Orthosorus* Straight Sorus Netted Dusky Brown Ocher Algae

SPECIES *atomarius*

GENUS *Pachydictyon* Thick Netted Dusky Brown Ocher Algae

SPECIES *coriaceum, paniculatum, polycladum*

GENUS *Padina* Padin Netted Dusky Brown Ocher Algae
SPECIES *antillarum, arborescens, australis, boergesenii, boryana, caulescens, concrescens, crispata, distromatica, dubia, durvillaei, elegans, fraseri, glabra, gymnospora, haitiensis, jonesii, melemele, minor, moffittiana, pavonica, perindusiata, phasiana, plumbea, profunda, ryukyuensis, sanctae-crucis, somalensis, stipitata, thivyi, zonata*
VARIETIES *pavonica tenuis*
GENUS *Padinopsis* Lesser Padin Netted Dusky Brown Ocher Algae
SPECIES *adriatica*
GENUS *Scoresbyella* Scoresby Netted Dusky Brown Ocher Algae
SPECIES *profunda*
GENUS *Spatoglossum* Spat Tongue Netted Dusky Brown Ocher Algae
SPECIES *asperum, australasicum, chapmanii, chaudhrianum, crassum, dichotomum, ecuadoreanum, flabelliforme, latum, macrodontum, pacificum, schmittii, schroederi, solierii, variabile, veleroae, vietnamense*
GENUS *Stoechospermum* Stoech Netted Dusky Brown Ocher Algae
SPECIES *marginatum*
GENUS *Stypopodium* Stipe Pod Netted Dusky Brown Ocher Algae
SPECIES *australasicum, flabelliforme, laciniatum, multipartitum, schimperi, shameelii, tubruqense, zonale*
GENUS *Taonia* Tao Netted Dusky Brown Ocher Algae
SPECIES *australasica, lacheana, lennebackeriae, pseudociliata*
GENUS *Zonaria* Zone Netted Dusky Brown Ocher Algae
SPECIES *atomaria, aureomarginata, ciliata, crenata, diesingiana, farlowii, flabellata, harveyana, inarticulata, tenuis, spiralis, stipitata, subarticulata, tournefortii, tournefortii, turneriana, turneriana angusta, zonalis, zonata*
VARIETIES *atomaria ciliata*

FAMILY 7 *Scytothamnaceae* Scyte Dusky Brown Ocher Algae
GENUS *Scytothamnus* Scyte Dusky Brown Ocher Algae
SPECIES *australis, fasciculatus*
GENUS *Asteronema* Star Thread Scyte Dusky Brown Ocher Algae
SPECIES *australe, breviarticulatum, ferruginea, rhodochortonoides*
GENUS *Splachnidium* Splachnid Scyte Dusky Brown Ocher Algae
SPECIES *rugosum*
VARIETIES *rugosum minus*
GENUS *Stereocladon* Stereo Clad Scyte Dusky Brown Ocher Algae
SPECIES *rugulosus*

FAMILY 8 *Sporochnaceae* Pear Dusky Brown Ocher Algae
GENUS *Sporochnus* Pear Dusky Brown Ocher Algae
SPECIES *anomalus, apodus, bolleanus, comosus, dotyi, elsieae, moorei, pedunculatus, radiciformis, rostratus, stylosus*
GENUS *Austronereia* Australia Nere Pear Dusky Brown Ocher Algae
SPECIES *australis*
GENUS *Bellotia* Bellot Pear Dusky Brown Ocher Algae
SPECIES *eriophorum*
GENUS *Carpomitra* Carpel Miter Pear Dusky Brown Ocher Algae
SPECIES *costata, filiformis, inermis, longicarpa, madagascariensis, scoparia*
VARIETIES *costata dichotoma, costata mediterranea*
GENUS *Encyothalia* Ency Thal Pear Dusky Brown Ocher Algae
SPECIES *cliftonii*
GENUS *Nereia* Nere Pear Dusky Brown Ocher Algae
SPECIES *filiformis, intricata, lophocladia, tropica*
GENUS *Perisporochnus* Perispore Pear Dusky Brown Ocher Algae
SPECIES *regalis*
GENUS *Perithalia* Perid Thal Pear Dusky Brown Ocher Algae
SPECIES *capillaris, caudata*
GENUS *Sporochnema* Thread Pear Dusky Brown Ocher Algae
SPECIES *tomentosum*
GENUS *Tomaculopsis* Tomacule Pear Dusky Brown Ocher Algae
SPECIES *herbertiana*

FAMILY 9 *Tilopteridaceae* Tile Wing Dusky Brown Ocher Algae
GENUS *Tilopteris* Tile Wing Dusky Brown Ocher Algae
SPECIES *mertensii*
GENUS *Haplospora* Single Spore Tile Wing Dusky Brown Ocher Algae
SPECIES *globosa*
GENUS *Phaeosiphoniella* Siphon Tile Wing Dusky Brown Ocher Algae
SPECIES *cryophila*

ORDER 2 *Ectocarpales* Ectocarp Dusky Brown Ocher Algae

FAMILY 1 *Ectocarpaceae* Ectocarp Dusky Brown Ocher Algae
GENUS *Ectocarpus* Ectocarp Dusky Brown Ocher Algae
SPECIES *acanthophorus, acuto-ramulis, acutus, adriaticus, affinis, balakrishnanii, barbadensis, bombycinus, borealis, bracchiolus, brevicellularis, bryantii, caespitulus, capensis, caspicus, chantransioides, chapmanii, cladosiphoniae, commixtus, confervoides, confusiophyllus, congregatus, constanciae, corticulatus, cryptophilus, cymosus, cystophylloides, dellowianus, denudatus, dictyoptericola, dimorphus, distortus, divergens, ellipticus, erectus, exiguus, exilis, falklandicus, fasciculatus, fenestroides, flagelliferus, flagelliformis, fructosus, fulvescens, fungiformis, fusiformis, gonodioides, hamulosus, hancockii, heterocarpus, hornericola, humilis, isopodicola, kellneri, kjellmanioides, laminariae, laurenciae, lepasicola, lutescens, macrocarpus, microscopicus, minor, mitchelloides, monzensis, multifurcus, myurus, natans, niigatensis, oblongatus, obovatus, obtusocarpus, obtusosus, parvulus, parvus, pectenis, plasticola, plumosus, polysiphoniae, pumilus, radicans, ramellosus, ramentaceus, rotundato-apicalis, rudis, rufulus, rufus, sadoensis, sargassicaulinis, sargassiphyllus, saxatilis, scytosipohae, shiiyaensis, shimokitaensis, siliculosus, simpliciusculus, simulans, sonorensis, sphaericus, strigosus, subulatus, tamarinii, taoniae, tappiensis, tasshaensis, tenellus, tomentosus, trichophorus, tsugaruensis, van-bosseae, variabilis, venetus, virescens, vungtauensis, zonariae, zosterae*
VARIETIES *acutus acutus, acutus haplogloiae, confervoides longipes, confervoides subulatus, confervoides variabilis, fasciculatus abbreviatus, fasciculatus congestus, fasciculatus macrosporus, fasciculatus polyrhizus, fasciculatus pycnocarpus, fasciculatus refractus, siliculosus adriaticus, siliculosus arctus, siliculosus crouaniorum, siliculosus dasycarpus, siliculosus divergens, siliculosus elongatus, siliculosus megacarpus, siliculosus parvus, siliculosus pygmaeus, siliculosus subulatus, siliculosus venetus, simpliciusculus vitiensis*
GENUS *Bodanella* Bodan Ectocarp Dusky Brown Ocher Algae
SPECIES *lauterbornii*
GENUS *Heribaudiella* Heribaud Ectocarp Dusky Brown Ocher Algae
SPECIES *fluviatilis*
GENUS *Ishige* Ishige Ectocarp Dusky Brown Ocher Algae
SPECIES *okamurae, sinicola*
GENUS *Kuckuckia* Kuckuck Ectocarp Dusky Brown Ocher Algae
SPECIES *catalinae, kylinii, spinosa*
GENUS *Mesospora* Mesospore Ectocarp Dusky Brown Ocher Algae
SPECIES *negrosensis, van-bosseae*

FAMILY 2 *Acinetosporaceae* Acinet Ectocarp Dusky Brown Ocher Algae
GENUS *Acinetospora* Acinet Ectocarp Dusky Brown Ocher Algae
SPECIES *crinita, nicholsoniae*
GENUS *Feldmannia* Feldmann Acinet Ectocarp Dusky Brown Ocher Algae
SPECIES *acuminata, battersiides, chitonicola, columellaris, cylindrica, enhali, filifera, formosana, hemisphaerica, indica, irregularis, kanyakumariensis, kjellmanii, lebelii, padinae, paradoxa, paradoxoides, rhizoidea, simplex, vaughanii, zeylanica*
VARIETIES *hemisphaerica minor, paradoxa caespitula, paradoxa cyclindrica, paradoxa mucosa*
GENUS *Geminocarpus* Gemini Acinet Ectocarp Dusky Brown Ocher Algae
SPECIES *austrogeorgiae, geminatus*
GENUS *Hincksia* Hincks Acinet Ectocarp Dusky Brown Ocher Algae
SPECIES *andamanensis, bhimlipatnamensis, clavata, dalmatica, fenestrata, fuscata, geniculata, granulosa, hauckii, hincksiae, mitchelliae, onslowensis, ovata, prolifera, rallsiae, sandriana, saundersii, secunda, sordida, terminalis, thyrsoidea*
VARIETIES *hincksiae australis, hincksiae californica, mitchelliae jamaicensis, mitchelliae neilii, mitchelliae parvus, sandriana baltica*
GENUS *Pogotrichum* Pogo Hairy Acinet Ectocarp Dusky Brown Ocher Algae
SPECIES *filiforme, setiforme, yezoense*

FAMILY 3 *Adenocystaceae* Gland Cyst Ectocarp Dusky Brown Ocher Algae
GENUS *Adenocystis* Gland Cyst Ectocarp Dusky Brown Ocher Algae
SPECIES *utricularis*
GENUS *Caepidium* Cepid Gland Cyst Ectocarp Dusky Brown Ocher Algae
SPECIES *antarcticum*
GENUS *Utriculidium* Bladder Gland Cyst Ectocarp Dusky Brown Ocher Algae
SPECIES *durvillei*

FAMILY 4 *Chordariaceae* Chord Ectocarp Dusky Brown Ocher Algae
GENUS *Chordaria* Chord Ectocarp Dusky Brown Ocher Algae
SPECIES *disessa, flagelliformis, flexuosa, gracilis, linearis, magellanica, nagaii*
VARIETIES *flagelliformis chordaeformis*
GENUS *Acrocytis* Acrocyte Chord Ectocarp Dusky Brown Ocher Algae
SPECIES *groenlandicus*
GENUS *Acrospongium* Tall Sponge Chord Ectocarp Dusky Brown Ocher Algae
SPECIES *ralfsioides*
GENUS *Acrothrix* Hairy Chord Ectocarp Dusky Brown Ocher Algae

SPECIES *gracilis, neglecta, pacifica*
GENUS *Acrotrichium* Lesser Hairy Chord Ectocarp Dusky Brown Ocher Algae
SPECIES *amphibolis*
GENUS *Actinema* Radial Chord Ectocarp Dusky Brown Ocher Algae
SPECIES *major, scutellum, subtilissimum*
GENUS *Adriogloia* Adriatic Glow Chord Ectocarp Dusky Brown Ocher Algae
SPECIES *adriatica*
GENUS *Ascocyclus* Sac Cyclic Chord Ectocarp Dusky Brown Ocher Algae
SPECIES *balticus, dichotomus, hypneae, orbicularis*
GENUS *Asperococcus* Asp Round Chord Ectocarp Dusky Brown Ocher Algae
SPECIES *attenuatus, bullosus, durvillei, ensiformis, fistulosus, norvegicus, scaber, tortilis*
VARIETIES *bullosus profundus, fistulosus confluens, fistulosus vermicularis*
GENUS *Botrytella* Berry Chord Ectocarp Dusky Brown Ocher Algae
SPECIES *micromora, pacifica, parva, reinboldii*
VARIETIES *micromora balticus*
GENUS *Buffhamia* Buffham Chord Ectocarp Dusky Brown Ocher Algae
SPECIES *speciosa*
GENUS *Chilionema* Claw Chord Ectocarp Dusky Brown Ocher Algae
SPECIES *hispanicum, ocellatum*
GENUS *Chnoospora* Chinospore Chord Ectocarp Dusky Brown Ocher Algae
SPECIES *bicanaliculata, implexa, minima, pannosa*
GENUS *Chordariopsis* Lesser Chord Ectocarp Dusky Brown Ocher Algae
SPECIES *capensis*
GENUS *Cladochroa* Branched Chord Ectocarp Dusky Brown Ocher Algae
SPECIES *chnoosporiformis*
GENUS *Cladosiphon* Branched Siphon Chord Ectocarp Dusky Brown Ocher Algae
SPECIES *contortus, cylindricus, filum, giraudyi, irregularis, mauritianus, mediterraneus, okamuranus, vermicularis, zosterae*
GENUS *Cladothele* Branched Thele Chord Ectocarp Dusky Brown Ocher Algae
SPECIES *decaisnei, striarioides*
GENUS *Clathrodiscus* Disc Chord Ectocarp Dusky Brown Ocher Algae
SPECIES *mandouli*
GENUS *Clavatella* Collared Chord Ectocarp Dusky Brown Ocher Algae
SPECIES *viridissima*
GENUS *Climacosorus* Climac Sorus Chord Ectocarp Dusky Brown Ocher Algae
SPECIES *mediterraneus*
GENUS *Coelocladia* Sky Branched Chord Ectocarp Dusky Brown Ocher Algae
SPECIES *arctica*
GENUS *Coilodesme* Coil Fiber Chord Ectocarp Dusky Brown Ocher Algae
SPECIES *arctica, bulligera, californica, corrugata, cystoseirae, fucicola, japonica, plana, polynampta, rigida*
VARIETIES *bulligera fucicola*
GENUS *Corycus* Cory Chord Ectocarp Dusky Brown Ocher Algae
SPECIES *lanceolatus, prolifera*
GENUS *Corynophlaea* Club Chord Ectocarp Dusky Brown Ocher Algae
SPECIES *crispa, cristata, cystophorae, flaccida, hamelii, umbellata*
GENUS *Cylindrocarpus* Cylindric Chord Ectocarp Dusky Brown Ocher Algae
SPECIES *microscopicus, rugosus*
GENUS *Dalmatogloia* Dalmation Glow Chord Ectocarp Dusky Brown Ocher Algae
SPECIES *bryozoi*
GENUS *Delamarea* Delamar Chord Ectocarp Dusky Brown Ocher Algae
SPECIES *attenuata*
GENUS *Dermatocelis* Dermal Chord Ectocarp Dusky Brown Ocher Algae
SPECIES *laminariae*
GENUS *Desmotrichum* Fiber Chord Ectocarp Dusky Brown Ocher Algae
SPECIES *plumosum, punctarioides*
GENUS *Dictyosiphon* Net Siphon Chord Ectocarp Dusky Brown Ocher Algae
SPECIES *chordarius, ekmanii, filiformis, foeniculaceus, hirsutus, macounii, sinicola, tenuis*
VARIETIES *foeniculaceus americanus, foeniculaceus filiformis*
GENUS *Elachista* Elach Chord Ectocarp Dusky Brown Ocher Algae
SPECIES *antarctica, australis, chondrii, claytoniae, coccophorae, confusicola, crassa, endarachnae, falcata, flaccida, fucicola, globulosa, intermedia, jabukae, lebelii, meridionalis, minutissima, mollis, neglecta, nigra, nipponica, okamurae, orbicularis, sadoensis, sargassicola, scutulata, stellaris, taeniiformis, tenuis, vellosa, zosterae*
VARIETIES *fucicola profunda, intermedia clavaeformis, intermedia profunda*
GENUS *Entonema* Entic Chord Ectocarp Dusky Brown Ocher Algae
SPECIES *penetrans, polycladum, subcorticale, tenuissimum*
GENUS *Eudesme* Fiber Chord Ectocarp Dusky Brown Ocher Algae

SPECIES *flavescens, virescens*
VARIETIES *virescens baltica*
GENUS *Flabellonema* Fan Chord Ectocarp Dusky Brown Ocher Algae
SPECIES *codii*
GENUS *Fosliea* Foslie Chord Ectocarp Dusky Brown Ocher Algae
SPECIES *curta*
GENUS *Giffordia* Gifford Chord Ectocarp Dusky Brown Ocher Algae
SPECIES *hainanensis, zhejiangensis*
GENUS *Giraudia* Giraud Chord Ectocarp Dusky Brown Ocher Algae
SPECIES *robusta, sphacelarioides*
GENUS *Gonodia* Gonod Chord Ectocarp Dusky Brown Ocher Algae
SPECIES *johnstonii, marchantae*
GENUS *Gononema* Gononeme Chord Ectocarp Dusky Brown Ocher Algae
SPECIES *pectinatum, ramosum*
GENUS *Gontrania* Gontran Chord Ectocarp Dusky Brown Ocher Algae
SPECIES *lubrica*
GENUS *Halodictyon* Haline Net Chord Ectocarp Dusky Brown Ocher Algae
SPECIES *mirabile*
GENUS *Halonema* Haline Thread Chord Ectocarp Dusky Brown Ocher Algae
SPECIES *subsimplex*
GENUS *Halorhipis* Haline Wicker Chord Ectocarp Dusky Brown Ocher Algae
SPECIES *winstonii*
GENUS *Halorhiza* Haline Rhizid Chord Ectocarp Dusky Brown Ocher Algae
SPECIES *vaga*
GENUS *Halothrix* Haline Hairy Chord Ectocarp Dusky Brown Ocher Algae
SPECIES *ambigua, coccophorae, ephemeralis, lumbricalis, rectiuscula, sadoensis, tortuosa*
GENUS *Hamelella* Hamel Chord Ectocarp Dusky Brown Ocher Algae
SPECIES *dermonematis, geminifructa*
GENUS *Hecatonema* Hecate Chord Ectocarp Dusky Brown Ocher Algae
SPECIES *aggregatum, enhali, floridanum, kjellmanii, lawsonii, liagorae, primarium, sargassicola, stewartensis, streblonematoides, terminale*
VARIETIES *terminale sauvageauii, terminale soluta*
GENUS *Herpodiscus* Herpes Disc Chord Ectocarp Dusky Brown Ocher Algae
SPECIES *durvilleae*
GENUS *Herponema* Herpes Chord Ectocarp Dusky Brown Ocher Algae
SPECIES *desmarestiae, graniferum, hormosirae, maculaeformis, minutum, solitarium, tortugense, valiantei, velutinum*
GENUS *Heterophycus* Variable Chord Ectocarp Dusky Brown Ocher Algae
SPECIES *plumosus*
GENUS *Heterosaundersella* Variable Saunders Chord Ectocarp Dusky Brown Ocher Algae
SPECIES *hattoriana*
GENUS *Hummia* Humm Chord Ectocarp Dusky Brown Ocher Algae
SPECIES *onusta*
GENUS *Hydroclathrus* Hydric Cage Chord Ectocarp Dusky Brown Ocher Algae
SPECIES *clathratus, stephanosorus, tenuis, tumulis*
GENUS *Isthmoplea* Isthmus Chord Ectocarp Dusky Brown Ocher Algae
SPECIES *rupincola, sphaerophora*
GENUS *Kuetzingiella* Kuetzing Chord Ectocarp Dusky Brown Ocher Algae
SPECIES *battersii, elachistaeformis, holmesii, maculans*
VARIETIES *battersii mediterranea*
GENUS *Kurogiella* Kurog Chord Ectocarp Dusky Brown Ocher Algae
SPECIES *saxatilis*
GENUS *Laminariocolax* Laminar Chord Ectocarp Dusky Brown Ocher Algae
SPECIES *aecidioides, draparnaldioides, eckloniae, macrocystis, tomentosoides*
VARIETIES *aecidioides pacificum, tomentosoides deformans, tomentosoides norvegicus, tomentosoides punctiformis*
GENUS *Laminarionema* Lesser Laminar Chord Ectocarp Dusky Brown Ocher Algae
SPECIES *elsbetiae*
GENUS *Leathesia* Leathes Chord Ectocarp Dusky Brown Ocher Algae
SPECIES *crassipilosa, difformioides, difformis, intermedia, japonica, monilicellulata, mucosa, nana, novae-zelandiae, primaria, pulvinata, sadoensis, saxicola, sphaerocephala, tsugaruensis, umbellata, verruculiformis, yezoensis*
VARIETIES *mucosa condensata*
GENUS *Leblondiella* Leblond Chord Ectocarp Dusky Brown Ocher Algae
SPECIES *densa*
GENUS *Leptonema* Thin Chord Ectocarp Dusky Brown Ocher Algae
SPECIES *arcticum*

GENUS *Leptonematella* Lesser Thin Chord Ectocarp Dusky Brown Ocher Algae
SPECIES *falklandica, fasciculata, neapolitana*
GENUS *Levringia* Levring Chord Ectocarp Dusky Brown Ocher Algae
SPECIES *atlantica, boergesenii, brasiliensis, decaryi, filiformis, natalensis, sordida*
GENUS *Liebmannia* Liebmann Chord Ectocarp Dusky Brown Ocher Algae
SPECIES *laccadivarum, leveillei*
GENUS *Litosiphon* Lite Siphon Chord Ectocarp Dusky Brown Ocher Algae
SPECIES *laminariae, setiformis, tenuis*
VARIETIES *laminariae major*
GENUS *Melanosiphon* Dark Siphon Chord Ectocarp Dusky Brown Ocher Algae
SPECIES *intestinalis*
VARIETIES *intestinalis tenuis*
GENUS *Mesogloia* Glow Chord Ectocarp Dusky Brown Ocher Algae
SPECIES *coccinea, hudsonii, lanosa, mediterranea, vermiculata*
GENUS *Mesogloiopsis* Lesser Glow Chord Ectocarp Dusky Brown Ocher Algae
SPECIES *tasmanica*
GENUS *Microcoryne* Lesser Club Chord Ectocarp Dusky Brown Ocher Algae
SPECIES *ocellata*
GENUS *Microspongium* Lesser Sponge Chord Ectocarp Dusky Brown Ocher Algae
SPECIES *alariae, gelatinosum, globosum, immersum, radians, tenuissimum*
VARIETIES *globosum affine*
GENUS *Microsporangium* Microspore Chord Ectocarp Dusky Brown Ocher Algae
SPECIES *kuckuckianum*
GENUS *Mikrosyphar* Syph Chord Ectocarp Dusky Brown Ocher Algae
SPECIES *pachymeniae, polysiphoniae, porphyrae, sphacelariae, zosterae*
GENUS *Myelophycus* Myel Chord Ectocarp Dusky Brown Ocher Algae
SPECIES *cavum, simplex*
GENUS *Myriactula* Myriad Chord Ectocarp Dusky Brown Ocher Algae
SPECIES *arabica, areschougii, caespitosa, chordae, clandestina, clavata, elongata, filiformis, fucorum, gracilis, haydenii, microscopica, minor, pulvinata, pusilla, rigida, rivulariae, rosarioides, sargassi, saromaensis, stellulata, vlastae*
GENUS *Myriocladia* Myriad Branch Chord Ectocarp Dusky Brown Ocher Algae
SPECIES *capensis, ekmani, lovenii, tomentosa*
GENUS *Myriogloea* Myriad Glow Chord Ectocarp Dusky Brown Ocher Algae
SPECIES *abbreviata, chilensis, grandis, intenstinalis, papenfussii, sciurus, simplex*
GENUS *Myriotrichia* Myriad Hair Chord Ectocarp Dusky Brown Ocher Algae
SPECIES *adriatica, canariensis, clavaeformis, occidentalis, repens*
GENUS *Nemacystus* Thread Cyst Chord Ectocarp Dusky Brown Ocher Algae
SPECIES *brandegeei, decipiens, erythraeus, flexuosus, hispanicus, howei, novae-zelandiae, subsimplex*
VARIETIES *flexuosus giraudyi*
GENUS *Omphalophyllum* Navel Chord Ectocarp Dusky Brown Ocher Algae
SPECIES *ulvaceum*
GENUS *Papenfussiella* Papenfuss Chord Ectocarp Dusky Brown Ocher Algae
SPECIES *extensa, gracilis, kuromo, lutea, tristanensis*
GENUS *Petrospongium* Rock Sponge Chord Ectocarp Dusky Brown Ocher Algae
SPECIES *berkeleyi, rugosum*
GENUS *Phaeostroma* Stroma Chord Ectocarp Dusky Brown Ocher Algae
SPECIES *bertholdii, endophyticum, longisetum, parasiticum, pusillum, pustulosum*
GENUS *Phaeostromatella* Lesser Stroma Chord Ectocarp Dusky Brown Ocher Algae
SPECIES *elegans*
GENUS *Phaeostrophion* Strophion Chord Ectocarp Dusky Brown Ocher Algae
SPECIES *irregulare*
GENUS *Phloeospora* Phloeospore Chord Ectocarp Dusky Brown Ocher Algae
SPECIES curta
GENUS *Phycocelis* Celis Chord Ectocarp Dusky Brown Ocher Algae
SPECIES *crouaniorum, foecunda, globosa, hispanica, major, ocellata*
VARIETIES *foecunda ramulosum, foecunda simplicissimum*
GENUS *Phycolapathum* Lapath Chord Ectocarp Dusky Brown Ocher Algae
SPECIES *debile*
GENUS *Pilocladus* Pilose Branch Chord Ectocarp Dusky Brown Ocher Algae
SPECIES *codicola*
GENUS *Platysiphon* Flat Siphon Chord Ectocarp Dusky Brown Ocher Algae
SPECIES *verticillatus*
GENUS *Pleurocladia* Pleural Branch Chord Ectocarp Dusky Brown Ocher Algae
SPECIES *lacustris, lucifuga*
VARIETIES *lacustris pacifica*
GENUS *Polycerea* Many Horn Chord Ectocarp Dusky Brown Ocher Algae

SPECIES *nigrescens, zostericola*
GENUS *Proselachista* Selachi Chord Ectocarp Dusky Brown Ocher Algae
SPECIES *taeniiformis*
GENUS *Protasperococcus* Asp Round Chord Ectocarp Dusky Brown Ocher Algae
SPECIES *myriotrichiiformis*
GENUS *Protectocarpus* Protected Carpel Chord Ectocarp Dusky Brown Ocher Algae
SPECIES *faeroensis, hecatonemoides, speciosus*
GENUS *Punctaria* Dotted Chord Ectocarp Dusky Brown Ocher Algae
SPECIES *chartacea, conglomerata, crispata, crouaniorum, expansa, fissalis, flaccida, glacialis, hesperia, kinoshitae, latifolia, lobata, mageshimensis, multifida, occidentalis, orbiculata, pilosa, plantaginea, projecta, stipitata, tenuis, tenuissima*
VARIETIES *latifolia angustifolia, latifolia laminarioides, latifolia lanceolata, plantaginea linearis, plantaginea rubescens*
GENUS *Rhadinocladia* Rhadin Branch Chord Ectocarp Dusky Brown Ocher Algae
SPECIES *cylindrica, farlowii*
GENUS *Saundersella* Saunders Chord Ectocarp Dusky Brown Ocher Algae
SPECIES *simplex*
GENUS *Sauvageaugloia* Sauvageau Glow Chord Ectocarp Dusky Brown Ocher Algae
SPECIES *griffithsiana, ikomae*
GENUS *Soranthera* Soral Flower Chord Ectocarp Dusky Brown Ocher Algae
SPECIES *ulvoidea*
VARIETIES *ulvoidea difformis*
GENUS *Sorocarpus* Soral Chord Ectocarp Dusky Brown Ocher Algae
SPECIES *pacificus*
GENUS *Spermatochnus* Ochnid Chord Ectocarp Dusky Brown Ocher Algae
SPECIES *paradoxus*
VARIETIES *paradoxus adriaticus*
GENUS *Sphaerotrichia* Sphere Hairy Chord Ectocarp Dusky Brown Ocher Algae
SPECIES *disessa, divaricata, firma, sadoensis*
GENUS *Spongonema* Thread Sponge Chord Ectocarp Dusky Brown Ocher Algae
SPECIES *tomentosum*
GENUS *Stictyosiphon* Stick Siphon Chord Ectocarp Dusky Brown Ocher Algae
SPECIES *adriaticus, arcticus, griffithsianus, laxus, lofotensis, lutarius, soriferus, tortilis*
VARIETIES *tortilis chordariaeformis*
GENUS *Stilophora* Stile Chord Ectocarp Dusky Brown Ocher Algae
SPECIES *attenuata, flanaganii, iranica, nodulosa, papillosa, tenella*
VARIETIES *tenella adriatica, tenella contorta*
GENUS *Stilopsis* Lesser Stile Chord Ectocarp Dusky Brown Ocher Algae
SPECIES *harveyana, lejolisii*
GENUS *Streblonema* Streble Chord Ectocarp Dusky Brown Ocher Algae
SPECIES *ambivalens, areschougii, breve, codiatile, codii, corymbiferum, desmarestiae, eudesmide, evagatum, fasciculatum, fasciculatum, gracilaricola, helophorum, inclusum, infestans, integratum, intestinum, investiens, invisibile, irregularis, maculans, minutissima, minutulum, myriocladiae, myrionematoides, nemastomae, pacificum, parasiticum, patagonicum, penetrale, porphyrae, reptans, rugosum, simplex, stilophorae, transfixum, tuberculi, turmale, turnerellae, vorax, zanardinii*
GENUS *Streblonemopsis* Lesser Streble Chord Ectocarp Dusky Brown Ocher Algae
SPECIES *irritans*
GENUS *Strepsithalia* Twisted Thal Chord Ectocarp Dusky Brown Ocher Algae
SPECIES *aemula, buffhamiana, curvata, leathesiae, liagorae, liebmanniae*
GENUS *Striaria* Striar Chord Ectocarp Dusky Brown Ocher Algae
SPECIES *attenuata*
GENUS *Stschapovia* Stschapov Chord Ectocarp Dusky Brown Ocher Algae
SPECIES *flagellaris*
GENUS *Suringariella* Suringar Chord Ectocarp Dusky Brown Ocher Algae
SPECIES *harveyana*
GENUS *Tinocladia* Tin Branch Chord Ectocarp Dusky Brown Ocher Algae
SPECIES *australis, crassa, falklandica, novae-zelandiae*
GENUS *Trachynema* Rough Chord Ectocarp Dusky Brown Ocher Algae
SPECIES *groenlandicum, mortenseni*
GENUS *Tremella* Tremor Chord Ectocarp Dusky Brown Ocher Algae
SPECIES *globulosa, natans, pisum*
GENUS *Trichocladia* Hairy Branch Chord Ectocarp Dusky Brown Ocher Algae
SPECIES *nostocoides*
GENUS *Ulonema* Ulo Chord Ectocarp Dusky Brown Ocher Algae
SPECIES *rhizophorum*
GENUS *Xanthosiphonia* Yellowish Siphon Chord Ectocarp Dusky Brown Ocher Algae
SPECIES *austrogeorgica, halliae*

GENUS *Zeacarpa* — Corn Chord Ectocarp Dusky Brown Ocher Algae
SPECIES *leiomorpha*
GENUS *Zosterocarpus* — Zoster Chord Ectocarp Dusky Brown Ocher Algae
SPECIES *abyssicolus, oedogonium, ocellata*

FAMILY 5 *Myrionemataceae* — Myriad Thread Ectocarp Dusky Brown Ocher Algae
GENUS *Myrionema* — Myriad Thread Ectocarp Dusky Brown Ocher Algae
SPECIES *acrochaetiae, balticum, compactum, compsonematoides, conchicola, corunnae, crustaceum, densum, dichotomum, feldmannii, fuegianum, furcatum, grateloupiae, inconspicuum, latipilosum, liechtensternii, macrocarpum, magnusii, myriodesmae, obscurum, padinae, papillosum, paradoxum, patagonicum, phyllophilum, primarium, ramulans, reptans, seriatum, setiferum, strangulans, terminale*
VARIETIES *corunnae filamentosum*
GENUS *Asterotrichia* — Star Myriad Thread Ectocarp Dusky Brown Ocher Algae
SPECIES *ocellata*
GENUS *Myrionemopsis* — Lesser Myriad Thread Ectocarp Dusky Brown Ocher Algae
SPECIES *prostratum*

FAMILY 6 *Pylaiellaceae* — Pyle Ectocarp Dusky Brown Ocher Algae
GENUS *Pylaiella* — Pyle Ectocarp Dusky Brown Ocher Algae
SPECIES *gardneri, littoralis, nana, penicilliformis, petaloniae, seriata, tenella, unilateralis, varia, washingtoniensis*
VARIETIES *littoralis firma-macrocarpa, littoralis fluviatilis, littoralis major, littoralis robusta, littoralis rupincola*
GENUS *Bachelotia* — Bachelot Pyle Ectocarp Dusky Brown Ocher Algae
SPECIES *antillarum, fulvescens*

FAMILY 7 *Ralfsiaceae* — Ralfs Ectocarp Dusky Brown Ocher Algae
GENUS *Ralfsia* — Ralfs Ectocarp Dusky Brown Ocher Algae
SPECIES *australis, bornetii, californica, ceylanica, confusa, endopluroides, expansa, fungiformis, hancockii, hesperia, hyalina, integra, lucida, minima, ovata, pacifica, pangoensis, pedicellata, pusilla, verrucosa*
GENUS *Basispora* — Base Spore Ralfs Ectocarp Dusky Brown Ocher Algae
SPECIES *africana*
GENUS *Hapalospongidion* — Sponge Ralfs Ectocarp Dusky Brown Ocher Algae
SPECIES *capitatum, gelatinosum, macrocarpum, pangoense, saxigenum, schmidtii, van-bosseae*
GENUS *Jonssonia* — Jonsson Ralfs Ectocarp Dusky Brown Ocher Algae
SPECIES *pulvinata*
GENUS *Lithoderma* — Rock Ralfs Ectocarp Dusky Brown Ocher Algae
SPECIES *antarcticum, fatiscens, fluviatilis, fontanum, piliferum*
VARIETIES *piliferum debile*
GENUS *Nemoderma* — Thread Ralfs Ectocarp Dusky Brown Ocher Algae
SPECIES *tingitanum*
GENUS *Petroderma* — Least Rock Ralfs Ectocarp Dusky Brown Ocher Algae
SPECIES *maculiforme, vietnamensis*
GENUS *Porterinema* — Porter Ralfs Ectocarp Dusky Brown Ocher Algae
SPECIES *fluviatile*
GENUS *Pseudolithoderma* — Lesser Rock Ralfs Ectocarp Dusky Brown Ocher Algae
SPECIES *adriaticum, australis, extensum, groenlandica, moreirae, nigra, paradoxum, roscoffense, rosenvingei, subextensum*

ORDER 3 *Fucales* — Painted Dusky Brown Ocher Algae

FAMILY 1 *Fucaceae* — Painted Dusky Brown Ocher Algae
GENUS *Fucus* — Painted Dusky Brown Ocher Algae
SPECIES *acicularis, aculeatus, albidus, arbuscula, areschougii, articulatus, atomarius, bastera, ceranoides, chalonii, cirrhosus, cottonii, crispatus, cristatus, dichotomus, distichus, divaricatus, evanescens, excisus, filicinus, filiformis, fimbriatus, flagellum, geniculatus, giganteus, gramineus, intermedius, lichenoides, membranaceus, membranifolius, menziesii, mytili, nereideus, osmunda, paniculatus, papillosus, polyphyllus, polypodioides, pygmaeus, rigidus,* ***serratus***(Medicinal Herb), *spermophorus, spiralis, stiriatus, tendo, teres, tetragonus, tremella, undulatus,* ***vesiculosus***(Medicinal Herb), *virsoides, zeylanicus*
VARIETIES *areschougii nanus, ceranoides glomerata, ceranoides harveyanus, ceranoides linearis, ceranoides ramosissima, crispatus brunneus, crispatus equalis, crispatus stellatus, crispatus virens, distichus edentatus-abbreviatus, evanescens angustus, evanescens edentatus, evanescens hesperius, filiformis gmelini, filiformis nanus, filiformis pylaisei, spiralis platycarpus, tremella lactuca, tremella porrifolia, tremella umbilicalis, vesiculosus compressus, vesiculosus foliaceus, vesiculosus gracillimus, vesiculosus linearis, vesiculosus mytili, vesiculosus vadorum, vesiculosus volubilis*
GENUS *Ascophyllum* — Sac Painted Dusky Brown Ocher Algae
SPECIES *nodosum*
VARIETIES *nodosum lusitanicum, nodosum mackayi*
GENUS *Hesperophycus* — Hesper Painted Dusky Brown Ocher Algae
SPECIES *californicus*
GENUS *Pelvetia* — Pelvet Painted Dusky Brown Ocher Algae

SPECIES *canaliculata, fastigiata*
VARIETIES *canaliculata acutiloba, fastigiata gracilis*
GENUS *Pelvetiopsis* Lesser Pelvet Painted Dusky Brown Ocher Algae
SPECIES *arborescens, limitata*
GENUS *Silvetia* Silvet Painted Dusky Brown Ocher Algae
SPECIES *babingtonii, compressa, siliquosa*
VARIETIES *compressa gracilis*
GENUS *Xiphophora* Xiph Painted Dusky Brown Ocher Algae
SPECIES *chondrophylla, gladiata*
VARIETIES *chondrophylla minor, gladiata novae-zelandiae*

FAMILY 2 *Cystoseiraceae* Cyst Painted Dusky Brown Ocher Algae
GENUS *Cystoseira* Cyst Painted Dusky Brown Ocher Algae
SPECIES *abies-marina, algeriensis, amentacea, baccata, barbata, barbatula, brachycarpa, compressa, corniculata, crassipes, crinita, crinitophylla, dubia, elegans, epiphytica, foeniculacea, funkii, geminata, granulata, hakodatensis, humilis, hyblaea, indica, jabukae, masoudii, mauritanica, mediterranea, montagnei, myrica, neglecta, nodicaulis, osmundacea, pelagosae, planiramea, platyclada, pycnoclada, sauvageauana, schiffneri, sedoides, selaginoides, senegalensis, setchellii, sicula, spinosa, squarrosa, susanensis, tamariscifolia, thysigera, trinodis, usneoides, wildpretii, wildprettii, zosteroides*
VARIETIES *amentacea spicata, amentacea stricta, barbata flaccida, barbata insularum, barbata repens, barbata tophuloidea, brachycarpa balearica, brachycarpa claudiae, compressa plana, compressa pustulata, corniculata laxior, crinita bosphorica, foeniculacea latiramosa, foeniculacea repens, foeniculacea tenuiramosa, humilis myriophylloides, jabukae tenuissima, mediterranea valiantei, myrica occidentalis, spinosa compressa, spinosa tenuior, tamariscifolia gibraltica, thysigera lepidium*
GENUS *Acrocarpia* Acrocarp Cyst Painted Dusky Brown Ocher Algae
SPECIES *paniculata, robusta*
GENUS *Axillariella* Axillar Cyst Painted Dusky Brown Ocher Algae
SPECIES *constricta*
GENUS *Bifurcaria* Bifurcated Cyst Painted Dusky Brown Ocher Algae
SPECIES *bifurcata, brassicaeformis, galapegensis*
GENUS *Bifurcariopsis* Lesser Bifurcated Cyst Painted Dusky Brown Ocher Algae
SPECIES *capensis*
GENUS *Blossevillea* Blosseville Cyst Painted Dusky Brown Ocher Algae
SPECIES *retorta, retroflexa*
GENUS *Carpoglossum* Tongue Cyst Painted Dusky Brown Ocher Algae
SPECIES *confluens*
GENUS *Caulocystis* Tailed Cyst Painted Dusky Brown Ocher Algae
SPECIES *cephalornithos, uvifera*
GENUS *Coccophora* Round Cyst Painted Dusky Brown Ocher Algae
SPECIES *langsdorfii*
GENUS *Cystophora* Lesser Cyst Painted Dusky Brown Ocher Algae
SPECIES *botryocystis, brownii, congesta, cuspidata, cymodocea, distenda, expansa, fibrosa, gracilis, grevillei, harveyi, intermedia, monilifera, moniliformis, pectinata, platylobium, polycystidea, racemosa, racemosa, retorta, retroflexa, scalaris, siliquosa, subfarcinata, tenuis, torulosa, xiphocarpa*
GENUS *Cystophyllum* Least Cyst Painted Dusky Brown Ocher Algae
SPECIES *hildebrandtii*
GENUS *Durvillaea* Durville Cyst Painted Dusky Brown Ocher Algae
SPECIES *antarctica, caepaestipes, chathamensis, potatorum, willana*
GENUS *Halidrys* Haline Cyst Painted Dusky Brown Ocher Algae
SPECIES *dioica, murmanica, siliquosa, volubilis*
GENUS *Himanthalia* Himan Thal Cyst Painted Dusky Brown Ocher Algae
SPECIES *elongata*
GENUS *Hormophysa* Bladder Cyst Painted Dusky Brown Ocher Algae
SPECIES *cuneiformis*
GENUS *Hormosira* Hormon Cyst Painted Dusky Brown Ocher Algae
SPECIES *banksii*
GENUS *Landsburgia* Landsburg Cyst Painted Dusky Brown Ocher Algae
SPECIES *myricifolia, quercifolia*
GENUS *Myagropsis* Myager Cyst Painted Dusky Brown Ocher Algae
SPECIES *myagroides*
GENUS *Myriodesma* Myriad Fiber Cyst Painted Dusky Brown Ocher Algae
SPECIES *calophyllum, harveyanum, integrifolium, leptophyllum, peronii, quercifolium, serrulatum, tuberosum*
VARIETIES *integrifolium serrulatum*
GENUS *Notheia* Nothe Cyst Painted Dusky Brown Ocher Algae
SPECIES *anomala*
GENUS *Platythalia* Flat Thal Cyst Painted Dusky Brown Ocher Algae
SPECIES *angustifolia, quercifolia*
GENUS *Scaberia* Scaber Cyst Painted Dusky Brown Ocher Algae

SPECIES *agardhii*

FAMILY 2 *Sargassaceae* Sargasso Painted Dusky Brown Ocher Algae

GENUS *Sargassum* Sargasso Painted Dusky Brown Ocher Algae

SPECIES *abbottiae, acinacifolium, acinaciforme, acinarium, agardhianum, agaviforme, albemarlense, albertisii, alternato-pinnatum, amabile, amaliae, ammophilum, anapense, angii, aquifolium, araii, armatum, asperifolium, assimile, autumnale, baccularia, balingasayense, banyuejiaoense, bataanense, beihaiense, belangeri, bermudense, bicorne, biserrula, biserrulioides, boreale, boryi, boveanum, brandegeei, bryantii, bulbiferum, calophyllum, camouii, capillare, capilliforme, capitatum, carpophyllum, cervicorne, cheirifolium, cinctum, cinereum, claviferum, concinnum, confusum, congkinhii, corderoi, cornigerum, cornutifructum, costatum, cotoense, crassifolium, crispifolium, cristaefolium, cristatum, cuneifolium, currimaoense, cylindrocarpum, cylindrocystum, cymosum, cystocarpum, cystophyllum, dazhouense, decipiens, densicystum, densifolium, densum, denticarpum, dentifolium, desfontainesii, desvauxii, distichum, divaricatum, diversifolium, dotyi, droserifolium, droserifolium spathulatum, dumosum, echinocarpum, ecuadoreanum, elegans, emarginatum, ensifolium, erinaceum, erumpens, esperi, euryphyllum, expansum, fallax, feldmannii, fergusonii, filicinum, filifolium, filipendula, fissifolium, flavicans, flavifolium, fluitans, fonanonense, frutescens, fruticulosum, fujianense, fuliginosoides, fulvellum, furcatum, fuscifolium,* ***fusiforme***(Medicinal Herb), *galapagense, gaudichaudii, gemmiphorum, giganteifolium, glaucescens, godeffroyi, gracile, gracillimum, graminifolium, granuliferum, grevillei, guangdongii, guardiense, hainanense, hawaiiensis, hemiphylloides, hemiphyllum, henslowianum, herbaceum, herklotsii, herporhizum, heterocystum, heteromorphum, hildebrandtii, hombronianum, horneri, hornschuchii, horridulum, horridum, howellii, huangluense, hystrix, ilicifolioides, ilicifolium, incanum, incisifolium, incisum, insulare, integerrimum, integrifolioides, intermedium, johnsonii, johnstonii, kashiwajimanum, kasyotense, kuetzingii, kushimotense, lacerifolium, lanceolatum, lapazeanuma, latifolium, laxifolium, leizhouense, lendigerum, leptopodum, liebmannii, ligulatum, linearifolium, longicaulis, longifructum, longivesiculosum, lophocarpum, lunense, macdougalii, maclurei, macrocarpum, mangarevense, marchantae, marginatum, mathiesonii, mcclurei, megalocystum, micracanthum, microceratium, microcystum, microphyllum, miyabei, muticum, myriocystum, nanshaense, naozhounse, natans, neurophorum, nigrifolioides, nigrifolium, nipponicum, notarisii, novae-hollandiae, nozhouense, obovatum, obtusifolium, odontocarpum, ohnoi, okamurae, oligocystum, oocyste, pachycarpum, pacificum, pallidum, palmeri, paniculatum, paradoxum, parvifolioides, parvifolium, parvivesiculosum, patens, peronii, persicum, phamhoangii, philippinense, phyllocystum, piluliferum, pinnatifidum, pinnatifolium, plagiophyllum, platycarpum, podacanthum, polyceratium, polycystum, polyphyllum, portierianum, primitivum, prismaticum, pseudocystocarpum, pseudolanceolatum, pteropleuron, pulchellum, pumilum, pusillum, pyriforme, qingdaoense, qinzhounse, qionghaiense, quinhonense, ramentaceum, ramifolium, rhizophorum, ringoldianum, rostratum, sagamianum, salicifolioides, saltii, samarense, sanyaense, scabridum, scabripes, scherzerianum, scopula, segii, serratifolium, setifolium, shandongense, shangchuanii, siliculosoides, siliquastrum, siliquosum, silvae, sinclairii, sinicola, skottsbergii, sociale, sonderi, spathulaefolium, spathulophyllum, spinifex, spinuligerum, squarrosum, steinitzii, stenophyllum, stolonifolium, subdroserifolium, subfalcatum, subrepandum, subspathulatum, subtilissimum, sullivanii, swartzii, symphyorhizoideum, tahitense, taiwanicum, telephifolium, templetonii, tenerrimum, tenuifolioides, tenuifolium, tenuissimum, teretifolium, thivyae, thunbergii, tilesii, torvum, tosaense, trichocarpum, trichophyllum, tristichum, turbinarioides, turbinatifolium, umezakii, vachellianum, varians, vaysierianum, velasquezii, verruculosum, vestitum, vigorosum, virgatum, vulgare, wakayamaense, wangii, weizhounse, wenchangense, wightii, wsizhouense, xishaense, yamadae, yamamotoi, yendoi, yezoense, yinggehaiense, yongxingense, yoshidae, zacae, zhangii*

VARIETIES *acinarium acinaris, acinarium crassiuscula, acinarium hildebrandtii, acinarium mauritianum, acinarium obockianum, acinarium salicifolium, aquifolium intercedens, aquifolium kuetzingianum, baccularia subcompressum, bermudense contractum, bermudense hillebrandtii, bermudense pinnatifidum, bermudense stagnalis, biserrula apiculiferum, biserrula ceylonense, biserrula goodiense, biserrula prionocarpum, biserrula rudolphianum, biserrula singapoorense, biserrula tranquebarense, boveanum aterrimum, boveanum somalense, capillare koenigii, capillare pseudocapillare, capillare tranquebarense, carpophyllum aristiferum, carpophyllum honomense, cinctum gracilentum, cinctum mixtum, cinctum thiebautii, cinereum berberifolium, cinereum lepidum, cinereum thunbergii, crassifolium samoensis, cristaefolium dubium, cristaefolium rotundatum, cristaefolium upolensis, cuneifolium adenense, cuneifolium integerrimum, cuneifolium kamaranense, cuneifolium steudneri, cylindrocystum faurotii, cylindrocystum obockianum, cymosum dichocarpum, cymosum esperi, cymosum gueinzii, cymosum lendigerum, cymosum nanum, cymosum preissii, cymosum scabriusculum, cystophyllum parcespinosa, diversifolium salicifolium, echinocarpum ambiguum, elegans ludwigii, fallax cygnorum, filifolium aciculare, filipendula contractum, filipendula laxum, filipendula montagnei, filipendula pinnatum, fissifolium furcaefolium, flavicans nudum, flavicans pervillei, furcatum diversifolium, furcatum humile, glaucescens ivanii, glaucescens wernerianum, gracile pseudogracile, gracile pseudogranuliferum, granuliferum dubiosum, hemiphyllum chinense, heterocystum timoriense, hombronianum ferax, hombronianum manilense, hombronianum plagiophylloides, hystrix buxifolium, hystrix spinulosum, ilicifolium acaraeocarpum, ilicifolium benkulense, ilicifolium compactum, ilicifolium conduplicatum, ilicifolium erythraeum, ilicifolium fauvelii, ilicifolium lasgoriense, ilicifolium microdon, ilicifolium montebellense, ilicifolium oocystoides, ilicifolium porosissimum, ilicifolium pseudospinulosum, ilicifolium pseudosquarrosum, ilicifolium venustum, incisifolium nulliporum, incisifolium suhrii, johnstonii gracile, johnstonii laxius, lanceolatum colomboense, lanceolatum nicobaricum, latifolium polycarpum, latifolium seychellense, latifolium zanzibaricum, lendigerum mombassaense, linearifolium serrulatum, marginatum anjerense, marginatum cannanorense, marginatum diotis, marginatum subplumbeum, microcystum dilatatum, microcystum euryphyllum, microcystum luzonense, microcystum microtis, myriocystum elegans, myriocystum grandifolium, myriocystum latiusculum, natans capillifolium, natans spinuligerum, novae-hollandiae subdistichum, obovatum pfeifferae, obovatum robillardii, obtusifolium chamberlainii, obtusifolium lendigersides, obtusifolium reichelii, odontocarpum lanceolatum, oligocystum bernardium, oligocystum ceylanicum, oligocystum maheense, oligocystum subflexuosum, oocyste bernardinum, oocyste chierchii, pacificum congestum, pacificum megaphyllum, pacificum rigidiusculum, parvifolium remissum, parvifolium runcinatum, patens rodgersianum, patens schizophyllum, piluliferum nhatrangense, plagiophyllum benkulense, plagiophyllum consimile, plagiophyllum hebetatum, plagiophyllum singapoorense, polyacanthum americanum, polyceratium ovatum, polycystum crinitum, polycystum festivum, polycystum intercedens, polycystum onustum, polycystum pergracile, polycystum proliferum, polycystum subecostatum, polyphyllum densa, polyphyllum fissifolium, portierianum lendigerioides, portierianum mombassiense, pulchellum gracilensis, pulchellum subspathulatum, ringoldianum coreanum, shandongense linearium, siliquastrum fuscescens, siliquastrum gardneri, siliquastrum tenue, siliquosum basilanicum, siliquosum bicornutum, siliquosum manipense, sinicola camouii, spathulaefolium cuneatum, squarrosum sumatranum, subrepandum figarianum, subrepandum turneri, telephifolium natalense, tenerrimum ambiguum, tenerrimum campbellianum, thunbergii latifolium, thunbergii nipponicum, vaysierianum acinarioides, vaysierianum assabiense, vaysierianum assarkense, vulgare angustifolium, vulgare diversifolium, vulgare foliosissimum, vulgare humile, vulgare indicum, vulgare latifolium, vulgare linearifolium, vulgare nanum, wightii sublineare, wightii submembranaceum*

GENUS *Acystis* — Cyst Sargasso Painted Dusky Brown Ocher Algae
SPECIES *heinii*
GENUS *Anthophycus* — Flower Sargasso Painted Dusky Brown Ocher Algae
SPECIES *longifolius*
GENUS *Carpophyllum* — Carpel Sargasso Painted Dusky Brown Ocher Algae
SPECIES *angustifolium, flexuosum, macrophyllum, maschalocarpum, plumosum*
VARIETIES *angustifolium filiforme*
GENUS *Cladophyllum* — Branched Sargasso Painted Dusky Brown Ocher Algae
SPECIES *schnetteri*
GENUS *Nizamuddinia* — Nizamudd Sargasso Painted Dusky Brown Ocher Algae
SPECIES *zanardinii*
GENUS *Oerstedtia* — Oerstedt Sargasso Painted Dusky Brown Ocher Algae
SPECIES *scalaris*
GENUS *Sargassopsis* — Lesser Sargasso Painted Dusky Brown Ocher Algae
SPECIES *filiformis*
GENUS *Turbinaria* — Turbin Sargasso Painted Dusky Brown Ocher Algae
SPECIES *condensata, conoides, costata, crateriformis, decurrens, denudata, elatensis, filamentosa, foliosa, gracilis, indica, kenyaensis, luzonensis, murrayana, ornata, papenfussii, parvifolia, tanzaniensis, trialata, tricostata, triquetra, turbinata*
VARIETIES *conoides laticuspidata, conoides retroflexa, ornata ecoronata, ornata evesiculosa, ornata hainanensis, ornata serrata, trialata capensis, tricostata weberae*

FAMILY 3 *Seirococcaceae* — Round Painted Dusky Brown Ocher Algae
GENUS *Seirococcus* — Round Painted Dusky Brown Ocher Algae
SPECIES *axillaris*
GENUS *Cystosphaera* — Cyst Round Painted Dusky Brown Ocher Algae
SPECIES *jacquinotii*
GENUS *Cytoseira* — (Synonym for *Seirococcus*)
GENUS *Marginariella* — Margin Round Painted Dusky Brown Ocher Algae
SPECIES *boryana, urvilliana*
GENUS *Phyllospora* — Spore Round Painted Dusky Brown Ocher Algae
SPECIES *comosa, menziesii*
GENUS *Scytothalia* — Scyte Thal Round Painted Dusky Brown Ocher Algae
SPECIES *dorycarpa, obscura*

ORDER 4 *Laminariales* — Laminar Dusky Brown Ocher Algae

FAMILY 1 *Laminariaceae* — Laminar Dusky Brown Ocher Algae
GENUS *Laminaria* — Laminar Dusky Brown Ocher Algae
SPECIES *abyssalis, agardhii, andersonii, angustata, belvisii, biruncinata, bongardiana, bracteata, brasiliensis, brongardiana, bullata, cichorioides, complanata, conica, coriacea, cornea, crassifolia, diabolica, digitata, discolor, ephemera, farlowii, fissilis, flagellum, flexicaulis, gunneri, hyperborea, japonica, lamourouxii, leptopoda, longicruris, longipedalis, longipes, longissima, lyrata, nigripes, ochotensis, ochroleuca, ophiura, pallida, palmata, papyrina, platymeris, polyphylla, pseudodigitata, punctata, religosa, remotifolia, repens, rodriguezii, ruprechtiana, ruprechtii, saccharina, sachalinensis, sarniensis, setchellii, sinclairii, solidungula, taeniata, turneri, viridissima, vittata, yendoana, yezoensis*
VARIETIES *agardhii angustissima, agardhii vittata, brongardiana palmata, diabolica longipes, digitata bifida, digitata elliptica, digitata ligulata, digitata lyrata, digitata palmata, digitata pseudosaccharina, hyperborea compressa, longipedalis diabolica, saccharina agardhi, saccharina viridissima, saccharina viridissima, saccharina vittata*
GENUS *Akkesiphycus* — Akkes Laminar Dusky Brown Ocher Algae

SPECIES *lubricum*
GENUS *Arthrothamnus* — Straight Thamnid Laminar Dusky Brown Ocher Algae
SPECIES *bifidus, kurilensis*
GENUS *Costaria* — Ribbed Laminar Dusky Brown Ocher Algae
SPECIES *costata*
VARIETIES *costata cuneata*
GENUS *Cymathere* — Cymathere Laminar Dusky Brown Ocher Algae
SPECIES *japonica, triplicata*
GENUS *Gigantea* — Giant Laminar Dusky Brown Ocher Algae
SPECIES *bulbosa*
GENUS *Halosiphon* — Halogen Siphon Laminar Dusky Brown Ocher Algae
SPECIES *tomentosus*
GENUS *Hedophyllum* — Hedon Laminar Dusky Brown Ocher Algae
SPECIES *sessile*
GENUS *Kjellmaniella* — Kjellman Laminar Dusky Brown Ocher Algae
SPECIES *crassifolia, gyrata*
GENUS *Phycocastanum* — Chestnut Laminar Dusky Brown Ocher Algae
SPECIES *bulbosum*
GENUS *Pleurophycus* — Pleural Laminar Dusky Brown Ocher Algae
SPECIES *gardneri*
GENUS *Polyschidea* — Schide Laminar Dusky Brown Ocher Algae
SPECIES *bulbosa*
GENUS *Saccharina* — Saccharin Laminar Dusky Brown Ocher Algae
SPECIES *plana*
GENUS *Streptophyllopsis* — Twisted Laminar Dusky Brown Ocher Algae
SPECIES *kuroshioense*
GENUS *Thalassiophyllum* — Thalass Laminar Dusky Brown Ocher Algae
SPECIES clathrum

FAMILY 2 *Alariaceae* — Alar Laminar Dusky Brown Ocher Algae
GENUS *Alaria* — Alar Laminar Dusky Brown Ocher Algae
SPECIES *angusta, cordata, crassifolia, crispa, dolichorhachis, eliptica, esculenta, fistulosa, flagellaris, grandifolia, linearis, marginata, nana, oblonga, paradisea, praelonga, pylaii, taeniata, tenuifolia, valida*
VARIETIES *esculenta angustifolia, esculenta latifolia, esculenta pinnatifida, esculenta polyphylla, marginata musaeformis*
GENUS *Ecklonia* — Ecklon Alar Laminar Dusky Brown Ocher Algae
SPECIES *biruncinata, brevipes, cava, exasperata, fastigiata, kurome, maxima, muratii, radiata, richardiana, stolonifera*
VARIETIES *radiata biruncinata*
GENUS *Eckloniopsis* — Lesser Ecklon Alar Laminar Dusky Brown Ocher Algae
SPECIES *radicosa*
GENUS *Egregia* — Greg Alar Laminar Dusky Brown Ocher Algae
SPECIES *laevigata, menziesii*
GENUS *Eisenia* — Eisen Alar Laminar Dusky Brown Ocher Algae
SPECIES *arborea, bicyclis, desmarestioides, galapagensis, masonii*
GENUS *Orgyia* — Orgy Alar Laminar Dusky Brown Ocher Algae
SPECIES *tetragona*
GENUS *Orimanthis* — Orim Flower Alar Laminar Dusky Brown Ocher Algae
SPECIES *foliacea, vesiculata*
GENUS *Pterygophora* — Winged Alar Laminar Dusky Brown Ocher Algae
SPECIES *californica*
GENUS *Undaria* — Undar Alar Laminar Dusky Brown Ocher Algae
SPECIES *crenata, pinnatifida, undarioides*
VARIETIES *pinnatifida elongata, pinnatifida vulgaris*
GENUS *Undariella* — Lesser Undar Alar Laminar Dusky Brown Ocher Algae
SPECIES *kurilensis, peterseniana*

FAMILY 3 *Chordaceae* — Chord Laminar Dusky Brown Ocher Algae
GENUS *Chorda* — Chord Laminar Dusky Brown Ocher Algae
SPECIES *filum, minuta, rigida*
VARIETIES *filum crassipes, filum thrix*
GENUS *Pseudochorda* — Lesser Chord Laminar Dusky Brown Ocher Algae
SPECIES *gracilis, nagaii*

FAMILY 4 *Lessoniaceae* — Lesson Laminar Dusky Brown Ocher Algae
GENUS *Lessonia* — Lesson Laminar Dusky Brown Ocher Algae
SPECIES *adamsiae, brevifolia, ciliata, corrugata, frutescens, fuscescens, laminarioides, nigrescens, repens, tholiformis, trabeculata, vadosa, variegata*
GENUS *Dictyoneuropsis* — Lesser Net Nerve Lesson Laminar Dusky Brown Ocher Algae

SPECIES *reticulata*
GENUS *Dictyoneurum* Net Nerve Lesson Laminar Dusky Brown Ocher Algae
SPECIES *californicum*
GENUS *Lessoniopsis* Lesser Lesson Laminar Dusky Brown Ocher Algae
SPECIES *littoralis*
GENUS *Macrocystis* Large Cyst Lesson Laminar Dusky Brown Ocher Algae
SPECIES *angustifolia, integrifolia, laevis, menziesii, pyrifera*
VARIETIES *pyrifera humboldtii*
GENUS *Nereocystis* Nere Cyst Lesson Laminar Dusky Brown Ocher Algae
SPECIES *luetkeana*
GENUS *Pelagophycus* Pelagic Lesson Laminar Dusky Brown Ocher Algae
SPECIES *porra*
GENUS *Postelsia* Postels Lesson Laminar Dusky Brown Ocher Algae
SPECIES *palmaeformis*

FAMILY 5 *Phyllariaceae* Phyllar Laminar Dusky Brown Ocher Algae
GENUS *Phyllaria* (Synonym for *Saccorhiza*)
GENUS *Haligenia* Halogen Phyllar Laminar Dusky Brown Ocher Algae
SPECIES *belvisii*
GENUS *Phyllariopsis* Lesser Phyllar Laminar Dusky Brown Ocher Algae
SPECIES *brevipes, purpurascens*
VARIETIES *brevipes pseudopurpurascens*
GENUS *Saccorhiza* Sac Rhizid Phyllar Laminar Dusky Brown Ocher Algae
SPECIES *dermatodea, polyschides*
VARIETIES *polyschides brevipes, polyschides gymnopoda, polyschides linearis, polyschides nodipes, polyschides punctata, polyschides turneri*

ORDER 5 *Scytosiphonales* Scyte Siphon Dusky Brown Ocher Algae

FAMILY 1 *Scytosiphonaceae* Scyte Siphon Dusky Brown Ocher Algae
GENUS *Scytosiphon* Scyte Siphon Dusky Brown Ocher Algae
SPECIES *canaliculatus, clathratus, complanatus, compressus, crispus, dotyi, echinatus, gracilis, lomentaria, percursus, ramulosus, templetonii, tenellus*
VARIETIES *clathratus uncinatus, compressus confervoideus, compressus crispatus, compressus nematodes, compressus trichodes, lomentaria cylindricus, lomentaria fistulosus, lomentaria gracilis*
GENUS *Analipus* Analip Scyte Siphon Dusky Brown Ocher Algae
SPECIES *filiformis, fusiformis, gunjii, japonicus*
GENUS *Colonella* Colon Scyte Siphon Dusky Brown Ocher Algae
SPECIES *kajrahatensis*
GENUS *Colpomenia* Colpomene Scyte Siphon Dusky Brown Ocher Algae
SPECIES *bullosa, durvillaei, durvillei, ecuticulata, mollis, nainativensis, peregrina, phaeodactyla, ramosa, sinuosa*
VARIETIES *sinuosa expansa, sinuosa firmior, sinuosa peregrina, tuberculata*
GENUS *Compsonema* Compsoneme Scyte Siphon Dusky Brown Ocher Algae
SPECIES *chordae, coccophorae, coniferum, dictyotoides, fructosum, gracile, immixtum, intercalare, intricatum, microspongium, minutum, nummuloides, oblongum, saxicolum, secundum, serpens, sessile, sporangiiferum, tenue*
VARIETIES *secundum terminale*
GENUS *Diplura* Diplural Scyte Siphon Dusky Brown Ocher Algae
SPECIES *simplex, simulans*
GENUS *Endarachne* Spider Scyte Siphon Dusky Brown Ocher Algae
SPECIES *binghamiae, biswasiana*
GENUS *Endopleura* Endopleural Scyte Siphon Dusky Brown Ocher Algae
SPECIES *aurea*
GENUS *Haplogloia* Glow Scyte Siphon Dusky Brown Ocher Algae
SPECIES *andersonii, kuckuckii, moniliformis*
GENUS *Hapterophycus* Hapter Scyte Siphon Dusky Brown Ocher Algae
SPECIES *echigoensis*
GENUS *Heterochordaria* Chord Scyte Siphon Dusky Brown Ocher Algae
SPECIES *abietina*
VARIETIES *abietina simplex*
GENUS *Heteroralfsia* Ralfs Scyte Siphon Dusky Brown Ocher Algae
SPECIES *saxicola*
GENUS *Ilea* Ile Scyte Siphon Dusky Brown Ocher Algae
SPECIES *bertolonii, clathrata, compressa, fulvescens, intestinalis, torta*
VARIETIES *clathrata uncinata, compressa crinita*
GENUS *Iyengaria* Iyengar Scyte Siphon Dusky Brown Ocher Algae
SPECIES *stellata*

GENUS *Jolyna* Jolyn Scyte Siphon Dusky Brown Ocher Algae
SPECIES *furcata, laminarioides*
GENUS *Petalonia* Petal Scyte Siphon Dusky Brown Ocher Algae
SPECIES *fascia, filiformis, zosterifolia*
VARIETIES *fascia caespitosa, fascia dictyotoides, fascia debilis, fascia pervulgata*
GENUS *Rosenvingea* Rosenvinge Scyte Siphon Dusky Brown Ocher Algae
SPECIES *antillarum, fastigiata, floridana, intricata, nhatrangensis, orientalis, sanctae-crucis*
VARIETIES *fastigiata major, sanctae-crucis scytosiphonoides*
GENUS *Sorapion* Sorapion Scyte Siphon Dusky Brown Ocher Algae
SPECIES *kjellmanii, simulans*
GENUS *Stragularia* Stragular Scyte Siphon Dusky Brown Ocher Algae
SPECIES *clavata, spongiocarpa*
VARIETIES *clavata laminariae*
GENUS *Symphyocarpus* Symphyony Scyte Siphon Dusky Brown Ocher Algae
SPECIES *strangulans*

ORDER 6 *Sphacelariales* Sphacelar Dusky Brown Ocher Algae

FAMILY 1 *Sphacelariaceae* Sphacelar Dusky Brown Ocher Algae
GENUS *Sphacelaria* Sphacelar Dusky Brown Ocher Algae
SPECIES *affinis, arctica, australis, axilaris, bertiana, biradiata, bornetii, brachygona, bracteata, brevicorne, caespitula, californica, capensis, carolinensis, carpoglossi, cervicornis, ceylanica, chorizocarpa, cirrosa, cornuta, didichotoma, divaricata, filicina, fusca, harveyana, implicata, indica, kovalamensis, lacustris, limicola, lutea, mexicana, mirabilis, multiplex, nana, nipponica, norrisii, novae-caledoniae, novae-hollandiae, plumigera, plumosa, plumula, pulvinata, racemosa, radicans, reinkei, reticulata, rigidula, sauvageaui, sessilis, solitaria, spuria, subfusca, sympodiocarpa, taitensis, tribuloides, tsengii, yamadae*
VARIETIES *cirrosa fusca, cirrosa meridionalis, filicina aestavalis, filicina hiemalis, radicans caespitosa, radicans elatior, recurva, rigidula princeps*
GENUS *Cladostephus* Branched Crown Sphacelar Dusky Brown Ocher Algae
SPECIES *ceratophyllum, clavaeforme, hariotii, hedwigioides, myriophyllum, setaceus, spongiosus*
VARIETIES *spongiosus verticillatus*
GENUS *Sphacella* Lesser Sphacelar Dusky Brown Ocher Algae
SPECIES *Sphacella*
FAMILY 2 *Stypocaulaceae* Stipe Sphacelar Dusky Brown Ocher Algae
GENUS *Stypocaulon* Stipe Sphacelar Dusky Brown Ocher Algae
SPECIES *durum, funiculare, paniculatum, scoparium*
VARIETIES *scoparium aestivalis, scoparium hiemalis, scoparium patentissima*
GENUS *Alethocladus* Branched Stipe Sphacelar Dusky Brown Ocher Algae
SPECIES *corymbosus*
GENUS *Halopteris* Winged Stipe Sphacelar Dusky Brown Ocher Algae
SPECIES *congesta, filicina, hordacea, novae-zelandiae, obovata, platycena, pseudospicata, ramulosa, virgata*
GENUS *Phloiocaulon* Tailed Stipe Sphacelar Dusky Brown Ocher Algae
SPECIES *foecundum, spectabile, squamulosum, suhrii*
GENUS *Ptilopogon* Beard Stipe Sphacelar Dusky Brown Ocher Algae
SPECIES *botryocladus*

CLASS 615.7 *Pinguiophyceae* Greasy Ocher Algae

ORDER 1 *Pinguiochrysidales* Golden Greasy Ocher Algae

FAMILY 1 *Pinguiochrysidaceae* Golden Greasy Ocher Algae
GENUS *Pinguiochrysis* Golden Greasy Ocher Algae
SPECIES *pyriformis*
GENUS *Glossomastix* Gloss Mastic Golden Greasy Ocher Algae
SPECIES *chrysoplasta*
GENUS *Phaeomonas* Dusky Monad Golden Greasy Ocher Algae
SPECIES *parva*
GENUS *Pinguiococcus* Round Golden Greasy Ocher Algae
SPECIES *pyrenoidosus*
GENUS *Polypodochrysis* Polypod Golden Greasy Ocher Algae
SPECIES *teissieri*

CLASS 615.8 *Raphidophyceae* Needle Ocher Algae

ORDER 1 *Raphidiales* Needle Ocher Algae

FAMILY 1 *Raphidiaceae*	Needle Ocher Algae
GENUS *Raphidium*	Needle Ocher Algae
SPECIES *falcatum, polymorphum*	
GENUS *Haramonas*	Hara Monad Needle Ocher Algae
SPECIES *dimorpha, viridis*	
GENUS *Heterosigma*	Sigma Needle Ocher Algae
SPECIES *akashiwo, carterae*	
VARIETIES *akashiwo sanguinea*	
GENUS *Raphidocelis*	Celis Needle Ocher Algae

SPECIES *contorta, danubiana, inclinata, pseudomucosa, rotunda, sigmoidea, subcapitata, turfosa, valida, van-goorii*

VARIETIES *contorta gracillima, inclinata serialis, van-goorii decussata*

GENUS *Rhapidomonas*	(Synonym for *Raphidium*)
GENUS *Raphidogloea*	Glow Needle Ocher Algae
SPECIES *interrupta*	
GENUS *Rickertia*	(Synonym for *Raphidium*)
GENUS *Thaumatomastix*	Thaumast Needle Ocher Algae
SPECIES *tripus*	
FAMILY 2 *Vacuolariaceae*	Vacuole Needle Ocher Algae
GENUS *Vacuolaria*	Vacuole Needle Ocher Algae
SPECIES *virescens*	
GENUS *Chattonella*	Chatton Vacuole Needle Ocher Algae

SPECIES *antiqua, globosa, harima, japonica, marina, minima, ovata, subsalsa, verruculosa*

GENUS *Fibrocapsa*	Fiber Capsule Vacuole Needle Ocher Algae
SPECIES *japonica*	
GENUS *Gonyostomum*	Gony Stoma Vacuole Needle Ocher Algae
SPECIES *latum, semen*	
GENUS *Hemieutreptia*	Half Trept Vacuole Needle Ocher Algae
SPECIES *antiqua*	
GENUS *Merotricha*	Sea Hairy Vacuole Needle Ocher Algae
SPECIES *bacillata*	

PHYLUM 616 *Prasinophyta* Unicell Green Algae

Latin: Prasinus green
Latin: Phyta phyton, plant
147 species [Guiry, Dec. 2004]

CLASS	Species
616.1 *Prasinophyceae* Unicell Green Algae	147

Phylum *Prasinophyta* Green Algae species are single cell organisms.

Other Classifications
In previous classification systems, Class *Prasinophyceae* was included in Phylum *Chlorophyta*. Scientist M. Guiry custodian of AlgaeBase www.algaebase.org raised the Class *Prasinophyceae* to the level of a Phylum on 14 Dec. 2004 because all other classes in Phylum *Chlorophyta* are multicell.

CLASS 616.1 *Prasinophyceae* Unicell Green Algae

ORDER 1 *Prasinales* Unicell Green Algae

FAMILY 1 *Prasinaceae* — Unicell Green Algae
GENUS *Pascheriella* — Pascher Unicell Green Algae
SPECIES *tetras*
GENUS *Platymonas* — Flat Unicell Green Algae
SPECIES *convolutae, subcordiformis, suecica, tetrahele*
GENUS *Saharifusa* — Saharifus Unicell Green Algae
SPECIES *arenacea, cucurbita, rochesterensis, saudiarabiae, striatogranulata*

FAMILY 2 *Chlorodendraceae* — Branched Unicell Green Algae
GENUS *Myochloris* — Mouse Branched Unicell Green Algae
SPECIES *collorhynchus*
GENUS *Prasinocladus* — Prasin Branched Unicell Green Algae
SPECIES *lubricus, malaysianus, marinus*

FAMILY 2 *Pterosphaeridiaceae* — Rock Sphere Unicell Green Algae
GENUS *Cymatiosphaera* — Cyme Rock Sphere Unicell Green Algae
SPECIES *craticula, globulosa, reticulosa*
GENUS *Dictyotidium* — Netted Rock Sphere Unicell Green Algae
SPECIES *craticula*

ORDER 2 *Dunaliellales* Dunal Unicell Green Algae

FAMILY 1 *Dunaliellaceae* — Dunal Unicell Green Algae
GENUS *Dunaliella* — Dunal Unicell Green Algae
SPECIES *acidophila, bardawil, bioculata, minuta, parva, peircei, polymorpha, primolecta, quartolecta, salina, tertiolecta*
GENUS *Hafniomonas* — Hafnion Dunal Unicell Green Algae
SPECIES *reticulata*

FAMILY 2 *Nephroselmidaceae* — Kidney Selm Dunal Unicell Green Algae
GENUS *Nephroselmis* — Kidney Selm Dunal Unicell Green Algae
SPECIES *discoidea, olivacea, pyriformis, rotunda, spinosa*
GENUS *Anticomonas* — (Synonym for *Pseudopedinomonas coccinea*)
GENUS *Argillamonas* — Argill Kidney Selm Dunal Unicell Green Algae
SPECIES *brasiliana*
GENUS *Fluitomonas* — Fluid Kidney Selm Dunal Unicell Green Algae
SPECIES *elliptica*
GENUS *Hiemalomonas* — Winter Kidney Selm Dunal Unicell Green Algae
SPECIES *frigida*
GENUS *Pseudopedinomonas* — Pedin Kidney Selm Dunal Unicell Green Algae
SPECIES *astigmata, coccinea, hiemalis, oculata*
GENUS *Sinamonas* — China Kidney Selm Dunal Unicell Green Algae
SPECIES *stagnalis*

FAMILY 3 *Polyblepharidaceae* — Eyelash Dunal Unicell Green Algae
GENUS *Polyblepharides* — Eyelash Dunal Unicell Green Algae
SPECIES *singularis*
GENUS *Asteromonas* — Star Eyelash Dunal Unicell Green Algae
SPECIES *gracilis*
GENUS *Aulacomonas* — Aulac Eyelash Dunal Unicell Green Algae
SPECIES *submarina*
GENUS *Chloronephris* — Kidney Eyelash Dunal Unicell Green Algae

SPECIES *pigra*
GENUS *Collodictyon* — Netted Eyelash Dunal Unicell Green Algae
SPECIES *sphaericum*
GENUS *Collonella* — Collon Eyelash Dunal Unicell Green Algae
SPECIES *kajrahatensis*
GENUS *Pedinomonas* — Pedin Eyelash Dunal Unicell Green Algae
SPECIES *major, minor, tuberculata*
GENUS *Phyllocardium* — Heart Eyelash Dunal Unicell Green Algae
SPECIES *complanatum*
GENUS *Polytomella* — Polytome Eyelash Dunal Unicell Green Algae
SPECIES *caeca, magna, papillata¸ parva¸ piriformis*
GENUS *Raciborskiella* — Raciborski Eyelash Dunal Unicell Green Algae
SPECIES *salina*
GENUS *Scherffelia* — Scherffel Eyelash Dunal Unicell Green Algae
SPECIES *dubia, pelagica*
GENUS *Scourfieldia* — Scourfield Eyelash Dunal Unicell Green Algae
SPECIES *caeca, complanata¸ cordiformis*
GENUS *Selenochloris* — Selen Eyelash Dunal Unicell Green Algae
SPECIES *lobata, stigmatophora*
GENUS *Spermatozopsis* — Spore Eyelash Dunal Unicell Green Algae
SPECIES *exsultans, similis*
GENUS *Spermipole* — Spore Pole Eyelash Dunal Unicell Green Algae
SPECIES *effusa*
GENUS *Spermosira* — Pitted Eyelash Dunal Unicell Green Algae
SPECIES *hallensis*
GENUS *Stephanoptera* — Crown Wing Eyelash Dunal Unicell Green Algae
SPECIES gracilis
GENUS *Sycamina* — Sycamine Eyelash Dunal Unicell Green Algae
SPECIES *nigrescens*
GENUS *Tetraselmis* — Tetrad Selm Eyelash Dunal Unicell Green Algae
SPECIES *alacris, apiculata, ascus, astigmatica, chuii, convolutae, cordiformis, desikacharyi, gracilis, hazeni, impellucida, inconspicua, levis, maculata, marina, micropapillata, rubens, striata, suecica, tetrabrachia,, tetrathele, verrucosa, wettsteinii*

ORDER 3 *Mamiellales* — Mamy Unicell Green Algae

FAMILY 1 *Mamiellaceae* — Mamy Unicell Green Algae
GENUS *Mamiella* — Mamy Unicell Green Algae
SPECIES *gilva*
GENUS *Bathycoccus* — Bottom Round Mamy Unicell Green Algae
SPECIES *prasinos*
GENUS *Dolichomastix* — Dolich Mastic Mamy Unicell Green Algae
SPECIES *nummulifera, tenuilepis*
GENUS *Mantoniella* — Manton Mamy Unicell Green Algae
SPECIES *antarctica, squamata*
GENUS *Maranhites* — Maranhite Mamy Unicell Green Algae
SPECIES *perplexus*
GENUS *Micromonas* — Micron Mamy Unicell Green Algae
SPECIES *pusilla*
GENUS *Ostreococcus* — Oyster Round Mamy Unicell Green Algae
SPECIES *tauri*

FAMILY 2 *Pycnococcaceae* — Dense Round Mamy Unicell Green Algae
GENUS *Pycnococcus* — Dense Round Mamy Unicell Green Algae
SPECIES *provasolii*
GENUS *Prasinococcus* — Prasin Dense Round Mamy Unicell Green Algae
SPECIES *capsulatus*
GENUS *Prasinoderma* — Prasin Dermal Dense Round Mamy Unicell Green Algae
SPECIES *coloniale*
GENUS *Pseudoscourfieldia* — Scourfield Dense Round Mamy Unicell Green Algae
SPECIES *marina*

ORDER 4 *Pyramimonadales* Pyram Unicell Green Algae

FAMILY 1 *Pyramimonadaceae* — Pyram Unicell Green Algae
GENUS *Pyramimonas* — Pyram Unicell Green Algae

SPECIES *amylifera, australis, chlorina, cyrtoptera, delicatula, disomata, formosa, grossii, inconstans, mucifera, obovata, octociliata, olivacea, orientalis, parkeae, propulsa, pseudoparkeae, reticulata, robusta, spinifera, tetrarhynchus, tychotreta, virginica*

GENUS *Gyromitus* — Gyre Miter Pyram Unicell Green Algae

SPECIES *cordiformis, distomatus*

GENUS *Halosphaera* — Halogen Sphere Pyram Unicell Green Algae

SPECIES *viridis*

GENUS *Kuzminia* — Kuzmin Pyram Unicell Green Algae

SPECIES *incerta*

GENUS *Pocillomonas* — Pocill Pyram Unicell Green Algae

SPECIES *flos-aquae*

GENUS *Prasinochloris* — Prasin Pyram Unicell Green Algae

SPECIES *sessilis*

GENUS *Protoaceromonas* — Primitive Sharp Pyram Unicell Green Algae

SPECIES *spinosa*

GENUS *Trichloridella* — Hairy Pyram Unicell Green Algae

SPECIES *paradoxa*

FAMILY 2 *Pterospermataceae* — Wing Spore Pyram Unicell Green Algae

GENUS *Pterosperma* — Wing Spore Pyram Unicell Green Algae

SPECIES *cristatum, marginatum, moebii, nationalis, polygonum, undulatum, vanhoeffenii*

GENUS *Pachysphaera* — Thick Sphere Wing Spore Pyram Unicell Green Algae

SPECIES *marshalliae, pelagica*

GENUS *Polyasterias* — Many Star Wing Spore Pyram Unicell Green Algae

SPECIES *problematica*

PHYLUM 617 *Rhodophyta* Red Algae

Latin: *Rhoda* red	**CLASSES**		**Species**
Latin: *Phyta* phyton, plant	617.1 *Bangiophyceae*	Bang Red Algae	
	617.2 *Compsopogonophyceae*	Beard Red Algae	
4100 species [Margulis & Schwartz, 1998]	617.3 *Florideophyceae*	Florid Red Algae	
6635 species [Guiry, Dec. 2004]	617.4 *Rhodellophyceae*	Rhodelle Red Algae	

Phylum *Rodophyta* Red Algae, rhodophyte, species grow symbiotically only on other Red Algae. They live at the edges of the sea on the shallow sea floor, on sea rocks, jetties, and pilings and are distributed worldwide. In the tropics, they are found on rocky shores. They range in size from microscopic to 3 feet long. They contain the pigments rhodopsin, allophycocyanina, phycocyanin, and phycoerythrin. Agar, the gelatin formed from a broth and used to grow bacteria in laboratories, is extracted from Red Algae. Polysaccharides from Red Seaweed Algae are used in the manufacture of ice cream, toothpaste, drugs, and cosmetics. The Dulce Red Seaweed Algae which grows in the coast of New England is eaten as a vegetable. Rhodophytes reproduce in alternating haploid sporophyte and diploid gametophyte generations.

CLASS 617.1 *Bangiophyceae* Bang Red Algae

ORDER 1 *Bangiales* Bang Red Algae

FAMILY 1 *Bangiaceae* Bang Red Algae
GENUS *Bangia* Bang Red Algae
SPECIES *atropurpurea, atrovirens, aureola, enteromorphioides, fulvescens, gloiopeltidicola, halymeniae, lutea, quadripunctata, rutilans, tanakai, tenuis, vermicularis, yamadae, zanardinii*
VARIETIES *atropurpurea brevisegmenta*
GENUS *Aspalatia* Aspal Bang Red Algae
SPECIES *andalousica*
GENUS *Bangiella* Lesser Bang Red Algae
SPECIES *atrovirens, crispa, fuscopurpurea, laminariae, mamillosa*
GENUS *Diploderma* Diploderm Bang Red Algae
SPECIES *tenuissimum*
GENUS *Histiona* Histion Bang Red Algae
SPECIES *velifera, zachariasii*
GENUS *Phyllona* Phyllon Bang Red Algae
SPECIES *amethystea, amplissima, angustinae, augustinae, bangiaeformis, carnea, coccinea, cordata, dentata, grayana, lactuca, lanceolata, minor, variegata, woolhouseae*
GENUS *Porphyra* Porphyry Bang Red Algae
SPECIES *abbottiae, acanthophora, adamsiae, aeodis, aestivalis, akasakae, amplissima, angusta, argentinensis, augustinae, autumnalis, bangiaeformis, birdiae, boryi, brumalis, bulbopes, cameronii, capensis, carnea, carolinensis, ceramicola, ceylanica, chauhanii, cordata, crassa, crispata, cucullata, cuneiformis, delicatula, dentata, denticulata, dentimarginata, dioica, drachii, drewiae, drewiana, endiviifolia, endiviifolium, fallax, fucicola, fuscopurpurea, gardneri, grateloupicola, grayana, guangdongensis, haitanensis, helenae, hiberna, hollenbergii, hospitans, inaequicrassa, indica, ionae, irregularis, ishigecola, kanakaensis, kanyakumariensis, kaspar, katadae, kinositae, koreana, kuniedae, kurogii, lacerata, laciniata, lanceolata, ledermannii, leucosticta, lilliputiana, linearis, lucasii, maculosa, marcosii, marginata, martensiana, microphylla, microphylla, miniata, minima, minor, monosporangia, moriensis, mumfordii, nereocystis, nobilis, norrisii, occidentalis, ochotensis, okamurae, okhaensis, oligospermatangia, onoi, papenfussii, pendula, perforata, plocamiestris, pseudoangusta, pseudocrassa, pseudolanceolata, pseudolinearis, pujalsiae, pulchra, punctata, purpurea, rediviva, reniformis, rizzinii, roseana, rosengurttii, saldanhae, schizophylla, segregata, seriata, sericea, smithii, spiralis, subcorticulata, suborbiculata, subtumens, tanegashimensis, tasa, tenera, tenuipedalis, tenuissima, thulaea, thuretii, torta, tristanensis, umbilicalis, variegata, vietnamensis, violacea, vulgaris, woolhouseae, yamadae, yezoensis*
VARIETIES *abbottiae fasciaria, acanthophora brasiliensis, amplissima crassa, amplissima elliptica, amplissima lanceolata, amplissima planiuscula, angusta sanrikuensis, brumalis oligospora, ceramicola vertebrata, cordata orbicularis, fallax conwayae, katadae hemiphylla, laciniata epiphytica, laciniata livida, laciniata novae-zelandiae, laciniata umbilicalis, laciniata umbilicalis, laciniata umbilicata, laciniata vulgaris, leucosticta atlantica, leucosticta coriacea, leucosticta elongata, leucosticta mediterranea, leucosticta suborbiculata, leucosticta undulata, miniata latior, ochotensis lanceolata, perforata patens, purpurea protensa, purpurea umbilicata, spiralis amplifolia, suborbiculata latifolia, tenera kjellmanii, tenera tamatsuensis, umbilicalis autumnalis, umbilicalis divisa, umbilicalis elongata, umbilicalis laciniata, umbilicalis laciniata, umbilicalis lanceolata, umbilicalis novae-zelandiae, umbilicalis pacifica, umbilicalis perforata, umbilicalis pudica, umbilicalis purpurea, umbilicalis rosea, umbilicalis sanguinea, umbilicalis simplex, umbilicalis vulgaris, vulgaris laciniata, vulgaris linearis, vulgaris minor, vulgaris umbilicata, yezoensis coreana, yezoensis narawaensis*

GENUS *Porphyrella* Lesser Porphyry Bang Red Algae
SPECIES *californica*
GENUS *Wildemania* Wildeman Bang Red Algae
SPECIES *amethystea, columbina*
VARIETIES *columbina decipiens, columbina kunthiana, columbina laingii*
GENUS *Zachariasia* Zacharias Bang Red Algae
SPECIES *epiphytica*

CLASS 617.2 *Compsopogonophyceae* Beard Red Algae

ORDER 1 *Compsopogonales* Beard Red Algae

FAMILY 1 *Compsopogonaceae* Beard Red Algae
GENUS *Compsopogon* Beard Red Algae
SPECIES *coeruleus, corticrassus, lividus, lusitanicum, minutus, prolificus, sparsus, tenellus*
VARIETIES *coeruleus catenatum*
GENUS *Boldia* Bold Beard Red Algae
SPECIES *angustata, erythrosiphon*
GENUS *Compsopogonopsis* Lesser Beard Red Algae
SPECIES *fruticosa, japonica*
GENUS *Rhodochaete* Bristle Beard Red Algae
SPECIES *parvula*

FAMILY 2 *Erythrotrichiaceae* Hairy Beard Red Algae
GENUS *Erythrotrichia* Hairy Beard Red Algae
SPECIES *ascendens, bangioides, bertholdii, carnea, foliiformis, hunterae, incrassata, investiens, japonica, kylinii, ligulata, parietalis, parksii, platyphylla, polymorpha, porphyroides, reflexa, regularis, rhizoidea, rosea, sargassicola, simplex, tetraseriata, tristanensis, vexillaris, welwitschii*
VARIETIES *parietalis majuscula, parksii minor*
GENUS *Erythrocladia* Branched Hairy Beard Red Algae
SPECIES *ectozoica, endophloea, irregularis, pinnata, polystromatica, violacea*
GENUS *Erythropeltis* Pelt Hairy Beard Red Algae
SPECIES *discigera*
VARIETIES *discigera flustrae*
GENUS *Membranella* Membrane Hairy Beard Red Algae
SPECIES *africana, nitens*
GENUS *Porphyropsis* Porphyry Hairy Beard Red Algae
SPECIES *coccinea, imperfecta, minuta, vexillaris*
VARIETIES *coccinea dawsonii*
GENUS *Porphyrostromium* Porphyry Stroma Hairy Beard Red Algae
SPECIES *boryanum, ciliare, obscurum*
GENUS *Sahlingia* Sahling Hairy Beard Red Algae
SPECIES *subintegra*
GENUS *Smithora* Smith Hairy Beard Red Algae
SPECIES *naiadum*
VARIETIES *naiadum australis, naiadum major, naiadum minor*

CLASS 617.3 *Florideophyceae* Florid Red Algae

ORDER 1 *Florideales* Florid Red Algae

FAMILY 1 *Florideaceae* Florid Red Algae
GENUS *Apona* Apon Florid Red Algae
SPECIES *barbata, ciliata, diaphana, elegans*
GENUS *Callonema* Callisto Thread Florid Red Algae
SPECIES *olivaceum*
GENUS *Chondria* Cartilage Florid Red Algae
SPECIES *acrorhizophora, angustissima, arcuata, armata, atropurpurea, baileyana, bernardii, boryana, bulbosa, californica, capensis, capillaris, capreolis, chejuensis, cnicophylla, coerulescens, collinsiana, compressa, cornuta, crassicaulis, curdieana, curvilineata, dangeardii, dasyphylla, decidua, decipiens, densa, expansa, flexicaulis, floridana, foliifera, fusifolia, hapteroclada, harveyana, hieroglyphica, hypoglossoides, incrassata, incurva, infestans, intertexta, lanceolata, lancifolia, leptacremon, littoralis, macrocarpa, mageshimensis, mairei, minutula, nidifica, oppositiclada, pellucida, platycladia, platyramea, polyrhiza, pumila, pygmaea, rainfordi, repens, riparia, ryukyuensis, scintillans, sedifolia, seticulosa, sibogae, simpliciuscula, stolonifera, subfasciculata, subsecunda, succulenta, transversalis, viticulosa*
VARIETIES *capillaris californica, capillaris divergens, capillaris minuta, capillaris uncinata, dasyphylla intermedia, dasyphylla stellata*
GENUS *Chondriopsis* (Synonym for *Chondria*)

GENUS *Commation* Commation Florid Red Algae
SPECIES *cryoporinum*
GENUS *Conchocelis* Conch Florid Red Algae
SPECIES *rosea*
GENUS *Donezella* Donez Florid Red Algae
SPECIES *delicata*
GENUS *Endosira* Endic Pitted Florid Red Algae
SPECIES *australis*
GENUS *Enigma* Enigma Florid Red Algae
SPECIES *aculeata*
GENUS *Eschara* Eschar Florid Red Algae
SPECIES *fistulosa, foliacea, verticillata*
GENUS *Gracilariocolax* Slender Colac Florid Red Algae
SPECIES *henriettae*
GENUS *Lenormandia* Lenormand Florid Red Algae
SPECIES *angustifolia, latifolia, marginata, muelleri, pardalis, spectabilis*
GENUS *Masloviporidium* Maslov Florid Red Algae
SPECIES *delicatum*
GENUS *Mastichonema* Mastic Thread Florid Red Algae
SPECIES *pluviale*
VARIETIES *pluviale kemmleri*
GENUS *Nullipora* Null Pore Florid Red Algae
SPECIES *polyphyllamea*
GENUS *Olivia* Olivia Florid Red Algae
SPECIES *botryoides*
GENUS *Perinema* Perin Thread Florid Red Algae
SPECIES *sibogae*
GENUS *Pihiella* Pihi Florid Red Algae
SPECIES *liagoraciphila*
GENUS *Quadricilia* Quad Ciliate Florid Red Algae
SPECIES *rotundata*
GENUS *Rhododiplobia* Diplobe Florid Red Algae
SPECIES *cor-margartiae*
GENUS *Sonderophycus* Sonder Florid Red Algae
SPECIES *australis*

FAMILY 2 *Balbianiaceae* Balbian Florid Red Algae
GENUS *Balbiania* Balbian Florid Red Algae
SPECIES *investiens, meiospora*
GENUS *Rhododraparnaldia* Draparnald Balbian Florid Red Algae
SPECIES *oregonica*

FAMILY 3 *Goniotrichaceae* Angled Hairy Florid Red Algae
GENUS *Asterocytis* Star Angled Hairy Florid Red Algae
SPECIES *echigoensis*
GENUS *Bangiopsis* Bang Angled Hairy Florid Red Algae
SPECIES *dumontioides, humpheryi, subsimplex*
GENUS *Chroodactylon* Finger Angled Hairy Florid Red Algae
SPECIES *depressum, ornatum, wolleanum*
VARIETIES *ornatum simplex*
GENUS *Colacodictyon* Colac Net Angled Hairy Florid Red Algae
SPECIES *reticulatum*

FAMILY 4 *Halymeniaceae* Halymene Florid Red Algae
GENUS *Halymenia* Halymene Florid Red Algae
SPECIES *abyssicola, actinophysa, amoena, angusta, asymmetrica, bermudensis, bifida, brasiliana, californica, cerratei, chiangiana, clathrata, cromwellii, curvicornis, dichotoma, dilatata, dubia, duchassaingii, durvillei, echinophysa, elongata, fasciata, fimbriata, flabellata, floresii, floridana, formosa, gardneri, gelinaria, hancockii, hollenbergii, hvari, imbricata, integra, jelinekii, kraftii, latifolia, lubrica, maculata, megaspora, microcarpa, muelleri, multifida, pennata, pinnulata, plana, platycarpa, pluribola, polyclada, porphyraeformis, porphyroides, pseudofloresii, pusilla, rosea, santamariae, schizymenioides, senegalensis, sinensis, spathulata, stipitata, templetonii, tenera, tenuispina, utriana, venusta, vinacea*
VARIETIES *elongata decipiens, elongata plana, floresii harveyana, floresii pinnata, floresii ulvoidea, polyclada aldabradensis*
GENUS *Acrodiscus* Disc Halymene Florid Red Algae
SPECIES *vidovichii*
GENUS *Aeodes* Aeode Halymene Florid Red Algae
SPECIES *digitata, dilatata, ecuadoreana, lanceolata, marginata, nitidissima, orbitosa, ulvoidea*
VARIETIES *lanceolata tenuis*

GENUS *Carpopeltis* Carpel Pelt Halymene Florid Red Algae
SPECIES *beckeri, bushiae, capitellata, decipiens, elata, maillardii, phyllophora, prolifera, spongeaplexus*
GENUS *Codiophyllum* Cody Halymene Florid Red Algae
SPECIES *flabelliforme, natalense*
GENUS *Cryptonemia* Cryptic Halymene Florid Red Algae
SPECIES *angustata, baileyi, bengryi, borealis, caliculata, chiangii, coriacea, crenulata, decolorata, decumbens, delicatula, denticulata, dichotoma, digitata, flabellifolia, hibernica, kallymenioides, latissima, limensis, lobata, lomation, luxurians, natalensis, nitophylloides, obovata, papenfussii, peruviana, rotunda, seminervis, semiprocumbens, tunaeformis, umbraticola, undulata, veleroae, wilsonii, yendoi*
VARIETIES *delicatula venezuelensis*
GENUS *Cyrtymenia* (Synonym for *Pachymenia*)
GENUS *Dermocorynus* Club Halymene Florid Red Algae
SPECIES *montagnei, occidentalis*
GENUS *Epiphloea* Epiphyte Halymene Florid Red Algae
SPECIES *bullosa*
GENUS *Estebania* Esteban Halymene Florid Red Algae
SPECIES *conjuncta*
GENUS *Gelinaria* Gelin Halymene Florid Red Algae
SPECIES *ulvoidea*
GENUS *Grateloupiocolax* Grateloupe Colac Halymene Florid Red Algae
SPECIES *colombiana*
GENUS *Isabbottia* Isometric Abbott Halymene Florid Red Algae
SPECIES *ovalifolia*
GENUS *Kintokiocolax* Kintok Colac Halymene Florid Red Algae
SPECIES *aggregato-cerantha*
GENUS *Lobocolax* Lobe Colac Halymene Florid Red Algae
SPECIES *deformans*
GENUS *Neoabbottiella* Modern Abbott Halymene Florid Red Algae
SPECIES *araneosa*
GENUS *Norrissia* Norris Halymene Florid Red Algae
SPECIES *setchellii*
GENUS *Pachymenia* Thick Halymene Florid Red Algae
SPECIES *carnosa, cornea, crassa, hieroglyphica, laciniata, lusoria, orbicularis, saxicola*
GENUS *Pachymeniopsis* Lesser Thick Halymene Florid Red Algae
SPECIES *pseudoelliptica*
GENUS *Phyllymenia* Phyllode Halymene Florid Red Algae
SPECIES *belangeri, papenfussii*
GENUS *Polyopes* Polyope Halymene Florid Red Algae
SPECIES *affinis, angustus, clarionensis, constrictus, hakalauensis, intricatus, lancifolius, ligulatus, microcarpus, polyideoides, prolifer, tasmanicus, tenuis, tosaensis*
GENUS *Prionitis* Prion Halymene Florid Red Algae
SPECIES *albemarlensis, angusta, articulata, australis, clarionensis, cornea, corymbifera, crispata, decipiens, divaricata, elata, filiformis, galapagensis, guaymasensis, hancockii, kinoensis, lanceolata, ligulata, linearis, lyallii, mexicana, nodifera, obtusa, patens, pterocladina, ramossissima, schmitziana, simplex, sternbergii, vietnamensis, wentii*
VARIETIES *filiformis delicatula*
GENUS *Sebdenia* Sebden Halymene Florid Red Algae
SPECIES *actinophysa, amoena, canariensis, dawsonii, dichotoma, flabellata, limensis, lindaueri, macaronesica, monardiana, okamurae, polydactlya, rodrigueziana, rubra, senegalensis, yamadae*
GENUS *Sinkoraena* China Korea Halymene Florid Red Algae
SPECIES *okamurae, tasmanica*
GENUS *Thamnoclonium* Thamnid Clone Halymene Florid Red Algae
SPECIES *dichotomum, lemannianum, procumbens, tissotii, treubii*
GENUS *Yonagunia* Yonagun Halymene Florid Red Algae
SPECIES *formosana, tenuifolia*
GENUS *Zanardinula* Zanardin Halymene Florid Red Algae
SPECIES *abbreviata, andersoniana, delicatula, lanceolata, mexicana, vizcainensis*
GENUS *Zymurgia* Zymurg Halymene Florid Red Algae
SPECIES *chondriopsidea*

FAMILY 5 *Hildenbrandiaceae* Hildenbrand Florid Red Algae
GENUS *Hildenbrandia* Hildenbrand Florid Red Algae
SPECIES *angolensis, arracana, chalikophila, crouaniorum, dawsonii, expansa, galapagensis, kerguelensis, lecannellieri, occidentalis, pachythallos, patula, rivularis, rubra, yessoensis*
VARIETIES *rubra fluviatilis, rubra rosea*
GENUS *Apophlaea* Apophlae Hildenbrand Florid Red Algae
SPECIES *lyallii, sinclairii*

FAMILY 6 *Plocamiaceae* Plocam Florid Red Algae

GENUS *Plocamium* Plocam Florid Red Algae

SPECIES *abnorme, affine, angustum, beckeri, brachiocarpum, brasiliense, cartilagineum, cirrhosum, concinnum, corallorhiza, cornutum, costatum, cruciferum, delicatulum, dilatatum, fimbriatum, froelichianum, fuscorubrum, glomeratum, hamatum, hookeri, leptophyllum, maxillosum, mertensii, microcladioides, minutum, oregonum, ovicornis, patagiatum, patens, preissianum, raphelisianum, recurvatum, rigidum, sandvicense, secundatum, serrulatum, suhrii, telfairiae, violacea*

VARIETIES *cartilagineum cartilagineum, cartilagineum coccineus, cartilagineum pacificum, rigidum tenuis, serrulatum pectinatum*

GENUS *Plocamiocolax* Colac Plocam Florid Red Algae

SPECIES *papenfussianus, pulvinata*

FAMILY 7 *Rhodogorgonaceae* Gorgon Florid Red Algae

GENUS *Rhodogorgon* Gorgon Florid Red Algae

SPECIES *ramosissima*

GENUS *Renouxia* Renoux Gorgon Florid Red Algae

SPECIES *antillana*

FAMILY 8 *Thoreaceae* Thor Florid Red Algae

GENUS *Thorea* Thor Florid Red Algae

SPECIES *bachmannii, brodensis, clavata, conturba, gaudichaudii, hispida, okadae, okaidai, prowsei, riekei, violacea, zollingeri*

GENUS *Nemalionopsis* Nemalion Thor Florid Red Algae

SPECIES *shawii, tortuosa*

VARIETIES *shawii carolinensis*

ORDER 2 *Batrachospermales* Tadpole Spore Florid Red Algae

FAMILY 1 *Batrachospermaceae* Tadpole Spore Florid Red Algae

GENUS *Batrachospermum* Tadpole Spore Florid Red Algae

SPECIES *abilii, aequinoctiale, aestivale, ambiguum, americanum, anatinum, androinvolucrum, angolense, annulatum, antipodites, antiquum, arcuatum, atrichum, atrum, australicum, azeredoi, bakarense, bambusinum, beraense, bharadwajii, boryanum, brasiliense, breutelii, breviarticulatum, caerulescens, campyloclonum, capense, capillaceum, carpocontortum, carpoinvolucrum, cayennense, confusum, crassiusculum, crispatum, curvatum, cylindrocellulare, dasyphyllum, debile, delicatulum, deminutum, densiverticillatum, densum, desikacharyi, diaphanum, diatyches, discorum, durum, elegans, equisetifolium, equisetoideum, faciferum, faroense, fennicum, ferreri, flagelliforme, fluitans, gelatinosum, gibberosum, gibbosum, globosporum, gombakense, gracillimum, guianense, gulbenkianum, guyanense, henriquesianum, heterocorticum, heteromorphum, hirosei, hypogynum, intortum, involutum, iriomotense, islandrinum, iyengarii, keratophytum, kraftii, kushiroense, kylinii, lamprogyne, latericium, lochmodes, loefgrenii, longiarticulatum, longipedicellatum, louisianae, luisitanicum, macrosporum, mahabaleshwarense, mahlacense, mascomi, microspermum, miniatum, nechochoense, nodiflorum, nonocense, nova-guineense, omobodoense, orthostichum, periplocum, procarpum, prominens, protimum, pseudocarpum, pseudogelatinosum, puiggarianum, pulchrum, pullum, ranuliferum, rivularioides, sinensis, skujae, skujanum, spermatiophorum, spermatoinvolucrum, stephensii, szechwanense, tabagatenense, tapirense, terawhiticum, theaquum, tiomanense, torridum, torsivum, tortuosum, transtaganum, trichocontortum, trichofurcatum, triste, turfosum, turgidum, ulandrium, verticillatum, virgato-decaisneanum, vittatum, vogesiacum, wattsii, woitapense, zeylanicum, zostericola*

VARIETIES *atrum puiggarianum, gelatinosum scopula, gelatinosum spermatoinvolucrum, procarpum americanum, tortuosum majus, triste chlora, triste coloratum*

GENUS *Batrachospermella* Lesser Tadpole Spore Florid Red Algae

SPECIES *moniliformis*

GENUS *Nothocladus* Branched Tadpole Spore Florid Red Algae

SPECIES *afroaustralis, lindaueri, nodosus*

GENUS *Psilosiphon* Psilum Siphon Tadpole Spore Florid Red Algae

SPECIES *scoparium*

GENUS *Sirodotia* Pitted Dot Tadpole Spore Florid Red Algae

SPECIES *ateleia, cirrhosa, delicatula, gardneri, goebelii, loefgrenii, polygama, segawae, sinica, suecica, yutakae*

VARIETIES *ateleia australis, suecica australis*

GENUS *Tuomeya* Tuomey Tadpole Spore Florid Red Algae

SPECIES *americana*

FAMILY 2 *Acrochaetiaceae* Bristle Florid Red Algae

GENUS *Acrochaetium* Bristle Florid Red Algae

SPECIES *acanthophora, actinocladium, affine, agardhiellae, alacranense, alariae, amahatanum, americanum, angustum, antillarum, arcuatum, armoricum, arnoldii, attenuatum, avrainvilleae, bahreinii, balliae, balticum, barbadense, battersianum, bengalicum, bisporum, blomquistii, boergesenii, brebneri, bulbosa, byssaceum, caesareae, caespitiforme, canariense, candelabrum, carneum, catenulatum, chaetomorphae, champiae, cheminii, chondriae, clandestinum, codicola, codii, colaconemoides, collopodum, compactum, concavum, corymbiferum, crassipes, curtum, cymopoliae, cytophagum, delisea, desmarestiae, dictyopteridis, discoideum, distichosporum, domerguei, dotyi, doumerquei, drewiae, duboscqii, dumontiae, dwarkense, eastwoodiae, efflorescens, effusum, elegans, emergens, endozoicum, entophyticum, epizooicum, erectum, ernothrix, extensum, fernandezianum, flexuosum, fuegiense, furcatum, galaxaurae, globulosum, gracile, grandis, grateloupiae, griffithsianum, gymnogongri, gynandrum, hallandicum, hamelii, hancockii, hauckii, hlulekaense, hormorhizum, howei, hoytii, humile, hummii, hypneae, imitator, immersum, inclusum, incrassatum, infestans, inkyui, interpositum, irregularis, islamii, iyengarii, jamaicense, japonicum, krusadii, kuckuckianum, kurilense, kurogii, kylini, laminariae, lanuginosum, lauterbachii, laxum, lenormandii, leptonema, leptonemoides, levringii, liagorae, liagoraefilum, liagoroides, lorrain-smithiae, luxurians, macropoda, macropus, macula, madinae, magnificum, maluinum, manorense, mauritianum, mediterraneum, microfilum, microscopicum, minimum, minutissimum, minutum, mirabilis, molinieri, molle, moniliforme, multisporum, naumannii, nemalii, netrocarpum, nitidulum, nitophylli, occidentale, opetigenum, pacificum, pallens, papenfussii, parvulum, penetrans, phacelorhizum, phuquocensis, plumosum, polysporum, porphyrae, posidoniae, pseudo-erectum, pubes, pulchellum, pulvereum, pulvinatum, punctatum, pusillum, radians, radiatum, ralfsii, reductum, reniforme, repens, reticulatum, robustum, roseolum, rosulatum, ryukyuensis, sagraeanum, sanctae-mariae, sancti-thomae, sargassicaulinus, sarmae, savianum, scapae, scinaiae, secundatum, seiriolanum, seriatum, sessile, sinicola, sparsum, spathoglossi, spongicola, stilophorae, strictum, stypopodiae, subimmersum, subpinnatum, subseriatum, subsimplex, subtilissimum, terminalis, trichogloeae, trifilum, tuticorinense, unifilum, unipes, vagum, vinculoides, virgatulum, yamadae*

VARIETIES *crassipes longisetum, efflorescens petrophila, extensum longicellulare, gracile vietnamense, hallandicum armoricum, hallandicum parvulum, immersum polysiphoniae, moniliforme mesogloiae, occidentale caespitosa, savianum agamum, rosulatum gallica, secundatum longiarticulata, trifilum schiffneri, vagum implicata, virgatulum crassitrichum, virgatulum luxurians, virgatulum opposita, virgatulum secundata, virgatulum tetrica*

GENUS *Audouinella* — Audouin Bristle Florid Red Algae

SPECIES *blumii, boryana, boweri, caesareae, callithamnionoides, chalybea, chnoosporae, clandestina, coccinea, compta, conchicola, conninea, curta, cylindrica, delisea, dictyotae, dotyi, dufourii, elegans, epispiculum, ernothrix, eugenea, extensa, funiformis, globosa, glomerata, godwardensis, hermannii, heterospora, hirsuta, indica, jogensis, jonsonii, kylinii, lanosa, macropa, macrospora, mahumetana, manxiana, monorhiza, mstinensis, nakamurai, naumannii, neozeelandica, occulta, parva, polyidis, pulvinata, pulvinatum, pygmaea, rongelapensis, sarmaii, seriaspora, serpens, sertulariae, siliculosa, sinensis, spinulosa, spongicola, streblocladiae, subtilis, tenella, variabilis, venezuelensis, welwitschii*

VARIETIES *chalybea brasiliensis, chalybea fuscoviolacea, chalybea leibleinii, chalybea major, chalybea marchica, chalybea maxima, chalybea muscicola, chalybea profunda, chalybea pulvinata, chalybea radians, chalybea thermalis, eugenea secundata, extensa longicellularis, hermannii alpina, hermannii dalmatica, hermannii expansa, hermannii fasciculata, hermannii garcilior, hermannii hercynica, hermannii kelseyi, hermannii pilosa, hermannii ramellosa, hermannii saxonica, hermannii subchalybea, pygmaea javanensis, pygmaea ramellifera*

GENUS *Colaconema* — Colac Thread Florid Red Algae

SPECIES *americanum, amphiroae, asparagopsis, bonnemaisoniae, caespitosum, chylocladiae, dasyae, daviesii, endophyticum, interpositum, macounii, membranaceum, monorhizum, nemalione, ophioglossa, panduripodium, pectinatum, proskaueri, rhizoideum, sarconemae, tetrasporum*

VARIETIES *daviesii royana, endophyticum tristansense, pectinatum cimbrica, rhizoideum patens*

GENUS *Kylinia* — Kylin Bristle Florid Red Algae

SPECIES *unilateralis*

GENUS *Rhodochorton* — Chorton Bristle Florid Red Algae

SPECIES *atlanticum, concrescens, corynosporoides, entozoicum, galaxaurae, kurilense, lepadicola, membranaceum, microscopicum, parkeri, purpureum, subsalsum, subsimplex, tetrasporum, velutinum*

VARIETIES *membranaceum macrocladum, purpureum kjellmani, purpureum typicum*

GENUS *Rhodochortonopsis* — Lesser Chorton Bristle Florid Red Algae

SPECIES *spongicola*

GENUS *Schmitziella* — Schmitz Bristle Florid Red Algae

SPECIES *cladophorae, endophloea*

FAMILY 3 *Balliaceae* — Ball Tadpole Spore Florid Red Algae

GENUS *Ballia* — Ball Tadpole Florid Red Algae

SPECIES *beckeri, callitricha, nana, pennoides, sertularioides, vestium*

GENUS *Balliella* — Least Ball Tadpole Florid Red Algae

SPECIES *cladoderma, crouanioides, grandis, hirsuta, pseudocorticata, repens, subcorticata*

GENUS *Balliopsis* — Lesser Ball Tadpole Spore Florid Red Algae

SPECIES *pinnulata, prieurii*

GENUS *Inkyuleea* — Inkyule Ball Tadpole Florid Red Algae

SPECIES *ballioides, beckeri, mariana*

FAMILY 4 *Lemaneaceae* — Leman Tadpole Spore Florid Red Algae

GENUS *Lemanea* — Leman Tadpole Spore Florid Red Algae

SPECIES *borealis, ciliata, condensata, deamii, fluviatilis, fucina, mamillosa, rigida, simplex, sinica, sudeica, thiryana, torulosa*

VARIETIES *fucina parva*

GENUS *Chantransia* — Chantran Leman Tadpole Spore Florid Red Algae

SPECIES *amethystea, arcta, ascosperma, atra, batrachospermi, beardslei, bergomensis, bichotoma, catenata, chiloensis, coccinea, coerulescens, compacta, crispata, dichlora, dichotoma, distans, distorta, flagellifera, fluviatilis, fontana, glaucida, gracilis, gracillima, hauckii, holsatica, hutchinsiae, incrustans, infusoria, intricata, marina, minima, minutissima, muscicola, nana, nigricans, nigrita, nodosa, nordstedtiana, pennata, polyrhiza, pulchella, pulvinata, radians, ramellosa, rigida, roseola, rupestris, scotica, sericea, subtilis, torulosa, trichotoma, vaucheriaeformis, vesciata, willei*
VARIETIES *incrustans pulvinata, pulchella chalybea*
GENUS *Paralemanea* Lesser Leman Tadpole Spore Florid Red Algae
SPECIES *annulata, brandegeei, californica, catenata, deamii, gardneri, grandis, mexicana, parishii, torulosa, tulensis*
GENUS *Pseudochantransia* Lesser Chantran Leman Tadpole Spore Florid Red Algae
SPECIES *lemaneae, thoreae, tuomeyae*
GENUS *Sacheria* Sacher Leman Tadpole Spore Florid Red Algae
SPECIES *ciliata*

ORDER 3 *Bonnemaisoniales* Bonne Maison Florid Red Algae

FAMILY 1 *Bonnemaisoniaceae* Bonne Maison Florid Red Algae
GENUS *Bonnemaisonia* Bonne Maison Florid Red Algae
SPECIES *asparagoides, australis, clavata, geniculata, hamifera, nootkana, spinescens*
VARIETIES *asparagoides irregularis*
GENUS *Asparagopsis* Asparagus Bonne Maison Florid Red Algae
SPECIES *armata, sanfordiana, svedelii, taxiformis*
VARIETIES *sanfordiana amplissima*
GENUS *Delisea* Delise Bonne Maison Florid Red Algae
SPECIES *compressa, elegans, flaccida, hypneoides, plumosa, pulchra*
GENUS *Falkenbergia* Falkenberg Bonne Maison Florid Red Algae
SPECIES *doubletii*
GENUS *Leptophyllis* Slender Bonne Maison Florid Red Algae
SPECIES *conferta*
GENUS *Pleuroblepharidella* Pleural Eyelash Bonne Maison Florid Red Algae
SPECIES *japonica*
GENUS *Ptilonia* Ptilon Bonne Maison Florid Red Algae
SPECIES *australasica, magellanica, mooreana, okadae, subulifera, willana*

FAMILY 2 *Naccariaceae* Naccar Bonne Maison Florid Red Algae
GENUS *Naccaria* Naccar Bonne Maison Florid Red Algae
SPECIES *antillana, corymbosa, hawaiiana, naccarioides, wiggii*
GENUS *Atractophora* Atract Naccar Bonne Maison Florid Red Algae
SPECIES *hypnoides*
GENUS *Reticulocaulis* Netted Stalk Naccar Bonne Maison Florid Red Algae
SPECIES *mucosissimus, obpyriformis*
GENUS *Rhododiscus* Disc Naccar Bonne Maison Florid Red Algae
SPECIES *carmelita*

ORDER 4 *Ceramiales* Ceram Florid Red Algae

FAMILY 1 *Ceramiaceae* Ceram Florid Red Algae
GENUS *Ceramium* Ceram Florid Red Algae
SPECIES *abyssale, adnuncum, aduncum, affine, amamiense, apiculatum, arborescens, arcticum, arenarium, areschougii, atlanticum, atrorubescens, aucklandicum, australe, avalonae, axillare, barbatum, bertholdii, bicorne, bisporum, boergesenii, borneense, botryocarpum, boydenii, brasiliense, brevizonatum, bulbosum, caespitosum, californicum, callipterum, camouii, capense, caudatum, centroceratiforme, ceratophyllum, chathamense, ciliatum, cimbricum, cingulatum, cingulum, circinatum, clarionense, clavaeforme, clavutatum, cliftonianum, codicola, codii, comptum, congestum, corniculatum, cornutum, crouanianum, cruciatum, cupulatum, danicum, dawsonii, derbesii, deslongchampsii, diaphanum, discorticatum, dorsiventrale, dozei, dumosertum, eatonianum, echionotum, elongatum, equisetoides, evermannii, excellens, filicula, fimbriatum, flaccidum, floridanum, forcipatum, fosliei, fruticulosum, furcatum, gaditanum, gaillonii, gardneri, giacconei, glanduliferum, graecum, guttatum, gymnogonium, hamatispinum, hanaense, hellenicum, heterospinum, hoodii, horridum, howellii, incospicuum, interruptum, involutum, irregulare, isogonum, japonicum, johnstonii, jolyi, kondoi, kororensis, krameri, ledermannii, lenticulare, lentiforme, leptocladum, leptosiphon, leptozonum, leutzelburgii, longissimum, loureiri, macilentum, macrotrichum, manorense, marshallense, maryae, mauritianum, mazatlanense, metcalfii, minutulum, monacanthum, multijugum, myriophyllum, nakamurae, nitens, obesum, obsoletum, ornatum, orthocladum, pacificum, pallidum, paniculatum, papenfussianum, parvulum, pedicellatum, penicillatum, pennatum, personatum, petitii, planum, pleurosporum, poeppigianum, procumbens, pseudostrictum, puberulum, punctiforme, pusillum, recissum, recticorticum, rescissum, rosenvingei, saviniae, scandinavicum, secundatum, seirosporum, septentrionale, serpens, shepherdii, shuttleworthianum, siliquosum, sinicola, skottsbergii, spinoso-pilosum, spyridioides, stichidosum, strobiliforme, subdichotomum, subtile,*

subverticillatum, suhrianum, tasmanicum, tenerrimum, tenue, tenuicorne, tenuicorticatum, tetricum, tranquillum, truncatum, uncinatum, ungulatum, uruguayense, vagans, variegatum, vatovai, verongiae, verrucosum, vestitum, vietnamense, villosum, virgatum, viscainoense, washingtoniense, womersleyi, zacae

VARIETIES *affine peninsularis, atlanticum norvegicum, brevizonatum caraibicum, ciliatum echinatum, ciliatum julaceum, ciliatum robustum, cimbricum flaccidum, cingulatum major, circinatum confluens, circinatum syntrophum, clavutatum inerme, deslongchampsii acrocarpum, deslongchampsii delicatum, deslongchampsii zostericolum, diaphanum capricornu, diaphanum decipiens, diaphanum elegans, diaphanum glabellum, diaphanum indicum, diaphanum minor, echionotum mediterraneum, forcipatum ciliatum, forcipatum glabellum, fruticulosum rubroides, gaditanum mediterraneum, leutzelburgii leutzelburgii, leutzelburgii australis, poeppigianum unilaterale, siliquosum acrocarpum, siliquosum elegans, siliquosum lophophorum, siliquosum minusculum, siliquosum zostericola, sinicola interruptum, sinicola johnstonii, stichidosum smithii, tenerrimum brevizonatum, tetricum pectinatum, virgatum virgatum, virgatum barbatum, virgatum corymbiferum, virgatum implexo-contortum, virgatum pedicellatum, virgatum secundatum, virgatum tenue*

GENUS *Acrothamnion* Tall Thamnion Ceram Florid Red Algae

SPECIES *butleriae, preissii*

GENUS *Acrothamniopsis* Lesser Tall Thamnion Ceram Florid Red Algae

SPECIES *eliseae*

GENUS *Aglaothamnion* Agla Thamnion Ceram Florid Red Algae

SPECIES *bipinnatum, boergesenii, callophyllidicola, caudatum, chadefaudii, chejuense, collinsii, cordatum, diaphanum, endovagum, fasciculatum, feldmanniae, felipponei, felliponei, flexibile, gaillonii, gallicum, halliae, herveyi, hookeri, monopodon, obstipum, okiense, oosumiense, priceanum, pseudobyssoides, roseum, sarcodiae, sepositum, tenuissimum, tripinnatum, uruguayense*

VARIETIES *tenuissimum divaricatum, tenuissimum mazoyerae, tenuissimum pygmaeum*

GENUS *Amoenothamnion* Sand Thamnion Ceram Florid Red Algae

SPECIES *lycopodioides, minimum, planktonicum*

GENUS *Anisoschizus* Anise Split Ceram Florid Red Algae

SPECIES *propaguli*

GENUS *Anotrichium* Hairy Ceram Florid Red Algae

SPECIES *barbatum, celianum, crinitum, elongatum, furcellatum, gracile, licmophorum, okamurae, planatum, secundum, subtile, tenue, towinna, yagii*

GENUS *Antarcticothamnion* Antarctic Thamnion Ceram Florid Red Algae

SPECIES *polysporum*

GENUS *Antithamnion* Anti Thamnion Ceram Florid Red Algae

SPECIES *adenocladellum, alternans, amamiense, amphigeneum, antillarum, armatum, baylesiae, biarmatum, callocladum, compactum, cristatum, cruciatum, decipiens, delicatulum, dendroideum, densum, diminuatum, divergens, eliseae, erucacladellum, gracilentum, hanovioides, heterocladum, hubbsii, imbricatum, kylinii, lherminieri, makroklonion, microptilum, nematocladellum, nigrescens, okiense, pectinatum, percurrens, piliferum, pinnafolium, pseudoarmatum, pterocladellum, pumilum, secundatum, subulatum, tanakae, tenuissimum, ternatum, ternirameum, tristicum, uncinatum, uniramosum, veleroae, verticale, villosum*

VARIETIES *cruciatum fragilissima, cruciatum radicans, diminuatum polyglandulum*

GENUS *Antithamnionella* Lesser Anti Thamnion Ceram Florid Red Algae

SPECIES *adnata, alternans, asymmetricum, atlantica, australis, basispora, bermudica, boergesenii, elegans, flagellata, floccosa, gardneri, graeffei, longicellulata, mcnabbii, multiramosa, nagaii, nigricans, rhodokraftia, schneideri, seriata, spirographidis, sublittoralis, tasmanica, ternifolia, tormentosa, verticillata*

VARIETIES *elegans decussata*

GENUS *Antrosphaera* Antrosphere Ceram Florid Red Algae

SPECIES *magnifica*

GENUS *Ardreanema* Ardrean Ceram Florid Red Algae

SPECIES *seriospora*

GENUS *Aristothamnion* Aristo Thamnion Ceram Florid Red Algae

SPECIES *callithamnioides, ramellifera*

GENUS *Ascocladium* Sac Branched Ceram Florid Red Algae

SPECIES *devoniense*

GENUS *Baldockia* Baldock Ceram Florid Red Algae

SPECIES *verticillata*

GENUS *Boreothamnion* Bore Thamnion Ceram Florid Red Algae

SPECIES *villosum*

GENUS *Bornetia* Bornet Ceram Florid Red Algae

SPECIES *binderiana, californica, repens, secundiflora, tenuis*

GENUS *Boryna* Boryn Ceram Florid Red Algae

SPECIES *axillaris, ciliata, elegans, forcipata, glabella, vinosa*

VARIETIES *ciliata denudata*

GENUS *Calliclavula* Callisto Club Ceram Florid Red Algae

SPECIES *trifurcata*

GENUS *Callidictyon* Callisto Netted Ceram Florid Red Algae

SPECIES *abyssorum*

GENUS *Callithamniella* Lesser Callisto Thamnion Ceram Florid Red Algae

SPECIES *capensis, flexilis, koreana, pacifica, silvae, tingitana*
GENUS *Callithamnion* Callisto Thamnion Ceram Florid Red Algae
SPECIES *acrospermum, acutum, aglaothamniodes, apicalis, apiculatum, arachnoideum, arborescens, arrawaricum, baileyi, biseriatum, bisporum, brachygonum, catalinense, caulescens, circinnatum, clavatum, codicola, colensoi, collabens, compactum, confertum, consanguineum, corymbosum, crispulum, cryptopterum, cystosirarum, debile, dudresnayi, echigoense, ecuadoreanum, ellipticum, epiphyticum, exiguum, fallax, felipponei, gailloni, gaudichaudii, gracile, grande, granulatum, griffithsioides, hirtellum, hypneae, japonicum, korffense, lanceolatum, lanosum, laricinum, lasioides, marshallense, mildbraedii, minimum, minutissimum, montagnei, multifidum, myurum, nipponicum, obstipum, octosporum, pacificum, parvulum, paschale, pedicellatum, pennula, perpusillum, pikeanum, piliferum, pinnatum, planum, propebyssoides, punicea, purpuriferum, pygmaeum, rabenhorstii, ramellosum, ramosissimum, reductum, repens, roseum, rupicolum, rupinicola, semicorticatum, shepherdii, soccoriense, spinescens, spinuliferum, stuposum, subsecundum, subverticillatum, tetragonum, tetricum, uruguayense, variabile, variegatum, veleroae, violaceum, virgulatum*
VARIETIES *bisporum australe, caulescens jamaicensis, corymbosum amphicarpum, corymbosum australis, corymbosum intricatum, pikeanum laxum, pikeanum pacificum, spinescens julaceum tetragonum divaricatum*
GENUS *Campylaephora* Field Ceram Florid Red Algae
SPECIES *crassa, hypnaeoides, japonica*
VARIETIES *crassa cymosa*
GENUS *Carpoblepharis* Carpel Fringed Ceram Florid Red Algae
SPECIES *ceylanica, curtissiae, flaccida, minima*
GENUS *Carpothamnion* Carpel Thamnion Ceram Florid Red Algae
SPECIES *gunnianum, molle*
GENUS *Centroceras* Central Horn Ceram Florid Red Algae
SPECIES *clavulatum, corallophiloides, cryptocanthum, distichum, inerme, internitens, japonicum, miniatum, minutum, secundum*
VARIETIES *cryptocanthum longiarticulatum*
GENUS *Centrocerocolax* Colac Central Horn Ceram Florid Red Algae
SPECIES *ubatubensis*
GENUS *Chaetoceras* Bristle Horn Ceram Florid Red Algae
SPECIES *pseudosimile*
GENUS *Compsothamnion* Adorned Thamnion Ceram Florid Red Algae
SPECIES *decompositum, gracillimum, thuyoides*
GENUS *Compsothamnionella* Lesser Adorned Thamnion Ceram Florid Red Algae
SPECIES *huismanii, sciadophila*
GENUS *Corallophila* Coral Ceram Florid Red Algae
SPECIES *apiculata, atlantica, bella, cinnabarina, eatoniana, howei, huysmansii, itonoi, kleiwegii, ptilocladioides, verongiae*
GENUS *Corynospora* (Synonym for *Monosporus*)
GENUS *Crouania* Crouan Ceram Florid Red Algae
SPECIES *attenuata, boergesenii, brunyana, capricornica, destriana, divaricata, ellisiae, francescoi, ischiana, iyengarii, mageshimensis, minutissima, mucosa, pleonospora, robbii, shepleyana, willae*
VARIETIES *attenuata bispora*
GENUS *Crouaniella* Lesser Crouan Ceram Florid Red Algae
SPECIES *laxitaxis*
GENUS *Crouanophycus* Least Crouan Ceram Florid Red Algae
SPECIES *mcnabbii*
GENUS *Cyclotrichium* Cyclic Hairy Ceram Florid Red Algae
SPECIES *gigas*
GENUS *Dasyphila* Daisy Ceram Florid Red Algae
SPECIES *cryptocarpa, plumarioides, preissii*
GENUS *Dasyptilon* Ptilon Daisy Ceram Florid Red Algae
SPECIES *harveyi, microptilum, pellucidum, ptilota*
GENUS *Dasythamniella* Thamnion Daisy Ceram Florid Red Algae
SPECIES *dasyura, latissima, plumigera, superbiens, wollastoniana*
GENUS *Delesseriopsis* Delesser Ceram Florid Red Algae
SPECIES *elegans*
GENUS *Desikacharyella* Desikachary Ceram Florid Red Algae
SPECIES *indica*
GENUS *Deucalion* Deucalion Ceram Florid Red Algae
SPECIES *levringii*
GENUS *Diapse* (Synonym for *Dasyptilon ptilota*)
GENUS *Diplothamnion* Diploid Thamnion Ceram Florid Red Algae
SPECIES *gordoniae, jolyi, tetrastichum*
VARIETIES *jolyi ecellulare*
GENUS *Dohrniella* Dohrn Ceram Florid Red Algae
SPECIES *antillana, nana, neapolitana*
VARIETIES *antillana brasiliensis*

GENUS *Drewiana* Drew Ceram Florid Red Algae
SPECIES *nitella*
GENUS *Elisiella* Elis Ceram Florid Red Algae
SPECIES *arbuscula, dispar*
GENUS *Episperma* Epiphyte Ceram Florid Red Algae
SPECIES *ciliatum, diaphanum, elegans*
GENUS *Episporium* Lesser Epiphyte Ceram Florid Red Algae
SPECIES *centroceratis*
GENUS *Eupogonium* Beard Ceram Florid Red Algae
SPECIES *spinellum*
GENUS *Euptilocladia* Ptilon Branched Ceram Florid Red Algae
SPECIES *magruderi, spongiosa, villosa*
GENUS *Euptilota* Ptilon Ceram Florid Red Algae
SPECIES *articulata, coralloidea, fergusonii, formosissima, krusadiensis, mooreana, pappeana*
GENUS *Gaillonia* Gaillon Ceram Florid Red Algae
SPECIES *lehmannii*
GENUS *Gattya* Gatty Ceram Florid Red Algae
SPECIES *obtusa, pinnella*
GENUS *Georgiella* George Ceram Florid Red Algae
SPECIES *confluens*
GENUS *Gongroceras* Gonger Horn Ceram Florid Red Algae
SPECIES *radicans*
GENUS *Gordoniella* Gordon Ceram Florid Red Algae
SPECIES *yonakuniensis*
GENUS *Grallatoria* Grall Ceram Florid Red Algae
SPECIES *reptans, tanakae*
GENUS *Griffithsia* Griffiths Ceram Florid Red Algae
SPECIES *antarctica, anthericephala, balara, bornetiana, capitata, caribaea, chilensis, coacta, confervoides, corallinoides, crassiuscula, devoniensis, elegans, genovefae, globulifera, grandis, gunniana, heteroclada, heteromorpha, japonica, metcalfii, monilis, multiramosa, okiensis, opuntioides, ovalis, pacifica, phyllamphora, pilalyea, pulvinata, radicans, rhizoidea, rhizophora, schousboei, subcylindrica, teges, tomo-yamadae, traversii, venusta, weber-van-bosseae*
VARIETIES *monilis monilis, monilis cincta, schousboei anastomosans, schousboei minor*
GENUS *Guiryella* Guiry Ceram Florid Red Algae
SPECIES *repens*
GENUS *Gulsonia* Gulson Ceram Florid Red Algae
SPECIES *annulata, ecorticata, nodulosa*
GENUS *Gymnophycus* Nude Ceram Florid Red Algae
SPECIES *hapsiphorus*
GENUS *Gymnothamnion* Nude Thamnion Ceram Florid Red Algae
SPECIES *elegans, nigrescens, pteroton*
VARIETIES *elegans bisporum*
GENUS *Haloplegma* Plegm Ceram Florid Red Algae
SPECIES *anweri, duperreyi, mexicanum, polyspore, preissii*
VARIETIES *duperreyi spinulosum*
GENUS *Halosia* Halose Ceram Florid Red Algae
SPECIES *eliseae*
GENUS *Halurus* Haline Tailed Ceram Florid Red Algae
SPECIES *equisetifolius, flosculosus*
VARIETIES *flosculosus irregularis, flosculosus sphaericus*
GENUS *Herpochondria* Herpetic Cartilage Ceram Florid Red Algae
SPECIES *corallinae, dentata, elegans, henriettae, kampenii, pygmaea*
GENUS *Heterothamnion* Variable Thamnion Ceram Florid Red Algae
SPECIES *episiliquosum, muelleri, platythaliae, sessile*
GENUS *Hirsutithallia* Hairy Thal Ceram Florid Red Algae
SPECIES *abaxialis, angustata, formosa, laricina, mucronata, tincta*
GENUS *Hollenbergia* Hollenberg Ceram Florid Red Algae
SPECIES *nigricans, subulata*
GENUS *Interthamnion* Inter Thamnion Ceram Florid Red Algae
SPECIES *attenuatum*
GENUS *Involucrana* Involucrate Ceram Florid Red Algae
SPECIES *crassa, meredithiana*
GENUS *Irtugovia* Irtugov Ceram Florid Red Algae
SPECIES *pacifica, shimamurana, spirographidis*
VARIETIES *pacifica uncinata*
GENUS *Lasiothalia* Woolly Thal Ceram Florid Red Algae
SPECIES *hirsuta*

GENUS *Laurenciophila* Laurence Ceram Florid Red Algae
SPECIES *minima*
GENUS *Lejolisia* Le Jolis Ceram Florid Red Algae
SPECIES *aegagropila, colombiana, exposita, mediterranea, pacifica, pusilla*
GENUS *Leptoklonion* Thin Clone Ceram Florid Red Algae
SPECIES *elongatum, fastigiatum*
GENUS *Liagorothamnion* Liagor Thamnion Ceram Florid Red Algae
SPECIES *mucoides*
GENUS *Lomathamnion* Loma Thamnion Ceram Florid Red Algae
SPECIES *capense, epicodii, humile*
GENUS *Lophothamnion* Crested Thamnion Ceram Florid Red Algae
SPECIES *hirtum*
GENUS *Macrothamnion* Large Thamnion Ceram Florid Red Algae
SPECIES *acanthophorum, pectenellum, pellucidum, secundum*
GENUS *Mazoyerella* Mazoyer Ceram Florid Red Algae
SPECIES *arachnoidea, australis, kraftii, krishnamurthyi*
GENUS *Medeiothamnion* Mede Thamnion Ceram Florid Red Algae
SPECIES *flaccidum, halurus, lyallii, norrissii, protensum, repens, sanctacrucense*
GENUS *Microcladia* Branched Ceram Florid Red Algae
SPECIES *alternata, borealis, californica, coulteri, exserta, gelidii, glandulosa, gloria-spei, moseleyi, novae-zelandiae, pinnata*
GENUS *Microgelidiopsis* Gelid Ceram Florid Red Algae
SPECIES *horrida*
GENUS *Microthamnion* Thamnion Ceram Florid Red Algae
SPECIES *elegans, interruptum, kuetzingianum*
VARIETIES *kuetzingianum strictissimum*
GENUS *Monosporus* Monospore Ceram Florid Red Algae
SPECIES *belangeri, herpesticus, indicus, inkyui, pedicellatus, sericatus*
VARIETIES *pedicellatus comosa, pedicellatus tenuis*
GENUS *Mortensenia* Mortensen Ceram Florid Red Algae
SPECIES *pulchra*
GENUS *Muellerena* Mueller Ceram Florid Red Algae
SPECIES *wattsii*
GENUS *Neomonospora* (Synonym for *Monosporus*)
GENUS *Neoptilota* Modern Ptilon Ceram Florid Red Algae
SPECIES *asplenioides, californica, densa, hypnoides*
GENUS *Nwynea* Nwyne Ceram Florid Red Algae
SPECIES *grandispora*
GENUS *Ochmapexus* Ochmapex Ceram Florid Red Algae
SPECIES *minimus*
GENUS *Ossiella* Ossy Ceram Florid Red Algae
SPECIES *pacifica*
GENUS *Papenfussia* Papenfuss Ceram Florid Red Algae
SPECIES *japonica*
GENUS *Perikladosporon* Branched Perion Ceram Florid Red Algae
SPECIES *percurrens*
GENUS *Perischelia* Claw Perion Ceram Florid Red Algae
SPECIES *glomulifera*
GENUS *Perithamnion* Perion Thamnion Ceram Florid Red Algae
SPECIES *ceramioides, densum, muelleri*
GENUS *Phlebothamnion* Phlebe Thamnion Ceram Florid Red Algae
SPECIES *gaillonii, lanceolatum*
GENUS *Platythamnion* Flat Thamnion Ceram Florid Red Algae
SPECIES *reversum*
VARIETIES *reversum laxum*
GENUS *Pleonosporium* Pleon Spore Ceram Florid Red Algae
SPECIES *boergesenii, borreri, callicladum, caribaeum, complanatum, constrictum, dichotomum, elongatum, filicinum, flexuosum, globuliferum, harveyanum, hirtum, intricatum, japonicum, kobayashii, mageshimense, mazeense, mexicanum, paternoster, pedicellatum, polymorphum, polystichum, pusillum, ramulosum, rhizoideum, segawae, squarrosum, squarrulosum, tohyamanum, vancouverianum, venustissimum, wuellerstorfianum, yagii*
VARIETIES *borreri elongatum*
GENUS *Pleuroblepharis* Pleural Fringed Ceram Florid Red Algae
SPECIES *japonica*
GENUS *Plumaria* Plum Ceram Florid Red Algae
SPECIES *pectinata, plumosa*
VARIETIES *pectinata tenerrima, plumosa tenerrimus*

GENUS *Plumariella* Lesser Plum Ceram Florid Red Algae
SPECIES *minima, yoshikawae*
GENUS *Plumariopsis* Least Plum Ceram Florid Red Algae
SPECIES *eatonii, peninsularis*
GENUS *Pseudocrouania* Lesser Crouan Ceram Florid Red Algae
SPECIES *ischiana*
GENUS *Pseudospora* Pseudospore Ceram Florid Red Algae
SPECIES *adriatica*
GENUS *Psilothallia* Psilum Thal Ceram Florid Red Algae
SPECIES *dentata, siliculosa, striata*
GENUS *Pterocladiopsis* Wing Branch Ceram Florid Red Algae
SPECIES *hirsuta*
GENUS *Pterothamnion* Winged Thamnion Ceram Florid Red Algae
SPECIES *aciculare, antarcticum, capense, confusum, crispum, cuspidatum, flexile, francisianum, heteromorphum, horridum, intermedium, latiramosum, manifestum, minutissimum, nodiferum, orbignianum, pectinatum, plumula, polyacanthum, polysporum, ramulentum, recurvatum, simile, spinescens, squarrulosum, tepocense, villosum, yezoense*
VARIETIES *plumula cryspum, plumula genuinum, plumula haplokladion, plumula laxum, plumula verticillatum*
GENUS *Ptilocladia* Ptilon Branched Ceram Florid Red Algae
SPECIES *australis, crouanioides, divaricata, gracilis, japonica, okiensis, pulchra, vestita, yueni*
GENUS *Ptilocladiopsis* Lesser Ptilon Branched Ceram Florid Red Algae
SPECIES *horrida*
GENUS *Ptilota* Ptilon Ceram Florid Red Algae
SPECIES *filicina, gunneri, hannafordii, phacelocarpoides, rhodocallis, serrata*
GENUS *Ptilothamnion* Ptilon Thamnion Ceram Florid Red Algae
SPECIES *cladophorae, codicolum, goukammae, occidentale, pluma, polysporum, pusillum, richardsii, rupicola, schmitzii, speluncarum, sphaericum, subsimplex*
GENUS *Ptilothamnionopsis* Lesser Ptilon Thamnion Ceram Florid Red Algae
SPECIES *lejolisea*
GENUS *Radiathamnion* Radial Thamnion Ceram Florid Red Algae
SPECIES *speleotis*
GENUS *Reinboldiella* Reinbold Ceram Florid Red Algae
SPECIES *filamentosa, robusta, schmitziana, warburgii*
GENUS *Rhipidothamnion* Wicker Thamnion Ceram Florid Red Algae
SPECIES *secundum*
GENUS *Rhodocallis* Callisto Ceram Florid Red Algae
SPECIES *elegans*
GENUS *Rhododictyon* Netted Ceram Florid Red Algae
SPECIES *bermudense*
GENUS *Scagelia* Scagel Ceram Florid Red Algae
SPECIES *americana, breviarticulata, pylaisaei*
VARIETIES *pylaisaei subnuda*
GENUS *Scageliopsis* Lesser Scagel Ceram Florid Red Algae
SPECIES *patens, strongylekystis, tsitsikammae*
GENUS *Scagelonema* Thread Scagel Ceram Florid Red Algae
SPECIES *parasitica*
GENUS *Scagelothamnion* Scagel Thamnion Ceram Florid Red Algae
SPECIES *pusillum*
VARIETIES *pusillum droebakense*
GENUS *Sciurothamnion* Squirrel Thamnion Ceram Florid Red Algae
SPECIES *stegengae*
GENUS *Seagriefia* Seagrief Ceram Florid Red Algae
SPECIES *fasciculifera*
GENUS *Seirospora* Seirospore Ceram Florid Red Algae
SPECIES *apiculata, byssoides, giraudyi, humilis, interrupta, occidentalis, orientalis, purpurea, sphaerospora, viridis*
GENUS *Shepleya* Shepley Ceram Florid Red Algae
SPECIES *australis, claviformis, elixithamnia, verticillata, wattsii*
GENUS *Spencerella* Spencer Ceram Florid Red Algae
SPECIES *australis*
GENUS *Spermothamnion* Spore Thamnion Ceram Florid Red Algae
SPECIES *antillarum, boergesenii, cymosum, echigoense, endophytica, flabellatum, gymnocarpum, inordinatum, investiens, irregulare, johannis, macromeres, miniatum, nonatoi, pinnatum, repens, snyderae attenuata, strictum, umamaheswarae*
VARIETIES *flabellatum disporum, investiens cidaricola, repens flagelliferum, repens monoica, repens sphaerica, repens subverticillatum, repens variabile*

GENUS *Sphondylothamnion* Sphondylid Thamnion Ceram Florid Red Algae
SPECIES *multifidum*
VARIETIES *multifidum distichum*
GENUS *Spongoclonium* Sponge Clone Ceram Florid Red Algae
SPECIES *australicum, brownianum, conspicuum, fasciculatum, orthocladum, pastorale, scopula*
VARIETIES *brownianum majus*
GENUS *Sporocanthus* Spiny Ceram Florid Red Algae
SPECIES *cristatus*
GENUS *Spyridia* Spyrid Ceram Florid Red Algae
SPECIES *alternans, ceramioides, clavata, cupressina, dasyoides, elongata, filamentosa, fusiformis, galeae, horridula, hypnoides, plumosa, squalida, tasmanica, tenuis, velasquezii*
VARIETIES *filamentosa refracta, hypnoides hypnoides, hypnoides berkeleyana, hypnoides complanata, hypnoides disticha, hypnoides disticha-inermis, hypnoides inermis*
GENUS *Spyridiocolax* Colac Spyrid Ceram Florid Red Algae
SPECIES *capixaba*
GENUS *Sympodothamnion* Sympodal Thamnion Ceram Florid Red Algae
SPECIES *leptophyllum*
GENUS *Syringocolax* Syringe Colac Ceram Florid Red Algae
SPECIES *macroblepharis*
GENUS *Tanakaella* Tanaka Ceram Florid Red Algae
SPECIES *itonoi, japonica, sericata*
GENUS *Tapinothrix* Tapin Thread Ceram Florid Red Algae
SPECIES *borneti, muscicola*
GENUS *Tetrathamnion* Tetrad Thamnion Ceram Florid Red Algae
SPECIES *arbuscula, densum, dispar, lineatum, myurum*
GENUS *Thamnocarpus* Carpel Thamnion Ceram Florid Red Algae
SPECIES *griffithsioides*
GENUS *Tiffaniella* Tiffany Ceram Florid Red Algae
SPECIES *apiculata, capitata, codicola, cymodoceae, feldmanniae, gorgonea, phycophilum, saccorhiza, schmitziana, snyderae, snyderiae, suyehioroi, tamamiru*
GENUS *Tokidaea* Tokida Ceram Florid Red Algae
SPECIES *chilkatensis, corticata, hirta, serrata*
GENUS *Trithamnion* Triad Thamnion Ceram Florid Red Algae
SPECIES *aculeatum, eubryani, eubryanii, gracilissimum, ovuligerum, tenellum, tetrapinnum, vulgare*
GENUS *Vickersia* Vickers Ceram Florid Red Algae
SPECIES *baccata*
GENUS *Warrenia* Warren Ceram Florid Red Algae
SPECIES *comosa*
GENUS *Wollastoniella* Wollaston Ceram Florid Red Algae
SPECIES *mucronata, myriophylloides*
GENUS *Wrangelia* Wrangel Ceram Florid Red Algae
SPECIES *abietina, anastomosans, argus, australis, balakrishnanii, bicuspidata, dumontii, elegantissima, galeae, hainanensis, nobilis, penicillata, plumosa, princeps, purpurifera, sceptrifera, sundaralingamii, tagoi, tanegana, tenuis, velutina*
VARIETIES *penicillata tenera*
GENUS *Zonariophila* Zoned Ceram Florid Red Algae
SPECIES *semiendophytica*

FAMILY 2 *Dasyaceae* Daisy Ceram Florid Red Algae
GENUS *Dasya* Daisy Ceram Florid Red Algae
SPECIES *anastomosans, antillarum, aperta, apiculata, atactica, baillouviana, baldockii, binghamiae, brasiliensis, capillaris, caraibica, ceramioides, cernua, clavigera, cliftonii, collabens, collinsiana, comata, corymbifera, crescens, crinita, crouaniana, cylindrica, divergens, echigoensis, echinata, elongata, extensa, flagellifera, frutescens, haffiae, haitiana, hapalathrix, harveyi, hookeri, hussoniana, hutchinsiae, iridescens, iyengarii, kamortensis, kraftii, kristeniae, longifila, magnei, malaccensis, minor, mollis, murrayana, naccarioides, ocellata, pachyclada, pallida, palmatifida, penicillata, punicea, quadrispora, ramosissima, rigescens, rigidula, roslyniae, schmidtiana, scoparia, scopulifera, sessilis, sinicola, spinuligera, stanfordiana, stanleyi, subtilis, tenuis, trichophora, velutina, villosa, wilsonis*
VARIETIES *hutchinsiae caespitosa, hutchinsiae minor, rigidula okiensis, sinicola abyssicola, sinicola californica*
GENUS *Amphisbetema* Dual Betem Daisy Ceram Florid Red Algae
SPECIES *indica*
GENUS *Baillouviana* Baillouv Daisy Ceram Florid Red Algae
SPECIES *kuetzingii*
GENUS *Colacodasya* Colac Daisy Ceram Florid Red Algae
SPECIES *australica, californica, inconspicua, sinicola*
GENUS *Dasyella* Lesser Daisy Ceram Florid Red Algae
SPECIES *gracilis*

GENUS *Dasyopsis* Least Daisy Ceram Florid Red Algae
SPECIES *palmatifida*
GENUS *Dasysiphonia* Siphon Daisy Ceram Florid Red Algae
SPECIES *chejuensis, concinna, doliiformis, okiensis*
GENUS *Dictyurus* Net Tailed Daisy Ceram Florid Red Algae
SPECIES *fenestratus, maldivensis, occidentalis, purpurascens*
GENUS *Dipterocladia* Dual Wing Branched Daisy Ceram Florid Red Algae
SPECIES *pinnatifolia, pulchella*
GENUS *Endogenes* Endogene Daisy Ceram Florid Red Algae
SPECIES *ceramii, polysiphoniae*
GENUS *Endogenia* Lesser Endogene Daisy Ceram Florid Red Algae
SPECIES *gracilaria*
GENUS *Eupogodon* Beard Daisy Ceram Florid Red Algae
SPECIES *geppii, grande, planus*
GENUS *Grateloupia* Grateloupe Daisy Ceram Florid Red Algae
SPECIES *acroidalea, acuminata, asiatica, aucklandica, carnosa, catenata, cerrosiana, comorinii, conferta, cosentinii, cuneifolia, cutleriae, dactylifera, dichotoma, divaricata, doryphora, elliptica, filicina, filiformis, furcata, gelatinosa, gibbesii, guajirae, hancockii, hawaiiana, howeii, imbricata, indica, intestinalis, kroneana, kurogii, lanceolata, latissima, lithophila, livida, longifolia, ovata, phuquocensis, pinnata, prolifera, prolongata, proteus, rojasii, schizophylla, scutellata, senegalensis, somalensis, sparsa, stipitata, subpectinata, subsimplex, turuturu, undulata, urvilleana, versicolor, violacea, wattii*
VARIETIES *doryphora linearis, filicina cirrhosa, filicina elongata, filicina horrida, filicina luxurians, filicina pectinata*
GENUS *Halydictyon* Haly Net Daisy Ceram Florid Red Algae
SPECIES *arachnoideum, australe, mirabile*
GENUS *Haplodasya* Single Daisy Ceram Florid Red Algae
SPECIES *tomentosa, urceolata*
GENUS *Heterosiphonia* Variable Siphon Daisy Ceram Florid Red Algae
SPECIES *arenaria, australis, berkeleyi, callithamnion, capensis, concinna, crassipes, crispa, crispella, curdieana, dubia, erecta, gibbesii, gunniana, hummii, japonica, lawrenciana, merenia, microcladioides, muelleri, multiceps, obscura, pellucida, plumosa, pulchra, punicea, rendlei, squarrosa, stuposa, subsecunda, tessellata, wrangelioides*
VARIETIES *crispella laxa*
GENUS *Nematophora* Filament Daisy Ceram Florid Red Algae
SPECIES *australis*
GENUS *Rhodoptilum* Ptilon Daisy Ceram Florid Red Algae
SPECIES *plumosum*
GENUS *Sinosiphonia* China Siphon Daisy Ceram Florid Red Algae
SPECIES *elegans*
GENUS *Tapeinodasya* Tapein Daisy Ceram Florid Red Algae
SPECIES *bornetii, etheliae*
GENUS *Thuretia* Thuret Daisy Ceram Florid Red Algae
SPECIES *australasica, bornetii, etheliae*

FAMILY 3 *Delesseriaceae* Delesser Ceram Florid Red Algae
GENUS *Delesseria* Delesser Ceram Florid Red Algae
SPECIES *aemula, appendiculata, belayevii, crassinervia, crozetii, decipiens, epiglossum, fuegiensis, intermedia, lancifolia, laurifolia, macloviana, minor, nereifolia, papenfussii, pontica, rubens, salicifolia, sanguinea, serrulata, subtilis*
GENUS *Abroteia* Abrote Delesser Ceram Florid Red Algae
SPECIES *orbicularis*
GENUS *Acrosorium* Tall Soral Delesser Ceram Florid Red Algae
SPECIES *acrospermum, amphiroae, ciliolatum, cincinnatum, decumbens, deformatum, flabellatum, fragile, maculatum, minus, odontophorum, okamurae, papenfussii, polyneurum, yendoi*
GENUS *Agarum* Agar Delesser Ceram Florid Red Algae
SPECIES *clathratum, fimbriatum, oharaense, pertusum*
VARIETIES *pertusum brassicaeformis, pertusum platyneurum*
GENUS *Aglaophyllum* Agla Delesser Ceram Florid Red Algae
SPECIES *reticulatum*
GENUS *Anisocladella* Anise Branched Delesser Ceram Florid Red Algae
SPECIES *serratodentata*
GENUS *Apoglossocolax* Colac Parsley Gloss Delesser Ceram Florid Red Algae
SPECIES *pusilla*
GENUS *Apoglossum* Parsley Gloss Delesser Ceram Florid Red Algae
SPECIES *gregarium, minimum, montagneanum, oppositifolium, ruscifolium, spathulatum, unguiculescens*
VARIETIES *ruscifolium minor*
GENUS *Arachnophyllum* Spider Delesser Ceram Florid Red Algae
SPECIES *confervaceum*

GENUS *Asterocolax* Star Colac Delesser Ceram Florid Red Algae
SPECIES *denticulata, erythroglossi, gardneri, hypophyllophila*
GENUS *Augophyllum* Auger Delesser Ceram Florid Red Algae
SPECIES *kentingii, marginifructum, wysorii*
GENUS *Austrofolium* Southern Frond Delesser Ceram Florid Red Algae
SPECIES *equatorianum, hassleri, howellii*
GENUS *Bartoniella* Barton Delesser Ceram Florid Red Algae
SPECIES *crenata*
GENUS *Botryocarpa* Berry Carpel Delesser Ceram Florid Red Algae
SPECIES *prolifera*
GENUS *Botryoglossum* Berry Gloss Delesser Ceram Florid Red Algae
SPECIES *cartilagineum, platycarpum*
VARIETIES *platycarpum linearis, platycarpum minor, platycarpum nana*
GENUS *Branchioglossum* Branch Tongue Delesser Ceram Florid Red Algae
SPECIES *bipinnatifidum, epiphyticum, minutum, nanum, prostratum, pygmaeum, spiniferum, undulatum, woodii*
GENUS *Brassierea* Brassier Delesser Ceram Florid Red Algae
SPECIES *vivens*
GENUS *Calloseris* Callisto Pitted Delesser Ceram Florid Red Algae
SPECIES *halliae*
GENUS *Caloglossa* Callisto Tongue Delesser Ceram Florid Red Algae
SPECIES *adhaerens, angustalata, apomeiotica, beccarii, bengalensis, continua, intermedia, leprieurii, monosticha, ogasawaraensis, postiae, saigonensis, stipitata, triclada*
VARIETIES *leprieurii hookeri*
GENUS *Calonitophyllum* Callisto Shiny Delesser Ceram Florid Red Algae
SPECIES *medium*
GENUS *Chauviniella* Chauvin Delesser Ceram Florid Red Algae
SPECIES *coriifolia, jadinii*
GENUS *Cladodonta* Branched Tooth Delesser Ceram Florid Red Algae
SPECIES *lyallii*
GENUS *Claudea* Claude Delesser Ceram Florid Red Algae
SPECIES *batanensis, elegans, multifida*
GENUS *Congregatocarpus* Congregated Carpel Delesser Ceram Florid Red Algae
SPECIES *pacificus*
GENUS *Crassilingua* Crass Lip Delesser Ceram Florid Red Algae
SPECIES *marginifera*
GENUS *Cryptopleura* Cryptic Pleural Delesser Ceram Florid Red Algae
SPECIES *callophylloides, crispa, crozieri, dichotoma, farlowiana, griffithsiana, hayamensis, imbricata, lobulifera, membranacea, peltata, ramosa, robusta, rosacea, ruprechtiana, violacea*
VARIETIES *ramosa uncinatum*
GENUS *Cumathamnion* Cuman Thamnion Delesser Ceram Florid Red Algae
SPECIES *sympodophyllum*
GENUS *Dicroglossum* Two Tongue Delesser Ceram Florid Red Algae
SPECIES *crispatulum*
GENUS *Dotyella* Doty Delesser Ceram Florid Red Algae
SPECIES *hawaiiensis, irregularis*
GENUS *Drachiella* Drach Delesser Ceram Florid Red Algae
SPECIES *heterocarpa, liaoii, minuta, spectabilis*
GENUS *Duckerella* Ducker Delesser Ceram Florid Red Algae
SPECIES *ferlusii*
GENUS *Erythroglossum* Tongue Delesser Ceram Florid Red Algae
SPECIES *balearicum, californicum, intermedium, laciniatum, latum, lusitanicum, minimum, pinnatum, sandrianum, schousboei, subcostatum, undulatissimum*
VARIETIES *laciniatum umbilicata*
GENUS *Frikkiella* Frikk Delesser Ceram Florid Red Algae
SPECIES *pseudoprostratum, searlesii*
GENUS *Gonimocolax* Gonad Colac Delesser Ceram Florid Red Algae
SPECIES *australis, corymbosa, roscoffensis*
GENUS *Gonimophyllum* Gonad Delesser Ceram Florid Red Algae
SPECIES *africanum, buffhamii, skottsbergii*
GENUS *Grinnellia* Grinn Delesser Ceram Florid Red Algae
SPECIES *americana, lanceolata*
GENUS *Halicnide* Halicnide Delesser Ceram Florid Red Algae
SPECIES *similans*
GENUS *Haraldia* Harald Delesser Ceram Florid Red Algae
SPECIES *australica, lenormandii, tenuis*
VARIETIES *lenormandii detonii*

GENUS *Haraldiophyllum* Lesser Harald Delesser Ceram Florid Red Algae
SPECIES *bonnemaisonii, erosum, infossum, mirabile, nottii, sinuosum*
VARIETIES *bonnemaisonii crassinerva*
GENUS *Hemineura* Half Neural Delesser Ceram Florid Red Algae
SPECIES *cruenta, frondosa, hassleri, howellii*
GENUS *Heterodoxia* Variable Dox Delesser Ceram Florid Red Algae
SPECIES *denticulata*
GENUS *Heteronema* Variable Filament Delesser Ceram Florid Red Algae *Heteronema*
SPECIES *abruptum, acus, acutissimum, diaphanum, discomorphum, distigmoides, globuliferum, hexagonum, klebsii, leptosomum, medusae, mutabile, nebulosum, ovale, polymorphum, proteus, sacculus, scaphurum, spirale, tortuosum, tremulum, trispira*
VARIETIES *hexagonum elegans*
GENUS *Hideophyllum* Hide Delesser Ceram Florid Red Algae
SPECIES *yezoense*
GENUS *Holmesia* Holmes Delesser Ceram Florid Red Algae
SPECIES *californica, capensis*
GENUS *Hymenena* Hymen Delesser Ceram Florid Red Algae
SPECIES *affinis, cuneifolia, curdieana, decumbens, durvillaei, endiviaefolia, falklandica, flabelligera, fryeana, kylinii, laciniata, multipartita, palmata, ruthenica, semicostata, setchellii, smithii, tenuis, variolosa, venosa*
GENUS *Hypoglossum* Hyponic Tongue Delesser Ceram Florid Red Algae
SPECIES *abyssicolum, androlamellare, annae, anomalum, armatum, attenuatum, barbatum, caloglossoides, dendroides, geminatum, guineense, harveyanum, heterocystideum, hypoglossoides, imperfectum, involvens, minimum, nipponicum, protendens, retusum, revolutum, rhizophorum, sagamianum, serratifolium, simulans, subsimplex, tenuifolium, wynnei*
VARIETIES *attenuatum abyssicolum, harveyanum fimbriatum, hypoglossoides penicillatum*
GENUS *Kurogia* Kurog Delesser Ceram Florid Red Algae
SPECIES *pulchra*
GENUS *Laingia* Laing Delesser Ceram Florid Red Algae
SPECIES *aleutica, hookeri*
GENUS *Marionella* Marion Delesser Ceram Florid Red Algae
SPECIES *prolifera, schmitziana*
GENUS *Martensia* Martens Delesser Ceram Florid Red Algae
SPECIES *australis, bibarii, denticulata, elegans, flabelliformis, formosana, fragilis, indica, jejuensis, lewisiae, martensii*
GENUS *Membranoptera* Membrane Wing Delesser Ceram Florid Red Algae
SPECIES *alata, denticulata, dimorpha, multiramosa, murrayi, platyphylla, robbeniensis, serrata, setchellii, spinulosa, tenuis, weeksiae*
GENUS *Microrhinus* Beak Delesser Ceram Florid Red Algae
SPECIES *carnosus*
GENUS *Mikamiella* Mikam Delesser Ceram Florid Red Algae
SPECIES *dentata, ruprechtiana*
GENUS *Myriogramme* Myriad Seaweed Delesser Ceram Florid Red Algae
SPECIES *alliaceum, caespitosa, carnea, cartilaginea, ciliata, costata, crispata, denticulata, dilabida, distromatica, divaricata, eckloniae, goaense, goaensis, gunniana, kerguelensis, kjellmanianum, kylinii, livida, multiloba, multilobata, nightingaliensis, okhaã«nsis, polyneura, prostrata, pulchra, quilonensis, repens, smithii, spectabilis, tristromatica, undulata, unistromaticum, variegata*
GENUS *Nancythalia* Nancy Thal Delesser Ceram Florid Red Algae
SPECIES *humilis*
GENUS *Neoholmesia* Modern Holmes Delesser Ceram Florid Red Algae
SPECIES *japonica, natalensis*
GENUS *Neohypophyllum* Modern Hyponic Delesser Ceram Florid Red Algae
SPECIES *middendorfii*
GENUS *Neuroglossum* Nerve Tongue Delesser Ceram Florid Red Algae
SPECIES *binderianum, delesseriae, multilobum*
GENUS *Nienburgia* Nienburg Delesser Ceram Florid Red Algae
SPECIES *andersoniana, pinnatifida, prolifera, serrata*
GENUS *Nienburgiella* Lesser Nienburg Delesser Ceram Florid Red Algae
SPECIES angustata
GENUS *Nitophyllum* Shiny Delesser Ceram Florid Red Algae
SPECIES *adhaerens, albidum, cincinnatum, circinnatum, confervaceum, crispum, delicatum, dentatum, divaricatum, dotyi, fallax, fimbriatum, flabellatum, galapagense, geminatum, hollenbergii, hommersandii, lenormandii, magontanum, marginale, marmoratum, medium, micropunctatum, northii, paessleri, peruvianum, pulchellum, punctatum, reptans, rotundum, stellato-corticatum, tongatense, tristromaticum, wilkinsoniae*
VARIETIES *lenormandii detonii*
GENUS *Nitospinosa* Shiny Spined Delesser Ceram Florid Red Algae
SPECIES *littledipensis, pristoidea, tasmanica*
GENUS *Okamurina* Okamura Delesser Ceram Florid Red Algae

SPECIES *pacifica*
GENUS *Ophiocythium* Snake Cythe Delesser Ceram Florid Red Algae
SPECIES *gracilipes*
GENUS *Pantoneura* Panton Nerve Delesser Ceram Florid Red Algae
SPECIES *angustissima, fabriciana, juergensii, plocamioides*
GENUS *Patulophycus* Patule Delesser Ceram Florid Red Algae
SPECIES *eclipes*
GENUS *Phitycolax* Phity Colac Delesser Ceram Florid Red Algae
SPECIES *inconspicuus*
GENUS *Phitymophora* Phity Delesser Ceram Florid Red Algae
SPECIES *amansioides, hypoglossum, linearis*
GENUS *Phycodrina* Lesser Dryad Delesser Ceram Florid Red Algae
SPECIES *elegans*
GENUS *Phycodrys* Dryad Delesser Ceram Florid Red Algae
SPECIES *amchitkensis, amplissima, antarctica, australasica, austrogeorgica, crenata, elegans, fimbriata, isabellae, lucasana, ovifolia, profunda, pulchra, quercifolia, radicosa, riggii, rossica, rubens, setchellii, simplex, sinuosa, vinogradovae*
VARIETIES *rubens aegagropila, vinogradovae moniliformis*
GENUS *Platyclinia* Flat Delesser Ceram Florid Red Algae
SPECIES *crenulata, crozieri, fuegiensis, fuscorubra, ramosa, stipitata, taylorii*
GENUS *Pollexfenia* Pollexfen Delesser Ceram Florid Red Algae
SPECIES *anacapensis, crispata, japonica, laciniata, lobata, minuta, pedicellata, tenella*
GENUS *Polycoryne* Many Club Delesser Ceram Florid Red Algae
SPECIES *compacta, radiata*
GENUS *Polyneura* Many Nerve Delesser Ceram Florid Red Algae
SPECIES *bonnemaisonii, denticulata, japonica, latissima, litterata, subtropica*
GENUS *Polyneurella* Lesser Many Nerve Delesser Ceram Florid Red Algae
SPECIES *hancockii*
VARIETIES *hancockii rhizoidea*
GENUS *Polyneuropsis* Least Many Nerve Delesser Ceram Florid Red Algae
SPECIES *stolonifera*
GENUS *Pseudobranchioglossum* Lesser Tongue Delesser Ceram Florid Red Algae
SPECIES *senegalense*
GENUS *Pseudogrinnellia* Lesser Grinn Delesser Ceram Florid Red Algae
SPECIES *barrattiae*
GENUS *Pseudolaingia* Lesser Laing Delesser Ceram Florid Red Algae
SPECIES *hancockii, larsenii*
GENUS *Pseudonitophylla* Lesser Shiny Delesser Ceram Florid Red Algae
SPECIES *deseadensis*
GENUS *Pseudophycodrys* Least Dryad Delesser Ceram Florid Red Algae
SPECIES *pacifica, phyllophora*
GENUS *Pteridium* Winged Delesser Ceram Florid Red Algae
SPECIES *proliferum*
GENUS *Radicilingua* Radicilingu Delesser Ceram Florid Red Algae
SPECIES *adriatica, reptans, thysanorhizans*
GENUS *Robea* Robe Delesser Ceram Florid Red Algae
SPECIES *costata*
GENUS *Schizoseris* Split Pitted Delesser Ceram Florid Red Algae
SPECIES *bombayensis, condensata, dichotoma, griffithsia, hymenena, multifoliata, perriniae, subdichotoma, tasmanica*
VARIETIES *condensata latiloba*
GENUS *Sorella* Soral Delesser Ceram Florid Red Algae
SPECIES *delicatula, pinnata, pulchra, repens*
GENUS *Sorellocolax* Soral Colac Delesser Ceram Florid Red Algae
SPECIES *stellaris*
GENUS *Sympodophyllum* Sympodal Delesser Ceram Florid Red Algae
SPECIES *reinboldii*
GENUS *Taenioma* Ribbon Delesser Ceram Florid Red Algae
SPECIES *nanum, perpusillum*
GENUS *Togula* Toga Delesser Ceram Florid Red Algae
SPECIES *jolla*
GENUS *Tokidadendron* Tokid Wood Delesser Ceram Florid Red Algae
SPECIES *kurilensis*
VARIETIES *kurilensis japonicum*
GENUS *Tsengiella* Tseng Delesser Ceram Florid Red Algae
SPECIES *spinulosa*
GENUS *Valeriemaya* Valerie May Delesser Ceram Florid Red Algae

SPECIES *geminata, maculata*
GENUS *Vanvoorstia* Vanvoorst Delesser Ceram Florid Red Algae
SPECIES *bennettiana, coccinea, incipiens, pocockiae, spectabilis*
GENUS *Womersleya* Womersley Delesser Ceram Florid Red Algae
SPECIES *monanthus*
GENUS *Yamadaphycus* Yamada Delesser Ceram Florid Red Algae
SPECIES *carnosus*
GENUS *Yendonia* Yendon Delesser Ceram Florid Red Algae
SPECIES *crassifolia, japonica*
GENUS *Yoshidaphycus* Yoshida Delesser Ceram Florid Red Algae
SPECIES *ciliatum*
GENUS *Zellera* Zeller Delesser Ceram Florid Red Algae
SPECIES *tawallina*
GENUS *Zinovaea* Zinova Delesser Ceram Florid Red Algae
SPECIES *acanthocarpa*

FAMILY 4 *Rhodomelaceae* Dark Ceram Florid Red Algae
GENUS *Rhodomela* Dark Ceram Florid Red Algae
SPECIES *confervoides, lycopodioides, muelleri, pinnata, preissii, sachalinensis, teres, vestita*
VARIETIES *lycopodioides tenuissima*
GENUS *Abbottella* Abbott Dark Ceram Florid Red Algae
SPECIES *concinna*
GENUS *Acanthophora* Spiny Dark Ceram Florid Red Algae
SPECIES *dendroides, muscoides, nayadiformis, pacifica, ramulosa, spicifera*
GENUS *Acrocystis* Cyst Dark Ceram Florid Red Algae
SPECIES *loriformis, multifida, nana*
GENUS *Adamsiella* Adams Dark Ceram Florid Red Algae
SPECIES *angustifolia, chauvinii, lorata, melchiori*
GENUS *Aiolocolax* Aiole Colac Dark Ceram Florid Red Algae
SPECIES *pulchellus*
GENUS *Alsidium* Alsid Dark Ceram Florid Red Algae
SPECIES *corallinum, cymatophilum, dasyphyllum, helminthochorton, pusillum*
GENUS *Amansia* Amans Dark Ceram Florid Red Algae
SPECIES *arabica, loriformis, mamillaris, multifida, pinnatifida, rhodantha, serrata*
GENUS *Amplisiphonia* Ample Siphon Dark Ceram Florid Red Algae
SPECIES *pacifica*
GENUS *Antarctocolax* Antarctic Colac Dark Ceram Florid Red Algae
SPECIES *lambii*
GENUS *Aphanocladia* Aphane Branched Dark Ceram Florid Red Algae
SPECIES *delicatula, ecorticata, pacifica, skottsbergii, stichidiosa*
GENUS *Ardissonea* Ardisson Dark Ceram Florid Red Algae
SPECIES *robusta*
GENUS *Ardissonula* Lesser Ardisson Dark Ceram Florid Red Algae
SPECIES *regularis*
GENUS *Benzaitenia* Benzaiten Dark Ceram Florid Red Algae
SPECIES *yenoshimensis*
GENUS *Beringiella* Bering Dark Ceram Florid Red Algae
SPECIES *fritilla, labiosa*
GENUS *Boergeseniella* Boergesen Dark Ceram Florid Red Algae
SPECIES *deludens, fruticulosa, martensiana, thuyoides*
GENUS *Bostrychia* Bostrych Dark Ceram Florid Red Algae
SPECIES *bipinnata, bispora, bryophila, calliptera, cornigera, disticha, flagellifera, fulcrata, glomerata, hamanae-tokidae, harveyi, hongkongensis, jelineckii, kaernbachii, lepida, leptoclada, mamillosa, monosiphonia, montagnei, moritziana, nigrescens, pannosa, pauperula, pilulifera, pinnata, radicans, scorpioides, sertularia, siliquis, simpliciuscula, sonderiana, tenella, tenuissima, vieillardii, wardii*
VARIETIES *hamanae-tokidae morii, harveyi chilensis, radicans hapteromanica, scorpioides montagnei, sertularia hiemalis, tenella denudata*
GENUS *Brongniartella* Brongniart Dark Ceram Florid Red Algae
SPECIES *australis, byssoides, elegans, mucronata*
GENUS *Bryocladia* Moss Branched Dark Ceram Florid Red Algae
SPECIES *cervicornis, cuspidata, densa, thwaitesii, thyrsigera*
GENUS *Bryothamnion* Moss Thamnion Dark Ceram Florid Red Algae
SPECIES *pacificum, seaforthii, triquetrum*
VARIETIES *seaforthii imbricatum*
GENUS *Chamaethamnion* Chamise Thamnion Dark Ceram Florid Red Algae
SPECIES *pocockiae, schizandra*
GENUS *Chiracanthia* Chir Spiny Dark Ceram Florid Red Algae
SPECIES *arborea*

GENUS *Chondrophycus* Cartilage Dark Ceram Florid Red Algae
SPECIES *articulatus, brandenii, capituliformis, carolinensis, cartilagineus, columellaris, concretus, corallopsis, cruciatus, dinhii, dotyi, flagelliferus, gemmiferus, intermedius, iridescens, jejuna, kangjaewonii, longicaulis, maris-rubri, palisadus, paniculatus, papillosus, parvipapillatus, patentirameus, perforatus, poiteaui, succisus, surculigerus, thuyoides, translucidus, tronoi, tumidus, undulatus, verticillatus, yamadanus*
VARIETIES *papillosus australasica, papillosus pacifica, papillosus thyrsoides, poiteaui gemmifera*
GENUS *Cladhymenia* Branched Hymen Dark Ceram Florid Red Algae
SPECIES *coronata, lenceifolia, lyallii, oblongifolia, pellucida*
GENUS *Cladurus* Branched Tail Dark Ceram Florid Red Algae
SPECIES *elatus*
GENUS *Cliftonaea* Clifton Dark Ceram Florid Red Algae
SPECIES *pectinata*
GENUS *Coeloclonium* Sky Clone Dark Ceram Florid Red Algae
SPECIES *debile, tasmanicum, umbellula, verticillatum*
GENUS *Colacopsis* Colac Dark Ceram Florid Red Algae
SPECIES *lophurellae, pulvinata, smitheniae, velutina*
GENUS *Ctenosiphonia* Comb Siphon Dark Ceram Florid Red Algae
SPECIES *hypnoides*
GENUS *Dasyclonium* Daisy Clone Dark Ceram Florid Red Algae
SPECIES *adiantiformis, bifurcatum, bipartitum, cuneifolium, flaccidum, harveyanum, incisum, ovalifolium, palmatifidum*
GENUS *Dawsoniella* Dawson Dark Ceram Florid Red Algae
SPECIES *bulborhiza*
GENUS *Dictyomenia* Netted Dark Ceram Florid Red Algae
SPECIES *angusta, ecostata, harveyana, sonderi, stephensonii, tridens*
GENUS *Digenea* Digene Dark Ceram Florid Red Algae
SPECIES *simplex, subarticulata*
GENUS *Digeneopsis* Lesser Digene Dark Ceram Florid Red Algae
SPECIES *subopaca*
GENUS *Diplocladia* Diploid Branched Dark Ceram Florid Red Algae
SPECIES *patersonis*
GENUS *Dipterocolax* Dual Wing Colac Dark Ceram Florid Red Algae
SPECIES *fernandezianus*
GENUS *Dipterosiphonia* Dual Wing Siphon Dark Ceram Florid Red Algae
SPECIES *australica, dendritica, heteroclada, prorepens, reversa, rigens, ringens*
VARIETIES *dendritica africana*
GENUS *Ditria* Ditriad Dark Ceram Florid Red Algae
SPECIES *expleta, reptans, zonaricola*
GENUS *Doxodasya* Dox Daisy Dark Ceram Florid Red Algae
SPECIES *bolbochaete, hirta, lanuginosa, lenormandiana*
GENUS *Echinosporangium* Urchin Spore Dark Ceram Florid Red Algae
SPECIES *semipennatum*
GENUS *Echinothamnion* Urchin Thamnion Dark Ceram Florid Red Algae
SPECIES *hookeri, hystrix, lyallii, mallardiae*
GENUS *Enantiocladia* Enanti BranchedDark Ceram Florid Red Algae
SPECIES *duperreyi, okamurae, prolifera, robinsonii*
GENUS *Endosiphonia* Endic Siphon Dark Ceram Florid Red Algae
SPECIES *gardineri, horrida, spinulosa*
GENUS *Enelittosiphonia* Enelitt Siphon Dark Ceram Florid Red Algae
SPECIES *hakodatensis, stimpsonii*
GENUS *Epiglossum* Epiphyte Tongue Dark Ceram Florid Red Algae
SPECIES *proliferum, smithiae*
GENUS *Erythrocystis* Cyst Dark Ceram Florid Red Algae
SPECIES *montagnei, saccata*
GENUS *Erythrostachys* Stack Dark Ceram Florid Red Algae
SPECIES *strobilifera*
GENUS *Exophyllum* Exophyte Dark Ceram Florid Red Algae
SPECIES *wentii*
GENUS *Fernandosiphonia* Fernand Siphon Dark Ceram Florid Red Algae
SPECIES *ecorticata, nana, unilateralis*
GENUS *Gonatogenia* Gonad Dark Ceram Florid Red Algae
SPECIES *subulata*
GENUS *Grammitella* Gramm Dark Ceram Florid Red Algae
SPECIES *guernisacii*
GENUS *Gredgaria* Gredgar Dark Ceram Florid Red Algae
SPECIES *maugeana*
GENUS *Halopithys* Haline Pithy Dark Ceram Florid Red Algae

SPECIES *incurva*
GENUS *Hawaiia* Hawaii Dark Ceram Florid Red Algae
SPECIES *trichia*
GENUS *Herpopteros* Herpetic Wing Dark Ceram Florid Red Algae
SPECIES *fallax*
GENUS *Herposiphonia* Herpetic Siphon Dark Ceram Florid Red Algae
SPECIES *akidoglossa, arcuata, bipinnata, brachyclados, caespitosa, calothrix, calva, ceratoclada, clavata, crassa, delicatula, dendroidea, didymosporangia, dubia, elongata, falcata, filifera, filipendula, fissidentoides, grandis, guineensis, heringii, hollenbergii, insidiosa, littoralis, monilifera, nuda, obscura, pacifica, paniculata, parca, parvula, pecten-veneris, pectinella, plumula, prorepens, rostrata, secunda, subdisticha, sulivanae, trichia, variabilis, versicolor, verticillata, vietnamica, xaymaca*
VARIETIES *dendroidea minor, pecten-veneris laxa, secunda densa, secunda tenella*
GENUS *Herposiphoniella* Lesser Herpetic Siphon Dark Ceram Florid Red Algae
SPECIES *plurisegmenta*
GENUS *Heterocladia* Variable Branched Dark Ceram Florid Red Algae
SPECIES *australis, caudata, umbellifera*
GENUS *Heterostroma* Variable Stroma Dark Ceram Florid Red Algae
SPECIES *nereidiis*
GENUS *Holotrichia* Hairy Dark Ceram Florid Red Algae
SPECIES *comosa*
GENUS *Husseya* Hussey Dark Ceram Florid Red Algae
SPECIES *rubra*
GENUS *Hutchinsia* Hutchins Dark Ceram Florid Red Algae
SPECIES *spinella*
GENUS *Janczewskia* Janczewsky Dark Ceram Florid Red Algae
SPECIES *australis, gardneri, hawaiiana, lappacea, meridionalis, moriformis, morimotoi, ramiformis, solmsii, tasmanica, teysmannii, verrucaeformis*
GENUS *Jantinella* Jantin Dark Ceram Florid Red Algae
SPECIES *sinicola, verrucaeformis*
GENUS *Kuetzingia* Kuetzing Dark Ceram Florid Red Algae
SPECIES *angusta, canaliculata, natalensis, pectinella*
GENUS *Laurencia* Laurence Dark Ceram Florid Red Algae
SPECIES *aldingensis, alsidioides, arbuscula, bolivarii, botrychioides, botryoides, brachyclados, brongniartii, caduciramulosa, calliclada, canariensis, capituliformis, caraibica, caspica, catarinensis, cervicornis, ceylanica, chilensis, chinensis, chondrioides, clarionensis, clavata, claviformis, coelenterata, columellaris, complanata, composita, congesta, coronopus, corymbosa, crustiformans, cymosa, decidua, decumbens, diegoensis, distichophylla, elata, epiphylla, estebaniana, fasciculata, fastigiata, filiformis, flexilis, flexuosa, foldatsii, forsteri, furcata, galstoffii, gardneri, glandulifera, glomerata, gracilis, hamata, hancockii, hongkongensis, humilis, indica, intricata, irieii, japonensis, johnstonii, lageniformis, lajolla, lata, laxa, longcaulis, majuscula, mariannensis, masonii, mcdermidiae, microcladia, minuscula, minuta, moretonensis, multibulba, nana, nangii, natalensis, nidifica, nipponica, nuda, obscura, obtusa, obtusiuscula, okamurai, oliveirana, omaezakiana, oophora, oppositocladia, pacifica, paniculata, pannosa, parvula, pedicularioides, peninsularis, pinnata, platyclada, ptychodes, pumila, pyramidalis, radicans, rigida, saitoi, shepherdii, silvae, similis, singaporensis, snackeyi, snyderiae, spicifera, spinulifera, striolata, subcolumnaris, subdisticha, subopposita, subsimplex, tasmanica, tenera, tenuissima, thuyoides, thyrsifera, tristicha, tropica, tuberculosa, uvifera, venusta, verruculosa, viridis, voragina*
VARIETIES *filiformis decussata, indica nidifica, minuta scammaccae, obtusa compacta, obtusa dendroidea, obtusa densa, obtusa divaricata, obtusa gracilis, obtusa laxa, obtusa mollissima, obtusa paniculata, obtusa pulvinata, obtusa pyramidica, obtusa racemosa, obtusa rigidula, obtusa snackeyi, obtusiuscula corymbifera, obtusiuscula laxa, paniculata patentiramea, pedicularioides queenslandica, tenuissima corymbulosa*
GENUS *Laurenciocolax* Laurence Colac Dark Ceram Florid Red Algae
SPECIES *polyspora*
GENUS *Lautosphaeria* Lauter Sphere Dark Ceram Florid Red Algae
SPECIES *monsfumosa*
GENUS *Lembergia* Lemberg Dark Ceram Florid Red Algae
SPECIES *allanii*
GENUS *Lenormandiopsis* Lenormand Dark Ceram Florid Red Algae
SPECIES *lorentzii, nozawae, parthasarathyi*
GENUS *Leptosiphonia* Thin Siphon Dark Ceram Florid Red Algae
SPECIES *schousboei*
GENUS *Leveillea* Leve Dark Ceram Florid Red Algae
SPECIES *comosa, jungermannioides, major, pectinata*
GENUS *Levringiella* Levring Dark Ceram Florid Red Algae
SPECIES *gardneri, microscopica*
GENUS *Lophocladia* Crest Branched Dark Ceram Florid Red Algae
SPECIES *japonica, kipukaia, kuetzingii, lallemandii, minima, trichoclados*
GENUS *Lophosiphonia* Crest Siphon Dark Ceram Florid Red Algae

SPECIES *adhaerens, capensis, cristata, hayashii, mexicana, obscura, prostrata, reptabunda*
VARIETIES *obscura major*
GENUS *Lophothalia* Crest Thal Dark Ceram Florid Red Algae
SPECIES *hormoclados, verticillata*
GENUS *Lophura* Crest Tailed Dark Ceram Florid Red Algae
SPECIES *tenuis*
GENUS *Lophurella* Lesser Crest Tailed Dark Ceram Florid Red Algae
SPECIES *caespitosa, christosphersenii, comosa, gaimardii, hookeriana, patula, periclados*
GENUS *Melanamansia* Melanamans Dark Ceram Florid Red Algae
SPECIES *daemelii, dietrichiana, fimbrifolia, glomerata, japonica, mamillaris, mitsuii, pumila, scalpellata, seagriefii*
GENUS *Melanothamnus* Melano Thamnion Dark Ceram Florid Red Algae
SPECIES *afaqhusainii, somalensis*
GENUS *Meridiocolax* Meridian Colac Dark Ceram Florid Red Algae
SPECIES *bracteata, narcissus, polysiphoniae*
GENUS *Metamorphe* Metamorph Dark Ceram Florid Red Algae
SPECIES *colensoi*
GENUS *Microcolax* Colac Dark Ceram Florid Red Algae
SPECIES *africanus, botryocarpa*
GENUS *Micropeuce* Peuce Dark Ceram Florid Red Algae
SPECIES *feredayae, glomerulifera, mucronata, sarcocaulon, setosus*
GENUS *Murrayellopsis* Murray Dark Ceram Florid Red Algae
SPECIES *dawsonii*
GENUS *Nanopera* Dwarf Dark Ceram Florid Red Algae
SPECIES *merrifieldiae*
GENUS *Neorhodomela* Modern Dark Ceram Florid Red Algae
SPECIES *aculeata, enomotoi, irtugoi, larix, munita, oregona*
VARIETIES *larix superlittoralis*
GENUS *Neosiphonia* Modern Siphon Dark Ceram Florid Red Algae
SPECIES *apiculata, decumbens, elongella, flavimarina, harlandii, harveyi, japonica, notoensis, poko, savatieri, sparsa, sphaerocarpa, teradomariensis, tongatensis, yendoi*
VARIETIES *harveyi olneyi japonica forfex, japonica savatieri, poko longii, sphaerocarpa distans, tongatensis upolensis*
GENUS *Neotenophycus* Modern Tenor Dark Ceram Florid Red Algae
SPECIES *ichthyosteus*
GENUS *Neurymenia* Neurymene Dark Ceram Florid Red Algae
SPECIES *fraxinifolia, nigricans*
GENUS *Odonthalia* Toothed Thal Dark Ceram Florid Red Algae
SPECIES *annae, corymbifera, dentata, floccosa, kamtschatica, kawabatae, lyallii, macrocarpa, ochotensis, setacea, uniseriata, washingtoniensis, yamadae*
VARIETIES *annae isomera, dentata angustata*
GENUS *Oligocladella* Small Branched Dark Ceram Florid Red Algae
SPECIES *boldinghii, prainii, pusilla*
GENUS *Onychocolax* Claw Colac Dark Ceram Florid Red Algae
SPECIES *polysiphoniae*
GENUS *Ophidocladus* Snake Branched Dark Ceram Florid Red Algae
SPECIES *simpliciusculus*
GENUS *Osmundaria* Lesser Osmund Dark Ceram Florid Red Algae
SPECIES *colensoi, fimbriata, melvillii, obtusiloba, oliveae, papenfussii, prolifera, serrata, volubilis*
GENUS *Osmundea* Osmund Dark Ceram Florid Red Algae
SPECIES *blinksii, crispa, hybrida, maggsiana, multibulba, osmunda, pelagiensis, pelagosae, pinnatifida, ramosissima, sinicola, spectabilis, splendens, truncata, verlaquei*
VARIETIES *pinnatifida angusta, pinnatifida littoralis, spectabilis diegoensis*
GENUS *Pachychaeta* Thick Bristle Dark Ceram Florid Red Algae
SPECIES *brachyarthra, cryptoclada*
GENUS *Periphykon* Periphyte Dark Ceram Florid Red Algae
SPECIES *beckeri, delesserioides*
GENUS *Perrinia* Perrin Dark Ceram Florid Red Algae
SPECIES *ericoides*
GENUS *Phaeocolax* Dusky Colac Dark Ceram Florid Red Algae
SPECIES *kajimurai*
GENUS *Picconiella* Piccon Dark Ceram Florid Red Algae
SPECIES *pectinata, plumosa*
GENUS *Pityophykos* Pity Dark Ceram Florid Red Algae
SPECIES *tasmanica*
GENUS *Placophora* Plaque Dark Ceram Florid Red Algae
SPECIES *binderi, japonica, monocarpa*

GENUS *Pleurostichidium* Pleural Stitch Dark Ceram Florid Red Algae
SPECIES *falkenbergii*
GENUS *Polysiphonia* Polysiphon Dark Ceram Florid Red Algae
SPECIES *abscissa, abscissoides, acuminata, adamsiae, adriatica, affinis, allochroa, amphibolis, anisogona, anomala, arachnoidea, arctica, aterrima, atlantica, atra, atricapilla, australiensis, azorica, bajacali, banyulensis, barbatula, beaudettei, beguinotii, bicornis, biformis, bifurcata, binneyi, blandii, boergesenii, boldii, breviarticulata, brevisegmenta, brodiei, caespitosa, callithamnioides, camerunensis, capucina, carettia, caspica, castagnei, castelliana, ceramiaeformis, cladorhiza, coacta, codicola, confusa, constricta, corymbosa, crassa, crassicollis, crassiuscula, curta, dasyoeformis, daveyae, decipiens, decussata, delicatula, denudata, derbesii, deusta, devoniensis, dichotoma, divergens, dotyi, dumosa, echigoensis, echinata, elongata, erythraea, exilis, fasciculata, fernandeziana, ferulacea, fibrata, fibrillosa, figariana, flabelliformis, flabellulata, flexella, flexicaulis, flocculosa, foeniculacea, foetidissima, forfex, fracta, fragilis, fucoides, funebris, furcellata, fuscorubens, gonatophora, gorgoniae, gracilis, guernisacii, hancockii, hapalacantha, haplodasyae, hassleri, havanensis, hawaiiensis, hemisphaerica, hendryi, herpa, hirta, hochstetteriana, homoia, howei, incompta, indigena, infestans, intricata, isogona, johnstonii, kampsaxii, kappannae, kieliana, kowiensis, lanosa, letestui, macounii, macrocarpa, marchantae, mollis, morisiana, morrowii, mottei, muelleriana, myriococca, namibiensis, nathanielii, neglecta, nhatrangense, nigra, nizamuddinii, nodulosa, nutans, opaca, ornata, orthocarpa, pacificapaniculata, paradoxa, parthasarathyi, parvula, pauperula, pentamera, perforans, pernacola, perriniae, platycarpa, plectrocarpa, polychroma, polyspora, porrecta, profunda, propagulifera, pseudovillum, pulvinata, purpurea, ramentacea, requienii, rhododactyla, rhunensis, rigidula, rubrorhiza, rudis, saccorhiza, sadoensis, sanguinea, scopulorum, senticulosa, sertularioides, setigera, shepherdii, simplex, simulans, sinicola, sonorensis, souriei, sparsa, sphaerospora, spinella, spinescens, spinosa, spinulifera, stricta, strictissima, stuposa, subtilissima, subulata, subulifera, succulenta, tapinocarpa, teges, tenerrima, tenuis, tenuistriata, tepida, tokidae, tripinnata, triton, tsudana, tuberosa, tuticorinensis, unguiformis, upolensis, urbana, urbanoides, utricularis, virgata, yonakuniensis*
VARIETIES *brodiei agardhii, brodiei densa, denudata leptura, divergens heteronema, echigoensis schuebelerii, elongata baltica, elongata denudata, fibrillosa griffithsiana, fibrillosa major, fibrillosa tenuissima, fucoides fucoides, fucoides affinis, havaniensis mucosa, hendryi compacta, hendryi deliquescens, hendryi gardneri, hendryi luxurians, johnstonii concinna, opaca intricata, opaca repens, , pacifica delicatula, pacifica determinata, pacifica distans, pacifica disticha, pacifica gracilis, pulvinata parvula, scopulorum villum, stricta intricata, stricta lepadicola, stricta patens, stricta roseola*
GENUS *Polyzonia* Polyzone Dark Ceram Florid Red Algae
SPECIES *elegans*
GENUS *Protokuetzingia* Primitive Kuetzing Dark Ceram Florid Red Algae
SPECIES *australasica, schottii*
GENUS *Pterochondria* Winged Cartilage Dark Ceram Florid Red Algae
SPECIES *woodii*
VARIETIES *woodii pygmaea*
GENUS *Pterosiphonia* Winged Siphon Dark Ceram Florid Red Algae
SPECIES *arctica, ardreana, baileyi, bipinnata, clevelandii, cloiophylla, complanata, concinna, dendroidea, fibrillosa, gardneri, gracilis, hamata, javanica, parasitica, pennata, pinnulata, spinifera, stangeri*
VARIETIES *bipinnata robusta, parasitica australis*
GENUS *Pycnothamnion* Dense Thamnion Dark Ceram Florid Red Algae
SPECIES *crustaceum*
GENUS *Rhodolachne* Lachne Dark Ceram Florid Red Algae
SPECIES *decussata, radicosa*
GENUS *Rhodomelopsis* Soft Dark Ceram Florid Red Algae
SPECIES *africana*
GENUS *Rodriguezella* Rodriguez Dark Ceram Florid Red Algae
SPECIES *bornetii, hongngai, ligulata, pennata, pinnata, rhizophylla, strafforelloi*
GENUS *Rytiphlaea* Rytiphl Dark Ceram Florid Red Algae
SPECIES *australis, semicristata, thuyoides, tinctoria*
VARIETIES *tinctoria horridula*
GENUS *Schizochlaenion* Split Chlaen Dark Ceram Florid Red Algae
SPECIES *rhodotrichum*
GENUS *Sonderella* Sonder Dark Ceram Florid Red Algae
SPECIES *linearis*
GENUS *Spirocladia* Spire Branched Dark Ceram Florid Red Algae
SPECIES *barodensis, hodgsoniae, loochooensis*
GENUS *Sporoglossum* Spore Tongue Dark Ceram Florid Red Algae
SPECIES *lophurellae*
GENUS *Stichothamnion* Stitch Thamnion Dark Ceram Florid Red Algae
SPECIES *antillarum, cymatophilum*
GENUS *Stictosiphonia* Stick Siphon Dark Ceram Florid Red Algae
SPECIES *arbuscula, gracilis, intricata, kelanensis, simpliciuscula, tangatensis, vaga*
VARIETIES *arbuscula gracilis, intricata fastigiata, intricata major, intricata minor*
GENUS *Streblocladia* Streble Branched Dark Ceram Florid Red Algae

SPECIES *atrata, camptoclada, collabens, corymbifera, glomerulata, tenuissima*
GENUS *Stromatocarpus* Stroma Carpel Dark Ceram Florid Red Algae
SPECIES *microscopicus, parasiticus*
GENUS *Symphyocladia* Symphyte Branched Dark Ceram Florid Red Algae
SPECIES *latissima, latiuscula, linearis, marchantioides, pennata, pumila*
GENUS *Tayloriella* Taylor Dark Ceram Florid Red Algae
SPECIES *abyssalis, dictyurus, divaricata, tenebrosa*
GENUS *Tiparraria* Tiparrari Dark Ceram Florid Red Algae
SPECIES *aurata*
GENUS *Tolypiocladia* Tolypion Branched Dark Ceram Florid Red Algae
SPECIES *calodictyon, condensata, glomerulata, penningtonensis*
GENUS *Trichidium* Hairy Dark Ceram Florid Red Algae
SPECIES *pedicellatum*
GENUS *Trigenea* Trigene Dark Ceram Florid Red Algae
SPECIES *australis*
GENUS *Tylocolax* Tyle Colac Dark Ceram Florid Red Algae
SPECIES *microcarpus*
GENUS *Ululania* Ulula Dark Ceram Florid Red Algae
SPECIES *stellata*
GENUS *Veleroa* Velero Dark Ceram Florid Red Algae
SPECIES *adunca, elongata, karuvalensis, magneana, subulata*
GENUS *Veryhachium* Veryhach Dark Ceram Florid Red Algae
SPECIES *gibbosum, globosum, longispinosum, martinum, subglobosum*
GENUS *Vestibulongum* Vestibule Dark Ceram Florid Red Algae
SPECIES *corlissi*
GENUS *Virgatasporites* Virgate Spore Dark Ceram Florid Red Algae
SPECIES *rudi*
GENUS *Vidalia* Vidal Dark Ceram Florid Red Algae
SPECIES *spiralis*
VARIETIES *spiralis cliftonii*
GENUS *Waldoia* Waldo Dark Ceram Florid Red Algae
SPECIES *antillana*
GENUS *Wilsonaea* Wilson Dark Ceram Florid Red Algae
SPECIES *dictyuroides*
GENUS *Womersleyella* Womersley Dark Ceram Florid Red Algae
SPECIES *kwazuluensis, pacifica, pacifica minor, setacea*
VARIETIES *pacifica minor*
GENUS *Wrightiella* Wright Dark Ceram Florid Red Algae
SPECIES *blodgettii, tumanowiczii*

FAMILY 5 *Sarcomeniaceae* Flesh Ceram Florid Red Algae
GENUS *Sarcomenia* Flesh Ceram Florid Red Algae
SPECIES *delesserioides*
GENUS *Cottoniella* Cotton Flesh Ceram Florid Red Algae
SPECIES *arcuata, filamentosa, fusiformis, libyensis, sanguinea*
VARIETIES *arcuata filamentosa, filamentosa algeriensis*
GENUS *Malaconema* Malac Thread Flesh Ceram Florid Red Algae
SPECIES *minimum, roeanum*
GENUS *Platysiphonia* Flat Siphon Flesh Ceram Florid Red Algae
SPECIES *caribaea, clevelandii, decumbens, delicata, hypneoides, intermedia, marginalis, mutabilis, parva, victoriae*
GENUS *Sarcotrichia* Hairy Flesh Ceram Florid Red Algae
SPECIES *dolichocystidea, tenera*

ORDER 5 *Corallinales* Coral Florid Red Algae

FAMILY 1 *Corallinaceae* Coral Florid Red Algae
GENUS *Corallina* Coral Florid Red Algae
SPECIES *aculeata, armata, berteroi, confusa, cuvieri, densa, elegans, elongata, frondescens, goughensis, hemisphaerica, hombronii, janioides, lobata, millegrana, muscoides, officinalis, panizzoi, pilulifera, pinnatifolia, polysticha, rigida, sachalinensis, tubulosa, vancouverensis*
VARIETIES *officinalis chilensis, officinalis flexilis, vancouverensis verticillata*
GENUS *Alatocladia* Alate Branched Coral Florid Red Algae
SPECIES *modesta*
GENUS *Amphiroa* Amphiro Coral Florid Red Algae
SPECIES *anastomosans, anceps, annulata, beauvoisii, brasiliana, brevianceps, breviarticulata, canaliculata, capensis, compressa, crassa, crosslandii, cryptarthrodia, cumingii, currae, dimorpha, drouetii, echigoensis, ephedraea, foliacea, fragilissima, franciscana, galapagensis, gracilis, hancockii, irregularis, itonoi, kuetzingiana,*

magdalenensis, mexicana, minutissima, misakiensis, nodulosa, pacifica, peninsularis, peruana, polymorpha, pusilla, rigida, subcylindrica, tasmanica, taylorii, tribulus, ungulata, valonioides, vanbossea, yendoi, zonata

VARIETIES *annulata pinnata, beauvoisii crassiuscula, compressa tenuis, crassa minuta, foliacea erecta, foliacea procumbens, fragilissima cuspidata, fragilissima cyathifera, franciscana robusta, rigida antillana*

GENUS *Archaeolithothamnion* Archaic Rock Thamnion Coral Florid Red Algae

SPECIES *affine, brevium, crosslandii, intermedium, lithothamnioides*

GENUS *Archaeolithothamnium* (Synonym for *Archaeolithothamnion*)

GENUS *Arthrocardia* Straight Heart Coral Florid Red Algae

SPECIES *anceps,carinata, corymbosa, duthieae, filicula, flabellata, lobata, palmata, setchellii, silvae, wardii*

VARIETIES *anceps modestum, flabellata australica*

GENUS *Austrolithon* Southern Rock Coral Florid Red Algae

SPECIES *intumescens*

GENUS *Boreolithon* Bore Rock Coral Florid Red Algae

SPECIES *van-heurckii*

GENUS *Bossea* Boss Coral Florid Red Algae

SPECIES *angustata, cooperi, gardneri, insularis, ligulata, pachyclada, sagittata*

GENUS *Bossiella* Lesser Boss Coral Florid Red Algae

SPECIES *californica, chiloensis, compressa, cretacea, dichotoma, orbignyana, plumosa*

VARIETIES *californica schmittii*

GENUS *Calliarthron* Callisto Straight Coral Florid Red Algae

SPECIES *latissimum, tuberculosum, yessoense*

GENUS *Cheilosporum* Lip Spore Coral Florid Red Algae

SPECIES *acutilobum, cultratum, elegans, proliferum, sagittatum, spectabile*

VARIETIES *cultratum debilis, cultratum minor, cultratum multifidum, sagittatum minor*

GENUS *Chiharaea* Chihara Coral Florid Red Algae

SPECIES *bodegensis*

GENUS *Chihsienella* Chihsien Coral Florid Red Algae

SPECIES *chihsienensis*

VARIETIES *chihsienensis anabaria*

GENUS *Choreonema* Chore Thread Coral Florid Red Algae

SPECIES *thuretii*

GENUS *Clathromorphum* Lattice Coral Florid Red Algae

SPECIES *circumscriptum, compactum, loculosum, nereostratum, obteculum, parcum, reclinatum, tubiforme, variabile*

VARIETIES *circumscriptum areolatum*

GENUS *Dermatolithon* Dermal Rock Coral Florid Red Algae

SPECIES *ascripticium, nodulosum, polycephalum, veleroae*

GENUS *Epilithon* Epiphyte Rock Coral Florid Red Algae

SPECIES *vallentinae*

GENUS *Exilicrusta* Exil Crust Coral Florid Red Algae

SPECIES *parva*

GENUS *Ezo* Ezo Coral Florid Red Algae

SPECIES *epiyessoense*

GENUS *Fosliella* Foslie Coral Florid Red Algae

SPECIES *affinis, atlantica, bermudensis, chamadoris, chamaedoris, fertilis, mediterranea, minuta, paschalis, sargassi, valida*

GENUS *Goniolithon* Gonad Rock Coral Florid Red Algae

SPECIES *acropetum, decutescens, dispalatum, solubile, versabile*

VARIETIES *dispalatum subsimplex*

GENUS *Haliptilon* Haline Ptilon Coral Florid Red Algae

SPECIES *attenuatum, cubense, gracile, mauritianum, paniculatum, polydactylum, roseum, splendens, squamatum, subulatum, virgatum*

VARIETIES *virgatum australe*

GENUS *Heteroderma* Variable Dermal Coral Florid Red Algae

SPECIES *corallinicola, gibbsii, nicholsii, parvicarpa, subtilissima*

GENUS *Hydrolithon* Hydric Rock Coral Florid Red Algae

SPECIES *arenicolum, boergesenii, boreale, breviclavium, cruciatum, cymodoceae, discoideum, farinosum, gardineri,improcerum, iyengarii, krusadiense, munitum, murakoshii, oligocarpum, onkodes, pellire, reinboldii, rupestre, rupestris, ruprestris, samoã«nse, sargassi, somaliae, superficiale, verrucosum*

VARIETIES *farinosum chalicodictyum, farinosum mauritiana, farinosum solmsiana, gardineri abbreviatum, improcerum tenue, onkodes subramosum*

GENUS *Jania* Jan Coral Florid Red Algae

SPECIES *adhaerens, affinis, antennina, arborescens, capillacea, decussatodichotoma, fastigiata, intermedia, iyengarii, longiarthra, longifurca, micrarthrodia, minuta, nipponica, novae-zelandiae, pacifica, parva, prolifera, pumila, pusilla, pygmaea, radiata, rubens, santae-marthae, subpinnata, tenella, ungulata, verrucosa, yenoshimensis*

VARIETIES *micrarthrodia tenuissima, rubens africana, rubens concatenata, rubens corniculata, rubens plumula, tenella zacae, ungulata brevior, verrucosa tenuior*

GENUS *Kvaleya* Kvaley Coral Florid Red Algae

SPECIES *epilaeve*

GENUS *Leptophytum* Thin Coral Florid Red Algae

SPECIES *acervatum, adeyi, bisporum, bornetii, coulmanicum, elatum, ferox, foecundum, foveatum, laeve*

GENUS *Leptorima* Lesser Thin Coral Florid Red Algae

SPECIES *nivea, oculata, undulata*

GENUS *Lesueuria* Lesueur Coral Florid Red Algae

SPECIES *minderiana*

GENUS *Litholepis* Rock Thin Coral Florid Red Algae

SPECIES *sonorensis*

GENUS *Lithophyllum* Rock Coral Florid Red Algae

SPECIES *absimile, acanthinum, accola, acrocamptum, albanense, alternans, amplostratum, aninae, applicatum, atalayense, azorum, bahrijense, bamleri, bipartitum, bracchiatum, brachiatum, byssoides, calcareum, canescens, chamberlainianum, complexum, congestum, conspectum, crassum, crodelioides, crouaniorum, cuneatum, cystoseirae, decussatum, dentatum, diguetti, dispar, divaricatum, duckerae, erosum, esperi, falkandcidum, fasciculatum, flavescens, geometricum, glangeaudii, gracile, grandiusculum, grumosum, hancocki, hermaphroditum, hibernicum, imitans, impressum, incrustans, insipidum, intermedium, ippolitoi, irregulare, irvineanum, johansenii, kotschyanum, laeve, leptothalloideum, lichenare, lithophylloides, lobatum, margaritae, mediocre, mgarrense, mutabile, neoatalayense, neofarlowii, nitorum, okamurae, orbiculatum, pallescens, papillosum, pinguinense, polycarpum, polycephalum, polyclonum, prelichenoides, proboscideum, prototypum, punctatum, pygmaeum, racemus, reesei, retusum, rileyi, sancti-georgei, shioense, sigi, simile, skottsbergii, stephensonii, stictaeforme, subantarcticum, tarentinum, tasmanicum, tedeschii, tortuosum, tumidulum, veleroae, vickersiae, whidbeyense, yessoense*

VARIETIES *calcareum eunana, congestum pseudodentatum, decussatum planiuscula, fasciculatum compressa, fasciculatum eunana, incrustans flabellata, incrustans harveyi, incrustans subdichotomum, kotschyanum affine, kotschyanum subplicatum, mediocre tasmanica, proboscideum delicatulum, pygmaeum geminostratum, stictaeforme cabiochae, stictaeforme expansum, tortuosum crassa, tortuosum decumbens, tortuosum undulosa*

GENUS *Lithoporella* Rock Pore Coral Florid Red Algae

SPECIES *atlantica, bermudensis, indica, melobesioides, sauvageaui*

VARIETIES *melobesioides varians*

GENUS *Lithothamnion* Rock Thamnion Coral Florid Red Algae

SPECIES *aculeiferum, adeyi, album, andrussowii, angolense, antarcticum, asperulatum, aucklandicum, australe, betieri, brasiliense, breviaxe, californicum, canariense, capense, cardinellense, corallioides, cottonii, coulmanicum, crassiusculum, crassum, crispatum, diguetii, ectocarpon, esperi, flavescens, fornicatum, fuegianum, giammarinoi, gibbosum, giganteum, glaciale, grade, grande, granuliferum, guadalupense, hamelii, heterocladum, heteromorphum, indicum, intermedium, islei, japonicum, kerguelenum, lemoineae, magnum, maldivicum, mangini, microsporum, minervae, montereyicum, muelleri, murmanicum, neglectum, nitidum, nodulosum, norvegicum, obtectulum, occidentale, pacificum, pauciporosum, peleense, philippii, phymatodeum, pocillum, praefruticulosum, proliferum, proponttidis, ruptile, scabiosum, sejunctum, sonderi, spissum, tophiforme, tusterense, ungeri, valens, validum, variabile, vescum, volcanum*

VARIETIES *australe americana, australe brachiatum, australe minutulum, australe subtile, australe tualensis, australe ubianum, capense eckloniae, corallioides flabelligerum, corallioides subsimplex, crispatum alternans, fornicatum apiculatum, fornicatum connatum, gibbosum crassum, glaciale boreale, glaciale rosea, indicum subtile, occidentale effusa*

GENUS *Lithothrix* Rock Thread Coral Florid Red Algae

SPECIES *aspergillum*

GENUS *Marginisporum* Margin Spore Coral Florid Red Algae

SPECIES *aberrans, crassissima, declinata*

GENUS *Mastophora* Mastigote Coral Florid Red Algae

SPECIES *laevis, pacifica, rosea, variegata*

GENUS *Mastophoropsis* Lesser Mastigote Coral Florid Red Algae

SPECIES *canaliculata*

GENUS *Matanomadhia* Matan Madhi Coral Florid Red Algae

SPECIES *indica, ovata*

GENUS *Melobesia* Dark Obese Coral Florid Red Algae

SPECIES *confervoides, galapagensis, grandiuscula, leptura, marginata, mediocris, melobesioides, membranacea, polycarpa, polystromatica, rosanoffii, tasmanica, tomitaroi*

GENUS *Mesophyllum* Mesocene Coral Florid Red Algae

SPECIES *aemulans, aleuticum, alternans, arcticum, brachycladum, canariense, caraiensis, commune, conchatum, crassiusculum, crispescens, curtum, cystocarpideum, ectocarpon, ehrmannii, engelhartii, erubescens, expansum, floridanum, fuegianum, funafutiense, imbricatum, incertum, incisum, inconspicuum, koritzae, lamellatum, laxum, lichenoides, macedonis, macroblastum, mesomorphum, nitidum, ornatum, printzianum, pulchrum, rupestre, sancti-dionysii, schmitzii, siamense, simulans, superpositum, syntrophicum, syrphetodes, tenue, vancouveriense*

VARIETIES *expansum agariciforme*
GENUS *Metagoniolithon* Meta Gonad Rock Coral Florid Red Algae
SPECIES *chara, charoides, gallioides, radiatum, stelliferum*
VARIETIES *chara dichotomum*
GENUS *Metamastophora* Meta Mastigote Coral Florid Red Algae
SPECIES *flabellata*
GENUS *Millepora* Mille Pore Coral Florid Red Algae
SPECIES *coriacea, fasciculata*
GENUS *Multisiphonia* Multisiphon Coral Florid Red Algae
SPECIES *hemicirculis, nanshanensis*
GENUS *Neogoniolithon* Modern Gonad Rock Coral Florid Red Algae
SPECIES *accretum, acropedum, affine, brassica-florida, caribaeum, ceylonense, clavacymosum, crassifructum, desikacharyi, dispalatum, erosum, ganesanii, hirtum, krusadii, mamillare, mamillosum, megalocystum, misakiense, oblimans, orotavicum, palkensei, propinquum, pudumadamense, rhizophorae, rugulosum, scabridum, setchellii, spectabile, strictum, tenuicrustaceum, tiruchendurense, trichotomum, varibaile*
VARIETIES *brassica-florida galapagense, brassica-florida ptichoides, caribaeum boreale, spectabile intermedia, spectabile nana, strictum nanum*
GENUS *Parachaetetes* Lesser Bristle Coral Florid Red Algae
SPECIES *reticulatus*
GENUS *Paracolonnella* Lesser Colon Coral Florid Red Algae
SPECIES *laohudingensis*
GENUS *Paraconophyton* Lesser Cone Coral Florid Red Algae
SPECIES *inconspicuum*
GENUS *Paulsilvella* Paul Silve Coral Florid Red Algae
SPECIES *antiqua, huveorum*
GENUS *Perispermon* Perispore Coral Florid Red Algae
SPECIES *hermaphroditum*
GENUS *Phymatolithon* Tumor Rock Coral Florid Red Algae
SPECIES *adeyi, brunneum, calcareum, investiens, laevigatum, lamii, lenormandii, masonianum, muricatum, notatum, purpureum, repandum, tenue, tenuissimum*
VARIETIES *calcareum compressum, calcareum crassa, calcareum norvegicum, calcareum palmatifidum, calcareum squarrulosum, calcareum subsimplex, calcareum subvalida, lenormandii squamulosa*
GENUS *Plectoderma* Plaited Dermal Coral Florid Red Algae
SPECIES *minus*
GENUS *Pneophyllum* Modern Coral Florid Red Algae
SPECIES *amplexifrons, confervicolum, conicum, coronatum, elegans, extensum, fragile, japonicum, keatsii, limitatum, lobescens, mauritianum, myriocarpum, submersiporum, subplanum, zonale, zostericola*
VARIETIES *zostericola tasmanicum*
GENUS *Porolithon* Pore Rock Coral Florid Red Algae
SPECIES *aequinoctiale, antillarum, castellum, colliculosum, conicum, craspedium, improcerum, oligarpum, orbiculatum, pachydermum, sonorense*
VARIETIES *pachydermum nexilis*
GENUS *Pseudogymnosolen* Nude Solitary Coral Florid Red Algae
SPECIES *mopanyuensis*
GENUS *Pseudolithophyllum* Lesser Rock Coral Florid Red Algae
SPECIES *consociatum, decipiens, discoideum, fosliei, fuegianum, irregulare, lemoinei, mannarense, mildbraedii, muricatum, neofarlowii, pambanense, panbamense, ptychoides, setchellii, tasmanicum, tuticorinense, whidbeyense*
GENUS *Rhizolamellia* Rhizid Lamella Coral Florid Red Algae
SPECIES *colli*
GENUS *Serraticardia* Serrated Heart Coral Florid Red Algae
SPECIES *macmillanii, maxima*
GENUS *Spongites* Sponge Coral Florid Red Algae
SPECIES *africanus, decipiens, discoidea, discoideus, fruticulosus, hyperellus, impar, sporolithon, sulawesiensis, tunicatus, yendoi*
VARIETIES *fruticulosus corymbiformis*
GENUS *Synarthrophyton* Straight Coral Florid Red Algae
SPECIES *chejuensis, eckloniae, magellanicum, munimentum, patena, robbenense, schielianum, schmitzii*
GENUS *Tenarea* Tenare Coral Florid Red Algae
SPECIES *adhaerens, carpophylli, erecta*
GENUS *Tielingella* Tieling Coral Florid Red Algae
SPECIES *tielingensis*
GENUS *Titanoderma* Titan Dermal Coral Florid Red Algae
SPECIES *corallinae, laminariae, mediterraneum, pustulatum, ramosissima, trochanter*
VARIETIES *pustulatum confine, pustulatum macrocarpum*
GENUS *Yamadaea* Yamada Coral Florid Red Algae
SPECIES *americana, melobesioides*

FAMILY 2 *Sporolithaceae* Spore Rock Coral Florid Red Algae
GENUS *Sporolithon* Spore Rock Coral Florid Red Algae
SPECIES *africanum, australasicum, brevium, crassiramosum, dimotum, durum, episoredion, episporum, erythraeum, glangeaudii, howei, indicum, lemoinei, liberum, molle, pacificum, ptychoides, schmidtii, sibogae, stefaninii*
VARIETIES *ptychoides mollis*
GENUS *Heydrichia* Heydrich Spore Rock Coral Florid Red Algae
SPECIES *groeneri, homalopasta, woelkerlingii*

ORDER 6 *Gelidiales* Arctic Florid Red Algae

FAMILY 1 *Gelidiaceae* Arctic Florid Red Algae
GENUS *Gelidium* Arctic Florid Red Algae
SPECIES *abbottiorum, aculeatum, affine, allanii, amamiense,* ***amansii***(Culinary Herb), *amboniense, americanum, anthonini, applanatum, arborescens, arbuscula, arenarium, asperum, australe, bipectinatum, canariensis, cantabricum, capense,* ***cartilagineum***(Agar Herb), *caulacantheum, ceramoides, chilense, claviferum, coarctatum, concinnum, corneum, coronadense, corrigerum, coulteri, crinale, deciduum, decompositum, decumbensum, delicatulum, divaricatum, elegans, elminense, fasciculatum, filicinum, flaccidum, floridanum, foliaceum, foliosum, galapagense, hancockii, heterocladum, hildenbrandtii, howei, hypnosum, hystrix, inagakii, inflexum, intertextum, isabelae, japonicum, johnstonii, kintaroi, latiusculum, lingulatum, linoides, longipes, lucasii, macnabbianum, madagascariense, maggsiae, maidenii, masudai, mcnabbianum, microdentatum, microdon, microdonticum, microphyllum, microphysa, micropterum, minusculum, multifidum, musciforme, nanum, nova-granatensis, nudifrons, omanense, pacificum, planiusculum, pluma, pristoides, proliferum, pteridifolium, pulchellum, pulchrum, pulvinatum, purpurascens, pusillum, reediae, regulare, reptans, rex, rigens, robustum, samoense, sclerophyllum, secundatum, senegalensis, serrulatum, sinicola, spathulatum, spinosum, subfastigiatum, tenue, torulosum, tsengii, usmanghanii, vagum, venetum, versicolor, vietnamense, vittatum, yamadae, zollingeri*
VARIETIES *aculeatum abnorme, cartilagineum setaceum, corneum abnorme, corneum aculeatum, corneum caespitosum, corneum claviger, corneum confertum, corneum flexuosum, corneum laciniatum, corneum lepadicola, corneum pectinatum, corneum setaceum, corneum simplex, corneum subrigidum, crinale corymbosum, crinale luxurians, crinale perpusillum, crinale platycladum, crinale polycladum, pulchellum claviger, pulchellum setaceum, pulchellum supradecompositum, pulvinatum parvissimum, pusillum cylindricum, pusillum mucronatum, pusillum pacificum, pusillum pakistancium, pusillum pulvinatum, pusillum simplex, spinosum confertum, spinosum elongatum, spinosum flexuosum, spinosum hystrix, spinosum laciniata, spinosum luxurians, vittatum laceratum*
GENUS *Acanthodiacrodium* Double Spine Arctic Florid Red Algae
SPECIES *aclinum, arburense, hexagonum, mamillatum*
GENUS *Acanthopeltis* Spiny Pelt Arctic Florid Red Algae
SPECIES *hirsuta, japonica, longiramulosa*
GENUS *Beckerella* Becker Arctic Florid Red Algae
SPECIES *biserrata, helenae, hildebrandtii, mediterranea, pectinata, pinnatifida, pterocladioides, rumpii, scalaramosa*
GENUS *Capreolia* Caper Arctic Florid Red Algae
SPECIES *implexa*
GENUS *Chaetangium* Bristle Angled Arctic Florid Red Algae
SPECIES *chilense, dichotomum, ornatum*
GENUS *Porphyroglossum* Porphyry Tongue Arctic Florid Red Algae
SPECIES *zollingeri*
GENUS *Pterocladia* Wing Branched Arctic Florid Red Algae
SPECIES *heteroplatos, lucida, mcnabbiana, media, musciformis,* ***pinnata***(Agar Herb), *rectangularis, tripolitana*
GENUS *Pterocladiella* Lesser Wing Branched Arctic Florid Red Algae
SPECIES *bartlettii, bulbosa, caerulescens, caespitosa, caloglossoides, capillacea, melanoidea, minima, nana, tenuis*
VARIETIES *bartlettii musciformis, capillacea densa, melanoidea filamentosa, melanoidea gracilis*
GENUS *Ptilophora* Ptilon Arctic Florid Red Algae
SPECIES *diversifolia, irregularis, prolifera, rhodoptera, spissa, subcostata*
GENUS *Yatabella* Yataba Arctic Florid Red Algae
SPECIES *hirsuta*

FAMILY 2 *Gelidiellaceae* Lesser Arctic Florid Red Algae
GENUS *Gelidiella* Lesser Arctic Florid Red Algae
SPECIES *acerosa, antipae, calcicola, diuens, feldmannii, hancockii, indica, ligulata, lubrica, machrisiana, mexicana, myrioclada, nigrescens, pannosa, ramellosa, rigidiuscula, rigiduscula, sanctarum, setacea, taylori, tinerfensis, trinitatensis, womersleyana*
VARIETIES *acerosa radicans*
GENUS *Acrocarpus* Tall Carpel Lesser Arctic Florid Red Algae
SPECIES *spathulatus*

ORDER 7 *Gigartinales* Gigartin Florid Red Algae

FAMILY 01 *Gigartinaceae* Gigartin Florid Red Algae
GENUS *Gigartina* Gigartin Florid Red Algae
SPECIES *alveata, ancistroclada, angulata, brachiata, bracteata, clavifera, compressa, crassicaulis, decipiens, densa, dilitata, disticha, divaricata, fissa, imperialis, insignis, isiforme, jardinii, kroneana, laingii, macrocarpa, minima, muelleriana, multidichotoma, ornithorhynchos, pachymenioides, paxillata, pinnata, pistillata, polycarpa, pumila, recurva, runcinata, simplex, skottsbergii, sonderi, spinosa, tysonii, wehliae*
VARIETIES *clavifera pseudopistillata, pistillata pectinata*
GENUS *Blinksia* Blinks Gigartin Florid Red Algae
SPECIES *californica*
GENUS *Catenellopsis* Chain Gigartin Florid Red Algae
SPECIES oligarthra
GENUS *Chondracanthus* Cartilage Spine Gigartin Florid Red Algae
SPECIES *acicularis, armatus, canaliculatus, chamissoi, chapmanii, chauvinii, corymbiferus, elegans, exasperatus, glomeratus, harveyanus, intermedius, johnstonii, macdougalii, spinosus, squarrulosus, teedei, tenellus, tepidus*
VARIETIES *acicularis compressa, chauvinii javanica, teedei lusitanicus*
GENUS *Chondrus* Cartilage Gigartin Florid Red Algae
SPECIES *albemarlensis, canaliculatus,* ***crispus***(Medicinal Herb), *elatus, elongatus, giganteus, griffithsiae, hancockii, nipponicus, ocellatus, pinnulatus, uncinalis, verrucosus, yamadae, yendoi*
VARIETIES *crispus aegagropilus, crispus densus, crispus lutescens, crispus membranaceus, crispus uncinatus, crispus virens, giganteus flabellatus, ocellatus parvus*
GENUS *Corynocystis* Club Cyst Gigartin Florid Red Algae
SPECIES *prostrata*
GENUS *Corynomorpha* Club Gigartin Florid Red Algae
SPECIES *clavata, prismatica*
GENUS *Cruoria* Cruory Gigartin Florid Red Algae
SPECIES *areschougii, cruoriaeformis, indica, middendorfii, pacifica, pellita, sachalinensis*
GENUS *Cubiculosporum* Cubic Spore Gigartin Florid Red Algae
SPECIES *koronicarpus*
GENUS *Gainia* Gain Gigartin Florid Red Algae
SPECIES *mollis*
GENUS *Haemescharia* Blood Schar Gigartin Florid Red Algae
SPECIES *hennedyi, polygyna*
GENUS *Hummbrella* Hummbrel Gigartin Florid Red Algae
SPECIES *hydra*
VARIETIES *hydra tentaculata*
GENUS *Iridaea* Iridescent Gigartin Florid Red Algae
SPECIES *angustinae, ciliata, convoluta, cordata, elongata, explanata, gigantea, hetercocca, lanceolata, latissima, mawsonii, oblongifructa, parvula, pinnata, platyna, remuliformis, ruprechtiana, tuberculosa, undulosa*
VARIETIES *cordata ligulata*
GENUS *Iridophycus* Lesser Iridescent Gigartin Florid Red Algae
SPECIES *agardhianum, ciliatum, undulosum*
GENUS *Mazzaella* Mazza Gigartin Florid Red Algae
SPECIES *affinis, californica, capensis, cobinae, convoluta, coriacea, denticulata, flaccida, hancockii, hemisphaerica, japonica, laminarioides, leptorhynchos, linearis, membranacea, minima, oregona, parksii, parva, phyllocarpa, rosea, sanguinea, splendens, volans*
VARIETIES *laminarioides papillata*
GENUS *Nizymenia* Nizymen Gigartin Florid Red Algae
SPECIES *australis, conferta, furcata*
GENUS *Ostiophyllum* Calcic Gigartin Florid Red Algae
SPECIES *sonderopeltae*
GENUS *Petrocelis* Rock Gigartin Florid Red Algae
SPECIES *anastomosans, ascendens, middendorfii*
GENUS *Polyides* Poly Gigartin Florid Red Algae
SPECIES *rotundus*
GENUS *Rhodoglossum* Tongue Gigartin Florid Red Algae
SPECIES *alcicorne, diffusum, digitatum, gigartinoides, latissimum, minimum, oweniae, pulchra, revolutum*
GENUS *Rissoella* Risso Gigartin Florid Red Algae
SPECIES *verruculosa*
GENUS *Sarcothalia* Flesh Thal Gigartin Florid Red Algae
SPECIES *atropurpurea, circumcincta, crassifolia, crispata, decipiens, dichotoma, insidiosa, lanceata, lapathifolia, livida, marginifera, papillosa, radula, scutellata, stiriata*
GENUS *Tichocarpus* Hairy Carpel Gigartin Florid Red Algae
SPECIES *crinitus*
GENUS *Wurdemannia* Wurdemann Gigartin Florid Red Algae

SPECIES *miniata*

FAMILY 02 *Acrosymphytaceae* Tall Symphyte Gigartin Florid Red Algae
GENUS *Acrosymphyton* Tall Symphyte Gigartin Florid Red Algae
SPECIES *miniata, brainardii, caribaeum, firmum, purpuriferum, taylorii, tenax*
GENUS *Acrosymphytonema* Lesser Tall Symphyte Gigartin Florid Red Algae
SPECIES *breemaniae*

FAMILY 03 *Acrotylaceae* Tall Tyle Gigartin Florid Red Algae
GENUS *Acrotylus* Tall Tyle Gigartin Florid Red Algae
SPECIES *australis*
GENUS *Amphiplexia* Amphiplex Tall Tyle Gigartin Florid Red Algae
SPECIES *hymenocladioides, racemosa*
GENUS *Antrocentrum* Hollow Center Tall Tyle Gigartin Florid Red Algae
SPECIES *nigrescens*
GENUS *Claviclonium* Club Clone Tall Tyle Gigartin Florid Red Algae
SPECIES *ovatum*
GENUS *Hennedya* Hennedy Tall Tyle Gigartin Florid Red Algae
SPECIES *crispa*
GENUS *Ranavalona* Ranavalon Tall Tyle Gigartin Florid Red Algae
SPECIES *duckerae*
GENUS *Reinboldia* Reinbold Tall Tyle Gigartin Florid Red Algae
SPECIES *polycarpa*

FAMILY 04 *Ahnfeltiaceae* Areschoug Gigartin Florid Red Algae
GENUS *Ahnfeltia* Ahnfelt Gigartin Florid Red Algae
SPECIES *elongata, fastigiata, gigartinoides, plicata, setacea, svensonii, tobuchiensis, yinggehaiensis*
GENUS *Ahnfeltiopsis* Lesser Ahnfelt Gigartin Florid Red Algae
SPECIES *catenata, chnoosporoides, complicata, concinna, densa, devoniensis, divaricatus, durvillei, fastigiata, flabelliformis, furcellata, gigartinoides, glomerata, gracilis, guangdongensis, hainanensis, humilis, intermedius, leptophylla, linearis, masudai, okamurae, pacifica, paradoxa, polyclada, pusilla, pygmaea, quinhonensis, serenei, smithii, triquetrifolia, vermicularis, yamadae*
VARIETIES *durvillei implicata, furcellata patens*

FAMILY 05 *Areschougiaceae* Areschoug Gigartin Florid Red Algae
GENUS *Areschougia* Areschoug Gigartin Florid Red Algae
SPECIES *congesta, ligulata, stuartii*
GENUS *Agardhiella* Agardh Areschoug Gigartin Florid Red Algae
SPECIES *coulteri, floridana, mexicana, ramossissima, subulata*
GENUS *Anatheca* Without Theca Areschoug Gigartin Florid Red Algae
SPECIES *montagnei*
GENUS *Betaphycus* Beta Areschoug Gigartin Florid Red Algae
SPECIES *gelatinum, philippinensis, speciosum*
GENUS *Callophycus* Callisto Areschoug Gigartin Florid Red Algae
SPECIES *condominius, costatus, densus, dorsifer, harveyanus, laxus, oppositifolius, serratus, tridentifer*
GENUS *Erythroclonium* Clone Areschoug Gigartin Florid Red Algae
SPECIES *angustatum, corallinum, muelleri, sedoides, sonderi*
GENUS *Eucheuma* Cheum Areschoug Gigartin Florid Red Algae
SPECIES *amakusaensis, arnoldii, cartilagineum, chondriforme, condriforme, crassum, crustiforme, deformans, denticulatum, edule, horridum, isiforme, johnstonii, jugatum, kraftianum, leeuwenii, nodulosum, odontophorum, perplexum, platycladum, serra, uncinatum, xishaensis*
VARIETIES *arnoldii alcyonida, horridum radicans isiforme denudatum, odontophorum mauritianum*
GENUS *Euryomma* Broad Areschoug Gigartin Florid Red Algae
SPECIES *platycarpa*
GENUS *Flahaultia* Flahault Areschoug Gigartin Florid Red Algae
SPECIES *appendiculata, palmata, tegetiformans*
GENUS *Gardneriella* Gardner Areschoug Gigartin Florid Red Algae
SPECIES *tubifera*
GENUS *Kappaphycus* Kappa Areschoug Gigartin Florid Red Algae
SPECIES *alvarezii, cottonii, inermis, interme, procrusteanum, striatum*
VARIETIES *alvarezii ajakii-assii, alvarezii alvarezii, alvarezii tambalangii, cottonii erecta*
GENUS *Lukinia* Lukin Areschoug Gigartin Florid Red Algae
SPECIES *dissecta*
GENUS *Melanema* Dark Thread Areschoug Gigartin Florid Red Algae
SPECIES *dumosum*
GENUS *Meristiella* Merist Areschoug Gigartin Florid Red Algae
SPECIES *echinocarpum, florigera, gelidium, schramii*
GENUS *Meristotheca* Merist Theca Areschoug Gigartin Florid Red Algae
SPECIES *coacta, decumbens, fergusonii, floridana, papulosa, procumbens, senegalense, tobagensis*
GENUS *Neoagardhiella* Modern Agardh Areschoug Gigartin Florid Red Algae

SPECIES *baileyi*
GENUS *Notophycus* — Notable Areschoug Gigartin Florid Red Algae
SPECIES *fimbriatus*
GENUS *Opuntiella* — Prickly Pear Areschoug Gigartin Florid Red Algae
SPECIES *californica, ornata, parva*
GENUS *Placentophora* — Placenta Areschoug Gigartin Florid Red Algae
SPECIES *colensoi*
GENUS *Reticulobotrys* — Netted Berry Areschoug Gigartin Florid Red Algae
SPECIES *catalinae*
GENUS *Rhabdonia* — Rod Areschoug Gigartin Florid Red Algae
SPECIES *clavigera, coccinea, decumbens, natalensis, verticillata*
VARIETIES *verticillata linearis*
GENUS *Sarcodiotheca* — Flesh Dual Theca Areschoug Gigartin Florid Red Algae
SPECIES *caribaea, dichotoma, divaricata, ecuadoreana, elongata, furcata, gaudichaudii, linearis, taylorii, tenuis*
GENUS *Sarconema* — Flesh Thread Areschoug Gigartin Florid Red Algae
SPECIES *filiforme, gracilarioides, scinaioides*
GENUS *Solieria* — Solier Areschoug Gigartin Florid Red Algae
SPECIES *anastomosa, chordalis, dichotoma, dura, filiformis, jaasundii, pacifica, robusta, tenuis*
GENUS *Tikvahiella* — Tikvah Areschoug Gigartin Florid Red Algae
SPECIES *candida*
GENUS *Turnerella* — Turner Areschoug Gigartin Florid Red Algae
SPECIES *mertensiana, pennyi*

FAMILY 06 *Calosiphonaceae* — Callisto Siphon Gigartin Florid Red Algae
GENUS *Calosiphonia* — Callisto Siphon Gigartin Florid Red Algae
SPECIES *dalmatica, vermicularis*
GENUS *Schmitzia* — Schmitz Callisto Siphon Gigartin Florid Red Algae
SPECIES *evanescens, hiscockiana, japonica, neapolitana, sanctae-crucis*

FAMILY 07 *Caulacanthaceae* — Tail Spine Gigartin Florid Red Algae
GENUS *Caulacanthus* — Tail Spine Gigartin Florid Red Algae
SPECIES *rigidus, salifugus, ustulatus*
VARIETIES *ustulatus fastigiatus*
GENUS *Catenella* — Chain Tail Spine Gigartin Florid Red Algae
SPECIES *caespitosa, fusiformis, impudica, nipae, subumbellata*
GENUS *Catenellocolax* — Chain Colac Tail Spine Gigartin Florid Red Algae
SPECIES leeuwenii
GENUS *Feldmannophycus* — Feldmann Tail Spine Gigartin Florid Red Algae
SPECIES *rayssiae*
GENUS *Heringia* — Hering Tail Spine Gigartin Florid Red Algae
SPECIES *mirabilis*
GENUS *Montemaria* — Montemar Tail Spine Gigartin Florid Red Algae
SPECIES *horridula*
GENUS *Taylorophycus* — Taylor Tail Spine Gigartin Florid Red Algae
SPECIES *filiformis, laxa*

FAMILY 08 *Choreocolacaceae* — Chore Colac Gigartin Florid Red Algae
GENUS *Choreocolax* — Chore Colac Gigartin Florid Red Algae
SPECIES *macronema, polysiphoniae, rabenhorstii, rhodymeniae, tumidus*
GENUS *Bostrychiocolax* — Bostrych Colac Chore Colac Gigartin Florid Red Algae
SPECIES *australis*
GENUS *Dawsoniocolax* — Dawson Colac Chore Colac Gigartin Florid Red Algae
SPECIES *bostrychiae*
GENUS *Harveyella* — Harvey Chore Colac Gigartin Florid Red Algae
SPECIES *mirabilis*
GENUS *Leachiella* — Leach Chore Colac Gigartin Florid Red Algae
SPECIES *pacifica*

FAMILY 09 *Crossocarpaceae* — Cross Carpel Gigartin Florid Red Algae
GENUS *Crossocarpus* — Cross Carpel Gigartin Florid Red Algae
SPECIES *lamuticus*
GENUS *Ionia* — Ion Cross Carpel Gigartin Florid Red Algae
SPECIES *cornu-cervi*
GENUS *Velatocarpus* — Veil Cross Carpel Gigartin Florid Red Algae
SPECIES *kurilensis, ochotensis, pustulosus*

FAMILY 10 *Cystocloniaceae* — Cyst Clone Gigartin Florid Red Algae
GENUS *Cystoclonium* — Cyst Clone Gigartin Florid Red Algae
SPECIES *difficile, purpureum*
VARIETIES *purpureum dendroideum*

GENUS *Acanthococcus* Spine Round Cyst Clone Gigartin Florid Red Algae
SPECIES *antarcticus*
GENUS *Austroclonium* Southern Cyst Clone Gigartin Florid Red Algae
SPECIES *charoides*
GENUS *Calliblepharis* Callisto Fringed Cyst Clone Gigartin Florid Red Algae
SPECIES *celatospora, ciliata,fimbriata, jolyi, jubata, planicaulis*
VARIETIES *ciliata ligulata, fimbriata filiformis*
GENUS *Craspedocarpus* Fringed Carpel Cyst Clone Gigartin Florid Red Algae
SPECIES *blepharicarpus, erosus, humilis, jolyi, ramentaceus, tenuifolius, venosus*
GENUS *Erythronaema* Thread Cyst Clone Gigartin Florid Red Algae
SPECIES *ceramioides*
GENUS *Fimbrifolium* Fringed Frond Cyst Clone Gigartin Florid Red Algae
SPECIES *dichotomum, spinulosum*
GENUS *Gloiophyllis* Glow Cyst Clone Gigartin Florid Red Algae
SPECIES *barkeriae*
GENUS *Peltasta* Pelt Cyst Clone Gigartin Florid Red Algae
SPECIES *australis*
GENUS *Reptataxis* Reptile Cyst Clone Gigartin Florid Red Algae
SPECIES *rhizophora*
GENUS *Rhodophyllis* Lesser Cyst Clone Gigartin Florid Red Algae
SPECIES *centrocarpa, divaricata, gracilarioides, madeirensis, membranacea, multipartita, peltata, reptans, strafforelloi, volans*
VARIETIES *divaricata werneri, peltata lacunosa*
GENUS *Stictosporum* Stick Spore Cyst Clone Gigartin Florid Red Algae
SPECIES *nitophylloides*

FAMILY 11 *Dicranemataceae* Dual Thread Gigartin Florid Red Algae
GENUS *Dicranema* Dual Thread Gigartin Florid Red Algae
SPECIES *cincinnalis, revolutum, rosaliae*
GENUS *Tenaciphyllum* Tenacious Dual Thread Gigartin Florid Red Algae
SPECIES *lobatum, rotundilobum, xishaensis*
GENUS *Tylotus* Tyle Ear Dual Thread Gigartin Florid Red Algae
SPECIES *indicus, lichenoides, obtusatus*

FAMILY 12 *Dumontiaceae* Dumont Gigartin Florid Red Algae
GENUS *Dumontia* Dumont Gigartin Florid Red Algae
SPECIES *alaskana, canariensis, clava, contorta, cornuta, simplex, triquetra*
GENUS *Constantinea* Constantine Dumont Gigartin Florid Red Algae
SPECIES *rosa-marina, simplex, sitchensis, subulifera*
GENUS *Cryptosiphonia* Cryptic Siphon Dumont Gigartin Florid Red Algae
SPECIES *woodii*
GENUS *Dasyphloea* Daisy Flow Dumont Gigartin Florid Red Algae
SPECIES *insignis*
GENUS *Dilsea* Dilse Dumont Gigartin Florid Red Algae
SPECIES *carnosa, integra, socialis*
GENUS *Dudresnaya* Dudresnay Dumont Gigartin Florid Red Algae
SPECIES *australis, babbittiana, bermudensis, canariensis, canescens, capricornica, colombiana, crassa, cupressina, georgiana, hawaiiensis, hudsonii, japonica, kuroshioensis, littleri, minima, okiensis, patula, puertoricensis, rubens, verticillata*
GENUS *Farlowia* Farlow Dumont Gigartin Florid Red Algae
SPECIES *compressa, conferta, mollis*
GENUS *Gibsmithia* Gibsmith Dumont Gigartin Florid Red Algae
SPECIES *dotyi, hawaiiensis, larkumii, womersleyi*
GENUS *Hyalosiphonia* Hyaline Siphon Dumont Gigartin Florid Red Algae
SPECIES *caespitosa*
GENUS *Kraftia* Kraft Dumont Gigartin Florid Red Algae
SPECIES *dichotoma*
GENUS *Leptocladia* Thin Branched Dumont Gigartin Florid Red Algae
SPECIES *binghamiae, laxa, peruviana*
GENUS *Masudaphycus* Masuda Dumont Gigartin Florid Red Algae
SPECIES *aenigmatosus, irregulare*
GENUS *Neodilsea* Modern Dilse Dumont Gigartin Florid Red Algae
SPECIES *araneosa, borealis, crispata, longissima, natashae, tenuipes, yendoana*
GENUS *Orculifilum* Orcule Dumont Gigartin Florid Red Algae
SPECIES *denticulatum*
GENUS *Pikea* Pike Dumont Gigartin Florid Red Algae
SPECIES *californica, robusta, yoshizakii*
GENUS *Rhodopeltis* Pelt Dumont Gigartin Florid Red Algae
SPECIES *australis, borealis*

GENUS *Sarcophyllis* — Flesh Dumont Gigartin Florid Red Algae
SPECIES *lobata*
GENUS *Thuretellopsis* — Thuret Dumont Gigartin Florid Red Algae
SPECIES *peggiana*
GENUS *Weeksia* — Weeks Dumont Gigartin Florid Red Algae
SPECIES *coccinea, digitata, fryeana, howellii, reticulata*

FAMILY 13 *Endocladiaceae* — Endic Branched Gigartin Florid Red Algae
GENUS *Endocladia* — Endic Branched Gigartin Florid Red Algae
SPECIES *muricata, vernicata, yasudae*
GENUS *Gloiopeltis* — Glow Pelt Endic Branched Gigartin Florid Red Algae
SPECIES *complanata, furcata, minuta, tenax*
VARIETIES *furcata ventricosa*

FAMILY 14 *Furcellariaceae* — Furco Gigartin Florid Red Algae
GENUS *Furcellaria* — Furco Gigartin Florid Red Algae
SPECIES *lumbricalis*
VARIETIES *lumbricalis tenuior*
GENUS *Halarachnion* — Haline Spider Furco Gigartin Florid Red Algae
SPECIES *ligulatum, parvum, pusillum*
VARIETIES *ligulatum aciculare, ligulatum dichotomum, ligulatum latifolium, ligulatum ramentaceum*
GENUS *Neurocaulon* — Nerve Stalk Furco Gigartin Florid Red Algae
SPECIES *foliosum, japonicum, thiebautii*

FAMILY 15 *Gloiosiphoniaceae* — Glow Siphon Gigartin Florid Red Algae
GENUS *Gloiosiphonia* — Glow Siphon Gigartin Florid Red Algae
SPECIES *capillaris, verticillaris*
GENUS *Cruoriopsis* — Cruory Glow Siphon Gigartin Florid Red Algae
SPECIES *danica, ensisae, hauckii, maris-rubri, reinboldii*
GENUS *Gloeophycus* — Lesser Glow Siphon Gigartin Florid Red Algae
SPECIES *koreanum*
GENUS *Peleophycus* — Pellet Glow Siphon Gigartin Florid Red Algae
SPECIES *multiprocarpium*
GENUS *Plagiospora* — Pelagic Spore Glow Siphon Gigartin Florid Red Algae
SPECIES *gracilis*
GENUS *Schimmelmannia* — Schimmelmann Glow Siphon Gigartin Florid Red Algae
SPECIES *bollei, dawsonii, elegans, frauenfeldii, plumosa, schousboei, venezuelensis*
GENUS *Thuretella* — Thuret Glow Siphon Gigartin Florid Red Algae
SPECIES *schousboei*

FAMILY 16 *Hypneaceae* — Mossy Gigartin Florid Red Algae
GENUS *Hypnea* — Mossy Gigartin Florid Red Algae
SPECIES *anastomosans, arbuscula, arenaria, boergesenii, bryoides, cenomyce, ceramioides, charoides, chordacea, coccinea, compressa, confervoides, congesta, cornuta, divaricata, divergens, ecklonii, esperi, filiformis, flagelliformis, flexicaulis, flexuosa, fruticulosa, furnariana, hamulosa, intricata, japonica, johnstonii, krugiana, marchantae, multicornis, musciformis, nidifica, nidulans, nigrescens, pannosa, pectinella, purpurascens, ramentacea, reptans, rosea, rugulosa, saidana, spicifera, spinella, stellulifera, tenuis, unilateralis, ustulata, valentiae, variabilis, viridis, volubilis, wurdemannii, yamadae*
VARIETIES *cenomyce tenuis, charoides delicatula, musciformis cervicornis, musciformis esperi, musciformis fasciculata, musciformis foliosissima, musciformis hippurioides, musciformis horrida, musciformis intricata, musciformis nuda, musciformis pygmaea, musciformis spinulosa, nigrescens tenuior, valentiae gardneri*
GENUS *Hypneocolax* — Colac Mossy Gigartin Florid Red Algae
SPECIES *stellaris*
VARIETIES *stellaris orientalis*

FAMILY 17 *Kallymeniaceae* — Kallymen Gigartin Florid Red Algae
GENUS *Kallymenia* — Kallymen Gigartin Florid Red Algae
SPECIES *agardhii, antarctica, berggrenii, callophylloides, crassiuscula, cribrogloea, cribrosa, feldmannii, guaymasensis, lacerata, lacinifolia, latiloba, limminghii, morelii, multiloba, norrisii, oligonema, pacifica, patens, perforata, perpusillum, pertusa, reniformis, requienii, rosea, rubra, sagamiana, schizophylla, schmitzi, sessilis, setchellii, spathulata, spinosa, stipitata, tasmanica, tenuifolia, thompsonii, veleroae, westii*
VARIETIES *reniformis cuneata*
GENUS *Austrophyllis* — Southern Kallymen Gigartin Florid Red Algae
SPECIES *alcicornis, harveyana*
GENUS *Beringia* — Bering Kallymen Gigartin Florid Red Algae
SPECIES *castanea*
GENUS *Callocolax* — Callisto Colac Kallymen Gigartin Florid Red Algae
SPECIES *acicularis, fungiformis, neglectus*
GENUS *Callophyllis* — Callisto Kallymen Gigartin Florid Red Algae

SPECIES *acrocarpa, adhaerens, adnata, angustifrons, atrosanguinea, beringensis, calliblepharoides, centrifuga, cervicornis, crassifolia, crenulata, crispata, cristata, decumbens, depressa, dissecta, divaricata, edentata, elongata, fastigiata, filicina, flabellulata, gardneri, haenophylla, hayamensis, heanophylla, japonica, johnstonii, laciniata, laingiana, lambertii, lecomtei, ligulata, linearis, linguata, mageshimensis, marginifructa, megalocarpa, microdonta, nana, obtusifolia, odonthalioides, okamurae, ornata, palmata, papulosa, phyllohaptera, pinnata, platyna, plumosa, radula, rangiferina, rhynchocarpa, stenophylla, tenera, thompsonii, variegata, violacea*

VARIETIES *rangiferina constrictus, rhynchocarpa incisa, violacea epiphytica*

GENUS *Cirrulicarpus* — Cirrule Carpel Kallymen Gigartin Florid Red Algae

SPECIES *carolinensis, gmelinii, nanus, polycoelioides, ruprechtianum*

GENUS *Ectophora* — Ectophore Kallymen Gigartin Florid Red Algae

SPECIES *depressa*

GENUS *Erythrophyllum* — Least Kallymen Gigartin Florid Red Algae

SPECIES *delesserioides, splendens*

GENUS *Euhymenia* — Hymen Kallymen Gigartin Florid Red Algae

SPECIES *schizophylla*

GENUS *Euthora* — Thor Kallymen Gigartin Florid Red Algae

SPECIES *tristanensis*

GENUS *Glaphyrymenia* — Glaphyry Kallymen Gigartin Florid Red Algae

SPECIES *pustulosa*

GENUS *Hommersandia* — Hommersand Kallymen Gigartin Florid Red Algae

SPECIES *palmatifolia*

GENUS *Hormophora* — Hormophore Kallymen Gigartin Florid Red Algae

SPECIES *australasica*

GENUS *Kallymeniopsis* — Lesser Kallymen Gigartin Florid Red Algae

SPECIES *lacera, oblongifructa, verrucosa*

GENUS *Meredithia* — Meredith Kallymen Gigartin Florid Red Algae

SPECIES *californica, microphylla*

GENUS *Nereoginkgo* — Nere Ginkgo Kallymen Gigartin Florid Red Algae

SPECIES *adiantifolia*

GENUS *Polycoelia* — Polysky Kallymen Gigartin Florid Red Algae

SPECIES *laciniata, van-hoevellii*

GENUS *Pugetia* — Puget Kallymen Gigartin Florid Red Algae

SPECIES *delicatissima, fragilissima, harveyana, kylinii, porphyroidea*

GENUS *Rhizopogonia* — Rhizid Beard Kallymen Gigartin Florid Red Algae

SPECIES *asperata*

GENUS *Thamnophyllis* — Thamnion Kallymen Gigartin Florid Red Algae

SPECIES *discigera, lacerata, laingii, pocockiae*

FAMILY 18 *Mychodeaceae* — Mych Gigartin Florid Red Algae

GENUS *Mychodea* — Mych Gigartin Florid Red Algae

SPECIES *acanthymenia, aciculare, australis, carnosa, chamaedoris, compressa, disticha, episcopalis, gracilaria, hamata, marginifera, pusilla, ramulosa, schrammii, spinulifera*

GENUS *Mychodeophyllum* — Lesser Mych Gigartin Florid Red Algae

SPECIES *papillitectum*

FAMILY 19 *Nemastomataceae* — Thread Stoma Gigartin Florid Red Algae

GENUS *Nemastoma* — Thread Stoma Gigartin Florid Red Algae

SPECIES *canariense, coliforme, comosum, confusum, constrictum, densum, dichotomum, feredayae, foliaceum, gelatinosum, laciniatum, levringii, lividum, marginatum, minor*

VARIETIES *constrictum longitrichogynum*

GENUS *Adelophycus* — Adel Thread Stoma Gigartin Florid Red Algae

SPECIES *corneus*

GENUS *Itonoa* — Itono Thread Stoma Gigartin Florid Red Algae

SPECIES *marginifera*

GENUS *Predaea* — Preda Thread Stoma Gigartin Florid Red Algae

SPECIES *aurora, bisporifera, feldmannii, goffiana, huismanii, incraspeda, japonica, kraftiana, kuroshiensis, kuroshioensis, laciniosa, masonii, ollivieri, pusilla, subpeltata, tenuis, tokidae, tumescens, weldii*

VARIETIES *feldmannii indica, pusilla alboranensis*

GENUS *Predaeopsis* — Lesser Preda Thread Stoma Gigartin Florid Red Algae

SPECIES *japonica*

GENUS *Tsengia* — Tseng Thread Stoma Gigartin Florid Red Algae

SPECIES *bairdii, comosa, feredayae, laingii, lanceolata, lancifolia, nakamurae, pulchra*

VARIETIES *nakamurae membranaceum*

FAMILY 20 *Peyssonneliaceae* — Peysson Gigartin Florid Red Algae

GENUS *Peyssonnelia* — Peysson Gigartin Florid Red Algae

SPECIES *adriatica, armorica, asiatica, atropurpurea, balanicola, bicolor, boergesenii, bornetii, boudouresquei, calcea, capensis, caulifera, clarionensis, conchicola, coriacea, coxalocalcificata,, crispata,*

dawsonii, de-zwaanii, delicata, distenda, dubyi, evae, foliosa, foveolata, guadalupensis, hairii, hancockii, hariotii, harveyana, hongii, immersa, inamoena, indica, intermedia, involvens, japonica, johansenii, lemoinei, luciparensis, luzonensis, magdalenae, magna, mariodinalis, maris-rubri, mariti, meridionalis, mexicana, nitida, nordstedtii, novae-hollandiae, obbesii, obscura, orientalis, pacifica, polymorpha, profunda, rara-avis, replicata, rosa-marina, rosenvingei, rubra, rugosa, rumoiana, simulans, splendens, squamaria, stoechas, thomassinii, valentinii
VARIETIES *calcea frutescens, obscura bombayensis, polymorpha gardineri, rosa-marina saxicola*
GENUS *Chevaliericrusta* Chevalier Crust Peysson Gigartin Florid Red Algae
SPECIES *polynesiae*
GENUS *Cruoriella* Cruory Peysson Gigartin Florid Red Algae
SPECIES *de-zwaanii, fissurata, mexicana*
GENUS *Ethelia* Ethel Peysson Gigartin Florid Red Algae
SPECIES *biradiata, mexicana, vanbosseae*
GENUS *Metapeyssonnelia* Lesser Peysson Gigartin Florid Red Algae
SPECIES *corallepida, feldmannii*
GENUS *Polystrata* Many Layer Peysson Gigartin Florid Red Algae
SPECIES *compacta, dura, fosliei*
GENUS *Pulvinia* Powdered Peysson Gigartin Florid Red Algae
SPECIES *epiphytica*
GENUS *Ramicrusta* Branch Crust Peysson Gigartin Florid Red Algae
SPECIES *nanhaiensis*
GENUS *Riquetophycus* Riquet Peysson Gigartin Florid Red Algae
SPECIES *polypus*
GENUS *Sonderopelta* Sonder Pelt Peysson Gigartin Florid Red Algae
SPECIES *coriacea*
GENUS *Squamaria* Scaly Peysson Gigartin Florid Red Algae
SPECIES *vulgaris*

FAMILY 21 *Phacelocarpaceae* Bundle Carpel Gigartin Florid Red Algae
GENUS *Phacelocarpus* Bundle Carpel Gigartin Florid Red Algae
SPECIES *alatus, apodus, complanatus, japonicus, neurymenioides, oligacanthus, peperocarpos, sessilis, tortuosus, tristichus*
GENUS *Phacelofimbria* Fringed Bundle Carpel Gigartin Florid Red Algae
SPECIES *minor*

FAMILY 22 *Phyllophoraceae* Frond Gigartin Florid Red Algae
GENUS *Phyllophora* Frond Gigartin Florid Red Algae
SPECIES *abyssalis, ahnfeltioides, antarctica, crispa, fimbriata, gelidioides, herediae, membranifolia, pseudoceranoides, sicula, submaritima*
GENUS *Actinococcus* Radial Round Frond Gigartin Florid Red Algae
SPECIES *chiton*
GENUS *Besa* Bes Frond Gigartin Florid Red Algae
SPECIES *papillaeformis, stipitata*
GENUS *Carpococcus* Carpel Round Frond Gigartin Florid Red Algae
SPECIES *linearis, perforatus*
GENUS *Ceratocolax* Horn Colac Frond Gigartin Florid Red Algae
SPECIES *hartzii*
GENUS *Coccotylus* Round Tyle Frond Gigartin Florid Red Algae
SPECIES *orientalis, truncatus*
GENUS *Erythrodermis* Dermal Frond Gigartin Florid Red Algae
SPECIES *haematis, pacifica, traillii*
GENUS *Gymnogongrus* Nude Gonger Frond Gigartin Florid Red Algae
SPECIES *antarcticus, carnosus, chiton, crenulatus, crustiforme, dilatatus, disciplinalis, filiformis, furcatus, gregarius, griffithsiae, guadalupensis, hancockiii, johnstonii, martinensis, melanothrix, minutus, nigricans, palmettoides, patens, platyphyllus, sinicola, smithii, tenuis, tetrasporifer, torulosus, turquetii*
VARIETIES *griffithsiae galapagensis*
GENUS *Mastocarpus* Mastigote Carpel Frond Gigartin Florid Red Algae
SPECIES *jardinii, pacificus, papillatus, stellatus, yendoi*
VARIETIES *stellatus acutus*
GENUS *Oncotylus* Onc Tyle Frond Gigartin Florid Red Algae
SPECIES *furcellatus*
VARIETIES *furcellatus minor*
GENUS *Ozophora* Ozo Frond Gigartin Florid Red Algae
SPECIES *clevelandii, latifolia, norrisii*
GENUS *Petroglossum* Rock Tongue Frond Gigartin Florid Red Algae
SPECIES *pacificum, parvum, undulatum*
GENUS *Schottera* Schotter Frond Gigartin Florid Red Algae
SPECIES *nicaensis*
GENUS *Stenogramme* Thin Frond Gigartin Florid Red Algae

SPECIES *interrupta, leptophylla, phyllophoroides*

FAMILY 23 *Rhizophyllidaceae* Rhizid Gigartin Florid Red Algae
GENUS *Rhizophyllis* Rhizid Gigartin Florid Red Algae
SPECIES *dentata*
GENUS *Contarinia* Contarine Rhizid Gigartin Florid Red Algae
SPECIES *magdae, okamurae, pacifica, peyssonneliaeformis, squamariae*
GENUS *Nesophila* Modern Rhizid Gigartin Florid Red Algae
SPECIES *hoggardii*
GENUS *Ochtodes* Ochto Rhizid Gigartin Florid Red Algae
SPECIES *crockeri, secundiramea*
GENUS *Portieria* Portier Rhizid Gigartin Florid Red Algae
SPECIES *dichotoma, harveyi, hornemannii, hornemannioides, japonica, spinulosa*
GENUS *Rhizophyllidium* Lesser Rhizid Gigartin Florid Red Algae
SPECIES *oedogonii*

FAMILY 24 *Sarcodiaceae* Flesh Gigartin Florid Red Algae
GENUS *Sarcodia* Flesh Gigartin Florid Red Algae
SPECIES *cuneifolia, dentata, dichotoma, flabellata, gattyae, grandifolia, marginata, montagneana, multifida, palmata*
VARIETIES *montagneana mauritiana*
GENUS *Chondrymenia* Cartilage Flesh Gigartin Florid Red Algae
SPECIES *lobata*
GENUS *Trematocarpus* Tremor Carpel Flesh Gigartin Florid Red Algae
SPECIES *acicularis, affinis, concinnus, dichotomus, flabellatus, fragilis, fragilis, papenfussii, pygmaeus*
GENUS *Trematosphaeridium* Tremor Sphere Flesh Gigartin Florid Red Algae
SPECIES *decoratum, zairense*

FAMILY 25 *Schizymeniaceae* Split Gigartin Florid Red Algae
GENUS *Schizymenia* Split Gigartin Florid Red Algae
SPECIES *apoda, binderi, dawsonii, dubyi, ecuadoreana, epiphytica, johnstonii, novae-zelandiae, obliqua, pacifica, violacea*
GENUS *Platoma* Flat Split Gigartin Florid Red Algae
SPECIES *abbottianum, ardreanum, cyclocolpa, heteromorphum, incrassata, izunosimensis, tenuis*
GENUS *Titanophora* Titan Split Gigartin Florid Red Algae
SPECIES *incrustans, marianensis, mauritiana, palmata, pikeana, pulchra, submarina, weberae*
GENUS *Wetherbeella* Wetherbe Split Gigartin Florid Red Algae
SPECIES *foliosa*

FAMILY 26 *Sphaerococcaceae* Sphere Gigartin Florid Red Algae
GENUS *Sphaerococcus* Sphere Gigartin Florid Red Algae
SPECIES *aculeata, coronopifolius, densus, domingensis, dumosus, gelidium, multipartitus, norvegicus, platyphyllus, rhizophylloides*
VARIETIES *coronopifolius pennata, densus fruticosus, multipartitus lichenoides*
GENUS *Thaumasia* Thaumas Sphere Gigartin Florid Red Algae
SPECIES *flavia*

ORDER 8 *Gracilariales* Graceful Florid Red Algae

FAMILY 1 *Gracilariaceae* Graceful Florid Red Algae
GENUS *Gracilaria* Graceful Florid Red Algae
SPECIES *abbottiana, aculeata, aggregata, ambigua, andersonii, apiculata, apiculifera, arcuata, armata, articulata, ascidiicola, attenuata, bailiniae, bangmeiana, beckeri, birdiae, blodgettii, brevis, bursa-pastoris, caerensis, camerunensis, canaliculata, capensis, cervicornis, changii, chilensis, chondroides, chorda, chouae, cliftonii, comosa, conferta, corallicola, corniculata, coronopifolia, corticata, crispata, crockeri, cuneata, cuneifolia, curtissiae, damaecornis, dawsonii, debilis, dendroides, denticulata, dichotomoflabellata, disputabilis, disticha, divaricata, domingiensis, dotyi, dumosa, dura, ecuadoreanus, epihippisora, fanii, firma, fisheri, flabelliforme, flagelliformis, foliifera, fruticosa, gigartinoides, gigas, glomerata,* ***gracilis***(Agar Herb), *guaymasensis, hainanensis, halogenea, hancockii, harveyana, hauckii, heteroclada, hikkaduwensis, incrustata, incurvata, indica, irregularis, johnstonii, kanyakumariensis, kilakkaraiensis, lacerata, latifroms, longa, longirostris, mammillaris, manilaensis, mannariensis, maramae, marcialana, mayae, megaspora, mexicana, microdendron, millardetii, minor, minuta, mixta, multifurcata, multipartita, occidentalis, opuntia, ornata, pachydermatica, pacifica, palmettoides, papenfussii, paradoxa, parvispora, percurrens, perplexa, peruana, pinnata, prolifera, protea, pudumadamensis, pulvinata, punctata, pygmaea, ramisecunda, ramulosa, reptans, rhodocaudata, rhodymenioides, robusta, rubra, rubrimembra, salicornia,salzmannii, secunda, secundata, shimodensis, sinicola, skottsbergii, sordida, spinigera, spinuligera, spinulosa, srilankia, stellata, stipitata, sublittoralis, subsecundata, subtilis, sullivanii, symmetrica, tenuifolia, tenuifrons, tenuistipitata, tepocensis, textorii, tikvahiae, truncata, tsudae, turgida, tuticorinensis, urvillei, vanbosseae, veloroae, venezuelensis, vermiculata, vermiculophylla, vieillardii, vieirae, vivesii, vivipara, yamamotoi, yinggehaiensis*

VARIETIES *arcuata attenuata, arcuata snackeyi, arcuata snackeyi-rhizophora, bursa-pastoris lyra, canaliculata conglomerata, chorda exilis, corticata cylindrica, corticata linearis, corticata ramalinoides, dura fruticosa, foliifera aeruginosa, foliifera granatea, gracilis ecorticata, gracilis procerrima, gracilis ramulosa,* ***gracilis verrucosa***(Agar Herb), *millardetii crenulata, millardetii exposita, millardetii latifolia, millardetii linearifolia, multipartita crispa, multipartita elongata, salicornia simplicior, secundata pseudoflagellifera, tenuistipitata tenuistipitata, tenuistipitata liui, textorii cunninghamii, textorii tenuis, vermiculophylla zhengii*

GENUS *Congracilaria* — Conger Graceful Florid Red Algae

SPECIES *babae*

GENUS *Corallopsis* — Coral Graceful Florid Red Algae

SPECIES *excavata*

GENUS *Curdiea* — Curdie Graceful Florid Red Algae

SPECIES *angusta, angustata, balthazar, codioides, coriacea, crassa, flabellata, irvineae, obesa, racovitzae, recovitzae*

GENUS *Gorgonosphaeridium* — Gorgon Sphere Graceful Florid Red Algae

SPECIES *frequens, suecicum*

GENUS *Gracilariophila* — Least Graceful Florid Red Algae

SPECIES *deformans, gardneri, infidelis, oryzoides, setchellii, sibogae*

VARIETIES *setchellii aggregata*

GENUS *Gracilariopsis* — Lesser Graceful Florid Red Algae

SPECIES *bailiniae, confervoides, costaricensis, heteroclada, lemaneiformis, longissima, megaspora, nganii, panamensis, phanthietensis, rhodotricha, robusta, tenuifrons*

GENUS *Hydropuntia* — Hydric Prickly Pear Graceful Florid Red Algae

SPECIES *caudata, cornea, corymbiata, crassissima, divergens, edulis, eucheumatoides, fisheri, multifurcata, pauciramosa, percurrens, preissiana, ramulosa, rangiferina, secunda, tsudae, usneoides, vanbosseae*

VARIETIES *edulis lemania, edulis taenioides*

GENUS *Melanthalia* — Dark Thal Graceful Florid Red Algae

SPECIES *abscissa, concinna, obtusata, polydactylis*

GENUS *Plocaria* — Plocar Graceful Florid Red Algae

SPECIES *virescens*

GENUS *Polycavernosa* — Polycavern Graceful Florid Red Algae

SPECIES *divergens, domingensis, fisheri, percurrens, tsudae, vanbosseae*

GENUS *Sarcocladia* — Flesh Branch Graceful Florid Red Algae

SPECIES *rhizophora*

FAMILY 2 *Pterocladiophilaceae* — Rock Branched Graceful Florid Red Algae

GENUS *Pterocladiophila* — Rock Branched Graceful Florid Red Algae

SPECIES *hemisphaerica*

GENUS *Gelidiocolax* — Arctic Colac Rock Branched Graceful Florid Red Algae

SPECIES *christianae, deformans, desikacharyi, lyndae, mammillata, margaritoides, microsphaericus, pustulata, suhriae, verruculata*

GENUS *Holmsella* — Holms Rock Branched Graceful Florid Red Algae

SPECIES *australis, pachyderma*

ORDER 9 *Nemaliales* — Thread Florid Red Algae

FAMILY 1 *Galaxauraceae* — Galaxy Ear Thread Florid Red Algae

GENUS *Galaxaura* — Galaxy Ear Thread Florid Red Algae

SPECIES *apiculata, articulata, barbata, capensis, cohaerens, comans, contigua, cyclindrica, diesingiana, distenta, divaricata, falcata, fasciculata, filamentosa, hawaiiana, hystrix, intermedia, kjellmanii, lenta, magna, marginata, obtusata, papillata, plicata, ramulosa, rugosa, striata, stupocaula, verprecula, vietnamensis, yamadae*

VARIETIES *marginata linearis, obtusata major*

GENUS *Actinotrichia* — Radial Hairy Galaxy Ear Thread Florid Red Algae

SPECIES *fragilis, robusta*

GENUS *Gloiophloea* — Glow Flow Galaxy Ear Thread Florid Red Algae

SPECIES *articulata, chinensis, rosea, scinaioides*

GENUS *Nothogenia* — Nothogenic Galaxy Ear Thread Florid Red Algae

SPECIES *cornea, erinacea, fastigiata, ovalis, pseudosaccata, pulvinata*

GENUS *Pseudoscinaia* — Lesser Scina Galaxy Ear Thread Florid Red Algae

SPECIES *snyderiae*

GENUS *Scinaia* — Scina Galaxy Ear Thread Florid Red Algae

SPECIES *aborealis, articulata, australis, bengalica, berggrenii, boergesenii, brasiliensis, canaliculata, capensis, caribaea, carnosa, complanata, confusa, fascicularis, firma, flabellata, furcata, furcellata, halliae, hatei, hormoides, howensis, huismanii, incrassata, interrupta, japonica, johnstoniae, latifrons, minima, moniliformis, okamurae, okiensis, proliferata, pseudojaponica, pseudomoniliformis, saifullahii, salicornioides, setchellii, shameelii, tokidae, tsinglanensis, undulata, verae*

VARIETIES *complanata intermedia, furcellata scandinavica*

GENUS *Tricleocarpa* — Tricle Carpel Galaxy Ear Thread Florid Red Algae

SPECIES *cylindrica, fragilis*
GENUS *Whidbeyella* Whidbey Galaxy Ear Thread Florid Red Algae
SPECIES *cartilaginea*

FAMILY 3 *Liagoraceae* Liagor Thread Florid Red Algae
GENUS *Liagora* Liagor Thread Florid Red Algae
SPECIES *abbottiae, albicans, articulata, australasica, bella, boergesenii, boninensis, brachyclada, californica, canariensis, ceranoides, cladonioides, clavata, daemelii, dendroidea, distenta, divaricata, donaldiana, doridis, engleriana, filiformis, fragilis, galaxauroides, galtsofii, gymnarthron, hainanensis, haitanensis, harveyana, hawaiiana, hirta, howensis, indica, japonica, julieae, maderensis, magniinvolucra, mannarensis, mauritiana, megagyna, norrissiae, papenfussii, pectinata, perennis, pikeana, pinnata, preissii, pseudorobusta, robusta, rubra, samaensis, segawae, sinensis, subarticulata, tenuis, tetrasporifera, tsengii, turneri, valida, vaughanii, viscida, voeltzkowii, wilsoniana*
VARIETIES *ceranoides leprosa, ceranoides pulverulenta, turneri coarctata*
GENUS *Akalaphycus* Akala Liagor Thread Florid Red Algae
SPECIES *liagoroides, setchelliae*
GENUS *Cumagloia* Cum Glow Liagor Thread Florid Red Algae
SPECIES *andersonii*
GENUS *Cylindraxis* Cylindric Liagor Thread Florid Red Algae
SPECIES *rotundatus*
GENUS *Dermonema* Dermal Liagor Thread Florid Red Algae
SPECIES *abbottiae, divaricatum, pulvinatum, virens*
GENUS *Dotyophycus* Doty Liagor Thread Florid Red Algae
SPECIES *abbottiae, corymbosus, pacificum, yamadae*
GENUS *Ganonema* Gano Liagor Thread Florid Red Algae
SPECIES *codii, farinosum, helminthaxis, lubrica, lubricum*
GENUS *Gloiotrichus* Glow Hairy Liagor Thread Florid Red Algae
SPECIES *fractalis*
GENUS *Helminthocladia* Helminth Branched Liagor Thread Florid Red Algae
SPECIES *agardhiana, andersonii, australis, beaugleholei, calvadosii, densa, dotyi, macrocephala, nizamuddinii, reyesii, rhizoidea, senegalensis, simplex, sreeramului, stackhousei, yendoana*
VARIETIES *calvadosii comorinensis, calvadosii indica*
GENUS *Helminthora* Helminth Liagor Thread Florid Red Algae
SPECIES *australis, divaricata, furcellata, guadelupensis, lindaueri, nipponica, stricta*
GENUS *Helmyton* Helmyton Liagor Thread Florid Red Algae
SPECIES *glomeratum, spirale*
GENUS *Izziella* Izzy Liagor Thread Florid Red Algae
SPECIES *orientalis*
GENUS *Liagoropsis* Lesser Liagor Thread Florid Red Algae
SPECIES *schrammii*
GENUS *Nemalion* Nemalion Liagor Thread Florid Red Algae
SPECIES *amoenum, attenuatum, cari-cariense, divaricatum, helminthoides, hudsonii, longicolle, perpusillum, vermiculare*
GENUS *Patenocarpus* Paten Carpel Liagor Thread Florid Red Algae
SPECIES *paraphysiferus*
GENUS *Stenopeltis* Thin Pelt Liagor Thread Florid Red Algae
SPECIES *gracilis*
GENUS *Trichogloea* Hairy Glow Liagor Thread Florid Red Algae
SPECIES *herveyi, javensis, lubrica, requienii, subnuda*
GENUS *Trichogloeopsis* Lesser Hairy Glow Liagor Thread Florid Red Algae
SPECIES *hawaiiana, mucosissima, pedicellata*
GENUS *Yamadaella* Yamada Liagor Thread Florid Red Algae
SPECIES *caenomyce*

ORDER A *Palmariales* Palmar Florid Red Algae

FAMILY 1 *Palmariaceae* Palmar Florid Red Algae
GENUS *Palmaria* Palmar Florid Red Algae
SPECIES *callophylloides, decipiens, georgica, hecatensis, marginicrassa, mollis, moniliformis,* ***palmata***(Dulse Dye Herb), *stenogona*
VARIETIES *palmata prolifera*
GENUS *Devaleraea* Devaler Palmar Florid Red Algae
SPECIES *compressa, microspora, ramentacea, yendoi*
GENUS *Halosacciocolax* Colac Halogen Sac Palmar Florid Red Algae
SPECIES *kjellmanii*
GENUS *Halosaccion* Halogen Sac Palmar Florid Red Algae
SPECIES *americanum, firmum, fistulosum, fucicolum, glandiforme, hydrophorum, intestinalis, lepechini, minjaii, saccatum, scopula, tilesii*

VARIETIES *firmum fibrillosa*
GENUS *Neohalosacciocolax* Modern Halogen Sac Palmar Florid Red Algae
SPECIES *aleutica*

FAMILY 2 *Rhodophysemataceae* Bladder Palmar Florid Red Algae
GENUS *Rhodophysema* Bladder Palmar Florid Red Algae
SPECIES *africana, elegans, feldmannii, georgei, nagaii, odonthaliae*
GENUS *Coriophyllum* Leather Bladder Palmar Florid Red Algae
SPECIES *expansum, setchellii*
GENUS *Meiodiscus* Disc Bladder Palmar Florid Red Algae
SPECIES *concrescens, spetsbergensis*
VARIETIES *spetsbergensis rupicola*
GENUS *Pseudorhododiscus* Lesser Disc Bladder Palmar Florid Red Algae
SPECIES *nipponicus*
GENUS *Reingardia* Reingard Bladder Palmar Florid Red Algae
SPECIES *laminaricola*
GENUS *Rhodophysemopsis* Lesser Bladder Palmar Florid Red Algae
SPECIES *hyperborea, laminariae*

FAMILY 3 *Rhodothamniellaceae* Thamnion Palmar Florid Red Algae
GENUS *Rhodothamniella* Thamnion Palmar Florid Red Algae
SPECIES *floridula, rhizoidea*
GENUS *Camontagnea* Camontagne Thamnion Palmar Florid Red Algae
SPECIES *hirsuta, oxyclada*

ORDER B *Rhodymeniales* Meny Florid Red Algae

FAMILY 1 *Rhodymeniaceae* Meny Florid Red Algae
GENUS *Rhodymenia* Meny Florid Red Algae
SPECIES *adnata, arborescens, ardissonei, caespitosa, californica, callophylloides, cinnabarina, coacta, coespitosella, corallina, crozetii, cuneata, dawsonii, decumbens, delicatula, dichotoma, dissecta, divaricata, erythraea, eyreana, flabellifolia, foliifera, hainanensis, halymenioides, hancockii, holmesii, howeana, indica, intricata, javanica, lanceolata, leptophylla, ligulata, linearis, liniformis, lobata, natalensis, novaezelandica, novazelandica, obtusa, pacifica, parva, peruviana, phyllodes, prolificans, prostrata, pseudopalmata, reniformis, rosea, schmittii, setchellii, skottsbergii, sonderi, stenoglossa, subantarctica, sympodiophyllum, verrucosa*
VARIETIES *californica attenuata, leptophylla minima, pseudopalmata acutifolia, pseudopalmata bahusiensis, pseudopalmata bifida, pseudopalmata crassiuscula, pseudopalmata divaricata, pseudopalmata filiformis, pseudopalmata flabelliformis, pseudopalmata fuscopurpurea, sonderi constricta*
GENUS *Asteromenia* Star Meny Florid Red Algae
SPECIES *peltata*
GENUS *Acanthogonyaulax* Spiny Gonad Aulac Meny Florid Red Algae
SPECIES *spinifera*
GENUS *Botryocladia* Berry Branched Meny Florid Red Algae
SPECIES *adhaerens, bahamensis, beaudettei, beckeriana, boergesenii, botryoides, bullosa, canariensis, caraibica, chiajeana, connexa, darwinii, ebriosa, feldmannii, fernandeziana, ganesanii, guaymasensis, guineensis, hancockii, kuckuckii, lawsonii, leptopoda, madagascariensis, microphysa, monoeca, neushulii, occidentalis, papenfussiana, paucivesicaria, pienaarii, pseudodichotoma, pyriformis, senegalensis, shanksii, skottsbergii, sonderi, spinulifera, tenuissima, uvarioides, wynnei*
VARIETIES *leptopoda luxurians, pseudodichotoma datilensis*
GENUS *Cephalocystis* Head Cyst Meny Florid Red Algae
SPECIES *furcellata, leucobotrys*
GENUS *Chamaebotrys* Chamise Berry Meny Florid Red Algae
SPECIES *boergesenii,erectus, lomentariae*
VARIETIES *boergesenii minimum*
GENUS *Chrysymenia* Golden Meny Florid Red Algae
SPECIES *acicularis, agardhii, clavellosa, dichotoma, dickieana, digitata, enteromorpha, glebosa, grandis, halymenioides, kaernbachii, littleriana, lobata, nodulosa, okamurae, ornata, pinnulata, planifrons, polydactyla, procumbens, ventricosa, wrightii*
GENUS *Coelarthrum* Celestial Straight Meny Florid Red Algae
SPECIES *albertisii, cliftonii, decumbens, mortensenii, muelleri, opuntia*
GENUS *Coelothrix* Celestial Hairy Meny Florid Red Algae
SPECIES *indica, irregularis*
GENUS *Cordylecladia* Club Branched Meny Florid Red Algae
SPECIES *erecta, guiryi, peasiae*
GENUS *Cryptarachne* Cryptic Spider Meny Florid Red Algae
SPECIES *polyglandulosa, reticulate*
GENUS *Epymenia* Epy Meny Florid Red Algae
SPECIES *capensis, elongata, falklandica, flabellata, marginifera, obtusa, variolosa, wilsonis*
VARIETIES *obtusa minor*

GENUS *Erythrocolon* Colon Meny Florid Red Algae
SPECIES *podagricum*
VARIETIES *podagricum minus*
GENUS *Gastroclonium* Gastric Clone Meny Florid Red Algae
SPECIES *clavatum, compressum, cylindricum, iyengarii, ovatum, pacificum, parvum, pygmaeum, reflexum, subarticulatum, trichodes, xishaensis*
GENUS *Gloiosaccion* Glow Sac Meny Florid Red Algae
SPECIES *brownii*
VARIETIES *brownii furcatum*
GENUS *Irvinea* Irvine Meny Florid Red Algae
SPECIES *ardreana*
GENUS *Leptosomia* Thin Soma Meny Florid Red Algae
SPECIES *rosea*
GENUS *Maripelta* Sea Pelt Meny Florid Red Algae
SPECIES *atlantica, rotata*
GENUS *Microphyllum* Micron Meny Florid Red Algae
SPECIES *borneense*
GENUS *Rhodymeniocolax* Colac Meny Florid Red Algae
SPECIES *austrina, botryoides*
GENUS *Sparlingia* Sparling Meny Florid Red Algae
SPECIES *pertusa*

FAMILY 2 *Champiaceae* Champ Meny Florid Red Algae
GENUS *Champia* Champ Meny Florid Red Algae
SPECIES *affinis, bifida, caespitosa, ceylanica, chathamensis, compressa, disticha, echigoensis, expansa, feldmannii, gigantea, globulifera, indica, insignis, irregularis, japonica, kotschyana, laingii, lumbricalis, minuscula, novae-zelandiae, parvula, plumosa, recta, salicornioides, somalensis, spathulata, stipitata, taironensis, tripinnata, vieillardii, viridis, womersleyi, zonata, zostericola*
VARIETIES *compressa scindica. novae-zelandiae dolchipoda, parvula amphibolis, parvula prostrata, parvula vaga*
GENUS *Champiocolax* Colac Champ Meny Florid Red Algae
SPECIES *lobatus, sarae*
GENUS *Chylocladia* Branched Champ Meny Florid Red Algae
SPECIES *capensis, grandis, pelagosae, perpusilla, phalligera, rigens, verticillata*
VARIETIES *verticillata bistratosa, verticillata breviarticulata, verticillata squarrosum, verticillata unistratosa*
GENUS *Coeloseira* Celestial Pitted Champ Meny Florid Red Algae
SPECIES *parva, trichodes*
GENUS *Dictyothamnion* Netted Thamnion Champ Meny Florid Red Algae
SPECIES *saltatum*

FAMILY 3 *Faucheaceae* Fauch Meny Florid Red Algae
GENUS *Fauchea* Fauch Meny Florid Red Algae
SPECIES *crispa, fryeana, furcellata, galapagensis, hassleri, laciniata, leptophylla, madagascariensis, microspora, mollis, procumbens, profunda, repens, sefferi, setchellii, spinulosa, stipitata, xishaensis*
VARIETIES *galapagensis pygmaea*
GENUS *Agardhinula* Agardh Fauch Meny Florid Red Algae
SPECIES *browneae*
GENUS *Cenacrum* Cenacre Fauch Meny Florid Red Algae
SPECIES *subsutum*
GENUS *Erythrymenia* Lesser Fauch Meny Florid Red Algae
SPECIES *minuta, obovata*
GENUS *Faucheocolax* Colac Fauch Meny Florid Red Algae
SPECIES *attenuata*
GENUS *Faucheopsis* Least Fauch Meny Florid Red Algae
SPECIES *coronata*
GENUS *Fryeella* Frye Fauch Meny Florid Red Algae
SPECIES *gardneri*
GENUS *Gloiocladia* Glow Branched Fauch Meny Florid Red Algae
SPECIES *atlantica, australe, australis, blomquistii, conjuncta, fruticulosa, furcata, halymenioides, iyoensis, japonica, mauritiana, minutula, polycarpa, ramellifera, robillardii, rubrispora, saccata, speciosa, subdichotoma*
GENUS *Gloiocolax* Glow Colac Fauch Meny Florid Red Algae
SPECIES *novae-zelandiae*
GENUS *Gloioderma* Glow Dermal Fauch Meny Florid Red Algae
SPECIES *atlanticum, minutulum, rubrisporum, saccatum, subdichotomum*
GENUS *Gloiodermatopsis* Lesser Glow Dermal Fauch Meny Florid Red Algae
SPECIES *setchellii*
GENUS *Halichrysis* Haline Golden Fauch Meny Florid Red Algae
SPECIES *coalescens, depressa, japonica, micans, thivyae*
GENUS *Hymenocladia* Hymen Branched Fauch Meny Florid Red Algae

SPECIES *chondricola, conspersa, dactyloides, divaricata, filiformis, kallymenioides, linearis, sanguinea, usnea*
VARIETIES *chondricola elongata*
GENUS *Hymenocladiopsis* — Lesser Hymen Branched Fauch Meny Florid Red Algae
SPECIES *crustigena*
GENUS *Leptofauchea* — Thin Fauch Meny Florid Red Algae
SPECIES *anastomosans, brasiliensis, pacifica, rhodymenioides*
GENUS *Minium* — Lead Fauch Meny Florid Red Algae
SPECIES *parvum*
GENUS *Sciadophycus* — Sciad Fauch Meny Florid Red Algae
SPECIES *expansus, stellatus*
GENUS *Webervanbossea* — Weber-van Bosse Fauch Meny Florid Red Algae
SPECIES *kaliformis, splachnoides, tasmanensis*

FAMILY 4 *Lomentariaceae* — Loment Meny Florid Red Algae
GENUS *Lomentaria* — Loment Meny Florid Red Algae
SPECIES *acicularis, amplexans, articulata, australis, baileyana, caespitosa, caseae, catenata, chylocladiella, clavaeformis, clavellosa, compressa, corallicola, diffusa, drouetii, ercegovicii, firma, flaccida, gracillima, hakodatensis, intertexta, jabukae, lubrica, mauritiana, monochlamydea, okamurae, orcadensis, pinnata, pyramidalis, rawitsheri, saxigena, secunda, squarrosa, strumosa, subdichotoma, tenera, tenerrima, umbellata, uncinata, verticillata*
VARIETIES *articulata linearis, clavellosa conferta, clavellosa reducta, firma compressa, uncinata filiformis*
GENUS *Binghamia* — Bingham Loment Meny Florid Red Algae
SPECIES *californica, forkii*
GENUS *Binghamiopsis* — Lesser Bingham Loment Meny Florid Red Algae
SPECIES *caespitosa*
GENUS *Ceratodictyon* — Horn Net Loment Meny Florid Red Algae
SPECIES *hachijoense, spongiosum*
GENUS *Gelidiopsis* — Arctic Loment Meny Florid Red Algae
SPECIES *capitata, hachijoensis, intricata, pannosa, planicaulis, repens, scoparia, tenuis, variabilis*
VARIETIES *pannosa exiguum*
GENUS *Semnocarpa* — Semnon Carpel Loment Meny Florid Red Algae
SPECIES *corynephora, minuta*
GENUS *Stirnia* — Stirn Loment Meny Florid Red Algae
SPECIES *prolifera*

CLASS 617.4 *Rhodellophyceae* Rhodell Red Algae

ORDER 1 *Porphyridiales* Porphyry Rhodell Red Algae

FAMILY 1 *Porphyridiaceae* — Porphyry Rhodell Red Algae
GENUS *Rhodella* — Rhodell Red Algae
SPECIES *maculata, violacea*
GENUS *Porphyridium* — Porphyry Rhodell Red Algae
SPECIES *aerugineum, purpureum, sordidum*
GENUS *Chaos* — Chaos Porphyry Rhodell Red Algae
SPECIES *nobile, sanguinarius*
GENUS *Chroothece* — Tinged Theca Porphyry Rhodell Red Algae
SPECIES *littorinae, mobilis, richteriana, rupestris*
GENUS *Dixoniella* — Dixon Porphyry Rhodell Red Algae
SPECIES *grisea*
GENUS *Flintiella* — Flint Porphyry Rhodell Red Algae
SPECIES *sanguinaria*
GENUS *Rhodosorus* — Sorus Porphyry Rhodell Red Algae
SPECIES *marinus*
GENUS *Vanhoeffenia* — Vanhoeffen Porphyry Rhodell Red Algae
SPECIES *antarctica*

FAMILY 2 *Phragmonemataceae* — Fragment Thread Porphyry Rhodell Red Algae
GENUS *Phragmonema* — Fragment Thread Porphyry Rhodell Red Algae
SPECIES *sordidum*
GENUS *Empselium* — Empsel Fragment Thread Porphyry Rhodell Red Algae
SPECIES *rubrum*
GENUS *Glauconema* — Glaucous Fragment Thread Porphyry Rhodell Red Algae
SPECIES *ramosa*
GENUS *Goniotrichiopsis* — Gonad Hairy Fragment Thread Porphyry Rhodell Red Algae
SPECIES *sublittoralis*
GENUS *Kyliniella* — Kylin Fragment Thread Porphyry Rhodell Red Algae

SPECIES *latvica*
GENUS *Neevea* — Neeve Fragment Thread Porphyry Rhodell Red Algae
SPECIES *repens*
GENUS *Rufusia* — Rufous Fragment Thread Porphyry Rhodell Red Algae
SPECIES *pilicola*

ORDER 2 *Stylonematales* Thread Rhodell Red Algae

FAMILY 1 *Stylonemataceae* — Thread Rhodell Red Algae
GENUS *Stylonema* — Thread Rhodell Red Algae
SPECIES *alsidii, alsidii strictum, cornu-cervi, reniforme, subcoeruleum*
VARIETIES *alsidii strictum*

SUBKINGDOM 62 *Fungi* Fungus

	PHYLUMS		**Species**	**Species**
Latin: Fungi fungus				
	621 *Ascomycota*	Sac Fungus	2414 [K]	30000 [M]
53,600 species [Margulis & Schwartz, 1998]	622 *Basidiomycota*	Club Fungus	9318 [K]	22500 [M]
32,853 [Kirk, Sept. 2006]	623 *Zygomycota*	Multinucleate Fungus	1121 [K]	1100 [M]

Subkingdom *Fungi* (Fungus) species are nucleus organisms that form chitinous resistant propagules and chitinous cell walls. All Subkingdom *Fungi* species lack undilipodia (i.e. they are amastigote) and are immotile at all stages of their life cycle. Because of their total immotility, they are placed into the Plant Kingdom. However, Fungus species are also closely related to the Animal Kingdom. Both Fungus and Animals have chitin as the main component of cell walls. Chitin is also the main component of Arthropod exoskeletons. Fungus species are an intermediate form having both Plant and Animal characteristics.

Non-motile Prefungus Protophyte (Protophyta) → **Fungus**

Motile Prefungus Protozoan (Protozoa) → **Fungus**

Other Classifications

In the Fusillo Classification System, the former Phylum *Glomeromycota* is the Class *Glomeromycetes* in the fungus Phylum *Zygomycota*.

PHYLUM 621 *Ascomycota* Sac Fungus

	CLASSES		**Species**
Latin: *Asco* sac Latin: *Mycota* mucus			
	621.1 *Ascomycetes*	Sac Fungus	
	621.2 *Ascomussomycetes*	Sac Mushroom	
30,000 species [Margulis & Schwartz, 1998]	621.2 *Lichinomycetes*	Lichen	
22,414 species [Kirk, Sept. 2006]	621.3 *Mucoromycetes*	Mold	
	621.4 *Hemiascomycetes*	Sugar Yeast	

Phylum *Ascomycota* Sac Fungus speces develop sac shaped spore bearing structures called ascocarps. There are three types of ascocarps; **(1) Cleistothecium Ascocarp, (2) Perithecium Ascocarp, and (3) Apothecium Ascocarp.**

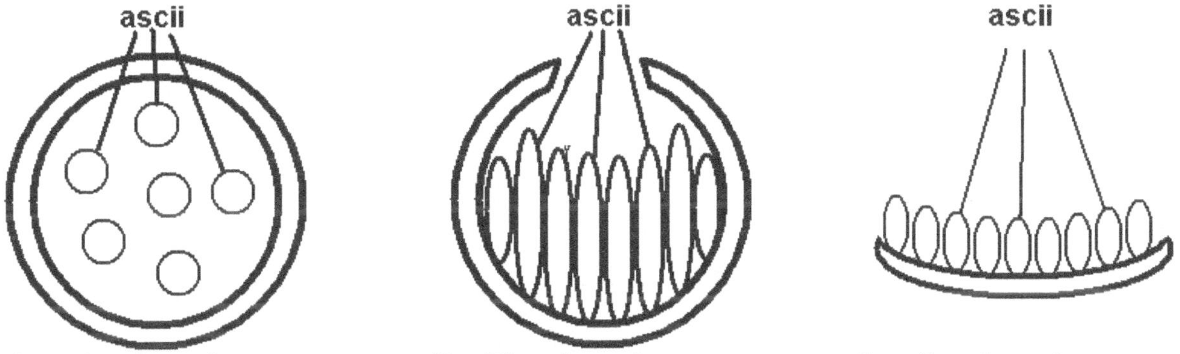

Cleistothecium Ascocarp **Perithecium Ascocarp** **Apothecium Ascocarp**

The **Cleistothecium Ascocarp** is characteristic of molds such as *Penicillium*. It is a spherica ascocarp that completely encloses the developing asci.

The **Perithecium Ascocarp** is characteristic of molds such as *Hypomyces* Mushroom Pimple Hypocreal Mold. It is globe or flask shaped ascocarp with an opening at the top and sometimes a neck.

The **Apothecium Ascocarp** is characteristic of cup fungus such as Peziza Cup Peziz Sac Fungus. It is flat to saucer to cup shaped. Many Sac Fungus species produce an Apothecium.

Phylum *Ascomycota* Sac Fungus species, ascomycotes, are either a saprobes, symbiotes, or, in the case of lichens, photosynthezers. Ascomycotes live on land (with the exception of *Verrucaria serpuloides* which is submerged in seawater). Phylum *Ascomycota* Sac Fungus species do not have a stem or base.

A saprobe excretes extra cellular enzymes which break down dead organic matter. The dead organic matter is then absorbed using thin threads (hypha). Saprobe ascomycotes live on insect carcasses and plant debris. Macromolecules of cellulose, lignin, and collagen are digested, releasing nutrient compounds of carbon, nitrogen, and phosphorous.

A symbiote extends the thin threads (hypha) directly into the host tissue and extracts nutrients from a host cell. Symbiote ascomycotes are parasites on algae, molds, and fungi.

Class *Ascomycetes* species include yeasts, chanterelles, lorchels, morels, and truffles.

Class *Lichinomycetes* Lichen Sac Fungus is the product of a symbiotic union of Class *Ascomycotes* species and photosynthetic bacteria species of the Phylum *Cyanobacteria*, or photosynthetic algae species of Phylum *Chlorophyta. Trebouxia* and *Pseudotrebouxia* chlorophytes and *Nostoc* cyanobacteria are most often the symbiotic partners of lichen. Lichen herbs synthesize lichenic pigments which are used as natural textile dyes.

Lichens release small spore fragments (soredia) consisting of one or more algal cells surrounded by fungal filaments (mycelia). The spore fragments are distributed on air currents. Lichens are found in Antarctica, the Sub-Arctic tundra, high mountains, the tropics, and in Canada, Alaska and Siberia boreal old growth forests. They commonly grow on tree bark or rocks. Lichens die in the presence of air pollution.

In the Genus listings the suffix *Brya* means that the lichen grows on moss; the suffix *Ligna* means that the lichen grows on wood, the suffix *Drysia* means that the lichen grows on trees; the suffix *Barca* means that the lichen grows on bark; the suffix *Sola* means that the lichen grows in soil; the suffix *Saxa* means that the lichen grows on rock.

Class *Mucoromycetes* includes blue-green molds. Ergot Hypocreal Sac Mold Herb *Claviceps purpurea-medica* produces a medicinal drug that is used to treat migraine headaches and uterus hemorrhage. Ergot Hypocreal Sac Mold Herb *Claviceps purpurea-aluca* produces a hallucinogenic drug. Poison Ergot Hypocreal Sac Mold *Claviceps purpurea-toxica* is poisonous to humans and domesticated animals. Sugar Yeast molds are single celled and do not have thin threads (hypha). Sugar Yeasts ferment the sugars glucose, fructose, and sucrose into ethyl alcohol and are used to make wine and beer. Elsinore Sac Mold *Elsinoe* causes disease in cirtrus, raspberry, and avocado plants. White Vaginal Saccharine Sugar Yeast *Candida albicans* causes moniliasis; a disease of the female vagina and thrush; a mucus membrane disease. Lung Cyst Sac Mold *Pneumocystis* causes fatal pneumonia in humans with compromised immune systems. Penicillin is a clear liquid produced on the top of Penicillin Asp Gill Sac Mold Herb *Penicillium chrysogenum.* Penicillin is an antibiotic. An antibiotic is a low molecular weight compound, secreted by a living bacteria or mold, which inhibits the growth of microbes including bacteria, protozoans, and protophytes. The sea mold *Cephalosporin* produces the antibiotic cephalosporin. The bacteria *Streptomyces* produces the antibiotic streptomycin.

Lichen Dye Herb Plants

Dye Species	Synonym	Acid	Boiling Water	Ammonia Method
Alectoria sarmentosa	—	Alectoronic Acid	bright orange	taupe violet
Arctoparmelia centrifuga	*Xanthoparmelia centrifuga*	Usnic Acid, Atranorin, Alectoronic Acid	red brown	—
Bryocaulon divergens	*Cornicularia divergens*	Olivetoric Acid	orange	—
Bryoria fremontii	—	Vulpinic Acid	tan brown	green
Bryoria fuscescens	—	Fumarprotocetraric Acid	yellow	—
Bulbothrix coronata	*Pseudocyphellaria coronata*	Lecanoric Acid	?	?
Caloplaca trachyphylla	—	Anthraquinones	—	red purple
Candelariella vitellina	—	Calycin	yellow	—
Cetraria ericetorum	—	Lichesterinic Acid	—	maize tan
Cetraria islandica	—	Fumarprotocetraric Acid	yellow	—
Cetraria islandica	—	Protolichesterinic Acid	brown	red
Cetraria islandica	—	Lichesterinic Acid	brown	rusty brown
Cetrariella delisei	*Cetraria delisei*	Gyrophoric Acid	—	fuchsia
Cladina portentosa	*Cladonia implexa*	Perlatolic Acid	leaf green	pink
Cladina rangiferina	*Cladonia rangiferina*	Fumarprotocetraric Acid, Atranorin	red brown	—
Cladina stellaris	*Cladonia alpestris*	Perlatolic Acid, Usnic Acid	yellow	—
Cladonia cristatella	—	Barbatic Acid, Didymic Acid	light green	—
Cladonia pyxidata	—	Fumarprotocetraric Acid	gold	—
Pertusaria maritima	*Coccotrema maritimum*	Stictid Acid, Constictid Acid	—	olive
Dactylina arctica	—	Gyrophoric Acid	—	fuchsia
Dermatocarpon miniatum	—	No Lichen Acids	olive brown	—
Dibaeis baeomyces	*Baeomyces fungoides*	Baeomycesic Acid	—	pale salmon

Lichen Dye Herb Plants

Dye Species	Synonym	Acid	Boiling Water	Ammonia Method
Diploschistes scruposus	*Urceolaria scruposa*	Atranorin, Lecanoric Acid	—	cherry
Evernia divaricata	—	Divaricatic Acid	—	green
Evernia punastri	—	Evernic Acid, Atranorin	light orange	voilet brown
Everniastrum catawbiense	—	Gyrophoric Acid	pink	pink
Flavocetraria culcullata	*Cetraria cucullata*	Protolichesterinic Acid	—	red purple
Flavocetraria nivalis	*Cetraria nivalis*	No Lichen Acids	—	violet
Flavoparmelia caperata	*Parmelia caperata*	Caperatic Acid	yellow	—
Flavopunctelia praesignis	*Parmelia caperata incorrupta*	Lecanoric Acid	?	—
Flavopunctelia soredica	*Parmelia ulophyllodes*	Usnic Acid, Lecanoric Acid	—	red
Hypogymnia physodes	—	Physodalic Acid	gold	—
Hypogymnia physodes	—	Physodic Acid	orange brown	—
Hypogymnia physodes	—	Protocetraric Acid	green	—
Kaernefeltia californica	*Cornicularia californica* *Cetraria californica*	Lichesterinic, Protolichesterinic Acid	—	light brown
Lasallia papulosa	*Lasallia pustulata papulosa*	Anthraquinones	—	pink
Lasallia papulosa	*Umbilicaria papulosa*	Anthraquinones	—	magenta
Lasallia pensylvanica	*Umbilicaria pensylvanica*	Anthraquinones		
Lasallia pustulata	—	Anthraquinones	—	red
Lasallia pustulata	*Umbilicaria pustulata*	Anthraquinones	—	purple
Lecanora muralis	—	Usnic Acid, Zeorin	tawny olive	—
Letharia columbiana	—	Vulpinic Acid, Atranorin	brilliant yellow	—
Letharia vulpina	—	Vulpinic Acid, Atranorin	brilliant lemon	—
Lobaria pulmonaria	—	Stictic Acid, Norstictic Acid	apricot	—
Lobaria spathulata		Lobaric Acid	light red	—
Melanelia fuliginosa	*Melanelia glabratula*	Lecanoric Acid	—	violet purple
Melenelia fuliginosa	*Parmelia glabratula*	Lecanoric Acid	—	gray red
Melanelia fuliginosa	*Parmelia fuliginosa*	Lecanoric Acid	—	violet purple
Melanelia olivacea	*Parmelia olivacea*	Fumarprotocetraric Acid	tan	—
Melanelia stygia	—	Fumarprotocetraric, Caperatic Acid	tan	—
Nephroma expallidum	—	Nephroarctin	brown	—
Nephroma helveticum	—	Zeorin	orange brown	dark brown
Ochrolechia pallescens	*Lecanora pallescens*	Gyrophoric, Olivetoric Acid,	—	—
Ochrolechia parella?	—	Orcinol Depsides?	—	red purple
Ochrolechia tartarea	*Lecanora tartarea*	Gyrophoric Acid	—	red
Ophioparma lapponica	*Haematomma lapponica*	Hypothamnolic Acid	—	purple red
Ophioparma ventosa	*Haematomma ventosum*	Divaricatic Acid	orange brown	deep magenta
Parmelia clavulifera?	—	β Orcinol Depsidones?	copper, rust	
Parmelia omphalodes	—	Salazinic Acid	orange red brown	—
Parmelia saxatilis	—	Salazinic Acid	bronze	—
Parmelia sulcata	—	Salazinic Acid	red brown	magenta
Parmotrema arnoldii	*Parmelia arnoldii*	Anatorin, Alectoronic Acid	gold	—
Parmotrema chinense	*Parmelia perlata*	Stictic Acid	russet orange	magenta pink
Parmotrema perforatum	—	Alectoronic Acid	—	red
Parmotrema tinctorum	—	Lecanoric Acid	gold	red
Peltigera aphthosa	—	Gyrophoric Acid, Zeorin	—	pink
Peltigera canina	—	No Lichen Acids	gold	—
Peltigera venosa	—	Zeorin, Tenuiorin	—	purple pink
Pertusaria corallina	*Phyllospora corallina?*	Atranorin, Argopsin, Pannarin, Zeorin?	—	red purple
Pertusaria pseudocorallina	*Isidium westringii*	Norstictic Acid?	—	orange pink

Lichen Dye Herb Plants

Dye Species	Synonym	Acid	Boiling Water	Ammonia Method
Physcia americana	—	Atranorin, Triterpenes	yellow	—
Placopsis gelida	—	Gyrophoric Acid	—	?
Platismatia glauca	—	Caperatic Acid, Atranorin	apricot yellow	—
Pseudevernia consocians	—	Lecanoric Acid, Atranorin	orange	pink
Pseudevernia furfuracea	—	Lecanoric Acid, Atranorin	vivid orange	—
Pseudocyphellaria crocata	—	Stictic Acid, Calycin	?	?
Punctelia subflava rudecta	*Parmelia rudecta*	Lecanoric Acid	gold	pink
Punctelia subflava rudecta	—	Lecanoric Acid	gold	pink
Ramalina farinacea	—	Protocetraric, Salazinic Acid	yellow	pink brown
Ramalina siliquosa	*Ramalina scopulorum?*	Orsinols?	yellow	red
Rimelia reticulata	*Parmelia reticulata*	Salazinic Acid	—	pink
Roccella babingtonii	—	Lecanoric Acid, Rocellic Acid	—	deep purple
Roccella fuciformis	—	Lecanoric Acid, Rocellic Acid	—	crimson
Roccella montagnei	—	Lecanoric Acid, Rocellic Acid	—	purple
Roccella phycopsis	—	Lecanoric Acid, Rocellic Acid	—	crimson
Roccella tinctoria	—	Lecanoric Acid, Rocellic Acid	—	purple
Sphaerophorus fragilis	—	Hypothamnolic Acid	—	red
Sphaerophorus globosus	—	Hypothamnolic Acid	—	red
Stereocaulon alpinum paschale	—	Lobaric Acid	rust	—
Stereocaulon tomentosum	—	Stictic Acid	orange	—
Teloschistes flavicans	—	Anthraquinones	yellow orange	—
Teloschistes velifera?	—	Anthraquinones?	—	purple pink
Umbilicaria americana	—	Gyrophoric Acid	—	purple magenta
Umbilicaria coriacea	*Pertusaria coriacea?*	Depsides, Depsidones, Xanthones?	dark brown	—
Umbilicaria deusta	*Gyrophora deusta*	Gyrophoric Acid	—	purple magenta
Umbilicaria hyperborea	—	Gyrophoric Acid	—	purple magenta
Umbilicaria lyngei	*Agrophora lyngei*	Gyrophoric Acid	brown	violet brown
Umbilicaria mammulata	—	Gyrophoric Acid		
Umbilicaria meuhlenbergii	*Actinogyra meuhlenbergii*	Gyrophoric Acid	brown	red pink
Umbilicaria meuhlenbergii	*Actinogyra meuhlenbergii*	Gyrophoric Acid	brown	red pink
Umbilicaria phaea	—	Gyrophoric Acid	—	purple magenta
Umbilicaria polyphylla	—	Gyrophoric, Umbilicaric Acid	—	purple magenta
Umbilicaria polyrrhiza	—	Gyrophoric Acid	—	purple magenta
Umbilicaria proboscidea	—	Gyrophoric Acid	—	purple magenta
Umbilicaria torreyfacta	—	Gyrophyoric Acid, Stictic Acid	brown	purple magenta
Umbilicaria vellea	—	Gyrophoric Acid	—	purple magenta
Umbilicaria virginis	*Omphalodiscus virginis*	Gyrophoric Acid, Norstictic	—	purple

Lichen Dye Herb Plants

Dye Species	Synonym	Acid	Boiling Water	Ammonia Method
		Asic		magenta
Usnea filipendula	*Usea dasypoga*	Salazinic Acid	orange brown	—
Usnea glabrata	—	Usnic Acid	—	gray green yellow
Usnea hirta	—	Usnic Acid	—	purple
Usnea articulata?	—	Usnic Acid?	ocher gold	—
Usnea filipendula barbata	Old Man's Beard Lichen	Barbatolic Acid	orange	blue purple
Basidia diffracta?	*Usnea diffracta?*	Usnic Acid, Atronin?	green	—
Usnea rubrotincta	*Usnea rubescens Cheanothecopsis rubescens*	Usnic, Vulpinic Acid?	gold orange	—
Usnea subfloridans	—	Salzinic Acid	gray olive	—
Caloplaca lactea? Julella lactea? Thomasella lactea?	*Variolaria lactea*	Lactic Acid?	—	pink
Variolaria orcina?	—	Orcinol?	—	purple
Vulpicida canadensis	*Cetraria canadensis*	Vulpinic Acid	green yellow	—
Vulpicida tilesii	*Cetraria tilesii*	Vulpinic Acid	bright yellow	—
Xanthoparmelia conspersa	—	Stictic Acid	red brown	drab apricot
Xanthoparmelia lineola	—	Salzinic Acid	russet	—
Ochrolechia tartarea?	*Xanthoparmelia tartarica*	Gyrophoric Acid?	copper	—
Xanthoria elegans	—	Anthraquinones	—	mauve
Xanthoria ectanea?	—	Anthraquinones?	—	slate blue
Xanthoria ectanoides?	—	Anthraquinones?	—	blue
Xanthoria parietina	—	Parietin	corn yellow	blue violet
Xanthoria polycarpa	—	Anthraquinones	—	blue

CLASS 621.1 *Ascomycetes* Sac Fungus

ORDER 01 *Ascomycetales* Sac Fungus

FAMILY 01 *Ascomycetaceae* — Sac Fungus
GENUS *Abropelta* — Delicate Pelt Sac Fungus
SPECIES *fusarioides*
GENUS *Abyssomyces* — Abyssal Sac Fungus
SPECIES *hydrozoicus*
GENUS *Acanthoderma* — Spiny Dermal Sac Fungus
SPECIES *memecyli*
GENUS *Acanthorus* — Spiny Sac Fungus
SPECIES *maranhensis*
GENUS *Acarellina* — Acarelline Sac Fungus
SPECIES *psidii*
GENUS *Acarocybellina* — Cybe Acarospore Sac Fungus
SPECIES *arengae*
GENUS *Acarocybiopsis* — Lesser Cybe Sac Fungus
SPECIES *cubitaensis*
GENUS *Acaropeltis* — Pelt Acarospore Sac Fungus
SPECIES *costaricensis*
GENUS *Achoropeltis* — Pelt Achorospore Sac Fungus
SPECIES *modesta*
GENUS *Aciesia* — Acies Sac Fungus
SPECIES *xylopiae*
GENUS *Acinula* — Acinule Sac Fungus
SPECIES *candicans*
GENUS *Acremoniella* — Acremon Sac Fungus
SPECIES *atra, cedrelae, lutzi, rugulosa, velata, verrucosa*
GENUS *Acroconidiellina* — Acro Cone Sac Fungus
SPECIES *chloridis, loudetiae, tropaeoli, urtiagae*
GENUS *Acrodictyopsis* — Lesser Acro Net Sac Fungus

SPECIES *lauri, patelii*
GENUS *Acrodictys* Acro Net Sac Fungus
SPECIES *elaeidis*
GENUS *Acrodontiella* Lesser Acro Tooth Sac Fungus
SPECIES *fallopiae*
GENUS *Acrodontium* Acro Tooth Sac Fungus
SPECIES *abietis, antarcticum, crateriforme, echinulatum, hydnicola, intermissum, salmoneum, simplex, virellum*
GENUS *Acrophialophora* Acro Phial Sac Fungus
SPECIES *fusispora, levis*
GENUS *Acrospeira* Acro Speir Sac Fungus
SPECIES *mirabilis*
GENUS *Acrostaurus* Acro Crosshatch Sac Fungus
SPECIES *turneri*
GENUS *Acrotheciella* Lesser Acro Theca Sac Fungus
SPECIES *javanica*
GENUS *Acrothecium* Acro Theca Sac Fungus
SPECIES *delicatulum, nigrum, robustum*
GENUS *Actinocladium* Radial Branched Sac Fungus
SPECIES *rhodosporum*
GENUS *Actinodochium* Radial Doch Sac Fungus
SPECIES *concinnum, jenkinsii, uniseptata*
GENUS *Acumispora* Cum Spore Sac Fungus
SPECIES *biseptata, phragmospora, uniseptata*
GENUS *Aegeritoidea* Aegerite Sac Fungus
SPECIES *foliicola*
GENUS *Aenigmatomyces* Enigmatic Sac Fungus
SPECIES *ampullisporus*
GENUS *Aenigmatospora* Enigmatic Spore Sac Fungus
SPECIES *pulchra*
GENUS *Agaricodochium* Agaric Doch Sac Fungus
SPECIES *camelliae*
GENUS *Agarwalia* Agarwal Sac Fungus
SPECIES *terricola*
GENUS *Agarwalomyces* Lesser Agarwal Sac Fungus
SPECIES *indicus*
GENUS *Agrabeeja* Agrabeej Sac Fungus
SPECIES *kavakapriya*
GENUS *Akanthomyces* Spiny Sac Fungus
SPECIES *aculeatus, aranearum, gracilis, johnsonii, novoguineensis*
GENUS *Alatosessilispora* Winged Sessile Spore Sac Fungus
SPECIES *bibrachiata*
GENUS *Alatospora* Winged Spore Sac Fungus
SPECIES *acuminata, crassipes, flagellata, pulchella*
GENUS *Albophoma* White Foam Sac Fungus
SPECIES *amanashiensis*
GENUS *Albosynnema* White Thread Sac Fungus
SPECIES *elegans, filicicola*
GENUS *Aleurodomyces* Floury Sac Fungus
SPECIES *signoretii*
GENUS *Algonquinia* Algonquin Sac Fungus
SPECIES *inflata*
GENUS *Aliquandostipite* Aliquando Stipe Sac Fungus
SPECIES *khaoyaiensis, sunyatsenii*
GENUS *Allantophoma* Diverse Foam Sac Fungus
SPECIES *endogenospora*
GENUS *Allonema* Diverse Thread Sac Fungus
SPECIES *roseum*
GENUS *Alloneottiosporina* Diverse Modern Spore Sac Fungus
SPECIES *carolinensis, infundibularis*
GENUS *Allothyrina* Diverse Thyre Sac Fungus
SPECIES *serjaniae*
GENUS *Allothyriopsis* Lesser Diverse Thyre Sac Fungus
SPECIES *landolphiae*
GENUS *Alpakesa* Alpakes Sac Fungus
SPECIES *nolinae, taiwanensis, uniseptata, yuccaefolia*
GENUS *Alysidiopsis* Alys Sac Fungus

SPECIES *foliicola, lignicola, pipsissewae, yunnanensis*
GENUS *Amallospora* Soft Spore Sac Fungus
SPECIES *dacrydion*
GENUS *Amblyosporium* Ambly Spore Sac Fungus
SPECIES *botrytis, coprophiloides, spongiosum*
GENUS *Ambrosiaemyces* Lesser Ambrosia Sac Fungus
SPECIES *zeylanicus*
GENUS *Ambrosiella* Ambrosia Sac Fungus
SPECIES *brunnea, ferruginea, gnathotrichi, hartigii, ips, macrospora, sulcati, tingens, xylebori*
GENUS *Amerodiscosiellina* American Disc Sac Fungus
SPECIES *annonacearum*
GENUS *Ameromassaria* Massar Sac Fungus
SPECIES *japonica*
GENUS *Amerosporiella* Lesser American Spore Sac Fungus
SPECIES *tristis*
GENUS *Amerosporiopsis* Least American Spore Sac Fungus
SPECIES *gaubae*
GENUS *Amerosporium* American Spore Sac Fungus
SPECIES *atrum, concinnum, congregatum, platense, polynematoides*
GENUS *Amoenodochium* Sand Doch Sac Fungus
SPECIES *humicola*
GENUS *Amoenomyces* Sand Sac Fungus
SPECIES *catenosporus*
GENUS *Amorphotheca* Amorphic Theca Sac Fungus
SPECIES *resinae*
GENUS *Ampelomyces* Ampel Sac Fungus
SPECIES *quercinus, quisqualis*
GENUS *Amphichaetella* Dual Bristle Sac Fungus
SPECIES *echinata*
GENUS *Amphophialis* Dual Phial Sac Fungus
SPECIES *concinna*
GENUS *Amphoropycnium* Dual Dense Sac Fungus
SPECIES *paracatuense, philippinense*
GENUS *Amphorothecium* Dual Theca Sac Fungus
SPECIES *occultum*
GENUS *Ampliotrema* Ample Tremor Sac Fungus
SPECIES *auratum*
GENUS *Ampulliferina* Ampule Sac Fungus
SPECIES *lauri*
GENUS *Amylis* Amyl Sac Fungus
SPECIES *memorabilis*
GENUS *Amylocarpus* Amyl Carpel Sac Fungus
SPECIES *encephaloides*
GENUS *Anaphysmene* Bladderless Sac Fungus
SPECIES *cupressi, heraclei*
GENUS *Anarhyma* Anarhyme Sac Fungus
SPECIES *rhododendri*
GENUS *Anastomyces* Resurrection Sac Fungus
SPECIES *microsporus*
GENUS *Ancoraspora* Ankora Spore Sac Fungus
SPECIES *cubensis*
GENUS *Ancorasporella* Lesser Ankora Spore Sac Fungus
SPECIES *mexicana*
GENUS *Angiopomopsis* Angiopome Sac Fungus
SPECIES *lophostoma*
GENUS *Anguillospora* Serpent Spore Sac Fungus
SPECIES *aquatica, crassa, curvula, filiformis, flagellifera, furtiva, fustiformis, gigantea, lanceolata, longissima, pseudolongissima, rosea*
GENUS *Anguillosporella* Lesser Serpent Spore Sac Fungus
SPECIES *coryli, vermiformis*
GENUS *Ankistrocladium* Ancestor Branched Sac Fungus
SPECIES *fuscum*
GENUS *Annellodentimyces* Toothed Annello Sac Fungus
SPECIES *antennatuloideus*
GENUS *Annellodochium* Doch Annello Sac Fungus
SPECIES *ramulisporum*
GENUS *Annellolacinia* Lace Annello Sac Fungus

SPECIES *dinemasporioides, pandanicola*
GENUS *Annellophorella* Annello Sac Fungus
SPECIES *ellisii*
GENUS *Annellophragmia* Fragment Annello Sac Fungus
SPECIES *coonoorensis*
GENUS *Annellospermosporella* Seed Spore Annello Sac Fungus
SPECIES *meliolinae*
GENUS *Anthina* Flower Sac Fungus
SPECIES *flammea*
GENUS *Anthopsis* Lesser Flower Sac Fungus
SPECIES *catenata, deltoidea, microspora*
GENUS *Anungitea* Anungite Sac Fungus
SPECIES *brevicatenata, continua, fragilis, heterospora, rhabdospora, uniseptata*
GENUS *Anungitopsis* Lesser Anungite Sac Fungus
SPECIES *intermedia, speciosa, triseptata*
GENUS *Aphanocladium* Aphane Branched Sac Fungus
SPECIES *album, spectabile*
GENUS *Aphanofalx* Aphane Falc Sac Fungus
SPECIES *irregularis, mali*
GENUS *Aplosporella* Aplospore Sac Fungus
SPECIES *acaciae, celtidis, hedericola, leucaenicola, terricola, vivanii*
GENUS *Apodothina* Apodothine Sac Fungus
SPECIES *pringlei*
GENUS *Aporellula* Poreless Sac Fungus
SPECIES *erigerontis*
GENUS *Aposporella* Apospore Sac Fungus
SPECIES *elegans*
GENUS *Aquamarina* Aquamarine Sac Fungus
SPECIES *speciosa*
GENUS *Arachnomyces* Spider Sac Fungus
SPECIES *minimus, nitidus, sulphureus*
GENUS *Arachnophora* Lesser Spider Sac Fungus
SPECIES *fagicola*
GENUS *Arborillus* Arbor Sac Fungus
SPECIES *llimonae*
GENUS *Arcispora* Arc Spore Sac Fungus
SPECIES *bisagittaria*
GENUS *Arcuadendron* Arc Wood Sac Fungus
SPECIES *ovatum, triangulare*
GENUS *Argentinomyces* Argentine Sac Fungus
SPECIES *naviculisporus*
GENUS *Argopericonia* Argo Pericon Sac Fungus
SPECIES *elegans*
GENUS *Aristastoma* Awn Stoma Sac Fungus
SPECIES *oeconomicum*
GENUS *Arnaudina* Arnaud Sac Fungus
SPECIES *manaosensis*
GENUS *Arnoldiella* Arnold Sac Fungus
SPECIES *robusta*
GENUS *Aropsiclus* Aropsicle Sac Fungus
SPECIES *junci*
GENUS *Arthrobotryomyces* Lesser Straight Berry Sac Fungus
SPECIES *amazonensis*
GENUS *Arthrobotryum* Straight Berry Sac Fungus
SPECIES *stilboideum*
GENUS *Arthrocladium* Straight Branched Sac Fungus
SPECIES *caudatum*
GENUS *Arthrocristula* Straight Cristule Sac Fungus
SPECIES *hyphenata*
GENUS *Arthrographis* Straight Graphic Sac Fungus
SPECIES *alba, cuboidea, kalrae, lignicola, pinicola, slphurea*
GENUS *Arthropsis* Straight Sac Fungus
SPECIES *cirrhata, hispanica, microsperma, truncata*
GENUS *Arthrowallemia* Straight Wallem Sac Fungus
SPECIES *formosa*
GENUS *Articulophora* Articulate Sac Fungus
SPECIES *variispora*

GENUS *Arxiella* Arc Sac Fungus
SPECIES *lunata, terrestris*
GENUS *Asbolisiomyces* Bowl Sac Fungus
SPECIES *ingae*
GENUS *Ascomauritiana* Mauritania Sac Fungus
SPECIES *lignicola*
GENUS *Ascochyta* Chytrid Sac Fungus
SPECIES *abelmoschi, achlydis, acori, agrostidis, alismatis, allii-cepae, anisomera, anthoxanthi, aquilegiae, ari, armoraciae, avenae, bambusina, bohemica, boltshauseri, brachypodii, brassicae, bryoniae, calystegiae, capsici, caricae, carvi, caulina, cheiranthi, chenopodii, chenopodii, clematidina, corticola, cucumeris, cynosuricola, cypericola, dahliicola, daturae, deformis, desmazieri, dianthi, dipsaci, doronici, ducis-aprutii, ducometii, epilobii, equiseti, eriobotryae, fabae, ferulae, festucae, galeopsidis, glaucii, gossypii, gracilispora, graminea, graminicola, grossulariae, heraclei, hordei, hordeicola, humuli, hyalospora, impatientis, iridis, irpina, juglandis, kabatiana, kleinii, lamiorum, lathyri, lentis, leptospora, ligustri, lupini, menyanthis, mercurialis, metulispora, microspora, missouriensis, moelleriana, nigripycnidiicola, obiones, oleae, opuli, orobi, paspali, philadelphi, piricola, pisi, plantaginicola, plantaginis, polemonii, primulae, psammae, pteridis, pyrethri, quadriguttulata, rhodesii, scotinospora, senecionicola, senecionis, sodalis, sojicola, solanicola, sonchi, sorghi, sorosii, stellariae, straminea, subalpina, symphoricarpi, syringae, tarda, tenerrima, teretiuscula, trifolii, ulicis, viciae, vincae, violae, vulgaris*
VARIETIES *avenae confusa, hordei hordei, hordei americana, hordei europaea, leptospora acuta, leptospora major, leptospora minor, leptospora septorioides, mercurialis autumnalis, subalpina penniseti, ulicis caledoniensis, viciae viciae, viciae pannonicae, viciae villosae, vulgaris symphoricarpi*
GENUS *Ascochytula* Least Chytrid Sac Fungus
SPECIES *atriplicis*
GENUS *Ascochytulina* Lesser Chytrid Sac Fungus
SPECIES *deflectens, pini-acicola, smilacina*
GENUS *Ascocodinaea* Codin Sac Fungus
SPECIES *polyporicola, stereicola*
GENUS *Ascocorticiellum* Corticle Sac Fungus
SPECIES *vermisporum*
GENUS *Ascominuta* Minute Sac Fungus
SPECIES *lignicola*
GENUS *Ascorhiza* Rhizid Sac Fungus
SPECIES *leguminosarum*
GENUS *Ascosubramania* Subraman Sac Fungus
SPECIES *melanographoides*
GENUS *Asperisporium* Asp Spore Sac Fungus
SPECIES *caricae, mikaniae, mikaniigena, sequoiae*
GENUS *Astelechia* Star Lichen Sac Fungus
SPECIES *longissima, radians*
GENUS *Asteroconium* Star Cone Sac Fungus
SPECIES *nothopegiae, saccardoi*
GENUS *Asteromyces* Star Sac Fungus
SPECIES *cruciatus*
GENUS *Asterosporium* Star Spore Sac Fungus
SPECIES *asterospermum, betulinum, hoffmannii, strobilorum*
GENUS *Atrosetaphiale* Dark Bristle Phial Sac Fungus
SPECIES *flagelliformis*
GENUS *Aulospora* Aulospore Sac Fungus
SPECIES *epimyces*
GENUS *Aurantiosacculus* Orange Sac Fungus
SPECIES *eucalypti*
GENUS *Avesicladiella* Bird Branched Sac Fungus
SPECIES *britannica, microsperma*
GENUS *Avettaea* Avetta Sac Fungus
SPECIES *alcornii, philippinensis, salvadorae*
GENUS *Azbukinia* Azbukin Sac Fungus
SPECIES *ferruginea*
GENUS *Azureothecium* Azure Theca Sac Fungus
SPECIES *australiense*
GENUS *Bachmanniomyces* Bachmann Sac Fungus
SPECIES *uncialicola*
GENUS *Bacillispora* Rod Spore Sac Fungus
SPECIES *aquatica*
GENUS *Bactridium* Bacteria Sac Fungus
SPECIES *acutum, flavum*
GENUS *Bactrodesmiella* Bacteria Fiber Sac Fungus

SPECIES *masonii*
GENUS *Baculospora* Bacule Spore Sac Fungus
SPECIES *pellucida*
GENUS *Badarisama* Badar Sac Fungus
SPECIES *sojae*
GENUS *Bahuchashaka* Bahuchashaka Sac Fungus
SPECIES *graminis*
GENUS *Bahukalasa* Bahukal Sac Fungus
SPECIES *samala*
GENUS *Bahusutrabeeja* Bahusutrabeej Sac Fungus
SPECIES *dwaya*
GENUS *Balaniopsis* Lesser Balan Sac Fungus
SPECIES *africana*
GENUS *Balanium* Balan Sac Fungus
SPECIES *stygium*
GENUS *Bartalinia* Bartalin Sac Fungus
SPECIES *bischofiae, pistacina, robillardoides, terricola*
GENUS *Basramyces* Basra Sac Fungus
SPECIES *marinus*
GENUS *Beauveriphora* Beauver Sac Fungus
SPECIES *sporothricis*
GENUS *Beccopycnidium* Becco Dense Sac Fungus
SPECIES *palmicola*
GENUS *Beejadwaya* Beejadway Sac Fungus
SPECIES *bispora*
GENUS *Belemnospora* Belem Spore Sac Fungus
SPECIES *epiphylla, pinicola, triseptata, verruculosa*
GENUS *Bellulicauda* Bell Tail Sac Fungus
SPECIES *dialii*
GENUS *Beltrania* Beltran Sac Fungus
SPECIES *africana, querna, rhombica*
GENUS *Beltraniopsis* Lesser Beltran Sac Fungus
SPECIES *esenbeckiae, tanzaniensis*
GENUS *Beniowskia* Beniowsky Sac Fungus
SPECIES *sphaeroidea*
GENUS *Benjpalia* Benjpal Sac Fungus
SPECIES *sundara*
GENUS *Beverwykella* Beverwyke Sac Fungus
SPECIES *pulmonaria*
GENUS *Bibanasiella* Biban Sac Fungus
SPECIES *ramispora*
GENUS *Biflagellosporella* Biflagellate Spore Sac Fungus
SPECIES *amazonensis*
GENUS *Biflua* Biflu Sac Fungus
SPECIES *physasca*
GENUS *Bilgramia* Bilgram Sac Fungus
SPECIES *indica*
GENUS *Bispora* Bispore Sac Fungus
SPECIES *antennata, betulina, christiansenii, effusa, fusca, novae-zelandiae*
GENUS *Bisporostilbella* Stilbe Bispore Sac Fungus
SPECIES *fusca*
GENUS *Bisseomyces* Bisse Sac Fungus
SPECIES *havanensis*
GENUS *Bitunicostilbe* Bitunic Stilbe Sac Fungus
SPECIES *clavata, linderae*
GENUS *Bizzozeriella* Bizzozer Sac Fungus
SPECIES *phyllogena*
GENUS *Blarneya* Blarney Sac Fungus
SPECIES *hibernica*
GENUS *Blasdalea* Blasdale Sac Fungus
SPECIES *disciformis*
GENUS *Blastacervulus* Cervule Blastid Sac Fungus
SPECIES *eucalypti*
GENUS *Blastobotrys* Berry Blastid Sac Fungus
SPECIES *arbuscula, aristata, capitulata, elegans, gigas, nivea, proliferans*
GENUS *Blastocatena* Chain Blastid Sac Fungus
SPECIES *pulneyensis*

GENUS *Blastoconium* Cone Blastid Sac Fungus
SPECIES *glumarum, tropicum*
GENUS *Blastodictys* Net Blastid Sac Fungus
SPECIES hibisci
GENUS *Blastofusarioides* Fused Blastid Sac Fungus
SPECIES *fusca*
GENUS *Blastophorella* Lesser Blastid Sac Fungus
SPECIES *smithii*
GENUS *Blastophorum* Blastid Sac Fungus
SPECIES *fusarioides, pini, truncatum, uniseptatum*
GENUS *Bondiella* Bond Sac Fungus
SPECIES *palmicola*
GENUS *Bonordeniella* Bonorden Sac Fungus
SPECIES *aspera*
GENUS *Bostrichonema* Bostrich Thread Sac Fungus
SPECIES *alpestre*
GENUS *Botryoderma* Dermal Berry Sac Fungus
SPECIES *lateritium, rostratum*
GENUS *Botryodiplodia* Diploid Berry Sac Fungus
SPECIES *acicola, circinans, excelsa, fraxini, gallae, hypodermia, juniperina, manihoticola, manihotis, palmarum, penzigii, pyrenophora, ricinicola, saccharina, ulmicola*
GENUS *Botryomonilia* Monad Berry Sac Fungus
SPECIES *scheeleae*
GENUS *Botryosporium* Spore Berry Sac Fungus
SPECIES *diffusum, longibrachiatum, pulchrum, pyramidale*
GENUS *Brachiosphaera* Sphere Short Sac Fungus
SPECIES *jamaicensis, tropicalis*
GENUS *Brachyconidiella* Cone Short Sac Fungus
SPECIES *monilispora*
GENUS *Brachydesmiella* Fiber Short Sac Fungus
SPECIES *anthostomelloidea, biseptata, caudata, orientalis, verrucosa*
GENUS *Brachysporiellina* Lesser Spore Short Sac Fungus
SPECIES *pulneyensis*
GENUS *Brachysporium* Spore Short Sac Fungus
SPECIES *bloxamii, breve, britannicum, dingleyae, graminis, masonii, nigrum, obovatum, oosporum*
GENUS *Briania* Brian Sac Fungus
SPECIES *fruticetum*
GENUS *Briosia* Briose Sac Fungus
SPECIES *ampelophaga*
GENUS *Brycekendrickia* Brycekendrick Sac Fungus
SPECIES *indica*
GENUS *Bulbilopycnis* Dense Bulb Sac Fungus
SPECIES *caffra*
GENUS *Bulbocatenospora* Chain Spore Bulb Sac Fungus
SPECIES *complanata*
GENUS *Bullaserpens* Bullet Serpent Sac Fungus
SPECIES *bignoniacearum*
GENUS *Byssotheciella* Bysse Theca Sac Fungus
SPECIES *eucalyptina, tiliae*
GENUS *Cacumisporium* Cacum Spore Sac Fungus
SPECIES *capitulatum, spooneri*
GENUS *Cadophora* Cadophore Sac Fungus
SPECIES *malorum*
GENUS *Calcarispora* Calcic Spore Sac Fungus
SPECIES *hiemalis*
GENUS *Calcarisporiella* Least Calcic Spore Sac Fungus
SPECIES *thermophila*
GENUS *Calcarisporium* Lesser Calcic Spore Sac Fungus
SPECIES *arbuscula, ovalisporum*
GENUS *Calceispora* Greater Calcic Spore Sac Fungus
SPECIES *hachijoensis, subglobosa*
GENUS *Callistospora* Callisto Spore Sac Fungus
SPECIES *gaubae*
GENUS *Calosphaeriopsis* Callisto Sphere Sac Fungus
SPECIES *huberiana*
GENUS *Calycidium* Calyce Sac Fungus
SPECIES *cuneatum, polycarpum*

GENUS *Calyptrozyma* Calypter Zyme Sac Fungus
SPECIES *arxii*
GENUS *Camarographium* Camaro Graphic Sac Fungus
SPECIES *stephensii*
GENUS *Camaropycnis* Camaro Dense Sac Fungus
SPECIES *libocedri*
GENUS *Camposporium* Field Spore Sac Fungus
SPECIES *antennatum, cambrense, hyalinum, japonicum, marylandicum, pellucidum*
GENUS *Camptomeris* Field Sea Sac Fungus
SPECIES *leucaenae*
GENUS *Campylospora* Lesser Field Spore Sac Fungus
SPECIES *chaetocladia, filicladia, parvula*
GENUS *Cancellidium* Cancellate Sac Fungus
SPECIES *applanatum*
GENUS *Candelabrum* Candelabra Sac Fungus
SPECIES *brocchiatum, japonense, spinulosum*
GENUS *Candelosynnema* Candle Sac Fungus
SPECIES *ranunculosporum*
GENUS *Capitorostrum* Cap Beak Sac Fungus
SPECIES *asteridiellae, cocoës*
GENUS *Capnobotryella* Cap Berry Sac Fungus
SPECIES *renispora, sessilispora*
GENUS *Capnocheirides* Cap Cheir Sac Fungus
SPECIES *rhododendri*
GENUS *Capsicumyces* Capsicum Sac Fungus
SPECIES *delicatus*
GENUS *Carmichaelia* Carmichael Sac Fungus
SPECIES *lignicola*
GENUS *Carnia* Carn Sac Fungus
SPECIES *tabebuiae*
GENUS *Carrismyces* Carris Sac Fungus
SPECIES *proliferatus*
GENUS *Casaresia* Casares Sac Fungus
SPECIES *sphagnorum*
GENUS *Catenophora* Chain Sac Fungus
SPECIES *garryae, pruni, yuccae*
GENUS *Catenophoropsis* Lesser Chain Sac Fungus
SPECIES *eucalypticola*
GENUS *Catenospegazzinia* Spegazzini Chain Sac Fungus
SPECIES *elegans, pulchra*
GENUS *Catenosubulispora* Ulispore Chain Sac Fungus
SPECIES *honiaraensis*
GENUS *Cenococcum* Ceno Round Sac Fungus
SPECIES *geophilum, graniforme*
GENUS *Centrospora* Central Spore Sac Fungus
SPECIES *clavata*
GENUS *Cephaliophora* Head Sac Fungus
SPECIES *irregularis, longispora, muscicola, navicularis, tropica, uniformis*
GENUS *Cephalotrichum* Hairy Head Sac Fungus
SPECIES *stemonitis*
GENUS *Ceratopycnidium* Horn Dense Sac Fungus
SPECIES *baccharidicola, citricola*
GENUS *Ceratosporella* Horn Spore Sac Fungus
SPECIES *compacta, deviata, novae-zelandiae, stipitata*
GENUS *Cercosporula* Cerco Spore Sac Fungus
SPECIES *corticola*
GENUS *Cerebella* Cerebellum Sac Fungus
SPECIES *andropogonis*
GENUS *Cesatiella* Cesat Sac Fungus
SPECIES *lancastriensis*
GENUS *Chaetantromycopsis* Antro Bristle Sac Fungus
SPECIES *bambusae*
GENUS *Chaetasbolisia* Bowl Bristle Sac Fungus
SPECIES *falcata*
GENUS *Chaetendophragmia* Fragment Bristle Sac Fungus
SPECIES *britannica*
GENUS *Chaetoblastophorum* Blastid Bristle Sac Fungus

SPECIES *ingramii*
GENUS *Chaetodiplodina* Diploid Bristle Sac Fungus
SPECIES *cubensis, graminicola*
GENUS *Chaetomastia* Mastigote Bristle Sac Fungus
SPECIES *canescens, hispidula*
GENUS *Chaetomella* Mellow Bristle Sac Fungus
SPECIES *acutiseta, cinnamomea, circinoseta, horrida, oblonga, raphigera*
GENUS *Chaetonectrioides* Nectar Bristle Sac Fungus
SPECIES *malaysiana*
GENUS *Chaetopeltaster* Pelt Star Bristle Sac Fungus
SPECIES *japonicus*
GENUS *Chaetophoma* Foam Bristle Sac Fungus
SPECIES *foeda, quercifolia*
GENUS *Chaetopsis* Bristle Sac Fungus
SPECIES *canovae, grisea*
GENUS *Chaetopyrena* Fire Bristle Sac Fungus
SPECIES *hederae-helicis, hesperidium, penicillata*
GENUS *Chaetospermopsis* Lesser Spore Bristle Sac Fungus
SPECIES *boninensis*
GENUS *Chaetospermum* Spore Bristle Sac Fungus
SPECIES *carneum, chaetosporum*
GENUS *Chaetosphaeronema* Sphere Bristle Sac Fungus
SPECIES *hispidulum, larense*
GENUS *Chaetospora* Lesser Spore Bristle Sac Fungus
SPECIES *quezelii*
GENUS *Chalara* Chalar Sac Fungus
SPECIES *affinis, africana, angustata, aotearoa, aspera, aurea, austriaca, brevipes, brevispora, bulbosa, constricta, crassipes, cylindrica, cylindrosperma, dennisii, ellisii, fungorum, fusidioides, hughesii, hyalina, inflatipes, insignis, kendrickii, laevis, longipes, microchona, microspora, neocaledoniae, novae-zelandiae, ovoidea, parvispora, phaeospora, populi, pteridina, ramosa, rhynchophialis, sessilis, setosa, spiralis, tubifera, ungeri, urceolata, vaccinii*
GENUS *Chalarodendron* Wood Chalar Sac Fungus
SPECIES *fuscum*
GENUS *Chalarodes* Lesser Chalar Sac Fungus
SPECIES *bisetis, obconica*
GENUS *Characonidia* Cone Charming Sac Fungus
SPECIES *ingae*
GENUS *Charomyces* Charming Sac Fungus
SPECIES *amphimelas, ramosus*
GENUS *Chaunopycnis* Chaun Dense Sac Fungus
SPECIES *alba, ovalispora*
GENUS *Cheilaria* Claw Sac Fungus
SPECIES *agrostidis*
GENUS *Cheiromoniliophora* Monad Cheir Sac Fungus
SPECIES *elegans*
GENUS *Cheiromyceopsis* Cheir Sac Fungus
SPECIES *echinulata*
GENUS *Cheirospora* Spore Cheir Sac Fungus
SPECIES *botryospora*
GENUS *Chiastospora* Spore Chiast Sac Fungus
SPECIES *parasitica*
GENUS *Chikaneea* Chikanee Sac Fungus
SPECIES *holleroniae*
GENUS *Chithramia* Chither Branched Sac Fungus
SPECIES *elegantissima*
GENUS *Chlamydomyces* Cloak Sac Fungus
SPECIES *palmarum, sympodialis*
GENUS *Chlamydopsis* Lesser Cloak Sac Fungus
SPECIES *proliferans*
GENUS *Chlamydorubra* Red Cloak Sac Fungus
SPECIES *verrucosa*
GENUS *Chmelia* Chmel Sac Fungus
SPECIES *slovaca*
GENUS *Choanatiara* Choan Sac Fungus
SPECIES *gracilis, lunata*
GENUS *Choreospora* Chore Spore Sac Fungus
SPECIES *tibellii*

GENUS *Chromosporium* Chromic Spore Sac Fungus
SPECIES *lateritium*
GENUS *Chrysalidopsis* Lesser Golden Sac Fungus
SPECIES *trevoae*
GENUS *Chryseidea* Golden Sac Fungus
SPECIES *africana*
GENUS *Chuppia* Chupp Sac Fungus
SPECIES *sarcinifera*
GENUS *Ciferriopeltis* Ciferri Pelt Sac Fungus
SPECIES *domingensis*
GENUS *Ciliochora* Ciliate Chore Sac Fungus
SPECIES *calabrica*
GENUS *Ciliochorella* Lesser Ciliate Chore Sac Fungus
SPECIES *mangiferae*
GENUS *Ciliospora* Ciliate Spore Sac Fungus
SPECIES *albida*
GENUS *Ciliosporella* Lesser Ciliate Spore Sac Fungus
SPECIES *selenospora, tuberculiformis*
GENUS *Circinoconis* Cone Circin Sac Fungus
SPECIES *paradoxa*
GENUS *Circinotrichum* Hairy Circin Sac Fungus
SPECIES *falcatisporum, maculiforme, olivaceum, papakurae, ponmudiense*
GENUS *Cirrenalia* Coiled Sac Fungus
SPECIES *basiminuta, donnae, lignicola, macrocephala*
GENUS *Cirrosporium* Tendril Spore Sac Fungus
SPECIES *novae-zelandiae*
GENUS *Civisubramaniania* Civi Sac Fungus
SPECIES *eucalypti*
GENUS *Cladochasiella* Chase Branched Sac Fungus
SPECIES *divergens*
GENUS *Cladoconidium* Cone Branched Sac Fungus
SPECIES *articulatum*
GENUS *Cladosterigma* Sterigm Branched Sac Fungus
SPECIES *fusisporum*
GENUS *Clathrosporium* Lattice Spore Sac Fungus
SPECIES *compactum*
GENUS *Clauzadeomyces* Clauzad Sac Fungus
SPECIES *verrucosus*
GENUS *Clavariana* Club Sac Fungus
SPECIES *aquatica*
GENUS *Clavatospora* Club Spore Sac Fungus
SPECIES *filiformis, longibrachiata*
GENUS *Cleistophoma* Closed Foam Sac Fungus
SPECIES *dryina*
GENUS *Cleistothelebolus* Closed Thele Bowl Sac Fungus
SPECIES *nipigonensis*
GENUS *Clohesyomyces* Clohesy Sac Fungus
SPECIES *aquaticus*
GENUS *Clypeopycnis* Dense Clype Sac Fungus
SPECIES *aeruginascens, lepidospermatis, punithalingamii*
GENUS *Clypeoseptoria* Septor Clype Sac Fungus
SPECIES *intricata, rockii, sparothospermi*
GENUS *Clypeostagonospora* Stag Spore Clype Sac Fungus
SPECIES *muehlenbergiae*
GENUS *Clypeostroma* Stroma Clype Sac Fungus
SPECIES *hemisphaericum*
GENUS *Codonmyces* Codon Sac Fungus
SPECIES *lecanorae*
GENUS *Colemaniella* Coleman Sac Fungus
SPECIES *ossoorii*
GENUS *Coleophoma* Cole Foam Sac Fungus
SPECIES *crateriformis, cylindrospora, empetri, fusiformis, maculans, maculiformis, oleae, rhododendri*
GENUS *Collematospora* Cole Spore Sac Fungus
SPECIES *venezuelensis*
GENUS *Colletoconis* Collet Cone Sac Fungus
SPECIES *aecidiophila*
GENUS *Comatospora* Tuft Spore Sac Fungus

SPECIES *suttonii*
GENUS *Complexipes* Complex Sac Fungus
SPECIES *moniliformis, tuberosus*
GENUS *Condylospora* Condylospore Sac Fungus
SPECIES *flexuosa, gigantea, spumigena*
GENUS *Coniambigua* Ambiguous Cone Sac Fungus
SPECIES *phaeographidis*
GENUS *Conioscypha* Scyph Cone Sac Fungus
SPECIES *hoehnelii, japonica, lignicola, varia*
GENUS *Conioscyphopsis* Lesser Scyph Cone Sac Fungus
SPECIES *australiensis*
GENUS *Coniosporium* Spore Cone Sac Fungus
SPECIES *ilicinum*
GENUS *Coniothecium* Theca Cone Sac Fungus
SPECIES *chomatosporum, complanatum, conglutinatum, effusum, silaceum*
GENUS *Coniothyrina* Thyre Cone Sac Fungus
SPECIES *agaves*
GENUS *Conjunctospora* Conjunctive Spore Sac Fungus
SPECIES *vesiculifera*
GENUS *Conostoma* Stoma Cone Sac Fungus
SPECIES *oenocarpi*
GENUS *Consetiella* Bristle Cone Sac Fungus
SPECIES *solida*
GENUS *Cooksonomyces* Cookson Sac Fungus
SPECIES *banksiae*
GENUS *Coprotiella* Copper Sac Fungus
SPECIES *gongylospora*
GENUS *Cordana* Cord Sac Fungus
SPECIES *boothii, crassa, ellipsoidea, inaequalis, musae, pauciseptata, solitaria*
GENUS *Coremiella* Corem Sac Fungus
SPECIES *cubispora*
GENUS *Cornucopiella* Cornucopia Sac Fungus
SPECIES *fusispora, mirabilis*
GENUS *Cornutispora* Spore Horn Sac Fungus
SPECIES *ciliata, lichenicola, limaciformis, triangularis*
GENUS *Cornutostilbe* Stilbe Horn Sac Fungus
SPECIES *medusa*
GENUS *Coryneliella* Club Sac Fungus
SPECIES *consimilis*
GENUS *Corynesporina* Lesser Club Spore Sac Fungus
SPECIES *elegans*
GENUS *Corynesporopsis* Club Spore Sac Fungus
SPECIES *antillana, biseptata, cylindrica, inaequiseptata, indica, isabelicae, quercicola, rionensis, uniseptata*
GENUS *Craneomyces* Crane Sac Fungus
SPECIES *verrucosus*
GENUS *Cremasteria* Cremaster Sac Fungus
SPECIES *cymatilis*
GENUS *Crinigera* Lily Sac Fungus
SPECIES *maritima*
GENUS *Crinitospora* Lily Spore Sac Fungus
SPECIES *pulchra*
GENUS *Crucellisporium* Crucell Spore Sac Fungus
SPECIES *africanum, selaginellae*
GENUS *Crustodiplodina* Crust Diploid Sac Fungus
SPECIES *lichenoides*
GENUS *Cryptoascus* Cryptic Sac Fungus
SPECIES *graminis, oligosporus*
GENUS *Cryptocoryneopsis* Lesser Cryptic Club Sac Fungus
SPECIES *umbraculiformis*
GENUS *Cryptocoryneum* Cryptic Club Sac Fungus
SPECIES *bigenimum, condensatum, hysterioides, rilstonii*
GENUS *Cryptomela* Cryptic Mellow Sac Fungus
SPECIES *acutispora, typhae*
GENUS *Cryptomycina* Least Cryptic Sac Fungus
SPECIES *pteridis*
GENUS *Cryptophiale* Cryptic Phial Sac Fungus
SPECIES *kakombensis, manifesta, minor, udagawae*

GENUS *Cryptosporium* Cryptic Spore Sac Fungus
SPECIES *betulinum, caricis, ceuthosporoides, crypticum, minimum, neesii, tami*
GENUS *Cryptostroma* Cryptic Stoma Sac Fungus
SPECIES *corticale*
GENUS *Cryptovalsa* Cryptic Vals Sac Fungus
SPECIES *elevata, protracta, suaedicola*
GENUS *Cryptumbellata* Cryptic Umbel Sac Fungus
SPECIES *terricola*
GENUS *Cubasina* Cuba Sac Fungus Culcitaln
SPECIES *albofusca*
GENUS *Culcitalna* Culcitaln Sac Fungus
SPECIES *achraspora*
GENUS *Culicidospora* Culic Spore Sac Fungus
SPECIES *aquatica, gravida*
GENUS *Culicinomyces* Culicin Sac Fungus
SPECIES *bisporalis, clavisporus*
GENUS *Cumulospora* Cumulo Spore Sac Fungus
SPECIES *marina*
GENUS *Curvicladium* Curved Branch Sac Fungus
SPECIES *cigneum*
GENUS *Curvidigitus* Curved Finger Sac Fungus
SPECIES *daphniphylli*
GENUS *Curvulariopsis* Curved Sac Fungus
SPECIES *cymbisperma*
GENUS *Cyanopyrenia* Cyan Fire Sac Fungus
SPECIES *japonica*
GENUS *Cyclomarsonina* Cyclic Marson Sac Fungus
SPECIES *cedrelae*
GENUS *Cylindrocarpostylus* Cylindric Carpel Sac Fungus
SPECIES *gregarius*
GENUS *Cylindrodendrum* Cylindric Wood Sac Fungus
SPECIES *album*
GENUS *Cylindrosympodium* Cylindric Sympodial Sac Fungus
SPECIES *variabile*
GENUS *Cyphellophora* Cypher Sac Fungus
SPECIES *laciniata, pluriseptata, vermispora*
GENUS *Cytoplacosphaeria* Plaque Sphere Cyto Sac Fungus
SPECIES *rimosa*
GENUS *Cytosphaera* Sphere Cyto Sac Fungus
SPECIES *mangiferae*
GENUS *Cytosporella* Least Spore Cyto Sac Fungus
SPECIES *chamaeropis*
GENUS *Cytosporina* Spore Cyto Sac Fungus
SPECIES *citriperda, crataegi, ludibunda*
GENUS *Cytosporium* Lesser Spore Cyto Sac Fungus
SPECIES *pittospori*
GENUS *Cytotriplospora* Triple Spore Cyto Sac Fungus
SPECIES *pini*
GENUS *Dactuliochaeta* Finger Bristle Sac Fungus
SPECIES *glycines*
GENUS *Dactuliophora* Lesser Finger Sac Fungus
SPECIES *tarrii*
GENUS *Dactylaria* Finger Sac Fungus
SPECIES *acicularis, affinis, biseptata, candidula, chrysosperma, clavata, congregata, dasguptae, dimorpha, dimorphospora, endophytica, flagellata, fusiformis, gallopava, higginsii, humicola, inaequilatera, irregularis, junci, lanosa, lepida, lunata, mitrata, monticola, obtriangularia, parvispora, purpurella, sahelensis, scolecospora, sporexamorpha, zapatensis*
GENUS *Dactylifera* Least Finger Sac Fungus
SPECIES *australiensis*
GENUS *Dactylosporium* Finger Spore Sac Fungus
SPECIES *macropus*
GENUS *Darwiniella* Darwin Sac Fungus
SPECIES *gracilis*
GENUS *Davisoniella* Davison Sac Fungus
SPECIES *eucalypti*
GENUS *Dawsomyces* Lesser Daws Sac Fungus
SPECIES *mirabilis, subinvisibilis*

GENUS *Dawsophila* Daws Sac Fungus
SPECIES *callichroma, polycarpa, pygmaea*
GENUS *Debaryella* Debary Sac Fungus
SPECIES *gracilis, hyalina*
GENUS *Deichmannia* Deichmann Sac Fungus
SPECIES *verrucispora*
GENUS *Deightoniella* Deighton Sac Fungus
SPECIES *arundinacea, torulosa*
GENUS *Delortia* Delort Sac Fungus
SPECIES *palmicola*
GENUS *Dematium* Demat Sac Fungus
SPECIES *echinobotryum, nigrum*
GENUS *Dendrodochium* Wood Chi Sac Fungus
SPECIES *citrinum, epistomum, hymenuloides, pinastri*
GENUS *Dendrodomus* Wood Dome Sac Fungus
SPECIES *annulatus*
GENUS *Dendrophoma* Wood Foam Sac Fungus
SPECIES *caespitosa, eumorpha, faginea, juglandina, mahoniae, pleurospora, poarum*
GENUS *Dendrospora* Wood Spore Sac Fungus
SPECIES *erecta, fastuosa, fusca, juncicola, microerecta, moniliformis, nana, polymorpha, tenella, torulosa*
GENUS *Dendrosporium* Lesser Wood Spore Sac Fungus
SPECIES *lobatum*
GENUS *Denticularia* Toothed Sac Fungus
SPECIES *mangiferae*
GENUS *Dentocircinomyces* Toothed Circin Sac Fungus
SPECIES *occidentalis*
GENUS *Derexia* (Synonym for *Neozygites lecanii*)
GENUS *Descalsia* Descals Sac Fungus
SPECIES *cruciata*
GENUS *Desertella* Desert Sac Fungus
SPECIES *globulifera*
GENUS *Dexhowardia* Dex Howard Sac Fungus
SPECIES *tetraspora*
GENUS *Diademospora* Diad Spore Sac Fungus
SPECIES *ramigera*
GENUS *Dicholobodigitus* Dual Lobe Finger Sac Fungus
SPECIES *niger*
GENUS *Dichotomophthora* Dichotomy Phthoral Sac Fungus
SPECIES *lutea, portulacae*
GENUS *Dichotomophthoropsis* Lesser Dichotomy Phthoral Sac Fungus
SPECIES *nymphaearum, safeeulaensis*
GENUS *Dicoccum* Dual Round Sac Fungus
SPECIES *asperum, fimicola*
GENUS *Dictyoarthrinium* Straight Net Sac Fungus
SPECIES *rabaulense, sacchari*
GENUS *Dictyocatenulata* Chain Net Sac Fungus
SPECIES *alba*
GENUS *Dictyochaetopsis* Bristle Net Sac Fungus
SPECIES *antillana, apicalis, dingleyae, elegantissima, filamentosa, glauconigra, gonytrichoides, hamata, intermedia, maharashtrensis, menisporoides, pahangensis*
GENUS *Dictyochorina* Chore Net Sac Fungus
SPECIES *arundinellae, portoricensis*
GENUS *Dictyopolyschema* Polyscheme Net Sac Fungus
SPECIES *pirozynskii*
GENUS *Dictyorostrella* Beak Net Sac Fungus
SPECIES *caricicola*
GENUS *Dictyospiropes* Spire Net Sac Fungus
SPECIES *ziziphi*
GENUS *Dictyosporium* Spore Net Sac Fungus
SPECIES *alatum, bulbosum, digitatum, elegans, foliicola, heptasporum, pelagicum, toruloides*
GENUS *Didymochaetina* Dual Bristle Sac Fungus
SPECIES *galactiae*
GENUS *Didymopsis* Dual Sac Fungus
SPECIES *helvellae*
GENUS *Didymosporium* Dual Spore Sac Fungus
SPECIES *striola*
GENUS *Diederichia* Diederich Sac Fungus

SPECIES *pseudeverniae*
GENUS *Diehliomyces* Diehl Sac Fungus
SPECIES *microsporus*
GENUS *Digitodesmium* Finger Fiber Sac Fungus
SPECIES *elegans*
GENUS *Digitodochium* Finger Doch Sac Fungus
SPECIES *rhodoleucum*
GENUS *Digitoramispora* Finger Branch Spore Sac Fungus
SPECIES *caribensis, excentrica*
GENUS *Digitosarcinella* Finger Sarcin Sac Fungus
SPECIES *caseariae*
GENUS *Dimastigosporium* Dual Mastigote Spore Sac Fungus
SPECIES *musimonum*
GENUS *Dinemasporium* Dual Thread Spore Sac Fungus
SPECIES *cytosporoides, decipiens, fimeti*
GENUS *Diplocladiella* Dual Branched Sac Fungus
SPECIES *scalaroides*
GENUS *Diplodiella* Diploid Sac Fungus
SPECIES *brassicae, fibricola*
GENUS *Diplolaeviopsis* Diploid Smooth Sac Fungus
SPECIES *ranula*
GENUS *Diplosporonema* Diploid Spore Thread Sac Fungus
SPECIES *delastrei*
GENUS *Discella* Disc Sac Fungus
SPECIES *abnormis, castanea, platani, platyspora*
GENUS *Dischloridium* Chloric Disc Sac Fungus
SPECIES *keniense, laeense, microsporum, naromoruense*
GENUS *Discocolla* Collar Disc Sac Fungus
SPECIES *nectrioides, pyrina*
GENUS *Discogloeum* Glow Disc Sac Fungus
SPECIES *veronicae*
GENUS *Discomycetoidea* Mycote Disc Sac Fungus
SPECIES *aequatorialis*
GENUS *Discosia* Osian Disc Sac Fungus
SPECIES *aquatica, artocreas, julia, strobilina, vagans*
GENUS *Discosiellina* Lesser Osian Disc Sac Fungus
SPECIES *longiciliata*
GENUS *Discosphaerina* Sphere Disc Sac Fungus
SPECIES *cytisi, fagi, fulvida, miribelii*
GENUS *Discosporina* Lesser Spore Disc Sac Fungus
SPECIES *carpinicola, commutata, deplanata, sulphurea*
GENUS *Discosporium* Spore Disc Sac Fungus
SPECIES *acaciae, eucalypti, eugeniae, tremuloides*
GENUS *Dissitimurus* Timur Diss Sac Fungus
SPECIES *exedrus*
GENUS *Dissoacremoniella* Acremon Diss Sac Fungus
SPECIES *silvatica*
GENUS *Ditangifibula* Dual Angle Fiber Sac Fungus
SPECIES *dikaryotae*
GENUS *Dolichousnea* Dolich Sac Fungus
SPECIES *trichodeoides*
GENUS *Domingoella* Domingo Sac Fungus
SPECIES *arctostaphyli*
GENUS *Dontuzia* Dontuz Sac Fungus
SPECIES *psychotriae*
GENUS *Dothideodiplodia* Doth Diploid Sac Fungus
SPECIES *agropyri*
GENUS *Dothioropsis* Doth Sac Fungus
SPECIES *corni*
GENUS *Drepanospora* Drepan Spore Sac Fungus
SPECIES *viridis*
GENUS *Dryosphaera* Dryad Sphere Sac Fungus
SPECIES *navigans, tropicalis*
GENUS *Dualomyces* Dual Sac Fungus
SPECIES *bambusicola, taiwanensis*
GENUS *Duosporium* Dual Spore Sac Fungus
SPECIES *cyperi*

GENUS *Dwayabeeja* Dwayabeej Sac Fungus
SPECIES *aethiopica, cubensis, graminis, sundara*
GENUS *Dwayalomella* Dwayalom Sac Fungus
SPECIES *setulosa, vaccinii*
GENUS *Dwibahubeeja* Dwibahubeej Sac Fungus
SPECIES *indica*
GENUS *Dwiroopella* Dwiro Sac Fungus
SPECIES sundara
GENUS *Dyrithium* Dyrith Sac Fungus
SPECIES *lividum*
GENUS *Echinocatena* Urchin Chain Sac Fungus
SPECIES *arthrinioides*
GENUS *Echinochondrium* Uchin Cartilage Sac Fungus
SPECIES *pulchrum*
GENUS *Ectostroma* Ectostrome Sac Fungus
SPECIES *iridis, liriodendri*
GENUS *Eiona* Eion Sac Fungus
SPECIES *tunicata*
GENUS *Elachopeltella* Elach Pelt Sac Fungus
SPECIES *rubescens, sapindacearum*
GENUS *Eladia* Elad Sac Fungus
SPECIES *saccula, sparsa, striatispora*
GENUS *Elegantimyces* Elegant Sac Fungus
SPECIES *sporidesmiopsis*
GENUS *Eleutheromyces* Eleutheron Sac Fungus
SPECIES *subulatus*
GENUS *Enantioptera* Enant Wing Sac Fungus
SPECIES *tetra-alata*
GENUS *Endocalyx* Calyx Endic Sac Fungus
SPECIES *cinctus, indumentum, melanoxanthus*
VARIETIES *melanoxanthus melanoxanthus, melanoxanthus grossus*
GENUS *Endoconospora* Cone Spore Endic Sac Fungus
SPECIES *cerastii, indica*
GENUS *Endocoryneum* Club Endic Sac Fungus
SPECIES *loculosum, quercus*
GENUS *Endomelanconium* Dark Cone Endic Sac Fungus
SPECIES *microsporum, nanum, phoenicicola, pini*
GENUS *Endophragmia* Fragment Endic Sac Fungus
SPECIES *dimorphospora, taxi*
GENUS *Endoxylina* Xyline Endic Sac Fungus
SPECIES *pini*
GENUS *Engelhardtiella* Engelhardt Sac Fungus
SPECIES *alba*
GENUS *Engyodontium* Engy Tooth Sac Fungus
SPECIES *album, aranearum, geniculatum, parvisporum, rectidentatum*
GENUS *Enridescalsia* Enry Descals Sac Fungus
SPECIES *speciosa*
GENUS *Entoderma* Entic Dermal Sac Fungus
SPECIES *colletosporium*
GENUS *Epaphroconidia* Epaph Cone Sac Fungus
SPECIES *hawksworthii*
GENUS *Ephelidium* Ephel Sac Fungus
SPECIES *heardense*
GENUS *Epicladonia* Epiphyte Branched Sac Fungus
SPECIES *sandstedei, simplex, stenospora*
GENUS *Epicoccospora* Epiphyte Round Spore Sac Fungus
SPECIES *nepalensis*
GENUS *Epicoccum* Epiphyte Round Sac Fungus
SPECIES *diversisporum, effusum, equiseti, granulatum, neglectum, nigrum*
GENUS *Epigloea* Epiphyte Glow Sac Fungus
SPECIES *bactrospora, biciliata, filifera, grummannii, intermedia, medioincrassata, pleiospora, renitens, soleiformis, tortuosa, urosperma*
GENUS *Episporogoniella* Epiphyte Spore Gonad Sac Fungus
SPECIES *muscigena*
GENUS *Eremascus* Eremasc Sac Fungus
SPECIES *albus, fertilis*
GENUS *Eschatogonia* Eschat Gonad Sac Fungus

SPECIES *marivelensis*
GENUS *Escovopsis* Escov Sac Fungus
SPECIES *aspergilloides, weberi*
GENUS *Esdipatilia* Esdipat Sac Fungus
SPECIES *indica*
GENUS *Esfandiariomyces* Esfandiar Sac Fungus
SPECIES *insignis*
GENUS *Esteya* Estey Sac Fungus
SPECIES *vermicola*
GENUS *Everhartia* Everhart Sac Fungus
SPECIES *lignatilis*
GENUS *Evlachovaea* Evlachova Sac Fungus
SPECIES *kintrischica*
GENUS *Excipularia* Excipular Sac Fungus
SPECIES *fusispora*
GENUS *Exochalara* Exochalar Sac Fungus
SPECIES *guadalcanalensis, longissima*
GENUS *Exosporium* Exospore Sac Fungus
SPECIES *stilbaceum, stylobatum, sultanae, tiliae*
GENUS *Extrusothecium* Extra Theca Sac Fungus
SPECIES *caffrum*
GENUS *Favomicrosporon* Fava Microspore Sac Fungus
SPECIES *pinettii*
GENUS *Favostroma* Fava Stroma Sac Fungus
SPECIES *crypticum*
GENUS *Fenestroconidia* Window Cone Sac Fungus
SPECIES *caloplacae*
GENUS *Filosporella* Filament Spore Sac Fungus
SPECIES *annelidica, fistucella, longispora, sympodialis, versimorpha*
GENUS *Fissuricella* Fissure Sac Fungus
SPECIES *filamenta*
GENUS *Flabellocladia* Fan Branched Sac Fungus
SPECIES *gigantea*
GENUS *Flabellospora* Fan Spore Sac Fungus
SPECIES *acuminata, amphibia, tetracladia*
GENUS *Flagellospora* Flagella Spore Sac Fungus
SPECIES *curvula, fusarioides, minuta, penicillioides*
GENUS *Flavobathelium* Yellow Bottom Sac Fungus
SPECIES *epiphyllum*
GENUS *Flosculomyces* Floscule Sac Fungus
SPECIES *floridaensis, trilobatus*
GENUS *Fluviostroma* Stream Stroma Sac Fungus
SPECIES *wrightii*
GENUS *Fontanospora* Fontana Spore Sac Fungus
SPECIES *eccentrica, fusiramosa*
GENUS *Foveostroma* Foveal Stroma Sac Fungus
SPECIES *drupacearum*
GENUS *Friedmanniomyces* Friedmann Sac Fungus
SPECIES *endolithicus*
GENUS *Frigidispora* Frigid Spore Sac Fungus
SPECIES *colnensis*
GENUS *Fujimyces* Fuji Sac Fungus
SPECIES *oödes*
GENUS *Fumago* Fumago Sac Fungus
SPECIES *salicina*
GENUS *Funicularius* Funnel Sac Fungus
SPECIES *triseriatus*
GENUS *Fusariella* Fused Sac Fungus
SPECIES *aegyptiaca, atrovirens, bizzozeriana, candida, concinna, formosana, hughesii, indica, intermedia, obstipa, sarniensis*
GENUS *Fusariopsis* Lesser Fused Sac Fungus
SPECIES *derrienii*
GENUS *Fusicatena* Fused Chain Sac Fungus
SPECIES *macrospora*
GENUS *Fusichalara* Fused Chalar Sac Fungus
SPECIES *clavatispora, minuta*
GENUS *Fusoma* Fused Soma Sac Fungus

SPECIES *biseptatum, triseptatum*
GENUS *Gabarnaudia* Gabarnaud Sac Fungus
SPECIES *betae, cucumeris, fimicola, humicola, tholispora*
GENUS *Gampsonema* Gamps Thread Sac Fungus
SPECIES *exile*
GENUS *Gamsia* Gams Sac Fungus
SPECIES *aggregata, simplex*
GENUS *Gamsylella* Gamsyl Sac Fungus
SPECIES *arcuata, gephyropaga, lobata, parvicollis, phymatopaga, robusta*
GENUS *Gangliophora* Ganglia Sac Fungus
SPECIES *casuarinae*
GENUS *Geastrumia* Ground Star Sac Fungus
SPECIES *polystigmatis*
GENUS *Gelatinocrinis* Gelatin Lily Sac Fungus
SPECIES *roseus*
GENUS *Gelatinopycnis* Gelatin Dense Sac Fungus
SPECIES *hartigiana*
GENUS *Gelatinosporium* Gelatin Spore Sac Fungus
SPECIES *magnum*
GENUS *Gemmulina* Gem Sac Fungus
SPECIES *botryosa*
GENUS *Genicularia* (Synonym for *Geniculifera*)
GENUS *Geniculifera* Genicule Sac Fungus
SPECIES *bogoriensis, cystosporia, effusa, paucispora*
GENUS *Gibbago* Gibbago Sac Fungus
SPECIES *trianthemae*
GENUS *Gilmaniella* Gilman Sac Fungus
SPECIES *bambusae, humicola, macrospora, multiporosa, punctiformis, subornata, thermophila*
GENUS *Giulia* Giule Sac Fungus
SPECIES *tenuis*
GENUS *Glabrotheca* Smooth Theca Sac Fungus
SPECIES *aciculispora*
GENUS *Glaphyriopsis* Glaph Sac Fungus
SPECIES *brunnea, hyalina*
GENUS *Glarea* Glare Sac Fungus
SPECIES *lozoyensis*
GENUS *Glioannellodochium* Ring Doch Glio Sac Fungus
SPECIES *australiense*
GENUS *Glioblastocladium* Blastid Branched Glio Sac Fungus
SPECIES *australiense*
GENUS *Gliocephalis* Head Glio Sac Fungus
SPECIES *hyalina, pulchella*
GENUS *Gliomastix* Mastigote Glio Sac Fungus
SPECIES *cerealis, fusigera, guttuliformis, inflata, luzulae, musicola, novae-zelandiae*
VARIETIES *murorum murorum, murorum felina*
GENUS *Glischroderma* Glisch Dermal Sac Fungus
SPECIES *cinctum*
GENUS *Globuliroseum* Globule Rosy Sac Fungus
SPECIES *cocciniae*
GENUS *Gloeocercospora* Cerco Spore Glow Sac Fungus
SPECIES *sorghi*
GENUS *Gloeocoryneum* Club Glow Sac Fungus
SPECIES *angamosense, cinereum hawaiiense*
GENUS *Goidanichiella* Goidan Claw Sac Fungus
SPECIES *barronii*
GENUS *Gonatobotryum* Gonad Berry Sac Fungus
SPECIES *apiculatum, fuscum, maculicola, parasiticum*
GENUS *Gonatophragmiella* Gonad Fragment Sac Fungus
SPECIES *didymospora*
GENUS *Goosiella* Lesser Goos Sac Fungus
SPECIES *helicospora*
GENUS *Goosiomyces* Goos Sac Fungus
SPECIES *digitatus*
GENUS *Grallomyces* Grall Sac Fungus
SPECIES *portoricensis*
GENUS *Graphiothecium* Graphic Theca Sac Fungus
SPECIES *phyllogenum*

GENUS *Graphium* Graphic Sac Fungus
SPECIES *album, aureum, calicioides, comatrichoides, erubescens, fissum, flexuosum, fragrans, fructicola, grovei, pallescens, penicillioides, putredinis, pyrinum, rigidum, rubrum, silanum, subulatum, tectonae, xanthocephalum*
GENUS *Groveolopsis* Grove Sac Fungus
SPECIES *pandani*
GENUS *Gymnodochium* Nude Doch Sac Fungus
SPECIES *fimicola*
GENUS *Gymnoxyphium* Nude Xyph Sac Fungus
SPECIES *malum, splendidum*
GENUS *Gyoerffyella* Gyoerffy Sac Fungus
SPECIES *biappendiculata, craginiformis, entomobryoides, gemellipara, myrmecophagiformis, rotula, speciosa, tricapillata*
GENUS *Gyrostroma* Gyre Stroma Sac Fungus
SPECIES *austroamericanum, missouriense, sinuosum*
GENUS *Gyrothrix* Gyre Hairy Sac Fungus
SPECIES *circinata, hughesii, microsperma, pediculata, podosperma, verticiclada, verticillata*
GENUS *Habrostictis* Haber Stick Sac Fungus
SPECIES *rubra*
GENUS *Halysiomyces* Halys Sac Fungus
SPECIES *saxatilis*
GENUS *Hansfordia* Hansford Sac Fungus
SPECIES *caricis, parasitica, pulvinata, triumfettae*
GENUS *Hansfordiopeltopsis* Hansford Pelt Sac Fungus
SPECIES *amazonensis*
GENUS *Hapalosphaeria* Single Sphere Sac Fungus
SPECIES *deformans*
GENUS *Haplariopsis* Single Sac Fungus
SPECIES *fagicola*
GENUS *Haplobasidion* Single Basal Sac Fungus
SPECIES *lelebae, thalictri*
GENUS *Hapsidascus* Haps Dasc Sac Fungus
SPECIES *hadrus*
GENUS *Haptocara* Car Hapteron Sac Fungus
SPECIES *latirostrum*
GENUS *Haptocillium* Ciliate Hapteron Sac Fungus
SPECIES *balanoides, campanulatum, glocklingiae, rhabdosporium, sinense, sphaerosporum, zeosporum*
GENUS *Haptospora* Spore Hapteron Sac Fungus
SPECIES *appendiculata, endoparasitica, tribrachispora*
GENUS *Harmoniella* Harmon Sac Fungus
SPECIES *chrysocephalus*
GENUS *Harpographium* Harp Graphic Sac Fungus
SPECIES *congestum, fasciculatum*
GENUS *Harposporium* Harp Spore Sac Fungus
SPECIES *anguillulae, arthrosporum, botuliforme, bysmatosporum, crassum, cylindrosporum, helicoides, leptospira, lilliputanum, microsporum, sicyodes*
GENUS *Harziella* Harz Sac Fungus
SPECIES *capitata*
GENUS *Hawksworthiana* Hawksworth Sac Fungus
SPECIES *peltigericola*
GENUS *Heimiodora* Heim Odor Sac Fungus
SPECIES verticillata
GENUS *Helhonia* Helhon Sac Fungus
SPECIES *rhamnigena*
GENUS *Helicoceras* Horn Helix Sac Fungus
SPECIES *oryzae*
GENUS *Helicofilia* Filament Helix Sac Fungus
SPECIES *madrasensis*
GENUS *Helicogoosia* Goos Helix Sac Fungus
SPECIES *paradoxa*
GENUS *Helicorhoidion* Rhoidion Helix Sac Fungus
SPECIES *dimorphum, irregulare*
GENUS *Helicosingula* Singular Helix Sac Fungus
SPECIES *leucadendri*
GENUS *Helicosporina* Spore Helix Sac Fungus
SPECIES *veronae*
GENUS *Helicothyrium* Thyre Helix Sac Fungus

SPECIES *ryukyuense*
GENUS *Helicoubisia* Oubis Helix Sac Fungus
SPECIES *coronata*
GENUS *Helicoön* Helix Sac Fungus
SPECIES *chlamydosporum, ellipticum, fuscosporum, pluriseptatum, reticulatum, richonis*
GENUS *Heliocephala* Head Helix Sac Fungus
SPECIES *proliferans*
GENUS *Heliscella* Lesser Helisc Sac Fungus
SPECIES *stellata, stellatacula*
GENUS *Heliscina* Helisc Sac Fungus
SPECIES *antennata, campanulata*
GENUS *Helminthophora* Helminth Sac Fungus
SPECIES *tenera, uniseptata*
GENUS *Hemibeltrania* Half Beltran Sac Fungus
SPECIES *echinulata, malaysiana*
GENUS *Hemicorynesporella* Half Club Spore Sac Fungus
SPECIES *guadalcanalensis*
GENUS *Hemidothis* Half Doth Sac Fungus
SPECIES *plurivora*
GENUS *Hendersoniopsis* Henderson Sac Fungus
SPECIES *thelebola*
GENUS *Henicospora* Henic Spore Sac Fungus
SPECIES *coronata, cylindroclavata, longissima, minor, queenslandica*
GENUS *Herreromyces* Herrer Sac Fungus
SPECIES *catenosporus*
GENUS *Heterocephalum* Variable Head Sac Fungus
SPECIES *aurantiacum, taiense*
GENUS *Heteroconium* Variable Cone Sac Fungus
SPECIES *chaetospira, solaninum*
VARIETIES *chaetospira pini*
GENUS *Heteroseptata* Variable Septic Sac Fungus
SPECIES *bacillospora*
GENUS *Hexacladium* Six Branch Sac Fungus
SPECIES *corynephorum*
GENUS *Himantia* Himant Sac Fungus
SPECIES *nitens*
GENUS *Hinoa* Hino Sac Fungus
SPECIES *bambusicola, sasae*
GENUS *Hoehneliella* Hoehnel Sac Fungus
SPECIES *perplexa*
GENUS *Holubovaniella* Holubovan Sac Fungus
SPECIES *elegans, gracilis*
GENUS *Hormiactella* Lesser Radial Hormo Sac Fungus
SPECIES *asetosa, fusca*
GENUS *Hormiactis* Radial Hormo Sac Fungus
SPECIES *alba, candida, fimicola, ontariensis*
GENUS *Hormiscioideus* Disc Hormo Sac Fungus
SPECIES *filamentosus*
GENUS *Hormococcus* Round Hormo Sac Fungus
SPECIES *conorum*
GENUS *Hormographiella* Lesser Graphic Hormo Sac Fungus
SPECIES *aspergillata, candelabrata, verticillata*
GENUS *Hormographis* Graphic Hormo Sac Fungus
SPECIES *ramirezii*
GENUS *Hormomyces* Hormo Sac Fungus
SPECIES *peniophorae*
GENUS *Hortaea* Horta Sac Fungus
SPECIES *werneckii*
GENUS *Hughesinia* Hughes Sac Fungus
SPECIES *chusqueae*
GENUS *Humicola* Humis Dwelling Sac Fungus
SPECIES *alopallonella, brevis, brunnea, fuscoatra, grisea, hyalothermophila, insolens, minima, nigrescens, parvispora*
VARIETIES *brunnea brunnea, brunnea africana, fuscoatra fuscoatra, fuscoatra longispora, grisea grisea, grisea thermoidea*
GENUS *Humicolopsis* Lesser Humis Dwelling Sac Fungus
SPECIES *cephalosporioides*

GENUS *Hyalobelemnospora* Belemn Spore Hyaline Sac Fungus
SPECIES *amazonica*
GENUS *Hyalocladium* Branched Hyaline Sac Fungus
SPECIES *moubasheri*
GENUS *Hyalodendron* Wood Hyaline Sac Fungus
SPECIES *lignicola*
GENUS *Hyalohelicomina* Helix Hyaline Sac Fungus
SPECIES *deutziae*
GENUS *Hyalopleiochaeta* Pleated Bristle Hyaline Sac Fungus
SPECIES *delicata*
GENUS *Hyalosynnema* Thread Hyaline Sac Fungus
SPECIES *multiseptatum*
GENUS *Hyalotia* Hyalote Sac Fungus
SPECIES *viridis*
GENUS *Hyalotiella* Lesser Hyalote Sac Fungus
SPECIES *transvalensis*
GENUS *Hydrometrospora* Hydric Metric Spore Sac Fungus
SPECIES *symmetrica*
GENUS *Hymenella* Hymen Sac Fungus
SPECIES *vulgaris*
GENUS *Hymenobia* Obe Hymen Sac Fungus
SPECIES *aporea*
GENUS *Hymenopsis* Lesser Hymen Sac Fungus
SPECIES *argentinensis, cudraniae, trochiloides, typhae*
GENUS *Hymenula* Least Hymen Sac Fungus
SPECIES *cerealis, rubella, stictoidea, vulgaris*
GENUS *Hyphodiscosioides* Disc Hypha Sac Fungus
SPECIES *tambopataensis*
GENUS *Hyphopolynema* Many Hypha Sac Fungus
SPECIES *australe*
GENUS *Hyphospora* Spore Hypha Sac Fungus
SPECIES *agavacearum*
GENUS *Hyphothyrium* Thyre Hypha Sac Fungus
SPECIES *pallidum*
GENUS *Hyphozyma* Zyme Hypha Sac Fungus
SPECIES *lignicola, roseonigra, sanguinea, variabilis*
VARIETIES *variabilis odora, variabilis variabilis*
GENUS *Hysteropycnis* Hyster Dense Sac Fungus
SPECIES *admixta, confluens, fraxini, globularis, occulta*
GENUS *Idiocercus* Idion Cercid Sac Fungus
SPECIES *australis, pirozynskii*
GENUS *Imimyces* Imi Sac Fungus
SPECIES *hollowayensis*
GENUS *Inesiosporium* Ines Spore Sac Fungus
SPECIES *longispirale, mauiense*
GENUS *Infundibura* Funnel Sac Fungus
SPECIES *adhaerens*
GENUS *Ingoldia* Ingold Sac Fungus
SPECIES *craginiformis, tricapillata*
GENUS *Inifatiella* Inifat Sac Fungus
SPECIES *nectandrae*
GENUS *Intercalarispora* Intercal Spore Sac Fungus
SPECIES *nigra*
GENUS *Isaria* Isar Sac Fungus
SPECIES *arachnophila, citrina, clavata, friesii, intricata, leprosa, puberula, tax, umbrina*
GENUS *Isariella* Lesser Isar Sac Fungus
SPECIES *auerswaldiae, candida*
GENUS *Isomunkia* Isomunk Sac Fungus
SPECIES *pulvinula*
GENUS *Isthmolongispora* Isthmus Long Spore Sac Fungus
SPECIES *ampulliformis, lanceata, minima, variabilis*
GENUS *Isthmophragmospora* Isthmus Fragment Spore Sac Fungus
SPECIES *verruculosa*
GENUS *Isthmotricladia* Isthmus Three Branch Sac Fungus
SPECIES *britannica, gombakiensis, verruculosum*
GENUS *Ityorhoptrum* Ityorhopter Sac Fungus
SPECIES *verruculosum*

GENUS *Janetia* Janet Sac Fungus
SPECIES *longispora*
GENUS *Jerainum* Jerain Sac Fungus
SPECIES *triquetrum*
GENUS *Johnstonia* Johnston Sac Fungus
SPECIES *colocasiae*
GENUS *Jubispora* Juby SporeSac Fungus
SPECIES *acaciae*
GENUS *Junctospora* Junction Spore Sac Fungus
SPECIES *pulchra*
GENUS *Kabatina* Kabat Sac Fungus
SPECIES *abietis, juniperi, mahoniae, populi, thujae*
GENUS *Kalchbrenneriella* (Synonym for *Torula cyanescens*)
GENUS *Kaleidosporium* Kaleidoscope Spore Sac Fungus
SPECIES *fenestratum*
GENUS *Kamatia* Kamat Sac Fungus
SPECIES *indica*
GENUS *Kameshwaromyces* Kameshwar Sac Fungus
SPECIES *globosus*
GENUS *Kananascus* Kanan Sac Fungus
SPECIES *koorchalomagnatus, verrucisporus*
GENUS *Keissleriomyces* Keissler Sac Fungus
SPECIES *sandstedeanus*
GENUS *Ketubakia* Ketubak Sac Fungus
SPECIES *indica*
GENUS *Kiliophora* Kili Sac Fungus
SPECIES *fusispora, ubiensis*
GENUS *Kionocephala* Cion Head Sac Fungus
SPECIES *catenulata*
GENUS *Kionochaeta* Cion Bristle Sac Fungus
SPECIES *aristata, ivoriensis, keniensis, malaysiana, ramifera, spissa, virtuosa*
GENUS *Kirramyces* Kirr Sac Fungus
SPECIES *hebes*
GENUS *Knufia* Knuf Sac Fungus
SPECIES *cryptophialidica*
GENUS *Kodonospora* Codon Spore Sac Fungus
SPECIES *tetracolumnaris*
GENUS *Kontospora* Kont Spore Sac Fungus
SPECIES *halophila*
GENUS *Koorchalomella* Koorchalom Sac Fungus
SPECIES *oryzae*
GENUS *Koralionastes* Coral Star Sac Fungus
SPECIES *angustus, ellipticus, giganteus, ovalis, violaceus*
GENUS *Korunomyces* Korun Sac Fungus
SPECIES *terminaliae, zapatensis*
GENUS *Kostermansinda* Kostermansind Sac Fungus
SPECIES *magna*
GENUS *Kostermansindiopsis* Lesser Kostermansind Sac Fungus
SPECIES *theobromae*
GENUS *Kramasamuha* Kramasamuh Sac Fungus
SPECIES *sibika*
GENUS *Kreiseliella* Kreisel Sac Fungus
SPECIES *typhae*
GENUS *Kumanasamuha* Kumanasamuh Sac Fungus
SPECIES *arakuensis, kalakadensis, novae-zelandica, sundara*
GENUS *Kylindria* (Synonym for *Cylindrotrichum*)
GENUS *Kyphophora* Cypher Sac Fungus
SPECIES *avicenniae*
GENUS *Lacazia* Lacaz Sac Fungus
SPECIES *loboi*
GENUS *Lacellina* Lace Sac Fungus
SPECIES *graminicola*
GENUS *Lacellinopsis* Lesser Lace Sac Fungus
SPECIES *sacchari*
GENUS *Lactydina* Lactic Dual Sac Fungus
SPECIES *tylopaga*
GENUS *Laeviomyces* Smooth Sac Fungus

SPECIES *opegraphae, pertusariicola*
GENUS *Lagenomyces* Lagen Sac Fungus
SPECIES *marginalis*
GENUS *Lambdasporium* Lambda Spore Sac Fungus
SPECIES *viridense*
GENUS *Lamproconium* Lamp Cone Sac Fungus
SPECIES *desmazieri*
GENUS *Laocoön* Laocoön Sac Fungus
SPECIES *paradoxus*
GENUS *Lappodochium* Lap Doch Sac Fungus
SPECIES *langeniforme*
GENUS *Lasmeniella* Lasmen Sac Fungus
SPECIES *dendritica*
GENUS *Latericonis* Wide Cone Sac Fungus
SPECIES *obscura*
GENUS *Lateriramulosa* Wide Branch Sac Fungus
SPECIES *bi-inflata, minitriangularia, quadriradiata, uni-inflata*
GENUS *Laterispora* Wide Spore Sac Fungus
SPECIES *brevirama*
GENUS *Lauriomyces* Laury Sac Fungus
SPECIES *bellulus, catenatus, heliocephalus, pulcher, ventricosus*
GENUS *Lautospora* Lauter Spore Sac Fungus
SPECIES *gigantea, simillima*
GENUS *Leandria* Leander Sac Fungus
SPECIES *momordicae*
GENUS *Lecythophora* Lecyth Sac Fungus
SPECIES *hoffmannii, lignicola, mutabilis*
GENUS *Leightoniomyces* Leighton Sac Fungus
SPECIES *phillipsii*
GENUS *Lemkea* Lemke Sac Fungus
SPECIES *sphaerospora*
GENUS *Lemonniera* Lemonnier Sac Fungus
SPECIES *aquatica, centrosphaera, cornuta, filiformis, pseudofloscula, terrestris*
GENUS *Lentescospora* Lente Sac Fungus
SPECIES *submarina*
GENUS *Leohumicola* Lion Humis Sac Fungus
SPECIES *minima*
GENUS *Lepidopterella* Butterfly Sac Fungus
SPECIES *palustris*
GENUS *Leptodermella* Thin Dermal Sac Fungus
SPECIES *incarnata*
GENUS *Leptodiscella* Thin Disc Sac Fungus
SPECIES *africana, japonica*
GENUS *Leptodontidium* Thin Tooth Sac Fungus
SPECIES *aciculare, beauverioides, boreale, camptobactrum, cubense, elatius, irregulare, obscurum, orchidicola, quercuum*
VARIETIES *elatius elatius, elatius ovalisporum*
GENUS *Leptosporina* Thin Spore Sac Fungus
SPECIES *aciculospora*
GENUS *Leptostromella* Thin Stroma Sac Fungus
SPECIES *graminicola, hysterioides, juncina, pteridina, septorioides*
GENUS *Leptothyrium* Thin Thyre Sac Fungus
SPECIES *alneum, asterinum, castaneae, clypeosphaerioides, gentianicola, ilicinum, melaleucum, osmundae, pinastri, protuberans*
GENUS *Libertiella* Libert Sac Fungus
SPECIES *malmedyensis*
GENUS *Linodochium* Lined Doch Sac Fungus
SPECIES *album, hyalinum*
GENUS *Listeromyces* Lister Sac Fungus
SPECIES *insignis*
GENUS *Lobatopedis* Lobe Pedal Sac Fungus
SPECIES *elegans, foliicola*
GENUS *Lomachashaka* Lomachashak Sac Fungus
SPECIES *africana, cynodontis, kera, sundara*
GENUS *Lunulospora* Moon Spore Sac Fungus
SPECIES *curvula*
GENUS *Luzfridiella* Luzfrid Sac Fungus

SPECIES *insignis*
GENUS *Lylea* Lyle Sac Fungus
SPECIES *catenulata, palmicola, tetracoila*
GENUS *Macrodiplodia* Large Diploid Sac Fungus
SPECIES *ulmi*
GENUS *Macrodiplodiopsis* Lesser Large Diploid Sac Fungus
SPECIES *desmazieri*
GENUS *Macrohilum* Large Hilo Sac Fungus
SPECIES *eucalypti*
GENUS *Macrophomina* Large Foam Sac Fungus
SPECIES *phaseolina, pseudeverniae*
GENUS *Magdalaenaea* Magdalaena Sac Fungus
SPECIES *monogramma*
GENUS *Mammaria* Nipple Sac Fungus
SPECIES *echinobotryoides*
GENUS *Mammariopsis* Lesser Nipple Sac Fungus
SPECIES *variospora*
GENUS *Mapletonia* Mapleton Sac Fungus
SPECIES *falcata*
GENUS *Margaritispora* Margarite Spore Sac Fungus
SPECIES *aquatica, monticola*
GENUS *Mariannaea* Marianna Sac Fungus
SPECIES *camptospora, clavispora, elegans, superimposita*
VARIETIES *elegans elegans, elegans punicea*
GENUS *Marisolaris* Sea Sun Sac Fungus
SPECIES *ansata*
GENUS *Martinellisia* Martin Sac Fungus
SPECIES *indica*
GENUS *Masoniomyces* Mason Sac Fungus
SPECIES *claviformis*
GENUS *Massalongina* Lesser Massalong Sac Fungus
SPECIES *aquilina*
GENUS *Massariothea* Massar Sac Fungus
SPECIES *attenuata, botulispora, paspali, scotica, shawiae, similis, themedae, triseptata*
GENUS *Mastigosporium* Mastigote Spore Sac Fungus
SPECIES *album, deschampsiae, kitzebergense, muticum, rubricosum*
GENUS *Matruchotia* Matruchot Sac Fungus
SPECIES *riograndensis, taquarae*
GENUS *Matsushimaea* Matsushima Sac Fungus
SPECIES *fasciculata, fertilis*
GENUS *Matsushimomyces* Lesser Matsushima Sac Fungus
SPECIES *indicus*
GENUS *Mauginiella* Maugin Sac Fungus
SPECIES *scaettae*
GENUS *Medeolaria* Medeole Sac Fungus
SPECIES *farlowii*
GENUS *Medusamyces* Medusa Sac Fungus
SPECIES *lunulosporus*
GENUS *Megacapitula* Large Cap Sac Fungus
SPECIES *villosa*
GENUS *Megaster* Large Star Sac Fungus
SPECIES *curvicornis, longicornis*
GENUS *Melanocephala* Dark Head Sac Fungus
SPECIES *australiensis, cubensis, cupulifera, manuensis, triseptata*
GENUS *Melanographium* Dark Graphic Sac Fungus
SPECIES *citri*
GENUS *Melanophoma* Dark Foam Sac Fungus
SPECIES *karroo*
GENUS *Melanosella* Dark Sac Fungus
SPECIES *mors-apis*
GENUS *Melophia* Mellow Sac Fungus
SPECIES *phyllachoroidea*
GENUS *Memnoniella* Memnon Sac Fungus
SPECIES *echinata, levispora, zingiberis*
GENUS *Menidochium* Meny Chi Sac Fungus
SPECIES *formosum*
GENUS *Menisporopsis* Meny Spore Sac Fungus

SPECIES *heterosetulosa, ludoviciana, novae-zelandiae, pirozynskii, theobromae*
GENUS *Menoidea* Meny Sac Fungus
SPECIES *abietis*
GENUS *Mercadomyces* Mercado Sac Fungus
SPECIES *camagueyensis*
GENUS *Merugia* Merug Sac Fungus
SPECIES *palicoureae*
GENUS *Mesniera* (Synonym)
GENUS *Metazythiopsis* Meta Zyther Sac Fungus
SPECIES *halepensis*
GENUS *Microcyclephaeria* Cyclepher Sac Fungus
SPECIES *palmicola*
GENUS *Microdiplodia* Diploid Sac Fungus
SPECIES *conigena, hedericola, junci, microsporella, miniuscula, minuta, myriospora, nissoliae, perpusilla, pittospororum, subtecta*
GENUS *Microdiscula* Disc Sac Fungus
SPECIES *phragmitis*
GENUS *Microhilum* Hil Sac Fungus
SPECIES *oncoperae*
GENUS *Micropera* Opera Sac Fungus
SPECIES *caespitosa, cotoneastri, padina, sorbi*
GENUS *Micropustulomyces* Pustule Sac Fungus
SPECIES *mucilaginosus*
GENUS *Microsphaeropsis* Sphere Sac Fungus
SPECIES *hellebori, ulmicola*
GENUS *Microstella* Star Sac Fungus
SPECIES *pluvioriens*
GENUS *Microtheliopsis* Thele Sac Fungus
SPECIES *uleana*
GENUS *Microtyle* Tyle Sac Fungus
SPECIES *bergii*
GENUS *Microxyphiella* Lesser Xyph Sac Fungus
SPECIES *hibiscifolia*
GENUS *Microxyphiopsis* Least Xyph Sac Fungus
SPECIES *byrsonimae, psidii*
GENUS *Microxyphium* Xyph Sac Fungus
SPECIES *aciculiforme, citri, leptospermi, purpuraefaciens, theae*
GENUS *Micula* Micule Sac Fungus
SPECIES *alni*
GENUS *Milospium* Milo Sac Fungus
SPECIES *graphideorum, lacoizquetae*
GENUS *Miniancora* Mini Ankora Sac Fungus
SPECIES *allisoniensis*
GENUS *Minutoexcipula* Minute Excipulate Sac Fungus
SPECIES *tuckerae, tuerkii*
GENUS *Minutophoma* Minute Foam Sac Fungus
SPECIES *chrysophthalmae*
GENUS *Mirandina* Miranda Sac Fungus
SPECIES *arnaudii, breviphora, corticola*
GENUS *Miratio* Mirat Sac Fungus
SPECIES *nectrioides*
GENUS *Mirimyces* Lesser Mirat Sac Fungus
SPECIES *pulcher*
GENUS *Mixtoconidium* Mixed Cone Sac Fungus
SPECIES *canariense*
GENUS *Moniliella* Monad Sac Fungus
SPECIES *acetoabutans, mellis, pollinis, suaveolens, tomentosa*
VARIETIES *suaveolens nigra, tomentosa pollinis*
GENUS *Monilochaetes* Least Monad Bristle Sac Fungus
SPECIES *infuscans*
GENUS *Monochaetia* Monad Bristle Sac Fungus
SPECIES *camelliae, carissae, cryptomeriae, dimorphospora, karstenii, lutea, monochaeta, saccardoi*
GENUS *Monochaetinula* Lesser Monad Bristle Sac Fungus
SPECIES *ampelophila, caffra, ceratoniae, geoffroeana, sterculiae, terminaliae*
GENUS *Monodidymaria* Monad Dual Maria Sac Fungus
SPECIES *canadensis, chenopodii, equiseti, scirpina, viticis*
GENUS *Monodisma* Mock Monad Sac Fungus

SPECIES *fragilis*
GENUS *Monosporiella* Lesser Monospore Sac Fungus
SPECIES *meliolicola*
GENUS *Monotrichum* Monad Hairy Sac Fungus
SPECIES *commelinae*
GENUS *Mucobasispora* Mucus Basal Spore Sac Fungus
SPECIES *tarikii*
GENUS *Mucosetospora* Mucus Bristle Spore Sac Fungus
SPECIES *gallica*
GENUS *Muirella* Muir Sac Fungus
SPECIES *alascensis*
GENUS *Multicladium* Multibranched Sac Fungus
SPECIES *digitatum*
GENUS *Multipatina* Multipatina Sac Fungus
SPECIES *citricola*
GENUS *Mycobacillaria* Rod Sac Fungus
SPECIES *simplex*
GENUS *Mycocentrodochium* Central Chi Sac Fungus
SPECIES *curvisporum*
GENUS *Mycocentrospora* Central Spore Sac Fungus
SPECIES *acerina, angulata, aquatica, cladosporioides, clavata*
GENUS *Mycochlamys* Cloak Sac Fungus
SPECIES *macrospora*
GENUS *Mycofalcella* Sickle Sac Fungus
SPECIES *calcarata*
GENUS *Mycospraguea* Sprague Sac Fungus
SPECIES *campicola*
GENUS *Mycosylva* Sylvan Sac Fungus
SPECIES *clarkii, reticulata, setosa*
GENUS *Mycotribulus* Tribule Sac Fungus
SPECIES *mirabilis*
GENUS *Mycousteria* Uster Sac Fungus
SPECIES *ellipsoideispora, morulinae*
GENUS *Myriellina* Myriel Sac Fungus
SPECIES *cydoniae, imperatae*
GENUS *Myriodontium* Myriad Tooth Sac Fungus
SPECIES *keratinophilum*
GENUS *Myrmecomyces* Ant Sac Fungus
SPECIES *annellisiae*
GENUS *Myrotheciastrum* Myriad Theca Star Sac Fungus
SPECIES *salvadorae*
GENUS *Mystrosporium* Mystic Spore Sac Fungus
SPECIES *adustum*
GENUS *Myxocephala* Head Sac Fungus
SPECIES *albida*
GENUS *Myxocyclus* Cyclic Sac Fungus
SPECIES *cenangioides, polycystis*
GENUS *Myxophora* Myxophore Sac Fungus
SPECIES *amerospora, apotheciicola, leptogiophila, ovalispora, placynthii, tetraspora, tjibodensis*
GENUS *Myxosporella* Lesser Myxospore Sac Fungus
SPECIES *schini*
GENUS *Myxosporium* Myxospore Sac Fungus
SPECIES *belludum, griseum, incarnatum, phaeosorum, pubescens, rosae, roumeguerei*
GENUS *Myxothyrium* Thyre Sac Fungus
SPECIES *leptideum*
GENUS *Nadsoniella* Nadson Sac Fungus
SPECIES *nigra*
GENUS *Naemosphaera* Thread Sphere Sac Fungus
SPECIES *acerina*
GENUS *Naemospora* Thread Spore Sac Fungus
SPECIES *crocea, croceola, microspora, strobi*
GENUS *Nagrajia* Nagraj Sac Fungus
SPECIES *havanensis*
GENUS *Nagrajomyces* Lesser Nagraj Sac Fungus
SPECIES *dictyosporus*
GENUS *Naiadella* (Synonym for *Classicula fluitans*)
GENUS *Nalanthamala* Nalanthamal Sac Fungus

SPECIES *madreeya, squamicola*
GENUS *Nanoschema* Nannite Scheme Sac Fungus
SPECIES *elaeocarpi*
GENUS *Naranus* Naran Sac Fungus
SPECIES *cryptomeriae*
GENUS *Nattrassia* Nattrass Sac Fungus
SPECIES *mangiferae*
GENUS *Navaneethospora* Navaneeth Spore Sac Fungus
SPECIES *cheiromycoides*
GENUS *Nematogonum* Gonad Thread Sac Fungus
SPECIES *aurantiacum, ferrugineum, mycophilum*
GENUS *Nematographium* Graphic Thread Sac Fungus
SPECIES *hippotrichoides, leucophlaeum, saccardoi, stilboideum, strictum*
GENUS *Nematospora* Spore Thread Sac Fungus
SPECIES *coryli*
GENUS *Neoalpakesa* Modern Alpakes Sac Fungus
SPECIES *poae*
GENUS *Neoarbuscula* Modern Tree Sac Fungus
SPECIES *eugeniae*
GENUS *Neofuckelia* Modern Fuckel Sac Fungus
SPECIES *pinicola*
GENUS *Neohendersonia* Modern Henderson Sac Fungus
SPECIES *congoensis, kickxii, piriformis*
GENUS *Neoheteroceras* Modern Variable Horn Sac Fungus
SPECIES *flageoletii*
GENUS *Neojohnstonia* Modern Johnston Sac Fungus
SPECIES *colocasiae*
GENUS *Neolamya* Modern Lamy Sac Fungus
SPECIES *peltigerae*
GENUS *Neomelanconium* Modern Dark Cone Sac Fungus
SPECIES *deightonii, gelatosporum*
GENUS *Neopericonia* Modern Pericon Sac Fungus
SPECIES *indica*
GENUS *Neophoma* Modern Foam Sac Fungus
SPECIES *graminella*
GENUS *Neoplaconema* Modern Plaque Thread Sac Fungus
SPECIES *cymbiforme, napelli*
GENUS *Neosporidesmium* Modern Spore Fiber Sac Fungus
SPECIES *maestrense*
GENUS *Neottiospora* Modern Otto Spore Sac Fungus
SPECIES *caricina, coprophila*
GENUS *Neottiosporina* Lesser Modern Otto Spore Sac Fungus
SPECIES *apoda, asymmetrica, australiensis, clavata, cylindrica, infundibularis, masonii, paspali, phragmiticola, sorghicola*
GENUS *Neta* Net Sac Fungus
SPECIES *angliae, compacta, gibbosospora, gigaspora, lignicola, patuxentica, quadriguttata, salmonicolor, tropicalis*
GENUS *Nidulispora* Nidule Spore Sac Fungus
SPECIES *quadrifida*
GENUS *Nigromacula* Black Spotted Sac Fungus
SPECIES *hypotrachynae, uniseptata*
GENUS *Nomuraea* Nomura Sac Fungus
SPECIES *anemonoides, atypicola, rileyi*
GENUS *Nothostrasseria* False Strasser Sac Fungus
SPECIES *dendritica*
GENUS *Nummospora* Money Spore Sac Fungus
SPECIES *hexamera*
GENUS *Nusia* Nus Sac Fungus
SPECIES *scheeleae*
GENUS *Nyctalospora* Night Spore Sac Fungus
SPECIES *compacta*
GENUS *Nypaella* Nypa Sac Fungus
SPECIES *frondicola*
GENUS *Obryzum* (Synonym for *Leptogium, Sphaerulina*)
GENUS *Obstipipilus* Obstipipil Sac Fungus
SPECIES *malabaricus*
GENUS *Obstipispora* Obstipipil Spore Sac Fungus

SPECIES *chewaclensis*
GENUS *Oceanites* Ocean Sac Fungus
SPECIES *scuticella*
GENUS *Ochroconis* Ocher Cone Sac Fungus
SPECIES *gamsii*
GENUS *Odontodictyospora* Tooth Net Spore Sac Fungus
SPECIES *comasii*
GENUS *Ojibwaya* Ojibway Sac Fungus
SPECIES *perpulchra*
GENUS *Omega* Omega Sac Fungus
SPECIES *coenobiticum*
GENUS *Ommatosporella* Ommate Spore Sac Fungus
SPECIES *burseracearum*
GENUS *Oncopodiella* Onc Pod Sac Fungus
SPECIES *hungarica, hyperparasitica, trigonella*
GENUS *Oncospora* Onc Spore Sac Fungus
SPECIES *pinastri*
GENUS *Oncostroma* Onc Stroma Sac Fungus
SPECIES *toddaliae*
GENUS *Onychocola* Claw Sac Fungus
SPECIES *canadensis*
GENUS *Onychophora* Claw Bearing Sac Fungus
SPECIES *coprophila*
GENUS *Oramasia* Oramas Sac Fungus
SPECIES *hirsuta*
GENUS *Orbimyces* Orb Sac Fungus
SPECIES *spectabilis*
GENUS *Orcadia* Orcad Sac Fungus
SPECIES *ascophylli*
GENUS *Ornatispora* Ornate Spore Sac Fungus
SPECIES *frondicola, gamsii, palmicola, punctata, taiwanensis*
GENUS *Orphanocoela* Orphan Sky Sac Fungus
SPECIES *calamagrostidis, kranzii, maydis*
GENUS *Ostracodermidium* Ostracod Dermal Sac Fungus
SPECIES *serpentinum*
GENUS *Ovadendron* Oval Wood Sac Fungus
SPECIES *sulphureo-ochraceum*
GENUS *Ozonium* Ozone Sac Fungus
SPECIES *aureum*
GENUS *Paliphora* Paliphore Sac Fungus
SPECIES *aurea, intermedia, porosa*
GENUS *Pantospora* Pantospore Sac Fungus
SPECIES *guazumae*
GENUS *Papilionospora* Papilion Spore Sac Fungus
SPECIES *aspergilloides*
GENUS *Pappimyces* Pappus Sac Fungus
SPECIES *hastatus*
GENUS *Papulaspora* Papule Spore Sac Fungus
SPECIES *appendicularis, aspera, aspergilliformis, brachiata, byssina, candida, coprophila, equi, halima, immersa, irregularis, pallidula, pannosa, parasitica, pisicola, rubida, sepedonioides*
GENUS *Papulosa* Lesser Papule Spore Sac Fungus
SPECIES *amerospora*
GENUS *Paraaoria* Para Aori Sac Fungus
SPECIES *himalayana*
GENUS *Paraarthrocladium* Para Straight Branch Sac Fungus
SPECIES *amazonense*
GENUS *Paraceratocladium* Para Horn Branch Sac Fungus
SPECIES *silvestre*
GENUS *Parachinomyces* Para China Sac Fungus
SPECIES *acroconidiellinae*
GENUS *Paracostantinella* Para Ribbed Sac Fungus
SPECIES *pulchra*
GENUS *Paracryptophiale* Para Cryptic Phial Sac Fungus
SPECIES *kamaruddinii*
GENUS *Paradactylaria* Para Finger Sac Fungus
SPECIES *elegans*
GENUS *Paradactylella* Lesser Para Finger Sac Fungus

SPECIES *peruviana*
GENUS *Paradendryphiopsis* Para Wood Phial Sac Fungus
SPECIES *anomala, cambrensis, laxa*
GENUS *Paradictyoarthrinium* Para Net Straight Sac Fungus
SPECIES *diffractum*
GENUS *Paradidymobotryum* Para Dual Berry Sac Fungus
SPECIES *oblongum*
GENUS *Paradischloridium* Para Disc Chloric Sac Fungus
SPECIES *schaffrei*
GENUS *Paradiscula* Para Disc Sac Fungus
SPECIES *spuria*
GENUS *Paraepicoccum* Para Epiphyte Round Sac Fungus
SPECIES *amazonense*
GENUS *Parafulvia* Para Fulvous Sac Fungus
SPECIES *indica*
GENUS *Parahyalotiopsis* Para Hyalote Sac Fungus
SPECIES *borassi*
GENUS *Paramassariothea* Para Massar Sac Fungus
SPECIES *falcata*
GENUS *Parapericonia* Para Pericon Sac Fungus
SPECIES *sessilispora*
GENUS *Paraphaeoisaria* Para Sphere Sac Fungus
SPECIES *alabamensis*
GENUS *Paraphialocephala* Para Phial Head Sac Fungus
SPECIES *himalayana*
GENUS *Parapithomyces* Para Pith Sac Fungus
SPECIES *brideliae, clitoriae*
GENUS *Parapleurotheciopsis* Para Pleural Theca Sac Fungus
SPECIES *ilicina, inaequiseptata*
GENUS *Pararobillarda* Para Robillard Sac Fungus
SPECIES *caffra*
GENUS *Parastenella* Para Stencil Sac Fungus
SPECIES *magnoliae*
GENUS *Parasympodiella* Para Sympodial Sac Fungus
SPECIES *clarkii, laxa*
GENUS *Paspalomyces* Paspal Sac Fungus
SPECIES *aureus*
GENUS *Patellina* Patella Sac Fungus
SPECIES *caesia, diaphana*
GENUS *Peethasthabeeja* Peethasthabeej Sac Fungus
SPECIES *krishna*
GENUS *Pellionella* Pellion Sac Fungus
SPECIES *tetonensis*
GENUS *Peltasterinostroma* Pelt Star Stroma Sac Fungus
SPECIES *rubi*
GENUS *Perelegamyces* Pereleg Sac Fungus
SPECIES *parviechinulatus*
GENUS *Periconia* Pericon Sac Fungus
SPECIES *algeriana, atra, atropurpurea, britannica, byssoides, cambrensis, circinata, clitoriae, cookei, curta, digitata, echinochloae, funerea, genistae, glaucocephala, glyceriicola, hispidula, laminella, lateralis, macrospinosa, manihoticola, minutissima, narsapurensis, paludosa, prolifica, sahariana, saraswatipurensis, shyamala, typhicola*
VARIETIES *atropurpurea microspora*
GENUS *Periconiella* Lesser Pericon Sac Fungus
SPECIES *araliacearum, ilicis, saccharicola, velutina*
GENUS *Periola* Periole Sac Fungus
SPECIES *tomentosa*
GENUS *Perisporium* Perispore Sac Fungus
SPECIES *arundinis, kunzei, princeps*
GENUS *Pestalozziella* Lesser Pestalozzi Sac Fungus
SPECIES *artocarpi, parva, subsessilis*
GENUS *Pestalozzina* Pestalozzi Sac Fungus
SPECIES *uniseptata*
GENUS *Petrakia* Petrak Sac Fungus
SPECIES *irregularis*
GENUS *Petrakiella* Lesser Petrak Sac Fungus
SPECIES *insignis*

GENUS *Petrakiopsis* Least Petrak Sac Fungus
SPECIES *elegans*
GENUS *Peyronelina* Peyronel Sac Fungus
SPECIES *glomerulata*
GENUS *Phacostromella* Lens Stroma Sac Fungus
SPECIES *coronata*
GENUS *Phaeoacremonium* Dusky Acremon Sac Fungus
SPECIES *aleophilum, angustius, inflatipes, mortoniae, rubrigenum, viticola*
GENUS *Phaeoannellomyces* Dusky Ringed Sac Fungus
SPECIES *elegans*
GENUS *Phaeococcomyces* Dusky Round Sac Fungus
SPECIES *catenatus, exophialae, nigricans*
GENUS *Phaeocytostroma* Dusky Cytose Stroma Sac Fungus
SPECIES *ambiguum, megalosporum, plurivorum, sacchari*
GENUS *Phaeodactylium* Dusky Finger Sac Fungus
SPECIES *alpiniae*
GENUS *Phaeodothiopsis* Dusky Doth Sac Fungus
SPECIES *zollingeri*
GENUS *Phaeohiratsukaea* Dusky Hiratsuka Sac Fungus
SPECIES *expansa*
GENUS *Phaeolabrella* Dusky Lip Sac Fungus
SPECIES *eryngiicola*
GENUS *Phaeomoniella* Dusky Monad Sac Fungus
SPECIES *chlamydospora*
GENUS *Phaeomonostichella* Dusky Monad Stitch Sac Fungus
SPECIES *symploci*
GENUS *Phaeophloeosporella* Dusky Flow Spore Sac Fungus
SPECIES *ekebergiae*
GENUS *Phaeosclera* Dusky Hard Sac Fungus
SPECIES *dematioides*
GENUS *Phaeostagonospora* Dusky Stag Spore Sac Fungus
SPECIES *nolinae*
GENUS *Phaeostalagmus* Dusky Stalagmite Sac Fungus
SPECIES *cyclosporus, peregrinus, tenuissimus*
GENUS *Phaeotheca* Dusky Theca Sac Fungus
SPECIES *fissurella, triangularis*
GENUS *Phaeotrichoconis* Dusky Hairy Cone Sac Fungus
SPECIES *crotalariae*
GENUS *Phalangispora* Phalange Spore Sac Fungus
SPECIES *constricta, nawawii*
GENUS *Phialea* Phial Sac Fungus
SPECIES *littoralis, strobilina*
GENUS *Phialemonium* Monad Phial Sac Fungus
SPECIES *curvatum, dimorphosporum, obovatum*
GENUS *Phialoarthrobotryum* Straight Berry Phial Sac Fungus
SPECIES *triseptatum*
GENUS *Phialocephala* Head Phial Sac Fungus
SPECIES *compacta, dimorphospora, fluminis, fusca, humicola, phycomyces, scopiformis, truncata, virens, xalapensis*
GENUS *Phialocladus* Branched Phial Sac Fungus
SPECIES *zsoltii*
GENUS *Phialocorona* Crowned Phial Sac Fungus
SPECIES *pleomorpha*
GENUS *Phialomyces* Lesser Phial Sac Fungus
SPECIES *macrosporus, striatus*
GENUS *Phialophaeoisaria* Sphere Phial Sac Fungus
SPECIES *spongiosa*
GENUS *Phialophora* Greater Phial Sac Fungus
SPECIES *alba, asteris, atrovirens, benthica, brachyconia, brevicollaris, bubakii, caliciformis, chrysanthemi, cinerescens, clavispora, cyanescens, cyclaminis, dacryospora, dancoi, decumbens, fasciculata, fastigiata, foetens, geniculata, gregata, hyalina, intermedia, japonica, lagerbergii, luteoviridis, melinii, mustea, olivacea, oxyspora, parasitica, pedrosoi, radicicola, repens, richardsiae, verrucosa, zeicola*
VARIETIES *asteris helianthi, radicicola radicicola, radicicola graminicola*
GENUS *Phialophorophoma* Foam Phial Sac Fungus
SPECIES *litoralis*
GENUS *Phialophoropsis* Least Phial Sac Fungus
SPECIES *cambrensis*

GENUS *Phialostele* Stele Phial Sac Fungus
SPECIES *scytopetali*
GENUS *Phialotubus* Tube Phial Sac Fungus
SPECIES *microsporus*
GENUS *Phomatosporella* Foam Spore Sac Fungus
SPECIES *albomaculans*
GENUS *Phragmocephala* Head Fragment Sac Fungus
SPECIES *atra, cookei, elegans, elliptica, glanduliformis, prolifera, stemphylioides*
GENUS *Phragmonaevia* Mole Fragment Sac Fungus
SPECIES *caulincola, hysterioides, peltigerae*
GENUS *Phragmospathulella* Spathe Fragment Sac Fungus
SPECIES *matsushimae*
GENUS *Phragmotrichum* Hairy Fragment Sac Fungus
SPECIES *rivoclarinum*
GENUS *Phyllocelis* Frond Sac Fungus
SPECIES *clibadii¸oyedaeae*
GENUS *Phyllopyrenia* Fire Frond Sac Fungus
SPECIES *macquariensis*
GENUS *Physalidiella* Bladder Sac Fungus
SPECIES *elegans, matsushimae*
GENUS *Physalidiopsis* Lesser Bladder Sac Fungus
SPECIES *carbonis*
GENUS *Pinatubo* Pinatubo Sac Fungus
SPECIES *oryzae*
GENUS *Piperivora* Piperivore Sac Fungus
SPECIES *nodosa*
GENUS *Piricauda* Pire Stalk Sac Fungus
SPECIES *arcticoceanorum, pelagica*
GENUS *Piricaudilium* Lesser Pire Stalk Sac Fungus
SPECIES *lobatum*
GENUS *Pirispora* Pire Spore Sac Fungus
SPECIES *trisetosa*
GENUS *Pirostoma* Pire Stoma Sac Fungus
SPECIES *circinans, viridisporum*
GENUS *Pithosira* Pith Pitted Sac Fungus
SPECIES *sydowii*
GENUS *Pittostroma* Pitted Stroma Sac Fungus
SPECIES *abietinum*
GENUS *Plectophomella* Plaited Foam Sac Fungus
SPECIES *concentrica, nypae, visci*
GENUS *Plectronidiopsis* Plaited Sac Fungus
SPECIES *chilensis*
GENUS *Pleiochaeta* Plethora Bristle Sac Fungus
SPECIES *setosa*
GENUS *Plenocatenulis* Double Chain Sac Fungus
SPECIES *magnoliae, osmanthi*
GENUS *Pleocouturea* Plethora Couture Sac Fungus
SPECIES *castagnei*
GENUS *Pleocryptospora* Plethora Cryptic Spore Sac Fungus
SPECIES *bambusae*
GENUS *Plesiospora* Plesiospore Sac Fungus
SPECIES *globosa*
GENUS *Pleurocolla* Pleural Collar Sac Fungus
SPECIES *compressa, tiliae*
GENUS *Pleurodesmospora* Pleural Fiber Spore Sac Fungus
SPECIES *coccorum*
GENUS *Pleuropedium* Pleural Pedal Sac Fungus
SPECIES *multiseptatum, tricladioides*
GENUS *Pleurophoma* Pleural Foam Sac Fungus
SPECIES *arabidis, cava, latvica, phyllachorivora, pleurospora, rubi, salicina, socialis*
GENUS *Pleurophomopsis* Lesser Pleural Foam Sac Fungus
SPECIES *lignicola*
GENUS *Pleurophragmium* Pleural Fragment Sac Fungus
SPECIES *acutum, capense, hippotrichoides*
GENUS *Pleurotheciopsis* Lesser Pleural Theca Sac Fungus
SPECIES *bramleyi, pusilla, setiformis, sylvestris, tropicalis, websteri*
GENUS *Pleurothecium* Pleural Theca Sac Fungus

SPECIES *magnum, malayense, pulneyense, recurvatum*
GENUS *Pochonia* Pochon Sac Fungus
SPECIES *bulbillosa, chlamydosporia, chlamydosporia catenulata, chlamydosporia chlamydosporia, goniodes, humicola, microbactrospora, rubescens, suchlasporia, suchlasporia catenata, suchlasporia suchlasporia*
GENUS *Pocillopycnis* Pocil DenseSac Fungus
SPECIES *umensis*
GENUS *Podoconis* Pod Cone Sac Fungus
SPECIES *megasperma*
GENUS *Podosporiella* Lesser Pod Spore Sac Fungus
SPECIES *verticillata*
GENUS *Podosporium* Pod Spore Sac Fungus
SPECIES *beccarianum, elongatum*
GENUS *Podotara* Pod Tara Sac Fungus
SPECIES *pilophoriformis*
GENUS *Polybulbophiale* Many Bulb Phial Sac Fungus
SPECIES *palmicola*
GENUS *Polycladium* Many Branch Sac Fungus
SPECIES *equiseti*
GENUS *Polycytella* Many Cyte Sac Fungus
SPECIES *hominis*
GENUS *Polylobatispora* Many Lobe Spore Sac Fungus
SPECIES *deltoidea, quinquecornuta*
GENUS *Polyschema* Polyscheme Sac Fungus
SPECIES *chambalensis, clavulata, congolensis, indica, larviformis, olivacea, sagarii, terricola*
GENUS *Polyscytalina* Lesser Many Scyte Sac Fungus
SPECIES *grisea*
GENUS *Polyscytalum* Many Scyte Sac Fungus
SPECIES *berkeleyi, fagicola, fecundissimum, fuegianum, fungorum, gracilisporum, griseum, hareae, pini, pustulans, sericeum, truncatum, verrucosum*
GENUS *Polysynnema* Many Thread Sac Fungus
SPECIES tropicum
GENUS *Pontogeneia* Pontogene Sac Fungus
SPECIES *calospora, codiicola, cubensis, enormis, erikae, padinae, valoniopsidis*
GENUS *Porocladium* Pore Branched Sac Fungus
SPECIES *aquaticum*
GENUS *Porosubramaniania* Lesser Pore Branched Sac Fungus
SPECIES *malabarica, moniliformis*
GENUS *Porrectotheca* Porrect Theca Sac Fungus
SPECIES *divaricata*
GENUS *Potamomyces* Water Sac Fungus
SPECIES *armatisporus*
GENUS *Protocucurbitaria* Cucumber Sac Fungus
SPECIES *ribicola*
GENUS *Psammina* Sand Sac Fungus
SPECIES *bommeriae, inflata, lobariae, simplex, stipitata*
GENUS *Pseudoanguillospora* Lesser Snake Spore Sac Fungus
SPECIES *gracilis, stricta*
GENUS *Pseudoaristastoma* Lesser Awn Stoma Sac Fungus
SPECIES *indicum*
GENUS *Pseudobasidiospora* Lesser Basal Spore Sac Fungus
SPECIES *caroliniana*
GENUS *Pseudobotrytis* Lesser Berry Sac Fungus
SPECIES *bisbyi, terrestris*
GENUS *Pseudocanalisporium* Lesser Channel Spore Sac Fungus
SPECIES *circumfecundum*
GENUS *Pseudocenangium* Lesser Angled Sac Fungus
SPECIES *succineum*
GENUS *Pseudocercosporidium* Lesser Cerco Spore Sac Fungus
SPECIES *carotae*
GENUS *Pseudochaetosphaeronema* Lesser Bristle Sphere Sac Fungus
SPECIES *larense*
GENUS *Pseudochuppia* Lesser Chupp Sac Fungus
SPECIES *sarcinispora*
GENUS *Pseudocladosporium* Lesser Branched Spore Sac Fungus
SPECIES *hachijoense*
GENUS *Pseudoclathrosphaerina* Lesser Lattice Sphere Sac Fungus

SPECIES *evamariae*
GENUS *Pseudoconium* Lesser Cone Sac Fungus
SPECIES *pallescens*
GENUS *Pseudocytoplacosphaeria* Lesser Cytoplaque Sphere Sac Fungus
SPECIES *coniicola*
GENUS *Pseudodictyosporium* Lesser Net Spore Sac Fungus
SPECIES *wauense*
GENUS *Pseudodidymaria* Lesser Dual Marine Sac Fungus
SPECIES *clematidis, wyethiae*
GENUS *Pseudodiplodia* Lesser Diploid Sac Fungus
SPECIES *ruticola*
GENUS *Pseudodiscula* Lesser Disc Sac Fungus
SPECIES *endogenospora, mali*
GENUS *Pseudofuscophialis* Lesser Fuscous Phial Sac Fungus
SPECIES *lignicola*
GENUS *Pseudogibellula* Lesser Gibe Sac Fungus
SPECIES *formicarum*
GENUS *Pseudogliophragma* Lesser Glow Fragment Sac Fungus
SPECIES *indica*
GENUS *Pseudographiella* Lesser Graphic Sac Fungus
SPECIES *indica, rhizomorpharum, variiseptata*
GENUS *Pseudographium* Least Graphic Sac Fungus
SPECIES *persicae*
GENUS *Pseudohansfordia* Lesser Hansford Sac Fungus
SPECIES *irregularis, mycophila*
GENUS *Pseudohendersonia* Lesser Henderson Sac Fungus
SPECIES *proteae*
GENUS *Pseudolachnea* Lesser Lachne Sac Fungus
SPECIES *hispidula*
GENUS *Pseudomeria* Lesser Merial Sac Fungus
SPECIES *mucosa*
GENUS *Pseudomicrodochium* Lesser Doch Sac Fungus
SPECIES *aciculare, candidum, cylindricum, lauri, suttonii*
GENUS *Pseudoneottiospora* Lesser Modern Otte Sac Fungus
SPECIES cunicularia, patouillardii
GENUS *Pseudopapulaspora* Lesser Pap Spore Sac Fungus
SPECIES *kendrickii*
GENUS *Pseudopatellina* Lesser Patella Sac Fungus
SPECIES *conigena*
GENUS *Pseudopeltistroma* Lesser Pelt Stroma Sac Fungus
SPECIES kyusyuense
GENUS *Pseudophloeosporella* Lesser Flow Spore Sac Fungus
SPECIES *dioscoreae*
GENUS *Pseudophragmotrichum* Lesser Fragment Hairy Sac Fungus
SPECIES *cubense*
GENUS *Pseudoramularia* Lesser Branched Sac Fungus
SPECIES *ponapensis, ruwenzoriensis*
GENUS *Pseudorhizopogon* Lesser Rhizid Beard Sac Fungus
SPECIES *sudae*
GENUS *Pseudorobillarda* Lesser Robillard Sac Fungus
SPECIES *agrostidis, aquatica, bambusae, indica, jaczewskii, magna, muehlenbergiae, peltigerae, phragmitis, setariae, sojae, texana*
GENUS *Pseudoschizothyra* Lesser Split Thyre Sac Fungus
SPECIES *thaungii*
GENUS *Pseudoseptoria* Lesser Septic Sac Fungus
SPECIES *bromigena, donacicola, donacis, everhartii, obtusa, stomaticola, usneae*
GENUS *Pseudostilbella* Lesser Stilbe Sac Fungus
SPECIES *euphorbiae*
GENUS *Pseudostracoderma* Lesser Ostracod Dermal Sac Fungus
SPECIES *terricola*
GENUS *Pseudotaeniolina* Lesser Ribbon Sac Fungus
SPECIES *convolvuli*
GENUS *Pseudotracylla* Lesser Tracy Sac Fungus
SPECIES *dentata, falcata*
GENUS *Pteromycula* Winged Sac Fungus
SPECIES *clypeiformis*
GENUS *Pterulopsis* Lesser Winged Sac Fungus

SPECIES *dummeri*
GENUS *Pterygosporopsis* Lesser Winged Spore Sac Fungus
SPECIES *fragilis, rhododendri*
GENUS *Pulchromyces* Pretty Sac Fungus
SPECIES *fimicola*
GENUS *Pullospora* Pull Spore Sac Fungus
SPECIES *macrospora, tetrachaeta*
GENUS *Pulvinotrichum* Powder Hair Sac Fungus
SPECIES *album, capitatum*
GENUS *Pycnidioarxiella* Arc Dense Sac Fungus
SPECIES *indica*
GENUS *Pycnodactylus* Finger Dense Sac Fungus
SPECIES *maranhensis*
GENUS *Pycnoharknessia* Harkness Dense Sac Fungus
SPECIES *pakistanica*
GENUS *Pycnomoreletia* Morelet Dense Sac Fungus
SPECIES *aristata, congoensis*
GENUS *Pycnopleiospora* Pleiospore Dense Sac Fungus
SPECIES *fagi*
GENUS *Pycnopsammina* Sand Dense Sac Fungus
SPECIES *lobariae*
GENUS *Pycnostysanus* Stysan Dense Sac Fungus
SPECIES *azaleae*
GENUS *Pycnothera* Theran Dense Sac Fungus
SPECIES *cordiae, ixorae*
GENUS *Pyrenothrix* Fire Hair Sac Fungus
SPECIES *nigra*
GENUS *Pyripnomyces* Pyripnon Sac Fungus
SPECIES *maranhensis*
GENUS *Quadracaea* Quad Sac Fungus
SPECIES *mediterranea*
GENUS *Quadricladium* Quad Branched Sac Fungus
SPECIES *aquaticum*
GENUS *Quasidiscus* Quasi Disc Sac Fungus
SPECIES *simmondsii*
GENUS *Quezelia* Quezel Sac Fungus
SPECIES *hoggariana*
GENUS *Rachidicola* Rachis Sac Fungus
SPECIES *palmae*
GENUS *Radiatispora* Radial Spore Sac Fungus
SPECIES *yunnanensis*
GENUS *Raizadenia* Raiz Gland Sac Fungus
SPECIES *garhwalensis*
GENUS *Ramakrishnanella* Ramakrishna Sac Fungus
SPECIES *indica*
GENUS *Ramaraomyces* Branched Sac Fungus
SPECIES *corticola*
GENUS *Ramichloridium* Branched Chloric Sac Fungus
SPECIES *anceps, apiculatum, cerophilum, ellipticum, epichloes, fasciculatum, mackenziei, musae, obovoideum, pini, schulzeri, subulatum, torvi, verrucosum*
VARIETIES *schulzeri schulzeri, schulzeri flexuosum, schulzeri tritici*
GENUS *Ramoconidiifera* Branched Cone Sac Fungus
SPECIES *verrucosa*
GENUS *Refractohilum* Refracted Hil Sac Fungus
SPECIES *achromaticum, galligenum, pluriseptatum*
GENUS *Reichlingia* Reichling Sac Fungus
SPECIES *leopoldii*
GENUS *Remersonia* Remerson Sac Fungus
SPECIES *thermophila*
GENUS *Repetophragma* Repet Fragment Sac Fungus
SPECIES *aburiense, afrormosiae, biseptatum, cambrense, cubense, dennisii, filiattenuatum, indicum, subulatum, wroblewskii, zambiense*
GENUS *Retiarius* Retiar Sac Fungus
SPECIES *bovicornutus, superficiaris*
GENUS *Retroconis* Retro Cone Sac Fungus
SPECIES *fusiformis*
GENUS *Rhabdoclema* Rod Clem Sac Fungus

SPECIES *chusqueae, olyrae*
GENUS *Rhexoampullifera* Ampule Rhex Sac Fungus
SPECIES *fagi*
GENUS *Rhexographium* Graphic Rhex Sac Fungus
SPECIES *fimbriasporum*
GENUS *Rhexoprolifer* Prolific Rhex Sac Fungus
SPECIES *variabilis*
GENUS *Rhinocladium* Snout Branched Sac Fungus
SPECIES *lesnei, pulchrum*
GENUS *Rhinotrichella* Lesser Snout Hair Sac Fungus
SPECIES *globulifera*
GENUS *Rhinotrichum* Snout Hair Sac Fungus
SPECIES *aureum, lanosum, niveum, repens, thwaitesii*
GENUS *Rhizohypha* Rhizid Hypha Sac Fungus
SPECIES *limodori*
GENUS *Rhizophila* Rhizid Sac Fungus
SPECIES *marina*
GENUS *Rhizopycnis* Rhizid Dense Sac Fungus
SPECIES *vagum*
GENUS *Rhizosphaerina* Rhizid Sphere Sac Fungus
SPECIES *variabilis*
GENUS *Rhodesiopsis* Rhodes Sac Fungus
SPECIES *gelatinosa*
GENUS *Rhombostilbella* Rhombic Stilbe Sac Fungus
SPECIES *crus-pavonis, rosae*
GENUS *Rhynchodiplodia* Beaked Diploid Sac Fungus
SPECIES *ursusiformae*
GENUS *Rhynchosporium* Beaked Spore Sac Fungus
SPECIES *orthosporum, secalis*
GENUS *Rhynchostoma* Beaked Stoma Sac Fungus
SPECIES *minutum*
GENUS *Rhynchnostrigula* Beaked Strigule Sac Fungus
SPECIES *papillata*
GENUS *Rileya* Riley Sac Fungus
SPECIES *piceae*
GENUS *Robillarda* Robillard Sac Fungus
SPECIES *sessilis*
GENUS *Rogergoosiella* Roger Goos Sac Fungus
SPECIES *roystoneicola*
GENUS *Roigiella* Roig Sac Fungus
SPECIES *lignicola*
GENUS *Rosulomyces* Rosule Sac Fungus
SPECIES *arthrosporioides*
GENUS *Rotaea* Rota Sac Fungus
SPECIES *flava*
GENUS *Rotiferophthora* Rotifer Phthoral Sac Fungus
SPECIES *attenuata, barronii, cylindrospora, gallinova, lunatospora, torquatospora*
GENUS *Rubikia* Rubik Sac Fungus
SPECIES *samsonii*
GENUS *Ruggieria* Ruggier Sac Fungus
SPECIES *glaucescens*
GENUS *Rutola* Rue Sac Fungus
SPECIES *graminis*
GENUS *Saccardoella* Saccardo Sac Fungus
SPECIES *mangrovei, transylvanica*
GENUS *Sadasivania* Sadasivan Sac Fungus
SPECIES *aristidae, bhustha, girisa*
GENUS *Saliastrum* Saline Star Sac Fungus
SPECIES *myrtilli*
GENUS *Sanjuanomyces* San Juan Sac Fungus
SPECIES *elegans*
GENUS *Saprochaete* Sapric Bristle Sac Fungus
SPECIES *saccharophila*
GENUS *Saprophragma* Sapric Fragment Sac Fungus
SPECIES *acerosum*
GENUS *Sarbhoyomyces* Sarbhoy Sac Fungus
SPECIES *biguttulatus*

GENUS *Sarcinomyces* Sarcin Sac Fungus
SPECIES *crustaceus, petricola, phaeomuriformis*
GENUS *Sarcopodium* Flesh Pod Sac Fungus
SPECIES *circinatum, circinosetiferum, roseum, tortuosum, vanillae*
GENUS *Sarcopyrenia* Flesh Fire Sac Fungus
SPECIES *beckhausiana, cylindrospora, gibba*
VARIETIES *gibba geisleri*
GENUS *Sarocladium* Saro Branched Sac Fungus
SPECIES *mycophilum, oryzae*
GENUS *Satchmopsis* Satchmo Sac Fungus
SPECIES *brasiliensis, sacciformis*
GENUS *Scenomyces* Scene Sac Fungus
SPECIES *perplexans*
GENUS *Sceptrifera* Scepter Sac Fungus
SPECIES *pulchra*
GENUS *Schiffnerula* Schiffner Sac Fungus
SPECIES *pulchra*
GENUS *Schizotrichum* Split Hair Sac Fungus
SPECIES *conocarpi*
GENUS *Schroeteria* Schroeter Sac Fungus
SPECIES *delastrina*
GENUS *Sclerococcum* Hard Round Sac Fungus
SPECIES *montagnei, normandinae, simplex, sphaerale*
GENUS *Sclerographiopsis* Hard Graphic Sac Fungus
SPECIES *dalbergiae*
GENUS *Sclerotiella* Hard Sac Fungus
SPECIES *indica*
GENUS *Scolecobasidiella* Lesser Scolec Basal Sac Fungus
SPECIES *avellanea, tropicalis*
GENUS *Scolecobasidium* Scolec Basal Sac Fungus
SPECIES *anellii, arenarium, cateniphorum, constrictum, excentricum, fusarioideum, fusiforme, humicola, longiphorum, salinum, terreum, tropicum, tshawytschae, verruculosum*
GENUS *Scolecodochium* Scolec Doch Sac Fungus
SPECIES *malayense*
GENUS *Scolecotheca* Scolec Theca Sac Fungus
SPECIES *cornuta*
GENUS *Scopularia* Scope Sac Fungus
SPECIES *corsicana*
GENUS *Scopulariella* Lesser Scope Sac Fungus
SPECIES *vaccinii*
GENUS *Scorpiosporium* Scorpio Spore Sac Fungus
SPECIES *minutum, rangiferinum*
GENUS *Scothelius* Scot Sun Sac Fungus
SPECIES *guianensis*
GENUS *Scotiosphaeria* Scot Sphere Sac Fungus
SPECIES *endoxylinae*
GENUS *Scutisporus* Scute Spore Sac Fungus
SPECIES *brunneus*
GENUS *Scytalidium* Scyte Sac Fungus
SPECIES *acidophilum, album, aurantiacum, circinatum, flavobrunneum, hyalinum, indonesiacum, infestans, japonicum, multiseptatum, muscorum, terminale, thermophilum, uredinicola, vaccinii*
GENUS *Seimatoantlerium* Seim Antler Sac Fungus
SPECIES *nepalense, tepuiense*
GENUS *Seimatosporiella* Seim Spore Sac Fungus
SPECIES *salvadorae*
GENUS *Seimatosporiopsis* Lesser Seim Spore Sac Fungus
SPECIES *salvadorae*
GENUS *Selenodriella* Selen Sac Fungus
SPECIES *fertilis, inaequilaterospora, indica, intermedia, perramosa*
GENUS *Selenosporella* Selen Spore Sac Fungus
SPECIES *curvispora, cymbiformis*
GENUS *Selenosporopsis* Lesser Selen Spore Sac Fungus
SPECIES *elegans*
GENUS *Selenozyma* Selen Zyme Sac Fungus
SPECIES *intestinalis, peltata*
GENUS *Septocyta* Septocyte Sac Fungus
SPECIES *ruborum, visci-britannicae*

VARIETIES *ruborum macrospora*
GENUS *Septodochium* Septic Doch Sac Fungus
SPECIES *triseptatum*
GENUS *Septogloeum* Septic Glow Sac Fungus
SPECIES *carthusianum, equiseti, gillii, glycines, oxysporum*
GENUS *Septomyrothecium* Septic Theca Sac Fungus
SPECIES *uniseptatum*
GENUS *Septopatella* Septic Patella Sac Fungus
SPECIES *septata*
GENUS *Septosporium* Septic Spore Sac Fungus
SPECIES *bulbotrichum*
GENUS *Septotrullula* Septic Trull Sac Fungus
SPECIES *bacilligera*
VARIETIES *bacilligera cambrica*
GENUS *Servazziella* Servazzi Sac Fungus
SPECIES *longispora*
GENUS *Sesquicillium* Sesquicil Sac Fungus
SPECIES *asymmetricum, buxi, candelabrum, impariphiale, microsporum, pseudosetosum, setosum*
GENUS *Sessiliospora* Sessile Spore Sac Fungus
SPECIES *bicolor*
GENUS *Setolibertella* Bristle Libert Sac Fungus
SPECIES *solani*
GENUS *Setophiale* Bristle Phial Sac Fungus
SPECIES *unisetulata*
GENUS *Setosporella* Bristle Spore Sac Fungus
SPECIES *mahmoudii*
GENUS *Setosynnema* Bristle Thread Sac Fungus
SPECIES *isthmosporum*
GENUS *Seychellomyces* Seychelles Sac Fungus
SPECIES *hexagonus*
GENUS *Shawiella* Shaw Sac Fungus
SPECIES *grevilleae*
GENUS *Shrungabeeja* Shrungabeej Sac Fungus
SPECIES *vadirajensis*
GENUS *Sirococcus* Pitted Round Sac Fungus
SPECIES *conigenus, strobilinus*
GENUS *Sirodesmium* Pitted Fiber Sac Fungus
SPECIES *diversum, olivaceum*
GENUS *Siroplacodium* Pitted Plaque Sac Fungus
SPECIES *atrum, caulocarpum, longisporum, shastense, umbelliferarum*
GENUS *Sirothyriella* Pitted Thyre Sac Fungus
SPECIES *pini-austriacae*
GENUS *Sirozythiella* Pitted Zyther Sac Fungus
SPECIES *sydowiana*
GENUS *Sitochora* Sit Chore Sac Fungus
SPECIES *ellipsospora*
GENUS *Solicorynespora* Soil Club Spore Sac Fungus
SPECIES *aterrima, calophylli, garciniae, kendrickii, litchii, obclavata, pseudolmediae, zapatensis*
GENUS *Solosympodiella* Soil Sympodal Sac Fungus
SPECIES *queenslandica, rigididentata*
GENUS *Soloterminospora* Soil Terminal Spore Sac Fungus
SPECIES *caffra*
GENUS *Sorosporella* Soral Spore Sac Fungus
SPECIES *uvella*
GENUS *Spegazzinia* Spegazzi Sac Fungus
SPECIES *intermedia, lobulata, ornata, sundara, tessarthra, xanthorrhoeae*
GENUS *Speiropsis* Speir Sac Fungus
SPECIES *pedatospora, scopiformis*
GENUS *Spermospora* Sperm Spore Sac Fungus
SPECIES *avenae, ciliata, lolii, poagena, subulata*
GENUS *Spermosporina* Lesser Sperm Spore Sac Fungus
SPECIES *aricola*
GENUS *Sphaeridium* Sphere Sac Fungus
SPECIES *candidum, miniatum, vitellinum*
GENUS *Sphaeriothyrium* Sphere Thyre Sac Fungus
SPECIES *filicinum, praecastrense*
GENUS *Sphaerocybe* Sphere Cybe Sac Fungus

SPECIES *concentrica*
GENUS *Sphaeronaema* Sphere Thread Sac Fungus
SPECIES *aemulans, allahabadense, colorata, conicum, indicum, leucoconium, pruinosum, rosea, rostratum, sesami, stellatum*
GENUS *Sphaeropsis* Lesser Sphere Sac Fungus
SPECIES *candollei, conspersa, glandulosa, hedericola, hyalina, menispora, necatrix, parca, ralfsii, sapinea, sumachi, thecicola, tumefaciens, visci*
GENUS *Sphaerulomyces* Least Sphere Sac Fungus
SPECIES *coralloides*
GENUS *Sphondylocephalum* Sphondylon Head Sac Fungus
SPECIES *verticillatum*
GENUS *Spicaria* Spicule Sac Fungus
SPECIES *griseola*
GENUS *Spicellum* Lesser Spicule Sac Fungus
SPECIES *roseum*
GENUS *Spinulospora* Spiny Spore Sac Fungus
SPECIES *pucciniiphila*
GENUS *Spiropes* Spire Sac Fungus
SPECIES *chusqueae*
GENUS *Spirosphaera* Spire Sphere Sac Fungus
SPECIES *beverwijkiana, carici-graminis, dimorpha, floriformis, keratinophila, lignicola, minuta, verruculosa*
GENUS *Spondylocladiella* Lesser Spondylon Branched Sac Fungus
SPECIES *botrytioides*
GENUS *Spondylocladiopsis* Least Spondylon Branched Sac Fungus
SPECIES *cupulicola*
GENUS *Spondylocladium* Spondylon Branched Sac Fungus
SPECIES *australe*
GENUS *Sporendocladia* Spore Endic Branched Sac Fungus
SPECIES *bactrospora, castaneae, foliicola, fumosa, ivoriensis, kionochaetoides, truncata*
GENUS *Sporendonema* Spore Endic Thread Sac Fungus
SPECIES *casei, purpurascens*
GENUS *Sporidesmiella* Lesser Spore Fiber Sac Fungus
SPECIES *claviformis, coronata, cuneiformis, hyalosperma, longissima, parva*
VARIETIES *hyalosperma hyalosperma, hyalosperma novae-zelandiae*
GENUS *Sporidesmium* Spore Fiber Sac Fungus
SPECIES *adscendens, altum, anglicum, atrum, aturbinatum, australiense, bombacis, cambrense, campiniae, cladii, clarkii, cookei, coronatum, cymbispermum, doliiforme, ehrenbergii, ensiforme, eupatoriicola, flexum, folliculatum, fusiforme, goidanichii, hormiscioides, ilicinum, larvatum, leptosporum, lobatum, multiseptatum, paludosum, pedunculatum, polymorphum, pseudobambusae, pyriforme, rubi, salinum, socium, tropicale, uvariicola, vagum*
GENUS *Sporocybe* Spore Cybe Sac Fungus
SPECIES *alternata, borzinii, brassicicola, hybrida, lobulata, nigrella, rhopaloides*
GENUS *Sporophiala* Spore Phial Sac Fungus
SPECIES *cookei, indica*
GENUS *Sporophora* Spore Bearing Sac Fungus
SPECIES *toruliformis*
GENUS *Sporotrichopsis* Spore Hair Sac Fungus
SPECIES *terrestris*
GENUS *Stachybotrys* Stack Berry Sac Fungus
SPECIES *alternans, chartarum, cylindrospora, dichroa, elegans, kampalensis, longispora, microspora, nephrospora, nilagirica, oenanthes, parvispora, sansevieriae, theobromae*
GENUS *Stachycoremium* Stack Corem Sac Fungus
SPECIES *parvulum*
GENUS *Stachylidium* Stack Sac Fungus
SPECIES *bicolor, extorre*
GENUS *Stagonosporopsis* Stag Spore Sac Fungus
SPECIES *curtisii, equiseti, salicorniae*
GENUS *Staheliella* Stahel Sac Fungus
SPECIES *nodosa*
GENUS *Staninwardia* Staninward Sac Fungus
SPECIES *breviuscula*
GENUS *Stanjemonium* Stanjemon Sac Fungus
SPECIES *fuscescens, grisellum, ochroroseum, spectabile*
GENUS *Staphylotrichum* Staph Hair Sac Fungus
SPECIES *coccosporum*
GENUS *Stauronema* Crosshatch Thread Sac Fungus
SPECIES *cruciferum, indicum, pingue, platense, sacchari, spinificis*

GENUS *Stellifraga* — Stella Fragment Sac Fungus
SPECIES *cladoniicola*
GENUS *Stellomyces* — Stella Sac Fungus
SPECIES *suidafrikanus*
GENUS *Stenocarpella* — Thin Carpel Sac Fungus
SPECIES *macrospora, maydis*
GENUS *Stenocephalopsis* — Thin Head Sac Fungus
SPECIES *subalutacea*
GENUS *Stenocladiella* — Thin Branched Sac Fungus
SPECIES *neglecta*
GENUS *Stenospora* — Thin Spore Sac Fungus
SPECIES *uredinicola*
GENUS *Stephembruneria* — Stephembrun Sac Fungus
SPECIES *elegans*
GENUS *Sterigmatobotrys* — Sterigmate Berry Sac Fungus
SPECIES *macrocarpa*
GENUS *Stevensomyces* — Stevens Sac Fungus
SPECIES *palmae*
GENUS *Stevensonula* — Stevenson Sac Fungus
SPECIES *ciliata*
GENUS *Stigmatea* — Stigma Sac Fungus
SPECIES *aegopodii, ostruthii, pelvetiae, polygoni, sylvatica*
GENUS *Stilbospora* — Stilbe Spore Sac Fungus
SPECIES *macrosperma, meridionalis, proteae*
GENUS *Strasseria* — Strasser Sac Fungus
SPECIES *geniculata*
GENUS *Strasseriopsis* — Lesser Strasser Sac Fungus
SPECIES *tsugae*
GENUS *Stratiphoromyces* — Layered Sac Fungus
SPECIES *brunneisporus*
GENUS *Striosphaeropsis* — Striped Sphere Sac Fungus
SPECIES *mirabilis*
GENUS *Stromatopogon* — Stroma Beard Sac Fungus
SPECIES *baldwinii*
GENUS *Stygiomyces* — Stygid Sac Fungus
SPECIES *galliformis*
GENUS *Stylaspergillus* — Aspergill Sac Fungus
SPECIES *laxus*
GENUS *Subicularium* — Subic Sac Fungus
SPECIES *reticulatum*
GENUS *Subramaniomyces* — Lesser Raman Sac Fungus
SPECIES *fusisaprophyticus*
GENUS *Subulispora* — Lesser Ulispore Sac Fungus
SPECIES *africana, britannica, cylindrospora, elegantissima, malaysiana, minima, procurvata, rectilineata, variabilis*
GENUS *Sutravarana* — Sutravaran Sac Fungus
SPECIES *samala*
GENUS *Suttoniella* — Sutton Sac Fungus
SPECIES *eriobotryae, gaubae, ixorae*
GENUS *Suttonina* — Lesser Sutton Sac Fungus
SPECIES *guatemaltica*
GENUS *Syamithabeeja* — Syamithabeej Sac Fungus
SPECIES *girisa*
GENUS *Symbiotaphrina* — Symbiote Aphrine Sac Fungus
SPECIES *buchneri, kochii*
GENUS *Symphysos* — Symphysal Sac Fungus
SPECIES *sapindaceifolii*
GENUS *Sympodiella* — Sympodial Sac Fungus
SPECIES *acicola, foliicola, multiseptata, roystoneae*
GENUS *Sympodiocladium* — Sympodial Branched Sac Fungus
SPECIES *frondosum*
GENUS *Sympodioclathra* — Sympodial Lattice Sac Fungus
SPECIES *globosa*
GENUS *Sympodiophora* — Lesser Sympodial Sac Fungus
SPECIES *viridigrisea*
GENUS *Sympodioplanus* — Sympodial Flat Sac Fungus
SPECIES *capensis*

GENUS *Synchronoblastia* Synchronoblast Sac Fungus
SPECIES *crypta*
GENUS *Syncladium* Synon Branched Sac Fungus
SPECIES *nietneri*
GENUS *Synglioc ladium* Synglion Branched Sac Fungus
SPECIES *acridiorum, aranearum, cleoni, intricatum, tetanopsis*
GENUS *Synnemaseimatoides* Synon Thread Seim Sac Fungus
SPECIES *nipponica*
GENUS *Synnematomyces* Synon Thread Sac Fungus
SPECIES *capitatus*
GENUS *Synnemellisia* Synon Thread Ellis Sac Fungus
SPECIES *hyalospora*
GENUS *Taeniolina* Ribbon Sac Fungus
SPECIES *centaurii, scripta*
GENUS *Talekpea* Talekpe Sac Fungus
SPECIES *foeticia*
GENUS *Talpapellis* Talpapel Sac Fungus
SPECIES *peltigerae*
GENUS *Tamsiniella* Tamsin Sac Fungus
SPECIES *labiosa*
GENUS *Tandonella* Lesser Tandon Sac Fungus
SPECIES *cubensis*
GENUS *Tandonia* Tandon Sac Fungus
SPECIES *dargentiana*
GENUS *Tarsodisporus* Tarsal Spore Sac Fungus
SPECIES *paraensis*
GENUS *Tawdiella* Tawdy Sac Fungus
SPECIES *doliiformis*
GENUS *Taxomyces* Taxon Sac Fungus
SPECIES *andreanae*
GENUS *Tectacervulus* Tecton Stag Sac Fungus
SPECIES *mahoniae*
GENUS *Temerariomyces* Temer Sac Fungus
SPECIES *acutulus*
GENUS *Teratosperma* Terat Spore Sac Fungus
SPECIES *microsporum, sclerotivorum*
GENUS *Termitariopsis* Termite Sac Fungus
SPECIES *cavernosa*
GENUS *Tetrabrachium* Tetrad Short Sac Fungus
SPECIES *elegans*
GENUS *Tetrabrunneospora* Tetrad Brown Spore Sac Fungus
SPECIES *ellisii*
GENUS *Tetrachaetum* Tetrad Bristle Sac Fungus
SPECIES *elegans*
GENUS *Tetracladium* Tetrad Branched Sac Fungus
SPECIES *apiense, breve, furcatum, marchalianum, palmatum, setigerum*
GENUS *Tetracoccosporium* Tetrad Round Spore Sac Fungus
SPECIES *aerium, paxianum*
GENUS *Tetrameronycha* Tetrad Claw Sac Fungus
SPECIES *bonaerensis*
GENUS *Tetraposporium* Tetrad Spore Sac Fungus
SPECIES *asterinearum, ravenelii*
GENUS *Textotheca* Text Theca Sac Fungus
SPECIES *caffra*
GENUS *Thallospora* Thallus Spore Sac Fungus
SPECIES *aspera*
GENUS *Thamnogalla* Thamnion Gallinule Sac Fungus
SPECIES *crombiei*
GENUS *Thaptospora* Thaptospore Sac Fungus
SPECIES *atrobrunnea*
GENUS *Thermomyces* Thermal Sac Fungus
SPECIES *ibadanensis, lanuginosus, stellatus, verrucosus*
GENUS *Thirumalacharia* Thirumalach Sac Fungus
SPECIES *curcumae*
GENUS *Thozetella* Thozet Sac Fungus
SPECIES *cristata, cubensis, effusa, tocklaiensis*
GENUS *Thyridaria* Thyre Sac Fungus

SPECIES *macrostomoides, minima, rubronotata*
GENUS *Thyridella* Lesser Thyre Sac Fungus
SPECIES *canadensis, colliculus, subrostrata*
GENUS *Thyronectria* Nectar Thyre Sac Fungus
SPECIES *balsamea, berolinensis, denigrata, lamyi*
GENUS *Thyrostroma* Stroma Thyre Sac Fungus
SPECIES *compactum*
GENUS *Thyrostromella* Lesser Stroma Thyre Sac Fungus
SPECIES *myriana*
GENUS *Thysanophora* Thysan Sac Fungus
SPECIES *asymmetrica, canadensis, glaucoalbida, longispora, penicillioides, striatispora, taiwanensis, taxi, verrucosa*
GENUS *Tiarosporellivora* Tiara Spore Sac Fungus
SPECIES *caricina*
GENUS *Ticogloea* Tico Glow Sac Fungus
SPECIES *guttulata*
GENUS *Tilakiopsis* Tilak Sac Fungus
SPECIES *trishulantha*
GENUS *Tirisporella* Tiri Spore Sac Fungus
SPECIES *beccariana*
GENUS *Titaeopsis* Tit Sac Fungus
SPECIES *ugandae*
GENUS *Titaeospora* Tit Spore Sac Fungus
SPECIES *equiseti*
GENUS *Tomenticola* Tomentose Sac Fungus
SPECIES *trematis*
GENUS *Torpedospora* Torpedo Spore Sac Fungus
SPECIES *ambispinosa, radiata*
GENUS *Torula* Torule Sac Fungus
SPECIES *abbreviata, caligans, cyanescens, deospora, divaricata, expansa, fusca, herbarum, ligniperda, lucifuga, ndjilensis, ovalispora, pulveracea, sporendonema, terrestris*
VARIETIES *herbarum quaternella*
GENUS *Torulomyces* Greater Torule Sac Fungus
SPECIES *lagena*
GENUS *Torulopsiella* Least Torule Sac Fungus
SPECIES *fumaginea, pseudogyroceras*
GENUS *Torulopsis* Lesser Torule Sac Fungus
SPECIES *gropengiesseri, versatilis*
GENUS *Toxosporiella* Toxic Spore Sac Fungus
SPECIES *bactrodesmioides*
GENUS *Trematophoma* Tremor Foam Sac Fungus
SPECIES *lignicola*
GENUS *Tremellina* Tremor Sac Fungus
SPECIES *pyrenophila*
GENUS *Tretophragmia* Fragment Treton Sac Fungus
SPECIES *longispora, nilgirensis*
GENUS *Tretopileus* Pile Treton Sac Fungus
SPECIES *indicus, opuntiae, sphaerophorus*
GENUS *Tretospeira* Speir Treton Sac Fungus
SPECIES *ugandensis*
GENUS *Tretovularia* Vular Treton Sac Fungus
SPECIES *villiana*
GENUS *Triacutus* Triad Cuticle Sac Fungus
SPECIES *subcuticularis*
GENUS *Triadelphia* Triadelphia Sac Fungus
SPECIES *diversa, heterospora, loudetiae, pulvinata, romanica, stilboidea, uniseptata*
GENUS *Triblidium* Trible Sac Fungus
SPECIES *caliciforme, crarense, octosporum*
GENUS *Tribolospora* Triad Bowl Spore Sac Fungus
SPECIES *sycopsidis*
GENUS *Tricellula* Triple Cell Sac Fungus
SPECIES *aquatica, aurantiaca, botryosa, curvata, inaequalis, patens*
GENUS *Trichobotrys* Hairy Berry Sac Fungus
SPECIES *effusa, pannosa*
GENUS *Trichocladium* Hairy Branched Sac Fungus
SPECIES *achrasporum, asperum, brosimi, canadense, englandense, ismailiense, lobatum, macrosporum, opacum, pyriforme*

GENUS *Trichoconis* Hairy Cone Sac Fungus
SPECIES *echinophila, lichenicola*
GENUS *Trichodochium* Hairy Doch Sac Fungus
SPECIES *pirozynskii*
GENUS *Trichopeltulum* Hairy Pelt Sac Fungus
SPECIES *pulchellum*
GENUS *Trichoseptoria* Hairy Septic Sac Fungus
SPECIES *maydina*
GENUS *Trichosporonoides* Hairy Spore Sac Fungus
SPECIES *madida, megachiliensis, nigrescens, oedocephalis, spathulata*
GENUS *Trichothecium* Hairy Theca Sac Fungus
SPECIES *campaniforme, domesticum, inaequale, parvum, piriferum, plasmoparae, roseum*
GENUS *Tricladiopsis* Triple Branched Sac Fungus
SPECIES *flagelliformis, foliosa*
GENUS *Tridentaria* Trident Sac Fungus
SPECIES *carnivora, glossopaga, implicans, subuliphora*
GENUS *Trimmatostroma* Trim Stroma Sac Fungus
SPECIES *abietis, betulinum, cordae, elginense, salicis, scutellare*
GENUS *Tripoconidium* Tripe Cone Sac Fungus
SPECIES *aphanophagum*
GENUS *Triscelophorus* Triple Hard Sac Fungus
SPECIES *acuminatus, magnificus, monosporus*
GENUS *Triscelosporium* Triple Hard Spore Sac Fungus
SPECIES *verrucosum*
GENUS *Trisulcosporium* Triple Furrowed Spore Sac Fungus
SPECIES *acerinum*
GENUS *Tritirachium* Tritirach Sac Fungus
SPECIES *cinnamomeum, dependens, heimii, isariae, oryzae*
VARIETIES *heimii heimii, heimii griseum*
GENUS *Trochophora* Wheel Sac Fungus
SPECIES *fasciculata*
GENUS *Tromeropsis* Tromer Sac Fungus
SPECIES *microtheca*
GENUS *Troposporella* Lesser Tropospore Sac Fungus
SPECIES *fumosa, monospora*
GENUS *Troposporium* Tropospore Sac Fungus
SPECIES *album*
GENUS *Troposporopsis* Least Tropospore Sac Fungus
SPECIES *atroapicis, rigidospora*
GENUS *Trullula* Trull Sac Fungus
SPECIES *melanochlora, olivascens*
GENUS *Tryssglobulus* Tryss Globule Sac Fungus
SPECIES *aspergilloides*
GENUS *Tuberculispora* Tubercule Spore Sac Fungus
SPECIES *jamaicensis*
GENUS *Tumularia* Tumor Sac Fungus
SPECIES *aquatica*
GENUS *Tunicago* Tunic Sac Fungus
SPECIES *triodiae, uniolae*
GENUS *Tunstallia* Tunstall Sac Fungus
SPECIES *aculeata*
GENUS *Tympanosporium* Tympan Spore Sac Fungus
SPECIES *parasiticum*
GENUS *Uberispora* Uber Spore Sac Fungus
SPECIES *heteroseptata, simplex, tropicalis*
GENUS *Ubrizsya* Ubrizsy Sac Fungus
SPECIES *oxycocci*
GENUS *Ulocoryphus* Ulocoryph Sac Fungus
SPECIES *mastigophorus*
GENUS *Umbellidion* Umbel Sac Fungus
SPECIES *radulans*
GENUS *Uncispora* Uncispore Sac Fungus
SPECIES *harroldiae*
GENUS *Urohendersonia* Tailed Henderson Sac Fungus
SPECIES *platensis, stipae, subterranea*
GENUS *Urohendersoniella* Lesser Tailed Henderson Sac Fungus
SPECIES *mastigospora*

GENUS *Uvarispora* Uvar Spore Sac Fungus
SPECIES *lignicola*
GENUS *Vagnia* Vagni Sac Fungus
SPECIES *cirriformis*
GENUS *Valsaria* Vals Sac Fungus
SPECIES *anserina, bambusae, capronii, fulvopruinata, insitiva, parmularia, petrakii, rubricosa*
GENUS *Vanakripa* Vanakrip Sac Fungus
SPECIES *gigaspora, parva*
GENUS *Vanbeverwijkia* Vanbeverwijki Sac Fungus
SPECIES *spirospora*
GENUS *Vanterpoolia* Vanterpool Sac Fungus
SPECIES *tsugae*
GENUS *Variocladium* Variable Branched Sac Fungus
SPECIES *giganteum, rangiferinum*
GENUS *Vasculomyces* Vascule Sac Fungus
SPECIES *xanthosomae*
GENUS *Velutipila* Velvet Pile Sac Fungus
SPECIES *poeltii*
GENUS *Ventrographium* Ventral Graphic Sac Fungus
SPECIES *nervicola*
GENUS *Venustisporium* Venus Spore Sac Fungus
SPECIES *chelyoforme*
GENUS *Venustocephala* Venus Head Sac Fungus
SPECIES *aequatorialis*
GENUS *Venustosynnema* Venus Thread Sac Fungus
SPECIES *ciliatum*
GENUS *Veramyces* Vera Sac Fungus
SPECIES *manuensis*
GENUS *Verdipulvinus* Verdant Powder Sac Fungus
SPECIES *nolinae*
GENUS *Vermiculariopsiella* Worm Sac Fungus
SPECIES *immersa*
GENUS *Vermispora* Worm Spore Sac Fungus
SPECIES *fusarina, fusarioides*
GENUS *Veronaea* Verona Sac Fungus
SPECIES *botryosa, caricis, carlinae, compacta, constricta, musae, simplex*
GENUS *Veronaella* Lesser Verona Sac Fungus
SPECIES *indica*
GENUS *Verrucariella* Verrucose Sac Fungus
SPECIES *capparicola*
GENUS *Verrucophragmia* Verrucose Fragment Sac Fungus
SPECIES *splendens*
GENUS *Verticicladiella* Vertic Branched Sac Fungus
SPECIES *humicola*
GENUS *Verticimonosporium* Vertic Monad Spore Sac Fungus
SPECIES *diffractum, ellipticum*
GENUS *Vesicladiella* Vesicle Branched Sac Fungus
SPECIES *capitata*
GENUS *Vestigium* Vestige Sac Fungus
SPECIES *felicis*
GENUS *Vialaea* Vial Sac Fungus
SPECIES *insculpta, minutella*
GENUS *Virgaria* Virgar Sac Fungus
SPECIES *nigra*
GENUS *Virgariella* Lesser Virgar Sac Fungus
SPECIES *atra, globigera, ovoidea*
GENUS *Virgatospora* Virgate Spore Sac Fungus
SPECIES *echinofibrosa*
GENUS *Vizella* (Synonym for *Blasdalea*)
GENUS *Vleugelia* Vleugel Sac Fungus
SPECIES *betulina*
GENUS *Volucrispora* Volucrate Spore Sac Fungus
SPECIES *graminea, graminea*
GENUS *Volutina* Volute Sac Fungus
SPECIES *concentrica*
GENUS *Vouauxiella* Lesser Vouaux Sac Fungus
SPECIES *caloplacae, lichenicola, pithospora, uniseptata, verrucosa*

GENUS *Vouauxiomyces* Vouaux Sac Fungus
SPECIES *ramalinae, santessonii, truncatus*
GENUS *Wallemia* Wallem Sac Fungus
SPECIES *sebi*
GENUS *Waydora* Waydor Sac Fungus
SPECIES *typica*
GENUS *Websteromyces* Webster Sac Fungus
SPECIES *verruculosum*
GENUS *Weufia* Weuf Sac Fungus
SPECIES *tewoldei*
GENUS *Wiesneriomyces* Wiesner Sac Fungus
SPECIES *laurinus*
GENUS *Wilsoniella* Wilson Sac Fungus
SPECIES *ovalis*
GENUS *Wojnowicia* Wojnowic Sac Fungus
SPECIES *hirta*
GENUS *Wolkia* Wolk Sac Fungus
SPECIES *decolorans*
GENUS *Xanthoriicola* Yellow Sac Fungus
SPECIES *physciae*
GENUS *Xenidiocercus* Strange Dual Rear Antenna Sac Fungus
SPECIES *macarangae, macrospora, pyriformis*
GENUS *Xenobotrytis* Strange Berry Sac Fungus
SPECIES *acaducospora*
GENUS *Xenogliocladiopsis* Strange Glow Branched Sac Fungus
SPECIES *eucalyptorum*
GENUS *Xenoheteroconium* Strange Variable Cone Sac Fungus
SPECIES *bicolor*
GENUS *Xenoplaca* Strange Plaque Sac Fungus
SPECIES *aequatoriensis*
GENUS *Xepicula* Xepicule Sac Fungus
SPECIES *leucotricha, leucotrichoides*
GENUS *Xepiculopsis* Lesser Xepicule Sac Fungus
SPECIES *graminea, perpulchra*
GENUS *Xeroconium* Xeric Cone Sac Fungus
SPECIES *boreale*
GENUS *Xiambola* Xiambol Sac Fungus
SPECIES *mirabilis*
GENUS *Xiphomyces* Xiph Sac Fungus
SPECIES *caricis, sacchari*
GENUS *Xylochia* Chi Wood Sac Fungus
SPECIES *stegonsporioides*
GENUS *Xyloglyphis* Glyph Wood Sac Fungus
SPECIES *striola*
GENUS *Xylogone* Gonad Wood Sac Fungus
SPECIES *sphaerospora*
GENUS *Xylohypha* Hypha Wood Sac Fungus
SPECIES *curta, ferruginosa, nigrescens, ortmansiae, pinicola*
GENUS *Xylomyces* Wood Sac Fungus
SPECIES *aquaticus*
GENUS *Yalomyces* Yalon Sac Fungus
SPECIES *pulcher*
GENUS *Ybotromyces* Bottom Sac Fungus
SPECIES *caespitosus*
GENUS *Yinmingella* Yin Ming Sac Fungus
SPECIES *mitriformis*
GENUS *Yuccamyces* Yucca Sac Fungus
SPECIES *canadensis, cubensis, luteus, pilosus, purpureus*
GENUS *Yunnania* Yunnan Sac Fungus
SPECIES *penicillata*
GENUS *Zakatoshia* Zakatoshi Sac Fungus
SPECIES *erikssonii, hirschiopori*
GENUS *Zalerion* Zaler Sac Fungus
SPECIES *arboricola, maritima*
GENUS *Zanclospora* Zancle Spore Sac Fungus
SPECIES *austroamericana, brevispora, indica, mystica, novae-zelandiae, stellata*
GENUS *Zebrospora* Zebra Spore Sac Fungus

SPECIES *bicolor*
GENUS *Zelandiocoela* Zealand Sky Sac Fungus
SPECIES *ambigua*
GENUS *Zeloasperisporium* Zelon Asp Spore Sac Fungus
SPECIES *hyphopodioides*
GENUS *Zelopelta* Zelon Pelt Sac Fungus
SPECIES *thrinacospora*
GENUS *Zetesimomyces* Zetes Sac Fungus
SPECIES *setibicolor*
GENUS *Zetiasplozna* Zeti Splozna Sac Fungus
SPECIES *caffra, cordylines, heteromorpha, thuemenii, unicolor*
GENUS *Zevadia* Zevad Sac Fungus
SPECIES *peroccidentalis*
GENUS *Zignoina* Zignoin Sac Fungus
SPECIES *subcorticalis*
GENUS *Zinzipegasa* Zinzipegas Sac Fungus
SPECIES *argentinensis*
GENUS *Zopheromyces* Zopher Sac Fungus
SPECIES *fasciatus*
GENUS *Zunura* Zuni Tailed Sac Fungus
SPECIES *appendiculata*
GENUS *Zygodesmus* Zygote Fiber Sac Fungus
SPECIES *fuscus*
GENUS *Zygosporium* Zygospore Sac Fungus
SPECIES *chartarum, deightonii, echinosporum, geminatum, gibbum, majus, masonii, minus, mycophilum, oscheoides, tuberculatum, verticillatum*
GENUS *Zythia* Zyther Sac Fungus
SPECIES *pinastri*
GENUS *Zyxiphora* Zyxiphore Sac Fungus
SPECIES *gorakhpurensis*

FAMILY 02 *Annulatascaceae* Ring Sac Fungus
GENUS *Annulatascus* (Synonym for *Cataractispora*)
GENUS *Aquasphaeria* Aquatic Ring Sac Fungus
SPECIES *dimorphospora*
GENUS *Cataractispora* Cataract Spore Ring Sac Fungus
SPECIES *appendiculata, aquatica, bipolaris, viscosa*
GENUS *Cataractispora* Horn Ring Sac Fungus
SPECIES *crinigera, fragilis, lampadophora, mycophila, ordinata, rhenana*
GENUS *Ceratostomella* Horn Stoma Ring Sac Fungus
SPECIES *dubia, investita*
GENUS *Diluviicola* Diluvial Ring Sac Fungus
SPECIES *capensis*
GENUS *Rhamphoria* Rhamphore Ring Sac Fungus
SPECIES *bevanii, pyriformis, separata, tympanidispora*
GENUS *Torrentispora* Torrent Spore Ring Sac Fungus
SPECIES *fibrosa*
GENUS *Vertexicola* Vertex Ring Sac Fungus
SPECIES *caudatus*

FAMILY 03 *Apiosporaceae* Parsley Spore Sac Fungus
GENUS *Apiospora* Parsley Spore Sac Fungus
SPECIES *camptospora, imperatae, montagnei, setosa*
GENUS *Arthrinium* Arthrine Parsley Spore Sac Fungus
SPECIES *aureum, caricicola, cuspidatum, euphorbiae, francii, fuckelii, georgii, hispanicum, hysterium, javii, marii, mediterranei, morthieri, phaeospermum, philipii, puccinioides, sacchari, serenense, sphaerospermum, sporophleum, urticae*
VARIETIES *aureum crassa, aureum minor*
GENUS *Cordella* Cord Parsley Spore Sac Fungus
SPECIES *clarkii*
GENUS *Lasiobertia* Lasiobert Parsley Spore Sac Fungus
SPECIES *africana*
GENUS *Pteroconium* Wing Cone Parsley Spore Sac Fungus
SPECIES *intermedium*
GENUS *Scyphospora* Scyph Parsley Spore Sac Fungus
SPECIES *hysterina*

FAMILY 04 *Ascosphaeraceae* Sphere Sac Fungus
GENUS *Ascosphaera* Sphere Sac Fungus

SPECIES *acerosa, apis, asterophora, atra, celerrima, cinnamomea, columbrina, fusiformis, larvis, major, naganensis, osmophila, pollenicola, proliperda, xerophila*
GENUS *Arrhenosphaera* Arrhen Sphere Sac Fungus
SPECIES *craneae*
GENUS *Bettsia* Betts Sphere Sac Fungus
SPECIES *alvei*

FAMILY 05 *Aspidotheliaceae* Asp Thele Sac Fungus
GENUS *Aspidothelium* Asp Thele Sac Fungus
SPECIES *cinerascens, fugiens, scutellicarpum, verruculosum*
GENUS *Musaespora* Banana Spore Asp Thele Sac Fungus
SPECIES *coccinea, corticola, epiphylla, gigas, kalbii, kassamensis, multilocularis*

FAMILY 06 *Asterinaceae* Star Sac Fungus
GENUS *Asterina* Star Sac Fungus
SPECIES *veronicae*
GENUS *Allothyrium* Thyre Star Sac Fungus
SPECIES *marcgraviae*
GENUS *Aulographina* Aulophyte Graphic Star Sac Fungus
SPECIES *eucalypti, pinorum*
GENUS *Lembosia* Lembos Star Sac Fungus
SPECIES *luzulae*
GENUS *Lembosina* Lesser Lembos Star Sac Fungus
SPECIES *aulographoides, gontardii*
GENUS *Leveillella* Leveille Star Sac Fungus
SPECIES *drimydis*
GENUS *Macowaniella* Macowan Star Sac Fungus
SPECIES *congesta, myrsinicola*
GENUS *Morenoina* Morenoin Star Sac Fungus
SPECIES *arundinariae, chamaecyparidis, clarkii, epilobii, festucae, fimbriata, minuta, paludosa, phragmitis, pteridicola, rhododendri, websteri*
GENUS *Parasterinopsis* Lesser Star Sac Fungus
SPECIES *caesalpiniae, orchidacearum, sersalisiae*
GENUS *Pirozynskia* Pirozynsky Star Sac Fungus
SPECIES *farriae, shoemakeri*
GENUS *Symphaster* Symphyte Star Sac Fungus
SPECIES *areolata, gesneraceae*
GENUS *Thyriopsis* Thyre Star Sac Fungus
SPECIES *halepensis*
GENUS *Triposporium* Tripe Spore Star Sac Fungus
SPECIES *elegans*
GENUS *Yamamotoa* Yamamoto Star Sac Fungus
SPECIES *bomjardinensis, carludovicae, guarapensis*

FAMILY 07 *Batistiaceae* Batista Sac Fungus
GENUS *Batistia* Batista Sac Fungus
SPECIES *annulipes*
GENUS *Acrostroma* Tall Stroma Batista Sac Fungus
SPECIES *annellosynnema*
GENUS *Batistinula* Lesser Batista Sac Fungus
SPECIES *gallesiae*
GENUS *Batistospora* Spore Batista Sac Fungus
SPECIES *septentrionalis*

FAMILY 08 *Boliniaceae* Bolin Sac Fungus
GENUS *Bolinia* (Synonym for *Camarops*)
GENUS *Apiocamarops* Parsley Camaro Bolin Sac Fungus
SPECIES *alba, cryptocellula*
GENUS *Camaropella* Lesser Camaro Bolin Sac Fungus
SPECIES *pugillus*
GENUS *Camarops* Camaro Bolin Sac Fungus
SPECIES *lutea, microspora, petersii, polysperma, tubulina*
GENUS *Catabotrys* Chain Berry Bolin Sac Fungus
SPECIES *deciduum*
GENUS *Endoxyla* Endic Wood Bolin Sac Fungus
SPECIES *cirrhosa, laevirostris, parallela, rostrata, vestita*
VARIETIES *vestita varvicensis*
GENUS *Linostomella* Linostome Bolin Sac Fungus
SPECIES *sphaerosperma*
GENUS *Neohypodiscus* Modern Hyponic Disc Bolin Sac Fungus

SPECIES *cerebrinus, irradians, rickii*
GENUS *Pseudovalsaria* Lesser Vals Bolin Sac Fungus
SPECIES *ferruginea*

FAMILY 09 *Calosphaeriaceae* Calico Sphere Sac Fungus
GENUS *Calosphaeria* Calico Sphere Sac Fungus
SPECIES *barbirostris, cyclospora, dryina, parasitica, princeps, pulchella, pusilla, wahlenbergii*
GENUS *Jattaea* Jatta Calico Sphere Sac Fungus
SPECIES *faginea, microtheca*
GENUS *Kacosphaeria* Kaco Calico Sphere Sac Fungus
SPECIES *antarctica*
GENUS *Pachytrype* Thick Trype Calico Sphere Sac Fungus
SPECIES *graphidioides, princeps*
GENUS *Phaeoisaria* Dusky Calico Sphere Sac Fungus
SPECIES *clavulata, clematidis, curvata, glauca*
GENUS *Phragmocalosphaeria* Fragment Calico Sphere Sac Fungus
SPECIES *piskorzii*
GENUS *Pleurostoma* Pleural Stoma Calico Sphere Sac Fungus
SPECIES *candollei, minimum, ootheca, vibratile*

FAMILY 10 *Coryneliaceae* Club Sac Fungus
GENUS *Bicornispora* Dual Club Sac Fungus
SPECIES *exophiala*
GENUS *Caliciopsis* Calcic Club Sac Fungus
SPECIES *orientalis, pinea, pseudotsugae*
GENUS *Coryneliopsis* Lesser Club Sac Fungus
SPECIES *antarctica, cupulifera*
GENUS *Coryneliospora* Spore Club Sac Fungus
SPECIES *fructicola, rapaneicola*
GENUS *Fitzpatrickella* Fitzpatrick Club Sac Fungus
SPECIES *operculata*
GENUS *Lagenulopsis* Lagenule Club Sac Fungus
SPECIES *bispora*

FAMILY 11 *Eremomycetaceae* Sand Sac Fungus
GENUS *Eremomyces* Sand Sac Fungus
SPECIES *bilateralis*
GENUS *Rhexothecium* Rhex Theca Sand Sac Fungus
SPECIES *globosum*

FAMILY 12 *Erysiphaceae* Blister Siphon Sac Fungus
GENUS *Erysiphe* Blister Siphon Sac Fungus
SPECIES *adunca, akebiae, aquilegiae, artemisiae, astragali, australiana, azaleae, baeumleri, betae, biocellata, brunneopunctata, buhrii, catalpae, circaeae, coluteae, convolvuli, cotini, cruciferarum, depressa, echinopis, euonymi, euphorbiae, flexuosa, friesii, galeopsidis, galii, guarinonii, hedwigii, heraclei, howeana, hyperici, knautiae, limonii, lonicerae, lycopsidis, lythri, magnicellulata, mayorii, montagnei, necator, ornata, palczewskii, pisi, prunastri, rayssiae, russellii, sordida, sparsa, symphoricarpi, syringae, thermopsidis, thesii, tortilis, ulmariae, urticae, valerianae, vanbruntiana, verbasci*
VARIETIES *adunca adunca, adunca regularis, aquilegiae aquilegiae, aquilegiae ranunculi, euonymi japonici, ornata ornata, ornata europaea, pisi pisi, pisi cruchetiana, pisi medicaginis-lupulinae, pisi medicaginis-sativae, pisi viciae-sativae, vanbruntiana vanbruntiana, vanbruntiana sambuci-racemosae*
GENUS *Arthrocladiella* Straight Branch Blister Siphon Sac Fungus
SPECIES *mougeotii*
GENUS *Blumeria* Blumer Blister Siphon Sac Fungus
SPECIES *graminis*
VARIETIES *graminis avenae, graminis tritici*
GENUS *Brasiliomyces* Brazil Blister Siphon Sac Fungus
SPECIES *cylobalanopsidis, entadae, kumaonensis, malachrae, malvastri, setosus, trina*
GENUS *Golovinomyces* Golovin Blister Siphon Sac Fungus
SPECIES *cichoracearum, cynoglossi, orontii*
VARIETIES *cichoracearum cichoracearum, cichoracearum fischeri, cichoracearum latisporus*
GENUS *Leveillula* Leveille Blister Siphon Sac Fungus
SPECIES *taurica, verbasci*
GENUS *Microsphaera* Sphere Blister Siphon Sac Fungus
SPECIES *alphitoides , begoniae , berberidis , diffusa , divaricata , grossulariae , hypophylla , penicillata , platani , pseudacaciae , trifolii , trifolii trifolii, trifolii intermedia*
GENUS *Neoerysiphe* Modern Blister Siphon Sac Fungus
SPECIES *chelones, cumminsiana, galeopsidis, galii, geranii*
GENUS *Oidium* Hypha Spore Blister Siphon Sac Fungus

SPECIES *abortifaciens, anacardii, arachidis, candicans, caricae-papayae, carpini, chartarum, chrysanthemi, concentricum, cyparissiae, ericinum, heveae, hortensiae, lauracearum, lycopersici, mangiferae, neolycopersici, pullorum, sesami*

GENUS *Oospora* — Oospore Blister Siphon Sac Fungus

SPECIES *colorans, crustacea, fusca, guerciana, humi, porriginis, sulphurea, sulphurella, tenuis, uvarum, variabilis*

GENUS *Ovulariopsis* — Ovule Blister Siphon Sac Fungus

SPECIES *papayae, passiflorae*

GENUS *Phyllactinia* — Radial Blister Siphon Sac Fungus

SPECIES *fraxini, guttata, mali*

GENUS *Podosphaera* — Pod Sphere Blister Siphon Sac Fungus

SPECIES *aphanis, balsaminae, clandestina, dipsacacearum, euphorbiae, ferruginea, fugax, fusca, helianthemi, leucotricha, macularis, mors-uvae, myrtillina, pannosa, parietariae, plantaginis, polemonii, spiraeae, thalictri, tridactyla, volkartii, xanthii*

VARIETIES *clandestina clandestina, clandestina aucupariae, euphorbiae hirtae*

GENUS *Queirozia* — Queiroz Blister Siphon Sac Fungus

SPECIES *turbinata*

GENUS *Sawadaea* — Sawada Blister Siphon Sac Fungus

SPECIES *bicornis, tulasnei*

GENUS *Sphaerotheca* — Sphere Theca Blister Siphon Sac Fungus

SPECIES *castagnei, epilobii, fuliginea*

GENUS *Streptopodium* — Twisted Pod Blister Siphon Sac Fungus

SPECIES *bonariense*

GENUS *Uncinula* — Hooked Blister Siphon Sac Fungus

SPECIES *clandestina*

FAMILY 13 *Glomerellaceae* — Glome Sac Fungus

GENUS *Glomerella* — Glome Sac Fungus

SPECIES *amenti, banisteriae, cingulata, glycines, graminicola, guevinae, magna, major, miyabeana, montana, pezizoidea, phacidiomorpha, phomoides, phyllanthi, rufomaculansvaccinii, septospora, tucumanensis*

VARIETIES *cingulata cingulata, cingulata camelliae, cingulata migrans, cingulata minor*

GENUS *Haplothecium* — Single Theca Glome Sac Fungus

SPECIES *caulicola, coptidis, dioscoreae, guianense*

FAMILY 14 *Halosphaeriaceae* — Halogen Sphere Sac Fungus

GENUS *Halosphaeria* — Halogen Sphere Sac Fungus

SPECIES *appendiculata, mediosetigera, pileata, quadri-remis, salina, spartinae, torquata, tubulifera*

VARIETIES *mediosetigera mediosetigera, mediosetigera grandispora*

GENUS *Aniptodera* — Anipt Dere Halogen Sphere Sac Fungus

SPECIES *fusiformis*

GENUS *Anisostagma* — Anise Stag Halogen Sphere Sac Fungus

SPECIES *rotundatum*

GENUS *Appendichordella* — Appendage Chord Halogen Sphere Sac Fungus

SPECIES *amicta*

GENUS *Arenariomyces* — Sand Halogen Sphere Sac Fungus

SPECIES *trifurcatus*

GENUS *Bovicornua* — Bovine Horn Halogen Sphere Sac Fungus

SPECIES *intricata*

GENUS *Buxetroldia* — Buxetrold Halogen Sphere Sac Fungus

SPECIES *bisaccata*

GENUS *Carbosphaerella* — Carbon Sphere Halogen Sphere Sac Fungus

SPECIES *leptosphaerioides, pleosporoides*

GENUS *Ceriosporopsis* — Wax Spore Halogen Sphere Sac Fungus

SPECIES *calyptrata, cambrensis, circumvestita, halima, longissima*

GENUS *Chadefaudia* — Chadefaud Halogen Sphere Sac Fungus

SPECIES *balliae, corallinarum, gymnogongri, marina, polyporolithi, schizymeniae*

GENUS *Clavariopsis* — Club Halogen Sphere Sac Fungus

SPECIES *aquatica, bulbosa*

GENUS *Corallicola* — Coral Halogen Sphere Sac Fungus

SPECIES *nana*

GENUS *Corollospora* — Corollo Spore Halogen Sphere Sac Fungus

SPECIES *comata, cristata, intermedia, maritima, trifurcata*

GENUS *Cucullosporella* — Hood Spore Halogen Sphere Sac Fungus

SPECIES *mangrovei*

GENUS *Etheirophora* — Etheir Halogen Sphere Sac Fungus

SPECIES *bijubata, blepharospora, unijubata*

GENUS *Fluviatispora* — Fluviate Spore Halogen Sphere Sac Fungus

SPECIES *reticulata, tunicata*

GENUS *Haligena* — Lesser Halogen Sphere Sac Fungus

SPECIES *elaterophora, unicaudata, viscidula*

GENUS *Iwilsoniella*	Iwilson Halogen Sphere Sac Fungus
SPECIES *rotunda*	
GENUS *Kohlmeyeriella*	Kohlmeyer Halogen Sphere Sac Fungus
SPECIES *tubulata*	
GENUS *Lanspora*	Lan Spore Halogen Sphere Sac Fungus
SPECIES *coronata*	
GENUS *Lignincola*	Wood Halogen Sphere Sac Fungus
SPECIES *laevis*	
GENUS *Limacospora*	Slug Spore Halogen Sphere Sac Fungus
SPECIES *sundica*	
GENUS *Moana*	Moan Halogen Sphere Sac Fungus
SPECIES *turbinulata*	
GENUS *Naïs*	Naïs Halogen Sphere Sac Fungus
SPECIES *inornata*	
GENUS *Naufragella*	Naufrag Halogen Sphere Sac Fungus
SPECIES *delmarensis, spinibarbata*	
GENUS *Nautosphaeria*	Nauticle Halogen Sphere Sac Fungus
SPECIES *cristaminuta*	
GENUS *Nimbospora*	Nimbo Spore Halogen Sphere Sac Fungus
SPECIES *bipolaris, effusa, octonae*	
GENUS *Nohea*	Nohe Halogen Sphere Sac Fungus
SPECIES *umiumi*	
GENUS *Ophiodeira*	Serpent Halogen Sphere Sac Fungus
SPECIES *monosemeia*	
GENUS *Phaeonectriella*	Dusky Cornicle Halogen Sphere Sac Fungus
SPECIES *lignicola*	
GENUS *Remispora*	Remy Spore Halogen Sphere Sac Fungus
SPECIES *hamata, maritima, stellata, trullifera*	
GENUS *Sigmoidea*	Sigmoid Halogen Sphere Sac Fungus
SPECIES *aurantiaca, marina, prolifera*	
GENUS *Thalassogena*	Thalass Halogen Sphere Sac Fungus
SPECIES *sphaerica*	
GENUS *Trailia*	Trail Halogen Sphere Sac Fungus
SPECIES *ascophylli*	
GENUS *Trichomaris*	Hairy Sea Halogen Sphere Sac Fungus
SPECIES *invadens*	
GENUS *Tunicatispora*	Tunic Spore Halogen Sphere Sac Fungus
SPECIES *australiensis*	
GENUS *Varicosporina*	Varicose Spore Halogen Sphere Sac Fungus
SPECIES *prolifera, ramulosa*	
FAMILY 15 *Hypsostromataceae*	Hypso Stroma Sac Fungus
GENUS *Hypsostroma*	Hypso Stroma Sac Fungus
SPECIES *caimitalense, saxicola*	
GENUS *Manglicola*	Mangled Hypso Stroma Sac Fungus
SPECIES *guatemalensis, samuelsii*	
FAMILY 16 *Lulworthiaceae*	Lulworth Sac Fungus
GENUS *Lulworthia*	Lulworth Sac Fungus
SPECIES *floridana, fucicola, halima, longirostris, medusa, opaca, purpurea, rufa, salina, submersa*	
GENUS *Lindra*	Linder Lulworth Sac Fungus
SPECIES *inflata*	
FAMILY 17 *Meliolaceae*	Meliole Sac Fungus
GENUS *Meliola*	Meliole Sac Fungus
SPECIES *clavulata, ellisii, tetradeniae*	
GENUS *Appendiculella*	Appendage Meliole Sac Fungus
SPECIES *calostroma*	
GENUS *Asteridium*	Star Meliole Sac Fungus
SPECIES *juniperinum*	
GENUS *Ceratospermopsis*	Horn Spore Meliole Sac Fungus
SPECIES *cupaniae, xylopiae*	
GENUS *Cryptomeliola*	Cryptic Meliole Sac Fungus
SPECIES *moiana, natans, orbicularis*	
GENUS *Endomeliola*	Endic Meliole Sac Fungus
SPECIES *dingleyae*	
GENUS *Ophioirenina*	Serpent Renine Meliole Sac Fungus
SPECIES *theae*	

GENUS *Pauahia* — Pauah Meliole Sac Fungus
SPECIES *sideroxyli*

FAMILY 18 *Monascaceae* — Monad Sac Fungus
GENUS *Monascus* — Monad Sac Fungus
SPECIES *aurantiacus, eremophilus, floridanus, kaoliang, pilosus, purpureus, ruber*
GENUS *Basipetospora* — Basal Petospore Monad Sac Fungus
SPECIES *chlamydospora, vesicarum*
GENUS *Xeromyce* — Xeric Monad Sac Fungus
SPECIES *bisporus*

FAMILY 18 *Orbiliaceae* — Orb Sac Fungus
GENUS *Orbilia* — Orb Sac Fungus
SPECIES *acuum, alnea, arundinacea, aurantiorubra, auricolor, coccinella, comma, cruenta, curvatispora, cyathea, delicatula, epipora, eucalypti, euonymi, fimicoloides, leucostigma, luteorubella, luzularum, pellucida, rectispora, retrusa, sarraziniana, subinflatula, vinosa, xanthostigma*
GENUS *Arthrobotrys* — Straight Berry Orb Sac Fungus
SPECIES *amerospora, arthrobotryoides, botryospora, brochopaga, cladodes, clavispora, conoides, cylindrospora, dactyloides, dendroides, entomopaga, foliicola, irregularis, javanica, musiformis, oligospora, oviformis, perpasta, polycephala, pyriformis, robusta, scaphoides, superba, vermicola*
VARIETIES *cladodes cladodes, cladodes macroides*
GENUS *Dactylella* — Lesser Finger Orb Sac Fungus
SPECIES *acrochaeta, arnaudii, atractoides, candida, copepodii, dasguptae, gampsospora, heptameres, implexa, lysipaga, mammillata, megalobrocha, microaquatica, oviparasitica, oxyspora, ramosa, rhopalota, spermatophaga*
GENUS *Dactylium* — Finger Orb Sac Fungus
SPECIES *obovatum, tenellum, tenuissimum*
GENUS *Dicranidion* — Dicran Orb Sac Fungus
SPECIES *fragile, gracilis, inaequalis*
GENUS *Duddingtonia* — Duddington Orb Sac Fungus
SPECIES *flagrans*
GENUS *Dwayaangam* — Dwayaangam Orb Sac Fungus
SPECIES *cornuta*
GENUS *Hyalinia* — Hyaline Orb Sac Fungus
SPECIES *dilutella, leightonii, rosella, rubella, rubicola, subdiaphana, succinea, tumidula*
VARIETIES *dilutella dilutella, dilutella smyrnii, leightonii lignicola*
GENUS *Monacrosporium* — Monad Acrospore Orb Sac Fungus
SPECIES *asthenopagum, bembicodes, cionopagum, cystosporum, doedycoides, drechsleri, elegans, ellipsosporum, eudermatum, fusiforme, globosporum, guizhouense, haptotylum, iridis, leptosporum, megalosporum, multiforme, mutabile, oxysporum, polybrochum, psychrophilum, reticulatum, rutgeriense, salinum, sarcopodioides, subtile, tentaculatum, thaumasium*

FAMILY 19 *Phyllobatheliaceae* — Bathel Sac Fungus
GENUS *Phyllobathelium* — Bathel Sac Fungus
SPECIES *nigrum*
GENUS *Phyllocratera* — Crater Bathel Sac Fungus
SPECIES *papuana*

FAMILY 20 *Piedraiaceae* — Pied Sac Fungus
GENUS *Piedraia* — Pied Sac Fungus
SPECIES *quintanilhae, sarmentoi*
GENUS *Trichosporum* — Hairy Spore Pied Sac Fungus
SPECIES *fuscum, murinum, symbioticum*

FAMILY 21 *Pyxidiophoraceae* — Pyxie Sac Fungus
GENUS *Pyxidiophora* — Pyxie Sac Fungus
SPECIES *asterophora, caulicola, grovei, marchalii, microspora, petchii, spinulorostrata*
GENUS *Mycorhynchidium* — Lesser Beak Pyxie Sac Fungus
SPECIES *saccatum*
GENUS *Mycorhynchus* — Beak Pyxie Sac Fungus
SPECIES *brunneocapitatus*
GENUS *Rhynchonectria* — Cornicle Beak Pyxie Sac Fungus
SPECIES *longispora*

FAMILY 22 *Polystomellaceae* — Polystome Sac Fungus
GENUS *Polystomella* — Polystome Sac Fungus
SPECIES *granulosa*
GENUS *Asteromellopsis* — Star Polystome Sac Fungus
SPECIES *insculpta*
GENUS *Dothidella* — Doth Polystome Sac Fungus
SPECIES *apiculata, australis, basirufa, confluens, depazeoides, marginata, millepunctata, oleandrina*

GENUS *Munkiella* — Munk Polystome Sac Fungus
SPECIES *caa-guazu*

FAMILY 23 *Pseudoperisporiaceae* — Perion Spore Sac Fungus
GENUS *Bryochiton* — Moss Chiton Perion Spore Sac Fungus
SPECIES *heliotropicus, microscopicus, monascus, perpusillus*
GENUS *Bryomyces* — Moss Perion Spore Sac Fungus
SPECIES *caudatus, doppelbaurorum, gymnomitrii, hemisphaericus, jungermanniae, microcarpus, nigrescens, scapaniae, velenovskyi*
VARIETIES *microcarpus microcarpus, microcarpus polytrichophilus, microcarpus rhacomitri*
GENUS *Epibryon* — Epiphyte Moss Perion Spore Sac Fungus
SPECIES *arachnoideum, bryophilum, bubakii, casaresii, chorisodontii, conductrix, craspedum, cryptosphaericum, dawsoniae, deceptor, diaphanum, dicrani, elegantissimum, endocarpum, eremita, filiforme, hepaticicola, hypophyllum, intercapillare, intercellulare, interlamellare, intracellulare, leucobryi, lichenicola, maculosum, marsupidii, metzgeriae, muscicola, notabile, odontophilum, parvipunctum, pedinophylli, perrumpens, plagiochilae, pogonati-urnigeri, polyphagum, polysporum, scapaniae, solorinae, tripartitum, turfosorum*
VARIETIES *casaresii casaresii, casaresii frullaniae*
GENUS *Epipolaeum* — Epiphyte Pole Perion Spore Sac Fungus
SPECIES *longisetosum*
GENUS *Lizonia* — Lizon Perion Spore Sac Fungus
SPECIES *baldinii, emperigonia, sexangularis*
GENUS *Nematostoma* — Thread Stoma Perion Spore Sac Fungus
SPECIES *parasiticum*
GENUS *Pododimeria* — Pod Dimer Perion Spore Sac Fungus
SPECIES *andina, gallica, gelatinosa, juniperi*
GENUS *Raciborskiomyces* — (Synonym for *Epipolaeum*)
GENUS *Stellosetifera* — Star Bristle Perion Spore Sac Fungus
SPECIES *malaysiana*
GENUS *Toroa* — Toro Perion Spore Sac Fungus
SPECIES *dimerosporioides, saurauiae*
GENUS *Wentiomyces* — Went Perion Spore Sac Fungus
SPECIES *lichenicola, peltigericola, sibiricus*
VARIETIES *lichenicola lichenicola, lichenicola bouteillei*

FAMILY 24 *Roesleriaceae* — Roesler Sac Fungus
GENUS *Roesleria* — Roesler Sac Fungus
SPECIES *subterranea*
GENUS *Roeslerina* — Lesser Roesler Sac Fungus
SPECIES *media*

FAMILY 25 *Saccardiaceae* — Heart Sac Fungus
GENUS *Angatia* — Angat Heart Sac Fungus
SPECIES *brasiliensis, eugeniae, palmarum, rondoniensis, thwaitesii*
GENUS *Ascolectus* — Lector Heart Sac Fungus
SPECIES *albus*
GENUS *Cyanodiscus* — Cyan Disc Heart Sac Fungus
SPECIES *glabrescens, occidentalis*
GENUS *Dictyonella* — Netted Heart Sac Fungus
SPECIES *alangii, circinata, dictyospora, erysiphoides, mirabilis, scabra, tabebuiae*
GENUS *Epibelonium* — Epiphyte Belon Heart Sac Fungus
SPECIES *gaeumannii*
GENUS *Pseudodiscus* — Lesser Disc Heart Sac Fungus
SPECIES *nootkatensis*
GENUS *Rivilata* — Rivil Heart Sac Fungus
SPECIES *ius*

FAMILY 26 *Seuratiaceae* — Seurat Sac Fungus
GENUS *Seuratia* — Seurat Sac Fungus
SPECIES *globifera, maunauluana, millardetii*
GENUS *Atichia* — Atich Seurat Sac Fungus
SPECIES *glomerulosa*

FAMILY 27 *Spathulosporaceae* — Spath Spore Sac Fungus
GENUS *Spathulospora* — Spath Spore Sac Fungus
SPECIES *adelpha, antarctica, calva, lanata, phycophila*
GENUS *Hispidicarpomyces* — Hairy Carpel Spath Spore Sac Fungus
SPECIES *galaxauricola*
GENUS *Retrostium* — Beak Spath Spore Sac Fungus
SPECIES *amphiroae*

FAMILY 28 *Testudinaceae* — Turtle Sac Fungus
GENUS *Testudina* — Turtle Sac Fungus
SPECIES *terrestris*
GENUS *Eremodothis* — Sand Doth Turtle Sac Fungus
SPECIES *angulata*
GENUS *Lepidosphaeria* — Thin Sphere Turtle Sac Fungus
SPECIES *nicotiae*
GENUS *Neotestudina* — Modern Turtle Sac Fungus
SPECIES *cunninghamiae, rosatii*
GENUS *Pseudophaeotrichum* — Dusky Hair Turtle Sac Fungus
SPECIES *sudanense*
GENUS *Ulospora* — Ulospore Turtle Sac Fungus
SPECIES *bilgramii*

FAMILY 29 *Thelebolaceae* — Thele Bowl Sac Fungus
GENUS *Thelebolus* — Thele Bowl Sac Fungus
SPECIES *caninus, crustaceus, microsporus, nanus, polysporus, psychrophilus, stercoreus, terrestris*
VARIETIES *crustaceus maritimus, crustaceus myriadeus*
GENUS *Antarctomyces* — Antarctic Thele Bowl Sac Fungus
SPECIES *psychotrophicus*
GENUS *Ascozonus* — Zoned Thele Bowl Sac Fungus
SPECIES *crouanii, cunicularius, leveilleanus, monascus, parvisporus, solms-laubachii, subhirtus, woolhopensis*
GENUS *Caccobius* — Cacco Thele Bowl Sac Fungus
SPECIES *minusculus*
GENUS *Coprobolus* — Copper Thele Bowl Sac Fungus
SPECIES *poculiformis*
GENUS *Ryparobius* — Rypar Thele Bowl Sac Fungus
SPECIES *brunneus, dubius, pachyascus*
VARIETIES *dubius lagopi*
GENUS *Streptotheca* — Twisted Theca Thele Bowl Sac Fungus
SPECIES *psychrophila*

FAMILY 30 *Thyridiaceae* — Thyre Sac Fungus
GENUS *Thyridium* — Thyre Sac Fungus
SPECIES *vestitum*
GENUS *Balzania* — Balzan Thyre Sac Fungus
SPECIES *platensis*
GENUS *Leucocrea* — Light Thyre Sac Fungus
SPECIES *nivea*
GENUS *Thyronectroidea* — Nectroid Thyre Sac Fungus
SPECIES *chrysogramma*

FAMILY 31 *Zopfiaceae* — Zopf Sac Fungus
GENUS *Zopfia* — Zopf Sac Fungus
SPECIES *rhizophila*
GENUS *Caryospora* — Cary Spore Zopf Sac Fungus
SPECIES *callicarpa*
GENUS *Celtidia* — Celtid Zopf Sac Fungus
SPECIES *duplicispora*
GENUS *Didymocrea* — Dual Zopf Sac Fungus
SPECIES *sadasivanii*
GENUS *Halotthia* — Halotth Zopf Sac Fungus
SPECIES *posidoniae*
GENUS *Zopfiofoveola* — Foveole Zopf Sac Fungus
SPECIES *punctata*

ORDER 02 *Anamorphales* Anamorph Sac Fungus

FAMILY 1 *Coelomycetaceae* — Carved Anamorph Sac Fungus
GENUS *Cystotrichiopsis* — Cyst Carved Anamorph Sac Fungus
SPECIES *salvadorae*
GENUS *Tetranacriella* — Tetrad Carved Anamorph Sac Fungus
SPECIES *papillata*

FAMILY 2 *Hyphomycetaceae* — Hypha Anamorph Sac Fungus
GENUS *Catenulifera* — Chain Hypha Anamorph Sac Fungus
SPECIES *rhodogena*
GENUS *Phaeobotrys* — Dusky Berry Hypha Anamorph Sac Fungus
SPECIES *acutisporum*

GENUS *Pleurovularia* Pleural Vular Hypha Anamorph Sac Fungus
SPECIES *polliniae*

ORDER 03 *Chaetothyriales* Bristle Sac Fungus

FAMILY 1 *Chaetothyriaceae* Bristle Sac Fungus
GENUS *Chaetothyrium* Bristle Sac Fungus
SPECIES *javanicum*
GENUS *Biciliopsis* Biciliate Bristle Sac Fungus
SPECIES *leptogiicola*
GENUS *Ceramothyrium* Ceram Thyre Bristle Sac Fungus
SPECIES *carniolicum*
GENUS *Merismella* Merism Bristle Sac Fungus
SPECIES *amazonensis, concinna, gracilenta, oligomera, proxima*
GENUS *Yatesula* Yates Bristle Sac Fungus
SPECIES *calami, sabalidis*

FAMILY 2 *Herpotrichiellaceae* Herpetic Bristle Sac Fungus
GENUS *Herpotrichiella* (Synonym for *Capronia*)
GENUS *Ardhachandra* Ardhachander Herpetic Bristle Sac Fungus
SPECIES *cristaspora*
GENUS *Berlesiella* Berles Herpetic Bristle Sac Fungus
SPECIES *nigerrima*
GENUS *Capronia* Capron Herpetic Bristle Sac Fungus
SPECIES *acutiseta, coronata, dactylotricha, epimyces, fungicola, inconspicua, macrotricha, mansonii, moravica, munkii, normandinae, parasitica, pilosella, pleiospora, pulcherrima, semi-immersa, sexdecimspora, villosa*
GENUS *Cladophialophora* Branched Phial Herpetic Bristle Sac Fungus
SPECIES *arxii, bantiana, boppii, brevicatenata, carrionii, devriesii*
GENUS *Exophiala* Exo Phial Herpetic Bristle Sac Fungus
SPECIES *alcalophila, angulospora, castellanii, dermatitidis, dopicola, jeanselmei, mansonii, mesophila, moniliae, negronii, pisciphila, psychrophila, salmonis, spinifera*
VARIETIES *jeanselmei jeanselmei, jeanselmei heteromorpha, jeanselmei lecanii-corni*
GENUS *Rhinocladiella* Snout Branched Herpetic Bristle Sac Fungus
SPECIES *aquaspersa, atrovirens, compacta, ellisii, indica, phaeophora*

ORDER 04 *Coronophorales* Crown Sac Fungus

FAMILY 1 *Coronophoraceae* Crown Sac Fungus
GENUS *Coronophora* Crown Sac Fungus
SPECIES *angustata, annexa, gregaria*
GENUS *Sphaeria* Sphere Crown Sac Fungus
SPECIES *acuminata, convolvulicola, cornicola, desmazieri, epimyces, exilis, funicola, graminicola, ochraceopallida, pallor, quercina, ribicola, vagans*

FAMILY 2 *Chaetosphaerellaceae* Hairball Crown Sac Fungus
GENUS *Chaetosphaerella* Hairball Crown Sac Fungus
SPECIES *fusca, fusispora, phaeostroma*
GENUS *Diplosporium* Diploid Spore Hairball Crown Sac Fungus
SPECIES *flavum*
GENUS *Veramycina* Vera Hairball Crown Sac Fungus
SPECIES *elegans*

FAMILY 2 *Nitschkiaceae* Nitschki Crown Sac Fungus
GENUS *Nitschkia* Nitschki Crown Sac Fungus
SPECIES *brevispina, broomeana, collapsa, confertula, cupularis, grevillei, parasitans*
GENUS *Acanthonitschkea* Spiny Nitschki Crown Sac Fungus
SPECIES *tristis*
GENUS *Bertia* Bert Nitschki Crown Sac Fungus
SPECIES *moriformis*
VARIETIES *moriformis multiseptata*
GENUS *Biciliosporina* Biciliate Spore Nitschki Crown Sac Fungus
SPECIES *karwarensis*
GENUS *Enchnoa* Enchno Nitschki Crown Sac Fungus
SPECIES *infernalis, lanata*
GENUS *Lasiosphaeriopsis* Woolly Sphere Nitschki Crown Sac Fungus
SPECIES *salisburyi, supersparsa*
GENUS *Loranitschkia* Lora Nitschki Crown Sac Fungus
SPECIES *viticola*
GENUS *Rhagadostoma* Berry Stoma Nitschki Crown Sac Fungus

SPECIES *lichenicola, rugosum*

ORDER 05 *Diaporthales* Dual Porthe Sac Fungus

FAMILY 1 *Diaporthaceae* Dual Porthe Sac Fungus
GENUS *Diaporthe* Dual Porthe Sac Fungus
SPECIES *acerina, acus, adunca, aesculicola, ailanthi, alleghaniensis, alnea, americana, arctii, aucubae, baccharidis, beckhausii, capsici, carpini, celastrina, chailletii, cinerascens, circumscripta, citri, coneglanensis, conjuncta, coramblicola, crataegi, crinigera, crustosa, cryptica, culta, decorticans, delitescens, desmazieri, detrusa, dulcamarae, epilobii, eres, eumorpha, fibrosa, furfuracea, garryae, genistae, hederae, helianthi, hickoriae, hippophaës, ilicis, importata, impulsa, inaequalis, incarcerata, incrustans, insignis, insularis, intermedia, juglandina, laschii, leycesteriae, ligulata, mahoniae, manihotis, medusaea, megalospora, meliloti, melonis, minuscula, mori, neilliae, nobilis, nodosa, nomurai, obsoleta, oncostoma, orobanches, orthoceras,oxyspora, pardalota, perexigua, perniciosa, phaseolorum, phillyreae, picea, pinophylla, pithya, pulchra, pulla, pungens, pustulata, putator, retecta, revellens, rhoina, rhois, rudis, ryckholtii, samaricola, sarmenticia, sarothamni, scandens, scobina, skimmiae, sorbariae, sorbicola, spiculosa, stictostoma, striiformis, strumella, syngenesia, tamaricina, tortuosa, toxica, tulasnei, vaccinii, varians, veronicae, vexans, viburni, vincae, viticola, woodii, woolworthii*
VARIETIES *arctii achilleae, arctii artemisiae, mahoniae foliicola, orthoceras senecionis, perniciosa pruni, phaseolorum phaseolorum, phaseolorum caulivora, phaseolorum meridionalis, phaseolorum sojae, picea obiones, strumella longispora, viburni spiraeicola*
GENUS *Anisomycopsis* Anise Dual Porthe Sac Fungus
SPECIES *rosae*
GENUS *Caudospora* Stalk Spore Dual Porthe Sac Fungus
SPECIES *taleola*
GENUS *Diatractium* Dual Atract Dual Porthe Sac Fungus
SPECIES *cordianum, ingae*
GENUS *Keinstirschia* Keinstirsch Dual Porthe Sac Fungus
SPECIES *megalosperma*
GENUS *Pedumispora* Pedal Spore Dual Porthe Sac Fungus
SPECIES *rhizophorae*
GENUS *Pseudocryptosporella* Cryptic Spore Dual Porthe Sac Fungus
SPECIES *polylepidis*
GENUS *Pseudothis* Doth Dual Porthe Sac Fungus
SPECIES *coccodes*
GENUS *Septomazzantia* Septic Mazzant Dual Porthe Sac Fungus
SPECIES *bicchiana, epitypha*
GENUS *Stioclettia* Stioclett Dual Porthe Sac Fungus
SPECIES *luzulina*

FAMILY 2 *Gnomoniaceae* Gnome Dual Porthe Sac Fungus
GENUS *Gnomonia* Gnome Dual Porthe Sac Fungus
SPECIES *alni-viridis, alpina, amoena, amygdalinae, ariae, betulina, borealis, campylostyla, cerastis, chamaemori, cingulata, comari, curvirostra, dispora, euphorbiacea, fimbriata, fimicola, gei, gei-montani, geranii, gnomon, graphis, intermedia, ischnostyla, leptostyla, longirostris, manihotis, nerviseda, nervisequa, padicola, papuana, pruni, pseudoamoena, radicicola, rhododendri, riparia, robertiani, rosae, rostellata, rubi, salina, sanguisorbae, setacea, tetraspora*
GENUS *Gnomoniella* Lesser Gnome Dual Porthe Sac Fungus
SPECIES *avellanae, nana, rubicola, tubiformis*

FAMILY 3 *Magnaporthaceae* Magnum Dual Porthe Sac Fungus
GENUS *Magnaporthe* Magnum Dual Porthe Sac Fungus
SPECIES *grisea, oryzae, poae, rhizophila, salvinii*
GENUS *Buergenerula* Buergen Magnum Dual Porthe Sac Fungus
SPECIES *biseptata, spartinae, typhae*
GENUS *Clasterosphaeria* Claster Sphere Magnum Dual Porthe Sac Fungus
SPECIES *cyperi*
GENUS *Clasterosporium* Claster Spore Magnum Dual Porthe Sac Fungus
SPECIES *caricinum, caulicola, coccolobae, fasciculare, flexum*
GENUS *Gaeumannomyces* Gaeumann Magnum Dual Porthe Sac Fungus
SPECIES *caricis, cylindrosporus, graminis, leptosporus*
VARIETIES *graminis graminis, graminis avenae, graminis tritici*
GENUS *Herbampulla* Herb Ampule Magnum Dual Porthe Sac Fungus
SPECIES *crassirostris*
GENUS *Juncigena* Rush Magnum Dual Porthe Sac Fungus
SPECIES *adarca*
GENUS *Mycoleptodiscus* Thin Disc Magnum Dual Porthe Sac Fungus
SPECIES *coloratus, geniculatus, lateralis, lunatus, sphaericus, terrestris, unilateralis, variabilis*
GENUS *Omnidemptus* Omnidempt Magnum Dual Porthe Sac Fungus

SPECIES *affinis*
GENUS *Ophioceras* Serpent Wax Magnum Dual Porthe Sac Fungus
SPECIES *leptosporum*
GENUS *Pseudohalonectria* Halogen Cornicle Magnum Dual Porthe Sac Fungus
SPECIES *adversaria, aomoriensis, eubenangeensis, falcata, lignicola, longirostrum, lutea, palmicola, phialidica*
GENUS *Pyricularia* Pear Magnum Dual Porthe Sac Fungus
SPECIES *borealis, higginsii, juncicola, lauri, parasitica, setariae, submersa*
FAMILY 3 *Melanconidaceae* Dark Cone Dual Porthe Sac Fungus
GENUS *Melanconis* Dark Cone Dual Porthe Sac Fungus
SPECIES *aceris, alni, apocrypta, chrysostroma, corni, decoraensis, everhartii, flavovirens, juglandis, modonia, nigrospora, ostryae, pallida, spodiaea, stilbostoma, xanthostroma*
VARIETIES *juglandis caryae*
GENUS *Melanconium* Lesser Dark Cone Dual Porthe Sac Fungus
SPECIES *apiocarpum, atrum, betulinum, bicolor, castaneae, hederae, juglandinum, pandani, sphaeroideum, stromaticum*
GENUS *Allantoporthe* Allanto Dark Cone Dual Porthe Sac Fungus
SPECIES *decedens, tessella*
GENUS *Apiosporopsis* Parsley Spore Dark Cone Dual Porthe Sac Fungus
SPECIES *carpinea*
GENUS *Calospora* Callisto Spore Dark Cone Dual Porthe Sac Fungus
SPECIES *arausiaca, undulata*
GENUS *Coniella* Least Dark Cone Dual Porthe Sac Fungus
SPECIES *australiensis, castaneicola, diplodiella, fragariae, fusiformis, granati, macrospora, musaiaensis, petrakii*
VARIETIES *musaiaensis musaiaensis, musaiaensis hibisci*
GENUS *Coryneum* Club Dark Cone Dual Porthe Sac Fungus
SPECIES *betulinum, castaneicola, disciforme, elevatum, foliicola, japonicum, kunzei, longistipitatum, microstictum, neesii, opacum, terrophilum*
GENUS *Cyclodomella* Cyclic Dark Cone Dual Porthe Sac Fungus
SPECIES *nigra*
GENUS *Dicarpella* Dual Carpel Dark Cone Dual Porthe Sac Fungus
SPECIES *dryina*
GENUS *Ditopella* Dual Topell Dark Cone Dual Porthe Sac Fungus
SPECIES *ditopa, farcta, fusispora, vizeana*
GENUS *Fremineavia* Fremine Bird Dark Cone Dual Porthe Sac Fungus
SPECIES *berkeleyi*
GENUS *Hapalocystis* Single Cyst Dark Cone Dual Porthe Sac Fungus
SPECIES *berkeleyi, bicaudata*
GENUS *Harknessia* Harkness Dark Cone Dual Porthe Sac Fungus
SPECIES *americana, eucalypti, hawaiiensis, leucospermi, renispora, uromycoides*
GENUS *Hercospora* Herco Spore Dark Cone Dual Porthe Sac Fungus
SPECIES *tiliae*
GENUS *Hypophloeda* Hyponic Flow Dark Cone Dual Porthe Sac Fungus
SPECIES *rhizospora*
GENUS *Massariovalsa* Massar Vals Dark Cone Dual Porthe Sac Fungus
SPECIES *californica, caudata, megalospora, sudans*
GENUS *Mastigosporella* Mastigote Spore Dark Cone Dual Porthe Sac Fungus
SPECIES *hyalina, nyssae*
GENUS *Mebarria* Mebarr Dark Cone Dual Porthe Sac Fungus
SPECIES *thujina*
GENUS *Melogramma* Marked Dark Cone Dual Porthe Sac Fungus
SPECIES *campylosporum, elongatum, spiniferum*
GENUS *Murogenella* Murogenic Dark Cone Dual Porthe Sac Fungus
SPECIES *terrophila*
GENUS *Phragmoporthe* Fragment Dark Cone Dual Porthe Sac Fungus
SPECIES *conformis*
GENUS *Pilidiella* Pilid Dark Cone Dual Porthe Sac Fungus
SPECIES *quercicola*
GENUS *Prosthecium* Primitive Theca Dark Cone Dual Porthe Sac Fungus
SPECIES *auctum, platanoidis*
GENUS *Prostratus* Prostrate Dark Cone Dual Porthe Sac Fungus
SPECIES *cyclobalanopsidis*
GENUS *Pseudovalsa* Vals Dark Cone Dual Porthe Sac Fungus
SPECIES *convergens, fusca, hapalocystis, lanciformis, longipes, stylospora, umbonata*
GENUS *Pseudovalsella* Lesser Vals Dark Cone Dual Porthe Sac Fungus
SPECIES *thelebola*

GENUS *Rabenhorstia* — Rabenhorst Dark Cone Dual Porthe Sac Fungus
SPECIES *tiliae*
GENUS *Schizoparme* — Split Parmel Dark Cone Dual Porthe Sac Fungus
SPECIES *straminea*
GENUS *Sydowiella* — Sydow Dark Cone Dual Porthe Sac Fungus
SPECIES *ambigua, depressula, fenestrans, juncina*
GENUS *Tubakia* — Tubak Dark Cone Dual Porthe Sac Fungus
SPECIES *castanopsis, japonica, rubra, subglobosa*
GENUS *Uleoporthe* — Ule Dark Cone Dual Porthe Sac Fungus
SPECIES *orbiculata*
GENUS *Wehmeyera* — Wehmeyer Dark Cone Dual Porthe Sac Fungus
SPECIES *acerina*
GENUS *Wuestneiopsis* — Wuestne Dark Cone Dual Porthe Sac Fungus
SPECIES *georgiana*

FAMILY 4 *Valsaceae* — Vals Dual Porthe Sac Fungus
GENUS *Valsa* — Vals Dual Porthe Sac Fungus
SPECIES *abietis, abrupta, ambiens, auerswaldii, ceratophora, ceratosperma, ceuthospora, cypri, eugeniae, germanica, intermedia, japonica, kunzei, laurocerasi, mali, nivea, oxystoma, paulowniae, pini, platani, pustulata, querna, salicina, sordida, syringae*
GENUS *Amphicytostroma* — Cyclic Stroma Vals Dual Porthe Sac Fungus
SPECIES *tiliae*
GENUS *Amphiporthe* — Least Vals Dual Porthe Sac Fungus
SPECIES *hranicensis, leiphaemia, raveneliana*
GENUS *Anisogramma* — Anise Marked Vals Dual Porthe Sac Fungus
SPECIES *anomala, apiospora, virgultorum*
GENUS *Apiognomonia* — Parsley Gnome Vals Dual Porthe Sac Fungus
SPECIES *errabunda, erythrostoma, petiolicola, supraseptata, veneta*
GENUS *Apioplagiostoma* — Parsley Pelagic Stoma Vals Dual Porthe Sac Fungus
SPECIES *aceriferum, carpinicola*
GENUS *Apioporthe* — Parsley Vals Dual Porthe Sac Fungus
SPECIES *phomospora, vepris*
GENUS *Aporhytisma* — Aporhytism Vals Dual Porthe Sac Fungus
SPECIES *urticae*
GENUS *Asteroma* — Star Vals Dual Porthe Sac Fungus
SPECIES *alneum, betulae, corni, coryli, impressum, labes, microspermum, padi, prunellae, veronicae*
GENUS *Chadefaudiomyces* — Chadefaud Vals Dual Porthe Sac Fungus
SPECIES *indicus*
GENUS *Chondroplea* — Cartilage Vals Dual Porthe Sac Fungus
SPECIES *populea*
GENUS *Cryphonectria* — Cryphon Cornicule Vals Dual Porthe Sac Fungus
SPECIES *cubensis, gyrosa, havanensis, parasitica*
GENUS *Cryptascoma* — Comma Cryptic Vals Dual Porthe Sac Fungus
SPECIES *bisetula*
GENUS *Cryptoderis* — Dere Cryptic Vals Dual Porthe Sac Fungus
SPECIES *caricina*
GENUS *Cryptodiaporthe* — Cryptic Vals Dual Porthe Sac Fungus
SPECIES *aculeans, aesculi, aubertii, castanea, corni, galericulata, hystrix, lebiseyi, lirella, populea, pyrrhocystis, robergeana, salicella, salicina*
GENUS *Cryptospora* — Spore Cryptic Vals Dual Porthe Sac Fungus
SPECIES *intexta, sowerbyi*
GENUS *Cryptosporella* — Lesser Spore Cryptic Vals Dual Porthe Sac Fungus
SPECIES *compta, hypodermia, platanigera, umbrina*
GENUS *Cytodiplospora* — Diploid Cytospore Vals Dual Porthe Sac Fungus
SPECIES *castaneae, tiliae*
GENUS *Cytospora* — Cytospore Vals Dual Porthe Sac Fungus
SPECIES *acaciae, aquifolii, carbonacea, carphosperma, cedri, ceratophora, corni, corylicola, decorticans, dubyi, eriobotryae, eucalypticola, euonymi, eutypelloides, hariotii, hendersonii, ilicina, jasmini, juglandina, laburni, lantanae, lauri, leucosperma, lonicerae, magnoliae, microspora, minuta, mougeotii, myricae-gales, occulta, orbicularis, oxyacanthae, personata, platani, pruinosa, prunorum, punica, ribis, sacchari, salicis, sambuci, sarothamni, subclypeata, tamaricis, taxi, tritici, vaccinii, vitis*
GENUS *Diaporthopsis* — Lesser Vals Dual Porthe Sac Fungus
SPECIES *angelicae, nigrella, pantherina, urticae*
GENUS *Diplodina* — Diploid Vals Dual Porthe Sac Fungus
SPECIES *acerina, bufonia, caricis, castaneae, coloradensis, equiseti, eurhododendri, galii, semi-immersa, watsoniana*
GENUS *Diplosclerophoma* — Hard Diploid Vals Dual Porthe Sac Fungus
SPECIES *ceratoniae*

GENUS *Discula* Disc Vals Dual Porthe Sac Fungus
SPECIES *betulina, campestris, caricis, cytosporea, destructiva, junci, kriegeriana, pyri*
GENUS *Ditopellina* Ditopell Vals Dual Porthe Sac Fungus
SPECIES *saccardiana*
GENUS *Endothia* Endic Doth Vals Dual Porthe Sac Fungus
SPECIES *fluens, radicalis*
GENUS *Hypospilina* Hyponic Vals Dual Porthe Sac Fungus
SPECIES *bifrons, pustula*
GENUS *Kapooria* Kapoor Vals Dual Porthe Sac Fungus
SPECIES *musarum*
GENUS *Laestadia* Laestad Vals Dual Porthe Sac Fungus
SPECIES *epilobii, faginea, iridis, periclymeni, polygonorum, pseudoplatani*
GENUS *Leptosillia* Thin Sill Vals Dual Porthe Sac Fungus
SPECIES *notha*
GENUS *Leucostoma* Light Stoma Vals Dual Porthe Sac Fungus
SPECIES *curreyi, kunzei, persoonii, translucens*
VARIETIES *persoonii armeniacae, persoonii avium, persoonii cerasi, persoonii insititiae, persoonii mahaleb, persoonii oeconomicae, persoonii persicae*
GENUS *Linospora* Line Spore Vals Dual Porthe Sac Fungus
SPECIES *ceuthocarpa, saligna, vulgaris*
GENUS *Maculatipalma* Spotted Palm Vals Dual Porthe Sac Fungus
SPECIES *frondicola*
GENUS *Mamiania* Mamy Vals Dual Porthe Sac Fungus
SPECIES *ostryae*
GENUS *Mamianiella* Lesser Mamy Vals Dual Porthe Sac Fungus
SPECIES *coryli*
GENUS *Mazzantia* Mazzant Vals Dual Porthe Sac Fungus
SPECIES *deplanata, galii, gougetiana, napelli, niesslii, sepium*
GENUS *Ophiognomonia* Serpent Gnome Vals Dual Porthe Sac Fungus
SPECIES *melanostyla*
GENUS *Ophiovalsa* Serpent Vals Dual Porthe Sac Fungus
SPECIES *betulae, corylina, suffusa*
GENUS *Paravalsa* Parasite Vals Dual Porthe Sac Fungus
SPECIES *indica*
GENUS *Phomopsis* Foam Vals Dual Porthe Sac Fungus
SPECIES *abdita, abietina, abutilonis, agrostidis, airae, albicans, anacardii, aquilariae, araucariae, asparagi, asphodelina, asteriscus, aucupariae, bougainvilleae, brachyceras, buddleiae, cacti, calystegiae, cancri, caricae-papayae, carnea, caryophylli, casuarinae, caulographa, cedrelae, cirsii, cocoës, coffeae, conii, cordylines, corni, corticis, cotoneastri, cruciferae, dauci, demissa, diachenii, digitalis, dioscoreae, diospyri, dipsaci, durandiana, ebulina, elaeagni, elaeidis, elliptica, emicis, ericaceana, eryngiicola, eucalypti, eugeniae, eupatoriicola, euphorbiae, extorris, foeniculi, folliculicola, fukushii, ganjae, gardeniae, glandicola, heveae, hieracii, hordei, hydrangeae, hyperici, hysteriola, ipomoeae, iridis, irregularis, japonica, juglandina, juniperivora, landolphiae, laurella, lavandulae, leptostromiformis, ligulata, linearis, lirellata, liriodendri, longicolla, lonicerae, lysimachiae, magnoliicola, malvacearum, mangiferae, manilkarae, mendax, microspora, morphaea, muelleri, musicola, oblita, obscurans, occulta, oncostoma, oryzae, oryzae-sativae, osmanthi, oxalina, padina, pallida, perniciosa, perseae, petiolorum, phoenicicola, pittospori, platanoidis, polygonorum, prunorum, pungens, putator, quercella, quercicola, quercina, radula, ramealis, ranojevicii, rhois, rudis, rusci, salicina, sambucina, sclerotioides, scobina, sedi, silenes, skimmiae, sophorae, sorbicola, stellariae, stictica, subordinaria, tecomae, theae, theicola, tinea, tritici, verbasci, viticola, xanthii*
VARIETIES *aquilariae nigra, leptostromiformis occidentalis, prunorum amygdali*
GENUS *Plagiostoma* Pelagic Stoma Vals Dual Porthe Sac Fungus
SPECIES *alneum, arnstadtiense, bavaricum, devexum, euphorbiae, inclinatum, lugubre, tormentillae*
GENUS *Pleuroceras* Pleural Wax Vals Dual Porthe Sac Fungus
SPECIES *groenlandicum, pseudoplatani*
GENUS *Septomyxa* Septic Vals Dual Porthe Sac Fungus
SPECIES *corni, fagicola, tulasnei*
GENUS *Sillia* Sill Vals Dual Porthe Sac Fungus
SPECIES *ferruginea*
GENUS *Skottsbergiella* Skottsberg Vals Dual Porthe Sac Fungus
SPECIES *diaporthoides*
GENUS *Stegophora* Stegophore Vals Dual Porthe Sac Fungus
SPECIES *ulmea*
GENUS *Valsella* Lesser Vals Dual Porthe Sac Fungus
SPECIES *adhaerens, amphoraria, clypeata, polyspora, salicis*
GENUS *Valseutypella* Typical Vals Dual Porthe Sac Fungus
SPECIES *multicollis, tristicha*
GENUS *Winterella* Winter Vals Dual Porthe Sac Fungus

SPECIES *cinctula*
GENUS *Xenotypa* Strange Vals Dual Porthe Sac Fungus
SPECIES *aterrima*

ORDER 06 *Dothideales* Doth Sac Fungus

FAMILY 1 *Dothideaceae* Doth Sac Fungus
GENUS *Dothidea* Doth Sac Fungus
SPECIES *berberidis, frangulae, funesta, genistae, insculpta, johnstonii, mezerei, muelleri, rugodisca, sambuci*
GENUS *Ascagilis* Agile Doth Sac Fungus
SPECIES *bipolaris*
GENUS *Auerswaldia* Auerswald Doth Sac Fungus
SPECIES *scabies*
GENUS *Bactrodesmium* Bacteroid Fiber Doth Sac Fungus
SPECIES *abruptum, arnaudii, atrum, betulicola, biformatum, cedricola, cubense, curvatum, esheri, hebridense, longisporum, obliquum, obovatum, pallidum, spilomeum, submoniliforme*
VARIETIES *obliquum obliquum, obliquum suttonii*
GENUS *Bagnisiella* Bagnis Doth Sac Fungus
SPECIES *australis, examinans*
GENUS *Bahusakala* Bahusaka Doth Sac Fungus
SPECIES *australiensis, longispora, niditermitae, olivaceonigra, ramosa*
GENUS *Belizeana* Belize Doth Sac Fungus
SPECIES *tuberculata*
GENUS *Biatriospora* Biatriospore Doth Sac Fungus
SPECIES *marina*
GENUS *Botryohypoxylon* Berry Hypha Sharp Doth Sac Fungus
SPECIES *amazonense*
GENUS *Bryopelta* Pelt Moss Doth Sac Fungus
SPECIES *variabilis*
GENUS *Bryorella* Moss Doth Sac Fungus
SPECIES *acrogena, complanata, compressa, conspecta, crassitecta, cryptocarpa, erumpens, gregaria, marginis, punctiformis, retiformis, semi-immersa*
GENUS *Bryosphaeria* Sphere Moss Doth Sac Fungus
SPECIES *brevicollis, bryophila, cinclidoti, echinoidea, epibrya, megaspora, pohliae, quinqueseptata, setifera*
GENUS *Bryostroma* Stroma Moss Doth Sac Fungus
SPECIES *axillare, bryi, guttulatum, halosporum, necans, rhacomitrii, trichostomi*
GENUS *Bryothele* Thele Moss Doth Sac Fungus
SPECIES *mira*
GENUS *Capillataspora* Capillary Spore Doth Sac Fungus
SPECIES *corticola*
GENUS *Caprettia* Caprett Doth Sac Fungus
SPECIES *amazonensis, nyssaegenoides*
GENUS *Catulus* Catule Doth Sac Fungus
SPECIES *aquilonius*
GENUS *Cercidospora* Weaver's Shuttle Spore Doth Sac Fungus
SPECIES *cladoniicola, decolorella, epipolytropa, lichenicola, macrospora, parva, punctillata, stereocaulorum, ulothii, verrucosaria*
GENUS *Cerodothis* Wax Doth Sac Fungus
SPECIES *aurea*
GENUS *Chaetoscutula* Bristle Scute Doth Sac Fungus
SPECIES *juniperi*
GENUS *Coccoidella* Round Doth Sac Fungus
SPECIES *euglypta, scutula*
GENUS *Colensoniella* Colenson Doth Sac Fungus
SPECIES *torulispora*
GENUS *Comminutispora* Minute Spore Doth Sac Fungus
SPECIES *agavacearum*
GENUS *Dangeardiella* Dangeard Doth Sac Fungus
SPECIES *fusiformis, macrospora*
GENUS *Daruvedia* Daruved Doth Sac Fungus
SPECIES *bacillata*
GENUS *Dichomera* Dichomer Doth Sac Fungus
SPECIES *gemmicola, mutabilis, rhamnicola*
GENUS *Didymopleella* Dual Pleiad Doth Sac Fungus
SPECIES *cladii*
GENUS *Dilophospora* Dual Crest Spore Doth Sac Fungus

SPECIES *alopecuri*
GENUS *Dolabra* Dolab Doth Sac Fungus
SPECIES *nepheliae*
VARIETIES *propinquus brachysporum*
GENUS *Gloeodiscus* Glow Disc Doth Sac Fungus
SPECIES *nigrorufus*
GENUS *Grandigallia* Grand Gall Doth Sac Fungus
SPECIES *dictyospora*
GENUS *Griggsia* Griggs Doth Sac Fungus
SPECIES *cyathea*
GENUS *Heleiosa* Heleiose Doth Sac Fungus
SPECIES *barbatula*
GENUS *Heterosphaeriopsis* Variable Sphere Doth Sac Fungus
SPECIES *fulvodisca*
GENUS *Homostegia* Homospore Cover Doth Sac Fungus
SPECIES *magnoliae, piggotii*
GENUS *Hormonema* Stimulating Thread Doth Sac Fungus
SPECIES *macrosporum, merioides, prunorum, schizolunatum*
GENUS *Hyalosphaera* Hyaline Sphere Doth Sac Fungus
SPECIES *ciliata, miconiae, pulchella*
GENUS *Hypobryon* Hyponic Moss Doth Sac Fungus
SPECIES *bicolor, heterotropum, insigne, perforans, poeltii, validum*
GENUS *Iledon* Iledo Doth Sac Fungus
SPECIES *versicolor*
GENUS *Jahnula* Jahn Doth Sac Fungus
SPECIES *aquatica*
GENUS *Karschia* Karsch Doth Sac Fungus
SPECIES *bloxamii, epiphorbia, talcophila, thalloides*
GENUS *Lanatosphaera* Woolly Sphere Doth Sac Fungus
SPECIES *dimerellae, porinicola*
GENUS *Leptospora* Thin Spore Doth Sac Fungus
SPECIES *rubella*
GENUS *Leveillina* Leve Doth Sac Fungus
SPECIES *arduinae, endocrypta*
GENUS *Licopolia* Licopol Doth Sac Fungus
SPECIES *franciscana, keniensis*
GENUS *Lidophia* Lid Doth Sac Fungus
SPECIES *graminis*
GENUS *Lineolata* Linear Doth Sac Fungus
SPECIES *rhizophorae*
GENUS *Loculohypoxylon* Purse Hyponic Sharp Doth Sac Fungus
SPECIES *grandineum*
GENUS *Megaloseptoria* Mega Septic Doth Sac Fungus
SPECIES *mirabilis*
GENUS *Microsphaeropsis* Lesser Sphere Doth Sac Fungus
SPECIES *arundinis, callista*
GENUS *Monodictys* Monad Net Doth Sac Fungus
SPECIES *anaptychiae, antiqua, asperospora, austrina, capensis, castaneae, cellulosa, fluctuata, fuliginosa, glauca, lepraria, levis, melanopa, paradoxa, putredinis, striata*
VARIETIES *antiqua compactum*
GENUS *Monographos* Monad Graphic Doth Sac Fungus
SPECIES *fuckelii*
GENUS *Mucomassaria* Massar Doth Sac Fungus
SPECIES *maxima*
GENUS *Muellerites* Mueller Doth Sac Fungus
SPECIES *juniperi*
GENUS *Mycocryptospora* Cryptic Spore Doth Sac Fungus
SPECIES *anthostomoides*
GENUS *Mycoglaena* Glen Doth Sac Fungus
SPECIES *acuminans, myricae*
GENUS *Mycothyridium* Thyre Doth Sac Fungus
SPECIES *lividum*
GENUS *Myriostigmella* Myriad Stigma Doth Sac Fungus
SPECIES *guatteriae*
GENUS *Neopeckia* Modern Peck Doth Sac Fungus
SPECIES *fulcita*
GENUS *Omphalospora* Navel Spore Doth Sac Fungus

SPECIES *himantia, melaena, stellariae*
GENUS *Otthia* Otth Doth Sac Fungus
SPECIES *lisae, spiraeae*
GENUS *Paraliomyces* Paral Doth Sac Fungus
SPECIES *lentifer*
GENUS *Passeriniella* Passerine Doth Sac Fungus
SPECIES *obiones*
GENUS *Phaeotomasellia* Dusky Thomas Doth Sac Fungus
SPECIES *ruwenzorensis*
GENUS *Philobryon* Moss Doth Sac Fungus
SPECIES *anuliferum*
GENUS *Phycorella* Algae Loving Doth Sac Fungus
SPECIES *scytonematis*
GENUS *Physalosporopsis* Bladder Spore Doth Sac Fungus
SPECIES *rhizophoricola*
GENUS *Placosphaeria* Plaque Sphere Doth Sac Fungus
SPECIES *campanulae*
GENUS *Plagiostromella* Pelagic Stroma Doth Sac Fungus
SPECIES *berberidis, indica, leptodermidis, pleurostoma*
GENUS *Pleostigma* Pleomorphic Stigma Doth Sac Fungus
SPECIES *fibrincolum, jungermannicola, lithospermi, malacodermum, marchicum, moelleriellae, piceae, quercinum, xenochaeta*
GENUS *Pleotrichiella* Pleomorphic Hairy Doth Sac Fungus
SPECIES *australiensis*
GENUS *Polystomellopsis* Polystome Doth Sac Fungus
SPECIES *mirabilis*
GENUS *Pseudomorfea* Morf Doth Sac Fungus
SPECIES *coffeae*
GENUS *Pseudopleospora* Pleomorphic Spore Doth Sac Fungus
SPECIES *galiorum, petrakii, ruthenica*
GENUS *Pteridiospora* Wing Spore Doth Sac Fungus
SPECIES *scoriadea, spinosispora*
GENUS *Racovitziella* Racovitz Doth Sac Fungus
SPECIES *endostromatica*
GENUS *Rosellinula* Rosellin Doth Sac Fungus
SPECIES *frustulosae, haplospora, kalbii*
GENUS *Rosenscheldia* Rosenscheld Doth Sac Fungus
SPECIES *abundans*
GENUS *Roumegueria* Roumeguer Doth Sac Fungus
SPECIES *goudotii*
GENUS *Roussoellopsis* Rousso Doth Sac Fungus
SPECIES *japonica, macrospora, tosaensis*
GENUS *Salsuginea* Salt Doth Sac Fungus
SPECIES *ramicola*
GENUS *Scirrhia* Scirrh Doth Sac Fungus
SPECIES *agrostidis, aspidiorum, castagnei, osmundae, rimosa, setariae*
GENUS *Sclerophoma* Hard Foam Doth Sac Fungus
SPECIES *entoxylina, eustomatis*
GENUS *Scolecobonaria* Scolex Bonar Doth Sac Fungus
SPECIES *filiformis, lithocarpi*
GENUS *Semifissispora* Fissure Spore Doth Sac Fungus
SPECIES *elongata, fusiformis, rotundata*
GENUS *Septonema* Sceptic Thread Doth Sac Fungus
SPECIES *crispulum, dendryphioides, fasciculare, leptaleum, ochraceum, pseudobinum, secedens, spilomeum*
GENUS *Septoriella* Sceptic Doth Sac Fungus
SPECIES *junci*
GENUS *Setomelanomma* Bristle Melanomma Doth Sac Fungus
SPECIES *holmii*
GENUS *Shiraia* Shira Doth Sac Fungus
SPECIES *bambusicola*
GENUS *Stigmina* Stigma Doth Sac Fungus
SPECIES *anacardii, carpophila, dura, eucalypti, glomerulosa, juniperina, longispora, millettiae, palmivora, platani, sapii, tinea*
GENUS *Stuartella* Stuart Doth Sac Fungus
SPECIES *carlylei*
GENUS *Stylodothis* Style Doth Sac Fungus

SPECIES *indica, puccinioides*
GENUS *Systremma* Tremor Doth Sac Fungus
SPECIES *bullata*
GENUS *Teichospora* Teich Spore Doth Sac Fungus
SPECIES *deflectens, ignavis*
GENUS *Teratoschaeta* Terat Schaete Doth Sac Fungus
SPECIES *rondoniensis*
GENUS *Thalassoascus* Thalass Doth Sac Fungus
SPECIES *cystoseirae, lessoniae, tregoubovii*
GENUS *Thelenidia* Thelen Doth Sac Fungus
SPECIES *monosporella*
GENUS *Tilakiella* Tilak Doth Sac Fungus
SPECIES *indica*
GENUS *Tomeoa* Tome Doth Sac Fungus
SPECIES *viridicolliculosa*
GENUS *Tyrannosorus* Tyrannosorus Doth Sac Fungus
SPECIES *pinicola*
GENUS *Vestergrenia* Vestergren Doth Sac Fungus
SPECIES *cestri, multipunctata*
GENUS *Vizellopsis* Vizell Doth Sac Fungus
SPECIES *grevilleae*
GENUS *Westea* West Doth Sac Fungus
SPECIES *banksiae*
GENUS *Xylopezia* Xylopez Doth Sac Fungus
SPECIES *hemisphaerica*
GENUS *Yoshinagella* Yoshinag Doth Sac Fungus
SPECIES *japonica, phyllostachydis*

FAMILY 2 *Botryosphaeriaceae* Berry Sphere Doth Sac Fungus
GENUS *Botryosphaeria* Berry Sphere Doth Sac Fungus
SPECIES *abietina, acaciae, cocoicola, corticis, dothidea, festucae, hyperborea, obtusa, pseudotsugae, quercuum, rhodina, ribis, stevensii, subglobosa, tsugae*
VARIETIES *ribis chromogena, ribis pyricola*
GENUS *Amarenomyces* Amar Berry Sphere Doth Sac Fungus
SPECIES *ammophilae*
GENUS *Auerswaldiella* Auerswald Berry Sphere Doth Sac Fungus
SPECIES *puccinioides, winteri*
GENUS *Diplodia* Diploid Berry Sphere Doth Sac Fungus
SPECIES *alni, amphisphaerioides, buxi, caespitosa, cajani, celtidis, confluens, consors, coryli, cowdellii, cydoniae, faginea, hederae, henriquesii, herbarum, ilicicola, inquinans, juglandis, lantanae, laurina, ligustri, lonicerae, magnoliae, malorum, mamillana, medicaginis, melaena, oleae, opuli, orae-maris, palmicola, paupercula, phoenicum, platanicola, populina, pruni, ramulicola, rhamni, rhododendri, rosarum, roumeguerei, rubi, rudis, saccardiana, salicina, salicorniae, sambucina, sarmentorum, sarothamni, spegazzinia, subtecta, sycina, syringae, taxi, tecta, tiliae, ulicis, viciae, vulgaris, warburgiana*
VARIETIES *saccardiana anglica*
GENUS *Discochora* Disc Chore Berry Sphere Doth Sac Fungus
SPECIES *pini*
GENUS *Dothidotthia* Dotth Berry Sphere Doth Sac Fungus
SPECIES *celtidis, ramulicola*
GENUS *Dothiorella* Lesser Berry Sphere Doth Sac Fungus
SPECIES *candollei, coronillae, fraxinea, gregaria, latitans, olearum, opuntiae, populina, pyrenophora, reniformis, smilacina, spartiicola, ulmi, zeae*
VARIETIES *pyrenophora salicis*
GENUS *Fusicoccum* Fused Berry Sphere Doth Sac Fungus
SPECIES *amygdali, cedrelae, depressum, fibrosum, juglandinum, luteum, macrosporum, noxium, obtusulum, pini, populi, quercinum, quercus, tingens, ulmi*
GENUS *Granulodiplodia* Granular Diploid Berry Sphere Doth Sac Fungus
SPECIES *granulosella*
GENUS *Guignardia* Guignard Berry Sphere Doth Sac Fungus
SPECIES *aesculi, bidwellii, bulgarica, calami, capsici, caricicola, citricarpa, cocoicola, dioscoreae, echinophila, eupatorii, foeniculata, haydenii, heveae, istriaca, latemarensis, mangiferae, manihoti, miribelii, musae, perpusilla, philoprina, pinastri, psidii, punctoidea, rhamni, sansevieriae, traversoana, vaccinii*
GENUS *Lasiodiplodia* Woolly Diploid Berry Sphere Doth Sac Fungus
SPECIES *theobromae*
GENUS *Leptoguignardia* Thin Guignard Berry Sphere Doth Sac Fungus
SPECIES *onobrychidis*
GENUS *Macrophoma* Foam Berry Sphere Doth Sac Fungus

SPECIES *collabens, coronillae-emeri, falconeri, flaccida, fraxini, gymnogongri, lageniformis, mangiferae, oleandrina, paeoniae, passifloricola, phlei, samaricola, strobi, sugi, taxi, theicola*
GENUS *Phomatosphaeropsis* Lesser Foam Berry Sphere Doth Sac Fungus
SPECIES *pinicola*
GENUS *Phyllosticta* Stick Berry Sphere Doth Sac Fungus
SPECIES *abietis, aceris, acetosae, ajugae, angulata, arbuti, artocarpina, asperulae, asplenii, batatas, beaumarisii, berberidis, bolleana, brassicae, briardii, buxina, cajani, camelliae, campestris, capitalensis, circumscissa, cirsii, colocasiae, colocasiicola, concentrica, congesta, cordylines, coryli, cotoneastri, crataegicola, cucurbitacearum, dioscoreicola, elettariae, eugeniae, filipendulae, flevolandica, forsythiae, fraxinicola, gei, ghaesembillae, hypoglossi, japonica, jasmini, lauri, lonicerae, lutetiana, lychnidina, magnoliae, magnoliae, mahoniae, mahoniicola, marchantiae, mercurialis, minima, monogyna, nicotianae, opuli, owaniana, phaseolina, pirolae, polygonorum, populina, primulicola, pseudacori, pyrolae, rhamni, rhamnicola, rhododendri, rhodorae, richardiae, roboris, rosae, ruborum, ruscicola, ruscigena, saccardoi, sambucicola, sapotae, scrophulariae, sechii, solitaria, sorbi, spinarum, straminella, syringae, telopeae, teucrii, thalictri, tiliae, tormentillae, typhina, ulmaria, ulmi, ulmicola, valerianae, viburni, violae, wistariae, zingiberis*
GENUS *Phyllostictina* Lesser Stick Berry Sphere Doth Sac Fungus
SPECIES *ilicis*

FAMILY 3 *Dothioraceae* Lesser Doth Sac Fungus
GENUS *Dothiora* Lesser Doth Sac Fungus
SPECIES *cannabinae, elliptica, europaea, laureolae, phaeosperma, rhamni-alpinae, ribesia, schizospora, sphaerioides, taxicola, thujae*
GENUS *Aureobasidium* Golden Basal Lesser Doth Sac Fungus
SPECIES *microstictum, pullulans*
VARIETIES *pullulans pullulans, pullulans aubasidani, pullulans melanogenum*
GENUS *Botryochora* Berry Chore Lesser Doth Sac Fungus
SPECIES *nigra*
GENUS *Delphinella* Delphin Lesser Doth Sac Fungus
SPECIES *abietis, strobiligena*
GENUS *Dothichiza* Chiz Lesser Doth Sac Fungus
SPECIES *foveolaris, pimprina, turgida*
GENUS *Kabatia* Kabat Lesser Doth Sac Fungus
SPECIES *periclymeni*
GENUS *Kabatiella* Lesser Kabat Lesser Doth Sac Fungus
SPECIES *caulivora, lini, microsticta, nigricans, zeae*
GENUS *Metasphaeria* Metasphere Lesser Doth Sac Fungus
SPECIES *acorella, anarithma, anisometra, ashwelliana, avenae, cetrariicola, clypeosphaerioides, cumana, empetri, graminum, hederae, helicicola, medicaginis, mosana, persistens, rimularum, rubida, rustica, sepincola, thwaitesii, tritorulosa*
VARIETIES *cumana macrospora*
GENUS *Pringsheimia* Pringsheim Lesser Doth Sac Fungus
SPECIES *chamaecyparidis, euphorbiae, karelii, sepincola, smilacis*
GENUS *Pullularia* Pull Lesser Doth Sac Fungus
SPECIES *prototropha*
GENUS *Saccothecium* Theca Lesser Doth Sac Fungus
SPECIES *sepincola*
VARIETIES *sepincola abbreviata, sepincola valde-evoluta*
GENUS *Sydowia* Sydow Lesser Doth Sac Fungus
SPECIES *polyspora*
VARIETIES *polyspora microspora*
GENUS *Yoshinagaia* Yoshinaga Lesser Doth Sac Fungus
SPECIES *quercus*

FAMILY 5 *Mycoporaceae* Pore Sac Fungus
GENUS *Mycoporum* Pore Sac Fungus
SPECIES *antecellens, compositum, eschweileri, lacteum*
GENUS *Cyrtidula* Curved Pore Sac Fungus
SPECIES *hippocastani, major, quercus*
GENUS *Dermatina* (Synonym for *Mycoporum*)
GENUS *Endococcus* Endic Round Pore Sac Fungus
SPECIES *alpestris, apiciicola, brachysporus, caudisporus, exerrans, macrosporus, parietinarius, propinquus, rugulosus, verrucosporus*
GENUS *Mycoporellum* Lesser Pore Sac Fungus
SPECIES *microspermum, obscurum, perexiguum*
GENUS *Mycoporopsis* Least Pore Sac Fungus
SPECIES *sorenocarpa*

FAMILY 6 *Planistromellaceae* Flatbed Doth Sac Fungus
GENUS *Planistromella* Flatbed Doth Sac Fungus

SPECIES *acervata, nolinifoliorum, parryi, torsifoliorum, uniseptata, yuccifoliorum, yuccigena*
GENUS *Loratospora* Loratospore Flatbed Doth Sac Fungus
SPECIES *aestuarii*
GENUS *Piptarthron* Piptarthron Flatbed Doth Sac Fungus
SPECIES *crassisporum, dasylirionis, limbatum, macrosporum, pluriloculare, uniloculare, yuccae*

ORDER 07 *Helotiales* Helot Sac Fungus

FAMILY 01 *Helotiaceae* Helot Sac Fungus
GENUS *Helotium* Helot Sac Fungus
SPECIES *aspegrenii, chloropodium, consobrinum, cribrosum, cruentatum, fulvum, immarginatum, improvisum, lutisedum, macrosporum, nubilipes, politum, pulchellum, rubens, rubescens, rubicola, schimperi, subcorticale, sulphuratum, tenuiculum, terrigenum, tetra-ascosporum, tuba*
GENUS *Adelodiscus* Adel Disc Helot Sac Fungus
SPECIES *philippinensis*
GENUS *Algincola* Algae Dwelling Helot Sac Fungus
SPECIES *quercina*
GENUS *Allophylaria* All Helot Sac Fungus
SPECIES *basalifusca, campanuliformis, clavuliformis, crystallifera, macrospora, subliciformis*
GENUS *Ambrodiscus* Amber Disc Helot Sac Fungus
SPECIES *pseudotsugae*
GENUS *Articulospora* Articulate Spore Helot Sac Fungus
SPECIES *angulata, antipodes, atra, moniliformis*
GENUS *Ascocalyx* Calyx Helot Sac Fungus
SPECIES *abietis*
GENUS *Ascoclavulina* Lesser Club Helot Sac Fungus
SPECIES *sakaii*
GENUS *Ascoconidium* Cone Helot Sac Fungus
SPECIES *castaneae, tsugae*
GENUS *Ascocorticium* Cortical Helot Sac Fungus
SPECIES *anomalum*
GENUS *Ascocoryne* Club Helot Sac Fungus
SPECIES *cylichnium, sarcoides, solitaria*
GENUS *Ascotremella* Tremor Helot Sac Fungus
SPECIES *faginea*
GENUS *Austrocenangium* Southern Cenang Helot Sac Fungus
SPECIES *australe, nanum*
GENUS *Bagnisimitrula* Bagnis Mitre Helot Sac Fungus
SPECIES *saccardoa*
GENUS *Banksiamyces* Banks Helot Sac Fungus
SPECIES *katerinae, maccannii, macrocarpus, toomansis*
GENUS *Belonioscyphella* Belon Scyph Helot Sac Fungus
SPECIES *hypnorum*
GENUS *Bioscypha* Scyph Helot Sac Fungus
SPECIES *cyatheae, pteridicola*
GENUS *Bisporella* Bispore Helot Sac Fungus
SPECIES *citrina, confluens, fuscocincta, lactea, pallescens, resinicola, scolochloae, subpallida, sulfurina*
GENUS *Bloxamia* Bloxam Helot Sac Fungus
SPECIES *bohemica, leucophthalma, truncata*
GENUS *Bryoglossum* Moss Tongue Helot Sac Fungus
SPECIES *gracile, rehmii*
GENUS *Bryophytomyces* Moss Helot Sac Fungus
SPECIES *sphagni*
GENUS *Bryoscyphus* Moss Scyph Helot Sac Fungus
SPECIES *aestivalis, atromarginatus, conocephali, dicrani, lichenicola, marchantiae, turbinatus*
GENUS *Calycella* Lesser Calyx Helot Sac Fungus
SPECIES *calonoides, ferruginea, induementicola, lenticularis, monilifera, ochracea, sublenticularis, subsessilis, terrestris, uliginosa*
GENUS *Carneopezizella* Flesh Peziz Helot Sac Fungus
SPECIES *salicicola*
GENUS *Cenangium* Cenang Helot Sac Fungus
SPECIES *acicola, acuum, coryli, ferruginosum, fuliginosum, graddonii, leoninum, sarothamni*
GENUS *Chlorencoelia* Chloric Celestial Helot Sac Fungus
SPECIES *torta, versiformis*
GENUS *Chlorociboria* Chloric Bory Helot Sac Fungus
SPECIES *aeruginascens, aeruginosa*
GENUS *Chloroscypha* Chloric Scyph Helot Sac Fungus

SPECIES *chloromela, cryptomeriae, sabinae, seaveri*
GENUS *Claussenomyces* Claussen Helot Sac Fungus
SPECIES *allenii, atrovirens, canariensis, clavatus, dacrymycetoideus, olivaceus, prasinulus, salicicola*
GENUS *Clinterium* Clinter Helot Sac Fungus
SPECIES *obturatum*
GENUS *Coryne* Club Helot Sac Fungus
SPECIES *albida, aquatica,* ***atrovirens***(Edible), *dubia, turbinata*
GENUS *Costantinella* Costantine Helot Sac Fungus
SPECIES *cristata, micheneri, terrestris*
GENUS *Criserosphaeria* Criser Sphere Helot Sac Fungus
SPECIES *phyllostictoides*
GENUS *Crocicreas* Saffron Helot Sac Fungus
SPECIES *amenti, complicatum, coronatum, culmicola, cyathoideum, dolosellum, furvum, gramineum, megalosporum, pallidum, rufescens, sellingensis, spicarum, starbaeckii, stramineum, subhyalinum, tomentosum*
VARIETIES *cyathoideum cyathoideum, cyathoideum cacaliae, cyathoideum epilobii, cyathoideum pteridicola, cyathoideum solani, gramineum ammophilae, gramineum incertellum, megalosporum gramineum*
GENUS *Crumenula* Crumen Helot Sac Fungus
SPECIES *urceoliformis*
GENUS *Crumenulopsis* Lesser Crumen Helot Sac Fungus
SPECIES *pinicola, sororia*
GENUS *Cryptocline* Cryptic Inclined Helot Sac Fungus
SPECIES *abietina, arctostaphyli, cinerascens, conigena, cyclaminis, exobasidioides, taxicola*
GENUS *Cyathicula* Cyath Helot Sac Fungus
SPECIES *culmicola, littoralis, rubescens*
GENUS *Cystodendron* Cyst Wood Helot Sac Fungus
SPECIES *dryophilum*
GENUS *Cystopezizella* Cyst Peziz Helot Sac Fungus
SPECIES *venceslai*
GENUS *Cystotricha* Cyst Hairy Helot Sac Fungus
SPECIES *striola*
GENUS *Dawsicola* Daws Helot Sac Fungus
SPECIES *neglecta*
GENUS *Dencoeliopsis* Wood Celestial Helot Sac Fungus
SPECIES *betulicola, johnstonii*
GENUS *Dendrostilbella* Wood Stilbe Helot Sac Fungus
SPECIES *byssina, chlorina, macrospora, mycophila*
GENUS *Dictyonia* Netted Helot Sac Fungus
SPECIES *lophirae, pouroumae*
GENUS *Dimorphospora* Dimorphic Spore Helot Sac Fungus
SPECIES *foliicola*
GENUS *Discinella* Disc Helot Sac Fungus
SPECIES *boudieri, exidiiformis, margarita, menziesii, minutissima, purpurascens*
VARIETIES *boudieri spadicea*
GENUS *Durella* Hard Helot Sac Fungus
SPECIES *atrocyanea, carestiae, commutata, connivens, lecideola, macrospora, melanochlora, suecica*
GENUS *Echinodiscus* Urchin Disc Helot Sac Fungus
SPECIES *lesdainii*
GENUS *Encoeliopsis* Entic Celestial Helot Sac Fungus
SPECIES *bresadola*
GENUS *Episclerotium* Epiphyte Hard Helot Sac Fungus
SPECIES *sclerotiorum, sclerotipus*
GENUS *Erikssonopsis* Eriksson Helot Sac Fungus
SPECIES *ericae*
GENUS *Fluminispora* River Spore Helot Sac Fungus
SPECIES *ovalis*
GENUS *Gelatinodiscus* Gelatin Disc Helot Sac Fungus
SPECIES *flavidus*
GENUS *Gelatinopsis* Gelatin Helot Sac Fungus
SPECIES *ericetorum*
GENUS *Geniculospora* Genicule Spore Helot Sac Fungus
SPECIES *inflata*
GENUS *Gloeopeziza* Glow Peziz Helot Sac Fungus
SPECIES *crozalsii*
GENUS *Gloeotinia* Glow Tin Helot Sac Fungus
SPECIES *granigena*
GENUS *Godronia* Godron Helot Sac Fungus
SPECIES *callunigera, cassandrae, fuliginosa, ledi, ribis, spiraeae, uberiformis, urceolus*

VARIETIES *cassandrae vaccinii*
GENUS *Gorgoniceps* Gorgon Helot Sac Fungus
SPECIES *aridula, boltonii, charnwoodensis, micrometra, viridula*
GENUS *Grahamiella* Graham Helot Sac Fungus
SPECIES *dryadis, variabile*
GENUS *Gremmeniella* Gremmen Helot Sac Fungus
SPECIES *abietina, laricina*
GENUS *Grimmicola* Grimm Helot Sac Fungus
SPECIES *parasiticus*
GENUS *Grovesia* Groves Helot Sac Fungus
SPECIES *pulchella*
GENUS *Grovesiella* Lesser Groves Helot Sac Fungus
SPECIES *abieticola*
GENUS *Helicodendron* Helix Wood Helot Sac Fungus
SPECIES *articulatum, coniferarum, cumbriense, foliicola, fractum, fuscum, giganteum, hyalinum, intestinale, longitubulosum, luteoalbum, multicatenulatum, multiseptatum, paradoxum, triglitziense, websteri, westerdijkae*
GENUS *Heteropatella* Variable Patella Helot Sac Fungus
SPECIES *antirrhini, lacera, umbilicata, valtellinensis*
GENUS *Heterosphaeria* Variable Sphere Helot Sac Fungus
SPECIES *alpestris, linariae, lojkae, ovispora, patella, pulsatillae, veratri*
GENUS *Heyderia* Heyder Helot Sac Fungus
SPECIES *abietis, agariciphila, americana, pusilla, sclerotiorum, sclerotipus*
GENUS *Holmiodiscus* Holm Disc Helot Sac Fungus
SPECIES *filipendulae*
GENUS *Hymenoscyphus* Hymen Scyph Helot Sac Fungus
SPECIES *albidus, albopectum, albopunctus, bryophilus, calyculus, carmichaelii, carpinicola, caudatus, ciborioides, clavatus, consobrinus, cortisedus, electrinus, epiphyllus, equisetinus, fagineus, fibrillosus, flavofuscescens, foliicola, friesii, fructigenus, geogenum, gregarius, herbarum, humuli, ianthium, imberbis, immutabilis, infarciens, laetus, limonium, lutescens, magnificus, menthae, myrtilli, nitidulus, perplexus, phascoides, phyllogenus, phyllophilus, pileatus, populneus, procerus, renisporus, repandus, rhizophilus, rhodoleucus, rhytidiadelphi, roburneus, robustior, rokebyensis, rumicis, salicellus, scutula, serotinus, sparsus, splendens, subferrugineus, sublateritius, subpallescens, subularis, symphoricarpi, syringicolor, tamaricis, tetracladius, tubus, vasaensis, vernus, vitellinus, vitigenus*
VARIETIES *albidus aesculi, epiphyllus acarius, fructigenus coryli, humuli minusculus, procerus excavatus, scutula album, scutula lysimachiae, scutula rubi, scutula rudbeckiae, scutula solani, tetracladius angulata*
GENUS *Idriella* Idry Helot Sac Fungus
SPECIES *australiensis, cagnizarii, couratarii, desertorum, lunata, rhododendri, tropicalis, uncinospora, variabilis*
GENUS *Ionomidotis* Violet Midas Helot Sac Fungus
SPECIES *fulvotingens*
GENUS *Jacobsonia* Jacobson Helot Sac Fungus
SPECIES *glauca*
GENUS *Lagerheima* Lagerheim Helot Sac Fungus
SPECIES *sphaerospora*
GENUS *Lasseria* Lasser Helot Sac Fungus
SPECIES *chrysophthalma*
GENUS *Lemalis* Lemal Helot Sac Fungus
SPECIES *aurea*
GENUS *Libartania* Libart Helot Sac Fungus
SPECIES *laserpitii, phragmiticola, themedae*
GENUS *Loramyces* Lora Helot Sac Fungus
SPECIES *juncicola, macrosporus*
GENUS *Llimoniella* Llimon Helot Sac Fungus
SPECIES *neglecta*
GENUS *Masseea* Masse Helot Sac Fungus
SPECIES *quisquiliarum*
GENUS *Merodontis* Seaside Toothed Helot Sac Fungus
SPECIES *tenella*
GENUS *Metapezizella* Meta Peziz Helot Sac Fungus
SPECIES *phyllachorivora*
GENUS *Micraspis* Asp Helot Sac Fungus
SPECIES *strobilina, tetraspora*
GENUS *Microglossum* Tongue Helot Sac Fungus
SPECIES *olivaceum, viride*
GENUS *Micropodia* Pod Helot Sac Fungus
SPECIES *arenula, filicea, pteridina*
GENUS *Midotiopsis* Midas Helot Sac Fungus

SPECIES *bambusicola, jamaicensis*
GENUS *Mniaecia* Mindful Helot Sac Fungus
SPECIES *nivea*
GENUS *Mollisinopsis* Mollisin Helot Sac Fungus
SPECIES *aurantiaca, filicis, indica*
GENUS *Muscicola* Fly Helot Sac Fungus
SPECIES *dubia*
GENUS *Mycosphaerangium* Sphere Angled Helot Sac Fungus
SPECIES *magnisporum, tetrasporum, tiliae*
GENUS *Myriodiscus* Myriad Disc Helot Sac Fungus
SPECIES *sparassoides*
GENUS *Mytilodiscus* Mytil Disc Helot Sac Fungus
SPECIES *alnicola*
GENUS *Nipterella* Nipter Helot Sac Fungus
SPECIES *duplex, parksii, tsugae*
GENUS *Nothomitra* False Mitre Helot Sac Fungus
SPECIES *cinnamomea, kovalii, sinensis*
GENUS *Ombrophila* Omber Helot Sac Fungus
SPECIES *ambigua, baeumleri, decolorans, helotioides, nigripes, obstricta, purpurea, rudis, violacea*
GENUS *Pachydisca* Thick Disc Helot Sac Fungus
SPECIES *belonea, brunnea, fibuliformis, fusispora, helotioides, ochracea, quisquiliaris, sclerotioides*
GENUS *Parencoelia* Paren Celestial Helot Sac Fungus
SPECIES *andina, biparasitica, kalmiae, myriostylidis*
GENUS *Parorbiliopsis* Orb Helot Sac Fungus
SPECIES *extumescens, luzulae-silvaticae, minuta, salicis, samarae*
GENUS *Patinellaria* Patina Helot Sac Fungus
SPECIES *sanguinea*
GENUS *Pestalopezia* Pestal Peziz Helot Sac Fungus
SPECIES *brunneopruinosa, rhododendri, tsugae*
GENUS *Pezizella* Peziz Helot Sac Fungus
SPECIES *albosanguinea, albula, alniella, chionea, conorum, cyathiformis, discreta, eburnea, epithallina, eriophori, fagi, glareosa, hughesii, incerta, leucostigma, milliaris, minutissima, muscicola, myriadea, nigrocorticata, parasitica, parilis, polytrichi, pulcherrima, pulvinata, punctoidea, roburnea, rubescens, subtilis, turgidella, vulgaris*
VARIETIES *eburnea filicis, pulvinata lignicola*
GENUS *Phaeangellina* Dusky Angled Helot Sac Fungus
SPECIES *empetri*
GENUS *Phaeohelotium* Dusky Helot Sac Fungus
SPECIES *extumescens, flexuosum, geogenum, italicum, lilacinum, monticola, nobile, purpureum, subcarneum, trabinellum, umbilicatum*
GENUS *Phaeopyxis* Dusky Pyxie Helot Sac Fungus
SPECIES *australis, carniolica, punctum, varia*
GENUS *Phragmiticola* Fragment Helot Sac Fungus
SPECIES *rhopalospermum*
GENUS *Pocillum* Pocil Helot Sac Fungus
SPECIES *cesatii, needhamii*
GENUS *Polydiscidium* Many Disc Helot Sac Fungus
SPECIES *martynii*
GENUS *Pragmopora* Fragment Pore Helot Sac Fungus
SPECIES *amphibola, bacillifera, pithya*
GENUS *Pragmopycnis* Fragment Dense Helot Sac Fungus
SPECIES *pithya*
GENUS *Pseudohelotium* Lesser Helot Sac Fungus
SPECIES *alaunae, pineti*
GENUS *Pseudopeltis* Pelt Helot Sac Fungus
SPECIES *filicum*
GENUS *Pseudopezicula* Lesser Peziz Helot Sac Fungus
SPECIES *tetraspora, tracheiphila*
GENUS *Rhizoscyphus* Rhizid Scyph Helot Sac Fungus
SPECIES *ericae, monotropae*
GENUS *Rhizothyrium* Rhizid Thyre Helot Sac Fungus
SPECIES *abietis, parasiticum*
GENUS *Rhodesia* Rhodesia Helot Sac Fungus
SPECIES subtecta
GENUS *Rhymbocarpus* Rhyme Carpel Helot Sac Fungus
SPECIES *cruciatus, neglectus, pubescens*
GENUS *Sarcoleotia* Flesh Lion Helot Sac Fungus

SPECIES *turficola*
GENUS *Scleroderris* Tough Leather Helot Sac Fungus
SPECIES *dennisii*
GENUS *Scutularia* Scute Helot Sac Fungus
SPECIES *citrina, littoralis, vermifera*
GENUS *Septopezizella* Septic Peziz Helot Sac Fungus
SPECIES *oreadum*
GENUS *Skyathea* Skyathe Helot Sac Fungus
SPECIES *hederae*
GENUS *Spathulariopsis* Spathe Helot Sac Fungus
SPECIES *velutipes*
GENUS *Sporonema* Spore Thread Helot Sac Fungus
SPECIES *hiemale*
GENUS *Stamnaria* Stamnar Helot Sac Fungus
SPECIES *persoonii*
VARIETIES *persoonii equiseti*
GENUS *Starbaeckia* Starbaeck Helot Sac Fungus
SPECIES *pseudotryblidioides*
GENUS *Strossmayeria* Strossmayer Helot Sac Fungus
SPECIES *atriseda, basitricha, viridiatra*
GENUS *Symphyosira* Symphony Pitted Helot Sac Fungus
SPECIES *rosea*
GENUS *Symphyosirinia* Lesser Symphony Pitted Helot Sac Fungus
SPECIES *angelicae, galii, heraclei*
GENUS *Tatraea* Tatra Helot Sac Fungus
SPECIES *macrospora*
GENUS *Thindiomyces* Thind Helot Sac Fungus
SPECIES *epiphyllus*
GENUS *Topospora* Top Spore Helot Sac Fungus
SPECIES *andromedae, myrtilli, proboscidea, uberiformis*
GENUS *Trichohelotium* Hairy Helot Sac Fungus
SPECIES *pineum, rosae*
GENUS *Tricladium* Triple Branched Helot Sac Fungus
SPECIES *angulatum, anomalum, attenuatum, brunneum, castaneicola, caudatum, chaetocladium, curvisporum, fallax, giganteum, gracile, indicum, intermedium, malaysianum, minutum, palmiforme, patulum, robustum, splendens, terrestre, varium*
GENUS *Tympanis* Drum Helot Sac Fungus
SPECIES *abietina, amelanchieris, aucupariae, confusa, conspersa, diospyri, fasciculata, hansbroughiana, hypopodia, laricina, ligustri, malicola, piceae, piceina, pitya, prunicola, pseudotsugae, saligna, spermatiospora, truncatula, tsugae*
GENUS *Unguiculariopsis* Claw Helot Sac Fungus
SPECIES *groenlandiae, ilicincola, lesdainii, lettaui, manriquei, refractiva, thallophila*
GENUS *Varicosporium* Variable Spore Helot Sac Fungus
SPECIES *delicatum, elodeae, giganteum, helicosporum, macrosporum, trimosum*
GENUS *Velutarina* Velvet Helot Sac Fungus
SPECIES *juniperi, rufo-olivacea*
GENUS *Weinmannioscyphus* Weinmann Scyph Helot Sac Fungus
SPECIES *messerschmidii*
GENUS *Xylogramma* Wood Gramm Helot Sac Fungus
SPECIES *longum, pallidum*
GENUS *Zugazaea* Zugaza Helot Sac Fungus
SPECIES *agyrioides*

FAMILY 02 *Bulgariaceae* Bulgaria Helot Sac Fungus
GENUS *Bulgaria* Bulgaria Helot Sac Fungus
SPECIES *inquinans, rufa*
GENUS *Bulgariella* Lesser Bulgaria Helot Sac Fungus
SPECIES *pulla*
GENUS *Neobulgaria* Modern Bulgaria Helot Sac Fungus
SPECIES *lilacina, premnophila, pura, undata*
VARIETIES *pura pura, pura foliacea*

FAMILY 03 *Cudoniaceae* Cudon Helot Sac Fungus
GENUS *Cudonia* Cudon Helot Sac Fungus
SPECIES *circinans, confusa*
GENUS *Cudoniella* Cudon Helot Sac Fungus
SPECIES *acicularis, buissonii, clavus, junciseda, rubicunda, tenuispora*
VARIETIES *clavus clavus, clavus grandis*
GENUS *Pachycudonia* Thick Cudon Helot Sac Fungus

SPECIES *constrictospora, monticola, spathulata*
GENUS *Spathularia* Spathe Cudon Helot Sac Fungus
SPECIES *flavida*

FAMILY 04 *Dermateaceae* Dermal Helot Sac Fungus
GENUS *Dermea* Dermal Helot Sac Fungus
SPECIES *ariae, balsamea, bicolor, cerasi, hamamelidis, libocedri, padi, piceina, pinicola, prunastri, tulasnei*
GENUS *Dermatea* Lesser Dermal Helot Sac Fungus
SPECIES *umbrina*
GENUS *Actinoscypha* Radial Scyph Dermal Helot Sac Fungus
SPECIES *scirpicola*
GENUS *Angelina* Angel Dermal Helot Sac Fungus
SPECIES *rufescens*
GENUS *Ascluella* Clue Dermal Helot Sac Fungus
SPECIES *symplocina*
GENUS *Atropellis* Dark Pell Dermal Helot Sac Fungus
SPECIES *piniphila*
GENUS *Belonium* Belon Dermal Helot Sac Fungus
SPECIES *cladii, excelsius, graddonii, incurvatum, nigromaculatum, psammicola, quercinum*
GENUS *Belonopsis* Lesser Belon Dermal Helot Sac Fungus
SPECIES *asteroma, ebudensis, filispora, graminea, guestphalicum, iridis, juncicola, junciseda, litoralis, mediella, obscura, pallens, retincola*
GENUS *Blumeriella* Blumer Dermal Helot Sac Fungus
SPECIES *haddenii, jaapii, kerriae*
GENUS *Bulbomollisia* Bulb Soft Dermal Helot Sac Fungus
SPECIES *radiata, striata, stromatica*
GENUS *Calloria* Callor Dermal Helot Sac Fungus
SPECIES *galiorum, neglecta*
GENUS *Calycellinopsis* Calyce Dermal Helot Sac Fungus
SPECIES *xishuangbanna*
GENUS *Catinella* Chain Dermal Helot Sac Fungus
SPECIES *olivacea*
GENUS *Cejpia* Cejp Dermal Helot Sac Fungus
SPECIES *hystrix*
GENUS *Cenangella* Central Angel Dermal Helot Sac Fungus
SPECIES *pinastri*
GENUS *Chaetonaevia* Bristle Mole Dermal Helot Sac Fungus
SPECIES *nannfeldtii, petasitis, ulmicola*
GENUS *Chlorosplenium* Chloric Splene Dermal Helot Sac Fungus
SPECIES *cenangium, viridulum*
GENUS *Coleosperma* Celestial Spore Dermal Helot Sac Fungus
SPECIES *lacustre*
GENUS *Coronellaria* Crown Dermal Helot Sac Fungus
SPECIES *amaena*
GENUS *Crustomollisia* Crust Soft Dermal Helot Sac Fungus
SPECIES *roburnea*
GENUS *Cryptohymenium* Cryptic Hymen Dermal Helot Sac Fungus
SPECIES *pycnidiophorum*
GENUS *Cryptosporiopsis* Cryptic Spore Dermal Helot Sac Fungus
SPECIES *balsameae, coryli, diplodioides, diversispora, fasciculata, grisea, inaequalis, longispora, platanicola, pruinosa, radicicola, tarraconensis, versiformis*
GENUS *Cryptosympodula* Cryptic Sympod Dermal Helot Sac Fungus
SPECIES *appendiculata*
GENUS *Cylindrosporium* Cylindric Spore Dermal Helot Sac Fungus
SPECIES *niveum, pseudoplatani*
GENUS *Dennisiodiscus* Dennis Disc Dermal Helot Sac Fungus
SPECIES *pinicola, prasinus*
GENUS *Dibeloniella* Dual Belon Dermal Helot Sac Fungus
SPECIES *citrinella*
GENUS *Diplocarpa* Lesser Dual Carpel Dermal Helot Sac Fungus
SPECIES *bloxamii*
GENUS *Diplocarpon* Dual Carpel Dermal Helot Sac Fungus
SPECIES *earlianum, impressum, mespili, rosae, saponariae*
VARIETIES *earlianum fragariae, earlianum tormentillae*
GENUS *Diplonaevia* Dual Mole Dermal Helot Sac Fungus
SPECIES *circinata, exigua, paulula, seriata*
GENUS *Discocurtisia* Disc Curtis Dermal Helot Sac Fungus
SPECIES *arundinariae*

GENUS *Discohainesia* Disc Haines Dermal Helot Sac Fungus
SPECIES *oenotherae*
GENUS *Drepanopeziza* Drepan Peziz Dermal Helot Sac Fungus
SPECIES *populi-albae, populorum, punctiformis, ribis, salicis, schoenicola, sphaeroides, tremulae, triandrae*
VARIETIES *ribis grossulariae*
GENUS *Duebenia* Dueben Dermal Helot Sac Fungus
SPECIES *compta*
GENUS *Durandiella* Durand Dermal Helot Sac Fungus
SPECIES *fraxini, gallica, pseudotsugae, seriata*
GENUS *Eupropolella* Propole Dermal Helot Sac Fungus
SPECIES *arundinariae, britannica, celata, vaccinii*
GENUS *Excipula* Excipule Dermal Helot Sac Fungus
SPECIES *chaetostroma, macrotricha*
GENUS *Felisbertia* Felisbert Dermal Helot Sac Fungus
SPECIES *melastomacearum*
GENUS *Gloeosporidiella* Lesser Glow Spore Dermal Helot Sac Fungus
SPECIES *nobilis, turgida*
GENUS *Gloeosporium* Glow Spore Dermal Helot Sac Fungus
SPECIES *alpinum, cylindrospermum, heveae, kaki, labes, limetticola, pedemontanum, pseudophoma, solani, theae-sinensis, trifolii, venetum, vogelianum*
GENUS *Graddonia* Graddon Dermal Helot Sac Fungus
SPECIES *coracina*
GENUS *Haglundia* Haglund Dermal Helot Sac Fungus
SPECIES *elegantior, penyardensis, perelegans, rubra, sarmentorum*
GENUS *Hainesia* Haines Dermal Helot Sac Fungus
SPECIES *marsdeniae, pertusariae, rubi*
GENUS *Hysteronaevia* Hyster Mole Dermal Helot Sac Fungus
SPECIES *fimbriata, olivacea*
GENUS *Hysteropezizella* Hyster Peziz Dermal Helot Sac Fungus
SPECIES *diminuens, foecunda, hebridensis, lyngei, prahliana, pusilla, rehmii*
VARIETIES *prahliana orcadensis*
GENUS *Hysterostegiella* Hyster Cover Dermal Helot Sac Fungus
SPECIES *crassomarginata, dowardensis, dumeti, fenestrata, lauri, quercea, valvata*
GENUS *Laetinaevia* Bright Mole Dermal Helot Sac Fungus
SPECIES *carneoflavida, luzulae, marina, minutissima, pustulata*
GENUS *Leptotrochila* Thin Runner Dermal Helot Sac Fungus
SPECIES *cerastiorum, lugubris, medicaginis, porri, prunellae, radians, ranunculi, repanda, sanguisorbae, verrucosa*
GENUS *Marssonina* Marsson Dermal Helot Sac Fungus
SPECIES *betulae, daphnes, dispersa, kriegeriana, salicigena, sorbi*
GENUS *Merostictis* Parted Stick Dermal Helot Sac Fungus
SPECIES *hypopyrrha*
GENUS *Microgloeum* Glow Dermal Helot Sac Fungus
SPECIES *pruni*
GENUS *Micropeziza* Least Peziz Dermal Helot Sac Fungus
SPECIES *cornea, karstenii, poae*
GENUS *Mollisia* Soft Dermal Helot Sac Fungus
SPECIES *amenticola, aquosa, caesia, caespiticia, caricina, casaresiae, chionea, cinerascens, cinerea, cinerella, cinereo-olivascens, cirsiicola, clavata, coerulans, conigena, culmina, dactyligluma, dextrinospora, discolor, erumpens, escharodes, euphrasiae, fallax, fallens, fungorum, fuscoparaphysata, fuscostriata, heterosperma, humidicola, hydrophila, incrustata, jugosa, juncina, leucosphaeria, ligni, lurida, lychnidis, lycopi, macrosperma, melaleuca, minutissima, mutabilis, myricariae, nigrescens, olivascens, orcadensis, palustris, plantaginis, poaeoides, populi, rabenhorstii, ramealis, rehmii, sordidula, spectabilis, sphaeroides, stellata, stromatica, stromaticola, subcorticalis, tenuispora, teucrii, typhae, uda, undulatodepressula, ventosa*
VARIETIES *cinerea cinerea, cinerea grisella, cinerea minutella, cinerella cinerella, cinerella caespitosa, discolor longispora, sordidula sordidula, sordidula major*
GENUS *Mollisiopsis* Lesser Soft Dermal Helot Sac Fungus
SPECIES *dennisii, lachnoideis, lanceolata, lobkovicensis, quercina, subantarctica, subcinerea*
GENUS *Monostichella* Monad Stitch Dermal Helot Sac Fungus
SPECIES *coryli, robergei, salicis*
GENUS *Naevala* Lesser Mole Dermal Helot Sac Fungus
SPECIES *hysteropezizoides, marginata, perexigua, ribis, vleugelii*
GENUS *Naeviella* Least Mole Dermal Helot Sac Fungus
SPECIES *paradoxa, poeltiana, volkartiana*
GENUS *Naeviopsis* Mole Dermal Helot Sac Fungus
SPECIES *arctica, caricis-brizoidis, carneola, carneopallida, epilobii, jenensis, montana, primulae, pusilla, rhodiolae, saticis, simulans, tithymalina, tornensis*

GENUS *Neofabraea* Modern Fabre Dermal Helot Sac Fungus
SPECIES *alba, krawtzewii, malicorticis, perennans*
GENUS *Nimbomollisia* Nimbo Soft Dermal Helot Sac Fungus
SPECIES *eriophori*
GENUS *Niptera* Nipter Dermal Helot Sac Fungus
SPECIES *ambigua, exsiliens, lacustris, melanophaea, melatephra, melatephroides, muelleri-argoviensis, myriadea, pilosa, pulla, ramincola, stockii, subbiatorina, submelaena, trichophoricola, umbelliferarum, umbonata*
VARIETIES *lacustris caricina*
GENUS *Obscurodiscus* Obscure Disc Dermal Helot Sac Fungus
SPECIES *myricae*
GENUS *Obtectodiscus* Obtect Disc Dermal Helot Sac Fungus
SPECIES *aquaticus, nectrioides*
GENUS *Ocellaria* Lesser Eyespot Dermal Helot Sac Fungus
SPECIES *masseeana*
GENUS *Oculimacula* Eyespot Dermal Helot Sac Fungus
SPECIES *acuformis, yallundae*
GENUS *Patellariopsis* Patella Dermal Helot Sac Fungus
SPECIES *atrovinosa, carnea, clavispora, dennisii, indica*
GENUS *Patinella* Patina Dermal Helot Sac Fungus
SPECIES *abietina, macrospora, rubrotingens*
GENUS *Pezicula* Peziz Dermal Helot Sac Fungus
SPECIES *acericola, alni, alnicola, amoena, aurantiaca, betulae, carnea, carpinea, cinnamomea, conigena, corticola, coryli, corylina, dennisii, eucrita, frangulae, hamamelidis, houghtonii, livida, myrtillina, ocellata, paradoxa, plantarium, pruinosa, rhododendri, rubi, scoparia, sepium, sporulosa, subcarnea*
GENUS *Phacidiella* Lens Dermal Helot Sac Fungus
SPECIES *asperulina*
GENUS *Phaeonaevia* Dusky Mole Dermal Helot Sac Fungus
SPECIES *monilispora*
GENUS *Phlyctema* Leaf Dermal Helot Sac Fungus
SPECIES *caulium, johnstonii*
GENUS *Pilidium* Soft Hair Dermal Helot Sac Fungus
SPECIES *acerinum, carbonaceum, concavum*
GENUS *Pirottaea* Pirotta Dermal Helot Sac Fungus
SPECIES *bresadolae, brevipila, caesiella, exilispora, inopinata, lamii, nigrostriata, paupercula, plantaginis, senecionis, strigosa, symphyti, veneta*
VARIETIES *bresadolae bresadolae, bresadolae bartsiae*
GENUS *Ploettnera* Ploettner Dermal Helot Sac Fungus
SPECIES *belladonnae, caeruleoviridis, conglomerata, exigua, hyperici, solidaginis*
GENUS *Podophacidium* Pod Lens Dermal Helot Sac Fungus
SPECIES *xanthomelum*
GENUS *Pseudonaevia* Lesser Mole Dermal Helot Sac Fungus
SPECIES *caricina*
GENUS *Pseudopeziza* Lesser Peziz Dermal Helot Sac Fungus
SPECIES *alismatis, calthae, jasiones, medicaginis, svalbardensis, trifolii*
GENUS *Pyrenopeziza* Fire Peziz Dermal Helot Sac Fungus
SPECIES *arctii, arenivaga, artemisiae, arundinacea, atrata, benesuada, betulicola, betulina, brassicae, carduorum, chailletii, chamaenerii, commoda, compressula, depressuloides, digitalina, ebuli, eryngii, foliicola, fuckelii, fuscescens, galii, galii-veri, hystrix, inornata, karstenii, kolaensis, laricina, lonicerae, lychnidis, lycopincola, maculans, maculata, mercurialidis, mercurialis, millegrana, morthieri, nervicola, oenanthes, osiliensis, pastinacae, petiolaris, plicata, polygoni, pulveracea, revincta, riccia, rubi, salicis, solidaginis, subplicata, thalictri, tini, umbellatorum, urticicola*
VARIETIES *atrata asparagi, atrata oenanthes, laricina microsperma, plicata conicola*
GENUS *Schizothyrioma* Split Thyre Dermal Helot Sac Fungus
SPECIES *aterrimum, ptarmicae*
GENUS *Scutobelonium* Scute Belon Dermal Helot Sac Fungus
SPECIES *amorilens*
GENUS *Scutomollisia* Scute Soft Dermal Helot Sac Fungus
SPECIES *clavata, contraria, fimbriomarginata, integromarginata, lanceata, leptoderma, microspora, morvernensis, operculata, pallido-ochracea, papillata, punctum, russea, stenospora*
GENUS *Spilopodia* Spotted Pod Dermal Helot Sac Fungus
SPECIES *melanogramma, nervisequa, simulata*
GENUS *Spilopodiella* Lesser Spotted Pod Dermal Helot Sac Fungus
SPECIES *arxii*
GENUS *Tapesia* Tapes Dermal Helot Sac Fungus
SPECIES *callunae, cinerella, evilescens, fusca, knieffii, lividofusca, rosae, strobilicola, toruloides, villosa*
VARIETIES *rosae prunicola*
GENUS *Trichodiscus* Hairy Disc Dermal Helot Sac Fungus

SPECIES *diversipilus, virescentulus*
GENUS *Trichosporiella* Hairy Spore Dermal Helot Sac Fungus
SPECIES *cerebriformis, flavificans, multisporum, ornithopoda, paludigena*
GENUS *Trochila* Runner Dermal Helot Sac Fungus
SPECIES *craterium, ilicina, laurocerasi*
GENUS *Waltonia* Walton Dermal Helot Sac Fungus
SPECIES *pinicola*

FAMILY 05 *Geoglossaceae* Earth Tongue Helot Sac Fungus
GENUS *Geoglossum* Earth Tongue Helot Sac Fungus
SPECIES *atropurpureum, barlae, cookeanum, difforme, elongatum, fallax, glabrum, glutinosum, nigritum, peckianum, simile, starbaeckii, uliginosum, umbratile, viscosum, vleugelianum*
GENUS *Maasoglossum* Maas Earth Tongue Helot Sac Fungus
SPECIES *verrucosporum*
GENUS *Phaeoglossum* Dusky Earth Tongue Helot Sac Fungus
SPECIES *zeylanicum*
GENUS *Thuemenidium* Thuemen Earth Tongue Helot Sac Fungus
SPECIES *atropurpureum, berteroi, hookeri*
GENUS *Trichoglossum* Hairy Earth Tongue Helot Sac Fungus
SPECIES *hirsutum, rasum, tetrasporum, variabile, walteri*
VARIETIES *hirsutum hirsutum, hirsutum capitatum, hirsutum irregulare*

FAMILY 06 *Hemiphacidiaceae* Half Lens Helot Sac Fungus
GENUS *Hemiphacidium* Half Lens Helot Sac Fungus
SPECIES *convexum, longisporum, planum*
GENUS *Didymascella* Dual Mask Half Lens Helot Sac Fungus
SPECIES *chamaecyparidis, tetraspora, thujina*
GENUS *Fabrella* Fabre Half Lens Helot Sac Fungus
SPECIES *tsugae*
GENUS *Korfia* Korf Half Lens Helot Sac Fungus
SPECIES *tsugae*
GENUS *Meria* Parted Half Lens Helot Sac Fungus
SPECIES *laricis, parkeri*
GENUS *Rhabdocline* Rod Inclined Half Lens Helot Sac Fungus
SPECIES *pseudotsugae, weirii*
GENUS *Sarcotrochila* Flesh Runner Half Lens Helot Sac Fungus
SPECIES *alpina, balsameae, macrospora, piniperda*

FAMILY 07 *Hyaloscyphaceae* Hyaline Scyph Helot Sac Fungus
GENUS *Hyaloscypha* Hyaline Scyph Helot Sac Fungus
SPECIES *albohyalina, aureliella, britannica, candidata, carpinacea, daedaleae, effugiens, epiporia, farinacea, flaveola, fuckelii, herbarum, hyalina, intacta, iridina, leuconica, mirabilis, paludosa, pellucida, priapi, quercicola, secalina, vitreola, zalewskii*
VARIETIES *albohyalina albohyalina, albohyalina spiralis, albohyalina tigillaris, fuckelii fuckelii, fuckelii alniseda, hyalina setosa, leuconica leuconica, leuconica bulbopilosa*
GENUS *Albotricha* White Hairy Hyaline Scyph Helot Sac Fungus
SPECIES *acutipila, albidorosea, albotestacea, ammophilae*
VARIETIES *acutipila laetior*
GENUS *Arachnopeziza* Spider Peziz Hyaline Scyph Helot Sac Fungus
SPECIES *aranea, aurata, aurelia, candidofulva, cornuta, eriobasis, obtusipila, trabinelloides*
GENUS *Asperopilum* Asp Downy Hyaline Scyph Helot Sac Fungus
SPECIES *juncicola*
GENUS *Austropezia* Southern Peziz Hyaline Scyph Helot Sac Fungus
SPECIES *samuelsii*
GENUS *Belonidium* Belon Hyaline Scyph Helot Sac Fungus
SPECIES *caricincola, melatephroides, minutissimum, pruinosum, sulphureum*
GENUS *Betulina* Birch Hyaline Scyph Helot Sac Fungus
SPECIES *fuscostipitata, hirta*
GENUS *Brunnipila* Brown Downy Hyaline Scyph Helot Sac Fungus
SPECIES *calyculiformis, clandestina*
VARIETIES *calyculiformis latebricola*
GENUS *Calycellina* Calyce Hyaline Scyph Helot Sac Fungus
SPECIES *asperipila, calycelloides, caricina, chlorinella, citricola, indumenticola, juniperina, leucella, microspis, myriadea, ochracea, phalaridis, populina, punctata, rivelinensis, spiraeae*
GENUS *Calycina* Lesser Calyce Hyaline Scyph Helot Sac Fungus
SPECIES *conorum*
GENUS *Calyptellopsis* Calypter Hyaline Scyph Helot Sac Fungus
SPECIES *reticulata*
GENUS *Ciliolarina* Ciliate Hyaline Scyph Helot Sac Fungus

SPECIES *corcontica, laetifica, laricina, ligniseda, neglecta, pinicola*
GENUS *Cistella* Cyst Hyaline Scyph Helot Sac Fungus
SPECIES *acuum, atenuipilus, citrina, coarctata, crassipila, dentata, fugiens, geelmuydenii, hungarica, lagenipilus, mali, perparvula, spicicola*
GENUS *Cistellina* Lesser Cyst Hyaline Scyph Helot Sac Fungus
SPECIES *stereicola*
GENUS *Clathrosphaerina* Lattice Sphere Hyaline Scyph Helot Sac Fungus
SPECIES *zalewskii*
GENUS *Clavidisculum* Club Disc Hyaline Scyph Helot Sac Fungus
SPECIES *caricis, crassipilum, graminicola, karstenii, kriegerianum, microsporum*
VARIETIES *caricis minor*
GENUS *Dasyscyphella* Lesser Daisy Hyaline Scyph Helot Sac Fungus
SPECIES *castaneicola*
GENUS *Dasyscyphus* Daisy Hyaline Scyph Helot Sac Fungus
SPECIES *abscondita, aureus, bullii, castaneus, comitissae, crassipilus, elegantulus, ericinus, flavofuligineus, fragariastri, globuligerus, horridulus, leucostomus, mughonicola, occultus, perplexus, pini, pudibundus, pulchricolor, roridus, serinellus, sesleriae, stigmellus, sydowii, tricolor, viridulus*
GENUS *Dematioscypha* Dermal Hyaline Scyph Helot Sac Fungus
SPECIES *dematiicola, richonis*
VARIETIES *richonis olivacea*
GENUS *Dimorphotricha* Dimorphic Hairy Hyaline Scyph Helot Sac Fungus
SPECIES *australis*
GENUS *Discocistella* Disc Cyst Hyaline Scyph Helot Sac Fungus
SPECIES *glyceriae, grevillei*
GENUS *Echinula* Urchin Hyaline Scyph Helot Sac Fungus
SPECIES *asteriadiformis*
GENUS *Eriopezia* Woolly Peziz Hyaline Scyph Helot Sac Fungus
SPECIES *caesia*
GENUS *Fuscolachnum* Fuscous Downy Hyaline Scyph Helot Sac Fungus
SPECIES *boreale, dumorum, inopinatum, labradoricum, misellum, pteridis*
GENUS *Fuscoscypha* Fuscous Hyaline Scyph Helot Sac Fungus
SPECIES *acicularum*
GENUS *Goidanichiella* Goidanich Hyaline Scyph Helot Sac Fungus
SPECIES *scopula, sphaerospora*
GENUS *Graddonidiscus* Graddon Disc Hyaline Scyph Helot Sac Fungus
SPECIES *coruscatus*
GENUS *Hamatocanthoscypha* Hooked Thorn Hyaline Scyph Helot Sac Fungus
SPECIES *laricionis, uncipila*
GENUS *Haplographium* Single Graphic Hyaline Scyph Helot Sac Fungus
SPECIES *bicolor*
GENUS *Hegermila* Hegermil Hyaline Scyph Helot Sac Fungus
SPECIES *andina, crassispora, octopartita, vermispora*
GENUS *Hyalacrotes* Tall Hyaline Scyph Helot Sac Fungus
SPECIES *hamulata, lupini*
GENUS *Hyalopeziza* Peziz Hyaline Scyph Helot Sac Fungus
SPECIES *alni, ciliata, corticicola, millepunctata, nectrioidea, niveocincta, pygmaea, spinicola, trichodea*
GENUS *Hyalotricha* Hairy Hyaline Scyph Helot Sac Fungus
SPECIES *salicicola*
GENUS *Hydrocina* Hydric Hyaline Scyph Helot Sac Fungus
SPECIES *chaetocladia*
GENUS *Hyphodiscus* Hypha Disc Hyaline Scyph Helot Sac Fungus
SPECIES *gemmarum, gregarius, hymeniophilus, otanii, theiodeus, viridipilosus*
GENUS *Incrucipulum* Incru Downy Hyaline Scyph Helot Sac Fungus
SPECIES *ciliare, sulphurellum, virtembergense*
GENUS *Incrupila* Lesser Incru Downy Hyaline Scyph Helot Sac Fungus
SPECIES *aspidii, melatheja, viridipilosa*
GENUS *Lachnellula* Lesser Downy Hyaline Scyph Helot Sac Fungus
SPECIES *abietis, arida, calyciformis, calycina, cervina, flavovirens, fuckelii, fuscosanguinea, hyalina, laricis, occidentalis, pseudofarinacea, resinaria, subtilissima, suecica, tuberculata, willkommii*
GENUS *Lachnum* Downy Hyaline Scyph Helot Sac Fungus
SPECIES *abnorme, acerinum, acutipilum, aeruginellum, apalum, barbatum,bicolor, brevipilosum, brevipilum, britzelmayrionum, callimorphum, calycioides, capitatum, caricis, carneolum, cerinum, clandestinum, clavigerum, clavisporum, coeruleoalbum, controversum, corticale, cruciferum, crystallinum, deflexum, diminutum, discolor, distinguendum, elongatisporum, fasciculare, fuscescens, humuli, imbecille, laburni, lachnoides, luteovinosum, luzulinum, mollissimum, nipponicum, nudipes, palearum, papyraceum, pseudolachnum, pteridialis, pulchellum, pulverulentum, pygmaeum, radotinense, rehmii, rhodoleucum, rhytismatis, salicariae, sesleriae, soppittii, spadiceum, subnudipes, sulphureum, tenuissimum, trapeziforme, virgineum*

VARIETIES *barbatum pellita, bicolor rubi, carneolum longisporum, fuscescens fuscescens, fuscescens fagicola, imbecille cladii, nudipes minor, palearum niger, virgineum selecti*
GENUS *Lasiobelonium* Woolly Belon Hyaline Scyph Helot Sac Fungus
SPECIES *lonicerae, mollissimum, nazarovae, nidulum, variegatum*
GENUS *Microscypha* Lesser Hyaline Scyph Helot Sac Fungus
SPECIES *ellisii, enrhiza, grisella, muelleri*
VARIETIES *ellisii eriophori*
GENUS *Mollisina* Soft Hyaline Scyph Helot Sac Fungus
SPECIES *acerina, alba, flava, oedema, rubi*
GENUS *Otwaya* Otway Hyaline Scyph Helot Sac Fungus
SPECIES *verruculospora*
GENUS *Parachnopeziza* Spider Peziz Hyaline Scyph Helot Sac Fungus
SPECIES *alba, aquilinella, bambusae, guangxiensis, miniopsis, sinensis, triseptata, variabilis*
GENUS *Perrotia* Perrot Hyaline Scyph Helot Sac Fungus
SPECIES *distincta, flammea*
GENUS *Pezizellaster* Star Peziz Hyaline Scyph Helot Sac Fungus
SPECIES *radiostriatus, serratus*
GENUS *Phaeoscypha* Dusky Hyaline Scyph Helot Sac Fungus
SPECIES *cladii*
GENUS *Phialina* Phial Hyaline Scyph Helot Sac Fungus
SPECIES *lachnobrachya, pseudopuberula, separabilis, ulmariae*
GENUS *Pilatia* Pilate Hyaline Scyph Helot Sac Fungus
SPECIES *foliorum*
GENUS *Pithyella* Pithy Hyaline Scyph Helot Sac Fungus
SPECIES *erythrostigma, hypnorum*
GENUS *Polaroscyphus* Polar Hyaline Scyph Helot Sac Fungus
SPECIES *spetsbergianus*
GENUS *Polydesmia* Many Fiber Hyaline Scyph Helot Sac Fungus
SPECIES *fructicola, pruinosa*
GENUS *Proliferodiscus* Prolific Disc Hyaline Scyph Helot Sac Fungus
SPECIES *alboviridis, caeruleoalbus, caeruleoniger, dingleyae, earoleucus, inspersus, pulveraceus*
GENUS *Protounguicularia* Claw Hyaline Scyph Helot Sac Fungus
SPECIES *barbata, brevicapitata, monoseptata, vandae, variepilosa*
VARIETIES *barbata barbata, barbata resinacea*
GENUS *Pseudaegerita* Aegerite Hyaline Scyph Helot Sac Fungus
SPECIES *corticalis, foliicola, matsushimae, viridis*
GENUS *Psilachnum* Psilum Downy Hyaline Scyph Helot Sac Fungus
SPECIES *acutum, asemum, auranticolor, chrysostigmum, granulosellum, inquilinum, lateritioalbum, micaceum, pteridigenum, rubrotinctum, striatum, tami, nympharum*
GENUS *Psilocistella* Psilum Cyst Hyaline Scyph Helot Sac Fungus
SPECIES *nympharum*
GENUS *Pubigera* Pubescent Hyaline Scyph Helot Sac Fungus
SPECIES *subvillosula*
GENUS *Ramulispora* Branched Spore Hyaline Scyph Helot Sac Fungus
SPECIES *alloteropsis, cerealis, sorghi, sorghicola, zonata*
GENUS *Rodwayella* Rodway Hyaline Scyph Helot Sac Fungus
SPECIES *citrinula, myricae, sessilis*
GENUS *Scutoscypha* Scute Hyaline Scyph Helot Sac Fungus
SPECIES *fagi*
GENUS *Tapesina* Tapes Hyaline Scyph Helot Sac Fungus
SPECIES *griseovitellina*
GENUS *Trichopeziza* Hairy Peziz Hyaline Scyph Helot Sac Fungus
SPECIES *episphaeria, viridula*
GENUS *Unguicularia* Clawed Hyaline Scyph Helot Sac Fungus
SPECIES *aspera, asterostoma, carestiana, costata, dilatopilosa, hamatopilosa, incarnatina*
GENUS *Unguiculariella* Least Clawed Hyaline Scyph Helot Sac Fungus
SPECIES *bhutanica*
GENUS *Unguiculella* Lesser Clawed Hyaline Scyph Helot Sac Fungus
SPECIES *eurotioides, foliicola, hamulata, microspora, robergei*
VARIETIES *robergei coelomyceticola*
GENUS *Urceolella* Urn Hyaline Scyph Helot Sac Fungus
SPECIES *appressipila, arundinis, crispula, curvipila, iridis, papillaris*
GENUS *Venturiocistella* Ventura Cyst Hyaline Scyph Helot Sac Fungus
SPECIES *diversipila, gaylussaciae, heterotricha, japonica, pini, ulicicola, venturioides*

FAMILY 08 *Leotiaceae* Helot Sac Mushroom
GENUS *Leotia* Slippery Cap Helot Sac Mushroom
SPECIES *atra, chlorocephala, infundibuliformis, lubrica*(Edible), *nana*

GENUS *Acleistia* Open Helot Sac Mushroom
SPECIES *alniella*
GENUS *Gelatinipulvinella* Gelatin Powder Helot Sac Mushroom
SPECIES *astraeicola*
GENUS *Geocoryne* Earth Club Helot Sac Mushroom
SPECIES *exogloea, variispora*
GENUS *Pezoloma* Peziz Helot Sac Mushroom
SPECIES *bhutanicum, ciliifera, fergussonii, griseum, iodocyanescens, iodopedis, kathiae, laricina, petiolatum, rhodocarpa, websteri*
GENUS *Sphagnicola* Peat Moss Dwelling Helot Sac Mushroom
SPECIES *marchantiae*

FAMILY 09 *Phacidiaceae* Lens Helot Sac Fungus
GENUS *Phacidium* Lens Helot Sac Fungus
SPECIES *abietinum, betulinum, carbonaceum, coniferarum, coronatum, falconeri, fennicum, gracile, infestans, lacerum, multivalve, pini, pusillum, vaccinii*
GENUS *Allantophomopsis* Membrane Foam Lens Helot Sac Fungus
SPECIES *abietina, cytisporea, fusiformis, lycopodina, peregrina, pusilla, robusta*
GENUS *Apostrasseria* Separated Strasser Lens Helot Sac Fungus
SPECIES *lunata*
GENUS *Ceuthospora* Ceuth Spore Lens Helot Sac Fungus
SPECIES *australis, buxi, calderae, eucalypti, euonymi, feurichii, foliicola, foliicola, innumera, lauri, laurocerasi, liriodendri, mahoniae, molleriana, phacidioides, phyllosticta, rhododendri*
GENUS *Coma* Comma Lens Helot Sac Fungus
SPECIES *circularis*
GENUS *Lophophacidium* Crest Lens Helot Sac Fungus
SPECIES *dooksii, hyperboreum*

FAMILY 10 *Rutstroemiaceae* Rue Stroma Helot Sac Fungus
GENUS *Rutstroemia* Rue Stroma Helot Sac Fungus
SPECIES *bolaris, bulgarioides, calopus, conformata, elatina, fruticeti, gallincola, henningsiana, hercynica, juniperi, lindaviana, maritima, microspora, myricae, paludosa, petiolorum, plana, rhenana, rubi, sydowiana*
GENUS *Dicephalospora* Two Head Spore Rue Stroma Helot Sac Fungus
SPECIES *calochroa, damingshanica, pinglongshanica, rufocornea*
GENUS *Lambertella* Lambert Rue Stroma Helot Sac Fungus
SPECIES *corni-maris, hicoriae, himalayensis, pruni, subrenispora, tubulosa, viburni*
GENUS *Lanzia* Lanzi Rue Stroma Helot Sac Fungus
SPECIES *coracina, cuniculi, echinophila, luteovirescens, stellariae, vacini*
GENUS *Poculum* Pocule Rue Stroma Helot Sac Fungus
SPECIES *firmum*
GENUS *Scleromitrula* Hard Mitre Rue Stroma Helot Sac Fungus
SPECIES *calthicola, spiraeicola*

FAMILY 11 *Sclerotiniaceae* Shell Helot Sac Fungus
GENUS *Sclerotinia* Shell Helot Sac Fungus
SPECIES *antarctica, asari, borealis, brunneocarpa, bulborum, filipes, heterocarpa, homoeocarpa, longisclerotialis, matthiolae, minor, muscorum, ricini, sativa, sclerotiorum, serica, spermophila, sulcata, trifoliorum*
VARIETIES *trifoliorum fabae, trifoliorum minor*
GENUS *Acarosporium* Acarospore Shell Helot Sac Fungus
SPECIES *sympodiale*
GENUS *Amphobotrys* Dual Berry Shell Helot Sac Fungus
SPECIES *ricini*
GENUS *Asterocalyx* Star Calyx Shell Helot Sac Fungus
SPECIES *ecuadorenis, mirabilis*
GENUS *Botryotinia* Lesser Berry Shell Helot Sac Fungus
SPECIES *allii, calthae, convoluta, draytonii, ficariarum, fuckeliana, globosa, moricola, narcissicola, pelargonii, polyblastis, porri, ranunculi, sphaerosperma, squamosa*
GENUS *Botrytis* Berry Shell Helot Sac Fungus
SPECIES *allii, anthophila, argillacea, arisaemae, bifurcata, carnea, cinerea, citrina, convallariae, cryptomeriae, curta, densa, diospyri, elliptica, epiphylla, fabae, galanthina, hyacinthi, isabellina, lateritia, lutescens, macrospora, paeoniae, pyramidalis, tulipae*
VARIETIES *cinerea sclerotiophila*
GENUS *Ciboria* Cibor Shell Helot Sac Fungus
SPECIES *acerina, aestivalis, alni, amentacea, americana, aschersoniana, batschiana, betulae, carunculoides, caucus, cistophila, dumbirensis, juncorum, rufofusca, seminicola, viridifusca*
GENUS *Ciborinia* Lesser Cibor Shell Helot Sac Fungus
SPECIES *bresadolae, camelliae, candolleana, ciborium, erythronii, foliicola, hirtella*
GENUS *Cristulariella* Crested Shell Helot Sac Fungus

SPECIES *depraedans*
GENUS *Dumontinia* Dumont Shell Helot Sac Fungus
SPECIES *tuberosa*
GENUS *Elliottinia* Elliott Shell Helot Sac Fungus
SPECIES *kerneri*
GENUS *Encoelia* Entic Celestial Shell Helot Sac Fungus
SPECIES *fascicularis, fimbriata, fuckelii, furfuracea, glaberrima, glauca, petrakii, populnea, siparia, tiliacea*
GENUS *Grovesinia* Groves Shell Helot Sac Fungus
SPECIES *pruni, pyramidalis*
GENUS *Haplaria* Haplar Shell Helot Sac Fungus
SPECIES *ochracea*
GENUS *Lambertellinia* Lambert Shell Helot Sac Fungus
SPECIES *scutuloides*
GENUS *Martininia* Martin Shell Helot Sac Fungus
SPECIES *panamaensis*
GENUS *Mitrula* Mitre Shell Helot Sac Fungus
SPECIES *alba, microspora, paludosa, sclerotipus, serpentina*
GENUS *Mitrulinia* Lesser Mitre Shell Helot Sac Fungus
SPECIES *ushuaiae*
GENUS *Moellerodiscus* Moeller Disc Shell Helot Sac Fungus
SPECIES *advenulus, lentus, tenuistipes*
GENUS *Monilia* Monad Shell Helot Sac Fungus
SPECIES *brunnea, carbonaria, laxa, polystroma, pruinosa*
GENUS *Monilinia* Lesser Monad Shell Helot Sac Fungus
SPECIES *aucupariae, baccarum, cydoniae, demissa, fructicola, fructigena, johnsonii, laxa, linhartiana, megalospora, mespili, oxycocci, padi, seaveri, urnula, vaccinii-corymbosi*
VARIETIES *laxa mali*
GENUS *Moserella* Moser Shell Helot Sac Fungus
SPECIES *radicicola*
GENUS *Myrioconium* Myriad Cone Shell Helot Sac Fungus
SPECIES *ambiens*
GENUS *Myriosclerotinia* Myriad Shell Helot Sac Fungus
SPECIES *curreyana, dennisii, juncifida, scirpicola, sulcatula, vahliana*
GENUS *Ovulinia* Oval Shell Helot Sac Fungus
SPECIES *azaleae, perplexa*
GENUS *Phaeangella* Dusky Angled Shell Helot Sac Fungus
SPECIES *ulicis*
GENUS *Poculinia* Pocule Shell Helot Sac Fungus
SPECIES *nothofagi*
GENUS *Pseudociboria* Lesser Cibor Shell Helot Sac Fungus
SPECIES *umbrina*
GENUS *Pycnopeziza* Dense Peziz Shell Helot Sac Fungus
SPECIES *pachyderma, sejournei, sympodialis*
GENUS *Rhacodiella* Beacon Shell Helot Sac Fungus
SPECIES *castaneae*
GENUS *Sclerocrana* Crane Shell Helot Sac Fungus
SPECIES *atra*
GENUS *Seaverinia* Seaver Shell Helot Sac Fungus
SPECIES *geranii*
GENUS *Septotis* Septic Ear Shell Helot Sac Fungus
SPECIES *podophyllina*
GENUS *Streptobotrys* Twisted Berry Shell Helot Sac Fungus
SPECIES *arisaematis, caulophylli, streptothrix*
GENUS *Streptotinia* Twisted Shell Helot Sac Fungus
SPECIES *arisaemae, caulophylli*
GENUS *Stromatinia* Stroma Shell Helot Sac Fungus
SPECIES *gladioli, narcissi, rapulum, smilacinae*
GENUS *Torrendiella* Torrent Shell Helot Sac Fungus
SPECIES *andina, brevisetosa, cannibalensis, ciliata, clelandii, dingleyae, eucalypti, grisea, guangxiensis, madsenii, quintocentenaria, setulata*
GENUS *Valdensia* Valdens Shell Helot Sac Fungus
SPECIES *heterodoxa*
GENUS *Valdensinia* (Synonym for *Valdensia*)
GENUS *Verrucobotrys* Verrucose Berry Shell Helot Sac Fungus
SPECIES *geranii*
GENUS *Zoellneria* Zoellner Shell Helot Sac Fungus
SPECIES *rosarum*

FAMILY 12 *Vibrisseaceae* Vibrisse Helot Sac Fungus
GENUS *Vibrissea* Vibrisse Helot Sac Fungus
SPECIES *decolorans, fergussonii, flavovirens, guernisacii, leptospora, margarita, microscopica, sporogyra, truncorum, turbinata*
GENUS *Anavirga* Virgo Vibrisse Helot Sac Fungus
SPECIES *dendromorpha, laxa, vermiformis*
GENUS *Apostemidium* Separated Stem Vibrisse Helot Sac Fungus
SPECIES *fiscellum, norvegicum, vibrisseoides*
VARIETIES *fiscellum submersum*
GENUS *Chlorovibrissea* Chloric Vibrisse Helot Sac Fungus
SPECIES *bicolor, melanochlora, phialophora, tasmanica*
GENUS *Leucovibrissea* Light Vibrisse Helot Sac Fungus
SPECIES *obconica*

ORDER 08 *Hysteriales* Hysteria Sac Fungus

FAMILY 1 *Hysteriaceae* Hysteria Sac Fungus
GENUS *Hysterium* Hysteria Sac Fungus
SPECIES *acuminatum, angustatum, curvatum, elongatum, hyalinum, insidens, pulicare*
GENUS *Acrogenospora* Acrogenic Spore Hysteria Sac Fungus
SPECIES *gigantospora, novae-zelandiae, ovalis, setiformis, sphaerocephala, subprolata*
GENUS *Encephalographa* Encephalograph Hysteria Sac Fungus
SPECIES *cerebrinella*
GENUS *Farlowiella* Farlow Hysteria Sac Fungus
SPECIES *australis, carmichaeliana*
GENUS *Gloniella* Least Glone Hysteria Sac Fungus
SPECIES *adianti, moliniae*
GENUS *Gloniopsis* Lesser Glone Hysteria Sac Fungus
SPECIES *biformis, brevisaccata, curvata, gerardiana, praelonga*
GENUS *Glonium* Glone Hysteria Sac Fungus
SPECIES *circumserpens, graphicum, lineare, nitidum, parvulum, simulans, stellatum, varium*
GENUS *Graphyllium* Lesser Graphic Hysteria Sac Fungus
SPECIES *pentamerum*
GENUS *Hysterographium* Graphic Hysteria Sac Fungus
SPECIES *fraxini, minutum, mori, vulvatum*
GENUS *Hysteropatella* Patella Hysteria Sac Fungus
SPECIES *clavispora, elliptica, prostii*
GENUS *Pseudoscypha* Scyph Hysteria Sac Fungus
SPECIES *abietis*

FAMILY 2 *Mytilinidiaceae* Mytiline Hysteria Sac Fungus
GENUS *Mytilinidion* Mytiline Hysteria Sac Fungus
SPECIES *acicola, decipiens, gemmigenum, mytilinellum, rhenanum, scolecosporum*
GENUS *Actidium* Radial Mytiline Hysteria Sac Fungus
SPECIES *hysterioides*
GENUS *Camaroglobulus* Camaro Globule Mytiline Hysteria Sac Fungus
SPECIES *resinae*
GENUS *Glyphium* Glyph Mytiline Hysteria Sac Fungus
SPECIES *elatum*
GENUS *Lophium* Crested Mytiline Hysteria Sac Fungus
SPECIES *dolabriforme, elegans, mytilinum*
GENUS *Ostreichnion* Ostre Ichnion Mytiline Hysteria Sac Fungus
SPECIES *curtisii, nova-caesariense, sassafras*
GENUS *Peyronelia* Peyron Mytiline Hysteria Sac Fungus
SPECIES *arbuscula, ramosa, rudis, sirodesmioides*
GENUS *Quasiconcha* Conch Mytiline Hysteria Sac Fungus
SPECIES *reticulata*
GENUS *Taeniolella* Ribbon Mytiline Hysteria Sac Fungus
SPECIES *alta, beschiana, breviuscula, delicata, faginea, phaeophysciae, plantaginis, pulvillus, punctata, rolfii, rudis, stilbospora, subsessilis*

ORDER 09 *Microthyriales* Thyre Sac Fungus

FAMILY 1 *Microthyriaceae* Thyre Sac Fungus
GENUS *Microthyrium* Thyre Sac Fungus
SPECIES *ciliatum, cytisi, fagi, gramineum, ilicinum, inconspicuum, lauri, macrosporum, microscopicum, pinophyllum, versicolor*
VARIETIES *ciliatum ciliatum, ciliatum hederae, cytisi cytisi, cytisi ulicis, cytisi ulicis-gallii*

GENUS *Actinomyxa*	Radial Thyre Sac Fungus
SPECIES *australiensis*	
GENUS *Actinopeltis*	Radial Pelt Thyre Sac Fungus
SPECIES *sordidula*	
GENUS *Arnaudiella*	Arnaud Thyre Sac Fungus
SPECIES *genistae*	
GENUS *Asterinella*	Star Thyre Sac Fungus
SPECIES *stuhlmannii*	
GENUS *Aulographum*	Aulograph Thyre Sac Fungus
SPECIES *hederae, maculare*	
GENUS *Byssopeltis*	Bysso Pelt Thyre Sac Fungus
SPECIES *maranhensis*	
GENUS *Chaetothyriopsis*	Bristle Thyre Sac Fungus
SPECIES *adianti*	
GENUS *Cirsosina*	Thistle Thyre Sac Fungus
SPECIES *calami, rhododendri*	
GENUS *Cirsosiopsis*	Lesser Thistle Thyre Sac Fungus
SPECIES *violacescens*	
GENUS *Cyclotheca*	Cyclic Theca Thyre Sac Fungus
SPECIES *sordidula*	
GENUS *Dictyoasterina*	Netted Star Thyre Sac Fungus
SPECIES *conopharyngiae*	
GENUS *Ellisiodothis*	Ellis Doth Thyre Sac Fungus
SPECIES *rehmiana*	
GENUS *Hidakaea*	Hidaka Thyre Sac Fungus
SPECIES *tumidula*	
GENUS *Hugueninia*	Huguen Thyre Sac Fungus
SPECIES *freycinetiae*	
GENUS *Muyocopron*	Muyo Copron Thyre Sac Fungus
SPECIES *heleocharidis*	
GENUS *Pachythyrium*	Thick Thyre Sac Fungus
SPECIES *parasiticum*	
GENUS *Phragmaspidium*	Fragment Asp Thyre Sac Fungus
SPECIES *corruscans, manaosense, viniferae*	
GENUS *Sapucchaka*	Sapucchaka Thyre Sac Fungus
SPECIES *madreeya*	
GENUS *Seynesiella*	Seynes Thyre Sac Fungus
SPECIES *juniperi*	
GENUS *Seynesiopeltis*	Seynes Pelt Thyre Sac Fungus
SPECIES *tetraplasandrae*	
GENUS *Trichothyriella*	Least Hairy Thyre Sac Fungus
SPECIES *quercigena*	
GENUS *Trichothyrina*	Lesser Hairy Thyre Sac Fungus
SPECIES *filicum*	
GENUS *Trichothyriomyces*	Hairy Thyre Sac Fungus
SPECIES *notatus*	
GENUS *Trichothyriopsis*	Greater Hairy Thyre Sac Fungus
SPECIES *alineum, densa, juruana, sexspora*	
FAMILY 2 *Leptopeltidaceae*	Thin Pelt Thyre Sac Fungus
GENUS *Leptopeltis*	Thin Pelt Thyre Sac Fungus
SPECIES *filicina, holmorum, litigiosa, pteridis*	
GENUS *Dothiopeltis*	Doth Thin Pelt Thyre Sac Fungus
SPECIES *arunci*	
GENUS *Leptopeltopsis*	Lesser Thin Pelt Thyre Sac Fungus
SPECIES *nebulosa*	
GENUS *Phacidina*	Lens Thin Pelt Thyre Sac Fungus
SPECIES *gracilis*	
GENUS *Staibia*	Staibe Thin Pelt Thyre Sac Fungus
SPECIES *connari*	
FAMILY 3 *Micropeltidaceae*	Pelt Thyre Sac Fungus
GENUS *Micropeltis*	(Synonym for *Stomiopeltis*)
GENUS *Armata*	Armed Pelt Thyre Sac Fungus
SPECIES *macrospora*	
GENUS *Bonaria*	Bonar Pelt Thyre Sac Fungus
SPECIES *lithocarpi, phyllanthi*	
GENUS *Dictyopeltella*	Netted Pelt Thyre Sac Fungus
SPECIES *consimilis, domingensis*	

GENUS *Mitopeltis* — Mite Pelt Thyre Sac Fungus
SPECIES *chilensis*
GENUS *Muricopeltis* — Hardpoint Pelt Thyre Sac Fungus
SPECIES *brasiliensis*
GENUS *Stomiopeltis* — Mouth Pelt Thyre Sac Fungus
SPECIES *betulae, cupressicola, dryadis, juniperina, pinastri*

FAMILY 4 *Schizothyriaceae* — Split Thyre Sac Fungus
GENUS *Schizothyrium* — Split Thyre Sac Fungus
SPECIES *pomi, speireum*
GENUS *Mycerema* — Cerem Split Thyre Sac Fungus
SPECIES *chandleri, vochysiacearum*
GENUS *Myriangiella* — Myriad Angle Split Thyre Sac Fungus
SPECIES *arcuata, costaricensis, orbicularis, roupalae, sabaleos*
GENUS *Orthobellus* — Straight Bell Split Thyre Sac Fungus
SPECIES *lauracearum, leguminosarum*
GENUS *Plochmopeltis* — Plochmon Pelt Split Thyre Sac Fungus
SPECIES *ellisii, graminicola, intricata, rodriguezii, roupalae*

ORDER 10 *Mycocaliciales* Calico Sac Fungus

FAMILY 1 *Mycocaliciaceae* — Calico Sac Fungus
GENUS *Mycocalicium* — Calico Sac Fungus
SPECIES *subtile*
GENUS *Asterophoma* — Star Foam Calico Sac Fungus
SPECIES *mazaediicola*
GENUS *Chaenothecopsis* — Gaping Theca Calico Sac Fungus
SPECIES *caespitosa, debilis, epithallina, haematopus, kalbii, nana, nigra, nigropedata, nivea, parasitaster, pusilla, pusiola, retinens,* rubescens(Dye Herb), *sagenidii, savonica, subparoica, tasmanica, vainioana, viridialba, viridireagens*
GENUS *Coniocybopsis* — (Synonym for *Mycocalicium*)
GENUS *Phaeocalicium* — Dusky Calico Sac Fungus
SPECIES *compressulum, populneum, praecedens*
GENUS *Stenocybe* — Thin Cybe Calico Sac Fungus
SPECIES *nitida, pullatula, septata*

FAMILY 2 *Sphinctrinaceae* — Sphincter Calico Sac Fungus
GENUS *Sphinctrina* — Sphincter Calico Sac Fungus
SPECIES *anglica, leucopoda, tubiformis, turbinata*
GENUS *Sphinctrinella* — (Synonym for *Sphinctrina*)
GENUS *Pyrgidium* — Fire Sphincter Calico Sac Fungus
SPECIES *montellicum*

ORDER 11 *Pezizales* Peziz Sac Fungus

FAMILY 1 *Pezizaceae* — Cup Peziz Sac Fungus
GENUS *Peziza* — Cup Peziz Sac Fungus
SPECIES *albopileata,* ***ammophila*****(Poison)**, *ampelina, ampliata, amplissima, apiculata, arvernensis, asterigma, atrovinosa, aurata, azureoides,* ***badia*****(Poison** Raw, Edible Well Cooked), *badioconfusa, badiofusca, boltonii, bovina, brunneoatra, buxea, cerea, chavetiae, chlorophaea, clavariarum, cocotina, congrex, depressa, domiciliana, echinospora,echinulata, epixyla, fimeti, flavida, gerardii, granularis, granulosa, helotioides, hortensis, howsei, indiscreta, isabellina, jungermanniae, labessiana, lignyota, limnaea, lividula, luculenta, macrospora, megalochondra, merdae, michelii, micropus, moravecii, moseri, natrophila, nidulariformis, ninguis, nivea, nuda, ochracea, ostracoderma, palustris, petersii, planoumbilicata, praetervisa, prosthetica, proteana, pseudoammophila, pseudoviolacea, pudica, rapula, recedens, repanda, ripensis, saccardiana, sanguinea, saniosa, sepiatra, solfatera, sterigmatizans, subcitrina, subisabellina, subrepanda, subumbrina, subviolacea,* ***succosa*****(Poison** Raw, Edible Well Cooked), *succosella, sylvatica, trachycarpa, udicola, vacini, varia, variecolor,* ***vesiculosa*****(Poison** Raw, Edible Well Cooked), *violacea, violascens*
VARIETIES *ampliata costifera, ampliata palustris, echinospora assimilata,* ***echinospora umbrina*****(Poison** Raw, Edible Well Cooked), *granulosa boudieri, isabellina isabellina, isabellina ianthina, ninguis ninguis, ninguis fortoulii, proteana proteana, proteana sparassoides*
GENUS *Amylascus* — Amyl Cup Peziz Sac Fungus
SPECIES *herbertianus, tasmanicus*
GENUS *Boudiera* — Boudier Cup Peziz Sac Fungus
SPECIES *acanthospora, areolata*
GENUS *Caloscypha* — Callisto Scyph Cup Peziz Sac Fungus
SPECIES *fulgens*
GENUS *Chromelosporium* — Chromic Spore Cup Peziz Sac Fungus
SPECIES *carneum, fulvum, glomerulosum, macrospermum, ochraceum, terrestre, tuberculatum*

GENUS *Cleistoiodophanus* Closed Odophane Cup Peziz Sac Fungus
SPECIES *conglutinatus*
GENUS *Coprotus* Copper Cup Peziz Sac Fungus
SPECIES *albidus, argenteus, aurora, glaucellus, granuliformis, lacteus, luteus, niveus, ochraceus, rhyparobioides, sexdecimsporus*
GENUS *Desmazierella* Desmazier Cup Peziz Sac Fungus
SPECIES *acicola*
GENUS *Discoitis* Disc Cup Peziz Sac Fungus
SPECIES *maturescens, **venosa***(**Poison** Raw, Edible Well Cooked)
VARIETIES *venosa reticulata*
GENUS *Filicupula* Filament Cup Peziz Sac Fungus
SPECIES *suboperculata*
GENUS *Fischerula* Fischer Cup Peziz Sac Fungus
SPECIES *macrospora, subcaulis*
GENUS *Galactinia* (Synonym for *Peziza*)
GENUS *Hapsidomyces* Hapsid Cup Peziz Sac Fungus
SPECIES *venezuelensis*
GENUS *Hydnobolites* Truffle Bolete Cup Peziz Sac Fungus
SPECIES *cerebriformis*
GENUS *Iodophanus* Red Violet Cup Peziz Sac Fungus
SPECIES *carneus, testaceus*
GENUS *Iodowynnea* Wynne Cup Peziz Sac Fungus
SPECIES *auriformis*
GENUS *Karstenella* Karsten Cup Peziz Sac Fungus
SPECIES *vernalis*
GENUS *Kimbropezia* Kimbro Cup Peziz Sac Fungus
SPECIES *campestris*
GENUS *Microeurotium* Rot Cup Peziz Sac Fungus
SPECIES *albidum*
GENUS *Muciturbo* Mucid Turbo Cup Peziz Sac Fungus
SPECIES *reticulatus, truncatus, verrucosus*
GENUS *Mycoarctium* Arctic Cup Peziz Sac Fungus
SPECIES *ciliatum*
GENUS *Mycoclelandia* Cleland Cup Peziz Sac Fungus
SPECIES *arenacea, bulundari*
GENUS *Oedocephalum* Oedipus Head Cup Peziz Sac Fungus
SPECIES *elegans, glomerulosum, laeticolor, lineatum, lunzinense, nicotianae, pallidum*
VARIETIES *glomerulosum candidum, glomerulosum cantonense*
GENUS *Pachyella* Thick Cup Peziz Sac Fungus
SPECIES *babingtonii, celtica, rhizinoides, violaceonigra*
GENUS *Pachyphloeus* Thick Flow Cup Peziz Sac Fungus
SPECIES *citrinus, conglomeratus, melanoxanthus*
GENUS *Plicaria* Pleated Cup Peziz Sac Fungus
SPECIES *carbonaria, endocarpoides, leiocarpa, radula*
GENUS *Pseudascozonus* Scozon Cup Peziz Sac Fungus
SPECIES *racemosporus*
GENUS *Ramgea* Ramge Cup Peziz Sac Fungus
SPECIES *annulispora*
GENUS *Rhizina* Rhizid Cup Peziz Sac Fungus
SPECIES *undulata*
GENUS *Rhodopeziza* Red Cup Peziz Sac Fungus
SPECIES *tuberculata*
GENUS *Sarcosphaera* Flesh Sphere Cup Peziz Sac Fungus
SPECIES *coronaria*
VARIETIES *coronaria crassa*(**Poison**), ***coronaria eximia***(**Poison** Raw, Edible Unripe Well Cooked)
GENUS *Sphaerozone* Sphere Zoned Cup Peziz Sac Fungus
SPECIES *echinulatum, ellipsosporum, ostiolatum, tulasnei*
GENUS *Sporocephalum* Spore Head Cup Peziz Sac Fungus
SPECIES *peniophorae*
GENUS *Strobiloscypha* Strobe Scyph Cup Peziz Sac Fungus
SPECIES *keliae*
GENUS *Urceolaria* Urn Cup Peziz Sac Fungus
SPECIES *scotica, tenuicorticata*

FAMILY 2 *Ascobolaceae* Bowl Peziz Sac Fungus
GENUS *Ascobolus* Bowl Peziz Sac Fungus
SPECIES *albidus, americanus, asininus, barbatus, behnitziensis, brassicae, brunneus, carbonarius, carletonii, cervinus, ciliatus, crenulatus, crosslandii, degluptus, denudatus, elegans, epimyces, equinus, foliicola,*

geophilus, hansenii, hawaiiensis, immersus, leveillei, lignatilis, lineolatus, macrosporus, mancus, masseei, michaudii, minutus, perplexans, purpurascens, rhytidosporus, roseopurpurascens, sacchariferus, stercorarius, stictoideus, viridis, winteri

GENUS *Ascophanus* Visible Bowl Peziz Sac Fungus

SPECIES *bresadolae, cinerellus, consociatus, globosopulvinatus, glumarum, minutellus, minutissimus, subfuscus*

GENUS *Cubonia* Cubon Bowl Peziz Sac Fungus

SPECIES *bulbifera*

GENUS *Lasiobolidium* Lesser Woolly Bowl Peziz Sac Fungus

SPECIES *orbiculoides, spirale*

GENUS *Lasiobolus* Woolly Bowl Peziz Sac Fungus

SPECIES *cuniculi, diversisporus, macrotrichus, microsporus, oligotrichus, ruber*

GENUS *Ochotrichobolus* Ocher Hairy Bowl Peziz Sac Fungus

SPECIES *polysporus*

GENUS *Rhizostilbella* Rhizid Stilbe Bowl Peziz Sac Fungus

SPECIES *hibisci*

GENUS *Saccobolus* Lesser Bowl Peziz Sac Fungus

SPECIES *beckii, caesariatus, chenocopricus, citrinus, depauperatus, dilutellus, glaber, globuliferellus, obscurus, quadrisporus, truncatus, versicolor*

GENUS *Thecotheus* Theca Bowl Peziz Sac Fungus

SPECIES *agranulosus, cinereus, keithii, pelletieri, rivicola*

GENUS *Trichobolus* Hairy Bowl Peziz Sac Fungus

SPECIES *sphaerosporus, zukalii*

FAMILY 3 *Ascodesmidaceae* Fiber Cup Peziz Sac Fungus

GENUS *Ascodesmis* Fiber Cup Peziz Sac Fungus

SPECIES *microscopica, nana, nigricans, porcina, sphaerospora, volutelloides*

GENUS *Eleutherascus* Liberty Fiber Cup Peziz Sac Fungus

SPECIES *cristatus, lectardii, peruvianus, tuberculatus*

FAMILY 4 *Pyronemataceae* Fire Cup Peziz Sac Fungus

GENUS *Pyronema* Fire Cup Peziz Sac Fungus

SPECIES *domesticum, glaucum, marianum, omphalodes, piggotii*

GENUS *Aleuria* Fairy Fire Cup Peziz Sac Fungus

SPECIES ***aurantia***(Commercial Edible), *cestrica, luteonitens, mellea, petaloidea, reperta, rhenana*

GENUS *Anthracobia* Coal Fire Cup Peziz Sac Fungus

SPECIES *macrocystis, maurilabra, melaloma, nitida, subatra, uncinata*

GENUS *Aparaphysaria* Bladder Fire Cup Peziz Sac Fungus

SPECIES *aparaphysata, doelloi*

GENUS *Ascorhizoctonia* Rhizid Octon Fire Cup Peziz Sac Fungus

SPECIES *intermedia*

GENUS *Boubovia* Boubov Fire Cup Peziz Sac Fungus

SPECIES *ascoboloides, luteola, nicholsonii, ovalispora, subprolata, vermiphila*

GENUS *Byssonectria* Silky Cornicle Fire Cup Peziz Sac Fungus

SPECIES *fusispora, tetraspora*

GENUS *Chalazion* Chalaz Fire Cup Peziz Sac Fungus

SPECIES *erinaceum, helveticum, sociabile*

GENUS *Cheilymenia* Cheilymen Fire Cup Peziz Sac Fungus

SPECIES *cadaverina, ciliata, cinnabarina, erecta, fibrillosa, fimicola, luteopallens, oligotricha, pulcherrima, raripila, rubra, stercorea, subhirsuta, theleboloides, vitellina*

GENUS *Ciliaria* Ciliate Fire Cup Peziz Sac Fungus

SPECIES *binominata, caudata, confusa, hirtococcinea, schizospora*

VARIETIES *hirtococcinea fulvococcinea*

GENUS *Coprobia* Dung Fire Cup Peziz Sac Fungus

SPECIES *granulata*

GENUS *Dictyocoprotus* Netted Dung Fire Cup Peziz Sac Fungus

SPECIES *mexicanus*

GENUS *Eoaleurina* Floury Fire Cup Peziz Sac Fungus

SPECIES *foliicola*

GENUS *Fimaria* Fimar Fire Cup Peziz Sac Fungus

SPECIES *cervaria, equina, hepatica, leporum, porcina, theioleuca, trochospora*

GENUS *Flavoscypha* Yellow Scyph Fire Cup Peziz Sac Fungus

SPECIES *cantharella, phlebophora*

VARIETIES *cantharella minor*

GENUS *Galeoscypha* Gale Scyph Fire Cup Peziz Sac Fungus

SPECIES *pileiformis*

GENUS *Genea* Gene Fire Cup Peziz Sac Fungus

SPECIES *hispidula, klotzschii, papillosa, sphaerica, verrucosa*

GENUS *Geopora* Pine Truffel Fire Cup Peziz Sac Fungus

SPECIES *arenicola, arenosa, cervina, foliacea, semi-immersa, sepulta, sumneriana, tenuis*
GENUS *Geopyxis* Acorn Fire Cup Peziz Sac Fungus
SPECIES *carbonaria, carnea, majalis, pinetorum, pulchra*
GENUS *Hiemsia* Hiems Fire Cup Peziz Sac Fungus
SPECIES *cleistocarpa, pseudoampezzana*
GENUS *Humaria* Humar Fire Cup Peziz Sac Fungus
SPECIES *haemastigma, hemisphaerica, ollaris, violascens*
GENUS *Hypotarzetta* Hyponic Tarzett Fire Cup Peziz Sac Fungus
SPECIES *insignis*
GENUS *Inermisia* Unarmed Fire Cup Peziz Sac Fungus
SPECIES *pilifera*
GENUS *Jafnea* (Synonym for *Jafneadelphus*)
GENUS *Jafneadelphus* Jafne Adelphi Fire Cup Peziz Sac Fungus
SPECIES *amethystinus*
GENUS *Kotlabaea* Kotlab Fire Cup Peziz Sac Fungus
SPECIES *deformis*
GENUS *Lachnea* Downy Fire Cup Peziz Sac Fungus
SPECIES *argillacea, caerulea*
GENUS *Lamprospora* Lamp Spore Fire Cup Peziz Sac Fungus
SPECIES *annulata, astroidea, campylopodis, carbonicola, crechqueraultii, crouanii, dicranellae, dictydiola, faroensis, minuta, modesta, polytrichi, seaveri, wrightii*
VARIETIES *crechqueraultii macrantha*
GENUS *Lathraeodiscus* Hidden Disc Fire Cup Peziz Sac Fungus
SPECIES *arcticus*
GENUS *Lazuardia* Lazuard Fire Cup Peziz Sac Fungus
SPECIES *lobata*
GENUS *Leucoscypha* Light Scyph Fire Cup Peziz Sac Fungus
SPECIES *alpestris, erminea, fossulae, leucotricha, ricciae*
GENUS *Luciotrichus* Lucy Hairy Fire Cup Peziz Sac Fungus
SPECIES *lasioboloides*
GENUS *Marcelleina* Marcelle Fire Cup Peziz Sac Fungus
SPECIES *atroviolacea, persoonii, rickii*
GENUS *Melastiza* False Eyelash Fire Cup Peziz Sac Fungus
SPECIES *chateri, contorta, cornubiensis, flavorubens, scotica*
GENUS *Miladina* Milad Fire Cup Peziz Sac Fungus
SPECIES *lecithina*
GENUS *Neottiella* Modern Otte Fire Cup Peziz Sac Fungus
SPECIES *crozalsiana, hetieri, ithacaensis, microspora, rutilans, rutilans ericetorum, vivida*
GENUS *Nothojafnea* Lesser Jafne Fire Cup Peziz Sac Fungus
SPECIES *cryptotricha, thaxteri*
GENUS *Octospora* Octal Spore Fire Cup Peziz Sac Fungus
SPECIES *axillaris, carbonigena, coccinea, convexula, crosslandii, humosa, leucoloma, melina, musci-muralis, neglecta, roxheimii, rubens, rustica*
GENUS *Octosporella* Lesser Octal Spore Fire Cup Peziz Sac Fungus
SPECIES *hemicrypta, jungermanniarum, ornithocephala, perforata, ptilidii, radulae, urosperma*
GENUS *Orbicula* Orb Fire Cup Peziz Sac Fungus
SPECIES *parietina*
GENUS *Otidea* Ear Fire Cup Peziz Sac Fungus
SPECIES *alutacea, apophysata, austica, bufonia, cochleata, grandis,* ***leporina***(Edible), ***onotica***(Edible), *platyspora, violacea*
VARIETIES *cochleata umbrina*
GENUS *Otideopsis* Lesser Ear Fire Cup Peziz Sac Fungus
SPECIES *kaushalii, yunnanensis*
GENUS *Oviascoma* Oval Fire Cup Peziz Sac Fungus
SPECIES *paludosum*
GENUS *Parascutellinia* Lesser Scute Fire Cup Peziz Sac Fungus
SPECIES *carneosanguinea*
GENUS *Paurocotylis* Tailed Embryo Fire Cup Peziz Sac Fungus
SPECIES *pila*
GENUS *Phaedropezia* Dusky Fire Cup Peziz Sac Fungus
SPECIES *epispartia*
GENUS *Pseudaleuria* Lesser Fairie Fire Cup Peziz Sac Fungus
SPECIES *quinaultiana*
GENUS *Pseudocollema* Collem Fire Cup Peziz Sac Fungus
SPECIES *cartilagineum*
GENUS *Pseudombrophila* Ombrophile Fire Cup Peziz Sac Fungus
SPECIES *deerrata, merdaria, misturae, pluvialis, ramosa*

GENUS *Pseudotis* (Synonym for *Otidea, Sowerbyella*)
GENUS *Psilopezia* Psilum Fire Cup Peziz Sac Fungus
SPECIES *nummularia*
GENUS *Pulvinula* Powdered Fire Cup Peziz Sac Fungus
SPECIES *archeri, carbonaria, cinnabarina, convexella, miltina*
GENUS *Pustularia* Pustule Fire Cup Peziz Sac Fungus
SPECIES *patavina*
GENUS *Pyropyxis* Pyxie Fire Cup Peziz Sac Fungus
SPECIES *rubra*
GENUS *Ramsbottomia* Ramsbottom Fire Cup Peziz Sac Fungus
SPECIES *asperior, lamprosporoidea, macracantha*
GENUS *Rhizoblepharia* Rhizid Eyelash Fire Cup Peziz Sac Fungus
SPECIES *jugispora, neotropica*
GENUS *Rhodoscypha* Red Scyph Fire Cup Peziz Sac Fungus
SPECIES *ovilla*
GENUS *Rhodotarzetta* Red Tarzett Fire Cup Peziz Sac Fungus
SPECIES *rosea*
GENUS *Scutellinia* Eyelash Fire Cup Peziz Sac Fungus
SPECIES *armatospora, barlae, cejpii, citrina, crinita, crucipila, decipiens, erinaceus, kerguelensis, legaliae, macrospora, minor, minutella, mirabilis, olivascens, paludicola, patagonica, pseudotrechispora, scutellata, setosa, subhirtella, superba, trechispora, umbrorum*
VARIETIES *minutella pallida, scutellata terrigena*
GENUS *Sepultaria* Sepultar Fire Cup Peziz Sac Fungus
SPECIES *geaster*
GENUS *Sowerbyella* Sowerby Fire Cup Peziz Sac Fungus
SPECIES *crassisculpturata, pallida, radiculata, radiculata petaloidea*
GENUS *Sphaerosoma* Sphere Soma Fire Cup Peziz Sac Fungus
SPECIES *echinulatum*
GENUS *Sphaerospora* Sphere Spore Fire Cup Peziz Sac Fungus
SPECIES *saccata*
GENUS *Sphaerosporella* Lesser Sphere Spore Fire Cup Peziz Sac Fungus
SPECIES *brunnea, hinnulea*
GENUS *Spooneromyces* Spooner Fire Cup Peziz Sac Fungus
SPECIES *asperula, laeticolor*
GENUS *Stephensia* Stephens Fire Cup Peziz Sac Fungus
SPECIES *arenivaga, bombycina, bynumii, crocea, shanorii, sumatrana, varia*
GENUS *Svrcekomyces* Svrcek Fire Cup Peziz Sac Fungus
SPECIES *pallidus*
GENUS *Tarzetta* Tarzett Fire Cup Peziz Sac Fungus
SPECIES *catinus, cupularis, gaillardiana, scotica, spurcata*
GENUS *Tricharina* Hairy Fire Cup Peziz Sac Fungus
SPECIES *ascophanoides, cretea, gilva, groenlandica, hiemalis, ochroleuca, praecox*
VARIETIES *praecox praecox, praecox intermedia*
GENUS *Trichophaea* Hairy Dusky Fire Cup Peziz Sac Fungus
SPECIES *abundans, albospadicea, bulbocrinita, gregaria, hemisphaerioides, hybrida, leucothecioides, lojkaeana, minuta, pseudogregaria, sublivida, woolhopeia*
VARIETIES *gregaria intermedia*
GENUS *Trichophaeopsis* Lesser Hairy Dusky Fire Cup Peziz Sac Fungus
SPECIES *bicuspis, paludosa*
GENUS *Warcupia* Warcup Fire Cup Peziz Sac Fungus
SPECIES *terrestris*
VARIETIES *terrestris minuta*
GENUS *Wilcoxina* Wilcox Fire Cup Peziz Sac Fungus
SPECIES *mikolae, rehmii*
VARIETIES *mikolae tetraspora*

FAMILY 5 *Sarcoscyphaceae* Elf Cup Peziz Sac Fungus
GENUS *Sarcoscypha* Elf Cup Peziz Sac Fungus
SPECIES *austriaca*, ***coccinea***(Edible), *dudleyi, jurana, macaronesica*, ***occidentalis***(Edible)
GENUS *Aurophora* Golden Elf Cup Peziz Sac Fungus
SPECIES *dochmia*
GENUS *Geodina* Earth Elf Cup Peziz Sac Fungus
SPECIES *guanacastensis*
GENUS *Kompsoscypha* Komp Scyph Elf Cup Peziz Sac Fungus
SPECIES *chudei, phyllogena, waterstonii, zizyphi*
GENUS *Microstoma* Stoma Elf Cup Peziz Sac Fungus
SPECIES *floccosum, protracta*
GENUS *Molliardiomyces* (Synonym for *Sarcoscypha*)

GENUS *Nanoscypha* Nannite Scyph Elf Cup Peziz Sac Fungus
SPECIES *bella, denisonii, euspora, macrospora, orissaensis, pulchra, striatispora, tetraspora, waterstonii*
GENUS *Pithya* Pithy Elf Cup Peziz Sac Fungus
SPECIES *cupressi, vulgaris*
GENUS *Thindia* Thind Elf Cup Peziz Sac Fungus
SPECIES *cupressi*

FAMILY 6 *Sarcosomataceae* Stalk Cup Peziz Sac Fungus
GENUS *Sarcosoma* Stalk Cup Peziz Sac Fungus
SPECIES *latahense*
GENUS *Conoplea* Cone Stalk Cup Peziz Sac Fungus
SPECIES *fusca, mangenotii*
GENUS *Galiella* Gale Stalk Cup Peziz Sac Fungus
SPECIES *rufa*
GENUS *Korfiella* Korf Stalk Cup Peziz Sac Fungus
SPECIES *karnika*
GENUS *Plectania* Plectan Stalk Cup Peziz Sac Fungus
SPECIES *melaena, melastoma, nannfeldtii*
GENUS *Pseudoplectania* Lesser Plectan Stalk Cup Peziz Sac Fungus
SPECIES *nigrella, sphagnophila, vogesiaca*
GENUS *Strumella* Strum Stalk Cup Peziz Sac Fungus
SPECIES *griseola*
GENUS *Urnula* Urn Stalk Cup Peziz Sac Fungus
SPECIES *helvelloides*
GENUS *Verticicladium* Vertic Branched Stalk Cup Peziz Sac Fungus
SPECIES *apicale*

FAMILY 7 *Terfeziaceae* Terfez Cup Peziz Sac Fungus
GENUS *Terfezia* Terfez Cup Peziz Sac Fungus
SPECIES *arenaria*
GENUS *Eoterfezia* Lesser Terfez Cup Peziz Sac Fungus
SPECIES *parasitica, perlucida*
GENUS *Loculotuber* Locule Tuber Terfez Cup Peziz Sac Fungus
SPECIES *gennadii*

ORDER 12 *Phyllachorales* Choral Sac Fungus

FAMILY 1 *Phyllachoraceae* Choral Sac Fungus
GENUS *Phyllachora* Choral Sac Fungus
SPECIES *aberiae, abortiva, abri, abutilonis, abyssinica, acaciae, acaenae, acalyphae, accedens, acervulata, actinodaphnes, acutispora, adolphiae, advena, aegiphilae, aegopodii, aequatoriensis, afra, africana, afzeliae, agharkarii, ajrekarii, alamoi, aliena, alnicola, aloëtica, alpiniae, alyxiae, amaniensis, amazonensis, ambigua, ambrosiae, americana, amica, ammophilae, amphibola, amphididyma, amphigena, amplexicaulis, amyridicola, amyridis, anacardiarum, anceps, andamanica, andicola, andropogonis, angelicae, angustispora, annonaceae, annonicola, annulata, annuliformis, anomala, antarctica, anthephorae, anthistiriae, apiahyna, apoensis, applanata, araliae, araliarum, aravalliensis, ardisiae, arechavaletae, aristidae, arthraxonis, arthrostylidii, arundinis, asclepiadis, aspidea, aspideoides, asterigena, asteromorpha, astronii, ateleiae, atrofigurans, atroinquinans, atromaculans, bakeriana, balakrishnanii, balansae, baldensis, bambusae, bambusina, banahaensis, banisteriae, banksiae, baphispora, barnadesiae, barringtoniae, barringtoniicola, bauhiniae, baumii, beaumontii, begoniae, bella, bersamae, betulae-nanae, biareolata, biguttulata, bischofiae, blanquillo, blepharoneuri, bogotensis, bontocensis, bourreriae, brachyspora, brachystegiae, brachystemonis, brasiliensis, brenesii, brevicarpa, brittoniana, bromi, brosimi, buchenaviae, buddleiae, buettneriae, bulbosa, bullata, bullulata, burgessiae, butleri, caesalpiniae, caffra, calami, callistemonis, calycophylli, canafistulae, canaliculata, canarii, canavaliae, cannabis, cantonensis, capensis, capparis, caricis, carpodini, caseariae, casimiroae, cassiae, castaneae, catervaria, catesbyana, cayennensis, cecropiae, celastri, centothecae, cepae, chalybaea, chenopodii, chimonobambusae, chusqueae, chusqueana, cibotii, ciferrii, cinerascens, cinerea, cinnae, cinnamomi, circinans, circinata, citharexyli, cladii-glomerati, clavata, clibadii, clusiae, clypeata, coccolobae, coicis, colombiensis, coloradensis, compositae, concentrica, concinna, condigna, conferta, congregata, conica, connari, connarina, consociata, conspicua, conspurcata, contigua, coorgiana, copaiferae, copeyensis, corallina, costaericae, costaricensis, coutareae, coutoubeae, crotonicola, crotonis, crustacea, cucurbitacearum, culmicola, curvulispora, cymbispora, cynodontis, cyperi, cyperina, dactylidis, dalibardae, dallasensis, danthoniae, davillae, dawei, decaisneana, delicatula, demersa, deminuta, dendritica, dendrocalami, denigrans, derridicola, detecta, deusta, deviata, devriesii, didyma, diehlii, digitariicola, dimeriae, dimorphocalycicola, diocleicola, dioscoreae, diospyri, dischidiae, dispersa, disseminata, distinguenda, dolgei, dolichogena, dolichospora, dombeyae, domingensis, dominicana, dothideoides, drypeticola, dubia, duplex, durantae, ectophytica, effigurata, effusa, egenula, ehrenbergii, ehretiae, ehrhartae, ekmaniana, elaeocarpi, elegans, elettariae, eleusines, elliptica, elmeri, elyonuri, emarginata, embeliae, engelhardtiae, entadae, epicampis, epicladii, episphaeria, eriochloae, erythroxyli, erythroxylina, escalloniae, espeletiae, eucalypti, eugeniae, eupatorii, euphorbiaceae, euryae, evernia, exanthematica, exasperans, excelsior, eximia, exostemae, explanata, fallax, fatiscens, feijoae, fici,*

ficicola, ficuum, filicina, filicum, fimbristylidicola, fimbristylidis, flabella, flavocincta, flemingiae, fluminensis, forsteroniae, fraseriana, freycinetiae, frigoris, fructicola, fructigena, fuscescens, fusicarpa, fusispora, galavisii, gaylussaciae, genipae, gentianae, gentilis, glaziovii, globispora, glochidii, glochidiicola, gloriana, glycinicola, glycosmidis, goeppertiae, gomphandrae, gondarensis, gorakhpurensis, gordoniae, gouaniae, goyazensis, gracillima, graminicola, graminis, grammica, granulosa, gratissima, greciana, grevilleae, grewiae, guaduae, guanacastica, guatteriae, guavira, guazumae, gudalurensis, gynericola, gynoxidis, hakeicola, halstediana, hauturu, heimii, helvetica, hendrickxii, henningsii, heraclei, heteropteridis, heterotrichi, hibisci, hibiscicola, hieronymi, himalayana, holci-fulvi, howardiana, hoyosensis, huallagensis, huberi, hugoniae, huigraensis, hymenachnei, hyssopi, icacoreae, idahoensis, impatientis, imperaticola, incarcerata, inclusa, inconspicua, incrustans, indica, indocalami, inelegans, infectoria, infesta, infuscans, ingicola, inimica, insueta, interstitialis, ipirangae, irregularis, ischaemi, isonandrae, ixorae, jacarandae, japonica, javanica, juglandicola, julocrotonis, junci, juruensis, kaernbachii, kamalii, kambakkamensis, kanarensis, kanyakumariana, karwarensis, kellermanii, keralensis, kerniana, klotzschiani, kniphofiae, koondrookensis, kwantungensis, kylei, kyllingae, lacrimiformis, lactea, laeviuscula, lagerheimiana, lagunensis, lamprothea, lapponica, lasiacis, lauracearum, lauricola, laurinearum, leeae, leeicola, leersiae, lehmanniana, leopoldensis, lepida, leptasca, leptocarydii, leptochloae, leptoderridis, leptospermi, leptostromoidea, leptotaeniae, leptotheca, lespedezae, leucospila, leveilleana, liebenbergii, lindmannii, lineola, litseae, litseicola, lonchotheca, longinaviculata, loudetiae, lucida, lundiae, lungusaensis, lunulata, luteomaculata, luzulae, lyonsiae, mabae, mabicola, macarangae, machaeriicola, macroloculata, macrospora, maculans, macularum, maculata, maculicola, madeirensis, madhucae, magnificens, malabarensis, malloti, malloticola, malvavisci, manaosensis, manuka, maprouneae, maquilingensis, marginalis, marisci-sieberiani, massinii, mauriae, maydis, mayepeae, mayorii, medellinensis, megalospora, megastroma, meibomiae, melaleucae, melanoplaca, melaspilea, melastomacearum, melastomatis, melatephra, meliae, melianthi, melicicola, meliosmae, menispermi, menothea, meridensis, merrillii, metastelmae, michelii, miconiicola, microcenta, microchita, micropeltoidea, microsperma, microspora, microstegia, microstroma, microtheles, millettiae, mindoensis, minuta, minutissima, mirandina, miryensis, miscanthi, miscanthidii, missouriensis, monanthochloes, monensis, monninae, montserratis, morthieri, mouririae, muehlenbergiae, muelleri, musae, musicola, mutisiae, myrciae, myrrhinii, myrsinicola, myrtincola, mysorensis, naqsii, nectandrae, nectandricola, negeriana, neolitseae, nepenthidis, nervicida, nervisequens, nervisequia, neurophila, nidulans, nigrescens, nitens, nitidissima, nitidula, njalensis, noackii, noblei, nodicola, notabilis, novoguineensis, nuttalliana, oblongispora, occultans, ochnacearum, ochnae, ocoteae, ocoteicola, olivascens, olyrae, opaca, opiferae, opposita, orbicula, orbicularis, orbiculata, orbis, oreodaphnes, ornans, oryzopsidis, ospinae, oxyspora, oyedaeae, pachinensis, paludicola, panamensis, panici, pappiana, pappophori, paraguaya, paralabatiae, paramo-nigra, parberyi, paritii-tiliacei, parvicapsa, parvula, paspalicola, paulliniae, peglerae, penicillata, pennellii, pennisetina, peregrina, perisporioides, perlata, permeans, permixta, permutata, pernettyae, perotidis, perplexans, perseae, perversa, pestis-nigra, petitmenginii, peucedani, phalaridis, phanerae, phlogis, phoebes, phoebicola, phyllanthophila, phylloplaca, phyllostachydis, physocarpi, phytolaccae, picea, pichinchae, picramniae, piperacearum, pipericola, piptadeniicola, piptocarphae, pittieri, pittospori, placida, plantaginis, platyelliptica, podagrariae, pogonatheri, polemonii, polygalae, polygonati, polypogonis, polytocae, pomigena, populi, potentillae, pouteriae, prataprajii, premnae, pressa, pretoriae, proteae, pseudes, pseudostromatica, psychotriae, pterocarpi, pterolobii, pterospermi, pulchra, puncta, punctifaciens, punctiformis, pusilla, pustulata, pycrei, pygei, quadrospora, qualeae, queenslandica, quercus, ramamurthyi, ramonensis, ramosa, ramosii, randiae, rapaneicola, ravenalae, reducta, renealmiae, repens, reticulata, rhamni, rhois, rhopographoides, rhynchosporae, rhytismoides, rickiana, ricksecker, rimulosa, rostellispora, rostkoviae, roupalae, roupalina, roureae, rouxii, rubefaciens, rudgeae, ruelliae, ruizii, ruprechtiae, saccardiana, sacchari, sageretiae, salaciae, samanensis, sancta, sarcomphali, sasae, sassafras, saurauiae, saurauiicola, scanica, scapincola, schimae, schizolobiicola, schoenicola, schotiae, schweinfurthii, scirpi, scleriae, scleriicola, scolopiae, scutiformis, secunda, securidacae, selenospora, serjaniae, serjaniicola, sesseae, setariicola, shettyi, shiraiana, shivasii, sideroxylonis, sikkimensis, silphii, simabae-cedronis, simabicola, simplex, sinensis, sinik-lagaraik, sintenisii, smilacicola, smilacis, socia, solidaginum, sordida, sororcula, spartinae, spatholobi, spegazzinii, sphaerosperma, sphaerospora, spicatae, spissa, sporoboli, stearnii, stena, stenocarpa, stenospora, stephaniae, stevensii, stewartii, stigmodes, stipata, strelitziae, subbrachyspora, subcircinans, subcuticularis, subopaca, subrepens, subtropica, swartziae, swieteniae, sycomori, sydowii, sylvatica, symploci, symplocicola, syntherismae, tabebuiae, tabernaemontanae, tachirensis, tamoyae, tandonii, tanensis, tapirirae, taruma, tecleae, tehonis, teneriffae, tenuis, tephrosiae, tequendamensis, terminaliae, tessariae, tetrantherae, tetrasperma, tetraspora, tetrasporicola, thanatophora, themedae, therophila, thiruvananthapurica, thwaitesii, tijucensis, tiliae, tipae, tirolensis, tjangkorreh, tjapukaiensis, tonduzii, tonkinensis, toroi, torrubiae, tragiae, travancorica, tricholaenae, tricuspidis, tripsacina, tritici-gracilis, triumfettae, trivialis, trophis, tropicalis, trujillensis, truncatispora, tumatumariana, tupi, uberata, ugandensis, ulcerata, ulei, umbilicata, uppalii, urbaniana, urophylla, urvilleana, usteriana, ustulata, vallecaucana, valsiformis, valsispora, velatispora, ventilaginis, veraguensis, verbesinae, vernicosa, vernoniae, vernoniicola, verrucosa, vesicata, vetiveriicola, victoriensis, viequesensis, vilis, vinosa, viridulocincta, vismiae, viticicola, viticola, vitis, viventis, vochysiae, vossiae, vulgata, weirii, whetzelii, winteri, wismarensis, woodiana, wrightiana, xanthii, xylopiae, xylosmatis, xylostei, yapensis, yassensis, yatesii, zanthoxyli, zanthoxylicola, zygopetali

VARIETIES *acaciae acaciae, acaciae enterolobii, acaciae leonensis, acaciae parkiae, acaciae pusaethae, acaciae subpapillata, banksiae banksiae, banksiae westaustraliensis, bauhiniae bauhiniae, bauhiniae australiensis, bauhiniae longispora, brachystegiae brachystegiae, brachystegiae innumerella, bromi bromi, bromi lloydii, bromi poae-nemoralis, callistemonis callistemonis, callistemonis langdonii, callistemonis similis, caricis caricis, caricis brasiliensis, circinata circinata, circinata sanguinea, crotonis crotonis, crotonis parvula, cyperi cyperi, cyperi obtusata, dolichogena dolichogena, dolichogena atroinquinans, dolichogena circinata, dolichogena phaseoli, dolichogena phaseolina, dolichogena pumila, dolichogena samoensis, eriochloae eriochloae, eriochloae colombiensis, fici albae, fici asperrimae, fici beechyanae, fici dekdekenae, fici fulvae, fici gibbosae, fici heterophyllae, fici hispidae, fici hochstetteri, fici minahassae, fici obscurae, fici pumilae, fici septicae, fici variolosae, fici wightianae, gentilis gentilis, gentilis calyptranthis, graminis graminis, graminis bambusae, graminis beckerae-polystachyae, graminis cynodonticola, graminis oplismeni, graminis panicicola, grevilleae grevilleae, grevilleae clelandii, hakeicola hakeicola, hakeicola cuttacuttae, hakeicola tasmaniensis, hauturu hauturu, hauturu lanceshawii, hauturu rekohu, leeae elatae, leucospila leucospila, leucospila destituta, magnificens magnificens, magnificens apeibae, melanoplaca melanoplaca, melanoplaca veratri, melastomatis candidae, microcenta microcenta, microcenta graphica, millettiae millettiae, millettiae ghanica, myrciae rostratae, nidulans nidulans, nidulans senecionis, nitens nitens, nitens diocleae, nitens galactiae, nitens isthmea, panici proliferi, pestis-nigra pestis-nigra, pestis-nigra caracaensis, phyllanthophila phyllanthophila, phyllanthophila egregia, puncta puncta, puncta clemensiana, puncta dalbergiicola, puncta lonchocarpi, puncta perforans, puncta puiggarii, randiae randiae, randiae aculeatae, roupalina roupalina, roupalina macrospora, sacchari spontanei, schizolobiicola schizolobiicola, schizolobiicola peltophori, sylvatica sylvatica, sylvatica brasiliensis, truncatispora truncatispora, truncatispora macrocarpa, urbaniana urbaniana, urbaniana halonospora, usteriana usteriana, usteriana piliformis, velatispora velatispora, velatispora hillianae, vernicosa vernicosa, vernicosa papulosa, vernoniicola vernoniicola, vernoniicola microspora, whetzelii whetzelii, whetzelii macrospora, yapensis yapensis, yapensis luzonensis, yapensis luzonensis-rhytismoides, yapensis pongamiae*

GENUS *Anisochora* — Anise Choral Sac Fungus

SPECIES *chardonii*

GENUS *Apiosphaeria* — Apiose Sphere Choral Sac Fungus

SPECIES *guaranitica, indica, nipponica, topographica*

GENUS *Bagnisiopsis* — Bagnis Choral Sac Fungus

SPECIES *coptidis, domingensis, eucalypti, maculans, moquileae, moricola, praestans, qualeae, rhamni, rhoina*

GENUS *Brobdingnagia* — Brobdingnag Choral Sac Fungus

SPECIES *eucalypticola, nigeriensis*

VARIETIES *nigeriensis nigeriensis, nigeriensis corneri*

GENUS *Camarotella* — Camaro Choral Sac Fungus

SPECIES *acrocomiae, costaricensis, symploci, triphasiae*

GENUS *Catacauma* — Catacomb Choral Sac Fungus

SPECIES *cabalii, contractum, costaricense, cubense, davillae, dussiae, egenulum, eugeniicola, garciae, makilingianum, microplacum, nigerrimum, nipponicum, paramoense, robinsonii, serra-nigrae, strychni, zanthoxyli*

GENUS *Clypeotrabutia* — Clype Trabute Choral Sac Fungus

SPECIES *meridensis*

GENUS *Coccodiella* — Round Choral Sac Fungus

SPECIES *advena, andicola, arundinariae, banksiae, bullosa, capparis, depressa, leandrae, machaerii, melastomatis, miconiae, miconiicola, minuta, myrtacearum, neurophila, nuda, peribebuyensis, polymorpha, puttemansii, toledoi, translucens*

VARIETIES *polymorpha pauciseta*

GENUS *Coccostroma* — Round Stroma Choral Sac Fungus

SPECIES *cynometrae, trattinickiae*

GENUS *Coccostromopsis* — Lesser Round Stroma Choral Sac Fungus

SPECIES *bambusae, chamaedoreae, diplothemii, palmicola*

GENUS *Colletotrichum* — Collet Hairy Choral Sac Fungus

SPECIES *acutatum, allii, ampelinum, boninense, capsici, caricae, caudatum, chlorophyti, circinans, coccodes, coffeanum, corchori, crassipes, curcumae, curvatum, dematium, destructivum, dianellae, fragariae, fuscum, gloeosporioides, gnaphalii, godetiae, gossypii, guaraniticum, helicis, higginsianum, holci, incarnatum, indicum, kahawae, liliacearum, lindemuthianum, linicola, lupini, mahoniae, malvarum, melicoccae, musae, neriicola, nigrum, nymphaeae, orbiculare, orchidearum, phaseolorum, phyllachoroides, piperis, pisi, psidii, solidaginis, sublineolum, tabacum, taiwanense, trichellum, trifolii, truncatum, typhae, xanthorrhoeae*

VARIETIES *acutatum hake, acutatum pineum, capsici cyamopsidicola, coffeanum virulans, dematium clitoriicola, dematium minus, dematium spinaceae, gloeosporioides alatae, gloeosporioides minus, gossypii cephalosporioides, lupini setosum*

GENUS *Deshpandiella* — Deshpand Choral Sac Fungus

SPECIES *jambolana*

GENUS *Diachora* — Diad Choral Sac Fungus

SPECIES *erebia, fennica, lathyri, onobrychidis*

GENUS *Discomycopsella* — Disc Choral Sac Fungus

SPECIES *bambusae*

GENUS *Endodothella* — Endic Doth Choral Sac Fungus

SPECIES *anacardiacearum, andropogonis, dispar, fallaciosa, istrica, kanarensis, leptosperma, litseae, lonchocarpicola, mararyensis, martinii, placentiformis, platensis, renealmiae, rickii, scirpina, tapirirae, tetraspora, tosensis, vismiae*

GENUS *Endotrabutia* Endic Trabute Choral Sac Fungus
SPECIES *tequendamensis*
GENUS *Erikssonia* Eriksson Choral Sac Fungus
SPECIES *carissae, melastomacearum, protii, pulchella, spatholobi*
GENUS *Fremitomyces* Fremite Choral Sac Fungus
SPECIES *mahe, punctatus*
GENUS *Geminispora* Gemini Spore Choral Sac Fungus
SPECIES *derridis, mimosae*
GENUS *Gibellina* Gibell Choral Sac Fungus
SPECIES *cerealis, concentrica, rehmiana*
GENUS *Haloguignardia* Halogen Guignard Choral Sac Fungus
SPECIES *cystoseirae, decidua, irritans, longispora, oceanica, tumefaciens*
GENUS *Helochora* Helon Choral Sac Fungus
SPECIES *hypertropha*
GENUS *Imazekia* Imazek Choral Sac Fungus
SPECIES *ryukyuensis*
GENUS *Isothea* Isometric Tea Choral Sac Fungus
SPECIES *chilensis, imazekii, nyssae, rhytismoides*
GENUS *Lindauella* Lindau Choral Sac Fungus
SPECIES *pyrenocarpoidea*
GENUS *Linochora* Line Choral Sac Fungus
SPECIES *rhododendri*
GENUS *Lohwagia* Lohwag Choral Sac Fungus
SPECIES *intermedia, kessleriana, verruciformis*
GENUS *Maculatifrondes* Spotted Frond Choral Sac Fungus
SPECIES *aequatoriensis*
GENUS *Malthomyces* Maltha Choral Sac Fungus
SPECIES *calamigena, coimbatoricus*
GENUS *Mangrovispora* Mangrove Spore Choral Sac Fungus
SPECIES *irregularis, pemphii*
GENUS *Ophiodothella* Serpent Doth Choral Sac Fungus
SPECIES *arengae, atromaculans, balansae, bignoniacearum, calami, calamicola, caseariae, cuervoi, cyclobalanopsidis, edax, ferruginea, fici, floridana, galophila, lagerstroemiae, leptospora, leucospila, liebenbergii, longispora, neurophila, orchidearum, palmicola, panamensis, paraguariensis, sydowii, syzygii, tithoniae, trichocarpa, ulei, vaccinii*
GENUS *Orphnodactylus* Orphan Finger Choral Sac Fungus
SPECIES *kalmiae*
GENUS *Oswaldina* Oswald Choral Sac Fungus
SPECIES *icarahyensis*
GENUS *Oxodeora* Oxode Choral Sac Fungus
SPECIES *petrakii*
GENUS *Palmomyces* Palm Choral Sac Fungus
SPECIES *montaneus*
GENUS *Parberya* Parbery Choral Sac Fungus
SPECIES *arxii, kosciuskoa*
GENUS *Periaster* Perimeter Star Choral Sac Fungus
SPECIES *strongylodontis*
GENUS *Phaeochorella* Dusky Choral Sac Fungus
SPECIES *artocarpi, ciliata, clypeata, machaerii, parinarii, zonata*
GENUS *Phaeotrabutia* Dusky Trabute Choral Sac Fungus
SPECIES *isabelae, smilacis*
GENUS *Phragmocarpella* Fragment Carpel Choral Sac Fungus
SPECIES *fusispora, ichnanthi, sasae*
GENUS *Phragmocauma* Fragment Caum Choral Sac Fungus
SPECIES *bambusae, kolowratiae, smilacis*
GENUS *Phycomelaina* Algae Loving Melain Choral Sac Fungus
SPECIES *laminariae*
GENUS *Phylleutypa* Typical Choral Sac Fungus
SPECIES *dioscoreae, kalmiae, wittrockii*
GENUS *Phyllocrea* Offspring Choral Sac Fungus
SPECIES *quitensis*
GENUS *Physalosporina* Bladder Spore Choral Sac Fungus
SPECIES *elymi*
GENUS *Placostroma* Plaque Stroma Choral Sac Fungus
SPECIES *diplothemii, ischaemi, lorentzianum, sporoboli*
GENUS *Plectosphaera* Plaited Sphere Choral Sac Fungus

SPECIES *actinodaphnes, amphidyma, atractina, banksiae, clarae-bonae, cryptomeriae, eleocharidis, embeliae, euryae, flagellariae, fragariae, ingae, kobresiae, memecyli, microsticta, polypodii, populina, quadraspora, rhynchosporae, rhytismophila, rosae, salicis, sanguinea, selaginellae, selenipediifolii, spartii, yakusimensis*

GENUS *Plectosporium* Plaited Spore Choral Sac Fungus

SPECIES *alismatis*

GENUS *Polystigma* Many Stigma Choral Sac Fungus

SPECIES *adenostomatis, amygdalinum, apophlaeae, bumeliae, caraganae, caulicola, deformans, eugeniae, fulvum, halimodendri, haraeanum, indicum, mahatbaleshwarense, melastomatum, orbiculatum, phyllodii, pusillum, rubrum, rubrum, sonneratiae, uniloculare, variipila, volkartianum, xanthostroma*

GENUS *Polystigmina* Lesser Many Stigma Choral Sac Fungus

SPECIES *pallescens*

GENUS *Pseudothiella* Lesser Doth Choral Sac Fungus

SPECIES *hirtellae*

GENUS *Pterosporidium* Wing Spore Choral Sac Fungus

SPECIES *rhizomorphae, rhizophorae*

GENUS *Puiggarina* Puiggar Choral Sac Fungus

SPECIES *amphilobii, atroinquinans, costaricensis, formosa, manaosensis, melastomaticola, morindae, myiocoproides, pirigena*

GENUS *Rehmiodothis* Rehm Doth Choral Sac Fungus

SPECIES *eucalypti, eugeniae, inaequalis, indica, lamulera, ocoteae, osbeckiae*

GENUS *Retroa* Retro Choral Sac Fungus

SPECIES *centrolobii, dimorphandrae*

GENUS *Schizochora* Split Choral Sac Fungus

SPECIES *calocarpa, elmeri, stenosperma*

GENUS *Schizotrichella* Lesser Split Choral Sac Fungus

SPECIES *lunata*

GENUS *Scolecoccoidea* Scolex Round Choral Sac Fungus

SPECIES *costaricensis*

GENUS *Scolecodothis* Scolex Doth Choral Sac Fungus

SPECIES *castaneae, hypophylla, kamatii*

VARIETIES *kamatii kamatii, kamatii macrospora*

GENUS *Sphaerodothis* Sphere Doth Choral Sac Fungus

SPECIES *arengae, balansae, borassi, circumscripta, consociata, danthoniae, densa, eremochloae, fourcroyae, gaultheriae, guilielmae, luquillensis, merianiae, merrillii, pirifera, raoi, schweinfurthii, tetracerae, trinitensis, viticifoliae*

GENUS *Stigmatula* Lesser Stigma Choral Sac Fungus

SPECIES *applanata, armandii, astragali, erysiphoides, francevilleana, gregaria, mucosa, submaculans, sutherlandiae*

GENUS *Stigmochora* Stigma Choral Sac Fungus

SPECIES *albiziae, chloroxyli, controversa, deightonii, natalensis, parkiae, simarubae, ulei, variegata*

GENUS *Stromaster* Stromaster Choral Sac Fungus

SPECIES *tuberculatus*

GENUS *Telimena* Telimen Choral Sac Fungus

SPECIES *bakeri, domingensis, ecastophylli, erythrinae, graminella, graminis, haraeana, panici, pleioblasti, rhoina*

GENUS *Telimenella* Lesser Telimen Choral Sac Fungus

SPECIES *bicincta, elaeicola, gangraena, phacidioidea*

GENUS *Telimenochora* Least Telimen Choral Sac Fungus

SPECIES *abortiva*

GENUS *Tolediella* Toledo Choral Sac Fungus

SPECIES *fusispora*

GENUS *Trabutiella* Lesser Trabute Choral Sac Fungus

SPECIES *ichnanthi*

GENUS *Tribulatia* Trabute Choral Sac Fungus

SPECIES *appendicospora*

GENUS *Vermicularia* Worm Choral Sac Fungus

SPECIES *leptosperma*

GENUS *Vitreostroma* Vitreous Stroma Choral Sac Fungus

SPECIES *desmodii*

VARIETIES *desmodii desmodii, desmodii asiaticum, desmodii lespedezae*

GENUS *Woronichina* Woronich Choral Sac Fungus

SPECIES *tranzschelii*

GENUS *Zimmermanniella* Zimmermann Choral Sac Fungus

SPECIES *trispora*

FAMILY 2 *Phaeochoraceae* Dusky Choral Sac Fungus

GENUS *Phaeochora* Dusky Choral Sac Fungus

SPECIES *indaya, livistonae, sphaerotheca, steinheilii*
GENUS *Cocoicola* Coco Dusky Choral Sac Fungus
SPECIES *californica., cylindrospora, fusispora, livistonicola, piperata*
GENUS *Phaeochoropsis* Lesser Dusky Choral Sac Fungus
SPECIES *diplothemiifolii, mucosa, neowashingtoniae, palmicola*
GENUS *Serenomyces* Seren Dusky Choral Sac Fungus
SPECIES *mauritiae, palmae, phoenicis, virginiae*

ORDER 13 *Pleosporales* Pleospore Sac Fungus

FAMILY 01 *Pleosporaceae* Pleospore Sac Fungus
GENUS *Pleospora* Pleospore Sac Fungus
SPECIES *addubitans, alfalfae, ambigua, androsaces, anthyllidis, aquatica, armeriae, bardanae, betae, brachyspora, calvescens, comata, cytisi, dianthi, dichondrae, discors, donacina, epilobii, equiseti, eustegia, fructicola, globosa, gracilariae, halophila, helvetica, herbarum, hydrophila, incerta, infectoria, iqbalii, junci, laminariana, laricina, leptogiicola, leptosphaerulinoides, lignicola, lycopersici, njegusensis, palustris, papaveracea, pellita,pelvetiae, penicillus, phaeocomoides, phaeospora, pomorum, punctata, richtophensis, rubelloides, rubicunda, scirpicola, sparganii, spartii-juncei, spartinae, straminis, subalpina, submersa, subriparia, tarda, triglochinicola, valesiaca, verecunda, vitalbae, welwitschiae*
VARIETIES *herbarum armeriae, herbarum caulicola, herbarum lactucum, herbarum medicaginis, pellita cirsiicola*
GENUS *Allewia* Allew Pleospore Sac Fungus
SPECIES *proteae*
GENUS *Alternaria* Alternate Pleospore Sac Fungus
SPECIES *acalyphicola, alternata, anagallidis, angustiovoidea, aragakii, arbusti, argyranthemi, atrocariis, bataticola, brassicae, brassicicola, burnsii, calendulae, capsici-annui, carthami, cassiae, cheiranthi, chlamydospora, chrysanthemi, cichorii, cinerariae, citri, citri, conjuncta, crassa, cucumerina, dauci, dennisii, dianthi, dianthicola, dichondrae, eichhorniae, euphorbiicola, gaisen, geophila, gossypina, grisea, grossulariae, gypsophilae, hawaiiensis, helianthi, helianthicola, helianthinficiens, iridis, japonica, kikuchiana, leucanthemi, limaciformis, limicola, lini, linicola, longipes, longissima, macrospora, mali, malvae, maritima, metachromatica, molesta, mouchaccae, multirostrata, nelumbii, oregonensis, padwickii, panax, passiflorae, peglionii, pellucida, petroselini, poonensis, porri, radicina, ramulosa, raphani, resedae, rhadina, ricini, ricini, saparva, scorzonerae, senecionis, sesami, sesamicola, smyrnii, solani, sonchi, sonchi, steviae, tagetica, tenuissima, tomato, triticicola, triticimaculans, triticina, tropica, violae, vitis, zinniae*
VARIETIES *anagallidis anagallidis, anagallidis linariae, cucumerina cucumerina, cucumerina cyamopsidis, tenuissima godetiae*
GENUS *Bipolaris* Bipolar Pleospore Sac Fungus
SPECIES *australis, brizae, buchloës, cactivora, clavata, coicis, colocasiae, crotonis, crustacea, cylindrica, eleusines, euchlaenae, halepensis, heveae, incurvata, indica, iridis, leersiae, luttrellii, microlaenae, micropus, miyakei, multiformis, nicotiae, novae-zelandiae, ovariicola, panici-miliacei, papendorfii, perotidis, sacchari, sorghicola, subpapendorfii, tropicalis, urochloae, zeae*
GENUS *Camarosporium* Camaro Pleospore Sac Fungus
SPECIES *aequivocum, ambiens, ariae, atriplicis, betulinum, dalmaticum, eucalypti, feurichii, ilicis, karstenii, kriegeria, laburni, lantanae, lauri, macrosporum, mirabile, nigricans, obiones, oreades, orni, passerinii, pini, pistaciae, propinquum, quaternatum, quercus, ribis, robiniae, rosae, roumeguerei, rubicola, salicinum, sarmenticium, spartii, suaedae, tamaricis, tiliae, virgiliae*
VARIETIES *pini conorum*
GENUS *Chlamydosporium* Cloak Pleospore Sac Fungus
SPECIES *submersum*
GENUS *Cochliobolus* Spiral Bowl Pleospore Sac Fungus
SPECIES *aberrans, akaii, akaiiensis, australiensis, bicolor, carbonum, chloridis, cymbopogonis, cynodontis, dactyloctenii, eleusines, ellisii, geniculatus, hawaiiensis, heteropogonis, heterostrophus, homomorphus, intermedius, kusanoi, lunatus, luttrellii, melinidis, microlaenae, miyabeanus, neergaardii, nisikadoi, nodulosus, ovariicola, peregianensis, perotidis, ravenelii, sativus, setariae, spicifer, stenospilus, tuberculatus, victoriae*
VARIETIES *lunatus aeria*
GENUS *Curvularia* Curved Pleospore Sac Fungus
SPECIES *affinis, andropogonis, borreriae, brachyspora, catenulata, clavata, comoriensis, crepinii, deightonii, falcata, fallax, gladioli, gudauskasii, harveyi, inaequalis, indica, intermedia, ischaemi, oryzae, ovoidea, penniseti, prasadii, protuberata, ramosa, richardiae, robusta, senegalensis, sorghina, spicata, trifolii, uncinata, verruciformis, verrucosa, verruculosa*
VARIETIES *trifolii gladioli*
GENUS *Decorospora* Decorated Pleospore Sac Fungus
SPECIES *gaudefroyi*
GENUS *Dendryphiella* Lesser Wood Pleospore Sac Fungus
SPECIES *infuscans, vinosa*
GENUS *Dendryphion* Wood Pleospore Sac Fungus
SPECIES *comosum, griseum, nanum, nodulosum*

GENUS *Didymella* Dual Mellow Pleospore Sac Fungus
SPECIES *aerospora, applanata, bryoniae, caricis, caulicola, cerastii, commanipula, cortaderiae, culmigena, eupyrena, exigua, fucicola, glacialis, graminicola, hebridensis, holosteae, hyphenis, hyporrhodia, ligulicola, lophospora, lycopersici, nigrella, operosa, pachyspora, phacae, phleina, planiuscula, prominula, proximella, purpurea, rubitingens, salicis, scotica, trifolii, urticicola, vincetoxici, vitalbina*
VARIETIES *culmigena endorhodia ligulicola ligulicola, ligulicola inoxydabilis*
GENUS *Drechslera* Drechsler Pleospore Sac Fungus
SPECIES *andersenii, apii, avenacea, biseptata, catenaria, dematioidea, euphorbiae, festucae, fugax, gigantea, multiformis, musae-sapientium, nicotiae, nobleae, ocella, pallida, patereae, phlei, poae, portulacae, salviniae, subpapendorfii, tripogonis, triseptata, zizaniae*
GENUS *Dwayamala* Dwayamal Pleospore Sac Fungus
SPECIES *prathilomaka*
GENUS *Embellisia* Embellis Pleospore Sac Fungus
SPECIES *abundans, allii, annulata, chlamydospora, conoidea, dennisii, didymospora, eureka, hyacinthi, indefessa, leptinellae, novae-zelandiae, oxalidicola, phragmospora, planifunda, proteae, telluster, thlaspis, tumida, verruculosa*
GENUS *Exserohilum* Exserohil Pleospore Sac Fungus
SPECIES *antillanum, curvatum, curvisporum, fusiforme, gedarefense, heteropogonicola, inaequale, macginnisii, oryzicola, paspali, protrudens*
GENUS *Falciformispora* Forked Pleospore Sac Fungus
SPECIES *lignatilis*
GENUS *Hendersonia* Henderson Pleospore Sac Fungus
SPECIES *aberrans, abietis, acicola, arcus, arundinacea, calamovilfae, culmicola, culmiseda, epicalamia, herpotricha, juncina, oleae, phragmitis, polycistis, sabaleos, sacchari, sambuci, sarmentorum, ulmi, vagans*
VARIETIES *sarmentorum lauri, sarmentorum mahoniae*
GENUS *Kirschsteiniothelia* Kirschstein Pleospore Sac Fungus
SPECIES *aethiops, maritima*
GENUS *Kriegeriella* Krieger Pleospore Sac Fungus
SPECIES *mediterranea, minuta, mirabilis, transiens*
GENUS *Leptosphaerulina* Thin Sphere Pleospore Sac Fungus
SPECIES *albulae, americana, arachidicola, argentinensis, australis, chartarum, hyperborea, myrtillina, nitida, peltigerae, personata, rupestris, trifoli*
GENUS *Lewia* Lew Pleospore Sac Fungus
SPECIES *ethzedia, eureka, infectoria, photistica, sauropi, scrophulariae*
GENUS *Macrospora* Lesser Large Pleospore Sac Fungus
SPECIES *scirpi*
GENUS *Macrosporium* Large Pleospore Sac Fungus
SPECIES *concinnum, ricini*
GENUS *Macroventuria* Large Venture Pleospore Sac Fungus
SPECIES *anomochaeta, wentii*
GENUS *Monascostroma* Monad Stroma Pleospore Sac Fungus
SPECIES *innumerosum*
GENUS *Mycodidymella* Lesser Dual Mellow Pleospore Sac Fungus
SPECIES *aesculi*
GENUS *Phoma* Foam Pleospore Sac Fungus
SPECIES *acetosellae, actaeae, aculeorum, acuta, acuum, adianticola, agnita, ajacis, alchemillae, aliena, allostoma, americana, ammophilae, andigena, anigozanthi, anserina, apiicola, aquilegiicola, aquilina, araucariae, armeriae-sibericae, arundinacea, asteriscus, astragali, aubrietiae, aurea, bakeriana, barbari, bellidis, bismarckii, cajani, capitulum, caricae-papayae, caricicola, cava, chamaenerii, chenopodiicola, chrysanthemicola, cirsii, clematidina, complanata, concentrica, coonsii, costaricensis, crataegi, crinicola, cruris-hominis, cucurbitacearum, cyanea, cytospora, dennisii, destructiva, deusta, dictamnicola, digitalis, dimorphospora, draconis, drobnjacensis, endorrhodia, enteroleuca, epitricha, eucalyptica, eupatorii, eupyrena, everniae, exigua, fallens, fimeti, flavescens, flavigena, foveata, frangulae, gallarum, gentianae, glaucispora, glomerata, grossulariae, haematocycla, hapalocystis, hedericola, henningsii, herbarum, heteroderae, heteromorphospora, hibernica, huancayensis, humicola, idaei, ilicis, incompta, insidiosa, insulana, intricans, jacquiniana, jolyana, labilis, lacustris, lecanorae, leguminum, leonuri, leveillei, libertiana, lichenis, lineolata, lini, lirelliformis, lobariae, longissima, lotivora, lupinella, lupini, macrocapsa, macrostoma, mali, mangiferae, medicaginis, microscopica, minutella, minutispora, minutula, multipora, multirostrata, mutica, myricae, narcissi, nebulosa, neglecta, negriana, nemophilae, nigricans, nigrificans, nigripycnidia, notha, nyctaginea, oculi-hominis, onagracearum, opuntiicola, ostiolata, paspali, pasteriscus, pedicularis, pedicularis, peltigerae, pereupyrena, phaseolina, phlomidis, physciicola, pimprina, pinodella, piperis, pirolae, plurivora, polemonii, polygramma, pomorum, poolensis, pratorum, prunicola, putaminum, radicina, radicis-callunae, radula, richardiae, riggenbachii, rubefaciens, rubella, rubiginosa, rumicicola, salicina, samararum, samarorum, saprophytica, saxifragarum, sclerotioides, scobina, selaginellicola, senecionis, septicidalis, siliquastrum, solieri, stagonosporioides, statices, strasseri, suaedae, subglomerata, subherbarum, subherbicola, sydowii, telephii, terricola, thujana, tracheiphila, trachelii, triticina, tropica, typharum, valerianae, valerianellae, vasinfecta, veronicicola, versabilis, viburnicola, violacea, violicola, visci, zantedeschiae, zingiberis, zopfii*

VARIETIES *acuta phlogis, andigena andigena, andigena crystalliniformis, enteroleuca enteroleuca, enteroleuca influorescens, exigua exigua, exigua diversispora, exigua foveata, exigua heteromorpha, exigua inoxydabilis, exigua lilacis, exigua linicola, exigua noackiana, exigua sambuci-nigrae, exigua viburni, gentianae sino-ornatae, herbarum herbarum, herbarum galiorum, herbarum lactaria, jolyana circinata, jolyana sahariensis, lirelliformis lirelliformis, lirelliformis weigeliae-roseae, macrostoma macrostoma, macrostoma incolorata, nyctaginea nyctaginea, nyctaginea boerhaaviae, ostiolata brunnea, pomorum pomorum, pomorum calorpreferens, poolensis poolensis, poolensis verbascicola, rubiginosa circumstipata*

GENUS *Pithomyces* Pithy Pleospore Sac Fungus

SPECIES *alabamensis, atro-olivaceus, chartarum, cynodontis, graminicola, karoo, maydicus, pulvinatus, quadratus, sacchari, terricola, valparadisiacus*

GENUS *Platysporoides* Flat Pleospore Sac Fungus

SPECIES *chartarum, cookei, crandallii, deflectens, donacis, multiseptata, patriniae, punctiformis, tirolensis, togwotiensis, vulgaris*

GENUS *Plenodomus* Double Dome Pleospore Sac Fungus

SPECIES *chelidonii, destruens, fuscomaculans, meliloti*

GENUS *Polyopeus* Many Pleospore Sac Fungus

SPECIES *recurvatus*

GENUS *Pseudocochliobolus* Spiral Shell Bowl Pleospore Sac Fungus

SPECIES *eragrostidis, pallescens, verruculosus*

GENUS *Pseudodidymella* Least Dual Mellow Pleospore Sac Fungus

SPECIES *fagi*

GENUS *Pseudoyuconia* Yucon Pleospore Sac Fungus

SPECIES *thalictri*

GENUS *Pyrenochaeta* Fire Bristle Pleospore Sac Fungus

SPECIES *dolichi, fallax, gardeniae, glycines, indica, lycopersici, mackinnonii, nobilis, oryzae, penniseti, quercina, romeroi, rubi-idaei, sacchari, terrestris, unguis-hominis*

VARIETIES *nobilis ilicis*

GENUS *Pyrenophora* Fire Pleospore Sac Fungus

SPECIES *avenae, bromi, chaetomioides, dictyoides, erythrospila, graminea, japonica, lolii, phaeocomes, seminiperda, teres, tetrarrhenae, tritici-repentis, typhicola, venturia*

GENUS *Rhopographus* Rhopal Graphic Pleospore Sac Fungus

SPECIES *filicinus*

GENUS *Setosphaeria* Seta Sphere Pleospore Sac Fungus

SPECIES *holmii, khartoumensis, minor, monoceras, pedicellata, prolata, rostrata, turcica*

GENUS *Stagonospora* Stag Pleospore Sac Fungus

SPECIES *anglica, aquatica, arenaria, arrhenatheri, atriplicis, bellunensis, bufonia, calystegiae, caricinella, caricis, cylindrica, dulcamarae, elegans, equiseti, equisetina, euonymi, foliicola, fragariae, gigaspora, graminella, heleocharidis, junciseda, lophiostoma, luzulae, macropus, macropycnidia, maculata, maritima, meliloti, neglecta, paludosa, pteridiicola, recedens, sacchari, samarorum, socia, sparganii, suaedae, subseriata, trimera, tussilaginis, typhoidearum, vexatula, vitensis*

VARIETIES *aquatica luzulicola, arenaria minor, neglecta colorata*

GENUS *Stemphylium* Stem Pleospore Sac Fungus

SPECIES *botryosum, callistephi, dendriticum, globuliferum, lancipes, loti, lycopersici, majusculum, maritimum, sarciniforme, solani, vesicarium, xanthosomatis*

GENUS *Teratosphaeria* Terat Sphere Pleospore Sac Fungus

SPECIES *microspora*

GENUS *Tremateia* Tremor Pleospore Sac Fungus

SPECIES *halophila*

GENUS *Ulocladium* Ulo Branched Pleospore Sac Fungus

SPECIES *alternariae, atrum, botrytis, chartarum, consortiale, cucurbitae, dauci, lanuginosum, manihoticola, microsporum, multiforme, obovoideum, oudemansii, septosporum, tuberculatum*

GENUS *Wettsteinina* Wettstein Pleospore Sac Fungus

SPECIES *carinthiaca, dryadis, engadinensis, eucarpa, macrotheca, mirabilis, niesslii, pachyasca*

GENUS *Zeuctomorpha* Zeuctomorph Pleospore Sac Fungus

SPECIES *arecae*

FAMILY 02 *Arthopyreniaceae* Fire Pleospore Sac Fungus

GENUS *Arthopyrenia* Fire Pleospore Sac Fungus

SPECIES *allogena, analepta, aphorisasa, atractospora, carneobrunneola, cerasi, cinchonae, cinereopruinosa, desistens, dirumpens, fallacior, fraxini, gravastella, grisea, macquariensis, nitescens, picea, planorbis, platypyrenia, punctiformis, rhyponta, salicis, saxicola, socialis, spilobola, stenotheca, subcerasi, subgregans, subpunctiformis, zostra*

GENUS *Leptorhaphis* Thin Needle Fire Pleospore Sac Fungus

SPECIES *atomaria, epidermidis, maggiana*

GENUS *Mycomicrothelia* Nipple Fire Pleospore Sac Fungus

SPECIES *atlantica, confusa, exigua, melanospora, miculiformis, obovata, subfallens, thelena, wallrothii*

GENUS *Naetrocymbe* (Synonym for *Arthopyrenia*)

GENUS *Semisphaeria* Semisphere Fire Pleospore Sac Fungus

SPECIES *sigmundii*
GENUS *Tomasellia* — Tomas Fire Pleospore Sac Fungus
SPECIES *diffusa, dispora, gelatinosa*

FAMILY 03 *Corynesporascaceae* — Club Pleospore Sac Fungus
GENUS *Corynesporasca* — Club Pleospore Sac Fungus
SPECIES *caryotae*
GENUS *Corynespora* — Lesser Club Pleospore Sac Fungus
SPECIES *biseptata, cambrensis, cassiicola, cespitosa, citricola, olivacea, proliferata, pruni, smithii, trichiliae*

FAMILY 04 *Cucurbitariaceae* — Cucumber Pleospore Sac Fungus
GENUS *Cucurbitaria* — Cucumber Pleospore Sac Fungus
SPECIES *acervata, aspegrenii, berberidis, bicolor, dulcamarae, elongata, euonymi, homalea, laburni, laurocerasi, naucosa, obducens, piceae, rhamni, ribis, rubefaciens, sorbi, staphula*
GENUS *Curreya* — Currey Cucumber Pleospore Sac Fungus
SPECIES *pityophila*
GENUS *Gibberidea* — Gibber Cucumber Pleospore Sac Fungus
SPECIES *visci*
GENUS *Rhytidiella* — Wrinkled Cucumber Pleospore Sac Fungus
SPECIES *baranyayi, beloniza, moriformis*
GENUS *Syncarpella* — Similar Carpel Cucumber Pleospore Sac Fungus
SPECIES *ribis*

FAMILY 05 *Dacampiaceae* — Dacamp Pleospore Sac Fungus
GENUS *Dacampia* — Dacamp Pleospore Sac Fungus
SPECIES *hookeri, rufescentis*
GENUS *Aaosphaeria* — Sphere Dacamp Pleospore Sac Fungus
SPECIES *arxii*
GENUS *Discothecium* — (Synonym for *Endococcus* & *Polycoccum*)
GENUS *Byssothecium* — Silky Theca Dacamp Pleospore Sac Fungus
SPECIES *circinans*
GENUS *Byssothecium* — Round Shield Dacamp Pleospore Sac Fungus
SPECIES *cladonema, epicrassum, hypocenomycis*
GENUS *Immotthia* — Immotth Dacamp Pleospore Sac Fungus
SPECIES *hypoxylon*
GENUS *Kalaallia* — Kalaal Dacamp Pleospore Sac Fungus
SPECIES *reactiva*
GENUS *Leptocucurthis* — Thin Cucumber Dacamp Pleospore Sac Fungus
SPECIES *quadrata*
GENUS *Moristroma* — Mor Stroma Dacamp Pleospore Sac Fungus
SPECIES *polysporum*
GENUS *Munkovalsaria* — Munk Vals Dacamp Pleospore Sac Fungus
SPECIES *donacina, rubra*
GENUS *Polycoccum* — Many Round Dacamp Pleospore Sac Fungus
SPECIES *arnoldii, crassum, dzieduszyckii, kerneri, marmoratum, microcarpum, microsticticum, minutulum, opulentum, peltigerae, pulvinatum, slaptoniense, sporastatiae, squamarioides, trypethelioides, vermicularium*
GENUS *Pseudonitschkia* — Nitschke Dacamp Pleospore Sac Fungus
SPECIES *parmotrematis*
GENUS *Pyrenidium* — Fire Dacamp Pleospore Sac Fungus
SPECIES *actinellum, hetairizans*
GENUS *Sinodidymella* — China Dual Mellow Dacamp Pleospore Sac Fungus
SPECIES *occidentalis, verrucosa*
GENUS *Weddellomyces* — Weddell Dacamp Pleospore Sac Fungus
SPECIES *epicallopisma, macrosporus, periphericus*

FAMILY 06 *Diademaceae* — Diad Pleospore Sac Fungus
GENUS *Diadema* — Diad Pleospore Sac Fungus
SPECIES *acutum, cinctum, curtum, hexamerum, obtusum, sieversiae, tetramerum*
GENUS *Clathrospora* — Lattice Diad Pleospore Sac Fungus
SPECIES *diplospora, elynae, permunda*
GENUS *Nimbya* — Nimby Diad Pleospore Sac Fungus
SPECIES *alternantherae, caricis, celosiae, crassoides, euphorbiicola, gomphrenae, heteroschemos, juncicola, major, pimpriana, rhapontici*

FAMILY 07 *Didymosphaeriaceae* — Double Round Pleospore Sac Fungus
GENUS *Didymosphaeria* — Double Round Pleospore Sac Fungus
SPECIES *anaxaea, arachidicola, arenaria, astragalina, celata, conoidea, dochmia, enormis, epidermidis, futilis, igniaria, massarioides, minuta, oblitescens, rubicola, syringae, tenebrosa, winteri*
GENUS *Asteromella* — Star Mellow Double Round Pleospore Sac Fungus
SPECIES *bellunensis, lupini, mali, platanoidis*
GENUS *Roussoella* — Rousso Double Round Pleospore Fungus

SPECIES *hysterioides, serrulata*
GENUS *Verruculina* — Verrucose Double Round Pleospore Sac Fungus
SPECIES *enalia*

FAMILY 08 *Fenestellaceae* — Window Pleospore Sac Fungus
GENUS *Fenestella* — Window Pleospore Sac Fungus
SPECIES *bipapillata, fenestrata, lycii, minor, salicis*
GENUS *Lojkania* — Lojkan Window Pleospore Sac Fungus
SPECIES *decorticata*

FAMILY 09 *Leptosphaeriaceae* — Thin Sphere Pleospore Sac Fungus
GENUS *Leptosphaeria* — Thin Sphere Pleospore Sac Fungus
SPECIES *abbreviata, acuta, acutiuscula, aggregata, agnita, albopunctata, allorgei, anceps, artemisiae, baggei, bellynckii, cesatiana, clara, clarkii, clavata, conferta, congesta, coniothyrium, cookei, crozalsii, cruenta, culmicola, culmifraga, densa, derasa, dioica, doliolum, dryadis, dumetorum, duplex, epilobii, eustomoides, fallaciosa, fuscella, galeopsidicola, galiorum, gloeospora, haematites, helminthospora, hispanica, korrae, lactuosa, ladina, lecanora, leucomelaria, libanotis, lindquistii, linearis, littoralis, lucina, luctuosa, lunariae, macrospora, maculans, marina, millefolii, modesta, nectrioides, nitschkei, octophragmia, ogilviensis, orae-maris, parmeliarum, parvula, pelagica, phormii, planiuscula, praetermissa, pratensis, purpurea, raphani, recutita, robusta, rubella, rubicunda, sacchari, scitula, senegalensis, septemcellulata, solani, submaculans, tamaricis, tompkinsii, trollii, uliginosa, clematidis-rectae*
VARIETIES *coniothyrium sporulosum, culmifraga propinqua, doliolum conoidea, doliolum doliolum, doliolum pinguicula, maculans scrophulariae, octophragmia octophragmia, octophragmia major*
GENUS *Coniothyrinula* — Lesser Cone Thyre Thin Sphere Pleospore Sac Fungus
SPECIES *clematidis-rectae*
GENUS *Coniothyrium* — Cone Thyre Thin Sphere Pleospore Sac Fungus
SPECIES *acaciae, ampelopsidis-hederaceae, bambusicola, batumense, boydeanum, carpaticum, cereale, clematidis-rectae, coffeae, concentricum, conicola, conoideum, conorum, crepinianum, cupressacearum, cydoniae, dispersellum, dracaenae, equiseti, ferrarisianum, fraxini, genistae, glomerulatum, hellebori, ilicinum, ilicis, insitivum, jasmini, juniperi, kallangurense, lavandulae, leguminum, lignorum, minitans, obiones, olivaceum, opuntiae, palmarum, palmicola, palmigenum, phormii, platani, populinum, psammae, pteridis, quercinum, rosarum, sarothamni, slaptoniense, sphaerospermum, sporulosum, tamaricis, tamarisci, terricola, ulmeum, wernsdorffiae*
VARIETIES *concentricum agaves, olivaceum visci, quercinum glandicola*
GENUS *Didymolepta* — Dual Thin Sphere Pleospore Sac Fungus
SPECIES *anomala*
GENUS *Ophiobolus* — Serpent Bowl Thin Sphere Pleospore Sac Fungus
SPECIES *acuminatus, culmorum, disseminans, erythrosporus, fruticum, fulgidus, intermedius, laminariae, littoralis, mathieui, ophioboloides, tenellus, troakei, typhae, ulnosporus, vulgaris*
GENUS *Parahendersonia* — Henderson Thin Sphere Pleospore Sac Fungus
SPECIES *dasylirii*
GENUS *Pleoseptum* — Septic Thin Sphere Pleospore Sac Fungus
SPECIES *yuccaesedum*

FAMILY 10 *Lophiostomataceae* — Crest Stoma Pleospore Sac Fungus
GENUS *Lophiostoma* — Crest Stoma Pleospore Sac Fungus
SPECIES *angustilabrum, appendiculatum, arundinis, bicuspidatum, caudatum, caulium, compressum, fibritectum, fuckelii, macrostomum, massarioides, microstomum, nucula, quadrinucleatum, semiliberum, ulicis, vagabundum, vagans, viridarium, winteri*
VARIETIES *angustilabrum crenatum, arundinis tritici, fuckelii fuckelii,fuckelii pulveraceum, fuckelii rubi, vagabundum rubidum*
GENUS *Acrocalymma* — Terminal Calyx Crest Stoma Pleospore Sac Fungus
SPECIES *medicaginis*
GENUS *Ascocratera* — Crater Crest Stoma Pleospore Sac Fungus
SPECIES *manglicola*
GENUS *Byssolophis* — Silky Crest Stoma Pleospore Sac Fungus
SPECIES *sphaerioides*
GENUS *Cilioplea* — Ciliate Crest Stoma Pleospore Sac Fungus
SPECIES *coronata*
GENUS *Entodesmium* — Entic Fiber Crest Stoma Pleospore Sac Fungus
SPECIES *niesslianum, rude*
VARIETIES *niesslianum viciae*
GENUS *Herpotrichia* — Serpent Hairy Crest Stoma Pleospore Sac Fungus
SPECIES *australis, callimorpha, diffusa, herpotrichoides, juniperi, macrotricha, rhenana, rhodosticta, striatispora*
GENUS *Keissleriella* — Keissler Crest Stoma Pleospore Sac Fungus
SPECIES *caudata, cladophila, culmifida, gallica, genistae, linearis, ocellata, pinicola, subalpina*
GENUS *Lophiosphaera* — Sphere Crest Stoma Pleospore Sac Fungus
SPECIES *ulicis, velata*

GENUS *Lophiotrema* Tremor Crest Stoma Pleospore Sac Fungus
SPECIES *curreyi*
GENUS *Massarina* Massar Crest Stoma Pleospore Sac Fungus
SPECIES *alni, aquatica, arundinacea, brunaudii, cisti, coffeicola, corni, corticola, eburnea, emergens, lacustris, microcarpa, microspora, operculicola, papulosa, rubi, tetraploa, tiliae, walkeri*
GENUS *Massariosphaeria* Massar Sphere Crest Stoma Pleospore Sac Fungus
SPECIES *autumnalis, grandispora, mosana, multiseptata, palustris, phaeospora, roumeguerei, scirpina, triseptata, typhicola*
GENUS *Muroia* Muro Crest Stoma Pleospore Sac Fungus
SPECIES *nipponica*
GENUS *Saccharicola* Sugar Crest Stoma Pleospore Sac Fungus
SPECIES *bicolor, taiwanensis*
GENUS *Tetraploa* Tetraploid Crest Stoma Pleospore Sac Fungus
SPECIES *aristata, ellisii*
GENUS *Trichometasphaeria* Hairy Sphere Crest Stoma Pleospore Sac Fungus
SPECIES *setulosa*
GENUS *Vaginatispora* Vagina Crest Stoma Pleospore Sac Fungus
SPECIES *aquatica*

FAMILY 11 *Melanommataceae* Melanoma Pleospore Sac Fungus
GENUS *Melanomma* Melanoma Pleospore Sac Fungus
SPECIES *aspegrenii, brachytele, fuscidulum, jenynsii, longicolle, medium, obliterans, panici-miliacei, pulvispyrius, pyriostictum, radicans, rhododendri, sanguinarium, spiniferum, subdispersum, vile*
GENUS *Anomalemma* Anomalous Melanoma Pleospore Sac Fungus
SPECIES *epochnii*
GENUS *Aposphaeria* Parsley Sphere Melanoma Pleospore Sac Fungus
SPECIES *populea, populina, pulviscula, sepulta, ulmicola*
GENUS *Astrosphaeriella* Star Sphere Melanoma Pleospore Sac Fungus
SPECIES *applanata, trochus*
GENUS *Asymmetricospora* Asymmetric Melanoma Pleospore Sac Fungus
SPECIES *calamicola*
GENUS *Bicrouania* Bicrouan Melanoma Pleospore Sac Fungus
SPECIES *maritima*
GENUS *Bimuria* Bimur Melanoma Pleospore Sac Fungus
SPECIES *novae-zelandiae*
GENUS *Byssosphaeria* Silky Sphere Melanoma Pleospore Sac Fungus
SPECIES *schiedermayeriana*
GENUS *Caryosporella* Cary Melanoma Pleospore Sac Fungus
SPECIES *rhizophorae*
GENUS *Javaria* Java Melanoma Pleospore Sac Fungus
SPECIES *samuelsii, shimekii*
GENUS *Karstenula* Karsten Melanoma Pleospore Sac Fungus
SPECIES *rhodostoma, shepherdiae*
GENUS *Ohleria* Ohler Melanoma Pleospore Sac Fungus
SPECIES *obducens, rugulosa*
GENUS *Pseudospiropes* Spire Melanoma Pleospore Sac Fungus
SPECIES *elaeidis, hachijoensis, hughesii, longipilus, lotorum, nodosus, obclavatus, rousselianus, simplex, subuliferus*
GENUS *Pseudotrichia* Hairy Melanoma Pleospore Sac Fungus
SPECIES *aurata, minor, mutabilis*
GENUS *Trematosphaeria* Tremor Sphere Melanoma Pleospore Sac Fungus
SPECIES *alpestris, britzelmayriana, clarkii, heterospora, hydrela, lunaria, megalospora, paradoxa, pertusa, tripartita*

FAMILY 12 *Montagnulaceae* Montagne Pleospore Sac Fungus
GENUS *Montagnula* Montagne Pleospore Sac Fungus
SPECIES *anthostomoides, dura, opulenta, phragmospora, rhodophaea, spartii, spinosella*
GENUS *Kalmusia* Kalmus Montagne Pleospore Sac Fungus
SPECIES *clivensis, sarothamni, stromatica, surrecta*
GENUS *Montagnella* Lesser Montagne Pleospore Sac Fungus
SPECIES *schweinitzii, tordillensis*
GENUS *Montagnellina* Least Montagne Pleospore Sac Fungus
SPECIES *stellaris*
GENUS *Paraphaeosphaeria* Dusky Sphere Montagne Pleospore Sac Fungus
SPECIES *michotii, pilleata, vectis*

FAMILY 13 *Parodiellaceae* Parody Pleospore Sac Fungus
GENUS *Parodiella* Parody Pleospore Sac Fungus
SPECIES *hedysari, pentanisiae*

GENUS *Pseudomeliola* — Mellow Parody Pleospore Sac Fungus
SPECIES *grammodes*

FAMILY 14 *Parodiopsidaceae* — Lesser Parody Pleonspore Sac Fungus
GENUS *Parodiops* — (Synonym for *Balladynocallia*)
GENUS *Alina* — Aline Lesser Parody Pleonspore Sac Fungus
SPECIES *jasmini*
GENUS *Balladynocallia* — Ball Dyno Callisto Lesser Parody Pleonspore Sac Fungus
SPECIES *amazonensis, glabra, magna*
GENUS *Dimeriella* — Lesser Dimer Lesser Parody Pleonspore Sac Fungus
SPECIES *sacchari*
GENUS *Dimerium* — Dimer Lesser Parody Pleonspore Sac Fungus
SPECIES *meliolicola*
GENUS *Pilgeriella* — Pilger Lesser Parody Pleonspore Sac Fungus
SPECIES *anacardii, perisporioides*
GENUS *Tretospora* — Treton Lesser Parody Pleonspore Sac Fungus
SPECIES *himalayana, indica, negrii, quaternata, shoreae, thitei, xeromphiicola*

FAMILY 15 *Phaeosphaeriaceae* — Dusky Sphere Pleospore Sac Fungus
GENUS *Phaeosphaeria* — Dusky Sphere Pleospore Sac Fungus
SPECIES *alpina, avenaria, berlesei, caricicola, caricinella, caricis, culmorum, dennisiana, epicalamia, eustoma, fuckelii, glyceriae-plicatae, graminis, herpotrichoides, insignis, juncicola, juncina, juncophila, licatensis, lindii, luctuosa, lutea, lycopodina, macrosporidium, marciensis, marram, microscopica, nardi, neomaritima, nigrans, nodorum, norfolcia, occulta, oreochloae, padellana, parvograminis, parvula, petkovicensis, phragmiticola, phragmitis, pleurospora, pontiformis, rousseliana, salebricola, setosa, silenes-acaulis, silvatica, sowerbyi, sparsa, spartinae, spartinicola, tofieldiae, triglochinicola, typhae, typharum, vagans, viridella, volkartiana*
VARIETIES *avenaria avenaria, avenaria triticae, nodorum hordei, typharum acori*
GENUS *Barria* — Barr Dusky Sphere Pleospore Sac Fungus
SPECIES *piceae*
GENUS *Blastostroma* — Blastid Stroma Dusky Sphere Pleospore Sac Fungus
SPECIES *aesculi*
GENUS *Eudarluca* — Darluc Dusky Sphere Pleospore Sac Fungus
SPECIES *caricis*
GENUS *Hadrospora* — Hadron Dusky Sphere Pleospore Sac Fungus
SPECIES *clarkii, fallax*
GENUS *Isthmosporella* — Isthmus Dusky Sphere Pleospore Sac Fungus
SPECIES *pulchra*
GENUS *Lautitia* — Lautite Dusky Sphere Pleospore Sac Fungus
SPECIES *danica*
GENUS *Mixtura* — Mixture Dusky Sphere Pleospore Sac Fungus
SPECIES *saginata*
GENUS *Mycopappus* — Pappus Dusky Sphere Pleospore Sac Fungus
SPECIES *alni*
GENUS *Neophaeosphaeria* — Modern Dusky Sphere Pleospore Sac Fungus
SPECIES *conglomerata, filamentosa, quadriseptata*
GENUS *Nodulosphaeria* — Nodule Dusky Sphere Pleospore Sac Fungus
SPECIES *aquilana, cadubriae, centaureae, cirsii, dolioloides, jaceae, spectabilis*
GENUS *Ophiosphaerella* — Serpent Dusky Sphere Pleospore Sac Fungus
SPECIES *agrostidis, herpotricha, korrae, narmari*
GENUS *Phaeodothis* — Doth Dusky Sphere Pleospore Sac Fungus
SPECIES *winteri*
GENUS *Phaeoseptoria* — Septic Dusky Sphere Pleospore Sac Fungus
SPECIES *airae, caricicola, eucalypti, musae, phalaridis, triticalina*
GENUS *Phaeosphaeriopsis* — Lesser Dusky Sphere Pleospore Sac Fungus
SPECIES *agavensis, glaucopunctata, nolinae, obtusispora*

FAMILY 16 *Phaeotrichaceae* — Dusky Hair Pleospore Sac Fungus
GENUS *Phaeotrichum* — Dusky Hair Pleospore Sac Fungus
SPECIES *benjaminii, cylindrosporum, setosum*
GENUS *Echinoascotheca* — Urchin Theca Dusky Hair Pleospore Sac Fungus
SPECIES *duplooformis*
GENUS *Trichodelitschia* — Hairy Delitsch Dusky Hair Pleospore Sac Fungus
SPECIES *bisporula*

FAMILY 17 *Pleomassariaceae* — Massar Pleospore Sac Fungus
GENUS *Pleomassaria* — Massar Pleospore Sac Fungus
SPECIES *holoschista, robiniae, siparia*
GENUS *Asteromassaria* — Star Massar Pleospore Sac Fungus
SPECIES *macrospora*

GENUS *Dendryphiopsis* Wood Massar Pleospore Sac Fungus
SPECIES *atra*
GENUS *Helminthosporium* Helminth Massar Pleospore Sac Fungus
SPECIES *apiculatum, asterinum, caladii, cylindricum, decacuminatum, delicatulum, densum, exasperatum, foveolatum, fusiforme, fusisporum, halodes, macilentum, microsorum, microsporum, minimum, palmigenum, parvum, rhopaloides, scolecoides, sechiicola, solani, triseptatum, turbinatum, velutinum*
GENUS *Peridiothelia* Envelope Nipple Massar Pleospore Sac Fungus
SPECIES *fuliguncta, grandiuscula, oleae*
GENUS *Prosthemium* Appendage Massar Pleospore Sac Fungus
SPECIES *asterosporum, betulinum, kentiae, palmatum, stellare, tiliae*
GENUS *Scolicosporium* Scolex Massar Pleospore Sac Fungus
SPECIES *typhae*
GENUS *Splanchnonema* Entrail Thread Massar Pleospore Sac Fungus
SPECIES *argus, carpini, foedans, loricatum, platani, pupula, quercicola*
GENUS *Stegonsporium* Cover Massar Pleospore Sac Fungus
SPECIES *cellulosum, ovatum, pyriforme*

FAMILY 18 *Sporormiaceae* Ormy Pleospore Sac Fungus
GENUS *Sporormia* Ormy Pleospore Sac Fungus
SPECIES *brassicae, fimetaria, lignicola, longipes, marchaliana, notarisii, pulchra, subticinensis*
GENUS *Chaetopreussia* Bristle Prussia Ormy Pleospore Sac Fungus
SPECIES *chadefaudii*
GENUS *Delitschia* Delitsch Ormy Pleospore Sac Fungus
SPECIES *bispora, consociata, didyma, furfuracea, marchalii, mesostenospora, pachylospora, patagonica, perpusilla, trichodelitschioides, trigonospora, winteri*
GENUS *Preussia* Prussia Ormy Pleospore Sac Fungus
SPECIES *aemulans, clavispora, flanaganii, fleischhakii, funiculata, globosa, isomera, planispora, terricola, typharum*
GENUS *Preussiella* Lesser Prussia Ormy Pleospore Sac Fungus
SPECIES *indica*
GENUS *Pycnidiophora* Dense Ormy Pleospore Sac Fungus
SPECIES *aurantiaca, dispersa*
GENUS *Sporormiella* Lesser Ormy Pleospore Sac Fungus
SPECIES *antarctica, australis, bipartis, borealis, corynespora, cymatomera, gigantea, grandispora, intermedia, isomera, lageniformis, leporina, megalospora, minima, minimoides, nigropurpurea, octomera, ovina, pascua, pilosa, pulchella, schadospora, similis, teretispora, vexans*
GENUS *Spororminula* Least Ormy Pleospore Sac Fungus
SPECIES *tenerifae*
GENUS *Westerdykella* Westerdyke Ormy Pleospore Sac Fungus
SPECIES *cylindrica, dispersa, multispora, nigra, ornata, purpurea, rapa-nuiensis*

FAMILY 19 *Tubeufiaceae* Tubeuf Pleospore Sac Fungus
GENUS *Tubeufia* Tubeuf Pleospore Sac Fungus
SPECIES *cerea, cylindrothecia, helicoma, helicomyces, paludosa, parvula, pezizula, trichella*
GENUS *Acanthophiobolus* Spiny Serpent Bowl Tubeuf Pleospore Sac Fungus
SPECIES *helicosporus*
GENUS *Acanthostigma* Spiny Stigma Tubeuf Pleospore Sac Fungus
SPECIES *ellisii, hederae, minutum, perpusillum, revocatum, scopulum*
GENUS *Acanthostigmella* Lesser Spiny Stigma Tubeuf Pleospore Sac Fungus
SPECIES *brevispina, genuflexa, orthoseta, pallida, pellucida, thaxteri*
GENUS *Acanthostigmina* Least Spiny Stigma Tubeuf Pleospore Sac Fungus
SPECIES *hebridensis*
GENUS *Allonecte* Bound Tubeuf Pleospore Sac Fungus
SPECIES *lagerheimii*
GENUS *Chaetocrea* Bristle Flesh Tubeuf Pleospore Sac Fungus
SPECIES *parasitica*
GENUS *Chaetosphaerulina* Bristle Sphere Tubeuf Pleospore Sac Fungus
SPECIES *bambusae, bambusicola, lignicola, vermicularispora, yasudai*
GENUS *Glaxoa* Glaxo Tubeuf Pleospore Sac Fungus
SPECIES *pellucida*
GENUS *Guelichia* Guelich Tubeuf Pleospore Sac Fungus
SPECIES *paradoxa*
GENUS *Helicoma* Helix Tubeuf Pleospore Sac Fungus
SPECIES *ambiens, chlamydosporum, dennisii, intermedium, irregulare, isiola, muelleri, muelleri, olivaceum, perelegans, stigmateum, vaccinii, violaceum*
GENUS *Helicomyces* Least Helix Tubeuf Pleospore Sac Fungus
SPECIES *bellus, roseus, scandens, tenuis, torquatus*
GENUS *Helicosporium* Lesser Helix Tubeuf Pleospore Sac Fungus

SPECIES *abuense, albidum, aureum, gracile, griseum, griseum, indicum, linderi, lumbricoides, lumbricopsis, pallidum, pannosum*
GENUS *Letendraea* Letendra Tubeuf Pleospore Sac Fungus
SPECIES *eurotioides, helminthicola, padouk*
GENUS *Letendraeopsis* Lesser Letendra Tubeuf Pleospore Sac Fungus
SPECIES *palmarum*
GENUS *Pendulispora* Pendant Tubeuf Pleospore Sac Fungus
SPECIES *venezuelanica*
GENUS *Podonectria* Pod Cornicle Tubeuf Pleospore Sac Fungus
SPECIES *aurantii, bambusicola, cicadellidicola, citrina, coccicola, coccorum, echinata, gahnia, larvispora, novae-zelandiae, tenuispora*
GENUS *Rebentischia* Rebentisch Tubeuf Pleospore Sac Fungus
SPECIES *unicaudata*
GENUS *Taphrophila* Tapher Tubeuf Pleospore Sac Fungus
SPECIES *argyllensis, cornu-capreoli*
GENUS *Thaxterina* Thaxter Tubeuf Pleospore Sac Fungus
SPECIES *multispora*
GENUS *Titaea* Titan Tubeuf Pleospore Sac Fungus
SPECIES *maxilliformis*
GENUS *Uredinophila* Blight Tubeuf Pleospore Sac Fungus
SPECIES *erinacea, tropicalis*
GENUS *Xenosporium* Strange Tubeuf Pleospore Sac Fungus
SPECIES *africanum, berkeleyi, boivinii, cubense, indicum, intermedium, larvale, mirabile, pirozynskii, pleurococcum, shoranoorense, subramanianii, thaxteri*

FAMILY 20 *Venturiaceae* Venture Pleospore Sac Fungus
GENUS *Venturia* Venture Pleospore Sac Fungus
SPECIES *aceris, anemones, atriseda, aucupariae, carpophila, centaureae, cephalariae, cerasi, chlorospora, comari, crataegi, ditricha, epilobii, eres, fraxini, geranii, helvetica, inaequalis, integra, lonicerae, macularis, maculiformis, mandshurica, minuta, nashicola, nitida, palustris, populina, potentillae, pyrina, rumicis, saliciperda, subcutanea, thwaitesii, tremulae, viennotii*
VARIETIES *tremulae grandidentatae, tremulae populi-albae*
GENUS *Acantharia* Spiny Venture Pleospore Sac Fungus
SPECIES *aterrima, chaetomoides, echinata, elegans, hamata, quercus-dilatatae, sinensis*
GENUS *Antennaria* Antenna Venture Pleospore Sac Fungus
SPECIES *laevigata, semiovata*
GENUS *Antennularia* Lesser Antenna Venture Pleospore Sac Fungus
SPECIES *rhododendri*
GENUS *Apiosporina* Bee Venture Pleospore Sac Fungus
SPECIES *morbosa*
GENUS *Arkoola* Arkool Venture Pleospore Sac Fungus
SPECIES *nigra*
GENUS *Atopospora* Atopic Venture Pleospore Sac Fungus
SPECIES *betulina*
GENUS *Caproventuria* Goat Venture Pleospore Sac Fungus
SPECIES *hanliniana, hystrioides*
GENUS *Coleroa* Colero Venture Pleospore Sac Fungus
SPECIES *alchemillae, chaetomium, circinans, rhynchosiae, robertiani*
GENUS *Dictyodochium* Netted Doch Venture Pleospore Sac Fungus
SPECIES *prinsepiae*
GENUS *Fusicladium* Fused Branch Venture Pleospore Sac Fungus
SPECIES *amelanchieris, orbiculatum*
GENUS *Gibbera* Gibber Venture Pleospore Sac Fungus
SPECIES *andromedae, conferta, elegantula, myrtilli, ramicola, rosae, salisburgensis, vaccinii*
GENUS *Lasiobotrys* Woolly Berry Venture Pleospore Sac Fungus
SPECIES *lonicerae*
GENUS *Lineostroma* Linear Stroma Venture Pleospore Sac Fungus
SPECIES *banksiae*
GENUS *Metacoleroa* Meta Colero Venture Pleospore Sac Fungus
SPECIES *dickiei*
GENUS *Phaeocryptopus* Dusky Cryptic Venture Pleospore Sac Fungus
SPECIES *gaeumannii, nudus*
GENUS *Phragmogibbera* Fragment Gibber Venture Pleospore Sac Fungus
SPECIES *xylariicola*
GENUS *Piggotia* Piggot Venture Pleospore Sac Fungus
SPECIES *astroidea, substellata, ulmi*
GENUS *Platychora* Flat Choral Venture Pleospore Sac Fungus
SPECIES *alni, ulmi*

GENUS *Pollaccia* Pollacci Venture Pleospore Sac Fungus
SPECIES *catenospora, radiosa*
GENUS *Protoventuria* Primitive Venture Pleospore Sac Fungus
SPECIES *alpina, arxii, barriae, engleriana, major, straussii, tetraspora*
GENUS *Pseudoparodia* Parody Venture Pleospore Sac Fungus
SPECIES *pseudopeziza*
GENUS *Pseudoparodiella* Lesser Parody Venture Pleospore Sac Fungus
SPECIES *vernoniae*
GENUS *Rhizosphaera* Rhizid Sphere Venture Pleospore Sac Fungus
SPECIES *kalkhoffii, macrospora, oudemansii, pini*
GENUS *Rosenscheldiella* Rosenscheld Venture Pleospore Sac Fungus
SPECIES *pulverulenta*
GENUS *Spilocaea* Spotted Venture Pleospore Sac Fungus
SPECIES *oleaginea, pyracanthae*
GENUS *Uleodothis* Ule Doth Venture Pleospore Sac Fungus
SPECIES *balansiana, munkii*
GENUS *Xenomeris* Strange Seaside Venture Pleospore Sac Fungus
SPECIES *abietis, acicola, alpina, arbuti, eucalypti, hemisphaerica, juniperi, nicholsonii, pruni, raetica, saccifolii*

ORDER 14 *Rhytismatales* Wrinkled Sac Fungus

FAMILY 1 *Rhytismataceae* Wrinkled Sac Fungus
GENUS *Rhytisma* Wrinkled Sac Fungus
SPECIES *acerinum, andromedae, asteris, decolorans, eucalypti, himalense, ilicis-canadensis, ilicis-latifoliae, itatiaiae, lagerstroemiae, prini, priscum, punctatum, salicinum, umbonatum, velatum*
GENUS *Bifusella* Dual Fused Wrinkled Sac Fungus
SPECIES *acuminata, cunninghamiicola, linearis, pini, saccata., superba*
GENUS *Bifusepta* Dual Fused Septic Wrinkled Sac Fungus
SPECIES *tehonii*
GENUS *Bivallum* Bivall Wrinkled Sac Fungus
SPECIES *diselmatis, heterosporum, microstrobi, pilgerodendri, podocarpi, zelandicum*
GENUS *Canavirgella* Cane Virgo Wrinkled Sac Fungus
SPECIES *banfieldii*
GENUS *Ceratophacidium* Horn Lens Wrinkled Sac Fungus
SPECIES *aristosporum*
GENUS *Cerion* Wax Wrinkled Sac Fungus
SPECIES *coccineum, leucophaeum*
GENUS *Coccomyces* Round Wrinkled Sac Fungus
SPECIES *agathidicola, annulatus, antillarum, araucariae, arbutifolius, arctostaphyli, atactus, basizonatus, binnaburrensis, bipartitus, boydii, canarii, castanopsis, cembrae, clavatus, clusiae, cocoës, concolor, consociatus, coronatus, croceus, crystalligerus, cunninghamii, cupressini, delta, dentatus, duplicarioides, ericae, farlowii, foliicola, fuscus, gaultheriae, globososimilis, globosus, heterophyllae, insignis, irretitus, kinabaluensis, kirkii, lauraceus, ledi, leptideus, leptosporus, libocedri, limitatus, longwoodicus, mertensianae, monticola, mucronatus, noosanus, palmicola, pampeanus, papillatus, paraphysincrustatus, parasiticus, parvulus, petersii, philippinus, phyllocladi, phyllocladicola, pseudotsugae, puiggarii, radiatus, sichuanensis, spegazzinii, strobi, strobilicola, tesselatus, triangularis, triseptatus, tumidus, tympanibaculi, tympanidiosporus, urceoloides, urceolus, venezuelae, vilis, wagnerianus, walkeri*
VARIETIES *dentatus castaneae, strobi strobi, strobi olympicus*
GENUS *Colpoma* Envelope Wrinkled Sac Fungus
SPECIES *agathidis, crispum, deustum, juniperi, ledi, morbidum, nothofagi, quercinum, styriacum*
GENUS *Criella* Crie Wrinkled Sac Fungus
SPECIES *austrocaledonica*
GENUS *Cryptomyces* Cryptic Wrinkled Sac Fungus
SPECIES *maximus*
GENUS *Cyclaneusma* Cyclic Floating Wrinkled Sac Fungus
SPECIES *minus, niveum*
GENUS *Darkera* Darker Wrinkled Sac Fungus
SPECIES *abietis, parca*
GENUS *Discocainia* Disc Cain Wrinkled Sac Fungus
SPECIES *arctica, laciniata, treleasei*
GENUS *Duplicaria* Duplicate Wrinkled Sac Fungus
SPECIES *antarctica, empetri*
GENUS *Duplicariella* Lesser Duplicate Wrinkled Sac Fungus
SPECIES *phyllodoces*
GENUS *Elytroderma* Membrane Dermal Wrinkled Sac Fungus
SPECIES *deformans, torres-juanii*

GENUS *Gelineostroma* Geline Stroma Wrinkled Sac Fungus
SPECIES *athrotaxis, swartii*
GENUS *Hypoderma* Hypodermic Wrinkled Sac Fungus
SPECIES *aceris, alborubrum, alpinum, bidwillii, bihospitum, borneense, campanulatum, carinatum, commune, cookianum, cordylines, corni, dryadis, dundasicum, eucalypti, gaultheriae, handelii, hansbroughii, hederae, ilicinum, labiorum-aurantiorum, liliense, mahuianum, obtectum, rubi, rufilabrum, sigmoideum, smilacis, sticheri, tillandsiae, tunicatum*
GENUS *Hypodermella* Lesser Hypodermic Wrinkled Sac Fungus
SPECIES *laricis, rhamni, symploci*
GENUS *Hypohelion* Hypha Helion Wrinkled Sac Fungus
SPECIES *parvum, scirpinum*
GENUS *Isthmiella* Isthmian Wrinkled Sac Fungus
SPECIES *abietis, crepidiformis, faullii, quadrispora*
GENUS *Karstenia* Karsten Wrinkled Sac Fungus
SPECIES *clematidis, corticoides, gregaria, idaei, inconspicua, lonicerae, macer, maydis, rubicola, sorbina*
GENUS *Laquearia* Laquear Wrinkled Sac Fungus
SPECIES *sphaeralis*
GENUS *Lasiostictella* Woolly Stick Wrinkled Sac Fungus
SPECIES *bambusae*
GENUS *Leptostroma* Thin Stroma Wrinkled Sac Fungus
SPECIES *donacinum, eupatorii, herbarum, juncacearum, luzulae, osmundicola, phragmitis, pteridis, spiraeae, spiraeinum*
GENUS *Lirula* Ridge Wrinkled Sac Fungus
SPECIES *abietis-concoloris, brevispora, exigua, japonica, macrospora, mirabilis, nervata, nervisequia, nervisequia nervisequia, nervisequia conspicua, punctata*
GENUS *Lophodermella* Lesser Crested Dermal Wrinkled Sac Fungus
SPECIES *arcuata, cerina, concolor, conjuncta, maureri, montivaga, morbida, orientalis, sulcigena*
GENUS *Lophodermium* Crested Dermal Wrinkled Sac Fungus
SPECIES *abietis, agathidis, airarum, alienum, alpinum, anhuiense, apiculatum, arundinaceum, atrum, aucupariae, australe, baculiferum, balaramii, borneoensis, brunneolum, camelliicola, canberrianum, caricinum, clusiae, conigenum, consociatum, crassum, croesicum, culmigenum, decorum, durilabrum, eriophori, eucalypti, festucae, foliicola, fusiforme, geniculosporum, gramineum, grandialpinum, hauturuanum, hedericola, herbarum, heteromelis, himalayense, inclusum, indianum, intermissum, intricatum, irregulare, juniperinum, kaikawakae, kinabaluense, kumaunicum, lacerum, lanceolatum, laricinum, lauri, licualae, litseae, litseaeoides, maculare, magellanicum, mahuianum, mangatepopense, mangiferae, medium, melaleucum, minor, molitoris, multimatricum, nanakii, neesii, nigrofactum, nitens, nitidum, nivale, nonramosum, orbiculare, orientale, oxycocci, paeoniae, petiolicola, petrakii, petriniae, phloxii, piceae, pinastri, pini-excelsae, pini-pumilae, prunicola, punctiforme, pyrolae, ramulosum, ravenelii, rectangulare, richeae, rubiicola, rubrum, sawadae, schweinitzii, seditiosum, sieglingiae, sphaerioides, staleyi, svalbardense, thujae, thujopsidis, tindalii, typhinum, uncinatum, unciniae, vagulum, versicolor*
GENUS *Lophomerum* Crested Seaside Wrinkled Sac Fungus
SPECIES *autumnale, clavuligerum, darkeri, ponticum, rhododendri, septatum*
GENUS *Macroderma* Large Dermal Wrinkled Sac Fungus
SPECIES *curtisii, uleanum*
GENUS *Marthamyces* Martha Wrinkled Sac Fungus
SPECIES *dendrobii, desmoschoeni, dracophylli, emarginatus, foliicola, panizzei, phacidioides, quadrifidus, quercifolius*
GENUS *Melasmia* Black Wrinkled Sac Fungus
SPECIES *empetri, ulmicola*
GENUS *Mellitiosporiella* Lesser Honeysweet Spore Wrinkled Sac Fungus
SPECIES *macrospora, pulchella*
GENUS *Mellitiosporium* Honeysweet Spore Wrinkled Sac Fungus
SPECIES *hysterinum, propolidoides, pteridinum*
GENUS *Meloderma* Melon Dermal Wrinkled Sac Fungus
SPECIES *desmazieri, dracophylli, richeae, sharmarum*
GENUS *Moutoniella* Mouton Wrinkled Sac Fungus
SPECIES *polita*
GENUS *Myriophacidium* Myriad Lens Wrinkled Sac Fungus
SPECIES *aphyophyllicum, corticola, quercinum, tridentatum*
GENUS *Naemacyclus* Thread Cyclic Wrinkled Sac Fungus
SPECIES *caulium, fimbriatus, griseus, korfii, macularis, steatopygioides*
GENUS *Neococcomyces* Modern Round Wrinkled Sac Fungus
SPECIES *rhododendri*
GENUS *Neophacidium* Modern Lens Wrinkled Sac Fungus
SPECIES *macrocarpum, quercus*
GENUS *Nothorhytisma* Another Wrinkled Sac Fungus
SPECIES *nahuelitae*

GENUS *Ocotomyces* Ocoto Wrinkled Sac Fungus
SPECIES *parasiticus*
GENUS *Parvacoccum* Small Round Wrinkled Sac Fungus
SPECIES *pini*
GENUS *Periperidium* Envelope Wrinkled Sac Fungus
SPECIES *acicola*
GENUS *Phacidiopycnis* Lens Dense Wrinkled Sac Fungus
SPECIES *padwickii, tuberivora*
GENUS *Phaeophacidium* Dusky Lens Wrinkled Sac Fungus
SPECIES *escalloniae*
GENUS *Ploioderma* Ploid Dermal Wrinkled Sac Fungus
SPECIES *cedri, hedgcockii, lethale, lowei, pedatum*
GENUS *Potebniamyces* Potebnia Wrinkled Sac Fungus
SPECIES *balsamicola, pyri*
VARIETIES *balsamicola balsamicola, balsamicola boycei*
GENUS *Propolidium* Lesser Bee Glue Wrinkled Sac Fungus
SPECIES *glaucum, mezerei*
GENUS *Propolis* Bee Glue Wrinkled Sac Fungus
SPECIES *angulosa, desmoschoeni, farinosa, lecanora, leonis, lugubris, phacidioides, pulchella*
GENUS *Pseudorhytisma* Lesser Wrinkled Sac Fungus
SPECIES *bistortae*
GENUS *Pureke* Pureke Wrinkled Sac Fungus
SPECIES *zelandicum*
GENUS *Soleella* Sole Wrinkled Sac Fungus
SPECIES *striiformis*
GENUS *Sporomega* Megaspore Wrinkled Sac Fungus
SPECIES *degenerans*
GENUS *Terriera* Terrier Wrinkled Sac Fungus
SPECIES *arundinacea, asteliae, brevis, cladophila, clithris, dracaenae, fourcroyae, fuegiana, javanica, latiascus, longissima, nematoidea, pandani, sacchari, samuelsii, stevensii*
VARIETIES *javanica javanica, javanica pandani*
GENUS *Therrya* Therry Wrinkled Sac Fungus
SPECIES *fuckelii, piceae, pini, pseudotsugae, tsugae*
VARIETIES *pini pini, pini mughicola*
GENUS *Tiarosporella* Tiara Spore Wrinkled Sac Fungus
SPECIES *graminis, madreeya, paludosa, tritici*
VARIETIES *graminis graminis, graminis karoo*
GENUS *Tryblidiopsis* Tryblid Wrinkled Sac Fungus
SPECIES *pinastri*
GENUS *Virgella* Virgo Wrinkled Sac Fungus
SPECIES *robusta*
GENUS *Vladracula* Vladrac Wrinkled Sac Fungus
SPECIES *annuliformis*
GENUS *Xyloschizon* Wood Split Wrinkled Sac Fungus
SPECIES *atratum, weirianum*
GENUS *Zeus* Zeus Wrinkled Sac Fungus
SPECIES *olympius*

FAMILY 2 *Ascodichaenaceae* Dual Gaping Wrinkled Sac Fungus
GENUS *Ascodichaena* Dual Gaping Wrinkled Sac Fungus
SPECIES *mexicana, rugosa*
GENUS *Dichaena* Lesser Dual Gaping Wrinkled Sac Fungus
SPECIES *strobilina*
GENUS *Macroallantina* Large Membrane Dual Gaping Wrinkled Sac Fungus
SPECIES *coccifera*
GENUS *Pseudophacidium* Lens Dual Gaping Wrinkled Sac Fungus
SPECIES *kumaonense, ledi, microspermum, piceae, rugosum*

ORDER 15 *Sordariales* Sordid Sac Fungus

FAMILY 1 *Sordariaceae* Sordid Sac Fungus
GENUS *Sordaria* Sordid Sac Fungus
SPECIES *alcina, arctica, arenicola, brevicollis, conoidea, destruens, fimicola, goundaensis, humana, islandica, lappae, mabokeensis, macrospora, minima, papyricola, polyspora, prolifica, sibutii, sparganiicola, superba, tenerifae, tomentoalba*
VARIETIES *fimicola canina, sparganiicola sparganiicola, sparganiicola velata*
GENUS *Ascotaiwania* Taiwan Sordid Sac Fungus

SPECIES *hsilio, hughesii, licualae, lignicola, mauritiana, mitriformis, pallida, palmicola, pennisetorum, persoonii, sawadae, wulai*

GENUS *Ascovaginospora* Vagina Spore Sordid Sac Fungus

SPECIES *stellipala*

GENUS *Asordaria* Lesser Sordid Sac Fungus

SPECIES *islandica, tenerifae*

GENUS *Bombardiella* Bombard Sordid Sac Fungus

SPECIES *caespitosa*

GENUS *Boothiella* Booth Sordid Sac Fungus

SPECIES *tetraspora*

GENUS *Carpoligna* Carpel Wood Sordid Sac Fungus

SPECIES *pleurothecii*

GENUS *Chrysonilia* Golden Nile Sordid Sac Fungus

SPECIES *crassa, sitophila, tetrasperma*

GENUS *Copromyces* Copper Sordid Sac Fungus

SPECIES *octosporus*

GENUS *Diplogelasinospora* Diploid Gelasin Spore Sordid Sac Fungus

SPECIES *grovesii, inaequalis, princeps*

GENUS *Effetia* Effet Sordid Sac Fungus

SPECIES *craspedoconidica*

GENUS *Fluminicola* River Sordid Sac Fungus

SPECIES *bipolaris*

GENUS *Garethjonesia* Gareth Jones Sordid Sac Fungus

SPECIES *lacunosispora*

GENUS *Gelasinospora* Gelasin Spore Sordid Sac Fungus

SPECIES *amorphoporcata, angusta, brasiliensis, brevispora, calospora, cerealis, cratophora, dictyophora, endodontia, fallaciosa, foveaconica, goundaensis, hapsidophora, heterospora, himalayensis, hippopotama, indica, inversa, kobi, longispora, macrospora, micropertusa, mirabilis, multiforis, novoguineensis, polytrema, pseudocalospora, pseudoreticulata, reticulata, retispora, saitoi, santi-florii, seminuda, sphaerospora, stellata, terricola, tetrasperma, udagawae, varians, zimbabweensis*

VARIETIES *calospora autosteira*

GENUS *Globosphaeria* Globe Sphere Sordid Sac Fungus

SPECIES *jamesii*

GENUS *Guanomyces* Guan Sordid Sac Fungus

SPECIES *polythrix*

GENUS *Isia* Isi Sordid Sac Fungus

SPECIES *anixielloides, neocaledoniensis*

GENUS *Melanocarpus* Black Carpel Sordid Sac Fungus

SPECIES *albomyces, oblatus, thermophilus*

GENUS *Monosporascus* Monospore Sordid Sac Fungus

SPECIES *cannonballus, eutypoides, ibericus, monosporus*

GENUS *Neurospora* Neurospore Sordid Sac Fungus

SPECIES *africana, crassa, discreta, dodgei, intermedia, lineolata, pannonica, sitophila, sublineolata, terricola, tetrasperma, toroi*

GENUS *Ophiostomella* Ophiostome Sordid Sac Fungus

SPECIES *rostellata*

GENUS *Phaeosporis* Dusky Spore Sordid Sac Fungus

SPECIES *melasperma*

GENUS *Plagiosphaera* Pelagic Sphere Sordid Sac Fungus

SPECIES *immersa*

GENUS *Reconditella* Recondite Sordid Sac Fungus

SPECIES *physconiarum*

GENUS *Rhexosporium* Rhex Spore Sordid Sac Fungus

SPECIES *terrestre*

GENUS *Roselliniella* Rosellin Sordid Sac Fungus

SPECIES *atlantica, cladoniae, microthelia, nephromatis*

GENUS *Roselliniomyces* Least Rosellin Sordid Sac Fungus

SPECIES *trichotheliorum*

GENUS *Roselliniopsis* Lesser Rosellin Sordid Sac Fungus

SPECIES *gelidaria, groedensis, tartaricola, tropica, ventosa*

GENUS *Savoryella* Savory Sordid Sac Fungus

SPECIES *appendiculata, aquatica, curvispora, fusiformis, grandispora, lignicola, limnetica, longispora, melanospora, paucispora, verrucosa*

GENUS *Scopinella* Scopin Sordid Sac Fungus

SPECIES *barbata, caulincola, gallicola, pleiospora, solani, sphaerophila*

GENUS *Wallrothiella* Wallroth Sordid Sac Fungus

SPECIES *melanostigmoides, subiculosa*

FAMILY 2 *Cephalothecaceae* Head Sheath Sordid Sac Fungus
GENUS *Cephalotheca* Head Sheath Sordid Sac Fungus
SPECIES *clarkii, kriegeria, purpurea, reniformis, savoryi, sulfurea*
GENUS *Albertiniella* Albertin Head Sheath Sordid Sac Fungus
SPECIES *polyporicola*
GENUS *Cryptendoxyla* Cryptic Endic Sharp Head Sheath Sordid Sac Fungus
SPECIES *hypophloia*

FAMILY 3 *Chaetomiaceae* Bristle Sordid Sac Fungus
GENUS *Chaetomium* Bristle Sordid Sac Fungus
SPECIES *anatolicum, anguipilium, angustispirale, angustum, apiculatum, arachnoides, aterrimum, atrobrunneum, aureum, barilochense, biapiculatum, biporatum, bostrychodes, brasiliense, britannicum, cancroideum, caprinum, carinthiacum, causiiforme, chiversii, circinatum, citrinum, cochlioides, convolutum, crispatum, cruentum, cuniculorum, cupreum, cymbiforme, deceptivum, deltosporum, distortum, dreyfussii, elatum, erectum, flavigenum, floriforme, funicola, fusiforme, fusisporum, fusum, gangligerum, gelasinosporum, globisporum, globosporum, globosum, gracile, griseum, hamadae, heterosporum, hexagonosporum, hispanicum, homopilatum, hungaricum, hyaloperidium, indicum, intricatum, irregulare, jabalpurense, jodhpurense, lentum, longicolleum, lucknowense, luteum, madrasense, malaysiense, mareoticum, medusarum, megalocarpum, microascoides, mollicellum, mollipilium, montblancense, muelleri, multispirale, murorum, nepalense, nigricolor, nozdrenkoae, oblatum, pachypodioides, paucipilum, perlucidum, piluliferum, purpurascens, quadrangulatum, raii, reflexum, repens, retardatum, reticulopilium, robustum, semenis-citrulli, senegalense, serpentinum, sphaerale, spinigerum, spinosum, spirale, spiralotrichum, spirilliferum, spirochaete, spirogyrum, strumarium, subaffine, subcircinatum, subspirale, subspirilliferum, subtorulosum, succineum, sulphureum, tetrasporum, thermophilum, tortuosum, torulosum, trigonosporum, trilaterale, umbonatum, uniapiculatum, uniporum, variosporum, variostiolatum, virescens, virescens thielavioideum, warcupii*
VARIETIES *bostrychodes bostrychodes, bostrychodes virescens, globosum arhizoides, globosum flavoviride, globosum griseum, globosum rectum, globosum rufipilum, thermophilum thermophilum, thermophilum coprophilum, thermophilum dissitum, trilaterale chiversii*
GENUS *Achaetomiella* Least Bristle Sordid Sac Fungus
SPECIES *fusispora*
GENUS *Achaetomium* Lesser Bristle Sordid Sac Fungus
SPECIES *globosum, macrosporum, sphaerocarpum, thermophilum, thielavioides*
GENUS *Aporothielavia* Pore Thiel Bird Bristle Sordid Sac Fungus
SPECIES *leptoderma*
GENUS *Botryotrichum* Berry Hairy Bristle Sordid Sac Fungus
SPECIES *atrogriseum, peruvianum, piluliferum*
GENUS *Chaetomidium* Mock Bristle Sordid Sac Fungus
SPECIES *arxii, cephalothecoides, fimeti, pilosum, subfimeti, trichorobustum, unciatum*
GENUS *Chaetomiopsis* Lesser Mock Bristle Sordid Sac Fungus
SPECIES *dinae*
GENUS *Corynascella* Lesser Club Bristle Sordid Sac Fungus
SPECIES *humicola, inquinata*
GENUS *Corynascus* Club Bristle Sordid Sac Fungus
SPECIES *digitulus, novoguineensis, sepedonium, setosus, similis, singapurensis, thermophilus*
GENUS *Emilmuelleria* Emil Mueller Bristle Sordid Sac Fungus
SPECIES *spirotricha*
GENUS *Farrowia* Farrow Bristle Sordid Sac Fungus
SPECIES *seminuda*
GENUS *Setiferotheca* Theca Bristle Sordid Sac Fungus
SPECIES *nipponica*
GENUS *Stellatospora* Star Spore Bristle Sordid Sac Fungus
SPECIES *terricola*
GENUS *Subramaniula* Branched Bristle Sordid Sac Fungus
SPECIES *irregularis, thielavioides*
GENUS *Thielavia* Thiel Bird Bristle Sordid Sac Fungus
SPECIES *appendiculata, arenaria, australiensis, basicola, coactilis, dacrydioides, fragilis, heterothallica, hyalocarpa, hyrcaniae, inaequalis, intermedia, kuwaitensis, microspora, minuta, ovispora, pallidospora, peruviana, subthermophila, terrestris, terricola, tortuosa, wareingii*
VARIETIES *terricola minor*

FAMILY 4 *Chaetosphaeriaceae* Sphere Bristle Sordid Sac Fungus
GENUS *Chaetosphaeria* Sphere Bristle Sordid Sac Fungus
SPECIES *abietis, acutata, anglica, aspergilloides, aterrima, barbicincta, bramleyi, brevispora, callimorpha, chalaroides, chloroconia, crustacea, cubensis, cupulifera, dingleyae, endophytica, fennica, fusiformis, inaequalis, innumera,lentomita, lignifrondentis, montana, myriocarpa, novae-zelandiae, ovoidea, pileoferruginea, preussii, pulchriseta, pulviscula, pygmaea, talbotii, tulasneorum, vermicularioides*
VARIETIES *inaequalis chloridioides, innumera botryoideum, innumera minutum, vermicularioides caudigerum, vermicularioides chlamydosporum*
GENUS *Catenularia* Chain Sphere Bristle Sordid Sac Fungus

SPECIES *piceae*
GENUS *Chloridium* — Chloric Sphere Bristle Sordid Sac Fungus
SPECIES *lignicola, pachytrachelum, paucisporum*
GENUS *Custingophora* — Custing Sphere Bristle Sordid Sac Fungus
SPECIES *olivacea, suidafrikana*
GENUS *Cylindrotrichum* — Cylindric Hairy Sphere Bristle Sordid Sac Fungus
SPECIES *clavatum, ellisii, hennebertii, oligospermum, triseptatum, zignoëllae*
GENUS *Dictyochaeta* — Netted Sphere Bristle Sordid Sac Fungus
SPECIES *assamica, coffeae, daphnioides, dendroidea, eucalypti, fertilis, hughesii, keniensis, malaysiana, parva, querna, ramulosetula, setosa, simplex, triseptata, vulgaris*
GENUS *Gonytrichum* — Gonad Hairy Sphere Bristle Sordid Sac Fungus
SPECIES *macrocladum, mirabile*
GENUS *Lentomita* — Lent Mite Sphere Bristle Sordid Sac Fungus
SPECIES *stylophora*
GENUS *Melanochaeta* — Melano Sphere Bristle Sordid Sac Fungus
SPECIES *aotearoae, hemipsila*
GENUS *Melanopsammella* — (Synonym for *Chaetosphaeria*)
GENUS *Menispora* — Menispore Sphere Bristle Sordid Sac Fungus
SPECIES *arnaudii, britannica, ciliata, lucida, minuta, tortuosa*
GENUS *Mesobotrys* — Middle Berry Sphere Bristle Sordid Sac Fungus
SPECIES *simplex*
GENUS *Porosphaerella* — Pore Sphere Bristle Sordid Sac Fungus
SPECIES *cordanophora, setosa*
GENUS *Porosphaerellopsis* — Lesser Pore Sphere Bristle Sordid Sac Fungus
SPECIES *sporoschismophora*
GENUS *Sporoschisma* — Spore Split Sphere Bristle Sordid Sac Fungus
SPECIES *juvenile, nigroseptatum, parcicuneatum, phaeocentron, uniseptatum*
GENUS *Sporoschismopsis* — Lesser Spore Split Sphere Bristle Sordid Sac Fungus
SPECIES *australiensis, caribensis, dingleyae, moravica, simmonsii*
GENUS *Striatosphaeria* — Striped Sphere Bristle Sordid Sac Fungus
SPECIES *codinaeaphora*
GENUS *Zignoella* — Zignoël Sphere Bristle Sordid Sac Fungus
SPECIES *collabens, dealbata, dolichospora, eutypoides, fallax, hysterioides, macrasca, morthieri, pachyspora, paecilostoma, rhodobapha, rhytidodes, seriata, slaptonensis*

FAMILY 5 *Coniochaetaceae* — Cone Bristle Sordid Sac Fungus
GENUS *Coniochaeta* — Cone Bristle Sordid Sac Fungus
SPECIES *angustispora, arxii, cypraeaspora, detonsa, discospora, ellipsoidea, extramundana, hansenii, kellermanii, leucoplaca, ligniaria, malacotricha, nepalica, nodulisporioides, ovata, perangusta, polymegasperma, polysperma, pulveracea, saccardoi, scatigena, sordaria, subcorticalis, tetraspora, trivialis, vagans, velutina, xylariispora*
GENUS *Ascotrichella* — Hairy Cone Bristle Sordid Sac Fungus
SPECIES *hawksworthii*
GENUS *Barrina* — Barr Cone Bristle Sordid Sac Fungus
SPECIES *polyspora*
GENUS *Coniochaetidium* — Lesser Cone Bristle Sordid Sac Fungus
SPECIES *boothii, ostreum, savoryi*
GENUS *Coniolaria* — Least Cone Bristle Sordid Sac Fungus
SPECIES *murandii*
GENUS *Ephemeroascus* — Ephemeral Cone Bristle Sordid Sac Fungus
SPECIES *verticillatus*
GENUS *Poroconiochaeta* — Pore Cone Bristle Sordid Sac Fungus
SPECIES *discoidea, punctulata*

FAMILY 6 *Lasiosphaeriaceae* — Sphere Woolly Sordid Sac Fungus
GENUS *Lasiosphaeria* — Sphere Woolly Sordid Sac Fungus
SPECIES *biformis, calva, canescens, caudata, dactylina, felina, glabrata, hirsuta, hispida, hispidula, lanuginosa, ovina, paucipilis, phyllophila, rhacodium, rugulosa, sorbina, spermoides, strigosa, trames*
VARIETIES *biformis terrestris*
GENUS *Anopodium* — Anopod Sphere Woolly Sordid Sac Fungus
SPECIES *ampullaceum, epile*
GENUS *Apiosordaria* — Parsley Sphere Woolly Sordid Sac Fungus
SPECIES *jamaicensis, longicaudata, microcarpa, otanii, rugosa, tenuilacunata, tetraspora, verruculosa, vestita*
VARIETIES *verruculosa maritima*
GENUS *Apodus* — Pod Sphere Woolly Sordid Sac Fungus
SPECIES *deciduus, oryzae*
GENUS *Arniella* — Lesser Arny Sphere Woolly Sordid Sac Fungus
SPECIES *echoura, polycercia*

GENUS *Arnium* Arny Sphere Woolly Sordid Sac Fungus
SPECIES *apiculatum, arizonense, caballinum, cervinum, hirtum, macrotheca, mendax, olerum, tomentosum*
GENUS *Biconiosporella* Dual Cone Sphere Woolly Sordid Sac Fungus
SPECIES *corniculata*
GENUS *Bombardia* Bombard Sphere Woolly Sordid Sac Fungus
SPECIES *bombarda*
GENUS *Bombardioidea* Lesser Bombard Sphere Woolly Sordid Sac Fungus
SPECIES *anartia, bombardioides, serignanensis, stercoris*
GENUS *Camptosphaeria* Field Sphere Woolly Sordid Sac Fungus
SPECIES *citrinella, clavispora, sulphurea, venezuelensis*
GENUS *Cercophora* Antenna Sphere Woolly Sordid Sac Fungus
SPECIES *ambigua, arenicola, caudata, coprophila, himalayensis, mirabilis, samala, septentrionalis, silvatica, sulphurella*
GENUS *Cladorrhinum* Branched Snout Sphere Woolly Sordid Sac Fungus
SPECIES *brunnescens, bulbillosum, foecundissimum, phialophoroides*
GENUS *Diffractella* Diffracted Sphere Woolly Sordid Sac Fungus
SPECIES *curvata*
GENUS *Echinopodospora* Urchin Pod Spore Sphere Woolly Sordid Sac Fungus
SPECIES *effusa, jamaicensis, sacchari, spinosa, terrestris, vermicularis, verruculosa*
GENUS *Emblemospora* Emblem Spore Sphere Woolly Sordid Sac Fungus
SPECIES *ditrema, monotrema*
GENUS *Endophragmiella* Endic Fragment Sphere Woolly Sordid Sac Fungus
SPECIES *africana, arranensis, aseptata, biseptata, boewei, boothii, cambrensis, cesatii, corticola, eboracensis, ellisii, fagicola, fallacia, hughesii, lauri, lignicola, naromoruensis, ovoidea, pallescens, pinicola, pulchra, resinae, rostrata, subolivacea, suttonii, taxi, uniseptata*
GENUS *Eosphaeria* Primeval Sphere Woolly Sordid Sac Fungus
SPECIES *uliginosa*
GENUS *Jugulospora* Jug Spore Sphere Woolly Sordid Sac Fungus
SPECIES *rotula*
GENUS *Lacunospora* Pitted Spore Sphere Woolly Sordid Sac Fungus
SPECIES *stercoraria*
GENUS *Mycomedusiospora* Medusa Spore Sphere Woolly Sordid Sac Fungus
SPECIES *flavida*
GENUS *Periamphispora* Perimeter Dual Spore Sphere Woolly Sordid Sac Fungus
SPECIES *phacelodes*
GENUS *Philocopra* Copper Loving Sphere Woolly Sordid Sac Fungus
SPECIES *discospora, pusilla*
GENUS *Phragmodiscus* Fragment Disc Sphere Woolly Sordid Sac Fungus
SPECIES *arundinariae, himalayensis*
GENUS *Pleurage* Lung Sphere Woolly Sordid Sac Fungus
SPECIES *dagobertii*
GENUS *Podospora* Pod Spore Sphere Woolly Sordid Sac Fungus
SPECIES *aloides, ampullacea, appendiculata, australis, austroamericana, austrohemisphaerica, bicolor, bifida, brasiliensis, carpinicola, castorinospora, cochleariformis, collapsa, comata, communis, conica, cupiformis, curvicolla, curvula, curvuloides, dagobertii, dakotensis, decidua, decipiens, didyma, ellisiana, excentrica, fibrinocaudata, fimbriata, fimicola, globosa, granulostriata, gwynne-vaughaniae, hirsuta, horridula, inaequalis, inflatula, inquinata, intestinacea, leporina, longicollis, minicauda, miniglutinans, myriospora, pauciseta, perplexens, platensis, pleiospora, praecox, prethopodalis, prolifica, pyriformis, serotina, setosa, tetraspora, unicaudata, vesticola*
GENUS *Pseudoproboscispora* Probiscis Spore Sphere Woolly Sordid Sac Fungus
SPECIES *aquatica*
GENUS *Schizothecium* Split Theca Sphere Woolly Sordid Sac Fungus
SPECIES *curvisporum, dubium, fimicola, glutinans, hispidulum, miniglutinans, nanum, pilosum, squamulosum*
GENUS *Spinulosphaeria* Spiny Sphere Woolly Sordid Sac Fungus
SPECIES *thaxteri*
GENUS *Strattonia* Stratton Sphere Woolly Sordid Sac Fungus
SPECIES *carbonaria, minor*
GENUS *Triangularia* Triangular Sphere Woolly Sordid Sac Fungus
SPECIES *angulispora, backusii, bambusae, batistae, mangenotii, striatispora, tanzaniensis*
GENUS *Zopfiella* Zopf Sphere Woolly Sordid Sac Fungus
SPECIES *attenuata, catalaunica, ebriosa, erostrata, inermis, karachiensis, latipes, leucotricha, longicaudata, macrospora, marina, ovina, pilifera, pleuropora, tabulata, tardifaciens, tetraspora*
GENUS *Zygospermella* Zygote Sphere Woolly Sordid Sac Fungus
SPECIES *insignis*

ORDER 16 *Trichosphaeriales* Hairball Sac Fungus

FAMILY 1 *Trichosphaeriaceae* Hairball Sac Fungus
GENUS *Trichosphaeria* Hairball Sac Fungus
SPECIES *barbula, caesia, crassipila, notabilis, pilosa, superficialis*
GENUS *Acanthosphaeria* Spiny Hairball Sac Fungus
SPECIES *punctillum, rostrupii*
GENUS *Arthropycnis* Dense Hairball Sac Fungus
SPECIES *praetermissa*
GENUS *Coniobrevicolla* Cone Hairball Sac Fungus
SPECIES *larsenii*
GENUS *Crassochaeta* Crass Bristle Hairball Sac Fungus
SPECIES *fusispora, nigrita*
GENUS *Cresporhaphis* Cree Spore Needle Hairball Sac Fungus
SPECIES *acerina, macrospora, muelleri, pinicola, rhoina, ulmi, wienkampii*
GENUS *Eriosphaeria* Woolly Hairball Sac Fungus
SPECIES *aggregata, investans, membranacea, subtomentosa, vermicularia*
GENUS *Khuskia* Khusk Hairball Sac Fungus
SPECIES *oryzae*
GENUS *Koorchaloma* Koorchal Hairball Sac Fungus
SPECIES *okamurae*
GENUS *Nigrospora* Black Spore Hairball Sac Fungus
SPECIES *musae, panici, sacchari, vietnamensis*
GENUS *Pseudogliomastix* Glow Mastigote Hairball Sac Fungus
SPECIES *protea*
GENUS *Setocampanula* Bristle Bell Hairball Sac Fungus
SPECIES *taiwanensis*

FAMILY 2 *Helminthosphaeriaceae* Worm Hairball Sac Fungus
GENUS *Helminthosphaeria* Worm Hairball Sac Fungus
SPECIES *clavariarum, corticiorum, hyphodermae*
GENUS *Diplococcium* Diploid Round Worm Hairball Sac Fungus
SPECIES *asperum, clarkii, lawrencei, mirabile, spicatum*
GENUS *Spadicoides* Spadix Worm Hairball Sac Fungus
SPECIES *atra, bina, grovei, obovata, ramosa, stoveri, verrucosa, xylogena*
GENUS *Tengiomyces* Teng Worm Hairball Sac Fungus
SPECIES *indicus*

ORDER 17 *Xylariales* Wood Sac Fungus

FAMILY 1 *Xylariaceae* Wood Sac Fungus
GENUS *Xylaria* Wood Sac Fungus
SPECIES *albisquamula, apiculata, arbuscula, bulbosa, carpophila, coremiifera, corniformis, cornu-damae, culleniae, digitata, duranii, enteroleuca, filiformis, fioriana, fraseri, friesii, furcata, guepini, hippotrichoides, hypoxylon, jaliscoensis, kaumanae, liquidambaris, longipes, magnoliae, mali, maraca, mellisii, michoacana, moliwensis, monstrosa, oxyacanthae, pisoniae, polymorpha, quercinophila, rhopaloides, sicula, subcoccophora, tumulosa, umbonata, vaporaria*
VARIETIES *furcata furcata, furcata hirsuta, magnoliae magnoliae, magnoliae microspora, sicula major*
GENUS *Acanthodochium* Spiny Doch Wood Sac Fungus
SPECIES *collodisculae*
GENUS *Adomia* Adom Wood Sac Fungus
SPECIES *avicenniae*
GENUS *Amphirosellinia* Amphir Sellin Wood Sac Fungus
SPECIES *evansii, fushanensis, nigrospora, quercina, tennesseensis*
GENUS *Amphisphaerella* Dual Sphere Wood Sac Fungus
SPECIES *dispersella, xylostei*
GENUS *Annulohypoxylon* Annular Hyponic Wood Sac Fungus
SPECIES *annulatum, archeri, atroroseum, bovei, cohaerens, discophorum, elevatidiscus, gombakense, hians, ilanense, leptascum, michelianum, microcarpum, minutellum, moriforme, multiforme, nothofagi, pouceanum, pseudostipitatum, purpureonitens, pyriforme, squamulosum, stygium, thouarsianum, truncatum, urceolatum*
VARIETIES *bovei bovei, bovei microspora, leptascum leptascum, leptascum macrosporum, moriforme moriforme, moriforme microdiscus, multiforme multiforme, multiforme alaskense, stygium stygium, stygium annulatum, thouarsianum thouarsianum, thouarsianum macrosporum*
GENUS *Anthostomella* Anther Stoma Wood Sac Fungus
SPECIES *acuminata, alchemillae, appendiculosa, applanata, aquatica, arenaria, atroalba, belalongensis, bipileatus, birima, brabeji, bruneiensis, caffrariae, calamicola, calligoni, caricis, castanopsis, chionostoma, clypeata, clypeoides, clypeosa, colligata, conorum, culmicola, cynaroides, cyperaceicola, daemonoropis, fausta,*

flagellariae, formosa, francisiae, frondicola, fuegiana, irregularispora, kapitiae, leptospora, leucospermi, licualicola, limitata, livistonicola, longa, longispora, lucens, lugubris, manawatua, meerensis, minutoides, miscanthea, myricae, nannorrhopis, notabilis, nypae, nypensis, nypicola, oblongata, okatina, oraniopsis, palmae, palmaria, pedemontana, petrinenesis, phaeosticta, phoenicis, pinangae, poecila, proteae, pullulans, punctulata, raphiae, rattanicola, reniformis, rhaphidophylli, rostrispora, rubicola, sabinianae, scirpi, scotina, semitecta, spartii, spiralis, spissitecta, taiwanensis, taiwaniana, theobromina, tomicoides, tomicum, torosa, trachycarpi, tumulosa, unguiculata, uniseriata, variabilis, xuanenensis, yushaniae, zongluensis

VARIETIES *formosa taxi*

GENUS *Appendicospora* — Appendage Spore Wood Sac Fungus

SPECIES *coryphae, hongkongensis*

GENUS *Appendixia* — Appendage Wood Sac Fungus

SPECIES *closterium*

GENUS *Ascotricha* — Hairy Wood Sac Fungus

SPECIES *amphitricha, bosei, chartarum, distans, erinacea, guamensis, lusitanica, novae-caledoniae, pusilla, xylina*

VARIETIES *chartarum orientalis*

GENUS *Ascovirgaria* — Virgo Wood Sac Fungus

SPECIES *occulta*

GENUS *Astrocystis* — Star Cyst Wood Sac Fungus

SPECIES *cyatheae, fimbriata, nypae, rudis, selangorensis, sinensis*

GENUS *Barrmaelia* — Barrmael Wood Sac Fungus

SPECIES *macrospora, moravica, oxyacanthae, picacea, pseudobombarda, sustenta*

GENUS *Basifimbria* — Basal Fringed Wood Sac Fungus

SPECIES *spinosa, succinea*

GENUS *Biporispora* — Bipore Spore Wood Sac Fungus

SPECIES *europaea*

GENUS *Biscogniauxia* — Cogniaux Wood Sac Fungus

SPECIES *africana, albosticta, ambiens, anceps, arima, atropunctata, baileyi, bartholomaei, capnodes, cinereolilacina, citriformis, communapertura, cylindrispora, dennisii, divergens, doidgeae, formosana, fuscella, grenadensis, kalchbrenneri, latirima, marginata, mediterranea, mucigera, nawawii, nothofagi, nummularia, philippinensis, pithodes, plana, plumbea, querna, repanda, reticulospora, schweinitzii, simplicior, sinuosa, uniapiculata, viscosicentra, waitpela, weldenii, zelandica*

VARIETIES *albosticta albosticta, albosticta orientalis, atropunctata atropunctata, atropunctata intermedia, atropunctata maritima, capnodes capnodes, capnodes limonispora, capnodes rumpens, capnodes theissenii, citriformis citriformis, citriformis macrospora, formosana formosana, formosana kentingensis, mediterranea mediterranea, mediterranea macrospora, mediterranea microspora, philippinensis philippinensis, philippinensis microspora, uniapiculata uniapiculata, uniapiculata indica, uniapiculata macrospora, weldenii weldenii, weldenii microspora*

GENUS *Calceomyces* — Calce Wood Sac Fungus

SPECIES *lacunosus*

GENUS *Camillea* — Camille Wood Sac Fungus

SPECIES *coroniformis, heterostoma, labiatirima*

VARIETIES *heterostoma heterostoma, heterostoma macrospora, heterostoma microspora*

GENUS *Cannonia* — Cannon Wood Sac Fungus

SPECIES *australis*

GENUS *Ceratocladium* — Horn Branched Wood Sac Fungus

SPECIES *microspermum*

GENUS *Chitonospora* — Chiton Spore Wood Sac Fungus

SPECIES *ammophila*

GENUS *Collodiscula* — Glue Disc Wood Sac Fungus

SPECIES *japonica*

GENUS *Creosphaeria* — Cree Sphere Wood Sac Fungus

SPECIES *sassafras*

GENUS *Cyanopulvis* — Cyan Powder Wood Sac Fungus

SPECIES *australiensis*

GENUS *Daldinia* — Daldin Wood Sac Fungus

SPECIES *bakeri, bambusicola, brachysperma, caldariorum, childiae, clavata, concentrica, cudonia, cuprea, eschscholzii, fissa, gelatinosa, graminis, grandis, lloydii, loculata, macrospora, mexicana, occidentalis, petriniae, sacchari, simulans, singularis, vernicosa*

VARIETIES *concentrica fraxineum*

GENUS *Diamantinia* — Diamond Wood Sac Fungus

SPECIES *citrina*

GENUS *Dicyma* — Dual Cyme Wood Sac Fungus

SPECIES *funiculosa, ovalispora, peruviana, vesiculifera*

GENUS *Discoxylaria* — Disc Wood Sac Fungus

SPECIES *myrmecophila*

GENUS *Distorimula* — Distal Rim Wood Sac Fungus

SPECIES *mexicana*
GENUS *Emarcea* Marked Wood Sac Fungus
SPECIES *castanopsidicola*
GENUS *Engleromyces* Engler Wood Sac Fungus
SPECIES *goetzei*
GENUS *Entoleuca* Entic Light Wood Sac Fungus
SPECIES *callimorpha, ellisii, mammata*
GENUS *Euepixylon* Pixylon Wood Sac Fungus
SPECIES *udum*
GENUS *Fasciatispora* Bound Spore Wood Sac Fungus
SPECIES *petrakii*
GENUS *Flagellosphaeria* Flagella Sphere Wood Sac Fungus
SPECIES *polytrichospora*
GENUS *Frondispora* Frond Spore Wood Sac Fungus
SPECIES *bicalcarata*
GENUS *Gigantospora* Giant Spore Wood Sac Fungus
SPECIES *gigaspora*
GENUS *Graphostroma* Graphic Stroma Wood Sac Fungus
SPECIES *platystoma*
GENUS *Guestia* Guest Wood Sac Fungus
SPECIES *gonetropospora*
GENUS *Hadrotrichum* Hadron Hairy Wood Sac Fungus
SPECIES *cedrelae, virescens*
GENUS *Halorosellinia* Halogen Sellin Wood Sac Fungus
SPECIES *oceanica*
GENUS *Helicogermslita* Helix Germ Slit Wood Sac Fungus
SPECIES *aucklandica, celastri, diversa, fleischhakii, gaudefroyi, gisbornia, johnstonii, mackenziei, valdiviensis*
GENUS *Hypocopra* Hyponic Copper Wood Sac Fungus
SPECIES *brefeldii, capillifera, equorum, fimeti, hypocoproides, merdaria, parvula, planispora, rostrata, stephanophora, stercoraria*
GENUS *Hypoxylon* Hyponic Sharp Wood Sac Fungus
SPECIES *aeruginosum, anthochroum, aucklandiae, begae, brevirimum, brevisporum, californicum, carneum, cercidicola, chathamense, chionostomum., cinnabarinum, commutatum, croceum, crocopeplum, curvirimum, dearnessii, diatrypeoides, dieckmannii, dingleyae, duranii, duriloculatum, erythrostroma, fendleri, ferrugineum, flavoargillaceum, fragiforme, fuhreri, fuscopurpureum, fuscum, fusoideosporum, gilbertsonii, haematostroma, howeanum, hughesii, hypomiltum, intermedium, investiens, jecorinum, julianii, karii, kretzschmarioides, laschii, laurus, lenormandii, lienhwacheense, lividicolor, lividipigmentum, macrocarpum, macrosporum, madangense, megalosporum, moellerianum, monticulosum, munkii, musceum, nicaraguense, nitens, notatum, nothofagicola, novemexicanum, ochraceum, olivicolor, olliforme, papillatum, parksianum, pelliculosum, perforatum, piceum, pilgerianum, placentiforme, polyporus, porphyreum, ravidoroseum, rectangulosporum, retpela, rickii, rubellum, rubigineoareolatum, rubiginosum, rugulosporum, rutilum, salicicola, samuelsii, sclerophaeum, shearii, subalbum, subcorticeum, subcrocopeplum, subgilvum, submonticulosum, subrutiloides, subrutilum, subticinense, symphyon, texcalense, ticinense, tinctor, tortisporum, trugodes, umbilicatum, undulatum, vandervekenii, venezuelense, vinosopulvinatum, vogesiacum*
VARIETIES *aeruginosum aeruginosum, aeruginosum macrosporum, hypomiltum hypomiltum, hypomiltum lavandulocinereum, rubiginosum rubiginosum, rubiginosum microsporum, shearii shearii, shearii minor, subgilvum subgilvum, subgilvum microsporum*
GENUS *Induratia* Indurate Wood Sac Fungus
SPECIES *apiospora*
GENUS *Jumillera* Jumiller Wood Sac Fungus
SPECIES *albida, cinerea, hawaiiensis, hypophlaea, mexicana, punctatobrunnea, viridis*
GENUS *Kretzschmaria* Kretzschmar Wood Sac Fungus
SPECIES *argentinensis, cetrarioides, clavus, curvirima, deusta, eriodendri, guyanensis, lucidula, macrosperma, megalospora, micropus, milleri, neocaledonica, pavimentosa, sandvicense, sigmoidirima, varians, zelandica, zonata*
GENUS *Kretzschmariella* Lesser Kretzschmar Wood Sac Fungus
SPECIES *culmorum*
GENUS *Leiosphaerella* Smooth Sphere Wood Sac Fungus
SPECIES *leonensis, succinea, vexata*
GENUS *Leprieuria* Leprieur Wood Sac Fungus
SPECIES *bacillum*
GENUS *Libertella* Libert Wood Sac Fungus
SPECIES *acerina, betulina, dissepta, heveae, opuli, parva, rosae, salicis*
GENUS *Lindquistia* Lindquist Wood Sac Fungus
SPECIES *indica*
GENUS *Lindquistomyces* Lesser Lindquist Wood Sac Fungus

SPECIES *antarcticus*
GENUS *Linocarpon* Lined Carpel Wood Sac Fungus
SPECIES *elaeidis, pandani*
GENUS *Lopadostoma* Lopad Stoma Wood Sac Fungus
SPECIES *hawaiianum, pouzarii*
GENUS *Manokwaria* Manokwar Wood Sac Fungus
SPECIES *notabilis*
GENUS *Melomastia* Mellow Mastigote Wood Sac Fungus
SPECIES *mastoidea*
GENUS *Monographella* Monograph Wood Sac Fungus
SPECIES *albescens, cucumerina, nivalis, stoveri*
VARIETIES *nivalis nivalis, nivalis neglecta*
GENUS *Nemania* Neman Wood Sac Fungus
SPECIES *aenea, albocincta, angusta, atropurpurea, beaumontii, bipapillata, carbonacea, caries, chestersii, chrysoconia, circostoma, confluens, creoleuca, diffusa, effusa, gwyneddii, illita, immersidiscus, kellermanii, latissima, macrocarpa, maculosa, maritima, memorabilis, minutula, nummularioides, pouzarii, quadrata, ravenelii, serpens, subaenea, venezuelensis*
VARIETIES *aenea aenea, aenea aureolutea, aenea macrospora, chestersii chestersii, chestersii microspora, serpens serpens, serpens colliculosa, serpens hydnicola*
GENUS *Nodulisporium* Nodule Spore Wood Sac Fungus
SPECIES *acervatum, africanum, cecidiogenes, cylindroconium, didymosporum, geochroum, gregarium, griseobrunneum, hinnuleum, honiaraense, indicum, melonis, tuberum, umbrinum, verrucosum*
GENUS *Nummularia* Money Wood Sac Fungus
SPECIES *gigas, succenturiata*
GENUS *Obolarina* Obolar Wood Sac Fungus
SPECIES *dryophila*
GENUS *Occultitheca* Occult Theca costaricensisWood Sac Fungus
SPECIES *costaricensis*
GENUS *Ophiorosellinia* Serpent Sellin Wood Sac Fungus
SPECIES *costaricensis*
GENUS *Oxydothis* Sharp Doth Wood Sac Fungus
SPECIES *pandanicola, sabalensis*
GENUS *Padixonia* Padixon Wood Sac Fungus
SPECIES *bispora*
GENUS *Paramphisphaeria* Dual Sphere Wood Sac Fungus
SPECIES *costaricensis*
GENUS *Pareutypella* Typical Wood Sac Fungus
SPECIES *nematoceras, sulcata*
GENUS *Penzigia* Penzig Wood Sac Fungus
SPECIES *berteri, compuncta*
GENUS *Phaeotrichosphaeria* Dusky Hairy Sphere Wood Sac Fungus
SPECIES *britannica*
GENUS *Phomatospora* Foam Spore Wood Sac Fungus
SPECIES *arenaria, argentinae, berkeleyi, coprophila, dinemasporium, endopteris, gelatinospora, hederae, ribesia, sphaerulina, striatispora*
GENUS *Pidoplitchkoviella* Pidoplitchkov Wood Sac Fungus
SPECIES *terricola*
GENUS *Podosordaria* Pod Sordid Wood Sac Fungus
SPECIES *jugoyasan, leporina, mexicana, pedunculata, tulasnei*
GENUS *Poroleprieuria* Pore Leprieur Wood Sac Fungus
SPECIES *rogersii*
GENUS *Poronia* Poron Wood Sac Fungus
SPECIES *punctata*
GENUS *Pulmosphaeria* Pulm Sphere Wood Sac Fungus
SPECIES *archontophoenicis*
GENUS *Pulveria* Powder Wood Sac Fungus
SPECIES *porrecta*
GENUS *Reticulosphaeria* Netted Sphere Wood Sac Fungus
SPECIES *indica*
GENUS *Rhopalostroma* Club Stroma Wood Sac Fungus
SPECIES *africanum, angolense, dennisii, gracile, hawksworthii, indica, indicum, kanyae, lekae, luzonense, sphaerocephalum*
GENUS *Rosellinia* Rosellin Wood Sac Fungus
SPECIES *abscondita, anthostomoides, aquila, arcuata, brevifissurata, britannica, bunodes, buxi, callosa, capetribulensis, coffeicola, communis, conferta, desmazieri, dingleyae, formosana, freycinetiae, gisbornia, helvetica, hughesii, insularis, johnstonii, laminariana, lignicola, limoniispora, mammiformis, mammoidea,*

mastoidea, minor, moroides, musispora, necatrix, nectrioides, nothofagi, novae-zelandiae, obliquata, palmae, papaverea, pepo, pyramidalis, quercina, rhopalostylidicola, samuelsii, thelena, tunicata

GENUS *Sabalicola* — Sabal Wood Sac Fungus
SPECIES *sabalensioides*

GENUS *Seynesia* — Seynes Wood Sac Fungus
SPECIES *livistonae*

GENUS *Spirodecospora* — Spiral Decospore Wood Sac Fungus
SPECIES *bambusicola, melnikii*

GENUS *Stilbohypoxylon* — Shining Hyponix Sharp Wood Sac Fungus
SPECIES *elaeicola, hypoxylinum, ignobile, immundum, macrosporum, minus, moelleri, novae-zelandiae, quisquiliarum, samuelsii, theissenii*

GENUS *Striatodecospora* — Striped Decospore Wood Sac Fungus
SPECIES *bambusae*

GENUS *Theissenia* — Theissen Wood Sac Fungus
SPECIES *cinerea, eurima, pyrenocrata*

GENUS *Triplicaria* — Triplicate Wood Sac Fungus
SPECIES *hypoxyloides*

GENUS *Urosporellopsis* — Urospore Wood Sac Fungus
SPECIES *taiwanensis*

GENUS *Vivantia* — Vivant Wood Sac Fungus
SPECIES *guadalupensis*

GENUS *Wawelia* — Wawel Wood Sac Fungus
SPECIES *argentea, microspora, octospora, regia*

GENUS *Whalleya* — Whalley Wood Sac Fungus
SPECIES *maculata, microplaca*

GENUS *Xylocladium* — Branched Wood Sac Fungus
SPECIES *clautriavii, claviforme*

GENUS *Xylocoremium* — Broom Wood Sac Fungus
SPECIES *flabelliforme*

FAMILY 2 *Amphisphaeriaceae* — Dual Sphere Wood Sac Fungus

GENUS *Amphisphaeria* — Dual Sphere Wood Sac Fungus
SPECIES *bufonia, culmicola, melanommoides, millepunctata, paedida, ulmicola, umbrina, ventosaria, vibratilis*

GENUS *Beltraniella* — Beltran Dual Sphere Wood Sac Fungus
SPECIES *amoena, clavatophora, humicola, odinae, pirozynskii, portoricensis*

GENUS *Blogiascospora* — Blog Spore Dual Sphere Wood Sac Fungus
SPECIES *marginata*

GENUS *Brachysporiella* — Short Spore Dual Sphere Wood Sac Fungus
SPECIES *gayana*

GENUS *Broomella* — Broom Dual Sphere Wood Sac Fungus
SPECIES *leptogiicola, montaniensis, vitalbae*

GENUS *Ceratosporium* — Horn Spore Dual Sphere Wood Sac Fungus
SPECIES *fuscescens, rilstonii*

GENUS *Coryneopsis* — Club Dual Sphere Wood Sac Fungus
SPECIES *canina, corni-albae, henriquesiana, microsticta, rubi, tamaricis*

GENUS *Cytoplea* — Cell Floating Dual Sphere Wood Sac Fungus
SPECIES *arundinacea, hederae*

GENUS *Diploceras* — Diploid Cherry Dual Sphere Wood Sac Fungus
SPECIES *hypericinum*

GENUS *Discostroma* — Disc Stroma Dual Sphere Wood Sac Fungus
SPECIES *corticola, stoneae, strobiligenum, tostum*

GENUS *Discostromopsis* — Lesser Disc Stroma Dual Sphere Wood Sac Fungus
SPECIES *stoneae*

GENUS *Edmundmasonia* — Edmund Mason Dual Sphere Wood Sac Fungus
SPECIES *pulchra*

GENUS *Ellurema* — Ellure Dual Sphere Wood Sac Fungus
SPECIES *indica*

GENUS *Hymenopleella* — Hymen Floating Dual Sphere Wood Sac Fungus
SPECIES *hippophaës*

GENUS *Lepteutypa* — Thin Dual Sphere Wood Sac Fungus
SPECIES *cupressi, hippophaës, podocarpi*

GENUS *Monotosporella* — Monotospore Dual Sphere Wood Sac Fungus
SPECIES *setosa, tuberculata*

GENUS *Neobroomella* — Modern Broom Dual Sphere Wood Sac Fungus
SPECIES *ciliata*

GENUS *Paracainiella* — Cain Dual Sphere Wood Sac Fungus
SPECIES *dryadis*

GENUS *Paradidymella* Mellow Dual Sphere Wood Sac Fungus
SPECIES *clarkii, holci*
VARIETIES *holci moliniae*
GENUS *Pestalosphaeria* Least Pestal Dual Sphere Wood Sac Fungus
SPECIES *elaeidis*
GENUS *Pestalotia* Pestal Dual Sphere Wood Sac Fungus
SPECIES *acaciae, algeriensis, bicolor, brassicae, caroliniana, cocculi, coffeae, diospyri, elasticae, eugeniae, heterocornis, juniperi, kawakamii, lignicola, longisetula, macrotricha, matildae, monochaeta, olivacea, pauciseta, pezizoides, pitospora, populi-nigrae, rhododendri, rosae, stevensonii, subcuticularis, thujae, torrendii, vaccinii, zonata*
GENUS *Pestalotiopsis* Lesser Pestal Dual Sphere Wood Sac Fungus
SPECIES *adusta, aeruginea, annulata, aquatica, baarnensis, breviseta, caudata, coffeae, crassiuscula, cruenta, disseminata, eupyrena, fici, filisetula, funerea, funereoides, glandicola, gracilis, guepinii, japonica, leucothoes, macrospora, maculans, maculiformans, mangiferae, mayumbensis, microspora, monochaetioides, nattrassii, neglecta, oxyanthi, paeoniae, palmarum, papposa, psidii, royenae, steyaertii, stictica, sydowiana, theae, triseta, versicolor, virgatula, westerdijkii*
VARIETIES *funerea discolor*
GENUS *Seimatosporium* Seam Spore Dual Sphere Wood Sac Fungus
SPECIES *alneum, anomalum, arbuti, caudatum, dilophosporum, discosioides, elegans, falcatum, foliicola, fusisporum, glandigenum, grevilleae, hakeae, lonicerae, macrospermum, mariae, parasiticum, ribis-alpini, rosarum, vaccinii*
GENUS *Seiridium* Silky Dual Sphere Wood Sac Fungus
SPECIES *cardinale, eucalypti, intermedium, papillatum*
GENUS *Truncatella* Truncated Dual Sphere Wood Sac Fungus
SPECIES *angustata, bella, conorum-piceae, hartigii, laurocerasi, truncata*

FAMILY 3 *Cainiaceae* Cain Wood Sac Fungus
GENUS *Cainia* Cain Wood Sac Fungus
SPECIES *desmazieri, graminis*
GENUS *Atrotorquata* Dark Torqued Cain Wood Sac Fungus
SPECIES *lineata*

FAMILY 4 *Clypeosphaeriaceae* Shield Sphere Wood Sac Fungus
GENUS *Clypeosphaeria* Shield Sphere Wood Sac Fungus
SPECIES *mamillana, perfidiosa, phillyreae*
GENUS *Apiorhynchostoma* Parsley Snout Stoma Shield Sphere Wood Sac Fungus
SPECIES *curreyi*
GENUS *Brunneiapiospora* Brown Parsley Spore Shield Sphere Wood Sac Fungus
SPECIES *aequatoriensis, australiensis, daemonoropis, deightoniella, javensis, jesseniae*
GENUS *Clypeophysalospora* Bladder Spore Shield Sphere Wood Sac Fungus
SPECIES *latitans*
GENUS *Duradens* Hard Shield Sphere Wood Sac Fungus
SPECIES *lignicola*

FAMILY 5 *Diatrypaceae* Diad Trype Wood Sac Fungus
GENUS *Diatrype* Diad Trype Wood Sac Fungus
SPECIES *berberidis, bullata, decorticata, disciformis, polycocca, sordida, spilomea, stigma, undulata*
GENUS *Anthostoma* Anther Stoma Diad Trype Wood Sac Fungus
SPECIES *amoenum, cubiculare, decipiens, denigrans, dryophilum, gastrinoides, gastrinum, melanotes, plowrightii, saprophilum, turgidum*
GENUS *Cryptosphaeria* Cryptic Sphere Diad Trype Wood Sac Fungus
SPECIES *eunomia, ligniota, moravica, ocellata, subcutanea*
VARIETIES *eunomia eunomia, eunomia fraxini, moravica hederae, moravica petrakii*
GENUS *Diatrypella* Lesser Diad Trype Wood Sac Fungus
SPECIES *cephalanthi, discoidea, exigua, favacea, pulvinata, quercina, rhois*
GENUS *Echinomyces* Urchin Diad Trype Wood Sac Fungus
SPECIES *echidna, obesa*
GENUS *Eutypa* Good Diad Trype Wood Sac Fungus
SPECIES *acerina, aspera, astroidea, crustata, flavovirens, hydnoidea, laevata, lata, leioplaca, leptoplaca, ludibunda, maura, polycocca, prorumpens, quercicola, scabrosa, sparsa, spinosa, subtecta, tetragona, ulicis, velutina*
GENUS *Eutypella* Lesser Good Diad Trype Wood Sac Fungus
SPECIES *acericola, alnifraga, alsophila, caricae, cerviculata, extensa, grandis, leprosa, microspora, parasitica, prunastri, quaternata, russodes, scoparia, sorbi, stellulata, tetraploa, virescens*
VARIETIES *quaternata minor*
GENUS *Fassia* Fass Diad Trype Wood Sac Fungus
SPECIES *scabrosa*
GENUS *Leptoperidia* Thin Envelope Diad Trype Wood Sac Fungus
SPECIES *applanata, asperrima, macropunctata, trifida*

GENUS *Peroneutypa* — Peron Good Diad Trype Wood Sac Fungus
SPECIES *corniculata*
GENUS *Quaternaria* — Quaternian Diad Trype Wood Sac Fungus
SPECIES *abnormis, dissepta*
GENUS *Rostronitschkia* — Rostronitschky Diad Trype Wood Sac Fungus
SPECIES *nervincola*

FAMILY 7 *Hyponectriaceae* — Hyponic Cornicle Wood Sac Fungus
GENUS *Hyponectria* — Hyponic Cornicle Wood Sac Fungus
SPECIES *buxi, cookeana*
GENUS *Amphorula* — Amphor Hyponic Cornicle Wood Sac Fungus
SPECIES *sachalinensis*
GENUS *Anisostomula* — Anise Stoma Hyponic Cornicle Wood Sac Fungus
SPECIES *areola*
GENUS *Aquadulciospora* — Aquadulce Spore Hyponic Cornicle Wood Sac Fungus
SPECIES *rhomboidia*
GENUS *Arwidssonia* — Arwidsson Hyponic Cornicle Wood Sac Fungus
SPECIES *empetri, loiseleuriae*
GENUS *Cainiella* — Cain Hyponic Cornicle Wood Sac Fungus
SPECIES *borealis, johansonii*
GENUS *Ceriophora* — Wax Bearing Hyponic Cornicle Wood Sac Fungus
SPECIES *palustris*
GENUS *Ceriospora* — Wax Spore Hyponic Cornicle Wood Sac Fungus
SPECIES *caudae-suis, polygonacearum*
GENUS *Chaetapiospora* — Bristle Parsley Spore Hyponic Cornicle Wood Sac Fungus
SPECIES *rhododendri*
GENUS *Chamaeascus* — Chamise Hyponic Cornicle Wood Sac Fungus
SPECIES *arcticus*
GENUS *Exarmidium* — Exarm Hyponic Cornicle Wood Sac Fungus
SPECIES *inclusum*
GENUS *Frondicola* — Frond Hyponic Cornicle Wood Sac Fungus
SPECIES *tunitricuspis*
GENUS *Microdochium* — Doch Hyponic Cornicle Wood Sac Fungus
SPECIES *bolleyi, caespitosum, dimerum,fusarioides, gracile, griseum, panattonianum, phragmitis, queenslandicum, sclerotiorum, tainanense, tripsaci*
VARIETIES *dimerum dimerum, dimerum nectrioides, dimerum pusillum, dimerum violaceum*
GENUS *Neoflageoletia* — Modern Flageolet Wood Sac Fungus
SPECIES *bambusina*
GENUS *Pellucida* — Pellucid Hyponic Cornicle Wood Sac Fungus
SPECIES *pendulina*
GENUS *Physalospora* — Bladder Spore Hyponic Cornicle Wood Sac Fungus
SPECIES *alpestris, anamalaiensis, aquatica, arctostaphyli, corni, empetri, ephedrae, eucalypti, euphorbiae, lonicerae, rhododendri, scirpi, vaccinii, vitis-idaeae*
VARIETIES *scirpi minus*
GENUS *Pseudomassaria* — Massar Hyponic Cornicle Wood Sac Fungus
SPECIES *chondrospora, corni, islandica, lycopodina, sepincoliformis, thistletonia, vaccinii*

CLASS 621.2 *Ascomussomycetes* Sac Mushroom

ORDER 1 *Pezizamussales* Peziz Sac Mushroom

FAMILY 1 *Discinaceae* — False Morel Peziz Sac Mushroom
GENUS *Discina* — False Morel Peziz Sac Mushroom
SPECIES ***ancilis***(Edible)
VARIETIES ***ancilis perlata***(Edible)
GENUS *Gymnohydnotrya* — Nude False Morel Peziz Sac Mushroom
SPECIES *australiana, echinulata, ellipsospora*
GENUS *Gyromitra* — Gyre False Morel Peziz Sac Mushroom
SPECIES *ambigua*(**Poison**), *brunnea*(**Poison**), *bubakii*, ***caroliniana***(Edible), *esculenta*(**Poison**), ***gigas***(Edible Unripe & Boiled), *infula*(**Poison**), *leucoxantha*
GENUS *Physomitra* — (Synonym for *Gyromitra*)
GENUS *Pseudorhizina* — Rhizid False Morel Peziz Sac Mushroom
SPECIES *californica*

FAMILY 2 *Helvellaceae* — Lorchel Peziz Sac Mushroom
GENUS *Helvella* — Lorchel Peziz Sac Mushroom

SPECIES ***acetabulum***(**Poison** Raw, Edible Well Cooked), ***albella***(Edible), *albipes*, ***atra***(Edible), *corium, costifera*, ***crispa***(Edible), *cupuliformis, dissingii*, ***elastica***(Edible), *ephippium, fusca*, ***lacunosa***(Edible), *latispora, leucomelaena, leucopus, macropus*, ***monachella***(Edible), *pezizoides, phlebophora, queletii, silvicola, solitaria*

GENUS *Acetabula* (Synonym for *Helvella*)

GENUS *Balsamia* Balsam Lorchel Peziz Sac Mushroom

SPECIES *fragiformis, platyspora, vulgaris*

GENUS *Barssia* Barss Lorchel Peziz Sac Mushroom

SPECIES *oregonensis, yezomontana*

GENUS *Cidaris* Cidar Lorchel Peziz Sac Mushroom

SPECIES ***caroliniana***(Edible)

GENUS *Cyathipodia* (Synonym for *Helvella*)

GENUS *Gyrocratera* (Synonym for *Hydnotrya*)

GENUS *Hydnotrya* Truffel Lorchel Peziz Sac Mushroom

SPECIES *confusa, cubispora, michaelis, tulasnei*

VARIETIES *michaelis sabuletorum*

GENUS *Macropodia* (Synonym for *Helvella*)

GENUS *Picoa* Pico Lorchel Peziz Sac Mushroom

SPECIES *juniperi, lefebvrei, melospora, pachyascus*

GENUS *Pindara* Pindar Lorchel Peziz Sac Mushroom

SPECIES *terrestris*

GENUS *Underwoodia* Underwood Lorchel Peziz Sac Mushroom

SPECIES *columnaris*

FAMILY 3 *Morchellaceae* Morel Peziz Sac Mushroom

GENUS *Morchella* Morel Peziz Sac Mushroom

SPECIES *ambigua*, ***angusticeps***(Edible), ***costata***(Edible), ***crassipes***(Edible), ***deliciosa***(Edible), ***elata***(Commercial Edible), ***esculenta***(Commercial Edible), ***hortensis***(Edible), *patula, smithiana, vaporaria*, ***vulgaris***(Commercial Edible)

VARIETIES ***elata purpurascens, esculenta rotunda, esculenta umbrina, vulgaris conica***

GENUS *Disciotis* Disc Morel Peziz Sac Mushroom

SPECIES *maturescens*, ***venosa***(Edible)

VARIETIES *venosa reticulata*

GENUS *Mitrophora* Mitre Morel Peziz Sac Mushroom

SPECIES *fusca, gigas, patula*, ***semilibera***(Edible)

VARIETIES ***semilibera hybrida***(Edible), ***semilibera rimosipes***(Edible)

GENUS *Ptychoverpa* Folded Bell Morel Peziz Sac Mushroom

SPECIES ***bohemica***(Edible)

GENUS *Verpa* Bell Morel Peziz Sac Mushroom

SPECIES ***conica***(Edible), ***digitaliformis***(Edible), *krombholzii, speciosa*

VARIETIES *conica conica, conica relhanii, krombholzii rufipes*

FAMILY 4 *Tuberaceae* Truffle Peziz Sac Mushroom

GENUS *Tuber* Truffle Peziz Sac Mushroom

SPECIES ***aestivum***(Edible), *albidum, bituminatum, bonnetii, borchii*, ***brumale***(Edible), *dryophilum, excavatum, ferrugineum, foetidum, macrosporum, maculatum*, ***magnatum***(Commercial Edible), ***melanosporum***(Commercial Edible), *nitidum, puberulum, rapaeodorum, rufum, scleroneuron*

GENUS *Choiromyces* Choir Truffle Peziz Sac Mushroom

SPECIES *alveolatus, meandriformis*

GENUS *Reddellomyces* Reddell Truffle Peziz Sac Mushroom

SPECIES *donkii, magnisporus, parvulosporus, westraliensis*

CLASS 621.3 *Lichinomycetes* Lichen

ORDER 1 *Lichinales* Lichen

FAMILY 01 *Lichinaceae* Lichen

GENUS *Lichina* Seaweed Lichen

SPECIES *antarctica, confinis, minutissima, pygmaea, tasmanica*, *willeyi*

GENUS *Anema* Anemic Lichen

SPECIES *dodgei*

GENUS *Cryptothele* Cryptic Nipple Lichen

SPECIES *granuliformis, permiscens, rhodosticta*

GENUS *Digitothyrea* Finger Thyre Lichen

SPECIES *divergens, polyglossa, rotundata*

GENUS *Edwardiella* Edward Lichen

SPECIES *mirabilis*

GENUS *Ephebe* Shag Lichen

SPECIES *americana, epheboides, fruticosa, hispidula, lanata, ocellata, perspinulosa, solida, tasmanica*

GENUS *Ephebeia* (Synonym for *Ephebe*)
GENUS *Euopsis* Good Lichen
SPECIES *granatina, pulvinata*
GENUS *Finkia* Fink Lichen
SPECIES *portoricensis*
GENUS *Forssellia* (Synonym for *Lichinella*)
GENUS *Gonohymenia* (Synonym for *Lichina*)
GENUS *Intralichen* Intra Lichen
SPECIES *christiansenii, lichenum*
GENUS *Kohlmeyera* (Synonym for *Turgidosculum*)
GENUS *Lemmopsis* Lesser Husk Lichen
SPECIES *arnoldiana, oblongans*
GENUS *Lempholemma* Husk Lichen
SPECIES *botryosum, chalazanum, cladodes, intricatum, isidioides, obliquepeltatum, polyanthes, radiatum, umbella, vesiculiferum*
GENUS *Lichinella* Licorice Jelly Lichen
SPECIES *americana, cribellifera, melamphylla, minnesotensis, nigritella, stipatula*
GENUS *Lichenobactridium* Rod Lichen
SPECIES *lecanorae, pertusariae*
GENUS *Lichenochora* Charming Lichen
SPECIES *aprica, bellemerei, clauzadei, coarctatae, collematum, constrictella, epidesertorum, epifulgens, epimarmorata, epinashii, gahavisukae, galligena, heppiae, inconspicua, lecidellae, mediterraneae, obscuroides, polycoccoides, pyrenodesmiae, sinapispermae, thallina, verrucicola, wasseri, weillii, xanthoriae*
GENUS *Lichenoconium* Cone Lichen
SPECIES *cargillianum, erodens, lecanorae, parasiticum, pyxidatae, usneae, xanthoriae*
GENUS *Lichenodiplis* Diple Lichen
SPECIES *lecanorae, lichenicola*
GENUS *Lichenopeltella* Lesser Pelt Lichen
SPECIES *alpestris, ammophilae, cetrariicola, coppinsii, cupularum, fimbriata, nigroannulata, norfolciana, palustris, peltigericola, pinophylla, ramalinae, salicis, santessonii*
GENUS *Lichenopuccinia* Puccin Lichen
SPECIES *poeltii*
GENUS *Lichenostella* Star Lichen
SPECIES *hypotrachynae*
GENUS *Lichenosticta* Stick Lichen
SPECIES *alcicornaria*
GENUS *Lichinodium* Odious Lichen
SPECIES *canadense, saxicola, sirosiphoideum*
GENUS *Mastodia* (Synonym for *Turgidosculum*)
GENUS *Mawsonia* Mawson Lichen
SPECIES *harrissonii*
GENUS *Melaspilea* Honey Lichen
SPECIES *amota, arthonioides, atroides, bagliettoana, cinerascens, congregans, congregantula, constrictella, cupularis, deformis, diplasiospora, epigena, gallowayi, gemella, granitophila, interjecta, leciographoides, lentiginosa, lentiginosula, leucina, maculosa, mangrovei, mesophlebia, microcarpa, ochrothalamia, octomera, opegraphoides, proximella, stellaris, tribuloides*
GENUS *Metamelaena* Dark Lichen
SPECIES *melambola*
GENUS *Omphalaria* (Synonym for *Lempholemma, Peccania*)
GENUS *Paulia* Paul Lichen
SPECIES *aldabrensis, caespitosa, pyrenoides*
GENUS *Peltula* Olive Pelt Lichen
SPECIES *bolanderi, clavata, cylindrica, deserticola, euploca, imbricata, impressa, langei, michoacanensis, obscurans, omphaliza, patellata, placodizans, radicata, richardsii, rodriguesii, sonorensis, tortuosa, zahlbruckneri*
GENUS *Phloeopeccania* Flow Peccant Lichen
SPECIES *australiensis*
GENUS *Phylliscidiopsis* Least Phyllode Lichen
SPECIES *abissinica*
GENUS *Phylliscidium* Lesser Phyllode Lichen
SPECIES *monophyllum*
GENUS *Phylliscum* Phyllode Lichen
SPECIES *demangeonii, tenue*
GENUS *Porocyphus* Pore Cypher Lichen
SPECIES *coccodes, kalbarriensis, kenmorensis, leptogiella, lichinelloides, rehmicus*
GENUS *Psorotichia* Scabies Lichen
SPECIES *hassei, diffundens, minuta, nigra, pyrenopsoides, schaereri, segregata, squamulosa*

GENUS *Pterygiopsis* Feather Lichen
SPECIES *atra, australiensis, convoluta, coracodiza, lacustris*
GENUS *Pyrenopsidium* (Synonym for *Pyrenopsis*)
GENUS *Pyrenopsis* Water Crust Lichen
SPECIES *australiensis, compacta, furfurea, fuscoatra, grumulifera, haemalella, haematina, impolita, lecidella, micrococca, phaeococca, phylliscella, polycocca, sanginea, subareolata, subfuliginosa, tasmanica, viridirufa*
GENUS *Rechingeria* (Synonym for *Lichinella*)
GENUS *Staurolemma* Crosshatch Lichen
SPECIES *fruticosum*
GENUS *Stromatella* Bed Lichen
SPECIES *bermudana*
GENUS *Synalissa* Lesion Lichen
SPECIES *ramulosa, symphorea*
GENUS *Thelignya* Theligny Lichen
SPECIES *lignyota*
GENUS *Thermutis* Thermal Lichen
SPECIES *velutina*
GENUS *Thermutopsis* Lesser Thermal Lichen
SPECIES *jamesii*
GENUS *Thrombium* Curdled Lichen
SPECIES *discordans, epigaeum, mongolicum, thelostomum*
GENUS *Thyrea* Thyre Lichen
SPECIES *confusa, girardii, pulvinata*
GENUS *Turgidosculum* Turgid Lichen
SPECIES *complicatulum, ulvae*
GENUS *Zahlbrucknerella* Zahlbruckner Lichen
SPECIES *calcarea, californica, fabispora*

FAMILY 02 *Agyriaceae* Agrarian Lichen
GENUS *Agyrium* Agrarian Lichen
SPECIES *rufum*
GENUS *Ainoa* Ainoan Agrarian Lichen
SPECIES *mooreana*
VARIETIES *mooreana brujeriana*
GENUS *Amylora* Amyl Agrarian Lichen
SPECIES *cervinocuprea*
GENUS *Anamylopsora* Anamyl Agrarian Lichen
SPECIES *pulcherrima*
GENUS *Coppinsia* Coppins Agrarian Lichen
SPECIES *minutissima*
GENUS *Elixia* Elixir Agrarian Lichen
SPECIES *flexella*
GENUS *Hafellnera* Hafellner Agrarian Lichen
SPECIES *parasemella*
GENUS *Lithographa* Script Agrarian Lichen
SPECIES *subantarctica, tesserata*
GENUS *Mosigia* (Synonym for *Rimularia*)
GENUS *Orceolina* Orceolin Agrarian Lichen
SPECIES *kerguelensis*
GENUS *Placopsis* Bullseye Agrarian Lichen
SPECIES *bicolor, brachyloba, clavifera, cribellans, effusa,* ***gelida***(Dye Herb), *lambii, macrophthalma, parellina, perrugosa, rhodocarpa, rhodophthalma, roseonigra*
VARIETIES *parellina microphylla*
GENUS *Placynthiella* Tarspot Agrarian Lichen
SPECIES *dasaea, hyporhoda, icmalea, oligotropha, uliginosa*
GENUS *Ptychographa* Folded Graphic Agrarian Lichen
SPECIES *xylographoides*
GENUS *Rimularia* Rimmed Agrarian Lichen
SPECIES *applanata, australis, badioatra, caeca, exigua, furvella, fuscosora, gibbosa, globulispora, globulosa, gyrizans, hepaticola, impavida, insularis, intercedens, limborina, mullensis, psephota, sphacelata, umbratilis*
VARIETIES *badioatra contractula*
GENUS *Saccomorpha* (Synonym for *Placynthiella*)
GENUS *Sarea* Sare Agrarian Lichen
SPECIES *difformis, resinae*
GENUS *Schaereria* False Map Agrarian Lichen
SPECIES *bullata, cinereorufa, corticola, endocyanea, fuscocinerea, xerophila*

VARIETIES *fuscocinerea fuscocinerea, fuscocinerea sorediata*
GENUS *Trapelia* — Disk Agrarian Lichen
SPECIES *coarctata, corticola, involuta, obtegens, placodioides*
GENUS *Trapeliopsis* — Mottled Disk Agrarian Lichen
SPECIES *aeneofusca, colensoi, congregans, flexuosa, gelatinosa, glaucolepidea, granulosa, percrenata, pseudogranulosa, viridescens, wallrothii*
GENUS *Xylographa* — Script Deadwood Agrarian Lichen
SPECIES *disseminata, hians, opegraphella, parallela, perminuta, trunciseda, vitiligo*
VARIETIES *parallela abietina, parallela laricicola, parallela minutula, parallela scaphoidea, vitiligo spilomatica*

FAMILY 03 *Gloeoheppiaceae* — Glow Pelt Lichen
GENUS *Gloeoheppia* — Glow Pelt Lichen
SPECIES *erosa, polyspora, rugosa*
GENUS *Gudelia* — Gudel Glow Pelt Lichen
SPECIES *mexicana*
GENUS *Pseudopeltula* — Lesser Glow Pelt Lichen
SPECIES *dicyanophora, heppioides, myriocarpa*

FAMILY 04 *Lepraricaceae* — Leper Lichen
GENUS *Lepraria* — Dust Leper Lichen
SPECIES *atlantica, cacuminum, caesioalba, crassissima, eburnea, elobata, incana, jackii, lobificans, neglecta, nivalis, nylanderiana, rigidula, sulphurella, umbricola, usnica*
VARIETIES *cacuminum angardiana, caesioalba borealis, caesioalba zonata, eburnea frigida, lobificans aeruginosa, lobificans finkii, nivalis crassissima, nivalis murorum*
GENUS *Botryolepraria* — Berry Leper Lichen
SPECIES *lesdainii*
GENUS *Cystocoleus* — Velvet Leper Lichen
SPECIES *ebeneus*
GENUS *Leprocaulon* — Cotton Thread Leper Lichen
SPECIES *albicans, arbuscula, gracilescens, microscopicum, pseudoarbuscula, subalbicans*
VARIETIES *microscopicum quisquiliare*
GENUS *Racodium* — Hair Leper Lichen
SPECIES *rupestre*

FAMILY 05 *Lichenotheliaceae* — Nipple Lichen
GENUS *Lichenothelia* — Nipple Lichen
SPECIES *calcarea, convexa, intermixta, metzleri, scopularia*
GENUS *Lichenostigma* — Stigma Nipple Lichen
SPECIES *elongata, maureri, rugosa*

FAMILY 06 *Patellariaceae* — Patella Lichen
GENUS *Patellaria* — Patella Lichen
SPECIES *abbotiana, fuscoater*
GENUS *Arthrorhaphis* — Joint Seam Dot Patella Lichen
SPECIES *aeruginosa, alpina, citrinella, grisea, muddii, vacillans*
VARIETIES *grisea grisea, grisea catolechioides*
GENUS *Baggea* — Bagge Patella Lichen
SPECIES *pachyascus*
GENUS *Endotryblidium* — Endic Tryble Patella Lichen
SPECIES *insculptum*
GENUS *Eutryblidiella* — Least Tryble Patella Lichen
SPECIES *hysterina*
GENUS *Gongylia* — (Synonym for *Arthrorhaphis*)
GENUS *Lahmiomyces* — Lahm Patella Lichen
SPECIES *piceae*
GENUS *Lecanidiella* — Lesser Lecan Patella Lichen
SPECIES *contortae*
GENUS *Lecanidion* — Lecan Patella Lichen
SPECIES *atratum, atroalbum, maurum*
GENUS *Lirellodisca* — Lyre Disc Patella Lichen
SPECIES *pyrenulispora*
GENUS *Murangium* — Murang Patella Lichen
SPECIES *sequoiae*
GENUS *Poetschia* — Poetsch Patella Lichen
SPECIES *cratincola*
GENUS *Rhizodiscina* — Rhizid Disc Patella Lichen
SPECIES *lignyota*
GENUS *Rhytidhysterium* — Wrinkled Bristle Patella Lichen
SPECIES *rufulum*

GENUS *Stratisporella* Layered Pore Patella Lichen
SPECIES *episemoides*
GENUS *Tryblidaria* Tryble Patella Lichen
SPECIES *melaxantha*
GENUS *Tryblidiella* Lesser Tryble Patella Lichen
SPECIES *fusca*

FAMILY 10 *Pertusariaceae* Perforated Lichen
GENUS *Pertusaria* Warty Perforated Lichen
SPECIES *aberrans, alaskensis, albescens, alboaspera, alpina,* ***amara***(Dye Herb), *amarescens, amblyogona, amnicola, aquilonia, arizonica, asiana, aspergilla, atra, atromaculata, barbatica, borealis, boweniana, bryontha, californica, carneopallida, ceylonica, chiodectonoides, cicatricosa, clarkeana, coccodes, commutata, complanata, coniophora, consanguinea, consocians, copiosa,* ***corallina***(Dye Herb), ***coriacea***(Dye Herb), *coronata, dactylina, dehiscens, dissita, disticha, doradorensis, elliptica, epacrospora, epixantha, errinundrensis, erubescens, erythrella, ewersii, excludens, expolita, flavicans, flavida, flavocorallina, flavoisidiata, floridana, follmanniana, geminipara, georgeana, gibberosa, glaucomela, globospora, globularis, glomerata, gundermanica, gymnospora, hadrocarpa, hakkodensis, hartmannii, hemisphaerica, hermaka, howeana, hutchinsiae, hymenea, hypothamnolica, hypoxantha, iners, injuneana, irregularis, isidiosa, jamesii, knightiana, lacerans, lacericans,* ***lactea***(Dye Herb), *lactescens, lavata, lecanina, leiocarpella, leioplaca, leioplacella, leucoplaca, leucosora, leucostigma, leucostomoides, leucothelia, limbata, lophocarpa, lordhowensis, macounii, macra, malabara,* ***maritima***(Dye Herb), *mattogrossensis, melaleucoides, melanochlora, melanospora, microstoma, miniatescens, miscella, modesta, monogona, montpittensis, multipuncta, muricata, nebulosa, neoscotica, nerrigensis, norfolkensis, norstictica, novae-hollandiae, novae-zelandiae, obruta, octomel, oculata, ophthalmiza, orarensis, ostiolata, pallida, papillata, paradoxica, paragibberosa, paratropa, paratuberculifera, patellifera, pertractata, pertusa, pertusella, petrophyes, pilosula, planaica, plittiana, pluripuncta, polythecia, porinella, praetermissa, propinqua, pruinifera, pseudococcodes,* ***pseudocorallina***(Dye Herb), *pseudodactylina, pseudothwaitesii, psoromica, puffina, pulchella, pupillaris, pustulata, pycnothelia, remota, rhexostoma, rigida, rubefacta, salebrosa, saltuensis, santamonicae, saximontana, scaberula, shenandoensis, sinusmexicana, sommerfeltii, sordida, stenhammari, subambigens, subamplicata, subcerussata, subdactylina, subisidiosa, sublacerans, subobducens, suboculata, subpertusa, subplanaica, subrigida, subventosa, sulcata, sydneyensis, tetrathalmia, texana, thamnolica, thiophaninica, thiospoda, thula, thwaitesii, trachyspora, trachythallina, trevethensis, trimera, trochisea, tropica, truncata, umbricola, undulata, urceolaria, valliculata, velata, ventosa, verdonii, verruculifera, virensica, vulpina, waghornei, wulfenioides, xanthodactylina, xanthodes, xanthonaria, xanthoplaca, xanthosorediata, xenismota, xylophyes, zeorina*
VARIETIES *albescens albescens, albescens corallina, albescens corallina-henrici, albescens discoidea, albescens scutellata, amara amara, amara faginea, amara flotowiana, amara isidiata, amara pulvinata, amara slesviciensis, aspergilla dealbata, aspergilla dealbescens, aspergilla leucosora, carneopallida protuberans, chiodectonoides inquinata, chiodectonoides nolens, clarkeana confusa, coccodes petraea, coccodes phymatodes, corallina chionea, corallina dealbata, corallina syncarpa, coronata isidiifera, dehiscens sekikaica, elliptica bispora, erubescens concava, erubescens perfida, excludens monogona, flavicans sulphurea, flavida lutescens, flavida variolosa, gibberosa anarithmetica, gibberosa graphidioides, gibberosa microspora, gibberosa microsporella, gibberosa moffatiana, gibberosa nitidula, gibberosa octospora, gibberosa trypetheliiformis, gibberosa virginea, gibberosa woollsiana, hadrocarpa asperata, hartmannii albinea, hartmannii muelleriana, hartmannii trypetheliiformis, hemisphaerica aspergilla, hemisphaerica hibernica, hemisphaerica speciosa, hutchinsiae panyrga, hymenea carnea, hymenea fallax, hymenea glabrescens, hymenea rupicola, hymenea wulfenii, knightiana whinrayi, lactescens spilomanthodes, leioplaca alpina, leioplaca pustulata, leioplacella confluens, lophocarpa arenacea, lophocarpa diffracta, lophocarpa glebosa, melaleucoides atropunctata, melanospora crassilabra, monogona brachyspora, monogona sorediata, monogona variolosa, multipuncta leptospora, ophthalmiza tenuescens, paradoxica tetraspora, pertusa chioneodes, pertusa colliculosa, pertusa fallax, pertusa leiotera, pertusa leioterella, pertusa rupestris, pertusella plicatula, pertusella straminea, petrophyes leucoxantha, pluripuncta gallica, polythecia constellata, polythecia fastigiata, polythecia sublactea, pseudocorallina ceuthocarpoides, pseudocorallina concreta, pseudocorallina microstictica, pseudocorallina westringii, pupillaris chloropolia, pupillaris planiuscula, pustulata melaleuca, subventosa deficiens, subventosa hypothamnolica, subventosa hypothamnolica-plicatula, subventosa paeminosa, subventosa sorediata, thiospoda leiotera, thiospoda minuta, thiospoda schizostomella, thwaitesii aphelospora, thwaitesii wilsonii, trimera trispora, truncata ascripta, truncata nothofagi, xanthoplaca persulphurata, xanthoplaca sulphurata*
GENUS *Coccotrema* Bud Volcano Perforated Lichen
SPECIES *citrinescens, cucurbitula, porinopsis*
GENUS *Megaspora* Megaspore Perforated Lichen
SPECIES *verrucosa*
GENUS *Melanaria* (Synonym for *Pertusaria*)
GENUS *Ochrolechia* Saucer Perforated Lichen
SPECIES *africana, androgyna, arborea, antillarum, apiculata, arborea, bryophaga, farinacea, frigida, gowardii, grimmiae, groenlandica, gyalectina, gyrophorica, inaequatula, inversa, juvenalis, laevigata, macrosperma, mexicana, microstictoides, montana, oregonensis,* ***pallescens***(Dye Herb), ***parella***(Dye Herb), *pseudopallescens, pseudotartarea, rhamni-purshianae, rhodoleuca, subathallina, subisidiata, subpallescens,*

subplicans, subrhodotropa, subviridis, szatalaensis, ***tartarea***(Dye Herb), *trachydactyla, trochophora, turneri, upsaliensis, weymouthii, xanthostoma*

VARIETIES *frigida frigida, frigida gonatodes, frigida grandiosa, frigida lapuënsis, parella parella, parella pallescens, parella phloeoleuca, subviridis bahusiensis, subviridis yasudae, szatalaensis pallescens, tartarea erichsenii, turneri alboflavescens*

GENUS *Pachyospora* (Synonym for *Megaspora*)
GENUS *Perforaria* (Synonym for *Coccotrema*)
GENUS *Porina* Pimple Perforated Lichen

SPECIES *adflata, aenea, africana, ahlesiana, albicera, aluniticola, amygdalina, aptrootii, athertonii, atlantica, atriceps, atrocoerulea, atropunctata, australiensis, austropacifica, bacillifera, bellendenica, blechnicola, borreri, bryophila, byssophila, canthicarpa, cestrensis, chlorotica, chloroticula, chrysophora, conica, constrictospora, coralloidea, corrugata, corruscans, crassa, curnowii, decrescens, deminuta, distans, dwesica, eminentior, epiphylla, exacta, exocha, farinosa, fluminea, fulvella, fulvula, ginzbergeri, grandis, guaranitica, guentheri, heterocarpa, heterospora, hibernica, howeana, hyperleptalea, impolita, impressa, insueta, interjungens, internigrans, kantvilasii, lectissima, leptalea, leptosperma, limbulata, limitata, linearis, longispora, lucida, malmei, mammillosa, mariae, mastoidea, mastoidella, meridionalis, minutissima, mirabilis, nigrofusca, nitidula, nucula, nuculastrum, ocellatoides, octomera, pallescens, papuensis, pelochroa, perminuta, pulla, rhaphidiophora, rhapidosperma, rhodostoma, riparia, rosei, rubentior, rudiuscula, rufula, salicina, semecarpi, similis, sphaerocephala, subapplanata, subargillacea, subnucula, sudetica, sylvatica, tahitiensis, tasmanica, terrae-reginae, tetracerae, tolgensis, trichothelioides, triseptata, ulceratula, virescens, whinrayi*

VARIETIES *borreri borreri, borreri leptospora, chlorotica chlorotica, chlorotica tenuifera, eminentior eminentior, eminentior sorediifera, guentheri guentheri, guentheri lucens, tetracerae persimilis*

GENUS *Pseudosagedia* (Synonym for *Porina*)
GENUS *Thamnochrolechia* Thamin Perforated Lichen
SPECIES *verticillata*
GENUS *Variolaria* (Synonym for *Pertusaria*)
GENUS *Zamenhofia* (Synonym for *Porina*)

FAMILY 11 *Trichotheliaceae* Hairy Nipple Lichen
GENUS *Trichothelium* Hairy Nipple Lichen
SPECIES *ake-assii, alboatrum, assurgens, horridulum, javanicum, meridionale, oceanicum*
VARIETIES *ake-assii epiphyllum, assurgens nanum, javanicum minus*
GENUS *Clathroporina* Lattice Hairy Nipple Lichen
SPECIES *confinis*
GENUS *Myeloconis* Myel Cone Hairy Nipple Lichen
SPECIES *erumpens, fecunda, guyanensis, parva*
GENUS *Phyllophiale* Saucer Hairy Nipple Lichen
SPECIES *alba*
GENUS *Polycornum* Many Horn Hairy Nipple Lichen
SPECIES *rubrofuscum*

FAMILY 12 *Strigulaceae* Bristle Lichen
GENUS *Strigula* Bristle Lichen

SPECIES *albicascens, americana, antillarum, australiensis, bermudana, concreta, confusa, decipiens, elixii, fossulicola, fractans, indutula, jamesii, janeirensis, macrocarpa, maculata, melanobapha, minutula, multipunctata, natalis, nemathora, nitidula, obducta, oceanica, orbicularis, phaea, phyllogena, platypoda, prasina, rupestris, smaragdula, stigmatella, subelegans, submuriformis, subtilissima, tagananae, taylorii, thelopsidoides, viridiseda, wilsonii*

VARIETIES *jamesii affinis, orbicularis complanata, smaragdula elatior, smaragdula elegans, smaragdula eumorpha, stigmatella stigmatella, stigmatella alpestris, tagananae lateralis*

GENUS *Phylloblastia* Blastid Bristle Lichen
SPECIES *dolichospora*
GENUS *Porinula* Pore Bristle Lichen
SPECIES *setifera*

FAMILY 13 *Thelenellaceae* Thelen Lichen
GENUS *Thelenella* Thelen Lichen

SPECIES *brasiliensis, hassei, inductula, kerguelena, larbalestieri, luridella, marginata, mawsonii, modesta, muscorum, sordidula, sychnogonioides, tasmanica, weberi*

VARIETIES *muscorum muscorum, muscorum octospora, sychnogonioides harrisii*

GENUS *Chromatochlamys* Chromic Claw Thelen Lichen
SPECIES *confusum*
GENUS *Julella* Jule Thelen Lichen
SPECIES *amygdalina, dispora, lactea, sericea, sublactea, vitrispora*
GENUS *Microglaena* (Synonym for *Thelenella*)
GENUS *Polyblastiopsis* Polyblast Thelen Lichen
SPECIES *coarctata, gregantula, nudata, pertusarioidea, tichospora, velata*
GENUS *Protothelenella* Primitive Thelen Lichen
SPECIES *corrosa, leucothelia, santessonii, sphinctrinoidella, sphinctrinoides*

VARIETIES *sphinctrinoides reducta*

FAMILY 14 *Umbilicariaceae* Tripe Lichen

GENUS *Umbilicaria* Tripe Lichen

SPECIES ***americana***(Dye Herb), *angulata, aprina, arctica, caroliniana, cinereorufescens,* ***coriacea***(Dye Herb), *crustulosa, cylindrica, decussata,* ***deusta***(Dye Herb), ***esculenta***(Edible), *grisea, havaasii, hirsuta,* ***hyperborea***(Dye Herb), *krascheninnikovii, lambii, leiocarpa,* ***lyngei***(Dye Herb), ***mammulata***(Dye Herb), ***muhlenbergii***(Edible), *nylanderiana,* ***phaea***(Dye Herb), ***polyphylla***(Dye Herb), ***polyrrhiza***(Dye Herb), ***proboscidea***(Dye Herb), *propagulifera, radicicula, scholanderi, spodochroa, subglabra,* ***torrefacta***(Dye Herb), ***vellea***(Dye Herb), ***virginis***(Dye Herb)

VARIETIES *coriacea rigida, cylindrica cylindrica, cylindrica crinita, cylindrica delisei, cylindrica denticulata, cylindrica denudata, cylindrica exasperata, cylindrica fimbriata, cylindrica laciniata, cylindrica tornata, deusta flocculosa, grisea murina, polyphylla anthracina, polyphylla congregata, polyphylla corrugata, polyphylla glabra, polyphylla lacera, polyphylla monophylla, polyphylla sulcata, polyrrhiza pellita, proboscidea corrugata, proboscidea deplicans, proboscidea exasperata, proboscidea fimbriata, proboscidea mesenteriformis, proboscidea pulla*

GENUS *Actinogyra* (Synonym for *Umbilicaria*)

GENUS *Agyrophora* (Synonym for *Umbilicaria*)

GENUS *Gyrophora* (Synonym for *Umbilicaria*)

GENUS *Lasallia* Toadskin Tripe Lichen

SPECIES ***papulosa***(Dye Herb), ***pensylvanica***(Dye Herb), ***pustulata***(Dye Herb)

FAMILY 15 *Verrucariaceae* Warty Lichen

GENUS *Verrucaria* Warty Lichen

SPECIES *acrotella, aethiobola, aethioboliza, amphibia, amylacea, applanata, aquatilis, aranensis, arctica, aspiciliicola, aucklandica, australiensis, austroalpina, baldensis, bryoctona, bubalina, buelliicola, bulbalina, bulgarica, caerulea, calciseda, calkinsiana, canella, collematodes, compacta, conturmatula, cootapatambensis, cyanea, dacryodes, degelii, devergescens, deversa, ditmarsica, divergens, dolosa, dufourii, durietzii, elaeina, elaeomelaena, fayettensis, finkiana, funckii, fuscella, fusconigrescens, glaucina, glaucovirens, halizoa, hochstetteri, howensis, hydrela, illinoisensis, inconstans, internigrescens, iovensis, knowlesiae, latebrosa, latericola, lecideoides, lobata, macrostoma, madida, margacea, marmorea, maura, mawsonii, melas, meridionalis, microsporoides, mucosa, mundula, muralis, murina, nigrescens, nigrescentoides, novomexicana, obnigrescens, obsoleta, ochrostoma, operculata, ossiseda, pachyderma, papillosa, parmigerella, phaeoderma, phaeosperma, phaeothelena, pinguicula, polysticta, praetermissa, prominula, pseudonigrescens, puncticulata, rheitrophila, ruderella, sandstedei, scabra, simplex, solicola, sorbicola, sordida, standfordii, striatula, subdiscreta, subglaucina, submersella, subtholocarpa, tasmanica, tessellata, tessellatula, tholocarpa, tuberculiformis, umbrinula, viridula, xyloxena, zamenhofiana*

VARIETIES *aethiobola acrotella, aethiobola aquilella, aethiobola catalepta, aethiobola cataleptoides, aethiobola fuscocinerascens, aethiobola hibernica, aethiobola internigrescens, aquatilis retecta, aucklandica tessellatuloidea, baldensis parmigera, baldensis sphinctrina, bryoctona psammophila, caerulea cineracea, degelii ceuthocarpa, ditmarsica erichsenii, ditmarsica santessonii, dolosa mutabilis, dufourii malhamensis, funckii degenerascens, funckii silicicola, fuscella glaucina, halizoa cribbii, halizoa friesiaca, halizoa mackenzie-lambii, hochstetteri carnea, hochstetteri dolomitica, hochstetteri foveolata, hochstetteri integra, hydrela hibernica, hydrela submersa, latericola granulosaria, latericola insularis, macrostoma macrostoma, macrostoma furfuracea, macrostoma furfuracea-tectorum, macrostoma griseorubens, macrostoma macrocarpa, macrostoma thrombioides, margacea leightonii, maura aractina, maura zschackeana, mucosa laetevirens, mundula deminuta, muralis rupestris, muralis submuralis, muralis virens, murina fugax, murina myriocarpa, murina parva, pinguicula peloclita, prominula scotina, rheitrophila kernstockii, sandstedei lorraine-smithiae, sandstedei ludovicinae, striatula continua, striatula microspora, viridula mauroides, xyloxena melaenella*

GENUS *Adelococcus* Adel Warty Lichen

SPECIES *alpestris, interlatens*

GENUS *Agonimia* Agony Warty Lichen

SPECIES *allobata, gelatinosa, globifera, octospora, papillata, repleta, tristicula*

GENUS *Anthracocarpon* Anthrax Warty Lichen

SPECIES *virescens*

GENUS *Awasthiella* Awasth Warty Lichen

SPECIES *indica*

GENUS *Catapyrenium* Earth Warty Lichen

SPECIES *andicola, boccanum, caeruleopulvinum, chilense, cinereum, daedaleum, globosum, granulosum, heppioides, lachneum, michelii, norvegicum, pilosellum, plumbeum, psoromoides, rufescens, schaereri, squamellum, squamulosum, tuckermanii*

GENUS *Clauzadella* Clauzade Warty Lichen

SPECIES *gordensis*

GENUS *Clavascidium* Collared Warty Lichen

SPECIES *antillarum, kisovense, liratum, umbrinum*

GENUS *Dermatocarpon* Stipple Back Warty Lichen

SPECIES *arnoldianum, deminuens, intestiniforme, leptophyllodes, leptophyllum, lorenzianum, luridum, meiophyllizum,* ***miniatum***(Dye Herb), *moulinsii, rivulorum, saxicola, vellereum*

VARIETIES *meiophyllizum leptophyllum, miniatum miniatum, miniatum complicatum*

GENUS *Diederimyces* Dieder Warty Lichen
SPECIES *fuscideae*
GENUS *Endocarpon* Stippled Warty Lichen
SPECIES *adscendens, adsurgens, aridum, crassisporum, helmsianum, lepidallum, macrosporum, pallidulum, pallidum, petrolepideum, pulvinatum, pusillum, robustum, rogersii, simplicatum, subnitescens, tortuosum*
VARIETIES *simplicatum bisporum*
GENUS *Endopyrenium* (Synonym for *Catapyrenium*)
GENUS *Haleomyces* Hale Warty Lichen
SPECIES *oropogonicola*
GENUS *Henrica* Henry Warty Lichen
SPECIES *ramulosa*
GENUS *Heterocarpon* Variable Fruit Warty Lichen
SPECIES *ochroleucum*
GENUS *Heteroplacidium* Variable Placid Warty Lichen
SPECIES *acarosporoides, acervatum, congestum, contumescens, divisum, endocarpoides, imbricatum, phaeocarpoides, podolepis*
GENUS *Hyalopyrenia* Hyaline Fire Warty Lichen
SPECIES *japonica*
GENUS *Involucropyrenium* Bract Ring Warty Lichen
SPECIES *sbarbaronis, terrigenum, tremniacense, waltheri*
GENUS *Kalbiana* Kalb Warty Lichen
SPECIES *brasiliensis*
GENUS *Lauderlindsaya* Lauder Lindsay Warty Lichen
SPECIES *acroglypta, borreri*
GENUS *Leucocarpia* Light Carpel Warty Lichen
SPECIES *biatorella*
GENUS *Leucocarpopsis* Lesser Light Carpel Warty Lichen
SPECIES *devensis*
GENUS *Macentina* Macent Warty Lichen
SPECIES *abscondita, stigonemoides*
GENUS *Melanotheca* Dark Theca Warty Lichen
SPECIES *achariana, cinnabarina, oxyspora, rubescens, subsimplex*
GENUS *Merismatium* Seaside Warty Lichen
SPECIES *discrepans, nigritellum*
GENUS *Muellerella* Mueller Warty Lichen
SPECIES *atricola, hospitans, lichenicola, melanostigma, polyspora, pygmaea, ventosicola*
VARIETIES *pygmaea pygmaea, pygmaea athallina*
GENUS *Mycophycias* Bladder Warty Lichen
SPECIES *apophlaeae, ascophylli*
GENUS *Neocatapyrenium* Modern Earth Warty Lichen
SPECIES *cladonioideum, latzelii, radicescens, rhizinosum*
GENUS *Normandina* Clam Warty Lichen
SPECIES *pulchella*
GENUS *Parmentaria* Rash Warty Lichen
SPECIES *australiensis, baileyi, chilensis, gregalis, grossa, interlatens, microspora, pallida, papillata, plana, subastroidea, subplana, toowoombensis, umbilicata*
VARIETIES *subastroidea subastroidea, subastroidea subsimplex*
GENUS *Phaeospora* Dusky Warty Lichen
SPECIES *arctica, exoriens, lemaneae, parasitica, rimosicola, supersparsa*
GENUS *Phaeosporobolus* Dusky Bowl Warty Lichen
SPECIES *alpinus, usneae*
GENUS *Placidiopsis* Lesser Stipple Scale Warty Lichen
SPECIES *cervinula, cinerascens, custnani, minor, pseudocinerea*
GENUS *Placidium* Stipple Scale Warty Lichen
SPECIES *lacinulatum, tuckermanii*
GENUS *Plurisperma* Many Spore Warty Lichen
SPECIES *dalbergiae*
GENUS *Pocsia* Pock Warty Lichen
SPECIES *borhidii, mucronata, septemseptata, triseptata*
GENUS *Polyblastia* Polyblast Warty Lichen
SPECIES *agraria, albida, australis, cruenta, cucurbitula, cupularis, deminuta, dermatodes, efflorescens, gothica, helvetica, hyperborea, inumbrata, melaspora, obsoleta, philaea, quartzina, sendtneri, septentrionalis, terrestris, theleodes, verrucosa, vouauxii, wheldonii*
VARIETIES *albida amota, cruenta henscheliana, cupularis bryophila, cupularis fuscoargillacea, cupularis intercedens, dermatodes schraderi, melaspora scotinospora, melaspora subinumbrata, terrestris subviridicans*
GENUS *Psoroglaena* Mangy Warty Lichen

SPECIES *cubensis, perminuta*
GENUS *Pyrenothamnia* (Synonym *for Placidium*)
GENUS *Pyrenulella* (Synonym for *Phaeospora*)
GENUS *Rhabdospora* Rod Striped Warty Lichen
SPECIES *acantophila, arabidicola, curva, hypochaeridis, inaequalis, lecanorae, lupini, pleosporoides, tanaceticola, visci*
GENUS *Sagediopsis* Sage Warty Lichen
SPECIES *aquatica, barbara, campsteriana, lomnitzensis*
GENUS *Scleropyrenium* Hard Fire Warty Lichen
SPECIES *japonicum, kurokawae*
GENUS *Staurothele* Pimple Warty Lichen
SPECIES *arctica, areolata, bacilligera, caesia, clopima, clopimoides, diffractella, discedens, drummondii, effigurata, elenkinii, fissa, frustulenta, geoica, guestphalica, hymenogonia, lecideoides, monicae, orispruinosa, pallidipora, polygonia, rufa, rugulosa, rupifraga, succedens, tenuissima, verruculosa*
VARIETIES *caesia ebbonensis, fissa catalepta, fissa glacialis, fissa umbrina, hymenogonia ventosa, rugulosa innata*
GENUS *Telogalla* Gall Warty Lichen
SPECIES *olivieri*
GENUS *Thelidium* Speck Warty Lichen
SPECIES *absconditum, areolatum, circumspersellum, decipiens, fontigenum, fumidum, heardense, impressum, incavatum, methorium, microbolum, microsporum, minimum, minutulum, nylanderi, papulare, pluvium, praevalescens, pyrenophorum, transsylvanicum, velutinum, zwackhii*
VARIETIES *decipiens immersum, decipiens leightonii, fumidum auruntii, incavatum auruntii, incavatum calcivorum, methorium aeniovinosum, microbolum cataractarum, minutulum acrotellum, minutulum aethioboloides, minutulum cataractarum, minutulum exile, minutulum hospitum, minutulum mesotropum, minutulum terrestre, papulare papulare, papulare sorediatum, pyrenophorum borreri, pyrenophorum explicatum, pyrenophorum subrimulatum, pyrenophorum viride, zwackhii cataractarum, zwackhii microcarpum, zwackhii sparsulum*
GENUS *Tichothecium* Hairy Theca Warty Lichen
SPECIES *gemmiferum*
GENUS *Trimmatothele* Trim Warty Lichen
SPECIES *glacialis, maritima, perquisita, umbellulariae, versipellis*

ORDER 2 *Arthoniales* Comma Lichen

FAMILY 1 *Arthoniaceae* Comma Lichen
GENUS *Arthonia* Dot Comma Lichen
SPECIES *accolens, aciniformis, albofarinosa, albofuscescens, albovirescens, aleuromela, almquistii, amoena, amylospora, anglica, anombrophila, apatetica, apotheciorum, apteropteridis, arthonioides, astericus, astroidestera, atlantica, atomaria, atrata, banksiae, bisepta, byssacea, caesia, caribaea, carneorufa, chiodectella, caudata, cinnabarina, cohabitans, complanata, conferta, conspersula, coronata, cupressina, cyanea, cytisi, delicatula, didyma, diffusa, diffusella, digitatae, diploiciae, dispersa, dispersula, eckfeldtii, elegans, endlicheri, epiphyscia, erubescens, erupta, excedens, excipienda, exilis, fissurina, floridana, fuliginosa, fuscopurpurea, galactitella, galactites, gelidae, glaucescens, glebosa, gracilenta, gracilior, gracillima, granosa, graphidicola, gyalectoides, hamamelidis, hypobela, ilicina, ilicinella, impallens, incarnata, interstes, intexta, invadens, lapidicola, lecanactidea, lecidella, lecideola, leptospora, leucastraea, leucopellaea, ligniaria, ligniariella, lividula, luridoalba, mediella, microsperma, microspermella, miserula, molendoi, muscigena, myriocarpella, neglectula, neoni, nigratula, nigrorufa, nymphaeoides, ochrocincta, ochrodiscodes, ochrolutea, palmulacea, patellulata, peltigerea, peraffinis, perminuta, phaeobaea, phaeophysciae, pinastri, platyspilea, polygramma, polymorpha, propinqua, pruinata, pruinosella, pruinosula, pseudocyphellariae, punctella, punctiformis, punctilliformis, pyrrhula, pyrrhuliza, quintaria, radiata, ravenelii, reniformis, rhoidis, rubella, rupicola, sagenidii, sexlocularis, sideria, spadicea, stellaris, stereocaulina, stictella, subcondita, subconveniens, subdiffusa, subdispuncta, subfuscicola, subgyrosa, subminutissima, subminutula, subrubella, tasmanica, teadescens, terrigena, tetramera, thelotrematis, thozetiana, torulosa, trilocularis, tuckermaniana, varians, vernans, vinosa, viridicans, willeyi, xylographica, zwackhii*
VARIETIES *anombrophila cinereopruinosa, apotheciorum clemens, apotheciorum galactinaria, apotheciorum subvarians, arthonioides aspersa, astroidestera armoricana, astroidestera macrospora, atlantica atlantica, atlantica dendritica, atlantica positiva, cinnabarina cinnabarina, cinnabarina anerythrea, cinnabarina astroidea, cinnabarina coccinea, cinnabarina concolor, cinnabarina cuspidans, cinnabarina kermesiana, cinnabarina marginata, cinnabarina nudata, cinnabarina purpurea, cinnabarina pruinata, didyma asperella, didyma atrofuscella, endlicheri lobata, excipienda hibernica, excipienda minutula, fuscopurpurea pelveti, intexta parasemoides, mediella hypnophila, muscigena leucodontis, phaeobaea paralia, pruinata impolita, punctiformis cembrina, punctiformis insinuata, punctiformis melantera, radiata astroidea, radiata cascarillae, radiata epipastoides, radiata opegraphina, radiata parallela, radiata swartziana, subconveniens ricasoliae, vinosa pineti*
GENUS *Allarthonia* (Synonym for *Arthonia*)
GENUS *Arthophacopsis* Lens Comma Lichen
SPECIES *parmeliarum*

GENUS *Arthothelium* Mouse Dot Comma Lichen
SPECIES *adveniens, albovirescens, ampliatum, cinereoargenteum, dictyosporum, distendens, gregarium, hallii, huegelii, infuscatum, interveniens, lirellans, macounii, macrothecum, microsporum, norvegicum, orbilliferum, polycarpum, pruinascens, pulverulentum, ruanum, sanguineum, spectabile, subcyrtodes, subspectabile, taediosum, velatius, violascens*
VARIETIES *interveniens ferax, macounii reagens, orbilliferum anastomosans, ruanum dispersum, ruanum rosaceum*
GENUS *Catarraphia* Catar Raphe Comma Lichen
SPECIES dictyoplaca
GENUS *Celidium* (Synonym for *Arthonia, Toninia*)
GENUS *Chrysothrix* Gold Dust Comma Lichen
SPECIES *candelaris, chlorina, chrysophthalma, flavovirens, xanthina*
GENUS *Coniarthonia* Cone Comma Lichen
SPECIES *wilmsiana*
GENUS *Cryptothecia* Cryptic Edge Comma Lichen
SPECIES *atropunctata, candida, eungellae, exilis, inexspectata, rubrocinctoides, scripta*
GENUS *Eremothecella* Sand Comma Lichen
SPECIES *calamicola, cyaneoides, macrocephala*
GENUS *Helicobolomyces* Helix Bowl Comma Lichen
SPECIES *lichenicola*
GENUS *Paradoxomyces* Paradox Comma Lichen
SPECIES *nymanii*
GENUS *Perigrapha* Perigraph Comma Lichen
SPECIES *superveniens*
GENUS *Sagenidiopsis* Sagen Comma Lichen
SPECIES *merrotsyi*
GENUS *Sipmania* Sipman Comma Lichen
SPECIES *peltata*
GENUS *Sporostigma* Stigma Comma Lichen
SPECIES *melaspora*
GENUS *Stirtonia* (Synonym for *Eremothecella*)
GENUS *Subhysteropycnis* Hyster Dense Comma Lichen
SPECIES *maculiformans*
GENUS *Tania* Tan Comma Lichen
SPECIES *lanosa*
GENUS *Tarbertia* Tarbert Comma Lichen
SPECIES *juncina*
GENUS *Wegea* Wege Comma Lichen
SPECIES *tylophorelloides*

FAMILY 2 *Opegraphaceae* Scribble Comma Lichen
GENUS *Opegrapha* Scribble Comma Lichen
SPECIES *agelaea, agelaeina, agelaeoides, areniseda, astraea, atra, bonplandii, brevis, calcarea, candida, cesareensis, corticola, cypressi, demutata, diaphoriza, dolomitica, ectolechiacearum, epiporina, foreaui, fumosa, glaucomaria, grossulina, gyrocarpa, hassei, herbarum, inalbescens, interveniens, intrusa, lacteella, lambinonii, lamyi, leptocarpoides, leptoplacella, leucina, leucoplaca, lithyrga, longissima, luzonensis, megagonidia, melanospila, microcycla, minutula, mougeotii, multipuncta, niveoatra, ochrocheila, parasitica, paraxanthodes, pertusariicola, physciaria, prosodea, puiggarii, pulvinata, quaternella, ravenelii, rotunda, rubefacta, rufescens, rupestris, saxicola, saxigena, sorediifera, sororiella, sphaerophoricola, staurothelicola, stellata, strigulae, subelevata, thelotrematis, umbellulariae, varia, vermicellifera, viridis, vulgata, xerica, zwackhii*
VARIETIES *areniseda actophila, atra arthonioidea, atra cerasi, atra denigrata, atra hapalea, atra nigrata, atra nimbosa, atra parallela, atra rimosa, atra stenocarpa, atra tenuior, bonplandii bonplandii, bonplandii abbreviata, calcarea agglomerata, calcarea atrula, calcarea chevallieri, calcarea conferta, calcarea confluens, calcarea decandollei, calcarea heteromorpha, cesareensis hysteriiformis, diaphoriza macquariensis, herbarum betulina, herbarum prosiliens, herbarum prosodeoides, herbarum protuberans, herbarum turneri, lithyrga ochracea, lithyrga steriza, mougeotii leightonii, niveoatra amphotera, niveoatra dubia, niveoatra subsiderella, ochrocheila atricolor, ochrocheila rubescens, parasitica monspeliensis, parasitica semicincta, prosodea taxicola, rufescens arthonioidea, rufescens contexta, rufescens fuscata, rufescens herpetica, rufescens rubella, rufescens siderella, rufescens subocellata, rupestris centrifuga, rupestris persoonii, rupestris saxatilis, saxigena conferta, varia varia, varia chlorina, varia diaphora, varia herbicola, varia heterocarpa, varia lichenoides, varia notha, varia pulicaris, varia rimalis, varia signata, varia tigrina, varia tridens, vermicellifera fuscella, vermicellifera hapaleoides, viridis involuta, viridis rubella, vulgata vulgata, vulgata cinerea, vulgata confluens, vulgata parallela, xerica pseudovaria*
GENUS *Bactrospora* Berry Scribble Comma Lichen
SPECIES *arthonioides, brevispora, brodoi, clementis, corticola, denticulata, dryina, granularis, homalotropa, macrospora, mesospora, metabola, micareoides, myriadea, paludicola, patellarioides, spiralis*
VARIETIES *homalotropa urceolatum*
GENUS *Chiodecton* Chi Scribble Comma Lichen

SPECIES *acarosporoides, colensoi, congestulum, effusum, farinaceum, flavovirens, heterogenum, leptosporum, macquariense, maculatum, malmei, montanum, mycelioides, ochroleucum, queenslandiae, sphaerale, stictathecium, sublaevigatum, subochroleucum*
VARIETIES *colensoi capense, colensoi grossum, colensoi hypoleucum, colensoi neohollandicum, colensoi ochraceofuscescens, colensoi perplexum, colensoi subdepressum, colensoi trypethelioides, colensoi velatum, colensoi virens, congestulum stromaticum, effusum endoleucum*
GENUS *Cresponea* Crespone Scribble Comma Lichen
SPECIES *apiculata, chloroconia, flava, flavescens, follmannii, leprieurii, leprieuroides, macrocarpoides, melanocheloides, pluriloculari, premnea, proximata*
GENUS *Enterographa* Graph Scribble Comma Lichen
SPECIES *anguinella, bella, compunctula, crassa, deslooveri, divergens, elaborata, elixii, hutchinsiae, lecanoroides, micrographa, pallidella, sorediata, subgelatinosa, subserialis, zonata*
GENUS *Lecanactis* Old Scribble Comma Lichen
SPECIES *abietina, akompsa, coniochlora, dilleniana, dubia, latebrarum, latispora, megaspora, neozelandica, platygraphoides, quassiae, salicina, subattigens, spermatospora, subabietina, sulphurea*
GENUS *Opgraphoidea* (Synonym for *Opegrapha*)
GENUS *Schismatomma* Split Tumor Scribble Comma Lichen
SPECIES *cretaceum, decolorans, glaucescens, graphidioides, niveum, occultum, pluriloculare, quercicola, rappii, rediunta, umbrinum, vernans*
VARIETIES *cretaceum virgineum, graphidioides abietinum, graphidioides farinosum, graphidioides pericleum, graphidioides rimatum, occultum nanocarpum*
GENUS *Sclerophyton* Hard Frond Scribble Comma Lichen
SPECIES *californica, conspicuum, inscriptum, occidentale, rostratum, seriale*
GENUS *Sclerophytonomyces* (Synonym for *Sclerophyton*)

FAMILY 3 *Roccellaceae* Orchil Comma Lichen
GENUS *Roccella* Orchil Comma Lichen
SPECIES ***babingtonii***(Dye Herb), *decipiens, difficilis, fimbriata,* ***fuciformis***(Dye Herb), ***montagnei***(Dye Herb), *peruensis,* ***phycopsis***(Dye Herb), ***tinctoria***(Dye Herb)
GENUS *Ancistroporella* Ancestor Orchil Comma Lichen
SPECIES *australiensis*
GENUS *Combea* (Synonym for *Roccella*)
GENUS *Dendrographa* False Orchil Comma Lichen
SPECIES *alectroides, leucophaea*
VARIETIES *leucophaea minor*
GENUS *Dendrographiella* Lesser False Orchil Comma Lichen
SPECIES *podosporioides*
GENUS *Dichosporidium* Dichotomy Spore Orchil Comma Lichen
SPECIES *boschianum*
GENUS *Dictyographa* Net Graph Orchil Comma Lichen
SPECIES *cinerea*
GENUS *Dirina* Fog Orchil Comma Lichen
SPECIES *catalinariae, insulae-howensis, massiliensis, paradoxa*
VARIETIES *massiliensis massiliensis, massiliensis sorediata, massiliensis sorediata-immersa, massiliensis sorediata-stenhammarii*
GENUS *Enterodictyon* Enteric Net Orchil Comma Lichen
SPECIES *velatum*
GENUS *Epiphora* Epiphyte Orchil Comma Lichen
SPECIES *encaustica*
GENUS *Erythrodecton* Red Orchil Comma Lichen
SPECIES *malacum*
GENUS *Feigeana* Feige Orchil Comma Lichen
SPECIES *socotrana*
GENUS *Follmanniella* Follmann Orchil Comma Lichen
SPECIES *scutellata*
GENUS *Graphidastra* Graph Star Orchil Comma Lichen
SPECIES *multiformis*
GENUS *Halographis* Halogen Graph Orchil Comma Lichen
SPECIES *runica*
GENUS *Hubbsia* Brittle Bag Orchil Comma Lichen
SPECIES *parishii*
GENUS *Lecanographa* Crust Orchil Comma Lichen
SPECIES *abscondita, amylacea, dimelaeonoides, grumulosa, hypothallinum, littoralis, lyncea, microcarpella, nothofagi, subdryophila*
VARIETIES *grumulosa hemisphaerica*
GENUS *Leciographa* Graph Orchil Comma Lichen
SPECIES *parasemoides*
GENUS *Mazosia* Mazos Orchil Comma Lichen

SPECIES *aptrootii, bambusae, melanophthalma, paupercula, phyllosema, quadriseptata, tomentifera*
GENUS *Pentagenella* Pentagon Orchil Comma Lichen
SPECIES *fragillima*
GENUS *Platygrapha* (Synonym for *Sigridea*)
GENUS *Platygraphopsis* Flat Graph Orchil Comma Lichen
SPECIES *interrupta*
GENUS *Peterjamesia* Peter James Orchil Comma Lichen
SPECIES *circumscripta*
VARIETIES *circumscripta circumscripta, circumscripta sorediatus*
GENUS *Plectocarpon* Plaited Orchil Comma Lichen
SPECIES *lichenum, sampaianae, scrobiculatae*
GENUS *Pulvinodecton* Powder Orchil Comma Lichen
SPECIES *kurzii, seychellensis*
GENUS *Roccellina* Peg Legged Rim Orchil Comma Lichen
SPECIES *conformis, exspectata, franciscana*
GENUS *Roccellodea* Lesser Orchil Comma Lichen
SPECIES *nigerrima*
GENUS *Sagenidium* Sage Orchil Comma Lichen
SPECIES *molle*
GENUS *Schizopeltis* Fog Fingers Orchil Comma Lichen
SPECIES *californica*
GENUS *Sigridea* Sigrid Orchil Comma Lichen
SPECIES *albella, californica, chloroleuca, glaucomoides, leptothallus*
GENUS *Simonyella* Simony Orchil Comma Lichen
SPECIES *variegata*
GENUS *Sporhaplus* Rod Spore Orchil Comma Lichen
SPECIES *rondoniensis*
GENUS *Streimannia* Streimann Orchil Comma Lichen
SPECIES *varieseptata*
GENUS *Syncesia* Synces Orchil Comma Lichen
SPECIES *myrticola*

ORDER 3 *Caliciales* Calico Lichen

FAMILY 1 *Caliciaceae* Calico Lichen
GENUS *Calicium* Calico Lichen
SPECIES *abietinum, adequatum, adspersum, atronitescens, chlorosporum, contortum, corynellum, diploellum, glaucellum, glebosum, hyperelloides, lenticulare, leucochlorum, parvum, quercinum, robustellum, salicinum, trabinellum, tricolor, victorianum, viride*
VARIETIES *abietinum curtum, adspersum roscidum, chlorosporum concinnum, chlorosporum glaucescens, chlorosporum queenslandiae, chlorosporum wilsonii, corynellum subsessile, glaucellum chlorodes, glaucellum claviculare, lenticulare amylocaule, lenticulare sphaerocephalum, lenticulare subquercinum, quercinum clavellum, quercinum curtiusculum, robustellum flavidum, robustellum pachypus, robustellum perbreve, robustellum stenosporum, robustellum validius, salicinum crustosum, salicinum hemiphaeum, salicinum hemiplodum, salicinum lichenoides, salicinum trachelinum, salicinum xylonellum, tricolor bulbosum, tricolor clarensis, tricolor cryptocroceum, tricolor deforme, tricolor minus, victorianum victorianum, victorianum desidiosum, victorianum obconicum, victorianum piperatum, viride baliolum, viride hyperellum*
GENUS *Acroscyphus* Tall Goblet Calico Lichen
SPECIES *sphaerophoroides*
GENUS *Cypheliopsis* (Synonym for *Cyphelium*)
GENUS *Cyphelium* Soot Calico Lichen
SPECIES *brachysporum, brunneum, inquinans, lucidum, marcianum, notarisii, pinicola, sessile, tigillare, trachylioides, tricinctum*
GENUS *Heterocyphelium* Variable Soot Calico Lichen
SPECIES *leucampyx*
GENUS *Microcalicium* Tiny Calico Lichen
SPECIES *ahlneri, arenarium, disseminatum*
GENUS *Sphinctrinopsis* Sphincter Calico Lichen
SPECIES *pertusariae*
GENUS *Texosporium* Texas Calico Lichen
SPECIES *santijacoba*
GENUS *Thelomma* Nipple Calico Lichen
SPECIES *santijacoba, californicum, carolinianum, mammosum, occidentale, ocellatum, santessonii*
GENUS *Tholurna* Urn Calico Lichen
SPECIES *dissimilis*

FAMILY 2 *Coniocybaceae* Pin Calico Lichen
GENUS *Coniocybe* Pin Calico Lichen

SPECIES *gracilescens, ochrocephala, rhodocephala*
GENUS *Chaenotheca* Stubble Pin Calico Lichen
SPECIES *brachypoda, brunneola, chlorella, chrysocephala, citriocephala, confusa, degelii, deludens, ferruginea, furfuracea, gracilenta, gracillima, hispidula, hygrophila, laevigata, phaeocephala, stemonea, subroscida, trichialis, xyloxena*
VARIETIES *chlorella carthusiae, chrysocephala filaris, chrysocephala melanocephala, ferruginea flavocitrina, ferruginea melanophaea, hispidula acicularis, stemonea viridis, trichialis aeruginosa, trichialis caerulescens, trichialis cinerea, trichialis elassospora, trichialis schaereri*
GENUS *Cybebe* (Synonym for *Chaenotheca gracilenta*)
GENUS *Sclerophora* Hard Pin Calico Lichen
SPECIES *farinacea, pallida, peronella, sanguinea*
VARIETIES *pallida nivea*

ORDER 4 *Capnodiales* Cap Lichen

FAMILY 1 *Capnodiaceae* Cap Lichen
GENUS *Capnodium* Cap Lichen
SPECIES *australe, citri, coffeae, elongatum, footii, salicinum, tiliae, walteri*
GENUS *Caldariomyces* Caldero Cap Lichen
SPECIES *fumago*
GENUS *Capnodendron* Wood Cap Lichen
SPECIES *trichomericola*
GENUS *Echinothecium* (Synonym for *Capnodium*)
GENUS *Hyaloscolecostroma* Hyaline Cap Lichen
SPECIES *rondoniense*
GENUS *Leptoxyphium* Thin Xiph Cap Lichen
SPECIES *fumago, graminum*
GENUS *Limaciniaseta* Lima Sharp Seta Cap Lichen
SPECIES *californica*
GENUS *Scolecoxyphium* Scole Xiph Cap Lichen
SPECIES *americanum, cirrhosum, formosum, fraserae*
GENUS *Scorias* Scoria Cap Lichen
SPECIES *spongiosa*
GENUS *Trichomerium* Hairy Cap Lichen
SPECIES *grandisporum*
GENUS *Tripospermum* Tripe Cap Lichen
SPECIES *camelopardus, myrti, pinophilum, prolongatum*

FAMILY 2 *Coccodiniaceae* Round Cap Lichen
GENUS *Coccodinia* (Synonym for *Dennisiella*)
GENUS *Dennisiella* Dennis Round Cap Lichen
SPECIES *babingtonii*
GENUS *Microxiphium* Xiph Round Cap Lichen
SPECIES *fagi*

FAMILY 3 *Euantennariaceae* Antenna Cap Lichen
GENUS *Euantennaria* (Synonym for Antennulata)
GENUS *Antennatula* Antenna Cap Lichen
SPECIES *pinophila*
GENUS *Hormisciella* Hormisc Antenna Cap Lichen
SPECIES *atra*
GENUS *Rasutoria* Rasutor Antenna Cap Lichen
SPECIES *abietis, pseudotsugae, terrieri, tsugae*
GENUS *Strigopodia* Bristle Pod Antenna Cap Lichen
SPECIES *resinae*
GENUS *Trichopeltheca* Hairy Pelt Theca Antenna Cap Lichen
SPECIES *asiatica, stevensii*
GENUS *Trichothallus* Hairy Thallus Antenna Cap Lichen
SPECIES *hawaiiensis*

FAMILY 4 *Metacapnodiaceae* Meta Cap Lichen
GENUS *Metacapnodium* Meta Cap Lichen
SPECIES *crassum, dennisii, dingleyae, fraserae, guava, juniperi, moniliforme, quinqueseptatum, smilacinum, spongiosum*
GENUS *Capnobotrys* Berry Meta Cap Lichen
SPECIES *dingleyae, splendens*
GENUS *Capnocybe* Cybe Meta Cap Lichen
SPECIES *fraserae, novae-zelandiae, spongiosa*
GENUS *Capnophialophora* Phial Meta Cap Lichen
SPECIES *fraserae, hughesii, pinophila*

GENUS *Capnosporium* — Lesser Meta Cap Lichen
SPECIES *moniliforme*
GENUS *Hormiokrypsis* — Hormion Crype Meta Cap Lichen
SPECIES *libocedri*

ORDER 5 *Cladoniales* — Cup Lichen

FAMILY 1 *Cladoniaceae* — Cup Lichen
GENUS *Cladonia* — Cup Lichen
SPECIES *abbreviatula, acervata, acuminata, adspersa, alaskana, albonigra, alinii, alpina, amauracraea, andereggii, angustata, anitae, apodocarpa, archeri, artuata, atlantica, attacta, bacillaris, baccilliformis, beaumontii, bellidiflora, bimberiensis, borbonica, borealis, boryi, botryocarpa, botrytes, buckii, brevis, caespiticia, callosa, calyciformis, capitellata, cariosa, carneola, caroliniana, celata, cenotea, cervicornis, chlorophaea, coccifera, complanata, coniocraea, conista, convoluta, corniculata, cornuta, corymbescens, crispata,* ***cristatella***(Dye Herb), *cucullata, cyanipes, cyanopora, cyathomorpha, dactylota, dahliana, decorticata, deformis, didyma, digitata, dimorpha, dimorphoclada, diversa, ecmocyna, enantia, extracorticata, fimbriata, firma, floerkeana, floridana, foliacea, fruticulosa, furcata, fuscofunda, glauca, glebosa, graciliformis, gracilis, granulans, homosekikaica, humilis, hypoxantha, imbricata, incrassata, kanewskii, krempelhuberi, kuringaiensis, labradorica, leporina, luteoalba, macilenta, macroceras, macrophylla, magyarica, mateocyatha, maxima, merochlorophaea, metacorallifera, monomorpha, multiformis, murrayi, nana, nanodes, neozelandica, nipponica, norvegica, nudicaulis, ochrochlora, pachycladodes, paeminosa, parasitica, perforata, perlomera, pertricosa, peziziformis, phyllophora, piedmontensis, pleurota, pocillum, polycarpa, polydactyla, poroscypha, praetermissa, prolifica, prostrata, pseudohondoensis, pseudomacilenta, psoromica, pulvinella,* ***pyxidata***(Dye Herb), *ramulosa, rangiformis, ravenelii, rei, rigida, robbinsii, santensis, sarmentosa, scabriuscula, schofeldii, simulata, singularis, sobolescens, southlandica, squamosa, staufferi, stereoclada, strepsilis, stricta, subcariosa, subcervicornis, subfurcata, subpityrea, subradiata, subsetacea, subsquamosa, subsubulata, subulata, sulcata, sulphurina, symphycarpia, tenerrima, tessellata, theirsii, thomsonii, transcendens, trassii, turgida, ustulata, verruculosa, vulcani, wainioi, weymouthii, zopfii*
VARIETIES *acuminata norrlinii, borbonica physodalica, callosa fragilissima, capitellata capitellata, capitellata amaurocraea, capitellata interhiascens, capitellata interhiascens-xanthoclada, capitellata squamatica, cervicornis cervicornis, cervicornis mawsonii, cervicornis pleuroclada, cervicornis pulvinata, cervicornis verticillata, chlorophaea asahinae, chlorophaea cryptochlorophaea, chlorophaea grayi, chlorophaea novochlorophaea, coccifera asotea, coccifera cornucopioides, convoluta endiviifolia, cornuta groenlandica, crispata crispata, crispata arborea, crispata carassensis, crispata cetrariiformis, crispata cetrariiformis-diffusa, crispata cetrariiformis-infundibulifera, crispata gracilescens, crispata japonica, didyma vulcanica, ecmocyna intermedia, ecmocyna occidentalis, enantia diffissa, fimbriata major, fimbriata nemoxyna, fimbriata simplex, firma nylanderi, floerkeana carcata, floerkeana intermedia, floerkeana trachypoda, furcata furcata, furcata cancellata, furcata narkodes, furcata palamaea, furcata pinnata, furcata subrangiformis, furcata tenuicaulis, gracilis elongata, gracilis turbinata, gracilis vulnerata, humilis humilis, humilis bourgeanica, humilis conistea, humilis conoidea, incrassata brebissonii, macilenta macilenta, macilenta cylindrica, macilenta flabellulata, macilenta ostreata, macilenta pityropoda, macilenta scabrosa, macilenta squamigera, macilenta subscyphifera, macilenta subulata, macilenta vermicularis, macrophylla alpicola, murrayi firma, ochrochlora ceratodes, ochrochlora lepidula, ochrochlora notabilis, ochrochlora pergracilis, ochrochlora spadicea, pertricosa filiformis, pertricosa peltastica, pertricosa pungens, peziziformis capitata, peziziformis leptophylla, phyllophora hypophylla, pleurota grandis, polydactyla polydactyla, polydactyla umbricola, praetermissa praetermissa, praetermissa ceratophylla, praetermissa modesta, pyxidata conchata, pyxidata neglecta, ramulosa anomaea, ramulosa foliolosa, ramulosa hololepis, ramulosa isignyi, ramulosa lamarkii, ramulosa leucocephala, ramulosa pachypoda, ramulosa pityrea, rangiformis aberrans, rangiformis foliosa, rangiformis klementii, rangiformis muricata, rei nemoxyna, rigida rigida, rigida acuta, rigida acuta-subsquamosula, rigida elegantula, rigida pulverulenta, sarmentosa campbelliana, sarmentosa johnstonii, scabriuscula adspersa, scabriuscula asperata, scabriuscula farinacea, scabriuscula gallica, scabriuscula gracillima, scabriuscula hians, scabriuscula pungens, scabriuscula recurva, scabriuscula virgulata, sobolescens clavulifera, squamosa squamosa, squamosa denticollis, squamosa phyllocoma, squamosa subsquamosa, stricta stricta, stricta hypophylla, stricta uliginosa, subcariosa polycarpoides, subcervicornis macrophyllodes, subcervicornis rappii, subpityrea formosana, subradiata balfourii, subulata subulata, subulata carassensis, subulata cornutoradiata, subulata radiata, subulata subantarctica, subulata subcornuta, sulcata sulcata, sulcata depleta, sulcata striata, sulcata wilsonii, sulphurina gonecha, ustulata subdigitata, ustulata tasmanica, weymouthii arrosa, zopfii destricta*
GENUS *Calathaspis* — Callisto Cup Lichen
SPECIES *devexa*
GENUS *Cladia* — Branched Cup Lichen
SPECIES *aggregata, corallaizon, deformis, dumicola, ferdinandii, fuliginosa, inflata, moniliformis, mutabilis, oreophila, retipora, schizopora, sullivani*
GENUS *Cladina* — Reindeer Cup Lichen
SPECIES *arbuscula, azorica, ciliata, confusa, conspicua, evansii, mediterranea, mitis,* ***portentosa***(Dye Herb), *pseudoevansii,* ***rangiferina***(Edible), *standstedei,* ***stellaris***(Dye Herb), *stygia, submitis, subtenuis, tasmanica, terrae-novae, uncialis*

VARIETIES *arbuscula arbuscula, arbuscula mitis, arbuscula scabrosa, arbuscula squarrosa, ciliata leucophaea, ciliata tenuis, ciliata tenuis-flavicans, ciliata tenuis-laxiuscula, ciliata tenuis-leucophaea, confusa confusa, confusa alpestris, confusa bicolor, confusa fallax, confusa leptoclada, confusa sylvatica, confusa sylvestris, mitis mitis, mitis stictica, portentosa portentosa, portentosa condensata, portentosa erinacea, portentosa impexa, portentosa laxiuscula, portentosa spumosa, portentosa subimpexa, uncialis uncialis, uncialis biuncialis, uncialis biuncialis-dicraea, uncialis pseudoparecha*

GENUS *Gymnoderma* Naked Skin Cup Lichen
SPECIES *lineare*
GENUS *Metus* Met Cup Lichen
SPECIES *conglomeratus*
GENUS *Muhria* Muhr Cup Lichen
SPECIES *urceolata*
GENUS *Myelorrhiza* Myel Rhizid Cup Lichen
SPECIES *antrea, jenjiana*
GENUS *Notocladonia* Notable Cup Lichen
SPECIES *cochleata, undulata*
GENUS *Pycnothelia* Gnome Finger Cup Lichen
SPECIES *caliginosa, papillaria*
GENUS *Ramalea* Ramal Cup Lichen
SPECIES *coilophylla, myriocladella, tribulosa*
GENUS *Squamella* Scaly Cup Lichen
SPECIES *spumosa*
GENUS *Thysanothecium* Thysan Theca Cup Lichen
SPECIES *hookeri, scutellatum*

FAMILY 2 *Baeomycetaceae* Beret Cup Lichen
GENUS *Baeomyces* Beret Cup Lichen
SPECIES *carneus, heteromorphus, placophyllus, rufus*
GENUS *Dibaeis* Earth Beret Cup Lichen
SPECIES *absoluta, arcuata,* ***baeomyces***(Dye Herb), *sorediata*
VARIETIES *arcuata fungoides, baeomyces roseus*
GENUS *Phyllobaeis* Phyllode Beret Cup Lichen
SPECIES *erythrella, imbricata, linearis, rhodochroa, rubescens*
GENUS *Thamnolia* White Worm Beret Cup Lichen
SPECIES *vermicularis*
VARIETIES *vermicularis vermicularis, vermicularis subuliformis, vermicularis taurica*

FAMILY 3 *Icmadophilaceae* Candy Cup Lichen
GENUS *Icmadophila* Candy Cup Lichen
SPECIES *ericetorum, splachnirima*
GENUS *Siphula* Fingers Candy Cup Lichen
SPECIES *australiensis, ceratites, comata, complanata, coriacea, dactyliza, decumbens, dissoluta, elixii, fastigiata, foliacea, fragilis, georginae, gracilis, jamesii, parhamii*
GENUS *Siphulella* Lesser Fingers Candy Cup Lichen
SPECIES *coralloidea*

ORDER 6 *Lecanorales* Rim Lichen

FAMILY 01 *Lecanoraceae* Rim Lichen
GENUS *Lecanora* Rim Lichen
SPECIES *achariana, achroa, acroides, actophila, agardhiana, aitema, alaskensis, alba, albella, albellula, albescens, alboflavida, aleutica, andina, andrewii, anoptiza, apochraeoides, applegatei, arafurensis, argentata, argentea, arnhemica, arthothelinella, atrella, atrocaesia, atromarginata, atrosulphurea, austro-oceanica, austrointumescens, austrosorediosa, austrotropica, barkmaniana, behringii, bicincta, bipruiniosa, boligera, brodoana, brouard, brownii, cadubriae, caesiorubella, caesiosora, caesiosulphurea, californica, campestris, canadensis, carpinea, casuarinophila, cateilea, cenisia, chionocarpoides, chlarotera, chlorophaeodes, cinereofusca, circumborealis, cladonioides, coilocarpa, collatolica, compallens, conferta, confusa, congesta, conizaeoides, contractuloides, coriensis, corysta, crenulata, cupressi, demissa, deplanans, discoensis, dispersa, dispersoareolata, dispersogranulata, ecorticata, elapheia, elatinoides, elixii, elmorei, epanora, epibryon, expallens, farinacea, farinaria, fibrosa, flavidofusca, flavidomarginata, flavopallida, flotoviana, flowersiana, formosa, formosula, frustulosa, fugiens, fuscescens, galactiniza, gangaleoides, garovaglii, geophila, gisleriana, glabrata, glaucopsina, granulifera, groenlandica, gyalectodes, handelii, helicopis, helva, horiza, hyalinescens, hybocarpa, hypocrocinoides, hypoptella, hypoptoides, imperfecta, impressa, imshaugii, insignis, interjecta, intricata, intumescens, iowensis, jamesii, kariana, kofae, laatokkaensis, lacteola, leproplaca, leprosa, leptacina, lividocinerea, lividolutea, louisianae, lugubris, luteovernalis, margaritula, margarodes, marginata, maxima, mayrhoferi, mazatzalensis, melacarpella, melaena, melanommata, mellea, meridionalis, microbola, microfusca, miculata, minutella, mobergiana, monticola, mughicola, mundula,* ***muralis***(Dye Herb), *neoalbomarginata, neoqueenslandica, nigromarginata, nordenskioeldii, novae-hollandiae, novae-semliae, novomexicana, oblutescens, ochroidea, olivaceopallida, opiniconensis, oraefrigidae, oreinoides, orosthea, pacifica,*

pangerangoensis, pannonica, parmelinoides, permutata, perplexa, persimilis, phaedropthalma, phaeocardia, phaeophora, phryganitis, pinguis, piniperda, placodiolica, planaica, plumosa, poliophaea, poluninii, polytropa, populicola, praecrenata, praepostera, pringlei, prolifera, pruinosa, pseudargentata, pseudistera, pseudodecorata, pseudogangaleoides, pseudomellea, pulicaris, queenslandica, quercicola, reagens, rhypoderma, rubicunda, rubina, rugosella, rupicola, rutilescens, salicicola, saligna, salina, sambuci, sarcopisoides, scotopholis, semitensis, sierrae, sipeana, sophodopsella, soralifera, sorediifera, sphaerospora, stenospora, stenotropa, straminea, strobilina, subaurea, subcarnea, subcoarctata, subimmergens, subimmersa, subintricata, sublivescens, subpiniperda, subpurpurea, subrugosa, subtartarea, subumbrina, sulfurescens, sulphurea, sulphurescens, swartzii, symmicta, tesselina, texana, thallophila, thysanophora, torrida, tristuiscula, tropica, umbrina, umbrosa, utahensis, valesiaca, varia, vegae, viriduloflava, weberi, willeyi, wilsonii, wisconsinensis, xanthosora, xanthostoma, xylophila, zosterae

VARIETIES *aitema sepincola, aitema symmicta, albella pallida, albella rubescens, albescens galactina, albescens urbana, albescens verrucosa, argentata subfuscata, caesiorubella australiensis, caesiorubella cancriformis, caesiorubella glaucomodes, caesiorubella leucoma, caesiorubella merrillii, caesiorubella pallida, caesiorubella pulverata, caesiorubella saximontana, caesiosora soredians, campestris campestris, campestris allophana, campestris dolomitica, campestris subfusca, carpinea angulosa, cenisia atrynea, chlarotera chlarona, chlarotera crassula, chlarotera distincta, chlarotera rugosa, chlarotera rugosella, cinereofusca appalachensis, cinereofusca degelii, conizaeoides conizaeoides, conizaeoides farinaria, conizaeoides pityrea, conizaeoides variola, crenulata exomila, demissa incusa, demissa subolivascens, ecorticata coriensis, elatinoides straminea, expallens conizaea, farinacea blanda, farinacea subglaucodea, farinacea subpallida, flavopallida glaucoflavens, flavopallida lineolata, formosa bullata, formosa nigroglomerata, frustulosa argopholis, frustulosa occidentalis, fugiens chlorophaeoides, geophila pachythallina, geophila superfluens, helicopis prosechoides, helicopis prosechoidiza, helva albellaria, helva alligata, horiza laevis, horiza parisiensis, intricata mutabilis, jamesii impudens, jamesii variolascens, leptacina intricata, margarodes continua, margarodes knightiana, marginata elata, melacarpella haywardiorum, muralis albomarginata, muralis diffracta, muralis versicolor, oreinoides hypospilota, oreinoides tesselina, orosthea sulphurea, pacifica tetraspora, piniperda glaucella, poliophaea spodophaea, polytropa alpigena, polytropa efflorescens, polytropa illosoria, populicola distans, praepostera schistina, pseudistera convexella, pseudistera glebularis, pseudistera subfuscella, pseudogangaleoides pseudogangaleoides, pseudogangaleoides verdonii, pulicaris chlarona, pulicaris pinastri, rubicunda sylvestris, rupicola rupicola, rupicola efflorens, rupicola efflorens-sorediata, rupicola glaucescens, rupicola glaucoma, rupicola inflexa, rupicola lindsayi, rupicola obscurata, rupicola sordida, saligna effusa, saligna sarcopis, soralifera intricata, sphaerospora grandinosa, subimmersa subimmersa, subimmersa laevissima, subimmersa ramboldii, sublivescens orosthea, swartzii karelica, swartzii subradiosa, umbrina hagenii, wilsonii wilsonii, wilsonii xanthophora, xylophila eyerdamii, xylophila laevis, xylophila riparia, zosterae hagenii, zosterae palanderi*

GENUS *Allophoron* — Allophore Rim Lichen
SPECIES *farinosum*
GENUS *Aphanopsis* — Inconspicuous Rim Lichen
SPECIES *coenosa*
GENUS *Arctopeltis* — Pelt Rim Lichen
SPECIES *thuleana*
GENUS *Auriculora* — Eared Rim Lichen
SPECIES *byssomorpha*
GENUS *Bartlettiella* — Bartlett Rim Lichen
SPECIES *fragilis*
GENUS *Bryodina* — Mossy Rim Lichen
SPECIES *rhypariza*
GENUS *Bryonora* — Moss Rim Lichen
SPECIES *castanea, curvescens, pruinosa, septentrionalis*
GENUS *Calvitimela* — Calvit Rim Lichen
SPECIES *aglaea, armeniaca*
GENUS *Carbonea* — Carbon Rim Lichen
SPECIES *aggregantula, assentiens, assimilis, atronivea, intrusa, latypizodes, montevidensis, phaeostoma, supersparsa, vitellinaria, vorticosa*
GENUS *Cladidium* — Branched Rim Lichen
SPECIES *bolanderi, thamnitis*
GENUS *Clauzadeana* — Clauzade Rim Lichen
SPECIES *macula*
GENUS *Corticifraga* — Corticle Fragment Rim Lichen
SPECIES *fuckelii, peltigerae*
GENUS *Corticiruptor* — Corticle Rupture Rim Lichen
SPECIES *abeloneae*
GENUS *Crocynia* — Cotton Rim Lichen
SPECIES *fragilis, gossypina, lanuginosa, pyxinoides*
GENUS *Diomedella* — Diomedia Rim Lichen
SPECIES *demersa, disjungenda*
GENUS *Edrudia* — Edrude Rim Lichen
SPECIES *constipans*

GENUS *Glaucomaria* (Synonym for *Lecanora*)
GENUS *Gypsoplaca* Gypsum Rim Lichen
SPECIES *macrophylla*
GENUS *Haematomma* Blood Tumor Rim Lichen
SPECIES *accolens, africanum, americanum, collatum, eremaeum, fenzlianum, flexuosum, infuscum, nothofagi, ochroleucum, pacificum, persoonii, puniceum, rufidulum, similis, sorediatum, stevensiae, subpuniceum*
VARIETIES *africanum bubalinum, collatum breviculum, eremaeum pruinosum, fenzlianum saxicola, ochroleucum ochroleucum, ochroleucum coccineum, ochroleucum leiphaemium, ochroleucum porphyrium, sorediatum leprarioides, sorediatum neglectum*
GENUS *Heterodea* Variable Rim Lichen
SPECIES *beaugleholei, muelleri*
GENUS *Lecanicillium* Cill Rim Lichen
SPECIES *acerosum, antillanum, aphanocladii, aranearum, attenuatum, dimorphum, evansii, fusisporum, lecanii, longisporum, muscarium, nodulosum, psalliotae, tenuipes*
GENUS *Lecanoropsis* (Synonym for *Lecanora*)
GENUS *Lecanostictopsis* Stick Rim Lichen
SPECIES *eucalypti, kamatii, noumeaensis, syzygii*
GENUS *Lecidella* Tile Disc Rim Lichen
SPECIES *alpestris, anomaloides, asema, buelliastrum, carpathica, chodatii, dimelaenophila, elaeochroma, enteroleucella, euphorea, flavosorediata, granulata, meiococca, parasema, patavina, placidensis, pulveracea, scabra, stigmatea, subauriculata, sublapicida, subviridis, viridans, wulfenii*
VARIETIES *anomaloides umbrosa, asema effugiens, asema elaeochromoides, asema polyantha, asema subincongrua, carpathica latypiza, elaeochroma elaeochroma, elaeochroma glomerulosa, elaeochroma olivacea, elaeochroma soralifera, scabra prasinula, sublapicida tristicula*
GENUS *Leproloma* Lobed Rim Lichen
SPECIES *diffusum, diffusum membranaceum, vouauxii*
VARIETIES *diffusum, diffusum chrysodetoides, membranaceum lanuginosum*
GENUS *Lopezaria* Lopez Rim Lichen
SPECIES *versicolor*
GENUS *Maronina* Maron Rim Lichen
SPECIES *australiensis, multifera*
GENUS *Marssoniella* Marsson Rim Lichen
SPECIES *carpetana, elegans, marssonii*
GENUS *Miltidea* Milt Rim Lichen
SPECIES *ceroplasta*
GENUS *Miriquidica* Miriquid Rim Lichen
SPECIES *atrofulva, complanata, deusta, garovaglii, griseoatra, intrudens, leucophaea, lulensis, nigroleprosa*
VARIETIES *complanata complanata, complanata sorediata, lulensis leucophaeoides, nigroleprosa nigroleprosa, nigroleprosa liljenstroemii*
GENUS *Mycoblastus* Blood Stain Rim Lichen
SPECIES *affinis, alpinus, caesius, campbellianus, fucatus, sanguinarius*
VARIETIES *sanguinarius sanguinarius, sanguinarius leprosus*
GENUS *Myrionora* Myrion Rim Lichen
SPECIES *albidula*
GENUS *Nimisiostella* Nimis Star Rim Lichen
SPECIES *lichenicola*
GENUS *Nesolechia* (Synonym for *Carbonea*)
GENUS *Notolecidea* Notable Rim Lichen
SPECIES *subcontinua*
GENUS *Parmularia* (Synonym for *Lecanora*)
GENUS *Pleopsidium* Cobblestone Rim Lichen
SPECIES *chlorophanum, flavum*
GENUS *Protoparmeliopsis* (Synonym for *Lecanora*)
GENUS *Psorinia* Mangy Dot Rim Lichen
SPECIES *conglomerata, lepidotella*
GENUS *Punctonora* Dot Rim Lichen
SPECIES *nigropulvinata*
GENUS *Pycnora* Dense Rim Lichen
SPECIES *sorophora, xanthococca*
GENUS *Pyrrhospora* Fire Spore Rim Lichen
SPECIES *arandensis, aurea, bullata, cinnabarina, elabens, laeta, quernea, rubiginans, sanguinolenta, subcinnabarina*
GENUS *Ramalinora* Linear Branch Rim Lichen
SPECIES *glaucolivida*
GENUS *Ramboldia* Rambold Rim Lichen
SPECIES *brunneocarpa, crassithallina, farinosa, insidiosa, petraeoides, plicatula, sorediata, stuartii, subnexa*

GENUS *Rhizoplaca* Posy Navel Rim Lichen
SPECIES *chrysoleuca, glaucophana, haydenii, marginalis, melanophthalma, peltata, subdiscrepans*
GENUS *Sagema* Sage Rim Lichen
SPECIES *potentillae*
GENUS *Schistophoron* Split Rim Lichen
SPECIES *tenue, variabile*
GENUS *Scoliciosporum* Shade Dot Rim Lichen
SPECIES *chlorococcum, compactum, curvatum, pruinosum, sarothamni, umbrinum*
GENUS *Stenhammarella* Stenhammar Rim Lichen
SPECIES *turgida*
GENUS *Strangospora* Strange Spore Rim Lichen
SPECIES *deplanata, microhaema, moriformis, ochrophora, pinicola*
GENUS *Timdalia* Timdal Rim Lichen
SPECIES *intricata*
GENUS *Traponora* Trapon Rim Lichen
SPECIES *asterella*
GENUS *Tylophoron* Tyle Rim Lichen
SPECIES *gibsonii, protrudens*
GENUS *Tylothallia* Tyle Thallus Rim Lichen
SPECIES *biformigera, pahiensis*
GENUS *Vainionora* Vain Rim Lichen
SPECIES *aemulans, flavovirens, pallidostraminea, stramineopallens, warmingii*
GENUS *Vezdaea* Vezda Rim Lichen
SPECIES *acicularis, aestivalis, cobria, dawsoniae, foliicola, leprosa, obscura, polyspora, retigera, rheocarpa, stipitata*

FAMILY 02 *Acarosporaceae* Cracked Rim Lichen
GENUS *Acarospora* Cracked Rim Lichen
SPECIES *admissa, aeruginosa, americana, anomala, applanata, arenacea, arenosa, asahinae, asperata, badiofusca, benedarensis, bullata, caesiofusca, californica, canadensis, carnegiei, cervina, cinereoalba, citrina, clauzadeana, coloradiana, complanata, desolata, dispersa, durietzii, elevata, fuscata, fuscescens, fuscorufa, gallica, geogena, glaucocarpa, hassei, heppii, heufleriana, hilaris, immersa, impressula, instrata, interjecta, interposita, interspersa, macrocarpa, macrospora, molybdina, negligens, nevadensis, nigromarginata, nitida, nitrophila, nodulosa, novae-hollandiae, obnubila, obpallens, oligospora, particularis, peltastica, pyrenopsoides, radicata, rhizobola, rufescens, rugulosa, saepincola, schleicheri, scotica, sinopica, smaragdula, socialis, sparsa, stapfiana, strigata, subrufula, succedens, superfusa, tasmanica, tenebrica, terricola, thamnina, thelococcoides, thermophila, tongleti, tusconensis, umbilicata, utahensis, veronensis, verruciformis, washingtonensis*
VARIETIES *cervina glaucocarpa, fuscata peliscypha, glaucocarpa depauperata, glaucocarpa percaenoides, heppii luteopruinosa, impressula atrata, impressula opaca, macrospora macrospora, macrospora albomarginata, macrospora incusa, macrospora marginata, macrospora murorum, macrospora squamulosa, nitrophila aequatula, nitrophila muddii, nitrophila normanii, nitrophila praeruptorum, nodulosa nodulosa, nodulosa ferdinandii, nodulosa reagens, schleicheri contigua, smaragdula amphibola, smaragdula fusca, smaragdula lesdainii, smaragdula murina, smaragdula rufescens, smaragdula scabrida, smaragdula scyphulifera, smaragdula sinopica, umbilicata umbilicata, umbilicata congredians, veronensis discreta, veronensis magnussonii*
GENUS *Endocena* Endocene Cracked Rim Lichen
SPECIES *informis*
GENUS *Glypholecia* Glyph Cracked Rim Lichen
SPECIES *scabra*
GENUS *Lithoglypha* Rock Glyph Cracked Rim Lichen
SPECIES *aggregata*
GENUS *Myriospora* (Synonym for *Acarospora*)
GENUS *Polysporina* Coal Dust Cracked Rim Lichen
SPECIES *cyclocarpa, ferruginea, lapponica, simplex, terricola, urceolata*
GENUS *Sarcogyne* Flesh Cracked Rim Lichen
SPECIES *anthrocarpa, bicolor, bolleana, californica, clavus, dakotensis, integra, kuleshanensis, magnusonii, novomexicana, oligospora, privigna, pruinosa, regularis, similis*
VARIETIES *privigna hypophaea, regularis pruinosa*

FAMILY 03 *Arctomiaceae* Arctic Tumor Rim Lichen
GENUS *Arctomia* Arctic Tumor Rim Lichen
SPECIES *delicatula, interfixa*
GENUS *Gregorella* Gregor Arctic Tumor Rim Lichen
SPECIES *humida*
GENUS *Pseudarctomia* Lesser Arctic Tumor Rim Lichen
SPECIES *microleuca*
GENUS *Wawea* Wawe Arctic Tumor Rim Lichen
SPECIES *fruticulosa*

FAMILY 04 *Bacidiaceae* Dot Rim Lichen

GENUS *Bacidia* Dot Rim Lichen

SPECIES *absistens, adastra, aggregatula, alaskensis, alutacea, arceutina, artyta, auerswaldii, augustinii, bagliettoana, beckhausii, biatorina, brandii, brouardii, caesiovirens, caligans, campalea, campbelliae, carneoglauca, chlorococca, circumspecta, conspicua, convexa, coralloidea, diffracta, effusa, entocosmensis, entodiaphana, flavens, fratruelis, friesiana, furfurella, fuscoviridis, granosa, helicospora, hemipolia, herbarum, heterochroa, hostheleoides, igniarii, illudens, incompta, insularis, ioessa, jacobi, killiasii, kingmanii, laurocerasi, leucocarpa, lividofusca, lividonigrans, macquariensis, medialis, microcarpa, micrommata, microphyllina, millegrana, modestula, multiseptata, mutabilis, nivalis, pallens, pallida, pallidonigrans, pammellii, phacodes, polychroa, quadrilocularis, raffii, ravenelii, reagens, rhodocardia, rubella, rudis, russeola, sabuletorum, salmonea, saxenii, saxicola, schweinitzii, scopulicola, septosior, siberiensis, stigmatella, subcircumspecta, subincompta, subproposita, subturgidula, suffusa, superbula, superula, trachona, triseptata, vallatula, verecundula, vermifera, vinicolor, viridescens, viridifarinosa*

VARIETIES *absistens intermissa, arceutina leightoniana, bagliettoana muscorum, beckhausii atriseda, beckhausii minuscula, beckhausii poliaena, beckhausii stenospora, carneoglauca antricola, carneoglauca byssoboliza, circumspecta quercicola, fuscoviridis albidocarnea, igniarii abbrevians, incompta incompta, incompta spissa, laurocerasi atrogrisea, millegrana millegrana, millegrana fusconigrescens, phacodes albescens, phacodes intermedia, polychroa acerina, polychroa fuscorubella, polychroa polysita, rubella luteola, rubella porriginosa, rubella rosella, subincompta affinis, vermifera hegetschweileri*

GENUS *Adelolecia* Adel Dot Rim Lichen

SPECIES *kolaensis, pilati*

GENUS *Bacidina* Jelly Dot Rim Lichen

SPECIES *aenea, apiahica, arnoldiana, assulata, californica, chloroticula, crystallifera, defecta, delicata, egenula, egenuloidea, inundata, neosquamulosa, mirabilis, pallidocarnea, phacodes, ramea, squamellosa, scutellifera, varia*

VARIETIES *arnoldiana arnoldiana, arnoldiana corticola, arnoldiana latebricola, egenula egenula, egenula epiphylla, inundata inundata, inundata allecta, phacodes phacodes, phacodes albescens, phacodes intermedia*

GENUS *Bacidiopsora* Spore Dot Rim Lichen

SPECIES *squamulosula*

GENUS *Boreoplaca* Bore Plaque Dot Rim Lichen

SPECIES *ultrafrigida*

GENUS *Catinaria* Cat Dot Rim Lichen

SPECIES *albocincta, atropurpurea, neuschildii, subcorallina*

GENUS *Compsocladium* Branch Dot Rim Lichen

SPECIES *archboldianum*

GENUS *Crustospathula* Crust Spoon Dot Rim Lichen

SPECIES *cartilaginea*

GENUS *Echidnocymbium* Urchin Cymb Dot Rim Lichen

SPECIES *speciosum*

GENUS *Frutidella* Bushy Dot Rim Lichen

SPECIES *caesioatra*

GENUS *Heppsora* Hepp Dot Rim Lichen

SPECIES *indica*

GENUS *Herteliana* Hertel Dot Rim Lichen

SPECIES *gagei, phaeops, taylorii*

GENUS *Japewia* Japew Dot Rim Lichen

SPECIES *carrollii, subaurifera, tornoënsis*

GENUS *Japewiella* Lesser Japew Dot Rim Lichen

SPECIES *tavaresiana*

GENUS *Jarmania* Jarman Dot Rim Lichen

SPECIES *tristis*

GENUS *Lecania* Crust Dot Rim Lichen

SPECIES *actaea, aipospila, atrynoides, baeomma, brunonis, californica, chlaronoides, chlorotiza, coeruleorubella, coerulescens, constricta, cuprea, cyrtella, cyrtellina, discreptans, dubitans, dudleyi, erysibe, fabacea, flavescens, fructigena, fuscella, groenlandica, hassei, heardensis, hutchinsiae, hyalina, inundata, johnstonii, molliuscula, muelleriana, naegelii, nylanderiana, olivacella, pepegospora, perproxima, polycycla, rabenhorstii, sanguinolenta, shastensis, suavis, subcaesia, subdispersa, subfuscula, sylvestris, toninioides, turicensis*

VARIETIES *aipospila maritima, aipospila spodophaeiza, baeomma caesia, baeomma rupicola, cuprea umbraticula, cyrtella sambucina, dubitans dimera, erysibe sorediata, fuscella syringea, hutchinsiae albocarnea, muelleriana subsquamosa, turicensis albariella, turicensis phaeoleucodes*

GENUS *Physcidia* Bladder Dot Rim Lichen

SPECIES *australasica*

GENUS *Schadonia* Schadon Dot Rim Lichen

SPECIES *alpina, fecunda*

GENUS *Solenopsora* Sole Psoral Dot Rim Lichen

SPECIES *candicans, crenata, elixiana, holophaea, liparina, tasmanica, vulturiensis*

GENUS *Speerschneidera* Rock Wool Dot Rim Lichen
SPECIES *euploca*
GENUS *Squamarina* Scaly Marine Dot Rim Lichen
SPECIES *cartilaginea, degelii, haysomii, lentigera*
GENUS *Tephromela* Ash Honey Dot Rim Lichen
SPECIES *aglaeida, arafurensis, armeniaca, atra, brisbanensis, bunyana, grumosa, isidiosa, korundensis, pertusarioides, testaceoatra*
VARIETIES *atra atra, atra torulosa*
GENUS *Tibellia* Tiby Dot Rim Lichen
SPECIES *dimerelloides*
GENUS *Waynea* Wayne Dot Rim Lichen
SPECIES *stoechadiana*

FAMILY 05 *Biatoraceae* Crust Dot Rim Lichen
GENUS *Biatora* Crust Dot Rim Lichen
SPECIES *albohyalina, anthracophila, botryosa, britannica, carneoalbida, chrysantha, cuprea, efflorescens, epixanthoides, flavopunctata, helvola, meiocarpa, porphyrospoda, pullata, sphaeroides, subduplex, tetramera, turgidula, vernalis*
VARIETIES *chrysantha epixanthoidiza, chrysantha gyrophorica, sphaeroides pilularis, tetramera obscurella, vernalis minor*
GENUS *Biatorella* Lesser Crust Dot Rim Lichen
SPECIES *australica, consanguinea, conspersa, cyphalea, desmaspora, floridensis, fossarum, hemisphaerica, plicata, rappii*
GENUS *Biatoridium* Mock Crust Dot Rim Lichen
SPECIES *delitescens, monasteriense*
GENUS *Biatorina* Least Crust Dot Rim Lichen
SPECIES *erysiboides, halophila, littoralis*
GENUS *Biatoropsis* False Crust Dot Rim Lichen
SPECIES *usnearum*
GENUS *Cliostomum* Clio Crust Dot Rim Lichen
SPECIES *corrugatum, flavidulum, griffithii, leprosum, tenerum, vitellinum*
VARIETIES *corrugatum graniforme, leprosum luteolum*
GENUS *Phyllopsora* Lace Crust Dot Rim Lichen
SPECIES *breviuscula, corallina, canoumbria, confusa, foliata, furfuracea, glabella, halei, parvifolia, parvifoliella, rosei, santensis, subhyalina*
VARIETIES *foliata foliata, foliata atrovirens, foliata subcorallina, parvifolia parvifolia, parvifolia fibrillifera, parvifolia granulosa, parvifolia subgranulosa*
GENUS *Piccolia* Tiny Crust Dot Rim Lichen
SPECIES *conspersa, haematina*

FAMILY 06 *Brigantiaeaceae* Brick Fire Dot Rim Lichen
GENUS *Brigantiaea* Brick Fire Dot Rim Lichen
SPECIES *fuscolutea, leucoxantha, lobulata, microcarpa, phaeomma, tricolor*
GENUS *Argopsis* Argo Brick Fire Dot Rim Lichen
SPECIES *megalospora*

FAMILY 07 *Candelariaceae* Candle Rim Lichen
GENUS *Candelaria* Candle Rim Lichen
SPECIES *concolor, crawfordii, fibrosa*
VARIETIES *concolor effusa*
GENUS *Caloplacopsis* (Synonym for *Candelina*)
GENUS *Candelariella* Goldspeck Candle Rim Lichen
SPECIES *antennaria, arctica, athallina, aurella, canadensis, citrina, coralliza, deflexa, dispersa, efflorescens, kuusamoensis, medians, placodizans, plumbea, reflexa, rosulans, spraguei, stenospora, subdeflexa, superdistans, terrigina,* ***vitellina***(Dye Herb), *xanthostigma, xanthostigmoides*
VARIETIES *aurella aurella, aurella epixantha, aurella smaragdula, aurella smaragdula-heidelbergensis, coralliza pulvinata, medians medians, medians steepholmensis, vitellina vitellina, vitellina corrusca, vitellina flavovirella, vitellina flavovirella-viridis, xanthostigma lutella*
GENUS *Candelina* Yoke Candle Rim Lichen
SPECIES *africana, mexicana, submexicana*
GENUS *Claurouxia* Clauroux Candle Rim Lichen
SPECIES *chalybeioides*

FAMILY 08 *Catillariaceae* Washed Rim Lichen
GENUS *Catillaria* Washed Rim Lichen
SPECIES *alba, alboflavicans, aphana, atomarioides, australica, banksiae, brisbanensis, bryophila, chalybeia, columbiana, contristans, croceella, cupressi, distorta, effugiens, endochroma, flavens, flavicans, franciscana, frenchiana, gilbertii, glauconigrans, globulosa, grossulina, heeri, jemtlandica, kelica, lenticularis, leptocheila, melaclina, melaclinoides, melaloma, minuta, modesta, muscicola, mycophila, nigroclavata,*

phaeoloma, picila, polycarpa, rimosa, rudolphii, scleroplaca, scotinodes, subfuscata, subintermixta, subnegans, subviridis, superflua, tasmanica, tenuilimbata, terrena, trachonoides, tristis, umbratilis, verrucosa
VARIETIES *chalybeia chalybeia, chalybeia chloropoliza, chalybeia chloropoliza-chloroscotina, contristans dufourei, lenticularis dolosa, lenticularis gagei, minuta arnoldii, minuta acrustacea, scotinodes confusior*
GENUS *Arthrosporum* Spider Washed Rim Lichen
SPECIES *populorum*
GENUS *Austrolecia* Southern Washed Rim Lichen
SPECIES *antarctica*
GENUS *Halecania* Halecan Washed Rim Lichen
SPECIES *alpivaga, australis, bryophila, elaeiza, giraltiae, lecanorina, micacea, ralfsii, rhypodiza, spodomela, tornensis, viridescens*
GENUS *Kiliasia* (Synonym for *Catillaria*)
GENUS *Sporastatia* Patch Washed Rim Lichen
SPECIES *polyspora, testudinea*
GENUS *Toninia* Bruised Blister Washed Rim Lichen
SPECIES *alutacea, arctica, aromatica, athallina, australiensis, australis, bullata, caeruleonigricans, candida, cinereovirens, coelestina, cumulata, diffracta, episema, fusispora, glaucocarpa, leucina, lutosa, massata, mesoidea, opuntioides, pennina, philippea, physaroides, plumbina, rosulata, ruginosa, scorigera, sculptuarata, sedifolia, squalescens, squalida, subdiffracta, subfuscae, submexicana , superioris , talparum, thiopsora, tristis, tumidula, verrucarioides, weberi*
VARIETIES *aromatica acervulata, aromatica turneri, bullata nitida, coelestina oribata, physaroides lurida, squalida cinereovirens, squalida havaasii, squalida squarrosa, thiopsora pulvinata, verrucarioides cervina, verrucarioides carbonacea, verrucarioides kolax*
GENUS *Wadeana* Wade Washed Rim Lichen
SPECIES *dendrographa, minuta*
GENUS *Xanthopsora* (Synonym *for Xanthopsorella*)
GENUS *Xanthopsorella* Yellow Washed Rim Lichen
SPECIES *texana*

FAMILY 09 *Coccocarpiaceae* Shell Rim Lichen
GENUS *Coccocarpia* Shell Rim Lichen
SPECIES *adnata, dissecta, domingensis, erythroxyli, glaucina, kerguelensis, palmicola, pruinosa, smaragdina, stellata*
VARIETIES *erythroxyli aurantiaca, erythroxyli ciliolata, erythroxyli leucorrhiza, erythroxyli mesomorpha, erythroxyli parmelioides, erythroxyli semiincisa, palmicola cronia, palmicola isidiophylla, palmicola isidiosa*
GENUS *Peltularia* Pelt Shell Rim Lichen
SPECIES *crassa*
GENUS *Spilonema* Hairball Shell Rim Lichen
SPECIES *dendroides, paradoxum, revertens*
GENUS *Steinera* Steiner Shell Rim Lichen
SPECIES *glaucella, werthii*

FAMILY 10 *Dactylosporaceae* Finger Rim Lichen
GENUS *Dactylospora* Finger Rim Lichen
SPECIES *amygdalariae, athallina, attendenda, australis, bloxamii, caledonica, deminuta, frigida, glaucomarioides, imperfecta, inquilina, lobariella, lurida, microspora, ophthalamizae, parasitica, parellaria, pertusaricola, purpurascens, saxatilis, scapanaria, stygia, urceolata*
GENUS *Leciographa* (Synonym for *Dactylospora*)
GENUS *Paruephaedria* Ephedra Finger Rim Lichen
SPECIES *heimerlii*

FAMILY 11 *Ectolechiaceae* Ecto Rim Lichen
GENUS *Ectolechia* (Synonym for *Sporopodium*)
GENUS *Badimia* Bay Ecto Rim Lichen
SPECIES *pallidula, polillensis*
GENUS *Badimiella* Lesser Bay Ecto Rim Lichen
SPECIES *pteridophila*
GENUS *Calopadia* Palmetto Dot Ecto Rim Lichen
SPECIES *foliicola, fusca, nymanii, perpallida, phyllogena, puiggarii, subcoerulescens, subfusca*
GENUS *Kantvilasia* Kantvilas Ecto Rim Lichen
SPECIES *hians*
GENUS *Lasioloma* Woolly Ecto Rim Lichen
SPECIES *arachnoideum, phycophilum, phycophorum, trichophorum*
GENUS *Loflammia* Flame Ecto Rim Lichen
SPECIES *epiphylla, gabrielis*
GENUS *Logilvia* Gilvan Ecto Rim Lichen
SPECIES *gilva*
GENUS *Lopadium* Urn Disc Ecto Rim Lichen

SPECIES *angustini, brisbanense, coralloideum, disciforme, dodgei, hepaticicola, lecanorellum, pezizoideum, pulchrum*

GENUS *Pseudocalopadia* — Lesser Palmetto Dot Ecto Rim Lichen

SPECIES *mira*

GENUS *Sporopodiopsis* — Lesser Spore Pod Ecto Rim Lichen

SPECIES *mortimeriana, sipmanii*

GENUS *Sporopodium* — Spore Pod Ecto Rim Lichen

SPECIES *aurantiacum, citrinum, flavescens, leprieurii, phyllocharis, xantholeucum*

GENUS *Tapellaria* — Tape Ecto Rim Lichen

SPECIES *bilimbioides, epiphylla, moelleri, nana, phyllophila*

FAMILY 12 *Heppiaceae* — Ruby Rim Lichen

GENUS *Heppia* — Ruby Rim Lichen

SPECIES *adglutinata, conchiloba, despreauxii, lutosa*

GENUS *Epiphloea* — Epiphyte Ruby Rim Lichen

SPECIES *terrena*

GENUS *Solorinaria* — (Synonym for *Heppia*)

FAMILY 13 *Hymeneliaceae* — Membrane Rim Lichen

GENUS *Hymenelia* — Membrane Rim Lichen

SPECIES *arctica, cyanocarpa, epulotica, heteromorpha, obtecta, prevostii, rhodopis*

VARIETIES *epulotica crustosa*

GENUS *Agrestia* — Vagabond Membrane Rim Lichen

SPECIES *hispida*

VARIETIES *hispida cyphellata*

GENUS *Aspicilia* — Sunken Disc Membrane Rim Lichen

SPECIES *albomarginata, albopruinosa, alboradiata, aliena, americana, annulata, anseris, artica, caesiocinerea, caesiopruinosa, calcarea, candida, cinerea, cingulata, composita, concinna, contigua, contorta, desertorum, disserpens, elevata, endochlora, epiglypta, esculenta*(Edible), *fimbriata, fruticulosa, grisea, heteroplaca, humboldtii, intermutans, karelica, laevata, laxula, leprosescens, lesleyana, limitata, mashiginensis, mastoidea, mazarina, moenium, myrinii, narssaquensis, nathoristii, nikrapensis, nordlandica, novae-semliae, pergibbosa, perradiata, pertusa, plicigera, polychroma, quartzitica, recedens, reptans, rolleana, rosulata, ryrkiapiae, simoënsis, sorediza, subcircinata, subdepressa, sublapponica, submersa, subplicigera, subradians, supertegens, tenuis, tuberculosa, verrucigera*

VARIETIES *caesiocinerea gibbosa, calcarea calcarea, calcarea caesioalba, calcarea lilliei, contorta contorta, contorta hoffmanniana, grisea insolata, laevata depressa, laevata lusca, laevata sylvatica, moenium excavata, recedens bohemica*

GENUS *Bouvetiella* — Bouvet Membrane Rim Lichen

SPECIES *pallida*

GENUS *Eiglera* — Eigler Membrane Rim Lichen

SPECIES *flavida*

GENUS *Ionaspis* — Watercolor Membrane Rim Lichen

SPECIES *alba, annularis, chrysophana, kerguelensis, lacustris, lavata*

VARIETIES *chrysophana suaveolens*

GENUS *Lobothallia* — Puffed Sunken Disc Membrane Rim Lichen

SPECIES *alphoplaca, melanaspis, praeradiosa, radiosa*

GENUS *Melanolecia* — Dark Membrane Rim Lichen

SPECIES *transitoria*

GENUS *Tremolecia* — Tremor Disc Membrane Rim Lichen

SPECIES *atrata*

FAMILY 14 *Lecideaceae* — Tile Rim Lichen

GENUS *Lecidea* — Tile Rim Lichen

SPECIES *aberrata, admiscens, advertens, ahlesii, albofescescens, amabilis, amniculensis, antiloga, aspera, atmoaria, atrobrunnea, atromorio, atroviridis, auriculata, bacidioides, baffiniana, berengeriana, brachyspora, brunneofusca, californica, calpodes, canorufescens, capensis, carnulenta, cascadensis, cellularis, cerarufa, cerviniicola, cetrariicola, cinerata, columnata, commaculans, confluens, confluentula, congesta, contigua, convexa, crassilabra, crisima, cruciaria, degelii, delincta, deminutula, demolita, despecta, diapensiae, diducens, doliiformis, dolodes, eckfeldtii, ecrustacea, ementiens, endomelaena, erythrophaea, exigua, extenuata, flavidolivens, flindersii, floridensis, fuliginosa, furfurosa, furva, furvonigrans, fuscoatra, fuscoatrula, glaucopholis, goniophiloides, granifera, haerjedalica, hassei, haysomii, hebescens, holopolia, homalodes, homosema, humilis, hypersporella, hypnorum, hypomela, hypopta, immarginata, inops, insulana, interjuncta, interstincta, interstita, intropallida, intrudens, katahdinensis, labradorica, lactea, lapicida, lepidastra, leprarioides, leptoboloides, leptoloma, leptolomoides, leucothallina, lichenicola, limosa, lithophila, louisianae, ludibunda, lurida, luteoatra, lygomma, lygommoides, lyngei, macquariensis, mammillana, mannii, marciensis, marylandensis, meiospora, melaphanoides, merrillii, micytho, minutula, moreliiensis, mucosa, multiflora, mutabilis, myriocarpella, nearingii, nemoralis, nylanderi, obluridata, occidentalis, ochroleuca, olivascens, pacifica, paddensis, pallidoatra, paraclitica, paupercula, peliaspis, pelochroa, phaeocarpa, phaeopelidna, phaeops, picea, plana, planetica, plebeja, polaris, polycocca, populina, porphyria, porphyrospoda, praenubila, praetermissa, promiscens, promiscua, promixta, protobacina, pruinosula, pulla, pumicicola, punctella, pycnocarpa, ramulosa, rivulorum, roseotincta, rufofusca, rugosa, russula, sanguineoatra, santaemonicae, sarcogynoides, scabrida, scrobiculata, scrupulosa, selenospora, shushanii, siderolithica, silacea, somphoterella, soredifera, sphaerella, steineri, subalgaea, subassentiens, subcaerulea, subcandida, subcinerascens, subduplex, sublimosa, sublygomma, subplana, subplebeja, subramosa, subrhagadiella, subsimilis, subspeirea, subtecta, subtilis, superjecta, swartzioidea, syncarpa, tenayucae, tenella, tenuissima, terrena, tessellata, theodori, torquens, tragorum, truckeei, turgidula, umbonata, vacciniicola, varians, violascens, virginiensis, werthii, xanthococcoides, xylogena*

VARIETIES *pycnocarpa pycnocarpa, pycnocarpa sorediata*

GENUS *Bahianora* — Bahi Tile Rim Lichen

SPECIES *poeltii*

GENUS *Cecidonia* — Cut Tile Rim Lichen

SPECIES *umbonella, xenophana*

GENUS *Hypocenomyce* — Cockle Shell Tile Rim Lichen

SPECIES *anthracophila, australis, caradocensis, castaneocinera, foveata, friesii, leucococca, praestabilis, scalaris*

GENUS *Lopacidia* — Lopac Tile Rim Lichen

SPECIES *multilocularis*

GENUS *Malcolmiella* — Malcolm Tile Rim Lichen

SPECIES *piperis, subgranifera*

GENUS *Nesolechia* — (Synonym for *Lecidea*)

GENUS *Rhizolecia* — Rhizid Tile Rim Lichen

SPECIES *hybrida*

GENUS *Steinia* — Stein Tile Rim Lichen

SPECIES *australis, geophana*

FAMILY 15 *Loxosporaceae* — Ragged Rim Lichen

GENUS *Loxospora* — Ragged Rim Lichen

SPECIES *cismonica, cyamidia, elatina, ochrophaea, pustulata, solenospora*

GENUS *Loxosporopsis* — Lesser Ragged Rim Lichen

SPECIES *corallifera*

FAMILY 16 *Megalariaceae* — Forest Dot Rim Lichen

GENUS *Megalaria* — Forest Dot Rim Lichen

SPECIES *brodoana, grossa, jemtlandica, laureri, melanotropa, pulverea*

GENUS *Tasmidella* — Tasm Forest Dot Rim Lichen

SPECIES *variabilis*

VARIETIES *variabilis variabilis, variabilis inactiva*

FAMILY 17 *Megalosporaceae* — Megalith Rim Lichen

GENUS *Megalospora* — Megalith Rim Lichen

SPECIES *melanodermia, occidentalis, pachycarpa, pachycheila, porphyritis, pulverata, queenslandica, reniformis, subtuberculosa, sulphurata, tuberculosa*

VARIETIES *melanodermia melanodermia, melanodermia galactocarpa, pulverata campylospora, sulphurata taitense*

GENUS *Austroblastenia* — Southern Blastid Megalith Rim Lichen

SPECIES *pauciseptata, pupa*

GENUS *Megaloblastenia* — Blastid Megalith Rim Lichen

SPECIES *flavidoatra, marginiflexa*

FAMILY 18 *Micareaceae* — Mica Dot Rim Lichen

GENUS *Micarea* — Mica Dot Rim Lichen

SPECIES *adnata, alabastrites, assimilata, austroternaria, bauschiana, botryoides, chlorosticta, cinerea, clavopycnidiata, contexta, coppinsii, crassipes* (Nail Fruit), *curvata, deminuta, denigrata, elachista, endocyanea, erratica, eximia, flagellispora, globulosella, hedlundii, incrassata, inquinans, intersociella, isabellina, lapillicola, leprosula, lignaria, lithinella, lutulata, lynceola, magellanica, marginata, melaena, melaenida, melanobola, micrococca, misella, mutabilis, myriocarpa, nigella, nitschkeana, olivacea, paratropa, parva, peliocarpa, polycarpella, prasina, prasinella, pseudomarginata, pycnidiophora, stipitata, subconfusa, submilliaria, subnigrata, subviridescens, sylvicola, synotheoides, ternaria, tuberculata, turfosa, viridiatra, viridileprosa, vulpinaris, xanthonica*

VARIETIES *alabastrites klementii, bauschiana semipallens, cinerea cinerea, cinerea tenuispora, globulosella bacidiella, lignaria lignaria, lignaria endoleuca, lignaria gomphillacea, lutulata polioides, lutulata umbrosa, lynceola excipulata, melaena globularis, paratropa subviolascens, peliocarpa trisepta, peliocarpa violacea, prasina sordidescens, subconfusa submoestula, submilliaria granulans, submilliaria subleprosula, subnigrata confusula, synotheoides longispora, turfosa verrucula*

GENUS *Helocarpon* — (Synonym for *Micarea crassipes, Trapeliopsis*)

GENUS *Psilolechia* — Bare Dust Mica Dot Rim Lichen

SPECIES *clavulifera, leprosa, lucida, purpurascens*

GENUS *Roccellinastrum* — Orchil Star Mica Dot Rim Lichen

SPECIES *flavescens, lagarostrobi, neglectum*

GENUS *Scutula* — Scute Mica Dot Rim Lichen

SPECIES *aggregata, cristata, dedicata, epiblastematica, epicladonia, krempelhuberi, miliaris, solorinaria, stereocaulorum*

FAMILY 19 *Pannariaceae* — Pan Rim Lichen

GENUS *Pannaria* — Mat Shingle Pan Rim Lichen

SPECIES *aenea, centrifuga, cheiroloba, conoplea, crenulata, crustata, decipiens, delicata, dichroa, dissecta, durietzii, elatior, elegantior, elixii, euphylla, fimbriata, fulvescens, globigera, hookeri, immixta, leproloma, lurida, mangroviana, mariana, microphyllizans, molkenboeri, obscura, pezizoides, ramosii, rubiginosa, sorediata, subcrustacea, tavaresii, tjibodensis, xanthorioides*

VARIETIES *conoplea pityrea, conoplea rubiginosa, crustata recedens, hookeri leucolepis, immixta flexuosa, lurida sublurida*

GENUS *Amphiloma* — (Synonym for *Leproloma*)

GENUS *Austrella* — Southern Pan Rim Lichen

SPECIES *arachnoidea*

GENUS *Degelia* — Degel Pan Rim Lichen

SPECIES *atlantica, calcicola, duplomarginata, durietzii, flabellata, gayana, ligulata, neozelandica, periptera, plumbea, rosulata, subcrustata*

GENUS *Erioderma* — Mouse Ears Felt Pan Rim Lichen

SPECIES *leylandii, mollissimum, pedicellatum, sorediatum*

VARIETIES *leylandii knightii*

GENUS *Fuscoderma* — Fuscous Pan Rim Lichen

SPECIES *limbatum*

GENUS *Fuscopannaria* — Fuscous Mat Shingle Pan Rim Lichen

SPECIES *ahlneri, granulans, ignobilis, leucophaea, leucosticta, leucostictoides, maritima, mediterranea, minor, praetermissa, pulveracea, sampaiana, saubinetii, subimmixta*

GENUS *Hosseusia* — Hosseus Pan Rim Lichen

SPECIES *gertrudiana*

GENUS *Kroswia* — Kroswian Pelt Pan Rim Lichen

SPECIES *crystallifera*

GENUS *Leioderma* — Tree Pelt Pan Rim Lichen

SPECIES *amphibolum, applanatum, duplicatum, erythrocarpum, glabrum, pycnophorum, sorediatum*

GENUS *Moelleropsis* — Moeller Pan Rim Lichen

SPECIES *nebulosa*

GENUS *Parmeliella* — Lead Shingle Pan Rim Lichen

SPECIES *aggregata, arctophila, brisbanensis, coerulescens, crossophylla, cyanolepra, diffracta, furfuracea, granulata, granulifera, gymnocheila, ligulata, mariana, nigrata, nigrocincta, palmatula, pannosa, papillata, parvula, polyphyllina, ruderatula, subtilis, testacea, thysanota, triptophylla*

VARIETIES *parvula jamesii, thysanota rubiginascens, triptophylla corallinoides, triptophylla incrassata*

GENUS *Protopannaria* — Primitive Pan Rim Lichen

SPECIES *pezizoides*

GENUS *Psoroma* — Shingle Pan Rim Lichen

SPECIES *asperellum, caliginosum, fruticulosum, geminatum, hirsutulum, hypnorum, macquariense, multifidum, pholidotoides, soccatum, sphinctrinum, tenue*

VARIETIES *caliginosum dispersum, hypnorum hypnorum, hypnorum paleaceum, pholidotoides crispellum, pholidotoides reticulatum*

GENUS *Psoromidium* — Lesser Shingle Pan Rim Lichen

SPECIES *aleuroides*

GENUS *Reinkella* — (Synonym for *Pannaria*)

GENUS *Santessoniella* — Santesson Pan Rim Lichen

SPECIES *pulchella, rugosa*

GENUS *Triclinum* — Tricline Pan Rim Lichen

SPECIES *cinchonarum*

FAMILY 20 *Phlyctidaceae* — Whitewash Rim Lichen

GENUS *Phlyctis* — Whitewash Rim Lichen

SPECIES *agelaea, argena, bifurcatus, cervicornis, cuneiformis, dichotomus, latifolius, polypodioides, speirea, subfistulosus, undulatus, willeyi*

VARIETIES *argena erythrosora*

GENUS *Phlyctella* — Lesser Whitewash Rim Lichen

SPECIES *andensis, subuncinata*

VARIETIES *subuncinata wilsonii*

GENUS *Phlyctidia* — Blemished Whitewash Rim Lichen

SPECIES *ludoviciensis*

GENUS *Phoracis* — Phorac Whitewash Rim Lichen

SPECIES *filicina*

GENUS *Psathyrophlyctis* — Thyre Whitewash Rim Lichen

SPECIES *serpentaria*

FAMILY 21 *Physciaceae* — Bladder Rim Lichen

GENUS *Physcia* Rosette Bladder Rim Lichen

SPECIES *adscendens, aipolia, alba, albata, albinea,* ***americana***(Dye Herb), *atrostriata, biziana, caesia, cascadensis, clementei, convexa, crispa, decorticata, dimidiata, dubia, duplicorticata, fragilescens, halei, integrata, jackii, leptalea, macquariensis, magnussoni, mexicana, millegrana, montana, neogaea, nubila, phaea, poncinsii, pseudospeciosa, pumilior, sorediosa, stellaris, subtilis, tenella, tribacia, tribacioides, undulata, verrucosa*

VARIETIES *adscendens anthelina, adscendens stellaris, aipolia cercidia, aipolia stellaris, albata alboplumbea, caesia subalbinea, caesia wainioi, clementei astroidea, dubia teretiuscula, jackii callosa, leptalea semipinnata, stellaris rosulata, tenella tenella, tenella marina, tenella marina-subobscura, tribacia tribacia, tribacia erosa, tribacia tenuis*

GENUS *Amandinea* Amandine Button Bladder Rim Lichen

SPECIES *coniops, dakotensis, efflorescens, endochroa, extenuata, insperata, lecideina, milliaria, petermannii, polyspora, punctata, subduplicata, subplicata, turgescens*

VARIETIES *punctata depauperata, punctata myriocarpa, punctata stigmatea*

GENUS *Anaptychia* Fringed Bladder Rim Lichen

SPECIES *bryorum, ciliaris, palmulata, runcinata, setifera, ulotrichoides*

VARIETIES *ciliaris ciliaris, ciliaris mamillata, ciliaris mamillata-melanosticta, ciliaris saxicola, runcinata aquila, runcinata caesiopruinosa, runcinata fusca*

GENUS *Australiaena* Australia Bladder Rim Lichen

SPECIES *streimannii*

GENUS *Awasthia* Awasth Bladder Rim Lichen

SPECIES *melanotricha*

GENUS *Buellia* Button Bladder Rim Lichen

SPECIES *adjuncta, aequata, aeruginascens, aeruginosa, aethalea, amblyogona, americana, amphidexia, arborea, arenaria, arnoldii, asterella, atroflavella, australica, badia, bolacina, calcariaecola, caloosensis, capitis-regnum, catasema, cinnabarina, coccinea, cretacea, curatellae, curtisii, desertorum, dialyta, dijiana, diplotommoides, disciformis, dispersa, elegans, elizae, endoleuca, epigaea, erubescens, excelsa, farinulenta, fertilis, fuliginosa, geophila, georgei, gerontoides, glomerulans, griseovirens, halonia, homophylia, hyperbolica, hypomelaena, imshaugiana, imshaugii, innata, insignis, inturgescens, jugorum, krempelhuberi, lactea, lacteoidea, lauri-cassiae, lecanoroides, lepidastra, leptocline, leptoclinoides, leucomela, lobata, mamillana, marginulata, maritima, mawsonii, microbola, molonglo, nantiana, nigra, notabilis, novomexicana, ocellata, oidalea, pachnidisca, papillata, perexigua, placodiomorpha, polyspora, protothallina, pruinosa, pulverea, pulverulenta, rappii, retrovertens, rimulosa, rosellotincta, rubifaciens, russa, sanguinolenta, saurina, saxorum, schaereri, semitensis, sequax, silicicola, smaragdula, spuria, stellulata, stigmaea, stillingiana, subalbula, subareolata, subcallispora, subcoronata, subdisciformis, submaritima, subrepleta, substellulans, succedens, talcophila, triphragmoides, tristiuscula, uberior, ventricosa, vernicoma, vilis, violaceofusca, wheeleri, wilsoniana*

VARIETIES *aethalea aethalea, aethalea baltica, aethalea impressula, aethalea sororioides, americana modesta, disciformis cinereoferruginea, disciformis vulgata, disciformis wilsonii, griseovirens betulina, homophylia homophylia, homophylia amphibola, homophylia emphytocarpa, krempelhuberi exilis, leptoclinoides saxicola, ocellata praeponens, ocellata verruculosa, polyspora punctata, retrovertens tusconensis, sequax abstracta, subdisciformis subdisciformis, subdisciformis meiosperma, subdisciformis ryssolea*

GENUS *Buelliella* Lesser Button Bladder Rim Lichen

SPECIES *minimula, physciicola, trypethelii*

GENUS *Buelliopsis* (Synonym for *Buellia*)

GENUS *Cratiria* Cratir Bladder Rim Lichen

SPECIES *obscurior, rutilantoides*

GENUS *Dermatiscum* (Synonym for *Dermiscellum*)

GENUS *Dermiscellum* Dermal Bladder Rim Lichen

SPECIES *catawbense*

GENUS *Dimelaena* Moon Glow Bladder Rim Lichen

SPECIES *australiensis, californica, elevata, oreina, radiata, suboriena, tenuis, thysanota, weberi*

GENUS *Diploicia* Lobed Button Bladder Rim Lichen

SPECIES *canescens*

VARIETIES *canescens canescens, canescens australasica*

GENUS *Diplotomma* Tumor Button Bladder Rim Lichen

SPECIES *alboatrum, ambiguum, chlorophaeum, chlorophaeum porphyricum, epipolium, murorum, nivalis, penichrum, pharcidium, venustum*

GENUS *Dirinaria* Medallion Bladder Rim Lichen

SPECIES *aegialita, applanata, applanata ochracea, aspera, batavica, complicata, confluens, confusa, consimilis, flava, frostii, leopoldii, melanoclina, minuta, papullifera, picta, purpurascens, subconfluens*

VARIETIES *applanata ochracea*

GENUS *Hafellia* Hafell Bladder Rim Lichen

SPECIES *bahiana, curatellae, demutans, dissa, forsteri, levieri, parastata, pleiotera, procellarum, pseudotetrapla, reagens, rechingeri, subcrassata, tetrapla*

VARIETIES *demutans callispora*

GENUS *Heterodermia* Fringed Bladder Rim Lichen

SPECIES *albicans, angustiloba, antillarum, appalachensis, appendiculata, barbifera, boryi, casarettiana, chilensis, chondroidea, comosa, crocea, dendritica, diademata, dissecta, domingensis, echinata, erinacea, flabellata, fragilissima, granulifera, gylactophylla, hypocaesia, hypoleuca, isidiophora, japonica, leucomela, microphylla, neoleucomelaena, obscurata, podocarpa, pseudospeciosa, rugulosa, sitchensis, speciosa, squamulosa, subcomosa, tremulans, tropica*

VARIETIES *japonica propagulifera*

GENUS *Hyperphyscia* Shadow Crust Bladder Rim Lichen

SPECIES *adglutinata, minor, pruinosa, syncolla*

GENUS *Mobergia* Moberg Bladder Rim Lichen

SPECIES *angelica, calculiformis*

GENUS *Phaeophyscia* Shadow Bladder Rim Lichen

SPECIES *adiastola, cernohorskyi, ciliata, constipata, decolor, endococcina, endococcinodes, endophoenicea, erythrocardia, fumosa, hirsuta, hirtella, hispidula, imbricata, insignis, kairamoi, leana, nigricans, orbicularis, pusilloides, rubropulchra, sciastra* VARIETIES *orbicularis hueana, orbicularis virella*

GENUS *Phaeorrhiza* Brown Fuzz Bladder Rim Lichen

SPECIES *nimbosa, sareptana*

GENUS *Physciella* Lesser Rosette Bladder Rim Lichen

SPECIES *chloantha, melanchra, nepalensis*

GENUS *Physciopsis* (Synonym for *Hyperphyscia*)

GENUS *Physconia* Frost Bladder Rim Lichen

SPECIES *detersa, distorta, elegantula, enteroxantha, farrea, grisea, isidiigera, kurokawae, leucoleiptes, muscigena, perisidiosa, pulverulenta, subpallida, thomsonii*

GENUS *Pyxine* Button Rosette Bladder Rim Lichen

SPECIES *albovirens, australiensis, berteriana, coccifera, cocoës, cognata, consocians, convexior, copelandii, coralligera, cylindrica, daedalea, desudans, elixii, endochrysina, fallax, farinosa, glaucovirescens, keralensis, microspora, minuta, nubila, petricola, plumea, pungens, retirugella, rugulosa, schmidtii, sorediata, subcinerea*

VARIETIES *berteriana endoxantha, berteriana meissneri, consocians physciaeformis, farinosa linearis, farinosa papuana, farinosa sorediosa, keralensis isidiolenta, microspora pyxinoides, petricola endoleuca, sorediata eschweileri, subcinerea caesiopruinosa, subcinerea chrysanthoides, subcinerea sorediosa*

GENUS *Rinodina* Pepper Spore Bladder Rim Lichen

SPECIES *adirondackii, americana, annulata, applanata, archaea, arctica, ascociscana, asperata, aspersa, athallina, atrocinerea, australiensis, beccariana, biloculata, bischoffii, blastidiata, bolanderi, bolodes, brattii, brouardii, cacuminum, calcarea, calcigene, castaneomela, castanmeloides, chrysomelaena, cinereovirens, colobinoides, coloradiana, compensata, confragosa, confragosula, confusa, conradii, constrictula, coriticola, dakotensis, darrowii, degeliana, destituta, dirinoides, disjuncta, dissa, dolichospora, efflorescens, elixii, endophragma, ericina, excrescens, filsonii, fimbriata, finkii, flavosoralifera, gennarii, glauca, granulans, granuligera, griseosoralifera, halei, hallii, herrei, hyberborea, inaequalis, immersa, interpolata, iowensis, isidioides, kentuckyensis, kozukensis, laevigata, lecanoides, lecanorina, lecideioides, lepida, luridata, luridescens, lycopodiicola, marysvillensis, metaboliza, milliaria, milvina, milvinodes, minutissima, mniaraea, moziana, mucrconatula, murrayi, obscura, occulta, ochrocea, oleae, olivaceobrunnea, orculariopsis, oregana, oxydata, pachysperma, papillata, parasitica, peloleuca, pennsylvanica, pityrea, polyspora, populicola, pyrina, pyriniformis, roboris, roscida, santae-monicae, septentrionalis, sheardii, silicicola, sophodes, stictica, striatitunicata, subbadioatra, subcrustacea, subminuta, subplumbea, subsophodes, substellulata, tasmanica, teichophila, tephraspis, thiomela, thujae, tibellii, trachytica, turfacea, vegasii, williamsii, xanthomelana, zwackhiana*

VARIETIES *beccariana subglaucescens, blastidiata teichophiloides, calcarea caesiella, confragosa metabolica, conradii intermedia, ericina madeirensis, gennarii demissa, gennarii lecideoides, gennarii salina, gennarii subexigua, immersa bischoffii, lecanorina ocellata, luridescens coniopta, mniaraea mniaraea, mniaraea cinnamomea, mniaraea mniaraeiza, oleae exigua, olivaceobrunnea archaeoides, oxydata biatorina, oxydata discolor, oxydata subarenaria, pityrea colobina, roboris roboris, roboris armeriicola, substellulata minutula, trachytica lecanorina*

GENUS *Rinodinella* Lesser Pepper Spore Bladder Rim Lichen

SPECIES *halophila*

GENUS *Sculptolumina* Sculpted Bladder Rim Lichen

SPECIES *japonica*

GENUS *Stigmatochroma* Stigma Colored Bladder Rim Lichen

SPECIES *epimarta*

GENUS *Tornabea* Tornabe Bladder Rim Lichen

SPECIES *scutellifera*

FAMILY 22 *Pilocarpaceae* Fuzzy Rim Lichen

GENUS *Pilocarpon* Fuzzy Rim Lichen

SPECIES *chlorinum, discordans, leucoblepharum*

GENUS *Bapalmuia* Bapalm Fuzzy Rim Lichen

SPECIES *buchananii, cacaotica, callichroa, confusa, costaricensis, halleana, ivoriensis, kakouettae, lafayetteana, lineata, marginalis, nigrescens, palmularis, rubicunda, sorediata, variratae, verrucosa*

GENUS *Byssoloma* Silky Fuzzy Rim Lichen

SPECIES *annuum, aurantiacum, confusum, dimerelloides, marginatum, meadii, microcarpum, murinum, pubescens, subdiscordans, subleucoblepharum, subundulatum, tricholomum, vanderystii* VARIETIES *annuum clauzadei, subdiscordans rotuliforme*

GENUS *Fellhanera* Fellhan Fuzzy Rim Lichen

SPECIES *aurantiaca, bouteillei, bullata, dominicana, endopurpurea, floridana, fuscatula, mastothallina, microdiscus, ochracea, parvula, rhapidophylli, semecarpi, subternella, subtilis, tasmanica, viridisorediata*

GENUS *Fellhaneropsis* Lesser Fellhan Fuzzy Rim Lichen

SPECIES *australiana, myrtillicola, vezdae*

GENUS *Pyriomyces* Fire Fuzzy Rim Lichen

SPECIES *protii*

FAMILY 23 *Porpidiaceae* Boulder Rim Lichen

GENUS *Porpidia* Boulder Rim Lichen

SPECIES *albocaerulescens, calcarea, carlottiana, cinereoatra, confluenta, contraponenda, crustulata, diversa, flavicunda, flavocaerulescens, flavocruenta, glaucophaea, grisea, herteliana, hydrophila, iowiana, islandica, macrocarpa, melinodes, musiva, nadvornikiana, ochrolemma, pachythallina, platycarpoides, rugosa, soredizodes, speirea, stephanodes, striata, subsimplex, superba, tahawasiana, thomsonii, tuberculosa, umbonifera, zeoroides*

VARIETIES *albocaerulescens albocaerulescens, albocaerulescens polycarpiza, macrocarpa macrocarpa, macrocarpa nigrocruenta, macrocarpa striata, platycarpoides cinereoatra, platycarpoides percontigua, superba superba, superba sorediata*

GENUS *Amygdalaria* Almond Boulder Rim Lichen

SPECIES *consentiens, continua, elegantior, haidensis, panaeola, pelobotryon, subdissentiens*

GENUS *Anamylospora* Enamel Boulder Rim Lichen

SPECIES *pulcherrima*

GENUS *Bellemerea* Sunken Disc Boulder Rim Lichen

SPECIES *alpina, cinereorufescens, cupreoatra, diamarta, pullata, sanguinea, subcandida, subsorediza*

GENUS *Bilimbia* Bilimbic Boulder Rim Lichen

SPECIES *lobulata, sabuletorum, sabulosa*

GENUS *Catarrhospora* Catarrh Boulder Rim Lichen

SPECIES *mira, splendida*

GENUS *Clauzadea* Clauzade Boulder Rim Lichen

SPECIES *chondrodes, immersa, metzleri, monticola*

GENUS *Farnoldia* Farnold Boulder Rim Lichen

SPECIES *hypocrita, jurana, micropis*

GENUS *Huilia* (Synonym for *Porpidia*)

GENUS *Immersaria* Submerged Boulder Rim Lichen

SPECIES *athroocarpa, carbonoidea*

GENUS *Koerberiella* Koerber Boulder Rim Lichen

SPECIES *johnsonii, wimmeriana*

GENUS *Labyrintha* Labyrinth Boulder Rim Lichen

SPECIES *implexa*

GENUS *Mycobilimbia* Twin Stem Dot Boulder Rim Lichen

SPECIES *fissuriseda, microcarpa*

GENUS *Paraporpidia* Lesser Boulder Rim Lichen

SPECIES *aboriginum, glauca, leptocarpa*

GENUS *Poeltiaria* Pelt Boulder Rim Lichen

SPECIES *coromandelica, corralensis, turgescens*

GENUS *Poeltidea* Lesser Pelt Boulder Rim Lichen

SPECIES *perusta*

GENUS *Schizodiscus* Split Disc Boulder Rim Lichen

SPECIES *afroalpinus*

GENUS *Stephanocyclos* Stephan Boulder Rim Lichen

SPECIES *henssenianus*

GENUS *Xenolecia* Strange Boulder Rim Lichen

SPECIES *spadicomma*

FAMILY 24 *Psoraceae* Scabby Rim Lichen

GENUS *Psora* Fish Scale Scabby Rim Lichen

SPECIES *californica, cerebriformis, crenata, crystallifera, decipiens, globifera, himalayana, icterica, luridella, montana, nipponica, pacifica, pseudorussellii, rubiformis, russellii, tenuifolia, tuckermanii, vellesiaca*

VARIETIES *crenata elegans, crystallifera testudinea, decipiens albomarginata*

GENUS *Lecidoma* Earth Crust Scabby Rim Lichen

SPECIES *demissum*

GENUS *Protoblastenia* Orange Dot Scabby Rim Lichen

SPECIES *calva, cyclospora, incrustans, lilacina, rupestris, siebenhaariana, terricola*

VARIETIES *cyclospora globulificans, rupestris irrubata*

GENUS *Protomicarea* Mica Scabby Rim Lichen

SPECIES *limosa*

GENUS *Psorula* — Lesser Fish Scale Scabby Rim Lichen
SPECIES *rufonigra*

FAMILY 25 *Ramalinaceae* — Cartilage Rim Lichen
GENUS *Ramalina* — Cartilage Rim Lichen
SPECIES *ahtii, almquistii, americana, arabum, australiensis, banzarensis, bistorta, caespitella, calicaris, canariensis, cannonii, capitata, celastri, chondrina, cochlearis, complanata, confirmata, cuspidata, dasypoga, dendriscoides, denticulata, dilacerata, disparata, dumeticola, exiguella, exilis,* ***farinacea***(Dye Herb), *fastigiata, filicaulis, fimbriata, fissa, fraxinea, glaucescens, inflata, lacera, leiodea, leptocarpha, litorea, luciae, menziesii, meridionalis, montagnei, nervulosa, obtusata, pacifica, paludosa, panizzei, peruviana, petrina, pollinaria, pollinaria humilis, polymorpha, portuensis, puberulenta, rigida, roesleri, scoparia,* ***siliquosa***(Dye Herb), *sinensis, sorediantha, stenospora, stevensiae, subamplicata, subfarinacea, subfraxinea, subleptocarpha, subpellucida, tenella, tenuis, tropica, unifolia, unilateralis, willeyi, whinrayi, zollingeri*
VARIETIES *australiensis anceps, australiensis angulosa, australiensis furcellata, australiensis myrioclada, australiensis usnea, calicaris canaliculata, calicaris subampliata, calicaris subfastigiata, canariensis baltica, canariensis ventricosa, capitata strepsilis, celastri ecklonii, celastri linearis, celastri membranacea, celastri ovalis, celastri ovalis-yemensis, celastri tenuissima, chondrina crinalis, chondrina thrausta, cuspidata armorica, cuspidata atlantica, cuspidata curnowii, cuspidata stenoclada, cuspidata variabilis, exiguella gracilenta, farinacea hypoprotocetrarica, farinacea multifida, farinacea pendulina, farinacea phalerata, farinacea reagens, farinacea rubescens, farinacea salazinica, fastigiata conglobata, fastigiata fenestrata, fastigiata minutula, fastigiata pellucida, fastigiata populina, fastigiata tortulosa, fissa brevissima, fissa platyna, fraxinea ampliata, fraxinea calicariformis, inflata inflata, inflata australis, inflata compacta, inflata geniculata, inflata gracilis, inflata intermedia, inflata knightiana, inflata olivacea, inflata perpusilla, inflata perpusilla-alba, inflata pusilla, inflata subgeniculata, inflata tasmanica, lacera duriaei, lacera evernioides, luciae tayloriana, peruviana dendroides, peruviana javanica, peruviana squarrosa, polymorpha emplecta, siliquosa cornuta, siliquosa crassa, siliquosa druidarum, siliquosa gracilescens, siliquosa incrassata, siliquosa rugosa, siliquosa scopulorum, siliquosa tuberculosa, subfarinacea angustissima, subfraxinea norstictica*
GENUS *Desmazieria* — (Synonym for *Ramalina*)
GENUS *Fistulariella* — (Synonym for *Ramalina*)
GENUS *Niebla* — Fog Cartilage Rim Lichen
SPECIES *cephalota, ceruchis, cerucoides, combeoides, flaccescens, homalea, isidiascens, laevigata, polymorpha, robusta*
GENUS *Ramalinopsis* — Lesser Cartilage Rim Lichen
SPECIES *mannii*
GENUS *Trichoramalina* — Hairy Cartilage Rim Lichen
SPECIES *crinita*

FAMILY 26 *Rhizocarpaceae* — Map Rim Lichen
GENUS *Rhizocarpon* — Map Rim Lichen
SPECIES *advenulum, alaxense, alpicola, amphibium, anaperum, anseris, athalloides, atroflavescens, badioatrum, bolanderi, caeruleoalbum, caesium, chioneum, cinereonigrum, cinereovirens, clausum, colludens, concentricum, cookeanum, copelandii, cumulatum, disporum, distinctum, effiguratum, eupetraeoides, eupetraeum, expallescens, ferax, furfurosum, geminatum, geographicum, grande, hensseniae, hochstetteri, inarense, infernulum, intermediellum, intersitum, jemptlandicum, johnstonii, kerguelense, lavatum, lecanorinum, macrosporum, mawsonii, ochrolechiae, microsporum, norvegicum, oederi, parvum, petraeum, polycarpoides, polycarpum, postumum, praebadium, pusillum, reductum, renneri, richardii, ridescens, rittokense, roridulum, saanaense, santessonii, simillimum, subgeminatum, sublavatum, submodestum, subpostumum, superficiale, tetramerum, timdalii, tinei, umbilicatum, vernicomoideum, viridiatrum*
VARIETIES *alpicola chionophilum, alpicola deludens, alpicola geronticum, alpicola oreites, badioatrum badioatrum, badioatrum incusum, copelandii cyclodes, disporum montagnei, distinctum ambiguum, distinctum orphinum, eupetraeoides atroalbescens, eupetraeoides chionophiloides, geminatum concretum, geographicum arcticum, geographicum atlanticum, geographicum diabasicum, geographicum frigidum, geographicum kittilense, geographicum lindsayanum, geographicum prospectans, geographicum riparium, geographicum vulgare, hochstetteri oceanicum, hochstetteri rufoatrum, infernulum infernulum, infernulum sylvaticum, lavatum obscuratum, lavatum roridulum, petraeum excentricum, petraeum perlutum, postumum lotum, reductum rubescens, richardii atlanticum, richardii atroalbum, richardii confervoides, richardii constrictum, superficiale superficiale, superficiale boreale, superficiale crystalligenum, superficiale splendidum, umbilicatum calcareum*
GENUS *Catolechia* — Cat Map Rim Lichen
SPECIES *wahlenbergii*
GENUS *Epilichen* — Downy Map Rim Lichen
SPECIES *scabrosus*
GENUS *Poeltinula* — Pelt Map Rim Lichen
SPECIES *cerebrina*

FAMILY 27 *Sphaerophoraceae* — Ball Rim Lichen
GENUS *Sphaerophorus* — Coral Ball Rim Lichen
SPECIES ***fragilis***(Dye Herb), ***globosus***(Dye Herb)
GENUS *Austropeltum* — Southern Pelt Ball Rim Lichen
SPECIES *glareosum*

GENUS *Bunodophoron* Bundle Ball Rim Lichen
SPECIES *australe, coomerense, diplotypum, flaccidum, formosanum, imshaugii, insigne, macrocarpum, melanocarpum, murrayi, notatum, patagonicum, ramuliferum, scrobiculatum, tibellii*
GENUS *Leifidium* Leif Ball Rim Lichen
SPECIES *tenerum*
GENUS *Neophyllis* Modern Ball Rim Lichen
SPECIES *melacarpa, pachyphylla*

FAMILY 28 *Stereocaulaceae* Foam Rim Lichen
GENUS *Stereocaulon* Snow Foam Rim Lichen
SPECIES *albicans, alpinum, apocalypticum, arcticum, arenarium, argus, botryosum, caespitosum, capitellatum, condensatum, coniophyllum, corticatulum, cumulatum, cymosum, dactylophyllum, delisei, depreaultii, evolutum, exalbidum, exutum, glareosum, grande, gregarium, groenlandicum, incrustatum, intermedium, leprocephalum, leucophaeopsis, microcarpum, myriocarpum, nanodes, octomerum, pileatum, plicatile, ramulosum, rivulorum, sasakii, saviczii, saxatile, spathuliferum, sterile, subcoralloides, symphycheilum, tennesseense,* ***tomentosum***(Dye Herb), *tornense, trachyphloeum, vesuvianum*
VARIETIES ***alpinum paschale***(Dye Herb), *alpinum tomentosum, corticatulum corticatulum, corticatulum humile, corticatulum leptaleum, corticatulum subcorticatum, dactylophyllum dactylophyllum, dactylophyllum coralloides, dactylophyllum occidentale, exutum subramulosum, pileatum pumilum, ramulosum compactum, ramulosum crebratum, ramulosum elegans, ramulosum exalbidum, ramulosum macquariensis, ramulosum macrocarpoides, ramulosum macrocarpum, ramulosum mixtum, ramulosum nudatum, ramulosum proximum, ramulosum pulvinare, ramulosum subcompressum, ramulosum submollescens, saxatile evolutoides, spathuliferum botryosum, spathuliferum dissolutum, trachyphloeum denudatum, vesuvianum vesuvianum, vesuvianum denudatum, vesuvianum nodulosum, vesuvianum nodulosum-capitatum, vesuvianum nodulosum-depressum, vesuvianum nodulosum-pulvinatum, vesuvianum nodulosum-umbonatum, vesuvianum symphycheileoides*
GENUS *Hertelidea* Hertel Foam Rim Lichen
SPECIES *botryosa, eucalypti, geophila, pseudobotryosa*
GENUS *Pilophorus* Matchstick Foam Rim Lichen
SPECIES *acicularis, clavatus, dovrensis, fibula, nigricaulis, robustus, strumaticus, vegae*
VARIETIES *strumaticus cereolus, strumaticus distans*

FAMILY 29 *Thelocarpaceae* Nipple Rim Lichen
GENUS *Thelocarpon* Nipple Rim Lichen
SPECIES *epibolum, hassei, impressellum, intermediellum, laureri, lichenicola, magnussonii, olivaceum, opertum, pallidum, robustum, saxicola, sphaerosporum, strasseri, superellum*
VARIETIES *epibolum epibolum, epibolum epithallinum, intermediellum intermixtulum, laureri epilithellum, laureri interceptum*
GENUS *Ahlesia* (Synonym for *Thelocarpon*)
GENUS *Melanophloea* Dark Nipple Rim Lichen
SPECIES *pacifica*
GENUS *Sarcosagium* Flesh Sage Nipple Rim Lichen
SPECIES *campestre*
VARIETIES *campestre campestre, campestre macrosporum*
GENUS *Thelocarpella* Lesser Nipple Rim Lichen
SPECIES *gordensis*

ORDER 7 *Ostropales* Oster Lichen

FAMILY 1 *Ostropaceae* Oster Lichen
GENUS *Ostropa* Oster Lichen
SPECIES *barbara*
VARIETIES *barbara cinerea*
GENUS *Papilionovela* Butterfly Oster Lichen
SPECIES *albothallina*
GENUS *Platygramme* Flat Oster Lichen
SPECIES *australiensis, muelleri*

FAMILY 2 *Asterothyriaceae* Star Thyre Oster Lichen
GENUS *Asterothyrium* Star Thyre Oster Lichen
SPECIES *decipiens, leucophthalium, rotuliforme*
GENUS *Epilithia* Epiphyte Rock Star Thyre Oster Lichen
SPECIES *cristata*
GENUS *Gyalidea* Earth Comfort Star Thyre Oster Lichen
SPECIES *diaphana, epiphylla, fritzei, hensseniae, hyalinescens, lecideopsis, rivularis, roseola, subscutellaris*
GENUS *Lopadiopsis* (Synonym for *Asterothyrium*)
GENUS *Microlychnus* Lamp Star Thyre Oster Lichen
SPECIES *epicorticis*
GENUS *Microspatha* Spoon Star Thyre Oster Lichen
SPECIES *glauca*

GENUS *Psorotheciopsis* Scabby Star Thyre Oster Lichen
SPECIES *patellarioides*
GENUS *Solorinella* Comfort Star Thyre Oster Lichen
SPECIES *asteriscus*

FAMILY 3 *Gomphillaceae* Nail Oster Lichen
GENUS *Gomphillus* Tooth Dot Nail Oster Lichen
SPECIES *americana, calycioides*
GENUS *Aulaxina* Furrowed Nail Oster Lichen
SPECIES *dictyospora, epiphylla, microphana, opegraphina, quadrangula*
GENUS *Bullatina* Blistered Nail Oster Lichen
SPECIES *aspidota*
GENUS *Calenia* Calen Nail Oster Lichen
SPECIES *bullatinoides, conspersa, graphidea, leptocarpa, phyllogena, subdepressa, thelotremella*
GENUS *Caleniopsis* Lesser Calen Nail Oster Lichen
SPECIES *laevigata, tetramera*
GENUS *Echinoplaca* Urchin Nail Oster Lichen
SPECIES *intercedens, epiphylla, leucotrichoides, pellicula, pellucida, tetrapla*
GENUS *Gyalectidium* Dimple Dot Nail Oster Lichen
SPECIES *australe, caucasicum, filicinum, flabellatum, imperfectum, microcarpum, novoguineense, verruculosum*
GENUS *Gyalideopsis* Dimple Tooth Nail Oster Lichen
SPECIES *anastomosans, athalloides, graminicola, lambinonii, muscicola, perlucida, piceicola, rogersii, rubescens, scotica, vainioi, verruculosa, williamsii, wirthii*
GENUS *Hippocrepidea* Horse Crepe Nail Oster Lichen
SPECIES *nigra*
GENUS *Paratricharia* Lesser Hairy Nail Oster Lichen
SPECIES paradoxa
GENUS *Rhexophiale* Rhex Nail Oster Lichen
SPECIES *rhexoblephara*
GENUS *Sagiolechia* Sage Nail Oster Lichen
SPECIES *protuberans*
GENUS *Tricharia* Hairy Nail Oster Lichen
SPECIES *albostrigosa, cretacea, fimbriata, guatemalensis, helminthospora, papillifera, plana, santessonii, vainioi, verrucosa, vezdae*

FAMILY 4 *Graphidaceae* Script Oster Lichen
GENUS *Graphidium* Script Oster Lichen
SPECIES *aphthosae, tesserata*
GENUS *Acanthothecis* Bristle Script Oster Lichen
SPECIES *gracilis*
GENUS *Anomalographis* Anomalous Script Oster Lichen
SPECIES *madeirensis*
GENUS *Carbacanthographis* Carbon Bristle Script Oster Lichen
SPECIES *hertelii*
GENUS *Cyclographina* Cyclic Script Oster Lichen
SPECIES *circumscissa, citri, cubana, hologlauca, holstii, icmadophiloides, leprevostii, lojkana, macgregorii, pervarians*
GENUS *Diorygma* Diory Script Oster Lichen
SPECIES *circumfusum, erythrellum, hieroglyphicum, junghuhnii, nothofagi, pruinosum, rufopruinosum, wilsonianum*
VARIETIES *junghuhnii mendax*
GENUS *Diplogramma* Dual Grass Script Oster Lichen
SPECIES *australiensis*
GENUS *Dyplolabia* Lip Script Oster Lichen
SPECIES *afzelii*
GENUS *Fissurina* Fissure Script Oster Lichen
SPECIES *comparilis, dumastii, glauca, globulifica, insidiosa, marginata, subcontexta, triticea*
GENUS *Glyphis* Blistered Script Oster Lichen
SPECIES *cicatricosa, cribrosa, montoensis, scyphulifera*
VARIETIES *cicatricosa cicatricosa, cicatricosa confluens, cicatricosa depauperata, cicatricosa intermedia*
GENUS *Graphina* Lesser Graphic Script Oster Lichen
SPECIES *abaphoides, abdita, acharii, anguina, antillarum, aquilonia, atrocelata, atronitens, babingtonii, borealis, celata, colliculosa, columbina, contorta, cypressi, daintreensis, dimidiata, elaiocarpa, fissofurcata, floridana, glaucoderma, gracilescens, gyridia, hiascens, incerta, incrustans, intertexta, laevigata, leprocarpa, leuconephala, lumbschii, marcescens, palmicola, parilis, pauciloculata, peplophora, pertenella, platycarpa, polyclades, psoromica, radicicola, repleta, rubens, ruiziana, salazinica, saxicola, scolectis, simulans, sophistica, sophisticascens, streblocarpa, streimannii, subaggregans, subnitida, subserpentina, substriatula, subtartarea, subvirginalis, tenuirima, virginea, xylophaga*

VARIETIES *anguina dendriticoides, anguina diffusa, anguina divaricata, anguina flexuosa, anguina inustula, anguina pulverulenta, anguina radiata, lumbschii lumbschii, lumbschii deficiens, platycarpa platycarpa, platycarpa recta, repleta repleta, repleta macrospora, repleta stictica*

GENUS *Graphis* Graphic Script Oster Lichen

SPECIES *albissima, albonitens, alboscripta, anfractuosa, angustata, aperiens, apertella, aphanes, assimilis, botryosa, caesiella, candidata, catherinae, centrifuga, chromothecia, crassilabra, descissa, desquamescens, dumastioides, duplicata, elegans, elixii, emersa, endoxantha, epimelaena, eulectra, glaucescens, grammatis, haleana, howeana, illiterata, immersella, immersicans, inamoena, insidiosa, intricata, inversa, kakaduensis, leptocarpa, leptoclada, leucopepla, librata, lineola, longula, lucifica, lumbricina, malacodes, mosquitensis, nakanishiana, nigririmis, nitida, paradoxica, pavoniana, persulcata, platycarpella, poitaeoides, propinqua, rimulosa, rufula, sayeri, scripta, semiaperta, stenotera, stipitata, streimannii, striatula, subamylacea, subelegans, subnitidula, subparilis, subregularis, subtenella, supracola, tenella, treubii, turbulenta, turgidula, vermifera, vinosa, xanthospora*

VARIETIES *duplicata duplicata, duplicata peruviana, elegans coacervata, elegans parallela, elegans petrina, elegans ramificans, elegans simplicior, elegans stellata, malacodes argopholis, nigririmis nigririmis, nigririmis deficiens, rimulosa rimulosa, rimulosa brachycarpa, rimulosa pulverulenta, scripta abietina, scripta cerasi, scripta diffracta, scripta diffusa, scripta divaricata, scripta eutypa, scripta flexuosa, scripta fraxini, scripta horizontalis, scripta limitata, scripta major, scripta minor, scripta minuta, scripta pulverulenta, scripta radiata, scripta recta, scripta serpentina, scripta spathea, scripta stellata, scripta tremulans, scripta typographica, scripta varia, striatula striatula, striatula sublaevis, turgidula turgidula, turgidula norstictica*

GENUS *Gymnographa* Nude Graphic Script Oster Lichen

SPECIES *eludens, heterospora, medusulina*

GENUS *Gyrostomum* (Synonym for *Glyphis scyphulifera*)

GENUS *Helminthocarpon* (Synonym for *Cyclographina*)

GENUS *Hemithecium* Half Theca Script Oster Lichen

SPECIES *chlorocarpoides, chrysenteron*

GENUS *Leiorreuma* Leiorreum Script Oster Lichen

SPECIES *exaltata, hypomelaenum*

GENUS *Medusulina* Medusa Script Oster Lichen

SPECIES *egenella, nitida, texana*

GENUS *Opegrapha* Ope Graphic Script Oster Lichen

SPECIES *vulgata*

GENUS *Phaeographina* Brick Script Oster Lichen

SPECIES *arechavaletae, asteroides, atromaculata, caesiopruinosa, caesioradians, contexta, echinocarpica, elixii, exilior, explicans, fuscescens, hadrospora, impudica, lamii, litoralis, montiscalvi, quassiaecola, wilsonii*

VARIETIES *arechavaletae banksiae*

GENUS *Phaeographis* Dusky Script Oster Lichen

SPECIES *arthonioides, australiensis, brasiliensis, ceratoides, colligata, dendritica, dentriticella, dendroides, elaeina, erumpens, haemitites, hypoglaucoides, inusta, leiogrammodes, lindigiana, lobata, lyellii, melanostalazans, mucronata, multicolor, nardiensis, necopinata, neotricosa, nornotatica, platycarpa, pseudomelana, punctiformis, sericea, smithii, subdividens, subfulgurata, subintricata, subinusta, subtigrina, tortuosa, tuberculifera*

VARIETIES *dendritica obtusa, dendritica sinensigrapha, mucronata cinerascens, mucronata inscripta, mucronata subcompulsa, smithii macularis, subintricata extenuata, subintricata intumescens*

GENUS *Phaeographopsis* Lesser Dusky Script Oster Lichen

SPECIES *indica*

GENUS *Sarcographa* Warty Script Oster Lichen

SPECIES *colliculosa, fenicis, intricans, kirtoniana, labyrinthica, medusula, medusulina, oculata, subtricosa, tricosa, verrucosa*

VARIETIES *labyrinthica actinota, subtricosa subtricosa, subtricosa pulverulenta*

GENUS *Sarcographina* Lesser Warty Script Oster Lichen

SPECIES *contortuplicata, cyclospora, farinulenta, glyphiza, gyrizans, maculata, sandwicensis*

GENUS *Thalloloma* Thallus Script Oster Lichen

SPECIES *subvelata*

GENUS *Thecaria* Theca Script Oster Lichen

SPECIES *montagnei, quassiaecola*

FAMILY 5 *Gyalectaceae* Dimple Oster Lichen

GENUS *Gyalecta* Dimple Oster Lichen

SPECIES *biformis, derivata, erythrozona, flotowii, foveolaris, friesii, geoica, herrei, jenensis, kukriensis, peziza, truncigena, ulmi*

VARIETIES *jenensis jenensis, jenensis cupularis, jenensis macrospora, jenensis marmorea, ulmi rubra*

GENUS *Belonia* Belony Dimple Oster Lichen

SPECIES *calcicola, incarnata, nidarosiensis, russula, uncinata*

GENUS *Bryophagus* Moss Eating Dimple Oster Lichen

SPECIES *gloeocapsa, minutissimus*

GENUS *Coenogonium* Pixie Hair Dimple Oster Lichen

SPECIES *botryosum, confervoides, curvulum, dilucidum, disjunctum, fallaciosum, flavicans, hypophyllum, implexum, interplexum, interpositum, linkii, lisowskii, missouriense, moniliforme, queenslandicum, subluteum, zonatum*

VARIETIES *interplexum acrocephalum*

GENUS *Cryptolechia* Hidden Dimple Oster Lichen

SPECIES *carneolutea, myriadella, nana, plurilocularis, saxatilis, stellaris, subincolorella, versicolor*

GENUS *Dimerella* Dimer Dimple Oster Lichen

SPECIES *lutea, pineti, pyrophthalma*

GENUS *Gyalectina* (Synonym for *Cryptolechia*)

GENUS *Microphiale* (Synonym for *Coenogonium, Dimerella*)

GENUS *Pachyphiale* Thick Phial Dimple Oster Lichen

SPECIES *carneola, fagicola, gyaizella*

VARIETIES *carneola cornea, fagicola corticola*

GENUS *Ramonia* Ramon Dimple Oster Lichen

SPECIES *ablephora, absconsa, azorica, chrysophaea, cupellina, dictyospora, eungellae, interjecta, leptospora, malmei, microspora, nigra, rappii, venezueliana*

VARIETIES *interjecta luteola*

GENUS *Secoligia* (Synonym for *Gyalecta*)

GENUS *Semigyalecta* Lesser Dimple Oster Lichen

SPECIES *paradoxa*

FAMILY 6 *Odontotremataceae* Tooth Tremor Oster Lichen

GENUS *Odontotrema* Tooth Tremor Oster Lichen

SPECIES *pertusariae*

GENUS *Bryodiscus* Moss Disc Tooth Tremor Oster Lichen

SPECIES *arctoalpinus, grimmiae, hepaticarum*

GENUS *Geltingia* Gelting Tooth Tremor Oster Lichen

SPECIES *associata*

GENUS *Lethariicola* Wolf Tooth Tremor Oster Lichen

SPECIES *cucularis*

GENUS *Odontura* Tailed Tooth Tremor Oster Lichen

SPECIES *rhaphidospora*

GENUS *Paralethariicola* Lesser Wolf Tooth Tremor Oster Lichen

SPECIES *aspiciliae*

GENUS *Paschelkiella* Paschelk Tooth Tremor Oster Lichen

SPECIES *pini*

GENUS *Pleospilis* Many Spot Tooth Tremor Oster Lichen

SPECIES *ascaridiella*

GENUS *Potriphila* Pot Tooth Tremor Oster Lichen

SPECIES *navicularis, neurogena*

GENUS *Rogellia* Rogell Tooth Tremor Oster Lichen

SPECIES *nectrioidea*

GENUS *Skyttea* Skytte Tooth Tremor Oster Lichen

SPECIES *buelliae, caesii, elachistophora, fusispora, gregaria, lecanorae, nitschkei, pyrenulae, spinosa, tephromelarum, viridis*

GENUS *Skyttella* Lesser Skytte Tooth Tremor Oster Lichen

SPECIES *mulleri*

GENUS *Spirographa* Spire Tooth Tremor Oster Lichen

SPECIES *fusisporella, vinosa*

GENUS *Stromatothecia* Bed Tooth Tremor Oster Lichen

SPECIES *nothofagi*

GENUS *Tryblis* Tryble Tooth Tremor Oster Lichen

SPECIES *arnoldii, signata*

GENUS *Xerotrema* Strange Tooth Tremor Oster Lichen

SPECIES *megalospora*

FAMILY 7 *Stictidaceae* Stick Oster Lichen

GENUS *Stictis* Stick Oster Lichen

SPECIES *arundinacea, elevata, elongatispora, friabilis, graminum, radiata, stellata*

GENUS *Absconditella* Departing Stick Oster Lichen

SPECIES *annexa, celata, delutula, lignicola, pauxilla, sphagnorum, trivialis*

GENUS *Acarosporina* Acarospore Stick Oster Lichen

SPECIES *microspora*

GENUS *Biostictis* Biote Stick Oster Lichen

SPECIES *chroodiscoides, psychotriae, puertoricensis, tjibodensis*

GENUS *Carestiella* Carest Stick Oster Lichen

SPECIES *socia*

GENUS *Conotrema* (Synonym for *Bactrospora*)

GENUS *Conotremopsis* Lesser Worm Stick Oster Lichen

SPECIES *weberiana*
GENUS *Cryptodiscus* Cryptic Disc Stick Oster Lichen
SPECIES *angulosus, foveolaris, microstomus, pallidus, rhopaloides*
GENUS *Cyanodermella* Cyan Stick Oster Lichen
SPECIES *candida, viridula*
GENUS *Delpontia* Delpont Stick Oster Lichen
SPECIES *pulchella*
GENUS *Eriospora* Woolly Stick Oster Lichen
SPECIES *leucostoma*
GENUS *Glomerobolus* Glome Bowl Sac Fungus
SPECIES *gelineus*
GENUS *Nanostictis* Dwarf Stick Oster Lichen
SPECIES *christiansenii, pseudocyphellariae*
GENUS *Petractis* Yoke Stick Oster Lichen
SPECIES *clausa, farlowii*
GENUS *Plejobolus* Plejo Bowl Stick Oster Lichen
SPECIES *arenarius*
GENUS *Propoliopsis* Propol Stick Oster Lichen
SPECIES *arengae*
GENUS *Robergea* Roberg Stick Oster Lichen
SPECIES *pupula*
GENUS *Schizoxylon* Split Stick Oster Lichen
SPECIES *berkeleyanum, ligustri, sepincola*
GENUS *Thelopsis* Lump Stick Oster Lichen
SPECIES *falveola, inordinata, isiaca, melathelia, rubella*
GENUS *Topelia* Topel Stick Oster Lichen
SPECIES *aperiens, brittonii, californica, heterospora, nimisiana, rosea*

FAMILY 8 *Thelotremataceae* Nipple Tremor Oster Lichen
GENUS *Thelotrema* Barnacle Nipple Tremor Oster Lichen
SPECIES *abrosellum, adjectum, africanum, aggregatum, ahtii, albidopallens, alboannuliforme, albomaculatum, allosporoides, amabilis, annuloelevatum, antoninii, argenteum, armellense, astroideum, attenuatum, australiense, berkeleyanum, bicuspidatum, biliranum, bisporum, byssoideum, californicum, cameroonensis, canarense, capense, carassense, catastictum, cinerellum, cinereovirens, clandestinum, colobicum, columellatum, compunctellum, conferendum, confertum, conformale, crateriforme, cubanum, cupulare, cyphelloides, decorticans, defectum, demersum, demissum, depressum, desquamescens, dignitosum, dilatatum, diminitum, dislaceratum, dissultum, dissutum, dodecamerum, dominicanum, eitenii, elmeri, endoxanthum, esslingeri, eximium, expallescens, expansum, exsertum, farinaceum, fissiporum, fissuratum, flavescens, floridense, foratum, formosanum, foveolare, galactizans, gibberulosum, glaucopallens, granatum, guadeloupense, guaiquinimae, halei, harmandii, hawaiiense, hiatum, homopastum, hypomelaenum, hypoprotocetraricum, imperfectum, inalbescens, indicum, inscalpens, integrellum, inturgescens, isidiatum, isidioides, isidiophorum, kalakkadense, kalarense, kamatii, khasianum, laceratulum, lacteum, laevius, lathraeum, lecanodeum, leioplacoides, lepademersum, lepadinum, lepadodes, leprieurii, leprocarpum, leprocarpoides, leucarpoides, leucocarpum, leucocheilum, leucohymenium, leucophthalmum, lirellizans, macrosporum, magnificum, manosporum, marginatum, megalosporum, meridense, mexicanum, minarum, mirabile, monospermum, monosporum, murinum, neei, nostalgicum, novae-zelandiae, obconicum, occlusum, osornense, papillosum, patens, pauperculum, petractoides, phlyctidioides, pidurutalagalum, piluliferum, piperis, platycarpoides, platycarpum, platysporum, poeltii, polythecium, porinoides, postpositum, profundum, pruinosum, pseudoexanthismocarpum, pseudoschizostomum, pulvereodiscum, puniceum, quitoënse, recurvum, refertum, rhamni-purshianae, rhodothecium, rimulosum, rockii, rudius, samaranum, santense, saxicola, scabiomarginatum, scabrosum, sendaiense, similans, simplex, sitianum, spondaicum, stellatum, stenosporum, stromatiferum, stylothecium, subcalvescens, subdenticulatum, subexpallescens, subgeminum, subgranulosum, subhiatum, subkeralense, suboccultum, subpatens, subsphaerosporum, subweberi, suecicum, tetrasporum, trypethelioides, tuberculiferum, umbonatum, umbratum, velatum, vernicosum, verrucorugosum, verruculosum, waasii, weberi, zenkeri*
VARIETIES *australiense microporum, lepadinum lepadinum, lepadinum inclusum, lepadinum rupestre, lepadinum scutelliforme, lepadodes lepadodes, lepadodes endochrysoides, leucophthalmum lacerata*
GENUS *Chapsa* Chaps Nipple Tremor Oster Lichen
SPECIES *indicum*
GENUS *Chroodiscus* Tinged Disc Nipple Tremor Oster Lichen
SPECIES *africanus, alborosellus, anomalus, asteliae, australiensis, australis, brasilianus, coccineus, lamelliferus, macrocarpus, megalophthalmus, minor, mirificus, neotropicus, submuralis, verrucosus*
VARIETIES *macrocarpus tasmanicus*
GENUS *Diploschistes* Crater Nipple Tremor Oster Lichen
SPECIES *actinostomus, aeneus, almbornii, arabiensis, badius, caesioplumbeus, candidissimus, conceptionis, diacapsis, diploschistoides, euganeus, farinosus, gypsaceus, gyrophoricus, hensseniae, hypoleucus, microsporus, muscorum, ocellatus,* ***scruposus***(Dye Herb), *sticticus, thunbergianus*

VARIETIES *actinostomus apertus, candidissimus calcareus, gypsaceus albissimus, muscorum bryophilus, muscorum lichenicola, muscorum parasitica, thunbergianus australasicus*

GENUS *Ingvariella* Ingvar Nipple Tremor Oster Lichen

SPECIES *bispora*

VARIETIES *bispora multispora*

GENUS *Leptotrema* Thin Nipple Tremor Oster Lichen

SPECIES *aemulans, aemulum, albocoronatum, argillaceum, calathiforme, endoxanthellum, lepadodes, lithophila, melanosporum, microglaenoides, neozelandicum, nitidulum, nuwarense, oxysporum, patulum, polycarpum, polyporum, schizoloma*

GENUS *Myriotrema* Volcano Nipple Tremor Oster Lichen

SPECIES *albidulum, albocinctum, anamalaiense, bahianum, barroense, calvescens, cinereoglaucescens, cinereum, clandestinum, columellatum, compunctum, concretum, configuratum, congestum, costaricense, craterellum, dactyliferum, deceptum, decorticatum, desquamans, elachistoteron, eminens, erodens, exile, extendens, fissurinum, flavolucens, fluorescens, foliicola, fragile, frondosum, glaucescens, glaucophaenum, glauculum, granulosum, halei, hartii, immersum, insigne, laeviusculum, lunense, mammiculum, mammillare, masonhalei, maximum, meghalayense, microphthalmum, microporum, microstomum, minutulum, minutum, multicavum, myriocarpon, myrioporoides, myrioporum, myriotremoides, neofrondosum, norsticticum, norstictideum, occultum, olivaceum, pachystomum, parvidiscum, peninsulae, permaculatum, pertusarioides, phaeosporum, polytretum, porinaceum, protoalbum, protocetraricum, pulverulentum, pycnoporellum, reclusum, rugiferum, santessonii, scabridum, schizostomum, secernendum, sembilanense, sordidescens, sphinctrinellum, squamuloides, steyermarkii, subcompunctum, subconforme, subcostaricense, subdactyliferum, subwrightii, tarmaguliense, terebrans, terebratulum, thwaitesii, trypaneoides, uniseptatum, urceolare, viride, viridialbum, vulcani, wightii, wrightii* VARIETIES *compunctum portoricense*

GENUS *Nadvornikia* Nadvornik Nipple Tremor Oster Lichen

SPECIES *hawaiiensis*

GENUS *Ocellularia* Volcano Eye Nipple Tremor Oster Lichen

SPECIES *actinota, agasthiensis, alba, albida, albo-olivacea, albogilva, albomaculata, albula, allospora, allosporiza, allosporoides, americana, amplior, andamanica, annulosa, anomorpha, antillensis, apayoënsis, arecae, ascidioidea, asiatica, astrolucens, auberiana, auberianoides, aurulenta, baileyi, bakoensis, bataana, bonplandii, bullata, caledoniensis, canara, canariana, capensis, carnea, carneodisca, carnosula, cavata, chionostoma, chiriquiensis, chonestoma, cinerascens, circumscripta, collativa, comparabilis, concentrica, concolor, conferta, confluens, conformis, confusa, conpsoromica, crassa, crispata, crocea, croceoisidiata, croceopora, cruentata, cryptotrema, dactyliza, decolorata, deformis, defossa, dehiscens, diacida, diffractella, diploschistoides, diplotrema, discoidea, discolor, dolichotata, domingensis, dominicana, elabens, emergens, emersa, endoleuca, endomelaena, epitrypa, erumpens, eumorpha, eurychades, exigua, exuta, fecunda, fissa, flavida, flavomedullosa, garoana, gentingensis, glaucoglyphica, glaziovii, globosa, glyphica, goniostoma, gracilis, grandis, granulatula, granulifera, granulosa, groenhartii, guptei, gymnocarpa, gyrostomoides, henatomma, holospora, inexpectata, interponenda, interposita, isidiifera, jamesii, jugalis, kanneliyensis, karnatakensis, keralensis, kinabaluensis, landronii, lankaensis, lathraea, latilabra, leiostoma, lepadinoides, leucina, leucomelaena, leucotrema, leucotylia, lirelliformis, lithophila, maculata, mahabalei, massalongii, mauretiana, megalostoma, meiosperma, meiospermoides, melanostoma, melanotremata, metaphorica, microascidia, microspora, monosporoides, mordenii, myrioporella, neocavata, neomasonhalei, neopertusariiformis, nigropuncta, nureliya, obovata, obscura, obturascens, octolocularis, orthomastia, panamensis, papillata, papuana, parvula, patwardhanii, percolumellata, perforata, persimilis, petrinensis, phaeotropa, phlyctidea, planaria, platychlamys, platystoma, pluripora, polillensis, polydisca, pomiformis, poncinsiana, praestans, pruinata, psathyroloma, psoromica, pulchra, punctulata, pycnophragmia, pyrenuloides, rassagala, recondita, retispora, rhabdospora, rhicnopora, rhodostroma, rimosa, ripleyi, sanfordiana, scolecospora, sinuosa, sorediata, stictica, sticticans, stictidea, straminea, subcavata, subemersa, sublilacina, submicrosporoides, subperforata, subpraestans, subsimilis, subtilis, tenuis, terebrata, texana, thelotrematoides, thryptica, trachodes, tuberculata, turgidula, udupiensis, umbilicata, verrucoisidiata, verrucomarginata, verrucosa, verruculosa, violacea, violacea violacea, violacea glauca, virens, viridipallens, viridis, wandoorensis, xantholeuca, xanthostroma, xanthostromiza, zeorina*

VARIETIES *alba alba, alba caesiascens, bonplandii bonplandii, bonplandii obliterata, perforata pauciseptatum, terebrata subeminescens*

GENUS *Phaeotrema* Dusky Nipple Tremor Oster Lichen

SPECIES *caracasanum, chlorosporum, consimile, deightonii, feeanum, fuscescens, grande, jamaicensis, microcarpa, picteanum, sticticum, uber*

GENUS *Platygrapha* Flat Nipple Tremor Oster Lichen

SPECIES congerens

GENUS *Pseudoramonia* Ramon Nipple Tremor Oster Lichen

SPECIES *richeae, stipitata*

GENUS *Reimnitzia* Reimnitz Nipple Tremor Oster Lichen

SPECIES *santensis*

GENUS *Rhabdodiscus* Striped Disc Nipple Tremor Oster Lichen

SPECIES *anomorphoides*

GENUS *Schistostoma* Split Mouth Nipple Tremor Oster Lichen

SPECIES *dehiscens*

GENUS *Topeliopsis* Topel Nipple Tremor Oster Lichen
SPECIES *acutispora, corticola, muscigena, rugosa, toensbergii, vezdae*
GENUS *Tremotylium* Tyle Nipple Tremor Oster Lichen
SPECIES *africanum, australiense, nitidulum, sierraleonense, sprucei*
GENUS *Urceolaria* (Synonym for *Diploschistes*)

ORDER 8 *Parmeliales* Shield Lichen

FAMILY 1 *Parmeliaceae* Shield Lichen
GENUS *Parmelia* Shield Lichen
SPECIES *crowii, cunninghamii, discordans, ernstiae, erumpens, fertilis, fraudans, glabra, hygrophila, neodiscordans, norcrambidiocarpa,* ***omphalodes***(Dye Herb), *protosignifera, protosulcata, pseudosulcata, pseudotenuirima, queenslandensis, salacinifera, salcrambidiocarpa,* ***saxatilis***(Dye Herb), *signifera, skultii, squarrosa, submontana,* ***sulcata***(Dye Herb), *tarkinensis, tenuirima, testacea*
VARIETIES *omphalodes caesiopruinosa, omphalodes glomellifera, omphalodes panniformis, omphalodes pinnatifida, omphalodes subconcentrica*
GENUS *Abrothallus* Rosary Shield Lichen
SPECIES *bertianus, bryoriarum, caerulescens, cetrariae, cladoniae, microspermus, parmeliarum, parmotrematis, peyritschii, prodiens, suecicus, usneae, welwitschii*
GENUS *Allantoparmelia* Grub Shield Lichen
SPECIES *almquistii, alpicola, siberica*
GENUS *Almbornia* Almborn Shield Lichen
SPECIES *azaniensis, cafferensis*
GENUS *Anzia* Black Foam Shield Lichen
SPECIES *americana, colpodes, isidiosa, minor, ornata, tianjarana*
GENUS *Arctoparmelia* Arctic Ring Shield Lichen
SPECIES ***centrifuga***(Dye Herb), *incurva, separata, subcentrifuga*
GENUS *Brodoa* Sausage Shield Lichen
SPECIES *intestiniformis, oroarctica*
GENUS *Bulborrhizina* Bulb Shield Lichen
SPECIES *africana*
GENUS *Bulbothrix* Eyelash Shield Lichen
SPECIES *apophysata, australiensis, bicornuta, confoederata,* ***coronata***(Dye Herb), *goebelii, isidiza, klementii, aevigatula, lordhowensis, microscopica, queenslandica, subtabacina, tabacina*
GENUS *Canomaculina* Mottled Ruffle Shield Lichen
SPECIES *conferenda, consors, erubescens, haitiensis, neotropica, pilosa, recipienda, reparata, subcaperata, subsumpta, subtinctoria*
GENUS *Canoparmelia* Lobed Shield Lichen
SPECIES *amazonica, aptata, caroliniana, corrugativa, crozalsiana, cryptochlorophaea, herveyensis, macrospora, martinicana, norpruinata, norsticticata, owariensis, pruinata, pustulescens, raunkiaeri, salicinifera, subarida, subtiliacea, texana*
GENUS *Cavernularia* Honeycomb Shield Lichen
SPECIES *hultenii, lophyrea*
GENUS *Cetrariastum* (Synonym for *Everniastrum*)
GENUS *Cetrelia* Giant Shield Lichen
SPECIES *alaskana, olivetorum*
VARIETIES *olivetorum cetrarioides, olivetorum chicitae, olivetorum monachorum*
GENUS *Cetreliopsis* Lesser Giant Shield Lichen
SPECIES *asahinae, endoxanthoides, laeteflava, papuae, rhytidocarpa*
GENUS *Chondropsis* (Synonym for *Xanthoparmelia*)
GENUS *Coelopogon* Hollow Beard Shield Lichen
SPECIES *abraxas, epiphorellus*
GENUS *Concamerella* Cam Shield Lichen
SPECIES *fistulata, pachyderma*
GENUS *Cornicularia* Bootstrap Shield Lichen
SPECIES *maritimum, normoerica, pocillarium*
GENUS *Coronoplectrum* Crown Plate Shield Lichen
SPECIES *namibicum*
GENUS *Dactylina* Finger Shield Lichen
SPECIES ***arctica***(Dye Herb), *beringica, madreporiformis, ramulosa*
GENUS *Dufourea* (Synonym for *Dactylina*)
GENUS *Eumitria* Mitre Shield Lichen
SPECIES *baileyi, pectinata*
VARIETIES *pectinata protocetrarica, pectinata squarrosa*
GENUS *Evernia* Oak Shield Lichen
SPECIES ***divaricata***(Dye Herb), *esoridiosa, mesomorpha, perfragilis,* ***prunastri***(Dye Herb)
VARIETIES *prunastri prunastri, prunastri arenaria, prunastri herinii, prunastri retusa, prunastri stictoceros*

GENUS *Everniastrum* Antler Shield Lichen
SPECIES ***catawbiense***(Dye Herb), *sorocheilum*
GENUS *Everniicola* Lesser Antler Shield Lichen
SPECIES *flexispora*
GENUS *Flavoparmelia* Green Shield Lichen
SPECIES *baltimorensis*, ***caperata***(Dye Herb), *concreta, diffractaica, euplecta, ferax, haysomii, haywardiana, helmsii, kantvilasii, marchantii, norfolkensis, proeuplecta, rutidota, scabrosina, secalonica, sorediens, springtonensis, succinprotocetrarica*
GENUS *Flavopunctelia* Speckleback Shield Lichen
SPECIES *darrowi, flaventior*, ***praesignis***(Dye Herb), ***soredica***(Dye Herb)
GENUS *Foraminella* (Synonym for *Parmeliopsis*)
GENUS *Himantormia* Himan Shield Lichen
SPECIES *lugubris*
GENUS *Hypogymnia* Tube Shield Lichen
SPECIES *apinnata, austerodes, australica, billardierei, duplicata, enteromorphoides, farinacea, heterophylla, imshaugii, inactiva, kosciuskoensis, krogiae, lugubris, metaphysodes, mollis, mundata, occidentalis, oceanica*, ***physodes***(Dye Herb), *pseudobitteriana, pulchrilobata, pulverata, rugosa, subcapitata, subphysodes, tasmanica, tubularis, tubulosa, turgidula, vittata*
VARIETIES *farinacea bitteriana, kosciuskoensis subobscura, lugubris lugubris, lugubris compactior, lugubris sublugubris, physodes labrosa, physodes platyphylla, physodes subtubulosa, physodes vittatoides, pulverata sorediosa, subphysodes subphysodes, subphysodes austerodioides, tubularis subteres*
GENUS *Hypotrachyna* Loop Shield Lichen
SPECIES *addita, adjuncta, baguioensis, booralensis, brevirhiza, britannica, costaricensis, croceopustulata, dahlii, densirhizinata, dentella, endochlora, ensifolia, exsecta, gondylophora, heterochroa, imbricatula, immaculata, laevigata, lividescens, novella, orientalis, oostingii, osseoalba, polydactyla, producta, prolongata, proserpinensis, pseudosinuosa, pulvinata, pustulifera, radiculata, reducens, revoluta, rockii, showmanii, sinuosa, subpustulifera, subsaxitilis, taylorensis, thysanota, virginica*
VARIETIES *heterochroa bahiana*
GENUS *Imshaugia* Starburst Shield Lichen
SPECIES *aleurites, evernica, placorodia*
GENUS *Karoowia* Karoow Shield Lichen
SPECIES *ralla, saxeti*
GENUS *Letharia* Wolf Shield Lichen
SPECIES ***columbiana***(Dye Herb), ***vulpina***(Dye Herb)
GENUS *Lethariella* Lesser Wolf Shield Lichen
SPECIES *canariensis, cashmeriana, cladonioides, flexuosa, intricata, mieheana, sinensis, smithii, togashii, zahlbruckneri*
GENUS *Melanelia* Camouflage Shield Lichen
SPECIES *agnata, albertana, commixta, culbertsonii, disjuncta, elegantula, exasperata, exasperatula*, ***fuliginosa***(Dye Herb), *fuscosorediata, glabra, glabroides, halei, hepatizon, infumata, laciniatula, multispora*, ***oliveacea***(Dye Herb), *olivaceoides, panniformis, piliferella, pseudoglabra, septentrionalis, sorediata*, ***stygia***(Dye Herb), *subargentifera, subaurifera, subelegantula, subglabra, subolivacea, tominii, trabeculata*
VARIETIES *disjuncta granulosa, fuliginosa fuliginosa, fuliginosa glabratula, fuliginosa laetevirens*
GENUS *Menegazzia* Flute Honeycomb Shield Lichen
SPECIES *aeneofusca, caesiopruinosa, caliginosa, castanea, confusa, conica, corrugata, elongata, enteroxantha, eperforata, fertilis, fissicarpa, globulifera, grandis, inactiva, jamesii, kantvilasii, minuta, myriotrema, neozelandica, norstictica, nothofagi, pertransita, platytrema, prototypica, sanguinascens, subbullata, subpertusa*, ***terebrata***(Dye Herb), *testacea, ultralucens, weindorferi*
VARIETIES *corrugata bullata, myriotrema retipora, pertransita endocitrina, subpertusa magellanica, terebrata pertusa*
GENUS *Myelochroa* Axle Bristle Shield Lichen
SPECIES *aurulenta, degelii, denegans, galbina, metarevoluta, obsessa*
GENUS *Namakwa* Namakwa Shield Lichen
SPECIES *aliphatica, exornata*
GENUS *Neofuscelia* Dusky Camouflage Shield Lichen
SPECIES *ahtii, atticoides, brunella, chiricahuensis, delisei, infrapallida, loxodes, occidentalis, pulla, pustulosa, subhosseana, verruculifera*
VARIETIES *verruculifera glomellifera*
GENUS *Neopsoromopsis* Scabby Shield Lichen
SPECIES *argentina*
GENUS *Neuropogon* Nerve Beard Shield Lichen
SPECIES *acromelanus, antarcticus, subcapillaris, trachycarpus*
GENUS *Nimisia* Nimis Shield Lichen
SPECIES *fuegiae*
GENUS *Nodobryoria* Foxtail Beard Shield Lichen
SPECIES *abbreviata, oregana, subdivergens*
GENUS *Omphalodiella* Navel Shield Lichen

SPECIES *patagonica*
GENUS *Omphalodium* (Synonym for *Omphalora*)
GENUS *Omphalora* Potato Chip Shield Lichen
SPECIES *arizonica*
GENUS *Ophioparma* Blood Spot Shield Lichen
SPECIES ***lapponica***(Dye Herb), *rubricosa*, ***ventosa***(Dye Herb)
GENUS *Pannoparmelia* Pan Shield Lichen
SPECIES *angustata, wilsonii*
GENUS *Paraparmelia* Least Shield Lichen
SPECIES *alabamensis, spodochroa*
GENUS *Parmelina* Fringed Shield Lichen
SPECIES *conlabrosa, endoleuca, euplectina, johnstoniae, labrosa, pallida, pastillifera, pseudorelicina, quercina, tiliacea, whinrayi*
VARIETIES *pseudorelicina stevensiana, quercina carporrhizans*
GENUS *Parmelinella* Lesser Fringed Shield Lichen
SPECIES *manipurensis, simplicior, versiformis, wallichiana*
GENUS *Parmelinopsis* Spined Shield Lichen
SPECIES *afrorevoluta, cryptochlora, horrescens, jamesii, minarum, neodamaziana, protocetrarica, spumosa, subfatiscens, swinscowii*
GENUS *Parmeliopsis* Bran Shield Lichen
SPECIES *ambigua, hyperopta, subambigua*
GENUS *Parmotrema* Scatter Rug Ruffle Shield Lichen
SPECIES *acrotrychum, adspersum*, ***arnoldii***(Dye Herb), *austrosinense*, ***chinense***(Dye Herb), *circinatum, convolutum, cooperi, crinitoides, crinitum, cristiferum, defectum, demethylmicrophyllinicum, dilatatum, dominicanum, eciliatum, elacinulatum, endosulphureum, eurysacum, fasciculatum, forsteri, gardneri, grayanum, hababianum, hypoleucinum, hypotropum, internexum, judithae, lobulascens, louisianae, maclayanum, madagascariaceum, margaritataum, mellissii, merrillii, michauxianum, mordenii, ochrocrinitum, overeemii, parahypotropum, permutatum, poolii, praesorediosum*, ***preperforatum***(Dye Herb), *pseudonilgherrense, queenslandense, rampoddense, reticulatum, rigidum, rimulosum, robustum, rubifaciens, saccatilobum, sancti-angelii, stuppeum, subrugatum, sulphuratum*, ***tinctorum***(Dye Herb), *ultralucens, xanthinum, zollingeri*
VARIETIES *chinense ciliata, chinense perlata, chinense trichotera, reticulatum leucosemothetum, reticulatum perforatum, reticulatum pseudovirens*
GENUS *Parmotremopsis* Lesser Scatter Rug Ruffle Shield Lichen
SPECIES *antillensis, phlyctina*
GENUS *Peccania* Peccant Shield Lichen
SPECIES *kansana, subnigra, texana*
GENUS *Phacopsis* Bundle Shield Lichen
SPECIES *huuskonenii, oxyspora, thallicola, vulpina*
GENUS *Placoparmelia* Plaque Shield Lichen
SPECIES *patagonica*
GENUS *Pleurosticta* Pleural Stick Shield Lichen
SPECIES *acetabulum*
GENUS *Protoparmelia* Chocolate Rim Shield Lichen
SPECIES *atriseda, badia, cupreoalbida, memnonia, montagnei, nephaea, ochrococca, oleagina, pulchra*
VARIETIES *badia picea, montagnei psarophana*
GENUS *Pseudephebe* Curly Rockwool Shield Lichen
SPECIES *minuscula, pubescens*
VARIETIES *pubescens subciliata*
GENUS *Pseudevernia* Least Antler Shield Lichen
SPECIES *cladonia*, ***consocians***(Dye Herb), ***furfuracea***(Dye Herb), *intensa*
VARIETIES *furfuracea furfuracea, furfuracea ceratea, furfuracea ceratea-olivetorina, furfuracea ericetorum*
GENUS *Pseudoparmelia* Least Shield Lichen
SPECIES *cubensis, floridensis, sphaerospora, uleana*
GENUS *Psiloparmelia* Nude Shield Lichen
SPECIES *arhizinosa, denotata, dichotoma, diffractaica, distincta, flavobrunnea, hypostictica, norstictica, pustulata, salazinica, sorediosa, subcrustosa*
GENUS *Psoromella* Scabby Honey Shield Lichen
SPECIES *pampana*
GENUS *Punctelia* Speckled Shield Lichen
SPECIES *appalachensis, bolliana, borreri, hypoluecites, missouriensis, nebulata, negata, perreticulata , pseudocoralloidea, punctilla, reddenda, semansiana, stictica, subalbicans, subflava, subpraesignis, subrudecta, ulophylla*
VARIETIES *subalbicans pallescens*, ***subflava rudecta***(Dye Herb)
GENUS *Relicina* Relic Shield Lichen
SPECIES *abstrusa, amphithrix, barringtonensis, circumnodata, clarkensis, conglutinata, connivens, demethylbarbatica, dentata, eximbricata, filsonii, kurandensis, limbata, planiuscula, ramboldii, ramosissima, relicinula, samoensis, schizospatha, subabstrusa, subcoronata, sublanea, subnigra, sydneyensis, terricrocodila*

VARIETIES *connivens butleri*
GENUS *Relicinopsis* Lesser Relic Shield Lichen
SPECIES *dahlii, intertexta, malaccensis, rahengensis, stevensii*
GENUS *Rimelia* Cracked Ruffle Shield Lichen
SPECIES *austrocetrata, cetrata, clavulifera, commensurata, diffractiaca, pustulata,* ***reticulata***(Dye Herb), *simulans, stictica, subisidiosa*
GENUS *Rimeliella* (Synonym for *Canomaculina*)
GENUS *Tuckermanopsis* Tuckerman Shield Lichen
SPECIES *chlorophylla*
GENUS *Tuckneraria* Tuckner Shield Lichen
SPECIES *ahtii, laureri, laxa, pseudocomplicata, togashii*
GENUS *Usnea* Beard Shield Lichen
SPECIES *aculeata, affinis, alboverrucata, alpina, amabilis, amblyoclada, angulata, arizonica, articulata, australis, californica, capillaris, catenulata, cavernosa, ceratina, chaetophora, ciliifera, cirrosa, condensata, confusa, cornuta, deformis, dimorpha, diplotypus, duriuscula, effusa, elata, elixii, endochrysia, erinacea, esperantiana, evansii,* ***filipendula***(Dye Herb), *finkii, flammea, flavocardia, florida, floriformis, fragilescens, fulvoreagens, furfurosula,* ***glabrata***(Dye Herb), *glabrescens, graciosa, hesperina, himantodes,* ***hirta***(Dye Herb), *inermis, intermedia, lapponica, leucosticta, longissima, maculata, madeirensis, mekista, merrillii, michauxii, mirabilis, molliuscula, montana, mutabilis, nidifica, nidulifera, occidentalis, oncodeoides, oncodes, pachyclada, pensylvanica, perplectata, perplexans, plicata, propagulifera, prostrata, pulvinata, punctulata, pycnoclada, ramillosa, ramulosissima, retifera, roseola, rubicunda,* ***rubrotincta***(Dye Herb), *rugulosa, sanguinea, scabiosa, scabrata, scabrida, scholanderi, silesiaca, silvatica, sphacelata, spinulifera, strigosa, stuppea, subalpina, subeciliata,* ***subfloridana***(Dye Herb), *subfusca, subhirta, subscabrosa, substerilis, sylvatica, tamborensis, taylorii, torulosa, trichodea, tristis, undulata, vainioi, variegata, variolosa, vrieseana, wasmuthii, wirthii, xanthopoga*
VARIETIES *angulata torquescens, articulata articulata, articulata intestiniformis, articulata mediterranea, ceratina incurviscens, cornuta constrictula, cornuta dasaea, cornuta inflata, cornuta intexta, cornuta subpectinata, filipendula dasypoga, filipendula fibrillosa, filipendula monstruosa, filipendula muricata, flammea scabrosa, flammea rupestris, fragilescens fragilescens, fragilescens mollis, fulvoreagens sorediifera, glabrata sorediifera, glabrescens compacta, glabrescens distincta, glabrescens extensa, glabrescens glabrella, glabrescens laricina, himantodes gracilior, himantodes neoguineensis, hirta comiformis, hirta minutissima, hirta villosa, inermis angulosa, molliuscula molliuscula, molliuscula bismolliuscula, molliuscula queenslandica, nidifica intercalaris, nidifica propinqua, nidifica straminea, pulvinata pectinata, pulvinata tropica, ramulosissima microcarpoides, ramulosissima pallida, rubicunda rubicunda, rubicunda protensa, rubicunda rubiginea, rubicunda spilota, rubicunda sublurida, rubrotincta arida, rubrotincta rubescens, scabrida scabrida, scabrida elegans, scabrida tayloriana, subfloridana comosa, subfloridana hirsutula, subfloridana melanopoda, subfloridana similis, subfloridana sordidula, torulosa torulosa, torulosa aurescens, torulosa glomerata*
GENUS *Xanthomaculina* Yellow Spot Shield Lichen
SPECIES *convoluta, frondosa, hottentotta*
GENUS *Xanthoparmelia* Yellow Shield Lichen
SPECIES *adusta, ajoensis, alexandrensis, alternata, amphixantha, amplexula, angustiphylla, antleriformis, aranaea, arapilensis, archeri, arida, aridella, atrobarbatica, atrocapnodes, aurifera, australasica, australiensis, austroalpina, austroconstrictans, ballingalliana, ballingallii, barbatica, barbellata, barthlottii, bellatula, bicolorans, bicontinens, biloelensis, blackdownensis, boonahensis, bourgeanica, brachinaensis, brattii, brunnthaleri, bungendorensis, burmeisteri, cafferensis, calida, californica, camshadalis, canobolasensis, caparidensis, centralis, cerussata, cheelii, chlorochroa, chudalupensis, colensoides, coloradensis, columbariensis, concomitans, condaminensis, congensis, congesta, conranensis, consociata,* ***conspersa***(Dye Herb), *constipata, convexa, convoluta, convolutella, cordillerana, cranfieldii, cravenii, crawfordensis, crespoae, cumberlandia, darlingensis, dayiana, delisiella, diadeta, dichotoma, dierythra, digitiformis, dissitifolia, donneri, dubitata, eilifii, elaeodes, eldridgei, elevata, elixii, epheboides, erosa, everardensis, examplaris, exillima, exuviata, ferruma, filarszkyana, filsonii, flavescentireagens, flindersiana, fracticollis, franklinensis, fumarprotocetrarica, fumigata, furcata, gerhardii, glabrans, glareosa, globulifera, gongylodes, graniticola, gregaria, hafellneri, heinarii, huachucensis, huttonii, hybridiza, hypoconstictica, hypoleia, hypoleiella, hypomelaena, hypomelaenoides, hypoprotocetrarica, hyposalazinica, idahoensis, imitatrix, imitratricoides, immutata, incantata, incerta, inconspicuella, incrustata, iniquita, inselbergia, isidiascens, isidiigera, isidiosa, isidiotegeta, jarmaniae, joranadia, juxtata, kalbarriensis, kimberleyensis, kondininensis, kosciuszkoensis, lavicola, laxencrustans, leppii, leucophaea,* ***lineola***(Dye Herb), *lipochlorochroa, lithophila, lithophiloides, louisii, loxodella, lumbschii, luminosa, luteonotata, lynii, maccarthyi, mannumensis, maricopensis, masonii, mayrhoferi, metaclystoides, metamorphosa, metastrigosa, mexicana, microcephala, microphyllizans, millerae, molliuscula, monadnockensis, mongaensis, montanensis, montezumensis, monticola, mougeotii, mougeotina, multipartita, murina, nana, nanoides, nashii, nebulosa, neochlorochroa, neoconspersa, neodelisei, neoglabrans, neomongaensis, neoquintaria, neorimalis, neotaractica, neotinctina, neotucsonensis, neowyomingica, nerrigensis, nigraoleosa, nigrocephala, nigropsomifera , nonreagens, norcapnodes, norchlorochroa, norconvoluta, norhypopsila, norpraegnans, norpumila, norstrigosa, nortegeta, notata, novomexicana, numinbahensis, oleosa, orchardii, pantherina, paraparmeliformis, parasitica, paratasmanica, parviloba, parvoclystoides, parvoincerta, perrugata, pertinax, petriseda, phillipsiana, piedmontensis, pinguiacida, planilobata, plittii, praegnans, prodomokosii, protomatrae, pseudoamphixantha, pseudohypoleia, psoromifera, pumila, purdieae, pustuliza, ralstoniana,*

rankinensis, remanella, remanens, remnantia, reptans, roderickii, rogersii, rubrireagens, rugulosa, rupestris, saginata, sammyi, sargentii, scabrosa, scabrosina, schistacea, schmidtii, scotophylla, segregata, semiviridis, separata, sleei, somloënsis, sorediata, spargenosa, squamariatella, stanthorpensis, stenophylloides, streimannii, stuartensis, stuartioides, stygiodes, subalpina, subbarbatica, subcrustacea, subcrustulosa, subdicipiens, subdistorta, suberadicata, subimitatrix, subincerta, sublineola, subloxodella, subluminosa, subnuda, subpigmentosa, subprolixa, subramigera, subrugulosa, subspodochroa, substenophylloides, substrigosa, subtropica, subverrucella, succedans, sulcifera, taractica, tasmanica, tegeta, terrestris, thamnoides, tinctina, toolbrunupensis, torulosa, trirosea, tropica, tusconensis, ustulata, vagans, valdeta, verdonii, verisidiosa, verrucella, verruciformis, versicolor, vicaria, vicariella, victoriana, waiporiensis, weberi, weberiella, willisii, wisangerensis, wyomingica, xanthofarinosa, xanthomelaena, xanthomelanoides, xerica, xerophila, yamblaensis, yowaensis, zonata

FAMILY 2 *Alectoriaceae* Hair Shield Lichen
GENUS *Alectoria* Witch Hair Shield Lichen
SPECIES *fallacina, imshaugii, lata, nigricans,* ***ochroleuca***(Edible), ***sarmentosa***(Dye Herb), *vancouverensis*
VARIETIES *nigricans geomoerensis, nigricans subchalybeiformis, nigricans thulensis, ochroleuca citrina, ochroleuca tenuior, sarmentosa sarmentosa, sarmentosa vexillifera*
GENUS *Bryocaulon* Fox Hair Shield Lichen
SPECIES ***divergens***(Dye Herb), *pseudosatoanum, satoanum*
GENUS *Bryopogon* (Synonym for *Bryoria*)
GENUS *Bryoria* Horse Hair Shield Lichen
SPECIES *bicolor, capillaris, carlottae, cervinula, chalybeiformis,* ***fremontii***(Edible), *friabilis, furcellata,* ***fuscescens***(Dye Herb), *glabra, implexa, lanestris, nadvornikiana, nitidula, oregana, pikei, pseudocapillaris, salazinica, simplicior, smithii, spiralifera, subcana, tenuis, tortuosa, trichodes*
VARIETIES *chalybeiformis prostratosteola, fuscescens fuscescens, fuscescens crispa, fuscescens pallida, fuscescens positiva, implexa pseudofuscescens, subcana haynaldii*
GENUS *Coelocaulon* (Synonym for *Bryocaulon*)
GENUS *Sulcaria* Grooved Horse Hair Shield Lichen
SPECIES *badia, isidiifera*

FAMILY 3 *Cetrariaceae* Heath Shield Lichen
GENUS *Cetraria* Iceland Heath Shield Lichen
SPECIES *aculeata, arenaria, australiensis, culbersonii,* ***ericetorum***(Dye Herb), ***islandica***(Edible), *kamczatica, laevigata, muricata, odontella, sepincola*
VARIETIES *aculeata sorediata, ericetorum crispa, ericetorum minor, ericetorum pustulata, ericetorum tenuifolia, islandica islandica, islandica crispiformis, islandica platyna, muricata hispida, muricata stuppea*
GENUS *Ahtiana* Candle Wax Heath Shield Lichen
SPECIES *aurescens, pallidula, sphaerosporella*
GENUS *Allocetraria* Palisade Heath Shield Lichen
SPECIES *madreporiformis, oakesiana*
GENUS *Arctocetraria* Arctic Heath Shield Lichen
SPECIES *andrejevii, nigricascens*
GENUS *Asahinea* Ground Rag Heath Shield Lichen
SPECIES *chrysantha, scholanderi*
GENUS *Cetrariella* Alpine Heath Shield Lichen
SPECIES ***delisei***(Dye Herb), *fastigiata*
GENUS *Esslingeriana* Tinted Rag Heath Shield Lichen
SPECIES *idahoensis*
GENUS *Flavocetraria* Snow Heath Shield Lichen
SPECIES ***cucullata***(Dye Herb), ***nivalis***(Dye Herb)
GENUS *Kaernefeltia* Thornbush Heath Shield Lichen
SPECIES ***californica***(Dye Herb), *merrillii*
GENUS *Masonhalea* Tumbleweed Heath Shield Lichen
SPECIES *richardsonii*
GENUS *Platismatia* Rag Heath Shield Lichen
SPECIES ***glauca***(Dye Herb), *herrei, lacunosa, norvegica, stenophylla, tuckermani*
VARIETIES *glauca coralloidea, glauca fallax, glauca fusca, glauca ulophylla*
GENUS *Tuckermannopsis* Wrinkled Heath Shield Lichen
SPECIES *americana, californica, chlorophylla, ciliaris, coralligera, fendleri, inermis, merrillii, orbata, platyphylla, sepincola, subalpina, weberi*
GENUS *Vulpicida* Sunshine Heath Shield Lichen
SPECIES ***canadensis***(Dye Herb), *juniperinus, pinastri,* ***tilesii***(Dye Herb), *tubulosus, viridis*

FAMILY 4 *Parmelariaceae* Lesser Shield Lichen
GENUS *Parmelaria* Lesser Shield Lichen
SPECIES *subthomsonii, thomsonii*
GENUS *Aldona* Aldon Lesser Shield Lichen
SPECIES *americana, minima, stella-nigra*
GENUS *Aldonata* Mock Aldon Lesser Shield Lichen

SPECIES *pterocarpi*
GENUS *Coccodothis* — Doth Lesser Shield Lichen
SPECIES *sphaeroidea*
GENUS *Cyclostomella* — Cyclic Lesser Shield Lichen
SPECIES *disciformis, oncophora*
GENUS *Dictyocyclus* — Netted Cyclic Lesser Shield Lichen
SPECIES *hydrangeae*
GENUS *Hemigrapha* — Graphic Lesser Shield Lichen
SPECIES *astericus, atlantica*
GENUS *Inocyclus* — Inner Cycle Lesser Shield Lichen
SPECIES *australiensis, blechni, calotheus, discoideus, dovyalidis, myrtacearum, psychotriae*
GENUS *Kentingia* — Kenting Lesser Shield Lichen
SPECIES *corticola*
GENUS *Parmulariopsella* — Mock Lesser Shield Lichen
SPECIES *burseracearum*
GENUS *Parmulina* — False Lesser Shield Lichen
SPECIES *exsculpta*
GENUS *Perischizon* — Split Lesser Shield Lichen
SPECIES *oleifolium*
GENUS *Polycyclus* — Many Cycle Lesser Shield Lichen
SPECIES *andinus, marginalis*
GENUS *Protothyrium* — Thyre Lesser Shield Lichen
SPECIES *borneense, salvadorae, tricalysiae, trichiliae*
GENUS *Rhagadolobium* — Lobed Lesser Shield Lichen
SPECIES *cucurbitacearum*
GENUS *Thallomyces* — Thallus Lesser Shield Lichen
SPECIES *oritis*

ORDER 9 *Peltigerales* — Pelt Lichen

FAMILY 1 *Peltigeraceae* — Pelt Lichen
GENUS *Peltigera* — Pelt Lichen
SPECIES ***aphthosa***(Dye Herb), *britannica*, ***canina***(Dye Herb), *cinnamomea, collina, degenii, didactyla, dilacerata, dolichorrhiza, elisabethae, evansiana, horizontalis, hymenina, kristinssonii, lactucifolia, lairdii, lepidophora, leptoderma, leucophlebia, malacea, membranacea, neckeri, neopolydactyla, pacifica, polydactylon, ponojensis, praetextata, pulverulenta, pusilla, retifoveata, rufescens, scabrosa, scabrosella, subhorizontalis, tereziana*, ***venosa***(Dye Herb)
VARIETIES *degenii nitens, didactyla erumpens, didactyla hazslinszkyi, didactyla spuria, dilacerata pellucida, dolichorrhiza dolichorrhiza, dolichorrhiza pseudocrispoides, horizontalis muscorum, kristinssonii occidentalis, leucophlebia variolosa, neckeri collina, neckeri crassoides, neckeri microcarpa, polydactylon polydactylon, polydactylon conjungens, polydactylon dissecta, praetextata innovans, rufescens incusa*
GENUS *Hydrothyria* — (Synonym for *Peltigera venosa*)
GENUS *Massalongia* — Lobed Pelt Lichen
SPECIES *carnosa, microphylliza*
GENUS *Nephroma* — Kidney Pelt Lichen
SPECIES *arcticum, australe, bellum, cellulosum*, ***expallidum***(Dye Herb), ***helveticum***(Dye Herb), *isidiosum, laevigatum, occultum, parile, resupinatum, rufum, silvae-veteris, tangeriense*
VARIETIES *australe antarcticum, cellulosum cellulosum, cellulosum isidioferum, laevigatum hibernicum, laevigatum lusitanicum, resupinatum tomentosum*
GENUS *Peltigeropsis* — Lesser Pelt Lichen
SPECIES *vainioi*
GENUS *Siphulastrum* — Siphon Star Pelt Lichen
SPECIES *cladinoides, granulatum, mamillatum, usneoides*
GENUS *Siphulina* — Siphon Pelt Lichen
SPECIES *neozelandica*
GENUS *Solorina* — Chocolate Chip Pelt Lichen
SPECIES *bispora, crocea, octospora, saccata, spongiosa*
VARIETIES *bispora bispora, bispora macrospora, bispora monospora*

FAMILY 2 *Collemataceae* — Jelly Pelt Lichen
GENUS *Collema* — Jelly Pelt Lichen
SPECIES *actinoptychum, auriforme, bachmanianum, callibotrys, callopismum, ceraniscum, coccophorum, confertum, conglomeratum, crispum, cristatum, dichotomum, durietzii, fasciculare, fecundum, flaccidum, fragile, fragrans, furfuraceum, fuscovirens, glaucophthalmum, glebulentum, implicatum, insulare, japonicum, kauaiense, laeve, latzelii, leptaleum, leucocarpum, limosum, multipartitum, nigrescens, novozelandicum, occultatum, parvum, polycarpon, pulcellum, pustulatum, quadriloculare, rugosum, ryssoleum, subconveniens, subflaccidum, subnigrescens, subparvum, subundulatum, tenax, texanum, thamnodes, undulatum*

VARIETIES *auriforme granosum, auriforme membranaceum, callopismum callopismum, callopismum rhyparodes, coccophorum atrum, coccophorum congestum, crispum crispum, crispum complicatum, crispum nudum, crispum graniforme, crispum metzleri, crispum metzleri-dispersum, crispum metzleri-monocarpum, cristatum cristatum, cristatum complicatum, cristatum granuliferum, cristatum gyrosum, cristatum hypergenum, cristatum marginale, cristatum marginale-jacobeaefolium, cristatum marginale-multifidum, cristatum marginale-papulosum, cristatum melaenum, cristatum minor, cristatum pinguescens, dichotomum cataclystum, dichotomum fluviale, fasciculare fasciculare, fasciculare aggregatum, fasciculare ascaridosporum, fasciculare colensoi, fasciculare microcarpum, flaccidum rupestre, fragrans microphyllum, fragrans terrulentum, fuscovirens lunaeforme, fuscovirens meizolobum, fuscovirens microlobum, fuscovirens tuniforme, glebulentum coralliferum, glebulentum furvellum, kauaiense limbatum, laeve laeve, laeve fimbriatum, laeve minus, laeve plumbeum, laeve senecionis, leptaleum gwytheri, leucocarpum petraeum, limosum glaucescens, quadriloculare quadriloculare, quadriloculare tasmaniae, subconveniens isidiosum, tenax tenax, tenax ceranoides, tenax concinnum, tenax confertum, tenax corallinum, tenax coronatum, tenax cristatulum, tenax pulposum, tenax vulgare, undulatum undulatum, undulatum granulosum, undulatum laureri*

GENUS *Collemodes* (Synonym for *Collema*)

GENUS *Leciophysma* Bladder Jelly Pelt Lichen

SPECIES *finmarkicum, furfurascens*

GENUS *Leightoniella* Leighton Jelly Pelt Lichen

SPECIES *zeylanensis*

GENUS *Leptogium* Skin Jelly Pelt Lichen

SPECIES *adpressum, appalachensis, arcticum, arsenei, asiaticum, austroamericanum, azureum, biatorinum, biloculare, brebissonii, britannicum, bullatulum, burgessii, burnetiae, byssinum, californicum, chloromelum, cochleatum, contortum, coralloideum, corniculatum, corticola, crenatulum, crispatellum, cyanescens, dactylinum, diffractum, digitatum, enkarodes, erectum, faciifictum, fallax, floridanum, furfuraceum, fusisporum, gelatinosum, granulans, hibernicum, hildenbrandii, hirsutum, hypotrachynum, imbricatum, intermedium, isidiosellum, javanicum, juniperinum, juressianum, laceroides, lichenoides, limbatum, marginellum, massiliense, menziesii, microstictum, milligranum, minutissimum, moluccanum, parculum, pecten, pellobatum, philorheuma, phyllocarpum, platynum, plicatile, poliophaeum, polycarpum, propaguliferum, pseudofurfuraceum, rivale, rivulare, rogersii, rugosum, saturninum, schraderi, sessile, stipitatum, subaridum, subtile, subtorulosum, tasmanicum, tenuissimum, teretiusculum, turgidum, victorianum, wilsonii*

VARIETIES *biatorinum cretaceum, biatorinum pusillum, brebissonii rubiginosum, byssinum amphineum, cochleatum tremelloides, corniculatum palmatum, corticola pulchellum, cyanescens caesium, diffractum placodiellum, gelatinosum scotinum, gelatinosum sinuatum, lichenoides fimbriatum, lichenoides lacerum, lichenoides lophaeum, lichenoides pulvinatum, phyllocarpum phyllocarpum, phyllocarpum turneri, plicatile hydrocharum, schraderi muscicola, tenuissimum humosum, teretiusculum microscopicum, turgidum andegavense, wilsonii denticulatum*

GENUS *Physma* Bladder Jelly Pelt Lichen

SPECIES *ahtianum, byrsaeum, byrsinum, cataractaecola, chilense, pseudoisidiatum*

GENUS *Ramalodium* Branched Jelly Pelt Lichen

SPECIES *succulentum*

GENUS *Synechoblastus* (Synonym for *Collema*)

FAMILY 3 *Lobariaceae* Lobed Pelt Lichen

GENUS *Lobaria* Lungwort Lobed Pelt Lichen

SPECIES *amplissima, discolor, hallii, hartmannii, isidiophora, kurokawae, oregana, patinifera, plurimiseptata, pseudopulmonaria, pseudoretigera,* ***pulmonaria***(Edible), *quercizans, ravenelii, retigera, rhaphispora, scrobiculata,* ***spathulata***(Dye Herb), *tenuis, virens*

VARIETIES *amplissima laciniata, discolor hypoleuca, discolor stictaeformis, pulmonaria hypomela, pulmonaria linita, retigera retigera, retigera subisidiosa, scrobiculata verrucosa*

GENUS *Dendriscocaulon* Thorn Lobed Pelt Lichen

SPECIES *dendriothamnodes, intricatulum, umhausense*

GENUS *Pseudocyphellaria* Speckle Belly Lobed Pelt Lichen

SPECIES ***anomala***(Dye Herb),, *anthraspis, ardesiaca, argyracea, aurata, bartlettii, billardierei, brattii, carpoloma, chloroleuca, clathrata, colensoi,* ***coronata***(Dye Herb),, ***crocata***(Dye Herb), *crocatoides, desfontainii, dissimilis, dozyana, faveolata, gilva, glabra, glaucescens, granulata, haywardiorum, insculpta, intricata, jamesii, lacerata, mooreana, multifida, neglecta, nermula, norvegica, pickeringii, poculifera, prolificans, punctillaris, ranierensis, rigida, rubella, rubrina, rudecta, sayeri, semilanata, sericeofulva, soredioglabra, stenophylla, sulphurea*

VARIETIES *argyracea sorediifera, bartlettii isidiosa, billardierei flotowiana, billardierei linearis, billardierei richardii, chloroleuca psilophylla, coronata durvillei, coronata endochrysea, coronata hirta, coronata orygmaea, faveolata cellulifera, faveolata physciospora, glabra conjungens, glabra delisea, glabra freycinetii, intricata thouarsii, multifida dissimulata, multifida subvariabilis, neglecta australiensis, neglecta coralloidea, neglecta crocata, neglecta esorediosa, neglecta isidiotyla, norvegica ecyphellata, sulphurea demutabilis, sulphurea karstenii*

GENUS *Sticta* Moon Crater Lobed Pelt Lichen

SPECIES *arctica, baileyi, beauvoisii, brevipes, caliginosa, camarae, canariensis, caperata, cyphellulata, diversa, duplolimbata, filix, flavocyphellata, fuliginosa, howei, hypopsiloides, latifrons, limbata, marginifera, martinii, myrioloba, pedunculata, rutilans, sayeri, stipitata, sublimbata, subtomentella, sylvatica, variabilis, weigelii, wrightii*

VARIETIES *brevipes suberecta, canariensis dufourii, diversa microphylla, howei macrophylla, stipitata shirleyana, subtomentella suborbicularis, weigelii appendiculata, weigelii quercizans*

FAMILY 4 *Placynthiaceae* — Cynthia Pelt Lichen

GENUS *Placynthium* — BlackThread Cynthia Pelt Lichen

SPECIES *asperellum, filiforme, flabellosum, garovaglii, lismorense, nigrum, pannariellum, petersii, pluriseptatum, stenophyllum, subradiatum, tantaleum, tremniacum*

VARIETIES *flabellosum adglutinatum, garovaglii caesium, pluriseptatum melanterum*

GENUS *Hertella* — Hert Cynthia Pelt Lichen

SPECIES *chilensis, subantarctica*

GENUS *Koerberia* — Koerber Cynthia Pelt Lichen

SPECIES *biformis, sonomensis*

GENUS *Leptochidium* — Vinyl Jelly Cynthia Pelt Lichen

SPECIES *albociliatum*

GENUS *Leptogidium* — (Synonym for *Polychidium*)

GENUS *Pilonema* — Downy Cynthia Pelt Lichen

SPECIES *heteromalla*

GENUS *Polychidium* — Woolly Bear Cynthia Pelt Lichen

SPECIES *contortum, dendriscum, muscicola*

GENUS *Vestergrenopsis* — Brownette Cynthia Pelt Lichen

SPECIES *elaeina, isidiata*

ORDER A *Pyrenulales* Rash Lichen

FAMILY 1 *Pyrenulaceae* — Rash Lichen

GENUS *Pyrenula* — Rash Lichen

SPECIES *actualis, adacta, annulata, anomala, aquila, aspistea, astroidea, baileyi, bicuspidata, cerina, chloroplaca, chlorospila, citriformis, cocoes, concatervans, confoederata, confinis, conspurcata, convexa, corticata, coryli, cruenta, cruentata, cubana, cuyabensis, decumbens, defossa, dermatodes, erumpens, falsaria, fibrata, finitima, galactina, immersa, indusiata, laevigata, leucostoma, lucifera, macounii, macrospora, macularis, maculata, mamillana, martinicana, mastophora, mastophorizans, melaleuca, michneri, microcarpa, microcarpoides, microtheca, nigrocincta, nitida, nitidans, nitidella, nitidula, occidentalis, occulta, ochraceoflava, ochraceoflavens, oleosa, oxysporiza, parvinuclea, personata, pinguis, plittii, porinoides, pseudobufonia, pyrenastroides, pyrenuloides, quassiaecola, ravenelii, rubrostoma, santensis, segregata, septicollaris, sexlocularis, subcongruens, subelliptica, subferruginea, tenuisepta, thelomorpha, velatior, warmingii, wetmorei, wheeleri*

VARIETIES *adacta adacta, adacta cinerascens, cruentata circumrubens, dermatodes achroocarpa, dermatodes glabratula, leucostoma libricola, mamillana kunthii, mamillana marginata, mastophora flaventior, occidentalis harrisii, occidentalis neglecta, occulta micromma, pinguis pinguis, pinguis emergens, pinguis emergens-punctella*

GENUS *Acrocordiella* — Tall Cord Rash Lichen

SPECIES *occulta*

GENUS *Anthracothecium* — Anthrax Rash Lichen

SPECIES *amphitropum, aurantiacum, australiense, doleschallii, macrosporum, nanum, oculatum, prasinum, sinapispermum, subglobosum, subvariolosum, thwaitesii, varians, variolosum*

GENUS *Asteroporum* — Star Pore Rash Lichen

SPECIES *punctuliforme, rimale*

GENUS *Bottaria* — (Synonym for *Pyrenula*)

GENUS *Celothelium* — Celo Rash Lichen

SPECIES *aciculiferum, ischnobelum*

GENUS *Clypeopyrenis* — Clype Rash Lichen

SPECIES *microsperma*

GENUS *Distopyrenis* — Distal Rash Lichen

SPECIES *americana*

GENUS *Eopyrenula* — Wart Rash Lichen

SPECIES *avellanae, grandicula, intermedia, leucoplaca, parvispora, septemseptata*

GENUS *Granulopyrenis* — Granular Rash Lichen

SPECIES *hymnothora*

GENUS *Lithothelium* — Rock Rash Lichen

SPECIES *austropacificum, cubanum, hyalosporum, illotum, macrosporum, obtectum, phaeosporum, quiescens, septemseptatum*

GENUS *Lyromma* — Lyre Rash Lichen

SPECIES *nectandrae*

GENUS *Mazaediothecium* — Maza Rash Lichen

SPECIES *album, rubiginosum*

GENUS *Melanothecopsis* — Dark Theca Rash Lichen

SPECIES *coactella, follmannii, sinensis, subpuncta*
GENUS *Parathelium* (Synonym for *Pyrenula*)
GENUS *Plagiocarpa* (Synonym for *Lithothelium*)
GENUS *Pleurotheliopsis* (Synonym for *Anthracothecium*)
GENUS *Porothelium* Pore Rash Lichen
SPECIES *album, arthonioides, melinostigma, rufofuscum*
GENUS *Pyrenastrum* (Synonym for *Pyrenula*)
GENUS *Pyrenowilmsia* Wilms Rash Lichen
SPECIES *ferruginosa*
GENUS *Pyrgillus* Gill Rash Lichen
SPECIES *cubanus, fuscus, indicus, javanicus*
GENUS *Rhaphidicyrtis* Striped Arch Rash Lichen
SPECIES *trichosporella*
GENUS *Splanchospora* Entrail Rash Lichen
SPECIES *ampullacea*
GENUS *Sulcopyrenula* Furrowed Rash Lichen
SPECIES *canellae-albae, staurospora*
GENUS *Xenus* Strange Rash Lichen
SPECIES *lithophylli*

FAMILY 2 *Massariaceae* Massar Rash Lichen
GENUS *Massaria* Massar Rash Lichen
SPECIES *anomia, gigaspora, inquinans*
GENUS *Aglaospora* Bright Massar Rash Lichen
SPECIES *profusa*
GENUS *Aigialus* Aigial Massar Rash Lichen
SPECIES *grandis, mangrovis, parvus, rhizophorae, striatispora*
GENUS *Mamillisphaeria* Nipple Massar Rash Lichen
SPECIES *dimorphospora*
GENUS *Navicella* Navel Massar Rash Lichen
SPECIES *pileata*

FAMILY 3 *Monoblastiaceae* Monoblast Rash Lichen
GENUS *Monoblastia* Monoblast Rash Lichen
SPECIES *borinquensis, buckii, cypressi, echinulospora, lutescens, palmicola, papillosa, pellucida, quisqueyana, rappii*
GENUS *Acrocordia* Tall Cord Monoblast Rash Lichen
SPECIES *cavata, conoidea, gemmata, macrospora, salweyi*
GENUS *Anisomeridium* Anise Monoblast Rash Lichen
SPECIES *albisedum, ambiguum, americanum, anisolobum, australiense, biforme, carinthiacum, consobrinum, distans, finkii, foliicola, juistense, leucochlorum, macrosporum, polypori, quaternarium, ranunculosporum, sanfordens, subnectendum, subnexum, subprostans, tarmarindi, terminatum, tuckerae, viridescens*
VARIETIES *polypori nyssogenum, polypori willeyanum*
GENUS *Ditremis* (Synonym for *Anisomeridium*)
GENUS *Microthelia* Nipple Monoblast Rash Lichen
SPECIES *biformis, verruculosa*
GENUS *Sarcinulella* Flesh Monoblast Rash Lichen
SPECIES *banksiae*

FAMILY 4 *Requienellaceae* Requiem Rash Lichen
GENUS *Requienella* Requiem Rash Lichen
SPECIES *seminuda*
GENUS *Lacrymospora* Teardrop Requiem Rash Lichen
SPECIES parasitica
GENUS *Mauritiana* Mauritania Requiem Rash Lichen
SPECIES *rhizophorae*
GENUS *Parapyrenis* Lesser Requiem Rash Lichen
SPECIES *aurora, conica, elongata, guayaci, lichenicola, maclurae, maritima, tecomatis*

FAMILY 5 *Trypetheliaceae* Blister Rash Lichen
GENUS *Trypethelium* Speckled Blister Rash Lichen
SPECIES *aeneum, ceratinum, eluteriae, epileucodes, floridanum, infuscatulum, medians, ochroleucum, oligocarpum, papillosum, scorites, subeluteriae, tropicum, variolosum, virens*
VARIETIES *eluteriae citrinum, epileucodes virginicum, floridanum annulare, tropicum tropicum, tropicum nigratum variolosum catervarium, variolosum nitidiusculum, virens exocanthum*
GENUS *Architrypethelium* Archaic Blister Rash Lichen
SPECIES *seminudum, uberinum*
GENUS *Astrothelium* Speckled Wart Blister Rash Lichen
SPECIES *cinnamomeum, confusum, galbineum, versicolor*

VARIETIES *galbineum ochrothelizu*
GENUS *Bathelium* Bath Blister Rash Lichen
SPECIES *madreporiforme, mastoideum*
GENUS *Campylothelium* Field Blister Rash Lichen
SPECIES *defossum, nitidum*
GENUS *Laurera* Laurer Blister Rash Lichen
SPECIES *chrysocarpa, cumingii, elatior, megasperma, meristospora, robusta, subdisjuncta*
GENUS *Megalotremis* Mega Tremor Blister Rash Lichen
SPECIES *biocellata, megalospora, pustulata, verrucosa*
GENUS *Ornatopyrenis* Ornate Blister Rash Lichen
SPECIES *muriformis, queenslandica*
GENUS *Pleurotrema* (Synonym for *Megalotremis*)
GENUS *Polymeridium* Suntan Blister Rash Lichen
SPECIES *albidum, albocinereum, catapastum, contendens, proponens, quinqueseptatum, subcinereum, sulphurescens*
GENUS *Pseudopyrenula* (Synonym for *Trypethelium*)
FAMILY 6 *Xanthopyreniaceae* Yellow Rash Lichen
GENUS *Xanthopyrenia* Yellow Rash Lichen
SPECIES *heardensis*
GENUS *Collemopsidium* Collem Yellow Rash Lichen
SPECIES *ostrearum, pelvetiae, pyrenuloides*
GENUS *Didymellopsis* Twin Yellow Rash Lichen
SPECIES *collematum, latitans, perigena, pulposi*
GENUS *Pyrenocollema* Quiver Yellow Rash Lichen
SPECIES *arenisedum, argilospilum, bryospilum, caesium, elegans, halodytes, monense, montanum, orustense, prospersellum, strontianense, subarenisedum, sublitorale, tasmanicum, trichotheciodes*
GENUS *Zwackhiomyces* Zwackh Yellow Rash Lichen
SPECIES *berengerianus, coepulonus, dispersus, euplocinus, immersae, lacustris, lecanorae, sphinctrinoides*

ORDER B *Teloschistales* Schist Lichen

FAMILY 1 *Teloschistaceae* Schist Lichen
GENUS *Teloschistes* Orange Bushy Schist Lichen
SPECIES *arcticus, californicus, chrysophthalmus, contortuplicatus, exilis, fasciculatus,* ***flavicans***(Dye Herb), *hypoglaucus, siebierianus, spinosus,* ***velifera***(Dye Herb), *xanthoroides*
VARIETIES *siebierianus excelsior, spinosus spinosus, spinosus subteres, velifera velifera, velifera fornicatus, velifera nodulosus*
GENUS *Apatoplaca* Plate Schist Lichen
SPECIES *oblongula*
GENUS *Blastenia* Blastid Schist Lichen
SPECIES *carnella, endochromoides, keroblasta, macquariensis, ochroleuca, pulcherrima, soredians*
GENUS *Caloplaca* Fire Dot Schist Lichen
SPECIES *adnexa, ahtii, albovariegata, alcarum, alociza, ammiospila, approximata, aractina, arenaria, arizonica, arnoldii, asserigena, atroalba, atrocyanescens, atroflava, atrosanguinea, aurantia, aurea, bassiae, bellemerei, bolacina, borealis, borreri, brattiae, britannica, brunneola, byrsonimae, caesiorufa, caesiorufella, californica, camptidia, carolinae, cascadensis, castellana, catalinae, celata, ceracea, cerina, cerinella, cerinelloides, chalybaea, chlorina, chrysodeta, chrysophthalma, cinnabarina, cinnamomea, cirrochroa, citrina, cladodes, clavigera, concilians, conjungens, coralloides, coronata, crenularia, crenulatella, cribrosa, crozetica, cupulifera, dakotensis, dalmatica, decipiens, depauperata, diphasia, diphyodes, diplacia, discolor, dispersa, durietzii, elixii, epiphyta, epithallina, epixantha, erythrosticta, eugyra, exsecuta, feracissima, ferruginea, ferrugineofusca, flavescens, flavogranulosa, flavorubescens, flavovirescens, floridana, fraudans, fraxinea, friesii, fulva, galactophylla, glaucescens, granularis, granulosa, grimmiae, groenlandica, haematites, hanneshertelii,* ***herbidella***(Dye Herb), *holocarpa, ignea, impolita, inclinans, inconspecta, insularis, intermedia, irrubescens, jungermanniae, lactea, lateritia, leptozona, litoricola, littorea, livida, lobulata, lucifuga, luteoalba, luteominia, macquariensis, marina, maritima, microloba, microphyllina, microthallina, modesta, montana, mooreae, neonii, neotropica, nivalis, novomexicana, obliterans, oblongula, obscurella, ochracea, ochrochroa, oregona, oxfordensis, parvula, paulsenii, peliophylla, pelodella, phaeantha, phaeocarpella, phlogina, pinicola, pollinii, polycarpa, quercicola, rosei, rubelliana, rubens, rubrofusca, ruderum, rugulosa, sarcopisoides, saxicola, saxifragarum, scopularis, scotoplaca, sideritis, sinapisperma, sipeana, sorocarpa, spaldingii, spitsbergensis, squamosa, stanfordensis, stantonii, suaedae, sublobulata, subnitida, subpyraceella, teicholyta, tetraspora, thallincola, tiroliensis, tolminii, tornoensis,* ***trachyphylla***(Dye Herb), *ulcerosa, ulmorum, urceolata, variabilis, verrucosa, verruculifera, vicaria, virescens, vitellinula, wetmorei, wrightii, xanthostigmoidea*
VARIETIES *alociza acrustacea, aractina fuscoatra, aractina viridirufa, arenaria erythrocarpa, arenaria lamprocheila, arenaria subpallida, atroflava submersa, atroflava turneriana, aurantia callopisma, cerina cerina, cerina anthracina, cerina anthracina-obscurata, cerina chloroleuca, cerina chloroleuca-stillicidiorum, cerina cyanolepra, cerina microcarpa, chlorina cyanophila, chlorina cyanopolia, chlorina isidiigera, cinnabarina cinnabarina, cinnabarina fulgens, cinnabarina opaca, cirrochroa cirrochroa, cirrochroa leprosa, citrina*

flavocitrina, citrina incrustans, crenularia festiva, dalmatica dolomiticola, dalmatica schaereri, dalmatica velana, depauperata vitellinella, ferruginea aurantiaca, flavescens brevilobata, flavescens heppiana, flavovirescens aurantiaca, flavovirescens inalpina, holocarpa lithophila, holocarpa pyracea, holocarpa pyrithroma, marina flavogranulata, microthallina obliterascens, ochracea tetrasticha, pollinii phaeocarpella, saxicola miniatula, saxicola murorum, saxicola pusilla, teicholyta albolutescens, teicholyta lallavei

GENUS *Cephalophysis* Head Bladder Schist Lichen
SPECIES *leucospila*
GENUS *Fulgensia* Sulfur Schist Lichen
SPECIES *bracteata, desertorum, fulgens, subbracteata*
VARIETIES *bracteata bracteata, bracteata alpina*
GENUS *Josefpoeltia* Joseph Poelt Schist Lichen
SPECIES *boliviensis, sorediosa*
GENUS *Leproplaca* Thin Plate Schist Lichen
SPECIES *chrysodeta, xantholyta*
GENUS *Letrouitia* Letrouit Schist Lichen
SPECIES *aureola, bifera, corallina, coralloidea, domingensis, flavocrocea, leprolyta, muralis, parabola, subvulpina, transgressa, vulpina*
GENUS *Polycaulonia* (Synonym for *Caloplaca*)
GENUS *Seirophora* Sirius Schist Lichen
SPECIES *magara*
GENUS *Xanthopeltis* Yellow Pelt Schist Lichen
SPECIES *rupicola*
GENUS *Xanthoria* Sunburst Schist Lichen
SPECIES *alaskana, borealis, calcicola, candelaria, concinna,* ***ectaneoides***(Dye Herb), ***elegans***(Dye Herb), *filsonii, fulva, hasseana, ligulata, papillifera,* ***parietina***(Dye Herb), ***polycarpa***(Dye Herb), *ramulosa, sorediata, subramulosa, ucrainica, ulophyllodes*
VARIETIES *calcicola congranulata, candelaria lychnea, candelaria oregana, parietina parietina, parietina convexa, parietina ectanea, ulophyllodes fallax*

FAMILY 2 *Fuscideaceae* Quilt Schist Lichen
GENUS *Fuscidea* Quilt Schist Lichen
SPECIES *arboricola, asbolodes, austera, australis, cyathoides, elixii, gothoburgensis, gyrodes, intercincta, iowensis, kochiana, lightfootii, lygaea, maccarthyi, mayrhoferi, mollis, poeltii, praeruptorum, pusilla, ramboldioides, recensa, subasbolodes, subfilamentosa, subreagens, thomsonii* VARIETIES *austera aggregata, austera taeniarum, australis australis, australis montana, cyathoides cyathoides, cyathoides sorediata, gothoburgensis maculosa, intercincta atlantica, intercincta decincta, kochiana albescens, lygaea periplaca, lygaea tenebrica, recensa arcuatula, recensa curvula*
GENUS *Hueidea* Hue Quilt Schist Lichen
SPECIES *australiensis*
GENUS *Lettauia* Lettau Quilt Schist Lichen
SPECIES *cladoniicola*
GENUS *Maronea* Maron Quilt Schist Lichen
SPECIES *carolinae, constans, polyphaea*
GENUS *Orphniospora* Orphan Spore Quilt Schist Lichen
SPECIES *moriopsis*
VARIETIES *moriopsis moriopsis, moriopsis atrata, moriopsis brunnea*
GENUS *Ropalospora* Comet Spore Quilt Schist Lichen
SPECIES *chlorantha, hibernica, lugubris, viridis*
VARIETIES *lugubris lugubris, lugubris sorediata*
GENUS *Sarrameana* Sarram Quilt Schist Lichen
SPECIES *albidoplumbea*

CLASS 621.4 *Mucoromycetes* Mold

ORDER 1 *Mucoromycetales* Mold

FAMILY 1 *Illosporaceae* Ill Spore Mold
GENUS *Illosporium* Ill Spore Mold
SPECIES *carneum, coccineum, olivatrum, roseum*
GENUS *Illosporiopsis* (Synonym for *Hobsonia*)
GENUS *Hobsonia* Hobson Ill Spore Mold
SPECIES *christiansenii, mirabilis*

FAMILY 2 *Myceliaceae* Mycel Mold
GENUS *Cheiromycella* Hand Mycel Mold
SPECIES *gyrosa, microscopica*
GENUS *Davisomycella* Davis Mycel Mold
SPECIES *ampla, asiatica, fragilis, hiratsukae, lacrimiformis, limitata, medusa, montana, ponderosae*

GENUS *Eleutheromycella* Eleuthsinian Mycel Mold
SPECIES *mycophila*
GENUS *Myceliasterila* (Synonym)
GENUS *Myceliophthora* Phthoral Mycel Mold
SPECIES *fergusii, hinnulea, lutea, thermophila, vellerea*
GENUS *Rehmiomycella* Rehm Mycel Mold
SPECIES *phoradendri*

FAMILY 3 *Myxotrichaceae* Hair Mold
GENUS *Myxotrichum* Hair Mold
SPECIES *aeruginosum, berkeleyi, cancellatum, chartarum, deflexum, herbariense, ochraceum, setosum, stipitatum*
GENUS *Byssoascus* Silky Hair Mold
SPECIES *striatisporus*
GENUS *Geomyces* Earth Hair Mold
SPECIES *asperulatus, auratus, pannorum, pulvereus, vinaceus*
GENUS *Malbranchea* Bad Branch Hair Mold
SPECIES *albolutea, arcuata, aurantiaca, chrysosporioidea, cinnamomea, circinata, dendritica, filamentosa, flava, flavorosea, flocciformis, fulva, graminicola, gypsea, multicolor, pulchella, sclerotica, setosa*
GENUS *Oidiodendron* Hypha Wood Hair Mold
SPECIES *ambiguum, cereale, chlamydosporicum, citrinum, echinulatum, flavum, griseum, maius, periconioides, pilicola, rhodogenum, scytaloides, setiferum, tenuissimum, truncatum*
GENUS *Pseudogymnoascus* Nude Hair Mold
SPECIES *alpinus, bhattii, roseus*

FAMILY 4 *Ophiostomataceae* Snake Mouth Mold
GENUS *Ophiostoma* Snake Mouth Mold
SPECIES *abiocarpum, adjuncti, aenigmaticum, ainoae, americanum, araucariae, bacillisporum, bicolor, bragantinum, brevicolle, brunneociliatum, canum, capense, clavatum, cuculatum, cucullatum, distortum, epigloeum, europhioides, flexuosum, francke-grosmanniae, galeiforme, grande, himal-ulmi, huntii, laricis, megalobrunneum, microsporum, multiannulatum, narcissi, neglectum, nigrocarpum, nigrum, novo-ulmi, olivaceum, penicillatum, perfectum, piceae, piceaperdum, polonicum, polyporicola, populinum, roraimense, rostrocoronatum, serpens, seticolle, sparsum, spinosum, splendens, tetropii, tremuloaureum, triangulosporum, ulmi, valdivianum, wageneri*
GENUS *Ceratocystiopsis* Horn Snake Mouth Mold
SPECIES *alba, collifera, crassivaginata, falcata, minima, minuta, pallidobrunnea, proteae, ranaculosa, retusi*
VARIETIES *minuta bicolor*
GENUS *Europhium* Greater Snake Mouth Mold
SPECIES *aureum, clavigerum, robustum, trinacriforme*
GENUS *Graphilbum* Graphic Snake Mouth Mold
SPECIES *pleomorphum*
GENUS *Kathistes* Kathay Snake Mouth Mold
SPECIES *analemmoides, calyculata, fimbriata*
GENUS *Klasterskya* Klastersky Snake Mouth Mold
SPECIES *acuum*
GENUS *Leptographium* Thin Graphic Snake Mouth Mold
SPECIES *abietinum, albopini, alethinum, antibioticum, brachiatum, douglasii, dryocoetidis, lundbergii, penicillatum, procerum, reconditum, terebrantis, wingfieldii*
GENUS *Raffaelea* Raffael Snake Mouth Mold
SPECIES *albimanens, ambrosiae, arxii, barbata, canadensis, hennebertii, montetyi, quercivora, santoroi, sulcati, tritirachium, variabilis*
GENUS *Sporothrix* Mammal Sporothricosis Snake Mouth Mold
SPECIES *alba, albicans, curviconia, cyanescens, foliorum, fungorum, ghanensis, guttuliformis, inflata, insectorum, inusitatiramosa, isarioides, nivea, pallida, ramosissima, ranii, schenckii, sclerotialis*
VARIETIES *schenckii schenckii, schenckii luriei*
GENUS *Spumatoria* Frothing Snake Mouth Mold
SPECIES *longicollis*
GENUS *Subbaromyces* Baro Snake Mouth Mold
SPECIES *aquaticus, splendens*

FAMILY 4 *Pneumocystidaceae* Lung Cyst Mold
GENUS *Pneumocystis* Lung Cyst Mold
SPECIES *carinii*

ORDER 2 *Eurotiales* Downy Mold

FAMILY 1 *Eurotiaceae* Greater Downy Mold
GENUS *Eurotium* Greater Downy Mold

SPECIES *amstelodami, appendiculatum, aridicola, chevalieri, costiforme, fimicola, glabrum, herbariorum, lateritium, medium, microsporum, niveoglaucum, parviverruculosum, spiculosum, tuberculatum, xerophilicum, xerophilum*

GENUS *Leiothecium* — Theca Greater Downy Mold

SPECIES *ellipsoideum*

FAMILY 2 *Pseudeurotiaceae* — Lesser Downy Mold

GENUS *Pseudeurotium* — Lesser Downy Mold

SPECIES *bakeri, desertorum, ovale, zonatum*

VARIETIES *ovale ovale, ovale milkoi*

GENUS *Connersia* — Conners Lesser Downy Mold

SPECIES *rilstonii*

GENUS *Leuconeurospora* — Light Lesser Downy Mold

SPECIES *pulcherrima*

GENUS *Pleuroascus* — Pleural Sac Lesser Downy Mold

SPECIES *nicholsonii*

FAMILY 3 *Trichocomaceae* — Downy Mold

GENUS *Trichocoma* — Downy Mold

SPECIES *paradoxa*

GENUS *Aspergillus* — Sprinkle Gill Downy Mold

SPECIES *acanthosporus, aculeatus, aeneus, allahabadii, alliaceus, alternatus, alutaceus, ambiguus, amylovorus, anthodesmis, appendiculatus, arenarius, asperescens, athecius, aurantiacus, aurantiobrunneus, auratus, aureolatus, aureolus, auricomus, auricularis, avenaceus, awamori, bicolor, biplanus, bisporus, bouffardi, brevipes, bridgeri, brunneouniseriatus, caelatus, caesiellus, caespitosus, campestris, candidus, carbonarius, carneus, carnoyi, cervinus, chevalieri, chrysellus, cinnamomeus, citrisporus, clavatoflavus, clavatonanica, clavatonanicus, clavatus, cleistominutus, conicus, conjunctus, coremiiformis, corrugatus, crassihyphae, cremeus, cristatus, crustosus, crystallinus, deflectus, dimorphicus, diversus, dorothicus, dubius, duricaulis, eburneocremeus, echinulatus, egyptiacus, elegans, ellipticus, elongatus, fennelliae, ficuum, fischeri, flaschentraegeri, flavescens, flavipes, flavofurcatus, flavus, floriformis, foetidus, foveolatus, fruticulosus, fumaricus, fumigatus, funiculosus, giganteus, glaber, glaucus, gorakhpurensis, gracilis, granulatus, granulosus, griseus, gymnosardae, halophilicus, helicothrix, hennebergii, heterocaryoticus, heteromorphus, heterothallicus, heyangensis, iizukae, implicatus, ingratus, insulicola, inuii, itaconicus, ivoriensis, janus, japonicus, kambarensis, kanagawaensis, lanosus, leporis, leucocarpus, longivesica, lucknowensis, malignus, malodoratus, manginii, melleus, mellinus, microcephalus, microcysticus, minimus, montevidensis, multicolor, murcuroides, nidulans, niger, nomius, nutans, ochraceoroseus, ochraceus, oryzae, ostianus, pallidus, panamensis, paradoxus, parasiticus, parvathecius, parvulus, penicilliformis, penicillioides, petrakii, peyronelii, proliferans, protuberus, pseudodeflectus, pseudoglaucus, pulverulentus, pulvinus, puniceus, quadricinctus, quadricingens, quadrilineatus, raperi, recurvatus, repens, restrictus, rhizopodus, robustus, roseus, ruber, rugulosus, salviicola, sclerotiorum, sepultus, silvaticus, sojae, sparsus, spathulatus, spectabilis, spelunceus, spinulosus, stellatus, stramenius, striatus, stromatoides, subolivaceus, subsessilis, sulphureus, sunderbanii, sydowii, tamarii, tardus, terreus, terricola, testaceocolorans, thomii, togoensis, tonophilus, tubingensis, umbrosus, unguis, unilateralis, ustus, varians, versicolor, violaceofuscus, violaceus, viridinutans, wentii, zhaoqingensis, zonatus*

VARIETIES *brunneouniseriatus nanus, chevalieri intermedius, fischeri glabra, fischeri spinosus, flavus flavus, flavus columnaris, flavus oryzae, foetidus acidus, foetidus pallidus, fumigatus acolumnaris, fumigatus ellipticus, fumigatus helvolus, janus janus,janus brevis, nidulans acristatus, nidulans dentatus, nidulans echinulatus, nidulans latus, niger niger, niger intermedius, niger nanus, niger phoenicis, oryzae oryzae, oryzae brunneus, oryzae effusus, parasiticus globosus, stellatus astellatus, sulphureus crassus, terreus africanus, terreus aureus, terreus globosus, terricola americanus, terricola indicus, versicolor rutilobrunneus*

GENUS *Byssochlamys* — Silky Cloak Downy Mold

SPECIES *fulva, nivea, verrucosa, zollerniae*

GENUS *Capsulotheca* — Capsule Theca Downy Mold

SPECIES *aspergilloides*

GENUS *Chromocleista* — Chromic Closed Downy Mold

SPECIES *cinnabarina, malachitea*

GENUS *Cladosarum* — Branched Downy Mold

SPECIES *olivaceum*

GENUS *Cristaspora* — Crest Spore Downy Mold

SPECIES *arxii*

GENUS *Dendrosphaera* — Wood Sphere Downy Mold

SPECIES *eberhardtii*

GENUS *Dichotomomyces* — Dichotomy Downy Mold

SPECIES *cejpii*

VARIETIES *cejpii spinosus*

GENUS *Elaphomyces* — Truffle Downy Mold

SPECIES *aculeatus, anthracinus, citrinus, granulatus, leucosporus, muricatus*

VARIETIES *granulatus cervinum*

GENUS *Emericella* — Meritorious Downy Mold

SPECIES *bicolor, cleistominuta, desertorum, falconensis, foeniculicola, navahoensis, nidulans, pluriseminata, purpurea, quadrilineata, rugulosa, similis, striata, subulata, undulata, variecolor, violacea*

VARIETIES *nidulans acristata*

GENUS *Erythrogymnotheca* Red Nude Theca Downy Mold

SPECIES *paucispora*

GENUS *Eupenicillium* Good Penicillin Downy Mold

SPECIES *angustiporcatum, cryptum, javanicum, javanicum lineolatum, javanicum meloforme, levitum, limosum, nepalense, parvum*

GENUS *Fennellia* Fennell Downy Mold

SPECIES *monodii, nivea*

GENUS *Geosmithia* Geosmith Downy Mold

SPECIES *cylindrospora, eburnea, emersonii, lavandula, malachitea, namyslowskii, putterillii, swiftii, viridis*

GENUS *Hamigera* Ham Downy Mold

SPECIES *reticulata*

GENUS *Hemicarpenteles* Carpenter Downy Mold

SPECIES *ornatum, paradoxum*

GENUS *Merimbla* Merimble Downy Mold

SPECIES *ingelheimense*

GENUS *Neosartorya* Sartory Downy Mold

SPECIES *delicata, fennelliae, glabra, hiratsukae, multiplicata, primulina*

GENUS *Olpitrichum* Olpe Downy Mold

SPECIES *macrosporum, patulum, sphaerosporum, tenellum*

GENUS *Paecilomyces* Paecil Downy Mold

SPECIES *aerugineus, amoeneroseus, byssochlamydoides, carneus, cateniannulatus, catenioblíquus, cinnamomeus, clavisporus, coleopterorum, cremeoroseus, dactylethromorphus, farinosus, fumosoroseus, ghanensis, inflatus, javanicus, lilacinus, marquandii, niphetodes, pascua, penicillatus, puntonii, simplex, smilanensis, tenuipes, variotii, victoriae, viridis*

VARIETIES *variotii antibioticus*

GENUS *Penicilliopsis* Lesser Penicillin Downy Mold

SPECIES *clavariiformis*

GENUS *Penicillium* Penicillin Downy Mold

SPECIES *abidjanum, aculeatum, adametzii, adametzioides, aethiopicum, albicans, albidum, alicantinum, alutaceum, anatolicum, angustiporcatum, aragonense, ardesiacum, arenicola, asturianum, atramentosum, atrosanguineum, atrovenetum, aurantioflammiferum, aurantiogriseum, bilaiae, brevicompactum, brevissimum, brunneum, buchwaldii, caerulescens, camemberti, canescens, capsulatum, carneum, caseifulvum, castellonense, catenatum, chalybeum, chermesinum, chrysogenum, citreonigrum, citrinum, coalescens, coeruleoviride, coeruleum, commune, coprobium, coprophilum, cordubense, corylophilum, corynephorum, cyaneum, cylindrosporum, daleae, dangeardii, decumbens, dendriticum, dierckxii, digitatum, dimorphosporum, discolor, diversum, dodgei, donkii, duclauxii, dupontii, echinulatum, emmonsii, erubescens, erythromellis, expansum, fasciculatum, fennelliae, flavidostipitatum, flavigenum, flavogenum, formosanum, fractum, freii, funiculosum, gaditanum, glabrum, gladioli, glandicola, glaucum, gossypii, gracilentum, graminicola, grancanariae, granulatum, griseofulvum, griseolum, griseoroseum, herqueri, heteromorphum, hirsutum, hispanicum, hordei, humuli, hypomycetis, ilerdanum, implicatum, incoloratum, indonesiae, inflatum, intermedium, inusitatum, isariiforme, islandicum, italicum, janczewskii, janthogenum, jensenii, kananaskense, kenraperi, klebahnii, kloeckeri, lacus-sarmientei, lagena, lanosum, lapatayae, lapidosum, lassenii, lehmanii, lignorum, lineatum, lividum, loliense, ludwigii, maclennaniae, madriti, majusculum, malacaense, mali, mariae-crucis, marneffei, megasporum, melanoconidium, melinii, meloforme, miczynskii, mimosinum, minioluteum, mirabile, moldavicum, molle, montanense, multicolor, murcianum, nalgiovense, neoechinulatum, nepalense, nilense, nodositatum, nodulum, novae-zelandiae, oblatum, ochrochloron, ochrosalmoneum, olivicolor, olsonii, onobense, ornatum, osmophilum, oxalicum, palitans, palmae, palmense, panamense, paneum, parmonense, patens, paxilli, phoeniceum, piceum, pinophilum, pinsaporum, pittii, polonicum, proteolyticum, pseudostromaticum, purpurescens, purpureum, purpurogenum, quercetorum, rademirici, radicum, raistrickii, raperi, rasile, resedanum, resticulosum, restrictum, reticulisporum, rolfsii, roqueforti, roseopurpureum, rubefaciens, rubidurum, rubrum, rugulosum, sabulosum, sajarovii, samsonii, scabrosum, sclerotigenum, sclerotiorum, selandiae, senticosum, severskii, shennonghianum, siamense, simplicissimum, sinaicum, smithii, solitum, sphaerum, spinulosum, spirillum, stolkiae, sublateritium, subtile, tarraconense, thomii, tricolor, turbatum, turolense, udagawae, ulaiense, valentinum, vanbeymae, variabile, variabile, vasconiae, velutinum, verrucosum, verruculosum, vinaceum, viridicatum, vulpinum, waksmanii, yugoslavicum, zacinthae, zonatum*

VARIETIES *abidjanum arvense, arenicola canadense, atrosanguineum manginii, aurantiogriseum album, aurantiogriseum aurantiocandidum, aurantiogriseum aurantiovirens, aurantiogriseum brunneoviolaceum, aurantiogriseum cyclopium, aurantiogriseum johanniolii, aurantiogriseum lanosocoeruleum, aurantiogriseum melanoconidium, aurantiogriseum neoechinulatum, aurantiogriseum puberulum, aurantiogriseum viridicatum, bilaiae lilacinoechinulatum, brevicompactum brunneostoloniferum, brevicompactum stoloniferum, brevicompactum volgaense, camemberti biforme, camemberti candidum, camemberti caseicola, canescens kapuscinskii, canescens raciborskii, canescens yarmokense, chermesinum indicum, chrysogenum chrysogenum, chrysogenum camerunense, chrysogenum dipodomyis, chrysogenum fulvescens, chrysogenum harmonense, chrysogenum meleagrinum, chrysogenum microsporum, chrysogenum notatum, citreonigrum cinereoatrum, citreonigrum citreoviride, citreonigrum gallaicum, citreonigrum hirayamae, citreonigrum katangense, citrinum baradicum, citrinum botryosum, citrinum sartoryi, citrinum steckii, commune album, commune flavoglaucum, commune lanosoviride, coprophilum coprophilum, coprophilum ribes-olens, corylophilum barcinonense, decumbens arabicum, decumbens glaucolanosum, decumbens meridianum, dierckxii atrovirens, dierckxii charlesii, dierckxii eben-bitarianum, dierckxii fellutanum, dierckxii sizovae, diversum aureum, dodgei brefeldianum, duclauxii clavigerum, echinulatum echinulatum, echinulatum discolor, emmonsii stipitatum, expansum crustaceum, expansum glaucum, funiculosum aurantiacum, funiculosum rubicundum, funiculosum varians, glabrum frequentans, glabrum terlikowskii, gladioli asperum, glandicola glandicola, glandicola confertum, glandicola glaucovenetum, glandicola mononematosum, griseofulvum dipodomyicola, griseofulvum urticae, griseoroseum concentricum, griseoroseum cyaneofulvum, herqueri coralligerum, herqueri luteocoeruleum, hirsutum hirsutum, hirsutum albocoremium, hirsutum allii, hirsutum corymbiferum, hirsutum venetum, indonesiae javanicum, indonesiae odoratum, italicum avellaneum, janczewskii granatense, janczewskii nigricans, jensenii godlewskii, klebahnii ehrlichii, kloeckeri wortmannii, lanosum kojigenum, lapidosum terrenum, lehmanii spiculisporum, megasporum asperosporum, megasporum echinosporum, megasporum giganteum, megasporum resinae, melinii damascenum, melinii estinogenum, melinii griseopurpureum, melinii radulatum, miczynskii michaelis, miczynskii pedemontanum, miczynskii shearii, miczynskii soppii, miczynskii syriacum, nilense egyptiacum, olivicolor ochraceum, phoeniceum idahoense, phoeniceum pusillum, pinophilum allahabadense, pinophilum korosum, raistrickii castellae, resedanum luteoaurantium, resticulosum japonicum, restrictum gorlenkoanum, restrictum kazachstanicum, restrictum kursanovii, restrictum striatisporum, roqueforti roqueforti, roqueforti conservandi, roqueforti gorgonzola, roseopurpureum vaccaeorum, rubrum rubrum, rubrum crateriforme, rugulosum phialosporum, rugulosum tardum, simplicissimum ciegleri, simplicissimum cremeogriseum, simplicissimum echinulonalgrovense, simplicissimum janthinellum, simplicissimum kabunicum, simplicissimum paraherquei, simplicissimum piscarium, simplicissimum pulvillorum, simplicissimum skrjabinii, solitum solitum, solitum carneolutescens, solitum crustosum, solitum crustosum-farinosum, solitum crustosum-pseudocasei, solitum crustosum-terrestre, solitum lanosogriseum, solitum melanochlorum, solitum patulum, solitum psittacinum, sphaerum rotundum, spinulosum abeanum, spinulosum nigricans, spinulosum spinuloramigenum, spinulosum trzebinskii, spirillum helicum, sublateritium ramusculum, thomii aurantioviolaceum, vanbeymae baarnense, velutinum fuscum, velutinum pinetorum, verrucosum casei, verrucosum gerundense, viridicatum olivinoviride, vulpinum claviforme, vulpinum ochraceum*

GENUS *Petromyces* Rock Downy Mold
SPECIES *albertensis*
GENUS *Polypaecilum* Polypaecil Downy Mold
SPECIES *botryoides, insolitum, pisce*
GENUS *Redaellia* Red Downy Mold
SPECIES *elegans*
GENUS *Sagenoma* Sagen Downy Mold
SPECIES *ryukyuense, viride*
GENUS *Sagenomella* Lesser Sagen Downy Mold
SPECIES *alba, bohemica, diversispora, griseoviridis, humicola, oligospora, sagenomatis, sclerotialis, striatispora, verticillata*
GENUS *Sarophorum* Sarophore Downy Mold
SPECIES *palmicola*
GENUS *Sterigmatocystis* (Synonym for *Aspergillus*)
GENUS *Talaromyces* Talar Downy Mold
SPECIES *austrocalifornicus, barcinensis, convolutus, derxii, eburneus, emersonii, emodensis, flavus, helicus, indigoticus, lanuginosus, leycettanus, luteus, macrosporus, muroii, spectabilis, subinflatus, tardifaciens, trachyspermus, ucrainicus, unicus*
VARIETIES *helicus boninensis, helicus major*
GENUS *Thermoascus* Thermal Sac Downy Mold
SPECIES *aurantiacus, crustaceus, thermophilus*

ORDER 3 Hypocreales Hypocreal Mold

FAMILY 1 *Hypocreaceae* Hypocreal Mold
GENUS *Hypocrea* Hypocreal Mold
SPECIES *argillacea, atrogelatinosa, aurantia, aureoviridis, ceramica, citrina, cornea, cupularis, dichromospora, gelatinosa, hunua, jecorina, lactea, lenta, lutea, muroiana, nigricans, pallida, pezizoides, pilulifera, placentula, poronioidea, protopulvinata, psychrophila, pulvinata, rufa, schweinitzii, semiorbis, spinulosa, splendens, strobilina, subsplendens, sulfurea, sulphurea, tawa, tremelloides* VARIETIES *aureoviridis macrospora, citrina citrina, citrina americana, rufa sterilis*
GENUS *Acremonium* Acremon Hypocreal Mold
SPECIES *acutatum, alabamense, alcalophilum, alternatum, antarcticum, apii, arxii, atrogriseum, bacillisporum, bactrocephalum, biseptum, blochii, borodinense, brachypenium, breve, brunnescens, byssoides, camptosporum, cavaraeanum, charticola, chrysogenum, crotocinigenum, cucurbitacearum, curvulum, cymosum, dichromosporum, diospyri, domschii, egyptiacum, exiguum, falciforme, flavum, furcatum, fusidioides, fusisporum, gamsii, glaucum, guillematii, hansfordii, hennebertii, hyalinulum, hypholomatis, implicatum,*

incoloratum, incrustatum, kiliense, lichenicola, lindtneri, lolii, longisporum, masseei, minutisporum, nectrioidea, nepalense, nigrosclerotium, ochraceum, olidum, persicinum, pinkertoniae, polychromum, potronii, psammosporum, pseudozeylanicum, psychrophilum, pteridii, radiatum, recifei, restrictum, rhabdosporum, roseogriseum, roseolum, rutilum, salmoneum, sclerotigenum, sordidulum, spicatum, spinosum, strictum, stromaticum, tectonae, thermophilum, tsugae, tubakii, typhinum, uncinatum, verruculosum, vitellinum, zeae, zeylanicum, zonatum

VARIETIES *typhinum typhinum, typhinum bulliforme, typhinum fasciculatum*

GENUS *Acrostalagmus* Stalag Hypocreal Mold

SPECIES *albus, aphidum, galeoides, luteoalbus*

GENUS *Aphysiostroma* Bladder Hypocreal Mold

SPECIES *stercorarium*

GENUS *Arachnocrea* Spider Hypocreal Mold

SPECIES *stipata*

GENUS *Battarrina* Battar Hypocreal Mold

SPECIES *inclusa*

GENUS *Bulbithecium* Bulb Theca Hypocreal Mold

SPECIES *hyalosporum*

GENUS *Cephalosporiopsis* Lesser Head Spore Hypocreal Mold

SPECIES *alpina*

GENUS *Cephalosporium* Head Spore Hypocreal Mold

SPECIES *cinnamomeum*

GENUS *Cladobotryum* Branched Berry Hypocreal Mold

SPECIES *amazonense, apiculatum, asterophorum, caribense, clavisporum, cubitense, fungicola, longiramosum, macrosporum, multiseptatum, mycophilum, penicillatum, pinarense, polypori, rubrobrunnescens, semicirculare, soroaense, sphaerocephalum, stereicola, tenue, verticillatum, virescens*

GENUS *Dialhypocrea* Dial Hypocreal Mold

SPECIES *puiggariana*

GENUS *Didymostilbe* Dual Stilbe Hypocreal Mold

SPECIES *aurantiospora, coffeae, matsushimae*

GENUS *Diplocladium* Dual Branched Hypocreal Mold

SPECIES *penicillioides*

GENUS *Emericellopsis* Meritorious Hypocreal Mold

SPECIES *donezkii, glabra, humicola, maritima, minima, mirabilis, pallida, robusta, sphaerospora, stolkiae, synnematicola, terricola*

VARIETIES *synnematicola magna, terricola glabra*

GENUS *Gibellulopsis* Gibel Hypocreal Mold

SPECIES *piscis*

GENUS *Gliocephalotrichum* Glow Hairy Head Hypocreal Mold

SPECIES *bulbilium, cylindrosporum, microchlamydosporum, ohiense, simplex*

GENUS *Gliocladium* Glow Branched Hypocreal Mold

SPECIES *album, ammoniphilum, aurifilum, catenulatum, cibotii, deliquescens, flavofuscum, flavum, luteolum, microspermum, nigrovirens, polyporicola, roseum, solani, vermoesenii, virens, viride*

GENUS *Halonectria* Halogen Cornicle Hypocreal Mold

SPECIES *milfordensis*

GENUS *Hapsidospora* Haphazard Spore Hypocreal Mold

SPECIES *irregularis*

GENUS *Hypocreopsis* Lesser Hypocreal Mold

SPECIES *lichenoides, rhododendri*

GENUS *Hypomyces* Mushroom Pimple Hypocreal Mold

SPECIES *aurantius, australis, baryanus, cervinus, chlorinigenus, chlorinus, chrysospermus, chrysostomus, dactylarioides, fulgens, hyalinus,* ***lactifluorum***(Edible), *lateritius, linkii, lithuanicus, miliarius, ochraceus, papulasporae, perniciosus, petchii, polyporinus, pseudopolyporinus, puertoricensis, rosellus, semitranslucens, spadiceus, stephanomatis, subiculosus, sympodiophorus, torminosus, tuberosus, tulasneanus, viridigriseus, viridis*

VARIETIES *chrysostomus porotheliiformis, ochraceus terrestris, subiculosus trichothecioides,* ***tulasneanus luteovirens***(Edible)

GENUS *Leucosphaerina* Light Sphere Hypocreal Mold

SPECIES *arxii, indica*

GENUS *Mycoarachis* Lesser Spider Hypocreal Mold

SPECIES *inversa*

GENUS *Mycogone* Gonad Hypocreal Mold

SPECIES *calospora, cervina, perniciosa, psilocybina, rosea*

GENUS *Myrothecium* Theca Hypocreal Mold

SPECIES *atroviride, bisetosum, carmichaelii, cinctum, gramineum, inundatum, lachastrae, masonii, prestonii, roridum, setiramosum, tongaense, verrucaria*

GENUS *Nigrosabulum* Black Sable Hypocreal Mold

SPECIES *ellipsoideum, globosum*

GENUS *Payosphaeria* Pay Sphere Hypocreal Mold
SPECIES *minuta*
GENUS *Penicillifer* False Penicillin Hypocreal Mold
SPECIES *pulcher*
GENUS *Phaeostilbella* Dusky Stilbe Hypocreal Mold
SPECIES *atra*
GENUS *Podostroma* Pod Stroma Hypocreal Mold
SPECIES *alutaceum, solmsii*
GENUS *Protocrea* Primitive Hypocreal Mold
SPECIES *delicatula, farinosa*
GENUS *Pseudohypocrea* Least Hypocreal Mold
SPECIES *citrinella*
GENUS *Rogersonia* Rogerson Hypocreal Mold
SPECIES *striolata*
GENUS *Sarawakus* Sarawak Hypocreal Mold
SPECIES *britannicus, fragilis, frustulosus, hexasporus, izawae, lycogaloides, roseus, sordidus, subtrachycarpus, succisus, trachycarpus*
GENUS *Selinia* Selin Hypocreal Mold
SPECIES *pulchra*
GENUS *Sepedonium* Sepedon Hypocreal Mold
SPECIES *ampullosporum, chalcipori, laevigatum, roseum*
GENUS *Septofusidium* Septic Fused Hypocreal Mold
SPECIES *berolinense, bifusisporum, elegantulum, herbarum, variabile*
GENUS *Sibirina* Sibir Hypocreal Mold
SPECIES *coriolopsicola, gamsii, orthospora, purpurea*
GENUS *Simplicillium* Simple Hypocreal Mold
SPECIES *lamellicola, lanosiniveum, obclavatum, wallacei*
GENUS *Sphaerostilbella* Sphere Stilbe Hypocreal Mold
SPECIES *aureonitens, berkeleyana, lutea, novae-zelandiae*
GENUS *Sporophagomyces* Spore Fragment Hypocreal Mold
SPECIES *chrysostomus, lanceolatus, moellerianus*
GENUS *Stephanoma* Stephan Hypocreal Mold
SPECIES *strigosum, tetracoccum*
GENUS *Stilbella* Stilbe Hypocreal Mold
SPECIES *acerina, aciculosa, albocitrina, annulata, bucidae, bulbicola, buquetii, buquetii, byssiseda, clavispora, clavulata, fimetaria, flavipes, fusca, jaapii, proliferans*
GENUS *Stromatocrea* Stroma Hypocreal Mold
SPECIES *cerebriformis*
GENUS *Tilachlidium* Tilach Hypocreal Mold
SPECIES *brachiatum, medietatis, subulatum*
GENUS *Tolypocladium* Tolyp Branched Hypocreal Mold
SPECIES *cylindrosporum, extinguens, geodes, lignicola, niveum, nubicola, parasiticum, tundrense*
GENUS *Trichoderma* Hairy Dermal Hypocreal Mold
SPECIES *aureoviride, citrinoviride, crassum, croceum, fasciculatum, fertile, ghanense, glaucum, hamatum, harzianum, inhamatum, koningii, longibrachiatum, minutisporum, oblongisporum, parceramosum, piluliferum, polysporum, pseudokoningii, pubescens, reesei, saturnisporum, spirale, strigosum, tomentosum, viride*
GENUS *Valsonectria* Vals Cornicle Hypocreal Mold
SPECIES *simpsonii*
GENUS *Verticillium* Whorled Hypocreal Mold
SPECIES *albo-atrum, biguttatum, botrytoides, calophylli, chlamydosporium, coccosporum, dahliae, distans, effusum, epimyces, epiphytum, falcatum, fungicola, griseum, hemipterigenum, incurvum, insectorum, intertextum, lecythidis, leptobactrum, lindauianum, longisporum, menisporoides, nanum, nigrescens, niveostratosum, nubilum, pseudohemipterigenum, rexianum, sacchari, seriatum, sulphurellum, terrestre, theobromae, tricorpus, vilmorinii*
VARIETIES *chlamydosporium chlamydosporium, chlamydosporium catenulatum, fungicola fungicola, fungicola aleophilum, fungicola flavidum*

FAMILY 2 *Bionectriaceae* Biotic Cornicle Hypocreal Mold
GENUS *Bionectria* Biotic Cornicle Hypocreal Mold
SPECIES *apocyni, aureofulva, byssicola, ochroleuca, tonduzii, tornata*
GENUS *Bryonectria* Moss Cornicle Hypocreal Mold
SPECIES *biseptata, callicarpa, cuneifera, hylocomii, phyllogena, praetermissa, racomitrii*
VARIETIES *cuneifera cuneifera, cuneifera jamaicensis*
GENUS *Clibanites* Cliban Cornicle Hypocreal Mold
SPECIES *paradoxa*
GENUS *Clonostachys* Clone Stack Cornicle Hypocreal Mold
SPECIES *compactiuscula*
GENUS *Heleococcum* Hele Round Cornicle Hypocreal Mold

SPECIES *alatosporum, aurantiacum, inapertum, japonense*
GENUS *Hydropisphaera* Hydric Sphere Cornicle Hypocreal Mold
SPECIES *arenula, arenuloides, boothii, cyatheae, dolichospora, erubescens, gigantea, haematites, hypoxantha, leucotricha, macrarenula, multiloculata, multiseptata, nymaniana, pachyderma, rufofusca, suffulta*
GENUS *Kallichroma* Callisto Chromic Cornicle Hypocreal Mold
SPECIES *glabrum, tethys*
GENUS *Lasionectria* Woolly Cornicle Hypocreal Mold
SPECIES *flavida*
GENUS *Mycocitrus* Citrus Cornicle Hypocreal Mold
SPECIES *phyllostachydis*
GENUS *Nectriella* Lesser Cornicle Hypocreal Mold
SPECIES *bloxamii, chrysites, consolationis, dacrymycella, exigua, laminariae, lophocoleae, luteola, minuta, paludosa, santessonii, versoniana*
GENUS *Nectriopsis* Least Cornicle Hypocreal Mold
SPECIES *broomeana, candicans, exigua, indigens, lecanodes, microthecia, oropensoides, parmeliae, rubefaciens, sporangiicola, squamulosa, tremellicola, tubariicola, violacea*
GENUS *Paranectria* Mock Cornicle Hypocreal Mold
SPECIES *affinis, oropensis, oropensis oropensis, oropensis parviseptata, superba*
GENUS *Paranectriella* False Cornicle Hypocreal Mold
SPECIES *arcuata, hemileiae, imperconspicua, juruana, meliolicola, miconiae, minuta*
GENUS *Peethambara* Peethambar Cornicle Hypocreal Mold
SPECIES *sundara*
GENUS *Peloronectriella* Pelor Cornicle Hypocreal Mold
SPECIES *sasae*
GENUS *Peristomialis* Peristome Cornicle Hypocreal Mold
SPECIES *berkeleyi, corynospora, dentifera*
GENUS *Pronectria* Greater Cornicle Hypocreal Mold
SPECIES *anisospora, echinulata, fissuriprodiens, pertusariicola, robergei, streimannii, tenacis, tenuispora, tincta, xanthoriae*
GENUS *Protocreopsis* Primitive Cornicle Hypocreal Mold
SPECIES *pertusa*
GENUS *Rhopalocladium* Club Branched Cornicle Hypocreal Mold
SPECIES *myxophilum*
GENUS *Roumegueriella* Roumeguer Cornicle Hypocreal Mold
SPECIES *rufula*
GENUS *Stephanonectria* Stephan Cornicle Hypocreal Mold
SPECIES *keithii*
GENUS *Stilbocrea* Stilbe Cornicle Hypocreal Mold
SPECIES *impressa, macrostoma*
GENUS *Trichonectria* Hairy Cornicle Hypocreal Mold
SPECIES *hirta, hyalocristata, rectipila*

FAMILY 3 *Ceratostomataceae* Horn Hypocreal Mold
GENUS *Ceratostoma* Horn Hypocreal Mold
SPECIES *notarisii, rhynchophora, spurium*
GENUS *Arxiomyces* Arx Horn Hypocreal Mold
SPECIES *campanulatus, vitis*
GENUS *Erostrotheca* Eros Theca Horn Hypocreal Mold
SPECIES *multiformis*
GENUS *Gibsonia* Gibson Horn Hypocreal Mold
SPECIES *phaeospora*
GENUS *Gonatobotrys* Gonad Berry Horn Hypocreal Mold
SPECIES *africana, flava, simplex*
GENUS *Harzia* Harz Horn Hypocreal Mold
SPECIES *acremonioides, velata, verrucosa*
GENUS *Melanospora* Dark Spore Horn Hypocreal Mold
SPECIES *aegyptiaca, asymmetrica, brevirostris, caprina, chionea, damnosa, exsola, fallax, fimbriata, fusispora, gigantea, inaequalis, karstenii, lagenaria, leucotricha, longisetosa, pegleri, phaseoli, vitrea, zamiae, zobelii*
GENUS *Microthecium* Theca Horn Hypocreal Mold
SPECIES *levitum*
GENUS *Persiciospora* Persic Spore Horn Hypocreal Mold
SPECIES *africana, japonica, masonii, moreaui*
GENUS *Pteridiosperma* Wing Spore Horn Hypocreal Mold
SPECIES *ciliatum, lenticulare*
GENUS *Pustulipora* Pustule Pore Horn Hypocreal Mold
SPECIES *corticola*
GENUS *Sphaeroderma* Sphere Dermal Horn Hypocreal Mold

SPECIES *sepultariae*
GENUS *Sphaerodes* Sphere Horn Hypocreal Mold
SPECIES *compressa, episphaericum, fimicola, micropertusa, retispora*
VARIETIES *fimicola ornata, retispora retispora, retispora inferior*
GENUS *Syspastospora* Syspast Horn Hypocreal Mold
SPECIES *parasitica*

FAMILY 4 *Clavicipitaceae* Ergot Hypocreal Mold
GENUS *Claviceps* Ergot Hypocreal Mold
SPECIES *africana, fusiformis, gigantea, grohii, junci, microcephala, nigricans, panicoidearum, paspali,* ***purpurea***(Medicinal Herb), *sorghi, viridis*
GENUS *Aschersonia* Ascherson Ergot Hypocreal Mold
SPECIES *aleyrodis, placenta*
GENUS *Atkinsonella* Atkinson Ergot Hypocreal Mold
SPECIES *hypoxylon*
GENUS *Atricordyceps* Atric Heart Ergot Hypocreal Mold
SPECIES *harposporifera*
GENUS *Balansia* Balans Ergot Hypocreal Mold
SPECIES *andropogonis, aristidae, claviceps, cyperi, henningsiana, oryzae-sativae, sclerotica, strangulans*
GENUS *Beauveria* Beauver Ergot Hypocreal Mold
SPECIES *amorpha, bassiana, brongniartii, caledonica, epigea, felina, velata, vermiconia*
GENUS *Byssostilbe* Silky Stilbe Ergot Hypocreal Mold
SPECIES *stilbigera*
GENUS *Cavimalum* Cavimal Ergot Hypocreal Mold
SPECIES *borneense, indicum*
GENUS *Cordycepioideus* Lesser Heart Ergot Hypocreal Mold
SPECIES *bisporus, octosporus*
GENUS *Cordyceps* Heart Ergot Hypocreal Mold
SPECIES *aphodii, australis, barberi, caloceroides, canadensis, capitata, clavulata, entomorrhiza, forquignonii, gracilis, incarnata, longisegmentis, memorabilis, militaris, ophioglossoides,* ***sinensis***(Medicinal Herb) *sphecocephala, tuberculata*
GENUS *Drechmeria* Drechmer Ergot Hypocreal Mold
SPECIES *coniospora, harposporioides*
GENUS *Echinodothis* Urchin Doth Ergot Hypocreal Mold
SPECIES *tuberiformis*
GENUS *Ephelis* Ephel Ergot Hypocreal Mold
SPECIES *oryzae*
GENUS *Epichloë* Epiphyte Green Ergot Hypocreal Mold
SPECIES *amarillans, baconii, clarkii, festucae, sylvatica, typhina*
GENUS *Epicrea* Epiphyte Ergot Hypocreal Mold
SPECIES *insignis*
GENUS *Gibellula* Gibe Ergot Hypocreal Mold
SPECIES *aranearum, clavulifera, leiopus, pulchra*
GENUS *Hirsutella* Hairy Ergot Hypocreal Mold
SPECIES *acridiorum, aphidis, brownorum, citriformis, cryptosclerotium, gregis, guyana, jonesii, kirchneri, necatrix, nodulosa, rhossiliensis, satumaensis, saussurei, strigosa, subramanianii, subulata, thompsonii, versicolor*
GENUS *Hymenostilbe* Hymen Stilbe Ergot Hypocreal Mold
SPECIES *muscarium, verrucosa*
GENUS *Hypocrella* Hyponic Ergot Hypocreal Mold
SPECIES *turbinata*
GENUS *Linearistroma* Linear Stroma Ergot Hypocreal Mold
SPECIES *lineare*
GENUS *Mycomalus* Bad Ergot Hypocreal Mold
SPECIES *bambusinus*
GENUS *Myriogenospora* Myriad Spore Ergot Hypocreal Mold
SPECIES *atramentosa*
GENUS *Neobarya* Bary Ergot Hypocreal Mold
SPECIES *aurantiaca, parasitica, peltigerae*
GENUS *Neocordyceps* Modern Heart Ergot Hypocreal Mold
SPECIES *kohyasanensis*
GENUS *Neotyphodium* Typhoid Ergot Hypocreal Mold
SPECIES *coenophialum*
GENUS *Podocrella* Pod Ergot Hypocreal Mold
SPECIES *poronioides*
GENUS *Polycephalomyces* Many Head Ergot Hypocreal Mold
SPECIES *ramosus, tomentosus*
GENUS *Torrubiella* Tor Ruby Ergot Hypocreal Mold

SPECIES *albolanata, albotomentosa, arachnophila, aranicida, rubra*
VARIETIES *arachnophila arachnophila, arachnophila pleiopus, arachnophila pulchra*
GENUS *Wakefieldiomyces* Wakefield Ergot Hypocreal Mold
SPECIES *peltatus*

FAMILY 5 *Nectriaceae* Cornicle Hypocreal Mold
GENUS *Nectria* Cornicle Hypocreal Mold
SPECIES *acrotyla, alata, aquifolii, atrofusca, augustoi, aurantiaca, aurantiicola, aureola, bactridioides, bainii, bicolor, borneensis, brassicae, calami, camerunensis, caulina, chaetopsinae, cinnabarina, cinnamomea, citrinoaurantia, coccinea, coccophila, coronata, coryli, curta, dealbata, decora, desmazieri, diploa, discophora, ditissima, diversispora, ellisii, epicallopisma, episphaeria, flavoviridis, freycinetiae, fuckeliana, funicola, fusispora, galligena, ganymede, gracilipes, grammicospora, grammicosporopsis, haematococca, hederae, humicola, illudens, inundata, inventa, ipomoeae, jatrophae, jungneri, kowhai, laetidisca, lagenae, lasiacidis, leptosphaeriae, lucifer, lugdunensis, macrostoma, magnoliae, magnusiana, mammoidea, manuka, mariannaeae, mauritiicola, modesta, muscivora, neogrammicospora, neomacrospora, orientalis, pallida, pallidula, pandani, penicillioides, pertusa, pertusoides, peziza, peziza, phaeodisca, phormiicola, pinea, pityrodes, platycephala, polyporina, pseudopeziza, pseudotrichia, pulcherrima, punicea, punicea ilicis, pyrosphaera, radicicola, ralfsii, ramulariae, rigidiuscula, rishbethii, rubrococca, rubropeziza, rugulosa, septospora, sesquicillii, setofusarii, sinopica, spirostriata, subquaternata, sylvana, tawa, tracosa, truncata, tuberculariformis, tuberculata, umbrina, urceolus, uredinicola, ventricosa, verrucosa, verrucospora, veuillotiana, vilior, viridigloea, viridispora, vulpina, westlandica, zealandica*
VARIETIES *chaetopsinae catenulatae, chaetopsinae penicillatae, haematococca brevicona, haematococca cucurbitae, haematococca fabae, haematococca phaseoli, haematococca piperis, haematococca pisi, mammoidea mammoidea, mammoidea rubi, mammoidea rugulosa, pityrodes saccharina, radicicola macroconidialis*
GENUS *Albonectria* White Cornicle Hypocreal Mold
SPECIES *albosuccinea, rigidiuscula, verrucosa*
GENUS *Allantonectria* All Anton Cornicle Hypocreal Mold
SPECIES *miltina*
GENUS *Allonectella* All Cornicle Hypocreal Mold
SPECIES *guaranitica, rubescens*
GENUS *Calonectria* Calisto Cornicle Hypocreal Mold
SPECIES *colhounii, crotalariae, gracilis, ilicicola, kyotensis,leightonii, minutissima, platasca, pyrochroa, quinqueseptata, spathulata, tessellata, theae, variabilis, xantholeuca*
VARIETIES *colhounii macroconidiale, kyotensis brasiliensis, kyotensis scoparium*
GENUS *Calostilbe* Calisto Stilbe Cornicle Hypocreal Mold
SPECIES *striispora*
GENUS *Calostilbella* Lesser Calisto Stilbe Cornicle Hypocreal Mold
SPECIES *calostilbe*
GENUS *Chaetopsina* Bristle Cornicle Hypocreal Mold
SPECIES *acutispora, fulva, ivoriensis, keniensis, unilateralis*
GENUS *Ciliciopodium* Cilique Pod Cornicle Hypocreal Mold
SPECIES *brevipes*
GENUS *Cosmospora* Cosmos Spore Cornicle Hypocreal Mold
SPECIES *purtonii, vilior, wegeliniana*
GENUS *Cyanochyta* Cyan Cornicle Hypocreal Mold
SPECIES *cyanogena*
GENUS *Cylindrium* Cylindric Cornicle Hypocreal Mold
SPECIES *aeruginosum, cordae, elongatum, flavovirens*
GENUS *Cylindrocarpon* Cylindric Carpel Cornicle Hypocreal Mold
SPECIES *aquaticum, candidum, carneum, congoense, decumbens, destructans, didymum, effusum, gracile, hederae, ianthothele, lichenicola, liriodendri, macroconidialis, macrosporum, magnusianum, musae, obtusisporum, olidum, orthosporum, ovatum, stilbophilum, tenue, theobromicola, vaginae*
VARIETIES *destructans destructans, destructans coprosmae, destructans crassum, destructans panacis, ianthothele ianthothele, ianthothele majus, ianthothele rugulosum, olidum olidum, olidum crassum, olidum suaveolens*
GENUS *Cylindrocladiella* Lesser Cylindric Branched Cornicle Hypocreal Mold
SPECIES *brevicollis, brevifilum, camelliae, elegans, lageniformis, mangiferae, novae-zelandiae, parva*
GENUS *Cylindrocladium* Cylindric Branched Cornicle Hypocreal Mold
SPECIES *avesticulatum, camelliae, candelabrum, citri, clavatum, heptaseptatum, lanceolatum, macrosporum, naviculatum, oumaiensis, ovatum, parvum, pauciramosum, penicilloides, peruvianum, pini, pteridis, spathiphylli*
GENUS *Dacryoma* Dacry Cornicle Hypocreal Mold
SPECIES *album*
GENUS *Dialonectria* Dial Cornicle Hypocreal Mold
SPECIES *flammeola*
GENUS *Fusarium* Fused Cornicle Hypocreal Mold

SPECIES *acutatum, anguioides, annulatum, anthophilum, aquaeductuum, arthrosporioides, babinda, bactridioides, begoniae, beomiforme, brachygibbosum, brassicae, brevicatenulatum, bugnicourtii, buharicum, bulbicola, buxicola, camptoceras, caucasicum, cavispermum, chlamydosporum, ciliatum, circinatum, coccidicola, coeruleum, compactum, concentricum, concolor, crookwellense, culmorum, de-tonianum, denticulatum, diversisporum, dlaminii, epistroma, eumartii, filiferum, flavum, flocciferum, fujikuroi, globosum, graminum, guttiforme, incarnatum, inflexum, lacertatum, lactis, larvarum, longipes, lunulosporum, macroceras, melanochlorum, merismoides, miscanthi, napiforme, neoceras, nisikadoi, nygamai, oxysporum, pallidoroseum, peltigerae, poae, polyphialidicum, proliferatum, pseudoanthophilum, pseudocircinatum, pseudonygamai, pyriforme, ramigenum, redolens, reticulatum, robustum, sacchari, sarcochroum, scirpisensu, sphaeriae, splendens, sporotrichioides, subglutinans, sublunatum, succisae, sulphureum, thapsinum, torulosum, trichothecioides, tumidum, venenatum*

VARIETIES *aquaeductuum medium, chlamydosporum chlamydosporum, chlamydosporum fuscum, larvarum rubrum, merismoides merismoides, merismoides acetilereum, merismoides chlamydosporale, merismoides crassum, merismoides violaceum, oxysporum aechmeae, oxysporum albedinis, oxysporum apii, oxysporum asparagi, oxysporum basilici, oxysporum batatas, oxysporum betae, oxysporum bouvardiae, oxysporum callistephi, oxysporum cannabis, oxysporum carthami, oxysporum cassiae, oxysporum cattleyae, oxysporum cepae, oxysporum chrysanthemi, oxysporum ciceris, oxysporum citri, oxysporum coffeae, oxysporum conglutinans, oxysporum crotalariae, oxysporum cubense, oxysporum cucumerinum, oxysporum cucurbitacearum, oxysporum cyclaminis, oxysporum dianthi, oxysporum elaeidis, oxysporum fabae, oxysporum fatshederae, oxysporum fragariae, oxysporum gladioli, oxysporum glycines, oxysporum hebes, oxysporum lagenariae, oxysporum lentis, oxysporum lilii, oxysporum lini, oxysporum lupini, oxysporum lycopersici, oxysporum matthiolae, oxysporum medicaginis, oxysporum melongenae, oxysporum melonis, oxysporum meniscoideum, oxysporum narcissi, oxysporum nicotianae, oxysporum niveum, oxysporum opuntiarum, oxysporum passiflorae, oxysporum perniciosum, oxysporum phaseoli, oxysporum pini, oxysporum pisi, oxysporum psidii, oxysporum radicis-lycopersici, oxysporum ranunculi, oxysporum raphani, oxysporum rauvolfiae, oxysporum redolens, oxysporum rhois, oxysporum sesami, oxysporum spinaciae, oxysporum tracheiphilum, oxysporum tuberosi, oxysporum tulipae, oxysporum vanillae, oxysporum vasinfectum, oxysporum zingiberi, proliferatum proliferatum, proliferatum minus, redolens dianthi, reticulatum reticulatum, reticulatum negundinis, sacchari elongatum, sporotrichioides sporotrichioides, sporotrichioides minus, sublunatum sublunatum, sublunatum elongatum*

GENUS *Fusidium* Lesser Fused Cornicle Hypocreal Mold
SPECIES *coccineum, griseum, heteronemum, viride*
GENUS *Fusidomus* Fused Dome Cornicle Hypocreal Mold
SPECIES *arcus, pulicaris*
GENUS *Fusisporium* Fused Spore Cornicle Hypocreal Mold
SPECIES *aurantiacum, bacilligerum, betae, foeni, georginae, roseolum*
GENUS *Gibberella* Humped Cornicle Hypocreal Mold
SPECIES *acervalis, acuminata, avenacea, baccata, buxi, cyanogena, flacca, fujikuroi, gordonii, imperatae, indica, intricans, moniliformis, pulicaris, stilboides, thapsina, tricincta, xylarioides, zeae*

VARIETIES *acuminata armeniacum, avenacea aywerte, avenacea nurragi, baccata buxi, baccata moricola, fujikuroi fujikuroi, fujikuroi intermedia, fujikuroi subglutinans, moniliformis intermedium, pulicaris coeruleum, pulicaris sambucinum*

GENUS *Gliocladiopsis* Glue Branched Cornicle Hypocreal Mold
SPECIES *irregularis, sumatrensis, tenuis*
GENUS *Glionectria* Glue Cornicle Hypocreal Mold
SPECIES *tenuis*
GENUS *Haematonectria* Blood Cornicle Hypocreal Mold
SPECIES *haematococca, illudens, ipomoeae, monilifera, termitum*
GENUS *Heliscus* Helix Cornicle Hypocreal Mold
SPECIES *submersus*
GENUS *Hyalofiorea* Hyaline Floral Cornicle Hypocreal Mold
SPECIES *ramosa*
GENUS *Lanatonectria* Woolly Cornicle Hypocreal Mold
SPECIES *flavolanata, flocculenta, mammiformis, raripila*
GENUS *Leuconectria* Light Cornicle Hypocreal Mold
SPECIES *clusiae*
GENUS *Metarhizium* Middle Rhizid Cornicle Hypocreal Mold
SPECIES *album, anisopliae, brunneum, flavoviride, guizhouense*
VARIETIES *flavoviride minus, flavoviride pemphigi*
GENUS *Neocosmospora* Modern Cosmos Spore Cornicle Hypocreal Mold
SPECIES *arxii, boninensis, parva, striata, tenuicristata, vasinfecta*
VARIETIES *vasinfecta africana*
GENUS *Neonectria* Modern Cornicle Hypocreal Mold
SPECIES *lucida, viridispora*
GENUS *Ophionectria* Snake Cornicle Hypocreal Mold
SPECIES *magniverrucosa, trichospora*
GENUS *Pleonectria* Pleomorphic Cornicle Hypocreal Mold

SPECIES pinicola
GENUS *Pseudonectria* Lesser Cornicle Hypocreal Mold
SPECIES *brongniartii, coronata, diparietispora, furfurella, pachysandricola, rousseliana*
GENUS *Psilonia* Psilum Cornicle Hypocreal Mold
SPECIES *arundinis*
GENUS *Pycnofusarium* Dense Fused Cornicle Hypocreal Mold
SPECIES *rusci*
GENUS *Rubrinectria* Rubric Cornicle Hypocreal Mold
SPECIES *olivacea*
GENUS *Stalagmites* Stalagmite Cornicle Hypocreal Mold
SPECIES *tumefaciens*
GENUS *Tetracytum* Tetracyte Cornicle Hypocreal Mold
SPECIES *lauri*
GENUS *Ticonectria* Tico Cornicle Hypocreal Mold
SPECIES *perianthii*
GENUS *Tubercularia* Tubercle Cornicle Hypocreal Mold
SPECIES *aurantiaca, lateritia, nigricans, perucinia, smaragdina, versicolor*
GENUS *Volutella* Volute Cornicle Hypocreal Mold
SPECIES *arundinis, buxi, ciliata, cinerescens, colletotrichoides, fructus, gilva, lini, melaloma, nivea, pachysandricola, ramkumarii, roseola, setosa, therryana*
VARIETIES *ciliata stipitata*
GENUS *Xenonectriella* Strange Cornicle Hypocreal Mold
SPECIES *leptaleae, lutescens, ornamentata, streimannii*
GENUS *Zythiostroma* Zyther Cornicle Hypocreal Mold
SPECIES *pinastri*

FAMILY 6 *Niessliaceae* Niessl Hypocreal Mold
GENUS *Niesslia* Niessl Hypocreal Mold
SPECIES *aemula, antarctica, cladoniicola, exigua, exilis, exosporioides, fuegiana, ilicifolia, lobariae, pulchriseta, subiculosa*
GENUS *Circinoniesslia* Thistle Niessl Hypocreal Mold
SPECIES *nectriae*
GENUS *Cryptoniesslia* Hidden Niessl Hypocreal Mold
SPECIES *setulosa*
GENUS *Malmeomyces* Chalk Clay Niessl Hypocreal Mold
SPECIES *pulchellus*
GENUS *Melanopsamma* Dark Sand Niessl Hypocreal Mold
SPECIES *borealis, pomiformis, pustula*
VARIETIES *borealis borealis, borealis minor*
GENUS *Miyakeomyces* Miyake Niessl Hypocreal Mold
SPECIES *bambusae*
GENUS *Monocillium* Monad Cilia Niessl Hypocreal Mold
SPECIES *arctiicola, bulbillosum, constrictum, dimorphosporum, indicum, loricatum, minutisporum, mucidum, nordinii, tenue, tenuissimum*
GENUS *Taiwanascus* Taiwan Sac Niessl Hypocreal Mold
SPECIES *tetrasporus*
GENUS *Trichosphaerella* Hairy Sphere Niessl Hypocreal Mold
SPECIES *ceratophora, decipiens*
GENUS *Valetoniella* Valet Niessl Hypocreal Mold
SPECIES *claviornata, crucipila, pauciornata*
GENUS *Valetoniellopsis* Lesser Valet Niessl Hypocreal Mold
SPECIES *laxa*

ORDER 4 *Laboulbeniales* Insect Parasite Mold

FAMILY 1 *Laboulbeniaceae* Insect Parasite Mold
GENUS *Laboulbenia* Insect Parasite Mold
SPECIES *anoplogenii, argutoris, cafii, casnoniae, clivinalis, cristata, dubia, eubradycelli, fasciculata, fennica, flagellata, giardi, gyrinidarum, inflata, leisti, marina, nebriae, notiophili, ophoni, pedicellata, polyphaga, pseudomasci, rougetii, stilicicola, subterranea, thaxteri, vulgaris*
VARIETIES *subterranea lecoareri*
GENUS *Acallomyces* Callisto Pyral Moth Parasite Mold
SPECIES *gyrophaenae, homalotae, platyolae*
GENUS *Acompsomyces* Adorned Gelech Moth Parasite Mold
SPECIES *stenichni*
GENUS *Amorphomyces* Amorphic Sphinx Moth Parasite Mold
SPECIES *biformis, falagriae, floridanus, hernandoi, minisculus, obliqueseptatus, ophioglossae, pronomaeae, rubescens, schistogeniae, stenusae, stipitatus*

GENUS *Apatelomyces* Apatel Zanole Moth Parasite Mold
SPECIES *ogmoceri*
GENUS *Apatomyces* Apate Wood Borer Beetle Parasite Mold
SPECIES *laboulbenioides*
GENUS *Aphanandromyces* Aphane Longhorn Beetle Parasite Mold
SPECIES *audisioi*
GENUS *Aporomyces* Perplexing Tachin Fly Parasite Mold
SPECIES *byrrhini, lutrochi, perpusillus, physemi, subulatus, szaboi, trinitatis, uniflagellatus*
GENUS *Arthrorhynchus* Straight Snout Insect Parasite Mold
SPECIES *cyclopodiae, eucampsipodae*
GENUS *Asaphomyces* Gad Fly Parasite Mold
SPECIES *tubanticus*
GENUS *Autophagomyces* Autophage Insect Parasite Mold
SPECIES *falcatus*
GENUS *Balazucia* Balaz Flesh Mite Parasite Mold
SPECIES *bilateralis, japonica*
GENUS *Benjaminiomyces* Benjamin Owlet Moth Parasite Mold
SPECIES *melanophthalmae, perpusilla, platensis, pumila*
GENUS *Botryandromyces* Berry Water Midge Parasite Mold
SPECIES *heteroceri, ornatus*
GENUS *Cainomyces* Cain Pyral Moth Parasite Mold
SPECIES *isomali*
GENUS *Camptomyces* Gall Gnat Parasite Mold
SPECIES *brunneomarginatus, europaeus, falcatus, guatemalensis, melanopus, recurvatus, subsigmoideus, sumatrae*
GENUS *Cantharomyces* Soldier Blister Beetle Parasite Mold
SPECIES *denigratus, italicus, orientalis, platystethi*
GENUS *Carpophoromyces* Phthoral Fruit Fly Parasite Mold
SPECIES *cybocephali*
GENUS *Chitonomyces* Chiton Insect Parasite Mold
SPECIES *melanurus, paradoxus*
GENUS *Clematomyces* Mild Grasshopper Parasite Mold
SPECIES *astenalis, calcaratus, indicus, pinophili, trochoideanus*
GENUS *Columnomyces* Fig Wax Scale Bug Parasite Mold
SPECIES *ptomaphagi*
GENUS *Compsomyces* Adorned Longhorn Beetle Parasite Mold
SPECIES *insignis, lestevi, macropoda, palameni, platensis, stilicopsis, verticillatus*
GENUS *Coreomycetopsis* Core Longhorn Beetle Parasite Mold
SPECIES *oedipus*
GENUS *Corylophomyces* Cory Crest Coat Beetle Parasite Mold
SPECIES *peyerimhoffii, reflexus, sarawakensis, sericoderi, weirii*
GENUS *Cucujomyces* Flat Coat Beetle Parasite Mold
SPECIES *bilobatus, celebensis, curtipes, cylindrocarpus, diplocoeli, elegans, elegantissimus, gonicoeli, intermedius, melanopus, phycophilus, reynoldsii, rotundatus, stipatus, lasiochili*
GENUS *Cupulomyces* Cupule Insect Parasite Mold
SPECIES *lasiochili*
GENUS *Dermapteromyces* Earwing Bug Parasite Mold
SPECIES *arimae, ctenophorus, labiae*
GENUS *Diandromyces* Diander Insect Parasite Mold
SPECIES *chilenus*
GENUS *Dichomyces* Dichotomy Tachin Fly Parasite Mold
SPECIES *inaequalis*
GENUS *Diclonomyces* Diclone Insect Parasite Mold
SPECIES *eumicrophilus, stilomedonis, subgaleatus*
GENUS *Dimeromyces* Dimer Moth Parasite Mold
SPECIES *balazucii, corynitis*
GENUS *Diphymyces* Diphyial Orb Web Spider Parasite Mold
SPECIES *appendiculatus, bidentatus, curvatus, niger, penicillifer, silphidarum, urbasoli*
GENUS *Diplomyces* Diploid Orb Web Spider Parasite Mold
SPECIES *atanygnathi, clavifer*
GENUS *Diplopodomyces* Diplopod Insect Parasite Mold
SPECIES *callipodos*
GENUS *Dipodomyces* Dipod Insect Parasite Mold
SPECIES *monstruosus, phloeocharidis*
GENUS *Distolomyces* Distol Insect Parasite Mold
SPECIES *forficulae*
GENUS *Ecteinomyces* Ectein Tachin Fly Parasite Mold

SPECIES *trichopterophilus*
GENUS *Enarthromyces* Rock Centipede Parasite Mold
SPECIES *indicus*
GENUS *Euhaplomyces* Trematode Worm Parasite Mold
SPECIES *ancyrophori*
GENUS *Euphoriomyces* Euphoria Scarab Beetle Parasite Mold
SPECIES *liodivorus*
GENUS *Gloeandromyces* Glow Insect Parasite Mold
SPECIES *nycteribiidarum, streblae*
GENUS *Haplomyces* Solitary Gall Gnat Parasite Mold
SPECIES *californicus, texanus, virginianus*
GENUS *Hesperomyces* Hesper Longhorn Beetle Parasite Mold
SPECIES *virescens*
GENUS *Histeridomyces* Hister Beetle Parasite Mold
SPECIES *acriti, europaeus, flagelliferus, ramosus, tishechkinii, venezolanus*
GENUS *Homaromyces* Homar Tussock Moth Parasite Mold
SPECIES *epieri*
GENUS *Hydraeomyces* Hydra Water Beetle Parasite Mold
SPECIES *halipli*
GENUS *Idiomyces* Idiope Orb Web Spider Parasite Mold
SPECIES *peyritschii*
GENUS *Kleidiomyces* Kleid Seed Longhorn Bug Parasite Mold
SPECIES *furcillatus, hoplandriae, venezolanus*
GENUS *Kruphaiomyces* Kruph Insect Parasite Mold
SPECIES *trochoidei*
GENUS *Limnaiomyces* Limnetic Cosmo Moth Parasite Mold
SPECIES *amphiopsis, hydrocharis, tropisterni*
GENUS *Majewskia* Majewsky Insect Parasite Mold
SPECIES *japonica*
GENUS *Meionomyces* Meionet Orb Web Spider Parasite Mold
SPECIES *astenalis, asteni, astenicola, dibelonitis, mychoceri, thaxterelli*
GENUS *Microsomyces* Tiny Gnaph Orb Web Spider Parasite Mold
SPECIES *psammoechi, telephani*
GENUS *Misgomyces* Misgolas Idiope Orb Web Spider Parasite Mold
SPECIES *dyschirii*
GENUS *Monandromyces* Monander Sea Squirt Parasite Mold
SPECIES *australis, elongatus, falcatus, longispinae, microveliae, neoalardi, polhemorum, protuberans, pseudoveliae, tenuistipitis, umbonatus*
GENUS *Monoicomyces* Moniognose Walking Stick Bug Parasite Mold
SPECIES *athetae, britannicus, fragilis, homalotae, nigrescens*
GENUS *Nanomyces* Tiny Weevel Beetle Parasite Mold
SPECIES *appendiculatus, fijianus, perpendicularis*
GENUS *Neohaplomyces* Modern Solitary Gall Gnat Parasite Mold
SPECIES *cubensis, medonalis, neomedonalis*
GENUS *Nycteromyces* Bat Tick Louse Fly Parasite Mold
SPECIES *streblidinus*
GENUS *Osoriomyces* Orso Orb Web Spider Parasite Mold
SPECIES *rhizophorus*
GENUS *Parvomyces* Small Musk Fly Parasite Mold
SPECIES *merophysiae*
GENUS *Peyritschiella* Peyritsch Tachin Fly Parasite Mold
SPECIES *biformis, furcifera, hybrida, protea, vulgata*
GENUS *Phalacrichomyces* Minute Marsh Beetle Parasite Mold
SPECIES *anomalus, normalis*
GENUS *Phaulomyces* Phaul Longhorn Beetle Parasite Mold
SPECIES *cis, corylophodis, denticulatus, dibelonetis, discolomidinus, ephistemi, euaestheti, leonensis, mediterraneus, octotemni, perparvus, ptilii, simplocariae, triramosus*
GENUS *Picardella* Picard Insect Parasite Mold
SPECIES *catalaunica, endogaea*
GENUS *Polyascomyces* Many Sac Insect Parasite Mold
SPECIES *trichophyae*
GENUS *Prolixandromyces* Prolix Insect Parasite Mold
SPECIES *corniculatus, lingulatus, rhinoceralis, tenuis, triandrus, veliae*
GENUS *Rhachomyces* Rhachis Tachin Fly Parasite Mold
SPECIES *canariensis, furcatus, lasiophorus, philonthinus*
GENUS *Rhadinomyces* Robber Fly Parasite Mold
SPECIES *cristatus, pallidus*

GENUS *Rhipidiomyces* Wedge Blister Beetle Parasite Mold
SPECIES *acriti*
GENUS *Rhizomyces* Rhizid Gall Gnat Parasite Mold
SPECIES *canzonerii, circinalis, confusus, cornutus, crispatus, ctenophorus, cucullatus, gibbosus, gracilis, kamerunus*
GENUS *Rickia* Rick Stonefly Parasite Mold
SPECIES *dendroiuli, hyperborea, zanettii*
GENUS *Rossiomyces* Ross Tachin Fly Parasite Mold
SPECIES *falcatus*
GENUS *Sandersoniomyces* Sanderson Insect Parasite Mold
SPECIES *divaricatus*
GENUS *Scalenomyces* Scalen Longhorn Beetle Parasite Mold
SPECIES *endogaeus*
GENUS *Scaphidiomyces* Shining Rove Beetle Parasite Mold
SPECIES *baeocerae, baeoceridii, platensis, scaphicoma*
GENUS *Scelophoromyces* Scelophore Insect Parasite Mold
SPECIES *osorianus*
GENUS *Siemaszkoa* Siemaszk Insect Parasite Mold
SPECIES *annae, fennica, flexa, ptenidii, pusillima, ramificans, valida*
GENUS *Smeringomyces* Smering Pholce Orb Web Spider Parasite Mold
SPECIES *anomalus, chaetophilus, lingulatus, trinitatis*
GENUS *Sphaleromyces* Slippery Tortric Moth Parasite Mold
SPECIES *lathrobii*
GENUS *Stichomyces* Stitch Jumping Orb Web Spider Parasite Mold
SPECIES *conosomatis*
GENUS *Stigmatomyces* Stigma Fruit Fly Parasite Mold
SPECIES *biformis, entomophilus, purpureus*
GENUS *Sugiyamaemyces* Sugiyama Insect Parasite Mold
SPECIES *oroussetii*
GENUS *Symplectromyces* Symplect Limon Fly Parasite Mold
SPECIES *vulgaris*
GENUS *Sympodomyces* Sympodial Insect Parasite Mold
SPECIES *pentacellularis*
GENUS *Synandromyces* Synander Insect Parasite Mold
SPECIES *amarygmae, floriformis, geniculatus, javanus, peltoidis, platydemae, psammoechi, sinensis, telephani, tomari*
GENUS *Tavaresiella* Tavares Insect Parasite Mold
SPECIES *hebri, majewskii, polhemi, santamariae*
GENUS *Teratomyces* Terat Fly Parasite Mold
SPECIES *actobii, philonothi*
GENUS *Tetrandromyces* Tetrander Insect Parasite Mold
SPECIES *arcuatus, bottegoi, brachidae, pallescens, terminalis, weirianus*
GENUS *Triainomyces* Triain Insect Parasite Mold
SPECIES *hollowayanus*
GENUS *Trochoideomyces* Trochoid Insect Parasite Mold
SPECIES *gracilicaulis*
GENUS *Troglomyces* Troglodyte Orb Web Spider Parasite Mold
SPECIES *manfredii*
GENUS *Zeugandromyces* Zeug Mosquito Parasite Mold
SPECIES *australis, orientalis, stilici*

FAMILY 2 *Ceratomycetaceae* Horn Insect Parasite Mold
GENUS *Ceratomyces* (Synonym for *Rhynchophoromyces*)
GENUS *Drepanomyces* Hook Tip Moth Horn Parasite Mold
SPECIES *malayanus*
GENUS *Eusynaptomyces* Eusyn Longhorn Beetle Parasite Mold
SPECIES *africanus, benjaminii, borealis, enochri, hydrobii*
GENUS *Helodiomyces* Marsh Beetle Horn Parasite Mold
SPECIES *elegans*
GENUS *Plectomyces* Plaited Longhorn Beetle Parasite Mold
SPECIES *gracilis*
GENUS *Rhynchophoromyces* Horn Snout Weevel Beetle Parasite Mold
SPECIES *amphiopsis, anacaenae, denticulatus, elephantinus, helophori, rostratus, sperchopsis, unguicola*
GENUS *Synaptomyces* Synapter Longhorn Beetle Parasite Mold
SPECIES *argentinus*
GENUS *Thaumasiomyces* Thaumas Longhorn Beetle Parasite Mold
SPECIES bipedalis, scabellularius, scarurus

FAMILY 3 *Euceratomycetaceae* Greater Horn Insect Parasite Mold

GENUS *Euceratomyces* — Greater Horn Insect Parasite Mold
SPECIES *terrestris*
GENUS *Cochliomyces* — Cochliar Palp Longhorn Beetle Parasite Mold
SPECIES *argentinensis, trinitatis*
GENUS *Colonomyces* — Colon Diad Cid Fly Parasite Mold
SPECIES *appendiculatus*
GENUS *Euzodiomyces* — Euzod Greater Horn Parasite Mold
SPECIES *lathrobii*
GENUS *Pseudoecteinomyces* — Ectein Greater Horn Parasite Mold
SPECIES *zuphiicola*

ORDER 5 *Microascales* — Sac Mold

FAMILY 1 *Microascaceae* — Sac Mold
GENUS *Microascus* — Sac Mold
SPECIES *albonigrescens, brevicaulis, caviariformis, cinereus, cirrosus, desmosporus, doguetii, giganteus, griseus, longirostris, manginii, pyramidus, senegalensis, singularis, trigonosporus*
VARIETIES *trigonosporus trigonosporus, trigonosporus macrosporus, trigonosporus terreus*
GENUS *Allescheria* — (Synonym for *Meria, Pseudallescheria, Thielavia*)
GENUS *Canariomyces* — Canary Sac Mold
SPECIES *notabilis*
GENUS *Cephalotrichum* — Hairy Head Sac Mold
SPECIES *microsporum, nanum, purpureofuscum*
GENUS *Doratomyces* — Dorat Sac Mold
SPECIES *asperulus, columnaris*
GENUS *Enterocarpus* — Enterocarp Sac Mold
SPECIES *grenotii*
GENUS *Graphiocladiella* — Graphic Branched Sac Mold
SPECIES *clavigera*
GENUS *Indiella* — (Synonym for *Pseudallescheria*)
GENUS *Kernia* — Kern Sac Mold
SPECIES *bifurcotricha, brachytricha, geniculotricha, hippocrepida, hyalina, macrospora, nitida, ovata, pachypleura, peruviana, retardata*
GENUS *Lophotrichus* — Crested Hairy Sac Mold
SPECIES *ampullus, bartlettii, incarnatus, macrosporus, martinii, plumbescens*
GENUS *Madurella* — Madura Mammal Parasite Sac Mold
SPECIES *grisea, mycetomi*
GENUS *Monosporium* — Monospore Mammal Parasite Sac Mold
SPECIES *venezuelense*
GENUS *Pachnodium* — Thick Node Sac Mold
SPECIES *canum*
GENUS *Petriella* — Petry Sac Mold
SPECIES *guttulata, lindforsii, musispora, setifera, sordida*
GENUS *Pithoascus* — Pith Sac Mold
SPECIES *exsertus, langeronii, nidicola, platysporus, schumacheri, stoveri*
GENUS *Pseudallescheria* — Allescher Sac Mold
SPECIES *africana, angusta, boydii, desertorum, ellipsoidea, fimeti, fusoidea*
GENUS *Scedosporium* — Scedospore Sac Mold
SPECIES *prolificans*
GENUS *Scopulariopsis* — Scope Sac Mold
SPECIES *acremonium, asperula, atra, brumptii, canadensis, candida, carbonaria, chartarum, coprophila, croci, flava, fusca, gracilis, halophilica, hibernica, koningii, murina, parva, sphaerospora*
GENUS *Sphaeronaemella* — Sphere Thread Sac Mold
SPECIES *fragariae, helvellae, mougeotii*
GENUS *Stysanus* — Stysan Sac Mold
SPECIES *cybosporus, fimetarius, fuscus, medius*
GENUS *Thielaviopsis* — Thielav Sac Mold
SPECIES *basicola*
GENUS *Trichurus* — Hairy Sac Mold
SPECIES *cylindricus, dendrocephalus, gorgonifer, spiralis, terrophilus*
GENUS *Viennotidia* — Vienna Sac Mold
SPECIES *fimicola, humicola*
GENUS *Wardomyces* — Ward Sac Mold
SPECIES *anomalus, columbinus, humicola, inflatus, moseri, ovalis, pulvinatus*
GENUS *Wardomycopsis* — Lesser Ward Sac Mold
SPECIES *humicola, inopinata*

FAMILY 2 *Ceratocystidaceae* — Horn Cyst Sac Mold

GENUS *Ceratocystis* Horn Cyst Sac Mold

SPECIES *abietina, adiposa, allantospora, angusticollis, arborea, autographa, brunnea, brunneociliata, cainii, clavata, clavigera, coerulescens, conicola, coronata, davidsonii, deltoideospora, eucastaneae, fagacearum, fimbriata, grandicarpa, introcitrina, ips, laricicola, minor, moniliformis, montia, nothofagi, obscura, olivacea, paradoxa, pilifera, pinicola, pluriannulata, prolifera, pseudoeurophioides, radicicola, stenoceras, subannulata, tenella, tetropii, virescens*

VARIETIES *coerulescens douglasii, paradoxa ramosissima*

GENUS *Chalaropsis* (Synonym for *Ceratocystis*)

GENUS *Gondwanamyces* Gondwana Horn Cyst Sac Mold

SPECIES *capensis, proteae*

FAMILY 3 *Chadefaudiellaceae* Chadefaud Sac Mold

GENUS *Chadefaudiella* Chadefaud Sac Mold

SPECIES *quezelii, thomasii*

GENUS *Faurelina* Faurel Chadefaud Sac Mold

SPECIES *elongata, fimigena*

ORDER 6 *Mycosphaerellales* Sphere Mold

FAMILY 1 *Mycosphaerellaceae* Sphere Mold

GENUS *Mycosphaerella* Sphere Mold

SPECIES *africana, allicina, allii-cepae, alnicola, angelicae, aquilina, arachidis, arbuticola, areola, aronici, ascophylli, aspidii, asteroma, atomus, aucupariae, berberidis, berkeleyi, brassicicola, brunneola, buxicola, calamagrostidis, cannabis, capronii, caricae, carinthiaca, caryigena, cerasella, chamaeropis, chrysanthemi, cirsii-arvensis, citri, clymenia, coffeicola, confusa, conglomerata, crataegi, cruciferarum, cruenta, cydoniae, dearnessii, deightonii, delegatensis, depazeiformis, dianthi, dryadicola, elodis, equiseti, eryngii, eugeniae, euphorbiae, fagi, fijiensis, filicum, fragariae, fraxini, gibsonii, gossypina, graminicola, groveana, harthensis, hedericola, henningsii, holci, horii, hyperici, idaeina, iridis, isariophora, juncaginearum, juniperi, juniperina, laricina, laricis-leptolepidis, latebrosa, ligustri, lineolata, linicola, longissima, macrospora, magnusiana, marksii, martagonis, microsora, mikaniae-micranthae, millepunctata, minima, minor, montellica, mori, musae, musicola, musicola, nawae, oblivia, octopetalae, oedema, osborniae, oxyacanthae, parkii, pedicularis, peregrina, perexigua, phacae-frigidae, phaseolorum, pini, pinodes, pirolae, pistaciarum, pistacina, plantaginis, podagrariae, pomi, populi, populicola, populorum, praecox, primulae, pruni-persicae, psammae, pseudomaculiformis, pseudoseptorioides, pteridis, punctiformis, pyri, rabiei, ranunculi, recutita, rhododendri, ribis, rosigena, rubella, rubi, sagedioides, salicorniae, scirpi-lacustris, sesami, silenes-acaulis, silenis, slaptoniensis, stellarinearum, stemmatea, stromatosa, suberosa, superflua, tabifica, tassiana, thais, tingens, trifolii, tussilaginis, typhae, ulmi, vaccinii, zeae-maydis*

GENUS *Cercoseptoria* Dual Antenna Septic Sphere Mold

SPECIES *theae*

GENUS *Cercospora* Dual Antenna Spore Sphere Mold

SPECIES *abelmoschi, althaeina, ampelopsidis, apii, ariminensis, armoraciae, asparagi, astragali, aurantia, berteroae, beticola, bizzozeriana, brachypoda, brassicicola, calendulae, calotropidis, canescens, cannae, cantuariensis, capsici, carbonacea, cardamines, caricis, carotae, carthami, chenopodii, cistinearum, comari, crepidis, depazeoides, diazu, duddiae, dulcamarae, elaeidis, erysimi, exosporioides, ferruginea, festucae, fusca, fusimaculans, handelii, hayi, insulana, italica, jasminicola, juncicola, kikuchii, krugiana, lactucae, ligustrina, lindericola, longipes, longissima, loti, malayensis, malvacearum, malvarum, malvicola, mangiferae, manihobae, medicaginis, meliae, melongenae, mercurialis, mikaniicola, moravica, moricola, murina, myrti, nicotianae, nigellae, olivascens, opuli, physospermi, piaropi, plantaginis, plumbaginea, pudicae, purpurea, radiata, rautensis, resedae, rhamni, rhizophorae, ricinella, rodmanii, rosicola, rubrotincta, scandens, sechii, sesami, setariae, smilacis, sojina, solani-melongenae, solanicola, sorghi, traversiana, unamunoi, violae, xanthosomatis, zeae-maydis, zebrina, zinniae, zonata*

GENUS *Cercosporella* Lesser Dual Antenna Spore Sphere Mold

SPECIES *acetosella, filicis-feminae, junci, mori, pantoleuca, primulae, rubi, virgaureae*

GENUS *Cercosporidium* Least Dual Antenna Spore Sphere Mold

SPECIES *bolleanum, bougainvilleae, gliricidiasis, punctum, uniseptatum*

GENUS *Cercostigmina* Dual Antenna Stigma Sphere Mold

SPECIES *protearum*

VARIETIES *protearum protearum, protearum leucadendri*

GENUS *Cladosporium* Branched Spore Sphere Mold

SPECIES *adianticola, aecidiicola, allii-porri, amoenum, antillanum, aphidis, argillaceum, brachormium, britannicum, carrionii, caryigenum, chlorocephalum, cladosporioides, colocasiae, coralloides, cucumerinum, effusum, elatum, epimyces, epiphyllum, ferrugineum, gallicola, gossypiicola, herbarum, humile, iridis, lacroixii, langeronii, laxicapitulatum, lignicola, macrocarpum, magnusianum, malorum, minourae, multigeniculatum, musae, nigrellum, nodulosum, orchidis, oxysporum, paeoniae, pallidum, perfragile, phlei, phyllachorae, porophorum, scillae, sphaerospermum, spongiosum, staurophorum, subsessile, syringae, tenuissimum, typharum, uniseptosporum, uredinicola, variabile, vignae*

VARIETIES *cladosporioides pisicola*

GENUS *Clypeispora* Clype Spore Sphere Mold

SPECIES *angustifoliorum*
GENUS *Colletogloeopsis* Collet Glow Sphere Mold
SPECIES *moelleriana, nubilosum*
GENUS *Cymadothea* Cyme Doth Sphere Mold
SPECIES *trifolii*
GENUS *Didymaria* Dual Sea Sphere Mold
SPECIES *linariae*
GENUS *Diplochorella* Diploid Solitary Sphere Mold
SPECIES *colensoi, tephrosia*
GENUS *Dissoconium* Twofold Cone Sphere Mold
SPECIES *aciculare, dekkeri*
GENUS *Dothistroma* Doth Stroma Sphere Mold
SPECIES *rhabdoclinis*
GENUS *Heterosporium* Variable Spore Sphere Mold
SPECIES *ornithogali*
GENUS *Hormodendrum* Chain Wood Sphere Mold
SPECIES *chaquense, olivaceum, viride*
GENUS *Lecanosticta* Lichen Stick Sphere Mold
SPECIES *gloeospora*
GENUS *Melanodothis* Dark Doth Sphere Mold
SPECIES *caricis*
GENUS *Microcyclus* Cyclic Sphere Mold
SPECIES *achalensis, amphimelaena, tinctoria, ulei*
GENUS *Micronectriella* Cornicle Sphere Mold
SPECIES *agropyri*
GENUS *Mycovellosiella* Vellose Sphere Mold
SPECIES *bellynckii, brachycarpa, cajani, concors, ferruginea, fulva, koepkei, murina, nattrassii, oryzae, perfoliata, phaseoli, puerariae, vaginae*
VARIETIES *cajani cajani, cajani indica, cajani trichophila*
GENUS *Ophiocarpella* Snake Carpel Sphere Mold
SPECIES *tarda*
GENUS *Ovularia* Ovule Sphere Mold
SPECIES *decipiens, duplex, farinosa, lamii, obliqua, sphaeroidea*
GENUS *Paracercospora* Parasite Dual Antenna Spore Sphere Mold
SPECIES *egenula*
GENUS *Passalora* Braided Sphere Mold
SPECIES *angelicae, bacilligera, depressa, galii, nodosa*
GENUS *Phacellium* Bundle Sphere Mold
SPECIES *alborosellum, carneum, episphaerium, mespili, rufibasis*
GENUS *Phaeoisariopsis* Dusky Sarian Sphere Mold
SPECIES *bataticola, griseola, liriodendri, magnoliae, robiniae, williamsiae*
GENUS *Phaeophleospora* Dusky Reed Sphere Mold
SPECIES *indica*
GENUS *Phaeoramularia* Dusky Branch Sphere Mold
SPECIES *angolensis, capsicicola, dissiliens, hachijoensis, kellermaniana, levieri, manihotis, periclymeni, punctiformis*
GENUS *Pharcidia* Beaming Sphere Mold
SPECIES *pelvetiae*
GENUS *Phloeospora* Bark Spore Sphere Mold
SPECIES *aceris, bupleuri, mimosae-pigrae, oxyacanthae, pseudoplatani, robiniae*
GENUS *Placocrea* Plaque Sphere Mold
SPECIES *pulchella*
GENUS *Pseudocercospora* Mock Dual Antenna Spore Sphere Mold
SPECIES *abelmoschi, cavarae, deightonii, fuligena, karaka, myrticola, nephelii, pachirae, pancratii, piperis, plunkettii, psidii, purpurea, salvadorae, subsessilis, unicolor, vitis*
GENUS *Pseudocercosporella* Lesser Mock Dual Antenna Spore Sphere Mold
SPECIES *aestiva, capsellae, oxyriae, scirpi*
GENUS *Ramularia* Branch Sphere Mold
SPECIES *acris, adoxae, aequivoca, agrestis, agrimoniae, ajugae, alismatis, alnicola, alpina, anagallidis, angelicae, anomala, anthrisci, aplospora, archangelicae, arenariae, ari, armoraciae, aromatica, arvensis, asteris, bellunensis, betae, beticola, bistortae, brunnea, bryoniae, calcea, cardamines, cardui, carneola, carthami, centaureae, centranthi, cerinthes, chaerophylli, chamaedryos, cicutae, circaeae, cirsii, cochleariae, coleosporii, collo-cygni, cupulariae, cyclaminicola, cynarae, cynoglossi, deusta, didyma, didymarioides, doronici, epilobiana, epilobii, epilobii-palustris, erodii, fatsiae, filaris, foeniculi, galegae, gei, geranii, glechomatis, haplospora, hellebori, heraclei, holci-lanati, hypochoeridis, inaequale, indica, interstitialis, keithii, kriegeriana, lactea, lamii, lamiicola, lapsanae, ligustrina, ludoviciana, lychnidicola, lysimachiarum, macrospora, maculiformis, magnusiana, meliloti, menthicola, montana, montana, myosotidis, nymphaearum, obducens, onobrychidis,*

ovata, parietariae, pastinacae, picridis, plantaginea, pratensis, primulae, primulana, pruinosa, purpurascens, pusilla, pygmaea, rhabdospora, rhei, rollandii, rosea, rubella, sabaudica, sambucina, saxifragae, scelerata, schulzeri, scolopendrii, scrophulariae, senecionis, septata, simplex, sparganii, spiraeae, stachydis, succisae, superflua, sylvestris, taraxaci, theicola, ulmariae, uredinis, urticae, valerianae, vallisumbrosae, variabilis, veronicae, violae, winteri

VARIETIES *filaris lappae, lamii minor, succisae knautiae*

GENUS *Readeriella* — Reader Sphere Mold

SPECIES *mirabilis*

GENUS *Selenophoma* — Selen Foam Sphere Mold

SPECIES *asterina, juncea, linicola, mahoniae, moravica*

GENUS *Septocylindrium* — Septic Cylindric Sphere Mold

SPECIES *acutum, leucum, viride*

GENUS *Septoria* — Septic Sphere Mold

SPECIES *acetosae, aciculosa, adanensis, adoxae, agropyrina, alnicola, alopecuri, alpicola, ammophilae, anemones, anthurii, apiicola, armeriae, armoraciae, arundinacea, astericola, atropurpurea, azaleae, bataticola, bellidis, betae, brissaceana, bromi, calendulae, cannabis, caricinella, caricis, carthami, castaneicola, cerastii, ceratoniae, cercidis, chamaecysti, chelidonii, chrysanthemella, cirsii, citri, citricola, clematidis, convolvuli, cornicola, crepidis, cruciatae, cucurbitacearum, cucurbitarum, cyclaminis, cymbalariae, dianthi, digitalis, dulcamarae, ekmaniana, epilobii, euonymi, exotica, ficariae, fragariae, galeopsidis, galiorum, gei, geranii, gerberae, gladioli, glaucis, glycines, hederae, helianthi, helianthicola, hippocastani, holci, hydrocotyles, hyperici, inconspicua, jasiones, lactucae, lactucae, lamii, lathyri, lavandulae, leontodontis, lepidii, leucanthemi, limonum, lineolata, lunariae, lychnidis, lycopersici, lycopi, lysimachiae, macropoda, matricariae, menthae, menyanthes, minuta, mougeotii, myricae, obesa, oenanthis, oenotherae, orchidearum, orobina, oudemansii, paeoniae, passerinii, passiflorae, pastinacae, petroselini, phlogis, pini-thunbergii, pistaciae, plantaginea, podagraria, polygonorum, posoniensis, potentillae, primulae, proteae, prunellae, pyrethri, quercicola, quercina, quevillensis, rhamni-catharticae, rhododendri, riparia, rosarum, rubiae, rumicum, salicicola, saponariae, scabiosicola, schnabliana, schoeni, scillae, scorodoniae, scrophulariae, scutellariae, semilunaris, senecionis-silvaticae, sii, silvicola, slaptonensis, slaptoniensis, smithiana, socia, sorbi, stachydis, stellariae, stellariae, tanaceti, taraxaci, thecicola, tormentillae, triseti, unedonis, urticae, veronicae, viciicola, vignae, villarsiae, violae-palustris, virgaureae*

VARIETIES *alopecuri alopecuri, alopecuri calamagrostidis, lycopersici malagutii*

GENUS *Sphaerella* — Lesser Sphere Mold

SPECIES *arbuti, brassicae, cinerascens, epistroma, erysiphina, fagicola, hieracii, nitidula, oblivia*

GENUS *Sphaerellothecium* — Theca Sphere Mold

SPECIES *araneosum, minutum, propinquellum*

GENUS *Sphaerulina* — Least Sphere Mold

SPECIES *dolichotera, dubiella, leightonii, mimosae-pigrae, myriadea, orae-maris, oryzina, pedicellata, polyspora, rehmiana*

GENUS *Stenella* — Thin Sphere Mold

SPECIES *adeniae, araguata, lythri*

GENUS *Stigmidium* — Stigma Sphere Mold

SPECIES *aggregatum, arthoniae, congestum, degelii, ephebes, epiramalina, fuscatae, gyrophorarum, hageniae, lecidellae, leucophlebiae, marinum, microspilum, mitchellii, mycobilimbiae, peltideae, pumilum, punctillum, rivulorum, schaereri, solorinarium, superpositum, tabacinae, xanthoparmeliarum*

GENUS *Thedgonia* — Thedgon Sphere Mold

SPECIES *bellocensis*

GENUS *Uwebraunia* — Uwebraun Sphere Mold

SPECIES *ellipsoidea, juvenis, lateralis*

ORDER 7 *Myriangiales* Myriad Angle Mold

FAMILY 1 *Myriangiaceae* — Myriad Angle Mold

GENUS *Myriangium* — Myriad Angle Mold

SPECIES *dolichosporum, duriaei, haraeanum*

GENUS *Anhellia* — Anhell Myriad Angle Mold

SPECIES *lantanae*

FAMILY 2 *Elsinoaceae* — Elsin Myriad Angle Mold

GENUS *Elsinoe* — Elsin Myriad Angle Mold

SPECIES *ampelina, annonae, australis, batatas, brasiliensis, fawcettii, jasmini, ledi, mangiferae, phaseoli, randii, rosarum, solidaginis, theae, veneta, ziziphi*

VARIETIES *fawcettii scabiosae*

GENUS *Butleria* — Butler Elsin Myriad Angle Mold

SPECIES *inaghatahani*

GENUS *Sphaceloma* — Sage Elsin Myriad Angle Mold

SPECIES *abutilonis, anacardii, arachidis, barleriicola, bidentis, coryli, cupaniae, embeliae, fagarae, fici, fici-caricae, flacourtiae, genipae, glycines, hederae, ichnocarpi, ixorae, krugii, lippiae, mattiroloanum, menthae, perseae, pongamiae, populi, rhois, ricini, semecarpi, siculum, symphoricarpi, terminaliae, violae*

ORDER 8 *Onygenales* Onyx Mold

FAMILY 1 *Onygenaceae* Onyx Mold
GENUS *Onygena* Onyx Mold
SPECIES *apus, corvina, equina, piligena*
GENUS *Amauroascus* Dark Sac Onyx Mold
SPECIES *albicans, aureus, kuehnii, niger, reticulatus, rwandensis, tropicalis, verrucosus, volatilis-patellis*
GENUS *Anixiopsis* Anix Onyx Mold
SPECIES *biplanata*
GENUS *Aphanoascus* Aphane Sac Onyx Mold
SPECIES *cinnabarinus, clathratus, cubensis, fulvescens, hispanicus, keratinophilus, multiporus, pinarensis, punsolae, reticulisporus, saturnoideus, verrucosus*
GENUS *Apinisia* Apinis Onyx Mold
SPECIES *graminicola, queenslandica*
GENUS *Arachnotheca* Spider Theca Onyx Mold
SPECIES *glomerata*
GENUS *Ascocalvatia* Sac Bald Onyx Mold
SPECIES *alveolata*
GENUS *Auxarthron* Auxarthron Onyx Mold
SPECIES *californiense, compactum, conjugatum, umbrinum, zuffianum*
GENUS *Bifidocarpus* Bifid Carpel Onyx Mold
SPECIES *cubensis*
GENUS *Byssoonygena* Silky Nygen Onyx Mold
SPECIES *ceratinophila*
GENUS *Chrysosporium* Golden Spore Onyx Mold
SPECIES *baduri, botryoides, carmichaelii, europae, filiforme, georgiae, globiferum, hispanicum, holmii, inops, kreiselii, littoralis, lobatum, lucknowense, medium, mephiticum, merdarium, minor, pannicola, pilosum, pseudomerdarium, pyriformis, queenslandicum, roseum, siglerae, sulfureum, synchronum, tropicum, undulatum, vallenarense, vespertilii, xerophilum, zonatum*
VARIETIES *globiferum articulatum, globiferum niveum, medium spissescens*
GENUS *Histoplasma* Histoplasm Onyx Mold
SPECIES *capsulatum, duboisii*
VARIETIES *capsulatum capsulatum, capsulatum farciminosum*
GENUS *Keratinophyton* Keratin Onyx Mold
SPECIES *terreum*
GENUS *Loboa* Lobo Onyx Mold
SPECIES *loboi*
GENUS *Monascella* Monad Sac Onyx Mold
SPECIES *botryosa*
GENUS *Nannizziopsis* Nannizzi Onyx Mold
SPECIES *vriesii*
GENUS *Neoxenophila* Modern Strange Onyx Mold
SPECIES *foetida*
GENUS *Pectinotrichum* Pectin Hairy Onyx Mold
SPECIES *llanense*
GENUS *Polytolypa* Many Tolyp Onyx Mold
SPECIES *hystricis*
GENUS *Renispora* Renal Spore Onyx Mold
SPECIES *flavissima*
GENUS *Shanorella* Shanor Onyx Mold
SPECIES *spirotricha*
GENUS *Spiromastix* Spiral Mastigote Onyx Mold
SPECIES *grisea, warcupii*
GENUS *Uncinocarpus* Hooked Carpel Onyx Mold
SPECIES *reesii, uncinatus*
GENUS *Xanthothecium* Yellow Theca Onyx Mold
SPECIES *peruvianus*
GENUS *Xynophila* Xyne Onyx Mold
SPECIES *mephitalis*

FAMILY 2 *Ajellomycetaceae* Ajell Onyx Mold
GENUS *Ajellomyces* Ajell Onyx Mold
SPECIES *crescens*
GENUS *Emmonsia* Emmons Ajell Onyx Mold
SPECIES *parva*
VARIETIES *parva parva, parva crescens*

FAMILY 3 *Arthrodermataceae* Ringworm Onyx Mold

GENUS *Arthroderma* Ringworm Onyx Mold
SPECIES *benhamiae, ciferrii, cuniculi, curreyi, flavescens, gertleri, gloriae, grubyi, insingulare, lenticulare, melis, multifidum, quadrifidum, simii, tuberculatum, uncinatum, vanbreuseghemii*
GENUS *Endodermophyton* (Synonym for *Trichophyton*)
GENUS *Ctenomyces* Comb Jelly Ringworm Onyx Mold
SPECIES *serratus*
GENUS *Epidermophyton* Epidermal Ringworm Onyx Mold
SPECIES *floccosum, stockdaleae*
VARIETIES *floccosum floccosum, floccosum nigricans*
GENUS *Grubyella* Gruby Ringworm Onyx Mold
SPECIES *aurea, ochoterenai*
GENUS *Keratinomyces* Keratin Ringworm Onyx Mold
SPECIES *ceretanicus, longifusus*
GENUS *Mycoderma* (Synonym for *Candida Paracoccidioides, Trichophyton, Trichosporon, Zymonema*)
GENUS *Microsporum* Mammal Ringworm Onyx Mold
SPECIES *amazonicum, audouinii, boullardii, distortum, duboisii, equinum, ferrugineum, fulvum, gallinae, gypseum, langeronii, nanum, praecox, ripariae, rivalieri*
VARIETIES *distortum canis, gypseum gypseum, gypseum vinosum*
GENUS *Nannizzia* Nannizzi Ringworm Onyx Mold
SPECIES *borellii, cajetana, cookiella, corniculata, fulva, grubyia, gypsea, incurvata, obtusa, ossicola, otae, persicolor, racemosa*
VARIETIES *cajetana cookei*
GENUS *Thallomicrosporon* Thallus Mammal Ringworm Onyx Mold
SPECIES *kuehnii*
GENUS *Trichophyton* Mammal Herpes Ringworm Onyx Mold
SPECIES *ajelloi, areolatum, balcaneum, batonrougei, bullosum, circonvolutum, concentricum, depressum, equinum, eriotrephon, fischeri, floriforme, fluviomuniense, glabrum, gourvilii, immergens, kanei, krajdenii, kuryangei, langeronii, megninii, mentagrophytes, pedis, phaseoliforme, proliferans, quinckeanum, radicosum, raubitschekii, rotundum, rubrum, schoenleinii, simii, soudanense, terrestre, thuringiense, tonsurans, vanbreuseghemii, verrucosum, violaceum, yaoundei*
VARIETIES *ajelloi ajelloi, ajelloi nanum, equinum equinum, equinum autotrophicum, gourvilii gourvilii, gourvilii intermedium, mentagrophytes mentagrophytes, mentagrophytes erinacei, mentagrophytes goetzii, mentagrophytes interdigitale, mentagrophytes nodulare, rubrum nigricans, tonsurans sulfureum, verrucosum verrucosum, verrucosum album, verrucosum autotrophicum, verrucosum discoides, verrucosum ochraceum, violaceum violaceum, violaceum indicum*
GENUS *Trichosporon* Intestine Parasite Ringworm Onyx Mold
SPECIES *asahii, beigelii, cutaneum, dulcitum, inkin, niger, sinense, sporotrichoides*

FAMILY 4 *Coccidioidaceae* Bud Onyx Mold
GENUS *Coccidioides* Insect Parasite Bud Onyx Mold
SPECIES *immitis*
GENUS *Paracoccidioides* Mammal Parasite Bud Onyx Mold
SPECIES *brasiliensis*
GENUS *Zymonema* Zyme Thread Mammal Parasite Bud Onyx Mold
SPECIES *dermatitidis*

FAMILY 5 *Gymnoascaceae* Nude Sac Onyx Mold
GENUS *Gymnoascus* Nude Sac Onyx Mold
SPECIES *alatosporus, arxii, boliviensis, demonbreunii, longitrichus, petalosporus, reessii, reticulatus, udagawae*
GENUS *Acitheca* Sharp Theca Nude Sac Onyx Mold
SPECIES *purpurea*
GENUS *Actinodendron* Radial Wood Nude Sac Onyx Mold
SPECIES *verticillatum*
GENUS *Arachniotus* Spider Nude Sac Onyx Mold
SPECIES *aurantiacus, candidus, flavoluteus, hebridensis, lanatus, littoralis, punctatus, ruber, verruculosus, volatilis-patellus*
GENUS *Gymnascella* Lesser Nude Sac Onyx Mold
SPECIES *citrina, confluens, dankaliensis, devroeyi, hyalinospora, marginispora, nodulosa*
GENUS *Gymnoascoideus* Least Nude Sac Onyx Mold
SPECIES *petalosporus*
GENUS *Leucothecium* Light Theca Nude Sac Onyx Mold
SPECIES *coprophilum, emdenii*
GENUS *Mallochia* Malloch Nude Sac Onyx Mold
SPECIES *echinulata*
GENUS *Orromyces* Orr Nude Sac Onyx Mold
SPECIES *spiralis*
GENUS *Petalosporus* Petal Spore Nude Sac Onyx Mold
SPECIES *afilamentosus*

GENUS *Pseudoarachniotus* Lesser Spider Nude Sac Onyx Mold
SPECIES *reticulatus, trochleosporus*

ORDER 9 *Taphrinales* Taphrin Mold

FAMILY 1 *Taphrinaceae* Taphrin Mold
GENUS *Taphrina* Taphrin Mold
SPECIES *alni, amentorum, americana, athyrii, betulina, bullata, caerulescens, californica, carnea, carpini, cerasi, communis, confusa, crataegi, dearnessii, deformans, farlowii, filicina, flavorubra, johansonii, maculans, mirabilis, nana, padi, polystichi, populi-salicis, populina, pruni, pruni-subcordatae, purpurascens, rhizophora, robinsoniana, sadebeckii, tormentillae, tosquinetii, ulmi, vestergrenii, virginica, wettsteiniana, wiesneri*
GENUS *Ascomyces* Sac Taphrin Mold
SPECIES *juglandis*

FAMILY 2 *Protomycetaceae* ArchaicTaphrin Mold
GENUS *Protomyces* ArchaicTaphrin Mold
SPECIES *lactucae-debilis, macrosporus, pachydermus*
GENUS *Burenia* Bueren ArchaicTaphrin Mold
SPECIES *inundata*
GENUS *Protomycopsis* Lesser ArchaicTaphrin Mold
SPECIES leucanthemi, phaseoli, purpureotingens
GENUS *Saitoella* Saito ArchaicTaphrin Mold
SPECIES *complicata*
GENUS *Taphridium* Least ArchaicTaphrin Mold
SPECIES *umbelliferarum*
GENUS *Volkartia* Volkart ArchaicTaphrin Mold
SPECIES *rhaetica*

CLASS 621.5 *Hemiascomycetes* Sugar Yeast

ORDER 1 *Saccharomycetales* Sugar Yeast

FAMILY 01 *Saccharomycetaceae* Sugar Yeast
GENUS *Saccharomyces* Sugar Yeast
SPECIES *bayanus, busse, capilitii, cerevisiae, diastaticus, heterogenicus, paradoxus, pasteurianus, subcutaneus, tumafaciens*
GENUS *Aciculoconidium* Acicular Cone Sugar Yeast
SPECIES *aculeatum*
GENUS *Ambrosiozyma* Ambrose Zyme Sugar Yeast
SPECIES *cicatricosa*
GENUS *Aphidomyces* Aphid Sugar Yeast
SPECIES *aphalarae-calthae, aphidis, drepanosiphi, psyllae-foersteri, sulcii*
GENUS *Arxiozyma* Arx Zyme Sugar Yeast
SPECIES *telluris*
GENUS *Ascoidea* Sac Sugar Yeast
SPECIES *africana, corymbosa, hylecoeti, rubescens*
GENUS *Candida* Candida Sugar Yeast
SPECIES *albicans, diddensiae, glabrata, guilliermondii, haemulonis, hylophila, krusei, lambica, maris, norvegica, parapsilosis, rugosa, saitoana, torresii, tropicalis, vini*
GENUS *Citeromyces* Citer Sugar Yeast
SPECIES *matritensis*
GENUS *Cyniclomyces* Cynicle Sugar Yeast
SPECIES *guttulatus*
GENUS *Debaryomyces* Debary Sugar Yeast
SPECIES *hansenii, kloeckeri, marama, polymorphus, vanrijiae*
GENUS *Dekkera* Dekker Sugar Yeast
SPECIES *anomala, bruxellensis, intermedia*
GENUS *Dolichoascus* Dolich Sac Sugar Yeast
SPECIES *schenckii*
GENUS *Entelexis* Entelex Sugar Yeast
SPECIES *magnoliae*
GENUS *Geotrichum* Earth Hairy Sugar Yeast
SPECIES *candidum, clavatum, decipiens, eriense, fermentans, fragrans, gracile, hirtum, klebahnii, loubieri, ludwigii, sericeum, terrestre*
VARIETIES *candidum candidum, candidum citri-aurantii*
GENUS *Hansenula* Hansen Sugar Yeast
SPECIES *silvicola, subpelliculosa*
GENUS *Helicogonium* Helix Gonad Sugar Yeast

SPECIES *mollisiophilum*
GENUS *Hormoascus* Chain Sac Sugar Yeast
SPECIES *ambrosiae*
GENUS *Hyphopichia* Hypha Pich Sugar Yeast
SPECIES *burtonii*
GENUS *Issatchenkia* Issatchenko Sugar Yeast
SPECIES *occidentalis, scutulata, terricola*
GENUS *Kluyveromyces* Kluyver Sugar Yeast
SPECIES *marxianus, thermotolerans*
VARIETIES *marxianus bulgaricus*
GENUS *Pachysolen* Thick Solen Sugar Yeast
SPECIES *tannophilus*
GENUS *Pachytichospora* Thick Tich Spore Sugar Yeast
SPECIES *transvaalensis*
GENUS *Parasaccharomyces* (Synonym for *Candida*)
GENUS *Pichia* Pich Sugar Yeast
SPECIES *angusta, anomala, delftensis, farinosa, fermentans, guilliermondii, jadinii, membranifaciens, minuta, onychis, pastoris, pini, subpelliculosa, vini*
GENUS *Saturnispora* Saturn Spore Sugar Yeast
SPECIES *ahearnii, dispora, saitoi, zaruensis*
GENUS *Sporopachydermia* Spore Pachyderm Sugar Yeast
SPECIES *cereana, lactativora, quercuum*
GENUS *Stephanoascus* Stephan Sac Sugar Yeast
SPECIES *ciferrii, farinosus*
GENUS *Sympodiomyces* Sympodial Sugar Yeast
SPECIES *parvus*
GENUS *Tetrapisispora* Tetrad Pea Spore Sugar Yeast
SPECIES *arboricola, iriomotensis, nanseiensis, phaffii*
GENUS *Torulaspora* Torule SporeSugar Yeast
SPECIES *delbrueckii*
GENUS *Trigonopsis* Trigon Sugar Yeast
SPECIES *variabilis*
GENUS *Tyridiomyces* Tyre Sugar Yeast
SPECIES *formicarum*
GENUS *Williopsis* Willie Sugar Yeast
SPECIES *mrakii*
GENUS *Yarrowia* Yarrow Sugar Yeast
SPECIES *lipolytica*
GENUS *Zygoascus* Zygote Sac Sugar Yeast
SPECIES *hellenicus*

FAMILY 02 *Cephaloascaceae* Head Sugar Yeast
GENUS *Cephaloascus* Head Sugar Yeast
SPECIES *albidus, fragrans*
GENUS *Ascocephalophora* Lesser Head Sugar Yeast
SPECIES *petasiformis*

FAMILY 03 *Dipodascaceae* Dipod Sac Sugar Yeast
GENUS *Dipodascus* Dipod Sac Sugar Yeast
SPECIES *aggregatus, albidus, ambrosiae, armillariae, australiensis, capitatus, geniculatus, ingens, macrosporus, spicifer, tetrasperma*
GENUS *Galactomyces* Galactic Dipod Sac Sugar Yeast
SPECIES *reessii*
GENUS *Magnusiomyces* Magnus Endic Sugar Yeast
SPECIES *magnusii*

FAMILY 04 *Endomycetaceae* Endic Sugar Yeast
GENUS *Endomyces* Endic Sugar Yeast
SPECIES *coprophilus, fibuliger, lactis, scopularum*
GENUS *Phialoascus* Phial Sac Endic Sugar Yeast
SPECIES *borealis*

FAMILY 05 *Eremotheciaceae* Dry Sugar Yeast
GENUS *Eremothecium* Dry Sugar Yeast
SPECIES *ashbyi, cymbalariae*
GENUS *Ashbya* (Synonym for *Nematospora*)
GENUS *Coccidiascus* Round Sac Dry Sugar Yeast
SPECIES *legeri*
GENUS *Holleya* Holley Dry Sugar Yeast
SPECIES *sinecauda*

GENUS *Nematospora* Thread Spore Dry Sugar Yeast
SPECIES *gossypii, phaseoli*

FAMILY 06 *Lipomycetaceae* Fat Sugar Yeast
GENUS *Lipomyces* Fat Sugar Yeast
SPECIES *starkeyi*
GENUS *Babjevia* Babjev Fat Sugar Yeast
SPECIES *anomala*
GENUS *Dipodascopsis* Dipod Sac Fat Sugar Yeast
SPECIES *tothii, uninucleata*
VARIETIES *uninucleata uninucleata, uninucleata wickerhamii*
GENUS *Kawasakia* Kawasaki Fat Sugar Yeast
SPECIES *arxii*
GENUS *Waltomyces* Walt Fat Sugar Yeast
SPECIES *lipofer*

FAMILY 07 *Metschnikowiaceae* Metschnikov Sugar Yeast
GENUS *Metschnikowia* Metschnikov Sugar Yeast
SPECIES *bicuspidata, pulcherrima, zobellii*
GENUS *Clavispora* Clave Spore Metschnikov Sugar Yeast
SPECIES *lusitaniae, opuntiae*

FAMILY 08 *Phaffomycetaceae* Phaff Sugar Yeast
GENUS *Phaffia* Phaff Sugar Yeast
SPECIES *rhodozyma*
GENUS *Komagataea* Komagata Phaff Sugar Yeast
SPECIES *pratensis*

FAMILY 09 *Saccharomycodaceae* Mycode Sugar Yeast
GENUS *Saccharomycodes* Mycode Sugar Yeast
SPECIES *ludwigii*
GENUS *Hanseniaspora* Hansen Spore Mycode Sugar Yeast
SPECIES *valbyensis*
GENUS *Wickerhamia* Wickerham Mycode Sugar Yeast
SPECIES *fluorescens*

FAMILY 10 *Saccharomycopsidaceae* Mycope Sugar Yeast
GENUS *Saccharomycopsis* Mycope Sugar Yeast
SPECIES *capsularis*
GENUS *Arthroascus* Arthropod Sac Mycope Sugar Yeast
SPECIES *javanensis*
GENUS *Botryoascus* Berry Sac Mycope Sugar Yeast
SPECIES *synnaedendrus*
GENUS *Guilliermondella* Guilliermond Mycope Sugar Yeast
SPECIES *selenospora*

ORDER 2 *Schizosaccharomycetales* Split Sugar Yeast

FAMILY 1 *Schizosaccharomycetaceae* Split Sugar Yeast
GENUS *Schizosaccharomyces* Split Sugar Yeast
SPECIES *japonicus, malidevorans, pombe*
VARIETIES *japonicus japonicus, japonicus versatilis*
GENUS *Octosporomyces* Octospore Split Sugar Yeast
SPECIES *octosporus*
GENUS *Zygosaccharomyces* Zygote Split Sugar Yeast
SPECIES *bailii, bisporus, microellipsoides, rouxii*

PHYLUM 622 *Basidiomycota* Club Fungus

Latin: *Basidio* base, basal
Latin: *Mycota* mucus

22,500 species [Margulis & Schwartz, 1998]
9,318 species [Kirk, Sept. 2006]

CLASSES	Species
622.1 *Basidiomycetes*	Club Fungus
622.2 *Agaricomycetes*	Gill Mushroom
622.3 *Boletomycetes*	Nongill Mushroom
622.4 *Gastromycetes*	Cavity Fungus
622.5 *Lichinobasidiomycetes*	Club Lichen
622.6 *Urediniomycetes*	Rust
622.7 *Ustilaginomycetes*	Smut

Phylum *Basidiomycota* Club Fungus species have a spore producing cell called a basidium. One basidum produces four basidiospores on a conical sterigma. When the basidiospores mature, they separate from the sterigma with a force that carries them away from the mushroom.

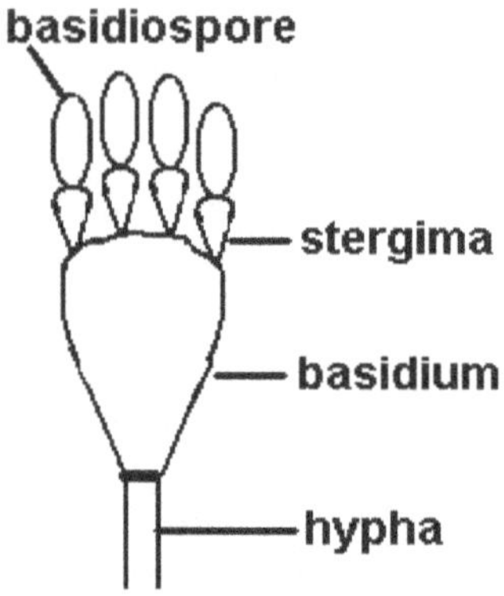

Phylum *Basidiomycota* Club Fungus species, basidiomycotes, are either **saprobes**, **symbiotes**, or, in the case of lichens, photosynthezers. Basidiomycotes live on land and have a stem or base called a thallus. A **saprobe** excretes extra cellular enzymes which break down dead organic matter. The dead organic matter is then absorbed using thin threads (hypha). Saprobe ascomycotes live on insect carcasses and plant debris. Macromolecules of cellulose, lignin, and collagen are digested, releasing nutrient compounds of carbon, nitrogen, and phosphorous. A **symbiote** extends the thin threads (hypha) directly into the host tissue and extracts nutrients from a host cell. Symbiote ascomycotes are parasites on algae, molds, and fungi.

Class *Basidiomycetes* Club Fungus includes the club fungus, coral fungus, jelly fungus, shelf fungus, pore fungus, and crust fungus. The Club Fungus *Erythromyces* produces the antibiotic erythromycin.

Class *Mussiriomycetes* Nongill Mushrooms includes the boletes, chanterelles, and spinecaps.

Class *Gastromycetes* includes bird nest fungus, earthstars, puffballs, and stinkhorns.

Phylum *Basidiomycota* Club Fungus Reclassification

Family	Order	Peterson	Species Fungorum	Class
Club Fungus *Basidiomycetaceae*	*Basidiomycetales*	—	*Hyphobasidiofera*	*Basidiomycetes*
Club Fungus *Basidiomycetaceae*	*Basidiomycetales*	—	~~*Mixiaceae*~~	*Basidiomycetes*
Club Fungus *Basidiomycetaceae*	*Basidiomycetales*	—	~~*Not Assigned*~~	*Basidiomycetes*
Horn Club Fungus *Ceratobasidiaceae*	*Basidiomycetales*	—	~~*Ceratobasidiales*~~	*Basidiomycetes*
Cyst Filament Club Fungus *Cystofilobasidiaceae*	*Basidiomycetales*	—	~~*Cystofilobasidiales*~~	*Basidiomycetes*
Filament Club Fungus *Filobasidiaceae*	*Basidiomycetales*	—	~~*Filobasidiales*~~	*Basidiomycetes*
Tule Club Fungus *Tulasnellaceae*	*Basidiomycetales*	—	~~*Tulasnellales*~~	*Basidiomycetes*
Coral Fungus *Clavariaceae*	*Clavariales*	*Clavariales*	~~*Agaricales*~~	*Basidiomycetes*
Branch Coral Fungus *Ramariaceae*	*Clavariales*	*Clavariales*	~~*Gomphaceae*~~	*Basidiomycetes*
Lace Coral Fungus *Sparassidaceae*	*Clavariales*	*Clavariales*	~~*Polyporales*~~	*Basidiomycetes*
Membrane Bristle Fungus *Hymenochaetaceae*	*Hymenochaetales*	—	*Hymenochaetales*	*Basidiomycetes*
Split Pore Membrane Bristle *Schizoporaceae*	*Hymenochaetales*	—	*Hymenochaetales*	*Basidiomycetes*
Pore Fungus *Polyporaceae*	*Polyporales*	*Polyporales*	*Polyporales*	*Basidiomycetes*
Pore Fungus *Polyporaceae*	*Polyporales*	—	*Globosomyces*	*Basidiomycetes*
Pore Fungus *Polyporaceae*	*Polyporales*	—	~~*Gloeophyllaceae*~~	*Basidiomycetes*
White Flesh Trellis Pore *Albatrellaceae*	*Polyporales*	—	*Polyporales*	*Basidiomycetes*
Bore Stereo Pore Fungus *Boreostereaceae*	*Polyporales*	—	*Polyporales*	*Basidiomycetes*

Family	Order	Peterson	Species Fungorum	Class
Cypher Pore FungusCyphellaceae	Polyporales	—	Polyporales	Basidiomycetes
Cyst Stereo Pore FungusCystostereaceae	Polyporales	—	Polyporales	Basidiomycetes
Epithelial Pore FungusEpitheliaceae	Polyporales	—	Polyporales	Basidiomycetes
Ox Tongue Pore FungusFistulinaceae	Polyporales	Polyporales	~~Agaricales~~	Basidiomycetes
Foam Pore FungusFomitopsidaceae	Polyporales	Polyporales	Polyporales	Basidiomycetes
Dye Pore FungusGanodermataceae	Polyporales	Polyporales	Polyporales	Basidiomycetes
Grass Pore FungusGrammotheleaceae	Polyporales	—	Polyporales	Basidiomycetes
Downy Pore FungusHapalopilaceae	Polyporales	—	Polyporales	Basidiomycetes
Hypha Dermal Pore FungusHyphodermataceae	Polyporales	—	Polyporales	Basidiomycetes
Seaside DownyPore FungusMeripilaceae	Polyporales	—	Polyporales	Basidiomycetes
Bristle Pore FungusPhanerochaetaceae	Polyporales	—	Polyporales	Basidiomycetes
Pod Scyph Pore FungusPodoscyphaceae	Polyporales	—	Polyporales	Basidiomycetes
Sistine Tremor Pore FungusSistotremataceae	Polyporales	—	Polyporales	Basidiomycetes
Tubule Lily Pore FungusTubulicrinaceae	Polyporales	—	Polyporales	Basidiomycetes
Strange Pore FungusXenasmataceae	Polyporales	—	Polyporales	Basidiomycetes
Groundwort CrustThelephoraceae	Thelephorales	Thelephorales	Thelephorales	Basidiomycetes
Corticle Wood CrustCorticiaceae	Thelephorales	Thelephorales	Thelephorales	Basidiomycetes
Wood CrustMeruliaceae	Thelephorales	Thelephorales	Thelephorales	Basidiomycetes
Tremor JellyTremellaceae	Tremellales	Tremellales	Tremellales	Basidiomycetes
Tremor JellyTremellaceae	Tremellales	Tremellales	~~Cuniculitremaceae~~	Basidiomycetes
Tremor JellyTremellaceae	Tremellales	Tremellales	Filobasidiella	Basidiomycetes
Tremor JellyTremellaceae	Tremellales	Tremellales	~~Hyaloriaceae~~	Basidiomycetes
Tremor JellyTremellaceae	Tremellales	Tremellales	~~Phragmoxenidiaceae~~	Basidiomycetes
Tremor JellyTremellaceae	Tremellales	Tremellales	~~Rhynchogastremataceae~~	Basidiomycetes
Tremor JellyTremellaceae	Tremellales	Tremellales	~~Sebacinaceae~~	Basidiomycetes
Tremor JellyTremellaceae	Tremellales	Tremellales	Sterigmatosporidium	Basidiomycetes
Tremor JellyTremellaceae	Tremellales	Tremellales	~~Syzygosporaceae~~	Basidiomycetes
Tremor JellyTremellaceae	Tremellales	Tremellales	~~Tremellodendropsidaceae~~	Basidiomycetes
Perplexed JellyAporpiaceae	Tremellales	Tremellales	~~Auriculariales~~	Basidiomycetes
Ear JellyAuriculariaceae	Tremellales	Tremellales	~~Auriculariales~~	Basidiomycetes
Tear JellyDacrymycetaceae	Tremellales	Tremellales	~~Dacrymycetales~~	Basidiomycetes
Tear JellyDacrymycetaceae	Tremellales	—	~~Cerinomycetaceae~~	Basidiomycetes
Warty JellyExidiaceae	Tremellales	—	Tremellales	Basidiomycetes
Pitted JellySirobasidiaceae	Tremellales	—	Tremellales	Basidiomycetes
BoleteBoletaceae	Boletales	Boletales	Boletales	Boletomycetes
BoleteBoletaceae	Boletales	Boletales	Endogonopsis	Boletomycetes
Lesser BoleteBoletinellaceae	Boletales	Boletales	Boletales	Boletomycetes
Slimecap BoleteGomphidiaceae	Boletales	Boletales	Boletales	Boletomycetes
Hollow Stalk BoleteGyroporaceae	Boletales	Boletales	Boletales	Boletomycetes
Chanterelle BoleteHygrophoropsidaceae	Boletales	Boletales	Boletales	Boletomycetes
Brimcap BoletePaxillaceae	Boletales	Boletales	Boletales	Boletomycetes
False Truffle BoleteRhizopogonaceae	Boletales	Boletales	Boletales	Boletomycetes
Slipperycap BoleteSuillaceae	Boletales	Boletales	Boletales	Boletomycetes
ChanterelleCantharaceae	Cantharellales	Cantharellales	Cantharellales	Boletomycetes
ChanterelleCantharaceae	Cantharellales	—	Stilbotulasnella	Boletomycetes
Simple ChanterelleAphelariaceae	Cantharellales	—	Cantharellales	Boletomycetes
Ball ChanterelleBotryobasidiaceae	Cantharellales	—	Cantharellales	Boletomycetes
Club ChanterelleClavilinaceae	Cantharellales	—	Cantharellales	Boletomycetes
Bolt ChanterelleGomphaceae	Cantharellales	Cantharellales	~~Phallales~~	Boletomycetes
SpinecapHydnaceae	Hydnales	Hydnales	~~Cantharellales~~	Boletomycetes
Pinecone SpinecapAuriscalpiaceae	Hydnales	Hydnales	~~Russulales~~	Boletomycetes
Banker SpinecapBankeraceae	Hydnales	Hydnales	~~Thelephorales~~	Boletomycetes
Hedgehog SpinecapHericiaceae	Hydnales	Hydnales	~~Russulales~~	Boletomycetes
Steccher SpinecapSteccherinaceae	Hydnales	Hydnales	~~Polyporales~~	Boletomycetes
Meadow Gill MushroomAgaricaceae	Agaricales	Agaricales	Agaricaceae	Agaricomycetes
Meadow Gill MushroomAgaricaceae	Agaricales	—	~~Gigaspermaceae~~	Agaricomycetes
Meadow Gill MushroomAgaricaceae	Agaricales	—	~~Hemigasteraceae~~	Agaricomycetes
Meadow Gill MushroomAgaricaceae	Agaricales	—	~~Niaceae~~	Agaricomycetes
Meadow Gill MushroomAgaricaceae	Agaricales	—	~~Physalacriaceae~~	Agaricomycetes
Meadow Gill MushroomAgaricaceae	Agaricales	—	Cribrospora	Agaricomycetes
Meadow Gill MushroomAgaricaceae	Agaricales	—	Disporotrichum	Agaricomycetes
Meadow Gill MushroomAgaricaceae	Agaricales	—	Phlebophyllum	Agaricomycetes

Family	Order	Peterson	Species Fungorum	Class
Bonnet Gill Mushroom*Bolbitaceae*	*Agaricales*	*Agaricales*	*Agaricales*	*Agaricomycetes*
Cap Gill Mushroom*Coprinaceae*	*Agaricales*	*Agaricales*	—	*Agaricomycetes*
Web Cap Gill Mushroom*Cortinariaceae*	*Agaricales*	*Agaricales*	*Agaricales*	*Agaricomycetes*
Pinkgill Gill Mushroom*Entolomataceae*	*Agaricales*	*Agaricales*	*Agaricales*	*Agaricomycetes*
Tallowgill Gill Mushroom*Hydnangiaceae*	*Agaricales*	*Agaricales*	*Agaricales*	*Agaricomycetes*
Waxycap Gill Mushroom*Hygrophoraceae*	*Agaricales*	*Agaricales*	*Agaricales*	*Agaricomycetes*
Parasol Gill Mushroom*Lepiotaceae*	*Agaricales*	*Agaricales*	*Agaricales*	*Agaricomycetes*
Pinwheel Gill Mushroom*Marasmiaceae*	*Agaricales*	*Agaricales*	*Agaricales*	*Agaricomycetes*
Cork Gill Mushroom*Mesophellaceae*	*Agaricales*	*Agaricales*	*Agaricales*	*Agaricomycetes*
Navelcap Gill Mushroom*Omphaliaceae*	*Agaricales*	—	*~~Omphalotaceae~~*	*Agaricomycetes*
Oyster Gill Mushroom*Pleurotaceae*	*Agaricales*	*Agaricales*	*Agaricales*	*Agaricomycetes*
Roofcap Gill Mushroom*Pluteaceae*	*Agaricales*	*Agaricales*	*Agaricales*	*Agaricomycetes*
Crumblecap Gill Mushroom*Psathyrellaceae*	*Agaricales*	—	*Agaricales*	*Agaricomycetes*
Wingcap Gill Mushroom*Pterulaceae*	*Agaricales*	—	*Agaricales*	*Agaricomycetes*
Splitgill Gill Mushroom*Schizophyllaceae*	*Agaricales*	*Agaricales*	*Agaricales*	*Agaricomycetes*
Ringstalk Gill Mushroom*Strophariaceae*	*Agaricales*	*Agaricales*	*Agaricales*	*Agaricomycetes*
Cavalier Gill Mushroom*Tricholomataceae*	*Agaricales*	*Agaricales*	*Agaricales*	*Agaricomycetes*
Typha Gill Mushroom*Typhulaceae*	*Agaricales*	*Agaricales*	*Agaricales*	*Agaricomycetes*
Brittle Gill Mushroom*Russulaceae*	*Russulales*	*Russulales*	*Russulales*	*Agaricomycetes*
Bondarzev Brittle Gill*Bondarzewiaceae*	*Russulales*	—	*Russulales*	*Agaricomycetes*
Urchin Brittle Gill Mushroom*Echinodontiaceae*	*Russulales*	—	*Russulales*	*Agaricomycetes*
Downy Brittle Gill Mushroom*Lachnocladiaceae*	*Russulales*	—	*Russulales*	*Agaricomycetes*
Pen Brittle Gill Mushroom*Peniophoraceae*	*Russulales*	—	*Russulales*	*Agaricomycetes*
Stereo Brittle Gill Mushroom*Stereaceae*	*Russulales*	*Russulales*	*Russulales*	*Agaricomycetes*
Cavity Fungus*Gasteraceae*	*Gasteromycetales*	—	*~~Gasterellaceae~~*	*Gasteromycetes*
Cavity Fungus*Gasteraceae*	*Gasteromycetales*	—	*~~Hybogasteraceae~~*	*Gasteromycetes*
Cavity Fungus*Gasteraceae*	*Gasteromycetales*	—	*~~Leucogastraceae~~*	*Gasteromycetes*
Cavity Fungus*Gasteraceae*	*Gasteromycetales*	—	*~~Melanogastraceae~~*	*Gasteromycetes*
Cavity Fungus*Gasteraceae*	*Gasteromycetales*	—	*~~Protogastraceae~~*	*Gasteromycetes*
Cone Cavity Fungus*Coniophoraceae*	*Gasteromycetales*	—	*~~Boletales~~*	*Gasteromycetes*
Membrane Cavity Fungus*Hymenogasteraceae*	*Gasteromycetales*	—	*~~Boletales~~*	*Gasteromycetes*
Bird Nest Cavity Fungus*Nidulariaceae*	*Gasteromycetales*		*~~Agaricales~~*	*Gasteromycetes*
Octavian Cavity Fungus*Octavianiaceae*	*Gasteromycetales*	—	*~~Boletales~~*	*Gasteromycetes*
Puffball*Lycoperdaceae*	*Lycoperdales*	*Lycoperdales*	*~~Agaricales~~*	*Gasteromycetes*
Earth Star Puffball*Geastraceae*	*Lycoperdales*	*Lycoperdales*	*~~Phallales~~*	*Gasteromycetes*
Earth Star Puffball*Geastraceae*	*Lycoperdales*	*~~Astraeaceae~~*	*~~Astraeaceae~~*	*Gasteromycetes*
Giant Puffball*Mycenastraceae*	*Lycoperdales*	*Lycoperdales*	*~~Agaricales~~*	*Gasteromycetes*
Pearock Puffball*Pisolithaceae*	*Lycoperdales*	*Lycoperdales*	*~~Boletales~~*	*Gasteromycetes*
Pod Stalk Puffball*Podaxaceae*	*Lycoperdales*	*Lycoperdales*	*~~Agaricales~~*	*Gasteromycetes*
Earthball Puffball*Sclerodermataceae*	*Lycoperdales*	*Lycoperdales*	*~~Boletales~~*	*Gasteromycetes*
Stalked Puffball*Tulostomataceae*	*Lycoperdales*	*Lycoperdales*	*~~Agaricales~~*	*Gasteromycetes*
Stinkhorn Horn Fungus*Phallaceae*	*Phallales*	*Phallales*	*Phallaceae*	*Gasteromycetes*
Lattice Horn Fungus*Calostomataceae*	*Phallales*	*Phallales*	—	*Gasteromycetes*
Stinkhorn Horn Fungus*Clathraceae*	*Phallales*	*Phallales*	—	*Gasteromycetes*
Uterus Horn Fungus*Hysterangiaceae*	*Phallales*	*Phallales*	*Phallales*	*Gasteromycetes*
Club Lichen*Lichinobasidiaceae*	*Lichinobasidiales*	—	*Ceratobasidiales*	*Lichinobasidiomycetes*
Club Lichen*Lichinobasidiaceae*	*Lichinobasidiales*	—	*Agaricales*	*Lichinobasidiomycetes*
Athel Club Lichen*Atheliaceae*	*Lichinobasidiales*	—	*Polyporales*	*Lichinobasidiomycetes*
Rust*Uredinaceae*	*Uredinales*	*Uredinales*	*~~Not Assigned~~*	*Urediniomycetes*
Rust*Uredinaceae*	*Uredinales*	—	*~~Agaricostilbales~~*	*Urediniomycetes*
Rust*Uredinaceae*	*Uredinales*		*Bensingtonia*	*Urediniomycetes*
Rust*Uredinaceae*	*Uredinales*		*~~Helicobasidiaceae~~*	*Urediniomycetes*
Rust*Uredinaceae*	*Uredinales*	—	*~~Heterogastridiaceae~~*	*Urediniomycetes*
Rust*Uredinaceae*	*Uredinales*	—	*~~Melampsoraceae~~*	*Urediniomycetes*
Rust*Uredinaceae*	*Uredinales*	—	*~~Mikronegeriaceae~~*	*Urediniomycetes*
Rust*Uredinaceae*	*Uredinales*	—	*~~Uncolaceae~~*	*Urediniomycetes*
Atract Rust*Atractiellaceae*	*Uredinales*	—	*~~Atractiellales~~*	*Urediniomycetes*
Chacon Rust*Chaconiaceae*	*Uredinales*	—	*Uredinales*	*Urediniomycetes*
Classic Rust*Classiculaceae*	*Uredinales*	—	*~~Classiculales~~*	*Urediniomycetes*
Coleospore Rust*Coleosporiaceae*	*Uredinales*	—	*Uredinales*	*Urediniomycetes*
Cronart Rust*Cronartiaceae*	*Uredinales*	—	*Uredinales*	*Urediniomycetes*
Phako Psoral Rust*Phakopsoraceae*	*Uredinales*	—	*Uredinales*	*Urediniomycetes*
Fragmented Rust*Phragmidiaceae*	*Uredinales*	—	*Uredinales*	*Urediniomycetes*

Family	Order	Peterson	Species Fungorum	Class
Pile Rust*Pileolariaceae*	*Uredinales*	—	*Uredinales*	*Urediniomycetes*
Flat Glow Rust*Platygloeaceae*	*Uredinales*	—	~~*Not Assigned*~~	*Urediniomycetes*
Puccini Rust*Pucciniaceae*	*Uredinales*	*Uredinales*	*Uredinales*	*Urediniomycetes*
Aster Puccini Rust*Pucciniastraceae*	*Uredinales*	—	*Uredinales*	*Urediniomycetes*
Pitted Puccini Rust*Pucciniosiraceae*	*Uredinales*	—	*Uredinales*	*Urediniomycetes*
Ravenel Rust*Raveneliaceae*	*Uredinales*	—	*Uredinales*	*Urediniomycetes*
Septic Rust*Septobasidiaceae*	*Uredinales*	—	~~*Septobasidiales*~~	*Urediniomycetes*
Sphere Fragmented Rust*Sphaerophragmiaceae*	*Uredinales*	—	~~*Raveneliaceae*~~	*Urediniomycetes*
Spore Bowl Rust*Sporidiobolaceae*	*Uredinales*	—	~~*Sporidiobolales*~~	*Urediniomycetes*
Pixie Tail Rust*Uropyxidaceae*	*Uredinales*	—	*Uredinales*	*Urediniomycetes*
Berry Rust*Microbotryaceae*	*Microbotryales*	—	*Microbotryales*	*Urediniomycetes*
Light Berry Rust*Leucosporidiaceae*	*Microbotryales*	—	*Microbotryales*	*Urediniomycetes*
Smut Berry Rust*Ustilentylomataceae*	*Microbotryales*	—	*Microbotryales*	*Urediniomycetes*
Smut*Ustilaginaceae*	*Ustilaginales*	*Ustilaginales*	*Ustilaginales*	*Ustilaginomycetes*
Smut*Ustilaginaceae*	*Ustilaginales*	—	~~*Clintamraceae*~~	*Ustilaginomycetes*
Smut*Ustilaginaceae*	*Ustilaginales*	—	~~*Cryptomycocolacales*~~	*Ustilaginomycetes*
Smut*Ustilaginaceae*	*Ustilaginales*	—	~~*Dermatosoraceae*~~	*Ustilaginomycetes*
Smut*Ustilaginaceae*	*Ustilaginales*	—	~~*Entorrhizales*~~	*Ustilaginomycetes*
Smut*Ustilaginaceae*	*Ustilaginales*	—	~~*Entylomatales*~~	*Ustilaginomycetes*
Smut*Ustilaginaceae*	*Ustilaginales*	—	~~*Farysiaceae*~~	*Ustilaginomycetes*
Smut*Ustilaginaceae*	*Ustilaginales*	—	~~*Geminaginaceae*~~	*Ustilaginomycetes*
Smut*Ustilaginaceae*	*Ustilaginales*	—	~~*Melanopsichiaceae*~~	*Ustilaginomycetes*
Smut*Ustilaginaceae*	*Ustilaginales*	—	~~*Microstromatales*~~	*Ustilaginomycetes*
Smut*Ustilaginaceae*	*Ustilaginales*	—	~~*Not Assigned*~~	*Ustilaginomycetes*
Smut*Ustilaginaceae*	*Ustilaginales*	—	~~*Websdaneaceae*~~	*Ustilaginomycetes*
Anthrax Smut*Anthracoideaceae*	*Ustilaginales*	—	*Ustilaginales*	*Ustilaginomycetes*
Cintract Smut*Cintractiaceae*	*Ustilaginales*	—	*Ustilaginales*	*Ustilaginomycetes*
Dossan Smut*Doassansiaceae*	*Ustilaginales*	—	~~*Doassansiales*~~	*Ustilaginomycetes*
Dossan Smut*Doassansiaceae*	*Ustilaginales*	—	~~*Doassansiopsidaceae*~~	*Ustilaginomycetes*
Dossan Smut*Doassansiaceae*	*Ustilaginales*	—	~~*Melaniellaceae*~~	*Ustilaginomycetes*
Dossan Smut*Doassansiaceae*	*Ustilaginales*	—	~~*Rhamphosporaceae*~~	*Ustilaginomycetes*
Glome Smut*Glomosporiaceae*	*Ustilaginales*	—	*Ustilaginales*	*Ustilaginomycetes*
Tillet Smut*Tilletiaceae*	*Ustilaginales*		~~*Tilletiales*~~	*Ustilaginomycetes*
Exo Smut*Exobasidiaceae*	*Exobasidiales*	—	*Exobasidiales*	*Ustilaginomycetes*
Exo Smut*Exobasidiaceae*	*Exobasidiales*	—	~~*Graphiolaceae*~~	*Ustilaginomycetes*
Exo Smut*Exobasidiaceae*	*Exobasidiales*	—	~~*Not Assigned*~~	*Ustilaginomycetes*
Short Exo Smut*Brachybasidiaceae*	*Exobasidiales*	—	*Exobasidiales*	*Ustilaginomycetes*
Cryptic Exo Smut*Cryptobasidiaceae*	*Exobasidiales*	—	*Exobasidiales*	*Ustilaginomycetes*
George Fischer Smut*Georgefischeriaceae*	*Georgefischeriales*	—	*Georgefischeriales*	*Ustilaginomycetes*
George Fischer Smut*Georgefischeriaceae*	*Georgefischeriales*	—	~~*Eballistraceae*~~	*Ustilaginomycetes*
George Fischer Smut*Georgefischeriaceae*	*Georgefischeriales*	—	~~*Gjaerumiaceae*~~	*Ustilaginomycetes*
Tillet George Fischer Smut*Tilletiariaceae*	*Georgefischeriales*	—	*Georgefischeriales*	*Ustilaginomycetes*
Tail Cyst Smut*Urocystaceae*	*Urocystales*	—	*Urocystales*	*Ustilaginomycetes*
Dark Tail Cyst Smut*Melanotaeniaceae*	*Urocystales*	—	*Urocystales*	*Ustilaginomycetes*

CLASS 622.1 *Basidiomycetes* Club Fungus

ORDER 1 *Basidiomycetales* Club Fungus

FAMILY 1 *Basidiomycetaceae* — Club Fungus
GENUS *Akenomyces* — Aken Club Fungus
SPECIES *costatus*
VARIETIES *costatus costatus, costatus enigmaticus*
GENUS *Arthrodochium* — Arthropod Doch Club Fungus
SPECIES *candidum*
GENUS *Arualis* — Arual Club Fungus
SPECIES *carolinensis*
GENUS *Atractocolax* — Atract Colac Club Fungus
SPECIES *pulvinatus*
GENUS *Cenangiomyces* — Cenang Club Fungus
SPECIES *luteus*
GENUS *Cerinosterus* — Cerin Steral Club Fungus

SPECIES *luteoalbus*
GENUS *Corticomyces* Corticle Club Fungus
SPECIES *xenasmatoides*
GENUS *Cruciger* Cross Club Fungus
SPECIES *lignatilis*
GENUS *Ellula* Ellule Club Fungus
SPECIES *guaduae*
GENUS *Fibulochlamys* Cloak Clasp Club Fungus
SPECIES *ferruginosa*
GENUS *Fibulocoela* Celestial Clasp Club Fungus
SPECIES *indica*
GENUS *Fibulotaeniella* Ribbon Clasp Club Fungus
SPECIES *canadensis*
GENUS *Geotrichopsis* Earth Hairy Club Fungus
SPECIES *mycoparasitica*
GENUS *Glomerulomyces* Clustered Club Fungus
SPECIES *fibulosus*
GENUS *Glutinoagger* Glutin Club Fungus
SPECIES *fibulatus*
GENUS *Hauerslevia* Hauerslev Club Fungus
SPECIES *pulverulenta*
GENUS *Heteromycophaga* Variable Phage Club Fungus
SPECIES *glandulosae*
GENUS *Hyphobasidiofera* Hypha Club Fungus
SPECIES malaysiana
GENUS *Itersonilia* Iterson Club Fungus
SPECIES *pastinacae, perplexans*
GENUS *Mixia* Mixed Club Fungus
SPECIES *osmundae*
GENUS *Mycaureola* Aureole Club Fungus
SPECIES *dilseae*
GENUS *Myriococcum* Myriad Round Club Fungus
SPECIES *praecox, thermophilum*
GENUS *Pagidospora* Pagidospore Club Fungus
SPECIES *amoebophila*
GENUS *Phymatotrichopsis* Tumor Hairy Club Fungus
SPECIES *omnivora*
GENUS *Piriformospora* Pear Spore Club Fungus
SPECIES indica
GENUS *Pneumatospora* Lung Spore Club Fungus
SPECIES *obcoronata*
GENUS *Pycnovellomyces* Dense Vellose Club Fungus
SPECIES *foliicola*
GENUS *Riessia* Riess Club Fungus
SPECIES *minima, naumovii, radicicola, semiophora*
GENUS *Riessiella* Lesser Riess Club Fungus
SPECIES *clavata, cylindrica*
GENUS *Spiniger* Spiny Club Fungus
SPECIES *curiosus, meineckellus*
GENUS *Sporotrichum* Spore Hairy Club Fungus
SPECIES *aurantiacum, azureum, bombycinum, caesiellum, candidum, chlorinum, chrysospermum, flavicans, flavissimum, flavovirens, griseum, laetum, lanatum, laxum, olivaceum, pruinosum, pulverulentum, roseolum, sporodochiale, tenue*
GENUS *Vanrija* Van Rija Club Fungus
SPECIES *ingeniosa*

FAMILY 2 *Ceratobasidiaceae* Horn Club Fungus
GENUS *Ceratobasidium* Horn Club Fungus
SPECIES *anceps, angustisporum, bicorne, calosporum, cereale, cornigerum, globisporum, obscurum, papillatum, pseudocornigerum, ramicola, setariae, sphaerosporum, stevensii, stridii*
GENUS *Aquathanatephorus* Aquatic Thanate Horn Club Fungus
SPECIES *pendulus*
GENUS *Cejpomyces* Cejp Horn Club Fungus
SPECIES *globosisporus*
GENUS *Ceratorhiza* Rhizid Horn Club Fungus
SPECIES *anacalospora, decidua, goodyerae-repentis*
GENUS *Heteroacanthella* Variable Spine Horn Club Fungus
SPECIES *acanthophysa, variabilis*

GENUS *Metabourdotia* Metabourdot Horn Club Fungus
SPECIES *tahitiensis*
GENUS *Oliveonia* Olive Horn Club Fungus
SPECIES *citrispora, fibrillosa, nodosa, pauxilla, termitophila*
GENUS *Oncobasidium* Onc Horn Club Fungus
SPECIES *theobromae*
GENUS *Scotomyces* Scotic Horn Club Fungus
SPECIES *subviolaceus*
GENUS *Thanatephorus* Thanate Horn Club Fungus
SPECIES *amygdalisporus, cucumeris, fusisporus, ochraceus, sterigmaticus, terrigenus*
VARIETIES *cucumeris ambigua, cucumeris anatolica, cucumeris cedri-deodarae, cucumeris fuchsiae, cucumeris hortensis, cucumeris lycopersici*
GENUS *Tofispora* Tof Spore Horn Club Fungus
SPECIES *biapiculatal, hebelomatospora, repetospora*
GENUS *Waitea* Waite Horn Club Fungus
SPECIES *circinata, nuda*

FAMILY 4 *Cystofilobasidiaceae* Cyst Filament Club Fungus
GENUS *Cystofilobasidium* Cyst Filament Club Fungus
SPECIES *bisporidii, capitatum, ferigula, infirmominiatum, lari-marini*
GENUS *Mrakia* Mrak Cyst Filament Club Fungus
SPECIES *frigida, gelida, nivalis, stokesii*
GENUS *Rhodozyma* Red Zyme Cyst Filament Club Fungus
SPECIES *montanae*

FAMILY 6 *Filobasidiaceae* Filament Club Fungus
GENUS *Filobasidium* Filament Club Fungus
SPECIES *capsuligenum, floriforme, uniguttulatum*
GENUS *Cryptococcus* Cryptic Round Filament Club Fungus
SPECIES *aerius, albidus, aquaticus, curvatus, diffluens, gastricus, laurentii, macerans, magnus, neoformans, skinneri*

FAMILY 7 *Tulasnellaceae* Tulasn Club Fungus
GENUS *Tulasnella* Tulasn Club Fungus
SPECIES *aggregata, albida, allantospora, anguifera, bifrons, bourdotii, brinkmannii, calospora, conidiata, cruciata, cystidiophora, danica, deliquescens, dissitispora, echinospora, eichleriana, falcifera, fuscoviolacea, guttulata, helicospora, hyalina, inclusa, interrogans, irregularis, pacifica, pallida, pallidocremea, papillata, permacra, pinicola, pruinosa, quasiflorens, rogersii, saveloides, thelephorea, tomaculum, vernicosa, violea, vitrea*
GENUS *Epulorhiza* Rhizid Tulasn Club Fungus
SPECIES *anaticula*
GENUS *Pseudotulasnella* Lesser Tulasn Club Fungus
SPECIES *guatemalensis*

ORDER 2 *Clavariales* Coral Fungus

FAMILY 1 *Clavariaceae* Coral Fungus
GENUS *Clavaria* Coral Fungus
SPECIES *acuta, alboglobospora, ardosiaca,* ***argillacea***(Edible)*, borealis, citrinorubra, corbierei, crosslandii, cupreicolor, echino-olivacea, echinobrevispora, echinonivosa, filiola, flavipes, flavopurpurea, fossicola, fragilis,* ***fumosa***(Edible)*, globospora, greletii, guilleminii, incarnata, krieglsteineri, macounii, maricola, megaspinosa, mima, muscula, musculospinosa, novozealandica, oronoensis, plumbeoargillacea,* ***purpurea***(Edible)*, redoleoalii, rosea, roseoviolacea, rubescens,* stegasauroides*, straminea, subacuta, subsordida, subviolacea, tenuipes, tortuosa, tuberculospora,* ***vermicularis***(Edible)*, ypsilonidia, zollingeri*
GENUS *Clavariadelphus* Adelphi Coral Fungus
SPECIES *americanus, caespitosus, cokeri, fasciculatus, flavidus, flavoimmaturatus, helveticus, himalayensis, lignicola,* ***ligula***(Edible)*, mirus, mucronatus, occidentalis, pallidoincarnatus,* ***pistillaris***(Edible)*, sachalinensis, subfastigiatus,* ***truncatus***(Edible)*, unicolor, xanthocephalus, yunnanensis*
GENUS *Clavicorona* Crown Coral Fungus
SPECIES *cristata, dichotoma, divaricata, mairei, microspora, piperata,* ***pyxidata***(Edible)*, taxophila, tuba, turgida*
GENUS *Clavulinopsis* Lesser Coral Fungus
SPECIES *alcicornis, amoena, archeri, arenicola, aurantiobrunnea, aurantiocinnabarina, boninensis, brevipes, candida, cinnamomea, citrinoalba, corallinorosacea, corniculata, depokensis, fruticula, fusiformis, griseola, hastula, helvola, hisingeri, laeticolor, ledermannii, lignicola, luteo-ochracea, luteoalba, luteonana, luticola, michelii, rufipes, semivestita, sibutiana, spiralis, subarctica, subtilis, sulcata, umbrinella, vernalis*
GENUS *Macrotyphula* Typhus Coral Fungus
SPECIES *contorta, defibulata,* ***fistulosa***(Edible)*, juncea, rhizomorpha, tremula*
GENUS *Multiclavula* Multiple Coral Fungus
SPECIES *clara, constans, coronilla, delicata, mucida, sinensis*

GENUS *Scytinopogon* Scyte Beard Coral Fungus
SPECIES *scaber*
GENUS *Setigeroclavula* Seta Coral Fungus
SPECIES *ascendens*

FAMILY 02 *Acrospermaceae* Acrospore Coral Fungus
GENUS *Acrospermum* Acrospore Coral Fungus
SPECIES *compressum, graminum, pallidulum*
GENUS *Oomyces* Oospore Acrospore Coral Fungus
SPECIES *carneoalbus*

FAMILY 2 *Ramariaceae* Branch Coral Fungus
GENUS *Ramaria* Branch Coral Fungus
SPECIES ***abietina***(Edible), *acrisiccescens, acutissima, aenea, africana, albidoflava, albocinerea, alborosea, altaica, ambigua, americana, amyloidea, anisata, anziana, apiahyna, apiculata, apiculata, araiospora, arcosuensis, argentea, armeniaca, articulotela, asiatica, atkinsonii, aurantiisiccescens,* ***aurea***(Edible), *aureofulva, aureorhiza, australiana, avellaneovertex, basirobusta, bonii, botrytis,* ***botrytis***(Edible), *botrytoides, bourdotiana, brevispora, brienzensis, broomei, brunneicontusa, brunneipes, brunneomaculata, bulbobasidiata, cacao, camelicolor, camellia, campestris, campoi, candida, canescens, capucina, celerivirescens, cettoi, cinereocarnea, cladoniae, clarobrunnea, claviramulata, conjuncta, coulterae, curta, cyaneigranosa, daucipes, decurrens, distinctissima, divaricata, dolomitica, echinovirens, eosanguinea, ephemeroderma, eryuanensis, eumorpha, fagetorum, fagicola, fennica, fennica, fistulosa, flaccida,* ***flava***(Edible), *flavescens, flaviceps, flavicolor, flavigelatinosa, flavoalba, flavobrunnescens, flavoides, flavomicrospora, flavosalmonicolor, flavosaponaria, flavoviridis, flavula, foetida, formosa, fragillima, fumosiavellanea, fuscobrunnea, gelatiniaurantia, gigantea, glaucoaromatica, gracilis, grandipes, grandis, griseobrunnea, grundii, guyanensis, gypsea, harrisonii, hemirubella, henriquesii, highlandensis, hilaris, himalayensis, holorubella, ignicolor, incognita, incongrua, indoyunnaniana, inedulis, inquinata, insignis, intimorosea, junquilleo-vertex, kewensis, kisantuensis, lacteobrunnescens, laeviformosoides, laevispora, largentii, leptoformosa, linearioides, longicaulis, longispora, longissimispora, lorithamnus, luteoflaccida, maculatipes, maculospora, madagascariensis, magnifica, magnipes, marrii, mediterranea, moelleriana, murrillii, mutabilis, myceliosa, nanispora, neoformosa, obtusissima, ochracea, ochraceosalmonicolor, ochrochlora, pallida, pallidissima, pallidosaponaria, pancaribbea, patagonica, perbrunnea, perfluopunicea, petersenii, polonica, polypus, primulina, prostrata, pseudobotrytis, pseudogracilis, pumila, pupulispora, pura, purpureopallida, purpurissima, pyrispora, rainierensis, rasilispora, raveneliana, reticulata, roellinii, rosella, rotundispora, rubella, rubiginosa, rubri-attenuipes, rubribrunnescens, rubricarnata, rubrievanescens, rubripermanens, rubrogelatinosa, samuelsii, sandaracina, sanguinea, sanguinipes, sardiniensis, schildii, sclerocarnosa, secunda, sesiana, sinoconjunctipes, solomonensis, somniculosa, spinulosa, strasseri, stricta, stricta, stuntzii, subaurantiaca, subbotrytis, subgelatinosa, subtilis, subviolacea, suecica, synaptopoda, terrea, testaceoflava, testaceoviolacea, thalliovirescens, thiersii, toxica, tridentina, tropicalis, tsugina, valdiviana, varians, velocimutans, verlotensis, vinaceipes, vinosimaculans, violaceibrunnea, watlingii, xanthosperma, zeppelinospora, zippelii*
GENUS *Austrogautieria* Southern Gautier Branch Coral Fungus
SPECIES *chlorospora, clelandii, costata, macrospora, manjimupana, rodwayi*
GENUS *Delentaria* Delentar Branch Coral Fungus
SPECIES *decurva*
GENUS *Gautieria* Gautier Branch Coral Fungus
SPECIES *morchelliformis*
GENUS *Kavinia* Kavin Branch Coral Fungus
SPECIES *alboviridis, globispora, himantia, salmonea, vivantii*
GENUS *Protogautieria* Primitive Gautier Branch Coral Fungus
SPECIES *lutea, substriata*
GENUS *Ramariopsis* Lesser Branch Coral Fungus
SPECIES *agglutinata, alutacea, asperulospora, asterella, avellanea, avellaneo-inversa, bicolor, biformis, californica, cinnamomipes, citrina, clavuligera, costaricensis, crocea, dealbata, flavescens, hibernica, holmskioldii, junquillea, kunzei, longipes, minutula, novae-hibernica, pseudosubtilis, pulchella, ramarioides, tenuicula, tenuiramosa, tortuosa, vestitipes*
GENUS *Ramaricium* Least Branch Coral Fungus
SPECIES *albo-ochraceum, alboflavescens, flavomarginatum*
GENUS *Rhodactina* Red Ray Branch Coral Fungus
SPECIES *himalayensis*

FAMILY 3 *Sparassidaceae* Lace Coral Fungus
GENUS *Sparassis* Lace Coral Fungus
SPECIES *brevipes,* ***crispa***(Edible Unripe Cooked), ***laminosa***(Edible), *nemecii, ramosa, spathulata*
VARIETIES ***crispa radicata, laminosa herbstii,*** *spathulata simplex*
GENUS *Bondarcevomyces* Bondarzev Lace Coral Fungus
SPECIES *taxi*

ORDER 3 *Hymenochaetales* Membrane Bristle Fungus

FAMILY 1 *Hymenochaetaceae* Membrane Bristle Fungus

GENUS *Hymenochaete* Membrane Bristle Fungus

SPECIES *aberrans, acanthophysata, adusta, agathicola, allantospora, americana, anomala, attenuata, barbata, berteroi, bispora, boidinii, boninensis, borbonica, borealis, burdsallii, carpatica, ceratophora, cervina, cervinoidea, cinnamomea, coffeana, colliculosa, contiformis, corrugata, cruenta, curtisii, damicornis, denticulata, depallens, dictator, digitata, dissimilis, dura, epichlora, episphaeria, escobarii, fasciculata, floridea, fomitoporioides, fuliginosa, fulva, fuscobadia, gigaspora, gillesii, gladiola, globispora, harpago, hauerslevii, innexa, intricata, jobii, konradii, lenta, leonina, lictor, lignosa, livens, longispora, luteobadia, macrospora, magnahypha, microspora, minuscula, mollis, mougeotii, murina, muroiana, nanospora, ochromarginata, opaca, palmicola, papyracea, patelliformis, paucisetosa, pellicula, pertenuis, pinnatifida, plurimaesetae, pratensis, proxima, pseudoadusta, reniformis, reticulata, rhabarberina, rheicolor, rigidula, rubiginosa, rufomarginata, ryvardenii, semistupposa, senatoumbrina, separabilis, separata, sordida, spathulata, sphaerospora, spreta, stratura, subferruginea, subpurpurascens, tabacina, tasmanica, tenuis, tomentelloidea, ungulata, unicolor, ustulata, vagans, vaginata, vallata, variegata, villosa, yasudai*

GENUS *Asterodon* Aster Membrane Bristle Fungus

SPECIES *ferruginosus*

GENUS *Aurificaria* Ear Fig Membrane Bristle Fungus

SPECIES *euphoriae, flammans, indica, luteoumbrina, poncei, shoreae*

GENUS *Clavariachaete* Coral Membrane Bristle Fungus

SPECIES *peckoltii, rubiginosa*

GENUS *Coltricia* Coltric Membrane Bristle Fungus

SPECIES *bambusicola, cinnamomea, confluens, duportii, focicola, fragilissima, hamata, montagnei, perennis, progressus, pyrophila, salpincta, sideroides, strigosa, strigosipes, subfastosa, truncicola*

VARIETIES *perennis cyathiformis, perennis fimbriatus, perennis infundibulum, perennis lejeunii, perennis leucoporus, perennis perfosssus, perennis subtomentosus*

GENUS *Coltriciella* Lesser Coltric Membrane Bristle Fungus

SPECIES *dependens, pusilla, tasmanica*

GENUS *Cyclomyces* Cyclic Membrane Bristle Fungus

SPECIES *fuscus, iodinus, setiporus*

GENUS *Erythromyces* Red Membrane Bristle Fungus

SPECIES *crocicreas*

GENUS *Fomitiporia* Foam Pore Membrane Bristle Fungus

SPECIES *mediterranea*

GENUS *Hydnochaete* Truffle Membrane Bristle Fungus

SPECIES *duportii, japonica, olivacea, peroxydata, saepiaria, tabacina, tabacinoides*

GENUS *Inonotus* Ino Membrane Bristle Fungus

SPECIES *albertinii, andersonii, chilanshanus, clemensiae, crocitinctus, crustosus, cuticularis, dryadeus, dryophilus, duostratosus, farlowii, flavidus, fulvomelleus, glomeratus, hamusetulus, hastifer, hispidus, juniperinus, micantissimus, mikadoi, munzii, nidus-pici, nodulosus, nothofagi, obliquus, ochroporus, patouillardii, pegleri, peristrophidis, pertenuis, porrectus, pseudoradiatus, pusillus, quercustris, radiatus, rheades, rickii, splitgerberi, subiculosus, tabacinus, tamaricis, texanus, triqueter, venezuelicus*

VARIETIES *hispidus spongiosus, hispidus velutinus, hispidus villosus*

GENUS *Onnia* Onni Membrane Bristle Fungus

SPECIES *circinata, orientalis, tomentosa*

GENUS *Phellinus* Cork Membrane Bristle Fungus

SPECIES *acontextus, adamantinus, adhaerens, allardii, andinus, apiahynus, appositus, aureobrunneus, baccharidis, badius, bambusinus, bicuspidatus, callimorphus, carteri, caryophylleus, caryophylli, cavicola, cesatii, chryseus, chrysoloma, cinchonensis, cinereus, coffeatoporus, conchatus,contiguus, crocatus, cyclobalanopsis, cylindrosporus, daedaliformis, dependens, discipes, erectus, everhartii, fastuosus, ferreus, ferrugineofuscus, ferrugineovelutinus, ferruginosus, flavomarginatus, fragrans, gilvus, grenadensis, griseoporus, hartigii, himalayensis, hippophaëicola, igniarius, jezoënsis, kamahi, kawakamii, laevigatus, linteus, lundellii, macgregorii, melanodermus, membranaceus, merrillii, minutiporus, neoquercinus, nigrolimitatus, nilgheriensis, nothofagi, noxius, occidentalis, overholtsii, pachyphloeus, palmicola, pectinatus, piceinus, pini, poeltii, pomaceus, populicola, prunicola, prunicola, pseudolaevigatus, pseudopunctatus, punctatiformis, punctatus, pusillus, reichingeri, repandus, resinaceus, rhabarbarinus, rhamni, rhytiphloeus, rimosus, robiniae, robustus, roseocinereus, rufus, sancti-georgii, sanfordii, sarcites, scleropileatus, senex, setulosus, sonorae, spiculosus, spinescens, sublamaensis, swieteniae, syringeus, tabaquilio, torulosus, tremulae, tricolor, tropicalis, umbrinellus, uncinatus, viticola, wahlbergii, weirianus, weirii, xeranticus*

VARIETIES *conchatus salicinus, igniarius nigricans, igniarius subresupinatus, pini cancriformans, pomaceus scutiformis, pomaceus tuberculosus, robustus abietis, viticola superficialis*

GENUS *Phylloporia* Pore Membrane Bristle Fungus

SPECIES *bibulosa, chrysites, ephedrae, fruticum, parasitica, psila, ribis, spathulata*

VARIETIES *ribis crataegi, ribis ulicis*

GENUS *Polystictus* Many Stick Membrane Bristle Fungus

SPECIES *ridleyi*

GENUS *Porodaedalea* Pore Daedale Membrane Bristle Fungus

SPECIES *niemelaei*

GENUS *Xanthochrous* Yellow Tinged Membrane Bristle Fungus

SPECIES *patouillardii*
VARIETIES *patouillardii patouillardii, patouillardii congoensis*
FAMILY 2 *Schizoporaceae* — Split Pore Membrane Bristle Fungus
GENUS *Schizopora* — Split Pore Membrane Bristle Fungus
SPECIES *carneolutea, cystidiata, flavipora, paradoxa*
GENUS *Basidioradulum* — Radial Split Pore Membrane Bristle Fungus
SPECIES *radula, tuberculatum*
GENUS *Echinoporia* — Urchin Split Pore Membrane Bristle Fungus
SPECIES *aculeifera, hydnophora*
GENUS *Grandinia* — (Synonym for *Hyphodontia*)
GENUS *Hyphodontia* — Hyphodont Split Pore Membrane Bristle Fungus
SPECIES *abieticola, adhaerispora, alba, alienata, aloha, altaica, alutacea, alutaria, apacheriensis, arguta, aspera, australis, barba-jovis, boninensis, borealis, brevidens, breviseta, byssoidea, candidissima, capitata, cineracea, crassispora, crustosa, crustosoglobosa, curvispora, detritica, efibulata, erastii, fimbriata, fimbriiformis, floccosa, gamundiae, gossypina, griseliniae, hallenbergii, halonata, hastata, hastifera, hjortstamii, juniperi, knysnana, lanata, latitans, macrescens, magnifica, microspora, mollis, nespori, nesporina, niemelaei, nothofagi, nudiseta, orasinusensis, pallidula, papillosa, pelliculae, poroideoefibulata, propinqua, pruni, pumilia, quercina, radula, rimosissima, sambuci, serpentiformis, setulosa, spathulata, stipata, subalutacea, subdetritica, subglobosa, subspathulata, taiwaniana, tenuicystidia, tetraspora, tomentosa, tropica, tuberculata, wrightii*
GENUS *Leucophellinus* — Light Cork Split Pore Membrane Bristle Fungus
SPECIES *hobsonii*
GENUS *Odontia* — (Synonym for *Hyphodontia*)
GENUS *Oxyporus* — Sharp Split Pore Membrane Bristle Fungus
SPECIES *borealis, bucholtzii, cervinogilvus, corticola, cuneatus, latemarginatus, lilaceus, obducens, pellicula, phellodendri, populinus, ravidus, similis, spiculifer, subflavus, subulatus*
GENUS *Paratrichaptum* — Hairy Split Pore Membrane Bristle Fungus
SPECIES *accuratum*
GENUS *Poriodontia* — Toothed Split Pore Membrane Bristle Fungus
SPECIES *subvinosa*
GENUS *Rogersella* — Rogers Split Pore Membrane Bristle Fungus
SPECIES *eburnea*

ORDER 4 *Polyporales* — Pore Fungus

FAMILY 01 *Polyporaceae* — Pore Fungus
GENUS *Polyporus* — Pore Fungus
SPECIES *admirabilis, alutaceus, alveolaris, anthracophilus, arcularius, aurantibrunneus, aurantioporus, australiensis, austroafricanus, badius, biskeletalis,* ***brumalis***(Edible), *carneus, castaneae, centroafricanus, cervinus, chocolatus, choseniae, ciliatus, corylinus, craterellus, doidgeae, expansus, ferruginipes, grammocephalus, guianensis, hypomelanus, incendiarius, indigenus, lateralis, lepideus, leprieurii, leptocephalus, ludovicianus, melaleucus, melanopus, meridionalis, mikawai, nigrocristatus, ostreiformis, perula, pervadens, philippinensis, platensis, pseudobetulinus, radicatus, rhizophilus, rugulosus, squamosus, tasmanicus, trametoides, tricholoma, tubaeformis, tuberaster, tumulosus, udus,* ***umbellatus***(Edible), *varius, violaceomaculatus*
VARIETIES *alveolaris mori, arcularius alveolarius, arcularius exasperatus, badius durus, badius perennis,* ***brumalis ciliatus, brumalis fasciculatus, brumalis fuscidulus, brumalis hypocrateriformis, brumalis umbilicatus,*** *ciliatus nummulariformis, leptocephalus nigripes, melanopus cyathoides, squamosus cellulosus, squamosus fagicola, squamosus juglandis, squamosus maculatus, squamosus maximus, squamosus michelii, squamosus polymorphus, squamosus rangiferinus, squamosus rostkovii, tumulosus tumulosus, tumulosus westraliensis,* ***umbellatus ramosissimus, umbellatus ramosus****, varius lateralis, varius nummularius*
GENUS *Antromycopsis* — Antro Pore Fungus
SPECIES *angustata, guzmanii, macrocarpa*
GENUS *Austrolentinus* — Southern Lentin Pore Fungus
SPECIES *tenebrosus*
GENUS *Bridgeoporus* — Bridge Pore Fungus
SPECIES *nobilissimus*
GENUS *Burgoa* — Burgo Pore Fungus
SPECIES *anomala, verzuoliana*
GENUS *Caloporus* — (Synonym for *Albatrellus, Boletopsis Buglossoporus, Ceriporia, Gloeoporus, Grifola, Laccocephalum, Meripilus, Meruliopsis, Postia, Rigidoporus, Scutiger*)
GENUS *Cerrena* — Cerren Pore Fungus
SPECIES *cystidiata, sclerodepsis, unicolor*
GENUS *Coriolopsis* — Lesser Turkeytail Pore Fungus
SPECIES *aspera, bataanensis, brunneoleuca, burchellii, byrsina, caperata, daedaloides, floccosa, gallica, helvola, occidentalis, polyzona, strumosa, taylorii, telfarii, tuberculata*
VARIETIES *gallica favus, gallica pileata*
GENUS *Coriolus* — Turkeytail Pore Fungus

SPECIES *azureus, fibula*
VARIETIES *azureus versicolor*
GENUS *Cryptoporus* Cryptic Pore Fungus
SPECIES *volvatus*
GENUS *Cystidiophorus* Cyst Pore Fungus
SPECIES *castaneus*
GENUS *Cystostiptoporus* Cyst Stiptic Pore Fungus
SPECIES *violaceocinerascens*
GENUS *Datronia* Datron Pore Fungus
SPECIES *decipiens, glabra, mollis, scutellata, scutellata, stereoides*
VARIETIES *scutellata noveboracensis*
GENUS *Dichomitus* Twin Mitre Pore Fungus
SPECIES *albidofuscus, anoectoporus, campestris, eucalypti, leucoplacus, newhookii, papuanus, squalens*
GENUS *Earliella* Earl Pore Fungus
SPECIES *scabrosa*
GENUS *Echinochaete* Urchin Bristle Pore Fungus
SPECIES *brachypora, cinnamomeosquamulosa, ruficeps, russiceps*
GENUS *Faerberia* Ferber Pore Fungus
SPECIES ***carbonaria***(Edible)
GENUS *Favolus* Favole Pore Fungus
SPECIES *tenuiculus*
GENUS *Fibroporia* Fiber Pore Fungus
SPECIES *radiculosa, vaillantii*
VARIETIES *vaillantii hybridus, vaillantii vaporarius*
GENUS *Flabellophora* Fan Pore Fungus
SPECIES *aurantiaca, brevipes, deceptiva, flaviporus, inconspicua, intertexta, kinabaluensis, latiporus, nana, obtorta, ochracea, parva, squamosa, subsimplex, superposita, variabilis, velutinosa*
GENUS *Fomes* Foam Pore Fungus
SPECIES *chaquensis, clelandii, extensus, fasciatus, fomentarius, fulvus, hemitephrus, meliae*
VARIETIES *fomentarius nigrescens*
GENUS *Funalia* Funal Pore Fungus
SPECIES *argentea, trogii*
GENUS *Fuscocerrena* Fuscous Cerren Pore Fungus
SPECIES *portoricensis*
GENUS *Globifomes* Glob Pore Fungus
SPECIES *graveolens*
GENUS *Globosomyces* Lesser Glob Pore Fungus
SPECIES *aggregatus*
GENUS *Gloeophyllum* Glow Pore Fungus
SPECIES *abietinum, carbonarium, concentricum, juniperinum, mexicanum, nigrozonatum, odoratum, protractum, sepiarium, striatum, subferrugineum, trabeum, vinaceobrunneum*
VARIETIES *odoratum annulatus, odoratum ceratophorus, odoratum polymorphus, trabeum communis*
GENUS *Grammothelopsis* Grass Thele Pore Fungus
SPECIES *incrusta, macrospora, puiggarii*
GENUS *Haploporus* Solitary Pore Fungus
SPECIES *odorus*
GENUS *Heliocybe* Helion Cybe Pore Fungus
SPECIES *sulcata*
GENUS *Hexagonia* Hexagon Pore Fungus
SPECIES *apiaria, dermatiphora, hirta, hydnoides, leprosa, niam-niamensis, nitida, pobeguinii, speciosa, umbrinella, variegata, velutina, vesparia, zambesiana*
VARIETIES *hirta favus, hydnoides fibrosus, hydnoides ursinus*
GENUS *Hirschioporus* Hirsch Pore Fungus
SPECIES *purpureus, quercinus, versatilis*
GENUS *Laccocephalum* Lakehead Pore Fungus
SPECIES *basilapidoides, hartmannii, mylittae, sclerotinum, tumulosum*
GENUS *Laetiporus* Bright Pore Fungus
SPECIES *baudonii, miniatus, persicinus, portentosus,* ***sulphureus***(Edible)
VARIETIES *sulphureus citrinus, sulphureus coriaceus, sulphureus imbricatus, sulphureus lingua-cervina, sulphureus ramosus, sulphureus tenax*
GENUS *Lentinus* Lentin Pore Fungus
SPECIES *anthocephalus, araucariae, badius, bertieri, caespiticola, concinnus, connatus, copulatus, courtetianus, crinitus, cyathiformis, fasciatus, fusipes, glabratus, hookerianus, lamelliporus, levis, martianoffianus, polychrous, prancei, punctaticeps, retinervis, sajor-caju, scleropus, similis, squamosus, squarrosulus, striatulus, strigosus, stuppeus, suavissimus, subdulcis, swartzii, tephroleucus, tigrinoides, tigrinus, tuber-regium, umbrinus, velutinus, villosus, zeyheri*
VARIETIES *badius calceoliformis, sajor-caju densifolius*

GENUS *Lenzites* Lenz Pore Fungus
SPECIES *betulinus, elegans, stereoides, styracina, warnieri*
GENUS *Leucopus* (Synonym for *Polyporus*)
GENUS *Lignosus* Lignose Pore Fungus
SPECIES *dimiticus, goetzii, rhinocerus, sacer*
GENUS *Loweporus* Lowe Pore Fungus
SPECIES *castaneus, fuscopurpureus, lividus, mollis, ochraceicinctus, pileoliferus, roseoalbus, rufescens, tephroporus*
GENUS *Macrohyporia* Hyponic Pore Fungus
SPECIES *dictyopora*
GENUS *Megasporoporia* Megaspore Pore Fungus
SPECIES *cavernulosa, hexagonoides, mexicana, setulosa*
GENUS *Melanoporella* Least Dark Pore Fungus
SPECIES *carbonacea*
GENUS *Melanoporia* Lesser Dark Pore Fungus
SPECIES *nigra*
GENUS *Melanopus* Dark Pore Fungus
SPECIES *badius, corylinus, dictyopus, guianensis, leprieurii, leptocephalus, radicatus, rhizophilus, squamosus* (Edible Wood Saddleback Nongill Mushroom), *tuberaster, varius* (Inedible Variable Nongill Mushroom)
VARIETIES *badius picipes* (Inedible Ringed Bay Nongill Mushroom), *corylinus tunetanus, dictyopus diabolicus, dictyopus infernalis, dictyopus nephridis, dictyopus pancheri, dictyopus rhizomorphus, dictyopus scabellus, guianensis marasmioides, leprieurii calyculus, leptocephalus elegans* (Inedible Elegant Thincap Nongill Mushroom), *tuberaster coronatus, tuberaster forquignonii* (Edible Forquignon Tuber Nongill Mushroom), *tuberaster lentus, varius noackianus, varius nummularius, varius melanopus, varius nummularius*
GENUS *Meruliporia* Merule Pore Fungus
SPECIES *incrassata*
GENUS *Microporellus* Lesser Tiny Pore Fungus
SPECIES *brunneus, chocolatus, clemensiae, dealbatus, fuliginosus, grandiporus, iguazuensis, inusitatus, labyrinthiformis, nigripes, obovatus, pahangensis, setigerus, subincarnatus, subumbonatus*
GENUS *Microporus* Tiny Pore Fungus
SPECIES *affinis, atroalbus, concinnus, incomptus, luteoceraceus, quarrei, vernicipes, xanthopus*
GENUS *Mollicarpus* Soft Carpel Pore Fungus
SPECIES *cognatus*
GENUS *Mycelithe* Rock Pore Fungus
SPECIES *fungifera*
GENUS *Navisporus* Navel Pore Fungus
SPECIES *africanus, floccosus, ortizii, sulcatus*
GENUS *Neolentinus* Modern Lentin Pore Fungus
SPECIES *adhaerens, cirrhosus, dactyloides, kauffmanii, lepideus, pallidus, papuanus, ponderosus, schaefferi, suffrutescens*
GENUS *Nigrofomes* Black Foam Pore Fungus
SPECIES *melanoporus*
GENUS *Nigroporus* Black Pore Fungus
SPECIES *durus, ussuriensis, vinosus, violaceus*
GENUS *Pachykytospora* Thick Cytospore Pore Fungus
SPECIES *alabamae, tuberculosa*
GENUS *Perenniporia* Perennial Pore Fungus
SPECIES *adnata, ahmadii, albocinnamomea, amylodextrinoidea, aurantiaca, centrali-africana, compacta, decurrata, detritus, dipterocarpicola, ellisiana, fergusii, ferruginea, fraxinea, fraxinophila, gomezii, hexagonoides, inflexibilis, isabellina, maackiae, malvena, martii, medulla-panis, micropora, minutissima, mundula, narymica, neofulva, ochroleuca, ohiensis, oviformis, pauciskeletalis, penangiana, phloiophila, piperis, roseoisabellina, rosmarini, semistipitata, stipitata, straminella, subacida, subcretacea, tenuis, tepeitensis, voeltzkowii*
VARIETIES *tenuis tenuis, tenuis pulchella*
GENUS *Phaeolus* Dusky Pore Fungus
SPECIES *schweinitzii* (Inedible Schweinitz Nongill Mushroom)
VARIETIES *schweinitzii sistotremoides, schweinitzii spadiceus*
GENUS *Phaeotrametes* Dusky Tramet Pore Fungus
SPECIES *decipiens*
GENUS *Piloporia* Pilose Pore Fungus
SPECIES *indica, sajanensis*
GENUS *Pocillaria* Poc Pore Fungus
SPECIES *reflexa*
GENUS *Podofomes* Pod Pore Fungus
SPECIES *corrugis, trogii*
GENUS *Polypilus* (Synonym for *Bondarzewia, Laetiporus, Meripilus, Polyporus*)
GENUS *Polyporellus* (Synonym for *Polyporus*)

GENUS *Poria* Greater Pore Fungus
SPECIES *chlorina, hypobrunnea, lindbladii, pavonina, sequoiae*
GENUS *Porodisculus* Disc Pore Fungus
SPECIES *pendulus*
GENUS *Poronidulus* Shiny Pore Fungus
SPECIES *conchifer*
GENUS *Pseudofavolus* Lesser Favole Pore Fungus
SPECIES *miquelii, nigrus, orinocensis*
GENUS *Pseudopiptoporus* Lesser Pipt Pore Fungus
SPECIES *devians*
GENUS *Ptychogaster* Folded Cavity Pore Fungus
SPECIES *corruscans, cubensis, effusus, rubescens*
GENUS *Pycnoporellus* Lesser Dense Pore Fungus
SPECIES *alboluteus, fulgens*
GENUS *Pycnoporus* Dense Pore Fungus
SPECIES *cinnabarinus, coccineus, puniceus, sanguineus*
VARIETIES *cinnabarinus miniatus, sanguineus ruber*
GENUS *Pyrofomes* Fire Foam Pore Fungus
SPECIES *albomarginatus, demidoffii, fulvoumbrinus, lateritius, perlevis, tricolor*
GENUS *Ryvardenia* Ryvarden Pore Fungus
SPECIES *campyla, cretacea*
GENUS *Skeletocutis* Skeleton Cut Pore Fungus
SPECIES *albocremea, alutacea, amorpha, australis, azorica, carneogrisea, friabilis, kuehneri, lenis, lilacina, nivea, niveicolor, nothofagi, novae-zelandiae, ochroalba, odora, papyracea, perennis, roseola, stellae, subincarnata, subvulgaris, tschulymica, vulgaris*
VARIETIES *vulgaris papyrae*
GENUS *Sparsitubus* Sparse Tube Pore Fungus
SPECIES *nelumbiformis*
GENUS *Stiptophyllum* Stiptic Pore Fungus
SPECIES *erubescens*
GENUS *Tinctoporellus* Tincture Pore Fungus
SPECIES *epimiltinus*
GENUS *Trametes* Tramet Pore Fungus
SPECIES *bresadolae, cervina, cingulata, cotonea, cubensis, cystidiata, decussata, discoidea, ectypa, ellipsospora, gibbosa, hirsuta, incerta, junipericola, lactinea, leonina, ljubarskii, lunispora, manilaensis, marianna, maxima, menziesii, meyenii, mimetes, modesta, morganii, nivosa, ochracea, ochroflava, pocas, pubescens,robiniophila, roseola, salebrosa, scopulosa, sinensis, socotrana, suaveolens, subectypus, tephroleuca, tyromycoides, unguliformis, versicolor, villosa, violacea*
VARIETIES *hirsuta nigromarginatus, hirsuta velutinus, menziesii convolutus, ochracea zonatus, pubescens velutinus, suaveolens suberosus*
GENUS *Trichaptum* Hairy Pore Fungus
SPECIES *abietinum, agglutinatum, basifuscum, biforme, byssogenum, ceraceicutis, favoloides, fumosoavellaneum, fuscoviolaceum, jackiae, lacunosum, laricinum, parvulum, perrottetii, sector, strigosum, subchartaceum, suberosum*
VARIETIES *abietinum purpurascens, fuscoviolaceum decipiens*
GENUS *Tyromyces* Tyrol Pore Fungus
SPECIES *albogilvus, amarus, atrostrigosus, caesioflavus, catervatus, chioneus, corniculatus, corticicola, duracinus, eberhardtii, fumidiceps, galactinus, hispidulinanus, humeana, hypocitrinus, immitis, incarnatus, inodermatus, interponens, kmetii, lacteus, languidus, leucospongia, levis, marianii, mediocris, merrittii, mexicanus, mollicaseus, nemorosus, ochraceicarneus, praeguttulatus, pulcherrimus, raduloides, sambuceus, setiger, singeri, squamulosus, stenomitis, subcaesius, sublacteus, subviridis, tephrus, wynnei, xuchilensis*
VARIETIES *chioneus candidus*
GENUS *Ungulina* (Synonym for *Fomes, Fomitopsis, Laricifomes*)
GENUS *Vanderbylia* Vanderbyl Pore Fungus
SPECIES *borneensis, peninsularis, subincarnata*
GENUS *Xerotus* Xerote Pore Fungus
SPECIES *afer, atrovirens, fuliginosus, glaucophyllus*

FAMILY 02 *Albatrellaceae* White Flesh Trellis Pore Fungus
GENUS *Albatrellus* White Flesh Trellis Pore Fungus
SPECIES *avellaneus, borneensis, cantharellus, cochleariformis,* ***confluens***(Edible), *congoensis, cristatus, ellisii, flettii,* ***ovinus***(Edible), *peckianus, pilosus, skamanius, syringae, tianschanicus, yasudae*
VARIETIES ***confluens artemidorus, confluens aurantius, ovinus carinthiacus, ovinus fragilis***
GENUS *Jahnoporus* Jahn White Flesh Trellis Pore Fungus
SPECIES *hirtus*
GENUS *Polyporoletus* Let White Flesh Trellis Pore Fungus
SPECIES *sublividus*
GENUS *Scutiger* Scute White Flesh Trellis Pore Fungus

SPECIES *oregonensis*
VARIETIES *oregonensis pes-caprae* (Edible Goatfoot Oregon Nongill Mushroom)

FAMILY 03 *Boreostereaceae* Bore Stereo Pore Fungus
GENUS *Boreostereum* Bore Stereo Pore Fungus
SPECIES *radiatum, vibrans*
GENUS *Columnocystis* Column Cyst Bore Stereo Pore Fungus
SPECIES *africana*
GENUS *Mycothele* Thele Bore Stereo Pore Fungus
SPECIES *disciformis*
GENUS *Veluticeps* Velvet Bore Stereo Pore Fungus
SPECIES abietina, *ambigua, berkeleyi, fimbriata, fusispora, heimii, pimeriensis, pini, tabacina*

FAMILY 04 *Cyphellaceae* Cypher Pore Fungus
GENUS *Cyphella* Cypher Pore Fungus
SPECIES *digitalis, ferruginea*
GENUS *Adustomyces* Dusty Cypher Pore Fungus
SPECIES *lusitanicus*
GENUS *Asterocyphella* Star Cypher Pore Fungus
SPECIES *floccosa, friesii, theiacantha*
GENUS *Globulicium* Globule Cypher Pore Fungus
SPECIES *hiemale*
GENUS *Gloeocorticium* Glow Cortic Cypher Pore Fungus
SPECIES *cinerascens*
GENUS *Halocyphina* Halogen Cypher Pore Fungus
SPECIES *villosa*
GENUS *Hyphoradulum* Hypha Ray Cypher Pore Fungus
SPECIES *conspicuum*
GENUS *Incrustocalyptella* Incrusted Calypter Cypher Pore Fungus
SPECIES *columbiana*
GENUS *Limnoperdon* Marsh Perdon Cypher Pore Fungus
SPECIES *incarnatum*
GENUS *Phaeoporotheleum* Dusky Thele Cypher Pore Fungus
SPECIES *bombycinum, revivescens*
GENUS *Pseudolagarobasidium* Lagar Cypher Pore Fungus
SPECIES *calcareum, leguminicola*
GENUS *Radulodon* Lesser Radial Cypher Pore Fungus
SPECIES *americanus, aneirinus, casearius, cirrhatinus, copelandii, erikssonii, licentii, subvinosus, venustus*
GENUS *Radulomyces* Radial Cypher Pore Fungus
SPECIES *confluens, fuscus, kamaaina, molaris, notabilis, poni, probatus, rickii*
VARIETIES *molaris stenodon*
GENUS *Radulum* (Synonym for *Radulodon*)
GENUS *Rhodoarrhenia* Red Arrhen Cypher Pore Fungus
SPECIES *pezizoidea*
GENUS *Sarcodontia* Flesh Tooth Cypher Pore Fungus
SPECIES *crocea, setosa*
GENUS *Seticyphella* Seta Cypher Pore Fungus
SPECIES *niveola, punctoidea, tenuispora*
GENUS *Sphaerobasidioscypha* Sphere Scyph Cypher Pore Fungus
SPECIES *citrispora, oberwinkleri*
GENUS *Thujacorticium* Tree Of Life Bark Cypher Pore Fungus
SPECIES *mirabile*
GENUS *Woldmaria* Woldmar Cypher Pore Fungus
SPECIES *filicina*

FAMILY 05 *Cystostereaceae* Cyst Stereo Pore Fungus
GENUS *Cystostereum* Cyst Stereo Pore Fungus
SPECIES *australe, heteromorphum, kenyense, murrayi, pini-canadense, stratosum*
GENUS *Cericium* Ceric Cyst Stereo Pore Fungus
SPECIES *luteoincrustatum*
GENUS *Crustomyces* Crust Cyst Stereo Pore Fungus
SPECIES *expallens, indecorus, subabruptus*
GENUS *Cystidiodontia* Tooth Cyst Stereo Pore Fungus
SPECIES *laminifera*
GENUS *Dentocorticium* Tooth Bark Cyst Stereo Pore Fungus
SPECIES *brasiliense, irregulare, sulphurellum, ussuricum*
GENUS *Parvobasidium* Small Cyst Stereo Pore Fungus
SPECIES *cretatum, lianaicola*
GENUS *Physodontia* Bladder Tooth Cyst Stereo Pore Fungus

SPECIES *lundellii*
GENUS *Pteridomyces* Wing Cyst Stereo Pore Fungus
SPECIES *bisporus, lacteus, roseolus*

FAMILY 06 *Epitheliaceae* Epiphyte Nipple Pore Fungus
GENUS *Epithele* Epiphyte Nipple Pore Fungus
SPECIES *citrispora, efibulata, fasciculata, interrupta, macarangae, malaiensis, nikau, ovalispora, typhae*
GENUS *Epithelopsis* Lesser Epiphyte Nipple Pore Fungus
SPECIES *bosei, fulva*

FAMILY 07 *Fistulinaceae* Ox Tongue Pore Fungus
GENUS *Fistulina* Ox Tongue Pore Fungus
SPECIES *africana, antarctica, brasiliensis, guzmanii,* ***hepatica***(Edible), *pallida, spiculifera*
VARIETIES *hepatica buglossum, hepatica bulliardii*
GENUS *Confistulina* Least Ox Tongue Pore Fungus
SPECIES *hepatica*
GENUS *Pseudofistulina* Lesser Ox Tongue Pore Fungus
SPECIES *radicata, sinensis*

FAMILY 08 *Fomitopsidaceae* Foam Pore Fungus
GENUS *Fomitopsis* Foam Pore Fungus
SPECIES *anhuiensis, arbitraria, avellanea, cajanderi, castanea, concava, cupreorosea, cytisina, dochmia, durescens, epileucina, euosma, feei, hainaniana, iberica, lignea, lilacinogilva, maire, minutispora, palustris, palustris, pinicola, pseudopetchii, quadrans, rhodophaea, rosea, rufolaccata, sanmingensis, sensitiva, spraguei, supina, widdringtoniae, zuluensis*
VARIETIES *pinicola fulvus, pinicola marginatus, pinicola semiovatus, pinicola ungulatus*
GENUS *Amylocystis* Amyl Cyst Foam Pore Fungus
SPECIES *lapponicus*
GENUS *Anomoporia* Anomalous Foam Pore Fungus
SPECIES *albolutescens, ambigua, bombycina, dumontii, flavissima, irpicoides, kamtschatica, myceliosa, vesiculosa*
GENUS *Auriporia* Ear Foam Pore Fungus
SPECIES *aurea, aurulenta, pileata*
GENUS *Buglossoporus* Bugloss Foam Pore Fungus
SPECIES *americanus, brunneiflavus, flavus, heritierae, magnus, malesianus, marmoratus, matangensis, pulvinus, rufescens*
VARIETIES *pulvinus quercinus*
GENUS *Daedalea* Daedale Foam Pore Fungus
SPECIES *africana, ambigua, dickinsii, flavida, imponens, langkawiensis, microsticta, moesta, quercina, sulcata*
GENUS *Daedaleopsis* Lesser Daedale Foam Pore Fungus
SPECIES *confragosa, nipponica, papyraceoresupinata, pergamenea*
VARIETIES *confragosa angustata*
GENUS *Donkioporia* Donk Foam Pore Fungus
SPECIES *expansa*
GENUS *Laricifomes* Larch Foam Pore Fungus
SPECIES *officinalis*
VARIETIES *officinalis agaricum, officinalis purgans*
GENUS *Osteina* Ostein Foam Pore Fungus
SPECIES *obducta*
GENUS *Parmastomyces* Parmast Foam Pore Fungus
SPECIES *corticola, deceptivus, glutinosus, kravtzevianus, mollissimus, taxi, umbrinus*
GENUS *Piptoporus* Fallen Foam Pore Fungus
SPECIES *betulinus, soloniensis*
VARIETIES *betulinus suberosus*
GENUS *Postia* Post Foam Pore Fungus
SPECIES *angulipora, balsamea, caesia, ceriflua, dissecta, floriformis, folliculocystidiata, fragilis, grata, guttulata, hibernica, inocybe, johnstonii, lateritia, leucomallella, lowei, luteocaesia, mappa, pelliculosa, perdelicata, placenta, ptychogaster, rancida, rennyi, septentrionalis, sericeomollis, simanii, stiptica, tephroleuca, undosa, venata, wakefieldiae*
VARIETIES *caesia candidus, placenta incarnatus, stiptica albidus*
GENUS *Spelaeomyces* Cave Foam Pore Fungus
SPECIES *heydenii*
GENUS *Wolfiporia* Wolf Foam Pore Fungus
SPECIES *cartilaginea, curvispora, dilatohypha, extensa, sulphurea*
VARIETIES ***extensa cocos***(Medicinal Herb)
GENUS *Xylostroma* Wood Tissue Foam Pore Fungus
SPECIES *giganteum*

FAMILY 09 *Ganodermataceae* Dye Pore Fungus

GENUS *Ganoderma* Dye Pore Fungus

SPECIES *adspersum, ahmadii, albomarginatum, applanatum, asperulatum, atrum, austrofujianense, bawanglingense, brownii, bruggemanii, calidophilum, cantharelloideum, capense, carnosum, cehengense, cervinum, chalceum, chenhaiense, colossus, corrugatum, crebrostriatum, cupreolaccatum, curtisii, daiqingshanense, dejongii, densizonatum, diaoluoshanense, donkii, dorsale, dussii, fassioides, fici, flexipes, formosanum, guinanense, guizhouense, hainanense, hoploides, impolitum, jianfenglingense, kosteri, kunmingense, lamaoense, leucocreas, leytense, limushanense, lobatoideum, lobatum,* ***lucidum***(Medicinal Herb), *luteicinctum, luteomarginatum, magniporum, manoutchehrii, mediosinense, meijangense, meredithiae, microsporum, miniatocinctum, mirabile, mirivelutinum, multipileum, multiplicatum, neojaponicum, nitidum, ochrolaccatum, oerstedii, orbiforme, oregonense, oroleucum, ostracodes, parviungulatum, pfeifferi, philippii, platense, puglisii, ramosissimum, ravenelii, resinaceum, reticulatosporum, rothwellii, rotundatum, sanmingense, sculpturatum, sessiliforme, shandongense, shangsiense, sichuanense, simaoense, sinense, stipitatum, stratoideum, sublucidum, subresinosum, subtornatum, tenue, testaceum, theaecolum, tibetanum, tornatum, trengganuense, triangulum, tropicum, trulla, trulliforme, ungulatum, valesiacum, vanheurnii, weberianum, williamsianum, xingyiense, zhenningense, zonatum*

VARIETIES *applanatum lipsiensis, lucidum annulatum, lucidum castaneus, lucidum crustatus, lucidum dimidiatus, lucidum flabelliformis, lucidum hemisphaericum, lucidum laccatus, lucidum rubellum, lucidum rugosus, lucidum verniceus, lucidum vernicosus, multiplicatum vitalii*

GENUS *Amauroderma* Dark Dye Pore Fungus

SPECIES *amoiense, auriscalpium, austrosinense, bataanense, boleticeum, bululoi, calcigenus, congregatum, conicum, conjunctum, dayaoshanense, ealaense, exile, fujianense, fuscoporia, guangxiense, infundibuliforme, insulare, jiangxiense, kwiluense, leptopus, leucosporum, longgangense, macrosporum, malesianum, parasiticum, perplexum, praetervisum, preussii, pudens, renidens, rude, rugosum, salisburiense, schomburgkii, secedens, solomonense, sprucei, subrugosum, unilaterum, wuzhishanense*

GENUS *Haddowia* Haddow Dye Pore Fungus

SPECIES *aëtii, longipes, neurospora*

GENUS *Humphreya* Humphrey Dye Pore Fungus

SPECIES *coffeata, eminii, endertii, lloydii*

GENUS *Lazulinospora* Lazulin Spore Dye Pore Fungus

SPECIES *wakefieldiae*

GENUS *Thermophymatospora* Thermal Nodule Dye Pore Fungus

SPECIES *fibuligera*

GENUS *Trachyderma* Rough Dermal Dye Pore Fungus

SPECIES *tsunodae*

FAMILY 10 *Grammotheleaceae* Grass Nipple Pore Fungus

GENUS *Grammothele* Grass Nipple Pore Fungus

SPECIES *bambusicola, delicatula, fuligo, hondurensis, lineata, ochraceus, pulchella, subargentea*

GENUS *Hymenogramme* Hymen Grass Nipple Pore Fungus

SPECIES *javensis*

GENUS *Porogramme* Lesser Grass Nipple Pore Fungus

SPECIES *graphica, lateritia, richeriae*

GENUS *Theleporus* Least Grass Nipple Pore Fungus

SPECIES *ajovalliensis*

FAMILY 11 *Hapalopilaceae* Downy Pore Fungus

GENUS *Hapalopilus* Downy Pore Fungus

SPECIES *africanus, albocitrinus, croceus, mutans, nidulans, phlebiiformis, placodes, rubescens, rutilans, salmonicolor, sibiricus*

GENUS *Aurantiporus* Orange Downy Pore Fungus

SPECIES *albidus, alborubescens, fissilis, pilotae*

GENUS *Bjerkandera* Bjerkander Downy Pore Fungus

SPECIES *adusta, fumosa*

VARIETIES *adusta carpineus, adusta concentricus, adusta crispus, adusta fuscoporus, adusta isabellinus, fumosa imberbis*

GENUS *Ceriporia* Wax Downy Pore Fungus

SPECIES *alachuana, camaresiana, excelsa, ferruginicincta, leptoderma, mellita, metamorphosa, otakou, purpurea, retamoana, reticulata, spissa, tarda, totara, viridans, xylostromatoides*

VARIETIES *spissa juglandinus*

GENUS *Ceriporiopsis* Lesser Wax Downy Pore Fungus

SPECIES *albonigrescens, balaenae, carnegiae, cerussata, consobrina, coprosmae, cremea, flavilutea, gilvescens, jelicii, lowei, mucida, obscura, portcrosensis, resinascens, rivulosa, subrufa, subsphaerospora, subvermispora, umbrinescens*

VARIETIES *resinascens pseudogilvescens*

GENUS *Climacocystis* Climax Cyst Downy Pore Fungus

SPECIES *borealis*

GENUS *Irpiciporus* Irpic Downy Pore Fungus

SPECIES *japonicus, tanakae*

GENUS *Ischnoderma* Ischnon Dermal Downy Pore Fungus

SPECIES *albotextum, benzoinum, brasiliense, porphyrites, resinosum, rosulatum, solomonense*
VARIETIES *, benzoinum rubiginosus, benzoinum velutinus, resinosum fuliginosus*
GENUS *Leptoporus* Thin Downy Pore Fungus
SPECIES *ellipsosporus, mollis*
VARIETIES *mollis substrigosus*
GENUS *Spongipellis* Sponge Pellet Downy Pore Fungus
SPECIES *caseosus, chubutensis, delectans, litschaueri, malicola, pachyodon, spumeus, unicolor*
VARIETIES *spumeus mongolicus*

FAMILY 12 *Hyphodermataceae* Hypha Dermal Pore Fungus
GENUS *Hyphoderma* Hypha Dermal Pore Fungus
SPECIES *acutocystis, acystidiatum, africanum, allantosporum, amoenum, anasaziense, anthracophilum, argillaceum, assimile, ayresii, baculorubrense, brunneocontextum, budingtonii, calcitrapa, capitatum, capitulatum, cinnamomeum, clavatum, clavigerum, comptum, cremeoalbum, cremeum, cristulatum, cryptocallimon, cylindrocystidiatum, definitum, densum, densustextum, deserticola, deviatum, echinocystis, effractum, etruriae, flagellatum, galactinum, gemmeum, gigasporum, granuliferum, guttuliferum, hjortstamii, iguazuense, incrustatissimum, inusitatum, involutum, lapponicum, leoninum, leptaleum, litschaueri, luridum, macedonicum, magnargillaceum, malenconii, medioburiense, microcystidium, moniliforme, multicystidiatum, multicystidium, mutatum, naiophila, neopuberum, obtusiforme, obtusum, occidentale, orphanellum, pallidum, pilisetum, pilosum, populneum, praetermissum, probatum, puberum, rimosum, rimulosum, roseocremeum, rubropallens, rubropunctatum, rude, sabinicum, scaevolae, setigerum, sibiricum, sikkimium, sphaeropedunculatum, sporulum, subclavigerum, subpraetermissum, subsetigerum, subsphaerosporum, tenue, transiens, tsugae, tuberculare, typhicola, utriculosum, variolosum, vulcanense*
GENUS *Aegerita* Aegerite Hypha Dermal Pore Fungus
SPECIES *candida, torulosa, webberi*
GENUS *Amaurohydnum* Dark Truffle Hypha Dermal Pore Fungus
SPECIES *flavidum*
GENUS *Amauromyces* Dark Hypha Dermal Pore Fungus
SPECIES *farinaceus, pallidus*
GENUS *Atheloderma* Athel Dermal Hypha Dermal Pore Fungus
SPECIES *mirabile, orientale*
GENUS *Brevicellicium* Short Hypha Dermal Pore Fungus
SPECIES *allantosporum, exile, mellinum, molle, olivascens, permodicum, uncinatum, viridulum, vulcanense*
GENUS *Bulbillomyces* Bulb Hypha Dermal Pore Fungus
SPECIES *farinosus*
GENUS *Cerocorticium* Wax Cortic Hypha Dermal Pore Fungus
SPECIES *calongei, canariense, molle, pseudomucidum, sulfureoisabellinum*
GENUS *Chrysoderma* Golden Hypha Dermal Pore Fungus
SPECIES *alboluteum*
GENUS *Conohypha* Cone Hypha Dermal Pore Fungus
SPECIES *albocremea, terricola*
GENUS *Coronicium* Crown Hypha Dermal Pore Fungus
SPECIES *alboglaucum, gemmiferum, proximum, thymicola*
GENUS *Cyanodontia* Cyan Tooth Hypha Dermal Pore Fungus
SPECIES *spathulata*
GENUS *Elaphocephala* Stag Head Hypha Dermal Pore Fungus
SPECIES *iocularis*
GENUS *Galzinia* Galzin Hypha Dermal Pore Fungus
SPECIES *cymosa, cystidiata, ellipsospora, forcipata, geminispora, incrustans, longibasidia, pedicellata, vesana*
GENUS *Hyphodontiastra* Tooth Star Hypha Dermal Pore Fungus
SPECIES *virgicola*
GENUS *Hyphodontiella* Tooth Hypha Dermal Pore Fungus
SPECIES *hauerslevii, multiseptata*
GENUS *Hypochnicium* Pear Hypha Dermal Pore Fungus
SPECIES *albostramineum, analogum, bombycinum, caucasicum, contiguum, cremicolor, cystidiatum, eichleri, erikssonii, geogenium, globosum, gomezii, longicystidiosum, lundellii, multiforme, pleurobasidiatum, punctulatum, sphaerosporum, stratosum, subrigescens, versatum, wakefieldiae*
VARIETIES *contiguum contiguum, contiguum laxum*
GENUS *Intextomyces* Intext Hypha Dermal Pore Fungus
SPECIES *aureus, contiguus, cystidiatus, umbrinus*
GENUS *Metulodontia* Met Tooth Hypha Dermal Pore Fungus
SPECIES *nivea*
GENUS *Odontiopsis* Lesser Tooth Hypha Dermal Pore Fungus
SPECIES *ambigua, hyphodontina*
GENUS *Pirex* Pirex Hypha Dermal Pore Fungus
SPECIES *concentricus*
GENUS *Subulicium* Ulic Hypha Dermal Pore Fungus

SPECIES *lautum, minus, rallum*
GENUS *Subulicystidium* Ulic Cyst Hypha Dermal Pore Fungus
SPECIES *brachysporum, cochleum, longisporum, meridense, naviculatum, nikau, perlongisporum*

FAMILY 13 *Meripilaceae* Meristic Downy Pore Fungus
GENUS *Meripilus* Meristic Downy Pore Fungus
SPECIES *applanatus, giganteus, maculatus, sumstinei, tropicalis, villosulus*
VARIETIES *giganteus acanthoides*
GENUS *Abortiporus* Abortive Meristic Downy Pore Fungus
SPECIES *biennis, fractipes, roseus*
VARIETIES *biennis distortus, biennis mesopodus*
GENUS *Amyloporia* (Synonym for *Antrodia*)
GENUS *Antrodia* Cavity Meristic Downy Pore Fungus
SPECIES *albida, albidoides, albobrunnea, alpina, aurantia, calceus, carbonica, cinnamomea, conchata, crassa, daedaleiformis, eutelea, ferox, gossypium, heteromorpha, hippophaës, infirma, juniperina, leucaena, macra, macrospora, malicola, mellita, oleracea, pini-cubensis, plicata, porothelioides, primaeva, pseudosinuosa, pulvinascens, ramentacea, rupamii, serialis, sinuosa, sitchensis, sordida, stratosa, subalbidoides, submalicola, subramentacea, variiformis, xantha*
VARIETIES *xantha pachymeres*
GENUS *Diacanthodes* Two Spine Meristic Downy Pore Fungus
SPECIES *fluminensis, griseus, novoguineensis*
GENUS *Grifola* Grifol Meristic Downy Pore Fungus
SPECIES *armeniaca, colensoi, eos, frondosa, platypora, rosularis, sordulenta*
VARIETIES *frondosa cristatus, frondosa intybaceus*
GENUS *Henningsia* Hennings Meristic Downy Pore Fungus
SPECIES *brasiliensis*
GENUS *Heteroporus* Variable Meristic Downy Pore Fungus
SPECIES *kinabaluensis, odoratus, sublacunosus, zonatus*
GENUS *Hydnopolyporus* Truffle Meristic Downy Pore Fungus
SPECIES *fimbriatus, hartmannii, palmatus*
GENUS *Physisporinus* Bladder Spore Meristic Downy Pore Fungus
SPECIES *vitreus*
GENUS *Rigidoporus* Rigid Meristic Downy Pore Fungus
SPECIES *adnatus, albiporus, amazonicus, aureofulvus, biokoensis, camschadalicus, crocatus, defibulatus, dextrinoideus, eminens, erectus, furcatus, hainanicus, hypobrunneides, incarnatus, incurvus, laetus, lineatus, longicystidius, microporus, moeszii, ochraceicinctus, parvulus, patellarius, pendulus, rigidus, sanguinolentus, subpileatus, sulphureus, trametoides, ulmarius, umbonatipes, undatus, vinaceus, vinctus*
VARIETIES *sanguinolentus carmichaelii*

FAMILY 14 *Phanerochaetaceae* Bristle Pore Fungus
GENUS *Phanerochaete* Bristle Pore Fungus
SPECIES *aculeata, affinis, albida, allantospora, andreae, arenata, arizonica, aurata, australis, avellanea, binucleospordida, brunnea, bubalina, burtii, cacaina, calotricha, cana, caucasica, chrysorhizon, chrysosporium, commixtoides, crescentispora, emplastra, exilis, filamentosa, galactites, globosa, hiulca, incrustans, intertexta, irpicoides, jose-ferreirae, laevis, leptoderma, luteoaurantiaca, magnoliae, martelliana, mauiensis, monomitica, parmastoi, percitrina, pseudomagnoliae, radicata, robusta, sacchari, salmoneolutea, salmonicolor, sanguinea, singularis, sordida, subceracea, subglobosa, subquercina, taiwaniana, tamariciphila, tuberculascens, tuberculata, velutina, viticola, xerophila*
VARIETIES *velutina cremicolor*
GENUS *Candelabrochaete* Candelabra Bristle Pore Fungus
SPECIES *adnata, africana, cirrata, dispar, eruciformis, langloisii, magnahypha, septocystidia, simulans, verruculosa*
GENUS *Ceraceomyces* Cerace Bristle Pore Fungus
SPECIES *americanus, borealis, cerebrosus, corymbatus, cremeo-ochraceus, crispatus, cystidiatus, microsporus, oligodontus, reidii, serpens, sublaevis, sulphurinus, tessulatus, variicolor, violascens*
GENUS *Erythricium* Red Bristle Pore Fungus
SPECIES *chaparralum, hypnophilum, laetum*
GENUS *Gyrophanopsis* Gyrophane Bristle Pore Fungus
SPECIES *polonensis, zealandica*
GENUS *Herpothallon* (Synonym for *Phanerochaete*)
GENUS *Hyphodermella* Hypha Dermal Bristle Pore Fungus
SPECIES *corrugata, maunakeaensis*
GENUS *Lopharia* Crest Bristle Pore Fungus
SPECIES *cinerascens, crassa, cystidiosa, papyrina, spadicea*
GENUS *Odonticium* Tooth Bristle Pore Fungus
SPECIES *australe, helgae, laxum, monfraguense, romellii*
GENUS *Phlebiopsis* Phlebe Bristle Pore Fungus
SPECIES *afibulata, darjeelingensis, gigantea, himalayensis, ravenelii, roumeguerei*
GENUS *Porostereum* Stereo Bristle Pore Fungus

SPECIES *medicum, novae-granatum, umbrinoalutaceum*
GENUS *Terana* Teran Bristle Pore Fungus
SPECIES *caerulea*

FAMILY 15 *Podoscyphaceae* Pod Scyph Pore Fungus
GENUS *Podoscypha* Pod Scyph Pore Fungus
SPECIES *aculeata, bolleana, brasiliensis, bubalina, caespitosa, corneri, cristata, elegans, fulvonitens, gillesii, glabrescens, involuta, mellisii, moelleri, moselei, multizonata, nitidula, nuda, ovalispora, parvula, petalodes, philippinensis, ravenelii, replicata, semiresupinata, thozetii, tomentipes, venustula, vespillonea, viridans*
VARIETIES *petalodes petalodes, petalodes rosulata*
GENUS *Aquascypha* Aquatic Pod Scyph Pore Fungus
SPECIES *hydrophora*
GENUS *Coralloderma* Coral Dermal Pod Scyph Pore Fungus
SPECIES *acroleucum, guzmanii*
GENUS *Cotylidia* Coty Pod Scyph Pore Fungus
SPECIES *aurantiaca, carpatica, diaphana, guttulata, harmandii, komabensis, marsicana, muscigena, pannosa, undulata*
VARIETIES *aurantiaca alba*
GENUS *Cymatoderma* Cyme Dermal Pod Scyph Pore Fungus
SPECIES *africanum, blumei, caperatum, dendriticum, elegans, fuscum, hainanense, infundibuliforme, pallens, plicatum, sclerotioides, venezuelae*
GENUS *Cyphellostereum* Cypher Stereo Pod Scyph Pore Fungus
SPECIES *laeve, muscicola, pusiolum, rivulorum*
GENUS *Granulobasidium* Granular Pod Scyph Pore Fungus
SPECIES *vellereum*
GENUS *Inflatostereum* Inflated Stereo Pod Scyph Pore Fungus
SPECIES *glabrum, radicatum*
GENUS *Pseudolasiobolus* Woolly Bowl Pod Scyph Pore Fungus
SPECIES *minutissimus*
GENUS *Stereopsis* Stereo Pod Scyph Pore Fungus
SPECIES *burtiana, cartilaginea, hiscens, humphreyi, malaiensis, mussooriensis, nigripes, radicans, raphiae, sparassoides, straminea, vitellina*

FAMILY 16 *Sistotremataceae* Sistine Tremor Pore Fungus
GENUS *Sistotrema* Sistine Tremor Pore Fungus
SPECIES *adnatum, alboluteum, albopallescens, athelioides, autumnale, biggsiae, binucleosporum, brinkmannii, citriforme, confluens, coroniferum, coronilla, dennisii, diademiferum, efibulatum, eximum, farinaceum, hamatum, heteronemum, hispanicum, hypogaeum, intermedium, muscicola, oblongisporum, octosporum, otagense, pistilliferum, populicola, porulosum, pteriphilum, pyrisporum, pyrosporum, raduloides, resinicystidium, sernanderi, suballantosporum, subtrigonospermum*
GENUS *Cristelloporia* Crested Sistine Tremor Pore Fungus
SPECIES *asperispora*
GENUS *Ingoldiella* Ingold Sistine Tremor Pore Fungus
SPECIES *fibulata, hamata*
GENUS *Minimedusa* Mini Medusa Sistine Tremor Pore Fungus
SPECIES *polyspora*
GENUS *Osteomorpha* Osteomorph Sistine Tremor Pore Fungus
SPECIES *fragilis*
GENUS *Paullicorticium* Paul Cortic Sistine Tremor Pore Fungus
SPECIES *delicatissimum, pearsonii*
GENUS *Repetobasidiellum* Lesser Repet Sistine Tremor Pore Fungus
SPECIES *fusisporum*
GENUS *Repetobasidium* Repet Sistine Tremor Pore Fungus
SPECIES *americanum, conicum, erikssonii, hastatum, intermedium, macrosporum, mirificum, vestitum, vile*
GENUS *Sistotremastrum* Star Sistine Tremor Pore Fungus
SPECIES *lateclavigerum, niveocremeum, suecicum*
GENUS *Sistotremella* Lesser Sistine Tremor Pore Fungus
SPECIES *cystidiolophora, hauerslevii, paullicorticioides, perpusilla*
GENUS *Sphaerobasidium* Sphere Sistine Tremor Pore Fungus
SPECIES *bisporum, subinvisibile*
GENUS *Trechispora* Trech Sistine Tremor Pore Fungus
SPECIES *alnicola, amianthina, bispora, byssinella, canariensis, candidissima, caucasica, clancularis, cohaerens, dimitica, farinacea, fastidiosa, granulifera, hymenocystis, incisa, invisitata, kavinioides, laevis, microspora, minima, mollusca, mutabilis, nivea, polygonospora, praefocata, silvae-ryae, sphaerocystis, sphaerospora, stellulata, stevensonii, tenuicula*

FAMILY 17 *Tubulicrinaceae* Tubule Lily Pore Fungus
GENUS *Tubulicrinis* Tubule Lily Pore Fungus

SPECIES *accedens, angustus, borealis, chaetophora, cinctoides, confusus, corneri, corticioides, evenii, globisporus, gloeocystidiatus, gracillimus, hamatus, hirtellus, incrassatus, inornatus, medius, meruensis, ovalisporus, propinquus, pseudoborealis, regificus, sceptrifer, sororius, strangulatus, subulatus, thermometrus, umbraculus*

GENUS *Leifia* — Leif Tubule Lily Pore Fungus
SPECIES *flabelliradiata*
GENUS *Litschauerella* — Litschauer Tubule Lily Pore Fungus
SPECIES *abietis, clematidis, gladiola, hastata*
GENUS *Tubulicium* — Tubule Lily Pore Fungus
SPECIES *dussii, vermiculare, vermiferum*

FAMILY 18 *Xenasmataceae* — Strange Pore Fungus
GENUS *Xenasma* — Strange Pore Fungus
SPECIES *parvisporum, praeteritum, pruinosum, pulverulentum, rimicola, tulasnelloideum*
GENUS *Aphanobasidium* — Aphane Strange Pore Fungus
SPECIES *aurobisporum, lloydii, sphaerosporum*
GENUS *Cunninghammyces* — Cunningham Strange Pore Fungus
SPECIES *fusisporus, umbonatus*
GENUS *Lepidomyces* — Thin Strange Pore Fungus
SPECIES *subcalceus*
GENUS *Phlebiella* — Phlebe Strange Pore Fungus
SPECIES *albida, allantospora, aurora, bourdotii, christiansenii, fibrillosa, filicina, gaspesica, grisella, paludicola, pseudotsugae*
GENUS *Xenasmatella* — Lesser Strange Pore Fungus
SPECIES *ardosiaca, insperata, nasti, romellii, vaga*
GENUS *Xenosperma* — Least Strange Pore Fungus
SPECIES *hexagonosporum, ludibundum, murrillii, pravum*

ORDER 5 *Thelephorales* Crust

FAMILY 1 *Thelephoraceae* — Groundwort Crust
GENUS *Thelephora* — Groundwort Crust
SPECIES *albidobrunnea, alta, anthocephala, arbuscula, atra, atrocitrina, aurantiotincta, borneensis, brunneoviolacea, caryophyllea, cerberea, cervicornis, cervina, congesta, crassitexta, cuticularis, cylindrica, dactylites, dentosa, erebia, fragilis, fucoides, ganbajun, gelidioides, griseozonata, intybacea, investiens, japonica, lutosa, magnifica, palmata, paraguayensis, pendens, penicillata, phyllophoroides, pseudoterrestris, ramarioides, robusta, scissilis, tenuis, terrestris, undulata, vialis*
GENUS *Amaurodon* — Dark Groundwort Crust
SPECIES *cyaneus, mustialaensis, viridis*
GENUS *Haplotrichum* — Single Hairy Groundwort Crust
SPECIES *chilense, curtisii, medium, rubiginosum, simile*
GENUS *Hydnodon* — Truffle Groundwort Crust
SPECIES *thelephorum*
GENUS *Lenzitopsis* — Lenzite Groundwort Crust
SPECIES *oxycedri*
GENUS *Phlyctibasidium* — Phlyct Groundwort Crust
SPECIES *polyporoideum*
GENUS *Polyozellus* — Poly Zeller Groundwort Crust
SPECIES ***multiplex***(Edible)
GENUS *Pseudotomentella* — Lesser Woolly Groundwort Crust
SPECIES *flavovirens, griseopergamacea, humicola, mucidula, nigra, tristis, vepallidospora*
GENUS *Skepperia* — Skepper Groundwort Crust
SPECIES *platensis*
GENUS *Tomentella* — Woolly Groundwort Crust
SPECIES *angulospora, asperula, atroarenicolor, atrocyanea, atrovirens, aurantiaca, badia, brevispina, brunneorufa, bryophila, calcicola, carbonaria, cinerascens, cinereoumbrina, clavigera, coerulea, crinalis, donkii, duemmeri, ellisii, epigaea, ferruginea, ferruginella, fragilis, fraseri, fungicola, fuscocinerea, fuscoferruginosa, galzinii, griseoumbrina, griseoviolacea, himalayana, indica, italica, juncicola, kentuckiensis, kootenaiensis, lapida, lateritia, lilacinogrisea, litschaueri, microspora, molybdaea, muricata, nitellina, olivascens, pellicularioides, pilatii, pilosa, pirolae, puberula, punicea, radiosa, scobinella, stuposa, subalpina, subcinerascens, subclavigera, subcorticioides, sublilacina, subtestacea, subvinosa, terrestris, testaceogilva, umbrinospora, variecolor, vesiculosa, viridescens, viridula*
GENUS *Tomentellago* — Mock Woolly Groundwort Crust
SPECIES *aeruginascens*
GENUS *Tomentellastrum* — Woolly Star Groundwort Crust
SPECIES *caesiocinereum*
GENUS *Tomentellina* — Least Woolly Groundwort Crust
SPECIES *fibrosa*

FAMILY 2 *Corticiaceae* Corticle Wood Crust
GENUS *Corticium* Corticle Wood Crust
SPECIES *appalachiense, boreoroseum, botryohypochnoideum, canfieldii, confine, conigenum, cremeoalbidum, durangense, efibulatum, floridense, gloeosporum, gramineum, griseoeffusum, invisum, koleroga, lilacinoroseum, lombardiae, meridioroseum, mississippiense, pini, quercicola, radiosum, roseocarneum, salmonicolor, utriculicum*
GENUS *Acantholichen* Spiny Lichen Wood Crust
SPECIES *pannarioides*
GENUS *Ambivina* Ambivalent Corticle Wood Crust
SPECIES *filobasidia*
GENUS *Amylobasidium* Amyl Corticle Wood Crust
SPECIES *tsugae*
GENUS *Corticirama* Branch Corticle Wood Crust
SPECIES *petrakii*
GENUS *Cytidia* Cytid Corticle Wood Crust
SPECIES *cristallifera, patelliformis, pezizoides, salicina, sarcoides, stereoides*
GENUS *Dendrophysellum* Wood Bladder Corticle Wood Crust
SPECIES *amurense*
GENUS *Dendrothele* Wood Nipple Corticle Wood Crust
SPECIES *acerina, alba, alliacea, amygdalispora, asterospora, biapiculata, bispora, candida, capitulata, commixta, corniculata, cyatheae, dryina, duthieae, globulispora, griseocana, incrustans, itihummensis, lepra, macrodens, macrospora, mangiferae, mexicana, microspora, moquiniarum, nivosa, pachysterigmata, pulvinata, sasae, seriata, strumosa, subfusispora, tetracornis*
GENUS *Laeticorticium* Bright Corticle Wood Crust
SPECIES *minnsiae, odontoides, piceinum, pinsapineum, polygonioides, pulverulentum, roseum, simplicibasidium, violaceum*
GENUS *Laetisaria* Bright Sar Corticle Wood Crust
SPECIES *agaves, arvalis, fuciformis*
GENUS *Licrostroma* Licron Stroma Corticle Wood Crust
SPECIES *subgiganteum*
GENUS *Limonomyces* Lemon Corticle Wood Crust
SPECIES *culmigenus, roseipellis*
GENUS *Lindtneria* Lindtner Corticle Wood Crust
SPECIES *alutacea, baboquivariensis, chordulata, flava, hydnoidea, lowei, panphyliensis, rugospora, thujatsugina*
GENUS *Matula* Mat Corticle Wood Crust
SPECIES *rompelii*
GENUS *Melzerodontia* Melzer Corticle Wood Crust
SPECIES *aculeata, udamentiens*
GENUS *Moniliopsis* Monad Corticle Wood Crust
SPECIES *anomala, foliicola, klebahnii*
GENUS *Mycolindtneria* Lesser Lindtner Corticle Wood Crust
SPECIES *flava, leucobryophila, pellicularis, pterospora, trachyspora*
GENUS *Nothocorticium* Nothing Corticle Wood Crust
SPECIES *patagonicum*
GENUS *Papyrodiscus* Papyrus Disc Corticle Wood Crust
SPECIES *ferrugineus*
GENUS *Punctularia* Dotted Corticle Wood Crust
SPECIES *atropurpurascens, strigosozonata*
GENUS *Rhizoctonia* Rhizid Octal Corticle Wood Crust
SPECIES *alpina, anaticula, anomala, callae, carotae, dichotoma, endophytica, ferruginea, floccosa, fragariae, fraxini, fumigata, globularis, lanuginosa, leguminicola, menthae, microsclerotia, mucoroides, muneratii, oryzae-sativae, pini-insignis, quercus, ramicola, repens, rubi, rubiginosa, silvestris, stahlii, tuliparum, versicolor*
VARIETIES *endophytica endophytica, endophytica filicata*
GENUS *Ripexicium* Ripexic Corticle Wood Crust
SPECIES *spinuliferum*
GENUS *Vuilleminia* Vuillemin Corticle Wood Crust
SPECIES *alni, comedens, corticola, coryli, cystidiata, maculata, megalospora, obducens, pseudocystidiata, subglobispora*

FAMILY 3 *Meruliaceae* Wood Crust
GENUS *Merulius* Wood Crust
SPECIES *aureus, tremellosus*
GENUS *Byssomerulius* Silky Wood Crust
SPECIES *pirottae*
GENUS *Ceraceohydnum* Cerace Truffle Wood Crust
SPECIES *brunneum*

GENUS *Chondrostereum* Cartilage Stereo Wood Crust
SPECIES *coprosmae, himalaicum, purpureum, vesiculosum*
GENUS *Climacodon* Climactic Wood Crust
SPECIES *annamensis, chlamydocystis, dubitativus, pulcherrimus, roseomaculatum, septentrionalis*
GENUS *Columnodontia* Column Wood Crust
SPECIES *resupinata*
GENUS *Crustoderma* Dermal Wood Crust
SPECIES *borbonicum, carolinense, corneum, dryinum, flavescens, gigacystidium, longicystidiatum, marianum, nakasoneae, opuntiae, resinosum, testatum*
GENUS *Cylindrobasidium* Cylindric Wood Crust
SPECIES *corrugum, eucalypti, evolvens, laeve, parasiticum, torrendii*
GENUS *Dacryobolus* Dacry Bowl Wood Crust
SPECIES *costratus, incarnatus, karstenii, phalloides, sudans*
GENUS *Gloeoporus* Glow Pore Wood Crust
SPECIES *carrii, chlorinus, citrinus, cremeoalbus, croceopallens, cystidiatus, dichrous, dimiticus, dolosus, friabilis, hispidus, nigrescens, pannocinctus, papuanus, pendens, phlebophorus, purpurascens, similis, subambiguus, subochraceus, subvinaceus, sulphureus, thelephoroides, umbrinus, vitellinus*
GENUS *Gloeostereum* Glow Stereo Wood Crust
SPECIES *incarnatum*
GENUS *Himantia* Himant Wood Crust
SPECIES *candida*
GENUS *Merulicium* Least Wood Crust
SPECIES *fusisporum*
GENUS *Meruliopsis* Lesser Wood Crust
SPECIES *bella, corium, hirtella, miniata, taxicola*
VARIETIES *corium purpurascens, taxicola rufus*
GENUS *Mycoacia* Sharp Wood Crust
SPECIES *aurea, austro-occidentalis, columellifera, fuscoatra, lutea, meridionalis, nothofagi, stenodon, subconspersa, uda*
VARIETIES *nothofagi australiensis*
GENUS *Mycoaciella* Lesser Sharp Wood Crust
SPECIES *bispora, hinnulea*
GENUS *Phlebia* Phlebe Wood Crust
SPECIES *acerina, albida, ardesiaca, argentea, argentina, argentinensis, brevispora, canadensis, caspica, celtidis, coccineofulva, cretacea, deflectens, diaphana, diffissa, donkii, femsjoeensis, firma, floridensis, formosana, griseoflavescens, griseolivens, hydnoidea, icterina, introversa, lacteola, leptospermi, lilascens, livida, lividina, ludoviciana, mellea, murrillii, nantahaliensis, nitidula, ochraceofulva, pallidolivens, pellucida, phlebioides, plumbea, queletii, radiata, rosea, rufa, ryvardenii, segregata, separata, serialis, sordida, subcretacea, subochracea, subserialis, subulata, totara, tremelloidea, tristis, verruculosa, weldeniana*
GENUS *Resinicium* Resinic Wood Crust
SPECIES *bicolor, chiricahuaense, furfuraceum, granulare, luteum*
GENUS *Scopuloides* Scope Wood Crust
SPECIES *hydnoides, leprosa, rimosa*
GENUS *Skvortzovia* Skvortzov Wood Crust
SPECIES furfurella

ORDER 6 *Tremellales* Jelly

FAMILY 1 *Tremellaceae* Tremor Jelly
GENUS *Tremella* Tremor Jelly
SPECIES *aurantia, callunicola, caloplacae, cetrariicola, cladoniae, coppinsii, encephala, exigua,* ***foliacea***(Edible), *globispora, hemifoliacea, hypogymniae, indecorata, invasa, karstenii, lichenicola, lobariacearum,* ***mesenterica***(Edible), *moriformis, normandinae, obscura, penetrans, pertusariae, phaeographidis, phaeophysciae, polyporina, protoparmeliae, sarniensis, simplex, steidleri, translucens, uliginosa, versicolor*
GENUS *Bullera* Buller Tremor Jelly
SPECIES *alba, armeniaca, crocea*
GENUS *Bulleromyces* Lesser Buller Tremor Jelly
SPECIES *albus*
GENUS *Christiansenia* (Synonym for *Syzygospora*)
GENUS *Cuniculitrema* Rabbit Tremor Jelly
SPECIES *polymorpha*
GENUS *Dictyotremella* Netted Tremor Jelly
SPECIES *novoguineensis*
GENUS *Epidochium* Epiphyte Doch Tremor Jelly
SPECIES *xylariae*
GENUS *Filobasidiella* Filament Tremor Jelly

SPECIES *depauperata, lutea*
GENUS *Holtermannia* Holtermann Tremor Jelly
SPECIES *coralloides, corniformis, damicornis, pinguis, prolifera, pulchella*
GENUS *Hyaloria* Hyaline Tremor Jelly
SPECIES *europaea, pilacre*
GENUS *Kockovaella* Kockova Tremor Jelly
SPECIES *calophylli, cucphuongensis, litseae, machilophila, phaffii, sacchari, schimae, vietnamensis*
GENUS *Neotremella* Modern Tremor Jelly
SPECIES *guzmanii*
GENUS *Phlogiotus* Funnel Tremor Jelly
SPECIES ***helvelloides***(Edible)
GENUS *Phragmoxenidium* Fragmented Tremor Jelly
SPECIES *mycophilum*
GENUS *Phyllopta* Opt Tremor Jelly
SPECIES *biparasitica*
GENUS *Rhynchogastrema* Beaked Tremor Jelly
SPECIES coronatum
GENUS *Sebacina* Seba Tremor Jelly
SPECIES *calcea, dimitica, epigaea, grisea, incrustans, umbrina*
GENUS *Sirotrema* Pitted Tremor Jelly
SPECIES *parvula, pusilla, translucens*
GENUS *Sterigmatosporidium* Sterigmate Tremor Jelly
SPECIES *polymorphum*
GENUS *Syzygospora* Joined Tremor Jelly
SPECIES *bachmannii, marasmoidea, mycophaga, norvegica, pallida, physciacearum, solida, sorana, subsolida, tumefaciens*
GENUS *Tremellodendropsis* Wood Tremor Jelly
SPECIES *clavulinoides, flagelliformis*
GENUS *Tremellostereum* Stereo Tremor Jelly
SPECIES *dichroum*
GENUS *Trimorphomyces* Trimorphic Tremor Jelly
SPECIES *papilionaceus*
GENUS *Tsuchiyaea* Tsuchiya Tremor Jelly
SPECIES *wingfieldii*

FAMILY 2 *Aporpiaceae* Perplexed Jelly
GENUS *Aporpium* Perplexed Jelly
SPECIES *caryae, dimidiatum, hexagonoides*
GENUS *Elmerina* Elmer Perplexed Jelly
SPECIES *cladophora, holophaea*

FAMILY 3 *Auriculariaceae* Ear Jelly
GENUS *Auricularia* Ear Jelly
SPECIES ***auricula***(Edible), *cornea, fuscosuccinea, mesenterica, polytricha, sordescens*
VARIETIES ***auricula judae, auricula judae-auricularis, auricula judae-lactea***, *mesenterica corrugata, mesenterica lobata, mesenterica tremelloides*
GENUS *Hirneola* (Synonym for *Auricularia*)
GENUS *Neotyphula* Modern Typha Ear Jelly
SPECIES *guianensis*
GENUS *Paraphelaria* Cork Ear Jelly
SPECIES *borneensis*

FAMILY 4 *Dacrymycetaceae* Tear Jelly
GENUS *Dacrymyces* Tear Jelly
SPECIES *australis, capitatus, chrysocomus, chrysospermus, confluens, dictyosporus, enatus, estonicus, intermedius, macnabbii, minor, novae-zelandiae, ovisporus, palmatus, punctiformis, stillatus, variisporus*
VARIETIES *chrysocomus microspora*
GENUS *Calocera* Coral Tear Jelly
SPECIES *cornea, furcata, glossoides, lutea, pallidospathulata, palmata, viscosa*
GENUS *Cerinomyces* Cerin Tear Jelly
SPECIES *aculeatus, albosporus, altaicus, ceraceus, grandinioides, lagerheimii*
GENUS *Dacryopinax* Pin Tear Jelly
SPECIES *aurantiaca, crenata, dennisii, elegans, foliacea, formosus, indacocheae, macrospora, martinii, maxidorii, petaliformis, spathularia, taibaishanensis, xizangensis, yungensis*
GENUS *Ditiola* Ditiole Tear Jelly
SPECIES *nuda, peziziformis, radicata*
GENUS *Guepiniopsis* Cup Tear Jelly
SPECIES *alpina, buccina*
GENUS *Heterotextus* Text Tear Jelly

SPECIES *luteus*
FAMILY 3 *Exidiaceae* Warty Jelly
GENUS *Exidia* Warty Jelly
SPECIES *glandulosa, nucleata, plana, recisa, repanda, saccharina, thuretiana, truncata*
GENUS *Basidiodendron* Wood Warty Jelly
SPECIES *caesiocinereum, cinereum, cremeum, eyrei, luteogriseum, pini, pulvereum, radians, rimosum, spinosum*
GENUS *Ceratosebacina* Horned Seba Warty Jelly
SPECIES *calospora, longispora*
GENUS *Ditangium* Ditang Warty Jelly
SPECIES *insigne*
GENUS *Efibulobasidium* Fibule Warty Jelly
SPECIES *albescens, rolleyi*
GENUS *Eichleriella* Eichler Warty Jelly
SPECIES *deglubens, leveilleana, spinulosa*
GENUS *Endoperplexa* Endic Perplexed Warty Jelly
SPECIES *dartmorica, enodulosa, subfarinacea*
GENUS *Exidiopsis* Lesser Warty Jelly
SPECIES *effusa, molybdea, strigosa*
GENUS *Fibulosebacea* Lesser Fibule Warty Jelly
SPECIES *strigosa*
GENUS *Flahaultia* Flahault Warty Jelly
SPECIES *hyalina*
GENUS *Gloeotromera* Glow Tromer Warty Jelly
SPECIES *alba*
GENUS *Guepinia* Guepin Tear Jelly
SPECIES ***helvelloides***(Edible)
VARIETIES ***helvelloides rufus***
GENUS *Gyrocephalus* (Synonym for *Guepinia*)
GENUS *Heterochaete* Variable Bristle Warty Jelly
SPECIES *gelatinosa*
GENUS *Heteroscypha* Variable Scyph Warty Jelly
SPECIES *applanata*
GENUS *Microsebacina* Seba Warty Jelly
SPECIES *fugacissima, microbasidia*
GENUS *Protodaedalea* Daedale Warty Jelly
SPECIES *hispida*
GENUS *Protomerulius* Wood Crust Warty Jelly
SPECIES *substuppeus*
GENUS *Pseudohydnum* Truffle Warty Jelly
SPECIES ***gelatinosum***(Edible)
GENUS *Renatobasidium* Renat Warty Jelly
SPECIES *notabile*
GENUS *Serendipita* Serendipity Warty Jelly
SPECIES *evanescens, inclusa, interna, invisibilis, orliensis, sigmaspora, vermifera*
GENUS *Stypella* Stype Warty Jelly
SPECIES *crystallina, dubia, glaira, grilletii, legonii, mirabilis, subgelatinosa, subhyalina, vermiformis*
GENUS *Tremellacantha* Tremor Spine Warty Jelly
SPECIES *sclerodontia*
GENUS *Tremellodendron* Tremor Wood Warty Jelly
SPECIES *cladonia, cuneatum*
GENUS *Tremellodon* (Synonym for *Pseudohydnum*)
FAMILY 4 *Sirobasidiaceae* Pitted Jelly
GENUS *Sirobasidium* Pitted Jelly
SPECIES *brefeldianum, compactum*
GENUS *Xenolachne* Lachne Pitted Jelly
SPECIES *flagellifera, longicornis*

CLASS 622.2 *Agaricomycetes* Gill Mushroom

ORDER 1 *Agaricales* Gill Mushroom

FAMILY 01 *Agaricaceae* Gill Mushroom
GENUS *Agaricus* Gill Mushroom
SPECIES *acheruntius, altipes,* ***arvensis***(Edible), ***augustus***(Edible), *benesii, bernardii, bisporus, bitorquis, bohusii, bresadolanus,* ***campestris***(Edible, Dye Herb), *clitocyboides, comtulus, connatus, cornucopioides,*

cupreobrunneus, depauperatus, devoniensis, dulcidulus, essettei, excellens, exsertus, fiardii, fissuratus, fontanae, friesii, fuscofibrillosus, gennadii, gilvus, glaucopus, gregarius, hondensis, impudicus, lachnopus, langei, lanipes, litoralis, luteomaculatus, lutosus, macrocarpus, moelleri, moellerianus, ***nivescens***(Edible), *osecanus, pampeanus, pequinii, perdicinus, phaeolepidotus, placomyces, platypus, porphyrizon, porphyrocephalus, praeclaresquamosus, putidus, radicatus, rodmanii, romagnesii, rufipes, rufotegulis,* ***silvaticus***(Edible), ***silvicola***(Edible), *subfloccosus, subgibbosus, subperonatus, subrufescens, subrutilescens, trutinatus,* ***urinascens***(Edible), *valdiviae, vaporarius, xanthodermus, xantholepis*

VARIETIES *altipes aestivalis, altipes albosericeus, altipes flavotactus,* ***arvensis exquisitus, arvensis leucotrichus, augustus perrarus,*** *benesii carolii, benesii squamuliferus, bernardii ingratus, bernardii maleolens, bisporus bisporus, bisporus albidus, bisporus avellaneus, bisporus brunnescens, bisporus cookeanus, bisporus hortensis, bitorquis edulis, bitorquis validus,* ***campestris campestris, campestris equestris, campestris fuscopilosellus, campestris isabellinus, campestris radicatus, campestris squamulosus, campestris villaticus,*** *comtulus niveolutescens, depauperatus deylii, devoniensis arenophilus, dulcidulus purpurellus, dulcidulus rubellus, fontanae niger, gennadii cellaris, impudicus koelerionensis, impudicus reae, impudicus variegatus, langei mediofuscus, litoralis maskae, litoralis spissicaulis, moelleri obscuratus, moelleri terricolor, moellerianus floccipes, osecanus parkensis, pequinii microsporus, placomyces placomyces, placomyces meleagris, porphyrizon purpurascens, praeclaresquamosus praeclaresquamosus, praeclaresquamosus terricolor,* ***silvaticus fagetorum, silvaticus haemorrhoidarius, silvaticus pallidus, silvaticus sanguinarius, silvaticus vinosobrunneus, silvicola abruptibulbus, silvicola flavescens, urinascens urinascens, urinascens albertii, urinascens crocodilinus, urinascens macrosporus, urinascens schaefferianus, urinascens stramineosquamulosus, urinascens stramineus, urinascens villaticus,*** *vaporarius pseudovillaticus, xanthodermus xanthodermus, xanthodermus griseus, xanthodermus lepiotoides, xanthodermus pseudocretaceus*

GENUS *Allopsalliota* — Psalliote Gill Mushroom
SPECIES *geesterani*
GENUS *Araneosa* — Spider Gill Mushroom
SPECIES *columellata*
GENUS *Attamyces* — Atta Gill Mushroom
SPECIES *bromatificus*
GENUS *Chamaemyces* — Chamise Gill Mushroom
SPECIES *fracidus*
GENUS *Clavogaster* — Club Cavity Gill Mushroom
SPECIES *novozelandicus*
GENUS *Constricta* — Constricted Gill Mushroom
SPECIES *africana*
GENUS *Cribrospora* — Crib Gill Mushroom
SPECIES *tulostomoides*
GENUS *Crucispora* — Cross Gill Mushroom
SPECIES *naucorioides, rhombisperma*
GENUS *Cystoagaricus* — Cyst Gill Mushroom
SPECIES *jujuyensis, sachaensis, strobilomyces, trisulphuratus*
GENUS *Disporotrichum* — Hairy Gill Mushroom
SPECIES *dimorphosporum*
GENUS *Endoptychum* — Endic Folded Gill Mushroom
SPECIES *agaricoides, arizonicum, depressum, kolya, melanosporum, moongum, wariatodes*
GENUS *Gasterellopsis* — Cavity Gill Mushroom
SPECIES *silvicola*
GENUS *Gigasperma* — Gigaspore Club Fungus
SPECIES *americanum, cryptica*
GENUS *Gymnogaster* — Nude Cavity Gill Mushroom
SPECIES *boletoides*
GENUS *Gyrophragmium* — (Synonym for *Longula*)
GENUS *Heinemannomyces* — Heinemann Gill Mushroom
SPECIES *splendidissima*
GENUS *Hiatulopsis* — Hiatule Gill Mushroom
SPECIES *amara, aureoflava*
GENUS *Holocotylon* — Cotylon Gill Mushroom
SPECIES *anomalum, brandegeeanum, texense*
GENUS *Hypogaea* — Hyponic Gill Mushroom
SPECIES *brunnea*
GENUS *Janauaria* — January Gill Mushroom
SPECIES *amazonica*
GENUS *Longula* — Long Gill Mushroom
SPECIES *texensis*
GENUS *Melanophyllum* — Dark Gill Mushroom
SPECIES *echinatum, eyrei, haematospermum*
GENUS *Metrodia* — Metrode Gill Mushroom

SPECIES *collybioides, excissa*
GENUS *Neosecotium* Modern Secot Gill Mushroom
SPECIES *africanum, macrosporum*
GENUS *Nia* Nia Gill Mushroom
SPECIES *vibrissa*
GENUS *Panaeolopsis* Pan Gill Mushroom
SPECIES *brasiliensis, nirimbii, obtusa, sanmartiniana*
GENUS *Phaeopholiota* Dusky Gill Mushroom
SPECIES *crinipellis*
GENUS *Phlebonema* Phlebe Thread Gill Mushroom
SPECIES *chrysotingens*
GENUS *Phlebophyllum* Phlebe Gill Mushroom
SPECIES *vitellinum*
GENUS *Phyllogaster* Cavity Gill Mushroom
SPECIES *pholiotoides*
GENUS *Physalacria* Bladder Gill Mushroom
SPECIES *aggregata, australiensis, bambusae, changensis, corticola, cryptomeriae, gillesii, indica, orientalis, orinocensis, pseudotropica, sasae, solomonensis, stilboidea*
GENUS *Psalliota* (Synonym for *Agaricus*)
GENUS *Pseudoauricularia* Ear Gill Mushroom
SPECIES *papuana*
GENUS *Rugosospora* Wrinkled Gill Mushroom
SPECIES *pseudorubiginosa*
GENUS *Schinzinia* Schinz Gill Mushroom
SPECIES *pustulosa*
GENUS *Schulzeria* Schulz Gill Mushroom
SPECIES *umkowaan*
GENUS *Singerina* Singer Gill Mushroom
SPECIES *indica*
GENUS *Smithiogaster* Smith Cavity Gill Mushroom
SPECIES *volvoagaricus*
GENUS *Termiticola* Termite Gill Mushroom
SPECIES *rubescens*
GENUS *Volvigerum* Volvo Gill Mushroom
SPECIES *conei*
GENUS *Xanthagaricus* Yellow Gill Mushroom
SPECIES *brunneolus, chrysosporus, erinaceus, flavidorufus, globisporus, gracilis, luteolosporus, myriostictus, nanus, rubescens, subepipastus, viridulus*

FAMILY 02 *Bolbitiaceae* Bonnet Gill Mushroom
GENUS *Bolbitius* Fairy Bonnet Gill Mushroom
SPECIES *coprophilus, mexicanus, reticulatus, titubans, variicolor, vitellinus*
VARIETIES *titubans olivaceus*
GENUS *Agrocybe* Earth Scale Bonnet Gill Mushroom
SPECIES *acericola,* ***aegerita***(Commercial Edible), *alnetorum, arvalis, attenuata, brunneola, cylindracea, dura, erebia, eresia, firma, molesta, paludosa, pediades, praecox, pusiola, putaminum, rivulosa, vervacti*
VARIETIES *arvalis tuberosa, brunneola ombrophila, pediades arenicola, pediades semiorbicularis, pediades subpediades, pediades temulenta, praecox gibberosa*
GENUS *Agrogaster* Cavity Bonnet Gill Mushroom
SPECIES *coneae*
GENUS *Conocybe* Ring Stalk Bonnet Gill Mushroom
SPECIES *aberrans, aeruginosa, albipes, ambigua, anthracophila, antipus, apala, aporos, appendiculata, arrhenii, aurea, blattaria, brachypodii, brunnea, brunneola, bulbifera, candida, crispa, crispella, cyanopus, dentatomarginata, digitalina, dumetorum, dunensis, echinata, exannulata, excedens, farinacea, filaris, fimetaria, fragilis, fuscimarginata, hadrocystis, hornana, incarnata, inocybeoides, intermedia, intrusa, juniana, kuehneriana, lateritia, lenticulospora, leucopus, macrocephala, macrospora, mairei, megalospora, merdaria, mesospora, moseri, murinacea, neoantipus, nivea, ochracea, pallidospora, percincta, pilosella, pinetorum, pinguis, plicatella, plumbeitincta, pubescens, pulchella, pygmaeoaffinis, rickeniana, rickenii, rostellata, rubiginosa, rugosa, sabulicola, semiglobata, siennophylla, siliginea, spicula, spiculoides, stercoraria, striaepes, subnuda, subovalis, subpubescens, sulcatipes, tenera, togularis, umbonata, utriformis, velutipes, vestita, vexans, watlingii*
VARIETIES *albipes lactea, albipes pseudocrispa, excedens excedens, excedens pseudomesospora, juniana sordescens*
GENUS *Cyttarophyllopsis* Cyttar Bonnet Gill Mushroom
SPECIES *cordispora*
GENUS *Hebeloma* Hebe Bonnet Gill Mushroom
SPECIES *aestivale, alpinum, ammophilum, anthracophilum, araneosa, arenosum, atrobrunneum, birrum, bruchetii, candidipes, cavipes, circinans, clavuligerum, crustuliniforme, cylindrosporum, dunense, edurum,*

fastibile, fragilipes, fusisporum, gigaspermum, gregarium, helodes, hetieri, hiemale, incarnatulum, ingratum, leucosarx, marginatulum, mesophaeum, microspora, pallidoluctuosum, populinum, psammophilum, pusillum, radicosum, sacchariolens, senescens, sinapizans, sinuosum, sordescens, spoliatum, sporadicum, syrjense, theobrominum, truncatum, vaccinum

VARIETIES *crustuliniforme microspermum, mesophaeum mesophaeum, mesophaeum crassipes*

GENUS *Naucoria* Naucor Bonnet Gill Mushroom

SPECIES *albotomentosa, alnetorum, amarescens, badia, bohemica, clavuligeroides, escharioides, luteolofibrillosa, pseudoamarescens, salicetorum, salicis, scolecina, silvaenovae, spadicea, sphagneti, striatula, suavis, subconspersa, subglobosa, tantilla*

GENUS *Panaeolina* Lesser Pan Bonnet Gill Mushroom

SPECIES *foenisecii*

GENUS *Panaeolus* Pan Bonnet Gill Mushroom

SPECIES *acuminatus, ater, atrobalteatus, campanulatus, cinctulus, cyanescens, fimicola, olivaceus, papilionaceus, retirugis, semiovatus, solidipes,* ***sphinctrinus***(Medicinal Herb), *subbalteatus, subfirmus*

VARIETIES *papilionaceus papilionaceus, papilionaceus parvisporus, semiovatus semiovatus, semiovatus phalaenarum*

GENUS *Pholiotina* Pholiote Bonnet Gill Mushroom

SPECIES *coprophila*

GENUS *Ptychella* Folded Bonnet Gill Mushroom

SPECIES *ochracea*

GENUS *Setchelliogaster* Setchell Bonnet Gill Mushroom

SPECIES *aurantius, australiensis, fragilis, rheophyllus, tenuipes, tetrasporus*

GENUS *Tubariopsis* Tube Bonnet Gill Mushroom

SPECIES *torquipes*

GENUS *Tympanella* Tympan Bonnet Gill Mushroom

SPECIES *galanthina*

GENUS *Wielandomyces* Wieland Bonnet Gill Mushroom

SPECIES *robustus*

FAMILY 03 *Coprinaceae* Cap Gill Mushroom

GENUS *Coprinus* Inky Cap Gill Mushroom

SPECIES *alopecius, amphibius, astroideus, bellulus, bipellis, brunnea, cinnamomeotinctus,* ***comatus***(Edible Unripe Cooked), *congregatus, coopertus, cordisporus, cortinatus, delicatulus, disseminatus, ephemeroides, foetidellus, galericuliformis, griseofoetidus, grossii, hexagonosporus, impatiens, lagopus, martinii, nemoralis, patouillardii, poliomallus, pseudoradiatus, pyrranthes, quadrifidus, saccharomyces, semilanatus, silvaticus, similis, stellatus, stercorarius, sterquilinus, truncorum, tuberosus*

VARIETIES *lagopus lagopus, lagopus vacillans, patouillardii patouillardii, patouillardii bisporus, phaeosporus*

GENUS *Coprinellus* Least Inky Cap Gill Mushroom

SPECIES *amphithallus, angulatus, aureogranulatus, bisporiger, bisporus, callinus, curtus, dilectus, domesticus, ellisii, ephemerus, flocculosus, heptemerus, heterosetulosus, hiascens, marculentus, micaceus, pellucidus, plagioporus, radians, sassii, sclerocystidiosus, subdisseminatus, subimpatiens, subpurpureus, truncorum, velatopruinatus, xanthothrix*

GENUS *Coprinopsis* Lesser Inky Cap Gill Mushroom

SPECIES *acuminata, ammophilae, argentea,* ***atramentaria***(Edible Unripe Cooked) *calospora, cinerea, cinereofloccosa, coniophora, cothurnata, echinospora, episcopalis, erythrocephala, filamentifer, friesii, gonophylla, herinkii, insignis, jonesii, kimurae, kriegIsteineri, kubickae, laanii, lagopides, luteocephala, macrocephala, myceliocephala, narcotica, nivea, ochraceolanata, pachyderma, pachysperma, phlyctidospora,* ***picacea***(Edible), *pseudofriesii, pseudonivea, pseudoradiata, psychromorbida, radiata, radicans, romagnesiana, sclerotiger, scobicola, semitalis, spelaiophila, spilospora, stangliana, stercorea, tigrinella, trispora, urticicola, utrifer, vermiculifer, verticillata, xantholepis, xenobia*

GENUS *Montagnites* Monday Cap Gill Mushroom

SPECIES *arenarius*

GENUS *Pseudocoprinus* (Synonym for *Coprinus*)

FAMILY 04 *Cortinariaceae* Web Cap Gill Mushroom

GENUS *Cortinarius* Web Cap Gill Mushroom

SPECIES *acetosus, acutus, adalbertii, agathosmus, ahsii, alcalinophilus, ammophilus, anfractus, angelesianus, aprinus, arquatus, arvinaceus, atropusillus, atrovirens, aureomarginatus, badiolilaceus, balaustinus, barbatus, basililaceus, basiroseus, bibulus, biformis, bivelus, bovinus, bulbosus, bulliardii, cagei, caledoniensis, californicus, callisteus, camphoratus, camptoros, cephalixus, chrysomallus, cotoneus, cumatilis, cupreorufus, cyaneus, cyanophyllus, cylindripes, damascenus, danili, decipiens, depressus, diabolicoides, diasemospermus, disjungendus, durus, emollitus, emunctus, erubescens, erythrinus, evernius, fervidus, flexipes, fulvescens, fulvosquamosus, fusisporus, galeobdolon, galerinoides, gausapatus, gentilis, heliotropicus, helvelloides, hemitrichus, herculeolens, hercynicus, heterosporus, hinnuloides, hoeftii, holophaeus, humicola, illuminus, imbutus, incisus, inconspicuus, iodioides, junghuhnii, laetissimus, laniger, lebretonii, lepidus, licinipes, limonius, livor, lucorum, magicus, malachius, myxo-anomalus, norvegicus, nothosaniosus, obtusus, orichalceus, paragaudis, parvannulatus, patibilis, pearsonii, percomis, pertristis, phaeophyllus, phaeopygmaeus, poecilopus, porphyropus, psammocephalus, pseudocandelaris, pseudovulpinus, quarciticus, rapaceus, redemitus, renidens, rheubarbarinus, riederi, rigens, rigidipes, rigidiusculus, rubicundulus, rufostriatus, safranopes, saniosus, saturninus, scandens, scotoides, simulatus, sommerfeltii, spilomeus, subargentatus, subbalaustinus, subtortus, subtorvus, suburaceus, suillus, sulphureus, tabacinus, terpsichores, testaceoviolaceus, testaceoviolascens, tofaceus, tophaceus, triformis, triumphans, turgidus, turmalis, umbrinolens, uraceus, valgus, variegatus, variiformis, venetus, venustus, vernus, vibratilis, violaceovelatus, violilamellatus, xantho-ochraceus, zosteroides*

VARIETIES *acutus striatulus, angelesianus strobilaceus, arquatus cookianus, arquatus xanthochrous, bibulus lilacinopusillus, bibulus pulchellus, bulliardii pseudocolus, cagei bicolor, camphoratus amethysteus, camphoratus hircinus, cinnamomeus luteus, cinnamomeus violaceus, cupreorufus orichalceus, decipiens decipiens, decipiens atrocoeruleus, decipiens insignis, durus balteatoclaricolor, erythrinus argyropus, evernius fragrans, flexipes flabellus, flexipes paleiferus, licinipes robustior, magicus subglaucopus, obtusus pseudostriatulus, orellanus rutilans, parvannulatus cedriolens, patibilis patibilis, patibilis scoticus, pearsonii cremeolaniger, pearsonii lanigeroides, poecilopus fuscopallens, rapaceus major, rigens duracinus, rigens raphanicus, saturninus cohabitans, scandens gracilis, spilomeus depauperatus, subargentatus argenteopileatus, subargentatus kauffmanianus, triformis triformis, triformis fuscopallens, triumphans crocolitus, valgus camurus, valgus fuliginosus, variegatus fulvidolilaceus, variegatus marginatus, venustus calopus, venustus traganulus, vulpinus albomarginatus, vulpinus rufoalbus, xantho-ochraceus langei*

GENUS *Anamika* — Anamik Web Cap Gill Mushroom

SPECIES *indica*

GENUS *Aroramyces* — Arora Web Cap Gill Mushroom

SPECIES *gelatinosporus, radiatus*

GENUS *Chromocyphella* — Chromic Cypher Web Cap Gill Mushroom

SPECIES *bryophyticola, burtii, galeata, muscicola, pinsapinea*

GENUS *Crepidotus* — Stump Foot Web Cap Gill Mushroom

SPECIES *applanatus, autochthonus, calolepis, carpaticus, cesatii, cinnabarinus, crocophyllus, epibryus, lanuginosus, lundellii, luteolus, maculans, maximus, mollis, muscigenus, subnidulans, subverrucisporus, variabilis, versutus*

VARIETIES *applanatus applanatus, applanatus putrigenus, applanatus subglobiger, autochthonus caspari, autochthonus fragilis, carpaticus wakefieldiae, cesatii cesatii, cesatii sphaerosporus, cesatii subsphaerosporus, epibryus commixtus, epibryus herbarum, epibryus hypnophilus, lundellii amygdalosporus, lundellii subtilis, mollis ralfsii, subverrucisporus bickhamensis, versutus bresadolae, versutus pubescens*

GENUS *Cribbea* — Cribb Web Cap Gill Mushroom

SPECIES *andina, gloriosa, lamellata, reticulata*

GENUS *Cuphocybe* — Cuph Cybe Web Cap Gill Mushroom

SPECIES *alborosea, calospora, ferruginea, melliolens, olivacea, phaeomyxa*

GENUS *Dermocybe* — Skin Web Cap Gill Mushroom

SPECIES *anomala, anthracina, bataillei, betuletorum, bolaris, canina, cinnabarina, cinnamomea, crocea, danica, huronensis, malicoria, olivaceofusca, orellana, pratensis, purpurea, rubella, sanguinea, semisanguinea, tubaria, uliginosa, zinzberatus*

VARIETIES *anomala azureovelata, anomala azurea, anomala epsomiensis, anomala lepidopa, anthracina purpureobadia, anthracina subanthracina, betuletorum raphanoides, cinnamomea lutea, cinnamomea violacea, crocea badia, crocea caerulea, huronensis olivacea, huronensis palustris, malicoria croceifolia, olivaceofusca schaefferi, orellana rutilans, purpurea phoenicea, rubella orellanoides, rubella speciosissima, sanguinea punicea, tubaria sphagneti, uliginosa concinna, uliginosa obtusus, uliginosa queletii*

GENUS *Episphaeria* — Round Web Cap Gill Mushroom

SPECIES *fraxinicola*

GENUS *Flammula* — (Synonym for *Gymnopilus, Pholiota*)

GENUS *Flammulaster* — Star Flamecap Web Cap Gill Mushroom

SPECIES *carpophilus, erinaceellus, ferrugineus, fusisporus, gracilis, granulosus, limulatus, muricatus, saliciphilus, siparius, speireoides, wieslandri*

VARIETIES *carpophilus carpophilus, carpophilus rhombosporus, carpophilus subincarnatus, limulatus limulatus, limulatus litus, limulatus novasilvensis*

GENUS *Galerina* — Skullcap Web Cap Gill Mushroom

SPECIES *allospora, ampullaceocystis, annulata, atkinsoniana, badipes, calyptrata, camerina, cephalotricha, cerina, cinctula, clavata, clavus, embolus, harrisonii, heimansii, heterocystis, hybrida, hypnorum, hypophaea, jaapii, laevis, marginata, mniophila, moelleri, nana, paludinella, paludosa, permixta, phillipsii, praticola, pseudocamerina, pseudocerina, pseudomniophila, pseudomycenopsis, pseudotundrae, pumila, salicicola, septentrionalis, sideroides, sphagnorum, stordalii, stylifera, subbadipes, subcerina, subclavata, subochracea, terrestris, tibiicystis, triscopa, uncialis, venenata, viscidula, vittiformis*

VARIETIES *allospora luteofulva, badipes cedretorum, camerina pseudobadipes, cerina longicystis, harrisonii antheliae, hypnorum calyptrospora, laevis graminea, marginata autumnalis, marginata unicolor, permixta cephalescens, pseudocamerina josserandii, pseudomycenopsis pseudopumila, pumila mycenopsis, stylifera indusiata, subcerina anglica, vittiformis muricellospora, vittiformis rubiginosa, vittiformis tetraspora*

GENUS *Gymnopilus* — Nude Flamecap Web Cap Gill Mushroom

SPECIES *aeruginosus, bellulus, decipiens, dilepis, flavus, fulgens, junonius, liquiritiae, obscurus, odini, peliolepis, penetrans, picreus, sapineus, stabilis, terrestris, validipes, ventricosus*

VARIETIES ***junonius spectabilis***(Medicinal Herb), *penetrans hybridus*

GENUS *Hebelomina* — Pie Web Cap Gill Mushroom

SPECIES *neerlandica*
GENUS *Inocybe* Fibercap Web Cap Gill Mushroom
SPECIES *acuta, adaequata, agardhii, albomarginata, albovelutipes, amethystina, appendiculata, arenicola, assimilata, asterosporus, aurea, auricoma, bongardii, calamistrata, calida, calospora, catalaunica, cervicolor, cincinnata, concinnula, cookei, corydalina, cryptocystis, curvipes, decipientoides, dulcamara, dunensis, duriuscula, erubescens, fibrosa, fibrosoides, flavella, flocculosa, fraudans, fuligineoatra, fulvella, furfurea, fuscidula, fuscomarginata, geophylla, giacomi, glabrescens, glabripes, glabrodisca, godeyi, grammata, griseolilacina, haemacta, heimii, hirtella, huijsmanii, humilis, hystrix, impexa, inodora, jacobi, lacera, langei, lanuginosa, leptocystis, leptophylla, maculata, malenconii, margaritispora, melanopoda, mimica, mixtilis, muricellata, napipes, nitidiuscula, oblectabilis, obscurobadia, ochroalba, olivaceobrunnea, pallida, paludinella, pelargonium, perlata, petiginosa, phaeodisca, piriodora, posterula, praetervisa, proximella, pruinosa, pseudoasterospora, pseudodestricta, pseudohiulca, pseudoreducta, pusio, putilla, rennyi, rimosa, salicis, sambucina, serotina, sindonia, soluta, splendens, squamata, squarrosa, striatorimosa, subcarpta, tabacina, taxocystis, tenebrosa, tjallingiorum, umboninota, umbratica, vaccina, vulpinella, whitei, xanthomelas*
VARIETIES *acuta striata, adaequata jurana, adaequata rhodiola, albomarginata ovalispora, albomarginata reducta, assimilata palaeotropica, assimilata umbrina, calamistrata hirsuta, calida brunneorufa, calospora gaillardii, cincinnata cincinnata, cincinnata conformata, cincinnata major, cincinnata obscuroides, cincinnata phaeocomis, corydalina corydalina, corydalina erinaceomorpha, corydalina erinaceomorpha-scabra, curvipes globocystis, curvipes lanuginella, dulcamara axantha, dulcamara delecta, dulcamara homomorpha, dulcamara squamosoannulata, dunensis heimiana, duriuscula trivialis, erubescens patouillardii, flavella xanthocephala, flocculosa flocculosa, flocculosa crocifolia, flocculosa gausapata, flocculosa geraniolens, flocculosa subtigrina, fraudans incarnata, fuscidula fuscidula, fuscidula bisporigera, fuscidula brunneoatra, fuscidula virgatula, geophylla geophylla, geophylla alba, geophylla clarkii, geophylla lilacina, geophylla lilacina-violacea, glabripes exiguispora, glabripes microspora, glabripes parvispora, glabrodisca decemgibbosa, godeyi boltonii, godeyi rickenii, godeyi rubescens, godeyi rufescens, haemacta rubra, heimii fibrillosa, hirtella hirtella, hirtella bispora, impexa maritima, jacobi rufoalba, lacera lacera, lacera helobia, lacera helobia-gracilis, lacera heterospora, lacera rhacodes, lacera subsquarrosa, lanuginosa lanuginosa, lanuginosa longicystis, lanuginosa ovatocystis, leptocystis hygrophila, leptophylla casimiri, margaritispora phaeosticta, nitidiuscula friesii, oblectabilis major, obscurobadia tenuicystidiata, posterula xanthodisca, pruinosa reidii, pseudoasterospora pseudoasterospora, pseudoasterospora microsperma, rennyi rennyi, rennyi major, rimosa alpina, rimosa argentata, rimosa curreyi, rimosa fastigiata, rimosa obsoleta, rimosa pseudofastigiata, rimosa schista, rimosa umbrinella, serotina devoniensis, serotina psammophila, sindonia commutabilis, sindonia kuehneri, sindonia mutica, soluta brevispora, splendens splendens, splendens phaeoleuca, tenebrosa atripes, umbratica commixta, umbratica infida, whitei pudica*
GENUS *Inoloma* Fiber Margin Web Cap Gill Mushroom
SPECIES *alboviolaceus, corrugatus, pholideus, traganum, violaceus, vulpinum*
VARIETIES *vulpinum albomarginatus, vulpinum rufoalbus*
GENUS *Kjeldsenia* Kjeldsen Web Cap Gill Mushroom
SPECIES *aureispora*
GENUS *Leucoinocybe* Light Fibercap Web Cap Gill Mushroom
SPECIES *lenta*
GENUS *Leucocortinarius* Creamcap Web Cap Gill Mushroom
SPECIES *bulbiger, odorifer*
GENUS *Melanomphalia* Dark Web Cap Gill Mushroom
SPECIES *nigrescens*
GENUS *Mycolevis* Levi Web Cap Gill Mushroom
SPECIES *siccigleba*
GENUS *Myxacium* Mucilage Web Cap Gill Mushroom
SPECIES *betulinum, collinitum, favrei, illibatum, livido-ochraceum, mucifluum, mucosum, ochroleucum, pluvium, pumilum, salor, stillatitium, triviale*
VARIETIES *collinitum muscigenum, illibatum subdelibutum, livido-ochraceum elatior, mucifluum pinicola, stillatitium integerrimum, stillatitium pseudosalor, triviale repandum*
GENUS *Nanstelocephala* Nanstel Web Cap Gill Mushroom
SPECIES *physalacrioides*
GENUS *Pellidiscus* Pellet Disc Web Cap Gill Mushroom
SPECIES *pallidus*
GENUS *Phaeocollybia* Dusky Colly Web Cap Gill Mushroom
SPECIES *arduennensis, festiva, jennyae, lugubris*
GENUS *Phaeogalera* Dusky Skullcap Web Cap Gill Mushroom
SPECIES *dissimulans, sphagneti, stagnina*
GENUS *Phaeoglabrotricha* Dusky Smooth Hair Web Cap Gill Mushroom
SPECIES *farinacea*
GENUS *Phaeomarasmius* Dusky Pinwheel Web Cap Gill Mushroom
SPECIES *erinaceus, rimulincola*
GENUS *Phaeosolenia* Dusky Solitary Web Cap Gill Mushroom
SPECIES *inconspicua, pelargonii, platensis*
GENUS *Phlegmacium* Mucus Web Cap Gill Mushroom

SPECIES *aleuriosmum, amoenolens, aureopulverulentum, aureoturbinatum, balteatoalbum, balteatocumatile, caesiocanescens, caerulescens, caesiostramineum, calochroum, cedretorum, citrinum, claricolor, coerulescentium, corrosum, corruscans, crassum, cumatile, cyanites, cyanopus, delibutus, dibaphus, dionysae, diosmus, elegantior, elegantissimum, fulvo-ochrascens, fulvoincarnatum, glaucopus, gracilior, infractum, largus, mairei, microspermum, mussivum, nanceiense, occidentale, odoratum, odorifer, olearioides, olidum, osmophorus, polymorphus, praestans, purpurascens, rickenianum, rufo-olivaceum, saginum, saporatum, scaurus, silvamonachorum, sodagnitum, spadicellum, splendens, subarquatum, tabulare, talum, urbicum, variicolor, varium, volvatum, xanthocephalum, xanthophyllum, calochroum coniferarum, calochroum parvum*

VARIETIES *aleuriosmum caroviolaceum, caerulescens caesiocyaneum, caesiostramineum amarescens, citrinum pseudosulphureum, citrinum sulphureum, claricolor subclaricolor, dibaphus nemorosus, elegantior quercus-ilicis, elegantior fulmineum, largus lividoviolaceus, mussivum russeoides, mussivum russeum, occidentale occidentale, occidentale obscurum, olearioides fulmineus, olearioides subfulgens, osmophorus evosmus, polymorphus multiformis, praestans berkeleyi, purpurascens largusoides, rufo-olivaceum testaceum, rufo-olivaceum vinosum, saginum subtriumphans, saginum validum, saporatum langei, saporatum lutescens, saporatum subturbinatus, scaurus herpeticum, scaurus herpeticum-fageticola, scaurus herpeticum-subvirentophyllus, tabulare decoloratum, talum melliolens, talum ochropallidum, variicolor marginatum, variicolor nemorense, varium decolorans*

GENUS *Pleurotellus* — Pleurote Web Cap Gill Mushroom

SPECIES *albellus, dictyorhizus, filicinus*

GENUS *Pseudogymnopilus* — Lesser Nude Flamecap Web Cap Gill Mushroom

SPECIES *pampeanus*

GENUS *Pyrrhoglossum* — Fire Tongue Web Cap Gill Mushroom

SPECIES *ferruginatum, hepatizon, holocrocinum, lilaceipes, macrosporum, pyrrhum, recedens, stipitatum, viriditinctum, yunnanense*

GENUS *Rapacea* — Rape Web Cap Gill Mushroom

SPECIES *mariae*

GENUS *Rozites* — Nitecap Web Cap Gill Mushroom

SPECIES ***caperatus***(Edible)

GENUS *Simocybe* — Simcap Web Cap Gill Mushroom

SPECIES *centunculus, haustellaris, reducta, sumptuosa*

VARIETIES *centunculus centunculus, centunculus laevigata, centunculus maritima, centunculus obscura*

GENUS *Stagnicola* — Stag Web Cap Gill Mushroom

SPECIES *perplexa*

GENUS *Stephanopus* — Stephan Web Cap Gill Mushroom

SPECIES *azureus, coerulea, stropharioides, trachyphyloeus, vilchensis*

GENUS *Telamonia* — Belted Web Cap Gill Mushroom

SPECIES *albovariegata, alnea, alnetorum, armillata, brunnea, casimiri, conica, hinnulea, ionophylla, roseipes, torva*

VARIETIES *alnetorum iliopodius, brunnea brunnea, brunnea glandicolor, casimiri hoffmannii, casimiri subsertipes, hinnulea radicata, ionophylla scutulata*

GENUS *Tubaria* — Scalecap Web Cap Gill Mushroom

SPECIES *albostipitata, confragosa, conspersa, dispersa, furfuracea, hiemalis, minutalis, pallidispora, praestans, pseudoconspersa, romagnesiana, umbrina*

FAMILY 05 *Entolomataceae* — Pinkgill Gill Mushroom

GENUS *Entoloma* — Pinkgill Gill Mushroom

SPECIES *abortivum, acidophilum, aethiops, albotomentosum, allospermum, ameides, anatinum, araneosum, argenteostriatum, atrocoeruleum, bisporigerum, bloxamii, brassicolens, byssisedum, caeruleoflocculosum, caeruleopolitum, caeruleum, calaminare, calthionis, carneogriseum, catalaunicum, cephalotrichum, cetratum, chlorophyllum, cocles, costatum, cruentatum, cuniculorum, cyaneoviridescens, dysthaloides, euchroum, excentricum, favrei, griseorubidum, henricii, hirtum, huijsmanii, indutoides, indutum, infula, inocybiforme, insidiosum, insolitum, inusitatum, inutile, jennyae, jubatum, kervernii, kristiansenii, langei, lanuginosipes, melanochroum, mougeotii, mougeotii murraii, mutabilipes, nausiosme, nigellum, nigroviolaceum, niphoides, occultopigmentatum, ochromicaceum, opacum, pallescens, parasiticum, phaeocyathus, placidum, plebeioides, plebejum, politoflavipes, politum, porphyrogriseum, pratulense, pseudocoelestinum, pseudoturci, pulvereum, pygmaeopapillatum, queletii, reae, reginae, romagnesii, rufocarneum, rugosum, rusticoides, salmoneum, sarcitum, scabropellis, scabrosum, sericeonitens, serrulatum, solstitiale, strictior, tenuipes, testaceum, tjallingiorum, transvenosum, triste, velenovskyi, ventricosum, verecundum, versatile, vinaceum, viridans, wynnei, xanthocaulon, xanthochroum*

VARIETIES *araneosum fulvostrigosum, bloxamii triste, infula infula, infula chlorinosum, mougeotii, mougeotii fuscomarginatum, plebejum erophilum, plebejum pyrenaicum, politum pernitrosum, rufocarneum andrianae, serrulatum atrides, vinaceum fumosipes*

GENUS *Calliderma* — Callisto Pinkgill Gill Mushroom

SPECIES *indigofera*

GENUS *Claudopus* — (Synonym for *Entoloma*)

GENUS *Clitopilus* — Depressed Cap Pinkgill Gill Mushroom

SPECIES *abortivus, cretatus, daamsii, hobsonii, omphaliiformis, passeckerianus, pinsitus,* ***prunulus***(Commercial Edible), *scyphoides*
VARIETIES *scyphoides scyphoides, scyphoides intermedius*
GENUS *Leptonia* Thin Pinkgill Gill Mushroom
SPECIES *subcyanula*
GENUS *Rhodocybe* Rosecap Pinkgill Gill Mushroom
SPECIES *caelata, fallax, gemina, griseospora, hirneola, melleopallens, nitellina, popinalis, roseoavellanea,* ***truncata***(Edible)
VARIETIES *gemina subvermicularis, popinalis lutetiana, popinalis mundula*
GENUS *Rhodocybella* Lesser Rosecap Pinkgill Gill Mushroom
SPECIES *rhododendri*
GENUS *Rhodogaster* Cavity Rose Pinkgill Gill Mushroom
SPECIES *calongei, chilensis*
GENUS *Rhodopaxillus* (Synonym for *Rhodocybe*)
GENUS *Rhodophyllus* Rosy Pinkgill Gill Mushroom
SPECIES *anthracinus, apiculatus, aprilis, asprellus, atromarginatus, babingtonii, caccabus, caesiocinctus, canosericeus, chalybaeus, chloropus, clandestinus, clypeatus, coelestinus, conferendus, corvinus, cuneatus, cuspidifer, cyanulus, depluens, dichrous, dysthales, elodes, exile, fernandae, formosus, fuscomarginatus, fuscotomentosus, griseocyaneus, griseoluridus, griseorubellus, hebes, hirtipes, hispidulus, ianthinus, incanus, incarnatofuscescens, juncinus, kuehnerianum, lampropus, lanicus, lividoalbus, lividocyanulus, longistriatus, lucidus, majaloides, mammosus, medius, minutus, myrmecophilus, neglectus, nitens, nitidus, olorinus, ortonii, pallens, papillatus, parkensis, pascuus, percandidus, pleopodius, poliopus, porphyrophaeus, prismatospermus, prunuloides, querquedula, rhodocalyx, rhodopolius, rhombisporus, roseus, sacchariolens, saundersii, scabiosus, sepium, sericatus, sericellus, sericeoides, sericeus, sinuatus, sodalis, sordidulus, sphagneti, sphagnorum, strigosissimus, subradiatus, tenellus, turbidus, turci, undatus, vernus*
VARIETIES *caccabus paludicola, chalybaeus chalybaeus, chalybaeus lazulinus, clypeatus pallidogriseum, coelestinus cruentatus, conferendus conferendus, conferendus pusillus, conferendus pusillus-xylophilus, conferendus rickenii, conferendus staurosporus, exile pyrospilus, fernandae psilopus, formosus fulvus, hebes leptopus, hebes obsoletus, incarnatofuscescens leptonipes, kuehnerianum sericoides, lanicus pusillus, longistriatus longistriatus, longistriatus majusculus, longistriatus sarcitulus, longistriatus sarcitulus-spurcifolius, minutus polymorphus, neglectus cancrinus, ortonii farinolens, percandidus omphaliiformis, pleopodius icterinus, pleopodius gracillimus, poliopus poliopus, poliopus discolor, poliopus parvisporigerum, prunuloides prunuloides, prunuloides inopiliforme, prunuloides obscurum, rhodopolius nidorosus, rhombisporus rhombisporus, rhombisporus floccipes, sericellus carneoalbus, sericellus decurrens, sericellus lutescens, sericeus sericeus, sericeus cinereo-opacum, sericeus nolaniformis, sinuatus sinuatus, sinuatus eulividum, sinuatus fertile, sinuatus lividum, sinuatus roseum, turbidus cordae, undatus undatus, undatus sericeonitidus, vernus cucullatus*
GENUS *Richoniella* Richon Pinkgill Gill Mushroom
SPECIES *afra, asterospora, leptoniispora, macrospora, pumila*

FAMILY 06 *Hydnangiaceae* Tallowgill Gill Mushroom
GENUS *Hydnangium* Hedgehog Tallowgill Gill Mushroom
SPECIES *carneum*
GENUS *Hemigaster* Half Bowl Tallowgill Gill Mushroom
SPECIES *candidus*
GENUS *Laccaria* Milky Tallowgill Gill Mushroom
SPECIES *amethystina, bicolor, fraterna, laccata, maritima, ochropurpurea, proxima, pumila, purpureobadia, tetraspora, tortilis*
VARIETIES *amethystina hudsonii, laccata affinis, laccata amethystea, laccata anglica, laccata farinacea, laccata moelleri, laccata pallidifolia, laccata rosella, laccata scotica, laccata subalpina, maritima trullisata, pumila altaica, pumila striatula*
GENUS *Maccagnia* Macca Tallowgill Gill Mushroom
SPECIES *carnica*
GENUS *Podohydnangium* Stump Foot Tallowgill Gill Mushroom
SPECIES *australe*

FAMILY 07 *Hygrophoraceae* Waxycap Gill Mushroom
GENUS *Hygrophorus* Ivorygill Waxycap Gill Mushroom
SPECIES *agathosmus, amarus,* ***arbustivus***(Edible), *bakerensis, calophyllus, camarophyllus,* ***chrysodon***(Edible), *cinereus,* ***discoxanthus***(Edible), ***eburneus***(Edible), *erubescens, flavodiscus, fulgeneus, gliocyclus, glutinosus,* ***hedrychii***(Edible), *hyacinthus,* ***hypothejus***(Edible), *inocybiformis, latitabundus,* ***lindtneri***(Edible), *lucorum,* ***marzuolus***(Commercial Edible), *mesotephrus, nemoreus, olivaceoalbus, penarius, persoonii, piceae,* ***poetarum***(Edible), *ponderatus,* ***pudorinus***(Edible), *purpurascens, pustulatus,* ***russula***(Edible), *sordidus, speciosus, subalpinus, tennesseensis, unicolor*
VARIETIES *agathosmus aureofloccosus, agathosmus cerasinus, camarophyllus caprinus,* ***chrysodon leucodon, eburneus eburneus, eburneus cossus, eburneus cossus-quercetorum, discoxanthus chrysaspis, discoxanthus discoideus, hedrychii melizeus, hypothejus aureus, hypothejus expallens,*** *latitabundus obesus,* ***lindtneri carpini, lindtneri leucophaeus****, persoonii limacinus, pustulatus tephroleucus*
GENUS *Camarophyllus* Archgill Waxycap Gill Mushroom

SPECIES *angustifolius, atropunctus, borealis, cantharellus, citrinovirens, colemannianus, flavipes, foetens, fornicatus, hymenocephalus, lacmus, nemoreus, pratensis, radiatus, russocoriaceus, schulzeri, virgineus*

VARIETIES *cantharellus lepida, citrinovirens brevispora, fornicatus clivalis, fornicatus lepidopus, fornicatus streptopus, lacmus subviolaceus, nemoreus leporinus, pratensis pallidus, pratensis pallidus-berkeleyanus, pratensis pallidus-ortonii, virgineus fuscescens, virgineus niveus, virgineus ochraceopallidus, virgineus roseipes, virgineus subradiata*

GENUS *Camarophyllopsis* Lesser Archgill Waxycap Gill Mushroom

SPECIES *darwinensis, deceptiva, dennisiana, hiemalis, micacea, microspora, phaeophylla, phaeoxantha, rugulosa, subfuscescens, tetraspora*

GENUS *Hygrocybe* Slender Waxycap Gill Mushroom

SPECIES *angustifolia, aurantia, aurantiosplendens, bresadolae, calciphila, calyptriformis, canescens, cantharellus, ceracea, chlorophana, citrinopallida, citrinovirens, coccinea, coccineocrenata, colemanniana, conica, conicoides, constrictospora, davisii, erinaceus, flavipes, fornicata, glutinipes, helobia, hondurensis, houghtonii, ingrata, insipida, intermedia, irrigata, lacma, laeta, lilacina, marchii, marginata, miniata, mucronella, nitrata, ovina, peckii, persistens, phaeococcinea,* ***pratensis***(Commercial Edible), *psittacina, punicea, quieta, reidii, roseiceps, russocoriacea, salicis-herbaceae, spadicea, splendidissima, subpapillata, substrangulata, tahquamenonensis, turunda, viola,* ***virginea***(Edible), *vitellina*

VARIETIES *calyptriformis calyptriformis, calyptriformis domingensis, calyptriformis nivea, ceracea vitellinoides, chlorophana aurantiaca, chlorophana euroflavescens, chlorophana flavescens, coccinea umbonata, conica conica, conica chloroides, conica cinereifolia, conica conicopalustris, conica nigrescens, conica olivaceonigra, conica pseudoconica, conica tristis, glutinipes glutinipes, glutinipes aurantioviscida, glutinipes citrina, glutinipes rubra, helobia miniatoalba, irrigata unguinosa, laeta laeta, laeta flava, miniata strangulata, mucronella reae, nitrata murinacea, ovina nitiosa, persistens persistens, persistens acutoconica, persistens aurantiolutescens, persistens constans, persistens cuspidata, persistens konradii, persistens konradii-silvatica, persistens konradii-subglobispora, persistens langei, persistens parapersistens, persistens pseudoconica, persistens subconica,* ***pratensis pratensis, pratensis pallida*** *psittacina psittacina, psittacina perplexa, psittacina perplexa-sciophana, punicea acutopunicea, quieta obrussea****, virginea virginea, virginea*** **fuscescens,** ***virginea niveus, virginea ochraceopallida,*** *vitellina luteolaeta*

GENUS *Limacium* Snail Waxycap Gill Mushroom

SPECIES *arbustivum, camarophyllum, capreolarium, carneipes, eburneum, hedrychii, hypothejus, karstenii*

VARIETIES *camarophyllum caprinum, eburneum eburneum, eburneum cossus, eburneum cossus-quercetorum, hedrychii melizeus, hypothejus aureus, hypothejus expallens*

GENUS *Limacella* Slimy Waxycap Gill Mushroom

SPECIES *delicata, guttata, illinita, ochraceolutea*

VARIETIES *delicata delicata, delicata glioderma, delicata vinosorubescens, delicata vinosorubescens-roseofloccosa, guttata lenticularis, guttata megalodactylus*

FAMILY 08 *Lepiotaceae* Parasol Gill Mushroom

GENUS *Lepiota* Parasol Gill Mushroom

SPECIES *aspera, bickhamensis, boertmannii, boudieri, brunneoincarnata, brunneolilacea, calcicola, castanea, cinerascens, cingulum, clypeolaria, cortinarius, coxheadii, cristata, cystophoroides, echinacea, echinella, efibulis, erminea, felina, forquignonii, fuscovinacea, glaucophylla, grangei, griseovirens, helveola, hymenoderma, hystrix, ignivolvata, jacobi, lilacea, locquinii, magnispora, medullata, nigromarginata, obscura, ochraceofulva, oreadiformis, parvannulata, perplexa, pseudoasperula, pseudolilacea, subalba, subgracilis, subincarnata, tomentella,* ***umbonata***(Edible), *xanthophylla*

VARIETIES *aspera acutesquamosa, aspera friesii, aspera furcata, boudieri acerina, boudieri fulvella, brunneoincarnata barlae, brunneoincarnata patouillardii, castanea rufidula, clypeolaria colubrina, clypeolaria ochraceosulfurescens, cortinarius cortinarius, cortinarius audreae, cristata felinoides, cristata pallidior, cristata subfelinoides, echinella echinella, echinella minuta, echinella rhodorhiza, echinella setulosa, erminea alba, forquignonii olivaceobrunnea, griseovirens poliochloodes, jacobi langei, magnispora ventriosospora, ochraceofulva cookei, oreadiformis laevigata, oreadiformis pratensis, oreadiformis rocabrunae, pseudolilacea pseudohelveola, subalba albosericea, subgracilis kuehneriana, subgracilis latispora, subgracilis minor, subgracilis ochraceosulfurescens, subgracilis wasseri, subincarnata josserandii*

GENUS *Chlorolepiota* Green Parasol Gill Mushroom

SPECIES *mahabaleshwarensis*

GENUS *Chlorophyllum* Greengill Parasol Gill Mushroom

SPECIES *abruptibulbum, agaricoides, alborubescens, brunneum, globosum, hortense, humei, mammillatum,* ***molybdites***(Edible Well Cooked), *neomastoideum, olivieri,* ***rhacodes***(Edible),*subfulvidiscum, subrhacodes*

VARIETIES ***rhacodes rhacodes, rhacodes bohemica, rhacodes brunnea, rhacodes hortensis, rhacodes olivieri***

GENUS *Cystolepiota* Cyst Parasol Gill Mushroom

SPECIES *adulterina, bucknallii, cystidiosa, hetieri, moelleri, ompnera, pseudogranulosa, pulverulenta, seminuda, sistrata*

GENUS *Endolepiotula* Endic Parasol Gill Mushroom

SPECIES *ruizlealii*

GENUS *Leucoagaricus* Smoothcap Parasol Gill Mushroom

SPECIES ***americanus***(Poison Raw, Edible Well Cooked), *barssii, brunneocingulatus, brunneus,* ***carneifolius***(Edible), *croceovelutinus, georginae, glabridiscus, gongylophorus, griseodiscus, ionidicolor,* *leucothites, marriagei, medioflavoides, melanotrichus, meleagris,* ***naucinus***(Edible), ***nympharum***(Edible), *pilatianus, purpureolilacinus, purpureolilacinus purpureorimosus, serenus, sericifer, subcretaceus, sublittoralis, tener, wichanskyi*

VARIETIES ***americanus bresadolae****, barssii macrorhizus, barssii pinguipes,* ***carneifolius densifolius****, croceovelutinus diversisporus, ionidicolor caeruleoviolaceus, leucothites holosericeus, melanotrichus melanotrichus, melanotrichus fuligineobrunneus, melanotrichus septentrionalis, pilatianus subrubens, sericifer sericatellus*

GENUS *Leucocoprinus* Lightcap Parasol Gill Mushroom

SPECIES ***badhamii***(Poison Raw, Edible Well Cooked)*, birnbaumii, brebissonii, cepistipes, cretaceus, cygneus, holospilotus, medioflavus, straminellus, tenellus*

VARIETIES *birnbaumii luteus, cepistipes cheimonoceps, cepistipes rorulentus, straminellus denudatus, straminellus gueguenii*

GENUS *Macrolepiota* Large Parasol Gill Mushroom

SPECIES ***excoriata***(Commercial Edible)*, fuligineosquarrosa, fuliginosa, gracilenta, konradii, mastoidea,* ***procera***(Commercial Edible)

VARIETIES *procera procera, procera pseudo-olivascens*

GENUS *Notholepiota* Counterfit Parasol Gill Mushroom

SPECIES *areolata*

GENUS *Parasola* Greater Parasol Gill Mushroom

SPECIES *auricoma, hemerobia, hercules, kuehneri, leiocephala, lilatincta, megasperma, misera, nudiceps, plicatilis, schroeteri*

GENUS *Phaeolepiota* Dusky Parasol Gill Mushroom

SPECIES *aurea*

VARIETIES *aurea spectabilis*

GENUS *Volvolepiota* Rolled Parasol Gill Mushroom

SPECIES *albida, brunnea*

FAMILY 09 *Marasmiaceae* Pinwheel Gill Mushroom

GENUS *Marasmius* Pinwheel Gill Mushroom

SPECIES *alliaceus, androsaceus, anomalus, bellipes, bulliardii, buxi, caricis, chordalis, cohaerens, collinus, corbariensis, cornelii, crinis-equi, curreyi, epidryas, epiphylloides, epiphyllus, favrei, fulvoferrugineus, graminum, hudsonii, inodorus, insititius, kroumirensis, limosus, minutus,* ***oreades***(Dye Herb)*, fulvoferrugineus, pallidocephalus, palmivorus, phalaricola, plicatulus, prasiosmus, pseudocaricis, pulcherripes, quercophilus, recubans, resinosus, rotula, sacchari, saccharinus, scorodonius, setosus, siccus, splachnoides, tenuiparietalis, thujinus, torquescens, wettsteinii, wynnei*

VARIETIES *anomalus epodius, anomalus littoralis, cohaerens ceratopus, crinis-equi repens, crinis-equi trichorhizus, epiphyllus plantaginis, insititius albipes, minutus capillipes, setosus eufoliatus, wynnei carpathicus, wynnei globularis, wynnei suaveolens*

GENUS *Amyloflagellula* Amyl Flagella Pinwheel Gill Mushroom

SPECIES *inflata, pseudoarachnoidea, pulchra, verrucosa*

GENUS *Anastrophella* Strophic Pinwheel Gill Mushroom

SPECIES *macrospora, subpeltata*

GENUS *Aphyllotus* Phyllote Pinwheel Gill Mushroom

SPECIES *campanelliformis*

GENUS *Armillaria* Bracelet Pinwheel Gill Mushroom

SPECIES *albolonaripes, borealis, calvescens, cepistipes, ectypa, fuscipes, gallica, gemina, heimii, hinnulea, laricina, limonea, luteobubalina, matsutake,* ***mellea***(Poison Raw, Edible Well Cooked)*, novae-zelandiae, obscura, ostoyae, robustum, sinapina, subannulatum, tabescens*

VARIETIES *gallica bulbosa, gallica lutea,* ***mellea glabra, mellea maxima, mellea minor, mellea sulphurea****, obscura polymyces, tabescens socialis*

GENUS *Armillariella* Lesser Bracelet Pinwheel Gill Mushroom

SPECIES *sparrei*

GENUS *Baeospora* Bael Pinwheel Gill Mushroom

SPECIES *myosura, myriadophylla*

GENUS *Calathella* Callisto Pinwheel Gill Mushroom

SPECIES *albolivida, columbiana, dichroa, digitiformis, ellisii, eruciformis, gayana, mangrovei*

GENUS *Campanella* Bell Pinwheel Gill Mushroom

SPECIES *boninensis, caesia, cucullata, eberhardtii, junghuhnii, purpureobrunnea*

GENUS *Cephaloscypha* Scyphcap Pinwheel Gill Mushroom

SPECIES *mairei, morlichensis*

GENUS *Chaetocalathus* Bristle Callisto Pinwheel Gill Mushroom

SPECIES *craterellus, eurysporus*

GENUS *Cymatellopsis* Cyme Pinwheel Gill Mushroom

SPECIES *ilmiana*

GENUS *Cyphellopsis* Cypher Pinwheel Gill Mushroom

SPECIES *anomala*

GENUS *Cyptotrama* Coincap Pinwheel Gill Mushroom

SPECIES *asprata, chrysopeplum, costesii, dennisii, depauperatum, deseynesiana, granulosa, hygrocyboides, lachnocephalum, macrobasidium, niveum, pauperum, platense, songolarum, verruculosum*
GENUS *Dactylosporina* Finger Pinwheel Gill Mushroom
SPECIES *macracantha, steffenii*
GENUS *Deigloria* Deiglor Pinwheel Gill Mushroom
SPECIES *amoena, modesta, pulchella, pulcherrima*
GENUS *Epicnaphus* Epic Naph Pinwheel Gill Mushroom
SPECIES *longispora, phalaropus, sphaerodermus*
GENUS *Favolaschia* Fava Lasch Pinwheel Gill Mushroom
SPECIES *alsophilae, andina, anemoenerosea, aulaxina, aurantiaca, auriscalpium, calocera, citrina, dumontii, dybowskyana, echinata, fendleri, furfurella, fusijanensis, heliconiae, lateritia, mainsii, meridae, moelleri, montana, nipponica, oligogloea, oligopora, pantherina, papuana, pezizoidea, phyllostachydis, pterigena, puberula, puiggarii, pygmaea, roseogrisea, rubra, sabalensis, selloana, singeriana, sprucei, subamyloidea, subceracea, teapae, varariotecta, violascens, volkensii, zenkeriana*
GENUS *Fissolimbus* Fissure Limbic Pinwheel Gill Mushroom
SPECIES *fallaciosus*
GENUS *Flagelloscypha* Flagella Scyph Pinwheel Gill Mushroom
SPECIES *abieticola, abruptiflagellata, christinae, dextrinoidea, donkii, faginea, fusispora, kavinae, libertiana, merxmuelleri, minutissima, montis-anagae, oblongispora, obovatispora, orthospora, pilatii, pseudopanacis, punctiformis, subnuda, tetraedrispora, venezuelae, virginea*
GENUS *Flammulina* Flamegill Pinwheel Gill Mushroom
SPECIES *fennae, ononidis, **velutipes***(Edible)
VARIETIES *ononidis pratensis, velutipes velutipes, velutipes lactea, velutipes lactea-alba*
GENUS *Gerronema* Gerron Pinwheel Gill Mushroom
SPECIES *albidum, josserandii, stevensonii*
GENUS *Glabrocyphella* Bald Cypher Pinwheel Gill Mushroom
SPECIES *ailanthi, bananae, brunneocrystallina, dermatoides, ellisiana, epileucina, filicicola, ohiensis, palmarum, rubescens, stercoraria, upplandensis*
GENUS *Hispidocalyptella* Hairy Calypter Pinwheel Gill Mushroom
SPECIES *australis*
GENUS *Hormomitaria* Hormon Miter Pinwheel Gill Mushroom
SPECIES *sulphurea, tenuipes*
GENUS *Lachnella* Downy Pinwheel Gill Mushroom
SPECIES *alboviolascens, disseminata, subiculosa, sulphurea, villosa*
GENUS *Lecanocybe* Lichencap Pinwheel Gill Mushroom
SPECIES *lateralis*
GENUS *Lentinula* Lentinule Pinwheel Gill Mushroom
SPECIES *aciculospora, boryana, edodes, guarapiensis, lateritia, novae-zelandiae, raphanica*
GENUS *Macrocystidia* Large Cyst Pinwheel Gill Mushroom
SPECIES *cucumis*
VARIETIES *cucumis latifolia*
GENUS *Manuripia* Manurip Pinwheel Gill Mushroom
SPECIES *bifida*
GENUS *Marasmiellus* Lesser Pinwheel Gill Mushroom
SPECIES *affixus, candidus, cocophilus, gossypinulus, humillimus, ornatissimus, ramealis, rosellus, scandens, tricolor, tropicalis, vaillantii*
GENUS *Merismodes* Merism Pinwheel Gill Mushroom
SPECIES *bresadolae, fasciculata*
GENUS *Micromphale* False Pinwheel Gill Mushroom
SPECIES *brassicolens, cauvetii, foetidum, perforans*
VARIETIES *brassicolens brassicolens, brassicolens pallidus, foetidum rufocarneum, foetidum venosum*
GENUS *Mucidula* (Synonym for *Xerula*)
GENUS *Nochascypha* Noch Scyph Pinwheel Gill Mushroom
SPECIES *dumontii, filicina, paraguayensis, stricta*
GENUS *Oudemansiella* Deep Root Pinwheel Gill Mushroom
SPECIES *canarii, melanotricha, mucida, platensis*
GENUS *Phaeodepas* Dusky Pinwheel Gill Mushroom
SPECIES *dennisii, nutans*
GENUS *Pseudotyphula* Typhus Pinwheel Gill Mushroom
SPECIES *ochracea*
GENUS *Rhizomarasmius* Rhizid Pinwheel Gill Mushroom
SPECIES *pyrrhocephalus, undatus*
GENUS *Rhodocollybia* Red Collybion Pinwheel Gill Mushroom
SPECIES ***butyracea***(Edible), *fodiens, maculata,* ***prolixa***(Edible)
VARIETIES ***butyracea butyracea, butyracea asema****, maculata maculata, maculata scorzonerea****, prolixa prolixa, prolixa distorta***
GENUS *Skepperiella* Skepper Pinwheel Gill Mushroom

SPECIES *cochlearis, merulioides, populi, spathularia*
GENUS *Stipitocyphella* Stipple Cypher Pinwheel Gill Mushroom
SPECIES *keniensis*
GENUS *Strobilurus* Strobe Tail Pinwheel Gill Mushroom
SPECIES *esculentus, stephanocystis, tenacellus*
GENUS *Stromatocyphella* Stroma Cypher Pinwheel Gill Mushroom
SPECIES *conglobata*
GENUS *Tetrapyrgos* Tetrad Pyrgose Pinwheel Gill Mushroom
SPECIES *alba, atrocyanea, austrochilensis, dendrophora, diplocystis, goniospora, nigripes, olivaceonigra, peullensis, reducta, simulans, stipitata, subcinerea, subdendrophora, tropicalis*
GENUS *Xerula* Strange Pinwheel Gill Mushroom
SPECIES *caussei, longipes, megalospora, pudens, radicata, xeruloides*
VARIETIES *caussei nigra, radicata marginata*

FAMILY 10 *Mesophelliaceae* Cork Gill Mushroom
GENUS *Mesophellia* (Synonym for *Malajczukia*)
GENUS *Andebbia* Debbie Cork Gill Mushroom
SPECIES *pachythrix*
GENUS *Gummiglobus* Gummy Globe Cork Gill Mushroom
SPECIES *agglutinosporus, joyceae*
GENUS *Malajczukia* Malajczuk Cork Gill Mushroom
SPECIES *amicorum, fusispora, ingratissima, karrialis, novae-zelandiae, spumoidea, tropica, viridigleba*
GENUS *Nothocastoreum* Castor Cork Gill Mushroom
SPECIES *cretaceum*

FAMILY 11 *Omphalinaceae* Navalcap Gill Mushroom
GENUS *Omphalina* Navelcap Gill Mushroom
SPECIES *arctica, brevibasidiata, chionophila, chlorocyanea, chromacea, demissa, epichysium, ericetorum, favrei, favrei brownii, fulvopallens, galericolor, gerardiana, griseopallida, kuehneri, luteovitellina, marchantiae, mutila, obatra, obscurata, oniscus, parvivelutina, peltigerina, philonotis, postii, pseudomuralis, pyxidata, rickenii, rigidipes, rivulicola, subhepatica, velutipes, wallacei*
VARIETIES *chlorocyanea viridis, ericetorum albida, ericetorum umbellifera, fulvopallens pseudoandrosacea, galericolor lilacinicolor, gerardiana fusconigra, luteovitellina flava, peltigerina cupulatoides, pyxidata muralis, rickenii cupulata, velutipes minor*
GENUS *Austroomphaliaster* Southern Navalcap Gill Mushroom
SPECIES *nahuelbutensis*
GENUS *Chrysomphalina* Golden Navalcap Gill Mushroom
SPECIES *chrysophylla, grossula*
GENUS *Clavomphalia* Collared Navalcap Gill Mushroom
SPECIES *yunnanensis*
GENUS *Cryptomphalina* Cryptic Navelcap Gill Mushroom
SPECIES *sulcata*
GENUS *Myxomphalia* Mucus Navalcap Gill Mushroom
SPECIES *maura*
VARIETIES *maura alba*
GENUS *Omphalia* Lesser Navalcap Gill Mushroom
SPECIES *pigmentata, tralucida*
GENUS *Omphaliaster* Star Navalcap Gill Mushroom
SPECIES *asterosporus, borealis, ianthinocystis, kyrtosporus, nauseodulcis, obolus, palustris*
GENUS *Omphalotus* Jack O Lantern Navalcap Gill Mushroom
SPECIES *illudens, mexicanus, olearius, olivascens, subilludens*
VARIETIES *olivascens indigo*
GENUS *Phytoconis* Cone Navalcap Gill Mushroom
SPECIES botryoides
GENUS *Pseudoomphalina* Least Navalcap Gill Mushroom
SPECIES *graveolens, pachyphylla*
GENUS *Semiomphalina* Half Navalcap Gill Mushroom
SPECIES *leptoglossoides*
GENUS *Xeromphalina* Strange Navalcap Gill Mushroom
SPECIES *campanella, cauticinalis, cornui, fellea*

FAMILY 12 *Pleurotaceae* Oyster Gill Mushroom
GENUS *Pleurotus* Oyster Gill Mushroom
SPECIES *abieticola, albidus, alocasiae, armeniascus, aureovillosus, australis, calyptratus, chrysorrhizus, citrinopileatus, columbinus, cornucopiae, craspedius, cyatheicola, cystidiosus, decipiens, djamor, dryinus, ëous,* ***eryngii***(Commercial Edible), *euosmus, favoloides, flabellatus, floridanus, hyacinthus, incarnatus, lactuosus, lampas, lazoi, lilaceilentus, lindquistii, musae, olivascens, omnivagus, ostreatoroseus,* ***ostreatus***(Commercial Edible), *penangensis, populinus, problematicus, pulmonarius, rattenburyi, roseolus, sapidus, smithii, staringii, subareolatus, submembranaceus, subviolaceus, viscidulus*

VARIETIES *albidus calvescens, albidus jacksonii, albidus laciniatocrenatus, cystidiosus formosensis, djamor djamor, djamor leptogrammus, djamor moselei, djamor ninguidus, djamor pacificus, djamor placentodes, djamor prometheus, djamor salmoneostramineus, djamor scabellus, djamor scabriusculus, dryinus acerinus, dryinus albertinii, dryinus corticatus, dryinus pometi, dryinus spongiosus, dryinus tephrotrichus,* ***eryngii elaeoselini*****,** ***eryngii ferulae*****,** ***eryngii fuscus eryngii nebrodensis*****,** ***ostreatus opuntiae, ostreatus salignus, ostreatus revolutus***

GENUS *Acanthocystis* (Synonyn for *Hohenbuehelia*)

GENUS *Agaricochaete* Bristle Oyster Gill Mushroom

SPECIES *hericium, indica, keniensis, mirabilis*

GENUS *Geopetalum* Earth Petal Oyster Gill Mushroom

SPECIES *album, brunnescens*

GENUS *Hohenbuehelia* Hohenbuehel Oyster Gill Mushroom

SPECIES *abietina, aciculospora, amazonica, angustata, approximans, atrocoerulea, aurantiocystis, auriscalpium, austrocedri, barbatula, campinaranae, culmicola, cyphelliformis, delasotae, espeletiae, fluxilis, grisea, heterosporica, horakii, hydrogeton, izonetae, leightonii, longipes, mastrucata, minutissima, mustialaensis, myxotricha, nigra, panelloides, pergelatinosa, petalodes, pinacearum, pinicola, recedens, reniformis, rickenii, sciadia, silvana, testudo, tropicalis*

VARIETIES *atrocoerulea albidotomentosa, culmicola culmicola, culmicola macrospora, grisea microspora, horakii grisea, petalodes petalodes, petalodes carbonaria, petalodes geogenia, geogenia queletii, rickenii tremula*

GENUS *Hypsizygus* Hyponic Viscose Oyster Gill Mushroom

SPECIES *tessulatus, ulmarius*

GENUS *Panellus* Panel Oyster Gill Mushroom

SPECIES *albifavolus, alutaceus, ambiguus, aureofactus, bambusarum, bambusifavolus, belangeri, brunneomaculatus, dichotomus, exiguus, fuscatus, glutinosus, hispidifavolus, intermedius, ligulatus, luminescens, magnus, megalosporus, melleo-ochraceus, microsporus, mitis, parvulus, patellaris, pauciporus, pendens, pusillus, pyruliferus, ringens, serotinus, stipticus, sublamelliformis, sublevatus, violaceofulvus*

VARIETIES *pusillus copelandii, pusillus gloeocystidiatus, stipticus albidotomentosus, stipticus farinaceus, stipticus lateralis*

GENUS *Panus* Fan Oyster Gill Mushroom

SPECIES *conchatus, domicola*

GENUS *Phyllotopsis* Fancap Oyster Gill Mushroom

SPECIES *nidulans*

GENUS *Pleurocybella* Manycap Oyster Gill Mushroom

SPECIES *porrigens*

GENUS *Pleurotopsis* Lesser Oyster Gill Mushroom

SPECIES *longinqua*

FAMILY 13 *Pluteaceae* Roofcap Gill Mushroom

GENUS *Pluteus* Roofcap Gill Mushroom

SPECIES *admirabilis, atromarginatus, aurantiorugosus, cervinus, chrysophaeus, cinereofuscus, dietrichii, ephebeus, exiguus, flavofulgineus, griseoluridus, hispidulus, inquilinus, insidiosus, leoninus, luctuosus, magnus, nanus, pellitus, petasatus, phlebophorus, plautus, podospileus, pouzarianus, pseudorobertii, robertii, romellii, rugosidiscus, salicinus, satur, semibulbosus, thomsonii, tomentosulus, umbrosus, villosus, washingtonensis*

VARIETIES *atromarginatus nigrofloccosus, atromarginatus tricuspidatus, aurantiorugosus caloceps, aurantiorugosus coccineus, cervinus cervinus, cervinus atricapillus, cervinus curtisii, cervinus eximius, chrysophaeus galeroides, chrysophaeus luteovirens, chrysophaeus xanthophaeus, cinereofuscus major, cinereofuscus olivaceus, dietrichii rimulosus, ephebeus lepiotoides, ephebeus murinus, ephebeus pearsonii, leoninus sororiatus, luctuosus marginatus, nanus griseopus, petasatus patricius, plautus alborugosus, plautus boudieri, plautus depauperatus, plautus dryophiloides, plautus gracilis, plautus granulatus, plautus punctipes, podospileus minutissimus, podospileus seticeps, romellii lutescens, romellii splendidus, salicinus beryllus, salicinus floccosus, satur pallescens, thomsonii cinereus*

GENUS *Amanita* Amanite Roofcap Gill Mushroom

SPECIES *amici, aspera, atkinsoniana, bisporigera, brunnescens,* ***caesarea***(Commercial Edible), *calyptrata, calyptroderma, chlorinosma, cinereoconia, cinereopannosa, citrina, cokeri, cothurnata, daucipes, eliae, farinosa, flavescens, flavoconia, flavorubescens, franchetii, frostiana, gemmata, inopinata, lepiotoides, longipes, malleata, microlepis, muscaria, mutabilis, nitida, olivaceogrisea, onusta, ovoidea, pachycolea, pantherina, peckiana, pelioma, phalloides, polypyramis, ponderosa, porphyria, psuedoporphyria, ravenelii, rhopalopus,* ***rubescens***(Poison Raw, Edible Well Cooked), *silvicola, singeri, smithiana, spissa, spreta,* ***strobiliformis***(Edible), *submembranacea, tenuifolia, tephrea,* ***vaginata***(Poison Raw, Edible Well Cooked), *verna, vernalis, virosa*

VARIETIES ***caesarea aurantia****, citrina citrina, citrina alba, citrina bulbosa, citrina mappa, flavescens contui, gemmata gemmata, gemmata adnata, gemmata exannulata, gemmata junquillea, muscaria muscaria, muscaria aureola, muscaria aureola-puella, muscaria circinnata, muscaria formosa, muscaria minor, muscaria regalis, muscaria regalis-emilii, muscaria regalis-umbrina, pantherina pantherina, pantherina robusta, phalloides alba, phalloides umbrina, phalloides viridis, porphyria recutita,* ***rubescens rubescens, rubescens alba, rubescens annulosulphurea, rubescens annulosulphurea-magnifica****, spissa ampla, spissa cariosa, spissa excelsa, spissa valida,* ***strobiliformis solitaria****, submembranacea submembranacea, submembranacea bispora,*

vaginata vaginata*, *vaginata alba*, *vaginata alba-fungites*, *vaginata albida*, *vaginata grisea*, *vaginata livida*, *vaginata plumbea*, *vaginata violacea, *verna virosa*

GENUS *Amanitopsis* Lesser Amanite Roofcap Gill Mushroom

SPECIES *argentea, battarrae, ceciliae, crocea, friabilis, fulva, lividopallescens, nivalis, pachyvolvata, punctata*

VARIETIES *argentea mairei, battarrae umbrinolutea, ceciliae inaurata, ceciliae strangulata*

GENUS *Aspidella* Asp Amanite Roofcap Gill Mushroom

SPECIES *echinocephala, vittadinii*

VARIETIES *echinocephala echinocephala, echinocephala aculeata*

GENUS *Torrendia* Torrend Roofcap Gill Mushroom

SPECIES *arenaria, pulchella*

GENUS *Volvaria* Sheath Roofcap Gill Mushroom

SPECIES *bakeri*, ***bombycina***(Commercial Edible), *diplasia, esculenta, gloiocephala, hypopithys, media, murinella, pusilla, surrecta, taylorii, volvacea*

VARIETIES *gloiocephala speciosa, hypopithys biloba, hypopithys plumulosa, hypopithys pubescentipes, pusilla parvula, surrecta loveiana, taylorii umbonata, volvacea heimii*

GENUS *Volvariella* Lesser Sheath Roofcap Gill Mushroom

SPECIES *caesiotincta, pubipes, reidii*

VARIETIES *reidii parvispora*

FAMILY 14 *Psathyrellaceae* Crumblecap Gill Mushroom

GENUS *Psathyrella* Crumblecap Gill Mushroom

SPECIES *ammophila, artemisiae, atrolaminata, badiovestita, bifrons, bipellis, borgensis, candolleana, canoceps, caput-medusae, castaneifolia, cernua, chondroderma, clivensis, conopilus, coprophila, corrugis, cotonea, dennyensis, dilineata, dunensis, fatua, fibrillosa, flexispora, friesii, frustulenta, fulvescens, fusca, gordonii, gossypina, gracilis, hirta, incerta, laevissima, leucotephra, longicauda, lutensis, maculata, marcescibilis, microrhiza, mucrocystis, multipedata, murcida, narcotica, noli-tangere, obtusata, ocellata, oedipus, olympiana, panaeoloides, pannucioides, pellucidipes, pennata, phegophila, piluliformis, ploddensis, polycystis, populina, prona, pseudobifrons, pseudocasca, pseudocorrugis, pseudogordonii, pseudogracilis, pygmaea, rigidipes, romseyensis, rostellata, rugocephala, sacchariolens, sarcocephala, solitaria, spadicea, spadiceogrisea, sphaerocystis, sphagnicola, spintrigera, spintrigeroides, stellata, stercoraria, tephrophylla, trepida, typhae, ulignicola, umbrina, vernalis, vinosofulva, vyrnwyensis*

VARIETIES *artemisiae artemisiae, artemisiae microspora, bipellis barlae, candolleana appendiculata, candolleana egenula, candolleana microlepidota, caput-medusae jerdonii, cernua papyracea, conopilus arata, conopilus subatrata, coprophila fimetaria, corrugis caudata, corrugis clavigera, corrugis gracilis, corrugis substerilis, corrugis vinosa, frustulenta cortinarioides, fulvescens fulvescens, fulvescens brevicystis, fulvescens trivialis, gossypina xanthocystis, hirta coprobia, marcescibilis fragilissima, marcescibilis lactea, microrhiza semivestita, microrhiza squamifera, obtusata obtusata, obtusata aberrans, olympiana amstelodamensis, pennata carbonicola, piluliformis hydrophila, piluliformis subpapillata, prona prona, prona albidula, prona cana, prona cana-atomata, prona orbitarum, prona subatomata, pygmaea consimilis, sacchariolens suavissima, solitaria rannochii, spadiceogrisea exalbicans, spadiceogrisea mammifera, spadiceogrisea utriformis, sphagnicola psathyroides, vernalis groegeri, vinosofulva utriformis*

GENUS *Drosophila* Drose Crumblecap Gill Mushroom

SPECIES *badia, spadicophylla*

GENUS *Gasteroagaricoides* Cavity Crumblecap Gill Mushroom

SPECIES *ralstoniae*

GENUS *Lacrymaria* Teardrop Crumblecap Gill Mushroom

SPECIES *glareosa, lacrymabunda, pyrotricha*

VARIETIES *lacrymabunda gracillima, lacrymabunda velutina*

GENUS *Macrometrula* Metric Crumblecap Gill Mushroom

SPECIES *rubriceps*

GENUS *Rhacophyllus* Beacon Crumblecap Gill Mushroom

SPECIES *lilacinus*

GENUS *Xerocoprinus* Inkycap Crumblecap Gill Mushroom

SPECIES *arenarius*

FAMILY 15 *Pterulaceae* Wingcap Gill Mushroom

GENUS *Pterula* Wingcap Gill Mushroom

SPECIES *bambusae, bresadolana, bromeliphila, brunneola, brunneosetosa, campoi, capillaris, caricis-pendulae, carpophila, complanata, cystidiata, decumbens, echo, epiphylla, epiphylloides, fluminensis, gordius, gracilis, intermedia, laxa, longispora, loretensis, moniliformis, navicula, palmicola, phyllodicola, plumosoides, pteridicola, pterulicioides, rigida, robusta, sclerotiicola, simplex, subulata, taxiformis, thindii, typhuloides, velutipes, verticillata, vinacea*

VARIETIES *subulata debilis, subulata multifida, taxiformis humilis*

GENUS *Actiniceps* Radial Wingcap Gill Mushroom

SPECIES *horrida, secunda*

GENUS *Allantula* Sac Membrane Wingcap Gill Mushroom

SPECIES *diffusa*

GENUS *Deflexula* Deflexed Wingcap Gill Mushroom

SPECIES *fascicularis, lilaceobrunnea, major, mangiformis, microspora, pacifica, pinnata, sprucei, sulcispora, ulmi*
GENUS *Dimorphocystis* Dimorphic Cyst Wingcap Gill Mushroom
SPECIES *laevis, subcapitata*
GENUS *Parapterulicium* Least Wingcap Gill Mushroom
SPECIES *octopodites, simplex, subarbusculum*
GENUS *Pterulicium* Lesser Wingcap Gill Mushroom
SPECIES *xylogenum*

FAMILY 16 *Schizophyllaceae* Splitgill Gill Mushroom
GENUS *Schizophyllum* Splitgill Gill Mushroom
SPECIES *amplum, commune, fasciatum, murrayi, radiatum, variabile*
VARIETIES *commune alneum, commune multifidum*
GENUS *Auriculariopsis* Ear Splitgill Gill Mushroom
SPECIES *albomellea*
GENUS *Henningsomyces* Henning Splitgill Gill Mushroom
SPECIES *candidus, mutabilis, patinaceus, puber, separatus*
GENUS *Rectipilus* Downy Splitgill Gill Mushroom
SPECIES *bavaricus, cistophilus, confertus, davidii, erubescens, fasciculatus, idahoensis, natalensis, sulphureus*

FAMILY 17 *Strophariaceae* Ringstalk Gill Mushroom
GENUS *Stropharia* Ringstalk Gill Mushroom
SPECIES *aeruginosa, albonitens, ambigua, aurantiaca, caerulea,* ***coronilla***(Edible), *halophila, hardii, hornemannii, inuncta, luteonitens, mammillata, melanosperma, percevalii,* ***rugosoannulata***(Edible), *semiglobata, squamosa, squamulosa, thrausta, umbonatescens*
VARIETIES *aeruginosa albocyanea, aeruginosa albocyanea-worthingtonii, coronilla obturata, hornemannii depilata, inuncta lundensis, inuncta pallida, inuncta upsaliensis,* ***rugosoannulata ferrii****, semiglobata stercoraria, squamulosa aeruginosa, squamulosa calolepis*
GENUS *Geophila* (Synonym for *Stropharia*)
GENUS *Hygrophana* (Synonym *for Stropharia*)
GENUS *Hypholoma* Fringed Margin Ringstalk Gill Mushroom
SPECIES *acutum, capnoides, dispersum, elongatipes, elongatum, epixanthum, ericaeoides, ericaeum, fasciculare, laeticolor, lapponicum, lateritium, marginatum, myosotis, polytrichi, radicosum, subericaeum, sublateritium, udum, xanthocephalum*
VARIETIES *fasciculare pusillum, lateritium pomposum, lateritium schaefferi, lateritium squamosum, lateritium vulgaris, udum fulvidulum*
GENUS *Kuehneromyces* Kuehner Ringstalk Gill Mushroom
SPECIES *leucolepidotus, lignicola, mutabilis, vernalis*
GENUS *Leratiomyces* Lerat Ringstalk Gill Mushroom
SPECIES *atrovirens, coccineus, similis, smaragdinus*
GENUS *Leucopholiota* Light Scalecap Ringstalk Gill Mushroom
SPECIES *decorosa*
GENUS *Melanotus* Dark Ringstalk Gill Mushroom
SPECIES *hartii, horizontalis, phillipsii, proteus*
GENUS *Mythicomyces* Mythic Ringstalk Gill Mushroom
SPECIES *corneipes*
GENUS *Naematoloma* (Synonym *for Stropharia*)
GENUS *Nivatogastrium* Snow Ringstalk Gill Mushroom
SPECIES *baylisianum, lignicola, nubigenum, sulcatum*
GENUS *Pholiota* Scalecap Ringstalk Gill Mushroom
SPECIES *abstrusa, adiposa, albocrenulata, alnicola, astragalina, aurivella, brunnescens, conissans, decussata, flammans, flavida, formosa, fulvozonata, gummosa, henningsii, heteroclita, hiemalis, highlandensis, jahnii, lenta, limonella, lubrica, lucifera, marginella, mixta, nameko, ochrochlora, pinicola, polychroa, populnea, pseudofascicularis, pusilla, scamba, spumosa, squarrosa, subochracea, squarrosoides, tuberculosa, veris, vernalis*
VARIETIES *albocrenulata fusca, alnicola alnicola, alnicola apicrea, alnicola aromatica, alnicola salicicola, alnicola salicicola-amara, aurivella lilacifolia, conissans graminis, conissans inaurata, conissans juncina, gummosa cookei, highlandensis carbonaria, highlandensis gigantea, highlandensis persicina, jahnii muelleri, limonella ceriferoides, lubrica groenlandica, mixta filia, mixta xanthophaea, populnea comosa, populnea destruens, squarrosa verruculosa, subochracea nematolomoides, tuberculosa curvipes*
GENUS *Pleuroflammula* Flamecap Ringstalk Gill Mushroom
SPECIES *ragazziana*
GENUS *Psilocybe* Smoothcap Ringstalk Gill Mushroom
SPECIES *apelliculosa, aztecorum, azurescens, bohemica, caerulescens, chionophila, coprophila, crobula,* ***cubensis***(Medicinal Herb), *cyanescens, fimetaria, inquilina, magica, merdaria, merdicola,* ***mexicana***(Medicinal Herb), *moelleri, montana, novae-zelandiae, phyllogena, pratensis, pseudobullacea, quebecensis, scobicola, semilanceata, strictipes, stuntzii, subaeruginascens, subcoprophila, subcubensis, subviscida, tampanensis, turficola, velifera, zapotecorum*

VARIETIES *inquilina muscorum, montana physaloides, phyllogena rhombispora, strictipes caerulescens, subviscida velata, subviscida velata-graminicola*
GENUS *Weraroa* Weraro Ringstalk Gill Mushroom
SPECIES *coprophila, cucullata, erythrocephala, nivalis, novae-zelandiae, patagonica, spadicea, virescens*
FAMILY 18 *Tricholomataceae* Cavalier Gill Mushroom
GENUS *Tricholoma* Cavalier Gill Mushroom
SPECIES *acerbum, aestuans, albobrunneum, album, apium,* ***argyraceum***(Edible), *arvernense, atrosquamosum, atroviolaceum, aurantium,* ***batschii***(Edible Well Cooked), *bresadolanum, bufonium,* ***caligatum***(Edible), *cingulatum,* ***colossus***(Edible), ***columbetta***(Edible), ***equestre***(Edible), *focale,* ***fulvum***(Edible), *gausapatum, goliath, goniospermum, groanense, hordum,* ***imbricatum***(Edible), *inamoenum, intermedia, lascivum, luridum,* ***magnivelare***(Edible), ***myomyces***(Edible), *nauseosum, orirubens, pardinum, pessundatum, platyphyllum, populinum,* ***portentosum***(Edible), *psammopus, radotinense,* ***robustum***(Edible), *salero,* ***saponaceum***(Edible), ***scalpturatum***(Edible), *sciodes, sejunctum, squarrulosum, stans, stiparophyllum, sudum, sulphurescens,* ***sulphureum***(Edible), *tigrinum, triste, unifactum, ustale, ustaloides,* ***vaccinum***(Edible), *venenata, vinaceogriseum, virgatum, viridifucatum, viridilutescens, zelleri*
VARIETIES *album raphanicum, apium helviodor,* ***argyraceum inocybeoides,*** *arvernense sejunctoides, atrosquamosum nigromarginatum,* ***batschii subannulatum, equestre auratum, equestre flavovirens, equestre populinum, fulvum flavobrunneum, fulvum nictitans,*** *inamoenum insigne, lascivum robustum,* ***magnivelare arenicola, magnivelare ponderosa, magnivelare murrillianum, myomyces bisporigerum, myomyces terreum,*** *orirubens horribile,* ***saponaceum saponaceum, saponaceum atrovirens, saponaceum lavedanum, saponaceum lavedanum-boudieri, saponaceum napipes, saponaceum squamosum, saponaceum squamosum-ardosiacum, saponaceum sulphurinum, scalpturatum chrysites,*** *stiparophyllum pseudoalbum, sulphurescens resplendens,* ***sulphureum sulphureum, sulphureum hemisulphureum, sulphureum pallidum****, ustale fulvellum*
GENUS *Amparoina* Amparo Cavalier Gill Mushroom
SPECIES *heteracantha, spinosissima*
GENUS *Arrhenia* Staminate Cavalier Gill Mushroom
SPECIES *auriscalpium, latispora, littoralis, lobata, retiruga, rustica, spathulata, sphagnicola, umbratilis*
GENUS *Arthrosporella* Arthrospore Cavalier Gill Mushroom
SPECIES *ditopa*
GENUS *Asproinocybe* Asp Cap Cavalier Gill Mushroom
SPECIES *brunneolilacina, lactifera, russuloides*
GENUS *Asterophora* Hazel Cavalier Gill Mushroom
SPECIES *agaricicola, lycoperdoides, mirabilis, parasitica*
VARIETIES *lycoperdoides agaricoides, lycoperdoides asterophora*
GENUS *Asterotus* Star Ear Cavalier Gill Mushroom
SPECIES *dealbatus*
GENUS *Biannulara* Double Ring Cavalier Gill Mushroom
SPECIES *imperialis, zelleri*
GENUS *Callistodermatium* Callisto Skin Cavalier Gill Mushroom
SPECIES *violascens*
GENUS *Calocybe* Callisto Cap Cavalier Gill Mushroom
SPECIES *alpestris, carnea, chrysenteron, constricta, fallax,* ***gambosa***(Commercial Edible), *ionides, juncicola, obscurissima, onychina, persicolor*
VARIETIES ***gambosa georgii***
GENUS *Calyptella* Calypter Cavalier Gill Mushroom
SPECIES *australis, bakeriana, bonairensis, campanula, capula, crateriformis, epibrya, flava, longipes, musae, nabambissoënsis, pteridophytorum, puiggarii, urbanii*
GENUS *Cantharellopsis* Least Chanterelle Cavalier Gill Mushroom
SPECIES *prescotii*
GENUS *Cantharellula* Lesser Chanterelle Cavalier Gill Mushroom
SPECIES *humicola, oregonensis, umbonata*
GENUS *Cantharocybe* Chanterelle Cap Cavalier Gill Mushroom
SPECIES *gruberi*
GENUS *Catathelasma* Cat Nipple Cavalier Gill Mushroom
SPECIES *evanescens, imperiale, singeri, ventricosum*
GENUS *Catatrama* Cat Tram Cavalier Gill Mushroom
SPECIES *costaricensis*
GENUS *Cellypha* Cellyph Cavalier Gill Mushroom
SPECIES *goldbachii, stictoidea*
GENUS *Cheimonophyllum* Cheimon Cavalier Gill Mushroom
SPECIES *candidissimum, roseum, stypticoides*
GENUS *Chromosera* Chromic Cavalier Gill Mushroom
SPECIES *cyanophylla*
GENUS *Clitocybe* Funnelcap Cavalier Gill Mushroom
SPECIES *agrestis, albirhiza, albofragrans, alexandri, amarescens, americana, angustissima, augeana, barbularum, bartelliae, brumalis, candicans, candida, catinus, clavipes, concava, costata, dealbata*(Poison),

diatreta, dicolor, diosma, ditopa, elegantula, eucalyptorum, festiva, festivoides, fragrans, frysica, fuscosquamula, ***geotropa***(Edible), ***gibba***(Edible), *glaucoalba, gracilipes, graveolens, harmajae, houghtonii, hydrogramma,* ***inornata***(Edible), *lateritia, lituus, marginella, martiorum, metachroa, morbifera, nebularis, nuda, odora, ornamentalis, phaeophthalma, phosphorea, phyllophila*(Poison), *pruinosa, rhizophora, rivulosa*(Poison), *robusta, ruderalis, sinopica, squamulosa, squamulosoides, suaveolens, subalutacea, subclavipes, subconnexa, subcordispora, subdryadicola, subinvoluta, subsalmonea, subspadicea, tarda, tenuissima, trulliformis, truncicola, umbilicata,* ***vermicularis***(Edible), *vibecina*

VARIETIES *agrestis graminicola, amarescens nitrophila, candicans dryadicola, candicans gallinacea, candicans tuba, clavipes comitialis, costata incilis, fragrans deceptiva, fragrans depauperata, frysica sericella, fuscosquamula exilis,* ***geotropa maxima gibba infundibuliformis, gibba membranacea****, gracilipes dionysae,* ***inornata zygophylla****, metachroa decembris, nebularis alba, odora alba, odora trogii, odora virens, odora viridis, phaeophthalma fritilliformis, phyllophila cerussata*(Poison), *phyllophila pithyophila*(Poison), *rivulosa minor, rivulosa neptunea, sinopica subsinopica, squamulosa sinopicoides, subcordispora pseudo-obbata, vibecina favrei, vibecina langei, vibecina orientalis*

GENUS *Clitocybula* Coincap Cavalier Gill Mushroom

SPECIES *abundans,* ***familia***(Edible), *lacerata*

GENUS *Collybia* Sooty Coincap Cavalier Gill Mushroom

SPECIES *cirrhata, cookei,* ***coracina***(Edible), *iocephala, luxurians, polyphylla, subnuda, tuberosa*

VARIETIES *cirrhata amanitae, tuberosa sclerotipes*

GENUS *Conchomyces* Conch Cavalier Gill Mushroom

SPECIES *bursiformis, verrucisporus*

GENUS *Crinipellis* Spider Pellet Cavalier Gill Mushroom

SPECIES *campanella, corticalis, minutula, scabella*

GENUS *Cynema* Cyneme Cavalier Gill Mushroom

SPECIES *alutacea*

GENUS *Cyphellocalathus* Cypher Callisto Cavalier Gill Mushroom

SPECIES *cecropiae*

GENUS *Cystoderma* Grainy Cap Cavalier Gill Mushroom

SPECIES *ambrosii, amianthinum, carcharias, chocoanum, cinnabarinum, fallax, granulosum, haematites, jasonis, lilacipes, simulatum, superbum*

VARIETIES *amianthinum album, amianthinum longisporum, amianthinum rugulosoreticulatum, cinnabarinum terrei*

GENUS *Delicatula* Delicate Cavalier Gill Mushroom

SPECIES *integrella, persimilis, pithyophila*

GENUS *Dendrocollybia* Wood Coincap Cavalier Gill Mushroom

SPECIES *racemosa*

GENUS *Dennisiomyces* Dennis Cavalier Gill Mushroom

SPECIES *fuscoalbus, glabrescentipes, griseus, lanzonii, rionegrensis*

GENUS *Dermoloma* Fringed Cavalier Gill Mushroom

SPECIES *cuneifolium, josserandii, magicum*

VARIETIES *cuneifolium cuneifolium, cuneifolium atrocinereum, cuneifolium fuscobrunneum, cuneifolium punctipes, josserandii josserandii, josserandii phaeopodium, josserandii pragense, josserandii pseudocuneifolium*

GENUS *Fayodia* Fayod Cavalier Gill Mushroom

SPECIES *anthracobia, bisphaerigera, gracilipes, pseudoclusilis*

GENUS *Filoboletus* Bolete Cavalier Gill Mushroom

SPECIES *gracilis*

GENUS *Flabellimycena* Fancap Cavalier Gill Mushroom

SPECIES *flava*

GENUS *Floccularia* Woolly Cavalier Gill Mushroom

SPECIES *albolanaripes, fusca, pitkinensis, rickenii, subcaligata*

VARIETIES *luteovirens, luteovirens luteovirens, luteovirens alba, luteovirens americana, luteovirens straminea*

GENUS *Gamundia* Gamund Cavalier Gill Mushroom

SPECIES *hygrocyboides, leucophylla, lonatii, metuloidigera, striatula, xerophila*

GENUS *Gymnopus* Nude Coincap Cavalier Gill Mushroom

SPECIES ***acervatus***(Edible), *alkalivirens, alpinus, aquosus, confluens,* ***dryophilus***(Edible), *erythropus, fagiphilus, fuscopurpureus,* ***fusipes***(Edible), *hariolorum, impudicus, ocior, oreadoides, peronatus, putillus, semihirtipes, terginus*

VARIETIES ***fusipes contorta, fusipes oedematopus****, ocior exsculptus*

GENUS *Haasiella* Haas Cavalier Gill Mushroom

SPECIES *venustissima*

GENUS *Hemimycena* Half Cavalier Gill Mushroom

SPECIES *angustispora, candida, cephalotricha, crispata, cucullata, cyphelloides, delectabilis, epichloë, gracilis, gypsella, hirsuta, lactea, mairei, mauretanica, ochrogaleata, pithya, pseudocrispula, pseudolactea, rickenii, subimmaculata, tortuosa*

GENUS *Hydropus* Hydric Cavalier Gill Mushroom

SPECIES *floccipes, scabripes, sphaerosporus, subalpinus, trichoderma*

GENUS *Lactocollybia* Milky Coincap Cavalier Gill Mushroom

SPECIES *angiospermarum*

GENUS *Lampteromyces* Lampter Cavalier Gill Mushroom

SPECIES *japonicus*

GENUS *Lepista* Wine Pitcher Cavalier Gill Mushroom

SPECIES ***caespitosa***(Edible), *densifolia, diemii, flaccida, glaucocana, inversa,* ***irina***(Edible), *multiformis,* ***nuda***(Poison Raw, Edible Well Cooked), *ovispora,* ***panaeolus***(Edible), *personata, piperata, pseudoectypa, pseudoirina, rickenii, sadica,* ***saeva***(Edible), *sordida, subaequalis*

VARIETIES ***caespitosa fasciculata***, *flaccida gilva,* ***nuda gracilis, nuda pruinosa***, *sordida ianthina, sordida lilacea, sordida obscurata*

GENUS *Leptoglossum* Thin Tongue Cavalier Gill Mushroom

SPECIES *omnivorum, polycephalum, tremellosum*

GENUS *Lyophyllopsis* Lesser Loosecap Cavalier Gill Mushroom

SPECIES *keralensis*

GENUS *Lyophyllum* Loosecap Cavalier Gill Mushroom

SPECIES *connatum, decastes, eustygium, favrei, fumosum, gangraenosum, hypoxanthum, infumatum, konradianum, leucophaeatum, loricatum, mephiticum, montanum, semitale, semitale, transforme, trigonosporum*

VARIETIES *decastes aggregatum, fumosum cinerascens, fumosum conglobatum, leucophaeatum fumatofoetens*

GENUS *Macrocybe* Large Cap Cavalier Gill Mushroom

SPECIES *crassa, gigantea, lobayensis, pachymeres, praegrandis, spectabilis, titans*

GENUS *Megacollybia* Large Coincap Cavalier Gill Mushroom

SPECIES *platyphylla*

GENUS *Melanoleuca* Dark & Light Cavalier Gill Mushroom

SPECIES *albifolia, alboflavida, atripes, brevipes, cinereifolia, cnista, cognata, exscissa, graminicola, grammopodia, humilis, iris, langei, melaleuca, nivea, oreina, polioleuca, polito-inaequalipes, rasilis, schumacheri, strictipes, stridula, subbrevipes, substrictipes, tabularis, turrita, verrucipes*

VARIETIES *albifolia leucophylla, cinereifolia maritima, cognata nauseosa, cognata robusta, exscissa exscissa, exscissa cinerascens, exscissa kuehneri, grammopodia macrocarpa, grammopodia obscura, oreina fragillima, oreina polioleuca, oreina pusilla, polioleuca vulgaris*

GENUS *Metulocyphella* Metule Cypher Cavalier Gill Mushroom

SPECIES *lanceolata, rostrata*

GENUS *Mniopetalum* Lead Petal Cavalier Gill Mushroom

SPECIES *distinctum, flabelliforme, miniatum, minutum*

GENUS *Moniliophthora* Decay Cavalier Gill Mushroom

SPECIES *perniciosa, roreri*

GENUS *Mycena* Fairy Helmet Cavalier Gill Mushroom

SPECIES *abramsii, acicula, aciculata, adonis, adscendens, aetites, alba, albidolilacea, alexandri, amicta, arcangeliana, arcifolia, aurantiomarginata, belliae, bulbosa, capillaripes, capillaris, chlorantha, cinerella, citricolor, citrinomarginata, clavicularis, clavularis, corynephora, crocata, cyanorrhiza, dasypus, diosma, epipterygia, epipterygioides, erubescens, fagetorum, filopes, flavescens, flavoalba, floridula, flos-nivium, galericulata, galopus, haematopus, hiemalis, inclinata, kuehneriana, latifolia, leaiana, leptocephala, lilaciflora, longiseta, maculata, megaspora, meliigena, metata, minutula, mirata, monticola, mucor, olida, olivaceomarginata, olivascens, overholtsii, parabolica, pearsoniana, pelianthina, phyllogena, picta, polyadelpha, polygramma, pseudocorticola, pterigena, pura, purpureofusca, renati, rorida, rosea, rosella, rubromarginata, sanguinolenta, septentrionalis, seynesii, silvae-nigrae, simia, smithiana, speirea, stipata, strobilicola, stylobates, texensis, tintinnabulum, truncosalicicola, urania, venustula, viridimarginata, vitilis, vitrea, vulgaris, xantholeuca, zephirus*

VARIETIES *abramsii praecox, aciculata codoniceps, adonis adonis, adonis coccinea, adonis rubella, adscendens tenerrima, aetites cinerea, aetites consimilis, aetites umbellifera, amicta iris, amicta caerulea, amicta mirabilis, arcangeliana olivascens, arcangeliana oortiana, aurantiomarginata elegans, capillaripes langei, capillaripes marginata, chlorantha pallida, citricolor flavida, citricolor tricolor, epipterygia epipterygia, epipterygia badiceps, epipterygia candida, epipterygia citrinella, epipterygia lignicola, epipterygia nutans, epipterygia plicatocrenata, epipterygia rubescens, epipterygia splendidipes, epipterygia viscosa, erubescens fellea, filopes amygdalina, filopes iodiolens, filopes pumila, flavescens sulphureomarginata, flavoalba luteoalba, galericulata galericulata, galericulata albida, galericulata berkeleyi, galericulata radicatella, galericulata rugosa, galericulata rugulosiceps, galopus galopus, galopus candida, galopus candida-alba, galopus nigra, galopus nigra-fusconigra, galopus nigra-leucogala, haematopus marginata, hiemalis epiphloea, hiemalis terrestris, inclinata calopus, leptocephala chlorinella, maculata alcalina, megaspora uracea, metata collariata, olivaceomarginata olivaceomarginata, olivaceomarginata avenacea, olivaceomarginata brunneomarginata, olivaceomarginata roseofusca, olivaceomarginata thymicola, polygramma candida, pura pura, pura alba, pura carnea, pura ianthina, pura lutea, pura multicolor, pura pseudopura, pura purpurea, pura roseoviolacea, pura violacea, renati flavipes, rosella rosea, rosella strobilina, sanguinolenta cruenta, speirea camptophylla, speirea tenuistipes, stylobates dilatata, stylobates torquata*

GENUS *Mycenella* Lesser Fairy Helmet Cavalier Gill Mushroom

SPECIES *bryophila, margaritispora, rubropunctata, salicina*

GENUS *Mycoalvimia* Alvim Cavalier Gill Mushroom
SPECIES *theobromicola*
GENUS *Nematoctonus* Thread Octal Cavalier Gill Mushroom
SPECIES *concurrens, hamatus, haptocladus, leiosporus, pachysporus, robustus, subreniformis, tripolitanius*
GENUS *Neoclitocybe* Modern Funnelcap Cavalier Gill Mushroom
SPECIES *euomphalos*
GENUS *Neonothopanus* Modern Pan Cavalier Gill Mushroom
SPECIES *nambi*
GENUS *Nothopanus* Pan Cavalier Gill Mushroom
SPECIES *hygrophanus, noctilucens*
GENUS *Nyctalis* (Synonym for *Asterophora*)
GENUS *Ossicaulis* Calcified Stalk Cavalier Gill Mushroom
SPECIES *lignatilis*
GENUS *Palaeocephala* Pale Cap Cavalier Gill Mushroom
SPECIES *cymatelloides*
GENUS *Peglerochaete* Pegler Bristle Cavalier Gill Mushroom
SPECIES *setiger*
GENUS *Pegleromyces* Pegler Cavalier Gill Mushroom
SPECIES *collybioides*
GENUS *Phaeomycena* Dusky Cavalier Gill Mushroom
SPECIES *albidula, aureophylla, fusca, indica, macrospora*
GENUS *Phaeotellus* Dusky Remote Cavalier Gill Mushroom
SPECIES *acerosus*
GENUS *Phyllotus* Ear Cavalier Gill Mushroom
SPECIES *hygrophanus, imbricatus*
GENUS *Physocystidium* Bladder Cyst Cavalier Gill Mushroom
SPECIES *cinnamomeum*
GENUS *Pleurella* Pleural Cavalier Gill Mushroom
SPECIES *ardesiaca*
GENUS *Porpoloma* Fat Fringe Cavalier Gill Mushroom
SPECIES *elytroides, metapodium, pes-caprae, spinulosum*
GENUS *Pseudoarmillariella* Lesser Bracelet Cavalier Gill Mushroom
SPECIES *ectypoides, fistulosa*
GENUS *Pseudobaeospora* Lesser Bael Cavalier Gill Mushroom
SPECIES *albidula, celluloderma, dichroa, laguncularis*
VARIETIES *dichroa dichroa, dichroa cystidiata*
GENUS *Pseudoclitocybe* Lesser Funnelcap Cavalier Gill Mushroom
SPECIES *cyathiformis, expallens, obbata*
GENUS *Resinomycena* Resin Fairy Helmet Cavalier Gill Mushroom
SPECIES *acadiensis, brunnescens, japonica, mirabilis, montana, pyrenaica, rhododendri, saccharifera*
GENUS *Resupinatus* Supine Cavalier Gill Mushroom
SPECIES *alboniger, algidus, applicatus, graminum, kavinii, physaroides, poriaeformis, striatulus, trichotis, unguicularis*
VARIETIES *algidus dendrocystis*
GENUS *Rhodopaxillus* (Synonym for *Lepista*)
GENUS *Rhodotus* Rosycap Cavalier Gill Mushroom
SPECIES *palmatus*
GENUS *Rickenella* Ricken Cavalier Gill Mushroom
SPECIES *fibula, mellea, pseudogrisella, swartzii*
GENUS *Rimbachia* Rimbach Cavalier Gill Mushroom
SPECIES *arachnoidea, bryophila, neckerae, paludosa, paradoxa*
GENUS *Ripartites* Riverbank Cavalier Gill Mushroom
SPECIES *metrodii, tricholoma*
VARIETIES *tricholoma panaeolus*
GENUS *Sinotermitomyces* China Termite Cavalier Gill Mushroom
SPECIES *carnosus, cavus, griseus, rugosiceps, taiwanensis*
GENUS *Squamanita* Scaly Cavalier Gill Mushroom
SPECIES *contortipes, odorata, paradoxa, pearsonii*
GENUS *Stanglomyces* Stangle Cavalier Gill Mushroom
SPECIES *taxophilus*
GENUS *Stigmatolemma* Stigma Lemma Cavalier Gill Mushroom
SPECIES *conspersum, hyalinum*
GENUS *Tephrocybe* Ash Cap Cavalier Gill Mushroom
SPECIES *ambusta, anthracophila, atrata, baeosperma, boudieri, cessans, confusa, ellisii, ferruginella, fuscipes, fusispora, impexa,* ***inolens***(Edible), *murina, osmophora, palustris, platypus, putida, rancida, striipilea, tylicolor*
GENUS *Termitomyces* Termite Cavalier Gill Mushroom

SPECIES *poonensis*
GENUS *Tilachlidiopsis* Tilachly Cavalier Gill Mushroom
SPECIES *piptadeniae*
GENUS *Tricholomopsis* Broad Cavalier Gill Mushroom
SPECIES *decora,* ***edodes***(Edible), ***rutilans***(Edible), *sulphureoides*
VARIETIES *rutilans variegata, sulphureoides ornata*
GENUS *Trogia* Trog Cavalier Gill Mushroom
SPECIES *cantharelloides, holochlora, icterina, inaequalis*
GENUS *Ugola* Ugol Cavalier Gill Mushroom
SPECIES *baryana, physaroides, praticola*

FAMILY 19 *Typhulaceae* Typha Gill Mushroom
GENUS *Typhula* Typha Gill Mushroom
SPECIES *abietina, anceps, buxi, candida, capitata, caricina, chamaemori, corallina, crassipes, culmigena, erumpens, erythropus, euphorbiae, filicina, fruticum, graminum, hedericola, hollandii, hyalina, idahoensis, ilicis, incarnata, ishikariensis, latissima, longispora, lutescens, luttrellii, megasperma, micans, mycophaga, olivascens, pachypus, pallens, pertenuis, phacorrhiza, piceicola, pteridicola, pungens, quisquiliaris, sclerotioides, setipes, spathulata, sphaeroidea, struthiopteridis, subsclerotioides, subulata, subvariabilis, thaxteri, thindii, todei, trifolii, tucumanensis, umbrina, uncialis, variabilis, vermiculata, virgata, viticola*
GENUS *Apiosporium* Bee Spore Typha Gill Mushroom
SPECIES *abietis, pinophilum*
GENUS *Lutypha* Lute Typha Gill Mushroom
SPECIES *sclerotiophila*
GENUS *Pistillaria* Pistil Typha Gill Mushroom
SPECIES *diaphana, petasitis, rhodocionidies, subuncialis, trispora*
VARIETIES *petasitis petasitis, petasitis curvispora*
GENUS *Pistillina* Lesser Pistil Typha Gill Mushroom
SPECIES *brunneola*
GENUS *Sclerotium* Hard Typha Gill Mushroom
SPECIES *aesculi, brassicae, bullatum, cacticola, cepivorum, coffeicola, complanatum, corrugatum, costantinii, crustuliforme, delphinii, denigrans, durum, fumigatum, glaucoalbidum, glucanicum, hyacinthi, hydrophilum, medullare, medullosum, muscorum, neglectum, oryzae-sativae, oryzicola, perniciosum, pustula, rhizodes, roseum, stercorarium, tectum, truncorum, tuliparum, wakkeri*

ORDER 2 *Russullales* Brittle Gill Mushroom

FAMILY 1 *Russulaceae* Brittle Gill Mushroom
GENUS *Russula* Brittle Gill Mushroom
SPECIES *abietum, acetolens, acriannulata, acrifolia, acriuscula, acuminata, adulterina,* ***adusta***(Edible), ***aeruginea***(Edible), *afronigricans, albella, albida, albidula, alboflocossa,* ***albonigra***(Edible), *albospissa, alnetorum,* ***alutacea***(Edible), *amarissima,* ***amethystina***(Edible), ***amoena***(Edible), *amoenicolor, amoenipes, amoenolens, anatina, anthracina, aquosa, atropurpurea, atrorubens, aurantiaca,* ***aurea***(Edible), *aurora,* ***azurea***(Edible), *badia, betularum, brevipes,* ***brunneoviolacea***(Edible), *caerulea, carminipes, carpini, cavipes, cerolens, cessans, chamiteae,* ***chloroides***(Edible), *cicatricata,* ***claroflava***(Edible), *clavipes, compacta, crassotunicata, cremeoavellanea, crustosa, cuprea, curtipes,* ***cyanoxantha***(Commercial Edible), *decipiens, decolorans,* ***delica***(Edible), ***densifolia***(Edible), *dissimilans, emetica, emeticicolor, exalbicans, faginea, farinipes, fellea, flavipes, flavispora, foetens, font-queri, fragilis, fragrantissima, fusconigra, fuscorubroides, gigasperma, gracillima, grata, graveolens,* ***grisea***(Edible), *helodes,* ***heterophylla***(Edible), *illota, inamoena,* ***innocua***(Edible), *insignis, intactior,* ***integra***(Edible), *intermedia,* ***ionochlora***(Edible), *laccata, laeta, langei, lateritia, lepidicolor,* ***lilacea***(Edible), *lilaceps, luteotacta, maculata, medullata, melitodes,* ***melliolens***(Edible), *melzeri, minutula,* ***mustelina***(Edible), *nana,* ***nauseosa***(Edible), ***nigricans***(Edible), ***nitida***(Edible), *nobilis, ochroleuca, odorata,* ***olivacea***(Edible), *oreina,* ***paludosa***(Edible), *parazurea, pascua, pectinatoides, pelargonia, persicina, poichilochroa, postiana, praetervisa, pseudo-olivascens, pseudoaeruginea, pseudoaffinis,* ***pseudodelica***(Edible), *pseudoimpolita,* ***pseudointegra***(Edible), *pseudopuellaris,* ***puellaris***(Edible), *puellula, pulchella, pulchrae-uxoris, pungens, queletii, raoultii, renidens, rhodella,* ***risigallina***(Edible), ***romellii***(Edible), ***rosea***(Edible), *roseicolor, rubens, ruberrima, rubroalba, rutila, sanguinaria, sardonia, scotica, sericatula, silvestris, solaris, sororia, sphagnophila, stenotricha, subfoetens, subrubens, subterfurcata, taeniospora, torulosa,* ***turci***(Edible), *unicolor, variata,* ***velenovskyi***(Edible), *venosopurpurea, ventricosipes, versatilis, versicolor,* ***vesca***(Edible), *veternosa,* ***vinosa***(Edible), *vinosobrunnea, vinosopurpurea, violacea,* ***violeipes***(Edible), ***virescens***(Edible), ***viscida***(Edible), *xenochlora,* ***xerampelina***(Edible), *zonatula, zvarae*
VARIETIES *alnetorum pumila, anthracina carneifolia, aquosa carminea, atropurpurea depallens, atropurpurea krombholzii, atropurpurea rubra, atropurpurea undulata,* ***aurea aurata,*** *aurora incarnata, aurora rosea, aurora velutipes, brunneoviolacea pseudoviolacea, cicatricata barlae, cicatricata fusca, claroflava constans, claroflava flava, cuprea urens,* ***cyanoxantha cutefracta, cyanoxantha pallida, cyanoxantha peltereaui, delica flavispora,*** *emetica clusii, emetica gregaria, fragilis fragilis, fragilis bataillei, fragilis fallax, fragilis knauthii, fragilis violascens, grata laurocerasi, graveolens gilvescens, graveolens megacantha, graveolens purpurata, graveolens purpurea, grisea palumbina, grisea pictipes,* ***heterophylla adusta, heterophylla livida, innocua smaragdina,*** *insignis depauperata,* ***integra integra, integra campestris,***

integra purpurella, *intermedia lundellii, intermedia mesospora, intermedia pulcherrima, laccata norvegica,* ***lilacea carnicolor, melliolens sapida,*** *melzeri laricina,* ***nauseosa fusca, nigricans elephantina, nitida venosa,*** *nobilis fageticola, nobilis mairei, ochroleuca citrina, ochroleuca granulosa, pectinatoides pectinata, persicina rubicunda, poichilochroa metachroa, pseudo-olivascens elaeodes, puellaris leprosa,* ***queletii***(Edible) *flavovirens,* ***risigallina armeniaca, risigallina chamaeleontina, risigallina lutea, risigallina luteorosella, risigallina ochracea, risigallina vitellina, rosea alba, rosea lactea, rosea lepida, rosea linnaei****, sanguinaria rosacea, sanguinaria sanguinea, sardonia chrysodacryon, sardonia citrina, sardonia drimeia, sardonia emeticiformis, sardonia flavovirens, sardonia mellina, sardonia viridis, silvestris emeticella, sororia consobrina, sororia intermedia, torulosa fuscorubra, vesca mitis, veternosa schiffneri,* ***vinosa obscura,*** *violacea violacea, violacea carneolilacina, violacea viridis,* ***violeipes violeipes, violeipes citrina, violeipes chlora, violeipes olivascens, violeipes punctata, viscida artesiana, viscida chrismantiae, viscida occidentalis, xerampelina xerampelina, xerampelina erythropus, xerampelina ochraceus****, zvarae armoricana, zvarae aurora*

GENUS *Cystangium* — Cyst Angled Brittle Gill Mushroom

SPECIES *balpineum, crichtonii, depauperatum, phymatodisporum, pineti, rodwayi, sessile*

GENUS *Gymnomyces* — Nude Brittle Gill Mushroom

SPECIES *xanthosporus*

GENUS *Lactarius* — Milkcap Brittle Gill Mushroom

SPECIES *acerrimus, acris, acutus, adhaerens, aerumbonatus, affinis, albocarneus, allardii, alnicola, alpinus, amarus, angustifolius, annulatoangustifolius, aquizonatus, argyllaceifolius, aspideus, atlanticus, atroviridis,* ***aurantiacus***(Edible), *aurantiifolius, aureifolius, auriolla, azonites, badiosanguineus, baliophaeus, barbatus, barrowsii, bertillonii, blennius, caeruleitinctus, caespitosus,* ***camphoratus***(Culinary Herb), *chelidonium, chromospermus, chrysorrheus, circellatus, citriolens, colorascens, controversus,* ***corrugis***(Edible), *cyathuliformis, deceptivus,* ***deliciosus***(Edible), *deterrimus, evosmus, fallax, flavidus, flexuosus, fluens, fuliginosus,* ***fulvissimus***(Edible), *fumosus, gerardii, glaucescens, glyciosmus, gossypinus, hatsudake,* ***helvus***(Culinary Herb), *hepaticus, hygrophoroides, hyphoinflatus, hysginoides, hysginus,* ***indigo***(Edible), *kauffmanii,* ***lacunarum***(Edible Well Cooked), *lanceolatus, lignyotus, lilacinus, luridus, maculatus, mairei, mammosus, mucidus, musteus, neuhoffii, obscuratus, olivaceoumbrinus, omphaliiformis, pallidus, paradoxus, payettensis, peckii, picinus, piperatus, populinus, porninsis, pseudoaffinis, pseudodeliciosus, pseudomucidus, pterosporus, pubescens, pyrogalus,* ***quieticolor***(Edible), ***quietus***(Edible), *repraesentaneus, resimus, romagnesii, rostratus, rubrilacteus, rubrocinctus,* ***rufus***(Culinary Herb), ***rugatus***(Edible), *ruginosus, salicis-reticulatae,* ***salmonicolor***(Edible), *scoticus, scrobiculatus,* ***semisanguifluus***(Edible), ***serifluus***(Edible), *sordidus, sphagneti, spinosulus,* ***subdulcis***(Edible), *subpalustris, subplinthogalus, subpurpureus, subumbonatus, subvellereus, tabidus, thyinos, torminosus, trivialis, turpis, uvidus, vellereus, vietus, vinaceorufescens,* ***violascens***(Edible), ***volemus***(Edible Well Cooked), *zonarioides, zonarius*

VARIETIES ***aurantiacus aurantiofulvus, aurantiacus mitissimus,*** *azonites albipes, bertillonii queletii, blennius virescens, blennius viridis, camphoratus cimicarius, camphoratus terrei, chrysorrheus theiogalus, flexuosus flexuosus, flexuosus roseozonatus, fluens albidopallens, fluens blennius, fulvissimus britannicus, fulvissimus subsericatus, glyciosmus impolitus, helvus aquifluus, helvus tomentosus, helvus tomentosus-marginatus,* ***lacunarum decipiens, lacunarum rubescens****, lilacinus lateritioroseus, mairei pearsonii, mairei zonatus, mammosus confusus, mammosus fuscus, obscuratus cyathula, obscuratus obnubilus, obscuratus radiatus, piperatus pergamenus, pubescens cilicioides,* ***quieticolor hemicyaneus****, romagnesii speciosus, rubrocinctus iners, rubrocinctus subsericeus, rufus exumbonatus, rufus mollis, salicis-reticulatae aspideoides,* ***salmonicolor salmoneus, salmonicolor subsalmoneus****, spinosulus violaceus, torminosus sublateritius, turpis plumbeus, vellereus vellereus, vellereus albivellus, vellereus hometii, vellereus velutinus, volemus ichoratus, volemus lactifluus, volemus oedematopus, volemus subrugatus, zonarius insulsus*

GENUS *Macowanites* — Macowan Brittle Gill Mushroom

SPECIES *tomentosa*

GENUS *Stephanospora* — Stephan Brittle Gill Mushroom

SPECIES *caroticola, corneri, flava, penangensis, redolens*

GENUS *Zelleromyces* — Zeller Brittle Gill Mushroom

SPECIES *australiensis, cinnabarinus, daucinus, gardneri, giennensis, gilkeyae, glabrellus, hispanicus, josserandii, major, malaiensis, meridionalis, oregonensis, ravenelii, sinensis, stephensii, striatus*

FAMILY 2 *Bondarzewiaceae* — Bondarzev Brittle Gill Mushroom

GENUS *Bondarzewia* — Bondarzev Brittle Gill Mushroom

SPECIES *berkeleyi, guaitecasensis, montana*

VARIETIES *montana mesentericus*

GENUS *Amylaria* — Amyl Bondarzev Brittle Gill Mushroom

SPECIES *himalayensis*

GENUS *Amylosporus* — Amyl Spore Bondarzev Brittle Gill Mushroom

SPECIES *bracei, campbellii, daedaliformis, ryvardenii, wrightii*

GENUS *Gloeodontia* — Glow Tooth Bondarzev Brittle Gill Mushroom

SPECIES *americana, columbiensis, discolor, pyramidata*

GENUS *Gloiodon* — Lesser Glow Tooth Bondarzev Brittle Gill Mushroom

SPECIES *nigrescens, occidentale, strigosus*

GENUS *Heterobasidion* — Variable Bondarzev Brittle Gill Mushroom

SPECIES *abietinum, annosum, araucariae, insulare, parviporum*

VARIETIES *annosum cryptarum*
GENUS *Rigidoporopsis* Rigid Pore Bondarzev Brittle Gill Mushroom
SPECIES *macrospora*
GENUS *Stecchericium* Steccher Bondarzev Brittle Gill Mushroom
SPECIES *abditum, isabellinum, rusticum, seriatum*
VARIETIES *seriatum fistulatum, seriatum westii*
GENUS *Wrightoporia* Wright Pore Bondarzev Brittle Gill Mushroom
SPECIES *avellanea, brunneo-ochracea, cinnamomea, cremea, cylindrospora, dimidiata, flava, gillesii, gloeocystidiata, gyropora, iobapha, lenta, micropora, neotropica, novae-zelandiae, ochrocrocea, perplexa, pouzarii, ramosa, rubella, subadusta, subrutilans, trametoides, tropicalis*

FAMILY 3 *Echinodontiaceae* Urchin Brittle Gill Mushroom
GENUS *Echinodontium* Urchin Brittle Gill Mushroom
SPECIES *ballouii, japonicum, tinctorium, tsugicola*
GENUS *Laurilia* Laurile Urchin Brittle Gill Mushroom
SPECIES *sulcata, taxodii*

FAMILY 4 *Lachnocladiaceae* Downy Brittle Gill Mushroom
GENUS *Lachnocladium* Downy Brittle Gill Mushroom
SPECIES *brasiliense, denudatum, flavidum, fulvum, molle, setulosum, subarticulatum, subpteruloides, ulei, zonatum*
GENUS *Asterostroma* Star Stroma Downy Brittle Gill Mushroom
SPECIES *apalum, cervicolor, cremeofulvum, laxum, medium, muscicola, persimile, pseudofulvum*
GENUS *Dichantharellus* Chanterelle Downy Brittle Gill Mushroom
SPECIES *brunnescens, malayanus*
GENUS *Dichopleuropus* Dual Downy Brittle Gill Mushroom
SPECIES *spathulatus*
GENUS *Dichostereum* Dual Stereo Downy Brittle Gill Mushroom
SPECIES *boreale, durum, effuscatum, granulosum, kenyense, orientale, pallescens, peniophoroides, ramulosum, rhodosporum, sordulentum*
GENUS *Scytinostroma* Scyte Stroma Downy Brittle Gill Mushroom
SPECIES *africanogalactinum, ahmadii, albocinctum, aluta, alutum, arachnoideum, caudisporum, corneri, crassum, crispulum, cystidiatum, decidens, duriusculum, fulvum, galactinum, hemidichophyticum, intextum, jacksonii, luteolum, mediterraneense, microspermum, neogalactinum, ochroleucum, odoratum, parvisporum, phaeosarcum, portentosum, praestans, protrusum, pseudopraestans, pulverulentum, renisporum, rhizomorpharum*
VARIETIES *intextum intextum, intextum magnisporum, ochroleucum luteum*
GENUS *Vararia* Varary Downy Brittle Gill Mushroom
SPECIES *abortiphysa, alticola, ambigua, amphithallica, athabascensis, aurantiaca, breviphysa, calami, callichroa, cinnamomea, cremea, cremeoavellanea, cunninghamii, dussii, ellipsospora, fibra, firma, fusispora, gallica, gillesii, gittonii, gomezii, gracilispora, hauerslevii, insolita, intricata, investiens, malaysiana, maremmana, mediospora, microphysa, minidichophysa, minispora, ochroleuca, parmastoi, pectinata, perplexa, phyllophila, pirispora, protrusa, racemosa, rhombospora, rosulenta, rugosispora, sigmatospora, sphaericospora, splendida, thujae, trinidadensis, tropica, ubatubensis, vassilievae, verrucosa*
VARIETIES *mediospora mediospora, mediospora makokouensis, minidichophysa amygdalispora*

FAMILY 5 *Peniophoraceae* Pen Brittle Gill Mushroom
GENUS *Peniophora* Pen Brittle Gill Mushroom
SPECIES *adjacens, albobadia, aurantiaca, bicornis, boidinii, bonariensis, borbonica, borbonica, bruneiensis, cinerea, confusa, coprosmae, crassitunicata, crustosa, decorticans, dipyrenosperma, duplex, elaeidis, ericina, erikssonii, exima, fasticata, fissilis, gabonensis, gilbertsonii, guadelupensis, halimi, incarnata, junipericola, laeta, laurentii, laxitexta, lilacea, limitata, lycii, macrocystidiata, malaiensis, malenconii, manshurica, meridionalis, molesta, monticola, multicystidiata, nuda, ovalispora, parvocystidiata, perexigua, piceae, pilatiana, pini, pithya, polygonia, proxima, pruinata, pseudonuda, pseudopini, pseudoversicolor, pusilla, quercina, reidii, rhodocarpa, robusta, rufa, rufomarginata, sacrata, scintillans, separans, septentrionalis, seymouriana, simulans, spathulata, sphaerocystidiata, subsalmonea, suecica, tamaricicola, taraguiensis, trigonosperma, unica, versicolor, versiformis, violaceolivida, viridis*
GENUS *Dendrophora* Wood Pen Brittle Gill Mushroom
SPECIES *erumpens*
GENUS *Duportella* Duport Pen Brittle Gill Mushroom
SPECIES *kuehneri, kuehneroides, miranda, pirispora, renispora, rhoica, tristicula*
GENUS *Entomocorticium* Entic Corticle Pen Brittle Gill Mushroom
SPECIES *dendroctoni*

FAMILY 6 *Stereaceae* Stereo Brittle Gill Mushroom
GENUS *Stereum* Stereo Brittle Gill Mushroom
SPECIES *antarcticum, armeniacum, bellum, complicatum, durum, gausapatum, hirsutum, hymenoglium, lobatum, magellanicum, ostrea, peculiare, pseudorimosum, rameale, reflexulum, rugosum, sanguinolentum, scutellatum, striatum, subtomentosum, sulphuratum, tjibodense, traplianum, vellereum*
VARIETIES *hirsutum auriformis*

GENUS *Acanthobasidium* Lesser Spiny Stereo Brittle Gill Mushroom
SPECIES *norvegicum, phragmitis*
GENUS *Acanthofungus* Spiny Stereo Brittle Gill Mushroom
SPECIES *ahmadii, rimosus, thoenii*
GENUS *Acanthophysium* Spiny Bladder Stereo Brittle Gill Mushroom
SPECIES *bisporum, buxicola, thoenii, tsugae*
GENUS *Aleurocystidiellum* Lesser Powder Cyst Stereo Brittle Gill Mushroom
SPECIES *subcruentatum*
GENUS *Aleurocystis* Powder Cyst Stereo Brittle Gill Mushroom
SPECIES *capensis, hakgallae, magnispora*
GENUS *Aleurodiscus* Powder Disc Stereo Brittle Gill Mushroom
SPECIES *aberrans, abietis, amorphus, apricans, atlanticus, aurantius, australiensis, berggrenii, botryosus, canadensis, cerussatus, coralloides, coronatus, cremicolor, croceus, delicatus, dendroideus, dextrinoideocerussatus, diffissus, disciformis, exasperatus, farlowii, fennicus, fruticetorum, gabonicus, gigasporus, grantii, ilexicola, jacksonii, lapponicus, laurentianus, limonisporus, lividocaeruleus, ljubarskii, macrocystidiatus, mesaverdensis, mirabilis, monilifer, oakesii, occidentalis, ochraceoflavus, parmuliformis, patelliformis, penicillatus, piceinus, roseoflavus, spiniger, subglobosporus, succineus, taxicola, tenuis, thujae, utahensis, vitellinus, wakefieldiae, weirii, zealandicus*
GENUS *Amylofungus* Lesser Amyl Stereo Brittle Gill Mushroom
SPECIES *corrosus, globosporus*
GENUS *Amylohyphus* Amyl Hypha Stereo Brittle Gill Mushroom
SPECIES *africanus*
GENUS *Amylosporomyces* Amyl Spore Stereo Brittle Gill Mushroom
SPECIES *camelicolor*
GENUS *Amylostereum* Amyl Stereo Brittle Gill Mushroom
SPECIES *areolatum, chailletii, ferreum, laevigatum*
GENUS *Boidinia* Boidin Stereo Brittle Gill Mushroom
SPECIES *cana, furfuracea, granulata, luteola, macrospora, permixta*
VARIETIES *furfuracea cinerellus*
GENUS *Chaetoderma* Bristle Stereo Brittle Gill Mushroom
SPECIES *luna*
GENUS *Conferticium* Crowded Stereo Brittle Gill Mushroom
SPECIES *insidiosum, ochraceum*
GENUS *Dextrinocystis* Dextrin Cyst Stereo Brittle Gill Mushroom
SPECIES *capitata*
GENUS *Gloeocystidiellum* Glow Cyst Stereo Brittle Gill Mushroom
SPECIES *aculeatum, aspellum, bisporum, clavuligerum, compactum, convolvens, cryptacanthum, fimbriatum, fistulatum, flammeum, formosanum, friesii, heimii, heimii, inconstans, kenyense, lactescens, laxum, leucoxanthum, luridum, marianum, moniliforme, peroxydatum, porosellum, porosum, purpureum, subasperisporum, sulcatum, tabacinum, triste, tropicale, turpe, wakullum, zawitense*
GENUS *Gloeocystidiopsis* Lesser Glow Cyst Stereo Brittle Gill Mushroom
SPECIES *salmonea*
GENUS *Gloiothele* Glow Nipple Stereo Brittle Gill Mushroom
SPECIES *citrinoidea, globosa*
GENUS *Hypochnus* (Synonym for *Boidinia*)
GENUS *Laxitextum* Lax Text Stereo Brittle Gill Mushroom
SPECIES *bicolor, lutescens*
GENUS *Megalocystidium* Megacyst Stereo Brittle Gill Mushroom
SPECIES *chelidonium, luteocystidiatum*
GENUS *Pseudoxenasma* Lesser Strange Stereo Brittle Gill Mushroom
SPECIES *verrucisporum*
GENUS *Scotoderma* Scotic Dermal Stereo Brittle Gill Mushroom
SPECIES *viride*
GENUS *Scytinostromella* Scyte Stroma Stereo Brittle Gill Mushroom
SPECIES *heterogenea, humifaciens, nannfeldtii, olivaceoalba*
GENUS *Vesiculomyces* Vesicule Stereo Brittle Gill Mushroom
SPECIES *citrinus, epitheloides*
GENUS *Xylobolus* Xylum Bowl Stereo Brittle Gill Mushroom
SPECIES *annosus, frustulatus, illudens, princeps, spectabilis, subpileatus*

CLASS 622.3 *Boletomycetes* Nongill Mushroom

ORDER 1 *Boletales* Bolete Nongill Mushroom

FAMILY 1 *Boletaceae* Bolete Nongill Mushroom
GENUS *Boletus* Bolete Nongill Mushroom

SPECIES ***aereus***(Commercial Edible), ***appendiculatus***(Commercial Edible), *armeniacus*, ***badius***(Edible), *betulicola, bicolor, byssinus, calopus, campestris, caucasicus, chrysenteron, cinnabarinus, cisalpinus, cookei, declivitatus,* ***edulis***(Commercial Edible), *fechtneri, ferrugineus, flavus, floridanus, fragrans, fraternus, frostii, immutatus, impolitus, legaliae, leucomelaena, lignatilis, longicurvipes, luridiformis, luridus, megalosporus, moravicus, nigricans, obscuratus, ornatipes, parvulus, persoonii,* ***pinicola***(Commercial Edible), *pinophilus, porosporus, projectellus, pruinatus, pseudoregius, pseudosulphureus, pulverulentus, queletii, radicans, radicatus, regius, reticulatus, retipes, rhodopurpureus, ripariellus, rubellus, rubropunctus, satanas, speciosus, subappendiculatus, subfraternus, subtomentosus, subvelutipes, torosus, variipes, xanthocyaneus, zelleri*

VARIETIES ***badius castaneus*** *calopus olivaceus, chrysenteron communis, edulis arcticus, edulis laevipes, edulis solidus, edulis trisporus, fechtneri pallescens, ferrugineus citrinovirens, ferrugineus spadiceus, impolitus suspectus, legaliae splendidus, luridiformis luridiformis, luridiformis discolor, luridiformis erythropus, luridus rubeolarius, luridus rubriceps, moravicus leonis, persoonii albus, queletii lateritius, queletii rubicundus, radicans albidus, radicans albus, radicans amarus, radicans pachypus, reticulatus aestivalis, rhodopurpureus xanthopurpureus, rubellus sanguineus, rubellus versicolor, subtomentosus marginalis, subtomentosus lanatus, subtomentosus leguei, subtomentosus striipes, subtomentosus xanthus*

GENUS *Afroboletus* — African Bolete Nongill Mushroom

SPECIES *azureotinctus, costatisporus, elegans, lepidellus, luteolus, pterosporus*

GENUS *Aureoboletus* — Golden Bolete Nongill Mushroom

SPECIES *auriporus, gentilis, novoguineensis, reticuloceps, subacidus, thibetanus*

VARIETIES *gentilis cramesinus, gentilis tenuipes*

GENUS *Austoboletus* — Southern Bolete Nongill Mushroom

SPECIES *betula*

GENUS *Boletellus* — Shagnet Bolete Nongill Mushroom

SPECIES *mirabilis*

GENUS *Boletopsis* — False Bolete Nongill Mushroom

SPECIES *atrata, grisea, leucomelaena, smithii, subcitrina*

GENUS *Buchwaldoboletus* — Buchwald Bolete Nongill Mushroom

SPECIES *hemichrysus, lignicola, sphaerocephalus*

VARIETIES *sphaerocephalus sulphureus*

GENUS *Chalciporus* — Chalk Pore Bolete Nongill Mushroom

SPECIES *piperatus, rubinellus*

GENUS *Chamonixia* — Chamise Nix Bolete Nongill Mushroom

SPECIES *albida, bispora, brevicolumna, caespitosa, caudata, mucosa, octorugosa, pachydermis, vittatispora*

GENUS *Endogonopsis* — Endogon Bolete Nongill Mushroom

SPECIES *sacramentarium*

GENUS *Gastroboletus* — False Truffle Bolete Nongill Mushroom

SPECIES *turbinatus*

GENUS *Gastroleccinum* — Cavity Scaber Stalk Bolete Nongill Mushroom

SPECIES *scabrosum*

GENUS *Gastrotylopilus* — Cavity Knob Bolete Nongill Mushroom

SPECIES *brunneus*

GENUS *Leccinum* — Scaber Stalk Bolete Nongill Mushroom

SPECIES *aerugineum, atrostipitatum, aurantiacum, crocipodium,* ***duriusculum***(Edible), *holopus, insigne, onychinum, palustre, pseudoscabrum, quercinum, rigidipes, roseofractum, roseotinctum, salicola, scabrum, schistophilum, umbrinoides, variicolor atrostellatum, variicolor, versipelle, vulpinum*

VARIETIES *aerugineum brunneogriseolum, aerugineum chlorinum, aerugineum cyaneobasi-leucum, aurantiacum populinum, aurantiacum rufum, crocipodium nigrescens, crocipodium tessellatus, holopus americanum, holopus nucatum, pseudoscabrum carpini, roseofractum pulchrum, scabrum scabrum, scabrum avellaneum, scabrum melaneum, scabrum molle, scabrum murinaceum, scabrum olivaceosum, scabrum subcinnamomeum, variicolor variicolor, variicolor bertauxii, versipelle floccopus, versipelle glandaceum, versipelle rufescens, versipelle testaceoscabrum*

GENUS *Phaeogyroporus* — Dusky Gyre Bolete Nongill Mushroom

SPECIES sudanicus

GENUS *Porphyrellus* — Chocolate Bolete Nongill Mushroom

SPECIES *arcularius, brumalis, ciliatus, porphyrosporus*

GENUS *Pseudoboletus* — Least Bolete Nongill Mushroom

SPECIES *parasiticus*

GENUS *Royoungia* — Roy Young Bolete Nongill Mushroom

SPECIES *boletoides*

GENUS *Setogyroporus* — Seta Gyre Bolete Nongill Mushroom

SPECIES *verus*

GENUS *Singeromyces* — Singer Bolete Nongill Mushroom

SPECIES *ferrugineus*

GENUS *Sinoboletus* — China Bolete Nongill Mushroom

SPECIES *duplicatoporus, fuscus, guizhouensis, maekawae, magniporus, magnisporus, tengii*

GENUS *Strobilomyces* — Strobe Bolete Nongill Mushroom

SPECIES *confusus, strobilaceus*
VARIETIES *strobilaceus cinereus, strobilaceus floccopus*
GENUS *Trachypus* Scaber Foot Bolete Nongill Mushroom
SPECIES *lepidus, leucophaeus*
GENUS *Tubiporus* Tube Pore Bolete Nongill Mushroom
SPECIES *appendiculatus, aereus, calopus, dupainii, eastwoodiae, impolitus, junquilleus, lupinus, luridus, pinicola, pulverulentus, purpureus, queletii, regius, reticulatus, rubellus, satanas, torosus*
VARIETIES *aereus edulis, calopus olivaceus, impolitus suspectus, luridus rubeolarius, luridus rubriceps, pinicola edulis, queletii lateritius, queletii rubicundus, reticulatus aestivalis, rubellus sanguineus, rubellus versicolor*
GENUS *Tubosaeta* Tube Seta Bolete Nongill Mushroom
SPECIES *alveolata, aureocystis, brunneosetosa, calocystis, goossensiae, heterosetosa*
GENUS *Tylopilus* Knob Bolete Nongill Mushroom
SPECIES *ballouii, chromapes, eximius, felleus, indecisis, plumbeoviolaceus, rubrobrunneus*
VARIETIES *felleus alutarius*
GENUS *Veloporphyrellus* Velvet Chocolate Bolete Nongill Mushroom
SPECIES *africanus, pantoleucus*
GENUS *Xanthoconium* Yellow Cone Bolete Nongill Mushroom
SPECIES *affine, chattoogaense, montaltoense, montanum, purpureum, separans, stramineum*
GENUS *Xerocomus* Dry Hair Bolete Nongill Mushroom
SPECIES *amarellus, armeniacus, badiorufus, badius, chrysenteron, cisalpinus, ferrugineus, moravicus, porosporus, pruinatus, ripariellus, russelii, subtomentosus*
VARIETIES *chrysenteron communis, ferrugineus ferrugineus, ferrugineus citrinovirens, ferrugineus spadiceus, moravicus leonis, pruinatus luteocarnosus, subtomentosus subtomentosus, subtomentosus lanatus, subtomentosus leguei, subtomentosus marginalis, subtomentosus striipes, subtomentosus xanthus*

FAMILY 2 *Gomphidiaceae* Slimecap Bolete Nongill Mushroom
GENUS *Gomphidius* Slimecap Bolete Nongill Mushroom
SPECIES ***glutinosus***(Edible), *gracilis, largus,* ***maculatus***(Edible), ***nigricans***(Edible), ***roseus***(Edible), *subroseus,* ***viscidus***(Edible)
VARIETIES ***maculatus cookei, viscidus giganteus, viscidus testaceus***
GENUS *Chroogomphus* Woolly Slimecap Bolete Nongill Mushroom
SPECIES *helveticus, leptocystis, ochraceus,* ***rutilus***(Edible), *tomentosus,* ***vinicolor***(Edible)
VARIETIES ***rutilus rutilus, rutilus britannicus, rutilus corallinus***
GENUS *Cystogomphus* Cyst Slimecap Bolete Nongill Mushroom
SPECIES *humblotii*
GENUS *Gomphogaster* Cavity Slimecap Bolete Nongill Mushroom
SPECIES *leucosarx*

FAMILY 3 *Gyroporaceae* Hollow Stalk Bolete Nongill Mushroom
GENUS *Gyroporus* Hollow Stalk Bolete Nongill Mushroom
SPECIES ***castaneus***(Edible), ***cyanescens***(Commercial Edible), *purpurinus*
VARIETIES *castaneus fulvidus, cyanescens constrictus, cyanescens lacteus*
GENUS *Rubinoboletus* Ruby Hollow Stalk Bolete Nongill Mushroom
SPECIES *ballouii, caespitosus, griseus, laetus, luteopurpureus, phaseolisporus, phlebopoides, reticulatus, rubinus, virescens*

FAMILY 4 *Paxillaceae* Brimcap Bolete Nongill Mushroom
GENUS *Paxillus* Brimcap Bolete Nongill Mushroom
SPECIES *ammoniavirescens, atraetopus, corrugatus, involutus, ionipus, psammophilus, rubicundulus*
VARIETIES *involutus excentricus*
GENUS *Austrogaster* Southern Cavity Brimcap Bolete Nongill Mushroom
SPECIES *baeospermum, marthae, novae-zelandiae*
GENUS *Austropaxillus* Southern Brimcap Bolete Nongill Mushroom
SPECIES *boletinoides, chilensis, contulmensis, infundibuliformis, macnabbii, muelleri, nothofagi, squarrosus, statuum*
VARIETIES *macnabbii aurantiacus*
GENUS *Boletinellus* (Synonym for *Gyrodon*)
GENUS *Gymnopaxillus* Nude Brimcap Bolete Nongill Mushroom
SPECIES *crubensis, morchelliformis*
GENUS *Gyrodon* Gyre Brimcap Bolete Nongill Mushroom
SPECIES *lividus, merulioides*
VARIETIES *lividus brachyporus, lividus sistotrema, lividus sistotremoides*
GENUS *Leucopaxillus* Funnel Brimcap Bolete Nongill Mushroom
SPECIES *gentianeus,* ***giganteus***(Edible), *kamatii, laterarius, macrocephalus, paradoxus, rhodoleucus,* ***tricolor***(Edible)
VARIETIES *paradoxus albissimus*
GENUS *Neopaxillus* Modern Brimcap Bolete Nongill Mushroom
SPECIES *bryogenus, echinospermus, echinosporus, plumbeus, reticulatus*

GENUS *Paragyrodon* — Lesser Gyre Brimcap Bolete Nongill Mushroom
SPECIES *sphaerosporus*
GENUS *Paxillogaster* — Cavity Brimcap Bolete Nongill Mushroom
SPECIES *luteus*
GENUS *Phylloporus* — Gill Brimcap Bolete Nongill Mushroom
SPECIES *pelletieri, rhodoxanthus*
VARIETIES *pelletieri paradoxus*
GENUS *Tapinella* — Tapin Brimcap Bolete Nongill Mushroom
SPECIES *atrotomentosa, panuoides*
VARIETIES *panuoides fagi, panuoides rubrosquamulosus*

FAMILY 5 *Rhizopogonaceae* — FalseTruffle Bolete Nongill Mushroom
GENUS *Rhizopogon* — FalseTruffle Bolete Nongill Mushroom
SPECIES *colossus, defectus, luteolus, luteorubescens, ochraceorubens, reticulatus, roseolus, villosulus, vinicolor, vulgaris*
VARIETIES *luteolus induratus, roseolus rubescens, vulgaris provincialis*
GENUS *Brauniellula* — Braun FalseTruffle Bolete Nongill Mushroom
SPECIES *albipes, crassitunicata, leucosarx, nancyae*
GENUS *Fevansia* — Fevans FalseTruffle Bolete Nongill Mushroom
SPECIES *aurantiaca*
GENUS *Truncocolumella* — Puffball FalseTruffle Bolete Nongill Mushroom
SPECIES *citrina, occidentalis, rubra*

FAMILY 6 *Suillaceae* — Slipperycap Bolete Nongill Mushroom
GENUS *Suillus* — Slipperycap Bolete Nongill Mushroom
SPECIES *acerbus, acidus, americanus, bellinii, borealis, bovinus, brevipes, caerulescens, cavipes, collinitus, cothurnatus, flavidus, granulatus, grevillei, lakei, luteus, neoalbidipes, pictus, pinorigidus, plorans, ponderosus, pseudobrevipes, pseudogranulatus, pungens, sibiricus, subaureus, subluteus, tomentosus, tridentinus, umbonatus, variegatus, viscidus*
VARIETIES *cavipes porosus, collinitus fluryi, flavidus pulchellus, granulatus lactifluus, grevillei annularius, grevillei badius, grevillei elegans, lakei landkammeri, tridentinus aurantiporus, viscidus aeruginascens, viscidus bresadolae, viscidus flavogriseus, viscidus flavogriseus-nueschii, viscidus laricinus*
GENUS *Boletinus* — (Synonym for *Suillus*)
GENUS *Fuscoboletinus* — Fuscous Slipperycap Bolete Nongill Mushroom
SPECIES *ochraceoroseus*
GENUS *Ixocomus* — Viscous Hair Slipperycap Bolete Nongill Mushroom
SPECIES *albidipes, americanus, bellinii, bovinus, flavidus, flavus, granulatus, grevillei, lakei, lignicola, luteus, mitis, placidus, sibiricus, subluteus, sulphureus, tridentinus, variegatus, viscidus*
VARIETIES *flavidus pulchellus, flavus elegans, granulatus lactifluus, granulatus leptopus, grevillei badius, grevillei elegans, lakei landkammeri, viscidus aeruginascens, viscidus bresadolae, viscidus flavogriseus, viscidus flavogriseus-nueschii, viscidus laricinus*
GENUS *Psiloboletinus* — Smooth Slipperycap Bolete Nongill Mushroom
SPECIES *lariceti*

ORDER 2 *Cantherellales* Chanterelle Nongill Mushroom

FAMILY 1 *Cantharellaceae* — Chanterelle Nongill Mushroom
GENUS *Cantharellus* — Chanterelle Nongill Mushroom
SPECIES *alboroseus, amethysteus, aurora, camphoratus,* ***cibarius***(Edible), ***cinereus***(Edible), ***cinnabarinus***(Edible), *crocelfolius, cyanescens, cyanoxanthus, densifolius, flavobrunneus, floridulus,* ***friesii***(Edible), *furfuraceus, guyanensis, ianthinoxanthus, ignicolor, insignis, isabellinus, kauffmanii, lateritius, lilacinus, longisporus, luteocomus, luteopunctatus,* ***lutescens***(Edible), *melanoxeros, miniatescens, minor, nothofagorum, odoratus, omphalinoides, pallidipes, persicinus, pseudofriesii, queletii, rhodophyllus, splendens, subalbidus, subdenticulatus, tenuis,* ***tubaeformis***(Edible), *umbriceps, xanthoscyphus*
VARIETIES *aurora xanthopus,* ***cibarius cibarius, cibarius albidus, cibarius alborufescens, cibarius bicolor, cibarius edulis, cibarius ferruginascens, cibarius flavipes, cibarius nanus, cibarius neglectus, cibarius pallens, cibarius pallidus, cibarius rufipes, cibarius salmoneus, cibarius umbrinus, cibarius vulgaris, cinereus hydrolyps****, subdenticulatus ignescens, subdenticulatus miniatus,* ***tubaeformis tubaeformis, tubaeformis infundibuliformis, tubaeformis luteolus, tubaeformis subramosus***
GENUS *Craterellus* — Crater Chanterelle Nongill Mushroom
SPECIES *aureus, caeruleofuscus,* ***cornucopioides***(Edible), *fallax, foetidus, peckii, verrucosus*
VARIETIES *aureus laetus,* ***cornucopioides ochrosporus***
GENUS *Goossensia* — Goossen Chanterelle Nongill Mushroom
SPECIES *cibarioides*
GENUS *Hygrophoropsis* — False Chanterelle Nongill Mushroom
SPECIES *aurantiaca, flabelliformis, fuscosquamula, laevis, macrospora, olida, pallida*
VARIETIES *aurantiaca lactea, aurantiaca nigripes, aurantiaca pallida, aurantiaca robusta, aurantiaca rufa*
GENUS *Parastereopsis* — Star Chanterelle Nongill Mushroom
SPECIES *borneensis*

GENUS *Pseudocraterellus* Lesser Crater Chanterelle Nongill Mushroom
SPECIES *alutaceus, calyculus, luteus, neotropicalis, pertenuis, sinensis, sinuosus, subundulatus*
GENUS *Stilbotulasnella* Stilbine Tulas Chanterelle Nongill Mushroom
SPECIES *conidiophora*

FAMILY 2 *Aphelariaceae* Simple Chanterelle Nongill Mushroom
GENUS *Aphelaria* Simple Chanterelle Nongill Mushroom
SPECIES *brunneola, complanata, deflectens, dendroides, portentosa, tropica, tuberosa*
GENUS *Phaeoaphelaria* Dusky Simple Chanterelle Nongill Mushroom
SPECIES *australiensis*
GENUS *Tumidapexus* Swollen Nap Simple Chanterelle Nongill Mushroom
SPECIES *ravus*

FAMILY 3 *Botryobasidiaceae* Ball Chanterelle Nongill Mushroom
GENUS *Botryobasidium* Ball Chanterelle Nongill Mushroom
SPECIES *ansosum, arachnoideum, asperulum, asterosporum, aureum, candicans, chilense, conspersum, curtisii, danicum, ellipsosporum, flavescens, grandinioides, grandisporum, intertextum, lacinisporum, laeve, latisporum, lembosporum, longisporum, medium, microverrucisporum, obtusisporum, olivaceum, pandani, parvisetosum, piliferum, pilosellum, pruinatum, robustius, sordidulum, sphaericosporum, stigmatisporum, subalbidum, subbotryosum, subcoronatum, sublaeve, tuberculisporum, tubulicystidium, vagum*
VARIETIES *intertextum angustisporum, pruinatum botryoideum, pruinatum coronatum, pruinatum laeve, pruinatum scabridum, vagum botryosum*
GENUS *Alysidium* Alyce Ball Chanterelle Nongill Mushroom
SPECIES *dubium, resinae*
VARIETIES *resinae microsporum*
GENUS *Botryohypochnus* Pock Mark Ball Chanterelle Nongill Mushroom
SPECIES *bisporus, bondarcevii, isabellinus*
GENUS *Suillosporium* Slipperycap Ball Chanterelle Nongill Mushroom
SPECIES *amygdalisporum, cystidiatum*

FAMILY 4 *Clavulinaceae* Club Chanterelle Nongill Mushroom
GENUS *Clavulina* Club Chanterelle Nongill Mushroom
SPECIES *alutaceosiccescens, amethystinoides, brunneocinerea, cartilaginea, cavipes, chondroides, cinerea, cirrhata, connata, copiosocystidiata, coralloides, decipiens, delicia, gallica, gigartinoides, gracilis, humilis, ingrata, leveillei, limosa, ornatipes, panurensis, puiggarii, purpurea, rugosa, samuelsii, septocystidiata, sprucei, subrugosa, tasmanica, urnigerobasidiata, vinaceocervina, viridula*
VARIETIES *cinerea cinerea, cinerea gracilis, coralloides cristata, coralloides incarnata, coralloides lappa, coralloides subrugosa, rugosa alcyonaria, rugosa canaliculata, rugosa fuliginea, rugosa herveyi*
GENUS *Clavulicium* Lesser Club Chanterelle Nongill Mushroom
SPECIES *pilatii, venosum*
GENUS *Membranomyces* Membrane Club Chanterelle Nongill Mushroom
SPECIES *delectabilis, spurius*

FAMILY 5 *Gomphaceae* Bolt Chanterelle Nongill Mushroom
GENUS *Gomphus* Bolt Chanterelle Nongill Mushroom
SPECIES *africanus, clavatus, crassipes, dingleyae,* ***floccosus***(Edible), *mamorensis, novae-zelandiae, pseudoclavatus, szechwanensis*
VARIETIES *clavatus truncatus, floccosus bonarii, floccosus canadensis, floccosus excavatus, floccosus wilsonii*
GENUS *Araeocoryne* Club Bolt Chanterelle Nongill Mushroom
SPECIES *elegans*
GENUS *Beenakia* Beenak Bolt Chanterelle Nongill Mushroom
SPECIES *dacostae, fricta, fuliginosa, informis, subglobospora*
GENUS *Ceratellopsis* Horn Bolt Chanterelle Nongill Mushroom
SPECIES *aciculata, aculeata, acuminata, asphodeli, corneri, equiseticola, graminicola, sagittiformis, terrigena*
GENUS *Chaetotyphula* Bristle Typha Bolt Chanterelle Nongill Mushroom
SPECIES *actiniceps, amylochaete, columbiana, gelatinosa, hyalina, montana, tetraspora*
GENUS *Gloeocantharellus* Glow Bolt Chanterelle Nongill Mushroom
SPECIES *corneri, lateritius*
GENUS *Lentaria* Lent Bolt Chanterelle Nongill Mushroom
SPECIES *afflata, boletosporioides, byssiseda, caribbeana, epichnoa, glaucosiccescens, micheneri, patouillardii, rionegrensis, surculus*
GENUS *Nevrophyllum* Turbin Bolt Chanterelle Nongill Mushroom
SPECIES *brunneum*
GENUS *Pseudogomphus* Lesser Bolt Chanterelle Nongill Mushroom
SPECIES *fragilissimus*
GENUS *Terenodon* Teren Bolt Chanterelle Nongill Mushroom
SPECIES *serenus*

ORDER 3 *Hydnales* — Spinecap Nongill Mushroom

FAMILY 1 *Hydnaceae* — Spinecap Nongill Mushroom
GENUS *Hydnum* — Spinecap Nongill Mushroom
SPECIES *albidum, albomagnum, ambustum, durieui, elatum, membranaceum stenodon, papyraceum,* ***repandum***(Edible Unripe Cooked), *rufescens,* ***umbilicatum***(Edible)
VARIETIES *ambustum crocidens,* ***repandum abietinum, repandum album, repandum aurantium, repandum bicolor, repandum bulbosum, repandum clandestinum, repandum diffractum, repandum flavidum, repandum medium, repandum pallidum, repandum washingtonianum****, rufescens sulcatipes*
GENUS *Corallofungus* — Coral Spinecap Nongill Mushroom
SPECIES *hatakeyamanus*
GENUS *Gloeomucro* — Glow Spinecap Nongill Mushroom
SPECIES *asahimontanus, chlorinus, flavus, limpidus, luteodiscus, yakusimensis*
GENUS *Nigrohydnum* — Black Spinecap Nongill Mushroom
SPECIES *nigrum*
GENUS *Phaeoradulum* — Dusky Spinecap Nongill Mushroom
SPECIES *guadelupense*

FAMILY 2 *Auriscalpiaceae* — Pinecone Spinecap Nongill Mushroom
GENUS *Auriscalpium* — Pinecone Spinecap Nongill Mushroom
SPECIES *barbatum, dissectum, luteolum, umbella, villipes, vulgare*
vulgare fechtneri
GENUS *Amylonotus* — Amyl Pinecone Spinecap Nongill Mushroom
SPECIES *tenuis*
GENUS *Gloeomyces* — Glow Pinecone Spinecap Nongill Mushroom
SPECIES *ginnsii, graminicola, moniliformis*
GENUS *Lentinellus* — Lens Pinecone Spinecap Nongill Mushroom
SPECIES *brunnescens, calyciformis, castoreus, cochleatus, crawfordii, flabelliformis, laurocerasi, micheneri, montanus, omphalomorphus, pulvinulus, semivestitus, tridentinus, ursinus, vulpinus*
VARIETIES *castoreus pusio, cochleatus inolens, cochleatus marcelianus, micheneri americanus, micheneri omphalodes, micheneri umbilicatus, pulvinulus hepatotrichus, pulvinulus hyracinus, pulvinulus pseudobarbatus, ursinus hispidus, ursinus luxurians, ursinus pelliculosa*
GENUS *Leptodon* — (Synonym for *Auriscalpium*)

FAMILY 3 *Bankeraceae* — Banker Spinecap Nongill Mushroom
GENUS *Bankera* — Banker Spinecap Nongill Mushroom
SPECIES *fuligineoalba, violascens*
VARIETIES *fuligineoalba macrodon, fuligineoalba reticulatus, violascens carnosa*
GENUS *Calodon* — (Synonym for *Hydnellum, Phellodon*)
GENUS *Hydnellum* — Toothed Banker Spinecap Nongill Mushroom
SPECIES *aurantiacum, auratile, caeruleum, chrysinum, coalitum, compactum, complicatum, complicatum carolinianum, concrescens, conigenum, cristatum, cruentum, crustulinum, cumulatum, cyanodon, cyanopodium, earlianum, ferrugineum, fraudulentum, frondosum, geogenium, gracilipes, multiceps, nigellum, papuanum, peckii, regium, scleropodium, scrobiculatum, septentrionale, singeri, spongiosipes, staurastrum, suaveolens, subzonatum*
VARIETIES *aurantiacum complectipes, aurantiacum floriforme, aurantiacum stohlii, caeruleum alachuanum, caeruleum cyaneotinctum, caeruleum ferrugipes, caeruleum inquinatum, compactum acre, compactum mirabile, compactum tuberculosum, concrescens fasciatum, concrescens parvum, concrescens queletii, concrescens spathulatum, concrescens subsuccosum, concrescens vespertilio, concrescens zonatum, concrescens zonatum, ferrugineum hybridum, ferrugineum pineticola, geogenium sulfureum, peckii diabolus, peckii rhizopes, scrobiculatum ferrugineoalbum, scrobiculatum sanguineofulvum, scrobiculatum testaceofulvum, scrobiculatum velutinum, spongiosipes nuttallii, suaveolens boreale, suaveolens gravis, suaveolens pullum*
GENUS *Phellodon* — Cork Banker Spinecap Nongill Mushroom
SPECIES *atratus, confluens, excentri-mexicanus, fibulatus, implicatus, melaleucus, niger, plicatus, putidus, rufipes, sinclairii, tenuis, tomentosus*
VARIETIES *confluens amicus, confluens vellereus, melaleucus alboniger, melaleucus brunneoroseus, melaleucus delicatus, melaleucus ellisianus, melaleucus graveolens, melaleucus pygmaeus, putidus cokeri, putidus hesleri, tomentosus connatum, tomentosus cyathiformis*
GENUS *Sarcodon* — Fleshy Banker Spinecap Nongill Mushroom
SPECIES *aglaosoma, amarescens, atroviridis, caliginosus, calvatus, conchyliatus, dissimulans, excentricus, fennicus, fuligineoviolaceus, fuscoindicus, glaucopus, harrisonii, humilis, ianthinus, illudens, imbricatus, lanuginosus, lepidus, leucopus, lundellii, martioflavus, praestans, procerus, quietus, regalis, rimosus, rutilus, scabrosus, squamosus, stereosarcinon, subfelleus, thwaitesii, underwoodii, ustalis, versipellis, wrightii*
VARIETIES *atroviridis bambusinus, atroviridis blackfordae, atroviridis fumosus, fuligineoviolaceus talpa, imbricatus aspratus, leucopus laevigatus, martioflavus armeniacus, stereosarcinon brevipes, underwoodii murrillii, underwoodii radicatus*

FAMILY 4 *Hericiaceae* — Hedgehog Spinecap Nongill Mushroom
GENUS *Hericium* — Hedgehog Spinecap Nongill Mushroom

SPECIES *abietis, alpestre, americanum, botryoides, cirrhatum, clathroides, coralloides, erinaceus*
VARIETIES *alpestre flagellum, alpestre stalactitium, cirrhatum diversidens, coralloides abietinum, coralloides caput-ursi, coralloides laciniatum, coralloides ramosum, coralloides reichii, erinaceus caput-medusae, erinaceus echinus, erinaceus grande, erinaceus hystrix, erinaceus quercinum, erinaceus unguiculatum*
GENUS *Dentipellis* Tooth Pellet Hedgehog Spinecap Nongill Mushroom
SPECIES *dissita, fragilis, parmastoi, separans*
VARIETIES *fragilis echinospora, separans isidioides, separans leptodon, separans subseparans*
GENUS *Dentipratulum* Tooth Meadow Hedgehog Spinecap Nongill Mushroom
SPECIES *bialoviesense*
GENUS *Dryodon* (Synonym for *Hericium*)
GENUS *Mucronella* Pointed Hedgehog Spinecap Nongill Mushroom
SPECIES *alba, albidula, bresadolae, calva, flava, polyporacea, pusilla, styriaca, subalpina*
VARIETIES *alba pendula, calva abnormis, calva aggregata*

FAMILY 5 *Steccherinaceae* Steccher Spinecap Nongill Mushroom
GENUS *Steccherinum* Steccher Spinecap Nongill Mushroom
SPECIES *adustum, aggregatum, alaskense, albidum, albofibrillosum, basibadium, bourdotii, ciliolatum, confragosum, cremeoalbum, ethiopicum, fimbriatum, galeritum, gilvum, labeosum, laeticolor, lanestre, lusitanicum, meridiochraceum, ochraceum, oreophilum, peckii, peruvianum, rawakense, reniforme, robustius, russum, scalare, scruposum, straminellum, subcrinale, tenue, willisii, zeylanicum*
VARIETIES *ochraceum rhois, reniforme morganii, subcrinale kavinae*
GENUS *Amethicium* Amethyst Steccher Spinecap Nongill Mushroom
SPECIES *rimosum*
GENUS *Antrodiella* Cavity Steccher Spinecap Nongill Mushroom
SPECIES *aurantilaeta, citrea, faginea, fissiliformis, flava, foliaceodentata, fragrans, genistae, hoehnelii, incrustans, induratus, multipileata, murrillii, onychoides, overholtsii, parasitica, rata, romellii, semisupina, subcrassa, thompsonii, versicutis, zonata*
GENUS *Australohydnum* Southern Steccher Spinecap Nongill Mushroom
SPECIES *dregeanum*
GENUS *Diplomitoporus* Diploid Mitre Steccher Spinecap Nongill Mushroom
SPECIES *crustulinus, flavescens, overholtsii, rimosus*
GENUS *Fibricium* Fiber Steccher Spinecap Nongill Mushroom
SPECIES *coriaceum, lapponicum, rude, subcarneum, subceraceum*
GENUS *Flaviporus* Yellow Pore Steccher Spinecap Nongill Mushroom
SPECIES *americanus, brownei, citrinellus, delicatus, hunuus, hydrophilus, liebmannii, minutisporus, stramineus, subhydrophilus, subundatus, venustus, xanthus*
GENUS *Flavodon* Yellow Tooth Steccher Spinecap Nongill Mushroom
SPECIES *flavus*
GENUS *Irpex* Barrow Steccher Spinecap Nongill Mushroom
SPECIES *consors, lacteus, vellereus*
VARIETIES *lacteus cinerascens, lacteus tulipiferae*
GENUS *Junghuhnia* Junghuhn Steccher Spinecap Nongill Mushroom
SPECIES *carneola, collabens, complicata, crustacea, fimbriatella, lacera, lacera, luteoalba, nitida, polycystidifera, pseudozilingiana, semisupiniformis, subfimbriata, undigera, zonata*
GENUS *Lamelloporus* Lamella Steccher Spinecap Nongill Mushroom
SPECIES *americanus*
GENUS *Metuloidea* Cone Steccher Spinecap Nongill Mushroom
SPECIES *tawa*
GENUS *Mycoleptodon* (Synonym for *Steccherinum*)
GENUS *Mycoleptodonoides* Thin Steccher Spinecap Nongill Mushroom
SPECIES *aitchisonii, pergamenea, pusilla, vassiljevae*
GENUS *Mycorrhaphium* Seam Steccher Spinecap Nongill Mushroom
SPECIES *stereoides*
GENUS *Odontia* Tooth Steccher Spinecap Nongill Mushroom
SPECIES *latemarginata*
GENUS *Stegiacantha* Cover Steccher Spinecap Nongill Mushroom
SPECIES *petaloides*

CLASS 622.4 *Gastromycetes* Cavity Fungus

ORDER 1 *Gastromycetales* Cavity Fungus

FAMILY 1 *Gastraceae* Cavity Fungus
GENUS *Gasterella* Lesser Cavity Fungus
SPECIES *luteophila*
GENUS *Gastrosporium* Spore Cavity Fungus
SPECIES *asiaticum, simplex*

GENUS *Hybogaster* Hybe Cavity Fungus
SPECIES *giganteus*
GENUS *Leucogaster* Light Cavity Fungus
SPECIES *liosporus, nudus*
GENUS *Melanogaster* Dark Cavity Fungus
SPECIES *ambiguus, broomeanus, intermedius*
GENUS *Protogaster* Primitive Cavity Fungus
SPECIES *rhizophilus*

FAMILY 2 *Coniophoraceae* Cone Cavity Fungus
GENUS *Coniophora* Cone Cavity Fungus
SPECIES *arida, cordensis, fusispora, hanoiensis, marmorata, merulioides, mollis, olivacea, opuntiae, prasinoides, puteana, submembranacea*
GENUS *Chrysoconia* Golden Cone Cavity Fungus
SPECIES *orthospora*
GENUS *Coniophoropsis* Lesser Cone Cavity Fungus
SPECIES *obscura*
GENUS *Corneromyces* Corner Cone Cavity Fungus
SPECIES *kinabalui*
GENUS *Gyrodontium* Gyre Tooth Cone Cavity Fungus
SPECIES *boveanum, sacchari, versicolor*
GENUS *Gyrophana* (Synonym *for Leucogyrophana, Serpula*)
GENUS *Jaapia* Jaapy Cone Cavity Fungus
SPECIES *argillacea, ochroleuca*
GENUS *Leucogyrophana* Light Gyre Cone Cavity Fungus
SPECIES *arizonica, lichenicola, luridochracea, mollusca, montana, olivascens, pinastri, pouzarii, pseudomollusca, pulverulenta, romellii, sororia, subtessulata*
GENUS *Podoserpula* Pod Serpentine Cone Cavity Fungus
SPECIES *pusio*
GENUS *Pseudomerulius* Intoxicating Cone Cavity Fungus
SPECIES *aureus, curtisii, elliottii*
GENUS *Serpula* Serpentine Cone Cavity Fungus
SPECIES *himantioides, lacrymans, tignicola*
VARIETIES *himantioides arboreus, lacrymans obliquus*

FAMILY 3 *Hymenogasteraceae* Membrane Cavity Fungus
GENUS *Hymenogaster* Membrane Cavity Fungus
SPECIES *arenarius, citrinus, griseus, hessei, luteus, muticus, olivaceus, sulcatus, tener, thwaitesii, vulgaris*
GENUS *Amogaster* Sand Membrane Cavity Fungus
SPECIES *viridigleba*
GENUS *Brauniella* Braun Membrane Cavity Fungus
SPECIES *alba*
GENUS *Descomyces* Truncated Membrane Cavity Fungus
SPECIES *albellus, albus, javanicus*
GENUS *Destuntzia* Destuntz Membrane Cavity Fungus
SPECIES *fusca, rubra, saylorii, solstitialis, subborealis*
GENUS *Marthanella* Martha Membrane Cavity Fungus
SPECIES *nidulosa*
GENUS *Mycoamaranthus* Amaranth Membrane Cavity Fungus
SPECIES *auriorbis*
GENUS *Protoglossum* Primitive Tongue Membrane Cavity Fungus
SPECIES *auriorbis, cribbiae, luteum, niveum, purpureum, violaceum, viscidum*
VARIETIES *luteum atratum*
GENUS *Quadrispora* Quadrospore Membrane Cavity Fungus
SPECIES *musispora, oblongispora*
GENUS *Thaxtergaster* Thaxter Membrane Cavity Fungus
SPECIES *pinguis*
GENUS *Timgrovea* Tim Grove Membrane Cavity Fungus
SPECIES *ferruginea, kwangsiensis, macrospora, reticulata, subtropica*

FAMILY 4 *Nidulariaceae* Bird Nest Cavity Fungus
GENUS *Nidularia* Bird Nest Cavity Fungus
SPECIES *confluens, deformis, farcta, pulvinata*
GENUS *Crucibulum* Crucible Bird Nest Cavity Fungus
SPECIES *laeve*
GENUS *Cyathus* Cup Bird Nest Cavity Fungus
SPECIES *africanus, berkeleyanus, bulleri, canna, earlei, helenae, julietae, limbatus, montagnei, olla, pallidus, pygmaeus, stercoreus, striatus*
GENUS *Mycocalia* Callisto Bird Nest Cavity Fungus

SPECIES *arundinacea, denudata, duriaeana, minutissima, reticulata, sphagneti*
GENUS *Nidula* Greater Bird Nest Cavity Fungus
SPECIES *candida, niveotomentosa*

FAMILY 5 *Octavianinaceae* Octavian Cavity Fungus
GENUS *Octaviania* Octavian Cavity Fungus
SPECIES *asterosperma*
GENUS *Sclerogaster* Hard Octavian Cavity Fungus
SPECIES *compactus, hysterangioides, lanatus*
GENUS *Wakefieldia* Wakefield Octavian Cavity Fungus
SPECIES *macrospora, striaespora*

ORDER 2 *Lycoperdales* Puffball

FAMILY 1 *Lycoperdaceae* Puffball
GENUS *Lycoperdon* Puffball
SPECIES *atropurpureum, caudatum, curtisii, decipiens, echinatum, echinulatum, ericaeum, flavotinctum, lambinonii, lividum, mammiforme, marginatum, molle, muscorum, nigrescens, oblongisporum, perlatum, pulcherrimum, pyriforme, spadiceum, umbrinum*
VARIETIES *caudatum pedicellatum, decipiens cepiforme, echinatum hoylei, lividum cervinum, lividum cookei, mammiforme velatum, nigrescens foetidum, nigrescens perlatum, perlatum perlatum, perlatum gemmatum, pyriforme tessellatum*
GENUS *Bovista* Globe Puffball
SPECIES *aestivalis, dermoxantha, limosa, minor, nigrescens, paludosa, pila, plumbea, polymorpha, pusilla, pusilliformis, tomentosa*
VARIETIES *aestivalis cepiformis, aestivalis furfuraceum, dermoxantha ericetorum, dermoxantha pusillum, nigrescens globosum, plumbea ovalispora*
GENUS *Bovistella* Lesser Globe Puffball
SPECIES *radicata*
GENUS *Calbovista* Callisto Globe Puffball
SPECIES *subsculpta*
GENUS *Calvatia* Giant Puffball
SPECIES *booniana, coelatum, cretacea, cyathiformis, fragilis, fumosa, gigantea, hesperia, leiospora, lepidophora, lilacina, maximum, pachyderma, polygonia, sculpta, sigillata, subcretacea, versispora*
VARIETIES *fragilis pilula*
GENUS *Disciseda* Case Puffball
SPECIES *candida, muelleri, subterranea*
GENUS *Gastropila* (Synonym for *Calvatia fragilis, fumosa, hesperia, subcretacea*)
GENUS *Glyptoderma* (Synonym for *Calvatia coelatum*)
GENUS *Handkea* Handke Puffball
SPECIES *excipuliformis*, ***utriformis***(Edible)
VARIETIES *excipuliformis elatum, excipuliformis saccatum*, ***utriformis bovista, utriformis caelatum, utriformis sinclairii***
GENUS *Langermannia* (Synonym for *Calvatia gigantia*)
GENUS *Mycenastrum* Giant Pasture Puffball
SPECIES *corium*
GENUS *Vascellum* Vase Puffball
SPECIES *pratense*
VARIETIES *pratense depressum, pratense hiemale*

FAMILY 2 *Geastraceae* Earth Star Puffball
GENUS *Geastrum* Earth Star Puffball
SPECIES *berkeleyi, campestre, corollinum, coronatum, elegans, fimbriatum, floriforme, fornicatum, lageniforme, limbatum, minimum, mirabilis, pectinatum, quadrifidum, rufescens, saccatum, schmidelii, striatum, striatum bryantii, triplex*
VARIETIES *corollinum mammosum, corollinum recolligens, elegans badium, elegans smithii, elegans umbilicatum, fimbriatum minor, fimbriatum sessile, pectinatum plicatum, pectinatum tenuipes, rufescens schaefferi, rufescens vulgatum, schmidelii nanum, triplex michelianum*
GENUS *Astraeus* Lesser Earth Star Puffball
SPECIES *hygrometricus*
VARIETIES *hygrometricus anglicus, hygrometricus stellatus, hygrometricus vulgaris*
GENUS *Geaster* (Synonym for *Geastrum*)
GENUS *Phialastrum* Phial Earth Star Puffball
SPECIES *barbatum*
GENUS *Radiigera* Radial Earth Star Puffball
SPECIES *paulensis*
GENUS *Sphaerobolus* Globe Earth Star Puffball
SPECIES *iowensis, stellatus*

VARIETIES *stellatus stellatus, stellatus brasiliensis, stellatus carpobolus, stellatus dentatus, stellatus tubulosus*

FAMILY 3 *Sclerodermataceae* Earthball Puffball
GENUS *Scleroderma* Earthball Puffball
SPECIES *albidum, areolatum, bovista, cepa, chevalieri, citrinum, columnare, dictyosporum, echinatum, flavidum, floridanum, hypogaeum, laeve, leptopodium, meridionale, michiganense, polyrhizum, reae, sinnamariense, stellatum, uruguayense, verrucosum*
VARIETIES *bovista fuscum, bovista texense, citrinum aurantiacum, polyrhizum geaster*
GENUS *Horakiella* Horak Earthball Puffball
SPECIES *clelandii*
GENUS *Myriostoma* Myriad Earthball Puffball
SPECIES *coliforme*
GENUS *Sclerangium* Star Earthball Puffball
SPECIES *polyrhhizon*
GENUS *Tremellogaster* Tremor Earthball Puffball
SPECIES *surinamensis*

FAMILY 4 *Tulostomataceae* Stalked Puffball
GENUS *Tulostoma* Stalked Puffball
SPECIES *brumale, melanocyclum, niveum, simulans, striatum*
VARIETIES *brumale mammosum, brumale pedunculatum*
GENUS *Battarrea* Flat Stalked Puffball
SPECIES *phalloides, stevensii*
GENUS *Chlamydopus* Cloak Stalked Puffball
SPECIES *meyenianus*
GENUS *Pisolithus* Pea Rock Stalked Puffball
SPECIES *arrhizus,* ***tinctorius***(Dye Herb)
VARIETIES *arrhizus arenarius, arrhizus capsuliferum, arrhizus olivaceus, arrhizus pisocarpium*
GENUS *Podaxis* Pod Stalked Puffball
SPECIES *argentinus, farlowii, longii, microporus, pistillari*
GENUS *Polysaccum* (Synonym for *Pisolithus*)
GENUS *Queletia* Quelet Stalked Puffball
SPECIES *mirabilis*
GENUS *Schizostoma* Split Stalked Puffball
SPECIES *laceratum, montellicum*

ORDER 2 *Phallales* Horn Fungus

FAMILY 1 *Phallaceae* Stinkhorn Horn Fungus
GENUS *Phallus* Penis Stinkhorn Horn Fungus
SPECIES *hadriani, impudicus, indusiatus*
VARIETIES *hadriani imperialis, hadriani iosmos, impudicus duplicata, impudicus foetidus, impudicus togatus*
GENUS *Aseroë* Aseroë Stinkhorn Horn Fungus
SPECIES *rubra*
GENUS *Chondrogaster* Cartilage Stinkhorn Horn Fungus
SPECIES *pachysporus*
GENUS *Dictyophora* (Synonym for *Phallus*)
GENUS *Gelopellis* Gel Stinkhorn Horn Fungus
SPECIES *macrospora, purpurascens, rufus, shanxiensis, thaxteri*
GENUS *Ileodictyon* Netted Stinkhorn Horn Fungus
SPECIES *cibarium, gracile*
GENUS *Itajahya* Itajahya Stinkhorn Horn Fungus
SPECIES *galericulata*
VARIETIES *galericulata typica*
GENUS *Kobayasia* Kobayasi Stinkhorn Horn Fungus
SPECIES *nipponica*
GENUS *Ligiella* Ligi Stinkhorn Horn Fungus
SPECIES *rodrigueziana*
GENUS *Lysurus* Lysure Stinkhorn Horn Fungus
SPECIES *cruciatus*
GENUS *Mutinus* Pointed Stinkhorn Horn Fungus
SPECIES *bambusinus, caninus, caninus inodorus, elegans, ravenelii*
GENUS *Neolysurus* Modern Lysure Stinkhorn Horn Fungus
SPECIES *arcipulvinus*
GENUS *Protubera* Protuberent Stinkhorn Horn Fungus
SPECIES *africana, borealis, brunnea, canescens, clathroidea, hautuensis, jamaicensis, maracuja, nipponica, nothofagi, parvispora, sabulonensis, termitum*

GENUS *Protuberella* Lesser Protuberent Stinkhorn Horn Fungus
SPECIES *borealis*

FAMILY 2 *Clathraceae* Lattice Horn Fungus
GENUS *Clathrus* Stinkhorn Lattice Horn Fungus
SPECIES *archeri, columnatus, crucibulum, olla, striatus, ruber*
VARIETIES *ruber cancellatus*
GENUS *Anthurus* Sponge Lattice Horn Fungus
SPECIES *muellerianus, gardneri*
VARIETIES *muellerianus aseroëformis*
GENUS *Calostoma* Pretty Mouth Lattice Horn Fungus
SPECIES *cinnibarina, lutescens*
GENUS *Pseudocolus* Dead Log Lattice Horn Fungus
SPECIES *fusiformis, schellenbergiae*
GENUS *Simblum* Meadow Lattice Horn Fungus
SPECIES *sphaerocepahum, texense*

FAMILY 3 *Hysterangiaceae* Uterus Horn Fungus
GENUS *Hysterangium* Uterus Horn Fungus
SPECIES *clathroides, coriaceum, nephriticum, thwaitesii*
GENUS *Boninogaster* Bonin Uterus Horn Fungus
SPECIES *phalloides*
GENUS *Circulocolumella* Column Uterus Horn Fungus
SPECIES *hahashimensis*
GENUS *Clathrogaster* Lattice Uterus Horn Fungus
SPECIES *beccarii, vulvarius*
GENUS *Gallacea* Gall Uterus Horn Fungus
SPECIES *avellanea, dingleyae, eburnea, scleroderma, violacea*
GENUS *Hallingea* Halling Uterus Horn Fungus
SPECIES *carneorosea, purpurea, violacea*
GENUS *Phallobata* Penis Bat Uterus Horn Fungus
SPECIES *alba*
GENUS *Phallogaster* Stinkhorn Uterus Horn Fungus
SPECIES *saccatus*
GENUS *Phlebogaster* Phlebe Uterus Horn Fungus
SPECIES *laurisylvicola, sinensis*
GENUS *Rhopalogaster* Club Uterus Horn Fungus
SPECIES *transversarium*

CLASS 622.5 *Lichinobasidiomycetes* Club Lichen

ORDER 1 *Lichinobasidiales* Club Lichen

FAMILY 1 *Lichinobasidiaceae* Club Lichen
GENUS *Lichenomphalia* Navelcap Mushroom Club Lichen
SPECIES *alpina, hudsoniana, umbellifera, velutina*
VARIETIES *velutina grisella*
GENUS *Marchandiomyces* Marchand Club Lichen
SPECIES *aurantiacus, corallinus*

FAMILY 2 *Atheliaceae* Athel Club Lichen
GENUS *Athelia* Athel Club Lichen
SPECIES *acrospora, alnicola, alutacea, andina, arachnoidea, bambusae, binucleospora, bombacina, coprophila, decipiens, epiphylla, fibulata, fuciformis, macularis, neuhoffii, nivea, ovata, phycophila, poeltii, pyriformis, repetobasidiifera, rolfsii, salicum, sibirica, singularis, subovata, tenuispora, teutoburgensis*
GENUS *Amphinema* Dual Thread Athel Club Lichen
SPECIES *angustispora, arachispora, byssoides, diadema*
GENUS *Amyloathelia* Amyl Athel Club Lichen
SPECIES *amylacea, aspera, crassiuscula*
GENUS *Amylocorticium* Amyl Cortic Athel Club Lichen
SPECIES *africanum, cebennense, cumminsii, indicum, pedunculatum, rhodoleucum, suaveolens, subincarnatum, subsulphureum*
GENUS *Athelicium* Least Athel Club Lichen
SPECIES stridii
GENUS *Athelidium* Mock Athel Club Lichen
SPECIES *aurantiacum*
GENUS *Athelopsis* Lesser Athel Club Lichen
SPECIES *baculifera, galzinii, glaucina, lacerata, lembospora, lunata, subinconspicua, virescens*
GENUS *Butlerelfia* Butler Elf Athel Club Lichen

SPECIES *eustacei*
GENUS *Byssocorticium* Silky Cortic Athel Club Lichen
SPECIES *atrovirens, californicum, efibulatum, lutescens, naviculare, neomexicanum, pulchrum*
GENUS *Byssoporia* Silky Athel Club Lichen
SPECIES *terrestris*
GENUS *Cora* Cora Athel Club Lichen
SPECIES *glabrata*
GENUS *Cristinia* Cristin Athel Club Lichen
SPECIES *artheniensis, brevicellularis, filia, gallica, helvetica, rhenana, sonorae*
GENUS *Dictyonema* Netted Thread Athel Club Lichen
SPECIES *interruptum, irpicinum, ligulatum, moorei, thelephora*
GENUS *Digitatispora* Finger Spore Athel Club Lichen
SPECIES *lignicola, marina*
GENUS *Fibulomyces* Fibule Athel Club Lichen
SPECIES *canadensis, crucelliger, fusoideus, mutabilis*
GENUS *Hypochnella* Hyponic Ochnon Athel Club Lichen
SPECIES *violacea*
GENUS *Hypochniciellum* Hyponic Ochnic Athel Club Lichen
SPECIES *cremeoisabellinum, iaganicum, molle, ovoideum, subillaqueatum*
GENUS *Irpicodon* Irpicodon Athel Club Lichen
SPECIES *pendulus*
GENUS *Leptosporomyces* Thin Spore Athel Club Lichen
SPECIES *adnatus, fuscostratus, galzinii, juniperinus, luteofibrillosus, montanus, mundus, raunkiaeri, septentrionalis*
GENUS *Lobulicium* Lobed Athel Club Lichen
SPECIES *occultum*
GENUS *Luellia* Lu Athel Club Lichen
SPECIES *cystidiata, furcata, recondita*
GENUS *Melzericium* Melzer Athel Club Lichen
SPECIES *bourdotii, rimosum, udicola*
GENUS *Mycostigma* Stigma Athel Club Lichen
SPECIES *aegeritoides*
GENUS *Piloderma* Pilose Dermal Athel Club Lichen
SPECIES *bicolor, byssinum, lanatum, lapillicola, olivaceum, sphaerosporum*
GENUS *Plicatura* Plicate Athel Club Lichen
SPECIES *nivea*
GENUS *Plicaturopsis* Lesser Plicate Athel Club Lichen
SPECIES *crispa, scarlatina*
GENUS *Taeniospora* Ribbon Spore Athel Club Lichen
SPECIES *gracilis, nasifera*
VARIETIES *gracilis gracilis, gracilis enecta*
GENUS *Tomentellopsis* Woolly Athel Club Lichen
SPECIES *bresadolana, echinospora, pallidoaurantiaca, pallidocitrina, pusilla, submollis, zygodesmoides*
GENUS *Tylospora* Tyle Spore Athel Club Lichen Column
SPECIES *asterophora, fibrillosa*

CLASS 622.6 *Urediniomycetes* Rust

ORDER 1 *Uredinales* Rust

FAMILY 01 *Uredinaceae* Rust
GENUS *Uredo* Rust
SPECIES *artocarpi, cajani, morvernensis, musae, oncidii, ranunculacearum*
GENUS *Aecidiconium* Hercules Cone Rust
SPECIES *barteti*
GENUS *Aecidium* Hercules Rust
SPECIES *mori*
GENUS *Bensingtonia* Bensington Rust
SPECIES *ciliata*
GENUS *Cerradoa* Enclosed Rust
SPECIES *palmaea*
GENUS *Coleopuccinia* Cole Puccini Rust
SPECIES *kunmingensis, sinensis*
GENUS *Cystopsora* Cyst Spore Rust
SPECIES *antidesmatis, notelaeae*
GENUS *Desmellopsis* Desmell Rust
SPECIES *aframomicola*

GENUS *Hemileia* Hemile Rust
SPECIES *coffeicola, vastatrix*
GENUS *Heterogastridium* Variable Cavity Rust
SPECIES *pycnidioideum*
GENUS *Hiratsukaia* Lesser Hiratsuka Rust
SPECIES *tsugae*
GENUS *Hiratsukamyces* Hiratsuka Rust
SPECIES *divinus, salaciicola*
GENUS *Intrapes* Intrap Rust
SPECIES *paliformis*
GENUS *Kweilingia* Kweiling Rust
SPECIES *bambusae*
GENUS *Masseeëlla* Massee Rust
SPECIES *narasimhanii*
GENUS *Mehtamyces* Mehta Rust
SPECIES *stereospermi*
GENUS *Melampsora* Dark Psoral Rust
SPECIES *allii-fragilis, allii-populina, amygdalinae, arctica, caprearum, epitea, euphorbiae, hypericorum, lini, medusae, populnea, salicis-albae, symphyti, vernalis*
VARIETIES *caprearum farinosa, epitea epitea, epitea alpina, epitea euonymi-capraearum, epitea hartigii, epitea laricis, epitea repentis, epitea repentis-orchidis, epitea reticulatae, epitea reticulatae-mixta, epitea ribesii-purpureae, epitea ribesii-viminalis, euphorbiae dulcis, euphorbiae helioscopiae, euphorbiae ricini, lini lini, lini liniperda, lini liniperda-major, medusae albertensis, populnea aecidioides, populnea laricis, populnea laricis-pentandrae, populnea laricis-populina, populnea laricis-tremulae, populnea pinitorqua, populnea rostrupii, salicis-albae vitellinae*
GENUS *Mikronegeria* Mikroneger Rust
SPECIES *alba, fagi*
GENUS *Milesia* Miles Rust
SPECIES *magnusiana*
GENUS *Peridiopsora* Perid Psoral Rust
SPECIES *adelocaryi*
GENUS *Phragmotelium* Fragmented Rust
SPECIES *barnardii, burmanicum, formosanum, mysorensis, okianum, pauciloculare, rubi-fraxinifolii, rubi-thunbergii, yoshinagai*
GENUS *Stakmania* Stakman Rust
SPECIES *indica*
GENUS *Sterigmatomyces* Sterigmate Rust
SPECIES *halophilus*
GENUS *Stomatisora* Stoma Soral Rust
SPECIES *geophilicola*
GENUS *Telomapea* Telomap Rust
SPECIES *inocarpi*
GENUS *Thirumalachariella* Thiru Malachar Rust
SPECIES *holwayi*
GENUS *Tunicopsora* Tunic Spore Rust
SPECIES *bagchii*
GENUS *Uncol* Uncol Rust
SPECIES *diazii*
GENUS *Uraecium* Urec Rust
SPECIES *africanum, argomuellerae, carpodini, cydistae, derridicola, extensum, guanacastensis, holwayi, leonense, lucumae, magnisporum, nothopegiae*
GENUS *Zymoxenogloea* Zymoxen Glow Rust
SPECIES *eriophori*

FAMILY 02 *Atractiellaceae* Atract Rust
GENUS *Atractiella* Atract Rust
SPECIES *brunaudiana, delectans, macrospora, muscigena, solani*
GENUS *Atractogloea* Glow Atract Rust
SPECIES *stillata*
GENUS *Chionosphaera* Chion Sphere Atract Rust
SPECIES *coppinsii, lichenicola*
GENUS *Fibulostilbum* Fiber Stilbe Atract Rust
SPECIES *phylaciicola*
GENUS *Pachnocybe* Pachnon Cap Atract Rust
SPECIES *albida, ferruginea, grisea*
GENUS *Phleogena* Phleogene Atract Rust
SPECIES *faginea*
GENUS *Stilbum* Stilbe Atract Rust

SPECIES *anomalum, buloloense, fasciculatum, macrosporum, piliforme, vaporarium*

FAMILY 03 *Chaconiaceae* Chacon Rust
GENUS *Chaconia* Chacon Rust
SPECIES *alutacea, brasiliensis, butleri, coaetanea, ingae, maprouneae*
GENUS *Achrotelium* Achrotel Chacon Rust
SPECIES *hemidesmi, ichnocarpi, lucumae, rhodesicum, urerae*
GENUS *Aplopsora* Aplon Psoral Chacon Rust
SPECIES *corni, hennenii, nyssae*
GENUS *Botryorhiza* Berry Rhizid Chacon Rust
SPECIES *hippocrateae*
GENUS *Ceraceopsora* Cerace Psoral Chacon Rust
SPECIES *elaeagni*
GENUS *Ceropsora* Cero Psoral Chacon Rust
SPECIES *piceae*
GENUS *Chrysocelis* Golden Chacon Rust
SPECIES *butleri, globbae, gymnostemmatis, lupini, muehlenbeckiae*
GENUS *Goplana* Goplan Chacon Rust
SPECIES *andina, aporosae, australis, cissi, concinna, dioscoreae, ecuadorica, espeletiae, indica, micheliae, mirabilis, ribis-andicolae*
GENUS *Olivea* Olive Chacon Rust
SPECIES *capituliformis, colebrookiana, fimbriata, isonandrae, petitiae, scitula, tectonae, viticis*

FAMILY 04 *Classiculaceae* Classic Rust
GENUS *Classicula* Classic Rust
SPECIES *fluitans*
GENUS *Jaculispora* Jacule Spore Classic Rust
SPECIES *submersa*

FAMILY 05 *Coleosporiaceae* Cole Spore Rust
GENUS *Coleosporium* Cole Spore Rust
SPECIES *asterum, ipomoeae, plumeriae, tussilaginis*
VARIETIES *tussilaginis calendulae, tussilaginis pulsatillae, tussilaginis senecionis-silvatici*
GENUS *Chrysomyxa* Golden Cole Spore Rust
SPECIES *abietis, arctostaphyli, empetri, ledi, ledicola, pyrolae*
VARIETIES *ledi rhododendri*
GENUS *Stilbechrysomyxa* Stilbe Golden Cole Spore Rust
SPECIES *himalensis, stilbae, succinea*

FAMILY 06 *Cronartiaceae* Cronart Rust
GENUS *Cronartium* Cronart Rust
SPECIES *coleosporioides, comandrae, comptoniae, flaccidum, himalayense, quercuum, ribicola, harknessii*
VARIETIES *flaccidum typica*
GENUS *Endocronartium* Endic Cronart Rust
SPECIES *harknessii, pini, sahoanum*

FAMILY 07 *Phakopsoraceae* Phako Psoral Rust
GENUS *Phakopsora* Phako Psoral Rust
SPECIES *gossypii, meibomiae, pachyrhizi, vitis*
GENUS *Aeciure* Aeciure Phako Psoral Rust
SPECIES *crotonis*
GENUS *Arthuria* Arthur Phako Psoral Rust
SPECIES *catenulata, columbiana, glochidii, tylophorae*
GENUS *Batistopsora* Batista Phako Psoral Rust
SPECIES *crucis-filii*
GENUS *Cerotelium* Cerotel Phako Psoral Rust
SPECIES *fici*
GENUS *Macabuna* Macabun Phako Psoral Rust
SPECIES *arrabideae, heterophragmae, malvicola, markhamiae, melanocephala, poae-sudeticae, qualeae, stereospermi, thaungae, uguressae, zizyphi*
GENUS *Monosporidium* Monospore Phako Psoral Rust
SPECIES *andrachnes, euphorbiae, machili, pavettae*
GENUS *Newinia* Newin Phako Psoral Rust
SPECIES *heterophragmatis, kigeliae, thaiana*
GENUS *Nothoravenelia* Ravenel Phako Psoral Rust
SPECIES *commiphorae, japonica*
GENUS *Phragmidiella* Fragmented Phako Psoral Rust
SPECIES *africana, heterophragmatis, markhamiae, ugandana*
GENUS *Physopella* Bladder Pellet Phako Psoral Rust
SPECIES *zeae*
GENUS *Scalarispora* Scalar Spore Phako Psoral Rust

SPECIES *hashiokai*
GENUS *Uredopeltis* — Pelt Phako Psoral Rust
SPECIES *dominicana*
GENUS *Uredostilbe* — Stilbe Phako Psoral Rust
SPECIES *pistila*

FAMILY 08 *Phragmidiaceae* — Fragmented Rust
GENUS *Phragmidium* — Fragmented Rust
SPECIES *acuminatum, bulbosum,fragariae, fusiforme, mucronatum, potentillae, rosae-pimpinellifoliae, rubi-idaei, sanguisorbae, tuberculatum, violaceum*
VARIETIES *bulbosum rubi, mucronatum subcorticium, mucronatum rosae, rubi-idaei gracilis*
GENUS *Arthuriomyces* — Arthur Fragmented Rust
SPECIES *peckianus*
GENUS *Frommeella* — Fromme Fragmented Rust
SPECIES *tormentillae*
GENUS *Gymnoconia* — Nude Cone Fragmented Rust
SPECIES *nitens*
GENUS *Hamaspora* — Ham Spore Fragmented Rust
SPECIES *acutissima, australis, dobremezii, engleriana, gedeana, hashiokai, longissima, nepalensis, okinawensis, ozeensis, rubi-sieboldii, sinica, sinica, taiwaniana, viennotii*
GENUS *Joerstadia* — Joerstad Fragmented Rust
SPECIES *alchemillae, aliena, keniensis, patouillardii*
GENUS *Kuehneola* — Kuehn Fragmented Rust
SPECIES *uredinis*
GENUS *Trachyspora* — Trachyspore Fragmented Rust
SPECIES *intrusa, keniensis, melospora, pentaphylleae, vestita, wurthii*
GENUS *Xenodochus* — Strange Doch Fragmented Rust
SPECIES *carbonarius*

FAMILY 09 *Pileolariaceae* — Pile Rust
GENUS *Pileolaria* — Pile Rust
SPECIES *terebinthi*
GENUS *Atelocauda* — Stalk Pile Rust
SPECIES *angustiphyllodia, bicincta, digitata, hyalospora, incrustans, koae*
VARIETIES *digitata phyllodiorum*
GENUS *Endoraecium* — Endic Raec Pile Rust
SPECIES *acaciae, hawaiiense*
GENUS *Uromycladium* — Tailed Pile Rust
SPECIES *tepperianum*

FAMILY 10 *Platygloeaceae* — Flat Glow Rust
GENUS *Platygloea* — Flat Glow Rust
SPECIES *crassa, herrerae, javanica, liui, singeri, tremelloidea*
GENUS *Achroomyces* — Achron Flat Glow Rust
SPECIES *arrhytidiae, henricii, insignis, lumbricifer, micrus, pachysterigmata*
GENUS *Aphelariopsis* — Aphelar Flat Glow Rust
SPECIES *kupemontis*
GENUS *Auriculoscypha* — Ear Scyph Flat Glow Rust
SPECIES *anacardiicola*
GENUS *Camptobasidium* — Field Flat Glow Rust
SPECIES *hydrophilum*
GENUS *Colacogloea* — Colac Flat Glow Rust
SPECIES *bispora, eniophorae*
GENUS *Crucella* — Cross Flat Glow Rust
SPECIES *subtilis*
GENUS *Cystobasidium* — Cyst Flat Glow Rust
SPECIES *fimetarium, sebaceum*
GENUS *Eocronartium* — Cronart Flat Glow Rust
SPECIES *muscicola*
GENUS *Helicobasidium* — Lesser Helix Flat Glow Rust
SPECIES *brebissonii, compactum, corticioides, longisporum, mompa*
GENUS *Helicogloea* — Helix Flat Glow Rust
SPECIES *angustispora, farinacea, globispora, globosa, graminicola, lagerheimii, musaispora, vestita*
GENUS *Herpobasidium* — Herpetic Flat Glow Rust
SPECIES *filicinum*
GENUS *Insolibasidium* — Insolar Flat Glow Rust
SPECIES *deformans*
GENUS *Kriegeria* — Krieger Flat Glow Rust
SPECIES *eriophori*

GENUS *Krieglsteinera* — Krieglstein Flat Glow Rust
SPECIES *lasiosphaeriae*
GENUS *Kryptastrina* — Cryptic Star Flat Glow Rust
SPECIES *inclusa*
GENUS *Mycogloea* — Mycote Flat Glow Rust
SPECIES *macrospora*
GENUS *Naohidea* — Naohide Flat Glow Rust
SPECIES *sebacea*
GENUS *Occultifur* — Occult Flat Glow Rust
SPECIES *corticiorum, interdendron, internus*
GENUS *Phyllogloea* — Lesser Flat Glow Rust
SPECIES *crassa, herrerae, javanica, liui, singeri, tremelloidea*
GENUS *Ptechetelium* — Ptechetel Flat Glow Rust
SPECIES *cyatheae*
GENUS *Spiculogloea* — Spicule Flat Glow Rust
SPECIES *minuta, occulta, subminuta*
GENUS *Tuberculina* — Tubercle Flat Glow Rust
SPECIES *maxima, persicina, sbrozzii*
GENUS *Zygogloea* — Zygote Flat Glow Rust
SPECIES *gemellipara*

FAMILY 11 *Pucciniaceae* — Puccini Rust
GENUS *Puccinia* — Puccini Rust
SPECIES *acetosae, achilleae, adoxae, aegopodii, albescens, albulensis, allii, angelicae, annularis, antirrhini, apii, arachidis, arenariae, asparagi, behenis, betonicae, bistortae, brachypodii, bulbocastani, bupleuri, buxi, calcitrapae, calthae, calthicola, campanulae, canaliculata, cancellata, caricina, carthami, chaerophylli, chaetochloae, chrysanthemi, chrysosplenii, cicutae, circaeae, cladii, clintonii, cnici, commutata, conclusa, conii, coronata, crepidicola, cyani, cynodontis, difformis, dioicae, distincta, dracunculina, elymi, epilobii, eriophori, eutremae, fergussonii, festucae, fragosoana, galii, gentianae, glechomatis, glomerata, graminis, helianthi, hemerocallidis, hieracii, hordei, horiana, hydrocotyles, hysterium, impatientis, iridis, kuehnii, kusanoi, lagenophorae, lapsanae, libanotidis, liliacearum, ljulinica, longicornis, longissima, luzulae, maculosa, magnusiana, major, malvacearum, mariana, melanocephala, menthae, microsora, moliniae, nemoralis, nitida, oahuensis, obscura, opizii, oxalidis, oxyriae, paludosa, pazschkei, pelargonii-zonalis, phragmitis, physospermi, pimpinellae, pittieriana, poarum, polemonii, polygoni-amphibii, polysora, pratensis, primulae, prostii, psidii, pulverulenta, punctata, punctiformis, purpurea, pygmaea, recondita, recondita triticina, ribis, rottboelliae, rufipes, rugulosa, saniculae, saxifragae, schedonnardii, schroeteri, scirpi, scorzonerae, septentrionalis, sessilis, smyrnii, soldanellae, sorghi, spegazzinii, striiformis, suaveolens, substriata, tanaceti, thesii, thymi, tumida, uliginosa, umbilici, urticata, variabilis, veronicae, verruca, vincae, violae, virgae, zoysiae*

VARIETIES *acetosae vaginalium, adoxae saxifragarum, albulensis veronicarum, allii blasdalei, allii mixta, allii phalaridis, allii porri, angelicae bullata, angelicae silai, annularis scorodoniae, arenariae circinans, arenariae dianthi, arenariae herniariae, arenariae lychnidearum, arenariae moehringiae, arenariae saginae, arenariae spergulae, arenariae stellariae, arenariae verrucosa, behenis silenes, bistortae angelicae, bistortae cari, bistortae conopodii, bistortae vivipari, brachypodii brachypodii, brachypodii arrhenatheri, brachypodii arrhenatheri-airae, brachypodii arrhenatheri-deschampsiae, brachypodii baryi, brachypodii poae-anthoxanthina, brachypodii poae-nemoralis, brachypodii poae-sudeticae, brachypodii poae-thalictri, bulbocastani bunii, bupleuri falcati, calcitrapae bardanae, calcitrapae bardanae-inquinans, calcitrapae cardui, calcitrapae cardui-pygnocephali, calcitrapae carlinae, calcitrapae compositarum, calcitrapae scabiosae, calcitrapae syngenesiarum, calthicola zopfii, campanulae rotundifoliae, caricina caricina, caricina grossulariae, caricina magnusii, caricina pringsheimiana, caricina ribesii-pendulae, caricina ribesii-pseudocyperi, caricina ribisnigri-lasiocarpae, caricina ribisnigri-paniculatae, caricina striola, carthami centaureae, chrysanthemi absinthii, chrysanthemi sinensis, clintonii sylvaticae, cnici andersonii, cnici cirsii, cnici cirsii-eriophori, cnici cirsii-lanceolati, cnici oleracei, cnici oleracei-asteris, cnici oleracei-cardui, cnici oleracei-chrysanthemi-leucanthemi, cnici oleracei-lemonnieriana, cnici oleracei-leucanthemi, cnici oleracei-millefolii, cnici oleracei-palustris, cnici oleracei-rudbeckiae, cnici oleracei-tripolii, commutata valerianae, coronata agrostidis, coronata alopecuri, coronata arrhenatheri, coronata avenae, coronata calamagrostis, coronata festucae, coronata holci, coronata lolii, coronata rhamni, difformis ambigua, dioicae arenariicola, dioicae dioicae, dioicae extensicola, dioicae schoeleriana, dioicae silvatica, elymi ammophilae, eriophori confinis, eutremae cochleariae, galii cruciatae, galii cruciatae-celakovskyana, galii verni, glomerata expansa, glomerata senecionis, graminis graminis, graminis albigensis, graminis anthistiriae, graminis avenae, graminis brizae-maximae, graminis cerealis, graminis dactylidis, graminis elymina, graminis favargeri, graminis graminicola, graminis graminicola-anthoxanthi, graminis graminicola-phlei-pratensis, graminis jubata, graminis linearis, graminis macrospora, graminis major, graminis megalopotamica, graminis oryzae, graminis secalis, graminis stakmanii, graminis tritici, graminis vilis, hemerocallidis funkiae, hieracii hieracii, hieracii cichorii, hieracii endiviae, hieracii flosculosorum, hieracii hypochaeridis, hieracii jaceae, hieracii leontodontis, hieracii picridis, hieracii piloselloidarum, hieracii taraxaci, hieracii tinctoriae, hordei hordei, hordei anomala, hordei fragosoi, hordei loliina, hordei murini, hordei schismi, hordei simplex, hordei simplex-straminis, hordei vulpiae, hordei vulpiae-myuri, hysterium sparsa, hysterium tragopogonis, impatientis argentata, impatientis noli-tangere, iridis truncata, lagenophorae erechtitis, lagenophorae terrieriana, luzulae oblongata, maculosa chondrillae, maculosa prenanthis, magnusiana epicaula, malvacearum sidae-rhombifoliae, menthae clinopodii, nemoralis aecidii-melampyri, nitida aethusae, nitida emondensis, nitida joerstadii, nitida petroselini, obscura maximae, pazschkei pazschkei, pazschkei jueliana, phragmitis trailii, poarum conspicua, polygoni-amphibii polygoni-amphibii, polygoni-amphibii convolvuli, pulverulenta tetragoni, punctata asperulae-cynanchicae, punctata asperulae-odoratae, punctata galiorum, punctiformis obtegens, purpurea andropogonis-hirti, purpurea prunicolor, purpurea sanguinea, pygmaea pygmaea, pygmaea ammophilina, recondita agropyrina, recondita agrostidis, recondita asperifolii, recondita borealis, recondita bromina, recondita bromina-symphyti, recondita dispersa, recondita holcina, recondita perplexans, recondita persistens, recondita secalina, recondita trisetina, rugulosa auloderma, rugulosa carniolica, rugulosa peucedani-parisiensis, schedonnardii hibisciata, schedonnardii muehlenbergiae, sessilis digraphidis, sessilis festucina, sessilis linearis, sessilis paradis, sessilis phalaridis, sessilis phalaridis-ari, sessilis phalaridis-orchidearum, sessilis winteriana, smyrnii olusatri, sorghi maydis, sorghi halepensis, sorghi zeae, striiformis striiformis, striiformis dactylidis, striiformis glumarum, substriata substriata, substriata penicillariae, substriata penicillariae-penniseti, substriata penicillariae-spicati, tanaceti artemisiae, tanaceti artemisiella, tanaceti discoidearum, tanaceti pyrethri, thymi caulincola, thymi ruebsaameni, thymi schneideri, tumida bunii, tumida umbelliferarum, umbilici cotyledonis, umbilici rhodiolae, urticata urticata, urticata acutae, urticata acutiformis, urticata biporula, urticata flaccae, urticata hirtae, urticata inflatae, urticata paniceae, urticata ripariae, urticata vesicariae, veronicae longifoliae, veronicae longifoliae-spicatae, verruca jaczewskii, vincae berkeleyi, violae aegra, violae depauperans, virgae aureae, virgae aureae-clandestina*

GENUS *Chrysella* — Golden Puccini Rust

SPECIES *mikaniae*

GENUS *Chrysocyclus* — Golden Cyclic Puccini Rust

SPECIES *cestri, mikaniae, senecionis*

GENUS *Cleptomyces* — Stolen Puccini Rust

SPECIES *lagerheimianus*

GENUS *Coleopucciniella* — Cole Puccini Rust

SPECIES *idei, simplex*

GENUS *Corbulopsora* — Corbule Psoral Puccini Rust

SPECIES *clemensiae, cumminsii, gravida*

GENUS *Cumminsiella* — Cummins Puccini Rust

SPECIES *mirabilissima*

GENUS *Endophyllum* — Endic Puccini Rust

SPECIES *euphorbiae-silvaticae, sempervivi*

GENUS *Gymnosporangium* — Nude Spore Puccini Rust

SPECIES *asiaticum, clavariiforme, clavipes, confusum, cornutum, globosum, juniperi-virginianae, sabinae, yamadae*

VARIETIES *sabinae juniperi*

GENUS *Kernella* — Kern Puccini Rust

SPECIES *lauricola*

GENUS *Kimuromyces* — Kimur Puccini Rust

SPECIES *cerradensis*

GENUS *Miyagia* — Miyagi Puccini Rust

SPECIES *anaphalidis, pseudosphaeria*

GENUS *Polioma* — Polio Puccini Rust

SPECIES *griseola*

GENUS *Uromyces* — Tailed Puccini Rust

SPECIES *acetosae, aecidiiformis, airae-flexuosae, anthyllidis, appendiculatus, ari-triphylli, armeriae, behenis, beticola, chenopodii, ciceris-arietini, colchici, coronatus, croci, dactylidis, dianthi, dolicholi, ervi, erythronii, excavatus, ficariae, gageae, galegae, gentianae, geranii, holwayi, inaequialtus, janiphae, junci, limonii, lineolatus, manihotis, minor, muscari, novissimus, pisi-sativi, polygoni, rumicis, salicorniae, scrophulariae, scutellatus, setariae-italicae, sparsus, striatus, tinctoriicola, transversalis, trifolii, tuberculatus, valerianae, viciae-fabae, vignae*

VARIETIES *aecidiiformis lilii, anthyllidis hippocrepidis, anthyllidis jaapianus, appendiculatus phaseoli, dactylidis festucae, dactylidis festucae-ranunculi, dactylidis graminum, dactylidis lycoctoni, dactylidis poae, dianthi caryophyllinus, dolicholi pamparum, geranii kabatianus, holwayi lilii, lineolatus maritimae, lineolatus scirpi, lineolatus scirpi-glaucis, lineolatus scirpi-oenanthi-crocati, muscari concentricus, muscari scillarum, novissimus hellerianus, pisi-sativi anglicae, pisi-sativi astragali, pisi-sativi corniculatae, pisi-sativi genistae-tinctoriae, pisi-sativi laburni, pisi-sativi lathyri, pisi-sativi loti, pisi-sativi onobrychidis, pisi-sativi phacae, pisi-sativi punctatus, pisi-sativi scoparii, pisi-sativi ulicis, polygoni avicularis, scrophulariae concomitans, setariae-italicae leptodermus, striatus medicaginis, tinctoriicola hybernae, transversalis freesiae, transversalis watsoniae, trifolii flectens, trifolii nerviphilus, trifolii repentis, trifolii repentis-fallens, tuberculatus excavatus, viciae-fabae viciae-fabae, viciae-fabae fallens, viciae-fabae globosa, viciae-fabae orobi*

GENUS *Zaghouania* — Zaghouan Puccini Rust

SPECIES *oleae, phillyreae*

FAMILY 12 *Pucciniastraceae* — Aster Puccini Rust

GENUS *Pucciniastrum* — Aster Puccini Rust

SPECIES *agrimoniae, americanum, areolatum, circaeae, epilobii, goeppertianum, goodyerae, guttatum, pyrolae*

VARIETIES *agrimoniae eupatoriae, areolatum padi, epilobii chamaenerii, epilobii chamaenerii-abieti, epilobii fuchsiae, epilobii pustulatum, guttatum galii*
GENUS *Hyalopsora* Hyaline Psoral Aster Puccini Rust
SPECIES *adianti-capilli-veneris, aspidiotus, polypodii*
GENUS *Melampsorella* Lesser Melam Psoral Aster Puccini Rust
SPECIES *caryophyllacearum, symphyti*
VARIETIES *caryophyllacearum cerastii*
GENUS *Melampsoridium* Melam Psoral Aster Puccini Rust
SPECIES *betulinum, hiratsukanum*
GENUS *Milesina* Miles Aster Puccini Rust
SPECIES *blechni, carpatorum, dieteliana, kriegeriana, murariae, scolopendrii, vogesiaca, whitei*
GENUS *Naohidemyces* Naohide Aster Puccini Rust
SPECIES *fujisanensis, vaccinii, vacciniorum*
GENUS *Uredinopsis* Lesser Aster Puccini Rust
SPECIES *americana, filicina*

FAMILY 13 *Pucciniosiraceae* Pitted Puccini Rust
GENUS *Pucciniosira* Pitted Puccini Rust
SPECIES *cornuta*
GENUS *Baeodromus* Baeodrome Pitted Puccini Rust
SPECIES *albertensis, californicus, eupatorii, holwayi, senecionis, tranzschelii*
GENUS *Ceratocoma* Horn Pitted Puccini Rust
SPECIES *guineensis, jacksoniae*
GENUS *Chardoniella* Chardon Pitted Puccini Rust
SPECIES *acuta, andina, capitata, gynoxidis*
GENUS *Cionothrix* Cion Thread Pitted Puccini Rust
SPECIES *praelonga*
GENUS *Dietelia* Dietel Pitted Puccini Rust
SPECIES *codiaei*
GENUS *Gambleola* (Synonym for *Pucciniosira*)
GENUS *Trichopsora* Hairy Psoral Pitted Puccini Rust
SPECIES *tournefortiae*

FAMILY 14 *Raveneliaceae* Ravenel Rust
GENUS *Ravenelia* Ravenel Rust
SPECIES *bottomleyae, caesalpiniae, cubensis*
GENUS *Allotelium* Allotel Ravenel Rust
SPECIES *mirabile*
GENUS *Anthomycetella* Anther Ravenel Rust
SPECIES *canarii*
GENUS *Apra* Apral Ravenel Rust
SPECIES *bispora*
GENUS *Cumminsina* Cummins Ravenel Rust
SPECIES *clavispora*
GENUS *Cystomyces* Cyst Ravenel Rust
SPECIES *costaricensis*
GENUS *Diabole* (Synonym for *Ravenelia cubensis*)
GENUS *Diabolidium* Diabolic Ravenel Rust
SPECIES *calliandrae*
GENUS *Diorchidiella* Orchid Ravenel Rust
SPECIES *australis, verlandii*
GENUS *Esalque* Esalque Ravenel Rust
SPECIES *holwayi*
GENUS *Hennenia* Hennen Ravenel Rust
SPECIES *ditelia*
GENUS *Kernkampella* Kernkamp Ravenel Rust
SPECIES *appendiculata, brevispora, breyniae, breyniae-patentis, coimbatorica, emblicae, kirganellae, phyllanthi*
GENUS *Lipocystis* (Synonym for *Ravenelia caesalpiniae*)
GENUS *Nyssopsora* Water Nymph Psoral Ravenel Rust
SPECIES *echinata*
GENUS *Spumula* Spumule Ravenel Rust
SPECIES *clemensiae, debilis, heteromorpha, quadrifida, serispora*
GENUS *Triphragmiopsis* Lesser Triple Fragment Ravenel Rust
SPECIES *isopyri, jeffersoniae, laricina*
GENUS *Triphragmium* Triple Fragment Ravenel Rust
SPECIES *filipendulae, ulmariae*
GENUS *Ypsilospora* Psilum Spore Ravenel Rust
SPECIES *africana, baphiae*

FAMILY 15 *Septobasidiaceae* Septic Rust
GENUS *Septobasidium* Septic Rust
SPECIES *accumbens, arachnoideum, carestianum, dictyodes, fissolobatum, lichenicola, pteruloides, rameale, rhabarbarinum, septobasidioides, siparium, spongium, stereoides, subcarbonaceum, sublilacinum, suffultum, thwaitesii, ussanguense*
GENUS *Coccidiodictyon* Round Net Septic Rust
SPECIES *inconspicuum*
GENUS *Glenospora* Glen Spore Septic Rust
SPECIES *graphii, lanuginosa*
GENUS *Johncouchia* John Couch Septic Rust
SPECIES *mangiferae*
GENUS *Uredinella* Lesser Septic Rust
SPECIES *coccidiophaga, spinulosa*

FAMILY 16 *Sporidiobolaceae* Spore Bowl Rust
GENUS *Sporidiobolus* Spore Bowl Rust
SPECIES *johnsonii, ruinenii, salmonicolor*
GENUS *Aessosporon* Aessospore Spore Bowl Rust
SPECIES *dendrophilum, salmonicolor*
GENUS *Ballistosporomyces* Ballistospore Spore Bowl Rust
SPECIES *ruber, xanthus*
GENUS *Cyrenella* Cyrene Spore Bowl Rust
SPECIES *elegans*
GENUS *Rhodotorula* Red Torus Spore Bowl Rust
SPECIES *acheniorum, bacarum, fragaria, glutinis, graminis, marina, minuta, mucilaginosa, pustula, rubra*
GENUS *Rogersiomyces* Rogers Spore Bowl Rust
SPECIES *okefenokeensis*
GENUS *Sporobolomyces* Lesser Spore Bowl Rust
SPECIES *holsaticus, roseus, tsugae*

FAMILY 17 *Uropyxidaceae* Tail Pyxie Rust
GENUS *Uropyxis* Tail Pyxie Rust
SPECIES *fraxini*
GENUS *Dipyxis* Double Tail Pyxie Rust
SPECIES *mexicana, viegasii*
GENUS *Leucotelium* Light Tail Pyxie Rust
SPECIES *cerasi, padi, pruni-persicae*
GENUS *Macruropyxis* (Synonym for *Uropyxis*)
GENUS *Mimema* Mime Tail Pyxie Rust
SPECIES *holwayi, venturae*
GENUS *Ochropsora* Ocher Psoral Tail Pyxie Rust
SPECIES *anemones, ariae, daisenensis, nambuana*
VARIETIES *ariae sorbi*
GENUS *Phragmopyxis* Fragmented Tail Pyxie Rust
SPECIES *acuminata, deglubens, leonensis, noelii*
GENUS *Poliomopsis* Polio Tail Pyxie Rust
SPECIES *thermopsidis*
GENUS *Sorataea* Sorat Tail Pyxie Rust
SPECIES *amiciae, arayatensis, baphiae, nephroidea, ostryoderridis, periodica*
VARIETIES *ostryoderridis insperata*
GENUS *Tranzschelia* Tranzschel Tail Pyxie Rust
SPECIES *anemones, pruni-spinosae*
VARIETIES *anemones fusca, anemones thalictri, pruni-spinosae pruni-spinosae, pruni-spinosae discolor*

ORDER 2 *Microbotryales* Berry Rust

FAMILY 1 *Microbotryaceae* Berry Rust
GENUS *Microbotryum* Berry Rust
SPECIES *dianthorum, lychnidis-dioicae, major, reticulatum, scabiosae, scorzonerae, silenes-inflatae, stellariae, violaceum*
GENUS *Liroa* Liron Berry Rust
SPECIES *emodensis*
GENUS *Reniforma* Kidney Berry Rust
SPECIES *strues*
GENUS *Sphacelotheca* Sphace Theca Berry Rust
SPECIES *andropogonis, destruens, erianthi, hydropiperis, macrospora, reiliana, schweinfurthiana, sorghi*
GENUS *Zundeliomyces* Zundel Berry Rust
SPECIES *polygoni*

FAMILY 2 *Leucosporidiaceae* Light Spore Berry Rust

GENUS *Leucosporidium* Light Spore Berry Rust
SPECIES *scottii*
GENUS *Sakaguchia* Sakaguchi Light Spore Berry Rust
SPECIES *dacryoidea*
FAMILY 3 *Ustilentylomataceae* Smut Berry Rust
GENUS *Ustilentyloma* Smut Berry Rust
SPECIES *brefeldii, fluitans, oreochloae, pleuropogonis*
GENUS *Aurantiosporium* Orange Spore Smut Berry Rust
SPECIES *marisci, subnitens*
GENUS *Bauerago* Bauer Smut Berry Rust
SPECIES *abstrusa, capensis, cyperi-lucidi, gardneri, vuyckii*
GENUS *Fulvisporium* Fulvous Spore Smut Berry Rust
SPECIES *restifaciens*

CLASS 622.7 *Ustilaginomycetes* Smut

ORDER 1 *Ustilaginales* Smut

FAMILY 1 *Ustilaginaceae* Smut
GENUS *Ustilago* Smut
SPECIES *affinis, anomala, avenaebistortarum bullata, cardui, cordae, crameri, crus-galli, cynodontis, duriaeana, echinata, filiformis, flosculorum, grandis, heufleri, hordei, hypodytes, intermedia, kuehneana, longissima, marina,* **maydis**(Edible Unripe), *nuda, parlatorei, pinguiculae, pustulata, scitaminea, serena, serpens, sphaerogena, striiformis, stygia, succisae, tragopogonis, trichophora, triodiae, tritici, utriculosa, vaillantii, vinosa*
VARIETIES , *avenae avenae, avenae jensenii, avenae kolleri, avenae levis, avenae nigra, avenae perennans, avenae rostrupii, , bistortarum candollei, bistortarum inflorescentiae, bistortarum marginalis, bistortarum ustilaginea, bullata bromivora, bullata patagonica, grandis typhoides, hordei carbo, hordei decipiens, hordei segetum, longissima longissima, longissima macrospora, longissima macrospora-davisii, maydis zeae-mays, nuda hordei, nuda tritici, nuda vavilovi, scitaminea sacchari, striiformis salveii, tragopogonis tragopogonis, tragopogonis pratensis, tragopogonis receptaculorum, vaillantii scillae*
GENUS *Clintamra* Clintamer Smut
SPECIES *nolinae*
GENUS *Cryptomycocolax* Cryptic Colac Smut
SPECIES *abnormis*
GENUS *Dermatosorus* Dermal Soral Smut
SPECIES *bulbostylidis, cyperi, eleocharidis, fimbristylidis, thirumalacharii*
GENUS *Endothlaspis* Endic Thlapse Smut
SPECIES *melicae, sorghi*
GENUS *Entorrhiza* Entic Rhizid Smut
SPECIES *aschersoniana, caricicola, casparyana, casparyanella, cypericola, fineranae, parvula, raunkiaeriana, scirpicola*
GENUS *Entyloma* Entic Tyle Smut
SPECIES *achilleae, arnoseridis, bellidis, calendulae, chrysosplenii, compositarum, dactylidis, dahliae, eryngii, fergussonii, ficariae, fuscum, gaillardianum, globigenum, helosciadii, henningsianum, hieracii, irregulare, linariae, matricariae, microsporum, nymphaeae, oryzae, ranunculi-repentis, serotinum, thirumalacharii, veronicae*
GENUS *Farysia* Farys Smut
SPECIES *thuemenii*
GENUS *Farysporium* Lesser Farys Smut
SPECIES *endotrichum*
GENUS *Franzpetrakia* Franz Petrak Smut
SPECIES *microstegii, okudairae*
GENUS *Geminago* Gemini Smut
SPECIES *nonveilleri*
GENUS *Juliohirschhornia* Julio Hirschhorn Smut
SPECIES *linderi*
GENUS *Kuntzeomyces* Kuntze Smut
SPECIES *ustilaginoideus*
GENUS *Macalpinomyces* MacAlpin Smut
SPECIES *chrysopogonis, effusus, eragrostiellae, eriachnes, loudetiae, nigritanae, sharmae, tanakae, trichopterygis, tripogonis, tristachyae, zonotriches*
GENUS *Malassezia* Malassez Smut
SPECIES *furfur, globosa, obtusa, pachydermatis, restricta, slooffiae, sympodialis*
GENUS *Melanopsichium* Melano Psich Smut
SPECIES *eleusinis, nepalense*
GENUS *Microstroma* Stroma Smut

SPECIES *album, juglandis, pithecolobii*
GENUS *Moesziomyces* Moeszi Smut
SPECIES *bullatus*
GENUS *Orphanomyces* Orphan Smut
SPECIES *arcticus, hungaricus, vankyi*
GENUS *Pseudozyma* Zyme Smut
SPECIES *antarctica, aphidis, flocculosa, fusiformata, prolifica, rugulosa, tsukubaensis*
GENUS *Restiosporium* Restion Spore Smut
SPECIES *lepidoboli, leptocarpi, meneyae, restionum*
GENUS *Rhombiella* Rhombic Smut
SPECIES *cardamines*
GENUS *Schizonella* Schizon Smut
SPECIES *cocconi, melanogramma*
GENUS *Sporisorium* Spore Soral Smut
SPECIES *cruentum, lepturi, ophiuri, puellare, sorghi, themedae-arguentis*
GENUS *Ustilaginoidea* Lesser Smut
SPECIES *virens*
GENUS *Websdanea* Websdane Smut
SPECIES *lyginiae*

FAMILY 2 *Anthracoideaceae* Anthrax Smut
GENUS *Anthracoidea* Anthrax Smut
SPECIES *arenariae, bigelowii, capillaris, caricis, caryophylleae, heterospora, karii, limosa, liroi, paniceae, pratensis, pseudirregularis, pulicaris, scirpi, subinclusa*
GENUS *Planetella* Plane Anthrax Smut
SPECIES *lironis*

FAMILY 3 *Cintractiaceae* Cintract Smut
GENUS *Cintractia* Cintract Smut
SPECIES *limitata, pulverulenta*
GENUS *Cintractiella* Lesser Cintract Smut
SPECIES *lamii*
GENUS *Heterotolyposporium* Variable Tolypospore Cintract Smut
SPECIES *lepidospermatis, piluliforme*
GENUS *Leucocintractia* Light Cintract Smut
SPECIES *affinis, leucoderma, leucodermoides, pachyderma, portus-argenti, scleriae*
GENUS *Stegocintractia* Stegid Cintract Smut
SPECIES *hyperborea, junci, lidii, luzulae, spadicea*
GENUS *Tolyposporium* Tolypospore Cintract Smut
SPECIES *ehrenbergii*
GENUS *Trichocintractia* Hairy Cintract Smut
SPECIES *utriculicola*
GENUS *Ustanciosporium* Ustance Spore Cintract Smut
SPECIES *gigantosporum, majus*

FAMILY 4 *Doassansiaceae* Dossan Smut
GENUS *Doassansia* Dossan Smut
SPECIES *alismatis, hygrophilae, limosellae, sagittariae*
GENUS *Doassansiella* Lesser Dossan Smut
SPECIES *aquatilis*
GENUS *Doassansiopsis* Least Dossan Smut
SPECIES *deformans, euryalis, furva, horiana, hydrophila, intermedia, limnocharidis, nymphaeae, nymphoides, ticonis*
GENUS *Doassinga* Inga Dossan Smut
SPECIES *callitrichis*
GENUS *Heterodoassansia* Variable Dossan Smut
SPECIES *callitrichis, hottoniae, morotiana, punctiformis, putkonenii*
GENUS *Melaniella* Melano Dossan Smut
SPECIES *oreophila, selaginellae*
GENUS *Nannfeldtiomyces* Nannfeldt Dossan Smut
SPECIES *anomalus, sparganii*
GENUS *Narasimhania* Narasimha Dossan Smut
SPECIES *alismatis*
GENUS *Pseudodermatosorus* Dermal Soral Dossan Smut
SPECIES *sagittariae*
GENUS *Pseudodoassansia* Mock Dossan Smut
SPECIES *obscura*
GENUS *Pseudotracya* Lesser Tracy Dossan Smut
SPECIES *otteliae*

GENUS *Rhamphospora* — Rhamphospore Dossan Smut
SPECIES *nymphaeae*
GENUS *Tracya* — Tracy Dossan Smut
SPECIES *hydrocharidis, lemnae*

FAMILY 5 *Glomosporiaceae* — Glome Smut
GENUS *Glomospora* — Glome Smut
SPECIES *empetri*
GENUS *Sorosporium* — Soral Glome Smut
SPECIES *heteropogonicola, indicum, saponariae*
GENUS *Thecaphora* — Theca Glome Smut
SPECIES *deformans, lathyri, seminis-convolvuli, solani, trailii*
GENUS *Thecaphorella* — Lesser Theca Glome Smut
SPECIES *antherarum*
GENUS *Tothiella* — Toth Glome Smut
SPECIES *thlaspeos*

FAMILY 6 *Tilletiaceae* — Tillet Smut
GENUS *Tilletia* — Tillet Smut
SPECIES *anthoxanthi, barclayana, caries, controversa, holci, horrida, indica, laevis, menieri, olida, setariae, sphaerococca*
VARIETIES *caries tritici*
GENUS *Conidiosporomyces* — Cone Spore Tillet Smut
SPECIES *ayresii, verruculosus*
GENUS *Erratomyces* — Erratic Tillet Smut
SPECIES *ajmeriensis, crotalariae, patelii, smithiae, thirumalacharii*
GENUS *Ingoldiomyces* — Ingold Tillet Smut
SPECIES *hyalosporus*
GENUS *Neovossia* — Voss Tillet Smut
SPECIES *opaca, sumatii*
GENUS *Oberwinkleria* — Oberwinkler Tillet Smut
SPECIES *anulata*

ORDER 2 *Exobasidiales* Exo Smut

FAMILY 1 *Exobasidiaceae* — Exo Smut
GENUS *Exobasidium* — Exo Smut
SPECIES *angustisporum, arescens, bisporum, camelliae, cylindrosporum, dubium, expansum, gracile, japonicum, juelianum, karstenii, myrtilli, oxycocci, pachysporum, pieridis-ovalifoliae, reticulatum, rhododendri, rostrupii, shiraianum, splendidum, sydowianum, symploci-japonicae, unedonis, vaccinii, vexans, yoshinagai*
VARIETIES *dubium magnusii, karstenii andromedae, myrtilli decolorans, vaccinii vaccinii, vaccinii caucasicum, vaccinii japonicum*
GENUS *Arcticomyces* — Arctic Exo Smut
SPECIES *warmingii*
GENUS *Endobasidium* — Endic Exo Smut
SPECIES *clandestinum*
GENUS *Graphiola* — Graphic Exo Smut
SPECIES *phoenicis*
GENUS *Laurobasidium* — Laurel Exo Smut
SPECIES *lauri*
GENUS *Sympodiomycopsis* — Sympodal Exo Smut
SPECIES *paphiopedili*

FAMILY 2 *Brachybasidiaceae* — Short Exo Smut
GENUS *Ceraceosorus* — Cerac Soral Short Exo Smut
SPECIES *bombacis*
GENUS *Dicellomyces* — Dicellular Short Exo Smut
SPECIES *scirpi*
GENUS *Exobasidiellum* — (Synonym for *Limonomyces, Helicogloea*)
GENUS *Kordyana* — Kordy Short Exo Smut
SPECIES *celebensis*
GENUS *Proliferobasidium* — Prolific Short Exo Smut
SPECIES *heliconiae*

FAMILY 3 *Cryptobasidiaceae* — Cryptic Exo Smut
GENUS *Botryoconis* — Berry Cone Cryptic Exo Smut
SPECIES *japonica*
GENUS *Coniodictyum* — Net Cone Cryptic Exo Smut
SPECIES *chevalieri*

ORDER 3 *Georgefischeriales* George Fischer Smut

FAMILY 1 *Georgefischeriaceae* George Fischer Smut
GENUS *Georgefischeria* George Fischer Smut
SPECIES *narasimhanii, riveae*
GENUS *Eballistra* Ballist George Fischer Smut
SPECIES *brachiariae, lineata, oryzae*
GENUS *Gjaerumia* Gjaerum George Fischer Smut
SPECIES *ossifragi*

FAMILY 2 *Tilletiariaceae* Tillet George Fischer Smut
GENUS *Tilletiaria* Tillet George Fischer Smut
SPECIES *anomala]*
GENUS *Phragmotaenium* Fragmented Ribbon Tillet George Fischer Smut
SPECIES indicum
GENUS *Tilletiopsis* Lesser Tillet George Fischer Smut
SPECIES *albescens, flava, fulvescens, lilacina, minor, pallescens, washingtonensis*

ORDER 4 *Urocystales* Tail Cyst Smut

FAMILY 1 *Urocystaceae* Tail Cyst Smut
GENUS *Urocystis* Tail Cyst Smut
SPECIES *agropyri, agrostidis, alopecuri, anemones, avenae-elatioris, bolivarii, bromi, cepulae, colchici, eranthidis, ficariae, filipendulae, fischeri, floccosa, gladiolicola, junci, melicae, occulta, poae, primulae, primulicola, ranunculi, sorosporioides, syncocca, trientalis, trollii, ulei, violae*
VARIETIES *agropyri campestris, agropyri tritici, anemones pompholygodes, cepulae colchici, cepulae magica, floccosa hellebori-viridis, occulta parallela, syncocca hepaticae-trilobae*
GENUS *Mundkurella* Mundkur Tail Cyst Smut
SPECIES *heptapleuri, kalopanacis, mohgaoensis, mossii*
GENUS *Vankya* Vanky Tail Cyst Smut
SPECIES *heufleri, ornithogali, vaillantii*

FAMILY 2 *Melanotaeniaceae* Dark Tail Cyst Smut
GENUS *Melanotaenium* Dark Tail Cyst Smut
SPECIES *ari, cingens, endogenum, hypogaeum, jaapii, lamii*
GENUS *Exoteliospora* Exotel Dark Tail Cyst Smut
SPECIES *osmundae*
GENUS *Yelsemia* Yelsem Dark Tail Cyst Smut
SPECIES *arthropodii, specularia*e

PHYLUM 623 *Zygomycota* Multinucleate Fungus

Latin: *Zygo* zygote, egg	CLASSES		Species
	623.1 *Zygomycetes*	Multinucleate Fungus	775 [K]
1,100 species	623.2 *Glomeromycetes*	Cluster Multinucleate Fungus	160 [K]
[Margulis & Schwartz] [M]	623.3 *Trichomycetes*	Hairy Multinucleate Fungus	186 [K]
1,121 species [Kirk, Sept. 2006] [K]			

Phylum *Zygomycota* Multinucleate Fungus species are either saprobes or symbiotes. Zygomycotes live on land. A saprobe excretes extra cellular enzymes which break down dead organic matter. The dead organic matter is then absorbed using thin threads (hypha). Saprobe zygomycotes live on insect carcasses and plant debris. Macromolecules of cellulose, lignin, and collagen are digested, releasing nutrient compounds of carbon, nitrogen, and phosphorous.

A symbiote extends the thin threads (hypha) directly into the host tissue and extracts nutrients from a host cell. Symbiote zygomycotes are parasites on algae, molds, and fungi.

Class 1 Zygomycetes Order 3 *Mucorales* species are saprobes. Species include the common Black Bread Mold *Rhizopus stolonifer*.

Class 1 Zygomycetes Order 4 Zoopagales species are symbiotes with animals. They invade amebas, protozoans, protophytes, nematode worms, and other small animals. They penetrate the surface of the small animal, coil inside it and violently debilitate it, killing it. From the remains emerge club-shaped reproductive cells (gametes).

Class 2 Glomeromycetes Order 1 *Glomerales* species are symbiiotes with plants. They attach themselves to the roots of plants (endomycorrhizal) and secrete hormones to stimulate the plant roots to branch. Their ancestor species existed 400 million years ago.

CLASS 623.1 *Zygomycetes* Multinucleate Fungus

ORDER 1 *Zygomycotales* Multinucleate Fungus

FAMILY 1 *Zygomycotaceae* — Multinucleate Fungus
GENUS *Basidiobolus* — Bowl Multinucleate Fungus
SPECIES *magnus, microsporus, ranarum*
GENUS *Densospora* — Dense Spore Multinucleate Fungus
SPECIES *nanospora, nuda, solicarpa, tubiformis*
GENUS *Enteroramus* — Enteric Branched Multinucleate Fungus
SPECIES *dimorphus*
GENUS *Nothadelphia* — Adelphia Multinucleate Fungus
SPECIES *mortierellicola*

FAMILY 2 *Dimargaritaceae* — Margarite Multinucleate Fungus
GENUS *Dimargaris* — Margarite Multinucleate Fungus
SPECIES *arida, bacillispora, cristalligena, oblongispora, simplex, verticillata, xerosporica*
GENUS *Dispira* — Spire Margarite Multinucleate Fungus
SPECIES *cornuta, implicata, parvispora, simplex*
GENUS *Tieghemiomyces* — Tieghem Margarite Multinucleate Fungus
SPECIES *californicus, parasiticus*

FAMILY 3 *Endogonaceae* — Endogon Multinucleate Fungus
GENUS *Endogone* — Endogon Multinucleate Fungus
SPECIES *flammicorona, lactiflua, macrocarpa, microcarpa, pisiformis*
GENUS *Peridiospora* — Peridiospore Endogon Multinucleate Fungus
SPECIES *reticulata, tatachia*
GENUS *Sclerogone* — Hard Endogon Multinucleate Fungus
SPECIES *eucalypti*
GENUS *Youngiomyces* — Young Endogon Multinucleate Fungus
SPECIES *aggregatus, carolinensis, multiplex, stratosus*

FAMILY 4 *Kickxellaceae* — Kickx Multinucleate Fungus
GENUS *Kickxella* — Kickx Multinucleate Fungus
SPECIES *alabastrina*
GENUS *Coemansia* — Coemans Kickx Multinucleate Fungus
SPECIES *aciculifera, bainieri, brasiliensis, breviramosa, ceylonensis, erecta, furcata, guatemalensis, interrupta, kamerunensis, linderi, mojavensis, nantahalensis, pectinata., reversa, scorpioidea, spiralis, thaxteri*
GENUS *Dipsacomyces* — Dip Sac Kickx Multinucleate Fungus

SPECIES *acuminosporus*

GENUS *Linderina*	Linder Kickx Multinucleate Fungus
SPECIES *macrospora, pennispora*	
GENUS *Martensella*	Lesser Martens Kickx Multinucleate Fungus
SPECIES *corticii*	
GENUS *Martensiomyces*	Martens Kickx Multinucleate Fungus
SPECIES *pterosporus*	
GENUS *Mycoëmilia*	Emily Kickx Multinucleate Fungus
SPECIES *scoparia*	
GENUS *Myconymphaea*	Nymph Kickx Multinucleate Fungus
SPECIES *yatsukahoi*	
GENUS *Ramicandelaber*	Candelabera Kickx Multinucleate Fungus
SPECIES *brevisporus, longisporus*	
GENUS *Spirodactylon*	Spire Finger Kickx Multinucleate Fungus
SPECIES *aureum*	
GENUS *Spiromyces*	Spire Kickx Multinucleate Fungus
SPECIES *aspiralis, minutus*	

ORDER 2 *Entomophthorales* Entic Phthoral Multinucleate Fungus

FAMILY 1 *Entomophthoraceae*	Entic Phthoral Multinucleate Fungus
GENUS *Entomophthora*	Entic Phthoral Multinucleate Fungus

SPECIES *aphrophorae, blunckii, brevinucleata, byfordii, byfordii, calliphorae, coleopterorum, erupta, exitialis, ferdinandii, ferdinandii, grandis, helvetica, israelensis, leyteensis, muscae, phalloides, philippinensis, planchoniana, pseudococci, richteri, rivularis, rivularis, scatophaga, schizophorae, simulii, simulii, sphaerosperma, syrphi, thripidum, trinucleata, weberi*

GENUS *Batkoa*	Batko Entic Phthoral Multinucleate Fungus
SPECIES *amrascae, apiculata, dysderci, gigantea, limoniae, major, papillata*	
GENUS *Completoria*	Complete Entic Phthoral Multinucleate Fungus
SPECIES *complens*	
GENUS *Culicicola*	Culic Entic Phthoral Multinucleate Fungus
SPECIES *chromaphidis, culicis*	
GENUS *Empusa*	Empuse Entic Phthoral Multinucleate Fungus
SPECIES *thaxteri*	
GENUS *Entomophaga*	Omnivore Entic Phthoral Multinucleate Fungus

SPECIES *aulicae, batkoi, bukidnonensis, calopteni, conglomerata, destruens, diprionis, grylli, kansana, lagriae, maimaiga, planchoniana, ptychopterae, pyriformis, saccharina, tabanivora, tenthredinis, tipulae, transitans*

GENUS *Erynia*	Eryn Entic Phthoral Multinucleate Fungus

SPECIES *aquatica, chironomi, conica, curvispora, delpiniana, erinacea, gigantea, gracilis, henricii, ovispora, plecopteri, radicans, rhizospora, sepulchralis, variabilis*

GENUS *Eryniopsis*	Lesser Eryn Entic Phthoral Multinucleate Fungus
SPECIES *caroliniana, lampyridarum, longispora*	
GENUS *Furia*	Furry Entic Phthoral Multinucleate Fungus

SPECIES *americana, creatonoti, crustosa, ellisiana, fujiana, ithacensis, montana, neopyralidarum, pieris, sciarae, shandongensis, triangularis, virescens, vomitoriae, zabri*

GENUS *Massospora*	Mass Spore Entic Phthoral Multinucleate Fungus

SPECIES *carinetae, cicadina, diceroproctae, diminuta, dorisianae, fidicinae, levispora, ocypetes, platypediae, spinosa, tettigatis*

GENUS *Orthomyces*	Ortho Entic Phthoral Multinucleate Fungus
SPECIES *aleyrodis*	
GENUS *Pandora*	Pandora Entic Phthoral Multinucleate Fungus

SPECIES *athaliae, bibionis, blunckii, borea, brahminae, bullata, cicadellis, dacnusae, delphacis, dipterigena, echinospora, formicae, gammae, gloeospora, kondoiensis, neoaphidis, nouryi, phalangicida, shaanxiensis, suturalis, uroleuconii*

GENUS *Strongwellsea*	Strongwell Entic Phthoral Multinucleate Fungus
SPECIES *castrans, magna*	
GENUS *Zoophthora*	Animal Entic Phthoral Multinucleate Fungus

SPECIES *anglica, anhuiensis, aphidis, arginis, autumnalis, bialoviezensis, brevispora, canadensis, crassispora, crassitunicata, elateridiphaga, falcata, forficulae, geometralis, giardii, humberi, ichneumonis, lanceolata, larvivora, miridis, nematoceris, obtusa, occidentalis, opomyzae, orientalis, pentatomis, petchii, phalloides, phytonomi, psyllae, radicans, tachypori, viridis*

FAMILY 2 *Ancylistaceae*	Ancylist Entic Phthoral Multinucleate Fungus
GENUS *Ancylista*	Ancylist Entic Phthoral Multinucleate Fungus
SPECIES *berdanii, closterii, miurii, netrii, pfeifferi*	
GENUS *Conidiobolus*	Conic Bowl Ancylist Entic Phthoral Multinucleate Fungus

SPECIES *adiaeretus, adieretus, antarcticus, bangalorensis, brefeldianus, chlamydosporus, coronatus, couchii, denaeosporus, firmipilleus, globuliferus, gonimodes, heterosporus, humicola, incongruus, iuxtagenitus, khandalensis, lachnodes, lamprauges, lichenicola, lobatus, megalotocus, mycophagus, mycophilus, nanodes, nodosus, obscurus, osmodes, paulus, polyspermus, polytocus, pumilus, rhysosporus, rugosus, stromoideus, terrestris, thromboides, undulatus*

VARIETIES *megalotocus indicus*

GENUS *Macrobiotophthora* — Macrobiotic Ancylist Entic Phthoral Multinucleate Fungus

SPECIES *vermicola, vimariensis*

FAMILY 3 *Meristacraceae* — Merist Entic Phthoral Multinucleate Fungus

GENUS *Meristacrum* — Merist Entic Phthoral Multinucleate Fungus

SPECIES *asterospermum, milkoi*

GENUS *Ballocephala* — Ballcap Merist Entic Phthoral Multinucleate Fungus

SPECIES *pedicellata, sphaerospora, verrucospora*

GENUS *Coelomomyces* — (Synonym for *Meristacrum*)

GENUS *Zygnemomyces* — Zygneme Merist Entic Phthoral Multinucleate Fungus

SPECIES *echinulatus, pendulatus*

FAMILY 4 *Neozygitaceae* — Modern Entic Phthoral Multinucleate Fungus

GENUS *Neozygites* — Modern Entic Phthoral Multinucleate Fungus

SPECIES *abacaridis, acaridis, adjarica, cinarae, cucumeriformis, floridana, fresenii, fumosa, heteropsyllae, lageniformis, lecanii, microlophii, parvispora, tanajoae, tetranychi, turbinata*

GENUS *Apterivorax* — Voraceous Modern Entic Phthoral Multinucleate Fungus

SPECIES *acaricida, sminthuri*

ORDER 3 *Mucorales* Furry Mold

Species of Order *Mucorales* cause a disease in humans called mucormycosis.

FAMILY 1 *Mucoraceae* — Furry Mold

GENUS *Mucor* — Furry Mold

SPECIES *adventitius, albus, aligarensis, amphibiorum, ardhlaengiktus, azygosporus, bacilliformis, bainieri, circinelloides, faisalabadensis, falcatus, flavus, fragilis, fuscus, genevensis, gigasporus, grandis, guilliermondii, hiemalis, inaequisporus, indicus, lausannensis, laxorrhizus, microsporus, minutus, mousanensis, mucedo, nanus, oblongiellipticus, oblongisporus, odoratus, piriformis, plasmaticus, plumbeus, prayagensis, psychrophilus, racemosus, ramosissimus, recurvus, saturninus, sinensis, strictus, subtilissimus, thermophilus, troglophilus, tuberculisporus, ucrainicus, variabilis, variosporus, virens, zonatus*

VARIETIES *aligarensis ovalisporus, bainieri tenuis, circinelloides circinelloides, circinelloides alternans, circinelloides ambiguus, circinelloides dubius, circinelloides griseocyanus, circinelloides griseoroseus, circinelloides janssenii, circinelloides janssenii-kursanovii, circinelloides janssenii-stagnalis, circinelloides janssenii-tenellus, circinelloides javanicus, circinelloides lusitanicus, circinelloides lusitanicus-griseolilacinus, circinelloides lusitanicus-jauchae, circinelloides lusitanicus-zeicolus, circinelloides mandshuricus, circinelloides prainii, circinelloides ramificus, flavus attenuatus, flavus macrosporus, flavus mephitis, flavus meridionalis, flavus petropolitanus, flavus schousboae, flavus sciurinus, fuscus bedrchani, fuscus echinosporus, fuscus griseobrunneus, fuscus intermedius, fuscus peacockensis, fuscus petrinsularis, fuscus ramiger, hiemalis hiemalis, hiemalis airensis, hiemalis alboflavus, hiemalis albogriseus, hiemalis albus, hiemalis allobrogensis, hiemalis alnicola, hiemalis aurantiacus, hiemalis corticola, hiemalis cumulicolus, hiemalis flavogriseus, hiemalis flavus, hiemalis formicogenus, hiemalis griseus, hiemalis intermedius, hiemalis luteus, hiemalis perennis, hiemalis sabulosus, hiemalis silvaticus, hiemalis toundrae, hiemalis vallesiacus, inaequisporus kaki, laxorrhizus laxorrhizus, laxorrhizus ovalisporus, microsporus cylindrosporus, mucedo bitabulatus, mucedo brevipes, mucedo coloratus, mucedo coprophilus, mucedo dependens, mucedo griseo-ochraceus, mucedo macrosporus, mucedo murorum, mucedo proliferus, mucedo robustus, mucedo rufescens, mucedo sphaerocephalus, mucedo vulgaris, odoratus rufescens, piriformis alboater, piriformis sphaerosporus, piriformis wosnessenskii, plasmaticus irkutensis, plasmaticus mucilagineus, plumbeus adriaticus, plumbeus brunneogriseus, plumbeus brunneus, plumbeus recurvus, plumbeus spinescens, plumbeus spinosus, plumbeus turfosus, racemosus racemosus, racemosus brunneus, racemosus chibinensis, racemosus christianensis, racemosus dimorphosporus, racemosus griseosporus, racemosus mediterraneus, racemosus oudemansii, racemosus pispekii, racemosus sphaerosporus, racemosus sphaerosporus-globosus, racemosus sphaerosporus-intermedius, racemosus sphaerosporus-levisporus, racemosus sphaerosporus-macrosporus, racemosus sphaerosporus-major, racemosus sphaerosporus-pyri, racemosus varians, recurvus recurvus, recurvus aromaticus, recurvus indicus, recurvus malaysiensis, strictus kanivcevii, subtilissimus nederlandicus, variosporus indicus*

GENUS *Absidia* — Absid Furry Mold

SPECIES *anomala, californica, coerulea, cuneospora, cylindrospora, fusca, glauca, griseola, heterospora, idahoensis, inflata, macrospora, parricida, pseudocylindrospora, psychrophilia, repens, spinosa*

VARIETIES *coerulea orchidis, coerulea saccardoi, coerulea tieghemii, cylindrospora cylindrospora, cylindrospora nigra, cylindrospora rhizomorpha, glauca asparagi, glauca paradoxa, glauca sphaerosporangioides, parricida parasitica, pseudocylindrospora brasiliensis, repens japonica, spinosa spinosa, spinosa azygospora, spinosa biappendiculata*

GENUS *Actinomucor* Ray Furry Mold
SPECIES *elegans, taiwanensis*
GENUS *Amylomyces* Amyl Furry Mold
SPECIES *rouxii*
GENUS *Benjaminiella* Benjamin Typha Furry Mold
SPECIES *multispora, poitrasii, youngii*
GENUS *Chaetocladium* Bristle Furry Mold
SPECIES *brefeldii, jonesii*
GENUS *Circinella* Coiled Furry Mold
SPECIES *angarensis, chinensis, lacrymispora, minor, mucoroides, muscae, rigida, simplex, umbellata*
GENUS *Hyphomucor* Hypha Furry Mold
SPECIES *assamensis*
GENUS *Mycotypha* Typha Furry Mold
SPECIES *africana, guadalcanalensis, indica, microspora*
GENUS *Parasitella* Parasite Furry Mold
SPECIES *parasitica*
GENUS *Philophora* Philophore Furry Mold
SPECIES *agaricina*
GENUS *Radiomyces* Radial Furry Mold
SPECIES *embreei, mexicanus, spectabilis*
GENUS *Rhizomucor* Rhizoid Furry Mold
SPECIES *endophyticus, miehei, pakistanicus, pusillus, tauricus, variabilis*
VARIETIES *miehei nainitalensis, pusillus parasiticus, pusillus septatus, variabilis variabilis, variabilis regularior*
GENUS *Rhizopodopsis* Rhizopod Furry Mold
SPECIES *javensis*
GENUS *Rhizopus* Rhizope Furry Mold
SPECIES *azygosporus, caespitosus, homothallicus, microsporus, oryzae, schipperae, sexualis, stolonifer*
VARIETIES *microsporus microsporus, microsporus chinensis, microsporus chinensis-bovinus, microsporus chinensis-liquifaciens, microsporus chinensis-pygmaeus, microsporus chinensis-thermosus, microsporus oligosporus, microsporus oligosporus-glaber, microsporus rhizopodiformis, microsporus rhizopodiformis-alpinus, microsporus rhizopodiformis-cohnii, microsporus rhizopodiformis-pusillus, oryzae oryzae, oryzae acetoinus, oryzae achlamydosporus, oryzae albus, oryzae angulisporus, oryzae araneosus, oryzae arrhizus, oryzae bahrnensis, oryzae batatas, oryzae betivorus, oryzae boreas, oryzae cambodja, oryzae chiuniang, oryzae chungkuoensis, oryzae delemar, oryzae formosaensis, oryzae fusiformis, oryzae hangchow, oryzae humilis, oryzae instriatis, oryzae isofermentans, oryzae japonicus, oryzae javanicus, oryzae kawasakiensis, oryzae konsho, oryzae liquifaciens, oryzae maydis, oryzae minimus, oryzae mochi, oryzae multiplicisporus, oryzae nodosus, oryzae norvegicus, oryzae peka, oryzae pseudochinensis, oryzae rugulosus, oryzae salebrosus, oryzae schizans, oryzae semarangensis, oryzae shanghaiensis, oryzae sontii, oryzae sterilis, oryzae suinus, oryzae tanekoji, oryzae thermosus, oryzae tonkinensis, oryzae tritici, oryzae trubini, oryzae usamii, oryzae verticillatum, sexualis sexualis, sexualis americanus, stolonifer stolonifer, stolonifer artocarpi, stolonifer lyococcos, stolonifer lyococcos-reflexus, stolonifer nigricans*
GENUS *Sporodiniella* Spore Furry Mold
SPECIES *umbellata*
GENUS *Syzygites* Joined Furry Mold
SPECIES *megalocarpus*
GENUS *Thermomucor* Heat Loving Furry Mold
SPECIES *indicae-seudaticae*
GENUS *Zygorhynchus* Beak Furry Mold
SPECIES *californiensis, exponens, heterogamus, japonicus, macrocarpus, moelleri, psychrophilus*
VARIETIES *exponens exponens, exponens polygonosporus, exponens smithii, heterogamus phosphoreus, macrocarpus hennebergii, moelleri albus, moelleri bernardii, moelleri circinelloides, moelleri dangeardii, moelleri griseocinereus, moelleri verruculosus, moelleri viridis, moelleri vuilleminii*

FAMILY 2 *Choanephoraceae* Funnel Furry Mold
GENUS *Choanephora* Funnel Furry Mold
SPECIES *cucurbitarum, infundibulifera*
VARIETIES *cucurbitarum americana, cucurbitarum heterospora, cucurbitarum mandshurica, cucurbitarum simsonii, infundibulifera infundibulifera, infundibulifera conjuncta, infundibulifera cunninghamiana*
GENUS *Blakeslea* Blakeslee Funnel Furry Mold
SPECIES *monospora, sinensis, trispora*
VARIETIES *trispora dichotoma, trispora tandonii*
GENUS *Gilbertella* Gilbert Funnel Furry Mold
SPECIES *hainanensis, persicaria*
VARIETIES *persicaria persicaria, persicaria indica*
GENUS *Poitrasia* Poitras Funnel Furry Mold
SPECIES *circinans*
VARIETIES *circinans indica, circinans prolifera*

FAMILY 3 *Cunninghamellaceae* — Cunningham Furry Mold
GENUS *Cunninghamella* — Cunningham Furry Mold
SPECIES *antarctica, blakesleeana, clavata, echinulata, elegans, homothallica, intermedia, phaeospora, polymorpha, vesiculosa*
VARIETIES *echinulata africana, echinulata bainieri, echinulata chibaensis, echinulata dalmatica, echinulata verticillata, echinulata japonica, echinulata ramosa, elegans batistae, elegans bertholletiae, phaeospora brunnea*
GENUS *Chlamydoabsidia* — Cloak Absid Cunningham Furry Mold
SPECIES *padenii*
VARIETIES *padenii dasguptae*
GENUS *Gongronella* — Gongron Cunningham Furry Mold
SPECIES *butleri, lacrispora*
VARIETIES *butleri urceolifera*
GENUS *Halteromyces* — Halter Cunningham Furry Mold
SPECIES *radiatus*
GENUS *Hesseltinella* — Hesseltin Cunningham Furry Mold
SPECIES *vesiculosa*

FAMILY 4 *Mortierellaceae* — Mortier Furry Mold
GENUS *Mortierella* — Mortier Furry Mold
SPECIES *acrotona, alliacea, alpina, ambigua, amoeboidea, angusta, antarctica, apiculata, armillariicola, bainieri, basiparvispora, beljakovae, biramosa, bisporalis, camargensis, candelabrum, capitata, chlamydospora, claussenii, clonocystis, cystojenkinii, dichotoma, echinosphaera, echinula, echinulata, elasson, elongata, elongatula, epicladia, epigama, exigua, fatshederae, fimbricystis, gamsii, gemmifera, globalpina, globulifera, histoplasmatoides, horticola, humilis, hyalina, hypsicladia, indohii, jenkinii, kuhlmanii, lignicola, longicollis, longigemmata, macrocystis, macrocystopsis, minutissima, mutabilis, nantahalensis, oligospora, parazychae, parvispora, pilulifera, polycephala, polygonia, pseudozygospora, pulchella, pusilla, reticulata, rishikesha, rostafinskii, sarnyensis, schmuckeri, sclerotiella, selenospora, simplex, strangulata, stylospora, tuberosa, turficola, umbellata, verrucosa, verticillata, wolfii, zonata, zychae, zychae brachyrhiza*
VARIETIES *alpina acuminata, alpina monospora, alpina renispora, alpina thaxteri, bisporalis decipiens, capitata vesiculosa, elongata debilis, exigua indica, exigua sterilis, gamsii spinosa, globulifera ericetorum, hyalina hygrophila, hyalina minor, hyalina minuta, hyalina subtilissima, lignicola sepedonioides, longicollis grandispora, minutissima minutissima, minutissima dubia, oligospora indica, parvispora gracilis, polycephala canina, polycephala crystallina, polycephala lemonnieri, polycephala raphani, polycephala vantieghemi, pulchella sossauensis, pusilla nodosa, pusilla stricta, verticillata marburgensis*
GENUS *Aquamortierella* — Aquatic Mortier Furry Mold
SPECIES *elegans*
GENUS *Dissophora* — Disson Mortier Furry Mold
SPECIES *decumbens, nadsonii, ornata*
GENUS *Gamsiella* — Gams Mortier Furry Mold
SPECIES *multidivaricata*
GENUS *Echinosporangium* — (Synonym for *Lobosporangium*)
GENUS *Lobosporangium* — Urchin Spore Mortier Furry Mold
SPECIES *transversale*
GENUS *Micromucor* — (Synonym for *Umbelopsis*)
GENUS *Modicella* — Modic Mortier Furry Mold
SPECIES *malleola, reniformis*
GENUS *Umbelopsis* — Umbel Mortier Furry Mold
SPECIES *angularis, autotrophica, dimorpha, fusiformis, gibberispora, isabellina, nana, ovata, ramanniana, swartii, versiformis, vinacea, westeae*
VARIETIES *versiformis roseonana, vinacea multispora*

FAMILY 5 *Phycomycetacea* — Algae Loving Furry Mold
GENUS *Phycomyces* — Algae Loving Furry Mold
SPECIES *blakesleeanus, microsporus, nitens*
VARIETIES *blakesleeanus piloboloides, nitens pirottianus, nitens spinulosus, nitens splendens*
GENUS *Spinellus* — Spined Algae Loving Furry Mold
SPECIES *chalybeus, fusiger, sphaerosporus*
VARIETIES *fusiger macrocarpus, fusiger rhombosporus*

FAMILY 6 *Pilobolaceae* — Pile Bowl Furry Mold
GENUS *Pilobolus* — Pile Bowl Furry Mold
SPECIES *crystallinus, lentiger, longipes, oedipus, roridus*
VARIETIES *crystallinus crystallinus, crystallinus hyalosporus, crystallinus kleinii, crystallinus kleinii-gracilis, crystallinus kleinii-hallensis, crystallinus kleinii-heterosporus, crystallinus kleinii-roseus, crystallinus ramosus, crystallinus schmidtii, crystallinus simplex, lentiger lentiger, lentiger borzianus, lentiger exiguus, lentiger geminatus, lentiger macrosporus, lentiger minutus, lentiger minutus-minor, lentiger minutus-morinii, lentiger sphaerosporus, oedipus argentinus, oedipus reticulatus, roridus roridus, roridus anisosporus, roridus microsporus, roridus umbonatus*

GENUS *Pilaira* Lesser Pile Bowl Furry Mold
SPECIES *anomala, caucasica, cesatii, dimidiata, moreaui*
VARIETIES *anomala saccardiana*
GENUS *Utharomyces* Uthar Pile Bowl Furry Mold
SPECIES *epallocaulus*
VARIETIES *epallocaulus indicus*

FAMILY 7 *Saksenaeaceae* Saksen Furry Mold
GENUS *Saksenaea* Saksen Furry Mold
SPECIES *vasiformis*
GENUS *Apophysomyces* Apophysal Saksen Furry Mold
SPECIES *elegans*

FAMILY 8 *Syncephalastraceae* Bristle Furry Mold
GENUS *Syncephalastrum* Starcap Bristle Furry Mold
SPECIES *monosporum, racemosum*
VARIETIES *monosporum monosporum, monosporum cristatum, monosporum pluriproliferum, racemosum cinereum, racemosum elegans, racemosum fuliginosum, racemosum javanicum, racemosum nigricans, racemosum paucisporum, racemosum verruculosum*
GENUS *Dichotomocladium* Dichotomy Bristle Furry Mold
SPECIES *elegans, floridanum, hesseltinei, robustum, sphaerosporum*
GENUS *Fennellomyces* Fennell Bristle Furry Mold
SPECIES *gigacellularis, heterothallicus, linderi, verticillatus*
GENUS *Mycocladus* Branched Bristle Furry Mold
SPECIES *blakesleeanus, corymbiferus, hyalospora, verticillata*
VARIETIES *blakesleeanus blakesleeanus, blakesleeanus atrospora, corymbiferus ramosus, corymbiferus regnieri, corymbiferus truchisi*
GENUS *Phascolomyces* Phascole Bristle Furry Mold
SPECIES *articulosus*
GENUS *Protomycocladus* Primitive Branched Bristle Furry Mold
SPECIES *faisalabadensis*
GENUS *Zychaea* Zych Bristle Furry Mold
SPECIES *mexicana*

FAMILY 9 *Thamnidiaceae* Thamnid Furry Mold
GENUS *Thamnidium* Thamnid Furry Mold
SPECIES *elegans*
VARIETIES *elegans arbuscula, elegans vantieghemii, elegans verticillatum*
GENUS *Backusella* Backus Thamnid Furry Mold
SPECIES *circina, ctenidia, lamprospora*
GENUS *Cokeromyces* Coker Thamnid Furry Mold
SPECIES *recurvatus*
GENUS *Dicranophora* Dicran Thamnid Furry Mold
SPECIES *fulva*
GENUS *Ellisomyces* Ellis Thamnid Furry Mold
SPECIES *anomalus*
GENUS *Helicostylum* Helix Thamnid Furry Mold
SPECIES *elegans, pulchrum*
GENUS *Kirkomyces* Kirk Thamnid Furry Mold
SPECIES *cordensis*
GENUS *Pirella* Pear Thamnid Furry Mold
SPECIES *circinans, naumovii*
GENUS *Thamnostylum* Lesser Thamnid Furry Mold
SPECIES *lucknowense, nigricans, piriforme, repens*

ORDER 4 *Zoopagales* Animal Phage Fungus

FAMILY 1 *Zoopagaceae* Animal Phage Fungus
GENUS *Zoopage* Animal Phage Fungus
SPECIES *atractospora, cladosperma, mitospora, nematospora, pachyblasta, phanera, tetraspora, thamnospira, toechospora, tryphera, virgispora*
GENUS *Acaulopage* Stalk Animal Phage Fungus
SPECIES *acanthospora, baculispora, bicornis, ceratospora, cercospora, crobylospora, dactylophora, dasyspora, dichotoma, gomphoclada, gyrinodes, hystricospora, ischnospora, lasiospora, longicornis, lophospora, macrospora, marantica, pectospora, rhaphidospora, rhicnospora, stenospora, tenuicornis, tetraceros, tigrina, trachyspora*
VARIETIES *tetraceros tetraceros, tetraceros longa*
GENUS *Basidiolum* Club Animal Phage Fungus
SPECIES *fimbriatum*
GENUS *Cystopage* Cyst Animal Phage Fungus

SPECIES *cladospora, ellipsospora, intercalaris, lateralis, sacciformis, sphaerospora, subtilis*
GENUS *Massartia* Massart Animal Phage Fungus
SPECIES *javanica*
GENUS *Stylopage* Lesser Animal Phage Fungus
SPECIES *anomala, araea, cephalote, grandis, hadra, haploe, leiohypha, lepte, minutula, rhabdoides, rhabdospora, rhicnacra, rhynchospora, scoliospora*
GENUS *Zoophagus* Least Animal Phage Fungus
SPECIES *cornus, insidians, tentaclum, tylopagus*

FAMILY 2 *Cochlonemataceae* Spiral Thread Animal Phage Fungus
GENUS *Cochlonema* Spiral Thread Animal Phage Fungus
SPECIES *agamum, bactrosporum, calosperma, cerasphorum, cylindricum, dolichosporum, euryblastum, explicatum, fusisporum, lineare, megalosomum, megaspirema, odontosperma, ozotum, pumilum, pygmaeum, symplocum, verrucosum*
GENUS *Amoebophilus* Ameba Spiral Thread Animal Phage Fungus
SPECIES *dangeardii, penardii, sicyosporus, simplex*
GENUS *Aplectosoma* Spurless Spiral Thread Animal Phage Fungus
SPECIES *microsporum*
GENUS *Bdellospora* Gum Spore Spiral Thread Animal Phage Fungus
SPECIES *helicoides*
GENUS *Endocochlus* Endic Spiral Thread Animal Phage Fungus
SPECIES *asteroides, binarius, brachysporus, gigas*
GENUS *Euryancale* Broad Spiral Thread Animal Phage Fungus
SPECIES *marsipioides, marsipospora, obliqua, phallospora, sacciospora*

FAMILY 3 *Helicocephalidaceae* Helix Head Animal Phage Fungus
GENUS *Helicocephalum* Helix Head Animal Phage Fungus
SPECIES *sarcophilum*
GENUS *Brachymyces* Short Helix Head Animal Phage Fungus
SPECIES *megasporus*
GENUS *Rhopalomyces* Club Helix Head Animal Phage Fungus
SPECIES *bennyi, candidus, elegans, magnus*
VARIETIES *elegans apiculatus*

FAMILY 4 *Piptocephalidaceae* Falling Head Animal Phage Fungus
GENUS *Piptocephalis* Falling Head Animal Phage Fungus
SPECIES *acephala, arrhiza, benjaminii, brijmohanii, cruciata, curvata, cylindrospora, debaryana, dichotomica, fimbriata, freseniana, fusispora, indica, lemonnieriana, lepidula, macrocephala, microcephala, minuta, pseudocephala, sphaerospora, tieghemiana, unispora, xenophila*
VARIETIES *indica indica, indica shantiniketna*
GENUS *Kuzuhaea* Kuzuh Falling Head Animal Phage Fungus
SPECIES *moniliformis*
GENUS *Syncephalis* Lesser Falling Head Animal Phage Fungus
SPECIES *adunca, annularis, asymmetrica, aurantiaca, basibulbosa, bispora, californica, cordata, cornu, curvata, depressa, fasciculata, furcata, fuscata, fusigera, glabra, hypogena, intermedia, leadbeateri, nana, nigricans, nodosa, obconica, parvula, pendula, penicillata, plumigaleata, pycnosperma, ramosa, rapacea, reflexa, sphaerica, tengi, tenuis, tetrathela, torpedospora, tranzschelii, truncata, ubatubensis, ventricosa, viridis, wynneae*

FAMILY 5 *Sigmoideomycetaceae* Sigma Animal Phage Fungus
GENUS *Sigmoideomyces* Sigma Animal Phage Fungus
SPECIES *dispiroides*
GENUS *Reticulocephalis* Netted Head Sigma Animal Phage Fungus
SPECIES *clathroides, gyrosus*
GENUS *Thamnocephalis* Thamnid Sigma Animal Phage Fungus
SPECIES *ovalispora, quadrupedata, sphaerospora*

CLASS 623.2 *Glomeromycetes* Cluster Multinucleate Fungus

ORDER 1 *Glomerales* Cluster Fungus

FAMILY 1 *Glomeraceae* Cluster Fungus
GENUS *Glomus* Cluster Fungus
SPECIES *aggregatum, ambisporum, arborense, arenarium, australe, avelingiae, bagyarasjii, boreale, botryoides, caesaris, caledonium, callosum, canadense, canum, citricola, claroideum, clarum, clavisporum, constrictum, convolutum, coronatum, corymbiforme, cuneatum, deserticola, diaphanum, dolichosporum, eburneum, epigaeum, etunicatum, fasciculatum, flavisporum, formosanum, fragile, fragilistratum, fuegianum, fulvum, geosporum, gibbosum, globiferum, glomerulatum, heterosporum, hoi, invermaium, kerguelense, laccatum, lamellosum, liquidambaris, luteum, macrocarpum, magnicaule, melanosporum, microaggregatum, microcarpum, minutum, monosporum, mortonii, mosseae, multicaule, multiforum, nanolumen, pallidum,*

pansihalos, pellucidum, przelewicense, pubescens, pulvinatum, pustulatum, radiatum, rubiforme, segmentatum, sinuosum, spinosum, taiwanense, tenebrosum, tenue, tortuosum, trimurales, verruculosum, vesiculiferum, viscosum, warcupii

GENUS *Paraglomus* — Lesser Cluster Fungus

SPECIES *brasilianum, occultum*

FAMILY 2 *Archaeosporaceae* — Archaic Cluster Fungus

GENUS *Archaeospora* — Archaic Cluster Fungus

SPECIES *gerdemannii, leptoticha, trappei*

GENUS *Geosiphon* — Earth Siphon Archaic Cluster Fungus

SPECIES *pyriformis*

ORDER 2 *Diversisporales* Diverse Spore Cluster Fungus

FAMILY 1 *Diversisporaceae* — Diverse Spore Cluster Fungus

GENUS *Diversispora* — Diverse Spore Cluster Fungus

SPECIES *spurca*

GENUS *Gerdemannia* — (Synonym for *Pacispora*)

GENUS *Pacispora* — Pac Diverse Spore Cluster Fungus

SPECIES *chimonobambusae, scintillans*

FAMILY 2 *Acaulosporaceae* — Stalkless Diverse Spore Cluster Fungus

GENUS *Acaulospora* — Stalkless Diverse Spore Cluster Fungus

SPECIES *capsicula, cavernata, delicata, denticulata, dilatata, elegans, excavata, foveata, gedanensis, koskei, lacunosa, laevis, longula, mellea, morrowiae, myriocarpa, nicolsonii, polonica, rehmii, rugosa, scrobiculata, spinosa, splendida, thomii, tuberculata, walkeri*

GENUS *Entrophospora* — Entrophic Stalkless Diverse Spore Cluster Fungus

SPECIES *baltica, colombiana, infrequens, kentinensis, schenckii*

FAMILY 3 *Gigasporaceae* — Giant Diverse Spore Cluster Fungus

GENUS *Gigaspora* — Giant Diverse Spore Cluster Fungus

SPECIES *albida, alboaurantiaca, decipiens, gigantea, margarita, ramisporophora, rosea*

GENUS *Scutellospora* — Scute Giant Diverse Spore Cluster Fungus

SPECIES *alborosea, arenicola, armeniaca, auriglоba, biornata, calospora, castanea, cerradensis, coralloidea, crenulata, dipapillosa, erythropa, fulgida, gilmorei, gregaria, hawaiiensis, heterogama, minuta, nigra, nodosa, pellucida, persica, projecturata, reticulata, rubra, savannicola, scutata, spinosissima, tricalypta, trirubiginopa, verrucosa, weresubiae*

CLASS 623.3 *Trichomycetes* Hairy Multinucleate Fungus

ORDER 1 *Trichomycetales* Hairy Multinucleate Fungus

FAMILY 1 *Trichomycetacea* — Hairy Multinucleate Fungus

GENUS *Spirogyromyces* — Hairy Multinucleate Fungus

SPECIES *vermicola*

FAMILY 2 *Asellariaceae* — Asell Hairy Multinucleate Fungus

GENUS *Asellaria* — Asell Hairy Multinucleate Fungus

SPECIES *armadillidii, aselli, caulleryi, gramenei, ligiae, unguiformis*

GENUS *Baltomyces* — Baltic Asell Hairy Multinucleate Fungus

SPECIES *styrax*

GENUS *Orchesellaria* — Penis Asell Hairy Multinucleate Fungus

SPECIES *lattesii, mauguioi, pelta, podurae*

ORDER 2 *Harpellales* Harp Hairy Multinucleate Fungus

FAMILY 1 *Harpellaceae* — Harp Hairy Multinucleate Fungus

GENUS *Harpella* — Harp Hairy Multinucleate Fungus

SPECIES *amazonica, leptosa, melusinae, meridionalis, tica*

GENUS *Carouxella* — Caroux Harp Hairy Multinucleate Fungus

SPECIES *coemeteriensis, scalaris*

GENUS *Ephemerellomyces* — Ephemeral Harp Hairy Multinucleate Fungus

SPECIES *aquilonius*

GENUS *Harpellomyces* — Lesser Harp Hairy Multinucleate Fungus

SPECIES *abruptus, eccentricus, montanus*

GENUS *Pseudoharpella* — Least Harp Hairy Multinucleate Fungus

SPECIES *arcolamylica*

GENUS *Stachylina* — Stack Harp Hairy Multinucleate Fungus

SPECIES *acutibasilaris, ceratopogonidarum, chironomidarum, euthena, grandispora, gravicaudata, jujuyensis, lentica, litoralis, longa, lotica, macrospora, magna, manicata, minima, minuta, nana, paludosa, paucispora, pedifer, penetralis, prolifica, queenslandiae, reflexa, robusta, stenospora, thaumaleidarum*

GENUS *Stachylinoides* — Lesser Stack Harp Hairy Multinucleate Fungus

SPECIES *arctata*

FAMILY 2 *Legeriomycetaceae* — Leger Harp Hairy Multinucleate Fungus

GENUS *Legeriomyces* — Leger Harp Hairy Multinucleate Fungus

SPECIES *aenigmaticus, ramosus, rarus*

GENUS *Allantomyces* — Sac Membrane Harp Hairy Multinucleate Fungus

SPECIES *caenidarum*

GENUS *Austrosmittium* — Southern Smitt Leger Harp Hairy Multinucleate Fungus

SPECIES *aussiorum, biforme, kiwiorum, norinsulare*

GENUS *Baetimyces* — Baetia Leger Harp Hairy Multinucleate Fungus

SPECIES *ancorae*

GENUS *Barbatospora* — Beard Spore Leger Harp Hairy Multinucleate Fungus

SPECIES *ambicaudata*

GENUS *Bojamyces* — Boja Leger Harp Hairy Multinucleate Fungus

SPECIES *repens, transfuga*

GENUS *Capniomyces* — Smokey Leger Harp Hairy Multinucleate Fungus

SPECIES *stellatus*

GENUS *Caudomyces* — Stalk Leger Harp Hairy Multinucleate Fungus

SPECIES japonicus

GENUS *Coleopteromyces* — Beetle Leger Harp Hairy Multinucleate Fungus

SPECIES *amnicus*

GENUS *Ejectosporus* — Ejecting Spore Leger Harp Hairy Multinucleate Fungus

SPECIES *magnus*

GENUS *Furculomyces* — Forked Leger Harp Hairy Multinucleate Fungus

SPECIES *boomerangus, septentrionalis, westraliensis*

GENUS *Gauthieromyces* — Gauthier Leger Harp Hairy Multinucleate Fungus

SPECIES *microsporus*

GENUS *Genistelloides* — Broom Leger Harp Hairy Multinucleate Fungus

SPECIES *amplispora, communis, helicoides, hibernus*

GENUS *Genistellospora* — Broom Spore Leger Harp Hairy Multinucleate Fungus

SPECIES *guanacastensis, homothallica, nubila, tepidaria, tropicalis*

GENUS *Glotzia* — Glotz Leger Harp Hairy Multinucleate Fungus

SPECIES *centroptili, coloradense, ephemeridarum, plecopterorum, stenospora, tasmaniensis*

GENUS *Graminella* — Grass Leger Harp Hairy Multinucleate Fungus

SPECIES *bulbosa, microspora*

GENUS *Graminelloides* — Lesser Grass Leger Harp Hairy Multinucleate Fungus

SPECIES *biconica*

GENUS *Lancisporomyces* — Lance Spore Leger Harp Hairy Multinucleate Fungus

SPECIES *vernalis*

GENUS *Legerioides* — Least Leger Harp Hairy Multinucleate Fungus

SPECIES *tumidus*

GENUS *Legeriosimilis* — Lesser Leger Harp Hairy Multinucleate Fungus

SPECIES *europaeus, tricaudata*

GENUS *Orphella* — Orphan Leger Harp Hairy Multinucleate Fungus

SPECIES *avalonensis, catalaunica, coronata, haysii, helicospora, hiemalis*

GENUS *Pennella* — Pen Leger Harp Hairy Multinucleate Fungus

SPECIES *angustispora, arctica, asymmetrica, grassei, hovassei, montana, simulii*

GENUS *Plecopteromyces* — Plaited Wing Leger Harp Hairy Multinucleate Fungus

SPECIES *patagoniensis*

GENUS *Pteromaktron* — Wing Maktron Leger Harp Hairy Multinucleate Fungus

SPECIES *protrudens*

GENUS *Simuliomyces* — Simulated Leger Harp Hairy Multinucleate Fungus

SPECIES *microsporus, spica*

GENUS *Smittium* — Smitt Leger Harp Hairy Multinucleate Fungus

SPECIES *aciculare, acutum, alpinum, angustum, annulatum, arcticum, arvernense, biforme, bisporum, brasiliense, brevisporum, bulbosporophorus, bullatum, caudatum, cellaspora, chironomi, coloradense, commune, compactum, culicis, culicisoides, culisitae, cylindrosporum, delicatum, dimorphum, dipterorum, elongatum, esteparum, fasciculatum, fastigatum, fecundum, fruticosum, gigasporum, gracilis, hecatei, heterosporum, imitatum, incrassatum, inexpectans, kansense, lentaquaticum, longisporum, macrosporum, megazygosporum, microsporum, minutisporum, morbosum, mucronatum, orthocladii, ouselii, paludis, parvum, pennelli, perforatum, phytotelmatum, precipitiorum, prostratum, pseudodimorphum, pusillum, rarum, rupestre, simulatum, simulii, tipulidarum, tronadorium, typhellum, urbanum*

VARIETIES *culisitae inopinatum, morbosum morbosum, morbosum rioplatense*

GENUS *Spartiella* — Spartan Leger Harp Hairy Multinucleate Fungus

SPECIES *animae, barbata*	
GENUS *Stipella*	Stipe Leger Harp Hairy Multinucleate Fungus
SPECIES *vigilans*	
GENUS *Tectimyces*	Tecton Leger Harp Hairy Multinucleate Fungus
SPECIES *leptophlebiidarum, robustus*	
GENUS *Trichozygospora*	Zygospore Leger Harp Hairy Multinucleate Fungus
SPECIES *chironomidarum*	
GENUS *Zygopolaris*	Zygopole Leger Harp Hairy Multinucleate Fungus
SPECIES *borealis, ephemeridarum*	

SUBKINGDOM 63 *Bryata* Moss

Latin: *Bryo* bryon, moss	PHYLUMS		Species	Species
	631 *Bryophyta*	Moss	13,372[C]	10000 [M]
16,100 species	632 *Anthocerophyta*	Hornwort	23 [R]	100 [M]
[Margulis & Schwartz] [M]	633 *Hepatophyta*	Liverwort	591 [R]	6000 [M]

13,986 species Crosby [C], Ruggiero [R]

Subkingom *Bryata* Moss plants along with algae, fungus, and lichens are non-vascular. All other plants are vascular. Vascular plants contain small cylindrical tubes (vascules) that conduct water and sap. Vascular plants have leaves, stems, and roots. Non-vascular plants have phyllodes (moss), fronds (ferns), a thallus (instead of a stem), and root-like rhizoids. Moss Plants reproduce with alternating haploid sporophyte and diploid gametophyte generations.

PHYLUM 631 *Bryophyta* Moss

Latin: *Bryo* bryon moss	CLASSES		Species
Latin: *Phyta* phyton, plant	631.1 *Bryopsida*	Moss	12,969 [C]
10,000 species	631.2 *Andreaeopsida*	Granite Moss	96 [C]
[Margulis & Schwartz, 1998]	631.3 *Sphagnopsida*	Peat Moss	307 [C]

13,372 species [Crosby, Jul. 2004]

Phylum *Bryophyta* Moss species have conducting cells to facilitate spore dispersal and follow the Aternating Gametophyte-Sporophyte Life Cycle. The male sperm (gamete) fertilizes the female egg (gamete) forming a zygote embryo. The zygote embryo matures into an adult plant. The female adult plant has a zygote embryo embedded inside. The embedded zygote embryo develops into a double chromosome (diploid) sporophyte. The sporophyte produces single chromosome (haploid) spores. The spores germinate and mature into male and female adult plants. The male and female adult plants have pollen sperm (gamete) and pistil eggs (gamete) respectively. The cycle then repeats.

Mosses are low growing plants that live in sub-tropical and tropical moist soil, marsh, swamp, bog, and shallow lake habitats. Mosses have a leafy gametophyte thallus, a stalk, and rhizioids. The rhizoids are root-like filaments which attach to the growing surface.

CLASS 631.1 *Bryopsida* Moss

ORDER 1 *Bryales* Moss

FAMILY 01 *Bryaceae* Moss

GENUS *Bryum* Moss

SPECIES *acuminatissimum, acutifolium, aequabile, afrocalophyllum, alandense, albomarginatum, albopulvinatum, albulanum, albulum, alexandri, algovicum, alpiniforme, alpinum, altaicum, alticaule, altisetum, amblyacis, amblyodon, amblyphyllum, amentirameum, amplirete, angustifolium, anomodon, apalodictyoides, apiculatum, appressifolium, appressum, arachnoideum, archangelicum, arcticum, argenteum, argentisetum, argyroglyphodon, aristatum, atenense, atropurpureum, atrovirens, auricomum, australe, austroturbinatum, austroventricosum, axel-blyttii, axillare, baeuerlenii, baldwinii, balticum, barnesii, baueri, bealeyense, bicolor, billetii, blandum, blindii, borellii, bornholmense, botterii, bourgeanum, brachycladulum, brachyneuron, brasiliense, brevicoma, brevicuspis, bulbiferum, bulbigerum, bulbillicaule, bullosum, cadetii, caespiticium, callistomum, calobryoides, calodictyon, calophyllum, campoanum, campylopodioides, canariense, canescens, capillatum, capituliforme, carbonicola, cellulare, cephalozioides, challaense, chlororhodon, chorizodontum, chryseum, chrysoneuron, chrysophyllum, cirrhiferum, clavatum, claviger, clintonii, coelophyllum, cognatum, colombi, coloradense, comense, congestiflorum, conoideo-operculatum, consimile, contortum, coronatum, corsicum, crassimucronatum, crassinervum, crassum, creberrimum, cremocarpum, cristatum, cristatum, crozetense, curvatum, cyathiphyllum, cyclophyllum, cylindrothecium, daenikeri, deciduum, decursivum, defractum, delitescens, demaretianum, densifolium, dentatum, denticulatinervium, depressum, diaphanum, dichotomum, dilatatum, dillenii, dimorphum, dissolutinerve, dixonii, dolichophyllum, donatii, dongolense, donianum, dunense, duplicatum, eatonii, ehlei, ekmanii, ekstamii, elegans, elegans, elegantulum, ellipsifolium, ellipticifolium, elwendicum, encalyptaceum, enisseense, eremaeum, ericoides, erythrocarpum, erythrocaulon, erythroloma, erythrotropis, euryphyllum, evanidinerve, fabronia, falcatum, felipponei, filicicola, flaccidifolium, flaccidum, flagellaceum, flagellans, flagellicoma, flavipes, flavituber, flavopallidum, fragile, francii, funckiioides, funkii, fuscescens, fuscomucronatum, fuscotomentosum, gamophyllum, garovaglii, gayanum, gemmascens, gemmatum, gemmiferum, gemmilucens, gemmiparum, geniculatum, germainii, gerwigii, gilliesii, gossypinum, goudotii, gracilisetum, graminifolium, gynostomoides, gyoerffianum, gypsophilum, haematoneurum, hagenii, handelii, hatcheri, hauthalii, hawaiicum, hedbergii, hieronymi, hioramii, hochreutineri, homalobolax, huillense, icodense, illecebraria, imbricatum, inaequale, incacorralis, incrassatolimbatum, innovans, insolitum, intermedium, intortulum, isleanum, jamaicense, joannis-meyeri, kerguelense, klinggraeffii, knowltonii, kunzei, laetum, laevifilum, laevigatum, lamii, lamprocarpum, lamprochaete, lamprocomum, lamprostegum, latifolium, latilimbatum, lawersianum, leonii, lepidum, leptocaulon, leptocladon, leptoglyphodon, leptoneurum, leptospeiron, leptotorquescens, leptotrichum, leucoglyphodon, leucophylloides, limbatum, limbifolium, lindbergii, lonchocaulon, lonchophyllum, lonchopus, longisetum, longisetum, lorentzii, ludovicae, lugubre, maceratum, macroblastum, macrodictyum, macrophyllum, macrosporum, mairei, maleteinorum, mamillatum, marginatum, marginatum, marratii, matto-grossense, mayrii, meesioides, megalacrion, melvilleanum, mendax, meruense, mesodon, micans, microcaespiticium, microcalophyllum, microcapillare, microchaeton, microcochleare, microimbricatum, micronitidum, mieheanum, mildeanum, miniatum, minii, minusculum, minutirosatum, minutissimum, mirabile, miserum, moei, mollifolium, molliusculum, mucronatum, mucronifolium, muehlenbeckii, multiflorum, murmanicum, nanoapiculatum, nanocapillare, nanophyllum, naviculare, neodamense, neodamense, nitidulum, nivale, norvegicum, nudicaule, obligogynum, obliquum, oblongum, obscurum, obtusidens, ochianum, oncophorum, oophyllum, orbiculatifolium, orientale, orthocladum, orthodontioides, orthopelma, orthotheciellae, orthothecium, osculatianum, oxycarpum, pabstianum, pachycladum, pachytheca, pallens, pallescens, pallidoviride, pamirense, pamiricomucronatum, pancheri, papuanum, paradoxum, paraguense, parvulum, patzeltii, pauperidens, pendulum, peralatum, perdecurrens, perlimbatum, perminutum, perrieri, phallus, picnoloma, piliferum, planiusculum, platyphyllum, pocsii, pohliaeforme, pootenianum, posthumum, prionotes, procerum, propium, prosatherum, provinciale, pseudoblandum, pseudocapillare, pseudointermedium, pseudomarginatum, pseudomicron, pseudopachytheca, pseudotriquetrum, pumilum, pungentifolium, purpurascens, purpuratum, purpureolucidum, purpureonigrum, pycnodictyum, pycnophyllum, pygmaeomucronatum, pygmaeum, pyrrhothrix, radiculosum, ramosum, rapaense, rectifolium, recurvulum, redboonii, reedii, remelei, renauldii, retusifolium, revolutum, rhexodon, rhizoblastum, rhypariocaulon, richardsii, riparioides, riparium, rivale, roscheri, rosulans, rothii, rotundifolium, rubens, rubescens, rubicundum, rubrocostatum, rubrolimbatum, ruderale, rudimentale, rufolimbatum, russulum, rutilans, sabuletorum, salakense, salinum, sandii, sauteri, savesii, savicziae, schleicheri, sclerodictyon, semirubrum, serotinum, sibericum, simii, simulans, sinense, skottsbergii, soboliferum, sordidum, spindleri, spinifolium, spininervium, splachnobryoides, splendidifolium, stellituber, stenocarpum, stenophyllum, stirtonii, subapiculatum, subargenteum, subcalophyllum, subclavatum, subelegans, subgracillimum, subhomalobolax, subleucophyllum, submucronatum, subneodamense, subnitidulum, subpercurrente, subsericeum, sullivanii, svihlae, synoicum, syriacum, taimyrense, taitae, taoense, tenue, tenuidens, tenuifolium, tenuisetum, tepintzensa, teres, terskeiense, tessellatum, thomasii, thomeanum, thomsonii, timmiostomoides, tisserantii, tjiburrumense, tophaceum, torquatum, trabutii, tristaniense, triste, truncorum, turbinatum, turgidum, uliginosum, urbanskyi, usambaricum, utriculatum, uvidum, valparaisense, vernicosum, vernum, veronense, versicolor, viguieri, violaceum, virescens, viridescens, voeltzkowii, vulcanicum, wagneri, wallaceanum, warneum, weigelii, winkelmannii, wrightii, zeballosicum*

GENUS *Acidodontium* — Acidic Tooth Moss

SPECIES *exaltatum, heteroneuron, integrifolium, lanceolatifolium, lonchotrachylon, longifolium, megalocarpum, pallidum, ramicola, rhamphostegium, seminerve, sprucei, subrotundum, trachyticola*

GENUS *Anomobryum* — Anomalous Moss

SPECIES *albo-imbricatum, alpinum, angustirete, astorense, auratum, brachymenioides, brachymeniopsis, bulbiferum, bullatum, clavicaule, compressulum, conicum, cygnicollum, drakensbergense, erectum, filinerve, gemmigerum, harriottii, humillimum, hyalinum, julaceum, kashmirense, laceratum, lanatum, leptostomoides, lusitanicum, marginatum, minutirete, nidificans, obtusatissimum, ochianum, ochii, parvifolium, pellucidum, polymorphum, polysetum, prostratum, pycnobaseum, schmidii, semiovatum, semireticulatum, sharpii, soquense, steerei, subnitidum, subrotundifolium, tereticaule, werthii, worthleyi, yasudae*

GENUS *Argyrobryum* — Agrarian Moss

SPECIES *virescens*

GENUS *Aulacomnium* — Aulac Omni Moss

SPECIES *androgynum, heterostichum, palustre, stolonaceum, turgidum*

GENUS *Brachymenium* — Short Crescent Moss

SPECIES *abyssinicum, acuminatum, alpinum, andersonii, angolense, angustatum, angustirete, apiculatum, brevicarpum, bryoides, bulbiferum, capitulatum, carionis, chilense, coarctatum, columbicum, commutatum, consimile, curvitheca, daguense, debilinerve, deceptivum, dicranoides, domingense, elgonense, erectum, erubescens, eurychelium, exile, exiloides, exoticosporum, fabronioides, flexifolium, glaucum, globirameum, gracile, himalayanum, hornschuchianum, huonii, immarginatum, indicum, jilinense, klotzschii, lanceolatum, leptophyllum, leucotrichum, longicolle, longidens, luteolum, madagassum, magellanicum, malachiticum, melanopyxis, mexicanum, meyenianum, micans, minutulum, morasicum, murale, muricola, myurella, nepalense, nietneri, nigrescens, niveum, notarisii, ovatum, paradoxum, patulum, pendulum, peraristatum, philonotula, preissianum, procerrimum, ptychothecium, pulchellum, pulchrum, radiculosum, regnellii, revolutum, rigidum, robertii, rubricarpum, saint-pierrei, semperlaxum, sikkimense, sinense, sipapoense, spathidophyllum, speciosum, speirocladum, spirale, spirifolium, standleyi, subuliferum, systylium, turgiditheca, turgidum, viviparum, walkeri, wattsii, wrightii*

GENUS *Catoscopium* — Cat Scope Moss

SPECIES *nigritum*

GENUS *Cylindrothecium* — Cylindric Theca Moss

SPECIES *akitense, attenuatum, chloroticum, flaccidum, herbaceum, nanocarpum, tosae*

GENUS *Dermatodon* — Dermal Tooth Moss

SPECIES *funkii*

GENUS *Dorcadion* — Dorcad Moss

SPECIES *pallidum*

GENUS *Entosthymenium* — Entic Thyme Moss

SPECIES *tristichum*

GENUS *Epipterygium* — Epiphyte Wing Moss

SPECIES *brasiliense, convalleum, immarginatum, koelzii, mandonii, mexicanum, obovatum, oparatense, puiggarii, tozeri, vanuatuicum, wrightii*

GENUS *Exodokidium* — Exodok Moss

SPECIES *donatii, subsymmetricum*
GENUS *Haplodontium* Single Tooth Moss
SPECIES *clavatum, diplodontum, purpusii, reticulatum, splendidum, subsplendidum, vilocense*
GENUS *Helicodontiadelphus* Helix Tooth Adelphi Moss
SPECIES *australiensis*
GENUS *Leersia* Leers Moss
SPECIES *extinctoria*
GENUS *Leptobryum* Thin Moss
SPECIES *pyriforme, sericeum, wilsonii*
GENUS *Leptostomum* Thin Stoma Moss
SPECIES *densum, depile, erectum, gerrardii, inclinans, macrocarpum, menziesii, millerianum, perfectum, splachnoideum*
GENUS *Leptotrichum* Thin Hair Moss
SPECIES *vaginans*
GENUS *Mielichhoferia* Mielichhofer Moss
SPECIES *acuminata, acutifolia, angustata, antisanensis, argentifolia, aristata, aristatula, assamica, auriseta, australis, badhwarii, borbonica, brachyclada, brotheri, bryocarpa, canescens, cardotii, castanea, chlorocarpa, clavitheca, coarctata, cuspidata, cuspidifera, demissa, elgonensis, elongata, fabronioides, filiramea, gracilis, grammocarpa, gymna, herzogii, himalayana, humipetens, hymenostomoides, immersidens, japonica, julacea, lahulensis, laxiretis, leptoclada, linderi, lonchocarpa, longipes, longiseta, lorentziana, macrocarpa, macrodonta, macrophylla, macrospora, manca, megalocarpum, microcarpa, microdonta, micropoma, mielichhoferiana, minutifolia, minutissima, modesta, ochracea, ovalis, pampae, paroica, patagonica, pilifera, plagiobrioides, plumosa, procera, punctulata, robusta, sanguinolenta, sasaokae, schmidii, secundifolia, seriata, seriola, serrae, serridens, sinensis, speluncae, striidens, subbasilaris, subcampylocarpa, subclavitheca, subminutifolia, subnuda, sullivanii, tehamensis, ulei, vardei, woodii*
GENUS *Mittenia* Mitten Moss
SPECIES *plumula*
GENUS *Mniobryoides* Lesser Lead Moss
SPECIES *degeneriae*
GENUS *Mniobryum* Lead Moss
SPECIES *alticaule, amplirete, aptychoides, aspillagae, bolivianum, brevisetum, linderi, pergracile, perlaxum, rubrum, ruwenzorense, sparsifolium, tortifolium*
GENUS *Orthodontium* Straight Tooth Moss
SPECIES *australe, denticulatum, gracile, inflatum, infractum, itacolumitis, lignicola, lineare, longisetum, loreifolium, pallens, pellucens, ruahinense, ruwenzorensis*
GENUS *Orthodontopsis* Lesser Straight Tooth Moss
SPECIES bardunovii
GENUS *Perssonia* Persson Moss
SPECIES *sanguinea*
GENUS *Phyllogonium* Phyllogon Moss
SPECIES *fulgens, viride, viscosum*
GENUS *Plagiobryum* Pelagic Moss
SPECIES *demissum, duthiei, giraldii, hultenii, japonicum, laxum, novaeseelandiae, piliferum, zierii*
GENUS *Poecilophyllum* Poecil Moss
SPECIES *macrodon*
GENUS *Pohlia* Pohl Moss
SPECIES *afrocruda, alba, aloysii-sabaudiae, ampullacea, andalusica, andrewsii, annotina, aristatula, atrothecia, australis, austroelongata, austropolymorpha, baldwinii, barbuloides, baronii, bequaertii, beringiensis, betulina, bolanderi, brachystoma, brevinervis, brevireticulata, brideliana, bulbifera, calopyxis, camptotrachela, cardotii, cavaleriei, chilensis, chitralensis, chrysoblasta, clavaeformis, clavicaulis, columbica, crassicostata, cruda, crudoides, cuspidata, debatii, drummondii, elatior, elongata, emergens, erecta, excelsa, excurrens, fauriei, filum, flexuosa, geniculata, grammocarpa, grammophylla, gromieri, hisae, humilis, inflexa, kenyae, korbiana, lacouturei, laticuspis, leptodontium, leptopoda, lescuriana, leucostoma, lonchochaete, longibracteata, longicollis, looseri, loriformis, ludwigii, lutescens, macleai, magnifica, marchica, mauiensis, melanodon, mielichhoferia, nemicaulon, nevadensis, novae-seelandiae, nudicaulis, nutans, nutanti-polymorpha, obtusifolia, ochii, oedoneura, oerstediana, orthocarpula, pacifica, papillosa, philonotula, plumella, procerrima, proligera, pseudobarbula, pseudodefecta, pseudogracilis, pseudophilonotula, revolvens, rhaetica, richardsii, rigescens, rostrata, rubella, rusbyana, saitoi, saprophila, schisticola, scotica, silvatica, sphagnicola, stewartii, subannulata, subcarnea, subcompactula, subleptopoda, tapintzense, tenuifolia, timmioides, trematodontea, tundrae, turgens, verrucosa, vexans, viridis, wahlenbergii, yunnanensis*
GENUS *Pseudopohlia* Lesser Pohl Moss
SPECIES *didymodontia, microstoma*
GENUS *Rhodobryum* Red Moss
SPECIES *andicola, andinoroseum, aubertii, beyrichianum, chilense, commersonii, confluens, domingense, giganteum, grandifolium, hieronymi, homalobolax, horizontale, humipetens, keniae, lato-cuspidatum, laxelimbatum, le-ratii, leucocanthum, longicaudatum, neelgheriense, ontariense, perspinidens, platense,*

preussii, pseudomarginatum, roseodens, roseolum, roseum, saprophilum, staudtii, subverticillatum, umbraculum, verticillatum

GENUS *Roellia* Roell Moss

SPECIES *roellii*

GENUS *Rosulabryum* Rosy Moss

SPECIES *albolimbatum, billarderi, campylothecium, capillare, epiphyticum, lamingtonicum, leptothrix, microrhodon, queenslandicum, subfasciculatum, subtomentosum, torquescens, tuberosum, wightii*

GENUS *Schistophyllum* Split Frond Moss

SPECIES *adianthoides, collinum, longifolium*

GENUS *Schistostega* Split Stegial Moss

SPECIES *pennata*

GENUS *Schizymenium* Schizymen Moss

SPECIES *ampullaceum, andinum, austrogeorgicum, basilare, bogotense, bolivianum, brevicaule, bryoides, campylocarpum, clavellatum, commixtum, costaricense, cratericola, cygnicollum, decurrens, dolichothecum, fulvonitens, fusiferum, gracilisetum, hariotii, hymenodontum, hymenostomum, javanicum, landii, latifolium, lindigii, linearicaule, linearifolium, madagassum, mildbraedii, multiflorum, nanum, nealiae, novoguineense, pectinatum, pleurogynum, pohlioideum, pontevedrense, pseudopohlia, pulchrum, pulvinatum, pusillum, ruwenzorense, schiedianum, schimperi, serratum, shevockii, skottsbergii, subglobosum, submacrodontum, subpohlioideum, weberbaueri*

GENUS *Skitophyllum* Skit Moss

SPECIES *longifolium*

GENUS *Synthetodontium* Synthetic Tooth Moss

SPECIES *pringlei*

GENUS *Tamariscella* Tamarisk Moss

SPECIES *cymbifoliola, promontorii*

GENUS *Tetracladium* Ten Branch Moss

SPECIES *komarovii*

GENUS *Webera* Weber Moss

SPECIES *alticaulis, ampliretis, carneoides, denticostata, julacea, leptoclada, luisieri, maderensis, merapicola, nudicaulis, pentasticha, striatidens, synoica, taylorii, ulei*

FAMILY 02 *Archidiaceae* Arch Moss

GENUS *Archidium* Arch Moss

SPECIES *acanthophyllum, acauloides, alternifolium, amplexicaule, andersonianum, birmannicum, brevinerve, capense, clarksonianum, clavatum, cubense, dinteri, donnellii, elatum, hallii, indicum, johannis-negrii, julaceum, julicaule, laterale, laxirete, microthecium, minus, minutissimum, muellerianum, ohioense, rehmannii, rothii, stellatum, subulatum, tenerrimum, thalliferum, wattsii, yunnanense*

GENUS *Pleuridium* Pleural Arch Moss

SPECIES *lamprropyxis, mexicanum, robinsonii*

FAMILY 03 *Bartramiaceae* Bartram Moss

GENUS *Bartramia* Bartram Moss

SPECIES *abbreviata, afro-ithyphylla, alaris, anacolioides, angustifolia, angustissima, aprica, auricola, austro-georgica, baldwinii, bogongia, borbonica, boulayi, brevifolia, brotheri, cacaltayae, capensis, circinnulata, compacta, conica, costa-ricensis, crassinervia, crispo-ithyphylla, defolians, delagoae, didymocarpa, dilatata, erwinii, faucium, fragilicuspis, fragilis, gathica, gigantea, glauca, glauca, hakonensis, halleriana, hampeana, hampei, henoni, heteromalla, humilis, inconspicua, incurva, inflata, ithyphylla, jungneri, leikipiae, leptodonta, longifolia, macounii, macowaniana, maderensis, madurensis, marionensis, mathewsii, mathildae, mauiensis, microbasis, microcarpa, microstoma, nevadensis, normanii, nothostricta, obscura, papulans, patens, pendula, peraristata, perpumila, perpusilla, pflanzii, pilicuspis, polytrichoides, pomiformis, potosica, prabaktiana, pruinata, pseudocrispata, pseudocryptopodium, pseudostricta, radicosa, rietmanniana, robusta, rufescens, rupta, ruvenzorensis, sobrina, sphaerica, spielhausii, squarrifolia, stenophylla, stricta, stricta, strictula, strumosa, subithyphylla, subpatens, subpellucida, subrobusta, subsymmetrica, subulata, thelioides, timmioides, tophacea, wedellii*

GENUS *Anacolia* Anacole Bartram Moss

SPECIES *aristifolia, aurescens, breutelii, cameruniae, laevisphaera, menziesii, scioana, webbii*

GENUS *Bartramidula* Lesser Bartram Moss

SPECIES *aristata, bartramioides, comosa, curta, cygnea, dispersa, erecta, fendleri, glaziovii, globosa, le-testui, nana, pusilla, roylei, tuerckheimii*

GENUS *Bartramiopsis* Least Bartram Moss

SPECIES *lescurii*

GENUS *Breutelia* Breutel Bartram Moss

SPECIES *aciphylla, affinis, anacolioides, angustiretis, anomala, aristifolia, arundinifolia, aureola, auriculata, austro-arcuata, azorica, baeuerlenii, borbonica, brachyphylla, brevifolia, breviseta, brittoniae, campbelliana, chrysea, chrysocoma, crassicaulis, dicranacea, diffracta, dominicensis, dumosa, eggersiana, elliptica, elongata, eugeniae, gnaphalea, gracillima, grandis, guilielmi-meyeri, harpophylla, hasskarliana, humbertii, incana, inclinata, integrifolia, jamaicensis, karsteniana, kilimandscharica, kinabaluensis, leptodontoides, lonchopelma, longicapsularis, luteola, macrocarpa, madagassa, maegdefraui, magdalenae, microdonta, minuta, mohriana, muelleri, muhavurensis, nigrescens, papuensis, pendula, perrieri, picardae, pilifera, plicata, polygastrica, popinqua, pseudophilonotis, reclinata, rhythidioides, robusta, roemeri, ryvardenii, scariosula, sciuroides, scoparia, scorpioides, secundifolia, squarrosa, stenodictyon, stricticaulis, stuhlmannii, subarcuata, subdisticha, submniocarpa, subplicata, substricta, subtomentosa, tabularis, tenuifolia, tomentosa, trianae, ulicina, viguieri, wainioi, witherheadii, yunnanensis*

GENUS *Bryobartramia* — Mock Bartram Moss

SPECIES *novae-valesiae*

GENUS *Conostomum* — Conic Stoma Bartram Moss

SPECIES *cleistocarpum, curvirostre, macrotheca, magellanicum, pentastichum, perpusillum, tetragonum*

GENUS *Fleischerobryum* — Fleischer Bartram Moss

SPECIES *longicolle, macrophyllum*

GENUS *Flowersia* — Flowers Bartram Moss

SPECIES *abyssinica, campylopus, setifolia, sinensis*

GENUS *Leiomela* — Leiomel Bartram Moss

SPECIES *aristifolia, bartramioides, brachyphylla, capillaris, deciduifolia, ecuadorensis, filifolia, fuscescens, lineata, lopezii, peruviana, piligera, subbrevifolia*

GENUS *Philonotis* — Philonote Bartram Moss

SPECIES *acicularis, acutissima, africana, amblystegioides, ampliretis, andina, angulata, angusta, angustifolia, arbusculacea, aristifolia, arnellii, asperrima, austrofalcata, baginsensis, bernoullii, boliviensis, brevicuspis, buckii, byssiformis, calcarea, caldensis, calomicra, capillaris, capillata, cernua, clavicaulis, conostomoides, coreensis, corticata, crenatula, curvata, curvula, dicranellacea, dregeana, elegantula, elongata, esquelensis, eurydictyon, evanescens, falcata, filiformis, filiramea, flaccidifolia, flavinervis, flexipes, fontana, fontanella, fontanoides, fragilicaulis, fragilicuspis, fugacissima, gardneri, gemmascens, glabrata, glaucescens, glomerata, gourdonii, gracilenta, gracilescens, gracillima, guyabayana, hansenii, hastata, hawaica, heleniana, helenica, hermannii, huallagensis, humilis, incrassata, jungneri, krausei, laeviuscula, lancifolia, laxitexta, leptocarpa, lignicola, litorea, longiseta, macrocarpa, macrodictya, macroglobus, marangensis, marchica, mauritiana, mbelabiensis, mercieri, microthamnia, minutifolia, mniobryoides, mollis, moritziana, muhlenbergii, myriocarpa, nanothecia, nanothecioidea, niam-niamiae, nigricans, nigroflava, operta, osculatiana, pallida, papillarioides, papillatomarginata, pechuelii, pellucidiretis, penicillata, perconferta, pergracilis, perigonialis, perlaxifolia, perpusilla, perrieri, pilifera, plana, platensis, platyneura, polymorpha, pomangium, pseudomollis, pugionifolia, pumila, pungens, pygmaeola, pyriformis, rificuspis, rigida, rufiflora, runcinata, ruwenzorensis, scabrifolia, schroederi, secunda, seriata, sharpiana, sikkimensis, simplicissima, slateri, soulii, sparsifolia, speirophylla, sphaerocarpa, spiralis, spongiosa, striata, striatula, strictiuscula, strictula, submarchica, subolescens, subrigida, subsimplex, subsphaericarpa, tenuis, thwaitesii, tjibodensis, tomentella, tortifolia, trachyphylla, trichodonta, trichophylla, tricolor, turneriana, uncinata, usambarica, vagans, vanderystii, vescoana, yezoana*

GENUS *Plagiopus* — Pelagic Bartram Moss

SPECIES *oederianus*

FAMILY 04 *Buxbaumiaceae* — Buxbaum Moss

GENUS *Buxbaumia* — Buxbaum Moss

SPECIES *aphylla, colyerae, himalayensis, javanica, minakatae, novae-zelandiae, piperi, punctata, symmetrica, tasmanica, thorsborneae, viridis*

GENUS *Diphyscium* — Diphysic Buxbaum Moss

SPECIES *chiapense, domingense, fasciculatum, fendleri, foliosum, fulvifolium, longifolium, mucronifolium, perminutum, pocsii, rhynchophorum, satoi, ulei*

GENUS *Muscoflorschuetzia* — Musk Florschuetz Buxbaum Moss

SPECIES *pilmaiquen*

GENUS *Theriotia* — Theriote Buxbaum Moss

SPECIES *kashmirensis, lorifolia*

FAMILY 05 *Fissidentaceae* — Fissure Moss

GENUS *Fissidens* — Fissure Moss

SPECIES *abbreviatus, acacioides, aciphyllus, acreanus, acutissimus, adianthoides, aeruginosus, afissidens, afropapillosus, aggestus, alatus, albolimbatus, alexandrinus, allionii, allisonii, allorgei, alomoides, amazonicus, amoenus, amplifolius, ampliretis, andicola, angustelimbatus, angustifolius, angustinervis, anisophyllus, annamensis, anomalus, antrophyi, aoraiensis, aphelotaxifolius, aporrocheilos, appalachensis, arboreus, arcticus, arcuatulus, arnigadhensis, arnoldii, asplenioides, asselii, autoicus, azoricus, badoglioi, badyinbarus, barretoi, basilaris, beccarii, beckettii, berterii, bessouensis, biformis, bilaspurense, bistratosus, blechnoides, bogoriensis, bogosicus, boninensis, borgenii, bourgaeanus, brachypus, brassii, braunii, brevidorsus, brevifolius, brevifrons, brevilingulatus, brevipes, bryoides, bushii, caespitosus, cagoui, calcicola, cambewarrae, canalae, capitatus, capitulatus, capriviensis, carnosus, celticus, ceylonensis, clebschii, closteri, coacervatus, collinus, commutatus, consociatus, crassinervis, crassipes, crenulatus, crispifolius, crispulus, crispus, cryptoneuron, cucullatus, curvatus, cuspidifer, cuynetii, cylindraceus, cylindrothecus, cyprius, dalamair, daltoniaefolius, danckelmanii, darntyi, darwinianus, dealbatus, decursivus, delicatulus, dendeliensis, dendrophilus, diaphanodontus, diaphanus, dietrichiae, diplodus, discolor, dissitifolius, diversifolius, dongensis, dubius, dumbeanus, duttonii, edamensis, elamellosus, elegans, elongatus, enervis, ernestii, erosulus, euryloma, excedens, excurrentinervis, exiguus, exilis, exsul, faniensis, fasciculatus, fautauae, filicicola, filiformis, firmus, fissicaulis, flabellatus, flabellulus, flaccidus, florschuetzii, fluitans, fontanus, formosanus, forsythii, fritzei, fujiensis, fuscoviridis, ganguleei, gardneri, gaultieri, geijskesii, geminiflorus, georgianus, geppii, gillianus, giraldii, gladiolus, glaucissimus, glossobryoides, gomae, goyazensis, gracilescens, gracilifolius, grainvillei, grandifolius, grandifrons, granulatus, griffithii, grossiretis, guangdongensis, guianensis, gymnocarpus, gymnogynus, gymnostomus, haconicus, hadii, hallianus, hampeanus, helenicus, henryae, herzogii, hildebrandtii, hillianus, hirsutus, hoeegii, hoei, hollianus, homomallulus, hookerioides, horizonticarpus, hornschuchii, humilis, hyalinus, hydropogon, hylogenes, hylophilus, hymenodon, hyophilus, idanreensis, imbricatus, inaequalilimbatus, inaequalis, inaequiretis, incertus, incisus, inclinis, incognitus, inconspicuus, incurvus, integerrimus, integrifolius, intromarginatus, involutus, iwatsukii, jaiorum, jansenii, jap-amabilis, javanicus, jeffreyi, jungermannioides, jungermanniopsis, kalimpongensis, karataviensis, karwarensis, kilaueae, kinabaluensis, kondoi, kosaninii, kurzii, lachmannii, lacouturei, lagenarius, lancifolius, laosianus, lateralioides, latifolius, lautokensis, laxetexturatus, laxitextus, lepidopiloides, leptocheilos, leptocladus, leptophyllus, le-testui, leucocinctus, leucodictyus, leucopteris, liliputanus, limbinervis, lindbergii, linearilimbatus, linearis, linearis, littlei, loennbergii, longevaginatus, longicaulis, longifolius, longisetus, luisierii, luteofuscus, luteo-limbatus, macleanus, macrobryoides, macrodus, macroglossus, macrosporoides, macrosporus, madecassus, marginatulus, mariei, marthae, maschalanthus, megalotis, menyhartii, metzgeria, michoacanus, microcarpus, microdictyon, micronesicus, minimus, minutifolius, minutipes, mobukensis, monguillonii, mooreae, multiflorus, nanobryoides, narbonensis, neglectus, neomagofukui, nigerianus, nigro-viridis, nobilis, nobreganus, nothotaxifolius, oblatus, obliquifolius, oblongifolius, obscurocostatus, obscurus, obtusatulus, obtusifolius, obtusissimus, obtuso-apiculatus, ocellatus, odontoloma, oediloma, oligistoloma, opacus, orishae, ornaticostatus, ornatus, osmundoides, ovatifolius, ovatus, pachyphyllus, pacificus, pallidinervis, pallidulus, pallidus, palmatus, palmifolius, parkii, pascuanus, patentifolius, patulifolius, pauperculus, pauperrimus, pechuelii, pellucidus, pellucinervis, peraculeatus, perangustus, perdecurrens, perfalcatus, perobtusus, perpellucidus, perplexans, persicus, perssonii, perumalensis, peruvianus, petrophilus, plagiothecioides, planifrons, platensis, platyneuros, plumosus, plumula, plurisetus, pocsii, pokhrensis, polyphyllus, polypodioides, polysetulus, porrectus, potieri, prionocheilos, prionodes, protonematicola, pseudoaldelphinus, pseudoceylonensis, pseudoclosteri, pseudo-eenii, pseudofirmus, pseudohollianus, pseudopallidus, pseudoplumosus, pseudorufescens, pugionifolius, pulchellus, punctulatus, purpureocaulis, pycnoglossus, pygmaeus, radicans, raiatensis, rambii, ramicola, ramiger, ramulosus, ranchiensis, ranuii, reflexus, reimersii, remotissimus, rigidulus, rivularis, robynsianus, rochensis, rosulatus, rufescens, rufinervis, rufulus, rupicola, ryukyuensis, santa-clarensis, saprophilus, sardous, saulensis, scalaris, scariosus, schiffneri, schusteri, schwabei, sciophyllus, scleromitrius, secundulus, sedgwickii, semicompletus, semilimbatus, semperfalcatus, sericeus, serrato-marginatus, serratus, serrulatus, seychellensis, shinii, siamensis, sigmocarpoides, socialis, somaliae, speluncae, splachnifolius, splachnoides, splendens, spurio-limbatus, steerei, stenophyllus, stenopteryx, stissotheca, stolonaceus, strictus, subangustus, subbasilaris, subbryoides, subdiscolor, subelamellosus, subelimbatus, subfirmus, subhumilis, sublimbatus, sublineaefolius, submarginatus, subnutans, subobtusatus, subobtusus, subpalmatus, subplanifrons, subpulchellus, subradicans, subramicola, subscleromitrius, subsessilis, subspathulatus, subulatifolius, subulatus, sufflatus, surumuensis, taiarapuensis, tapes, taxifolius, taylorii, teniolatus, terebrifolius, termitarum, terrae-reginae, terricola, teysmannianus, thorsbornei, thwaitesii, titalyanus, tonkinensis, tosaensis, townsendianus, usambaricus, valiae, vanzantenii, variolimbatus, ventanae, ventricosus, vesiculosus, victorialis, virens, vitreo-limbatus, vulcanicus, wageri, waiensis, waldheimii, wallisii, wattsii, weirii, welwitschii, wichurae, widgrenii, yamamotoi, yanoae, yasudae, yokohamensis, yucatanensis, zollingeri, zwickeyi*

GENUS *Conomitrium* — Conic Mitre Fissure Moss
SPECIES *arboreum, diplodus, dussianum, hemiloma, stissotheca*
GENUS *Moenkemeyera* — Moenke-Meyer Fissure Moss
SPECIES *alomifolia, australiensis, rarotongae*

FAMILY 06 *Hypnodendraceae* — Hypnic Wood Moss
GENUS *Hypnodendron* — Hypnic Wood Moss
SPECIES *arcuatum, auricomum, beccarii, brevipes, camptotheca, colensoi, comatulum, comatum, comosum, dendroides, diversifolium, flagelliferum, fuscomucronatum, junghuhnii, kerrii, marginatum, menziesii, microstictum, milnei, reinwardtii, samoanum, spininervium, subarborescens, subspininervium, tahiticum, vitiense*
GENUS *Braithwaitea* — Braithwait Hypnic Wood Moss
SPECIES *arborescens, sulcata*
GENUS *Mniodendron* — Lead Hypnic Wood Moss
SPECIES *nanum*

FAMILY 07 *Lembophyllaceae* — Lembic Moss
GENUS *Lembophyllum* — Lembic Moss
SPECIES *divulsum, orbiculatum*
GENUS *Camptochaete* — Field Bristle Lembic Moss
SPECIES *angustata, arbuscula, curvata, deflexa, excavata, fallax, fruticosa, leichardtii, porotrichoides, pulvinata, subporotrichoides*
GENUS *Dixonia* — Dixon Lembic Moss
SPECIES *thamnioides*
GENUS *Dolichomitra* — Dolich Mitre Lembic Moss
SPECIES *cymbifolia, robusta, speciosa*
GENUS *Dolichomitriopsis* — Lesser Dolich Mitre Lembic Moss
SPECIES *crenulata, diversiformis, obtusifolia, sasaokae*
GENUS *Fallaciella* — Deceptive Lembic Moss
SPECIES *gracilis, obusta*
GENUS *Fifea* — Fife Lembic Moss
SPECIES *aciphylla*
GENUS *Neobarbella* — Modern Beard Lembic Moss
SPECIES *amoena, comes*
GENUS *Porotrichodendron* — Pore Hairy Wood Lembic Moss
SPECIES *biforme, flavidulum, glaziovii, gracile, lindigii, longipes, mahahaicum, nitidum, porteri, robustum, substolonaceum, superbum*
GENUS *Porotrichopsis* — Pore Hairy Lembic Moss
SPECIES *flacca*
GENUS *Rigodium* — Rigode Lembic Moss

SPECIES *adpressum, brachypodium, implexum, penicilliferum, pseudothuidium, tamarix, toxarion*
GENUS *Tripterocladium* Three Wing Branch Lembic Moss
SPECIES *leucocladulum*

FAMILY 08 *Meesiaceae* Mees Moss
GENUS *Meesia* Mees Moss
SPECIES *hexasticha, longiseta, muelleri, novae-zealandia, triquetra, ulei, uliginosa*
GENUS *Amblyodon* Ambler Tooth Mees Moss
SPECIES *dealbatus*
GENUS *Neomeesia* Modern Mees Moss
SPECIES *paludella*
GENUS *Paludella* Marsh Mees Moss
SPECIES squarrosa

FAMILY 09 *Mniaceae* Lead Moss
GENUS *Mnium* Lead Moss
SPECIES *ambiguum, arcticum, arizonicum, blyttii, geniculatum, heterophyllum, hornum, ilicifolium, integrifolium, laevinerve, lycopodioides, marginatum, nipponense, orientale, perpusillum, punctatum, rutheanum, spinosum, spinulosum, stellare, sublimbatum, submarginatum, thomsonii*
GENUS *Cinclidium* Cinclid Lead Moss
SPECIES *arcticum, latifolium, stygium, subrotundum*
GENUS *Cyrtomnium* Curved Lead Moss
SPECIES *hymenophylloides, hymenophyllum*
GENUS *Leucolepis* Light Thin Lead Moss
SPECIES *acanthoneura*
GENUS *Orthomnion* Straight Lead Moss
SPECIES *bryoides, dilatatum, elimbatum, handelii, javense, loheri, noguchii, nudum, piliferum, yunnanense*
GENUS *Plagiomnium* Pelagic Lead Moss
SPECIES *acutum, affine, arbuscula, carolinianum, ciliare, confertidens, cordatum, cuspidatum, drummondii, elatum, elimbatum, ellipticum, insigne, integrum, japonicum, maximoviczii, medium, novae-zealandiae, rhynchophorum, rostratum, succulentum, tezukae, undulatum, venustum, vesicatum*
GENUS *Pseudobryum* Lesser Lead Moss
SPECIES *cinclidioides, speciosum*
GENUS *Rhizomnium* Rhizid Lead Moss
SPECIES *andrewsianum, appalachianum, glabrescens, gracile, hattorii, horikawae, magnifolium, nudum, parvulum, pseudopunctatum, punctatum, striatulum, tuomikoskii*
GENUS *Stellariomnium* (Synonym for *Mnium*)
GENUS *Trachycystis* Rough Cyst Lead Moss
SPECIES *flagellaris, microphylla, ussuriensis*

FAMILY 10 *Polytrichaceae* Many Hair Moss
GENUS *Polytrichum* Many Hair Moss
SPECIES *angustifolium, autranii, brachymitrium, brachypelma, breviceps, cataractarum, commune, croceum, cupreum, elegans, elongatum, glabrum, gracilisetum, hyperboreum, juniperellum, juniperinum, lycopodioides, microcephalum, micropysix, novae-hollandiae, obliquirostre, oerstedianum, pachyneuron, piliferum, polakowskyi, recurvatum, recurvipilum, sinense, stolonigerum, strictum, subcarinatum, subpilosum, subremotifolium, swartzii, tibetanum, tonkinense, trachythecium, transvaaliense, vaginatum*
GENUS *Alophosia* Alophose Many Hair Moss
SPECIES *azorica*
GENUS *Atrichopsis* Least Many Hair Moss
SPECIES *compressa*
GENUS *Atrichum* Lesser Many Hair Moss
SPECIES *altecristatum, androgynum, angustatum, anomalum, brevilamellatum, controversum, crispulum, crispum, cylindricum, flavisetum, hirtellum, longifolium, minutum, muelleri, nigricans, obtusulum, oerstedianum, polycarpum, pusillum, rhystophyllum, selwynii, subserratum, tenellum, undulatiforme, undulatum, undulatum, xanthopodum, yakushimense, yunnanense*
GENUS *Dawsonia* Dawson Many Hair Moss
SPECIES *beccarii, crispifolia, filicaulis, gigantea, grandis, insignis, lativaginata, longifolia, longiseta, papuana, polytrichoides, pulchra, pullei, robbinsii, superba*
GENUS *Dendroligotrichum* Wood Many Hair Moss
SPECIES *dendroides, squamosum*
GENUS *Hebantia* Hebant Many Hair Moss
SPECIES *rigida*
GENUS *Itatiella* Itati Many Hair Moss
SPECIES *ulei*
GENUS *Lyellia* Lyell Many Hair Moss
SPECIES *aspera, bifurcata, crispa, platycarpa*
GENUS *Meiotrichum* Lesser Many Hair Moss
SPECIES *lyallii*

GENUS *Notoligotrichum* Notable Wood Many Hair Moss
SPECIES *australe, bellii, crispulum, laxifolium, mexicanum, minimum, tapes, trichodon, tristaniense*
GENUS *Oligotrichum* Fewer Many Hair Moss
SPECIES *afrolaevigatum, aligerum, aristatulum, atrichopsis, austroaligerum, canaliculatum, cavallii, crispatissimum, crossidioides, denudatum, erosum, falcatum, glaciale, hercynicum, javanicum, laevifolium, minutum, nepalense, novae-guineae, obtusatum, obtusifolium, parallelum, riedelianum, semilamellatum, serratomarginatum, suzukii, tenuirostre, tristaniense, wageri*
GENUS *Philocrya* (Synonym for *Polytrichum*)
GENUS *Pogonatum* Bearded Many Hair Moss
SPECIES *aloides, belangeri, brachyphyllum, brasiliense, campylocarpon, camusii, capense, cirratum, comosum, congolense, contortum, convolutum, dentatum, fastigiatum, geheebii, gracilifolium, inflexum, iwatsukii, japonicum, lamellosum, marginatum, microphyllum, microstomum, minus, misimense, nanum, neesii, neglectum, neocaledonicum, nipponicum, norrisii, nudiusculum, otaruense, patulum, pensilvanicum, pergranulatum, perichaetiale, peruvianum, petelotii, philippinense, piliferum, procerum, proliferum, rufisetum, rutteri, semipellucidum, sinense, spinulosum, subfuscatum, subtortile, subulatum, tahitense, tortile, tubulosum, urnigerum, usambaricum, volvatum*
GENUS *Polytrichadelphus* Adelphi Many Hair Moss
SPECIES *abriaquiae, archboldii, aristatus, bolivianus, ciferrii, ciliatus, cuspidirostrum, ericoides, giganteus, grossidens, integrifolius, longisetus, magellanicus, mitis, nukahivensis, organensis, peruvianus, pseudopolytrichum, purpureus, rubescens, rubiginosus, subrubescens, trianae, umbrosus, valenciae*
GENUS *Polytrichastrum* Star Many Hair Moss
SPECIES *alpinum, altaicum, appalachianum, emodi, formosum, longisetum, ohioense, pallidisetum, papillatum, sexangulare, tenellum, torquatum, xanthopilum*
GENUS *Psilopilum* Smooth Many Hair Moss
SPECIES *aequinoctiale, cavifolium, gymnostomulum, laevigatum*
GENUS *Steereobryon* Steere Many Hair Moss
SPECIES *subulirostrum*

FAMILY 11 *Racopilacea* Racon Downy Moss
GENUS *Racopilum* Racon Downy Moss
SPECIES *africanum, angustistipulaceum, aristatum, ayresii, brevipes, buettneri, capense, cardotii, chevalieri, crassicuspidatum, crinitum, cuspidigerum, ellipticum, epiphyllosum, fernandezianum, ferriei, floridae, francii, gracile, gracillimum, intermedium, laxirete, leptocarpum, leptotapes, macrocarpum, madagassum, magnirete, marginatum, mauritianum, microdictyon, microides, microphyllum, mougeotianum, naumannii, niutensis, orthocarpioides, pacificum, pectinatum, penzigii, perrieri, plicatum, polythrincium, purpurascens, robustum, siamense, spectabile, speluncae, squarrifolium, strumiferum, thomeanum, tomentosum, ugandae, verrucosum*
GENUS *Powellia* Powell Racon Downy Moss
SPECIES *acutifolia, australis, breviseta, integra, involutifolia, macromitrioides, parvula, subelimbata*

FAMILY 12 *Rhizogoniaceae* Rhizid Gonad Moss
GENUS *Rhizogonium* Rhizid Gonad Moss
SPECIES *alpestre, distichum, graeffeanum, hattorii, lamii, lindigii, novae-hollandiae, pennatum, sublimbatum*
GENUS *Cryptopodium* Cryptic Pod Rhizid Gonad Moss
SPECIES *bartramioides*
GENUS *Goniobryum* Lesser Rhizid Gonad Moss
SPECIES *subaloma, subbasilare*
GENUS *Hymenodon* Hymen Tooth Rhizid Gonad Moss
SPECIES *aeruginosus, angustifolius, ovatus, pilifer, reggaeus, sphaerothecius*
GENUS *Hymenodontopsis* Lesser Hymen Tooth Rhizid Gonad Moss
SPECIES *stresemannii*
GENUS *Leptotheca* Thin Theca Rhizid Gonad Moss
SPECIES *boliviana, gaudichaudii*
GENUS *Mesochaete* Medium Bristle Rhizid Gonad Moss
SPECIES *taxiforme, undulata*
GENUS *Pyrrhobryum* Fire Rhizid Gonad Moss
SPECIES *bifarium, dozyanum, latifolium, mauritianum, medium, mnioides, novae-caledoniae, paramattense, pungens, setosum, spiniforme, vallis-gratiae*

FAMILY 13 *Seligeriaceae* Seliger Moss
GENUS *Seligeria* Seliger Moss
SPECIES *acutifolia, austriaca, brevifolia, calcarea, campylopoda, cardotii, careyana, diminuta, diversifolia, donniana, galinae, gymnostoma, irrigata, oelandica, paucifolia, polaris, pusilla, recurvata, subimmersa, trifaria, tristichoides*
GENUS *Blindia* Blind Seliger Moss
SPECIES *acuta, brachystegia, caespiticia, campylopodioides, capillifolia, contecta, dalatensis, dryptodontoides, gradsteinii, immersa, inundata, jamesonii, japonica, lewinskyae, magellanica, martinii, maxwellii, robusta, roerichii, seppeltii, sonsoniae, subtortifolia, torrentium, tortelloides*
GENUS *Brachydon* (Synonym for *Brachydontium*)
GENUS *Brachydontium* Short Tooth Seliger Moss

SPECIES *curvisetum, flexisetum, intermedium, microcarpum, notorogenes, olsonii, olympicum, polycarpum, trichodes*
GENUS *Hymenolomopsis* Hymen Seliger Moss
SPECIES *tolucensis*
GENUS *Trochobryum* Wheel Seliger Moss
SPECIES *carniolicum*

FAMILY 14 *Spiridentaceae* Spire Moss
GENUS *Spiridens* Spire Moss
SPECIES *armatus, balfourianus, camusii, capillifer, cheesmaniae, flagellosus, longifolius, perichaetialis, reinwardtii, vieillardii*
GENUS *Franciella* Francis Spire Moss
SPECIES *spiridentoides*

FAMILY 15 *Tetraphidaceae* Ten Tooth Moss
GENUS *Tetraphis* Ten Tooth Moss
SPECIES *geniculata, pellucida*
GENUS *Calomnion* Callisto Lead Ten Tooth Moss
SPECIES *brownseyi, ceramense, complanatum, denticulatum, iwatsukii, lillianiae, melanesicum, milleri, schistostegiellum*
GENUS *Tetrodontium* Greater Ten Tooth Moss
SPECIES *brownianum, ovatum, repandum*

ORDER 2 *Dicranales* Dicran Moss

FAMILY 1 *Dicranaceae* Dicran Moss
GENUS *Dicranum* Dicran Moss
SPECIES *albescens, bentzelii, boswellii, breviflagellare, brevisetum, calymperaceum, camptophyllum, chlorophyllosum, congestum, crispatum, crispulum, densifolium, expallidum, falcatum, falcatum, fallax, fuscescens, fuscescens, geniculatum, goughii, grimmioides, hawaiicoflexuosum, humile, incanum, latifolium, leucobasis, multicapsulare, nietneri, nigro-flavum, orthocarpum, pumilum, purpureum, rupestre, schleicheri, scopelliforme, setosum, sieberianum, strictipilum, strictisetum, subarctocarpum, subpungens, subscoparium, subsetosum, subulatum, subulifolium, subviride, sulcatum, tectorum, trachyphyllum, viridissimum, viridulum, yakushimense, acanthoneurum, acutifolium, arcuatipes, assamicum, atratum, bonjeanii, borbonicum, braunsiae, brevifolium, caesium, cheoi, condensatum, crassifolium, crispifolium, decumbens, deflexicaulon, delavayi, dilatinerve, diplospiniferum, dispersum, drummondii, dubium, eggersii, elongatum, filum, flagellare, fragilifolium, fragillimum, frigidum, fulvum, fuscescens, gonoi, gregoryi, groenlandicum, hamulosum, himalayanum, howellii, japonicum, jashii, johnstonii, kashmirense, klauteri, leiodontium, leioneuron, leucobryoides, linzianum, longicylindricum, lophoneuron, lorifolium, majus, mayrii, montanum, muehlenbeckii, nipponense, novaustrinum, obliquatum, ontariense, orthophylloides, orthophyllum, pallidisetum, papillidens, peruvianum, petrophylum, polysetum, psathyrum, pseudofalcatum, pseudoleucoloma, pseudorobustum, rectifolium, rhabdocarpum, rodriguezii, savatieri, schensianum, scoparium, scottianum, semperi, setifolium, spadiceum, speirophyllum, spurium, sumichrastii, symblepharoides, syrrhopodontoides, tauricum, thelinotum, thraustophyllum, toninii, transsylvanicum, truncicola, tubulifolium, undulatum, viride, yezomontanum*
GENUS *Anisodon* Anise Dicran Moss
SPECIES *acutirostris*
GENUS *Anisothecium* Anise Theca Dicran Moss
SPECIES *brachyangium, brachydontium, campylophyllum, capituligerum, clathratum, convolutum, crispum, cyrtodontum, elegans, fontanum, globuligerum, gracillimum, grevilleanum, hioramii, hookeri, horridum, humile, javanicum, laxirete, lorentzii, macrostomum, madagassum, molliculum, nicholsii, palustre, planinervium, pseudorufescens, pycnoglossum, recurvimarginatum, rotundatum, rufescens, rufipes, ruttneri, schreberianum, schreberianum, skottsbergii, spirale, staphylinum, submacrostomum, ugandae, vaginale, vaginatum, varium, yezoanum*
GENUS *Aongstroemia* Aong Stroma Dicran Moss
SPECIES *appressa, filiformis, fitzgeraldii, gayana, heteromalla, julacea, lamyi, longipes, orientalis, sauteri, subcompressa*
GENUS *Aongstroemiopsis* Lesser Aong Stroma Dicran Moss
SPECIES *julacea*
GENUS *Arctoa* Arctic Dicran Moss
SPECIES *anderssonii, fulvella, hyperborea, schistioides*
GENUS *Atractylocarpus* Dark Ctylocarp Dicran Moss
SPECIES *alpinus, alticaulis, brasiliensis, intermedius, longisetus, madagascariensis, nanus, neocaledonicus, novoguineensis, patagonicus*
GENUS *Bartleya* (Synonym for *Dicranum*)
GENUS *Braunfelsia* Braunfels Dicran Moss
SPECIES *aenea, angustifolia, dicranoides, edentula, enervis, longipes, muscicola, plicata*
GENUS *Brothera* Lesser Brother Dicran Moss
SPECIES *leana*
GENUS *Brotherobryum* Brother Dicran Moss

SPECIES *dekockii, macgregorii, sleumeri, undulatifolium*
GENUS *Bryohumbertia* Humbert Dicran Moss
SPECIES *filifolia, flavicoma, subcomosa*
GENUS *Bryotestua* Testue Dicran Moss
SPECIES *brevicuspis, longicuspis*
GENUS *Bryoxiphium* Xiph Dicran Moss
SPECIES *madeirense, mexicanum, norvegicum*
GENUS *Camptodontium* Field Tooth Dicran Moss
SPECIES *cryptodon, fallax*
GENUS *Campylopodiella* Least Field Pod Dicran Moss
SPECIES *crenulata, flagellacea, himalayana, stenocarpa*
GENUS *Campylopodium* Lesser Field Pod Dicran Moss
SPECIES *euchlorum, lineare, medium, phascoides*
GENUS *Campylopus* Field Pod Dicran Moss
SPECIES *acuminatus, aemulans, albescens, albidovirens, alienus, ambiguus, amboroensis, anderssonii, andreanus, angustiretis, arctocarpus, arcuatus, areodictyon, argyrocaulon, asperifolius, atlanticus, atrovirens, aureonitens, australis, austro-alpinus, austrostramineus, austrosubulatus, beauverdianus, belangeri, bicolor, boswellii, brassii, brevipilus, brunneus, bryotropii, cambouei, capillaceus, capitulatus, carolinae, catarractilis, catharinensis, cavifolius, chevalieri, chilensis, chlorophyllosus, circinatus, clavatus, cleefii, clemensiae, cockaynii, comatus, comosus, concolor, crateris, crispatulus, crispatus, crispifolius, cruegeri, cryptopodioides, cubensis, cucullatifolius, cuspidatus, cygneus, decaryi, densicoma, densifolius, denticuspis, dichrostris, dicranoides, dietrichiae, diversinervis, edithae, ericoides, exasperatus, exfimbriatus, extinctus, filifolius, flaccidus, flagellifer, flexuosus, flindersii, fragilis, fuscocroceus, fuscolutescens, gardneri, gastro-alaris, gemmatus, gemmiparus, geniculatus, geraensis, goughii, gracilis, grimmioides, griseus, guaitecae, hawaiicoflexuosus, hawaiicus, hensii, heterostachys, hildebrandtii, homalobolax, huallagensis, humilis, incertus, incrassatus, incurvatus, introflexus, itacolumitis, jamesonii, johannis-meyeri, jugorum, julaceus, julicaulis, kivuensis, lamellinervis, laxitextus, laxoventralis, longicellularis, luetzelburgii, luteus, macgregorii, madagassus, matarensis, megalotus, milleri, modestus, multicapsularis, nanophyllus, nepalensis, nietneri, nigro-flavus, nivalis, oblongus, obrutus, obtextus, occultus, oerstedianus, papuensis, pauper, perangustifolius, percurvatus, perexilis, perfalcatus, perichaetialis, perporosus, perpusillus, pilifer, pinfaensis, pittieri, praemorsus, praetermissus, pseudobicolor, purpureocaulis, pyriformis, rabenii, recurvus, reflexisetus, reflexus, refractus, richardii, rigens, robillardei, savannarum, schimperi, schmidii, scopelliformis, scoposum, searellii, sehnemii, sericeoides, serratus, setifolius, sharpii, shawii, sinensis, skottsbergii, smaragdinus, sphagnicola, spiralis, squalidus, stewartii, strictisetus, subannotinus, subarctocarpus, subcomatus, subcubitus, subcuspidatus, subfalcatus, subfragilis, subjugorum, subluteus, subnitens, subperichaetialis, subporodictyon, subtrachyblepharus, subulatus, subvirescens, surinamensis, tablasensis, taiwanensis, tallulensis, terebrifolius, thwaitesii, torrentis, trachyblepharon, trichophylloides, trivialis, trollii, tubulosus, uleanus, umbellatus, valerioi, validinervis, vesticaulis, wawraeanus, wheeleri, widgrenii, zygodonticarpus*
GENUS *Cecalyphum* Calyph Dicran Moss
SPECIES *fertile*
GENUS *Chorisodontium* Choris Tooth Dicran Moss
SPECIES *aciphyllum, burrowsii, dicranellatum, luridum, magellanicum, mittenii, nigricans, setaceum, speciosum, spegazzinii*
GENUS *Cladophascum* Branch Bundle Dicran Moss
SPECIES *gymnomitrioides*
GENUS *Cladopodanthus* Branch Pod Dicran Moss
SPECIES *heterophyllus, microcarpus, speciosus*
GENUS *Cnestrum* Nester Dicran Moss
SPECIES *alpestre, glaucescens, schisti*
GENUS *Cryptodicranum* Cryptic Dicran Moss
SPECIES *armittii*
GENUS *Cynodontium* Dog Tooth Dicran Moss
SPECIES *asperifolium, bruntonii, crassirete, fallax, gracilescens, guatemalense, jenneri, mexicanum, polycarpum, sinensifugax, strumiferum, strumulosum, subsphaericarpum, suecicum, tanganyikae, tenellum*
GENUS *Dichodontium* Two Toothed Dicran Moss
SPECIES *debile, flavescens, fulvescens, integrum, olympicum, pellucidum*
GENUS *Dicnemoloma* Dual Thread Fringed Dicran Moss
SPECIES *fraseri, incanum, sieberi, sieberianum, strictipilum*
GENUS *Dicranella* Lesser Dicran Moss
SPECIES *acroclada, affinis, afroexigua, alpina, amplexans, ampullacea, angustifolia, annamensis, apolensis, apophysatula, argentinica, ascensionica, aulacocarpa, austroexigua, austro-sinensis, baileyana, barbensis, barnesii, bicolor, bogotensis, boliviana, borbonica, brachyblepharis, breviseta, caldensis, callosa, cameruniae, canariensis, cardotii, cerviculata, circinata, coarctata, congolensis, consimilis, cratericola, crenulata, crinalis, densa, densifolia, denticulata, dietrichiae, dilatatinervis, ditissima, ditrichoides, divaricata, divaricatula, dussii, edentata, elata, erectotheca, euryphylla, eustegia, exigua, exilis, falcularia, fallax, fitz-geraldii, flaccidula, flavipes, fuegiana, fukienensis, fusca, glauca, glaziovii, globicarpa, gonoi, grandispora, gremmenii, griffithii, guadelupensis, guilleminiana, gymna, harrisii, hawaiica, heteromalla, heteromalla, hilariana, hillebrandii, hochreutineri, howei, ihsibae, infuscata, integrifolia, itatiaiae, juliformis, kenyae, kiushiana, kunzeana, lagunaria, leibergii, lenta, leptoneura, ligulata, ligulifolia, liliputana, lindigiana, linearifolia, longifolia, longirostris, lusitanica, lutaria, luteola, macromorpha, macrospora, madagassa, martiana, mayebarae, micro-divaricata, minuta, moenkemeyeri, muelleri, muralis, nanocarpa, nitida, nitidula, nodicoma, osculatiana, oshimae, pabstiana, pacifica, pallidiseta, papillinervis, papua-palustris, paraguensis, paucifolia, perrottetii, pertenella, pervilleana, polii, pomiformis, ponapensis, proscripta, pseudosubulata, puiggarii, pumila, pusilla, pyrrhotricha, quelpaertensis, remotifolia, reticulata, rigida, rigidula, rio-grandensis, rivalis,*

rufiseta, samoana, schmidii, schreberi, sclerophylla, sericea, setifera, stackhousiana, standleyi, stickinensis, stricticaulis, strumulosa, subangulata, subfalcularia, subpycnoglossa, subsecunda, subsubulata, subsulcata, subulata, sumatrana, temperata, tenax, tenuisetula, torrentium, tovariensis, trumpffii, ulei, usambarica, varia, weberbaueri

GENUS *Dicranodontium* Tooth Dicran Moss

SPECIES *asperulum, denudatum, didictyon, didymodon, filifolium, humile, insularum, papillifolium, porodictyon, pulchroalare, schwabei, sordidum, tanganyikae, tapes, tristaniense, uncinatum*

GENUS *Dicranoloma* Fringed Dicran Moss

SPECIES *angustatum, angustiflorum, angustifolium, arfakianum, argutum, assimile, atlanticum, australe, austrinum, austro-scoparium, bartramianum, bartramioides, billardieri, blumii, brachycarpum, brachysteleum, brasiliense, braunii, breviflagellare, brevifolium, brevisetum, burchardtii, calycinum, calymperaceum, capillare, capillifolioides, chilense, confusum, cutlackii, cylindropyxis, cylindrothecium, daymannianum, deplanchei, diaphanoneurum, dicarpoides, dicarpum, dusenii, elimbatum, eucamptodontoides, falklandicum, fasciatum, francii, geluense, gracile, grossialare, hariotii, havilandii, hollermayeri, imponens, kerguelense, kunkelii, macrodon, menziesii, microcarpum, muelleri, nelsonii, obesifolium, onraedtii, pancheri, perdecurrens, perichaetiale, perlongifolium, perremotifolium, perviride, platycaulon, platyloma, plicatum, plurisetum, punctulatum, reflexum, robustum, rufifolium, rugifolium, rugosum, serratum, setosum, singalangense, skottsbergii, speightii, steenisii, steyermarkii, striatulum, subassimile, subconfine, subenerve, subimponens, subpungens, subreflexifolium, subsetosum, sullivanii, trichopodum, wattsii*

GENUS *Dicranoweisia* Weis Dicran Moss

SPECIES *africana, alpina, antarctica, austro-crispula, brevifolia, brevipes, breviseta, brunnea, cirrata, contermina, crispula, falcifolia, fastigiata, flexipes, funiculipes, grimmiacea, indica, insularis, jugellifera, mackayi, macrospora, papuana, roellii, spenceri, subglobosa, tortifolia, turpis*

GENUS *Eucamptodontopsis* Lesser Field Tooth Dicran Moss

SPECIES *aberrans, brittoniae, mcphersonii, pilifera, tortuosa*

GENUS *Eustichia* Stitch Dicran Moss

SPECIES *longirostris*

GENUS *Holodontium* Whole Tooth Dicran Moss

SPECIES *trictum*

GENUS *Holomitriopsis* Lesser Whole Mitre Dicran Moss

SPECIES *laevifolia*

GENUS *Holomitrium* Whole Mitre Dicran Moss

SPECIES *angolense, antennatum, arboreum, aristatum, austroalpinum, azuayensis, borbonicum, calycinum, camptocarpum, comorense, crispulum, cylindraceum, densifolium, dietrichiae, diversirete, dubiosum, flexuosum, francii, glaziovii, gracilisetum, hawkinsii, hillieri, lepervanchei, longifolium, lutescens, moritzianum, mucronatum, nitidum, nodosum, obliquum, olfersianum, paraguense, perichaetiale, pervaginatum, pulchellum, rhyparodictyon, seramense, seticalycinum, seticalyx, sinuosum, stenobasis, subglobosum, subperichaetiale, terebellatum, tortuosum, undulatum, urvilleanum, vaginatum, williamsii, xolocotzianum*

GENUS *Hygrodicranum* Hydric Dicran Moss

SPECIES *bolivianum, falklandicum, herrerae*

GENUS *Kiaeria* Kiaer Dicran Moss

SPECIES *blyttii, falcata, glacialis, pumila, riparia, starkei*

GENUS *Kingiobryum* King Dicran Moss

SPECIES *paramicola*

GENUS *Macrodictyum* Large Net Dicran Moss

SPECIES *latifolium, proliferum, wrightii*

GENUS *Mesotus* Middle Ear Dicran Moss

SPECIES *celatus*

GENUS *Microcampylopus* Small Field Dicran Moss

SPECIES *khasianus, laevigatus, leucogaster*

GENUS *Microdus* Odious Dicran Moss

SPECIES *acutidentatus, assamicus, filicaulis, friedensis, miquelianus, picquenotii, rubrisetus, sinensis, subnitidulus, tutuilae, yuennanensis*

GENUS *Mitrobryum* Mitre Dicran Moss

SPECIES *koelzii*

GENUS *Muscoherzogia* Musk Herzog Dicran Moss

SPECIES *pauperculata*

GENUS *Nanomitriopsis* Nannite Mitre Dicran Moss

SPECIES *longifolia*

GENUS *Ochrobryum* Ocher Dicran Moss

SPECIES *gardneri, kurzianum, mittenii, nietneri, sakalavum, sessile, subulatum*

GENUS *Oncophorus* Onc Dicran Moss

SPECIES *crispifolius, fuegianus, gracilentus, luteovirens, rauei, virens, wahlenbergii*

GENUS *Oreas* Mountain Dicran Moss

SPECIES *martiana*

GENUS *Oreoweisia* Mountain Weis Dicran Moss

SPECIES *brasiliensis, brevidens, brevifolia, chilensis, delgadilloi, erosa, laxifolia, laxiretis, mexicana, mulahacenii, setschwanica, subintegra, torquescens, tunariensis, weisioides*

GENUS *Orthodicranum* (Synonym for *Dicranum*)
GENUS *Parisia* Paris Dicran Moss
SPECIES *ciliata, laevipila, neocaledonica*
GENUS *Pilopogon* Downy Beard Dicran Moss
SPECIES *africanus, guadalupensis, laevis, longirostratus, macrocarpus, peruvianus, schilleri, tiquipayae*
GENUS *Platyneuron* Flat Nerve Dicran Moss
SPECIES *praealtum*
GENUS *Pleurophascum* Multiple Bundle Dicran Moss
SPECIES *grandiglobum, occidentale*
GENUS *Pocsiella* Pocs Dicran Moss
SPECIES *hydrogonioides*
GENUS *Polymerodon* Polymer Tooth Dicran Moss
SPECIES *andinus*
GENUS *Pseudephemerum* Ephemeral Dicran Moss
SPECIES *nitidum*
GENUS *Pseudochorisodontium* Floss Silk Tooth Dicran Moss
SPECIES *conanenum, gymnostomum, hokinense, mamillosum, ramosum, setschwanicum*
GENUS *Pseudohyophila* Hyophile Dicran Moss
SPECIES *peruviana*
GENUS *Rhabdoweisia* Rod Weis Dicran Moss
SPECIES *africana, crenulata, crispata, fugax*
GENUS *Schistomitrium* Split Mitre Dicran Moss
SPECIES *apiculatum, brevi-apiculatum, mucronifolium, robustum, sparei*
GENUS *Schliephackea* Schliephacke Dicran Moss
SPECIES *meteorioides, prostrata*
GENUS *Sclerodontium* Hard Tooth Dicran Moss
SPECIES *clavinerve, pallidum*
GENUS *Sorapilla* Soral Pill Dicran Moss
SPECIES *papuana, sprucei*
GENUS *Sphaerothecium* Round Theca Dicran Moss
SPECIES *phascoideum, reconditum, subchlorophyllosum*
GENUS *Steyermarkiella* Steyermark Dicran Moss
SPECIES *anomalodictya*
GENUS *Symblepharis* Symbol Eyelash Dicran Moss
SPECIES *fragilis, gracillima, guizhouensis, krausei, lindigii, oncophoroides, reinwardtii, sinensis, tenuis, vaginata*
GENUS *Thysanomitrion* Thysan Mitre Dicran Moss
SPECIES *leioneuron, miserum, wainioi*
GENUS *Viridivellus* Green Velvet Dicran Moss
SPECIES *pulchellum*

FAMILY 2 *Bruchiaceae* Bruch Dicran Moss
GENUS *Bruchia* Bruch Dicran Moss
SPECIES *aurea, bolanderi, brevifolia, brevipes, carinata, carolinae, drummondii, eckloniana, flexuosa, fusca, hallii, hampeana, microspora, paricutinensis, queenslandica, ravenelii, sinensis, texana, uruguensis, vogesiaca*
GENUS *Eobruchia* Lesser Bruch Dicran Moss
SPECIES *bruchioides, ecuatoriana*
GENUS *Trematodon* Tremor Tooth Bruch Dicran Moss
SPECIES *ambiguus, amoenus, angolensis, assamensis, aureus, baileyi, boasii, bolivianus, borbonicus, brachyphyllus, brevicalyx, brevicarpus, brevicollis, brevifolius, brevirostris, brevisetus, capillifolius, cheesemanii, conformis, congolensis, crispifolius, curvicollis, decaryi, dentatus, divaricatus, felipponei, fendleri, flexipes, geniculatus, gymnostomus, hakusanensis, heterophyllus, hildebrandtii, hookeri, humilis, intermedius, intermixtus, kurzii, lacunosus, laetevirens, latinervis, lato-obtusus, le-testui, longescens, longicollis, longifolius, lozanoi, ludovicae, mackayi, mayebarae, mayottensis, microthecius, minutulus, mirabilis, montanus, nitidulus, norrisii, novae-hannoverae, nudus, palettifolius, papuensis, paradoxus, pascuanus, pauperifolius, perssonorum, pillansii, platybasis, puteensis, rapaensis, reticulatus, schmidii, semitortidens, setaceus, subambiguus, suberectus, subulosus, tisserantii, tonkinensis, uruguensis, usambaricus, vaginatus, victoriae, viguieri*

FAMILY 3 *Dicnemonaceae* Dual Thick Thread Dicran Moss
GENUS *Dicnemon* Dual Thick Thread Dicran Moss
SPECIES *aneitense, calycinum, cochlearifolium, cuspidatum, dixonianum, novae-guineae, pancheri, planifolium, robbinsii, semicryptum, seriatum, spathoideum, subpiliferum*
GENUS *Eucamptodon* Field Tooth Dual Thick Thread Dicran Moss
SPECIES *muelleri, perichaetialis, scalarirete*
GENUS *Synodontia* Tooth Bearing Dual Thick Thread Dicran Moss
SPECIES *connivens*

FAMILY 4 *Ditrichaceae* Dual Hair Dicran Moss
GENUS *Ditrichum* Dual Hair Dicran Moss

SPECIES *ambiguum, apophysatum, atlanticum, aureum, bogotense, brachycarpum, brachypodum, brevidens, brevirostre, brevisetum, brotherusii, buchananii, capillare, colijnii, conicum, cornubicum, crinale, crispatissimum, cylindricarpum, darjeelingense, difficile, divaricatum, elatum, ferrugineum, flexicaule, francii, gemmiferum, gracile, hallei, heteromallum, hookeri, hyalinocuspidatum, hyalinum, immersum, itatiaiae, javense, laxissimum, levieri, lewis-smithii, liliputanum, lineare, longisetum, luteum, macrorhynchum, madagassum, montanum, pallidum, pancheri, paulense, perporodictyon, plagiacron, plumbicola, punctulatum, pusillum, rhynchostegium, roivanenii, rufo-aureum, schimperi, sekii, sericeum, spinulosum, strictum, subaustrale, subcapillaceum, submersum, subrufescens, subulatum, tenuinerve, tortipes, tortuloides, ulei, validinervium, zonatum*

GENUS *Astomiopsis* — Stomaless Dual Hair Dicran Moss
SPECIES *amblyocalyx, exserta, julacea, magilliana, pacifica, sinensis, subulata*

GENUS *Austrophilibertiella* — Southern Philbert Dual Hair Dicran Moss
SPECIES *ditrichoidea, nitens*

GENUS *Bryomanginia* — Mangini Dual Hair Dicran Moss
SPECIES *saint-pierrei*

GENUS *Ceratodon* — Horn Tooth Dual Hair Dicran Moss
SPECIES *amblyocalyx, bryophilus, capillisetus, conicus, crassinervis, delicatulus, elimbatus, emodii, heterophyllus, inclinatus, indicus, mexicanus, minor, moravicus, novo-granatensis, plano-marginatus, purpureus, semilunaris, strictifolius, valdiviae, venezuelensis, vulcanicus*

GENUS *Cheilothela* — Claw Nipple Dual Hair Dicran Moss
SPECIES *chloropus, jaegeri*

GENUS *Chrysoblastella* — Golden Blastid Dual Hair Dicran Moss
SPECIES *chilensis*

GENUS *Cladastomum* — Branch Stoma Dual Hair Dicran Moss
SPECIES *robustum, ulei*

GENUS *Cleistocarpidium* — Closed Carpel Dual Hair Dicran Moss
SPECIES *palustre*

GENUS *Crumuscus* — Crumby Musk Dual Hair Dicran Moss
SPECIES *vitalis*

GENUS *Cygniella* — Long Neck Dual Hair Dicran Moss
SPECIES *sharpii*

GENUS *Distichium* — Two Rank Dual Hair Dicran Moss
SPECIES *asperrimum, austroinclinatum, brachyphyllum, brachystegium, brevifolium, brevisetum, bryoxiphioidium, capillaceum, crispatum, hagenii, inclinatum, lorentzii, remotifolium, vernicosum*

GENUS *Ditrichopsis* — Lesser Dual Hair Dicran Moss
SPECIES *clausa, gymnostoma*

GENUS *Eccremidium* — Hanging Dual Hair Dicran Moss
SPECIES *arcuatum, brisbanicum, exiguum, floridanum, minutum, pulchellum*

GENUS *Garckea* — Garcke Dual Hair Dicran Moss
SPECIES *abbreviata, elongata, flexuosa, mathieui, moenkemeyeri*

GENUS *Pleuridium* — Pleural Dual Hair Dicran Moss
SPECIES *acuminatum, andinum, arnoldii, aurantiacum, brachycaulon, californicum, costesii, curvisetum, denticulatum, ecklonii, holdridgei, japonicum, laxirete, lindigianum, longirostre, macrothecium, nervosum, papillosum, pappeanum, ravenelii, subenerve, subnervosum, subulatum, sullivantii, tenue, valetonii, venezuelanum, viride*

GENUS *Pringleella* — Pringle Dual Hair Dicran Moss
SPECIES *pleuridioides, sinensis, subulata*

GENUS *Pseudoditrichum* — Least Dual Hair Dicran Moss
SPECIES *mirabile*

GENUS *Rhamphidium* — Beak Dual Hair Dicran Moss
SPECIES *borinquense, brasiliense, brevifolium, crispifolium, dicranoides, dixonii, fendleri, laetum, laticuspe, madurense, montanus, novoguineensis, ovale, purpuratum, pygmaeolum, vaginatum, veitchii*

GENUS *Saelania* — Saelani Dual Hair Dicran Moss
SPECIES *glaucescens*

GENUS *Skottsbergia* — Skottsberg Dual Hair Dicran Moss
SPECIES *paradoxa*

GENUS *Strombulidens* — Strombule Dual Hair Dicran Moss
SPECIES *longirostris*

GENUS *Trichodon* — Tooth Dual Hair Dicran Moss
SPECIES *cylindricus, muricatus*

GENUS *Tristichium* — Three Stitch Dual Hair Dicran Moss
SPECIES *apodum, lorentzii, mirabile, sinense*

GENUS *Wilsoniella* — Wilson Dual Hair Dicran Moss
SPECIES *blindioides, bornensis, crispidens, decipiens, flaccida, hampeana, jardinii, karsteniana, subvaginans, tonkinensis*

FAMILY 5 *Leucobryaceae* — White Dicran Moss
GENUS *Leucobryum* — White Dicran Moss

SPECIES *acutifolium, aduncum, albescens, albicans, albidum, aneitense, angustum, antillarum, arfakianum, argentinicum, babetii, ballinense, bistratosum, boivinianum, boninense, boryanum, bowringii, brevifolium, calycinum, cambouei, cameruniae, candidum, chevalieri, chlorophyllosum, clavatum, comorense, congolense, crispum, cruegerianum, cucullatum, flavo-mucronatum, fouta-djalloni, fragile, galinonii, giganteum, glaucum, goyazense, gracile, gunnii, heterodictyon, homalophyllum, humillimum, imbricatum, incurvifolium, isleanum, javense, juniperoideum, laminatum, letestui, leucophanoides, macro-falcatum, madagassum, martianum, mayottense, microcarpum, molliculum, neocaledonicum, pachyphyllum, parvulum, perangustum, perrotii, polakowskyi, pseudomadagassum, pungens, rehmannii, rugosum, sanctae-mariae, sanctum, scabrum, seemannii, sericeum, sordidum, sphagnoides, strictum, subandinum, subchlorophyllosum, subobtusifolium, sumatranum, tahitense, tumidum, turgidulum, uleanum, wattsii*

GENUS *Leucoloma* — Fringed White Dicran Moss

SPECIES *aduncum, albocinctum, allorgei, amblyacron, ambreanum, amoene-virens, annamense, bauerae, bifidum, boivinianum, breviflagellare, brevioperculatum, brotheri, caespitulans, caldense, calymperaceum, candidulum, candidum, capillifolium, chlorophyllum, chrysobasilare, cinclidiotioides, circinale, circinatulum, cirrosulum, crosbyi, cruegerianum, cuneifolium, decaryi, delicatulum, dichelymoides, ecaudatum, entabeniense, fontinaloides, fuscifolium, garnieri, gracilescens, grandidieri, grimmioides, guineense, herzogii, holstii, humbertii, incanum, incrassatum, isleanum, itatiaiense, latifolium, leichhardtii, lepervancheri, longifolium, madagascariense, mafatense, malabarense, mariei, marojenziense, membranaceum, mittenii, molle, mosenii, nitens, normandii, ochrobasilare, okamurae, oncophorellum, pallidulum, perrieri, persecundum, perviride, phumiengse, pobeguinii, procerum, pumilum, pusillum, pygmaeum, rutenbergii, sanctae-mariae, scaberulum, scabricuspis, schelpei, schwaneckeanum, secundifolium, sericeum, serraticuspis, serrulatum, setosum, seychellense, siamense, sieberi, sieberianum, sinuosulum, sprengelianum, squarrosulum, strictipilum, subbiplicatum, subcespitulans, subchrysobasilare, subimmarginatum, subpungens, subsecundifolium, subsetosum, syrrhopodontoides, talazaccii, tanganyikae, taylorii, tenerum, tenuifolium, theriotii, thraustum, thuretii, tortellum, triforme, tuberculosum, villaumei, zeyheri, zuluense*

GENUS *Octoblepharum* — Octal Eyelash White Dicran Moss

SPECIES *africanum, albidum, ampullaceum, brevisetum, cocuiense, costatum, cylindricum, depressum, erectifolium, exiguum, leptoneuron, leucobryoides, pulvinatum, rhaphidostegium, stramineum, tatei*

GENUS *Paraleucobryum* — Lesser Fringed White Dicran Moss

SPECIES *enerve, longifolium, sauteri*

FAMILY 6 *Phyllodrepaniaceae* — Creeping Phyllode Dicran Moss

GENUS *Phyllodrepanium* — Creeping Phyllode Dicran Moss

SPECIES *falcifolium*

GENUS *Mniomalia* — Lead Bad Creeping Phyllode Dicran Moss

SPECIES *semilimbata, viridis*

ORDER 3 *Funariales* — Cord Moss

FAMILY 1 *Funariaceae* — Cord Moss

GENUS *Funaria* — Cord Moss

SPECIES *acicularis, aequidens, altissima, annulata, anomala, arctica, aristatula, berteroana, beyrichii, bogosica, bullata, cameruniae, capillaris, capillipes, chevalieri, chilensis, clavellata, commixta, convoluta, costesii, cuspidata, decaryi, delicatula, discelioides, eberhardtii, erectiuscula, euryloma, excurrentinervis, flava, flavicans, flaviseta, fritzei, fuegiana, fuscescens, grossidens, groutiana, hosseusii, hygrometrica, hygrometrica, imerinensis, incompleta, incurvifolia, inflata, japonica, kilimandscharica, koelzei, laxissima, leptopoda, linearidens, ludoviciae, luteo-limbata, macrocarpa, macrospora, maireana, mayottensis, meeseacea, mexicana, microstoma, minuticaulis, nepalensis, nubica, orizabensis, orthocarpa, orthopoda, perlaxa, pilifera, polaris, porteri, pulchella, pulchra, pulchricolor, ramulosa, renauldii, rhomboidea, riparia, robustior, saharae, sartorii, schnyderi, sickenbergeri, sinuato-limbata, sovatensis, subimmarginata, subleptopoda, submarginata, subplanifolia, succuleata, tenella, valdiviae, varia, wijkii*

GENUS *Aphanorrhegma* — Inconspicuous Fractured Cord Moss

SPECIES *serratum*

GENUS *Brachymeniopsis* — Short Meny Cord Moss

SPECIES *gymnostoma*

GENUS *Bryobeckettia* — Beckett Cord Moss

SPECIES *bartlettii*

GENUS *Corynotheca* — Club Theca Cord Moss

SPECIES *poeltii*

GENUS *Cygnicollum* — Long Neck Cord Moss

SPECIES *immersum*

GENUS *Discelium* — Disc Cord Moss

SPECIES *nudum*

GENUS *Entosthodon* — Entic Osthe Tooth Cord Moss

SPECIES *acaulis, acidotus, americanus, andicola, angustifolius, apiculatopilosus, apophysatus, aristatus, attenuatus, balansae, beccarii, bergianus, bolanderi, bonplandii, borbonicus, borneensis, buseanus, californicus, cartilagineus, chiloensis, clavatus, curviapiculatus, deserticola, diversinervis, dixonii, dozyanus, drummondii, duriaei, fascicularis, flexisetus, handelii, helmsii, hildebrandtii, holstii, hungaricus, italicus, jamesonii, kochii, krausei, laevis, latifolius, laxus, lepervanchei, limbatus, lindigii, marginatulus, mathewsii, mauritianus, mouretii, muhlenbergii, obtusatus, obtusifolius, obtuso-apiculatus, obtusus, ouropratensis, papillosus, perrottetii, pertenella, physcomitrioides, plagiothecius, planifolius, planoconvexus, productus, puiggarii, radians, rottleri, rubiginosus, rubrisetus, schimperi, serratus, smithhurstii, sonorae, spathulatus, subintegrus, subnudus, subpallescens, tucsoni, urceolatus, usambaricus, volkensii, wichurae, wigginsii*

GENUS *Funariella* — Lesser Cord Moss
SPECIES *curviseta*
GENUS *Goniomitrium* — Angled Mitre Cord Moss
SPECIES *acuminatum, africanum, seroi, speluncae*
GENUS *Loiseaubryum* — Louise Aubry Cord Moss
SPECIES *nutans*
GENUS *Nanomitriella* — Nannite Mitre Cord Moss
SPECIES *ciliata*
GENUS *Physcomitrella* — Lesser Physce Mitre Cord Moss
SPECIES *patens, readeri*
GENUS *Physcomitrellopsis* — Least Physce Mitre Cord Moss
SPECIES *africana*
GENUS *Physcomitrium* — Physce Mitre Cord Moss
SPECIES *acutifolium, arenicola, argentinicum, aubertii, badium, brevirostre, brevisetum, brisbanicum, bukobense, chlorodictyon, coarctatum, collenchymatum, coorgense, courtoisii, cupulare, cupuliferum, delicatulum, dilatatum, eurystomum, falcifolium, felipponei, flaccidum, flexifolium, germanillae, gibertii, hemisphaericum, herteri, hookeri, immersum, indicum, insigne, integrifolium, japonicum, lindmannii, longicolle, lorentzii, luteolum, mahui, martianovii, mirabile, niloticum, nodulifolium, orbignyanum, paraguense, perflaccidum, platyphylloides, puiggarii, pusillum, pygmaeum, pyriforme, repandum, rooperi, roseae, sahapurense, serrifolium, serrulatum, sinensi-sphaericum, spathulatum, sphaericum, subminutulum, subspathulatum, subsphaericum, sumatranum, sylvestre, thieleanum, umbonatum*
GENUS *Pyramidula* — Pyramid Cord Moss
SPECIES *tetragona*
GENUS *Steppomitra* — Steppe Mitre Cord Moss
SPECIES *hadacii*

FAMILY 2 *Ephemeraceae* — Ephemeral Cord Moss
GENUS *Ephemerum* — Ephemeral Cord Moss
SPECIES *aequinoctiale, apiculatum, argentinicum, asiaticum, capense, cohaerens, conicum, crassinervium, crassinervium, cristatum, diversifolium, fimbriatum, furcatum, homomallum, jamesii, ligulatum, longifolium, longifolium, namaquense, pachyneuron, papillosum, pechuelii, perminutum, piliferum, recurvifolium, rehmannii, rutheanum, serratum, sessile, spinulosum, stellatum, uleanum*
GENUS *Ephemeridium* — Lesser Ephemeral Cord Moss
SPECIES *papillosum*
GENUS *Micromitrium* — Mitre Ephemeral Cord Moss
SPECIES *densifolium, lacustre, megalosporum, perexiguum, subaequinoctale, synoicum, tenerum, thelephorothecum, wrightii*
GENUS *Nanomitrium* — (Synonym for *Micromitrium*)

FAMILY 3 *Gigaspermaceae* — Giant Spore Cord Moss
GENUS *Gigaspermum* — Giant Spore Cord Moss
SPECIES *mouretii, repens*
GENUS *Chamaebryum* — Chamise Giant Spore Cord Moss
SPECIES *pottioides*
GENUS *Costesia* — Costes Giant Spore Cord Moss
SPECIES *spongiosa*
GENUS *Lorentziella* — Lorentz Giant Spore Cord Moss
SPECIES *imbricata, uruguensis*
GENUS *Neosharpiella* — Modern Sharp Giant Spore Cord Moss
SPECIES *aztecorum, turgida*
GENUS *Oedipodiella* — Lesser Riddle Pod Cord Moss
SPECIES *australis*
GENUS *Oedipodium* — Riddle Pod Cord Moss
SPECIES *griffithianum*

FAMILY 4 *Splachnaceae* — Visceral Cord Moss
GENUS *Splachnum* — Visceral Cord Moss
SPECIES *adolphi-friederici, ampullaceum, austriacum, luteum, melanocaulon, pensylvanicum, resectum, rubrum, sphaericum, vasculosum, weberbaueri*
GENUS *Aplodon* — Aple Tooth Visceral Cord Moss
SPECIES *wormskioldii*
GENUS *Brachymitrion* — Short Mitre Visceral Cord Moss
SPECIES *cochabambae, immersum, jamesonii, laciniatum, moritzianum, pocsii*
GENUS *Kopononobryum* — Koponon Visceral Cord Moss

SPECIES *bengalense*
GENUS *Moseniella* — Mosen Visceral Cord Moss
SPECIES *brasiliensis, ulei*
GENUS *Splachnobryum* — Lesser Visceral Cord Moss
SPECIES *aquaticum, assamicum, crassinervium, gracile, limbatum, novae-guineae, obtusum, oorschotii, valdiviae, wiemansii*
GENUS *Tayloria* — Taylor Visceral Cord Moss
SPECIES *acuminata, alpicola, altorum, arenaria, borneensis, braithwaitiana, callophylla, chiapensis, dubyi, froelichiana, gunnii, hornschuchii, indica, isleana, jacquemontii, kilimandscharica, lingulata, longiseta, magellanica, mirabilis, nepalensis, novo-guinensis, octoblepharum, orthodonta, pseudoalpicola, purpurascens, recurvimarginata, reinerii, rubicaulis, rudimenta, rudolphiana, sandwicensis, scabriseta, serrata, solitaria, splachnoides, squarrosa, stenophysata, subglabra, tasmanica*
GENUS *Tetraplodon* — Tetrad Aple Tooth Visceral Cord Moss
SPECIES *angustatus, blyttii, caulescens, itatiaiae, mnioides, pallidus, paradoxus, tomentosus, urceolatus*
GENUS *Voitia* — Voit Visceral Cord Moss
SPECIES *grandis, hyperborea, nivalis*

ORDER 4 *Grimmiales* — Grimm Moss

FAMILY 1 *Grimmiaceae* — Grimm Moss
GENUS *Grimmia* — Grimm Moss
SPECIES *abyssinica, afroincurva, agassizii, ahmadiana, alpestris, alpicola, americana, angusta, anodon, anomala, apocarpa, apocarpa, arctica, arcuatifolia, arenaria, argyrotricha, atrata, australis, austrofunalis, bicolor, brachydictyon, brownii, caespiticia, camonia, campylotricha, compactula, crinita, crinitoleucophaea, curviseta, cyathocarpa, decipiens, dissimulata, donniana, elatior, elongata, erythraea, exquisita, funalis, fuscolutea, gebhardii, glauca, gracilis, grisea, haleakalae, hamulosa, handelii, hartmanii, humilis, immergens, incrassicapsulis, incurva, indica, involucrata, kansuana, khasiana, kidderi, laevidens, laevigata, latifolia, lawiana, leibergii, limbatula, lisae, longicapusula, longicaulis, longirostris, macroperichaetialis, macrotheca, madagassa, mairei, mammosa, mariniana, maritima, meridionalis, mesopotamica, mexicana, mixta, molesta, mollis, montana, muehlenbeckii, navicularis, nepalensis, nevadensis, nivalis, novae-zeelandiae, obtusata, obtusolinealis, ochyriana, olneyi, orbicularis, ovalis, paramattensis, percarinata, pilifera, pilosissima, plagiopodia, pseudoanodon, pulla, pulvinata, pulvinatula, pygmaea, ramondii, reflexidens, rigidissima, saxatilis, serrana, somervellii, stenobasis, stenophylla, stolonifera, subcallosa, subleucophaea, subtergestina, teretinervis, tergestina, torngakiana, torquata, tortuosa, trichophylla, triformis, trinervis, ungeri, unicolor, verticillatula, vulcanica, wilsonii*
GENUS *Aligrimmia* — Ali Grimm Moss
SPECIES *peruviana*
GENUS *Coscinodon* — Coscine Tooth Grimm Moss
SPECIES *arctolimnius, bolivianus, calyptratus, cribrosus, hartzii, humilis, patersonii, pilifer, yukonensis*
GENUS *Coscinodontella* — Lesser Coscine Tooth Grimm Moss
SPECIES *bryanii*
GENUS *Dryptodon* — Tree Wing Tooth Grimm Moss
SPECIES *obtusus, patens*
GENUS *Hydrogrimmia* — (Synonym for *Grimmia*)
GENUS *Indusiella* — Indus Grimm Moss
SPECIES *thianschanica*
GENUS *Jaffueliobryum* — Jaffuel Grimm Moss
SPECIES *arsenei, raui, williamsii, wrightii*
GENUS *Leucoperichaetium* — Light Bristle Fringed Grimm Moss
SPECIES *eremophilum*
GENUS *Racomitrium* — Racon Mitre Grimm Moss
SPECIES *aciculare, aduncoides, affine, afoninae, alare, albipiliferum, angustifolium, aquaticum, barbuloides, bartramii, brachypus, brevipes, brevisetum, canescens, carinatum, crispipilum, crispulum, crumianum, cucullatifolium, cucullatulum, curiosissimum, decurrens, depressum, dichelymoides, didymum, ellipticum, elongatum, emersum, ericoides, fasciculare, fuscescens, gracillimum, grimmioides, hespericum, heterostichum, himalayanum, japonicum, joseph-hookeri, kerguelense, laetum, laevigatum, lamprocarpum, lanuginosum, lawtonae, lepervanchei, lusitanicum, macounii, marginatum, membranaceum, microcarpum, minutum, muticum, nigritum, nitidulum, norrisii, obesum, obtusum, occidentale, ochraceum, orthotrichaceum, pachydictyon, pacificum, panschii, papeetense, patagonicum, pruinosum, pseudopatens, ptychophyllum, pygmaeum, rupestre, ryszardii, seychellarum, steerei, subrupestre, subsecundum, sudeticum, sulcipilum, valdon-smithii, varium, venustum, verrucosum, visnadiae, vulcanicola, vulcanicum*
GENUS *Schistidium* — Schist Grimm Moss
SPECIES *agassizii, alpicola, amblyophyllum, andreaeopsis, angustissimum, angustum, antarctici, apocarpum, celatum, chenii, chrysoneurum, cinclidodonteum, confusum, crassipilum, cribrodontium, cryptocarpum, cupulare, donatii, elegantulum, falcatum, fallax, flaccidum, frigidum, frisvollianum, gracile, grandirete, halinae, heterophyllum, hyalinocuspidatum, liliputanum, lineare, lorentzianum, maritimum, nodulosum, obtusifolium, occidentale, occultum, olivaceum, perichaetiale, poeltii, pulchrum, readeri, recurvum,*

rivulare, robustum, scabripes, scandicum, serratomucronatum, squamatulum, steerei, stirlingii, streptophyllum, subconfertum, subjulaceum, submuticum, subpraemorsum, syntrichiaceum, tenerum, torreyanum, trichodon, truncatoapocarpum, urnulaceum, venetum, yaulense

GENUS *Scouleria* — Scouler Grimm Moss

SPECIES *aquatica, marginata, patagonica*

GENUS *Tridontium* — Three Tooth Grimm Moss

SPECIES *tasmanicum*

FAMILY 2 *Ptychomitriaceae* — Folded Mitre Grimm Moss

GENUS *Ptychomitrium* — Folded Mitre Grimm Moss

SPECIES *aligrimmioides, angusticarpum, australe, barrii, chimborazense, crassinervium, crispatum, cucullatifolium, deltorii, dentatum, diexaratum, drummondii, exaratifolium, fauriei, fernandesianum, fluviatile, formosicum, gardneri, helenicum, hieronymi, incurvum, indicum, isoskelos, kiusiuense, laxifolium, lepidomitrium, ligulatum, lindmanii, linearifolium, longisetum, mamillosum, mittenii, mucronatum, muelleri, nigrescens, obtusifolium, papillosum, patens, polyphyllum, sellowianum, serratum, sinense, standleyi, subcrispatum, subcylindricum, subdentatum, tortula, vaginatum, viride, wilsonii, yulongshanum*

GENUS *Campylostelium* — Field Stele Folded Miter Grimm Moss

SPECIES *brachycarpum, pitardii, saxicola, strictum, venezuelanum*

GENUS *Glyphomitrium* — Glyph Folded Miter Grimm Moss

SPECIES *acuminatum, calycinum, crispifolium, daviesii, elatum, formosanum, humillimum, hunanense, minutissimum, warburgii*

GENUS *Ptychomitriopsis* — Lesser Folded Miter Grimm Moss

SPECIES *africana, aloinoides*

ORDER 5 *Hookeriales* — Hooker Moss

FAMILY 1 *Hookeriaceae* — Hooker Moss

GENUS *Hookeria* — Hooker Moss

SPECIES *acutifolia, commutata, contracta, dussii, janeirensis, lucens, magellanica, orbignyana, uliginosa*

GENUS *Achrophyllum* — Achromic Hooker Moss

SPECIES *anomalum, crassirete, dentatum, haesselianum, javense, magellanicum, quadrifarium, tenuinerve, bogotense*

GENUS *Adelothecium* — Adel Theca Hooker Moss

SPECIES *bogotense*

GENUS *Beeveria* — Beever Hooker Moss

SPECIES *distichophylloides*

GENUS *Brymela* — Honey Hooker Moss

SPECIES *acuminata, callicostelloides, cuspidata, fissidentoides, fluminensis, obtusifolia, parkeriana, rugulosa, tutezona, websteri*

GENUS *Bryobrothera* — Brother Hooker Moss

SPECIES *crenulata*

GENUS *Calyptrochaeta* — Calypter Bristle Hooker Moss

SPECIES *albescens, apiculata, asplenioides, boniana, brassii, brownii, cristata, deflexa, enervis, flexicollis, grandiretis, haitensis, isophylla, japonica, leptoloma, lucida, marginata, microblasta, mniadelphus, mollis, nutans, odontoloma, otwayensis, parviretis, pocsii, ramosa, remotifolia, rotundifolia, setigera*

GENUS *Canalohypopterygium* — Canal Needle Wing Hooker Moss

SPECIES *tamariscinum*

GENUS *Chaetomitriopsis* — Lesser Bristle Mitre Hooker Moss

SPECIES *glaucocarpa*

GENUS *Chaetomitrium* — Bristle Mitre Hooker Moss

SPECIES *acanthocarpum, aneitense, auriculatum, beccarii, borbonicum, borneense, brassii, callichrous, ciliatum, comorense, confertum, ctenidioides, darnaedii, densum, depressum, dusenii, elegans, elmeri, elongatum, fimbriatum, friedense, frondosum, hebridense, horridulum, integrifolium, laevifolium, laevisetum, lanceolatum, lancifolium, lauterbachii, leptopoma, macrohystrix, madangense, maryatii, nanohystrix, nervosum, orthorrhynchum, paleatum, papillifolium, perakense, perarmatum, perlaeve, philippinense, poecilophyllum, pseudoelongatum, recurvifolium, robbinsii, roemeri, schofieldii, seramense, setosum, sikkimense, smithii, speciosum, spinosum, sublaevisetum, tahitense, torquescens, volutum, vrieseanum, warburgii, weberi, werneri, wheeleri*

GENUS *Chaetophora* — Bristle Hooker Moss

SPECIES *perrinii*

GENUS *Crosbya* — Crosby Hooker Moss

SPECIES *nervosa, straminea*

GENUS *Curviramea* — Curved Branch Hooker Moss

SPECIES *mexicana*

GENUS *Cyathophorella* — Lesser Cyath Hooker Moss

SPECIES *adianthoides, aoyagii, densifolia, doii, grandistipulacea, japonica, philippinense*

GENUS *Cyathophorum* — Cyath Hooker Moss

SPECIES *densirete, japonicum, minus, philippinense, adiantum, africanum, bulbosum, hookerianum, parvifolium, spinosum, tahitense*

GENUS *Dendrocyathophorum* Wood Cyath Hooker Moss

SPECIES *decolyi*

GENUS *Dendrohypopterygium* Wood Needle Wing Hooker Moss

SPECIES *arbuscula, filiculiforme*

GENUS *Dimorphocladon* Dimorphic Branch Hooker Moss

SPECIES borneense

GENUS *Diploneuron* Diploid Nerve Hooker Moss

SPECIES *connivens*

GENUS *Distichophyllum* Two Rank Hooker Moss

SPECIES *aciphyllum, albomarginatum, angustifolium, angustissimum, apiculigerum, assimile, borneense, brevicuspis, capillatum, carinatum, catinifolium, cavaleriei, ceylanicum, cirratum, collenchymatosum, crispulum, cucullatum, cuspidatum, decolyi, denticulatum, dicksonii, dixonii, ellipticum, eremitae, evanidolimbatum, fasciculatum, fernandezianum, flaccidum, flavescens, fossombronioides, francii, freycinetii, fuegianum, gracile, gracilicaule, graeffeanum, grandifolium, griffithii, heterophyllum, humifusum, imbricatum, integerrimum, jungermannioides, kinabaluense, koghiense, krausei, leiopogon, leskeodontoides, limpidum, lingulatum, lixii, longicuspis, longipes, longobasis, lorianum, macropodum, madurense, maibarae, malayense, meizhiae, microcarpum, microcladum, minutum, mittenii, mniifolium, montagneanum, nadeaudii, nanospathulatum, nidulans, nigricaule, noguchianum, oblongum, obovatum, obtusifolium, osterwaldii, paradoxum, patagonicum, procumbens, pseudomalayense, pulchellum, pullei, rakotomariae, rigidicaule, rotundifolium, samoanum, santosii, scabrisetum, schmidtii, semimarginatum, sinense, spathulatum, stipitatifolium, subcuspidatum, subelimbatum, submersum, submucronatum, subnigricaule, succulentum, telmaphila, theriotianum, torquatifolium, tortile, turgidum, ulukahiense, undulatum, vitianum, wanianum*

GENUS *Elharveya* Harvey Hooker Moss

SPECIES *visicularioides*

GENUS *Ephemeropsis* Ephemeral Hooker Moss

SPECIES *tjibodensis, trentepohlioides*

GENUS *Eriopus* Woolly Hooker Moss

SPECIES *flexicaulis, lorifolius, pacificus, papillatus, perlimbatus, subremotifolius*

GENUS *Hemiragis* Half Rag Hooker Moss

SPECIES *aurea*

GENUS *Hookeriopsis* Lesser Hooker Moss

SPECIES *ambigua, angustiretis, armata, balazsii, bartramii, beyrichiana, borneensis, bowersiana, brachypelma, brunneophylla, cavifolia, corcovadensis, crosbyi, cuspidatissima, diversifolia, exesa, exigua, gabonensis, heterophylla, hornschuchiana, hydrophila, latifrondea, leiophylla, lepidopiloides, leucomioides, levieri, lonchopelma, luteorufescens, minutiretis, mittenii, negrensis, pachydictyon, pallidifolia, papillidioides, papillosula, perfulva, pocsii, ptari-tepuiensis, puiggarii, rhynchostegioides, rubens, saxicola, serrata, standleyi, staudtii, steyermarkii, subscabrella, thwaitesiana, venezuelensis, vesicularia, wichurae*

GENUS *Hypopterygium* Needle Wing Hooker Moss

SPECIES *didictyon, discolor, elatum, flavolimbatum, lutescens, nematosum, nivale, polythrix, rotulatum, sandwicense, sasaokae, speciosum, sylvaticum, tamarisci, thouinii, vriesei, wolfffhuegelii*

GENUS *Lepidopilidium* Thin Downy Hooker Moss

SPECIES *attenuatum, aureo-purpureum, brevisetum, brunneoleum, caudicaule, cespitosum, chenagonii, corbieri, cyrtostegium, devexum, divaricatum, entodontella, flexuosum, furcatum, gracilifrons, hanningtonii, isleanum, laevisetum, nitens, parvulum, plebejum, portoricense, purpurissatum, regnellii, rupestre, subdevexum, subdivaricatum, subrevolutum, theriotii, wainioi*

GENUS *Lopidium* Lope Hooker Moss

SPECIES *concinnum, nematosum, struthiopteris*

GENUS *Metadistichophyllum* Meta Two Rank Hooker Moss

SPECIES *rhizophorum*

GENUS *Philophyllum* Lesser Downy Hooker Moss

SPECIES *tenuifolium*

GENUS *Pilosium* Downy Hooker Moss

SPECIES *chlorophyllum*

GENUS *Pilotrichidium* Downy Hairy Hooker Moss

SPECIES *antillarum, callicostatum*

GENUS *Pterygophyllum* Wing Hooker Moss

SPECIES *balantii, chonoticum, flavum, fragile, jungermannioides, splachnifolium*

GENUS *Rhynchostegiopsis* Beak Stegid Hooker Moss

SPECIES *brasiliensis, carolae, costaricensis, flexuosa, tunguraguana*

GENUS *Sauloma* Saul Hooker Moss

SPECIES *tenella, tenuis*

GENUS *Schimperobryum* Schimper Hooker Moss

SPECIES *splendidissimum*

GENUS *Sclerohypnum* Hard Hypnic Hooker Moss

SPECIES *littorale*

GENUS *Stenodictyon* Thin Net Hooker Moss
SPECIES *bisodalense, pallidum, wrightii*
GENUS *Tetrastichium* Tetrad Stitched Hooker Moss
SPECIES *fontanum, virens*
GENUS *Thamniopsis* Thamnid Hooker Moss
SPECIES *cheiloneura, cruegeriana, diffusa, incurva, killipii, langsdorffii, pendula, purpureophylla, sinuata, stenodictyon, terrestris, undata, utacamundiana*
GENUS *Vesiculariopsis* Vesicle Hooker Moss
SPECIES *spirifolium*

FAMILY 2 *Callicostellaceae* Calico Star Hooker Moss
GENUS *Callicostella* Calico Star Hooker Moss
SPECIES *africana, aiomensis, apophysata, aquatica, armata, ascensionis, attenuata, beccariana, belangeriana, bernoullii, bisexualis, brevipes, caledonica, campbelliana, chevalieri, chionophylla, chlorina, chloroneura, ciliata, circinata, colombica, constricta, cruegeri, daltoniaecarpa, depressa, diatomophila, disticha, eberhardtiana, emarginatula, erosotruncata, fallax, fissidentella, frateri, gabonensis, galipanoana, guatemalensis, heterophylla, hondurensis, integrifolia, irrorata, jungermannioides, juruensis, kaernbachii, kreaniana, lacerans, laeviuscula, leptocladula, limosa, loriae, maclaudii, martiana, melanotheca, merkelii, mexicana, microcarpa, monofaria, mosenii, nukahivensis, oblongifolia, oerstediana, pallida, paludicola, papillata, papillosula, parvicellulata, paulensis, paupera, pellucida, perpallida, perpapillata, perrotii, pilotrichidioides, prabaktiana, pterygophylloides, pusilla, rivularis, rufescens, rugiseta, salaziae, saxatilis, scaberrima, scabripes, scabriseta, scabriuscula, sellowiana, seychellensis, spurio-pallida, strumulosa, subdepressa, subemarginatula, submicrocarpa, submonofaria, subsecunda, tenerrima, torrentium, tristis, usambarica, vesiculata, virens*
GENUS *Callicostellopsis* Lesser Calico Star Hooker Moss
SPECIES *meridiensis*
GENUS *Cyclodictyon* Cyclic Net Calico Star Hooker Moss
SPECIES *aciculifolium, aeruginosum, albatum, albicans, albicaule, allionii, amnigenum, angustirete, arsenei, aubertii, bakeri, benoistii, bescherellei, bicolor, bidentatum, blumeanum, bombonasicum, borbonicum, breve, brevifolium, brittonae, caespitosum, castaneum, chimborazense, chloroleucum, crassicaule, cuspidatum, delicatum, dixonianum, erubescens, fendleri, filicuspis, flexicuspis, glareosum, glaucifolium, hildebrandtii, humectatum, humile, immersum, iporangeanum, jagianum, kinabaluense, krauseanum, krebedjense, laete-virens, latifolium, laxifolium, lebrunii, leucomitrium, limbatum, longifrons, lorentzii, marginatum, maxonii, minarum, minus, mittenii, molliculum, nivale, obscurifolium, obscurum, olfersianum, pandurifolium, pergracile, perlimbatum, perrottetii, plicatulum, preussii, pusillum, recognitum, regnellianum, regnellii, richardsii, rivale, roridum, rubrisetum, shillicatense, spectabile, stephanii, subbrevifolium, sublimbatum, submarginatum, subobtusifolium, subtortifolium, tocoraniense, vallis-gratiae, varians, vesiculosum, viridissimum*
GENUS *Schizomitrium* Split Mitre Calico Star Hooker Moss
SPECIES *cirrhosum*
GENUS *Trachyxiphium* Thick Xiph Calico Star Hooker Moss
SPECIES *aduncum, glanduliferum, guadalupense, heteroicum, hypnaceum, pernutans, steerei, subfalcatum, tenue, vagum, variabile*

FAMILY 3 *Daltoniaceae* Dalton Hooker Moss
GENUS *Daltonia* Dalton Hooker Moss
SPECIES *androgyna, angustifolia, apiculata, aristata, aristifolia, armata, baldwinii, baumgartneri, bilimbata, brasiliensis, brevinervis, brevipedunculata, cardotii, constricta, contorta, cucullata, decolyi, dusenii, euryloma, forsythii, gemmipara, gomezii, gracilis, himalayensis, jamesonii, latimarginata, latolimbata, lindigiana, longicuspidata, longifolia, longinervis, macgregorii, macrotheca, marginata, mildbraedii, minor, minuta, mittenii, onraedtii, ovalis, pellucida, peruviana, plicata, pseudostenophylla, pulvinata, reticulata, robbinsii, rufescens, schiffneri, semitorta, sphaerica, splachnoides, stenoloma, stenophylla, subangustifolia, subapiculata, tenuifolia, tortifolia, trachyodonta, tristaniensis, tuberculosa*
GENUS *Actinodontium* Ray Tooth Dalton Hooker Moss
SPECIES *adscendens, dusenii, integrifolium, pygmaeum, rhaphidostegum, sprucei, standleyi, streptopogoneum*
GENUS *Amblytropis* Blunt Dalton Hooker Moss
SPECIES *gemmacea, hispidula, ovata, setosa*
GENUS *Benitotania* Benito Tan Dalton Hooker Moss
SPECIES *elimbata*
GENUS *Catharinea* China Dalton Hooker Moss
SPECIES *anomala, hirtella, lateralis, minuta, obtusula, schimperi, undulatiformis, xanthopoda, yunnanensis*
GENUS *Catharomnion* China Omni Dalton Hooker Moss
SPECIES *ciliatum*
GENUS *Crossomitrium* Cross Mitre Dalton Hooker Moss
SPECIES *acuminatum, epiphyllum, goebelii, patrisiae, phragmidiaceum, saprophilum, scabrisetum, sintenisii, tenellum, wallisii*
GENUS *Distichophyllidium* Two Rank Dalton Hooker Moss
SPECIES *africanum, antarense, jungermanniaceum, muticum, nymanianum*
GENUS *Helicoblepharum* Helix Eyelash Dalton Hooker Moss
SPECIES *brasiliense, daltoniaceum, fuscidulum, venustum*

GENUS *Lepidopilum* Thin Downy Dalton Hooker Moss

SPECIES *affine, amplirete, angustifrons, apollinairei, apophysatum, arcuatum, armatum, aurifolium, ballivianii, brevifolium, brevipes, caudatum, caviusculum, chloroneuron, convallium, crassisetum, curvifolium, curvirameum, cuspidans, diaphanum, erectiusculum, falcatulum, filiferum, filosum, flavescens, floresianum, frondosum, glaziovii, grevilleanum, hirsutum, inflexum, krauseanum, lastii, leiomitrium, leucomioides, longifolium, maculatum, mniaceum, mosenii, muelleri, nitidum, niveum, novae-guineae, ovalifolium, pallido-nitens, pectinatum, pergracile, permarginatum, phyllophilum, polytrichoides, rupestre, scabrisetum, stillicidiorum, subsubulatum, subulatum, surinamense, tenuifolium, tenuissimum, tortifolium, verrucipes, wallisii*

GENUS *Leskeodon* Leske Tooth Dalton Hooker Moss

SPECIES *acuminatus, andicola, aristatus, auratus, brevicuspidatus, caducifolius, cubensis, densiretis, dussii, longicaulis, mexicanus, minusculus, paisa, palmarum, pandurifolius, parvifolius, philippinensis, ponapensis, seramensis, wallisii*

GENUS *Leskeodontopsis* Lesser Leske Tooth Dalton Hooker Moss

SPECIES *pustulata*

GENUS *Pilotrichum* Downy Hair Dalton Hooker Moss

SPECIES *affine, andersonii, armatum, asperifolium, bipinnatum, compositum, corrugatum, cristatum, debile, denticulatum, divaricatum, evanescens, fendleri, husnotii, hypnoides, longicaule, lophophyllum, luciae, microthamnium, procerum, ramosissimum, rugifolium*

GENUS *Stenodesmus* Thin Fiber Dalton Hooker Moss

SPECIES *tenuicuspis*

GENUS *Symphyodon* Symphyte Tooth Dalton Hooker Moss

SPECIES *angustus, asper, complanatus, copelandii, echinatus, erinaceus, erraticus, gollanioides, imbricatifolius, oblongifolius, orientalis, perrottetii, pygmaeus, scaber, scabrisetus, weymouthioides, yuennanensis*

ORDER 6 *Hypnales* Hypne Moss

FAMILY 01 *Hypnaceae* Hypne Moss

GENUS *Hypnum* Hypne Moss

SPECIES *aemulans, alternans, aluminicola, amabile, andoi, austrostramineum, bambergeri, bicolor, bizotii, boryanum, brachythecium, calcicola, callichroum, campoanum, caperatum, cariosum, cataractarum, celatum, chrysogaster, circinale, cupressiforme, curvifolium, deflectens, densirameum, dieckei, eccremocarpum, emodi-fertile, fauriei, fertile, formianum, franchetti, freuchenianum, fuegianum, fujiyamae, gebhardii, gelidum, geminum, gracilirameum, hampei, hamulosum, hercynicum, heseleri, hildebrandtii, holmenii, imponens, imponentiforme, inflectens, intortoplicatum, jutlandicum, kushakuense, lamarckii, lutescens, macrogynum, moenkemeyeri, mukahivense, napoanum, nelsonii, nervosum, obsoletinerve, oldhamii, pallescens, patulum, pensilvanicum, plicaefolium, plicatulum, plumaeforme, pratense, procerrimum, pseudosplendens, punae, radiatum, recurvatum, resupinatum, revolutum, rigidum, riparioides, rotula, rubricaule, saitoi, sakuraii, sauteri, saxetorum, schiedeanum, setschwanicum, sikkimense, siphotheca, skottsbergii, speciosum, spegazzinii, subchrysogaster, subimponens, submolluscum, subpinnatum, subreptile, thouinii, tristoviride, uii, uncinulatum, usagarum, vaucheri, vernicosum, wilhelmense*

GENUS *Andoa* Ando Hypne Moss

SPECIES *berthelotiana*

GENUS *Breidleria* Breidler Hypne Moss

SPECIES *erectiuscula*

GENUS *Bryocrumia* Crum Hypne Moss

SPECIES *vivicolor*

GENUS *Bryosedgwickia* Sedgwick Hypne Moss

SPECIES *barlassinae*

GENUS *Buckiella* Buck Hypne Moss

SPECIES *draytonii, undulata*

GENUS *Callicladium* Calico Branch Hypne Moss

SPECIES *haldanianum*

GENUS *Caribaeohypnum* Caribe Hypne Moss

SPECIES *polypterum*

GENUS *Catagonium* Cat Angled Hypne Moss

SPECIES *brevicaudatum, complanatum, emarginatum, myurum, nitens, nitidum, serrulatum*

GENUS *Chrysohypnum* Golden Hypne Moss

SPECIES *camptorhynchum, cavifolium, demaretii, diminutivum, elegantulum, expallescens, frondosum, patens, salleanum, simorrhynchum, squarrosulum*

GENUS *Crepidophyllum* Sandel Hypne Moss

SPECIES *modestum*

GENUS *Ctenidiadelphus* Comb Adelphi Hypne Moss

SPECIES *plumularia, spinulosus*

GENUS *Ctenidium* Comb Hypne Moss

SPECIES *andoi, capillifolium, ceylanicum, distinguendum, elegantulum, floribundarioides, hastile, homalophyllum, hondurense, luzonense, lychnites, malacobolum, malacodes, molluscum, novoguineense, percrassum, pinnatum, plumicaule, polychaetum, pubescens, pulchellum, schofieldii, serratifolium, stellulatum*

GENUS *Cyathothecium* Cup Theca Hypne Moss

SPECIES *distichaceum*

GENUS *Dolichotheca* Long Pod Theca Hypne Moss

SPECIES *spinulosa*

GENUS *Echinodium* Urchin Hypne Moss

SPECIES *hispidum, prolixum, renauldii, setigerum, spinosum, umbrosum*

GENUS *Ectropodon* Ectropod Hypne Moss

SPECIES *urceolatus*

GENUS *Ectropotheciella* Lesser Ectropod Theca Hypne Moss

SPECIES *decrescens, distichophylla*

GENUS *Ectropotheciopsis* Least Ectropod Theca Hypne Moss

SPECIES *novoguineensis*

GENUS *Ectropothecium* Ectropod Theca Hypne Moss

SPECIES *aciculare, adnatum, aeruginosum, affine, afromolluscum, andoi, andrei, aneitense, angusto-textum, anisophyllum, annamense, antarense, aquaticum, arcuatum, arfakense, aubertii, aureo-crispum, aureum, aurifolium, barbelloides, bingerianum, bowiei, brachycarpum, brevifalcatum, brotheri, buitenzorgii, burmense, campaniforme, campanulatum, capillisetum, carolinarum, chamissonis, chenagonii, compressifolium, condensatum, costaricense, cupressoides, cyathothecioides, cyathothecium, cygnicollum, cylindricum, cyperoides, dafilae, dealbatum, decurrens, densum, dentatum, dentigerum, diffusum, diminutum, diversifolium, diversirete, dixonii, doliare, dolosum, drepanocladioides, drepanophyllum, duemmeri, eccremocladum, effusum, eleganti-pinnatum, engleri, epiphytum, erythrocaule, eurycladium, eurydictyon, falcifolium, falciforme, ferrugineum, fiorinae, flavo-viride, fluviatile, fosbergii, giganteum, goliathense, gunnii, haplocladum, howeanum, hyalinum, hypnoides, ichnotocladum, incubans, intertextum, intorquatum, isocladum, isophyllum, kerstanii, kinabaluense, kweichowense, laevigatum, lateriticola, laticuspis, latifolium, leiophyllum, leptochaeton, leptotapes, leucochloron, leucocladium, lonchocormus, longicapillare, longicaule, longicollum, longipedunculatum, luzoniae, mac-gregorii, macrobolax, malacoblastum, manii, mauritianum, meti, micholitzii, micronesiense, micro-verrucosum, minutum, molle, mollicellum, mollissimum, monumentorum, moritzii, nishimurii, nitidum, novae-guineae, novae-valesiae, obtusulum, occultum, ohosimense, ovalifolium, pacificum, pailotii, pallascens, palustre, papillosulum, papuanum, penangense, penzigianum, percomplanatum, perlatifolium, perminutum, perplicatum, perrevolutum, perrieri, perrotii, perscabrum, pinnatum, plicatum, plumigerum, plumosum, polyandrum, ponapense, protractulum, ptychofolium, pulchrum, ramuligerum, rechingeri, rectifolium, regulare, revolutum, rhynchostegioides, riparioides, robbinsii, robustum, rostellatum, rufulum, rupicola, sandwichense, saprophilum, savaicum, sericeum, serratum, seubertii, seychellarum, siamense, singapurense, sodale, sparsipilum, squarrifolium, stereodontoides, stigmophyllum, striatulum, subangense, subcallichroides, subdistichellum, subhaplocladum, sublaticuspis, sublaxirete, subsparsipilum, subulosum, sydneyense, tamatavense, tannaense, tapes, taxiforme, tophigerum, trichomitrium, triviale, turgidum, tutuilum, umbilicatum, urceolatum, valentinii, venustulum, vesicularioides, viridifolium, viridulum, vitiense, wainioi, wangianum, yasudae, zollingeri*

GENUS *Elmeriobryum* Elmer Hypne Moss

SPECIES *guatemalense, philippinense*

GENUS *Entodontella* Entic Tooth Hypne Moss

SPECIES *cameruniae*

GENUS *Eurohypnum* European Hypne Moss

SPECIES *leptothallum*

GENUS *Giraldiella* Giraldi Hypne Moss

SPECIES *levieri*

GENUS *Glossadelphus* Tongue Adelphi Hypne Moss

SPECIES *acutifolius, amboinensis, anisopterus, bilobatus, borneensis, congolensis, dimorphus, glossoides, guineensis, hermaphroditus, ivoreanus, julaceus, limnobioides, mauiensis, ogatae, ovatus, perplanicaulis, semiscabrus, similans, singalangensis, strictifolius, tahitensis, viridis, yakoushimae*

GENUS *Gollania* Gollan Hypne Moss

SPECIES *arisanensis, bartramiophila, benguetensis, clarescens, cylindricarpa, elbertii, homalothecioides, japonica, longii, neckerella, revoluta, robusta, ruginosa, schensiana, sinensis, splendens, taxiphylloides, tereticaulis, turgens, varians*

GENUS *Herzogiella* Herzog Hypne Moss

SPECIES *adscendens, boliviana, cylindricarpa, letestui, perrobusta, renitens, seligeri, striatella, turfacea*

GENUS *Holmgrenia* (Synonym for *Hypnum*)

GENUS *Homomallium* Homomorphic Mall Hypne Moss

SPECIES *adnatum, andoi, connexum, gollanii, homalostegium, incurvatum, japonico-adnatum, mexicanum, plagiangium, sharpii, simlaense, yuennanense*

GENUS *Hondaella* Honda Hypne Moss

SPECIES *brachytheciella, entodontea*

GENUS *Hyocomium* Pig Comb Hypne Moss

SPECIES *armoricum*

GENUS *Irelandia* Ireland Hypne Moss
SPECIES *robusticaulis*
GENUS *Isopterygium* Isometric Wing Hypne Moss
SPECIES *acuminatum, acutifolium, affusum, albescens, ambiguum, amblyocarpum, ambreanum, amoenum, andamanicum, annamense, antsihanakense, antunesii, applanatum, aptychopsis, aquaticum, arachnoideum, argillicola, argyroleucum, assamicum, asymmetricum, austrodenticulatum, bancanum, battakense, baurii, boivinii, borneense, brassii, brevicuspis, brevisetum, brownii, byssaceum, byssicaule, byssobolax, candidulum, chryseolum, citrinellum, clerophilum, combae, complanatulum, conangium, congoanum, courtoisii, cratericola, dewevrei, dubyi, elodes, exiguum, exile, fallax, filamentosum, fissum, fruticola, gabonense, gibbosulum, gracile, gracilisetum, gracillimum, gunnii, hasimotoi, hookeriophilum, howeanum, hygrophilum, intortum, ivoiriense, jamaicense, jelinkii, kilimandscharicum, latifolium, leiotheca, leptoblastum, leptoplumosum, leucophanes, leucopsis, lignicola, limatum, lioui, longicaule, longipes, longisetulum, longitheca, luteonitens, macoense, mbangae, meylanii, microplumosum, microthecium, minutirameum, minutum, molle, nanoglobum, nematosum, neocaledonicum, nervatulum, nitentivirens, norfolkianum, novoguinense, pallidulum, palmarum, paludigenum, pedunculatum, pendulum, perchlorosum, perminutum, perpallidum, perpusillum, phlyctitheca, planifolium, plumicaule, podorrhizum, praenitens, prasiellum, propaguliferum, pseudosubulatum, punctulatum, pygmaeocarpum, radicans, rubellum, samoanum, saporense, sapricola, sarasinii, saxense, serrulatum, strangulatum, strictirameum, subaptychopsis, subarachnoideum, subbrevisetum, subconangium, subcurvicolle, subglobosum, subleptoblastum, submicrothecium, subpinnatum, subsplendidulum, taxithelioides, taylorii, tenerifolium, tenerum, teretiusculum, tisserantii, tonkinense, tristaniense, undulatum, verruculosum, vestitum, vineale, welwitschii*
GENUS *Leiodontium* Smooth Tooth Hypne Moss
SPECIES *gracile, robustum, surculare*
GENUS *Leucomium* Light Hypne Moss
SPECIES *commixtum, crinitifolium, letestui, steerei, strumosum*
GENUS *Macrothamniella* Large Thamne Hypne Moss
SPECIES *novoguinensis, pilosula, robbinsii*
GENUS *Microctenidium* Comb Hypne Moss
SPECIES *assimile, leveilleanum*
GENUS *Microthamnium* Small Thamne Hypne Moss
SPECIES *longirostrum, pseudosubperspicuum, russellii, smithii*
GENUS *Mittenothamnium* Mitten Thamne Hypne Moss
SPECIES *andicola, aptychella, atroviride, bescherellei, brachythecioides, burelae, capillirameum, caudiforme, circinatum, ctenidioides, decurrens, deguchii, delicatulum, eurystomum, evardii, glaucissimum, heterostachys, hookerioides, horridulum, humile, hylophilum, incompletum, iporanganum, isopterygioides, jamesonii, lacouterei, langsdorffii, laxulum, lehmannii, leptoreptans, limosum, longoreptans, loriforme, macroblepharum, macrodontium, madagassum, megapelmatum, microthamnioides, micrurum, modestum, mucidum, mycostelium, nano-operculatum, nanopolymorphum, overlaetii, oxyrrhynchioides, oxystegum, pachythecium, perrieri, plinthophilum, plumosum, pseudoelegans, pseudoreptans, puiggarii, reduncum, reptans, rivulare, robustiusculum, rostratulum, rostratum, saproadelphus, scalpellifolium, sellovii, simorrhynchum, sordidum, stigmopyxis, subcampaniforme, subdiminutivum, submacrodontium, subobscurum, subpatens, substriatum, tamarisciforme, tamariscifrons, tapes, tenellum, trichocladon, trichopelmatum, versipoma, viscidulum, volvatum*
GENUS *Orthothecium* Straight Theca Hypne Moss
SPECIES *acuminatum, austrocatenulatum, chryseum, diminutivum, hyalopiliferum, intricatum, lapponicum, ovicarpum, rufescens, schlagintweitii, strictum*
GENUS *Phyllodon* Tooth Frond Hypne Moss
SPECIES *scutellifolius, truncatulus, truncatus*
GENUS *Platygyriella* Lesser Flat Gyre Hypne Moss
SPECIES *aurea, aureoides, densa, fragilifolia, kirtikarii, pringlei*
GENUS *Platygyrium* Flat Gyre Hypne Moss
SPECIES *fuscoluteum, repens, russulum, subrussulum*
GENUS *Podperaea* Pod Peran Hypne Moss
SPECIES *krylovii*
GENUS *Ptilium* Ptile Hypne Moss
SPECIES *crista-castrensis*
GENUS *Puiggariopsis* Puiggar Hypne Moss
SPECIES *aurifolia*
GENUS *Pylaisia* Pylais Hypne Moss
SPECIES *australis, brevirostris, brotheri, condensata, cristata, curviramea, extenta, falcata, intricata, latifolia, leptoclada, macrotis, nana, obtusa, polyantha, robusta, selwynii, steerei, subcircinata, subimbricata*
GENUS *Pylaisiella* Lesser Pylais Hypne Moss
SPECIES *frahmii, kunisawae*
GENUS *Rhacopilopsis* Rhace Downy Hypne Moss
SPECIES *trinitensis, variegata*
GENUS *Rhizohypnella* Rhyzid Hypne Moss
SPECIES *bartramii, sundaensis*

GENUS *Sharpiella* (Synonym for *Hypnum*)
GENUS *Stenotheciopsis* Thin Theca Hypne Moss
SPECIES *serrula*
GENUS *Stereodon* Stereo Tooth Hypne Moss
SPECIES *aureo-nitidus, bipinnatus, lilliei, subhamulosus, vivicola*
GENUS *Stereodontopsis* Lesser Stereo Tooth Hypne Moss
SPECIES *excavatam pseudorevoluta*
GENUS *Taxiphyllopsis* Lesser Yew Hypne Moss
SPECIES *iwatsukii*
GENUS *Taxiphyllum* Yew Hypne Moss
SPECIES *alternans, angustirete, aomoriense, arcuatum, autoicum, barbieri, cuspidifolium, densifolium, deplanatum, eberhardtii, fluitans, gabonense, giraldii, inundatum, isopterygioides, laevifolium, laxalare, ligulaefolium, moutieri, papuanum, pilosum, prostratum, punctulatum, ramivagum, schweinfurthii, serrulatum, splendescens, subretusum, taxirameum, torrentium, whittierianum, wissgrillii*
GENUS *Trachythecium* Thick Theca Hypne Moss
SPECIES *calcicola, dixonii, hebridense, le-testui, mamillatum, micropyxis, submammillosulum, tuberculatum, verrucosum*
GENUS *Vesicularia* Vesicle Hypne Moss
SPECIES *acinacifolia, albo-viridis, amboinensis, angustifolia, anisothecia, aperta, aquatilis, argentinica, ayresii, bescherellei, borealis, brachytheciopsis, buruensis, caloblasta, calodictyon, chlorotica, codonopyxis, combae, crassiramea, cruegeri, debilis, demangei, dubyana, ectropotheciopsis, elegantula, eligiana, eurycladia, eurydictyon, ferriei, firma, flaccida, flavo-viridis, fonticola, galerulata, glaucina, glaucissima, glaucopinnata, glaucula, glazioviana, golungensis, hainanensis, hamata, hapalyptera, hilliana, hygrobium, immunda, immutata, inflectens, inundata, ischyropteris, isopterygiiforme, janowskyi, kurzii, latiramea, lepervanchei, leucocladium, levieri, lonchocormus, longo-fluitans, loricatifolia, marginata, mayumbensis, miquelii, nanocarpa, nigeriana, nitidula, ochracea, oedicarpa, orbicifolia, oreadelphus, padangensis, paranahybae, paulensis, pelvifolia, perangusta, perpallida, perpinnata, perreticulata, perviridis, pilifera, plicaefolia, reimersiana, reticulata, rhynchostegiocarpa, rivalis, rodriguezii, rootii, sarcoblasta, scaturigina, setchellii, sigmangia, sigmatellopsis, slateri, soyauxii, squamatifolia, stillatitia, stillicidiorum, stramineola, strephomischos, subcalodictyon, subchlorotica, subenervis, subfuscescens, subpilicuspis, subsarcoblasta, subscaturiginosa, subsphaerica, suburceolata, succosa, tahitensis, tenaci-inserta, tenuatipes, tepida, terrestris, thermophila, thollonii, tjibodensis, tonkinensis, tophacea, trullifolia, vesicularis, vitiana*

FAMILY 02 *Amblystegiaceae* Blunt Cover Hypne Moss
GENUS *Amblystegium* Blunt Cover Hypne Moss
SPECIES *arvernense, byssoides, camisassae, campyliopsis, chalaropelma, cuspidarioides, excurrens, filicinum, formianum, gallicum, glaziovii, jaffuelii, juratzkanum, patenti-flexuosum, perminimum, pseudosubtile, rigescens, riparioides, rishiriense, saxatile, saxicola, serbicum, serpens, sparsile, speirophyllum, stricto-serpens, subtrifarium, subulatum, tenax, tenuifolium, versirete*
GENUS *Acrocladium* Tall Branch Blunt Cover Hypne Moss
SPECIES *auriculatum, chlamydophyllum, trichocladium*
GENUS *Amblystegiella* (Synonym for *Amblystegium*)
GENUS *Callialaria* Callialar Blunt Cover Hypne Moss
SPECIES *curvicaulis*
GENUS *Calliergidium* Calliergid Blunt Cover Hypne Moss
SPECIES *austrostramineum*
GENUS *Calliergon* Calliergon Blunt Cover Hypne Moss
SPECIES *cordifolium, giganteum, joveti-astii, megalophyllum, orbiculare-cordatum, richardsonii, subpapillosum*
GENUS *Calliergonella* Lesser Calliergon Blunt Cover Hypne Moss
SPECIES *cuspidata, complanata, lindbergii*
GENUS *Campyliadelphus* Field Adelphi Blunt Cover Hypne Moss
SPECIES *elodes, polygamus, stellatus*
GENUS *Campylium* Field Blunt Cover Hypne Moss
SPECIES *amblystegioides, annamense, calcareum, campylophylloides, cardotii, chrysophyllum, courtoisii, creperum, glaucocarpoides, gollanii, hispidulum, hygrophilum, lacerulum, laxifolium, longicuspis, pachytheca, porphyreticum, pseudochrysophyllum, pseudocomplexum, radicale, reichenbachianum, squarrosulum, subdecursivulum, tenerum, uninervium*
GENUS *Campylophyllum* Lesser Field Blunt Cover Hypne Moss
SPECIES *halleri, quisqueyanum, sommerfeltii*
GENUS *Conardia* Conard Blunt Cover Hypne Moss
SPECIES *compacta*
GENUS *Cratoneuron* Crate Nerve Blunt Cover Hypne Moss
SPECIES *commutatovirescens, filicinum, formianum, formosanum, longicostatum, punae, subcurvicaule, sulcatoirrigatum, tenerrimum*
GENUS *Cratoneuropsis* Lesser Crate Nerve Blunt Cover Hypne Moss
SPECIES *relaxa, subrelaxa*
GENUS *Donrichardsia* Don Richards Blunt Cover Hypne Moss

SPECIES *macroneuron*
GENUS *Drepanocladus* Sickle Branch Blunt Cover Hypne Moss
SPECIES *aduncus, andinus, arcticus, austro-aduncus, austro-fluitans, austrostramineus, brachiatus, frigidus, hallii, herthae, jamesii-macounii, latinervis, longifolius, perplicatus, revolvens, revolvens, secundifolius, sendtneri, sordidus, sparsus, stagnatus*
GENUS *Gradsteinia* Gradstein Blunt Cover Hypne Moss
SPECIES *andicola*
GENUS *Hamatocaulis* Hooked Stem Blunt Cover Hypne Moss
SPECIES *lapponicus, vernicosus*
GENUS *Hygroamblystegium* Hydric Blunt Cover Hypne Moss
SPECIES *austrofluviatile, calcareum, caudicaule, ecuadorense, excurrens, filum, fluviatile, formianum, luridum, noterophilum, punae, tenax*
GENUS *Hygrohypnum* Lesser Hydric Blunt Cover Hypne Moss
SPECIES *alpestre, alpinum, bestii, caussequei, choprae, closteri, cochlearifolium, coreanum, duriusculum, eugyrium, fontinaloides, hedbergii, laevigatum, luridum, molle, montanum, nairii, norvegicum, ochraceum, polare, purpurascens, smithii, styriacum, subeugyrium, tsurugizamicum*
GENUS *Hypnobartlettia* Bartlett Blunt Cover Hypne Moss
SPECIES *fontana*
GENUS *Koponenia* Koponen Blunt Cover Hypne Moss
SPECIES *holoneuron*
GENUS *Leptodictyum* Thin Net Blunt Cover Hypne Moss
SPECIES *humile, kurdicum, loeskei, mizushimae, riparium*
GENUS *Limprichtia* (Synonym for *Amblystegium*)
GENUS *Loeskypnum* Loesky Blunt Cover Hypne Moss
SPECIES *badium, wickesii*
GENUS *Ochyraea* Ocher Blunt Cover Hypne Moss
SPECIES *tatrensis*
GENUS *Orthotheciella* Straight Theca Blunt Cover Hypne Moss
SPECIES *varia*
GENUS *Palustriella* Marsh Blunt Cover Hypne Moss
SPECIES *commutata, decipiens*
GENUS *Pictus* Painted Blunt Cover Hypne Moss
SPECIES *scoticus*
GENUS *Platydictya* Flat Net Blunt Cover Hypne Moss
SPECIES *confervoides, fauriei, hattorii, jungermannioides, madurensis, minutissima, subtilis*
GENUS *Platylomella* Flat Honey Blunt Cover Hypne Moss
SPECIES *lescurii*
GENUS *Pseudocalliergon* Lesser Calliergon Blunt Cover Hypne Moss
SPECIES *angustifolium, brevifolium, lycopodioides, trifarium, turgescens*
GENUS *Sanionia* Sanion Blunt Cover Hypne Moss
SPECIES *georgicouncinata, orthothecioides, uncinata*
GENUS *Sarmenthypnum* (Synonym for *Amblystegium*)
GENUS *Sasaokaea* Sasaoka Blunt Cover Hypne Moss
SPECIES *aomoriensis*
GENUS *Sciaromiella* Sciarome Blunt Cover Hypne Moss
SPECIES *bartlettii*
GENUS *Sciaromiopsis* Lesser Sciarome Blunt Cover Hypne Moss
SPECIES *sinensis*
GENUS *Sciaromium* (Synonym for *Sciaromiella*)
GENUS *Scorpidium* Scorpion Blunt Cover Hypne Moss
SPECIES *cossonii, revolvens, scorpioides*
GENUS *Sinocalliergon* China Calliergon Blunt Cover Hypne Moss
SPECIES *satoi*
GENUS *Straminergon* Straw Blunt Cover Hypne Moss
SPECIES *stramineum*
GENUS *Vittia* Vitt Blunt Cover Hypne Moss
SPECIES *pachyloma, salina*
GENUS *Warnstorfia* Warnstorf Blunt Cover Hypne Moss
SPECIES *austrostraminea, exannulata, fluitans, luipichensis, procera, pseudosarmentosa, pseudostraminea, sarmentosa, trichophylla, tundrae*

FAMILY 03 *Brachytheciaceae* Short Sheath Hypne Moss
GENUS *Brachythecium* Short Sheath Hypne Moss
SPECIES *acuminatum, acutum, afroglareosum, albicans, allisonii, appleyardiae, athrocladum, austroglareosum, austrosalebrosum, austrostramineum, bellicum, bellii, biventrosum, bodinieri, bolanderi, borgenii, brachycarpum, brachycladum, brotheri, buchananii, calcareum, calliergonoides, campestre, camptothecioides, campylothallum, capillaceum, caucasicum, cavifolium, chakratense, chauvetii, chocayae, cirriphylloides, collinum, complanatum, comtifolium, conostomum, coreanum, coruscum, curvatulum,*

cymbifolium, decurrens, delicatulum, dicranoides, dieckei, digastrum, dovrense, duemmeri, ehrenbergii, emodi-glaucum, erythrorrhizon, falcatulum, falcatum, fendleri, filiforme, flagellare, flagellare, fontanum, formosanum, frigidum, gangulianum, garhwalense, garovaglioides, geheebii, gelidum, glaciale, glaciale, glareosum, gloriosum, hastile, hawaiicum, hedbergii, helminthocladum, holzingeri, hylotapetum, immersum, inconditum, indicopopuleum, indistinctum, kamounense, kashmirense, konoi, koponenii, krausei, kuroishicum, laetum, laevipes, laevivelutinum, lamprocarpum, latifolium, latinervium, laxifolium, leibergii, leiopodium, longicuspidatum, longidecurrens, longidens, longipes, lutescens, majusculum, microcollinum, microsericeum, mieheanum, mildeanum, mittenii, miyabei, moriense, nakazimae, negrii, nelsonii, nitidissimum, nitidulum, nivale, noguchii, noterophiloides, obsoletinerve, occidentale, oedipodium, olympicum, ornellanum, otaruense, pallescens, papillipes, papuense, paradoxum, pendulum, percurrens, perscabrum, piligerum, pinnatum, pinnirameum, plumosum, poadelphus, polyoicum, populeum, populeum, praelongum, preussii, procumbens, propinnatum, pseudopopuleum, pseudouematsui, pseudovelutinum, pulchellum, quelpaertense, ramicola, reflexum, rigens, rivulare, rivularoides, roccatii, rotaeanum, ruderale, rutabulum, ryanii, sakuraii, salebrosum, sapporense, satsumense, siamense, spectabile, spuriopopuleum, starkii, stricticalyx, subfalcatum, subfalcatum, sublaetum, subpilosum, subplicatum, subrutabulum, svihlae, tearapense, trachypodium, turgidum, turquetii, uematsui, ugandae, umbilicatum, valentinii, vallium, vanekii, vellereum, velutinum, venturii, venustum, vesiculariopsis, waziriense, weinmannii, woronowii, yakushimense, zanonii

GENUS *Aerolindigia* — Aerole Short Sheath Hypne Moss
SPECIES *capillacea*
GENUS *Brachytheciella* — Lesser Short Sheath Hypne Moss
SPECIES *stolonifera*
GENUS *Bryhnia* — Bryhn Short Sheath Hypne Moss
SPECIES *angustifolia, brachycladula, concavifolia, decurvans, graminicolor, hultenii, nepalensis, nitida, novae-angliae, pflanzii, polyclada, tenerrima, tokubuchii, trichomitria*
GENUS *Bryoandersonia* — Anderson Short Sheath Hypne Moss
SPECIES *illecebra*
GENUS *Bryostreimannia* — Streimann Short Sheath Hypne Moss
SPECIES *turgida*
GENUS *Camptothecium* — Field Short Sheath Hypne Moss
SPECIES *fallax, lutescens, naumannii, paulianum, pseudolutescens, subhumile*
GENUS *Catagoniopsis* — Catagon Short Sheath Hypne Moss
SPECIES *berteroana*
GENUS *Chamberlainia* — (Synonym for *Brachythecium*)
GENUS *Cirriphyllum* — Tendril Short Sheath Hypne Moss
SPECIES *andinum, brandegei, cirrosum, piliferum, pirottae, romanum, subnerve, tommasinii*
GENUS *Cratoneurella* — Crate Nerve Short Sheath Hypne Moss
SPECIES *uncinifolia*
GENUS *Eriodon* — Woolly Short Sheath Hypne Moss
SPECIES *conostomus*
GENUS *Eurhynchiella* — Lesser Beak Short Sheath Hypne Moss
SPECIES *acanthophylla, decurrens, semitorta, tenuinervis, toncolensis*
GENUS *Eurhynchium* — Beak Short Sheath Hypne Moss
SPECIES *acicladium, africanum, altaicum, angustirete, arbuscula, argenteum, asperipes, asperisetum, clinocarpum, coarctum, confertum, corralense, crassinervium, cuestarum, dawsonii, devexum, dives, dumosum, eustegium, flotowianum, fuegianum, glaciale, hercynicum, hians, integrifolium, jovet-astiae, kirishimense, laevisetum, latifolium, laxirete, meridionale, meyeri, nivium, oreganum, peyronelii, praelongum, pseudopiliferum, pulchellum, pumilum, remotifolium, rugisetum, savatieri, schleicheri, serpenticaule, serricuspis, speciosum, spiculosum, spinulinerve, sprengelii, squarrifolium, striatulum, striatum, suberectum, submegapolitanum, subsquarrulosum, substrigosum, tenuivagum, trichocladoides, yezoanum*
GENUS *Flabellidium* — Fan Short Sheath Hypne Moss
SPECIES *spinosum*
GENUS *Homalotheciella* — Lesser Flat Short Sheath Hypne Moss
SPECIES *sinensis, subcapillata, tenerrima*
GENUS *Homalothecium* — Lesser Flat Short Sheath Hypne Moss
SPECIES *aequatoriense, arenarium, aureum, fallax, gracillimum, incompletum, integerrimum, laevisetum, luteolum, lutescens, neckeroides, nevadense, nuttallii, philippeanum, sericeum*
GENUS *Isotheciopsis* — Lesser Isometric Short Sheath Hypne Moss
SPECIES *amoena*
GENUS *Isothecium* — Isometric Short Sheath Hypne Moss
SPECIES *algarvicum, alopecuroides, ceylonense, cristatum, hakkodense, holtii, marocanum, mertensii, myosuroides, myuroides, nervosum, obtusatulum, rigidissimum, semitortum, sericeum, sprucei, subdiversiforme, trichocladon*
GENUS *Kindbergia* — Kindberg Short Sheath Hypne Moss
SPECIES *kenyae, oedogonium*
GENUS *Lepyrodontopsis* — Thin Tooth Short Sheath Hypne Moss
SPECIES *trichophylla*
GENUS *Mandoniella* — Mandon Short Sheath Hypne Moss

SPECIES *spicatinervia*
GENUS *Myuroclada* Mouse Tail Branch Short Sheath Hypne Moss
SPECIES *maximowiczii, rotundifolia*
GENUS *Nobregaea* Nobrega Short Sheath Hypne Moss
SPECIES *latinervis*
GENUS *Okamuraea* Okamura Short Sheath Hypne Moss
SPECIES *micrangia*
GENUS *Orthorrhynchium* Straight Beak Short Sheath Hypne Moss
SPECIES *balanseanum, elegans*
GENUS *Oxyrrhynchium* Sharp Beak Short Sheath Hypne Moss
SPECIES *altisetum, confervoideum, laosianum, ovatum, porothamnioides, protractum, subasperum*
GENUS *Palamocladium* Open Branch Short Sheath Hypne Moss
SPECIES *congestum, euchloron, leskeoides, wilkesianum*
GENUS *Platyhypnidium* Flat Short Sheath Hypne Moss
SPECIES *aquaticum, austrinum, brotheri, esquirolii, grolleanum, intermedium, lusitanicum, mac-owanianum, muelleri, obscurum, patulifolium, peruviense, pringlei, riparioides, torrenticola, validum*
GENUS *Pleuropus* (Synonym for *Clitopilus, Hypsizygus, Panus, Rhodotus*)
GENUS *Pseudopleuropus* Pleural Short Sheath Hypne Moss
SPECIES *himalayanus, indicus, morrisonensis*
GENUS *Pseudisothecium* (Synonym for *Isothecium*)
GENUS *Rhynchostegiella* Lesser Beak Cover Sheath Hypne Moss
SPECIES *assamica, bequaertii, bourgaeana, brachypodia, brunelii, chilensis, curviseta, divaricatifolia, durieui, fabroniadelphus, hawaica, holstii, humillima, japonica, keniae, laeviseta, leiopoda, leptoneura, linderi, litorea, macilenta, menadensis, microtheca, mindorensis, novae-zealandiae, opacifolia, ovalifolia, papuensis, pendula, pseudoteesdalei, ramicola, sachensis, sakuraii, santosii, scabriseta, smithii, sublaevipes, sumatrana, tanneri, tenella, tenelliformis, teneriffae, tenuicaulis, trichophylla, vitiensis, vriesei, zeyheri*
GENUS *Rhynchostegium* Beak Cover Short Sheath Hypne Moss
SPECIES *acicula, alboviridum, ambiguum, aneuron, angustifolium, apophysatum, assumptionis, bequaertii, brachypterum, brachypyxis, brachythecioides, brevicuspis, brevirete, buluense, cacticola, calderi, campylocarpum, campylocladulum, cataractarum, celebicum, chrysophylloides, comorae, complanum, compridense, conchophyllum, confertum, contortulum, contractum, cylindritheca, dasyphyllum, distans, distratum, drepanocladioides, duthiei, fauriei, finitimum, fissidens, fissidentellum, fragilicuspis, gaudichaudii, georgianum, glaucovirescens, globipyxis, gracilipes, herbaceum, hercynicum, homaliocaulon, hookeri, hopfferi, horridum, huttonii, inaequale, inclinatum, inerme, irriguum, isopterygioides, javanicum, lamasicum, laxatum, leptomitophyllum, leptopteridium, leucodictyon, lindmanii, luteonitens, malmei, megapolitanum, microthamnioides, minutum, murale, muriculatum, nanopennatum, nanothecium, nigrescens, oblongifolium, obtusatum, omocrates, ovalfolium, oxyodon, pallenticaule, pallidifolium, pallidius, pampae, parvulum, patulum, pectinatum, pellucidum, philippinense, pinnicaule, plagiotheciella, planifolium, planiusculum, pseudodistans, raphidorrhynchum, revelstokense, rivale, robustum, rotundifolium, ruvenzorense, santaiense, sarcoblastum, savatieri, selaginellifolium, sellowii, semiscabrum, semitortulum, sinense, sparsirameum, strongylense, subbrachypterum, subconfertum, subenerve, submenadense, subperspicuum, subrectocarpum, subrotundum, subrusciforme, subserrulatum, subspeciosum, subtrachypterum, surrectum, taphrophilum, tenuifolium, tocaremae, trachynotum, trachypelma, trieblingii, tubaronense, ulicon, vagans, vitianum, volkensii*
GENUS *Schimperella* Schimper Short Sheath Hypne Moss
SPECIES *bello-intricata, rhynchostegioides*
GENUS *Scleropodiopsis* Lesser Hard Pod Short Sheath Hypne Moss
SPECIES *laxiretis*
GENUS *Scleropodium* Hard Pod Short Sheath Hypne Moss
SPECIES *australe, brachyphyllum, californicum, cespitans, coreense, julaceum, obtusifolium, subcaespitosum, touretii*
GENUS *Scorpiurium* Scorpion Tail Short Sheath Hypne Moss
SPECIES *circinatum, cucullatum, deflexifolium, sendtneri, spelaeorum*
GENUS *Steerecleus* Steerecle Short Sheath Hypne Moss
SPECIES *argute-serratus, beskeanus, scariosus, serrulatus*
GENUS *Stenocarpidiopsis* Thin Carpel Short Sheath Hypne Moss
SPECIES *salicicola*
GENUS *Tomentypnum* Tomentose Short Sheath Hypne Moss
SPECIES *falcifolium, nitens, paulianum*
GENUS *Trachybryum* Rough Short Sheath Hypne Moss
SPECIES *megaptilum*
GENUS *Unclejackia* Uncle Jack Short Sheath Hypne Moss
SPECIES *crispifolia, longisetula*

FAMILY 04 *Entodontaceae* Entic Tooth Hypne Moss
GENUS *Entodon* Entic Tooth Hypne Moss
SPECIES *abbreviatus, acutifolius, adyris, aeruginosus, akitensis, argyreus, attenuatus, beyrichii, brevirameus, brevirostris, brevisetus, buckii, caliginosus, calycinus, campipatrum, cernuus, challengeri, chloropus, chloroticus, cladorrhizans, columnaris, complanatulus, complicatus, conchophyllus, concinnus,*

corbieri, crassirameus, cupressiformis, curvatus, cymbifolius, denticuspis, diffusinervis, distans, divergens, diversinervis, dolichocucullatus, drepanoides, expallens, felicis, flaccidus, flavissimus, flavovirens, flexipes, floridanus, fulvonitens, geminidens, giraldii, gracilis, gracilisetus, hampeanus, herbaceus, incurvatus, jamesonii, kungshanensis, laetus, laxo-alaris, lindbergii, longicostatus, longifolius, luridus, luteonitens, mackaviensis, macropodus, madagassus, mechoacanus, micans, microcarpus, micropodus, morrisonensis, mosenii, motelayi, myurus, nanocarpus, nanoclimacium, natalensis, neglectus, nepalensis, novarae, obtusatus, ovicarpus, pallescens, pallidisetus, perplicatus, piovani, plumosus, propinquus, prorepens, pseudoseductrix, pulchellus, punctulatus, pylaisioides, rigidus, rubicundus, rufescens, rutenbergii, scabridens, scariosus, schensianus, schleicheri, seductrix, serrulatus, seubertianus, smaragdinus, splendidulus, stereophylloides, stewartii, subcuspidatus, subdiversinervis, suberythropus, subflexipes, subgeminidens, subplicatus, sullivantii, sullivantii, taiwanensis, tobaensis, tokyensis, tosae, transbaicalensis, truncatus, truncorum, usambaricus, verruculosus, virens, viridulus, vulcanicus, werneri, yunnanensis

GENUS *Erythrodontium* — Red Entic Tooth Hypne Moss

SPECIES *argentinicum, barteri, engleri, julaceum, lacoutourei, lamoruense, latifolium, longisetum, platygyrioides, rotundifolium, squarrosum, subbiforme, subjulaceum, vanderystii, warmingii*

GENUS *Mesonodon* — Meson Entic Tooth Hypne Moss

SPECIES *flavescens, regnellianus*

GENUS *Nanothecium* — Nannite Entic Tooth Hypne Moss

SPECIES *foreaui*

GENUS *Plagiotheciopsis* — Pelagic Entic Tooth Hypne Moss

SPECIES *oblonga*

GENUS *Pseudoscleropodium* — Hard Pod Entic Tooth Hypne Moss

SPECIES *purum*

GENUS *Pylaisiobryum* — Pylais Entic Tooth Hypne Moss

SPECIES *abyssinicum*

GENUS *Rozea* — Roze Entic Tooth Hypne Moss

SPECIES *andrieuxii, diversifolia, fulva, kenoyeri, pterogonioides, subjulacea*

GENUS *Trachyphyllum* — Rough Entic Tooth Hypne Moss

SPECIES *borgenii, carinatum, dusenii, gastrodes, inflexum, jeyporense, touwianum*

FAMILY 05 *Fabroniaceae* — Fabron Hypne Moss

GENUS *Fabronia* — Fabron Hypne Moss

SPECIES *abyssinica, affinis, anacamptodens, andina, angustifolia, argentinica, assamica, australis, balansae, bartramii, basilaris, beccarii, bizotii, brachyphylla, breutelii, campenonii, ciliaris, claviramea, congolensis, crassiretis, curvirostris, degeneri, dentata, eckloniana, fastigiata, filamentosa, flavinervis, gardneriana, garnieri, glauca, goetzei, goughii, guarapiensis, gueinzii, gymnostoma, hampeana, imbricata, jamesonii, julacea, kiusiana, lachenaudii, latidens, latifolia, leikipiae, longipila, lorentzii, macroblepharis, madurensis, marianna, matsumurae, microspora, minuta, minutissima, motelayi, niam-niamiae, nietneri, nivalis, obtusatula, papillidens, patentissima, perimbricata, pilifera, pobeguinii, pocsii, podocarpi, pusilla, ravenelii, rehmannii, rostrata, schensiana, schmidii, schwetschkeoides, scottiae, secunda, socotrana, subpolycarpa, tahitensis, tayloriana, trichophora, trichophylla, tucumanensis, victoriae, villaumii, wageri, zollingeri*

GENUS *Anacamptodon* — Spur Fabron Hypne Moss

SPECIES *africanus, amblystegioides, compactus, cubensis, fortunei, japonicus, latidens, marginatus, minimus, splachnoides, subulatus, validinervis, wattsii*

GENUS *Austinia* — Austin Fabron Hypne Moss

SPECIES *tenuinervis*

GENUS *Clasmatodon* — Clasmate Tooth Fabron Hypne Moss

SPECIES *parvulus*

GENUS *Dimerodontium* — Double Tooth Fabron Hypne Moss

SPECIES *acuminatum, africanum, aurescens, balansae, chlorophyllosum, kenyae, mendozense, ovatifolium, pellucidum, schnyderi*

GENUS *Helicodontium* — Helix Tooth Fabron Hypne Moss

SPECIES *acuminatum, capillare, chloronema, clarazii, complanatum, doii, fabroniopsis, formosicum, guineense, italicum, jacobi-felicis, kiusianum, lanceolatum, leptodontium, minutum, pendulum, pervirens, pseudolimnobium, rhyparobolax, sanctum, siambonense*

GENUS *Ischyrodon* — Pelvic Tooth Fabron Hypne Moss

SPECIES *lepturus, seriolus*

GENUS *Leptoischyrodon* — Thin Pelvic Tooth Fabron Hypne Moss

SPECIES *congoanus*

GENUS *Levierella* — Levier Fabron Hypne Moss

SPECIES *densiramea, fabroniacea, perserrata, subfabroniacea*

GENUS *Macgregorella* — Macgregor Fabron Hypne Moss

SPECIES *indica*

GENUS *Merrilliobryum* — Merrill Fabron Hypne Moss

SPECIES *fabronioides*

GENUS *Rhizofabronia* — Rhizid Fabron Hypne Moss

SPECIES *perpilosa, persoonii*

GENUS *Taxitheliella* — Yew Fabron Hypne Moss

SPECIES *richardsii*

FAMILY 06 *Hylocomiaceae* — Wood Tuft Hypne Moss
GENUS *Hylocomium* — Wood Tuft Hypne Moss
SPECIES *interruptum, splendens*
GENUS *Girgensohnia* — (Synonym for *Hylocomium*)
GENUS *Hylocomiastrum* — Star Wood Tuft Hypne Moss
SPECIES *himalayanum, pyrenaicum, umbratum*
GENUS *Leptocladiella* — Thin Branch Wood Tuft Hypne Moss
SPECIES *delicatula, flagellaris, psilura*
GENUS *Leptohymenium* — Thin Membrane Wood Tuft Hypne Moss
SPECIES *brachystegium, breutelii, dentatum, hokinense, longisetaceum, sharpii, stramineum, tenue*
GENUS *Loeskeobryum* — Loeske Wood Tuft Hypne Moss
SPECIES *brevirostre, cavifolium*
GENUS *Macrothamnium* — Large Thamne Wood Tuft Hypne Moss
SPECIES *hylocomioides, javense, leptohymenioides, macrocarpum*
GENUS *Neodolichomitra* — Long Pod Mitre Wood Tuft Hypne Moss
SPECIES *yunnanensis*
GENUS *Orontobryum* — Syria Wood Tuft Hypne Moss
SPECIES *hookeri*
GENUS *Pleurozium* — Pleuroze Wood Tuft Hypne Moss
SPECIES *schreberi*
GENUS *Pleuroziopsis* — Lesser Pleuroze Wood Tuft Hypne Moss
SPECIES *ruthenica*
GENUS *Rhytidiadelphus* — Wrinkled Adelphi Wood Tuft Hypne Moss
SPECIES *japonicus, loreus, printzii, squarrosus, triquetrus*
GENUS *Rhytidiopsis* — Lesser Wrinkled Wood Tuft Hypne Moss
SPECIES *robusta*
GENUS *Rhytidium* — Wrinkled Wood Tuft Hypne Moss
SPECIES *rugosum*

FAMILY 07 *Leskeaceae* — Leske Hypne Moss
GENUS *Leskea* — Leske Hypne Moss
SPECIES *angustata, australis, brevifolia, catenularia, consanguinea, filaria, filivaga, gracilescens, guineensis, integra, leskeoides, mauritiana, obscura, perstricta, plumaria, polenburgii, polycarpa, riparia, scabrinervis, subacuminata, subfiliramea, sublaevifolia, vaucheri*
GENUS *Bryohaplocladium* — (Synonym for *Leskea*)
GENUS *Bryonorrisia* — Norris Leske Hypne Moss
SPECIES *acutifolia*
GENUS *Chileobryon* — Chile Leske Hypne Moss
SPECIES *callicostelloides*
GENUS *Duthiella* — Duth Leske Hypne Moss
SPECIES *declinata, emodi, flaccida, formosana, japonica, lacustris, myuriiformis, pellucens, perpapillata, rivicola, robusta, speciosissima, wallichii*
GENUS *Fabronidium* — Fabron Leske Hypne Moss
SPECIES *espinosae, guatemaliense*
GENUS *Lescuraea* — Tail Leske Hypne Moss
SPECIES *arizonae, chilensis, darjeelingensis, glacialis, longipes, mutabilis, saxicola, secunda, serrata*
GENUS *Leskeadelphus* — Adelphi Leske Hypne Moss
SPECIES *bolivianus*
GENUS *Leskeella* — Lesser Leske Hypne Moss
SPECIES *incrassata, nervosa, pusilla, williamsii, zuluensis*
GENUS *Lindbergia* — Lindberg Leske Hypne Moss
SPECIES *brachyptera, brevifolia, duthiei, haplocladioides, japonica, koelzii, longinervis, magniretis, maritima, mexicana, ovata, patentifolia, pellicula, pseudoleskeoides, sinensis, taxifolia, viridis*
GENUS *Mamillariella* — Nipple Leske Hypne Moss
SPECIES *geniculata*
GENUS *Orthoamblystegium* — Blunt Cover Leske Hypne Moss
SPECIES *spuriosubtile*
GENUS *Pseudoleskea* — Least Leske Hypne Moss
SPECIES *andina, atricha, attenuata, baileyi, brachyclados, coreana, dispersa, falcifolia, incurvata, julacea, korjakorum, laevifolia, laevissima, leikipiae, leskeoides, obtusiuscula, papillarioides, patens, penzigii, perraldieri, plagiostoma, pseudoattenuata, radicosa, ramuligera, saviana, semensis, setschwanica, stenophylla, tonkinensis, trichodes, yuennanensis*
GENUS *Pseudoleskeella* — Mock Leske Hypne Moss
SPECIES *catenulata, glaberrima, laxiramea, mildeana, papillosa, rupestris, serpentinensis, tectorum*
GENUS *Pseudoleskeopsis* — False Leske Hypne Moss
SPECIES *artariae, bollei, claviramea, imbricata, mollicula, osterwaldii, tosana, unilateralis, zippelii*
GENUS *Rigodiadelphus* — Rigode Adelphi Leske Hypne Moss

SPECIES *arcuatus, robustus*
GENUS *Schwetschkea* Schwetschke Leske Hypne Moss
SPECIES *applanata, arachnoidea, brevipes, capillifolia, courtoisii, enervis, fabronioides, grateloupii, gymnostoma, incerta, javensis, laxa, peracuminata, pygmaea, schweinfurthii, sinica, sublaxa, usambarica*

FAMILY 08 *Myriniaceae* Myrin Hypne Moss
GENUS *Myrinia* Myrin Hypne Moss
SPECIES *brasiliensis, pulvinata*
GENUS *Juratzkaeella* Juratzka Myrin Hypne Moss
SPECIES *sinensis*
GENUS *Nematocladia* Thread Branch Myrin Hypne Moss
SPECIES *tesserata*
GENUS *Schwetschkeopsis* Schwetschke Myrin Hypne Moss
SPECIES *elongata, fabronia, formosana, robustula*

FAMILY 09 *Myuriaceae* Mouse Tail Hypne Moss
GENUS *Myurium* Mouse Tail Hypne Moss
SPECIES *acuminatum, alaris, borii, hochstetteri, warburgii*
GENUS *Eumyurium* Lesser Mouse Tail Hypne Moss
SPECIES *sinicum*
GENUS *Oedicladium* Riddle Branch Mouse Tail Hypne Moss
SPECIES *fragile, involutaceum, perplexum, praenitens, quinquefarium, rufescens, serricuspe, subnitens, tortifolium*
GENUS *Piloecium* Downy Mouse Tail Hypne Moss
SPECIES *pseudorufescens*

FAMILY 10 *Plagiotheciaceae* Pelagic Sheath Hypne Moss
GENUS *Plagiothecium* Pelagic Sheath Hypne Moss
SPECIES *berggrenianum, bicolor, cavifolium, ceylonense, cochleatum, cordifolium, corticola, curvifolium, dehradunense, denticulatum, drepanophyllum, entodontella, euryphyllum, falklandicum, formosicum, georgicoantarcticum, glossophylloides, handelii, herzogii, laetum, lamprostachys, lancifolium, latebricola, lonchochaete, lucidum, mauiense, membranosulum, mildbraedii, monbuttoviae, neckeroideum, nemorale, nitens, nitidifolium, noricum, novogranatense, obtusissimum, orthocarpum, ovalifolium, paleaceum, perminutum, piliferum, platyphyllum, rhynchostegioides, shinii, subglaucum, subulatum, succulentum, svihlae, tjuzenii, vesiculariopsis*
GENUS *Bardunovia* Bardunov Pelagic Sheath Hypne Moss
SPECIES *baicalensis*
GENUS *Isocladiella* Isometric Branch Pelagic Sheath Hypne Moss
SPECIES *surcularis*
GENUS *Isopterygiopsis* Isometric Wing Pelagic Sheath Hypne Moss
SPECIES *alpicola, muelleriana, pulchella*
GENUS *Plagiotheciella* (Synonym for *Plagiothecium*)
GENUS *Pseudotaxiphyllum* Yew Pelagic Sheath Hypne Moss
SPECIES *densum, distichaceum, elegans, falcifolium, fauriei, homomallifolium, laetevirens, maebarae, pohliaecarpum, richardsii*

FAMILY 11 *Pterigynandraceae* Winged Gynander Hypne Moss
GENUS *Pterigynandrum* Winged Gynander Hypne Moss
SPECIES *boscii, filiforme*
GENUS *Habrodon* Habron Tooth Winged Gynander Hypne Moss
SPECIES *kashmiriensis, perpusillus*
GENUS *Iwatsukiella* Iwatsuki Winged Gynander Hypne Moss
SPECIES *leucotricha*

FAMILY 12 *Sematophyllaceae* Sign Hypne Moss
GENUS *Sematophyllum* Sign Hypne Moss
SPECIES *aberrans, adnatum, affine, allinckxiorum, amblystegiocarpum, ampullulatum, aneuron, angusticuspis, angusticybeum, angustifolium, angustirete, aptychoides, auratum, aureonitens, aureosulphureum, aureoviride, beyrichii, bottinii, brachycarpum, brachycladulum, brachytheciiforme, bracteatum, brasiliense, brotheri, calamicola, campicola, capilliferum, cataractae, cellulosum, ceylonense, chlorocormum, chlorothecium, chrysostegum, circinicaule, cirrhifolium, coppeyi, crassirete, crassiusculum, curvirostre, cuspidatum, cuspidiferum, cyparissoides, danckelmannii, decumbens, demissum, dimorphum, dregei, dubium, elgonense, entodontoides, erythrocaulon, erythropodium, euryphyllum, fabronia, flaccidifolium, flagelliferum, flavidum, flavovesiculosum, fluminicola, fragilirostrum, francii, friesiorum, frullaniadelphus, fulvifolium, galipense, gracile, grandicellulosum, harpidioides, hawaiiense, helenicum, homomallum, humile, humillimum, implanum, incrassetum, incurvum, integrifolium, jollifii, kunkelii, laetevirens, laticuspis, latifolium, leucostomum, lithophilum, lonchophyllum, longinerve, longisetum, macrocytus, macrorhynchum, mandobboense, marylandicum, masafuerae, meiothecioides, micrangium, microcarpoides, microcladiellum, microcladioides, minutum, mittenianum, nanocephalum, nebulosum, nematocaulon, nigro-alare, obtusifolium, oedophysidium, orthocarpum, pacimoniense, parisii, parvifolium, parvulum, pennellii, perichaetiale, perrevolutum, phoeniceum, physaophyllos, plagiothecioides, pobeguinii, pulvinale, reitzii, replicatum,*

rivuletorum, robustulum, roridum, ruvenzorense, santae, scabriusculum, schimperi, scorpiurus, serandii, serifolium, serrulatum, simulans, sinuosulum, socotrense, sphaeropyxis, squarrosum, steerei, stenopyxidium, steyermarkii, sticticola, stylites, subbrachycarpum, subbrachytheciiforme, subcespitosum, subcochleatum, subconnivens, subcylindricum, subdepressum, subfulvum, subhomomallum, subhumile, subjulaceum, subpiliferum, subpinnatum, subrugifolium, subscabrellum, subscabrum, subsecundum, subsimplex, substrumulosum, succedaneum, swartzii, temperatum, tenerifolium, tenuicarpum, tequendamense, tonduzii, trachaelocarpum, tucumanense, uncinatum, wageri, warmingii, zuluense

GENUS *Acanthocladium* Spine Branch Sign Hypne Moss

SPECIES *brevifolium, tanytrichoides*

GENUS *Acanthodium* Spine Tooth Sign Hypne Moss

SPECIES *rigidum*

GENUS *Acanthorrhynchium* Spine Beak Sign Hypne Moss

SPECIES *altisetulum, grosso-papillatum, microcarpum, monostictum, papillatum, subintegrum, substigmosum*

GENUS *Acritodon* Acrite Tooth Sign Hypne Moss

SPECIES *nephophilus*

GENUS *Acroporium* Tall Pore Sign Hypne Moss

SPECIES *aciphyllum, adspersum, affine, angustum, antarense, baviense, brevicuspidatum, brevisetulum, caespitosum, cataractarum, catharinense, ceylonense, condensatum, consanguineum, convolutifolium, convolutum, denticulatum, depressum, dicranoides, diminutum, dixonii, downii, eburense, erythropodium, esmeraldicum, estrellae, exiguum, flexisetum, fuscoflavum, gracilescens, gracillimum, hermaphroditum, joannis-winkleri, laevifolium, lamprophyllum, letestui, longicaule, longicuspis, longirostre, macro-turgidum, megasporum, microcladum, microthecium, nietnerianum, novae-guineae, perserratum, platycladum, plicatum, pocsii, praelongum, prionophylax, procerum, procumbens, pungens, ramicola, ramuligerum, rhaphidostegioides, ridleyi, rigens, rufum, savesianum, secundum, seriatum, serricalyx, smallii, stellatum, stramineum, strepsiphyllum, subluxurians, vincensianum, warburgii*

GENUS *Anhymenium* Membraneless Sign Hypne Moss

SPECIES *polysetum*

GENUS *Allioniellopsis* Allion Sign Hypne Moss

SPECIES *cryphaeoides*

GENUS *Aptychella* Folded Sign Hypne Moss

SPECIES *borii, chilensis, chlorophyllosa, linearifolia, proligera, speciosa, tenuiramea*

GENUS *Aptychopsis* Lesser Folded Sign Hypne Moss

SPECIES *pungifolia, pycnodonta, pyrrophylla, subpungifolia*

GENUS *Brotherella* Brother Sign Hypne Moss

SPECIES *amblystegia, complanata, coreana, crassipes, curvirostris, deplanatula, dixonii, erythrocaulis, falcata, fauriei, filiformis, henonii, herbacea, indosinensis, longipes, lorentziana, luzonensis, mercieri, minutula, nictans, opaeodon, pallida, propinqua, recurvans, roellii, subarcuata*

GENUS *Chionostomum* Snow Stoma Sign Hypne Moss

SPECIES *baolocense, hainanensis, pinicola, rostratum*

GENUS *Clastobryella* Lesser Clastic Sign Hypne Moss

SPECIES *tenella*

GENUS *Clastobryophilum* Clastic Sign Hypne Moss

SPECIES *balansaeanum, bogoricum*

GENUS *Clastobryopsis* Least Clastic Sign Hypne Moss

SPECIES *brevinervis, muelleri, perdecurrens, planula, robusta*

GENUS *Clastobryum* Greater Clastic Sign Hypne Moss

SPECIES *caudatum, conspicuum, cuculligerum, dimorphum, epiphyllum, glabrescens, indicum, leucophyllum, novo-guinense, panchoi, pungens, scalare, spiculiferum*

GENUS *Colobodontium* Monkey Tooth Sign Hypne Moss

SPECIES *vulpinum*

GENUS *Donnellia* Donnell Sign Hypne Moss

SPECIES *commutata, lagenifera, lageniformis, laxiuscula, longidens, matutina, usambarica*

GENUS *Foreauella* Foreau Sign Hypne Moss

SPECIES *orthothecia*

GENUS *Gammiella* Gamma Sign Hypne Moss

SPECIES *ceylonensis, koningsbergeri, panchienii, pterogonioides, rugosa, tonkinensis*

GENUS *Hageniella* Hagen Sign Hypne Moss

SPECIES *assamica, isopterygioides, micans, sikkimensis*

GENUS *Heterophyllium* Variable Sign Hypne Moss

SPECIES *acunae, affine, albescens, albicans, bulbilliferum, henryi, humbertii, micro-alare, microcarpum, nematosum, scabrifolium, subpiligerum, tonkinense*

GENUS *Horridohypnum* Horrible Sign Hypne Moss

SPECIES *mexicanum*

GENUS *Hypnella* Lesser Sign Hypne Moss

SPECIES *ambrosia, diversifolia, guayanensis, leptorrhyncha, pallescens, pilifera, punctata, recurvula, sigmatelloides*

GENUS *Macrohymenium* Membrane Sign Hypne Moss
SPECIES *acidodon, incrassatum, muelleri, strictum*
GENUS *Mahua* Mahu Sign Hypne Moss
SPECIES *enervis*
GENUS *Mastopoma* Mast Pome Sign Hypne Moss
SPECIES *armitii, brevifolium, denticulatum, epiphyllum, laevisetum, loriforme, papillosum, parvidentatum, raphidostegioides, robinsonii, robustum, salaense, scabrifolium, subfiliferum, subrobustum, uncinifolium*
GENUS *Meiotheciella* Lesser Small Sheath Sign Hypne Moss
SPECIES *papillosa*
GENUS *Meiotheciopsis* Least Small Sheath Sign Hypne Moss
SPECIES *asiatica, burmensis*
GENUS *Meiothecium* Small Sheath Sign Hypne Moss
SPECIES *attenuatum, boryanum, chlorocladum, congolense, crassirete, crispum, fuscescens, greenwoodii, guineense, hamatulum, hamatum, intextum, jagorii, longisetum, madagascariense, mediopapillatum, microcarpum, papuanum, pendulum, pobeguinii, prunicola, rechingeri, revolubile, secundifolium, serrulatum, stratosum, submicrotheca, tenellum, tenuirete, turgidellum, urceolatum, wattsii*
GENUS *Palisadula* Palisade Sign Hypne Moss
SPECIES *chrysophylla, katoi*
GENUS *Papillidiopsis* Papilla Sign Hypne Moss
SPECIES *bruchii, complanata, koponenii, luxurians, macrosticta, mahensis, malesiana, papillata, ramulina, stissophylla*
GENUS *Paranapiacabaea* Nap Cab Sign Hypne Moss
SPECIES *paulista*
GENUS *Potamium* Water Sign Hypne Moss
SPECIES *sanctae-mariae*
GENUS *Pseudohypnella* Least Sign Hypne Moss
SPECIES *verrucosa*
GENUS *Pseudotrismegistia* Triad Smegist Sign Hypne Moss
SPECIES *undulata*
GENUS *Pterogonidium* Wing Angled Sign Hypne Moss
SPECIES *liliputanum, pulchellum*
GENUS *Pterogoniella* Lesser Wing Angled Sign Hypne Moss
SPECIES *undulata*
GENUS *Pterogoniopsis* Least Wing Angled Sign Hypne Moss
SPECIES *cylindrica*
GENUS *Pylaisiadelpha* Pylais Adelphi Sign Hypne Moss
SPECIES *brasiliensis, capillacea, drepanioides, duellii, sharpii, tenuirostris, yokohamae*
GENUS *Radulina* Radial Sign Hypne Moss
SPECIES *aequorea, borbonica, elegantissima, hamata, scaberula*
GENUS *Rhaphidorrhynchium* Beaked Sign Hypne Moss
SPECIES *amoenum, andinum, angustissimum, argyrophyllum, argyroviride, aurescens, berberidis, brachypus, brevi-horridum, callidioides, callidum, calliferum, cambouei, capilliforium, chichibuense, confertissimum, crispans, dallii, decurvifolium, dendroligotrichum, distantifolium, gabonense, geheebii, gueinzii, hermaphroditum, hyoji-suzukii, incurvum, lamprosericeum, leptophyllum, levieri, lignicola, marmellense, meiotocladium, noduliferum, olfersii, pallido-nitidum, percondensatum, prominulum, purpuripes, quaylei, rubricaule, rueckeri, sanguisetum, subfalcatulum, symbolax, tegeticola, tereticaule, trichostegum, tuloferum*
GENUS *Rhaphidostegium* Beaked Cover Sign Hypne Moss
SPECIES *chiquitanum, cochleatulum, condensatulum, conterminum, cylindrothecium, densirete, diaphanodictyon, dixonii, eurycystis, fulvoalare, guarayum, hoehnei, micropyxidium, nanopyxis, plumosum, rigescens, sebillei, subovale, undulatum*
GENUS *Rhaphidostichum* Beaked Stitch Sign Hypne Moss
SPECIES *acestrostegium, brevisetum, bunodicarpum, clemensiae, cucullifolium, dubium, eberhardtii, elegans, gracile, gunckelii, leeanum, leptocarpoides, leptocarpum, loriforme, luzonense, macromonostictum, pallidifolium, papuanum, piliferoides, piliferum, pullei, pustulatum, schwaneckeanum, subleptocarpum, subrevolutum, theliporum*
GENUS *Schroeterella* Schroeter Sign Hypne Moss
SPECIES *zygodonta*
GENUS *Struckia* Struck Sign Hypne Moss
SPECIES *argentata, zerovii*
GENUS *Syringothecium* Syringe Sheath Sign Hypne Moss
SPECIES *brasiliense, sprucei*
GENUS *Taxithelium* Yew Sign Hypne Moss
SPECIES *afroacuminulatum, alare, andongense, annandii, archboldii, argyrophyllum, bakeri, benguetiae, binsteadii, borneense, brassii, capillarisetum, capillipes, chloropterum, clastobryoides, compressicaule, concavum, confusum, convolutum, deningeri, diaphanum, dimorphophyllum, epapillosum, falcatulum, falcifolium, francii, friedense, glabratum, glabrisetum, glabriusculum, glaucophyllum, glaucovirens, gottscheanum, guineense, herpetium, hirtellum, homalophyllum, impellucidum, inerme, instratum, isocladioides,*

isocladum, juruense, kerianum, kuilui, kuniense, laeviusculum, latitruncatum, leptopunctatum, leptosigmatum, levieri, liukiuense, ludovicae, magnum, meiothecioides, merrillii, microthamnioides, mundulum, muscicola, neocaledonicum, nepalense, nigerianum, nitidulum, nossianum, novae-valesiae, novae-zeelandiae, oblongifolium, papillisetum, parvulum, perglabrum, perminutum, petrophilum, planissimum, planulum, planum, pluripunctatum, pobeguinii, polyandrum, portoricense, protensum, pseudoamoenum, ramicola, ramivagum, rhaphidostegioides, rhizophoreti, robinsonii, rotundatulum, samoanum, schmidtii, selenithecium, spathulifolium, spuriosubtile, stenosekos, subandinum, sublaevifolium, suboctodiceras, subrotundatulum, subsimilans, sumatranum, tenuisetum, tongense, trachaelocarpum, uematsui, ventrifolium, vernieri, voeltzkowii, wattsii, werneri, wewakense

GENUS *Timotimius* — Tim Ear Sign Hypne Moss

SPECIES *titanotus*

GENUS *Trichosteleum* — Hair Stele Sign Hypne Moss

SPECIES *adhaerens, albifolium, altisetum, angustifolium, angustirete, annamense, apiocarpum, aquatile, arrectum, asperifolium, asselii, bernoullianum, bolivarense, boschii, brachydictyon, brongersmae, carolinarum, chrysophyllum, complanatulum, confertulum, consanguineum, constrictum, cymbiforme, debettei, dicranelloides, dicranoides, dimorphum, elegantulum, everettii, fissum, flexuosa-hamatum, francii, fruticola, fuscescens, fusco-virescens, gabonense, glaucinum, glaziovii, grosse-papillosum, grosso-mamillosum, gunnii, hebridarum, heterophyllum, horridulum, humbertii, inflexifolium, insigne, inundatum, jaegeri, janeirense, jonesii, lachmanii, laevi-hamatum, le-ratii, le-testui, leviusculum, longisetulum, lorifolium, louisiadum, lutschianum, mammosum, mastopomatoides, microcalyx, microdontum, microstegium, mindanense, minutissimum, monostictum, montanum, neocaledonicum, novae-guineae, orthophyllum, papillosissimum, papillosum, patens, pendens, pendulum, perchlorosum, perfalcatum, permixtum, perrotii, pervilleanum, petrophilum, piliferum, pinnatum, pobeguinii, popokvilense, procerum, pseudo-acuminulatum, pseudomammosum, punctipapillosum, pusillum, pygmaeum, robbinsii, rubrisetum, ruficaule, saproxylophilum, sarapiquense, schlimii, sematophylloides, sentosum, singapurense, smithii, staudtii, stereodontoides, stictum, stigmosum, subcucullifolium, subdemissum, subinstratum, sublaeve, sublaevifolium, sublaevigatum, sublongisetum, subrhinophyllum, subscabrisetulum, subulatulum, taxithelioides, tisserantii, tortipilum, trachyamphorum, trachycystis, trachypyxis, turgidulum, usambaricum, verrucosum, vieillardii, vietnamense, vincentinum, wattsii*

GENUS *Trismegistia* — Triad Smegist Sign Hypne Moss

SPECIES *brachyphylla, brauniana, breviflagellosa, calderensis, complanatula, delicatula, dendroides, deningeri, densiretis, gracilicaulis, lancifolia, monodii, panduriformis, pedunculata, pulchella, rigidicaulis, squarrosa, strangei, valetonii*

GENUS *Trolliella* — Troll Sign Hypne Moss

SPECIES *euendostoma*

GENUS *Warburgiella* — Warburg Sign Hypne Moss

SPECIES *ambigua, bistrumosa, breviseta, celebensis, circinata, concavifolia, cupressinoides, cuspidatifolia, falcatula, filicuspis, hygrophila, isopterygioides, kjelbergii, leptocarpa, leptorhynchoides, leptorrhyncha, leucocytus, macrospora, nitens, pallida, perviridis, philippinensis, pycnophylla, subleptorrhyncha, subleptorrhynchoides, subpapuana, subpycnophylla, venemae*

GENUS *Wijkia* — Wijk Sign Hypne Moss

SPECIES *albescens, alboalaris, alstonii, annamensis, baculifera, bessonii, brevifolia, carlottae, ceylonensis, clarkii, clastobryoides, comosa, concavifolia, costaricensis, crossii, cuynetii, deflexifolia, dentigera, extenuata, filifera, filipendula, flagellifera, flagelliformis, gracilis, hornschuchii, jacobsonii, jungneri, laxa, laxitexta, lepida, letestui, macgregorii, madagassa, nivea, pendula, penicillata, pinnata, polymorpha, protensa, radiculosa, rutenbergii, sublepida, subnitida, surcularis, tanytricha, trichocolea, trichocoleoides*

GENUS *Wijkiella* — Lesser Wijk Sign Hypne Moss

SPECIES *kenyae*

FAMILY 13 *Stereophyllaceae* — Stereo Hypne Moss

GENUS *Stereophyllum* — Stereo Hypne Moss

SPECIES *andungense, aptychopsis, bremondii, brunnthaleri, cataractarum, confusum, favaroi, fulvum, guineense, jamaicense, laxirete, lepidopiloides, leucothallum, limnobioides, macrocarpum, moelleri, neocaledonicum, omalosekos, radiculosum, reclinatum, rigescens, subacuminatum, vitiense, woodii*

GENUS *Entodontopsis* — Entic Tooth Stereo Hypne Moss

SPECIES *anceps, angustiretis, contorte-operculata, leucostega, mexicana, nitens, panamensis, papillifera, pygmaea, radicalis, rhynchostegioides, rufescens, sedgwickii, setschwanica, tavoyensis, tenuinervis, wightii*

GENUS *Eulacophyllum* — Lake Stereo Hypne Moss

SPECIES *cultelliforme*

GENUS *Juratzkaea* — Juratzk Stereo Hypne Moss

SPECIES *argentinica, incisa, indica, seminervis*

GENUS *Sciuroleskea* — Squirrel Leske Stereo Hypne Moss

SPECIES *mittenii, xanthophylla*

GENUS *Stenocarpidium* — Thin Carpel Stereo Hypne Moss

SPECIES *leucodon*

FAMILY 14 *Thamnobryaceae* — Thamne Hypne Moss

GENUS *Thamnobryum* — Thamne Hypne Moss

SPECIES *alleghaniense, alopecurum, aneitense, angustifolium, arbusculosum, assimile, capense, caroli, cataractarum, ceylonense, confertum, coreanum, corticola, ellipticum, fasciculatum, fruticosum, grandirete, incurvum, ingae, latifolium, liesneri, macrocarpum, maderense, malgachum, microalopecurum, neckeroides, negrosense, pandum, parvulum, pendulirameum, planifrons, plicatulum, proboscideum, pumilum, rigidum, speciosum, sublatifolium, subseriatum, subserratum, tumidicaule, tumidum, vorobjovii*

GENUS *Thamnium* — Lesser Thamne Hypne Moss

SPECIES *afrum, crassipes, dolichomitrioides, fluviaticum, hildebrandtii, moenkemeyeri, novae-walesiae, planissimum, serpenticaule, suspectum, variifolium*

GENUS *Bryolawtonia* — Lawton Thamne Hypne Moss

SPECIES *vancouveriensis*

GENUS *Crassiphyllum* — Crass Thamne Hypne Moss

SPECIES *fernandesii*

GENUS *Porothamnium* — Pore Thamne Hypne Moss

SPECIES *arbusculans, baculiferum, bigelovii, comorense, comosum, elatulum, flagellatum, floridum, frahmii, gracile, gymnopodum, hildebrandti, leptopteris, leucocaulon, minutum, obliquifolium, panduraefolium, parvirete, protensum, scariosum, serricola, sparsiflorum, subexplanatum, tenuinerve, valdiviae, variifolioides, viguieri*

FAMILY 15 *Theliaceae* — Nipple Hypne Moss

GENUS *Thelia* — Nipple Hypne Moss

SPECIES *asprella, hirtella, lescurii*

GENUS *Fauriella* — Fauri Nipple Hypne Moss

SPECIES *petelotii, robustiuscula, tenerrima, tenuis*

GENUS *Myurella* — Mouse Tail Nipple Hypne Moss

SPECIES *acuminata, julacea, sibirica, tenerrima*

FAMILY 16 *Thuidiaceae* — Thuy Hypne Moss

GENUS *Thuidium* — Thuy Hypne Moss

SPECIES *aculeoserratum, allenii, alvarezianum, assimile, batakense, bipinnatulum, blepharophyllum, brachymenium, brachypyxis, brachythecium, brasiliense, breviacuminatum, carantae, caudicaule, chacoanum, complanum, costaricense, cylindrella, cymbifoliolum, cymbifolium, delicatulum, djuriense, eccremocarpum, exasperatum, glaucescens, granulatum, haplohymenium, himantophyllum, hygrophilum, inconspicuum, ise-sanctum, kanedae, komarovii, laeviusculum, lepidoziaceum, ligulifolium, loricalycinum, mattogrossense, nanophyllum, neocaledonicum, nutans, occultirete, ochraceum, orthocarpum, paraguense, patrum, pellucens, perscissum, peruvianum, pinnatulum, plumulosum, poeppigii, pristocalyx, promontorii, pseudodelicatulum, pseudoglaucinum, pseudoprotensum, pulvinatulum, purpureum, quadrifarium, recognitum, scabribracteatum, schiedeanum, serricola, siphotheca, subbifarium, subglaucinum, subgranulatum, subinvolvens, submicropteris, subserratum, subtamariscinum, subtilissimum, tamariscinum, tomentosum, torrentium, tuerckheimii, undulatifolium, unipinnatum, urceolatum, verrucipes, wrightii, yungarum*

GENUS *Abietinella* — Fir Thuy Hypne Moss

SPECIES *abietina, hystricosa*

GENUS *Actinothuidium* — Ray Thuy Hypne Moss

SPECIES *hookeri*

GENUS *Aequatoriella* — Equatorial Thuy Hypne Moss

SPECIES *bifaria*

GENUS *Anomodon* — Anomalous Tooth Thuy Hypne Moss

SPECIES *abbreviatus, attenuatus, dentatus, flagelliformis, giraldii, grandiretis, heteroideus, longifolius, longinervis, minor, pseudotristis, rostratus, rotundatus, rugelii, sieboldii, solovjovii, thraustus, tristis, viticulosus*

GENUS *Boulaya* — Boulay Thuy Hypne Moss

SPECIES *mittenii*

GENUS *Bryochenea* — Chen Thuy Hypne Moss

SPECIES *vestitissima*

GENUS *Bryonoguchia* — Noguchi Thuy Hypne Moss

SPECIES *molkenboeri*

GENUS *Claopodium* — Claopod Thuy Hypne Moss

SPECIES *aciculum, assurgens, bolanderi, crispifolium, fulvellum, gracillimum, kiusiense, pellucinerve, prionophyllum, rugulosifolium, undulatifolium, whippleanum*

GENUS *Cyrtohypnum* — Curved Thuy Hypne Moss

SPECIES *arzobispoae, atlanticum, brachythecium, cylindraceum, espinosae, frontinoae, intermedium, lepidoziaceum, pseudoinvolvens, sharpii, stevensii*

GENUS *Echinophyllum* — Urchin Thuy Hypne Moss

SPECIES *sachalinense*

GENUS *Haplocladium* — Single Branch Thuy Hypne Moss

SPECIES *angustifolium, discolor, fauriei, incurvum, jacquemontii, larminatii, microphyllum, parvulum, percapillatum, perparaphyllum, persistens, plumulosum, porphyropelma, schimperi, stratosum, strictulum, virginianum*

GENUS *Haplohymenium* — (Synonym for *Thuidium*)

GENUS *Garysmithia* — (Synonym for *Thuidium*)

GENUS *Helodium* — Helode Thuy Hypne Moss

SPECIES *blandowii, paludosum*

GENUS *Herpetineuron* — Serpent Nerve Thuy Hypne Moss

SPECIES *toccoae*

GENUS *Heterocladium* Variable Branch Thuy Hypne Moss
SPECIES *angustifolium, capillaceum, dimorphum, heteropterum, macounii, pilicuspis, procurrens, tenellum, wulfsbergii*
GENUS *Hylocomiopsis* Wood Tuft Thuy Hypne Moss
SPECIES *cylindracarpa, ovicarpa*
GENUS *Indothuidium* India Thuy Hypne Moss
SPECIES *kiasense*
GENUS *Inouethuidium* Inoue Thuy Hypne Moss
SPECIES *robbinsii*
GENUS *Leptocladium* Thin Branch Thuy Hypne Moss
SPECIES *sinense*
GENUS *Leptopterigynandrum* Thin Wing Gynander Thuy Hypne Moss
SPECIES *austro-alpinum, autoicum, brevirete, clavatum, decolor, filiforme, incurvatum, stricticaule, subintegrum, tenellum*
GENUS *Miyabea* Miyabe Thuy Hypne Moss
SPECIES *fruticella, nishimurae, rotundifolia, thuidioides*
GENUS *Orthothuidium* Straight Thuy Hypne Moss
SPECIES *curtisetum*
GENUS *Pelekium* Pelek Thuy Hypne Moss
SPECIES *bonianum, campanulatum, chenagonii, ciliatum, contortulum, erosifolium, fuscatum, gratum, intricatum, investe, involvens, leptocladum, microphyllum, minusculum, minutulum, muricatulum, pygmaeum, ramosissimum, ramusculosum, scabrosulum, schistocalyx, siphotheca, subbifarium, subpinnatum, synoicum, varians, velatum, versicolor, watanabei*
GENUS *Ptychodium* Folded Thuy Hypne Moss
SPECIES *plicatum*
GENUS *Rauia* (Synonym for *Rauiella*)
GENUS *Rauiella* Raui Thuy Hypne Moss
SPECIES *bornii, fujisana, lagoensis, niveo-calycina, praelonga, scita, siambonica, teretiuscula*
GENUS *Thuidiopsis* Least Thuy Hypne Moss
SPECIES *chilensis, dusenii, furfurosa, sparsa*

ORDER 7 *Isobryales* Isometric Moss

FAMILY 1 *Isobryaceae* Isometric Moss
GENUS *Lepyrodon* Thin Tooth Isometric Moss
SPECIES *australis, hexastichus, lagurus, parvulus, patagonicus, pseudolagurus, tomentosus*
GENUS *Regmatodon* Regma Tooth Isometric Moss
SPECIES *declinatus, gracillimus, nietneri, orthostegius, polysetus*
GENUS *Wardia* Ward Isometric Moss
SPECIES *hygrometrica*

FAMILY 2 *Cyrtopodaceae* Arch Pod Isometric Moss
GENUS *Cyrtopus* Arch Pod Isometric Moss
SPECIES *setosus*
GENUS *Bescherellia* Bescher Arch Pod Isometric Moss
SPECIES *elegantissima*

FAMILY 3 *Fontinalaceae* Fontinal Isometric Moss
GENUS *Fontinalis* Fontinal Isometric Moss
SPECIES *allenii, amblyphylla, antipyretica, bogotensis, bryhnii, chrysophylla, dalecarlica, dichelymoides, duriaei, flaccida, gigantea, howellii, hypnoides, involuta, kindbergii, lachenaudii, lescurii, macmillanii, microphylla, missourica, mollis, neomexicana, novae-angliae, patula, perfida, redfearnii, rigens, sphagnifolia, squamosa, subcarinata, sullivantii, utahensis, waghornei, welchiana*
GENUS *Brachelyma* Short Rye Fontinal Isometric Moss
SPECIES *subulatum*
GENUS *Dichelyma* Dual Rye Fontinal Isometric Moss
SPECIES *capillaceum, falcatum, japonicum, pallescens, sinense, uncinatum*

FAMILY 4 *Hydropogonaceae* Hydric Beard Isometric Moss
GENUS *Hydropogon* Hydric Beard Isometric Moss
SPECIES *fontinaloides*
GENUS *Hydropogonella* Lesser Hydric Beard Isometric Moss
SPECIES *gymnostoma*

FAMILY 5 *Prionodontaceae* Sawtooth Isometric Moss
GENUS *Prionodon* Sawtooth Isometric Moss
SPECIES *balboi, contortus, densus, fuscolutescens, hoffmannii, jamesonii, lycopodioides*
GENUS *Taiwanobryum* Taiwan Sawtooth Isometric Moss
SPECIES *speciosum*

FAMILY 6 *Ptychomniaceae* Folded Omni Isometric Moss

GENUS *Ptychomnion* Folded Omni Isometric Moss
SPECIES *aciculare, cygnisetum, densifolium, falcatulum, ptychocarpon*
GENUS *Cladomnion* Branch Folded Omni Isometric Moss
SPECIES *ericoides*
GENUS *Cladomniopsis* Lesser Branch Folded Omni Isometric Moss
SPECIES *crenato-obtusa*
GENUS *Dichelodontium* Dual Tooth Folded Omni Isometric Moss
SPECIES *nitidum*
GENUS *Glyphothecium* Glyph Sheath Folded Omni Isometric Moss
SPECIES *gracile, papillosum, pendulum, sciuroides*
GENUS *Hampeella* Hampe Folded Omni Isometric Moss
SPECIES alaris, pallens
GENUS *Tetraphidopsis* Tetrad Aphid Folded Omni Isometric Moss
SPECIES *pusilla*

FAMILY 7 *Rutenbergiaceae* Rutenberg Isometric Moss
GENUS *Rutenbergia* Rutenberg Isometric Moss
SPECIES *borbonica, cirrata, limbata, madagassa, prionodon*
GENUS *Neorutenbergia* Modern Rutenberg Isometric Moss
SPECIES *usagarae*

FAMILY 8 *Trachypodaceae* Rough Pod Isometric Moss
GENUS *Trachypus* Rough Pod Isometric Moss
SPECIES *appendiculatus, bicolor, humilis, longifolius, perplicatus*
GENUS *Bryowijkia* Wijk Rough Pod Isometric Moss
SPECIES *ambigua, madagassa*
GENUS *Diaphanodon* Diaphanous Tooth Rough Pod Isometric Moss
SPECIES *blandus, friisii, laevivittatus, procumbens*
GENUS *Pseudospiridentopsis* Spire Tooth Rough Pod Isometric Moss
SPECIES *horrida*
GENUS *Pseudotrachypus* Lesser Rough Pod Isometric Moss
SPECIES *convolvens, martinicensis, spiculatus, stevensii, wallichii*
GENUS *Trachypodopsis* Least Rough Pod Isometric Moss
SPECIES *auriculata, formosana, lancifolia, laxoalaris, normandii, serrulata*

ORDER 8 *Leucodontales* Light Tooth Moss

FAMILY 1 *Leucodontaceae* Light Tooth Moss
GENUS *Leucodon* Light Tooth Moss
SPECIES *alpinus, andrewsianus, angustiretis, assimilis, atrovirens, brachypus, canariensis, coreensis, cryptotheca, curvirostris, denticulatus, dracaenae, exaltatus, felipponei, flagellaris, flagelliformis, formosanus, giganteus, immersus, jaegerinaceus, julaceus, laxifolius, luteus, morrisonensis, nipponicus, pendulus, radicalis, rutenbergii, sapporensis, sciuroides, secundus, sinensis, sohayakiensis, sphaerocarpus, subgracilis, subulatus, temperatus, tibeticus, treleasei*
GENUS *Alsia* Alsian Light Tooth Moss
SPECIES *californica*
GENUS *Antitrichia* Hairless Light Tooth Moss
SPECIES *californica, curtipendula, kilimandscharica*
GENUS *Bestia* Best Light Tooth Moss
SPECIES *longipes*
GENUS *Climacium* Climax Light Tooth Moss
SPECIES *acuminatum, americanum, dendroides, japonicum*
GENUS *Dendroalsia* Wood Alsian Light Tooth Moss
SPECIES *abietina*
GENUS *Dozya* Dozy Light Tooth Moss
SPECIES *japonica*
GENUS *Eoleucodon* Greater Light Tooth Moss
SPECIES *pacificus*
GENUS *Felipponea* Felippe Light Tooth Moss
SPECIES *esquirolii, hollermayeri, montevidensis*
GENUS *Forsstroemia* Forsster Light Tooth Moss
SPECIES *canariensis, coronata, cryphaeoides, indica, japonica, lasioides, neckeroides, noguchii, producta, stricta, thomsonii, trichomitria, tripinnata*
GENUS *Leucodontopsis* (Synonym for *Leucodon*)
GENUS *Okamuraea* Okamura Light Tooth Moss
SPECIES *brachydictyon, brevipes, hakoniensis, plicata*
GENUS *Pterogonium* Wing Angled Light Tooth Moss
SPECIES *beyrichianum, gracile, peregrinum*

FAMILY 2 *Cryphaeaceae* Cryphe Light Tooth Moss

GENUS *Cryphaea* Cryphe Light Tooth Moss
SPECIES *acuminata, amurensis, apiculata, attenuata, brevipila, chlorophyllosa, clandestina, consimilis, filiformis, furcinervis, glomerata, gracillima, heteromalla, hygrophila, jamesonii, lanceolata, lorentziana, nervosa, obovatocarpa, omeiensis, orizabae, ovalifolia, parvula, patens, pilifera, protensa, pusilla, raddiana, ragazzii, ramosa, rhacomitrioides, rutenbergii, schiedeana, songpanensis, tenella*
GENUS *Cryphaeophilum* Least Cryphe Light Tooth Moss
SPECIES *molle*
GENUS *Cryphidium* Lesser Cryphe Light Tooth Moss
SPECIES *leucocoleum*
GENUS *Cyptodon* Cryptic Tooth Cryphe Light Tooth Moss
SPECIES *dilatatus, fasciculatus, gracilis, muelleri*
GENUS *Cyptodontopsis* Lesser Cryptic Tooth Cryphe Light Tooth Moss
SPECIES *leveillei*
GENUS *Dendrocryphaea* Wood Cryphe Light Tooth Moss
SPECIES *crassinervis, cuspidata, gorveana, lamyana, latifolia, lechleri, tasmanica*
GENUS *Dendropogonella* Wood Beard Cryphe Light Tooth Moss
SPECIES *rufescens*
GENUS *Monocryphaea* Monad Cryphe Light Tooth Moss
SPECIES *ravenelii*
GENUS *Pilotrichopsis* Downy Hair Cryphe Light Tooth Moss
SPECIES *dentata, ferruginea, robusta*
GENUS *Schoenobryum* Beautiful Cryphe Light Tooth Moss
SPECIES *blumenauianum, caripense, concavifolium, evanescens, henschenii, kunertii, paraguense, rubricaule*
GENUS *Sphaerotheciella* Sphere Sheath Cryphe Light Tooth Moss
SPECIES *koponenii, pachycarpa, pinnata, sinensis, sphaerocarpa*

FAMILY 3 *Hedwigiaceae* Hedwig Light Tooth Moss
GENUS *Hedwigia* Hedwig Light Tooth Moss
SPECIES *ciliata, detonsa, filiformis, stellata*
GENUS *Braunia* Braun Hedwig Light Tooth Moss
SPECIES *alopecura, arbuscula, attenuata, camptoclada, cirrhifolia, cochlearifolia, delavayi, divaricatula, entodonticarpa, exserta, laxifolia, liebmanniana, nephelogenes, obtusicuspis, plicata, reflexifolia, rupestris, schimperianum, secunda, squarrulosa, subincana, subplicata, tucumanensis*
GENUS *Hedwigidium* Lesser Hedwig Light Tooth Moss
SPECIES *glyphocarpum, integrifolium*
GENUS *Pararhacocarpus* Lesser Rhace Carpel Hedwig Light Tooth Moss
SPECIES *patagonicus*
GENUS *Pseudobraunia* Lesser Braun Hedwig Light Tooth Moss
SPECIES *californica*
GENUS *Rhacocarpus* Rhace Carpel Hedwig Light Tooth Moss
SPECIES *alpinus, apiculatus, chlorotus, excisus, inermis, purpurascens, rehmannianus, strictipilus*

FAMILY 4 *Leptodontiaceae* Thin Light Tooth Moss
GENUS *Leptodontium* Thin Light Tooth Moss
SPECIES *aggregatum, allorgei, araucarieti, brachyphyllum, capituligerum, erythroneuron, filicola, flexifolium, fuhrmannii, gambaragarae, gemmascens, handelii, interruptum, latifolium, longicaule, luteum, neesii, paradoxum, planifolium, proliferum, pungens, repens, rhynchophorum, saxicola, scaberrimum, stellaticuspis, stellatifolium, stoloniferum, styriacum, subintegrifolium, syntrichioides, tricolor, trifarium, viticulosoides, wallisii, zygodontoides*
GENUS *Caduciella* Wand Thin Light Tooth Moss
SPECIES *guangdongensis, mariei*
GENUS *Leptodon* Lesser Thin Light Tooth Moss
SPECIES *fuciformis, smithii*
GENUS *Leptodontiella* Least Thin Light Tooth Pott Moss
SPECIES *apiculata*
GENUS *Paraleptodontium* (Synonym for *Leptodontium*)
GENUS *Pseudocryphaea* Cryphe Thin Light Tooth Moss
SPECIES *domingensis*

FAMILY 5 *Meteoriaceae* Meteor Light Tooth Moss
GENUS *Meteorium* Meteor Light Tooth Moss
SPECIES *africanum, aongstromii, araucariophila, balansaeanum, buchananii, ciliaphyllum, crispifolium, cucullatum, denticulatum, deppei, elatipapilla, filiforme, flexicaule, gerale, illecebraria, illecebrinum, imbricatum, intricatum, latiphyllum, leuconeurum, medium, nigrescens, nitens, papillarioides, polytrichum, pseudoteres, regnellii, rigidum, riograndense, spathulifolium, squamidioides, striatum, subpolytrichum, taiwanense, teres, undulifolium, viride*
GENUS *Aerobryidium* Lesser Aero Meteor Light Tooth Moss
SPECIES *aureonitens, crispifolium, filamentosum, fuscescens*

GENUS *Aerobryopsis* Least Aero Meteor Light Tooth Moss
SPECIES *capensis, cirrifolia, cochlearifolia, deflexa, horrida, leptosigmata, longissima, membranacea, parisii, pulchricoma, spurio-convolvens, striatula, subdivergens, subleptostigmata, vitiana*
GENUS *Aerobryum* Aero Meteor Light Tooth Moss
SPECIES *speciosum*
GENUS *Ancistrodes* Ancestor Meteor Light Tooth Moss
SPECIES *genuflexa*
GENUS *Barbella* Beard Meteor Light Tooth Moss
SPECIES *amoena, angustifolia, biformis, capillaris, capillicaulis, chrysonema, compressiramea, conferva, elegantula, eusebii, flagellifera, futunensis, horridula, linearifolia, macroblasta, perpinnata, phyllogonioides, rufifolia, turgida*
GENUS *Barbellopsis* Lesser Beard Meteor Light Tooth Moss
SPECIES *trichophora*
GENUS *Chrysocladium* Golden Branch Meteor Light Tooth Moss
SPECIES *retrorsum*
GENUS *Cryptopapillaria* Cryptic Papilla Meteor Light Tooth Moss
SPECIES *chrysoclada, feae, fuscescens, helictophylla, penicillata*
GENUS *Floribundaria* Abundant Flower Meteor Light Tooth Moss
SPECIES *flaccida, floribunda, floribundula, glabrata, golae, implicata, intermedia, laxifolia, leptonema, pseudofloribunda, schenckii, setschwanica, subintegra, torquata, vaginans, walkeri*
GENUS *Lindigia* Lindig Meteor Light Tooth Moss
SPECIES *debilis*
GENUS *Meteoridium* Lesser Meteor Light Tooth Moss
SPECIES *remotifolium, tenuissimum*
GENUS *Meteoriopsis* Least Meteor Light Tooth Moss
SPECIES *auronitens, javensis, maroniensis, novoguinensis, pseudostraminea, reclinata, squarrosa, straminea, undulata*
GENUS *Neodicladiella* Two Branch Meteor Light Tooth Moss
SPECIES *pendula*
GENUS *Neonoguchia* Noguchi Meteor Light Tooth Moss
SPECIES *auriculata*
GENUS *Orthostichella* Straight Stitch Meteor Light Tooth Moss
SPECIES *auricularis, filamentosula, hexasticha, microcarpa, mucronatula, pandurifolia, versicolor, viridis*
GENUS *Papillaria* Papilla Meteor Light Tooth Moss
SPECIES *acinacifolia, araucarieti, bamforthiae, bazzaccoi, bipinnata, brasiliense, breviculifolia, brotheri, caldensis, callochlorosa, capillicuspis, catharinae, catharinensis, cordatifolia, crenifolia, crocea, densifolium, dilatata, filifunalis, flaccidula, flagellifera, flavolimbata, fulva, guarapiensis, hyalinotricha, laxifolia, longifolia, meteorioides, mosenii, myrtacearum, perauriculata, perichaetialis, pilifolia, pseudoappressa, ptychophylla, renauldii, scariosa, socia, subaongstroemiana, subintegra, tenella, tijucae, trachyblasta, viridata, wagneri, warszewiczii, zeloflexicaulis*
GENUS *Pilotrichella* Downy Hair Meteor Light Tooth Moss
SPECIES *allionii, ampullacea, angustifolia, attenuata, caldensis, calomicra, capillicaulis, communis, curvifrons, cuspidans, cyathipoma, dimorpha, filamentosula, flexilis, gracilicaulis, grimaldii, holstii, incurva, inflatifolia, isleana, latiramea, leptoclada, mascarenica, mauiensis, moenkemeyeri, muelleri, nudiramulosa, pachygaster, pachygastrella, pallidicaulis, perrobusta, phleoides, pilifolia, profusicaulis, pseudoimbricata, pulchella, quadrangularis, quitensis, ragazzii, rigida, sericea, sordido-viridis, squarrulosa, stuhlmannii, subpachygastrella, subpanduraefolia, subsemitorta, tenella, tenellula, turgidellacea, viridis, welwitschii*
GENUS *Pseudobarbella* Least Beard Meteor Light Tooth Moss
SPECIES *ancistrodes, attenuata, hattorii, laosiensis, laxifolia, levieri, propagulifera, validiramosa*
GENUS *Sinskea* Sinske Meteor Light Tooth Moss
SPECIES *flammea, phaea*
GENUS *Squamidium* Scaly Meteor Light Tooth Moss
SPECIES *angustifolium, brasiliense, cuspidatum, diversicoma, isocladum, leucotrichum, livens, macrocarpum, nigricans, ternstroemiae*
GENUS *Toloxis* Tolox Meteor Light Tooth Moss
SPECIES *borchgrevinkii, imponderosa, semitorta*
GENUS *Trachycladiella* Rough Branch Meteor Light Tooth Moss
SPECIES *aurea, sparsa*
GENUS *Tricholepis* (Synonym for *Meteorium*)
GENUS *Weymouthia* Weymouth Meteor Light Tooth Moss
SPECIES *cochlearifolia, mollis*
GENUS *Zelometeorium* Zelo Meteor Light Tooth Moss
SPECIES *allionii, ambiguum, patens, patulum*

FAMILY 6 *Neckeraceae* Necker Light Tooth Moss
GENUS *Neckera* Necker Light Tooth Moss
SPECIES *aequalifolia, americana, andina, andrei, angustifolia, aurescens, balansae, besseri, bhutanensis, borealis, brownii, caldensis, chilensis, complanata, coreana, crenulata, crispa, decurrens, denigricans, douglasii,*

ehrenbergii, emersa, falcata, fauriei, flexiramea, formosana, goughiana, hawaiico-pennata, hedbergii, heterophylla, himalayana, humilis, intermedia, konoi, laevigata, leichhardtii, leptodontea, nakazimae, neckeroides, nelloi, obtusifolia, oligocarpa, pachycarpa, pennata, perpinnata, platyantha, plumosa, polyclada, praelonga, puiggarii, pumila, pusilla, remota, rigida, rotundata, sanctae-catharinae, scabridens, semicrispa, setschwanica, spruceana, submacrocarpa, subuliformis, sundaensis, tenera, tjibodensis, undulatifolia, urnigera, valentiniana, villae-ricae, viridis, yezoana, yunnanensis

GENUS *Baldwiniella* — Baldwin Necker Light Tooth Moss
SPECIES *kealeensis*
GENUS *Bissetia* — Bisset Necker Light Tooth Moss
SPECIES *lingulata*
GENUS *Curvicladium* — Curved Branch Necker Light Tooth Moss
SPECIES *kurzii*
GENUS *Handeliobryum* — Handel Necker Light Tooth Moss
SPECIES *sikkimense*
GENUS *Himantocladium* — Himante Branch Necker Light Tooth Moss
SPECIES *cyclophyllum, formosicum, giulianettii, implanum, pacificum, plumula, submontanum, warburgii*
GENUS *Homalia* — Flat Necker Light Tooth Moss
SPECIES *glabella, lusitanica, pennatula, sakontala, trichomanoides, variifolia, webbiana*
GENUS *Homaliadelphus* — Flat Adelphi Necker Light Tooth Moss
SPECIES *sharpii, targionianus*
GENUS *Homaliodendron* — Flat Wood Necker Light Tooth Moss
SPECIES *crassinervium, dentatum, erosifolium, exiguum, flabellatum, gracile, handelii, intermedium, javanicum, ligulaefolium, microdendron, microphyllum, montagneanum, opacum, papillosum, paquei, piniforme, pinnatelloides, punctulatum, pygmaeum, rectifolium, scalpellifolium, spinosum, squarrulosum, stracheyanum, undulatum*
GENUS *Hydrocryphaea* — Hydric Cryphe Necker Light Tooth Moss
SPECIES *wardii*
GENUS *Isodrepanium* — Isometric Sickle Necker Light Tooth Moss
SPECIES *lentulum*
GENUS *Limbella* — Limbic Necker Light Tooth Moss
SPECIES *fryei, tricostata*
GENUS *Metaneckera* — Meta Necker Light Tooth Moss
SPECIES *menziesii*
GENUS *Neckeradelphus* — (Synonym for *Neckera*)
GENUS *Neckeropsis* — Lesser Necker Light Tooth Moss
SPECIES *andamana, beccariana, boiviniana, boniana, calcicola, chevalieri, crinita, darjeelingensis, disticha, exserta, fimbriata, foveolata, gracilenta, gracilis, inundata, lepineana, liliana, madecassa, moutieri, nanodisticha, nitidula, obtusata, pocsii, semperiana, spurio-truncata, submarginata, subtruncata, takahashii, touwii, undulata*
GENUS *Neomacounia* — Macoun Necker Light Tooth Moss
SPECIES *nitida*
GENUS *Noguchiodendron* — Noguchi Wood Necker Light Tooth Moss
SPECIES *sphaerocarpum*
GENUS *Pendulothecium* — Pendent Sheath Necker Light Tooth Moss
SPECIES *auriculatum, oblongifolium, punctatum*
GENUS *Pinnatella* — Pinnate Necker Light Tooth Moss
SPECIES *alopecuroides, ambigua, amblyphylla, anacamptolepis, calcutensis, foreauana, gollanii, kuehliana, limbata, makinoi, minuta, mixta, mucronata, robusta, taiwanensis, uroclada*
GENUS *Porotrichum* — Pore Hairy Necker Light Tooth Moss
SPECIES *amboroicum, atlanticum, bolivianum, brevifolium, caesium, caldense, cavifolium, chilense, crassipes, dentatum, elongatum, eurydictyon, expansum, filiferum, globiglossum, guatemalense, korthalsianum, krauseanum, lancifrons, liliputanum, longirostre, macropoma, madagassum, microthecium, mucronulatulum, mutabile, pennaefrondeum, plumosum, pobeguinii, pterops, quintasii, stipitatum, subangustifolium, subpennaeforme, substriatum, tenuinerve, thieleanum, usagarum*
GENUS *Touwia* — Touwi Necker Light Tooth Moss
SPECIES *laticostata*

FAMILY 7 *Pterobryaceae* — Winged Light Tooth Moss
GENUS *Pterobryon* — Winged Light Tooth Moss
SPECIES *arbuscula, australiense, densum, excelsum, flagelliferum, julaceum, pusillum, subarbuscula*
GENUS *Calyptothecium* — Calypter Sheath Winged Light Tooth Moss
SPECIES *acostatum, acutum, auriculatum, australinum, bescherellei, burmense, compressum, crispulum, duplicatum, extensum, hamatum, hookeri, integrifolium, nitidum, oxyphyllum, phyllogonioides, pinnatum, prainii, pterobryoides, ramosii, recurvulum, rhystotis, seminerve, squarrosulum, subacutum, subecostatum, subhumile, symphysodontoides, urvilleanum, wightii*
GENUS *Cryptogonium* — Cryptic Angle Winged Light Tooth Moss
SPECIES *phyllogonioides*
GENUS *Cryptoleptodon* — Cryptic Thin Tooth Winged Light Tooth Moss
SPECIES *acuminatus, longisetus, pluvinii, rigidulus*

GENUS *Cyrtopodendron* Cryptic Wood Winged Light Tooth Moss
SPECIES *vieillardii*
GENUS *Endotrichella* Lesser Endic Hairy Winged Light Tooth Moss
SPECIES *arundinifolia, assimilis, boutaniana, elmeri, fragarioides, gracilescens, irwinii, lepida, poilaneana, robinsonii, sarawakensis, secunda, serricuspis, straminea, wallisii*
GENUS *Endotrichellopsis* Least Endic Hairy Winged Light Tooth Moss
SPECIES laevifolia
GENUS *Endotrichum* Endic Hairy Winged Light Tooth Moss
SPECIES *densum*
GENUS *Euptychium* Fold Winged Light Tooth Moss
SPECIES *assimile, cuspidatum, dumosum, gunnii, mucronatum, papillosum, philippinense, pungens, robustum, setigerum, vitiense*
GENUS *Garovaglia* Garovagli Winged Light Tooth Moss
SPECIES *angustifolia, aristata, baeuerlenii, binsteadii, brassii, compressa, crispa, crispata, eberhardtii, elegans, lepida, luzonensis, mujuensis, plicata, plumosa, powellii, robbinsii, subelegans, tortifolia, zantenii*
GENUS *Henicodium* Henicode Winged Light Tooth Moss
SPECIES *geniculatum*
GENUS *Hildebrandtiella* Hildebrandt Winged Light Tooth Moss
SPECIES *cuspidans, endotrichelloides, guyanensis, longiseta, pachyclada*
GENUS *Horikawaea* Horikawa Winged Light Tooth Moss
SPECIES *dubia, nitida, redfearnii*
GENUS *Jaegerina* Jaeger Winged Light Tooth Moss
SPECIES *formosa, guatemalensis, luzonensis, plicata, retrosquarrosa, robillardii, scariosa, solitaria, stolonifera*
GENUS *Jaegerinopsis* Lesser Jaeger Winged Light Tooth Moss
SPECIES *brasiliensis*
GENUS *Meteoriella* Meteor Winged Light Tooth Moss
SPECIES *soluta*
GENUS *Micralsopsis* Micrals Winged Light Tooth Moss
SPECIES *complanata*
GENUS *Muellerobryum* Mueller Winged Light Tooth Moss
SPECIES *whiteleggei*
GENUS *Neolindbergia* Lindberg Winged Light Tooth Moss
SPECIES *cladomnioides, deningeri, rigida, robusta, rugosa, veloirae, vitiensis*
GENUS *Orthorrhynchidium* Straight Beak Winged Light Tooth Moss
SPECIES *planifrons*
GENUS *Orthostichidium* Straight Stitch Winged Light Tooth Moss
SPECIES *elongatum, involutifolium, subpendulum*
GENUS *Orthostichopsis* Lesser Straight Stitch Winged Light Tooth Moss
SPECIES *auricosta, debilinervis, latifolia, longinervis, madagassa, pinnatella, praetermissa, scaberula, schiffneri, strictula, subimbricata, sublivens, subtenuis, tenuis, tetragona, tijucae, tortipilis*
GENUS *Osterwaldiella* Osterwald Winged Light Tooth Moss
SPECIES *monostricta*
GENUS *Penzigiella* Penzig Winged Light Tooth Moss
SPECIES *cordata*
GENUS *Pireella* Pire Winged Light Tooth Moss
SPECIES *angustifolia, cymbifolia, falcifolia, filicina, formosana, guatemalensis, husnotiana, mariae, papillosula, pergemmescens, pohlii, pycnothallodes, trichomanoides*
GENUS *Pseudopterobryum* Least Winged Light Tooth Moss
SPECIES *laticuspis, tenuicuspis*
GENUS *Pterobryella* Lesser Winged Light Tooth Moss
SPECIES *brevi-acuminata, longifrons, papuensis, praenitens, rigida, speciosissima*
GENUS *Pterobryidium* Greater Winged Light Tooth Moss
SPECIES *australe*
GENUS *Pterobryopsis* Mock Winged Light Tooth Moss
SPECIES *acuminata, acutifolium, angustifolia, aurantia, breviflagellosa, burmensis, crassicaulis, crassiuscula, cucullatifolia, divergens, fasciculata, foulkesiana, gedehensis, hoehnelii, itahiae, kegeliana, longissima, mexicana, myurioides, orientalis, pilifolia, rehmannii, rubrinervis, scabriuscula, schmidii, setschwanica, stolonacea, subcrassicaulis, subcrassiuscula*
GENUS *Pulchrinodus* Pretty Winged Light Tooth Moss
SPECIES *inflatus*
GENUS *Renauldia* Renauld Winged Light Tooth Moss
SPECIES *baueri, chilensis, dusenii, hildebrandtielloides, lycopodioides, mexicana, paradoxica, patentissima, peruviana*
GENUS *Rhabdodontium* Rod Winged Light Tooth Moss
SPECIES *buftonii*
GENUS *Scabridens* Scabre Winged Light Tooth Moss

SPECIES *sinensis*
GENUS *Spiridentopsis* Spire Tooth Winged Light Tooth Moss
SPECIES *longissima*
GENUS *Symphysodon* Symphony Tooth Winged Light Tooth Moss
SPECIES *enervis, gunnii, langbianensis, longicuspis, neckeroides, pilifer, siamensis, squarrosus, symphysodontelloides, vitianus*
GENUS *Symphysodontella* Lesser Symphony Tooth Winged Light Tooth Moss
SPECIES *attenuatula, convoluta, cylindracea, involuta, laxissima, parvifolia, scaphidiophylla, siamensis, splendens, subulata, tortifolia*
GENUS *Trachyloma* Rough Winged Light Tooth Moss
SPECIES *concavifolium, dimorpha, diversinerve, indicum, planifolium, pycnoblastum, wattsii*

ORDER 9 *Orthotrichales* Straight Hair Moss

FAMILY 1 *Orthotrichaceae* Straight Hair Moss
GENUS *Orthotrichum* Straight Hair Moss
SPECIES *acuminatum, aequatoreum, affine, afrofastigiatum, alpestre, anaglyptodon, anomalum, araucarieti, arborescens, armatum, assimile, bartramii, bicolor, bolanderi, brassii, brotheri, callistomum, calvum, casasianum, compactum, consimile, consobrinum, crenatoerosum, crenulatum, crispifolium, cupulatum, cyathiforme, dasymitrium, delavayi, densum, denticulatum, diaphanum, elegans, elegantulum, elongatum, erosum, erpodiaceum, erubescens, euryphyllum, exiguum, fenestratum, firmum, flowersii, furcatum, gigantosporum, graphiomitrium, griffithii, gymnostomum, hallii, handiense, hawaiicum, heteromallum, hispanicum, holmenii, holzingeri, hooglandii, hookeri, hortense, hortoniae, ibericum, ibukiense, incrassatum, incurvomarginatum, iwatsukii, johnstonii, kaurinii, keeverae, laevigatum, latimarginatum, laxifolium, laxum, leikipiae, leiolecythis, leiothecium, letourneuxii, lewinskyae, liliputanum, ludificans, lyellii, macrocephalum, malacothecium, mandonii, meyenianum, microcarpum, mollissimum, notabile, obtusatum, obtusifolium, ohioense, oreophilum, pallens, pallens, pariatum, patens, pellucidum, penicillatum, perexiguum, philibertii, praemorsum, pseudostramineum, psychrophilum, pulchellum, pulchrum, pumilum, pungens, pusillum, pycnophyllum, pylaisii, revolutum, rhytiore, rivulare, robustum, rogeri, rupestre, sainsburyi, scanicum, schoddei, sharpii, shawii, shevockii, sinuosum, sordidum, spanotrichum, speciosum, spjutii, sprucei, steerei, stellatum, stramineum, strangulatum, striatum, subexsertum, subpumilum, subulatum, taiwanense, tasmanicum, tenellum, tenuicaule, tortidontium, tortifolium, trachymitrium, transvaaliense, tristriatum, truncatodentatum, truncatum, underwoodii, undulatifolium, urnaceum, urnigerum, vermiferum, vicarium, vittii, vladikavkanum*
GENUS *Amphidium* Dual Straight Hair Moss
SPECIES *aloysii-sabaudiae, brevifolium, clastophyllum, lapponicum, le-testui, mougeotii, papillosum, remotidens, tortuosum*
GENUS *Bryomaltaea* Malta Straight Hair Moss
SPECIES *obtusifolia*
GENUS *Cardotiella* Cardot Straight Hair Moss
SPECIES *appendiculata, elimbata, quinquefaria, renauldii, schlotheimiaeformis, secunda, subappendiculata*
GENUS *Ceuthotheca* Ceuth Sheath Straight Hair Moss
SPECIES *cryptocarpa*
GENUS *Codonoblepharon* Tooth Eyelash Straight Hair Moss
SPECIES *menziesii, pungens*
GENUS *Craspedophyllum* (Synonym for *Orthotrichum*)
GENUS *Desmotheca* Fiber Sheath Straight Hair Moss
SPECIES *apiculata, brachiata*
GENUS *Drummondia* Drummond Straight Hair Moss
SPECIES *mitelloides, obtusifolia, prorepens, pursellii, sinensis, thomsonii, turkestanica, ussuriensis*
GENUS *Euzygodon* Lesser Zygote Tooth Straight Hair Moss
SPECIES *sendtneri*
GENUS *Florschuetziella* Florschuetz Straight Hair Moss
SPECIES *scaberrima, steerei*
GENUS *Groutiella* Grout Straight Hair Moss
SPECIES *apiculata, chimborazensis, husnotii, laxotorquata, macrorrhyncha, maracaibensis, mucronifolia, obtusa, reesei, subgoniorrhyncha, thraustophylla, tomentosa, tuberculata, tumidula, wagneriana*
GENUS *Helicophyllum* Helix Straight Hair Moss
SPECIES *torquatum*
GENUS *Kleioweisiopsis* Kleioweis Straight Hair Moss
SPECIES *denticulata*
GENUS *Leiomitrium* Smooth Mitre Straight Hair Moss
SPECIES *plicatum*
GENUS *Leptodontiopsis* Thin Tooth Straight Hair Moss
SPECIES *elata, fragilifolia, macrocarpa, orientalis*
GENUS *Leratia* Lerat Straight Hair Moss
SPECIES *neocaledonica*
GENUS *Macrocoma* Comma Straight Hair Moss
SPECIES *abyssinica, brasiliensis, frigida, gastonyi, iwatsukii, lycopodioides, orthotrichoides, pulchella, sullivantii, tenuis*
GENUS *Macromitrium* Mitre Straight Hair Moss

SPECIES *acuminatum, acutirameum, acutissimum, adelphinum, adnatum, altum, amaniense, amboroicum, ancistrophyllum, angulatum, angulicaule, angulosum, angustifolium, annamense, anomodictyon, antarcticum, archboldii, archeri, argutum, astroideum, atratum, atroviride, attenuatum, aurantiacum, aurescens, aureum, austrocirrosum, belangeri, benguetense, bequaertii, bifasciculare, bifasciculatum, binsteadii, bistratosum, blumei, brachycarpum, brachypodium, braunii, brevicaule, brevihamatum, brevisetum, brevissimum, brevituberculatum, brotheri, caldense, caloblastoides, calocalyx, calomicron, calymperoideum, campoanum, cancellatum, cardotii, carinatum, carionis, cataractarum, catharinense, cavaleriei, chungkingense, cirrosum, clastophyllum, clavatum, clemensiae, comatum, complicatulum, concinnum, constrictum, contextum, courtoisii, crassirameum, crassiusculum, crenulatum, crinale, crispatulum, crosbyorum, cucullatulum, cuspidatum, cylindricum, cylindrothecium, densifolium, densum, diaphanum, dickasonii, dielsii, divaricatum, diversifolium, divortiarum, doeringianum, drewii, dubium, dusenii, echinatum, eckendorffii, ecrispatum, ellipticum, emarginatum, emersulum, eriomitrium, erubescens, evrardii, exsertum, falcatulum, fasciculare, fendleri, fernandezianum, ferriei, filicaule, flavopilosum, flexuosum, formosae, fortunatii, fragile, fragilicuspis, francii, frondosum, frustratum, fulgescens, fulvum, funicaule, funiforme, fuscescens, fusco-aureum, galipense, gebaueri, gigasporum, gimalacii, glabratum, glaziovii, globirameum, gracile, gracilipes, greenmanii, grossirete, guatemalense, gymnostomum, hamatum, handelii, harrisi, helmsii, hemitrichodes, herzogii, heterodictyon, hildebrandtii, himalayanum, hoehnei, holomitrioides, homaloblastum, hornschuchii, hortoniae, huigrense, hymenostomum, incrustatifolium, incurvifolium, intortifolium, involutifolium, japonicum, kinabaluense, koghiense, krausei, laevigatum, lanceolatum, laosianum, lauterbachii, lebomboense, leprieurii, leptocarpum, leratii, leratioides, levatum, ligulaefolium, ligulare, liliputanum, lingulatum, lomasense, lonchomitrioides, longibrachteatum, longicaule, longifolium, longipes, longipilum, longirostre, lorifolium, ludovicae, macrocomoides, macropelma, macrosporum, macrothele, marginatum, masafuerae, mauritianum, mcphersonii, megalocladon, megalosporum, melanostomum, melinii, menziesii, microcarpum, microstomum, minutum, mittenianum, moorcroftii, mosenii, muellerianum, nanothecium, negrense, nematosum, neo-caledonicum, nepalense, nietneri, nigricans, nitidum, noguchianum, norrisianum, nubigenum, oblongum, ochraceoides, ochraceum, okabei, onraedtii, orthophyllum, orthostichum, osculatianum, ousiense, ovale, pallidum, papillisetum, paridis, parvifolium, parvirete, patens, pellucidum, peraristatum, perdensifolium, perfragile, perichaetiale, perpusillum, perreflexum, pertriste, petelotii, picobonitum, pilicalyx, piliferum, pilosum, pinnulatum, platyphyllaceum, plicatum, podocarpi, polygonostomum, proliferum, prolongatum, prorepens, proximum, pseudofimbriatum, pseudoramentosum, pseudoserrulatum, pulchrum, pullenii, punctatum, pyriforme, quercicola, ramsayae, recurvifolium, refractifolium, regnellii, renauldii, repandum, retusulum, retusum, rhacomitrioides, rhaphidophyllum, rhizomatosum, richardii, rigbyanum, rimbachii, robinsonii, ruginosum, rugulosum, rusbyanum, saddleanum, salakanum, sanctae-mariae, sarasinii, savatieri, schiffneri, schmidii, sclerodictyon, scoparium, seemannii, sejunctum, semipapillosum, semperi, seriatum, serpens, sharpii, similirete, sinense, solitarium, soulae, speirostichum, st.-johnii, standleyi, stellulatum, stephanodictyon, stolonigerum, stoneae, streimannii, striatum, strictfolium, stricticuspis, subapiculatum, subbrevihamatum, subcirrhosum, subcrenulatum, subdiscretum, suberosulum, subhemitrichodes, subincurvum, sublaeve, subleptocarpum, sublongicaule, submucronifolium, subnitidum, subpaucidens, subperichaetiale, subpungens, subscabrum, subsessile, substrictifolium, subtortum, subulatum, subvillosum, sulcatum, swainsonii, syntrichophyllum, tahitisecundum, taiheizanense, taiwanense, taoense, tenax, thwaitesii, tocaremae, tongense, tortifolium, torulosum, tosae, trachypodium, trichophyllum, trinitense, trollii, tuberculatum, turgidum, tylostomum, ulophyllum, undatum, uraiense, urceolatulum, urceolatum, validum, venezuelense, vesiculatum, vesiculosum, viridissimum, vitianum, viticulosum, voeltzkowii, wattsii, weissioides, xenizon, yuleanum, zikanii, zimmermannii, zollingeri*

GENUS *Matteria* — Matter Straight Hair Moss
SPECIES *gracillima, papillosa*
GENUS *Microtheciella* — Sheath Straight Hair Moss
SPECIES *kerrii*
GENUS *Muelleriella* — Mueller Straight Hair Moss
SPECIES *angustifolia, atrata, aucklandica, crassifolia*
GENUS *Nyholmiella* — (Synonym for *Orthotrichum*)
GENUS *Orthomitrium* — Lesser Mitre Straight Hair Moss
SPECIES *schofieldii, tuberculatum*
GENUS *Pentastichella* — Pentad Stitch Straight Hair Moss
SPECIES *jaffuelii, pentasticha*
GENUS *Pleurorthotrichum* — Many Straight Hair Moss
SPECIES *chilense*
GENUS *Schlotheimia* — Schlotheim Straight Hair Moss
SPECIES *acutifolia, angulosa, angustata, appressifolia, appunii, araucarieti, argentinica, asperrima, atlantica, badiella, balfourii, bequaertii, boiviniana, brachyphylla, brachypodia, breviseta, brownii, calomitria, campbelliana, campylopus, capillaris, capillidens, chamissonis, clavata, compacta, congolensis, conica, crumii, cuspidifera, cyrtophylla, densifolia, dichotoma, elata, emarginato-pilosa, emergens, excorrugata, fasciculata, ferruginea, fornicata, foveolata, fragilifolia, funiformis, furcata, fuscoviridis, gauthieri, glauca, glaziovii, gracilescens, gracilis, grammocarpa, grandi-areolata, grevilleana, henscheniana, horridula, illecebra, jamesonii, juliformis, knightii, krausei, laete-virens, lancifolia, lasiomitria, lindmanii, linearifolia, longicaulis, mac-gregorii, malacophylla, microcarpa, microphylla, muelleri, nossi-beana, papuana, paraguensis, pauli, percuspidata, perrieri, perrotii, perserrata, pilomitria, pobeguinii, poecilodictyon, pseudoaffinis, puiggarii, pungens, pungentissima, recurvifolia, regnellii, rhystophylla, richardii, rigescens, robillardii, robusta, robusticuspis, rufo-aeruginosa, rufopallens, rugifolia, rugulosa, rusbyana, schweinfurthii, semidiaphana, serricalyx, spinulosa, squarrosa, striata, subfornicata, sublaxa, subsinuata, tecta, tenuiseta, tomentosa, torquata, tortilis, trichomitria, trichophora, trypanoclada, uncialis, ventrosa, wainioi, wallisii*
GENUS *Sehnemobryum* — Sehnem Straight Hair Moss
SPECIES *paraguense*
GENUS *Stoneobryum* — Stone Straight Hair Moss
SPECIES *bunyaense, mirum*
GENUS *Uleastrum* — Ule Star Straight Hair Moss

SPECIES *palmicola, paraguense*
GENUS *Ulota* Ulote Straight Hair Moss
SPECIES *angustilimbata, angustissima, aurantiaca, bellii, bellissima, breviseta, calvescens, carinata, cervina, coarctata, crispa, curvifolia, dixonii, drummondii, ecklonii, eurystoma, fernandeziana, fuegiana, fulva, germana, glabella, hattorii, hutchinsiae, japonica, laticiliata, lativentrosa, lobbiana, lutea, luteola, macrocalycina, macrocarpa, macrodontia, magellanica, megalospora, membranata, morrisonensis, novae-seelandiae, novo-guinensis, obtusiuscula, perbreviseta, perichaetialis, phyllantha, phyllanthoides, pilifera, pusilla, pycnophylla, pygmaeothecia, rehmannii, reptans, robusta, rubella, rufula, schmidii, splendida, tanganyikae, ventricosa, viridis, yakushimensis*
GENUS *Zygodon* Zygote Tooth Straight Hair Moss
SPECIES *acutifolius, altarensis, angustatus, apiculatus, barbuloides, bartramioides, brevipes, brevisetus, campylophyllus, cernuus, conoideus, corralensis, cylindricus, dentatus, dioicus, dixonii, ehrenbergii, erosus, fasciculatus, forsteri, forsteri, fragilis, fruticola, gracilis, gracillimus, hirsutus, hookeri, humilis, inermis, insularum, intermedius, johnstonii, laxifolius, leptobolax, leptocarpus, liebmannii, longicellularis, longisetus, lukasii, macrophyllus, magellanicus, microgemmaceus, microtheca, minutus, nivalis, novoguinensis, obovalis, ochraceus, oeneus, papillatus, parvulus, patrum, perichaetialis, perreflexus, peruvianus, petrophilus, pichinchensis, pilosulus, podocarpi, polyptychus, porteri, quitensis, ramulosus, reinwardtii, robustus, roccatii, rubrigemmius, rufescens, runcinatus, rupestris, schenckei, schimperi, semitortus, sendtneri, seriatus, setulosus, sibiricus, skujae, sordidus, squarrosus, stenocarpus, stresemannii, strictus, subrecurvifolius, subsquarrosus, tetragonostomus, theriotii, trichomitrius, tristis, venezuelensis, vestitus, viridissimus, wightii, yuennanensis*

FAMILY 2 *Erpodiaceae* Erpod Straight Hair Moss
GENUS *Erpodium* Erpod Straight Hair Moss
SPECIES *acrifolium, beccarii, ceylonicum, coronatum, cubense, distichum, domingense, glaziovii, grossirete, holstii, luzonense, madagassum, mangiferae, perrottetii, pringlei, schimperi, theriotii*
GENUS *Aulacopilum* Awl Downy Erpod Straight Hair Moss
SPECIES *abbreviatum, glaucum, hodgkinsoniae, japonicum, schaeferi, trichophyllum, tumidulum*
GENUS *Solmsiella* Solms Erpod Straight Hair Moss
SPECIES *biseriata, ceylonica, javanica*
GENUS *Venturiella* Venture Erpod Straight Hair Moss
SPECIES *sinensis*
GENUS *Wildia* Wild Erpod Straight Hair Moss
SPECIES *solmsiellacea*

FAMILY 3 *Rhachitheciaceae* Rhachis Sheath Straight Hair Moss
GENUS *Rhachithecium* Rhachis Sheath Straight Hair Moss
SPECIES *nipponicum, papillosum, perpusillum, welwitschii*
GENUS *Hypnodontopsis* Hypne Tooth Rhachis Sheath Straight Hair Moss
SPECIES *apiculatus, mexicana, spathulata*
GENUS *Jonesiobryum* Jones Rhachis Sheath Straight Hair Moss
SPECIES *cerradense, sphaerocarpum, termitarum*
GENUS *Rhachitheciopsis* Lesser Rhachis Sheath Straight Hair Moss
SPECIES *tisserantii*
GENUS *Tisserantiella* Tisserant Rhachis Sheath Straight Hair Moss
SPECIES *minutissima, pulchella*
GENUS *Ulea* Ule Rhachis Sheath Straight Hair Moss
SPECIES *yesensis*
GENUS *Zanderia* Zander Rhachis Sheath Straight Hair Moss
SPECIES *octoblepharis*

ORDER A *Pottiales* Pott Moss

FAMILY 1 *Pottiaceae* Pott Moss
GENUS *Pottia* Pott Moss
SPECIES *alpicola, altipes, appertii, caucasica, chubutensis, cucullata, flavipes, gemmifera, globosa, humillima, latifolia, latzii, ligularifolia, longifolia, namaquensis, naumanii, neocaledonica, notarisii, obliqua, pellata, pilifera, propagulifera, recurvifolia, scabrifolia, serrata, stenocarpa, tasmanica, thraustophylla, uleana*
GENUS *Acaulon* Stalkless Pott Moss
SPECIES *capense, casasianum, chrysacanthum, crassinervium, dertosense, eremicola, fontiquerianum, granulosum, integrifolium, leucochaete, longifolium, muticum, nanum, recurvatum, robustum, schimperianum, sphaericum, triquetrum, uleanum, vesiculosum*
GENUS *Aloina* Aloin Pott Moss
SPECIES *aloides, ambigua, apiculata, bifrons, brevirostris, catillum, collina, cornifolia, ericaefolia, hamulus, humilis, longirostris, macrorrhyncha, recurvipatula, rigida, roseae, sedifolia*
GENUS *Aloinella* Lesser Aloin Pott Moss
SPECIES *andina, boliviana, catenula, cucullatifolia, cucullifera, galeata, venezuelana*
GENUS *Anictangium* Anict Angled Pott Moss
SPECIES *orthotrichoides*
GENUS *Anoectangium* Anoecton Angled Pott Moss

SPECIES *abyssinicum, aestivum, afrocompactum, bellii, bicolor, borbonense, brachyphyllum, brotherusii, canadense, clarum, contortum, crassinervium, eukilimandscharicum, flexuosum, fuscum, haleakalae, hanningtonii, harttiae, herzogii, hobsonii, humblotii, hymenodontoides, imberbe, impressum, kashmiriense, keniae, lineare, madagassum, mafatense, magnirete, marinum, microphyllum, nigerianum, pallidicaule, papuanum, patagonicum, pflanzii, pleuroweisioides, rhaphidostegium, rivale, rufoviride, schimperi, sellae, shepherdae, sikkimense, spathulatum, stracheyanum, tapes, thomsonii, walkeri, weisioides, wilmsianum*

GENUS *Aschisma* Split Pott Moss

SPECIES *carniolicum, kansanum*

GENUS *Astomum* Stoma Pott Moss

SPECIES *borbonicum, intermedium, japonicum, lorentzii, minutum, mollifolium*

GENUS *Barbula* Beard Pott Moss

SPECIES *acrophylla, acuta, afrofontana, albicuspis, alpicola, altipapillosa, ambigua, amplexifolia, anastomosans, anceps, anoectangiacea, appressifolia, aquatica, arctoamericana, arcuata, aureola, austrogracilis, bagelensis, bicolor, bolleana, brachymenia, bulbiformis, calycina, calyculosa, capillipes, catenulata, chenia, chlorophana, chocayensis, confertifolia, congoana, conica, convoluta, coreensis, costaricensis, costata, costesii, crocea, crozalsii, cucullata, cylindrangia, declivium, dharvarensis, dioritica, dissita, dixoniana, dusenii, ehrenbergii, emarginata, enderesii, ericaefolia, eubryum, farriae, fendleri, fidelis, francii, frigida, funalis, furvofusca, fuscescens, fuscoviridis, geminata, geniculata, glaucescens, glaucula, gracilenta, gracilis, gracillima, grimmiacea, gymnostoma, hampeana, helvetica, hiroshii, hispaniolensis, hosseusii, hyalinobasis, hymenostylioides, inclinans, indica, integrifolia, isoindica, itzigsohnii, jacutica, javanica, juniperoidea, kiaerii, lamprocalyx, laureriana, lavardei, leiophylla, leucobasis, leucodontoides, linguaecuspis, lonchodonta, longicostata, lurida, macassarensis, macrorrhyncha, majuscula, malagana, marginans, marginatula, mauiensis, meidensis, microcalycina, munyensis, novae-caledoniae, novo-granatensis, novo-guinensis, occidentalis, omissa, orizabensis, pachyloma, perlinearis, pernana, pertorquescens, peruviana, pflanzii, planifolia, plebeja, potaninii, pseudo-ehrenbergii, pseudogracilis, pseudonigrescens, punae, purpurascens, pycnophylla, pygmaea, rechingeri, riograndensis, riparia, robbinsii, rothii, rubriseta, salisburiensis, sambakiana, santessonii, santiagensis, saxicola, scleromitra, semilimbata, semirosulata, seramensis, singkarakensis, somaliae, sordida, spathulifolia, speirostega, stenocarpa, subcaespitosa, subcalycina, subcernua, subcomosa, subdenticulata, subglaucescens, subgracilis, subgrimmiacea, subobtusa, subpellucida, subreflexifolia, subreplicata, subrigidula, subrufa, subruncinata, subtorquescens, sumatrana, svihlae, swartziana, tenuicoma, tenuirostris, thelimitria, tisserantii, tortelloides, translucens, trichomanoides, tuberculosa, umtaliensis, uncinicoma, unguiculata, unguiculatula, vaginata, vardei, ventanica, vulcanica, williamsii, wollei, zambesiaca, zennoskeanna*

GENUS *Bellibarbula* Bell Beard Pott Moss

SPECIES *kurziana, recurva*

GENUS *Bryoceuthospora* Ceuth Spore Pott Moss

SPECIES *aethiopica, mexicana*

GENUS *Bryoerythrophyllum* Red Pott Moss

SPECIES *alpigenum, andersonianum, binnsii, bolivianum, brachystegium, calcareum, caledonicum, campylocarpum, chimborazense, columbianum, compactum, ferruginascens, fuscinervium, gymnostomum, hostile, inaequalifolium, jamesonii, ligulare, machadoanum, noguchianum, recurvirostrum, rotundatum, rubrum, sharpii, wallichii, yunnanense*

GENUS *Calymperastrum* Calymper Star Pott Moss

SPECIES *latifolium*

GENUS *Calyptopogon* Calypter Beard Pott Moss

SPECIES *mnioides*

GENUS *Chenia* Chen Pott Moss

SPECIES *leptophylla, lorentzii, subobliqua*

GENUS *Cinclidotus* Cinclid Ear Pott Moss

SPECIES *acutifolius, aquaticus, bistratosus, confertus, danubicus, fontinaloides, herzogii, nyholmiae, pachyloma, pachylomoides, riparius*

GENUS *Crossidium* Cross Pott Moss

SPECIES *aberrans, apiculatum, crassinervia, davidai, deserti, desertorum, erosum, geheebii, laevipilum, laxefilamentosum, rosei, seriatum, spiralifolium, squamiferum, succulentum, woodii*

GENUS *Crumia* Crum Pott Moss

SPECIES *latifolia*

GENUS *Desmatodon* Fiber Tooth Pott Moss

SPECIES *brevicaulis, latifolius*

GENUS *Dialytrichia* Dialy Hair Pott Moss

SPECIES *mucronata*

GENUS *Didymodon* Double Tooth Pott Moss

SPECIES *aaronis, albicuspis, alticaulis, amblyophyllus, ampliretis, andreaeoides, anserinocapitatus, argentinicus, argentiniensis, asperifolius, australasiae, austroalpigena, barbulae, barbuloides, bartramii, berthoanus, bistratosus, brachyphyllus, brunneus, calycinus, camusii, canaliculatus, capitatus, cardotii, catenulatus, ceratodonteus, challaense, constrictus, contortus, cordatus, cuspidatus, deciduus, dubius, eckeliae, eckendorfii, epunctatus, erosodenticulatus, fallax, ferrugineus, filicaulis, fragilicuspis, funkii, gaochienii, giganteus, glaucoviridis, guineensis, gymnus, hampei, hastatus, haussknechtii, hedysariformis, herzogii, humboldtii, humidus, imperfectus, incrassatolimbatus, incrassatus, inundatus, jackvancei, japonicus, johansenii, juniperinus, laevigatus, lamyanus, leskeoides, lindigii, lingulatus, loeskei, lorentzianus, luehmannii, macrophyllus, mamillosus, marginatus, maschalogena, maximus, microthecius, minusculus, mittenii, montevidensis, moritzianus, murrayae, neesii, nevadensis, nicholsonii, nigrescens, norrisii, occidentalis, orbignyanus, pallidobasis, patagonicus, perexilis, perobtusus, planotophaceus, polycephalus, pruinosus, pulvinans, reflexus, reticulatus, revolutus, rigidulus, rivicola, rubiginosus, rufidulus, schilleri, sicculus, sinuosus, soaresii, spadiceus, spathulatolinearis, stenopyxis, stewartii, strictorubellus, subandreaeoides, subfontanus, subrevolutus, subtorquatus, subtriquetrus, subulatus, taylorii, tectorum, tenellus, tomaculosus, tophaceopsis, tophaceus, torquatus, trivialis, umbrosus, uruguayensis, vinealis, waymouthii, wisselii, wollei, xanthocarpus*

GENUS *Dolotortula* — Dole Tortule Pott Moss

SPECIES *mniifolia*

GENUS *Erythrobarbula* — (Synonym for *Erythrophyllum*)

GENUS *Erythrophyllastrum* — Red Star Pott Moss

SPECIES *andinum*

GENUS *Erythrophyllopsis* — Least Red Pott Moss

SPECIES *fuscula*

GENUS *Erythrophyllum* — Lesser Red Pott Moss

SPECIES *pulvinans, stenopyxis, yunnanense*

GENUS *Eucladium* — Branch Pott Moss

SPECIES *commutatum, tasmanicum, verticillatum*

GENUS *Ganguleea* — Gangule Pott Moss

SPECIES *angulosa*

GENUS *Geheebia* — (Synonym for *Pottia*)

GENUS *Gertrudiella* — Gertrude Pott Moss

SPECIES validinervis

GENUS *Globulinella* — Globule Pott Moss

SPECIES *benoistii, globifera*

GENUS *Grevilleanum* — (Synonym for *Pottia*)

GENUS *Gymnostomiella* — Lesser Nude Stoma Pott Moss

SPECIES *burmensis, erosulum, longinervis, monodii, vernicosa*

GENUS *Gymnostomum* — Nude Stoma Pott Moss

SPECIES *aeruginosum, bescherellei, bewsii, boreale, calcareum, carthusianum, chenii, foliosum, hymenostylioides, jacksharpii, lanceolatum, laxirete, lessonii, lingulatum, ludovicae, mosis, simplicissimum, splachnobryoides, subrigidulum, tenerrimum, unguiculatum, viridulum*

GENUS *Gyroweisia* — Gyre Weis Pott Moss

SPECIES *barbulacea, monterreia, reflexa, rohlfsiana, shansiensis, tenuis, yuennanensis*

GENUS *Hennediella* — Hennedy Pott Moss

SPECIES *acletoi, acutidentata, angustifolia, antarctica, arenae, bellii, densifolia, denticulata, diaguita, heimii, heteroloma, kunzeana, limbata, longipedunculata, longirostris, macrophylla, marginata, oedipodioides, polyseta, serrulata, stanfordensis, steereana*

GENUS *Hilpertia* — Hilpert Pott Moss

SPECIES *velenovskyi*

GENUS *Husnotiella* — Husnot Pott Moss

SPECIES *glossophylla*

GENUS *Hydrogonium* — Hydric Angled Pott Moss

SPECIES *dicranelloides, heterophyllum, hygrophilum, mamatkulovii, mussoorianum, patulifolium*

GENUS *Hymenostomum* — Membrane Stoma Pott Moss

SPECIES *aristatulum, densirete, fasciculatum, francii, goyazense, guineense, laxirete, leratii, noumeanum, obscurissimum, olivaceum, siamense, subrostellatum, sullivanii*

GENUS *Hymenostyliella* — Lesser Membrane Style Pott Moss

SPECIES *alata, calcarea, llanosii*

GENUS *Hymenostylium* — Membrane Style Pott Moss

SPECIES *congoanum, contextum, crassinervium, crispulum, dicranelloides, diversifolium, filiforme, grandirete, hildebrandtii, kunzeanum, papillinerve, recurvirostre, rigescens, scaturiginosum, sinense, subcrispulum, subrigidulum*

GENUS *Hyophila* — Hyophile Pott Moss

SPECIES *acuminata, acutifolia, acutipocsii, angustifolia, anoectangioides, apiculata, argentinica, ascensionis, assimilis, baginsensis, bartramiana, beruensis, bingeri, blanda, brevifolia, combae, congolensis, crispula, cuspidatissima, dendroides, elliptica, fouta-djalloni, glaucoviridis, grandiretis, grossidens, guaraya, holstii, incurva, integrifolia, involuta, involutifolia, javanica, kurziana, laete-virens, lanceolata, latifolia, lindigii, linguaeformis, loxorhyncha, mattogrossensis, microphylla, mollifolia, mosenii, neocaledonica, niam-niamiae, novae-seelandiae, nymaniana, ochracea, ovalifolia, pampaninii, paraguensis, parietalis, perpendiculata, perpusilla, procera, propagulifera, punctulata, regnellii, rubiginosa, sakalavensis, setschwanica, spathulata, streimannii, subacutiuscula, subcucullata, subflaccida, uleana, variegata, victoriae, viridula, vitiana, warmingii*

GENUS *Hyophilopsis* — (Synonym for *Hyophila*)

GENUS *Hyophiladelphus* — Hyophile Adelphi Pott Moss

SPECIES *agrarius*

GENUS *Hypodontium* Hyponic Tooth Pott Moss
SPECIES *dregei, pomiforme*
GENUS *Leptobarbula* Thin Beard Pott Moss
SPECIES *berica*
GENUS *Luisierella* Luisier Pott Moss
SPECIES *barbula*
GENUS *Merceya* Mercey Pott Moss
SPECIES *repandula*
GENUS *Microbryum* Minute Pott Moss
SPECIES *brevicaule, curvicolle, davallianum, floerkeanum, longipes, raddei, rectum, rufochaete, starkeanum, subplanomarginatum, tasmanicum, vlassovii, zealandiae*
GENUS *Mironia* Miron Pott Moss
SPECIES *crassicuspis, ehrenbergiana, stenotheca*
GENUS *Molendoa* Molendo Pott Moss
SPECIES *andina, cucullata, duthiei, fuegiana, hornschuchiana, kitaibelana, ogalalensis, platyphyllum, roylei, schliephackei, sendtneriana, seravschanica, subobtusifolia, taeniatifolia, warburgii*
GENUS *Neocardotia* (Synonym for *Pottia*)
GENUS *Neohyophila* (Synonym for *Pottia*)
GENUS *Neophoenix* Phoenix Pott Moss
SPECIES *matoposensis*
GENUS *Oxystegus* (Synonym for *Pottia*)
GENUS *Pharomitrium* (Synonym for *Pottia*)
GENUS *Phascopsis* Lesser Phasce Pott Moss
SPECIES rubicunda
GENUS *Phascum* Phasce Pott Moss
SPECIES *beaugleholei, calodictyon, cuynetii, galilaeum, laticostum, longipilum, nepalense, peraristatum, piptocarpum, readeranum*
GENUS *Plaubelia* Plaubel Pott Moss
SPECIES *involuta, sprengelii*
GENUS *Pleurochaete* Many Bristle Pott Moss
SPECIES *beccarii, ecuadoriensis, malacophylla, squarrosa*
GENUS *Pottiopsis* Lesser Pott Moss
SPECIES *caespitosa*
GENUS *Pseudocrossidium* Lesser Cross Pott Moss
SPECIES *apiculatum, austrorevolutum, carinatum, chilense, crinitum, elatum, excavatum, granulosum, hornschuchianum, leucocalyx, mendozense, obtusulum, pachygastrellum, perrevolutum, porphyreoneurum, replicatum, revolutum*
GENUS *Pseudosymblepharis* Symbol Eyelash Pott Moss
SPECIES *bombayensis, cavernarum, circinnatula, mauiensis, schimperiana, syrrhopodontoides*
GENUS *Pterygoneurum* Wing Angled Nerve Pott Moss
SPECIES *californicum, chotticum, compactum, crossidioides, kozlovii, lamellatum, mac-leanum, medium, ovatum, papillosum, squamosum, subsessile*
GENUS *Quaesticula* Quaesticle Pott Moss
SPECIES *navicularis*
GENUS *Reimersia* Reimer Pott Moss
SPECIES *inconspicua*
GENUS *Rhexophyllum* Rhex Pott Moss
SPECIES *subnigrum*
GENUS *Sagenotortula* Sagen Tortule Pott Moss
SPECIES *quitoensis*
GENUS *Saitobryum* Saito Pott Moss
SPECIES *lorentzii*
GENUS *Sarconeurum* Flesh Nerve Pott Moss
SPECIES *glaciale*
GENUS *Scopelophila* Scope Pott Moss
SPECIES *cataractae, infericola, ligulata*
GENUS *Sebillea* Sebill Pott Moss
SPECIES *brasiliensis*
GENUS *Semibarbula* Half Beard Pott Moss
SPECIES *ranuii*
GENUS *Serpotortella* Serpent Tortus Pott Moss
SPECIES *chenagonii, cyrtophylla*
GENUS *Stegonia* Cover Pott Moss
SPECIES *hyalinotrichum, latifolia, mouretii, pilifera*
GENUS *Stonea* Stone Pott Moss
SPECIES *oleaginosa*
GENUS *Streblotrichum* Streble Hairy Pott Moss

SPECIES *propaguliferum*
GENUS *Streptocalypta* Twisted Calypter Pott Moss
SPECIES *lorentziana, pulchriretis, santosii, tortelloides*
GENUS *Streptopogon* Twisted Beard Pott Moss
SPECIES *calymperes, cavifolius, clavipes, erythrodontus, lindigii, matudianus*
GENUS *Streptotrichum* Twisted Hair Pott Moss
SPECIES *ramicola*
GENUS *Syntrichia* Hairy Pott Moss
SPECIES *aculeata, alpestris, ammonsiana, amphidiacea, amplexa, anderssonii, andicola, antarctica, austro-africana, baileyi, bartramii, bipedicellata, bogotensis, bolanderi, brachyclada, brandisii, brevisetacea, cainii, calcicola, campestris, caninervis, cavallii, chisosa, ciliata, costesii, epilosa, filaris, flagellaris, fontana, fragilis, geheebiaeopsis, gemmascens, glabra, glacialis, gromschii, handelii, inermis, intermedia, jaffuelii, lacerifolia, laevipila, latifolia, leucostega, limensis, linguifolia, magellanica, magilliana, minor, mollis, norvegica, obtusissima, pagorum, papillosa, papillosissima, percarnosa, phaea, princeps, prostrata, pseudohandelii, pseudorobusta, pygmaea, ramosissima, reflexa, rigescens, robusta, rubella, rubra, ruralis, saxicola, scabrella, scabrinervis, schnyderi, serrata, serripungens, sinensis, socialis, subaristata, submontana, subpapillosa, subpapillosissima, virescens, viridula*
GENUS *Teniolophora* Tendril Pott Moss
SPECIES *fluviatile*
GENUS *Tetracoscinodon* Tetrad Coscine Tooth Pott Moss
SPECIES *irroratus*
GENUS *Tetrapterum* Tetrad Wing Pott Moss
SPECIES *brachypelma, cylindricum, lamprocarpum, lamprothecium, sullivanii, tetragonum, tetrapteroides, weymouthii*
GENUS *Timmia* Timmy Pott Moss
SPECIES *alataviensis, austriaca, megapolitana, norvegica, sibirica*
GENUS *Timmiella* Lesser Timmy Pott Moss
SPECIES *acaulon, anomala, barbuloides, brevidens, cameruniae, corniculata, crassinervis, diminuta, flexiseta, grosseserrata, pelindaba, subintegra, umbrosa*
GENUS *Tortella* Lesser Tortus Pott Moss
SPECIES *alpicola, aprica, brotheri, bryotropica, cirrifolia, contortifolia, cryptocarpa, cyrtobasis, dakinii, eckendorffii, flavovirens, fragilis, fragillima, fristedtii, fruchartii, germainii, goniospora, hosseusii, humilis, inclinata, inflexa, involutifolia, jugicola, kmetiana, knightii, lilliputana, limbata, lindmaniana, linearis, mooreae, nitida, novae-valesiae, perrufula, pilcomayica, pseudocaespitosa, rigens, rubripes, simplex, smithii, somaliae, subflavovirens, tortuosa, undulatifolia, vernicosa, walkeri, xanthocarpa*
GENUS *Tortula* Tortus Pott Moss
SPECIES *abyssinica, acaulon, aculeonervis, afanassievii, altipes, ambigua, amphidiifolia, ampliretis, appressa, areolata, argentinica, astoma, atrata, atrovirens, austroruralis, bauriana, berthoana, bidentata, bogosica, brevipes, breviseta, brevissima, brunnea, buchtienii, buyssonii, calcarea, californica, canescens, capillaris, cardotii, cernua, characodonta, chrysopila, chubutensis, chungtienia, cochlearifolia, contorta, crawfordii, crenata, cucullifolia, cuneifolia, cuspidatissima, deciduidentata, demawendica, emarginata, entosthodontacea, ericaefolia, eucalyptrata, felipponei, ferruginea, fragillima, gracilis, grandiretis, guepinii, israelis, kabir-khanii, kneuckeri, laevinervis, lanceola, laureri, lazarenkoi, leiostomoides, leptotheca, leucochlora, leucostoma, ligulata, lindbergii, lingulata, litorea, madagassa, marginata, maritima, minima, minor, minutirosula, modica, mucronifera, mucronifolia, muralis, murina, napoana, nevadensis, obscuriretis, obtusifolia, omissa, pallida, parramattana, paulsenii, perarmata, perpusilla, planicosta, planifolia, platyphylla, plinthobia, podocarpi, polycarpa, polylepidis, porteri, protobryoides, pseudolatifolia, pseudoprinceps, pugionata, pulvinatula, rallieri, randii, raucopapillosa, recurvata, revolutifolia, revolvens, rhodonia, sabinae, sainsburyana, santiagensis, santorinensis, savatieri, sinuata, solmsii, solomensis, sordida, splachnoides, squarripila, sublimbata, subrufa, subtranscaspica, subulata, systylia, thianschanica, thomsonii, tonkinensis, toutonii, trachyphylla, transcaspica, truncata, ucrainica, umbrosa, vahliana, viridipila, websteri, wilczekii, willisiana, wilsonii, xerophila, yuennanensis, zoddae*
GENUS *Trachycarpidium* Rough Carpel Pott Moss
SPECIES *brisbanicum, echinatum, lonchophyllum, tisserantii, verrucosum*
GENUS *Trachyodontium* Rough Tooth Pott Moss
SPECIES *zanderi*
GENUS *Trichostomopsis* Lesser Hairy Stoma Pott Moss
SPECIES *curvipes*
GENUS *Trichostomum* Hairy Stoma Pott Moss
SPECIES *abyssinicum, acutiusculum, aequatoriale, alpinum, apophysatulum, arboreum, arcticum, atrocaule, austrocrispum, bellii, borneense, brachydontium, brittonianum, calymperaceum, cardotii, carinatum, castaneum, clavinerve, compactulum, connivens, contractum, criotum, crispulum, deciduaefolium, distans, eckelianum, edentulum, elliottii, exulatum, fallaciosum, fallax, flavisetum, fragilifolium, gracillimum, hattorianum, hibernicum, hondurense, hyalinoblastum, imshaugii, incertum, insulare, involutum, kanieriense, knightii, lambii, laticostatum, leptocylindricum, leptotheca, lignicola, ligulaefolium, lillei, lindigii, lorifolium, mauiense, melanostomum, mildeanum, minutissimum, mitteneanum, muticum, orthodontum, ovatifolium, paludicola, pennequinii, perangustum, perinvolutum, perligulatum, perplexum, perpusillum, perrieri, persicum, philippinense, piliferum,*

planifolium, platyphyllum, plicatulum, pomangium, portoricense, prionodon, pulicare, pygmaeum, recurvifolium, rhodesiae, ruvenzorense, sinaloense, sinochenii, soulae, spirale, sporaphyllum, squarrosum, stanilandsii, subangustifolium, subcirrhatum, subconnivens, subintegrum, sublamprothecium, sweetii, tenuirostre, termitarum, tortelloides, tucumanense, unguiculatum, urceolare, usambaricum, villaumei, wagneri, weisioides, whittonii, williamsii, zanderi

GENUS *Triquetrella* — Triquet Pott Moss

SPECIES *arapilensis, californica, fragilis, papillata, patagonica, richardsiae, tasmanica, tristicha*

GENUS *Tuerckheimia* — Tuerckheim Pott Moss

SPECIES *guatemalensis, robusta, svihlae, valeriana*

GENUS *Uleobryum* — Ule Pott Moss

SPECIES *naganoi, occultum, peruvianum*

GENUS *Weisiopsis* — Lesser Weis Pott Moss

SPECIES *anomala, bahiensis, nigeriana, norrisii, oblonga, plicata*

GENUS *Weissia* — Weis Pott Moss

SPECIES *abbreviata, alianuda, andersoniana, argentinica, artocosana, atra, ayresii, balansae, balansaeana, bizotii, brachycarpa, brachypelma, brachypoma, breutelii, breviseta, canaliculata, condensa, contermina, controversa, cucullata, dieterlenii, diffidentia, edentula, erythrogona, exserta, fallax, felipponei, fornicata, ghatensis, glazioui, groenlandica, humicola, inoperculata, jamaicensis, kaikouraensis, krassavinii, kunzeana, latifolia, latiuscula, levieri, ligulaefolia, lineaefolia, longidens, longifolia, lorentzii, ludoviciana, macrospora, micacea, mittenii, muhlenbergiana, multicapsularis, neocaledonica, newcomeri, nitida, norkettii, novae-valisiae, obtusata, obtusifolia, occidentalis, occulta, opaca, ovalis, ovatifolia, papillosa, patagonica, patula, perpusilla, pilifera, platystegia, ricciae, riograndensis, rostellata, rutilans, semidiaphana, semiinvoluta, semipallida, sharpii, simplex, socotrana, splachnum, squarrosa, sterilis, subacaulis, submicacea, termitidarum, triumphans, tyrrhena, vardei, veviridis, welwitschii, willisiana*

GENUS *Weissiodicranum* — Weiss Dicran Pott Moss

SPECIES *insularum*

GENUS *Willia* — Willie Pott Moss

SPECIES *austroleucophaea, brachychaete, calobolax*

FAMILY 2 *Calymperaceae* — Calympere Pott Moss

GENUS *Calymperes* — Calympere Pott Moss

SPECIES *aeruginosum, afzelii, armatum, asperum, bartramii, beccarii, boninense, boulayi, conterminum, couguiense, crassinerve, derelictum, disciforme, erosum, fasciculatum, glaziovii, graeffeanum, guildingii, guineense, hispidum, hyophilaceum, lamellosulum, levyanum, linearifolium, linguatum, lonchophyllum, loucoubense, mangalorense, meyeri, mitrafugax, motleyi, mussoriense, nanum, nicaraguense, noakhalense, norkettii, othmeri, palisotii, pallidum, platyloma, porrectum, proligerum, pulvinatum, pungens, pygmaeum, robinsonii, rubiginosum, rupestre, sakaranae, schmidtii, serratum, smithii, strictifolium, subintegrum, subserratum, subulatum, tahitense, tenerum, thomeanum, venezuelanum, woodii*

GENUS *Arthrocormus* — Straight Corm Calympere Pott Moss

SPECIES *schimperi*

GENUS *Calymperopsis* — Lesser Calympere Pott Moss

SPECIES *disciformis, vietnamensis*

GENUS *Exodictyon* — Exo Net Calympere Pott Moss

SPECIES *dentatum, incrassatum, parkinsonii*

GENUS *Exostratum* — Exo Layer Calympere Pott Moss

SPECIES *asperum, blumii, scolopendrium, sullivantii*

GENUS *Leucophanes* — Light Transparent Calympere Pott Moss

SPECIES *angustifolium, candidum, glaucum, hildebrandtii, milleri, molleri, naumannii, octoblepharioides, pucciniferum, renauldii, rodriguezii, serratulum, seychellarum, unguiculatum, vitianum*

GENUS *Mitthyridium* — Mit Thyre Calympere Pott Moss

SPECIES *adpressum, cardotii, constrictum, crassum, fasciculatum, flavum, iwatsukianum, jungquilianum, leucoloma, luteum, megamorphum, micro-undulatum, obtusifolium, papuanum, perundulatum, repens, retusum, subluteum, undulatum, wallisii*

GENUS *Syrrhopodon* — Syrrhopod Calympere Pott Moss

SPECIES *africanus, albo-vaginatus, annotinus, aristifolius, armatispinosus, armatissimus, armatus, asper, autotomaius, bartramii, brasiliensis, chenii, ciliatus, circinatus, clavatus, confertus, crenulatus, croceus, cryptocarpus, cuneifolius, curticancellinatus, cymbifolius, cyrtacanthos, disciformis, elatus, elongatus, erubescens, fimbriatus, fissipapillatus, flammeonervis, flavifolius, flexifolius, fraserianus, gardneri, gaudichaudii, graminifolius, hainanensis, helicophyllus, hispidissimus, hongkongensis, hornschuchii, incompletus, involutus, isthmi, japonicus, kiiense, laevis, lamprocarpus, lanceolatus, langbianensis, leprieurii, ligulatus, loreus, lycopodioides, mahensis, mammillosus, mauritianus, meijeri, muelleri, obuduensis, orientalis, parasiticus, peguensis, perangustifolius, perarmatus, perdusenii, phragmidiaceus, pilulifer, planifolius, platycerii, prolifer, pulcher, rigidus, rodriguezii, rupestris, sarawakensis, semicircularis, semiliber, simmondsii, sinii, spiculosus, stenophyllus, steyermarkii, stoneae, strigosus, stuhlmannii, subdisciformis, terebellum, texanus, theriotii, tixieri, tjibodensis, tortilis, trachyphyllus, tristichus, viguieri, virgulicola, xanthophyllus*

GENUS *Thyridium* — Thyre Calympere Pott Moss

SPECIES *andamense*

FAMILY 3 *Encalyptaceae* — Calypter Pott Moss

GENUS *Encalypta* Calypter Pott Moss
SPECIES *affinis, alpina, apophysata, armata, asperifolia, brevicollis, brevipes, buxbaumioidea, capillata, ciliata, cuspidata, cylindrica, flowersiana, hedbergii, intermedia, labradorica, laevigata, leiocarpa, longicollis, microstoma, mutica, novae-valesiae, obovatifolia, obtusata, obtusifolia, peruviana, pilifera, pilosa, procera, rauhii, rhaptocarpa, sandwicensis, serbica, sibirica, sinica, spathulata, squarrosa, streptocarpa, subbrevicollis, tianschanica, tibetana, vittiana, vulcanica, vulgaris*
GENUS *Anacalypta* False Calypter Pott Moss
SPECIES *latifolia*
GENUS *Bryobrittonia* Britton Calypter Pott Moss
SPECIES *longipes*

CLASS 631.2 *Andreaeopsida* Granite Moss

ORDER 1 *Andreaeales* Granite Moss

FAMILY 1 *Andreaeaceae* Granite Moss
GENUS *Andreaea* Granite Moss
SPECIES *acuminata, acutifolia, alpestris, alpina, amblyophylla, angustifolia, apiculata, appendiculata, arachnoidea, arthuriana, asperula, atlantica, australis, barbuloides, bistratosa, blyttii, borbonica, brevipes, camerunensis, cockaynei, commutata, crassinervia, cucullata, densifolia, depressinervis, dissitifolia, erubescens, erythrodictyon, filamentosa, firma, flabellata, flexuosa, fragilis, frigida, fuegiana, gainii, gibbosa, hartmanii, heinemannii, huttonii, indica, javanica, julicaulis, karsteniana, kilimandscharica, lanceolata, latinervis, laxifolia, leiophylla, likiangensis, lorentziana, marginata, megistospora, microphylla, microvaginata, mildbraedii, mitchellii, morrisonensis, mutabilis, nitida, nivalis, novae-zealandiae, obovata, opaca, pachyphylla, parallela, patagonica, peruviana, pilifera, planinervia, pseudomutabilis, pseudosubulata, regularis, remotifolia, rigida, robusta, rothii, rupestris, schofieldiana, semisquarrosa, seriata, sinuosa, spurioalpina, squarrosofiliformis, striata, subfluitans, subremotifolia, subulata, taiwanensis, tasmanica, tsaratananae, tunariensis, turgescens, urophylla, vaginalis, vilocensis, viridis, vulcanica, wangiana, willii, wilsonii*
FAMILY 2 *Andreaeobryaceae* Lesser Granite Moss
GENUS *Andreaeobryum* Lesser Granite Moss
SPECIES *macrosporum*

CLASS 631.3 *Sphagnopsida* Peat Moss

ORDER 1 *Sphagnales* Peat Moss

FAMILY 1 *Sphagnaceae* Peat Moss
GENUS *Sphagnum* Peat Moss
SPECIES *aciphyllum, acutirameum, aequalipunctatum, aequiporosum, africanum, alaskense, alegrense, algentryi, amazonense, amazonicum, amoenoides, amoenum, andersonianum, angermanicum, angustifolium, annulatum, antarense, antioquiense, aongstroemii, arcticum, atroligneum, australe, austro-americanum, azuayense, balslevii, balticum, barclayae, bartlettianum, billbuckii, bocainense, boliviae, boomii, bourbonense, boyacanum, brachybolax, brachycaulon, breedlovei, brevirameum, buckianum, calymmatophyllum, capense, capillifolioides, capillifolium, capillifolium, carolinianum, centrale, ceylonicum, chevalieri, chi-chiense, chinense, cleefii, columniforme, compactum, condensatum, conflatum, connectens, contortulum, contortum, costae, cribriforme, crispatum, cristatum, crumii, cucullatum, cuculliforme, cundinamarcanum, curicuriariense, curvatulum, cuspidatulum, cuspidatum, cyclocladum, cyclophyllum, davidii, delamboyense, denticulatum, diblastoides, dicladum, dimorphophyllum, dissimile, divisum, dominii, dusenioides, ehyalinum, elenkini, engelii, ericetorum, exile, exquisitum, falcatulum, fallax, fimbriatum, fitzgeraldii, flaccidum, flavicomans, flexuosum, fontanum, frahmii, fraudulentum, funkiae, fuscovinosum, fuscum, garysmithii, geraisense, girgensohnii, globicephalum, gomezii, gordjogini, gracilescens, grasslii, griseum, guanabarae, guwassanense, hamiltonii, hampeanum, harleyi, hegewaldii, helenicum, henryense, hertelianum, herteri, homophyllum, huilense, imbricatum, imperforatum, incertum, incommodum, inretortum, inundatum, irwinii, isoviitae, itabense, itatiaiae, jensenii, juliforme, junghuhnianum, krylovi, kushiroense, laceratum, laegaardii, lankesteri, lapazense, laxirameum, laxiramosum, laxulum, lenense, lescurii, leucobryoides, lewisii, liesneri, limbatum, lindbergii, lojense, longicomosum, longistolo, luetzelburgii, luzonense, macrophyllum, maegdefraui, magellanicum, magistri, majus, matogrossense, mendocinum, meridense, microcuspidatum, microporum, minutulum, mirabile, mississippiense, molle, moronum, mosenii, multifibrosum, multiporosum, negrense, nepalense, nitidulum, nitidum, noryungasense, novae-caledoniae, novo-guineense, novo-zelandicum, obliquefibrosum, obtusum, olafii, oligoporum, orientale, ornatum, ovalifolium, ovatum, oxyphyllum, pacificum, pallens, palustre, papillosum, paranense, parcoramosum, pendulirameum, perforatum, perichaetiale, perrieri, personatum, planifolium, platyphylloides, platyphylloideum, platyphyllum, pluriporosum, poasense, portoricense, priceae, pseudomedium, pseudoramulinum, pulchrum, pulvinatum, pungifolium, pycnocladulum, pylaesii, quinquefarium, ramulinum, reclinatum, recurvum, reichardtii, richardsianum, rio-negrense, riparium, ripense, robinsonii, roraimense, roseum, rotundatum, rotundifolium, rubellum, rubiginosum, rubroflexuosum, rugense, russowii, rutenbergii, sancto-josephense, sanguinale, santanderense, schofieldii, schwabeanum, scorpioides, sehnemii, septatoporosum, septatum, sericeum, simplex, simplicicaulis, sipmanii, skyense, slooveri, sonsonense, sordidum, sparsum, splendens, squarrifolium, squarrosum, steerei, strictum, subacutifolium, subaequifolium, subbalticum, subditivum, subfalcatulum, subfulvum, subhomophyllum, submedium, subnitens, subobesum, subovalifolium, subrigidum, subrufescens, subsecundoides, subsecundum, subserratum, subtile, sucrei, sumapazense, tabuleirense, tenellum, tenerum, teres, torreyanum, tosaense, trinitense, triporosum, trirameum, troendelagicum, trollii, truncatum, tumidulum, tundrae, turgens, turgescens, typicum, uleanum, umbrosum, uruguayense, veresczagini, versiporum, violascens, viride, vitalii, wieboldtii, wilfii, wulfianum*

PHYLUM 632 *Anthocerophyta* Hornwort

Latin: *Antho* anther, flower	**CLASS**		**Species**
Latin *Cero* wax	632.1 *Anthocerotopsida*	Hornwort	100

Latin: *Phyta* phyton, plant
100 species [Margulis & Schwartz, 1998]
23 species [Ruggiero, Nov. 2006]

Phylum *Anthocerophyta* Hornworm Moss species have coils which facilitate zygote spore dispersal (gameophyte elators) and have an elongated horn shaped sporophyte. They live on tree trunks, cliffs, and muddy stream banks. Within a green horn shaped sporophyte thallus is a mucus filled cavity. Hornworts first appeared in the fossil record in the Cretaceous period 144 to 66 million years ago.

CLASS 632.1 *Anthocerotopsida* Hornwort

ORDER 1 *Anthocerotales* Hornwort

FAMILY 1 *Anthocerotaceae* Hornwort
GENUS *Anthoceros* Hornwort
SPECIES *adscendens, agrestis, donnelli, fusiformis, lamellatus, macounii, scarious*
VARIETIES *adscendens joorii, adscendens lescurii, adscendens megalosporus, adscendens olneyi, adscendens ravenelii, agrestis agrestis, agrestis cavernosus, agrestis constans, agrestis douinii, agrestis douinii-crispulus, agrestis husnotii, agrestis longicapsulus, agrestis multifidus, agrestis multilobulus, agrestis nagasakiensis, fusiformis pseudohusnotii, fusiformis stomatifer, lamellatus appalachianus, scarious eatonii, scarious grossilamellus, scarious meridionalis, scarious ohionensis, scarious reticulatus*
GENUS *Aspiromitus* (Synonym for *Anthoceros*)
GENUS *Carpoceros* (Synonym for *Phaeoceros*)
GENUS *Folioceros* (Synonym for *Phaeoceros*)
GENUS *Megaceros* Large Hornwort
SPECIES *aenigmaticus*
GENUS *Notothylas* Notable Thyle Hornwort
SPECIES *breutelii, orbicularis*
VARIETIES *orbicularis orbicularis, orbicularis decurva, orbicularis pseudotemperata*
GENUS *Phaeoceros* Dusky Hornwort
SPECIES *carolinianus, hallii, laevis, mohrii, pearsonii*
VARIETIES *carolinianus carolinianus, carolinianus arsenii, carolinianus atlanticus, carolinianus bilobata, carolinianus cubanus, carolinianus incrassatus, carolinianus japonicus, carolinianus koreanus, carolinianus kuhlmannii, carolinianus kuntzeanus, carolinianus laciniatus, carolinianus major, carolinianus miyoshianus, carolinianus moldavicus, carolinianus nordenskjoeldii, carolinianus occidentalis, carolinianus oreganus, carolinianus parvus, carolinianus planus, carolinianus subcostatus, carolinianus tenuis, hallii phymatodes, hallii sulcatus, laevis laevis, laevis aquatica, laevis beltrani, laevis carolinianus, laevis curnowii, laevis gussonei, laevis major, laevis miyakeanus, laevis polymorphus, mohrii microsporus, pearsonii bolanderi*
GENUS *Sewardiella* Seward Hornwort
SPECIES *tuberifera*
GENUS *Sphaerosporoceros* (Synonym for *Anthoceros*)

PHYLUM 633 *Hepatophyta* Liverwort

Latin: *Hepata* hepatic, liver	**CLASS**	**Species**
Latin: *Phyta* phyton, plant	633.1 *Hepatopsida* Liverwort	6000

6000 species [Margulis & Schwartz, 1998]
591 species [Ruggiero, Nov. 2006]

Phylum *Hepatophyta* Liverwort species have coils which facilitate zygote spore dispersal (gametophyte elators) and have a liver shaped gametophyte thallus. After algae, fungus, and lichen, Liverwort is the simplest type of plant. The thallus has flattened green lobes. There is no mucus filled cavity. They inhabit burned forests, stream banks, and new growth areas colonized by pioneer plants. Liverworts are the descendents of green algae.

CLASS 633.1 *Hepatopsida* Liverwort

ORDER 1 *Haplomitriales* Mitre Liverwort

FAMILY 1 *Haplomitriaceae* Mitre Liverwort
GENUS *Haplomitrium* Mitre Liverwort
SPECIES *hookeri*
VARIETIES *hookeri hookeri, hookeri cordae*

ORDER 2 *Jungermanniales* Jungermann Liverwort

FAMILY 01 *Jungermanniaceae* Jungermann Liverwort
GENUS *Jungermannia* Jungermann Liverwort
SPECIES *atrovirens, caespiticia, confertissima, crenuliformis, exsertifolia gracillima, hyalina, hyalinum, karl-meulleri, leiantha, polaris, pumila, pyriflora, sphaerocarpa, subulata*
VARIETIES *atrovirens tristis, confertissima bolanderi, confertissima danicola, , exsertifolia exsertifolia, exsertifolia cordifolia, exsertifolia pendletonii, karl-meulleri karl-muelleri, karl-meulleri heteroicum, karl-meulleri lanceolata, polaris schiffneri, pumila rivularis, pyriflora pyriflora, pyriflora innovata, pyriflora purpurea, sphaerocarpa amplexicaulis, sphaerocarpa nana, subulata amakawana*
GENUS *Acrobolbus* Tall Bulb Jungermann Liverwort
SPECIES *ciliatus*
GENUS *Anastrepta* Untwisted Jungermann Liverwort
SPECIES *orcadensis*
GENUS *Anastrophyllum* No Star Jungermann Liverwort
SPECIES *assimile, donianum, helleranum, michauxii, minutum, saxicola, sphenoloboides, tenue*
VARIETIES *minutum minutum, minutum grandis*
GENUS *Anthelia* No Nipple Jungermann Liverwort
SPECIES *julacea, juratzkana*
GENUS *Barbilophozia* Beard Jungermann Liverwort
SPECIES *atlantica, attenuata, barbata, binsteadii, cavifolia, floerkei, hatcheri, hyperborea, kunzeana, lycopodioides, quadriloba*
VARIETIES *quadriloba quadriloba, quadriloba collenchymatica, quadriloba glareosa*
GENUS *Chandonanthus* Chandon Jungermann Liverwort
SPECIES *cavallii, filiformis, hirtellus, setiformis*
GENUS *Chonecolea* Chone Jungermann Liverwort
SPECIES *doellingeri*
GENUS *Cryptocolea* Cryptic Jungermann Liverwort
SPECIES *imbricata*
GENUS *Goebeliella* Goebel Jungermann Liverwort
SPECIES *cornigera*
GENUS *Gymnocolea* Nude Jungermann Liverwort
SPECIES *acutiloba, inflata*
VARIETIES *inflata inflata, inflata heterostipa*
GENUS *Gyrothyra* Gyre Thyre Jungermann Liverwort
SPECIES *underwoodiana*
GENUS *Herzogiaria* Herzog Jungermann Liverwort
SPECIES *teres*
GENUS *Jamesoniella* Jameson Jungermann Liverwort
SPECIES *autumnalis, undulifolia*
VARIETIES *autumnalis autumnalis, autumnalis heterostipa, autumnalis nipponica*
GENUS *Lophozia* Crested Jungermann Liverwort

SPECIES *alboviridis, alpestris, ascendens, badensis, bantriensis, bicrenata, capitata, collaris, elongata, excisa, gillmanii, grandiretis, groenlandica, guttulata, heterocolpos, heteromorpha, hyperarctica, incisa, latifolia, laxa, longidens, obtusa, opacifolia, pellucida, perssonii, polaris, rubrigemma, rutheana, subapiculata, sudetica, ventricosa, wenzelii*

VARIETIES *badensis badensis, badensis apiculata, excisa excisa, excisa elegans, excisa infuscata, excisa succulenta, gillmanii gillmanii, gillmanii ciliolata, grandiretis grandiretis, grandiretis parviretis, heterocolpos heterocolpos, heterocolpos arctica, heterocolpos harpanthoides, longidens longidens, longidens arctica, polaris polaris, polaris sphagnorum, ventricosa ventricosa, ventricosa confusa, ventricosa grandiretis, ventricosa longiflora, ventricosa rigida, ventricosa silvicola, wenzelii wenzelii, wenzelii lapponica*

GENUS *Mastigophora* Mastigophore Jungermann Liverwort

SPECIES *caledonica, diclados, flagellifera, woodsii*

GENUS *Mesoptychia* Folded Jungermann Liverwort

SPECIES *sahlbergii*

GENUS *Mylia* Mylian Jungermann Liverwort

SPECIES *anomala, taylorii*

GENUS *Nardia* Nard Jungermann Liverwort

SPECIES *breidleri, compressa, geoscyphus, insecta, lescurii, scalaris*

VARIETIES *geoscyphus geoscyphus, geoscyphus bifida, scalaris scarlaris, scalaris botryoides*

GENUS *Odontoschisma* Tooth Split Jungermann Liverwort

SPECIES *denudatum, elongatum, gibbsiae, macounii, prostratum, sphagni*

VARIETIES *denudatum denudatum, denudatum laevissima*

GENUS *Plectocolea* Plaited Jungermann Liverwort

SPECIES *fossombronioides, obovata, evansii, rubra, subelliptica*

VARIETIES *evansii obscura*

GENUS *Pleurozia* Pleural Jungermann Liverwort

SPECIES *purpurea*

GENUS *Ptilidium* Ptile Jungermann Liverwort

SPECIES *californicum, ciliare, pulcherrimum*

GENUS *Radula* Radial Jungermann Liverwort

SPECIES *auriculata, australis, bolanderi, complanata, flaccida, mollis, obconica, obtusiloba, prolifera, sullivantii, tenax, voluta*

VARIETIES *obtusiloba obtusiloba, obtusiloba polyclada*

GENUS *Sphenolobopsis* Sphene Lobe Jungermann Liverwort

SPECIES *pearsonii*

GENUS *Tritomaria* Triton Sea Jungermann Liverwort

SPECIES *exsecta, exsectiformis, heterophylla, polita, quinquedentata, scitula*

VARIETIES *exsectiformis exsectiformis, exsectiformis arctica, polita polita, polita polymorpha, quinquedentata quinquedentata, quinquedentata grandiretis, quinquedentata turgida*

GENUS *Vetaforma* Vetaform Jungermann Liverwort

SPECIES *dusenii*

FAMILY 02 *Arnelliaceae* Arne Jungermann Liverwort

GENUS *Arnellia* Arne Jungermann Liverwort

SPECIES *fennica*

GENUS *Gongylanthus* Gongy Arne Jungermann Liverwort

SPECIES *ericetorum*

FAMILY 03 *Calypogejaceae* Callisto Beard Jungermann Liverwort

GENUS *Calypogeja* Callisto Beard Jungermann Liverwort

SPECIES *fissa, integristipula, muelleriana, neesiana, peruviana, sphagnicola, suecica, sullivantii, trichomanis*

VARIETIES *fissa fissa, fissa neogaea, muelleriana muelleriana, muelleriana blomquistii*

GENUS *Metacalypogeia* Meta Callisto Beard Jungermann Liverwort

SPECIES *cordifolia, schusterana*

FAMILY 04 *Cephaloziaceae* Cap Jungermann Liverwort

GENUS *Cephalozia* Cap Jungermann Liverwort

SPECIES *affinis, bicuspidata, bicuspidata bicuspidata, bicuspidata ambigua, bicuspidata lammersiana, bicuspidata otaruensis, catenulata, connivens, connivens connivens, connivens bifida, connivens compacta, lacinulata, leucantha, loitlesbergeri, lunulifolia, macounii, macrostachya, macrostachya macrostachya, macrostachya australis, pleniceps, pleniceps pleniceps, pleniceps caroliniana, pleniceps sphagnorum*

GENUS *Cephaloziella* Lesser Cap Jungermann Liverwort

SPECIES *arctica, artica, biloba, brinkmani, dentata, divaricata, elachista, gracillima, grimsulana, hampeana, hyalina, integerrima, massalongi, minima, patulifolia, phyllacantha, rappii, rubella, spinicaulis, subdentata, turneri, uncinata*

VARIETIES *divaricata divaricata, divaricata scabra, grimsulana grimsulana, grimsulana angustiloba, rubella rubella, rubella arctogena, rubella bifida, rubella elegans, rubella sullivantii, , uncinata uncinata, uncinata sphagnicola*

GENUS *Cladopodiella* Branch Pod Cap Jungermann Liverwort

SPECIES *fluitans, francisci*
GENUS *Cylindrocolea* — Cylindric Cap Jungermann Liverwort
SPECIES *obliqua, rhizantha*
GENUS *Fuscocephaloziopsis* — Fuscous Cap Jungermann Liverwort
SPECIES *pulvinata*
GENUS *Hygrobiella* — Hygrobe Cap Jungermann Liverwort
SPECIES *laxifolia*
GENUS *Metahygrobiella* — Meta Hygrobe Cap Jungermann Liverwort
SPECIES *acuminata*
GENUS *Nowellia* — Nowell Cap Jungermann Liverwort
SPECIES *curvifolia*
GENUS *Pleuroclada* — Many Branch Cap Jungermann Liverwort
SPECIES *albescens*
VARIETIES *albescens albescens, albescens islandica*
GENUS *Schofieldia* — Schofield Cap Jungermann Liverwort
SPECIES *monticola*
GENUS *Trabacellula* — Trabak Cap Jungermann Liverwort
SPECIES *tumidula*

FAMILY 05 *Chaetophyllopsaceae* — Bristle Jungermann Liverwort
GENUS *Chaetophyllopsis* — Bristle Jungermann Liverwort
SPECIES *whiteleggei*
GENUS *Chaetocolea* — Lesser Bristle Jungermann Liverwort
SPECIES *palmata*
GENUS *Herzogianthus* — Herzog Bristle Jungermann Liverwort
SPECIES *vaginatus*

FAMILY 06 *Geocalycaceae* — Earth Calyx Jungermann Liverwort
GENUS *Geocalyx* — Earth Calyx Jungermann Liverwort
SPECIES *graveolens*
GENUS *Chiloscyphus* — Claw Scyph Earth Calyx Jungermann Liverwort
SPECIES *gemmiparus, pallescens, polyanthos*
VARIETIES *pallescens pallescens, pallescens fragilis, polyanthos polyanthos, polyanthos rivularis*
GENUS *Harpanthus* — Harp Earth Calyx Jungermann Liverwort
SPECIES *drummondii, flotovianus, scutatus*
GENUS *Leptoscyphus* — Thin Scyph Earth Calyx Jungermann Liverwort
SPECIES *cuneifolius*
GENUS *Lophocolea* — Crest Earth Calyx Jungermann Liverwort
SPECIES *bidentata, cuspidata, gottscheoides, heterophylla, martiana, minor, muricata*
VARIETIES *cuspidata cuspidata, cuspidata alata*
GENUS *Pachyglossa* — Thick Tongue Earth Calyx Jungermann Liverwort
SPECIES *gottscheoides*

FAMILY 07 *Gymnomitriaceae* — Nude Mitre Jungermann Liverwort
GENUS *Gymnomitrion* — Nude Mitre Jungermann Liverwort
SPECIES *apiculatum, concinnatum, corralioides, laceratum, obtusum, pacificum*
GENUS *Anomomarsupella* — Anomalous Marsupial Nude Mitre Jungermann Liverwort
SPECIES *cephalozielloides*
GENUS *Marsupella* — Marsupial Nude Mitre Jungermann Liverwort
SPECIES *alpina, arctica, boeckii, bolanderi, brevissima, commutata, condensata, emarginata, funckii, paroica, revoluta, sparsifolia, sphacelata, ustulata, ustulata ustulata, ustulata sprucei*
VARIETIES *emarginata emarginata, emarginata aquatica, emarginata tubulosa*
GENUS *Prasanthus* — Prase Nude Mitre Jungermann Liverwort
SPECIES *suecicus*

FAMILY 08 *Herbertaceae* — Herbert Jungermann Liverwort
GENUS *Herbertus* — Herbert Jungermann Liverwort
SPECIES *aduncus, hawaiiensis, himalayanus, stramineus*
VARIETIES *aduncus aduncus, aduncus tenuis*
GENUS *Grollea* — Groll Herbert Jungermann Liverwort
SPECIES *antheliopsis*
GENUS *Triandrophyllum* — Triander Herbert Jungermann Liverwort
SPECIES *fernandeziense, heterophyllum, maegdefrauni, subtrifidum*

FAMILY 09 *Isotachidaceae* — Isometric Swift Jungermann Liverwort
GENUS *Isotachis* — Isometric Swift Jungermann Liverwort
SPECIES *lyallii*
GENUS *Eoisotachis* — Lesser Isometric Swift Jungermann Liverwort
SPECIES *stephanii*
GENUS *Neesioscyphus* — Nees Scyph Isometric Swift Jungermann Liverwort
SPECIES *carneus*

FAMILY 10 *Jubulaceae* — Jubul Jungermann Liverwort
GENUS *Jubula* — Jubul Jungermann Liverwort
SPECIES *pennsylvanica*
GENUS *Frullania* — Frullan Jubul Jungermann Liverwort
SPECIES *arietina, bolanderi, brasiliensis, brittoniae, californica, catalinae, chilcootiensis, cobrensis, cucullata, donnellii, eboracensis, franciscana, gymnotis, inflata, jackii, kunzei, oakesiana, obcordata, plana, riparia, selwyniana, squarrosa, tamarisci*
VARIETIES *tamarisci tamarisci, tamarisci asagrayana, tamarisci nisquallensis*
GENUS *Neohattoria* — Hattor Jubul Jungermann Liverwort
SPECIES *herzogii*
GENUS *Steerea* — Steere Jubul Jungermann Liverwort
SPECIES *clemensiana*

FAMILY 11 *Lejeuneaceae* — Lejeune Jungermann Liverwort
GENUS *Lejeunea* — Lejeune Jungermann Liverwort
SPECIES *autoica, blomquistii, caespitosa, calcicola, cardoti, cavifolia, cladogyna, flava, floridana, glaucescens, laetivirens, lamacerina, minutiloba, pililoba, ruthii, spiniloba, ulicina*
VARIETIES *calcicola calcicola, calcicola mexicana, glaucescens glaucescens, glaucescens acrogyna, glaucescens obsoleta, lamacerina lamacerina, lamacerina gemminata, minutiloba minutiloba, minutiloba heterogyna ulicina ulicina, ulicina bullata*
GENUS *Brachiolejeunea* — Short Lejeune Jungermann Liverwort
SPECIES *bahamensis, corticalis*
GENUS *Caudalejeunea* — Stalk Lejeune Jungermann Liverwort
SPECIES *lehmanniana*
GENUS *Ceratolejeunea* — Horn Lejeune Jungermann Liverwort
SPECIES *confusa, cubensis, laetefusca, rubiginosa*
GENUS *Cheilolejeunea* — Claw Lejeune Jungermann Liverwort
SPECIES *clausa, decidua, evansii, parvula, polyantha, rigidula*
VARIETIES *polyantha polyantha, polyantha caduciloba*
GENUS *Cololejeunea* — Cole Lejeune Jungermann Liverwort
SPECIES *biddlecomiae, cardiocarpa, contractiloba, cornutissima, diaphana, ephemeroides, macounii, minutissima, ornata, setiloba, sicaefolia, subcristata, tuberculata*
VARIETIES *minutissima minutissima, minutissima myriocarpa*
GENUS *Crossotolejeunea* — Cross Lejeune Jungermann Liverwort
SPECIES *bermudiana*
GENUS *Diplasiolejeunea* — Double Lejeune Jungermann Liverwort
SPECIES *rudolphiana*
GENUS *Drepanolejeunea* — Drepan Lejeune Jungermann Liverwort
SPECIES *appalachiana, sabaliana*
GENUS *Harpalejeunea* — Harp Lejeune Jungermann Liverwort
SPECIES *ovata, stricta*
GENUS *Hygrolejeunea* — Hygro Lejeune Jungermann Liverwort
SPECIES *alaskana*
GENUS *Leptolejeunea* — Thin Lejeune Jungermann Liverwort
SPECIES *elliptica*
GENUS *Leucolejeunea* — Light Lejeune Jungermann Liverwort
SPECIES *clypeata, conchifolia, unciloba, xanthocarpa*
GENUS *Lopholejeunea* — Crest Lejeune Jungermann Liverwort
SPECIES *muelleriana, subfusca*
GENUS *Mastigolejeunea* — Mastigote Lejeune Jungermann Liverwort
SPECIES *auriculata*
GENUS *Neurolejeunea* — Nerve Lejeune Jungermann Liverwort
SPECIES *breutelii*
GENUS *Ptychocoleus* — Fold Lejeune Jungermann Liverwort
SPECIES *heterophyllus*
GENUS *Rectolejeunea* — Rectangle Lejeune Jungermann Liverwort
SPECIES *berteroana, brittoniae, maxonii, phyllobola*
GENUS *Taxilejeunea* — Yew Lejeune Jungermann Liverwort
SPECIES *cladiophora, obtusangula, sharpii*

FAMILY 12 *Lepidolaenaceae* — Thin Jungermann Liverwort
GENUS *Lepidolaena* — Thin Jungermann Liverwort
SPECIES *clavigera*
GENUS *Gackstroemia* — Gack Thin Jungermann Liverwort
SPECIES *magellanica*
GENUS *Jubulopsis* — Jubul Thin Jungermann Liverwort
SPECIES *novae-zelandiae*

FAMILY 13 *Lepidoziaceae* — Slender Jungermann Liverwort

GENUS *Lepidozia* Slender Jungermann Liverwort
SPECIES *filamentosa, reptans, sandvicensis*
GENUS *Bazzania* Bazzan Slender Jungermann Liverwort
SPECIES *ambigua, denudata, nudicaulis, pearsonii, tricrenata, trilobata*
VARIETIES *trilobata trilobata, trilobata depauperata*
GENUS *Kurzia* Kurz Slender Jungermann Liverwort
SPECIES *crenacanthoidea, pauciflora, setacea, sylvatica*
GENUS *Mytilopsis* Mytile Slender Jungermann Liverwort
SPECIES *albifrons*
GENUS *Phycolepidozia* Phycote Slender Jungermann Liverwort
SPECIES *exigua*
GENUS *Telaranea* Telaran Slender Jungermann Liverwort
SPECIES *chaetophylla, nematodes*
GENUS *Zoopsis* Animal Slender Jungermann Liverwort
SPECIES *argentea*

FAMILY 14 *Plagiochilaceae* Pelagic Claw Jungermann Liverwort
GENUS *Plagiochila* Pelagic Claw Jungermann Liverwort
SPECIES *acanthophylla, appalachiana, arctica, aspleniformis, asplenioides, austini, caduciloba, columbiana, corniculata, diffusa, dubia, eurphyllon, floridana, hypnoides, invisus, ludoviciana, miradorensis, porelloides, satoi, schoefieldiana, semidecurrens, sharpii, sullivantii, undata, virginica*
VARIETIES *acanthophylla acanthophylla, acanthophylla japonica, arctica arctica, arctica intermedia, arctica subarctica, dubia dubia, dubia integrifolia, eurphyllon euryphyllon, eurphyllon echinata, miradorensis miradorensis, miradorensis convoluta, semidecurrens semidecurrens, semidecurrens alaskana, sharpii sharpii, sharpii yakusimensis, sullivantii sullivantii, sullivantii spinigera, undata undata, undata crispata, virginica virginica, virginica caroliniana, virginica euryphylla*
GENUS *Pedinophyllum* Pedin Pelagic Claw Jungermann Liverwort
SPECIES *interruptum*

FAMILY 15 *Porellaceae* Pore Jungermann Liverwort
GENUS *Porella* Pore Jungermann Liverwort
SPECIES *aboris-vitae, bolanderi, cordaeana, macroloba, navicularis, pinnata, platyphylla, platyphylloidea, roellii, swartziana, vernicosa*
GENUS *Ascidiota* Ascidiote Pore Jungermann Liverwort
SPECIES *blepharophylla*
VARIETIES *blepharophylla alaskana*

FAMILY 16 *Pseudolepicoleaceae* Lesser Thin Jungermann Liverwort
GENUS *Pseudolepicolea* Lesser Thin Jungermann Liverwort
SPECIES *fryei, quadrilaciniata*
GENUS *Blepharostoma* Eyelash Stoma Thin Jungermann Liverwort
SPECIES *arachnoideum, trichophyllum*
VARIETIES *trichophyllum trichophyllum, trichophyllum brevirete*
GENUS *Flufordiella* Fluford Thin Jungermann Liverwort
SPECIES *fuegiana*
GENUS *Isophyllaria* Isometric Thin Jungermann Liverwort
SPECIES *murrayana*
GENUS *Trichotemnoma* Hairy Thin Jungermann Liverwort
SPECIES *corrugatum*

FAMILY 17 *Scapaniaceae* Scapan Jungermann Liverwort
GENUS *Scapania* Scapan Jungermann Liverwort
SPECIES *aequiloba, americana, apiculata, bolanderi, brevicaulis, calcicola, carinthiaca, compacta, crassiretis, curta, cuspiduligera, degenii, glaucocephala, gymnostomophila, hyperborea, invisa, irrigua, kaurinii, ligulifolia, lingulata, massalongi, mucronata, nemorosa, obcordata, obscura, paludicola, paludosa, paradoxa, parvifolia, perssonii, pulcherrima, saxicola, scandica, serrulata, simmonsii, spitzbergensis, subalpina, tundrae, uliginosa, umbrosa, undulata*
VARIETIES *curta curta, curta grandiretis, curta isoloba, cuspiduligera cuspiduligera, cuspiduligera diphophyllopsis, degenii degenii, degenii dubia, irrigua irrigua, irrigua rufescens, lingulata lingulata, lingulata microphylla, mucronata mucronata, mucronata praetervisa, paludicola paludicola, paludicola rotundiloba, paludicola viridigemma, scandica scandica, scandica dimorpha, undulata undulata, undulata aequatiformis, undulata oakesii*
GENUS *Diplophyllum* Diploid Scapan Jungermann Liverwort
SPECIES *albicans, andrewsii, apiculatum, imbricatum, microdontum, obtusatum, obtusifolium, plicatum, taxifolium*
VARIETIES *apiculatum apiculatum, apiculatum taxifolioides, taxifolium taxifolium, taxifolium macrosticta, taxifolium mucronatum*
GENUS *Douinia* Douin Scapan Jungermann Liverwort
SPECIES *ovata*

FAMILY 18 *Schistochilaceae* Split Claw Jungermann Liverwort

GENUS *Paraschistochila* — Split Claw Jungermann Liverwort
SPECIES *pinnatifolia*
GENUS *Pluerocladopsis* — Many Branch Split Claw Jungermann Liverwort
SPECIES *simulans*

FAMILY 19 *Trichocoleaceae* — Hairy Jungermann Liverwort
GENUS *Trichocolea* — Hairy Jungermann Liverwort
SPECIES *tomentella*
GENUS *Eotrichocolea* — Good Hairy Jungermann Liverwort
SPECIES *polycantha*
GENUS *Neotrichocolea* — Modern Hairy Jungermann Liverwort
SPECIES *bisseti*
GENUS *Trichocoleopsis* — Lesser Hairy Jungermann Liverwort
SPECIES *saccatula*

ORDER 3 *Marchantiales* Marchant Liverwort

FAMILY 1 *Marchantiaceae* — Marchant Liverwort
GENUS *Marchantia* — Marchant Liverwort
SPECIES *domingensis, paleacea, polymorpha*
GENUS *Bucegia* — Bucege Marchant Liverwort
SPECIES *romanica*
GENUS *Carrpos* — Carpel Marchant Liverwort
SPECIES *sphaerocarpos*
GENUS *Conocephalum* — Conehead Marchant Liverwort
SPECIES *conicum, supradecompositum*
GENUS *Corsinia* — Corsin Marchant Liverwort
SPECIES *coriandrina, weddelli*
GENUS *Dumortiera* — Dumortier Marchant Liverwort
SPECIES *hirsuta*
VARIETIES *hirsuta hirsuta, hirsuta nepalensis*
GENUS *Lunularia* — Moon Marchant Liverwort
SPECIES *cruciata*
GENUS *Monosolenium* — Monad Sole Marchant Liverwort
SPECIES *tenerum*
GENUS *Oxymitra* — Sharp Mitre Marchant Liverwort
SPECIES *androgyna, paleacea*
GENUS *Preissia* — Preiss Marchant Liverwort
SPECIES *italica, quadrata*
GENUS *Targionia* — Targion Marchant Liverwort
SPECIES *hypophylla*

FAMILY 2 *Aytoniaceae* — Ayton Marchant Liverwort
GENUS *Asterella* — Ayton Marchant Liverwort
SPECIES *bolanderi, californica, elegans, gracilis, lindenbergiana, palmeri, saccata, tenella*
VARIETIES *elegans echinella*
GENUS *Cryptomitrium* — Cryptic Mitre Ayton Marchant Liverwort
SPECIES *tenerum*
GENUS *Mannia* — Mann Ayton Marchant Liverwort
SPECIES *californica, fragrans, pilosa, sibirica, triandra*
GENUS *Plagiochasma* — Pelagic Chasm Ayton Marchant Liverwort
SPECIES *cuneatum, rupestre, wrightii*
GENUS *Reboulia* — Reboul Ayton Marchant Liverwort
SPECIES *hemisphaerica*

FAMILY 3 *Cleveaceae* — Cleve Marchant Liverwort
GENUS *Athalamia* — Cleve Marchant Liverwort
SPECIES hyalina
GENUS *Peltolepis* — Pelt Thin Cleve Marchant Liverwort
SPECIES *quadrata*
GENUS *Sauteria* — Sauter Cleve Marchant Liverwort
SPECIES *alpina*

FAMILY 4 *Ricciaceae* — Riccian Marchant Liverwort
GENUS *Riccia* — Riccian Marchant Liverwort
SPECIES *albida, austini, beyrichiana, bifurca, californica, campbelliana, canaliculata, cavernosa, curtisii, dictyospora, donnellii, eldeeniae, fluitans, frostii, glauca, hirta, macallisteri, membranacea, nigrella, ozarkiana, sorocarpa, sullivantii, tenella, trichocarpa, violacea*
GENUS *Ricciocarpos* — Carpel Riccian Marchant Liverwort
SPECIES *natans*

ORDER 4 *Metzgeriales* Metzger Liverwort

FAMILY 1 *Metzgeriaceae* Metzger Liverwort
GENUS *Metzgeria* Metzger Liverwort
SPECIES *angusta, conjugata, crassipilis, furcata, leptoneura, myriopoda, pubescens, temperata, uncigera*
VARIETIES *crassipilis novicrassipilis, furcata flavovirens, furcata glabra, furcata planiuscula, furcata setigera, furcata ulvula, leptoneura leptoneura, leptoneura australis, leptoneura extensa, leptoneura fuscescens, leptoneura hamata, leptoneura hamatiformis, leptoneura linearis, leptoneura longipila, leptoneura nitida, leptoneura polychaeta, leptoneura recurva, myriopoda ciliifera, pubescens duricosta, temperata fruticulosa*
GENUS *Makinoa* Makino Metzger Liverwort
SPECIES *crispata*
GENUS *Metzgeriopsis* Lesser Metzger Liverwort
SPECIES *pusilla*
GENUS *Phyllothallia* Thallus Metzger Liverwort
SPECIES *fuegiana, nivicola*
GENUS *Steerella* Steere Metzger Liverwort
SPECIES *linearis*
GENUS *Verdoornia* Verdoorn Metzger Liverwort
SPECIES *succulenta*

FAMILY 2 *Allisoniaceae* Allison Metzger Liverwort
GENUS *Allisonia* Allison Metzger Liverwort
SPECIES *cockaynii*
GENUS *Calycularia* Calyx Allison Metzger Liverwort
SPECIES *compacta, crispula, golae, laxa, radiculosa*

FAMILY 3 *Aneuraceae* Aneural Metzger Liverwort
GENUS *Aneura* Aneural Metzger Liverwort
SPECIES *maxima, pinguis*
VARIETIES *pinguis pinguis, pinguis angustior*
GENUS *Cryptothallus* Cryptic Thallus Aneural Metzger Liverwort
SPECIES *mirabilis*
GENUS *Riccardia* Riccardo Aneural Metzger Liverwort
SPECIES *chamedryfolia, incurvata, jugata, latifrons, multifida, palmata, stricta*
VARIETIES *latifrons latifrons, latifrons arctica, multifida multifida, multifida synoica*

FAMILY 4 *Blasiaceae* Blasé Metzger Liverwort
GENUS *Blasia* Blasé Metzger Liverwort
SPECIES *pusilla*
GENUS *Cavicularia* Cavicle Blasé Metzger Liverwort
SPECIES *densa*

FAMILY 5 *Fossombroniaceae* Fossombron Metzger Liverwort
GENUS *Fossombronia* Fossombron Metzger Liverwort
SPECIES *alaskana, brasiliensis, foveolata, hispidissima, lamellata, longiseta, macounii, wondraczekii, zygospora*
VARIETIES *brasiliensis brasiliensis, brasiliensis lamellifera, brasiliensis leucorhiza, foveolata foveolata, foveolata cristula*
GENUS *Petalophyllum* Petal Fossombron Metzger Liverwort
SPECIES *ralfsii*

FAMILY 6 *Hymenophytaceae* Membrane Metzger Liverwort
GENUS *Hymenophyton* Membrane Metzger Liverwort
SPECIES *flabellatum*
GENUS *Podomitrium* Pod Mitre Membrane Metzger Liverwort
SPECIES *malaccense, phyllanthus*

FAMILY 7 *Pallaviciniaceae* Pallavic Metzger Liverwort
GENUS *Pallavicinia* Pallavic Metzger Liverwort
SPECIES *lyellii*
GENUS *Moerckia* Moerck Pallavic Metzger Liverwort
SPECIES *blyttii, hibernica*

FAMILY 8 *Pelliaceae* Pellet Metzger Liverwort
GENUS *Pellia* Pellet Metzger Liverwort
SPECIES *appalachiana, endiviifolia, epiphylla, megaspora, neesiana*
VARIETIES *endiviifolia endiviifolia, endiviifolia alpicola, neesiana neesiana, neesiana columbiana*
GENUS *Noteroclada* Branch Pellet Metzger Liverwort
SPECIES *confluens*

FAMILY 9 *Treubiaceae* Treube Metzger Liverwort
GENUS *Treubia* Treube Metzger Liverwort
SPECIES *insignis*

GENUS *Apotreubia* — Apo Treube Metzger Liverwort
SPECIES *nana*

ORDER 5 *Sphaerocarpales* Round Carpel Liverwort

FAMILY 1 *Sphaerocarpaceae* — Round Carpel Liverwort
GENUS *Sphaerocarpos* — Round Carpel Liverwort
SPECIES *cristatus, donnellii, drewei, hians, michelii, texanus*
GENUS *Geothallus* — Earth Thallus Round Carpel Liverwort
SPECIES *tuberosus*
GENUS *Riella* — Riel Round Carpel Liverwort
SPECIES *affinis, americana*

ORDER 6 *Takakiales* Takak Liverwort

FAMILY 1 *Takakiaceae* — Takak Liverwort
GENUS *Takakia* — Takak Liverwort
SPECIES *ceratophylla, lepidozioides*

SUBKINGDOM 64 *Filicata* (Fern)

Latin: *Filicata* filament, thread	**PHYLUMS**		**Species**	**Species**
i.e. narrow leaflets on ferns	641 *Lycopodiophyta*	Moss Fern	161 [R]	2000 [M]
12,117 species	642 *Pteridophyta*	Leaf Fern	931 [R]	12000 [M]
[Margulis & Schwartz, 1998] [M]				
1,114 species [Ruggiero, Nov. 2006]				

Boughton Cobb (see ISBN:0-618-39406-0) revised Fern Classification in 2005 by dividing Ferns into two main groups Moss Ferns (*Lycophyta*) and Leaf Ferns (*Monilophyta* also known as *Pteridophyta*). He further subdivided Moss Ferns into two subgroups; (1) Club/Fir Moss, and (2) Quillwort/Spike Moss. He further subdivided Leaf Ferns into five subgroups; (1) Horsetail, (2) Whisk Fern, (3) Succulent Fern, (4) Tropical Fern, and (5) Subtropical Fern.

In Fusillo Fern Classification, Cobb Leaf Fern subgroup (1) Horsetail and subgroup (2) Whisk Fern have been combined into one subgroup Primitive Fern (*Protofilicopsida*). In the Fusillo Fern Classification, Cobb Leaf Fern subgroup (3) Dryland Succulent Ferns and subgroup (4) Tropical Ferns have been combined into one subgroup Wetland, Dryland, & Tropical Fern (*Parafilicopsida).*

Fusillo Fern Classification

Domain *Plantae*	Plant
Kingdom 6 *Sporophyta*	Spore Plant
Subkingdom 4 *Filicata*	Fern
Phylum 1 *Lycopodiophyta*	**Moss Fern**
Class 1 *Lycopodiopsida*	Moss Fern
Order 1 *Lycopodiales*	Club Moss
Family 1 *Lycopodiaceae*	Club Moss
Order 2 *Huperziales*	Fir Moss
Family 1 *Huperziaceae*	Fir Moss
Class 2 *Isosellopsida*	Quillwort & Spike Moss
Order 1 *Isoetales*	Quillwort
Family 1 *Isoetaceae*	Quillwort
Order 2 *Sellaginellales*	Spike Moss
Family 1 *Sellaginellaceae*	Spike Moss
Phylum 2 *Pteridophyta*	**Leaf Fern**
Class 1 *Protofilicopsida*	Primitive Fern
Order 1 *Equistetales*	Horsetail
Family 1 *Equisetaceae*	Horsetail
Order 2 *Psilotales*	Whisk Fern
Family 1 *Psilotaceae*	Whisk Fern
Class 2 *Parafilicopsida*	Wetland, Tropical, & Dryland Fern
Order 1 *Hydropterales*	Wetland Fern
Family 1 *Azollaceae*	Mosquito Pond Fern
Family 2 *Parkeriaceae*	Water Sprite Fern
Family 3 *Marsileaceae*	Water Clover Fern
Family 4 *Salviniaceae*	Floating Water Fern
Order 2 *Marattales*	Tropical Fern
Family 1 *Marattaceae*	Tropical Fern
Order 3 *Ophioglossales*	Dryland Fern
Family 1 *Ophioglossaceae*	Succulent Fern
Class 4 *Filicopsida*	Subtropical Fern
Order 1 *Filicales*	Fern
Family 1 *Pteridaceae*	Fern
Family 2 *Hymenophyllaceae*	Filmy Fern
Family 3 *Osmundaceae*	Flowering Fern
Family 4 *Schizeaceae*	Curly Grass Fern
Order 2 *Polypodiales*	Polypody Fern
Family 01 *Polypodiaceae*	Polypody Fern
Family 02 *Aspleniaceae*	Spleenwort Fern
Family 03 *Blechnaceae*	Misorus Fern
Family 04 *Cyathaceae*	Tree Fern
Family 05 *Dennstaedtiaceae*	Hay Scented Fern
Family 06 *Dicksoniaceae*	Dickson Fern
Family 07 *Drypopteridaceae*	Wood Fern
Family 08 *Gleichenaceae*	Forked, Scrambling, & Umbrella Fern
Family 09 *Grammatidaceae*	Marked Fern
Family 10 *Thelypteridaceae*	Delicate Fern
Family 11 *Vittariaceae*	Shoestring Fern

PHYLUM 641 *Lycopodiophyta* Moss Fern

Latin: *Lycos* wolf	CLASS		Species
Latin: *Phyta* phyton, plant	643.1 *Lycopodiopsida*	Club Moss & Fir Moss	1000
1000 species	643.2 *Isosellopsida*	Quillwort & Spike Moss	
[Margulis & Schwartz, 1998]			
161 species [Ruggiero, Nov. 2006]			

Phylum *Lycopodiophyta* Clubmoss Fern, lycopod, species first appeared in the fossil record 400 million years ago. They are evergreen vascular spore plants. Glossy fronds are arranged in spirals or whorls and held close to the branches. They resemble miniature pine trees (1 inch to 4 inches high). Their spores have been used to coat pills and condoms. Their spores are also used to create "pink" fireworks. Lycopods reproduce in alternating haploid sporophyte and diploid gametophyte generations.

CLASS 643.1 *Lycopodiopsida* Clubmoss Fern

ORDER 1 *Lycopodiales* Club Moss & Fir Moss

FAMILY 1 *Lycopodiaceae* Club Moss
GENUS *Lycopodium* Club Moss
SPECIES *clavatum, lagopus, phlegmaria, venustulum, volubile*
VARIETIES *clavatum clavatum, clavatum laurentianum, clavatum subremotum, clavatum tristachyum, lagopus brevispicatum, lagopus integerrimum, lagopus megastachyon*
GENUS *Diphasium* (Synonym for *Lycopodium*)
GENUS *Lepidotis* (Synonym for *Lycopodiella*)
GENUS *Dendrolycopodium* Tree Club Moss
SPECIES *dendroideum, hickeyi, obscurum*
VARIETIES *dendroideum hybridum, hickeyi isophyllum*
GENUS *Lycopodiella* Bog Club Moss
SPECIES *alopecuroides, appressa, brucei, cernua, copelandii, inundata, margueritiae, prostrata, subappressa*
VARIETIES *appressa bigelovii, , cernua cernua, cernua crassifolia, cernua curvata, copelandii chapmanii, copelandii elongatum, copelandii robustum, prostrata pinnatum*
GENUS *Palhinhaea* (Synonym for *Lycopodiella*)
GENUS *Phlegmariurus* (Synonym for *Huperzia*)
GENUS *Phylloglossum* Tongue Club Moss
SPECIES *drummondii*
GENUS *Pseudolycopodiella* Slender Bog Club Moss
SPECIES *caroliniana*
VARIETIES *caroliniana caroliniana, caroliniana meridionalis*
GENUS *Spinulum* Bristly Club Moss
SPECIES *annotinum, canadense*
VARIETIES *annotinum annotinum, annotinum acrifolium, annotinum alpestre, annotinum dubium, canadense pungens*

ORDER 2 *Huperziales* Fir Moss

FAMILY 1 *Huperziaceae* Fir Moss
GENUS *Huperzia* Fir Moss
SPECIES *acerosa, appalachiana, aqualupiana, bartleyi, buttersii, chinensis, dichotoma, erubescens, filiformis, funiformis, haleakalae, helleri, hippuridea, linifolia, lucidula, mannii, montana, nutans, occidentalis, phyllantha, pithyoides, polytrichoides, porophila, reflexa, selago, serrata, sintenisii, somai, stemmermanniae, sulcinervia, taxifolia, tenuicaulis, wilsonii*
VARIETIES *acerosa portoricense, acerosa setaceum, acerosa verticillata, chinensis miyoshiana, lucidula reflexa, occidentalis tryonii, porophila patens, selago selago, selago densa, selago densa-appressa, selago densa-arctica, serrata serrata, serrata dentata*
GENUS *Diphasiastrum* Ground Cedar Fir Moss
SPECIES *alpinum, complanatum, digitatum, habereri, issleri, sabinifolium, sitchense, tristachyum, zeilleri*
VARIETIES *complanatum anceps, complanatum canadense, digitatum flabelliforme, habereri laurentianum, issleri decipiens, issleri pseudoalpinum, sabinifolium armatum, sabinifolium patens, sabinifolium sharonense, sabinifolium superfertile, tristachyum chamicyparissus, tristachyum patentifolium, zeilleri boreale, zeilleri elongatum, zeilleri gartonis*
GENUS *Urostachys* (Synonym for *Huperzia*)

CLASS 643.2 *Isosellopsida* Quillwort & Spikemoss

ORDER 1 *Isoetales* Quillwort

FAMILY 1 *Isoetaceae* Quillwort
GENUS *Isoetes* Quillwort
SPECIES *acadiensis, altonharvillii, appalachiana, bolanderi, boomii, brittonii, bruntonii, butleri, dodgei, eatonii, engelmannii, fairbrothersii, flaccida, foveolata, georgiana, harveyi, hawaiiensis, heterospora, hickeyi, howellii, hyemalis, jeffreyi, lacustris, lithophila, louisianensis, maritima, melanopoda, melanospora, nuttallii, occidentalis, orcuttii, piedmontana, prototypus, riparia, saccharata, tegetiformans, tenella, truncata, tuckermanii, valida, virginica*
VARIETIES *appalachiana georgiana, bolanderi bolanderi, bolanderi pygmaea, eatonii gravesii, engelmannii engelmannii, engelmannii gracilis, flaccida flaccida, flaccida alata, flaccida chapmanii, flaccida rigida, howellii minima, lacustris hieroglyphica, lacustris macrospora, maritima pseudotruncata, occidentalis flettii, occidentalis paupercula, occidentalis piperi, riparia riparia, riparia amesii, riparia amesii-reticulata, riparia canadensis, riparia palmeri, riparia robbinsii, tenella asiatica, tenella braunii, tenella echinospora, tenella hesperia, tenella muricata, tenella robusta, tenella savilei, tenella setacea, valida caroliniana*

ORDER 2 *Selaginellales* Spike Moss

FAMILY 1 *Selaginellaceae* Spike Moss
GENUS *Selaginella* Spike Moss
SPECIES *acanthonota, apoda, arbuscula, arenicola, arizonica, armata, asprella, bigelovii, braunii, cinerascens, cordifolia, deflexa, densa, douglasii, eatonii, eclipes, eremophila, flabellata, hansenii, heterodonta, kraussiana, krugii, laxifolia, lepidophylla, leucobryoides, ludoviciana, menziesii, mutica, neomexicana, oregana, ovifolia, peruviana, pilifera, plana, plumosa, rupestris, rupincola, sandvicensis, selaginoides, sibirica, stipitata, subcaulescens, substipitata, tenella, tortipila, uncinata, underwoodii, utahensis, viridissima, wallacei, watsonii, weatherbiana, willdenovii, wrightii*
VARIETIES *arenicola arenicola, arenicola riddellii, armata plagiochila, densa densa, densa engelmannii, densa scopulorum, densa standleyi, mutica mutica, mutica limitanea, mutica limitanea-texana, peruviana sheldonii, pilifera pringlei, plumosa stolonifera, sandvicensis parvula, selaginoides spinosa, subcaulescens sintenisii, substipitata portoricensis, tenella albonitens, underwoodii dolichotricha*

PHYLUM 642 *Pteridophyta* Leaf Fern

Latin: *Ptera* wing	**CLASS**		**Species**
Latin: *Phyta* plant	641.1 *Protofilicopsida*	Primitive Fern	
	641.2 *Parafilicopsida*	Wetland, Tropical, Dryland Fern	
	641.3 *Eufilicopsida*	Subtropical Fern	12000

12000 species [Margulis & Schwartz, 1998]
931 species [Ruggiero, Nov. 2006]

Members of the **Class Protofilicopsida Primitive Fern Order *Equisetales* Horsetail** have a reed like stalk with a thick elongated cylinder at the top similar to cat tail plants of the genus *Typha*. The elongated bulb has scales and resembles the bushy tail of a horse. They first appeared in the fossil record 300 million years ago. They are also referred to as scouring rush because their "horse tail" top was used to scour pots. They are poisonous to cattle and horses because they produce the enzyme thiaminase which destroys the vitamin thiamine.

Class *Protofilicopsida* Primitive Fern Order *Psilotales* Whisk Fern species are the only vascular plants that lack roots and leaves. They grow in rock crevices, on tree trunks and in soil in subtropical areas. They first appeared in the fossil record 400 million years ago. Hawaiian men used Whisk Fern spores to prevent irritation from loin cloths.

Class *Eufilicopsida* Fern species are vascular spore plants. Like algae, fungus, lichen, moss, club moss, and whisk ferns, they have only spores, they do not have seeds. They have finely divided fronds with many frondlets per frond. They grow in moist soil in a subtropical climate.

CLASS 642.1 *Protofilicopsida* Primitive Fern

ORDER 1 *Equisetales* Horsetail

FAMILY 1 *Equisetaceae* — Horsetail
GENUS *Equisetum* — Horsetail
SPECIES *arcticum, arvense, dycei, ferrissii, fluviatile, font-queri, hyemale, laevigatum, litorale, mackaii, nelsonii, palustre, pratense, ramosissimum, scirpoides, sylvaticum, telmateia, variegatum*
VARIETIES *arvense alpestre, arvense boreale, arvense calderi, arvense campestre, arvense riparium, ferrissii elatum, ferrissii intermedium, fluviatile limosum, hyemale affine, hyemale affine-californicum, hyemale affine-praealtum, hyemale affine-pseudohyemale, hyemale affine-robustum, laevigatum funstonii, laevigatum kansanum, mackaii jesupii, mackaii trachyodon, palustre americanum, palustre simplicissimum, sylvaticum multiramosum, sylvaticum pauciramosum, telmateia telmateia, telmateia braunii, telmateia braunii-maximum, variegatum variegatum, variegatum alaskanum, variegatum anceps*
GENUS *Hippochaete* — (Synonym for *Equisetum*)

ORDER 2 *Psilotales* Whisk Fern

FAMILY 1 *Psilotaceae* — Whisk Fern
GENUS *Psilotum* — Whisk Fern
SPECIES *complanatum, intermedium, nudum*
GENUS *Tmesipteris* — (Synonym for *Psilotum*)

CLASS 642.2 *Parafilicopsida* Wetland, Tropical, Dryland Fern

ORDER 1 *Hydropteridales* Wetland Fern

FAMILY 1 *Azollaceae* — Mosquito Pond Fern
GENUS *Azolla* — Mosquito Pond Fern
SPECIES *caroliniana, filiculoides, mexicana, pinnata*

FAMILY 2 *Marsileaceae* — Water Fern
GENUS *Marsilea* — Water Clover Fern
SPECIES *ancylopoda, hirsuta, macropoda, minuta, mollis, oligospora, polycarpa, quadrifolia, vestita, villosa*
VARIETIES *ancylopoda berteroi, ancylopoda mexicana, vestita vestita, vestita fournieri, vestita minuta, vestita mucronata, vestita tenuifolia, vestita uncinata*
GENUS *Pilularia* — Pillwort Water Fern
SPECIES *americana*
GENUS *Regnellidium* — Regnell Water Fern
SPECIES *diphyllum*

FAMILY 3 *Parkeriaceae* Water Sprite Fern
GENUS *Ceratopteris* Water Sprite Fern
SPECIES *pteridoides, richardii, thalictroides*
VARIETIES *thalictroides deltoidea, thalictroides siliquosa*

FAMILY 4 *Salviniaceae* Floating Water Fern
GENUS *Salvinia* Floating Water Fern
SPECIES *auriculata, biloba, herzogii, minima, molesta, natans*
VARIETIES *auriculata natans, auriculata rotundifolia*

ORDER 2 *Marattiales* Tropical Fern

FAMILY 1 *Marattiaceae* Tropical Fern
GENUS *Marattia* Potato Fern
SPECIES *douglasii, laevis*
VARIETIES *laevis kaulfussii*
GENUS *Angiopteris* Mule's Foot Tree Fern
SPECIES *evecta*
GENUS *Danaea* Dana Fern
SPECIES *elliptica, jamaicensis, jenmanii, nodosa, urbanii*

ORDER 3 *Ophioglossales* Dryland Fern

FAMILY 1 *Ophioglossaceae* Succulent Fern
GENUS *Ophioglossum* Adder's Tongue Fern
SPECIES *azoricum, californicum, concinnum, crotalophoroides, engelmannii, lusitanicum, nudicaule, pendulum, petiolatum, pusillum, reticulatum, vulgatum*
VARIETIES *crotalophoroides crotalophoroides, crotalophoroides nanum, nudicaule dendroneuron, nudicaule ellipticum, nudicaule minus, nudicaule mononeuron, nudicaule tenerum, pendulum pendulum, pendulum falcatum, pusillum alaskanum, pusillum pseudopodum, vulgatum pycnostichum*
GENUS *Botrychium* Grape Fern
SPECIES *acuminatum, ascendens, biternatum, boreale, campestre, crenulatum, dissectum, echo, gallicomontanum, hesperium* (Moonwort), *jenmanii, lanceolatum, lineare, lunaria* (Moonwort), *lunarioides* (Moonwort), *matricariifolium* (Moonwort), *minganense* (Moonwort), *montanum, mormo, multifidum, oneidense, pallidum, paradoxum, pedunculosum, pinnatum, pseudopinnatum, pumicola, robustum, rugulosum, simplex, spathulatum, subbifoliatum, virginianum, watertonense*
VARIETIES *biternatum tenuifolium, crenulatum dusenii, dissectum obliquum, dissectum oblongifolium, hesperium occidentale, jenmanii alabamense, lanceolatum lanceolatum, lanceolatum angustisegmentum, lunaria onondagense, multifidum californicum, multifidum coulteri, multifidum intermedium, multifidum matricariae, multifidum silaifolium, pinnatum obtusilobum, rugulosum ternatum, simplex simplex, simplex compositum, simplex laxifolium, simplex tenebrosum, virginianum europaeum*
GENUS *Cheiroglossa* Hand Fern
SPECIES *palmata*
GENUS *Helminthostachys* Snake Fern
SPECIES *zeylanica*

CLASS 642.3 *Filicopsida* Fern

ORDER 1 *Filicales* Fern

FAMILY 1 *Pteridaceae* Fern
GENUS *Pteris* Brake Fern
SPECIES *altissima, arborea, bahamensis, biaurita, cretica, deflexa, delchampsii, denticulata, ensiformis, excelsa, grandifolia, hillebrandii, irregularis, longifolia, lydgatei, multifida, mutilata, plumula, pungens, tremula, tripartita, vittata*
VARIETIES *arborea crassipes, cretica cretica, cretica albolineata, ensiformis victoriae, multifida serrulata, plumula quadriaurita*
GENUS *Acrostichum* Leather Fern
SPECIES *aureum, danaeifolium*
VARIETIES *danaeifolium excelsum*
GENUS *Adiantopsis* Lesser Maidenhair Fern
SPECIES *asplenioides, pedata, radiata, rupicola*
GENUS *Adiantum* Maidenhair Fern
SPECIES *aleuticum, anceps, capillus-veneris, caudatum, concinnum, deltoideum, dolosum, fragile, hirsutum, hispidulum, jordanii, latifolium, lucidum, macrophyllum, melanoleucum, obliquum, pedatum, petiolatum, pulverulentum, pyramidale, raddianum, tenerum, tetraphyllum, tracyi, trapeziforme, tricholepis, villosum, viridimontanum, vivesii, wilsonii*

VARIETIES *aleuticum calderi, aleuticum pedatum, aleuticum subpumilum, capillus-veneris modestum, capillus-veneris protrusum, fragile fragile, fragile rigidulum, pyramidale cristatum, raddianum cuneatum*

GENUS *Aleuritopteris* (Synonym for *Notholaena*)

GENUS *Anopteris* Moore Brake Fern

SPECIES *hexagona*

VARIETIES *hexagona hexagona, hexagona heterophylla, hexagona intermedia*

GENUS *Argyrochosma* Silver False Cloak Fern

SPECIES *dealbata, fendleri, incana, jonesii, limitanea, limitanea, limitanea, microphylla*

VARIETIES *fendleri cancellata, microphylla parvifolia*

GENUS *Aspidotis* Shield Lace Fern

SPECIES *californica, carlotta-halliae, densa*

VARIETIES *densa siliquosa*

GENUS *Astrolepis* Scaly Cloak Fern

SPECIES *cochisensis, integerrima, sinuata, windhamii*

VARIETIES *cochisensis cochisensis, cochisensis arizonica, cochisensis chihuahuaensis, sinuata sinuata, sinuata mexicana*

GENUS *Bommeria* Copper Fern

SPECIES *hispida*

GENUS *Cheilanthes* Lip Fern

SPECIES *aemula, alabamensis, arizonica, bonariensis, clevelandii, cooperae, covillei, decipiens, decora, eatonii, feei, fendleri, fibrillosa, gracillima, horridula, intertexta, kaulfussii, lanosa, lendigera, leucopoda, lindheimeri, microphylla, newberryi, notholaenoides, parishii, parryi, paupercula, pringlei, takeuchii, tomentosa, villosa, viscida, wootonii, wrightii, yavapensis*

VARIETIES *bonariensis aurea, eatonii castanea, gracillima aberrans, lanosa vestita*

GENUS *Coniogramme* Loulu Fern

SPECIES *pilosa*

GENUS *Cryptogramma* Rock Brake Fern

SPECIES *acrostichoides, cascadensis, crispa, sitchensis, stelleri*

GENUS *Doryopteris* Finger Fern

SPECIES *pedata*

GENUS *Eriosorus* Chestnuthair Fern

SPECIES *hispidulus*

GENUS *Gymnopteris* (Synonym for *Bommeria*)

GENUS *Hemionitis* Star Fern

SPECIES *palmata*

GENUS *Notholaena* False Cloak Fern

SPECIES *aliena, aschenborniana, californica, candida, copelandii, grayi, greggii, lemmonii, nealleyi, neglecta, schaffneri, standleyi, trichomanoides*

VARIETIES *californica californica, californica accessita, californica deserti, californica leucophylla, californica nigrescens, grayi grayi, grayi sonorensis, standleyi hookeri*

GENUS *Onychium* (Synonym for *Anopteris*)

GENUS *Pellaea* Cliff Brake Fern

SPECIES *andromedifolia, atropurpurea, brachyptera, breweri, bridgesii, cordifolia, gastonyi, glabella, glaciogena, intermedia, lyngholmii, mucronata, ovata, ternifolia*

VARIETIES *andromedifolia pubescens, andromedifolia rubens, cordifolia cardiomorpha, glabella glabella, glabella bushii, glabella missouriensis, glabella occidentalis, glabella occidentalis-nana, glabella simplex, glabella simplex-suksdorfiana, mucronata mucronata, mucronata californica, mucronata californica-compacta, ternifolia ternifolia, ternifolia arizonica, ternifolia villosa, truncata, viridis, wrightiana*

GENUS *Pentagramma* Five Mark Goldback & Silverback Fern

SPECIES *pallida, triangularis*

VARIETIES *triangularis maxonii, triangularis semipallida, triangularis semipallida-viridis, triangularis triangularis, triangularis viscosa*

GENUS *Pityrogramma* Powder Frond Goldback & Silverback Fern

SPECIES *austroamericana, calomelanos, chrysophylla, ebenea, mackenneyi, sulphurea, trifoliata*

VARIETIES *austroamericana aureoflava, chrysophylla gabrielae, chrysophylla subflexuosa, ebenea tartarea*

GENUS *Psilogramme* (Synonym for *Eriosorus*)

GENUS *Schizostege* (Synonym for *Pteris*)

GENUS *Trismeria* (Synonym for *Pityrogramma*)

FAMILY 2 *Hymenophyllaceae* Filmy Fern

GENUS *Hymenophyllum* Filmy Fern

SPECIES *asplenioides, axillare, brevifrons, decurrens, fendlerianum, fragile, fucoides, hirsutum, lanatum, lanceolatum, lineare, macrothecum, microcarpum, obtusum, paucicarpum, polyanthos, recurvum, sericeum, sieberi, tayloriae, tegularis, tunbrigense, undulatum, wrightii*

GENUS *Callistopteris* (Synonym for *Trichomanes*)

GENUS *Lecanium* (Synonym for *Hymenophyllum*)

GENUS *Mecodium* (Synonym for *Hymenophyllum*)

GENUS *Trichomanes* Soft Hair Fruit Dot Filmy Fern

SPECIES *alatum, angustifrons, bauerianum, boschianum, capillaceum, crinitum, crispum, cyrtotheca, davallioides, draytonianum, elegans, fimbriatum, holopterum, hookeri, hymenoides, hymenophylloides, intricatum, kapplerianum, krausii, lineolatum, membranaceum, minutum, osmundoides, ovale, padronii, petersii, pinnatum, polypodioides, punctatum, pusillum, radicans, rigidum, robustum, scandens, trigonum*

VARIETIES *punctatum punctatum, punctatum floridanum, punctatum sphenoides, trigonum trigonum, trigonum fimbriatum*

GENUS *Vandenboschia* Vandenbosch Filmy Fern

SPECIES *tubiflora*

FAMILY 3 *Osmundaceae* Flowering Fern

GENUS *Osmunda* Flowering Fern

SPECIES *cinnamomea, claytoniana, regalis, ruggii*

VARIETIES *cinnamomea cinnamomea, cinnamomea frondosa, cinnamomea glandulosa, cinnamomea imbricata, regalis spectabilis*

GENUS *Anemia* Anemic Flowering Fern

SPECIES *adiantifolia, cicutaria, coriacea, cuneata, hirsuta, hirta, mexicana, phyllitidis, portoricensis, underwoodiana, wrightii*

FAMILY 4 *Schizaeaceae* Curly Grass Fern

GENUS *Schizaea* Curly Grass Fern

SPECIES *digitata, fluminensis, poeppigiana, pusilla*

VARIETIES *fluminensis elegans*

GENUS *Actinostachys* Ray Curly Grass Fern

SPECIES *pennula*

VARIETIES *pennula digitatum, pennula germanii*

GENUS *Lophidium* (Synonym for *Schizaea*)

GENUS *Lygodium* Climbing Curly Grass Fern

SPECIES *cubense, japonicum, microphyllum, palmatum, venustum, volubile*

GENUS *Microschizaea* (Synonym for *Actinostachys*)

ORDER 2 *Polypodiales* Polypody Fern

FAMILY 01 *Polypodiaceae* Polypody Fern

GENUS *Polypodium* Polypody Fern

SPECIES *amorphum, appalachianum, attenuatum, californicum, calirhiza, cambricum, dissimile, dulce, glycyrrhiza, hesperium, jubiformis, lanceolatum, loriceum, pellucidum, saximontanum, scouleri, sibiricum, tobagense, triseriale, virginianum*

VARIETIES *amorphum montense, calirhiza intermedium, dissimile chnoodes, dulce sororium, glycyrrhiza occidentale, hesperium columbianum, pellucidum pellucidum, pellucidum knudsenii, pellucidum vulcanicum*

GENUS *Campyloneurum* Strap Fern

SPECIES *angustifolium, costatum, latum, phyllitidis, repens*

GENUS *Goniophlebium* (Synonym for *Polypodium*)

GENUS *Hymenolepis* (Synonym for *Polypodium*)

GENUS *Lepisorus* Weeping Fern

SPECIES *thunbergianus*

VARIETIES *thunbergianus lineare*

GENUS *Lophosoria* Diamond Leaf Fern

SPECIES *quadripinnata*

GENUS *Marginaria* (Synonym for *Pleopeltis*)

GENUS *Microgramma* Snake Fern

SPECIES *heterophylla, lycopodioides, piloselloides, reptans, vacciniifolia*

VARIETIES *reptans ciliatum*

GENUS *Microsorum* Monarch Fern

SPECIES *punctatum, scolopendria, spectrum*

VARIETIES *punctatum irregulare, scolopendria phymatodes*

GENUS *Neurodium* Ribbon Fern

SPECIES *lanceolatum*

GENUS *Niphidium* Graceful Fern

SPECIES *crassifolium*

GENUS *Paltonium* (Synonym for *Neurodium*)

GENUS *Pecluma* Rock Cap Fern

SPECIES *camptophyllaria, consimilis, dispersa, pectinata, plumula, ptilodon*

VARIETIES *camptophyllaria camptophyllaria, camptophyllaria abbreviata, ptilodon ptilodon, ptilodon caespitosum*

GENUS *Phlebodium* Golden Polypody Fern

SPECIES *aureum, decumanum, pseudoaureum*

GENUS *Phymatosaurus* Nodule Saurian Polypody Fern

SPECIES *grossus*

GENUS *Platycerium* (Synonym for *Polypodium*)

GENUS *Pleopeltis* Scaly Polypody Fern
SPECIES *astrolepis, polylepis, polypodioides, riograndensis, squamata*
VARIETIES *astrolepis revoluta, polylepis erythrolepis, polypodioides polypodioides, polypodioides michauxiana*
GENUS *Pyrrosia* (Synonym for *Polypodium*)

FAMILY 02 *Aspleniaceae* Spleenwort Fern
GENUS *Asplenium* Spleenwort Fern
SPECIES *abscissum, acuminatum, adiantum-nigrum, adulterinum, aethiopicum, alternifolium, auriculatum, auritum, biscaynianum, boydstoniae, bradleyi, caudatum, ceterach, cirrhatum, clermontiae, contiguum, cookii, corderoanum, cristatum, cuneatum, curtissii, dalhousiae, densum, ebenoides, enatum, exiguum, feei, flabellulatum, formosum, gravesii, herb-wagneri, heterochroum, heteroresiliens, hobdyi, horridum, inexpectatum, insiticium, juglandifolium, kaulfussii, kentuckiense, kokeense, laetum, lobulatum, macraei, monanthes, montanum, morganii, myriophyllum, neobrackenridgei, nidus, normale, obtusifolium, ocoense, palmeri, patens, pinnatifidum, platyneuron, plenum, plenum, polyodon, pseudoerectum, pteropus, pumilum, radicans, resiliens, rhizophyllum, rhomboidale, rhomboideum, ruta-muraria, rutaceum, salicifolium, schizophyllum, septentrionale, serra, serratum, sessilifolium, shawneense, trichomanes, trudellii, unilaterale, uniseriale, varians, vespertinum, virginicum, wherryi*
VARIETIES *acuminatum goldmanii, adiantum-nigrum andrewsii, aethiopicum furcatum, aethiopicum praemorsum, aethiopicum rhipidoneuron, auritum auritum, auritum bipinnatifidum, bradleyi stotleri, contiguum contiguum, contiguum hirtulum, curtissii subtile, enatum flexuosum, exiguum fontanum, exiguum glenniei, feei sanguinolentum, flabellulatum partitum, insiticium sphenotomum, juglandifolium integerrimum, kaulfussii lydgatei, kaulfussii meiotomum, macraei sphenolobum, macraei diplaziosorum, myriophyllum scalifolium, myriophyllum verecundum, normale pavonicum, platyneuron platyneuron, platyneuron bacculum-rubrum, platyneuron incisum, platyneuron proliferum, polyodon knudsenii, polyodon nitidulum, polyodon nitidulum-falcatum, polyodon sectum, polyodon subcaudatum, pumilum anthriscifolium, radicans alloeopteron, radicans crytopteron, rhomboidale sintenisii, rhomboideum insulare, ruta-muraria cryptolepis, ruta-muraria lanceolum, ruta-muraria lanceolum-ohionis, ruta-muraria subtenuifolium, schizophyllum nephelephyllum, trichomanes trichomanes, trichomanes dentatum, trichomanes quadrivalens, trichomanes ramosum, trichomanes ramosum-viride, virginicum mixtum*
GENUS *Asplenosorus* (Synonym for *Asplenium*)
GENUS *Camptosorus* (Synonym for *Asplenium*)
GENUS *Ceterach* (Synonym for *Asplenium*)
GENUS *Deparia* False Spleenwort Fern
SPECIES *acrostichoides, fenzliana, kaalaana, marginalis, petersonii, prolifera*
VARIETIES *acrostichoides thelypterioides, petersonii japonica*
GENUS *Diellia* Island Spleenwort Fern
SPECIES *erecta, falcata, laciniata, leucostegioides, mannii, pallida, unisora*
VARIETIES *erecta pumila*
GENUS *Phyllitis* Hart's Tongue Fern
SPECIES *scolopendrium*
VARIETIES *scolopendrium scolopendrium, scolopendrium americana, scolopendrium japonica*

FAMILY 03 *Blechnaceae* Midsorus Fern
GENUS *Blechnum* Midsorus Fern
SPECIES *brownei, divergens, fragile, francii, insularum, lanceola, lineatum, occidentale, polypodioides, serrulatum, spicant*
VARIETIES *fragile polypodioides, lanceola pyramidatum, occidentale minor, polypodioides unilaterale, serrulatum indicum, spicant spicant, spicant elongatum, spicant nipponicum*
GENUS *Anchistea* (Synonym for *Woodwardia*)
GENUS *Doodia* Hacksaw Fern
SPECIES *kunthiana, lyonii*
GENUS *Lomaria* (Synonym for *Stenochlaena*)
GENUS *Lomariopsis* Fringed Fern
SPECIES *amydrophlebia, kunzeana, sorbifolia*
GENUS *Lorinseria* (Synonym for *Woodwardia*)
GENUS *Sadleria* Plume Fern
SPECIES *cyatheoides, pallida, souleyetiana, squarrosa*
VARIETIES *pallida hillebrandii, squarrosa polystichoides, squarrosa unisora*
GENUS *Stenochlaena* Vine Fern
SPECIES *tenuifolia*
GENUS *Struthiopteris* (Synonym for *Blechnum*)
GENUS *Woodwardia* Chain Fern
SPECIES *areolata, fimbriata, radicans, virginica*
VARIETIES *fimbriata chamissoi*

FAMILY 04 *Cyatheaceae* Tree Fern
GENUS *Cyathea* Tree Fern

SPECIES *andina, arborea, armata, balanocarpa, borinquena, cooperi, cubensis, furfuracea, grandifolia, grevilleana, insignis, obtusa, parvula, pungens,spectabilis, tenera, tenuis, wilsonii, woodwardioides*
VARIETIES *andina escuquensis, grevilleana elegans, parvula aquilina, pungens procera, tenera brittoniana, woodwardioides araneosa*
GENUS *Alsophila* Alsophile Tree Fern
SPECIES *amintae, brooksii, bryophila, portoricensis*
VARIETIES *amintae dryopteroides*
GENUS *Cnemidaria* Deep Woods Tree Fern
SPECIES *horrida*
GENUS *Hemitelia* (Synonym for *Cnemidaria*)
GENUS *Nephelea* (Synonym for *Alsophila*)
GENUS *Trichopteris* Hairy Tree Fern
SPECIES *borinquena, pubescens*

FAMILY 05 *Dennstaedtiaceae* Hay Scented Fern
GENUS *Dennstaedtia* Hay Scented Fern
SPECIES *bipinnata, cicutaria, dissecta, flaccida, globulifera, obtusifolia, punctilobula*
VARIETIES *bipinnata adiantoides, cicutaria rubiginosa, obtusifolia ordinata*
GENUS *Anisosorus* (Synonym for *Lonchitis*)
GENUS *Davallia* (Synonym for *Odontosoria*)
GENUS *Histiopteris* Bat Wing Fern
SPECIES *incisa*
GENUS *Hypolepis* Bead Fern
SPECIES *flaccida, nigrescens, punctata, repens, tenerrima, tenuifolia, urbanii*
GENUS *Lindsaea* Necklace Fern
SPECIES *ensifolia, guianensis, lancea, portoricensis, quadrangularis, repens, stricta*
VARIETIES *quadrangularis antillensis, quadrangularis antillensis-montana, repens macraeana, stricta jamesoniiformis*
GENUS *Lindsaeosoria* Lesser Necklace Fern
SPECIES *flynnii*
GENUS *Lonchitis* Tomato Fern
SPECIES *hirsuta*
GENUS *Microlepia* Limp Leaf Fern
SPECIES *mauiensis, speluncae, strigosa*
VARIETIES *strigosa hirta, strigosa setosa*
GENUS *Odontosoria* Toothed Creeping Fern
SPECIES *aculeata, chinensis, clavata, jenmanii, scandens*
VARIETIES *scandens uncinella*
GENUS *Orthiopteris* (Synonym for *Saccoloma*)
GENUS *Pteridium* Bracken Fern
SPECIES *aquilinum, arachnoideum, caudatum*
VARIETIES *aquilinum champlainense, aquilinum decompositum, aquilinum latiusculum, aquilinum pseudocaudatum, aquilinum pubescens*
GENUS *Saccoloma* Soral Pouch Fern
SPECIES *domingense, inaequale*
GENUS *Sphenomeris* (Synonym for *Odontosoria*)

FAMILY 06 *Dicksoniaceae* Dickson Fern
GENUS *Dicksonia* Dickson Fern
SPECIES *menziesii*
GENUS *Cibotium* Man Fern
SPECIES *chamissoi, glaucum, nealiae*
VARIETIES *chamissoi hawaiense, chamissoi splendens, glaucum st.-johnii*

FAMILY 07 *Dryopteridaceae* Wood Fern
GENUS *Dryopteris* Wood Fern
SPECIES *acutidens, algonquinensis, arguta, australis, benedictii, boottii, burgessii, campyloptera, carthusiana, celsa, cinnamomea, clintoniana, crinalis, cristata, dowellii, expansa, filix-mas, fragrans, fuscoatra, glabra, goldiana, hawaiiensis, intermedia, leedsii, ludoviciana, marginalis, neowherryi, parvula, patula, pittsfordensis, podosora, rossii, sandwicensis, separabilis, slossoniae, subbipinnata, tenebrosa, triploidea, uliginosa, unidentata, wallichiana*
VARIETIES *campyloptera americana, campyloptera austriaca, carthusiana spinulosa, celsa atropalustris, celsa wherryi, clintoniana poyseri, expansa americana, expansa assimilis, expansa dilatata, fragrans fragrans, fragrans remotiuscula, glabra nuda, intermedia concordiana, triploidea fructuosa, wallichiana paleacea, wallichiana parallelogramma*
GENUS *Arachniodes* Spider Holly Fern
SPECIES *aspidioides, carvifolia, chaerophylloides, insularis, simplicior*
GENUS *Aspidium* (Synonym for *Nephrolepis*)
GENUS *Athyrium* Lady Fern

SPECIES *americanum, distentifolium, filix-femina, microphyllum*
VARIETIES *americanum alpestre, americanum distentifolium, americanum gaspense, filix-femina filix-femina, filix-femina angustum, filix-femina angustum-michauxii, filix-femina angustum-rubellum, filix-femina angustum-subtripinnatum, filix-femina asplenioides, filix-femina cyclosorum, filix-femina cyclosorum-boreale, filix-femina cyclosorum-californicum, filix-femina cyclosorum-elatius, filix-femina cyclosorum-sitchense, microphyllum baldwinii, microphyllum multisectum, microphyllum poiretianum, microphyllum vexans*
GENUS *Bolbitis* Bulb Creeping Fern
SPECIES *aliena, fluviatilis, heudelotii, nicotianifolia, pergamentacea, portoricensis, serratifolia, sinuata*
VARIETIES *portoricensis cladorrhizans, sinuata hydrophylla*
GENUS *Ctenitis* Lace Fern
SPECIES *hirta, honolulensis, latifrons, nemorosa, sloanei, squamigera, submarginalis*
VARIETIES *hirta hirta, hirta portoricana, sloanei ampla*
GENUS *Cyclopeltis* Lime Fern
SPECIES *semicordata*
GENUS *Cyrtomium* Net Vein Holly Fern
SPECIES *caryotideum, falcatum, fortunei*
GENUS *Cystopteris* Bladder Fern
SPECIES *bulbifera, douglasii, fragilis, illinoensis, laurentiana, montana, protrusa, reevesiana, tennesseensis, tenuis, utahensis, wagneri*
VARIETIES *fragilis angustata, fragilis dickieana, fragilis woodsioides, laurentiana huteri, reevesiana tenuifolia, tennesseensis simulans, tenuis mackayi*
GENUS *Diplazium* Twinsorus Fern
SPECIES *centripetale, cristatum, esculentum, expansum, grandifolium, hymenodes, kaalanum, lherminieri, lonchophyllum, maximum, meyenianum, molokaiense, plantaginifolium, pycnocarpon, riedelianum, sandwichianum, striatum, unilobum*
VARIETIES *cristatum arboreum, meyenianum arnottii, riedelianum verapax*
GENUS *Dryostichum* Stitch Wood Fern
SPECIES *singulare*
GENUS *Elaphoglossum* Tongue Fern
SPECIES *aemulum, alatum, apodum, chartaceum, crassifolium, crinitum, cubense, decoratum, eggersii, erinaceum, glabellum, herminieri, hirtum, latifolium, lingua, longifolium, martinicense, maxonii, muscosum, palmeri, parvisquameum, pellucidum, peltatum, petiolatum, procurrens, rigidum, serpens, simplex, smithii, spathulatum, wawrae*
VARIETIES *aemulum conforme, aemulum gorgoneum, alatum alatum, alatum crassicaule, alatum fauriei, alatum parvisquamaeum, crassifolium reticulatum, hirtum micans, martinicense underwoodianum, maxonii pteropus, pellucidum micradenium, petiolatum dussii, rigidum flaccidum, simplex firmum, spathulatum piloselloides, spathulatum pusillum*
GENUS *Fadyenia* Dotted Fern
SPECIES *hookeri*
GENUS *Filix* (Synonym for *Cystopteris*)
GENUS *Gymnocarpium* Oak Fern
SPECIES *achriosporum, appalachianum, brittonianum, disjunctum, dryopteris, heterosporum, intermedium, jessoense, robertianum*
VARIETIES *dryopteris linnaeana, jessoense parvulum, robertianum pumilum*
GENUS *Hemidictyum* Half Net Fern
SPECIES *marginatum*
VARIETIES *marginatum limbatum*
GENUS *Hymenodium* (Synonym for *Elaphoglossum*)
GENUS *Hypoderris* Hair Vein Fern
SPECIES *brownii*
GENUS *Lastreopsis* False Shield Fern
SPECIES *effusa*
GENUS *Leptochilus* (Synonym for *Bolbitis*)
GENUS *Lomagramma* Fringe Marked Fern
SPECIES *guianensis*
GENUS *Matteuccia* Ostrich Fern
SPECIES *struthiopteris*
VARIETIES *struthiopteris pensylvanica, struthiopteris pubescens*
GENUS *Maxonia* Climbing Wood Fern
SPECIES *apiifolia*
GENUS *Megalastrum* Jungle Fern
SPECIES *subincisa*
GENUS *Nephrodium* (Synonym for*Triplophyllum*)
GENUS *Nephrolepis* Kidney Sword Fern
SPECIES *averyi, biserrata, cordifolia, exaltata, falcata, hirsutula, laurifolia, multiflora, pectinata, rivularis, undulata*
VARIETIES *falcata furcans, undulata occidentalis*

GENUS *Nothoperanema* — Island Lace Fern
SPECIES *rubiginosum*
GENUS *Oleandra* — Oleander Fern
SPECIES *articulata*
VARIETIES *articulata nodosa*
GENUS *Olfersia* — Woodland Island Fern
SPECIES *cervina*
GENUS *Onoclea* — Sensitive Fern
SPECIES *sensibilis*
GENUS *Peltapteris* — (Synonym for *Elaphoglossum*)
GENUS *Phanerophlebia* — Vein Fern
SPECIES *auriculata, umbonata*
GENUS *Polybotrya* — (Synonym for *Olfersia*)
GENUS *Polystichopsis* — (Synonym for *Arachniodes*)
GENUS *Polystichum* — Christmas Holly Fern
SPECIES *acrostichoides, aleuticum, andersonii, bonseyi, braunii, calderonense, californicum, dudleyi, echinatum, hagenahii, haleakalense, hillebrandii, imbricans, kruckebergii, kwakiutlii, lemmonii, lonchitis, microchlamys, minutum, muricatum, platyphyllum, polystichiforme, potteri, rhizophyllum, scopulinum, setigerum, tsus-simense*
VARIETIES *acrostichoides acrostichoides, acrostichoides lonchitoides, acrostichoides schweinitzii, braunii purshii, echinatum triangulum, hagenahii marginale, imbricans imbricans, imbricans curtum, imbricans nudatum, lemmonii mohrioides, minutum incisoserratum, setigerum alaskense*
GENUS *Pteretis* — (Synonym for *Matteuccia*)
GENUS *Rumohra* — Iron Fern
SPECIES *adiantiformis*
GENUS *Tectaria* — Halberd Fern
SPECIES *abbreviatum, amesiana, cicutaria, coriandrifolia, estremeriana, fimbriata, gaudichaudii, heracleifolia, incisa, pedata, plantaginea, trifoliata*
VARIETIES *fimbriata lobata, fimbriata minima, incisa martinicensis, plantaginea plantaginea, plantaginea confluens*
GENUS *Triplophyllum* — Fountain Fern
SPECIES *funestum*
GENUS *Woodsia* — Cliff Fern
SPECIES *abbeae, alpina, appalachiana, cochisensis, glabella, gracilis, ilvensis, kansana, maxonii, mexicana, neomexicana, obtusa, phillipsii, plummerae, scopulina, tryonis*
VARIETIES *abbeae confusa, abbeae squammosa, alpina bellii, obtusa obtusa, obtusa occidentalis, oregana, oregana oregana, oregana cathcartiana, scopulina scopulina, scopulina laurentiana, scopulina lyallii*

FAMILY 08 *Gleicheniaceae* — Forked, Scrambling, & Umbrella Fern
GENUS *Gleichenia* — (Synonym for *Dicranopteris, Diplopterygium, Sticherus*)
GENUS *Dicranopteris* — Forked Fern
SPECIES *emarginata, flexuosa, linearis, pectinata, rubiginosa, sandwicensis*
VARIETIES *emarginata emarginata, emarginata inaequalis, linearis linearis, linearis maxima*
GENUS *Diplopterygium* — Scrambling Fern
SPECIES *glaucum, pinnatum*
GENUS *Hicriopteris* — (Synonym for *Diplopterygium*)
GENUS *Sticherus* — Umbrella Fern
SPECIES *bifidus, brevipubis, owhyensis, palmatus, remotus*
VARIETIES *remotus trachyrhizoma*

FAMILY 09 *Grammitidaceae* — Marked Polypody Fern
GENUS *Grammitis* — Marked Dwarf Polypody Fern
SPECIES *aspleniifolia, baldwinii, cultrata, forbesiana, hanekeana, hessii, hookeri, liogieri, mollissima, nimbata, sherringii, taxifolia, tenella, trichomanoides*
VARIETIES *hessii limbata*
GENUS *Adenophorus* — Kihi Fern
SPECIES *abbottiae, abietinus, carsonii, epigaeus, haalilioanus, hymenophylloides, montanus, oahuensis, periens, pinnatifidus, tamariscinus, tripinnatifidus*
VARIETIES *pinnatifidus sarmentosus*
GENUS *Amphoradenium* — (Synonym for *Adenophorus*)
GENUS *Cochlidium* — Snail Fern
SPECIES *jungens, rostratum, seminudum, serrulatum*
VARIETIES *jungens minus, serrulatum duale*
GENUS *Enterosora* — Three Forked Polypody Fern
SPECIES *trifurcata*
GENUS *Lellingeria* — Lellinger Polypody Fern
SPECIES *hartii, myosuroides, saffordii, suspensa*
VARIETIES *suspensa jubiformis*
GENUS *Melpomense* — Melpomene Dwarf Polypody Fern

SPECIES *anfractuosa, flabelliformis*
VARIETIES *flabelliformis taenifolium*
GENUS *Xiphopteris* (Synonym for *Lellingeria*)
GENUS *Zygophlebia* Octopus Fern
SPECIES *sectifrons*

FAMILY 10 *Thelypteridaceae* Maiden Fern
GENUS *Thelypteris* Maiden Fern
SPECIES *abdita, abrupta, angustifolia, augescens, balbisii, boydiae, brittonae, cheilanthoides, cordata, cyatheoides, decussata, deltoidea, dentata, exindusiata, forsteri, germaniana, globulifera, grandis, guadalupensis, hastata, hildae, hispidula, inabonensis, incesta, interrupta, invisa, kunthii, leptocladia, linkiana, namophila, nephrodioides, nevadensis, noveboracensis, oligocarpa, opposita, opulenta, ovata, palmeri, palustris, parasitica, pennata, piedrensis, pilosa, poiteana, puberula, quelpaertensis, reptans, resinifera, reticulata, retroflexa, rheophyta, rolandii, rudis, sancta, sclerophylla, serra, serrata, simulata, tetragona, truncata, verecunda, wailele, yaucoensis*
VARIETIES *balbisii sprengelii, cordata cordata, cordata imitata, dentata reducta, guadalupensis domingensis, hastata hastata, hastata heterodoxa, hispidula hispidula, hispidula inconstans, hispidula versicolor, hispidula versicolor-macilenta, interrupta gongylodes, interrupta totta, invisa oligophylla, kunthii macrorhizoma, kunthii normalis, kunthii saxatilis, kunthii unca, nevadensis oregana, ovata ovata, ovata harperi, ovata lindheimeri, palustris pubescens, palustris pubescens-haleana, patens, patens patens, patens scabriuscula, patens smithiana, pennata megalodus, pilosa pilosa, pilosa alabamensis, puberula puberula, puberula sonorensis, quelpaertensis limbosperma, quelpaertensis oreopteris, reptans reptans, reptans conformis, reptans tenera, retroflexa asplenioides, sancta portoricensis*
GENUS *Christella* (Synonym for *Thelypteris*)
GENUS *Cyclosorus* (Synonym for *Thelypteris*)
GENUS *Goniopteris* (Synonym for *Thelypteris*)
GENUS *Lastrea* (Synonym for *Thelypteris*)
GENUS *Leptogramma* (Synonym for *Thelypteris*)
GENUS *Macrothelypteris* Maiden Sword Fern
SPECIES *torresiana*
VARIETIES *torresiana setigera, torresiana uliginosa*
GENUS *Meniscium* (Synonym for *Thelypteris*)
GENUS *Oreopteris* (Synonym for *Thelypteris*)
GENUS *Phegopteris* Beech Fern
SPECIES *connectilis, hexagonoptera*
VARIETIES *connectilis polypodioides*
GENUS *Pneumatopteris* Air Fern
SPECIES *hudsoniana, sandwicensis*
VARIETIES *sandwicensis stegnogrammoides*
GENUS *Pseudophegopteris* False Beech Fern
SPECIES *keraudreniana*
GENUS *Sphaerostephanos* (Synonym for *Thelypteris*)

FAMILY 11 *Vittariaceae* Shoestring Fern
GENUS *Vittaria* Shoestring Fern
SPECIES *appalachiana, costata, elongata, graminifolia, lineata, remota*
VARIETIES *elongata rigida, graminifolia filifolia*
GENUS *Ananthacorus* (Synonym for *Vittaria*)
GENUS *Anetium* Treetrunk Fern
SPECIES *citrifolium*
GENUS *Antrophyum* Lineleaf Fern
SPECIES *cajenense, intramarginale, lanceolatum, lineatum, plantagineum, urbanii*
GENUS *Hecistopteris* Moss Fern
SPECIES *pumila*
GENUS *Polytaenium* (Synonym for *Antrophyum*)

Index

About the Author

Lawrence Fusillo was born in Greenbelt Maryland in 1949. While in elementary school, he was introduced to plant life, by his neighbor, Anna Belle Owens, who had a large garden in her back yard. His interest in biology continued in junior high school in a biology club, and in high school when he was a volunteer in a microbiology laboratory at the Veteran's Administration hospital in Washington DC. He received a B. A. in Physics from John's Hopkins University in 1971, and an M. S. in Applied Physics from the University of New Orleans in 1981.

He became interested in the names of colors in 1996, working at the Naval Air Development Center in Warminster Pennsylvania while developing software to display the colors of seafloor sediment. He found that most colors were named after plants that exhibited the color in the form of flowers. He began studying the relationship between plants, animals, minerals, and color names and discovered that the Linnaeus *Genus species* Latin classification system for plants, animals and minerals had not been updated since 1758.

After a 26 year career designing magnetometer, sonar and radar software for the US Navy first as a government employee (1972-1981) and later as a US Navy contractor (1982-1998), he designed air traffic control software as a contractor for the Federal Aviation Administration (1999-2001). He began background research for writing scientific books in May of 2001.

The "Fusillo Nature Classification System Genus & Higher Taxa" for plants, animals, and minerals with category names in English, is designed to replace the Linnaeus Classification System and all subsequent classification schemes. The categories Domain, Kingdom, Subkingdom, Phylum, Class, Order, and Family were assigned both in English and in Latin and the Latin Genus was assigned. In subsequent books in the series, the *Genus* and *species* categories were assigned both in English and in Latin. A companion CD "PC CD Fusillo Nature Classification System" contains an implementation of the Fusillo Nature Classification System in Excel Spreadsheet files (.xls) which can be imported into database programs.

Books by Lawrence J. Fusillo

1	**Fusillo Nature Classification System (FNCS) Genus & Higher Taxa 1st Ed.**	1 Jan 2008	ISBN:0-7414-4395-3
2	FNCS Nonliving Matter Species Checklist 1st Edition	1 Mar 2008	ISBN:0-7414-4459-3
3	FNCS Virus Species Checklist 1st Edition	1 Mar 2009	ISBN:0-7414-5214-6
4	FNCS Microbe Species Checklist 1st Edition	1 Sep 2009	ISBN:0-7414-0000-0
5	FNCS Spore Plant Species Checklist 1st Edition	1 May 2010	ISBN: 0-7414-5878-0
6	FNCS Seed Plant Species Checklist 1st Edition		
7	FNCS Orchid Species Checklist 1st Edition	1 Nov 2009	ISBN: 0-7414-5668-0
8	FNCS Herb, Vegetable, & Fruit Variety Checklist		
9	FNCS Cultivated Flower Variety Checklist 1st Edition		
10	FNCS Anna Belle Nature Trail Guide 1st Editon	1 Sep 2007	ISBN:0-7414-4201-9
11	FNCS Invertebrate Coral, Nettle, Sponge & Worm Species Checklist 1st Edition		
12	FNCS Invertebrate Arthropod Species Checklist 1st Edition		
13	FNCS Invertebrate Shellfish & Vertebrate Fish Species Checklist 1st Edition		
14	FNCS Vertebrate Species Checklist 1st Edition		

1 **Fusillo Knowledge Classification System (FKCS) 1st Edition**
2 FKCS PC Hardware Physics 1st Edition
3 FKCS PC Hardware & Operating System Advances 1st Edition
4 FKCS PC Hardware Components Catalog 1st Edition
5 FKCS PC Programmer's Guide 1st Edition
6 FKCS PC Color Language and Dictionary of Names 8th Edition
7 FKCS PC Programmer's Data Dictionary 1st Edition
8 FKCS PC User's Guide to Computer Books, Magazines & User Groups 1st Edition
9 FKCS PC User's Guide to Educational Software 1st Edition
10 FKCS PC User's Guide to Educational Web Sites Edition
11 FKCS Body Tuneup Organic Foods, Air Filter, & Exercise 1st Edition
12 FKCS Mind Tuneup Brain Foods and Mental Agility Exercises 1st Edition

Made in United States
Cleveland, OH
06 November 2026

47787688R00295